Dielectric Materials For Wireless Communication

Dielectric Materials for Wireless Communication

MAILADIL T. SEBASTIAN
National Institute for Interdisciplinary Science & Technology (NIIST),
Trivandrum, 695019, India

ELSEVIER
Amsterdam • Boston • Heidelberg • London • New York • Oxford
Paris • San Diego • San Francisco • Singapore • Sydney • Tokyo

Elsevier
Linacre House, Jordan Hill, Oxford OX2 8DP, UK
Radarweg 29, PO Box 211, 1000 AE Amsterdam, The Netherlands

First edition 2008

Library of Congress Cataloging-in-Publication Data
A catalog record for this book is available from the Library of Congress

British Library Cataloguing in Publication Data
A catalogue record for this book is available from the British Library

ISBN: 978-0-08-045330-9

For information on all Elsevier publications
visit our website at books.elsevier.com

Printed and bound by CPI Group (UK) Ltd, Croydon, CR0 4YY
Transferred to Digital Print 2011

Contents

Foreword

In this book Dr. Sebastian describes the current state of the art of what are now broadly described as microwave dielectric materials. The history of these materials stretches back to the late 19th century. In 1897 Lord Rayleigh described a dielectric waveguide and in 1909 Debye described dielectric spheres. It was not until 1939 that Richtmyer coined the term "Dielectric Resonator" when he suggested that a dielectric ring could confine high-frequency electromagnetic waves and hence form a resonator. Richtmyer also realized that an open resonator would resonate into free space and three quarters of a century later these ideas have spawned a multibillion dielectric antenna industry and dielectric resonator industry. Astonishingly, our lives have been completely transformed by the science of a handful of people.

Today, microwave dielectric materials are all-pervasive. Several people buy a new mobile phone every second of every day of every year. This book takes us to the heart of the science and it takes us through the science in a comprehensive manner. We learn about the key properties of relative permittivity, of dielectric loss and of temperature coefficients and we learn how the microstructure and chemistry of the dielectric is crucial in determining the key properties. We learn about the beginnings of the now huge dielectric resonator industry in the pioneering work of Hank O'Brian and Taki Negas on barium titanate compositions. Historically the book is faithful and we next learn about the zirconium titanates, finally ending up with the newer perovskites.

The amazingly forgiving properties of the perovskites, in terms of substitution, are described and the ability of these substitutions to affect all the key properties – the temperature coefficient, the dielectric loss and the relative dielectric constant. The book describes how one can tailor the dielectric properties of materials by judicious choice of substituent or dopant.

In the final chapters we see interesting information of specific materials such as titania and alumina as well as low sintering temperature materials that can be cofired with electrodes such as silver. Included in an appendix is the most comprehensive list of microwave dielectric materials, along with their key properties, that exists.

This book will serve a wide range of communities – from University students and tutors to industrial laboratories. The volume of information available is prodigious as a rapid glance of the contents indicates and this in combination with a truly comprehensive list of over a thousand references makes this book a most valuable source of information. Dr. Sebastian has worked in the area of microwave dielectrics for many years and has published extensively in this area. This book is a considerable achievement.

Professor Neil McN Alford FREng
Imperial College London

Acknowledgment

The subject matter presented in this book has been derived from several publications in addition to our own and I am grateful to many authors and publishers for allowing us to use their material. I am indebted to my doctoral students and postdoctoral fellows whose researches in this area immensely helped me to write this book. I am grateful to Dr. Tamura, Murata Manufacturing Company in Japan and Nokia in Finland for permitting me to use their images in the cover page of this book. I wish to thank Prof. Jerzy Krupka, Prof. Stanislav Kamba, Prof. Heli Jantunnen, Prof. Neil Alford, Prof. Hitoshi Ohsato, Prof. Roberto Moreira, Prof. P. Mohanan, Prof. V R K Murthy, Dr. R. Ratheesh, Dr. H. Sreemoolanathan and Dr. J. James for critically reading parts of the manuscript and giving useful suggestions. I am also thankful to Prof. T. K. Chandrasekhar, Director NIIST for his encouragement and to Mr. G. Subodh and Sumesh George for drawing some of the figures. I am also grateful to Prof. Neil Alford, Imperial College for writing the Foreword to this book.

M. T. Sebastian

CHAPTER ONE

INTRODUCTION

Microwave dielectric materials play a key role in global society with a wide range of applications from terrestrial and satellite communication including software radio, GPS, and DBS TV to environmental monitoring via satellites. In order to meet the specifications of the current and future systems, improved or new microwave components based on dedicated dielectric materials and new designs are required. The recent progress in microwave telecommunication, satellite broadcasting and intelligent transport systems (ITS) has resulted in an increasing demand for dielectric resonators (DRs), which are low loss ceramic pucks used mainly in wireless communication devices. With the recent revolution in mobile phone and satellite communication systems using microwaves as the carrier, the research and development in the field of device miniaturization has been one of the biggest challenges in contemporary Materials Science. This revolution is apparent on a daily basis in the ever increasing number of cell phone users. The recent advances in materials development has led to these revolutionary changes in wireless communication technology. Dielectric oxide ceramics have revolutionized the microwave wireless communication industry by reducing the size and cost of filter, oscillator and antenna components in applications ranging from cellular phones to global positioning systems. Wireless communication technology demands materials which have their own specialized requirements and functions. The importance of miniaturization cannot be overemphasized in any hand-held communication application and can be seen in the dramatic decrease in the size and weight of devices such as cell phones in recent years. This constant need for miniaturization provides a continuing driving force for the discovery and development of increasingly sophisticated materials to perform the same or improved function with decreased size and weight.

A DR is an electromagnetic component that exhibits resonance with useful properties for a narrow range of frequencies. The resonance is similar to that of a circular hollow metallic waveguide except for the boundary being defined by a large change in permittivity rather than by a conductor. Dielectric resonators generally consist of a puck of ceramic that has a high permittivity and a low dissipation factor. The resonant frequency is determined by the overall physical dimensions of the puck and the permittivity of the material and its immediate surroundings. The key properties required for a DR are high quality factor (Q), high relative permittivity (ε_r) and near zero temperature coefficient of resonant frequency (τ_f). An optimal DR that satisfies these three properties simultaneously is difficult to achieve in a particular material.

In the early microwave systems, bulk metallic cavities were used as resonators, but were huge and not integrated with microwave integrated circuits (MICs). On the other hand, stripline resonators have a poor quality factor and poor temperature stability resulting in the instability of the circuit. Hence the importance of DRs, which are easily integrated with MICs with low loss and with thermally stable frequency, especially at mm wavelengths. Most of the microwave-based device systems are located in the frequency range

Dielectric Materials for Wireless Communications
ISBN-13: 978-0-08-045330-9

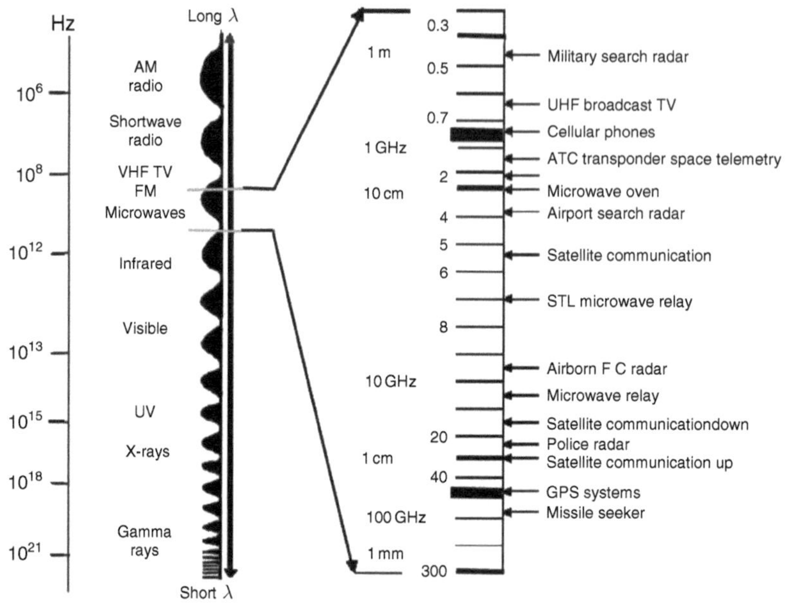

Figure 1.1 Microwave spectrum and applications.

300 MHz–300 GHz as shown in Figure 1.1. Technological improvements in DRs have contributed to considerable advancements in modern wireless communications. Ceramic DRs have the advantage of being more miniaturized as compared to traditional microwave cavities, and have a significantly higher quality factor. DRs have replaced cavity resonators in most microwave and millimeter-wave applications for reasons of cost, dimension, mass, stability, efficiency, tenability, ruggedness and ease of use. In addition, the temperature variation of the resonant frequency of DRs can be engineered to a desired value to meet circuit designer's requirements. Functioning as important components in communication circuits, DRs can create and filter frequencies in oscillators, amplifiers and tuners. In order to respond to the requirement for increased channel capacity in ground-based cellular and satellite communications, new devices with superior performance must be developed. The system performance is closely related to material properties. In microwave communications, DR filters are used to discriminate between wanted and unwanted signal frequencies from the transmitted and received signals. The desired frequency is extracted and detected to maintain a strong signal-to-noise ratio. For clarity, it is also critical that the wanted signal frequencies are not affected by seasonal temperature changes.

The low permittivity ceramics are used for millimeter-wave communication and also as substrates for microwave integrated circuits. The medium ε_r ceramics with permittivity in the range 25–50 are used for satellite communications and in cell phone base stations. The high ε_r materials are used in mobile phones, where miniaturization of the device is very important. For millimeter-wave and substrate application, a temperature-stable low permittivity and high Q (low loss) materials are required for high speed signal transmission with minimum attenuation.

The term "dielectric resonator" first appeared in 1939, when Richtmeyer of Stanford University showed that a suitably shaped dielectric piece can function as a microwave resonator [1]. However, it took more than 20 years to generate further interest on DRs and to test Richtmeyer's prediction experimentally. In 1953, Schlicke [2] reported on super high permittivity materials (~1000 or more) and their applications as capacitors at relatively low RF frequencies. In the early 1960s, Okaya and Barash from Columbia University rediscovered DRs while working on rutile single crystals [3, 4]. Okaya and Barash [3, 4] measured the permittivity and Q of TiO_2 single crystals at room temperature down to 50 K in the microwave frequency range, using the commensurate transmission line technique [4]. Later several authors developed methods for measuring the ε_r, quality factor (Q) and τ_f of DRs. These methods are discussed in Chapter 2. In the early 1960s, Cohen and his co-workers [5] from Rantec Corporation performed the first extensive theoretical and experimental evolution of DR. Rutile ceramics were used for the experiments that had an isotropic permittivity of about 100. The TiO_2 has a poor (+450 ppm/°C) stability of resonant frequency that prevented its commercial exploitation. The first microwave filter using TiO_2 ceramics was proposed by Cohen in 1968 [6, 7]. But this filter was not useful for practical applications because of its high τ_f. A real breakthrough in DR ceramic technology occurred in the early 1970s, when the first temperature-stable, low loss barium tetratitanate ($BaTi_4O_9$) ceramics were developed by Masse et al. [8]. Later, barium nanotitanate ($Ba_2Ti_9O_{20}$) with improved performance was reported by Bell Laboratories [9]. The next breakthrough came from Japan when Murata Manufacturing Company produced (Zr,Sn)TiO_4 ceramics [10, 11]. They offered adjustable compositions so that temperature coefficients could be varied between +10 and −10 ppm/°C. Later, in 1975, Wakino et al. realized the miniaturization of the DR-based filters and oscillators [12]. Since then extensive theoretical and experimental work and development of several DR materials has occurred. This early work resulted in the actual use of DRs as microwave components. Commercial production of DRs started in the early 1980s. The number of papers published and patents filed on the science and technology of DRs increased considerably over the years as shown in Figure 1.2. There are about 2300 low loss dielectric materials reported in the literature (see Appendix 2). More than 5000 papers have been published and over 1000 patents were filed on DR materials and devices. However, with only a limited number of useful dielectric ceramic materials to choose from, the electronic industry is constantly searching for new materials that are easily affordable for manufacture.

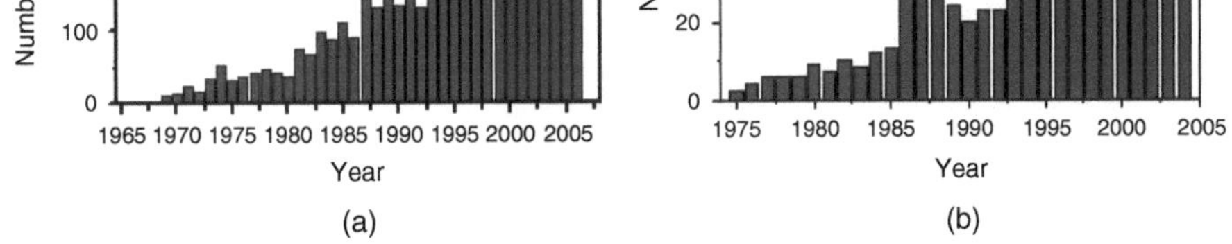

Figure 1.2 (a) The number of papers published on dielectric resonator materials and technology versus year of publishing (b) Number of patents filed versus year.

Richtmeyer [1] in 1939 theoretically predicted that a piece of dielectric with regular geometry and high ε_r can confine electromagnetic energy within itself, but still be prone to energy loss due to radiation. It was found that through total multiple internal reflections, a piece of high ε_r dielectric can confine microwave energy at a few discrete frequencies, provided the energy is fed in the appropriate direction (see Figure 1.3). If the transverse dimensions of the sample are comparable to the wavelength of the microwave, then certain field distributions or modes will satisfy Maxwell's equations and boundary conditions. The reflection coefficient approaches unity as ε_r approaches infinity. In the microwave frequency range, free space wavelength (λ_c) is in centimeters and hence the wavelength (λ_g) inside the dielectric will be in millimeters only when the value of ε_r is in the range 20–100. To get resonance, dimensions of the dielectric must be of the same order (in millimeters). Still larger ε_r gives higher confinement of energy, reduced radiation loss and better miniaturization. However, high ε_r will result in higher dielectric losses because of inherent material properties. When exposed to free space, a DR can also radiate microwave energy when it is fed suitably and can be used as efficient radiators, called Dielectric Resonator Antennas (DRA). A DR with finite values of ε_r prevents 100% reflection from the air/dielectric boundary and hence some field will always exist in the vicinity of the dielectric. This is of great advantage since it enables one to couple microwave power easily to the DR by matching the field pattern of the coupling elements to that of the DR. Figure 1.4 (a) and (b) illustrates the variation of electric and magnetic fields inside a dielectric ($Ca_5Nb_2TiO_{12}$ ceramic puck with $\varepsilon_r = 48$) kept inside a copper cavity and simulated using a three-dimensional transmission line matrix modeling method [13].

The size of a DR is considerably smaller than the size of an empty resonant cavity operating at the same frequency, provided the relative permittivity (ε_r) of the material is

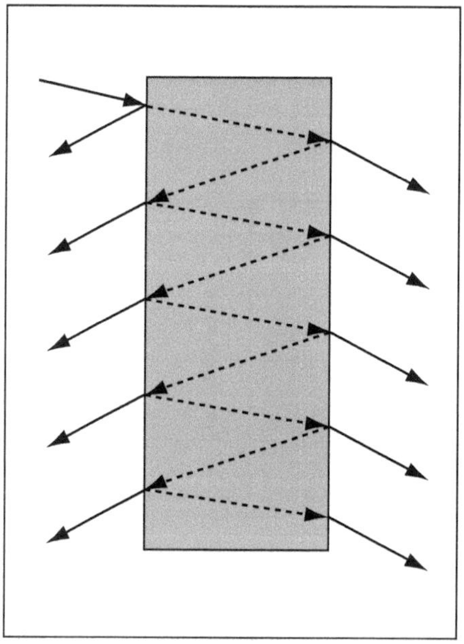

Figure 1.3 Schematic sketch of total multiple internal reflections in a high ε_r dielectric piece.

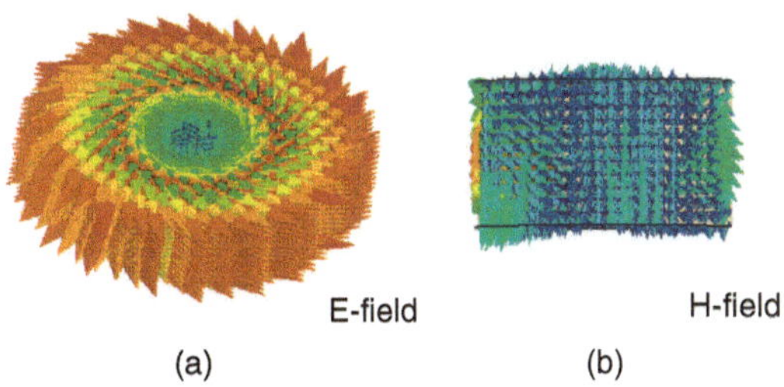

Plate 1 Variation of (a) electric and (b) magnetic fields of $TE_{01\delta}$ resonance mode of a $Ca_5Nb_2TiO_{12}$ ceramic resonator with $\varepsilon_r = 48$ (after Ref. [13]) (see Figure 1.4, p. 5).

Plate 2 Picture of dielectric ceramic packs developed at the author's laboratory (see Figure 1.5, p. 5).

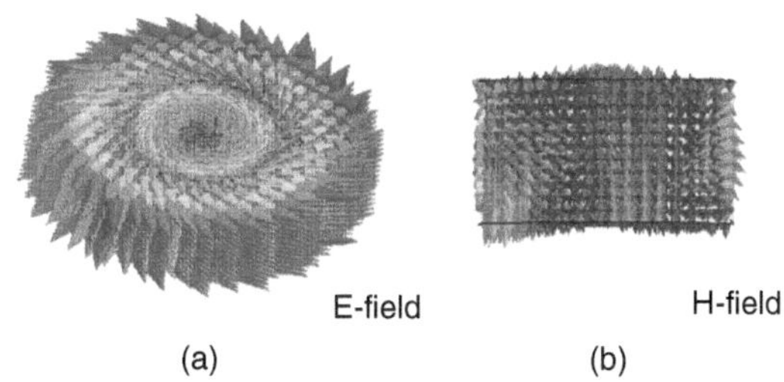

Figure 1.4 Variation of (a) electric and (b) magnetic fields of $TE_{01\delta}$ resonance mode of a $Ca_5Nb_2TiO_{12}$ ceramic resonator with $\varepsilon_r = 48$ (after Ref. [13]) (see Color Plate section).

substantially higher than unity. Higher ε_r shrinks overall circuit/device size proportional to $(1/\varepsilon_r)^{1/2}$. For example, a circuit is compressed by a factor of six when a high Q ceramic with $\varepsilon_r = 36$ is substituted for a high Q air cavity $\varepsilon_r = 1$. The shape of a DR is usually a solid cylinder but can also be tubular, spherical and parallelepiped. Figure 1.5 shows some of the low loss dielectric pucks made at the author's laboratory. A commonly used resonant mode of a cylindrical DR is $TE_{01\delta}$. At resonant frequency, electromagnetic fields inside a resonator store energy equally in electric and magnetic fields. When ε_r is about 40, more than 95% of the stored electric energy and over 60% stored magnetic energy are located within the dielectric cylinder. The remaining energy is distributed in the air around the resonator, decaying rapidly with distance away from the resonator boundary. The DR can be incorporated into a microwave network by exciting it with microstrip transmission lines, as shown in Figure 1.6. The distance between the resonator and the microstrip conductor determines the amount of coupling. In order to prevent losses due to radiation, the entire device is usually enclosed in a metallic shielding box.

High Q minimizes circuit insertion losses and can be used as a highly selective circuit. In addition, high Q suppresses the electrical noise in oscillator devices. Although several manufacturers may produce similar components for the same application, there are subtle differences in circuit design, construction and packaging. Since frequency drift of a device is a consequence of the overall thermal expansion of its unique combination of

Figure 1.5 Picture of dielectric ceramic packs developed at the author's laboratory (see Color Plate section).

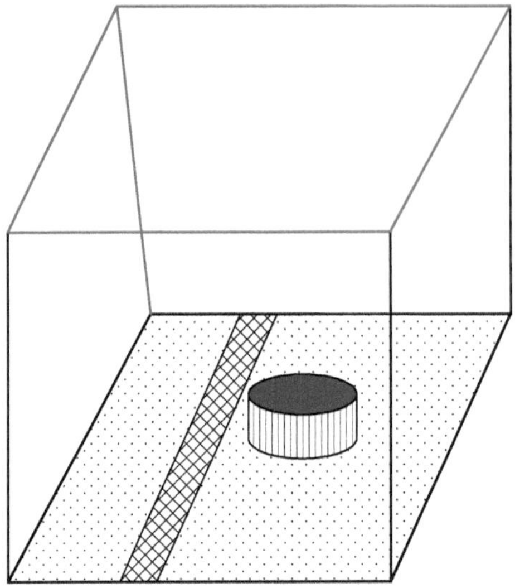

Figure 1.6 Dielectric resonator mounted on a microstrip.

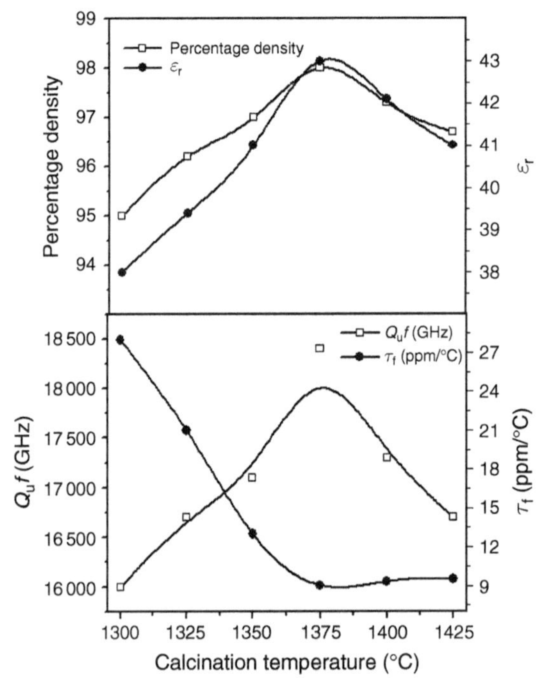

Figure 1.7 The variation of the relative density, ε_r, Qf and τ_f of $Ba(Sm_{1/2}Nb_{1/2})O_3$ ceramic versus calcination temperature. Sintering temperature 1550°C for 2 hours (after Ref. [16]).

construction materials, each design requires a slightly different τ_f for temperature compensation. Typically, ceramics with a specific τ_f in the range 15 to -15 ppm/°C are selected. In ceramic production, τ_f and ε_r specifications must be held in demanding tolerances ±1%.

Oxide ceramics are critical elements in these microwave circuits, and a full understanding of their crystal chemistry is fundamental to future development. Properties of microwave ceramics critically depend on several parameters, such as purity of starting materials, calcination temperature and duration, shaping method and sintering temperature and durations. Design of the heating/cooling schedule requires knowledge of formation mechanism of various phases in multicomponent systems. The starting powders must sinter to high density to get optimum electrical properties. Figures 1.7–1.10 show typical examples of the effect of calcination and sintering temperatures and their durations on the density and microwave dielectric properties of ceramic DRs. The microwave ceramics are usually optimized for the best density for which the ε_r and *Qf* are normally the best. Powders prepared by solid state methods require a higher sintering temperature as compared to those prepared by chemical methods. However, the sintering temperature can be lowered by using sintering aids. High *Qf* can be achieved only in ceramics with fine grains and/or homogeneous microstructures. The ceramics with discontinuous and excess grain growth exhibit poor performance. Incomplete densification is mainly due to discontinuous grain growth. Due to rapid and discontinuous grain growth, porosity is trapped in. If the distance from these trapped pores to the path of the fast material transport or grain boundaries is large, the rate of material transport to close

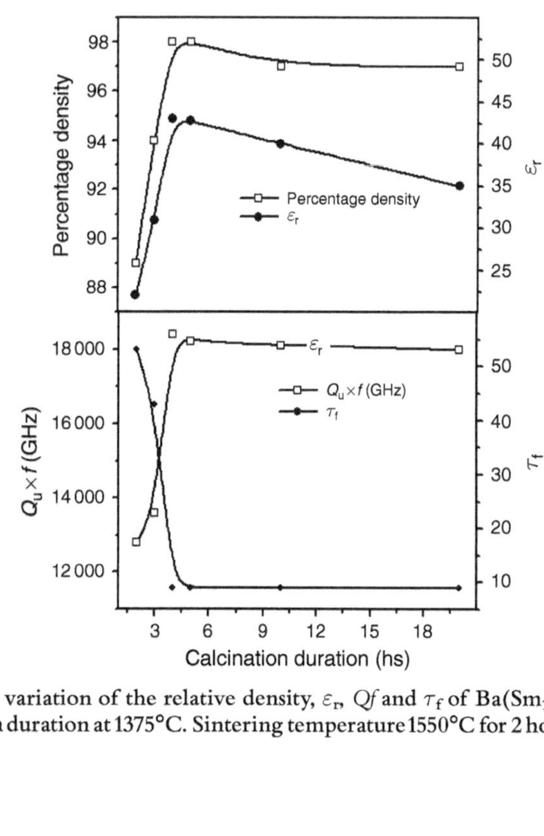

Figure 1.8 The variation of the relative density, ε_r, *Qf* and τ_f of $Ba(Sm_{1/2}Nb_{1/2})O_3$ ceramic versus calcination duration at 1375°C. Sintering temperature 1550°C for 2 hours (after Ref. [16]).

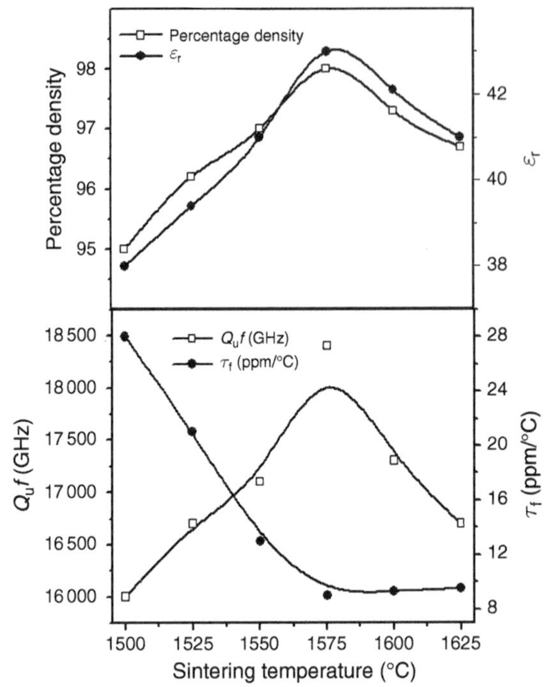

Figure 1.9 The variation of the relative density, ε_r, Qf and τ_f of $Ba(Sm_{1/2}Nb_{1/2})O_3$ ceramic versus sintering temperature (after Ref. [16]).

the pores will be limited. The materials will be porous and the dielectric and mechanical properties will be poor. Densification can be promoted by the use of additives. Additives aid fabrication by allowing densification to occur at lower temperatures in shorter times or by inhibiting discontinuous grain growth and allowing pore elimination to proceed to completion. Use of these additives is for most part empirical, and experimental verification of their role is impossible usually limited to the grain boundaries. At least two theories were advanced to explain the effect of these additives. (a) Liquid phase assistance. Additive with a lower MPt is used for densification by melting and coating the ceramic particles, rapid dissolution and transfer of the base material to fill the interparticle spaces, which leads to enhanced densification. The additive has to be fairly soluble in the base ceramic and several wt% of the additive is necessary to provide sufficient liquid phase to coat all the powder particles. (b) Solid solution effect. Addition of a solute with a different valency enhances bulk material transport due to the introduction of vacancies in the ceramic. These vacancies render the diffusion coefficient extrinsic by making thermal vacancy concentration insignificant up to a certain temperature. If added in sufficient quantity, it can significantly alter the rate of material transport through the solid phase. Sintering aids affect ceramics in many ways by changing the density, microstructure, defect structure and possibly crystal structure. These changes brought about by the sintering aids affect the resulting dielectric properties. The density ε_r, Q and τ_f are all affected by the additives. A higher relative density results in a higher ε_r. The selection of proper additives, their optimum quantity and optimum processing conditions are effective in enhancing the quality factor.

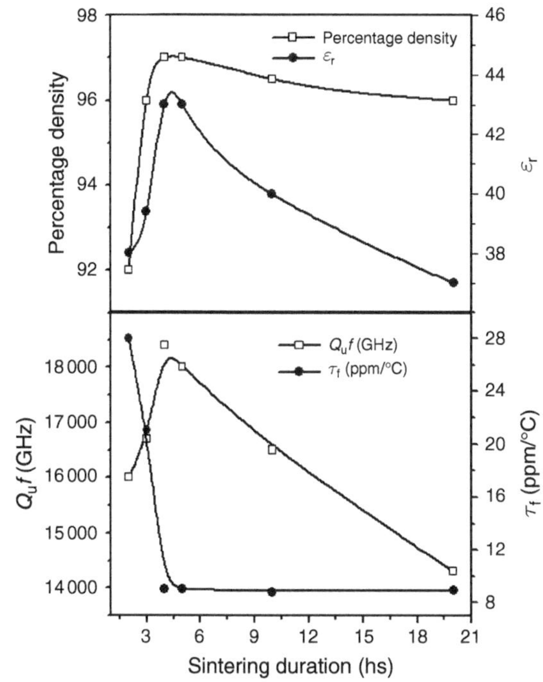

Figure 1.10 The variation of the relative density, ε_r, Qf and τ_f of $Ba(Sm_{1/2}Nb_{1/2})O_3$ ceramic versus sintering duration on sintering at 1575°C (after Ref. [16]).

Although the title of the book is *Dielectric Materials for Wireless Communication*, it may be noted that the materials discussed in this book are useful for several other applications such as capacitors and gate dielectrics. Although DRs are so promising in practical applications, surprisingly no reference or edited books summarizing the research results on low loss dielectric materials are available. The first book on DRs by Kajfezz and Guillon [14] and the book on DRA by Luk and Leung [15] essentially deal with electromagnetic theory and applications and only discuss DR materials in passing. During the last 25 years, scientists the world over have developed a large number of new low loss dielectric materials and improved the properties of existing materials. Actually a few thousand research papers have been published on the preparation, characterization and properties of DRs. However, the data of this multitude of very useful materials are scattered. Hence it is the purpose of this book to bring the data and science of these several useful materials together which will be of immense help to researchers and technologists the world over. The book describes the preparation, characterization and properties of important DR materials and how one can tailor the properties to meet the requirements of the design engineer.

The topics covered in the book includes factors affecting the dielectric properties, measurement of dielectric properties, important low loss dielectric material systems such as perovskites, tungsten bronze type materials, materials in $BaO–TiO_2$ system, $(Zr,Sn)TiO_4$, alumina, rutile, $A_nB_{n-1}O_{3n}$ type materials, LTCC, etc. The book also has a data table listing the reported low loss dielectric materials with properties and references arranged in the order of increasing permittivity.

Filling an existing need, Dielectric Materials for Wireless Communication is the first comprehensive book in this rapidly growing field.

REFERENCES

[1] R. D. Richtmeyer. Dielectric resonators. J. Appl. Phys. 15 (1939)391–398.

[2] H. M. Schlicke. Quasidegenerate modes in high ε_r dielectric cavities. J. Appl. Phys. 24 (1953)187–191.

[3] A. Okaya. The rutile microwave resonator. Proc. IRE 48 (1960) 1921.

[4] A. Okaya and L. F. Barash.The dielectric microwave resonator. Proc. IRE 50 (1962) 2081–2092.

[5] S. B. Cohen. Microwave band pass filters containing high Q dielectric resonators. IEEE Trans. Microw. Theory and Techniques MTT-16 (1968)1628–1629.

[6] S. B. Cohen. Microwave filters containing high Q dielectric resonators. IEEE MTT-Symp. Dig. (1965)49–53.

[7] S. B. Cohen. Microwave band pass filters containing high Q dielectric resonators. IEEE Trans. Microw. Theory and Techniques. MTT-33 (1985)586–592.

[8] D. J. Masse, R. A. Purcel, D. W. Ready, E. A. Maguire and C. P. Hartwig. A new low loss high K temperature compensated dielectric for microwave applications. Proc. IEEE 59 (1971)1628–1629.

[9] J. K. Plourde, D. F. Limn, H. M. O'Bryan, and J. Thomson Jr. $Ba_2Ti_9O_{20}$ as a microwave resonator. J. Am. Ceram. Soc. 58 (1975)418–420.

[10] K. Wakino, M. Katsube, H. Tamura, T. Nishikawa, and Y. Ishikawa. Microwave dielectric materials (Japanese). In IEEE Four Joint Cov. Rec. (1977) paper No. 235.

[11] K. Wakino, T. Nishikawa, H. Tamura, and Y. Ishikawa. Microwave band pass filters containing dielectric resonator with improved temperature stability and spurious response. IEEE MTT-S Int. Microw. Symp. Dig. (1975)63–65.

[12] K. Wakino, T. Nishikawa, S. Tamura, and Y. Ishikawa. Microwave band pass filters containing dielectric resonators with improved temperature stability and spurious response. IEEE MTT-S. Int. Symp. Dig. (1975)63–65.

[13] P. V. Bijumon, M. T. Sebastian, and P. Mohanan. Experimental investigations and three dimensional transmission line matrix simulation of $Ca_{5-x}A_xB_2TiO_{12}$ (A=Mg, Zn, Ni and Co; B=Nb and Ta) ceramic resonators. J. Appl. Phys. 98 (2005)124105.

[14] D. Kajfezz and P. Guillon (Eds). Dielectric Resonators, 2nd Edition. Noble Publishing Corporation, Atlanta, US (1998).

[15] K. M. Luk and K. W. Leung (Eds). Dielectric Resonator Antennas. Research Studies Press Ltd, Baldock, Hertfordshire, UK (2002).

[16] L. A. Khalam. The $A(B'_{1/2}B''_{1/2})O_3$ {A=Ba, Sr, Ca, Mg; B=Re, and B=Nb,Ta} microwave ceramics for wireless communications. Ph.D. Thesis, Kerala University, India (2007).

CHAPTER TWO

Measurement of Microwave Dielectric Properties and Factors Affecting them

Dielectric Resonators (DR) are dielectric bodies of high permittivity and high Q-factor that can be used as energy storage devices. Ceramic DRs are usually prepared in the form of cylindrical or rectangular pucks by the sintering process. They are much smaller in size compared to its metallic counterpart. The three important characteristics of an ideal DR are high relative permittivity (ε_r) for resonator applications and low ε_r for millimeter wave applications, low dielectric loss (loss tangent) and low coefficient of temperature variation of the resonant frequency (τ_f). These three aspects and different measurement methodology to measure them are elaborately discussed in the following sections.

2.1 Permittivity (ε_r)

The relative permittivity (ε_r) of the material shows its energy storing capacity when a potential is applied across it. It is related to the macroscopic properties like polarization or capacitance. For circuit miniaturization, usually one employs a high ε_r material. A high ε_r facilitates circuit miniaturization because the wavelength inside the DR is inversely proportional to the square root of its permittivity as given by the equation.

$$\lambda_d = \frac{\lambda_0}{\sqrt{\varepsilon_r}} \tag{2.1}$$

where λ_d is the wavelength in the dielectric, λ_0 is the wavelength in air (actually in vacuum). The dimension of the dielectric sample must be an integral multiple of half-wavelength in the dielectric to resonate in the simplest fundamental mode [1]. If that wavelength is reduced, then the physical dimensions of the resonator must be reduced as well. The permittivity of a material determines the relative speed that an electrical signal can travel in that material. A low permittivity will result in a high signal propagation speed. When microwaves enter a dielectric material, they are slowed down by a factor roughly equal to the square root of the permittivity which implies that the wavelength decreases by the same amount and the frequency is unaffected as shown in Figure 2.1.

By definition, the ε_r is related to the refractive index n by

$$\varepsilon_r = n^2 \tag{2.2}$$

As permittivity is frequency dependent, it is very rare that the square of the refractive index measured at optical frequencies is the same as permittivity measured at microwaves. This rule is applicable only if the same polarization processes are excited by both optical

Dielectric Materials for Wireless Communications
ISBN-13: 978-0-08-045330-9

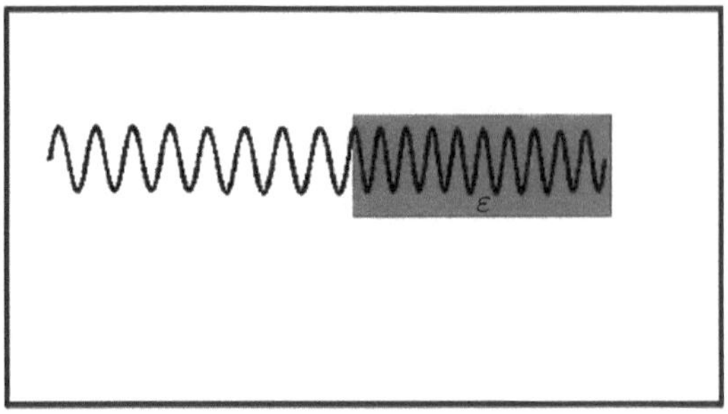

Figure 2.1 The wavelength is reduced by a factor of $\sqrt{\varepsilon}$ when the wave enters the dielectric.

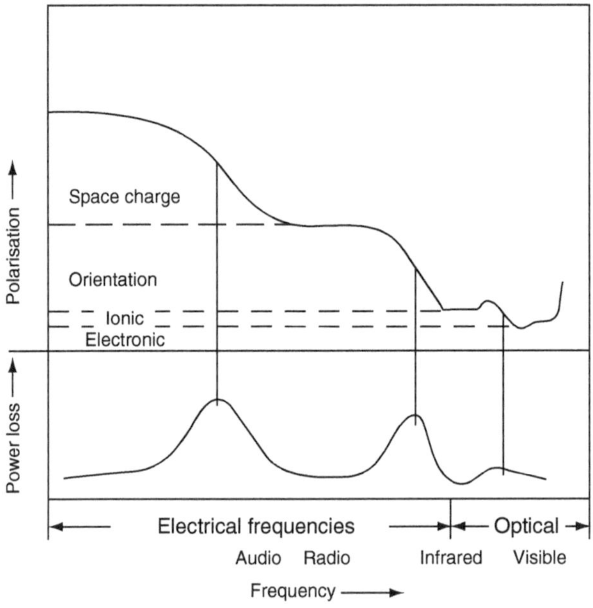

Figure 2.2 Frequency dependence of polarization processes and peak power losses (after Ref. [3]).

and microwave (or RF) frequencies, which is generally true only for elemental solid materials like diamond ($\varepsilon_r = 5.68$, $n^2 = 5.85$) or germanium ($\varepsilon_r = 16$, $n^2 = 16.73$) [2]. In other materials, this rule is not valid since dipolar polarization processes which occur at lower frequencies do not usually occur at higher optical frequencies. At microwave frequencies, ionic and electronic polarization mechanisms contribute predominantly to the net dipole moments and the permittivity as depicted in Figure 2.2.

2.2 Quality Factor (*Q*)

The dielectric loss tangent (tan δ) of a material denotes quantitatively dissipation of the electrical energy due to different physical processes such as electrical conduction, dielectric relaxation, dielectric resonance and loss from non-linear processes [4]. Origin of dielectric losses can also be considered as being related to delay between the electric field and the electric displacement vectors [5]. The total dielectric loss is the sum of intrinsic and extrinsic losses. Intrinsic dielectric losses are the losses in the perfect crystals which depend on the crystal structure and can be described by the interaction of the phonon system with the ac electric field. Gurevich and Tagantsev developed a complete theory of intrinsic dielectric losses [6, 7]. The ac electric field alters the equilibrium of the phonon system and the subsequent relaxation is associated with energy dissipation [6–8]. The phonon frequency is much higher than the microwave frequency. Hence the low frequency dielectric relaxation in the ideal lattice should be of an harmonic origin. As a result, energy of the field dissipates heat and the sample gets heated up. Gurevich and Tagantsev have reviewed [6] the theory of intrinsic losses. The intrinsic dielectric losses depend on the crystal symmetry, ac field frequency and temperature. These intrinsic losses fix the lower limit of losses in defect-free single crystals or ideal pure materials. Extrinsic losses are associated with imperfections in the crystal lattice such as impurities, microstructural defects, grain boundaries, porosity, microcracks, order–disorder, random crystallite orientation, dislocations, vacancies, dopant atoms etc. The extrinsic losses are caused by lattice defects and therefore can be in principle eliminated or reduced to the minimum by proper material processing. The losses due to different types of defects show different frequency and temperature dependence. The crystals belonging to different symmetry groups have very different temperature and frequency dependences of dielectric loss [6]. Manufacturers of dielectric ceramics often use the name "quality factor" for the reciprocal of the tan δ. One should carefully distinguish this quantity from the Q-factor of a resonator which is defined as

$$Q = 2\pi \frac{\text{maximum energy stored per cycle}}{\text{Average energy dissipated per cycle}}$$

The term "quality factor" is more commonly associated with microwave resonators. Quality factor, or Q, is a measure of the power loss of a microwave system. For the microwave resonator, losses can be of four types: (a) dielectric, (b) conduction, (c) radiation and (d) external [1]. The dielectric Q_d, conduction Q_c and radiation Q_r quality factors are given by

$$Q_d = 2\pi \frac{W_1}{P_d T} = \frac{\omega_0 W_1}{P_d},\ Q_c = \frac{\omega_0 W_1}{P_c},\ Q_r = \frac{\omega_0 W_1}{P_r} \tag{2.3}$$

where W_1 is the total stored electric energy in the resonator, ω_0 is the angular resonant frequency, P_d, P_c and P_r represent the power dissipated in the dielectric, conductor and radiation respectively and

$$T = \frac{2\pi}{\omega_0}$$

The unloaded quality factor Q_u is related to other Q-factors by

$$\frac{1}{Q_u} = \frac{1}{Q_d} + \frac{1}{Q_c} + \frac{1}{Q_r} \tag{2.4}$$

where $1/Q_d$ is dielectric loss, $1/Q_c$ the loss due to conductivity of the metallic plates and $1/Q_r$ is the loss due to radiation. Most resonant cavities are completely shielded, so there is no radiation effect, and that term can be ignored. The design of the shielding box or metal housing plays an important role in the final performance of the circuit. It affects insertion loss, spectral purity, temperature stability and spurious mode rejection.

In practice, external losses ($1/Q_{ext}$) arise due to coupling. To introduce an electromagnetic field in the resonator, microwave conducting probes are brought close to it (typically within millimeters of each other) – the higher the ε_r of the resonator, the closer the probe must be to it. The electromagnetic fields around them induce electromagnetic fields in the dielectric ceramic, and so they are coupled; however, the presence of conducting probes in the electromagnetic field lines of the resonator leads to additional loss. The total or loaded Q-factor is defined as [1]

$$\frac{1}{Q_L} = \frac{1}{Q_d} + \frac{1}{Q_c} + \frac{1}{Q_r} + \frac{1}{Q_{ext}} \tag{2.5}$$

where $1/Q_L$ is the total loss of the system and $1/Q_{ext}$ is the loss due to external coupling.

Q_L is determined experimentally from the shape of the resonance peak, as illustrated in Figure 2.3. Loaded quality factor refers to a resonator coupled with external circuit and its relationship with unloaded quality factor depends on the coupling coefficients. A peak occurs in the transmitted signal amplitude at the resonant frequency, and the peak has some finite breadth. A bandwidth (BW) is defined as the width of resonance curve at half power points (3 dB down from the peak). The peak frequency (resonant frequency) f divided by the 3 dB width is equal to Q_L.

The loaded Q_L is obtained from the measured resonant frequency f and half power (–3 dB) bandwidth Δf of TE_{011} mode resonance.

$$Q_L = \frac{f}{\Delta f} \tag{2.6}$$

The resonator BW is inversely proportional to the Q-factor. Thus high Q resonators have narrow BW. If the coupling is low (coupling coefficients << 1), then the unloaded

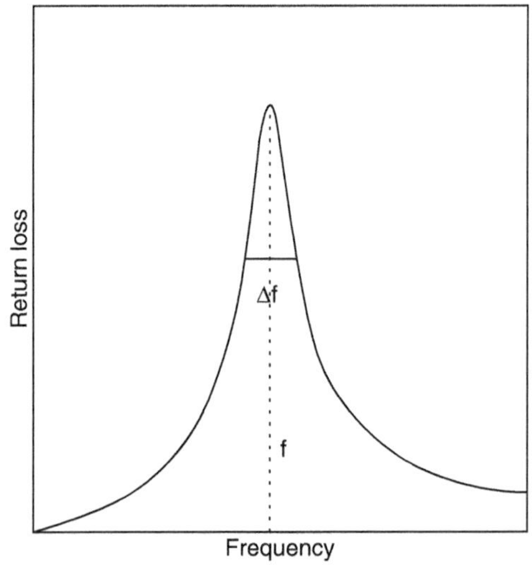

Figure 2.3 The $TE_{01\delta}$ resonant peak and associated parameters.

Q-factor is approximately equal to the loaded Q-factor. In practice, resonators are often used with adjustable couplings which allow such approximation without the need for measuring coupling coefficients. If conduction, radiation and external losses are negligible, then $Q_L = 1/\tan\delta$. When we measure the dielectric loss, at a particular frequency, we get the total loss tangent and we cannot distinguish the contributing factors.

If all conduction, radiation and external losses are negligible, then the loaded Q-factor depends on dielectric losses in the resonant structure. If the resonant structure contains several (N) dielectrics (one of them is sample under test) then the Q-factor due to dielectric losses is related to the dielectric losses in particular dielectric regions by the following formulae [5]:

$$\frac{1}{Q_d} = \sum_{i=1}^{N} P_{ei} \tan\delta_i \tag{2.7}$$

$$P_{ei} = \frac{\iiint_{vd} \varepsilon_i |E|^2 \mathrm{d}v}{\iiint_{vt} \varepsilon(v) |E|^2 \mathrm{d}v} \tag{2.8}$$

where P_{ei} is the electric energy filling factor for the ith dielectric region and $\tan\delta_i$ is the dielectric loss tangent for the ith dielectric region. V_d is the volume of the DR, v_t is the volume of the whole resonant structure, $\varepsilon(v)$ is the spatially dependent permittivity inside the whole resonant structure and ε_i is the permittivity of the ith dielectric region. For low loss dielectric materials, electric energy filling factors can be determined from incremental frequency rules that require computations of the derivative of the resonant frequency with respect to the real permittivities of the dielectric regions [9]. For cylindrical TE_0 mode resonant structures having axial symmetry, incremental frequency rule can also be used to determine geometric factor [1].

In most practical structures conductor losses appear and this must be accounted for in order to determine Q-factor due to dielectric losses (Q_d) from measured values of the loaded Q-factors

$$\frac{1}{Q_c} = \frac{R_s}{G} \tag{2.9}$$

The geometric factor G is defined as [5]

$$G = \omega \frac{\iiint_{vi} \mu_0 |H|^2 \mathrm{d}v}{\iint_S |H_t|^2 \mathrm{d}S} \tag{2.10}$$

where μ_0 is the permeability of the resonator.

The coupling losses ($1/Q_{ext}$) are usually determined experimentally and is given by

$$\frac{1}{Q_u} = \frac{1}{Q_d} + \frac{1}{Q_c} + \frac{1}{Q_r} = \frac{1/Q_L}{1 + \beta c_1 + \beta c_2} \tag{2.11}$$

where βc_1 and βc_2 are coupling coefficients for a resonator with two coupling ports. The radiation losses are negligible for a closed cavity. It should be mentioned here that the inverse of dielectric loss tangent which is commonly named as Q-factor due to dielectric losses is in most cases different from the Q-factor due to dielectric losses of a DR (Q_d). Even if we have resonant structure with one lossy dielectric and an air, there is a difference between those two quantities. This is because the total electric energy in the resonant structure is stored partly in the dielectric and partly in the air.

2.3 Measurement of Microwave Dielectric Properties

The measurement of dielectric properties plays a crucial role for characterizing the materials for different applications. Precisely measuring the material properties of a material is important to predict the system performance. Many methods have been developed and employed for measuring permittivity and permeability of materials. The most accurate measurement at high frequency can be done using the high frequency Q resonant cavity technique. However, the main disadvantage of the cavity method is that the measured results are applicable only over a narrow frequency band. Electrical properties of material over a wide range of frequencies can be measured with less accuracy using the transmission line methods. The transmission and reflection methods are used to evaluate the dielectric properties of medium and high loss materials as a function of frequency. The reader is referred to the recent reviews [5, 10] for the details of these techniques. In this book we restrict the discussions to measure the relative permittivity and loss tangents of low loss dielectric materials. These measurements are performed by resonance techniques which provide the highest accuracy of measurements for low loss dielectrics.

There are various methods which enable measurement of the quality factors of low loss dielectrics [10–17]. However, not all of them take into account practical effects introduced by a real measurement system. The practical effects include noise, crosstalk, coupling losses, transmission line delay and impedance mismatch. Inadequate accounting of these effects may lead to significant uncertainty in the measured Q-factor. The determination of the complex permittivity and other electromagnetic properties require precise measurements of the resonant frequencies and Q-factors. For some measurement techniques, these parameters have to be measured in the presence and absence of the test sample, for other techniques only once in the presence of the test sample. Once the resonant frequencies, Q-factors of the resonant structures and dimensions of the test samples are measured, computations have to be performed to obtain the ε_r and tan δ. It may be noted that as the ε_r increases, the resonant frequency decreases and as the dimensions of the sample decrease the resonant frequency increases.

2.3.1 Hakki and Coleman (Courtney) method

2.3.1.1 Permittivity

The complex permittivity of the material of DR is often measured by the method developed by Hakki and Coleman [11] and modified by Courtney [12], in which a cylindrical disc of material to be measured is inserted between two mathematically infinite conducting plates, as shown in Figure 2.4. The end plates are usually made of well-polished copper plates coated with silver or gold. Consider a circular cylindrical rod of relative permittivity ε_r, length L and diameter D placed between the end metal plates (See Figure 2.4). The diameter of the conducting plates should be much larger than that of the dielectric puck. The TE_{011} mode is normally used to make the measurements. The dielectric puck diameter-to-height ratio should be about two to get wide mode separation; so that the TE_{011} mode is not disturbed by other adjacent modes. If the dielectric material is isotropic then the characteristic equation for such a resonant structure for the TEoml mode is [11] given by

$$\alpha \frac{J_0(\alpha)}{J_1(\alpha)} = -\beta \frac{K_0(\beta)}{K_1(\beta)} \tag{2.12}$$

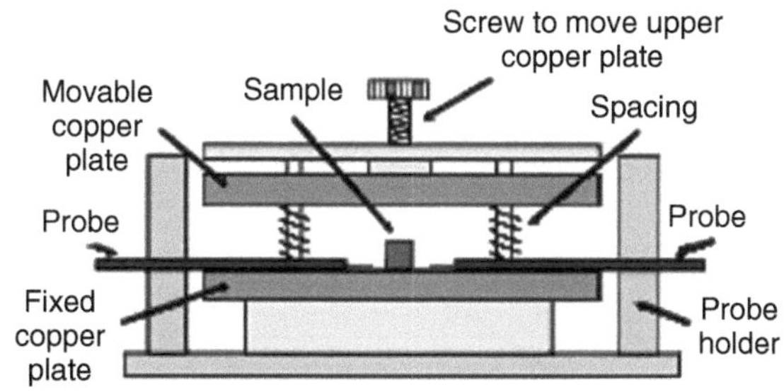

Figure 2.4 Schematic sketch of Courtney setup for measuring the dielectric constant under end shorted condition (after Ref. [12]).

where $J_o(\alpha)$ and $J_1(\alpha)$ are the Bessel functions of first kind of order zero and one. $K_0(\beta)$ and $K_1(\beta)$ are the modified Bessel functions of second kind of orders zero and one respectively. The parameters α and β depend on the geometry, the resonant wavelength inside the DR and dielectric properties. Thus

$$\alpha = \frac{\pi D}{\lambda_0}\left[\varepsilon_r - \left(\frac{l\lambda_0}{2L}\right)^2\right]^{1/2} \tag{2.13}$$

$$\beta = \frac{\pi D}{\lambda_0}\left[\left(\frac{l\lambda_0}{2L}\right)^2 - 1\right]^{1/2} \tag{2.14}$$

where

l is the longitudinal variations of the field along the axis,

L is the length of the DR,

D is the diameter of the DR and

λ_0 is the free space resonant wavelength.

The characteristic equation is a transcendental equation and hence a graphical solution or numerical iteration becomes necessary. Hakki and Coleman used a graphical solution to solve the characteristic equation. Corresponding to each value of β there is infinite number of (α_n) that solves the characteristic equation. Hakki and Coleman provided [11] a mode chart showing the variation of α as a function of β as shown in Figure 2.5. The characteristic equation and the resulting mode charts are universal as far as permittivity is concerned. Hence the use of mode charts to evaluate the permittivity of all materials is conceivable.

α can be easily evaluated from the curve-fitted values given by the following equations

$$X = 1 + (\beta - 1)/2$$

$$\alpha = -0.0091087 + x^2 + 0.050169x + 2.6084$$

The real part of the permittivity of the resonator can be calculated using the mode chart parameters (α_1 and β_1), the resonant frequency (f) and the dimensions of the dielectric puck using the equation

$$\varepsilon_r = 1 + \left[\frac{c}{\pi D f}\right]^2 \left(\alpha_1^2 + \beta_1^2\right) \tag{2.15}$$

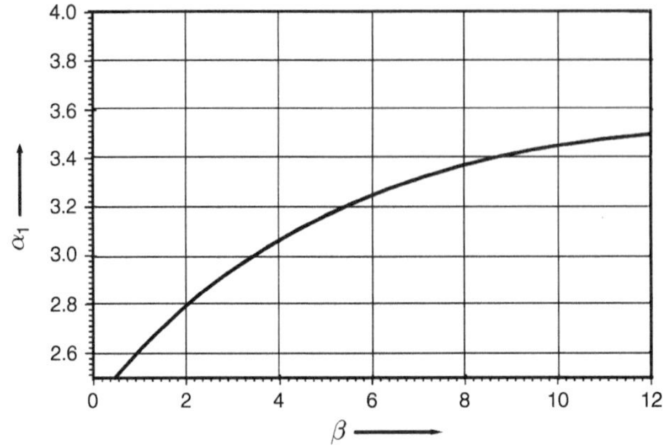

Figure 2.5 Mode chart (after Ref. [11]).

Hakki and Coleman [11] used an iris coupling from a waveguide to couple microwave to the DR. Later, Courtney [12] modified the method by using two horizontally oriented E-field probes for coupling microwave to the DR. This enabled to span a wide range of frequencies, since there is no cut-off frequency for coaxial lines. The TE_{011} mode is used for the measurements since this mode propagates inside the sample but is evanescent outside the DR sample. Therefore a large amount of electrical energy can be stored in high Q DRs [18]. In the end shorted condition the E field becomes zero close to the metal wall and electric energy vanishes in the air gap [13]. The TE and TM modes do not contain electric and magnetic fields in the axial (z) direction. The TE_{011} mode is chosen for measurement because for this mode only azimuthal component of the electric field exists and the error due to the air gap is practically eliminated [19]. For cylindrical resonator, TE and TM modes exist only if the azimuthal mode index $m = 0$, otherwise all other modes are hybrid, i.e. they have all six electromagnetic components. Hybrid modes are usually divided into two mode families: HE and TM. They are only occasionally used in measurements of dielectrics (uniaxially anisotropic crystals).

A network analyzer is used to measure the Q-factor of a DR. An identical rigid cable with connectors on both ends (THRU) having length equal to the sum of the length of the two-probe cable is required for calibration. The network analyzer is set for the S_{21} mode and the THRU cable is connected between port 1 and port 2 of the network analyzer. The start frequency, stop frequency and number of points are then set in the network analyzer. A THRU calibration is performed in the network analyzer. The THRU cable is removed and the jig cables are connected to conduct the measurements. The silver- or gold-coated plate is then placed on the top of the sample. The probe is kept very close to the DR to provide strong coupling. The network analyzer will display the transmission coefficient with frequency. Different modes are represented by the spikes in the spectrum. When the top plate is tilted, most of the spikes will move to the right side. There will be one spike which will not move to the right side. On further tilting, this spike will move to the left side. This is the TE_{011} mode. Now the maximum of this spike is found and the centre frequency is set. The span is reduced as much as possible so that it can display the resonance curve with −5 dB on both the sides. The distance between the two probes is increased to reduce the coupling so that the above curve is visible in the display. The resonant frequency and –3 dB width of the spectrum is noted. The maximum amplitude of S_{21} is also noted. This is the insertion loss of the system. The value of the coupling constant

$$\beta_c = 10^{\frac{-S_{21}}{20}}$$

The measured Q value is called the loaded quality factor (Q_L), since the measurement is done by external circuit (the network analyzer with coupling probes) which will load the resonator. However, under weak coupling the Q_L is obtained from Equation (2.6). The unloaded quality factor is given by

$$Q_u = Q_L/(1 - \beta_c)$$

Figure 2.6 shows the frequency response of the TE_{011} mode resonator. By knowing the diameter "D" and length "L" of the sample, β is calculated using Equation (2.14). From the mode chart, the value of α_1 corresponding to the β_1 value is noted. The permittivity ε_r is calculated using Equation (2.15). The sample should be well polished to avoid the error due to surface roughness. In this method the accuracy is limited to the accuracy of the measurement of the resonant frequency and the dimensions of the sample. The possible error in the measurement of permittivity is of the order of 0.3%. Such an error is possible when dimensional uncertainties of the samples are in the order of 0.15%. The advantages of this method are very simple measurement configuration and easy access for introducing and removing the test samples. This is one of the fairly accurate and the most frequently used techniques for measurement of permittivity and this method is proposed as one of the international standards IEC techniques [20] for measurements of the complex permittivity

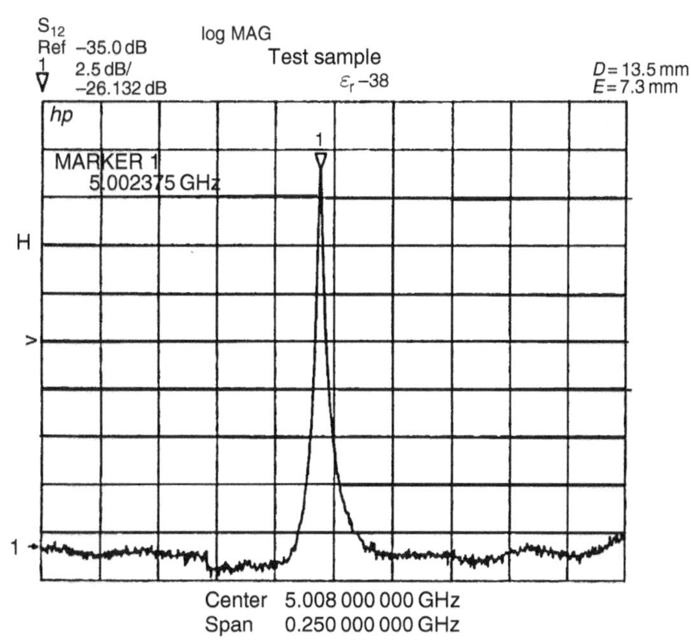

Figure 2.6 The TE_{011} resonance of a ceramic puck with $\varepsilon_r = 38$ under end shorted condition.

of low loss solids. Hennings and Schnabel [21] studied the reproducibility of the ε_r measured by the end shorted method of Courtney using 10 different samples prepared in a batch. Their results showed a maximum variation of 0.6% in ε_r. In the Courtney method, the ε_r is measured only at one resonant frequency corresponding to the TE_{011} mode. If one can identify other resonant modes, then it is possible to measure ε_r at other resonant frequencies. By using the resonant modes TE_{011}, TE_{021}, TE_{031} and TE_{041}, the ε_r of a sample can be measured in a range of frequencies.

If the distance between the end conducting plates is smaller than half wavelength corresponding to the resonant frequency, then the electromagnetic fields will be evanescent outside the dielectric in the air region [5]. This occurs when the aspect ratio (D/L) of the resonator sample is larger than certain minimum value which depends on the ε_r of the sample. The smaller the ε_r of the sample, the larger must be the aspect ratio of the sample to avoid the radiation losses. When the metal plates are separated by a distance larger than the half wavelength, then the TE_{011} mode has very low Q-factors mainly due to radiation losses. The radiation losses are usually not present for infinitely extended metal plates for the TE_{011} mode [5]. For materials with $\varepsilon_r \geq 10$, the aspect ratio, $D/L = 1.4$–1.8, is not suitable for measurements since low Q leaky modes TM_{210} and TM_{020} are close to the TE_{011} mode. For low $\varepsilon_r < 10$, the cut-off conditions require to use $D/L > 1.55$ [18]. The quasi-TM modes and TM modes are not suitable for ε_r measurements. This is due to the fact that a minute air gap between the dielectric sample and the metal plate considerably alter the resonant frequency which affects the accuracy of ε_r measurement [13, 22]. Moreover, these modes are leaky with low Q-factor [14].

2.3.1.2 Measurement of loss tangent

The quality factor can be measured by Hakki and Coleman end shorted method [10–13, 14, 17, 23, 24]. The quality factor measured by this method will be low since loss occurs due to the conducting plates and radiation effects. However, correction to conductor losses can be applied knowing the surface resistance of the conducting plates. The unloaded Q_u is obtained from the measured resonant frequency f and half power (–3 dB) bandwidth Δf of TE_{011} mode resonance given by Equation (2.6).

The tan δ can be calculated [10–13, 14, 17, 23, 24] from

$$\tan\delta = \frac{A}{Q_u} - BR_s \tag{2.16}$$

$$= A\left[\frac{1}{Q_u} - \frac{R_s}{A/B}\right] = \frac{1}{P_e}\left[\frac{1}{Q_u} - \frac{1}{Q_c}\right]$$

Q_c is given by Equation (2.9) and $G = A/B$ and $P_e = 1/A$

R_s is the surface resistance of the conducting plates and is given by

$$R_s = \sqrt{\frac{\pi f \mu}{\sigma}} \tag{2.17}$$

where σ is the conductivity of the conducting plates.

The permeability for a non-magnetic metal $\mu = 4\pi \times 10^{-7}$ H/m [17]

$$A = 1 + \frac{W}{\varepsilon_r} \tag{2.18}$$

$$B = \left(\frac{\lambda_0}{\lambda_g}\right)^3 \left(\frac{1+W}{30\pi^2 \varepsilon_r l}\right) \tag{2.19}$$

$$W = \frac{J_1^2(\alpha)}{K_1^2(\beta)} \frac{K_0(\beta)K_2(\beta) - K_1^2(\beta)}{J_1^2(\beta) - J_0(\alpha)J_2(\alpha)} \tag{2.20}$$

where λ_0 is the resonant wavelength.

$\lambda_g = 2L/l$ ($l = 1, 2, 3, \ldots$)

λ_g is the guiding wavelength of an infinitely long dielectric rod waveguide and W is the ratio of electric field energy stored outside to inside the rod [17, 18].

Thus knowing the value of σ, the tan δ can be obtained. Kobayashi and Tamura [17] has reported a method of measuring the value of R_s using two rod samples cut from a dielectric rod which have same diameters but different lengths. One of the rod for a TE_{01p} mode resonator is p times as long as the other for a TE_{011} mode resonator where $p \geq 2$. The modes have almost the same resonant frequency but differ in observed unloaded quality factors because of different conductor loss contributions in the two cases. Since both rods have the same tan δ, the following equation can be obtained

$$R_s = 30\pi^2 \left[\lambda_g/\lambda_0\right]^3 \frac{\varepsilon_r + W}{1+W} \frac{p}{p-1} \left[\frac{1}{Q_{ml}} - \frac{1}{Q_{mp}}\right] \tag{2.21}$$

where Q_{ml} and Q_{mp} are measured unloaded quality factors for TE_{011} and TE_{01p} modes respectively.

2.3.2 $TE_{01\delta}$ mode dielectric resonator method

When the Q of a DR sample is measured by the end shorted method of Hakki and Coleman, or Courtney, the measured Q-factor is affected by the conductor and radiation losses. However, these effects can be avoided by using the cavity method in which the DR is placed on a low loss single crystal quartz or teflon spacer inside the cavity. The quality factor (Q), permittivity (ε_r) and τ_f of the DRs can be measured using a transmission mode cavity proposed by Krupka et al. [5, 25]. The DR is placed inside a cylindrical metallic cavity usually made of copper and the inner surfaces are finely polished and usually gold or silver coated. The cavity is closed with a lid after loading the DR sample. Microwave is fed using loop coupling. The cavity has infinite number of modes, when excited with microwave spectrum of frequencies. Usually D/L ratio of 2–2.5 is maintained to get maximum mode separation and to avoid interference from other adjacent modes. The resonant frequency, quality factor and τ_f are dependent on the resonator surroundings. The electric field is symmetric with the geometry of the sample and the cavity, which helps to reduce the sources of loss due to cavity. In the $TE_{01\delta}$ cavity method, the field confinement is not complete in the z direction and hence TE_{011} mode is designated as $TE_{01\delta}$. As seen in Figure 2.7, the sample is isolated using a quartz spacer from the effects of losses due to the finite resistivity of the metallic cavity. Observe S_{21} or transmission characteristics of the $TE_{01\delta}$ resonant mode versus frequency which is displayed. One can assume that the unloaded Q-factor is equal to loaded Q-factor if the coupling is weak. Figure 2.8 shows the typical resonance spectra in reflection and transmission configuration of $Ba(Mg_{1/3}Ta_{2/3})O_3$ ceramic sample having $\varepsilon_r = 24$. The $TE_{01\delta}$ mode frequency is noted and the unloaded Q-factor is measured.

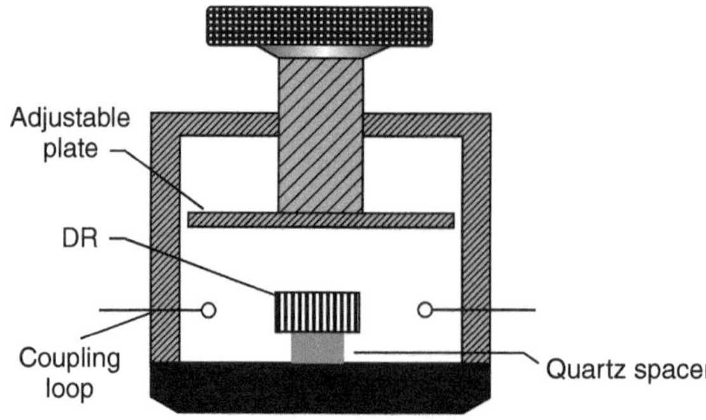

Figure 2.7 The cavity setup for the measurement of Q-factor.

After identifying the mode, the resonant frequency and 3 dB BW are determined by using the network analyzer. The network analyzer is then calibrated for full two port and S_{11} and S_{22} are measured at the resonant frequency. From this, the coupling coefficients βc_1 and βc_2 for the coupling ports are determined using the relations

$\beta c_1 = (1 - S_{11})/(S_{11} + S_{22})$ and $\beta c_2 = (1 - S_{22})/(S_{11} + S_{22})$, where S_{11} and S_{22} are the reflection coefficients of port 1 and port 2 in magnitude [26].

From the measured Q_L, Q_u can be calculated using the Equation (2.11).

Sometimes the desired mode (the $TE_{01\delta}$ one) may be close to other modes. In such cases the cavity volume can be slightly changed by rotating the top screw which moves the top plate up or down which separates the modes. The ability to tune the frequency is very useful for the identification of the desired resonant mode and to allow it to measure samples of various dimensions. Figure 2.9 shows a typical test fixture manufactured by QWED. Rigorous electromagnetic analysis must be performed to evaluate permittivity and the dielectric loss tangent of the sample under test. Rayleigh–Ritz method has been used in computer program for a typical test fixture manufactured by QWED.

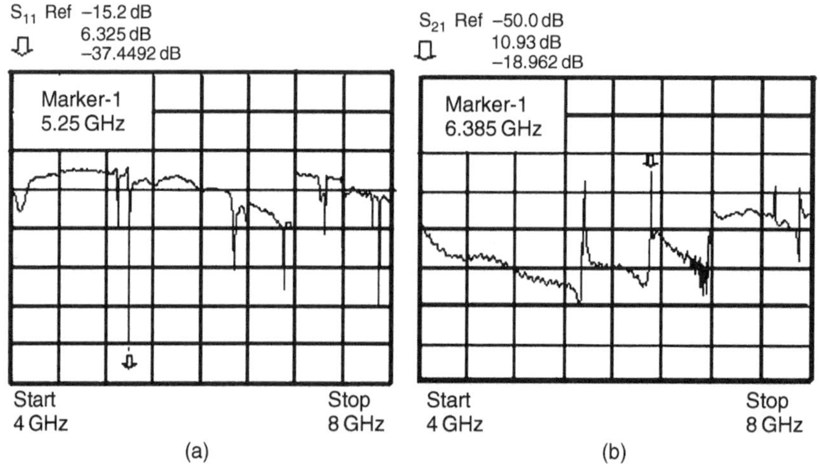

Figure 2.8 Microwave resonance spectra of $Ba(Mg_{1/3}Ta_{2/3})O_3$ ceramic with $\varepsilon_r = 24$ (a) reflection (b) transmission configuration.

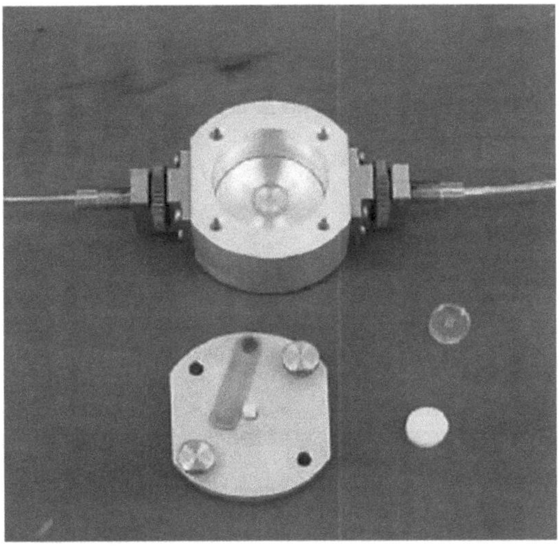

Figure 2.9 The Cavity manufactured by QWED for quality factor measurement (Courtesy, J Krupka QWED, Warsaw, Poland).

The $TE_{01\delta}$ mode method is one of the most accurate techniques for measuring loss tangent and permittivity of isotropic low loss materials [25, 27], although manufacturers of dielectric materials use it in simplified form of this technique only for measurements of the dielectric loss tangent. One can notice that neglecting all parasitic losses (that are small for $TE_{01\delta}$ mode if permittivity of the sample is large) and assuming that electric energy filling factor is equal to unity, the inverse of measured unloaded Q-factor is approximately equal to the dielectric loss tangent. Such assumptions are not valid for very low loss dielectric (in this case conductor losses must be rigorously taken into account) and for low permittivity materials (electric energy filling factor in the sample is substantially smaller than one). The cavity method using the $TE_{01\delta}$ mode has several advantages such as easy mode identification, small parasitic losses and lack of mode degeneracy [5]. However, the evaluation of ε_r and tan δ requires advanced numerical computations (this can be only done employing dedicated computer programs) because of the absence of exact solutions of Maxwell's equation. For this reason, Hakki and Coleman method is still used as it allows relatively easy determination of permittivity. The uncertainty in dielectric loss tangent using $TE_{01\delta}$ mode cavity method with optimized enclosure is of the order of $\pm 2 \times 10^{-6}$ or 0.03 tan δ (whichever is larger) and uncertainty of ε_r measurements is $\Delta\varepsilon/\varepsilon = \pm\Delta$ dim/dim (where dim = dimensions of the test sample). The measurement frequency depends on the size and permittivity of the test samples. Measurements at higher frequencies are possible by using smaller cavities and smaller test samples or by using several higher order quasi TE_{0nm} modes [28].

Valant et al. [29] reported that the presence of resonator support and the coupling loop perturb the electromagnetic field and this may lower the measured Q value and shift the resonant frequency. Hence in order to get the unloaded Q value, the test cavity should be large in size. However, on increasing the size of the test cavity, the resonant modes of the cavity move to lower frequencies. When ε_r of the DR is larger than about 20, the position of the $TE_{01\delta}$ mode of the DR is below the first test cavity mode

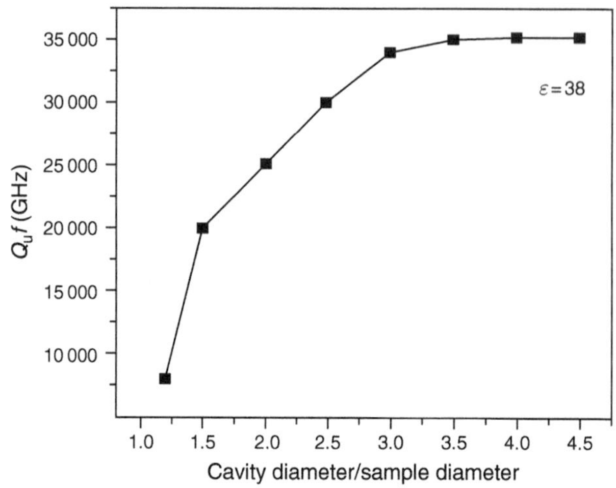

Figure 2.10 Variation of *Qf* with ratio of cavity diameter/sample diameter.

(TM_{010}). Valant et al. [29] made a detailed study of the influence of test cavity dimensions on the microwave dielectric properties of the ceramic puck. The electromagnetic field could penetrate into the conducting walls of the test cavity (skin effect). This lowers the measured Q value of the DR. Hence the test cavity size should be large enough to avoid the skin effect. Usually the dimensions of the test cavities are such that the $TE_{01\delta}$ mode of the DR is the lowest resonance and hence it can be easily identified. The quality factor decreases when the cavity diameter/puck diameter ratio is smaller than 3 as shown in Figure 2.10. The cavity normally used is 3–5 times the size of the test sample. The surface resistance of copper can be calculated from the quality factor of the TE_{011} resonance of the empty cavity to apply correction to the measured Q of the sample for the loss due to cavity walls [1].

One can measure the ε_r and tan δ at low frequencies by the parallel plate capacitor method using an LCR meter for new materials. This will give an approximate idea of ε_r and tan δ. This will in turn help to calculate the approximate resonant frequency of the DR using the following equation [30]

$$f \alpha \frac{1}{\nu_r \sqrt{\varepsilon_r}} \tag{2.22}$$

where ν_r is the volume of the resonating body.

The knowledge of the value of the resonant frequency further helps to know the size of the DR at a given frequency or the size of the cavity required to measure the Q-factor.

2.3.3 Measurement of quality factor by stripline excited by cavity method

In the microstripline excited cavity method, the DR is magnetically coupled to a 50 Ω microstripline as shown in Figure 2.11 along with the equivalent circuit [31]. The ratio

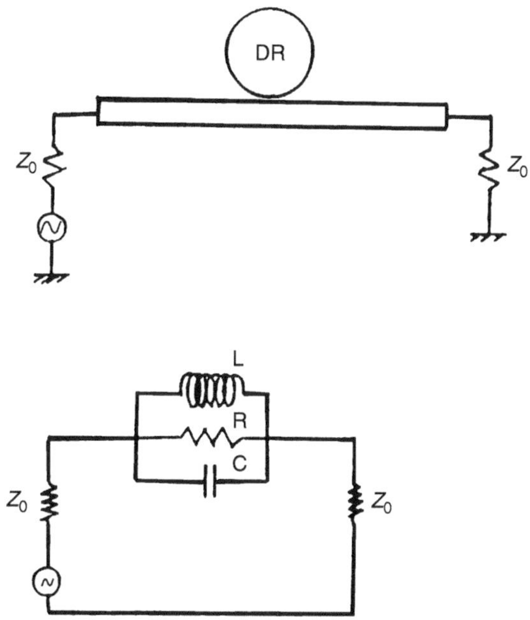

Figure 2.11 Schematic diagram of a DR coupled to a (a) microstripline and (b) equivalent circuit (after Ref. [31]).

of the resonator-coupled resistance R at the resonant frequency to the resistance external to the resonator is called the coupling factor β_c.

$$\beta_c = \frac{R}{R_{ext}} = \frac{S_{110}}{S_{210}} \tag{2.23}$$

where S_{110} and S_{210} are the real quantities representing the reflection and transmission coefficients respectively at the resonant frequency.

Under critical coupling ($\beta_c = 1$), the power dissipated in the external circuit is equal to the power dissipated in the resonator (P_d), which is equally divided into the power reflected to the generator ($P_r = S_{110}{}^2$) and the power transmitted to the load ($P_t = S_{210}{}^2$). In the shielded resonator configuration, from the conservation of energy, power dissipated in the resonator being given by

$$P_d = 1 - |S_{110}|^2 - |S_{210}|^2 \tag{2.24}$$

The coupling factor β_c is a function of the distance between the DR and the microstripline under fixed shielding conditions.

The expression for the unloaded voltage transmission coefficient S_{21u} derived by Khanna and Garault is given by [31]

$$S_{21u} = S_{210}\sqrt{\frac{2}{(1 + S_{210}{}^2)}} \tag{2.25}$$

S_{21u} corresponds to the voltage transmission coefficient of the unloaded resonator.

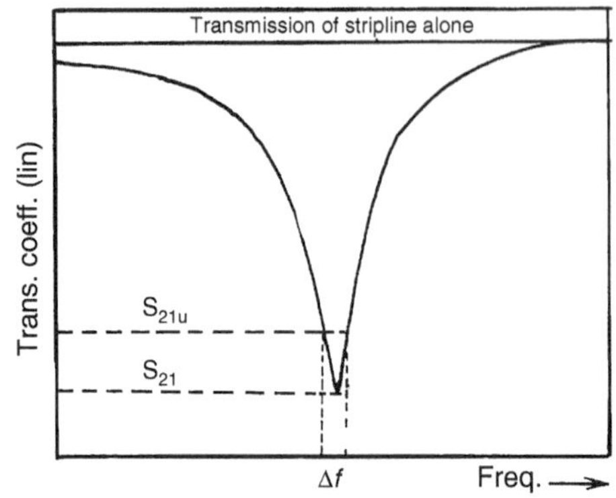

Figure 2.12 Typical resonant curve of a DR coupled to a microstripline used in determining the quality factor by the stripline method.

The frequencies f_1 and f_2 corresponding to S_{21u} given by Equation 2.25 is noted from the network analyzer (Figure 2.12). The difference in frequencies (f_2–f_1) is Δf. The frequency that corresponds to the peak of the S_{21} curve is resonant frequency f. Knowing the resonant frequency f and Δf, the unloaded quality factor, Q_u, is calculated using the Equation 2.6.

Figure 2.13 shows the experimental setup for the Q measurement by the Microstripline excited cavity method. A 50 Ω Microstripline of width 3 mm is etched on RT-Duroid 5880 ($\varepsilon_r \sim 2.2$ and thickness 1.9 mm) and kept at the bottom wall of rectangular cavity made of copper. The cavity is excited using 3.5 mm Microstrip edge connectors as shown in Figure 2.13. The DR is placed near the Microstripline. Resonance spectra are displayed in the network analyzer screen. Among them, the $TE_{01\delta}$ mode is identified. Set the central frequency as the resonant frequency and reduce the span to enhance the frequency resolution. Then network analyzer is calibrated for S_{21} THRU by connecting an identical microstrip transmission line used in the cavity. Then connect it to the cavity with DR and measure the resonant frequency f. The transmission coefficient S_{210} corresponding to f is

Figure 2.13 The experimental set-up for measuring quality factor by the stripline method. The DR is coupled to the stripline.

taken. The factor S_{21u} is calculated using Equation 2.25. From the width Δf corresponding S_{21u} and f, the unloaded quality factor is calculated.

2.3.4 Whispering gallery mode resonators

Dielectric resonators are normally operated using $TE_{01\delta}$, $TM_{01\delta}$ or $HE_{11\delta}$ modes measured by the end shorted Courtney, $TE_{01\delta}$ or stripline methods [11, 12, 23, 27]. However, the measured Q of these modes depends not only on the material loss tangent but also on the radiation and conductor losses of the cavity. Hence simple measurement of quality factor by Courtney method, $TE_{01\delta}$ or Microstripline methods is not sufficient to determine accurately the dielectric loss of low loss dielectric materials. Recently it has been established [32–39] that Whispering Gallery modes (WGMs) would confine the entire fields within the resonator which in turn give negligible radiation and conductor losses at microwave frequencies. The quality factor of WGM DR is limited only by the intrinsic losses in the dielectric material. Since the radiation and conduction losses are negligible, the measured WGM Q-factor is approximately equal to $1/\tan\delta$. In WGM resonators, most of the electromagnetic energy is confined to the dielectric near the perimeter of the air–dielectric interface which in turn reduces the radiation and conductor losses [34, 36]. One additional advantage of using WGM technique is that they allow measurements of two permittivity components of uniaxially anisotropic materials (several single crystals exhibit uniaxial anisotropy). Permittivity components can be determined from measurements of resonant frequencies and Q-factors of two modes belonging to different mode families employing rigorous numerical analysis, e.g. mode-matching. The electrical energy filling factors for E (quasi TM mode) and H (quasi TE mode) modes are given by [34, 37, 38]

$$P_{\varepsilon\perp} = 2\frac{\partial f}{\partial \varepsilon_\perp}\frac{\varepsilon_\perp}{f} \tag{2.26}$$

$$P_{\varepsilon ll} = 2\frac{\partial f}{\partial \varepsilon_{ll}}\frac{\varepsilon_{ll}}{f} \tag{2.27}$$

The dielectric loss tangent for the dielectric can be solved [37, 38] using the equation

$$Q_{(E)}^{-1} = \tan\delta(P_{\varepsilon_\perp} + P_{\varepsilon_{\text{ll}}}) + R_s/G_{(E)} \tag{2.28}$$

$$Q_{(H)}^{-1} = \tan\delta(P_{\varepsilon_\perp} + P_{\varepsilon_{\text{ll}}}) + R_s/G_{(H)} \tag{2.29}$$

where R_s is the surface resistance of the cavity enclosing the DR, ε_{ll} is the permittivity parallel to the anisotropic axis and $\varepsilon_\perp$ is the one perpendicular to it. The conductor losses decrease as the surface resistance becomes smaller and as the geometric factor (G) increases [38]. For WGMs the geometric factor G is significantly large that the effect of cavity can be ignored when compared to the loss tangent. The energy filling factors of DR for all these modes are close to unity. The modes which have high electric energy filling factor (WGM modes) will have the highest quality factors. The dimensions are relatively large even in the millimeter wavelength band. In effect, acting on these modes, radiation losses are negligible. This is an important feature of WGMs behind conventional TE or TM modes, the unloaded quality factor of which depends not only on the material loss tangent but also on the metallic shieldings in which they are enclosed.

The WGM method offers good suppression of spurious modes because the propagation constant along the z axis is very small and unwanted modes leak out axially. They offer a high level of integration. The WGM DRs are classified as either $WGE_{n,m,l}$, in which the electric field is essentially transverse, or $WGH_{n,m,l}$, for which the electric field is essentially axial. The integer n denotes the azimuthal variation, m radial variation and l, the axial ones. WGMs are periodic according to the azimuthal number, and the number of modes in a BW increases with the diameter of the DR. So for small diameter of the resonator, the frequency interval between two successive modes will be large. Dielectric resonators acting on their WGMs can be excited in different ways. In the low frequency range, one can use an electric or magnetic dipole. However, this type of excitation is stationary and travelling WGM cannot be excited. In the mm wavelength frequency region either dielectric image waveguides or microstrip transmission lines are used to excite travelling WGMs. The WGM method is commonly used to estimate the Q-factor of sapphire single crystal resonators.

2.3.5 Split post dielectric resonator

The split post dielectric resonator (SPDR) provides an accurate method for measuring the complex permittivity and loss tangent of substrates and thin films at a single frequency point in the frequency range of 1–20 GHz. In the SPDR method [40–44], the sample should be in the form of a flat rectangular piece or a sheet. The SPDR uses a particular resonant mode which has a specific resonant frequency depending on the resonator dimensions and the permittivity. This method does not have flexibility in the measurement frequency and dimensions as the samples need to be prepared in the form of thin sheets. In this method, flat samples of the test material are inserted through one of the open sides of the fixture. The laminar dielectric under test is placed between two low loss dielectric rods or resonators kept in a metallic enclosure as shown in Figure 2.14. The electric field in the resonator sample is parallel to the surface of the sample. Hence the test sample should have strictly parallel faces and the thickness of the sample should be less than the fixture air gap and the sample should have enough area to cover inside of the fixture. The air gap between the sample and the DR does not affect the accuracy of the measurement. The required thickness of the sample also depends on the ε_r of the material. Materials with high ε_r must have less thickness. Figure 2.14 schematically shows the SPDR. A pair of DRs and a metal enclosure of relatively small height are used in the construction of the SPDR fixture. This allows formation of an evanescent

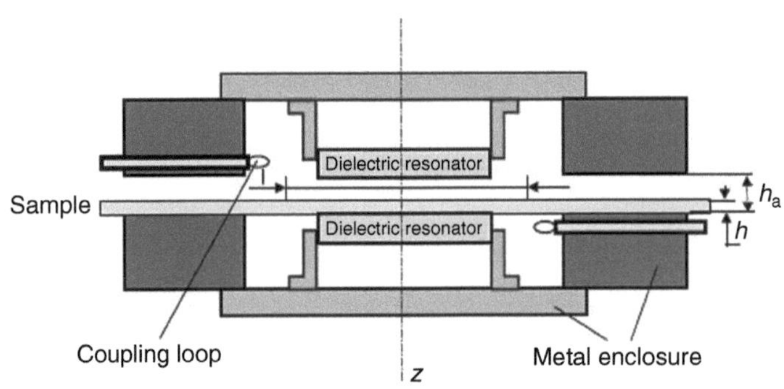

Figure 2.14 Schematic sketch of SPDR.

electromagnetic field, not only in the air gap between the DRs, but also in the cavity region for radii greater than the radius of the DR. This simplifies the numerical analysis and reduces possible radiation effects. Although different modes of the resonator can be identified and used for the microwave characterization, $TE_{01\delta}$ mode is preferable since this mode is insensitive to the presence of air gaps perpendicular to z-axis of the fixture. The thickness of the sample needs to be measured and is provided as a parameter to the software.

The complex permittivity can be calculated based on the rigorous electromagnetic modeling of the split post resonant structure using the Rayleigh–Ritz technique [41]. The real part of the complex permittivity can be computed from the measured resonant frequencies and thickness of the sample as an iterative solution of the following equation [41]

$$\varepsilon_r' = 1 + \frac{f_0 - f_s}{hf_0 K_\varepsilon(\varepsilon_r', h)} \tag{2.30}$$

where h is the thickness of the test sample, f_0 is the resonant frequency of the empty SPDR, f_s is the resonant frequency of the SPDR with the dielectric sample. K_ε is a function of ε'_r and has to be evaluated for a number of ε_r and using Rayleigh–Ritz technique [41]. Iterative procedure is used to evaluate subsequent values of ε_r and ε_r'' from Equation 2.30. The loss tangent of the test sample is calculated from the measured unloaded Q-factors of the SPDR with and without the dielectric sample based on

$$\tan\delta = \left(\frac{1}{Q_u} - \frac{1}{Q_{DR}} - \frac{1}{Q_c}\right) / P_e \tag{2.31}$$

where Q_c^{-1} and Q_{DR}^{-1} denote losses of the metallic and dielectric parts of the resonator respectively and P_e is the electric energy filling factor of the sample given by Equation (2.8).

Uncertainty of the permittivity measurements of a sample of thickness h can be estimated as $\Delta\varepsilon/\varepsilon = \pm(0.0015 + \Delta h/h)$ and uncertainty in loss tangent measurements $\Delta \tan\delta = 2 \times 10^{-5}$ or $\pm 0.03 \tan\delta$. In order to determine the complex permittivity of a sample, the resonant frequencies and Q-factors of the empty SPDR and the SPDR containing the test sample need to be measured. The SPDR uses a particular mode. This mode has a particular resonant frequency, depending on resonator dimensions and to a limited extent on the electrical properties of the test sample. Thus each SPDR is designed for a particular nominal frequency and the actual measurement is taken close to the nominal frequency. The nominal frequency determines the requirements for the size of the sample. For example, for SPDR of nominal frequency 5–6 GHz, the minimum sample size should be 30 × 30 mm and thickness 2.1 mm. QWED manufactures SPDRs with dedicated software for the evaluation of permittivity and loss tangents. Figure 2.15 shows the general view of SPDRs manufactured by QWED. SPDR has superior accuracy as compared to the reflection–transmission methods. The method is convenient, fast to measure low loss laminar dielectrics such as substrates or LTCC, printed circuit boards and even thin films and not suitable for DRs.

2.3.6 Cavity perturbation method

In the past, cavity perturbation technique was the only method available to obtain approximate solutions but its applications were limited not only to low permittivity samples but also to specific modes and specific samples. The cavity perturbation technique is widely used for the determination of the dielectric characteristics of thin

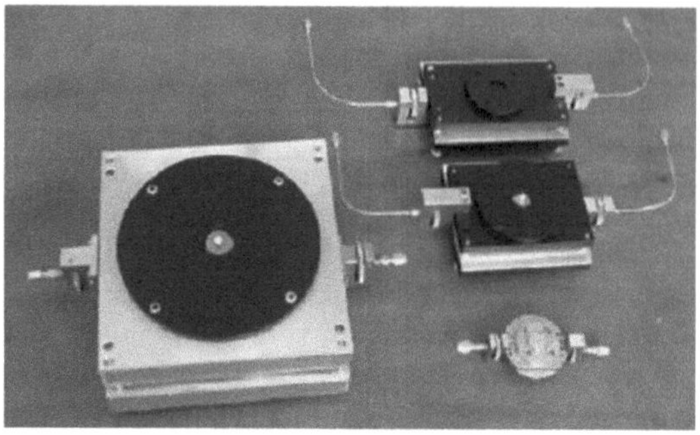

Figure 2.15 Photographs of SPDRs produced by QWED, Poland. (Courtesy J. Krupka, QWED, Warsaw Poland).

sheet samples of low and medium dielectric loss [10, 45]. In the cavity perturbation technique, a small piece of the material usually in the form of a disk or sheet is placed in a metallic resonant cavity operating in a known mode. The material characteristics are estimated from the shift in the resonant frequency and change in the Q of the system [46–49]. This technique was pioneered by Slater [49] and is a suitable method for measuring the dielectric properties of materials with permittivity less than 10. The cavity perturbation method is not a swept frequency measurement since the measurement frequencies are determined by the cavity as well as the dimensions of the sample. Hence it can be used only for discrete frequency measurements. In this method a rectangular waveguide (WG) with a small slot at the broader wall at the middle is used. The cavity is excited with optimum iris coupling, typically the diameter of the iris is equal to the shorter dimension of the waveguide (WG)/2.2, can be used for the measurement of dielectric properties of the samples. The resonant frequency and quality factor of the empty cavity is determined for different cavity modes. Then the thin sheet sample is inserted and positioned at the E-field antinode. If the sample is purely dielectric, the maximum electric field can be easily determined by simply moving the sample across the slit. The mode will shift to low frequency side and retraces from there. The sample is kept at the retracing position, this is the electric field maximum position. If it is slightly magnetic, the permittivity can be measured only for the odd modes by keeping it at the middle of the cavity. The new resonant frequency and Q of the sample is again measured. The complex permittivity of the sample is calculated [10, 45, 50] using the Equations (2.32–2.34).

$$\varepsilon_r' = 1 + \left[\frac{V_c(f_0 - f_s)}{2V_s f_s}\right] \tag{2.32}$$

$$\varepsilon_r'' = \frac{(\varepsilon' - 1)}{2\varepsilon'}\frac{f_s}{(f_0 - f_s)}\left[\frac{1}{Q_s} - \frac{1}{Q_0}\right] \tag{2.33}$$

$$\tan\delta = \frac{\varepsilon_r''}{\varepsilon_r'} \tag{2.34}$$

where f_0 is the resonant frequency of the empty cavity, f_s is the resonant frequency of the cavity with sample, V_c is the volume of the cavity and V_s is the volume of the sample,

Q_0 is the quality factor of the empty cavity and Q_s is the quality factor of the cavity with sample. The experimental error was found to be less than 2% in case of permittivity and 1.3% in the case of dielectric loss. Here also the measured Q_s and Q_0 can be corrected by measuring S_{11} and S_{22} as mentioned earlier by proper calibration of the network analyzer. It is better to use Waveguide TRL calibration for better accuracy at the ends of the waveguide to coaxial adaptor. Calibrate the network analyzer for full two port using TRL calibration. Now S_{11} and S_{22} are measured at the resonant frequency. As mentioned earlier, calculate the coupling coefficients for the two ports and find the unloaded quality factors Q_u and Q_s. The main advantage of this method is the easiness of determining the permittivity and loss using simple device with moderate accuracy.

2.3.7 TM_{ono} mode and re-entrant cavity method

The microwave dielectric properties can also be measured in the frequency range 2–10 GHz using the TM_{0n0} mode cavities with rod dielectric samples [51–53]. In the frequency range 50 MHz–2 GHz a re-entrant cavity method can be employed [51, 54, 55] to evaluate the dielectric properties. Both TM_{010} and re-entrant cavities are closed with a lid after insertion of the samples. For TM_{010} mode cavity, transcendental equation is known only if the height of the sample is equal to the height of the cavity. For re-entrant cavity, the exact solution for transcendental equation in a closed form does not exist. It can be solved using rigorous mode-matching methods [55]. Karpov [56] was the first who presented mode-matching numerical technique to solve Maxwell's equation for re-entrant cavity. Since the electric energy filling factor in re-entrant cavity is close to unity, resolution in the loss tangent measurement is of the order of 5×10^{-5}. A similar resolution in loss tangent is possible for the TM_{010} mode cavities, provided the test sample has a sufficiently large diameter. The uncertainty in real permittivity measurement is about 0.5–2% for TM_{010} mode cavity and about 1–3% for re-entrant cavity. The loss tangent and permittivity can be measured as a function of frequency by employing higher order TM_{0n0} modes.

2.3.8 TE_{01n} mode cavities

The TE_{01n} mode method is employed to measure the complex permittivity and Q-factor of low loss disc samples [57–59] as shown in Figure 2.16. The operating frequency range of TE_{01n} mode cavities is in the range of 8—40 GHz. The TE_{01n} mode cavities have a circumferential electric field distribution which is tangential to a cylindrical sample kept symmetrically in the cavity [5]. Hence the electric field is continuous across dielectric air interfaces which allow air gap omission without degradation of measurement uncertainties. The surface currents in the metal cavity walls are circumferential so that physical contact between the lateral surface and the cavity bottoms is not important. Figures 2.17 and 2.18 show the variation of resonant frequency and Q-factor due to dielectric losses versus imaginary part of permittivity for TE_{011} mode cylindrical cavity containing the dielectric sample. In the low loss dielectric region, the resonant frequency is smaller than that for the empty cavity (Figure 2.17) and resonant frequency shift (f_0–f) depends on the real permittivity and thickness of the sample. The quality factor due to dielectric loss in this region given by Equation (2.7) and depends linearly on the dielectric loss tangent.

In the low dielectric loss region, real part of the complex permittivity can be determined from the measured resonant frequency and the simplified transcendental

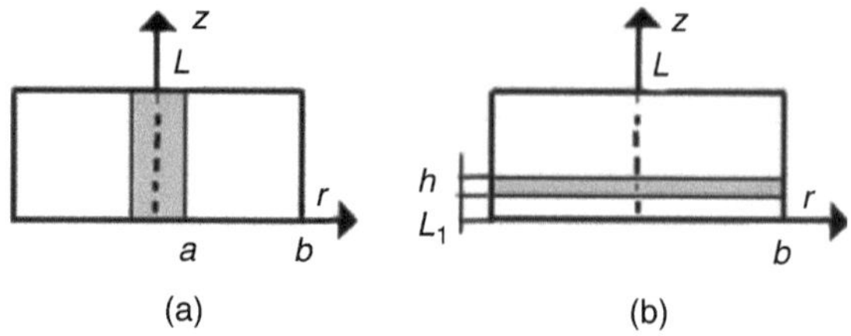

Figure 2.16 Cylindrical cavities containing (a) dielectric rod (b) dielectric disc.

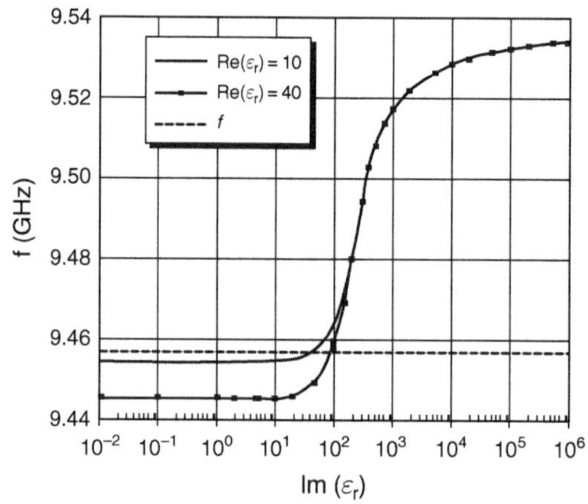

Figure 2.17 Variation of resonant frequency with imaginary part of permittivity for TE_{011} mode cylindrical cavity with dielectric disc sample. Dotted line corresponds to the TE_{011} mode frequency of the empty cavity (after Ref. [5]).

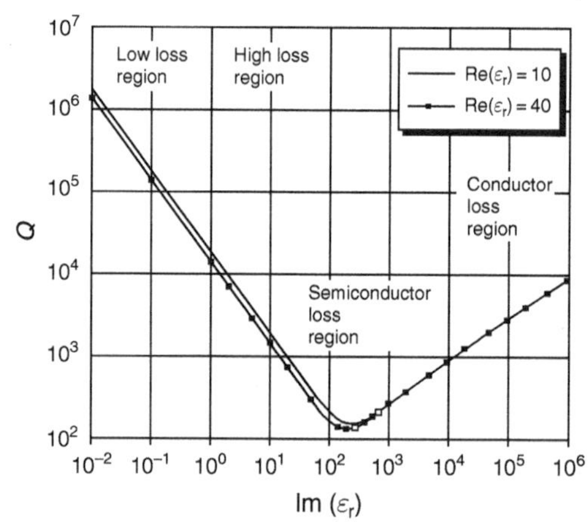

Figure 2.18 Variation of Q-factor due to dielectric losses as a function of imaginary part of permittivity for TE_{011} mode cylindrical cavity containing dielectric disc sample. (after Ref. [5]).

equation where both complex permittivity and complex angular frequency have imaginary parts equal to zero. Evaluation of the dielectric loss tangent requires determination of all parasitic losses and the electric energy filling factor in the sample. The surface resistance in a closed cavity can be determined from the measured Q-factor of the empty cavity and then scaling up to the frequency of the cavity containing the sample using the formula

$$R_s(\omega) = R_s(\omega_0)\sqrt{\frac{\omega}{\omega_0}} \tag{2.35}$$

where ω_0 is the resonant angular frequency of the empty cavity. Having known the surface resistance value and electromagnetic field distribution in the presence of sample under test, Q-factor due to conductor losses can be evaluated and accounted for determination of Q_d from measured value of Q_u.

If imaginary part of permittivity becomes very large (see Figure 2.18) then both the resonant frequency and the Q_d predominantly depend on imaginary part of the permittivity. In such cases only imaginary part of permittivity can be determined. When imaginary part of permittivity becomes small and the sample is thin (and therefore electric energy filling factor is small), then the Q-factor due to dielectric loss becomes very large and accuracy of measurements of dielectric loss tangent suffers. Measurements of low loss materials using the TE_{01n} mode cavities can be performed using thicker samples (thus enlarging electric energy filling factor) or by keeping the sample to the position where the electric field approaches the maximum (using low loss dielectric support). The ideal thickness of the sample is half wavelength or its multiple [5]. In both the above cases, the electric energy filling factors in the dielectric sample becomes relatively large and a higher resolution loss tangent measurement (5×10^{-5}) can be achieved. The uncertainty in the permittivity measured using TE_{01n} mode cavities is of the order of 0.5% [5]. At lower frequencies, <8 GHz, the dimensions of the TE_{01n} mode cavities and samples become too large for practical applications.

2.4 Estimation of Dielectric Loss by Spectroscopic Methods

The intrinsic losses, which are the lowest possible losses in a given compound, can be estimated by the combination of the far infrared and submillimeter (or THz) spectroscopy. This is possible because the intrinsic microwave losses which are fully determined by the multiphonon absorption of the ideal (but necessarily anharmonic) crystal and the polar lattice vibration (optic phonons) can be directly measured by means of infrared spectroscopy [6].

It was reported [60–62] already in the 1960s that far infrared reflection spectra $R(\nu) = |r(\nu).\exp(i\theta(\nu))|^2$ could be analyzed by the Kramers–Kronig relation and classical dispersion theory.

The K–K relationship between $r = \sqrt{R}$ and the phase of reflected wave Θ is given by [60]

$$\Theta(\nu) = \frac{2\nu}{\pi}\int_0^\infty \frac{\ln r(\nu')}{\nu'^2 - \nu^2}\,d\nu' \tag{2.36}$$

The K–K integration requires data from zero to infinite frequency. The accuracy of the analysis is affected by the extrapolation of experimental data taken in finite frequency range to zero and to infinite frequency. The combination of K–K and classical dispersion theory proves [63] a better and more reliable analysis.

From the reflected intensity R and the phase angle Θ, the optical parameters n and k can be determined using the following relationships [63]

$$n = \frac{1 - R}{1 + R - 2\sqrt{R}\cos\Theta} \tag{2.37}$$

$$k = \frac{2\sqrt{2R}\sin\Theta}{1 + R - 2\sqrt{2}R\cos\Theta} \tag{2.38}$$

$$\varepsilon' = n^2 - k^2 \tag{2.39}$$

$$\varepsilon'' = 2nk \tag{2.40}$$

where n is the refractive index,

ε' and ε'' are the real and imaginary parts of the permittivity and

k is the extinction coefficient.

The analysis involves four steps: (a) The K–K analysis allows to calculate phase angle $\Theta(\nu)$ spectrum from the experimental reflectivity spectrum $R(\nu)$. (b) Real and imaginary parts of complex index of refraction are calculated using Equations (2.37) and (2.38). (c) The real and imaginary parts of complex permittivity are calculated using Equations (2.39) and (2.40). (d) Finally the complex permittivity is fitted by means of a suitable model.

According to classical dispersion theory both the components of complex permittivity spectra $\varepsilon = \varepsilon' + i\varepsilon''$ can be expressed as a sum of quasi harmonic damped oscillators and ε' and ε'' are then given by [60]

$$\varepsilon'(\nu) = \varepsilon_\infty + \sum_j 4\pi\rho_j\nu_j^2 \frac{\nu_j^2 - \nu^2}{\left(\nu_j^2 - \nu^2\right)^2 + \gamma_j^2\nu^2} \tag{2.41}$$

$$\varepsilon''(\nu) = \sum_j 4\pi\rho_j\nu_j^2 \frac{\gamma_j\nu}{\left(\nu_j^2 - \nu^2\right)^2 + \gamma_j^2\nu^2} \tag{2.42}$$

The summation is over j resonances (i.e., oscillators describing phonons) in the spectrum. The dispersion parameters are

$4\pi\rho_j$ = strength of the oscillator ($= \Delta\varepsilon_j\nu_j^2$, where $\Delta\varepsilon_j$ marks the contribution of the jth mode to the static permittivity)

γ_j = width (phonon damping)

ν_j = resonant frequency (phonon eigen frequency).

It can be shown that at frequencies much lower than the phonon eigen frequencies ($\nu << \nu_j$, i.e., in microwave region) equations 2.41 and 2.42 can be written in the form

$$\varepsilon' = \varepsilon_\infty + 4\pi\rho_j \text{ for } \nu << \nu_j \tag{2.43}$$

$$\varepsilon''(\nu) \propto \nu \sum_j \frac{4\pi\rho_j\gamma_j}{\nu_j^2} \tag{2.44}$$

It means that the permittivity is independent of frequency and dielectric loss ε'' is proportional to frequency much below the phonon frequencies ν_j which lies in THz range.

ε_∞ is the permittivity caused by electronic polarization at higher frequencies. The ε_∞ can be obtained using the frequency-independent middle-infrared reflectivity by the equation [64]

$$\varepsilon_\infty = \frac{(1+\sqrt{R})^2}{(1-\sqrt{R})^2} \tag{2.45}$$

The normal incidence of reflectivity is given by Fresnels formula

$$R = \frac{(n-1)^2 + k^2}{(n+1)^2 + k^2} \tag{2.46}$$

The phonon parameters can be obtained by K–K analysis. Each peak in the plot of ε'' versus frequency obtained by the K–K analysis corresponds to a phonon frequency ν_j. γ_j is given by the frequency half-width of the ε_j'' peak. The oscillator strength is determined by

$$4\pi\rho_j = \frac{\gamma_j \varepsilon_j''}{\nu_j} \tag{2.47}$$

Complex permittivity can be calculated also without K–K analysis, just by direct fitting of the experimental reflectivity spectra using the formula in Equations 2.41 and 2.42 and

$$R = \left| \frac{(\sqrt{\varepsilon}-1)}{(\sqrt{\varepsilon}+1)} \right|^2 \tag{2.48}$$

Simulation process is done in the following way

(a) Fix ν_j and γ_j and adjust ρ_j to give the correct width of the reflection band.
(b) Fix ν_j and ρ_j and adjust γ_j to give the correct maximum reflectivity.
(c) ν_j is adjusted to align the fitted curve with the experimental.

Once the best fit is obtained, tan δ can be determined.

$$\tan\delta = \frac{\varepsilon''}{\varepsilon'} = \frac{\sum_j 4\pi\rho_j\nu_j^2 \frac{\gamma_j\nu}{(\nu_j^2-\nu^2)^2+\gamma_j^2\nu^2}}{\varepsilon_\infty + \sum_j 4\pi\rho_j\nu_j^2 \frac{\nu_j^2-\nu^2}{(\nu_j^2-\nu^2)^2+\gamma_j^2\nu^2}} \tag{2.49}$$

At low frequencies $\nu << \nu_j$ [63], we obtain

$$\tan\delta = \frac{\sum_j 4\pi\rho_j \frac{\gamma_j\nu}{\nu_j^2}}{\varepsilon_\infty + \sum_j 4\pi\rho_j} \tag{2.50}$$

Equation (2.50) shows that the tan δ is linearly frequency dependent.

The far infrared spectroscopy (FIR) and terahertz transmission spectroscopy [65, 66] are useful methods to estimate the intrinsic losses coming from multiphonon absorption. As early as 1962,Ruprecht and Bell reported experimentally that ε'' is proportional to the frequency well below the phonon frequencies ($f \leq 10^{12}$ Hz) [67]. It was also reported in 1962 [67, 68] that lattice absorption is the dominant loss mechanism at high frequencies. Hence several authors extrapolated the data at high frequencies obtained by FIR and

SMM to the microwave frequency range [66, 69]. Several authors [63, 64, 70–75] analyzed the FIR reflectivity data using K–K analysis and classical dispersion theory and estimated the permittivity and dielectric loss tangent. There is reasonable agreement between the ε_r obtained by the spectroscopic method and that measured by microwave methods. However, the loss tangent obtained by far infrared method was in some cases much smaller than that measured by microwave methods. In the measurement by microwave methods, the presence of porosity and crystal defects contribute to the loss factor. The microwave methods give the sum of intrinsic and extrinsic dielectric losses. The spectroscopic methods give the intrinsic loss factor. Thus it is possible by far infrared and SMM method to find the lowest loss factor in a particular material which in turn is helpful to optimize the ceramic preparation to get a low loss material and to reduce extrinsic losses. The relative influence of defects on losses decreases with increasing frequency [72]. Hence their influence cannot be measured in the optical frequency range. This enables one to estimate the order of magnitude of intrinsic microwave losses from IR reflectivity on new materials without much processing. Unlike the microwave losses, the IR reflectivity of dense ceramics is less sensitive to sample preparation (i.e., extrinsic losses contribute only slightly to IR spectra), so that with new materials it can be successfully and simply used for estimating the intrinsic microwave losses and dielectric constant [66, 70–72]. Nevertheless, it should be stressed that the accuracy of far-IR (FIR) experiment is rather limited. Therefore it is necessary to combine FIR reflectivity studies with SMM or THz measurements, which are more sensitive than FIR reflectivity on small dielectric losses below phonon frequencies.

The fundamental loss mainly corresponds to the absorption of energy quantum of the electromagnetic field $\hbar\omega$ ($\omega = 2\pi\nu$ is the ac field frequency) in collision process with thermal phonons, which have much higher energies [6, 76]. This large difference in energies makes it difficult to satisfy the conservation laws. Under such a situation, the absorption of $\hbar\omega$ of energy corresponds to three fundamental loss mechanisms, i.e. three quantum, four quantum and quasi–Debye mechanisms. The three quantum process involves an $\hbar\omega$ quantum and two phonons [6, 76–78]. The four quantum mechanism corresponds to the field quantum absorption processes involving three phonons. The application of a dc field to a centrosymmetric crystal breaks its central symmetry and therefore gives rise to quasi-Debye mechanism [6, 76, 77]. In centrosymmetric materials, the crystalline symmetry permits only three quantum and four quantum mechanisms. But in non-centrosymmetric materials all the three mechanisms are allowed and the quasi-Debye mechanism is dominant for the intrinsic loss. As the ε_r increases, the dielectric loss also increases. For intrinsic loss mechanism in cubic centrosymmetric crystals, the real ε' and imaginary ε'' are related by $\varepsilon'' \propto \varepsilon^x$, where $x = 2.5$–5 [77]. The role of intrinsic losses at microwave frequencies increases with increasing ε_r.

Petzelt and co-workers [66, 69–72, 75, 79, 80] at Prague made measurements on many microwave ceramic materials in the IR and SMM (submillimeter) range using Fourier transform infrared (FTIR) and Backward Wave Oscillator (BWO) and/or time-resolved THz spectroscopy and compared the extrapolated losses from THz to MW range (using $\varepsilon'' \propto \omega$) with those measured directly by microwave methods. It was found in many cases that the extrapolated loss factors were much lower than those measured by microwave methods. This apparently seems that the proportionality $\varepsilon'' \propto \omega$ is not satisfied. The microwave measurements are very sensitive to extrinsic loss depending on the ceramic processing whereas the spectroscopic methods are not (i.e., extrinsic losses contribute only slightly into dielectric losses in THz range). Thus the difference in loss factor between those extrapolated from SMM and the microwave methods is due to extrinsic loss factor.

The model predicts the simple proportionality for intrinsic losses $\varepsilon'' \propto \omega$ (or Qf is a constant) which is usually obeyed in well-processed ceramics. However, the extrapolation of the damped harmonic oscillator models down to the microwave range is questionable, because it assumes frequency-independent phonon damping, which is not necessary [79, 81]. The microscopic phonon transport theory [6, 79, 82] indicates that the simple damped oscillator model with frequency-independent damping is valid only in the vicinity of phonon eigen-frequencies, and the microwave frequencies lie 3–4 orders of magnitude below the phonon frequencies. Nevertheless, FIR was used frequently (together with THz spectroscopy) [72, 74, 80, 82] for estimation of intrinsic microwave dielectric losses, and it was shown that the linear extrapolation from THz to MW range works quite well, although there is no exact detailed theoretical reason for it.

As regards the temperature dependence of the dielectric losses, two phonon difference decay processes dominate at room temperature and near Debye temperature (T_D) in the microwave range far below the eigen frequencies and therefore $\varepsilon'' \propto \omega T^2$. At low temperatures ε'' dependence on temperature is much steeper which differs from the classical damped oscillator model $\varepsilon'' \propto \omega T$. The difference is attributed to the presence of extrinsic sources of losses which are strongly frequency dependent [66, 79].

Finally we should stress that to distinguish between the intrinsic and the extrinsic sources of losses, the $\varepsilon''(T, \omega)$ dependence should be measured in a broad range of temperature down to low temperatures and up to SMM or THz frequencies [80].

2.5 Factors Affecting Dielectric Losses

The tan δ is very sensitive to humidity [12, 83]. Hence the microwave measurements should be done in an air-conditioned room. The samples should be heated in an oven to remove the moisture before starting the experiments. For a low loss material, it should contain lowest possible concentration of dipoles and charge carriers with lowest possible mobility [84]. However, it is a fact that most technically important insulating materials are far from very pure and often contain deliberate or accidental admixtures of substances which are necessary for their processing. The disordered charge distributions in the crystal lattice also contribute to dielectric loss [84, 85]. Dielectric losses occur if the charge distribution in the crystal deviates from perfect periodicity. In 1964, Schlömann [84] reported that in ionic non-conducting crystals, the loss tangent increases when the ions are distributed disorderly in such a way that they break the periodic arrangement of atoms in the crystal. The loss tangent depends strongly on the spatial correlation between charge deviations. He reported that the loss tangent is negligible if the disordered charge distribution in the crystal maintains the charge neutrality within a short range of the order of lattice constant.

The intrinsic quality factor ($Q_i = 1/\tan \delta$) of any given material will vary with the frequency of measurements. For many materials the dielectric loss tangent almost linearly increases as the frequency increases. Hence often the intrinsic quality factor ($1/\tan \delta$) is reported as $Q_i f = f/\tan \delta$ (in GHz) since this value is the first approximation constant. Assumption that the value of $Q_i f$ is constant is satisfied the best for the well-densified ceramics at certain limited frequency range. In practice, samples measured at higher frequencies (5–12 GHz) always give higher $Q_i f$ values than the same material measured at lower frequencies of 1–3 GHz. This difference may be related to ceramic processing. The bigger samples resonating at lower frequencies statistically contain more imperfections than smaller ceramic pucks resonating at larger frequencies. The presence of porosity decreases the quality factor due to the presence of moisture in the pores. Hence porous samples show an increase in $Q_i f$ on warming up due to escape of trapped moisture.

Fundamental theory of intrinsic losses was given by Gurevich and Tagantsev [6, 7, 76]. These intrinsic losses set the lower limit of losses found in pure defect-free single crystals. Several phonon processes contribute to intrinsic losses in a dielectric, their importance depending on the ac field frequency, temperature range and symmetry of the crystal under consideration. The loss mechanisms are different for a crystal with a centre of symmetry and that without centre of symmetry.

Guvevich and Tagantsev [6, 7, 76] analyzed the lattice anharmonicity for various crystal symmetries and obtained numerical estimates of tan δ of ideal crystals.

For hexagonal symmetry and $T << T_D$ (T_D = Debye temperature)

$$\tan\delta = \frac{\eta\omega(kT)^5}{\varepsilon_r \rho v_s^5/\hbar^2 (kT_D)^2} \tag{2.51}$$

and for rhombohedral or cubic symmetry

$$\tan\delta = \frac{\eta\omega^2(kT)^4}{\varepsilon_r \rho v_s^5/\hbar (kT_D)^2} \tag{2.52}$$

η is a dimensionless anharmonicity parameter ranging between 1 and 100, ω is the angular frequency, k = Boltzman constant, T = absolute temperature, v_s = sound velocity, T_D = Debye temperature and ρ is the density.

For the hexagonal symmetry, the tan δ decreases rapidly with decrease of temperature as T^5 and in centrosymmetric crystals as T^4. For hexagonal sapphire, Braginsky et al. verified [8] experimentally the decrease of tan δ as a function of temperature in the range 3.5–300 K and frequency 9 GHz for three different sapphire samples of different defect concentration as shown in Figure 2.19. The curve (b) corresponds to a sample grown at

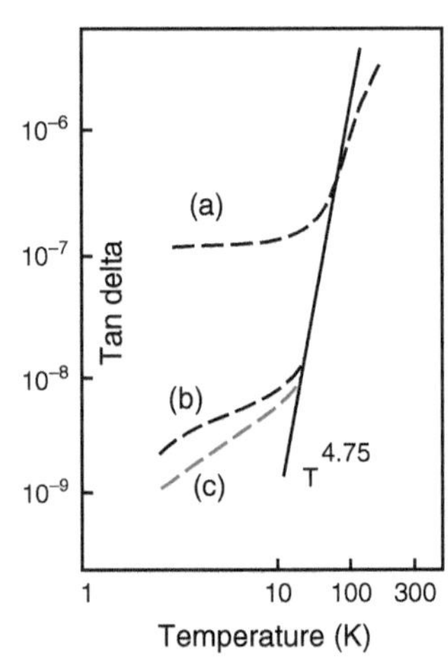

Figure 2.19 Variation of tan δ of three different sapphire crystal as a function of temperature on cooling (a) Crystal with disorientation (b) Crystal grown a the rate of 8 mm/h (c) Crystal grown at the rate of 4 mm/h (after Ref. [8]).

the rate of 8 mm/h and curve (c) to that at the rate of 4 mm/h. The curve (a) is for a sample which contained a block structure with a disorientation of 1°. This sample had the largest tan δ. The tan δ decreased with decreasing temperature. All the three samples showed a straight section in the temperature range 50–300 K, where tan $\delta \sim T^{\alpha}$ ($\alpha = 4.75$). Braginsky reported that this is the lower limit of the dielectric losses at a given frequency, which is independent of the variation of the defect structure and is of fundamental origin. At temperatures $T < 50$ K, the value of tan δ is greatly affected by the level of crystal perfection. Bragisnsky et al. [8] extrapolated from the experimentally obtained tan δ–temperature plot (Figure 2.19) to predict a tan $\delta \sim 5 \times 10^{-15}$ at 4 K at 10 GHz for a defect-free crystal. This corresponds to a Q value of about 2×10^{14} which can be considered as the upper limit of the quality factor in a perfect sapphire crystal.

There is no predictive theory to account for the microwave loss in dielectric ceramics. Hence the approach to find new DR materials is largely done by trial and error method involving preparing and testing a large number of samples. This is a laborious and time-consuming job. The quality factor is highly dependent not only on the intrinsic quality of the ceramic material, but also on the method of measurement, the measurement environment and the frequency at which the sample is measured. A given material sample may exhibit greatly differing Q values when tested in different test fixtures and environments which may vary in size, shape, conductor quality, coupling, type of sample support, ambient temperature and relative humidity.

2.6 Correction for Porosity

The presence of porosity in the sintered ceramic puck influences the ε_r. Hence the measured ε_r should be corrected for porosity to get the actual permittivity of the material. This can be performed using the following equation obtained by Penn et al. [86]

$$\varepsilon' = \varepsilon_m \left(1 - \frac{3P(\varepsilon_m - 1)}{2\varepsilon_m + 1}\right) \tag{2.53}$$

where ε_m is the permittivity of the material corrected for porosity, ε' is the experimentally obtained permittivity and P is the fractional porosity.

2.7 Calculation of Permittivity using Clausius Mossotti Equation

The permittivity ε_r can be calculated theoretically using Clausius–Mossotti equation for cubic or isotropic materials [87]

$$\frac{\varepsilon_r - 1}{\varepsilon_r + 1} = \left(\frac{4\pi}{3}\right)\left(\frac{\alpha_D}{V_m}\right) \tag{2.54}$$

For convenience we can rearrange to get

$$\varepsilon_r = \frac{3V_m + 8\pi\alpha_D}{3V_m - 4\pi\alpha_D} \tag{2.55}$$

where V_m is the molar volume and α_D is the sum of the dielectric polarizabilities of individual ions.

The V_m of the dielectric material can be obtained from X-ray diffraction studies. Shannon reported [88] on the applicability of the concept of additivity of molecular polarizabilities in oxides and fluorides and derived a set of 61 ion polarizabilities to be used to estimate mean permittivity of well-behaving compounds. This rule states that the molecular polarizabilities α_D of a complex material can be broken up into the molecular polarizabilities of simpler compounds by

$$\alpha_D(M_2M'O_4) = 2\alpha_D(MO) + \alpha_D(M'O_2) \tag{2.56}$$

where M are the cations. Furthermore, it is possible to break up the molecular polarizabilities of complex compounds in to ions according to

$$\alpha(M_2M'O_4) = 2\alpha(M^{2+}) + \alpha(M^{4+}) + 4\alpha(O^{2-}) \tag{2.57}$$

The dielectric polarizabilities of several ions are reported by Shannon [88] and are given in Figure 2.20. The calculated ε_r usually agree well with porosity-corrected experimental values for well-behaved ceramics [88]. It may be noted that deviations from calculated values can occur due to deviations from cubic symmetry, and also the fact that the sample is ceramic and not a single crystal. Presence of ionic or electronic conductivity, H_2O or CO_2 in channels, rattling of ions, presence of dipolar impurities or ferroelectric behavior also cause deviations from the calculated values [88]. The deviations in the reported values of dielectric polarizability can also affect the calculated values. Vineis et al. reported [89] that a more correct value of the dielectric polarizability of La is 4.82 instead of 6.12 reported by Shannon [88]. It may be noted that the CM equation used for the calculation of ε_r is highly sensitive to the value of the denominator and even a small error in determining the cell volume can significantly affect the calculated value of the permittivity. The ε_r depends on the dielectric polarizability of the constituent ions and the crystal structure. The Ba, Pb, and Bi have relatively high dielectric polarizabilities and the compounds based on them have relatively higher ε_r.

Ion dielectric polarisabilities in Å³ (modified after Shannon)

Li 1.20	Be 0.19											B 0.05	C	N	O 1.21	F 1.62	Ne
Na 1.80	Mg 1.32											Al 0.79	Si 0.87	P V1.22	S	Cl	Ar
K 3.83	Ca 3.16	Sc 2.81	Ti IV2.93	V 2.92	Cr II1.45	Mn II2.64	Fe II2.23	Co II1.65	Ni II1.23	Cu II2.11	Zn 2.04	Ga 1.5	Ge 1.63	As V1.72	Se	Br	Kr
Rb 5.29	Sr 4.24	Y 3.81	Zr 3.25	Nb 3.97	Mo	Tc	Ru	Rh	Pd	Ag	Cd	In 2.62	Sn	Sb III2.83	Te IV5.23	I	Xe
Cs 7.43	Ba 6.40	La 4.82	Hf	Ta 4.73	W	Re	Os	Ir	Pt	Au	Hg	Tl 7.28	Pb II6.58	Bi 6.12	Po	At	Rn
Lanthinide Series		Ce III6.15 IV3.94	Pr 5.32	Nd 5.01	Pm	Sm 4.74	Eu II4.83 III4.53	Gd 4.37	Tb 4.25	Dy 4.07	Ho 3.97	Er 3.81	Tm 3.82	Yb 3.58	Lu 3.64		
Actinide Series		Th 4.92	Pa	U IV4.45	Np	Pu	Am	Cm	Bk	Cf	Es	Fm	Md	No	Lr		

Figure 2.20 Dielectric polarizabilities of ions (after Ref. [88]).

Differentiating Equation (2.54) [90, 91]

$$\frac{1}{(\varepsilon_r - 1)(\varepsilon_r + 2)}\left(\frac{\delta\varepsilon_r}{\delta T}\right)_P = -\frac{1}{3V_m}\left(\frac{\delta V_m}{\delta T}\right)_P + \frac{V_m}{\alpha_D}\left(\frac{\delta\alpha_D}{\delta V_m}\right)_T \frac{1}{3V_m}\left(\frac{\delta V_m}{\delta T}\right)_P + \frac{1}{3\alpha_D}\left(\frac{\delta\alpha_D}{\delta T}\right)_{V_m}$$
$$= A + B + C \qquad (2.58)$$

The first term represents the thermal expansion and its effect produces a decrease in the concentration of the dipoles. The minus sign indicates that its contribution is negative. The number of polarizable ions per unit volume decreases as the temperature increases. The second term represents the possibility that the polarization will increase with increasing volume in which ions are able to move as the temperature increases. This contribution is positive unless $(\delta V/\delta T)_P$ becomes negative which is unlikely. The third term represents the direct dependence of polarizability on temperature, the volume remaining constant. In general this contribution is negative since ε_r decreases with increase of temperature. Bosman and Havinga [90] investigated the temperature and pressure dependence of ε_r of a number of cubic halides and oxides. They reported that for low ε_r materials the permittivity increased with increasing temperature whereas for high ε_r materials the permittivity decreased with temperature. The ε_r decreased with increase of pressure for all the materials they measured.

2.8 Measurement of Temperature Coefficient of Resonant Frequency (τ_f)

The temperature coefficient of resonant frequency τ_f is the parameter which indicates the thermal stability of the resonator. The τ_f indicates how much the resonant frequency drifts with changing temperature. The electronic device with microwave resonators requires τ_f values as close to zero as possible. Microwave circuits will normally have some low characteristics τ_f, so the resonator components which go into them are required to compensate for the inherent drift. For this reason, the τ_f values of resonators required are typically non-zero but with some low finite value. The origin of τ_f is related to linear expansion coefficient α_L which affects the resonator dimensions and its dielectric constant variation with temperature [1]. Mathematically the relationship is

$$\tau_f = -\alpha_L - \frac{\tau_\varepsilon}{2} \qquad (2.59)$$

where τ_ε is the temperature coefficient of the permittivity and α_L is the linear thermal expansion coefficient of the dielectric material which is usually positive.

In practice, Equation (2.59) is valid for 100% electric energy storage in the sample and the thermal expansion of the metal cavity enclosing the DR is negligible. For an ideal resonator the temperature coefficient of resonant frequency (τ_f) should be near to zero. Hence from Equation (2.59) for a zero τ_f, the τ_ε should have twice the value of α_L and should be negative. Since resonators are used in communication systems, temperature stability is an important factor and should be close to zero. For most of the electronic ceramic materials, α_L is about +10 ppm/°C indicating the significant influence of τ_ε on τ_f.

Experimentally τ_f is measured by following the drift in the resonant peak frequency f as the temperature is slowly varied. In order to measure τ_f, DR is kept end shorted between two copper plates under Courtney setup. This is then kept inside a temperature-controlled oven. The E-field probe is kept near the DR in such a way to get resonance. The TE_{011} mode is identified and the setup is then slowly heated (~1°C/min) in the range 25–80°C. The probe of the thermocouple is kept just inside the oven so that it does not disturb the resonant frequency. Shift of the resonant frequency as a result of heating in the reflection mode is noted using network analyzer when the temperature is steady. The variation of resonant frequency is plotted as a function of temperature. The τ_f is calculated from the slope of the curve using equation

$$\tau_f = \frac{f_{80} - f_{25}}{f_{25}(80 - 25)} = \frac{1}{f}\frac{\Delta f}{\Delta T} \tag{2.60}$$

The τ_f is expressed as parts per million per degree Celsius (ppm/°C)

The τ_f can also be measured by the cavity method used for measuring the quality factor. The thermal expansion of the cavity during heating limits the accuracy of the method. However, use of a cavity made of invar can minimize the inaccuracy. In fact thermal expansion of the cavity is negligibly small for WGMs and also for $TE_{01\delta}$ mode resonant structure if permittivity of the sample is large and the sample is situated away from all cavity walls. The τ_f is related to thermal expansion and relative permittivity by the Equation (2.59). The temperature coefficient of dielectric constant τ_ε is of considerable interest to users of dielectric substrates. This can be obtained by the parallel plate capacitor method using an LCR meter at low frequency (e.g., 1 MHz) and heating the sample. From the ε_r noted at different temperatures the τ_ε can be obtained. At microwave frequency, one can obtain τ_ε from the value of τ_f and α_L and using Equation (2.59).

Temperature coefficient of resonant frequency and Q-factor are quantities that characterize resonator and they depend not only on properties of ceramic sample but also on the properties of other parts of resonance system.

2.9 Tuning the Resonant Frequency

The resonant frequency of the resonator depends on the resonator surroundings, relative permittivity and the sample dimensions. The frequency of a particular resonator can be changed [92–95] by a tuning plate, dielectric plug or a dielectric disk as shown in Figure 2.21. The plate tuning leads to considerable decrease in the quality factor whereas the dielectric plug or dielectric disk tuning may decrease the quality factor only by less than 5%. If a metallic tuner is used, resonant frequency will increase as the tuner approaches the DR. If a dielectric tuner is used, resonant frequency will decrease as the tuner approaches the DR. A change in resonant frequency up to about 15% can be achieved by these methods. However, when using a metallic plate tuner, the tuning range may be limited to only a few percent in order to avoid serious degradation of Q-factor and ε_r. It is also possible to tune the frequency by changing the physical size of the DR by machining and reducing the thickness or diameter of the DR. It is also

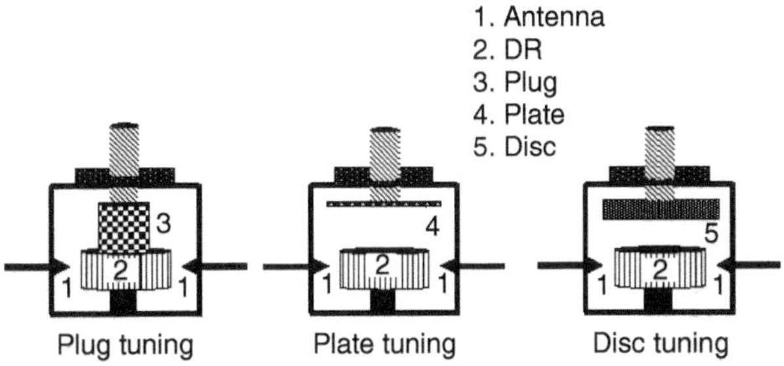

Figure 2.21 Schematic diagrams showing tuning of the resonant frequency (a) plug tuning (b) plate tuning (c) disc tuning.

possible to tune the frequency by increasing the size of the resonator by attaching a small piece of low loss ceramic to the DR using low loss glue. The resonators can be mounted to a ceramic substrate using a small drop of low loss adhesive such as cyanoacrylic. Recently Petrov and Alford [96] reported fast resonant frequency tuning in DRs by incorporating a thick film of $Ba_{1-x}Sr_xTiO_3$.

The QWED in Poland manufacture several types of test fixtures for the precise measurement of electromagnetic properties of materials at microwave frequencies. The test fixtures with software for extracting the relevant data are based on the expertise of Prof. Jerzy Krupka, who is well known for the precise measurement of microwave dielectric properties.

REFERENCES

[1] D. Kajfezz and P. Guillon. Dielectric Resonators. Noble Publishing Corporation, Tucker, Georgia, USA (1998).
[2] A. J. Moulson and J. M. Herbert. Electroceramics. Chapman and Hall, London (1990).
[3] V. Raghavan. Materials Science and Engineering. Prentice Hall India, New Delhi (1982).
[4] J. C. Burtfoot. Ferroelectrics: An Introduction to the Physical Principles. Van Nostrand-Reinbold, London (1967).
[5] J. Krupka. Frequency domain complex permittivity measurements at microwave frequencies. Meas. Sci. Technol. 16(2005)R1–R16.
[6] G. L. Gurevich and A. K. Tagantsev. Intrinsic dielectric loss in crystals. Adv. Phys. 40(1991) 719–767.
[7] V. L. Gurevich and A. K. Tagantsev. Intrinsic dielectric loss in crystals: Low temperature. Sov. Phys. JETP. 64(1986)142–151.
[8] V. B. Braginsky, V. S. Ilschenko, and Kh. S. Bagdassarov. Experimental observation of fundamental microwave absorption in high quality dielectric crystals. Appl. Phys. A. 120(1987)300–305.
[9] Y. Kobayashi, Y. Aoki, and Y. Kabe. Influence of conductor shields on the Q factor of a TE_0 dielectric resonator. IEEE Trans. Microw. Theory Tech. MTT-33(1985)1361–1366.
[10] L. F. Chen, C. K. Ong, C. P. Neo, V. V. Varadan, and V. K. Varadan. Microwave Electronics. Measurement and Materials Characterization. Wiley, Chichester, England (2004).
[11] B. W. Hakki and P. D. Coleman. A dielectric resonator method of measuring inductive capacities in the millimeter range. IRE Trans. Microw. Theory Tech. MTT-8(1960)402–410.
[12] W. E. Courtney. Analysis and evaluation of a method of measuring complex permittivity and permeability of microwave materials. IEEE Trans. Microw. Theory Tech. MTT-18(1970) 476–485.

[13] S. B. Cohn and K. C. Kelly. Microwave measurements of high dielectric constant materials. IEEE Trans. Microw. Theory Tech. MTT-14(1966)406–410.

[14] Y. Kobayashi and M. Katoh. Microwave measurement of dielectric properties of low loss materials by the dielectric rod resonator method. IEEE Microw. Theory Tech. MTT-33(1985) 586–592.

[15] E. L. Ginzton. Microwave Measurements. McGraw Hill, Boston (1957).

[16] E. J. Vanzura and J. E. Rogers. Proceedings of the IEEE Conference on Instrumentation and Measurement Technology, May 14–16, Atlanta (1991).

[17] Y. Kobayashi and H. Tamura. Round robin test on a dielectric resonator method for measuring complex permittivity at microwave frequency. ICIE Trans. Electronics. E77-C(1994)882–887.

[18] Y. Kobayashi and S. Tanaka. Resonant modes of dielectric rod resonator short circuited at both ends by parallel plate conducting plates. IEEE. MTT-28(1980)1077–1085.

[19] D. L. Rebsh, D. C. Webb, R. A. Moore, and J. D. Cawlishaw. Mode chart for accurate design of cylindrical dielectric resonators. IEEE Microw. Theory Tech. MTT-13(1965)468–469.

[20] IEC 61338, section 1–1, 1–2 and 1–3. (1996–2003). Wave guide type dielectric resonators. Part 1. General information and test conditions-section 3. Measurement method of complex permittivity for dielectric resonator materials at microwave frequency (www.iec.ch).

[21] D. Hennings and P. Schnabel. Dielectric characterization of $Ba_2Ti_9O_{20}$ type ceramics at microwave frequencies. Philips Res. Rep. 38(1983)295–311.

[22] P. Whelss and D. Kajfezz. The use of higher order resonant modes in measuring the dielectric constant of dielectric resonators. IEEE MTT-S. Symp. Dig.. St Louis. (1985)473–476.

[23] M. W. Pospieszalski. On the theory and applications of the dielectric post resonator. IEEE Microw. Theory Technol. MTT-25(1977)697–700.

[24] H. Takagi, N. Funnami, H. Tamura, and K. Akino. Measurement accuracy of complex permittivity in the dielectric rod resonator method. Jpn. J. Appl. Phys. 31(1992)3269–3271.

[25] J. Krupka, K. Derzakowski, B. Riddle, and J. B. Jarvis. A dielectric resonator for measurements of complex permittivity of low loss materials as a function of temperature. Meas. Sci. Technol. 9(1998)1751–1756.

[26] K. Leong, J. Mazierska, and J. Krupka. Measurement of unloaded Q factor of transmission mode dielectric resonators. IEEE MTTS-97 Symp. Dig. (1997)1639–1642.

[27] H. Takamura, H. Matsumoto, and K. Wakino. Low temperature properties of microwave dielectrics. Proc. 7th Meeting on Ferroelectric Materials and their Applications. Jpn. J. Appl. Phys. 28:Suppl. 2(1989) 21–23.

[28] J. Krupka, W.-T. Huang, and M.-J. Tung. Complex permittivity measurements of low loss microwave ceramics employing higher order quasi TE_{0np} modes excited in a cylindrical dielectric sample. Meas. Sci. Technol. 16(2005)1014–1020.

[29] M. Valant, D. Suvorov, and S. Macek. Measurement error analysis in the determination of microwave dielectric properties using cavity reflection method. Ferroelectrics. 176(1996)167–177.

[30] A. W. Scott. Understanding Microwaves. Wiley, New York (1993).

[31] A. P. S. Khanna and G. Garault. Determination of loaded, unloaded and external quality factors of a dielectric resonator coupled to a microstripline. IEEE Trans. Microw. Theory Tech. MTT-31(1983)261–264.

[32] J. R. Wait. Electromagnetic whispering gallery modes in a dielectric rod. Radio Sci. 2(1967) 1005–1017.

[33] D. Cros and P. Guillon. Whispering gallery dielectric resonator modes for w-band devices. IEEE Microw. Theory Tech. MTT-38(1990)1667–1674.

[34] J. Krupka, D. Cros, M. Aubourg, and P. Guillon. Study of whispering gallery modes in anisotropic single crystal dielectric resonators. IEEE Trans. Microw. Theory Tech. 42(1994) 56–61.

[35] R. A. Woode, E. N. Ivanov, M. E. Tobar, and D. G. Blair. Measurement of dielectric loss tangent of alumina at microwave frequencies at room temperature. Electron. Lett. 30(1994)2120–2122.

[36] E. N. Ivanov, D. G. Blair, and V. I. Kalinichev. Approximate approach to the design of shielded dielectric disk resonators with whispering gallery modes. IEEE Trans. Microw. Theory Tech. MTT-41(1993)632–637.
[37] M. E. Tobar and A. G. Mann. Resonant frequencies of high order modes in cylindrical anisotropic dielectric resonators. IEEE Trans. Microw. Theory Tech. 39(1991)2077–2082.
[38] J. Krupka, K. Derzakowski, A. Abramowicz, M. E. Tobar, and R. G. Geyer. Whispering gallery modes for complex permittivity measurements of ultra low loss dielectric materials. IEEE Microw. Theory Tech. 47(1999)752–759.
[39] J. Krupka, D. Cros, M. Aubourg, and P. Guillon. Study of whispering gallery modes in anisotripic single crystal dielectric resonators. IEEE Trans. Microw. Theory Tech. MTT-42(1994)56–61.
[40] J. Krupka, R. G. Geyer, J. Baker-Jarvis, and J. Ceremurga. Measurements of the complex permittivity of microwave circuit board substrates using split dielectric resonator and reentrant cavity techniques. Seventh Intl. Conference on Dielectric Materials, Measurements and Applications Conf. Publ. No. 430, Bath, UK (37–40 SPDR) (Sept. 1996) pp. 21–24.
[41] J. Krupka, A. P. Gregonry, O. C. Kochard, R. N. Clarke, B. Riddle, and J. Baker-Jarvis. Uncertainity of complex permittivity measurement by split post dielectric resonator techniques. J. Eur. Ceram. Soc. 21(2001)2673–2676.
[42] J. Krupka, S. Gabelich, K. Derzakowski, and B. M. Pierce. Comparison of split post dielectric resonator and ferrite disc resonator techniques for microwave permittivity measurements of polycrystalline ytterium iron garnet. Meas. Sci. Technol. 10(1999)1004–1008.
[43] G. Kent. An evanescent mode tester for ceramic dielectric substrates. IEEE Microw. Theory Tech. MTT-36(1988)1451–1454.
[44] J. Krupka. Precise measurement of complex permittivity of dielectric materials at microwave frequencies. Mater. Chem. Phys. 79(2003)195–198.
[45] A. Prakash, J. K. Vaid, and A. Mansingh. Measurement of dielectric parameters at microwave frequencies by cavity perturbation technique. IEEE Trans. Microw. Theory Tech. MTT-27 (1979)791–795.
[46] W. Von Auluck and J. H. Rowen. Measurement of dielectric and magnetic materials at microwave frequencies. Bell System Technical Journal. 36(1957)427–448.
[47] E. G. Spencer, R. C. Lecraw, and F. Reggia. Measurement of microwave dielectric constant and tensor permeabilities of ferrites. Proc. IRE 44(1956)790–800.
[48] J. O. Artman and P. E. Tannenwald. Measurement of permeability tensor in ferrites. Phys. Rev. 91(1953)1014–1015.
[49] J. Slater. Microwave electronics. Rev. Mod. Phys. 18(1946)441–512.
[50] D. Li, C. E. Free, E. G. Pitt, and P. G. Barnell. A simple method for accurate loss tangent measurements of dielectrics using a microwave resonant cavity. IEEE Microw. Wireless Components Lett. 11(2001)118–120.
[51] A. Brandt. Investigations of dielectrics at ultra high frequencies. GIFL, Moscow (in Russian) (1963).
[52] H. E. Bussy and L. A. Steinert. Exact solution for a gyromagnetic sample and measurement on a ferrite. IRE Trans. Microw. Theory Tech. 6(1958)72–76.
[53] P. O. Rinsman and T. Ohlsson. Theory for experiments with a TM_{020} applicator. J. Microw. Power. 10(1975)271–280.
[54] J. V. Parry. The measurement of permittivity and power factor of dielectrics at frequencies from 300 to 600 Mc/S. Proc. Inst. Electr. Engn. 98(1961)303–3111.
[55] A. Kaczkowski and A. Milewski. A high accuracy wide range measurement method for determination of complex permittivity in re-entrant cavity. Part A: theoretical analysis of the method. IEEE Trans. Theory Tech. MTT-28(1980)225–228.
[56] O. V. Karpov. On an absolute method of measurement of dielectric properties of a solid using shaped resonator. Sov. Phys. 1(1959)220–228.
[57] F. Horner, T. A. Taylor, R. Dunsmuir, J. Lamb, and W. Jackson. Resonance methods of dielectric measurement at centimeter wavelengths. J. IEEE. 93(1946)53–68.
[58] R. Cook, R. G. Jones, and C. B. Rosenberg. Comparison of cavity and open resonator measurement of permittivity and loss angle at 35 GHz. IEEE Trans. Instrum. Meas. 23(1974)438–442.

[59] U. Stumper. A TE_{01n} cavity resonator method to determine the complex permittivitty of low loss liquids at millimeter wave lengths. Rev. Sci. Instrum. 44(1973)165–169.
[60] W. G. Spitzer, R. C. Miller, D. A. Klienman, and L. E. Howarth. Far infrared dielectric dispersion in $BaTiO_3$, $SrTiO_3$ and TiO_2. Phys. Rev. 126(1962)1710–1721.
[61] W. G. Spitzer and D. A. Kleinman. Far infrared lattice bands in quartz. Phys. Rev. 121(1961) 1324–1335.
[62] C. H. Perry, D. J. Mc Carthy, and G. Rupprecht. Dielectric dispersion of some perovskite zirconates. Phys. Rev. 138(1965)A1537–1538.
[63] R. Kudesia, A. E. McHale, R. A. Condrate, and R. L. Snyder. Microwave characteristics and far infrared reflection spectra of zirconium tin titanate dielectrics. J. Mater. Sci. 28(1993) 5569–5575.
[64] K. Wakino, D. A. Sagala, and H. Tamura. Far infrared reflection spectra of $Ba(ZnTa)O_3$–$BaZrO_3$ dielectric resonator material. Proc. 6th Meeting on Ferroelectricity, Kobe 1985 (Japan). Jpn. J. Appl. Phys. 24: Suppl. 2(1985)1042–1044.
[65] D. S. Keiding, M. Van Exter, and, Ch.. Fattinger. Far-infrared domain spectroscopy with terahertz beams of dielectrics and semiconductors. J. Opt. Soc. Am. B7(1990)2006–2015.
[66] J. Petzelt, S. Kamba, G. V. Kozlov, and A. A. Volkov. Dielectric properties of microwave ceramics investigated by far infrared and submillimeterwave spectroscopy. Ferroelectrics. 176(1996)145–165.
[67] G. Ruprecht and R. O. Bell. Microwave losses in strontium titanate above the phase transition. Phys. Rev. 125(1962)1915–1920.
[68] B. D. Silverman. Microwave absorption in cubic strontium titanate. Phys. Rev. 125(1962)1921–1950.
[69] S. Kamba, J. Petzelt, E. Buixaderas, D. Haubrich, P. Vanek, P. Kuzel, I. N. Jawahar, M. T. Sebastian, and P. Mohanan. High frequency dielectric properties of $A_5B_4O_{15}$ microwave ceramics. J. Appl. Phys. 89(2001)3900–3906.
[70] R. Zurmühlen, J. Petzelt, S. Kamba, V. V. Voitsekhovskii, and N. Setter. Dielectric spectroscopy of $Ba(B'_{1/2}B''_{1/2})O_3$ complex perovskite ceramics: Correlation between ionic parameters and microwave dielectric properties. Part I – Infrared reflectivity study (10^{12}–10^{14}Hz). J. Appl. Phys. 77(1995)5341–5350.
[71] R. Zurmühlen, J. Petzelt, S. Kamba, G. Kozlov, A. Volkov, B. Gorshunov, D. Dube, A. Tagantsev, and N. Setter. Dielectric spectroscopy of $Ba(B(B'_{1/2}B''_{1/2})O_3$ complex perovskite ceramics: Correlation between ionic parameters and microwave dielectric properties. Part II.: Studies below the phonon eigen frequencies (10^2–10^{12}Hz). J. Appl. Phys. 77(1995) 5351–5364.
[72] J. Petzelt and N. Setter. Far infrared spectroscopy and origin of microwave losses in low loss ceramics. Ferroelectrics. 150(1993)89–102.
[73] K. Fukuda. R. Kitoh, I. Awai. Microwave characteristics of mixed phases of $BaTi_4O_9$–$BaPr_2Ti_4O_{12}$ ceramics. J. Mater. Sci. 30(1995)1208–1216.
[74] V. M. Ferreira, J. L. Baptista, S. Kamba, and J. Petzelt. Dielectric spectroscopy of $MgTiO_3$ based ceramics in the 10^9–10^{14}Hz region. J. Mater. Sci. 28(1993)5894–5900.
[75] J. Petzelt, S. Pacessova, J. Fousek, S. Kamba,V. Zelezny,V. Koukal, J. Schwarzbach, B. P. Gorshunov, G. V. Kozlav, and A. A. Volkov. Dielectric spectra of some ceramics for microwave applications in the range of 10^{10}–10^{14} Hz. Ferroelectrics. 93(1989)77–85.
[76] V. L. Gurevich. Kinetics of Phonon Systems. Nauk, Moscow (1980).
[77] A. K. Tagantsev, V. O. Sherman, K. F. Astafiev, J. Venkatesh, and N. Setter. Ferroelectric materials for microwave tunable applications. J. Electroceram. 11(2003)5–66.
[78] K. A. Subbhaswamy and D. L. Mills. Theory of microwave absorption in wide band-gap-insulators: The role of thermal phonon life times. Phys. Rev. B33(1986)4213–4220.
[79] A. K. Tagantsev, J. Petzelt and N. Setter. Relation between intrinsic microwave and sub-millimeter losses and permittivity in dielectrics. Solid State Commun. 87(1993)1117–1120.
[80] J. Petzelt and S. Kamba. Submillimeter and infrared response of microwave materials: Extraploation to microwave properties. Mater. Phys. Chem. 79(2003)175–180.

[81] N. Setter, E. L. Colla, I. M. Reaney, R. Zuhrmuhlen, and A. K. Tagantsev. Structure property relations in microwave dielectrics. Proc. Electroceram. R. Waser (Ed.), Vol. 1, Aachen, Germany (1994) pp. 11–16.
[82] K. Wakino. Miniaturization techniques of microwave components for mobile communication systems using low loss dielectrics. Ferroelectrics Rev. 2(2001)1–49.
[83] J. Molla, M. Gonzalez, R. Villa, and A. Ibara. Effect of humidity on microwave dielectric losses of porous alumina. J. Appl. Phys. 85(1999)1727–1730.
[84] E. Schlömann. Dielectric losses in ionic crystals with disordered charge distributions. Phys. Rev. 135(1964)A413–A419.
[85] V. S. Vinogradov. Fiz. Tverd. Tela. 2(1960)2622. English Translation. Sov. Phys. Solid State. 2(1961)2338.
[86] S. J. Penn, N. N. Mc N. Alford, A. Templeton, X. Wang, M. Xu, M. Rece, and K. Schrapel. Effect of porosity and grain size on the microwave dielectric properties of sintered alumina. J. Am. Ceram. Soc. 80(1997)1885–1888.
[87] H. Frohlichs. Theory of Dielectrics. Clarendon Press, Oxford (1950).
[88] R. D. Shannon. Dielectric polarizabilities of ions in oxides and fluorides. J. Appl. Phys. 73(1993)348–366.
[89] C. Vineis, P. K. Davies, T. Negas, and S. Bell. Microwave dielectric properties of hexagonal perovskites. Mater. Res. Bull. 31(1996)431–437.
[90] A. J. Bosman and E. E. Havinga. Temperature dependence of dielectric constants of cubic ionic compounds. Phys. Rev. 129(1963)1593–1600.
[91] E. L. Colla, I. M. Reaney, and N. Setter. Effect of structural changes in complex perovskites on the temperature coefficient of relative permittivity. J. Appl. Phys. 74(1993)3414–3425.
[92] T. Shen and K. A. Zaki. Tunable dielectric resonators with dielectric tuning disks. IEEE Microw. Theory Tech. MTT-48(2000)2439–2444.
[93] F. Hermandez- Gil and J. Perez-Martinez. Analysis of dielectric resonators with tuning screw and supporting structure. IEEE Microw. Theory Tech. MTT-33(1985)1453–1457.
[94] S.-W. Chen, K. A. Zaki, and R. G. West. Tunable temperature compensated dielectric resonators and filters. IEEE Microw. Theory Tech. MTT-38(1990)1046–1052.
[95] Y. M. Poplako, Y. V. Porkopenko, V. I. Molchanov, and A. Dogan. Frequency tunable microwave dielectric resonator. IEEE Microw. Theory Tech. MTT-49(2001)1020–1026.
[96] P. K. Petrov and Mc N. Alford. Tunable dielectric resonator with ferroelectric element. Electron. Lett. 37(2001)1066–1067.

CHAPTER THREE

Microwave Dielectric Materials in the BaO–TiO_2 System

3.1 Introduction

Barium titanate compounds in the BaO–TiO_2 system are archetypal electronic ceramics with a long history of technological applications in the ceramic capacitor industry. The Ti-rich region (>50% TiO_2) of the BaO–TiO_2 gives few compositions which exhibit [1–6] good temperature compensation with very good relative permittivity and low dielectric loss at radio frequencies. In 1947, Bunting and co-workers reported [1, 2] a low-loss temperature-compensated composition containing 83.3 mol% TiO_2 with $\varepsilon_r = 37$ at 1 MHz. In 1958, Jonker and Kwestroo [3] reported a promising combination of dielectric properties at 1 MHz for ternary compositions in the BaO–TiO_2–SnO_2/ZrO_2 system containing 18.5 mol% BaO and 4 mol% SnO_2/ZrO_2. During 1966–1971 Schwarzbach and Plocek [4] and Naumann et al. [5] and Masse et al. [6] reported temperature compensation in the binary phase at 80 mol% TiO_2 ($BaTi_4O_9$). Rase and Roy [7] published the first comprehensive phase equillibria study of the BaO–TiO_2 system. They reported the existence of compounds Ba_2TiO_4, $BaTiO_3$, $BaTi_2O_5$, $BaTi_3O_7$ and $BaTi_4O_9$. Crystal structures have been determined for Ba_2TiO_4 [8], $BaTiO_3$ [9, 10], $BaTi_2O_5$ [11] and $BaTi_4O_9$ [12]. Several subsequent investigations [3, 13] have presented contradictory data on compound identification and stability. Jonker and Kwestroo [3] reported the existence of $Ba_2Ti_9O_{20}$ and $Ba_2Ti_5O_{12}$ and questioned the formation of $BaTi_2O_5$. They reported that the new single-phase compounds are formed only when small amounts of SnO_2 and ZrO_2 were present. The X-ray diffraction pattern of $Ba_2Ti_9O_{20}$ reported by Schwarzbach and Plocek [4] is different from that reported by Jonker and Kwestroo [3]. Figure 3.1 shows the phase diagram of TiO_2-rich region of BaO–TiO_2 system [14]. It has been reported [15, 16] that the TiO_2-rich region in the BaO–TiO_2 system consists of four stable compounds such as $Ba_6Ti_{17}O_{40}$, $Ba_4Ti_{13}O_{30}$, $BaTi_4O_9$ and $Ba_2Ti_9O_{20}$. The first three compounds decompose peritectoidally at 1340°C, 1360°C and 1430°C respectively. O'Bryan and Thomson [15] reported that $Ba_2Ti_9O_{20}$ decomposes peritectoidally at 1300°C but later works [16, 17] established that the peritectoid reaction occurs above 1400°C. Several metastable compounds such as $BaTi_5O_{11}$, $BaTi_6O_{13}$, $BaTi_2O_5$ have also been reported. $BaTi_5O_{11}$ and $BaTi_2O_5$ also form as reaction intermediates during solid state reaction [18]. Tillmanns [19, 20] has reported the existence of $BaTi_5O_{11}$ and $BaTi_6O_{13}$ obtained by heating $BaTi_4O_9$ and quenching. In the following, we discuss the three important low loss dielectric ceramics $BaTi_4O_9$, $BaTi_5O_{11}$ and $Ba_2Ti_9O_{20}$ in detail.

Dielectric Materials for Wireless Communications
ISBN-13: 978-0-08-045330-9

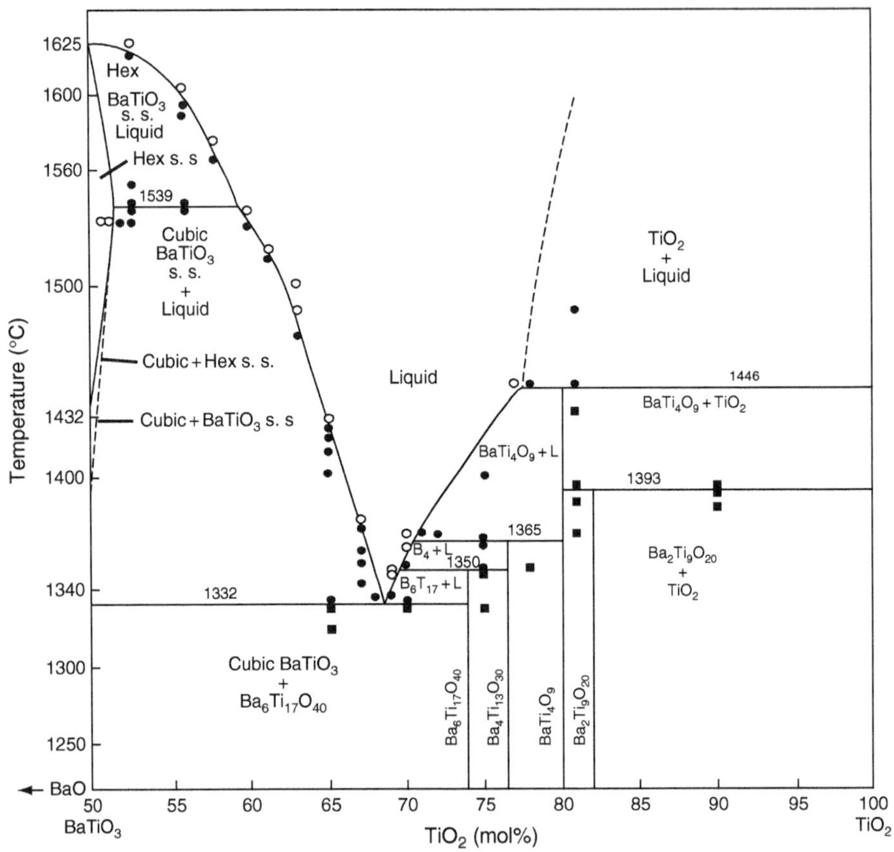

Figure 3.1 Phase diagram of BaTiO$_3$–TiO$_2$ system (after Ref. [14]).

3.2 BaTi$_4$O$_9$

The BaTi$_4$O$_9$ was first reported by Stratton [21] and Rase and Roy [7] as a TiO$_2$-rich BaO–TiO$_2$ compound. Mixtures of BaCO$_3$ and TiO$_2$ with >80 mol% TiO$_2$ initially form BaTi$_4$O$_9$ and Ba$_4$Ti$_{13}$O$_{40}$ which in turn reacts with TiO$_2$ to form BaTi$_4$O$_9$ [15]. Further reactions occur on prolonged heating at temperatures above 1200°C to form Ba$_2$Ti$_9$O$_{20}$. The BaTi$_4$O$_9$ appears in equilibrium with Ba$_2$Ti$_9$O$_{20}$ below 1420°C for compositions between 80 and 81.8 mol% TiO$_2$. Thus at high temperatures a sample containing 80 mol% TiO$_2$ contains two phases, i.e a small amount of Ba$_2$Ti$_9$O$_{20}$ in addition to BaTi$_4$O$_9$. At temperatures above 1428°C, BaTi$_4$O$_9$ melts incongruently producing TiO$_2$ and a liquid phase [7, 14–16, 22]. In Ba-doped TiO$_2$ samples which have been slowly cooled, exsolved Ba reacts with TiO$_2$ to form BaTi$_4$O$_9$ and Ba$_2$Ti$_9$O$_{20}$ surface phases [23].

The BaTi$_4$O$_9$ is prepared by the conventional solid state ceramic route by ball milling stoichiometric amounts of BaCO$_3$ and TiO$_2$ for about 24 hours [24, 25] and calcining at about 1100°C. It is then again ball milled and shaped and sintered at about 1350°C/2 h to

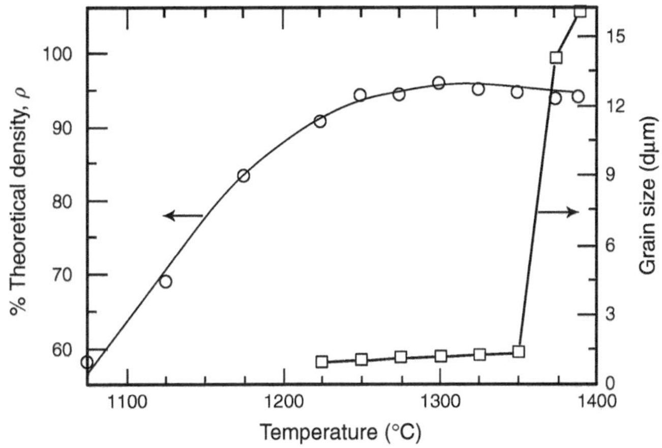

Figure 3.2 Variation of percent theoretical density and grain size as function of temperature. Two hours firing time (after Ref. [25]).

get low loss resonator pucks. Figure 3.2 shows the variation of percentage of theoretical density and grain size with sintering temperature [25]. Sintering at temperatures above 1350°C considerably increased the grain size. The $BaTi_4O_9$ ceramic showed about 95% of theoretical density on sintering at about 1350°C. It has been reported [26–28] that pure single-phase $BaTi_4O_9$ powder could not be formed below 1300°C by conventional synthesis method due to complexity in composition and structure. Normally $BaTi_4O_9$ is not formed even after heating the precursors at 1200°C/2 h. Usually at this temperature, $Ba_2Ti_9O_{20}$, $Ba_4Ti_{13}O_{30}$, and $BaTi_2O_5$ appear as secondary phases. The formation of secondary phase of $Ba_2Ti_9O_{20}$ is often been observed in $BaTi_4O_9$ [25]. Liou et al. [29] prepared $BaTi_4O_9$ by a reaction sintering method without calcination. $BaCO_3$ and TiO_2 in stoichiometric proportion ball milled for 12 hours and then pelletized and sintered at 1170°C/6 h showed up to 98% density. Mhaisalker et al. [25] reported that ball milling $BaCO_3$ and TiO_2 powders using alumina balls introduced Al_2O_3 impurities which reacted during sintering to form hollandite ($BaAl_2Ti_6O_{16}$) and degraded the dielectric properties. Sintering at high temperatures is also known to reduce Ti^{4+} to Ti^{3+} degrading the quality factor. Several authors [30–36, 137–140] reported synthesis of $BaTi_4O_9$ by wet chemical methods. Weng et al. [30] and Javadpour and Eror [35] prepared $BaTi_4O_9$ at low temperatures by a modified citrate route using $BaCO_3$, $Ti(OCH(CH_3)_2)_4$ starting materials in the presence of citric acid and ethylene glycol. It was then heated to get a polymerized glassy solid which was powdered and heat treated between 700 and 1300°C. Calcination at 750°C resulted in $BaTi_5O_{11}$ with secondary-phase $Ba_2Ti_9O_{20}$ whereas calcining at 800°C gave single-phase $BaTi_4O_9$. Samples sintered at 1250°C/3 h give 95% density with grains of 3–8 μm size. Li et al. [140] prepared $BaTi_4O_9$ by a modified pechini method, i.e. EDTA-citrate gel process. The starting materials were $Ba(NO_3)_2$, butyl titanate, EDTA and citric acid. The powder calcined at 800°C shown to be predominantly $BaTi_5O_{11}$, which on further heating to 1200°C transformed to pure $BaTi_4O_9$. Choy et al. [32–34] prepared nanometer-size $BaTi_4O_9$ powders by citrate route using $Ba(NO_3)_2$ and $TiCl_4$. Powder X-ray diffraction analysis confirmed the formation of crystalline $BaTi_4O_9$ on heating the citrate precursors at 1100°C/1 h. Purohit and Tyagi [37] showed that monophase $BaTi_4O_9$ powder can be

Table 3.1 Low loss dielectric compositions in the TiO_2-rich region of the BaO–TiO_2 system

Compound	Structure	Z	Crystal data	Density calculated	Reference
$BaTi_4O_9$	Orthorhombic *Pnmn*	2	$a = 14.53$ Å $b = 3.797$ Å $c = 6.294$ Å	4.522 g/cm^3	[7, 12, 17, 45]
$Ba_2Ti_9O_{20}$	Triclinic $P\bar{1}$	4	a = 7.471 Å b = 14.081 Å c = 14.344 Å $\alpha = 89.94$ $\beta = 79.43$ $\gamma = 84.45$	4.61 gm/cm^3	[3, 46]
$BaTi_5O_{11}$	Monoclinic *P2/n*	4	$A = 7.67$ Å $b = 14.02$ Å $c = 7.52$ Å $\beta = 98.55$	4.58 g/cm^3	15

prepared by autoigniting the citrate–nitrate precursor solution and $BaTi_4O_9$ ceramics with 96% density formed on sintering at 1200°C/6 h.

Very small single crystals of $BaTi_4O_9$ were made by quenching the melt having a ratio of 1:4 of BaO:TiO_2 [12, 38]. Larger crystals have been grown by the Float Zone technique [39, 40]. Single crystals with Ba partially substituted by Ca or Sr and Ti by Zr have also been reported [41]. $BaTi_4O_9$ is orthorhombic with *Pnmn* space group and the lattice parameters are given in Table 3.1 [11, 12, 42–44]. The $BaTi_4O_9$ is reported to be isostructural with KTi_3NbO_9 [43].

3.2.1 Microwave dielectric properties

It was Masse et al. [6], who first pointed out that $BaTi_4O_9$ is a suitable material for microwave dielectric resonator applications. Mhaisalkar et al. [24] reported that $BaTi_4O_9$ has ε_r of 37, *Qf* of 22 700 GHz and $\tau_f = +15$ ppm/°C. Several authors studied [24, 39–44, 47–51] the effect of dopants on the dielectric properties of $BaTi_4O_9$. The dielectric properties of $BaTi_4O_9$ with different dopants are given in Table 3.2. Mhaisalker et al. [24] studied the effect of dopants such as Mn, Sn, Zr, Ca, Sr, Pb on the properties of $BaTi_4O_9$. Secondary phase formation occurred for larger dopant levels. Doping with Mn, Sn, Pb significantly lowered the *Qf* whereas doping with Zr, Sr, Ca increased the *Qf*. However, very small amount (0.5 mol%) of Mn increased the quality factor but further addition lowered it. Doping has not much effect on ε_r and τ_f, although the *Qf* very much depends on it. The solubility of WO_3 in $BaTiO_3$ is extremely small (<0.1%) and the addition of WO_3 to $BaTiO_3$ increases the loss tangent of the ceramics [57]. However, BaO–4TiO_2–0.1WO_3 ceramic was found to possess excellent microwave properties [48, 50]. Several authors studied [48–50] the effect of WO_3 addition on the dielectric properties of $BaTi_4O_9$. The addition of WO_3 to the BaO–TiO_2 system results in multiple phases including $BaTi_4O_9$, $Ba_2Ti_9O_{20}$, $BaWO_4$ and TiO_2 [48]. Nishigaki et al. [48] found

Table 3.2 Microwave dielectric properties of $BaTi_4O_9$ ceramics

$BaTi_4O_9$ + dopants	Sintering temperature (°C)	ε_r	Qf (GHz)	τ_f (ppm/°C)	Reference
$BaTi_4O_9$	1300/2 h in O_2	37	22 700	15	[24]
$BaTi_4O_9$ + 0.5 mol% Mn	1300/2 h in O_2	37	30 400	10	[24]
$BaTi_4O_9$ + 5 mol% Ca	1300/2 h in O_2	37	28 000	15	[24]
$BaTi_4O_9$ + 2 mol% Sr	1300/2 h in O_2	38	24 000	20	[24]
$BaTi_4O_9$ + 6 mol% Zr	1300/2 h in air	37	28 000	20	[24]
$BaTi_4O_9$ + 2 mol% $BaWO_4$	1400/2 h in O_2	35	50 400	0	[48]
$BaTi_4O_9$ + ZnO–Ta_2O_5 + 0.1 wt% Mn	1275	36	45 000	0	[49]
0.62$BaTi_4O_9$–0.35ZnO–0.3Ta_2O_5 + 0.3 wt% Mn	1280/2 h	35.4	48 150	0.5	[47, 49]
0.62$BaTi_4O_9$–0.35ZnO–0.3Nb_2O_5 + 0.3 wt% Mn	1280/2 h	35.8	50 760	1.1	[47, 49]
0.615$BaTi_4O_9$–0.35ZnO–0.3Nb_2O_5 + 0.3 wt% Mn	1280/2 h	35.8	50 800	1.1	[47]
$BaTi_4O_9$ + 1 mol% Ta_2O_5	1280/2 h	–	36 000	–	[52]
BaO–4TiO_2–0.1WO_3	1360/in O_2	36	43 900	–	[50]
BaO–4TiO_2–0.1WO_3	1360/in air	36	44 500	–	[50]
BaO–4TiO_2–0.1WO_3	1400/2 h in O_2	35	50 400	–0.5	[48]
$BaTi_4O_9$ + 5 mol% Pb	1300/2 h in O_2	37	4000	–	[24]
$BaTi_4O_9$ + 4 mol% Sn	1300/2 h in O_2	36	7200	10	[24]
0.64$BaTi_4O_9$–0.36$BaPr_2Ti_4O_{12}$	–	–	9500	25	[53]
$BaTi_4O_9$ + 5 wt% ZnO–B_2O_3	900/2 h	33	25 000	6	[54]
$BaTi_4O_9$ + 1 wt% ZnO–B_2O_3	900/2 h	33	27 000	7	[54]

(Continued)

Table 3.2 (Continued)

$BaTi_4O_9$ + dopants	Sintering temperature (°C)	ε_r	Qf (GHz)	τ_f (ppm/°C)	Reference
$BaTi_4O_9$ –3 wt% MCAS glass	1200/4 h	34.6	42 050	14.2	[55]
$BaTi_4O_9$–citrate route	1250/10 h	36	50 500	16	[32, 33]
$BaTi_4O_9$–co-precipitation	1300/2 h	38	28 000	11	[56]
$BaTi_4O_9$–polymer precursor route	1250/3 h	35.6	42 000	12	[30]

that addition of a small amount of WO_3 lowered the τ_f of $BaTi_4O_9$ and increased Q. The BaO–4TiO$_2$–5yWO$_3$ for $y = 0.02$, Qf increased from 42 200 to 50 400 GHz as shown in Figure 3.3. The Q is saturated for y greater than 0.02. The addition of WO_3 leads to the formation of $Ba_2Ti_9O_{20}$ (with nearly zero τ_f) and $BaWO_4$ with negative τ_f. Hence

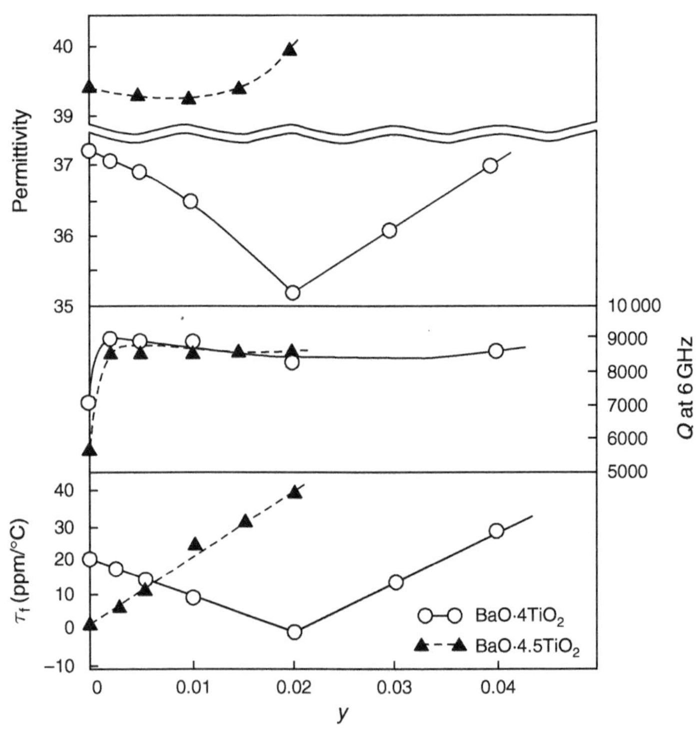

Figure 3.3 Microwave dielectric properties of BaO–4TiO$_2$–5yWO$_3$ and BaO–4.5TiO$_2$–5yWO$_3$ ceramics as a function of the amount of WO_3 (after Ref. [48]).

the presence of $BaWO_4$ improves the τ_f, and the formation of $Ba_2Ti_9O_{20}$ does not change the τ_f since it has nearly zero τ_f. EPMA studies of $BaTi_4O_9$ shows the presence of $BaTi_4O_9$, $Ba_2Ti_9O_{20}$ and $BaWO_4$. The sintered $BaWO_4$ has $\varepsilon_r = 8.2$, $Qf = 18\,000$ GHz and $\tau_f = -33$ ppm/°C. As the amount of WO_3 increased, the fraction $BaTi_4O_9$ decreased and $Ba_2Ti_9O_{20}$, $BaWO_4$ fractions increased. For $y > 0.025$, TiO_2 was also formed and the τ_f increased (see Figure 3.3). XRD and EPMA studies showed that addition of larger amount of WO_3 leads to the formation of TiO_2 in addition to $BaWO_4$. The τ_f depends on the amount of $BaWO_4$ and TiO_2. In general, addition of $BaWO_4$, WO_3, MnO_2, $ZnO–Ta_2O_5$, $WO_3–B_2O_3$ considerably improves the quality factor [24, 47–50]. Addition of $Ta_2O_5 < 1$ mol% in $BaTi_4O_9$ improves Qf. The Ta^{5+} ions reduce the number of oxygen vacancies due to the presence of impurities like Fe_2O_3

$$Ta_2O_5 \rightarrow 2Ta_{Ti} + 5O_o + 2e'$$

$$Fe_2O_3 \rightarrow 2Fe_{Ti} + 3O_o + V_{\ddot{O}}''$$

$$Fe_2O_3 + Ta_2O_5 \rightarrow 2Fe_{Ti} + 2Ta_{Ti} + 8O_O$$

Addition of $Ta_2O_5 > 1$ mol% would increase [48] the electron concentration as shown above since there are no electron compensators. Therefore formation of electrons due to Ta_2O_5 dopant is compensated by the reduction of Ti^{4+} to Ti^{3+} lowering the quality factor. At high temperature, Ti^{4+} in $BaTi_4O_9$ gets reduced to Ti^{3+}. The reduction product reacts with Ta_2O_5, forming $Ba(Ti^{3+}Ta^{5+})_2O_9$ since this compound with Ti^{3+} is isostructural with $BaTi_4O_9$. It can easily form a solid solution of $Ba(Ti^{4+})_{4-2x}[(Ti^{3+})_x(Ta^{5+})_x]O_9$ thus stabilizing the Ti^{3+} ions at room temperature. The Mn can trap electrons associated with intrinsic oxygen vacancies of titanates within the grains, e.g.

$$Mn_{Ti} + e' \rightarrow Mn'_{Ti}$$
$$Mn'_{Ti'} + e' \rightarrow Mn''_{Ti}$$

where Mn_{Ti}, Mn'_{Ti}, $Mn_{Ti''}$ are Mn^{4+}, Mn^{3+} and Mn^{2+} ions on a titanium site. Mn thus behaves as a compensator in defect equilibrium helping to maintain Ti^{4+} during cooling. It is well known that oxygen is lost from titanate systems during sintering in an atmosphere of low oxygen content [25, 58–61]. The formation of oxygen vacancies in titanates accompanies the formation of electrons which results in the reduction of Ti^{4+} into Ti^{3+} [25]. The presence of Ti^{3+} ions in $BaO–TiO_2$ systems is considered to be the reason for low quality factor. Firing pure $BaTi_4O_9$ ceramics in oxygen atmosphere decreased the loss by an order of magnitude [25]. Mn doping and heat treatment in oxygen atmosphere can eliminate Ti^{3+} ions [25, 62].

In order to obtain the desired dielectric properties, the mixing techniques of two low loss materials with positive and negative τ_f has become well known [63–67]. This approach is very ideal to produce desired properties by selecting the proper compounds but in general it is difficult to obtain a material with intermediate properties of such two-component systems because of the difficulty in retaining their individual properties at the sintering temperature. Fukuda et al. [53, 68] investigated the microwave dielectric properties of $BaTi_4O_9–BaPr_2Ti_4O_{12}$ mixture phases. The $(1-x)BaTi_4O_9–xBaPr_2Ti_4O_{12}$ prepared by conventional ceramic route and X-ray diffraction study showed only $BaTi_4O_9$ and $BaPr_2Ti_4O_{12}$ phases. SEM recorded from the sintered samples clearly shows two phases (Figure 3.4). The small white grains in Figure 3.4 are $BaPr_2Ti_4O_{12}$ and large grains are $BaTi_4O_9$. The TEM (Figure 3.5) revealed the interfacial behavior of the two phases $(1-x)BaTi_4O_9–xBaPr_2Ti_4O_{12}$ ($x = 0.36$).

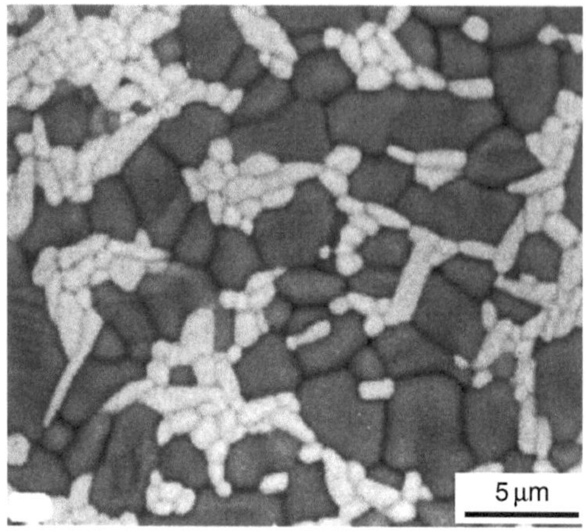

Figure 3.4 SEM photograph (1–x)$BaTi_4O_9$–$x$$BaPr_2Ti_4O_{12}$ ($x = 0.36$)(after Ref. [53], courtesy Springer Science and Business Media)).

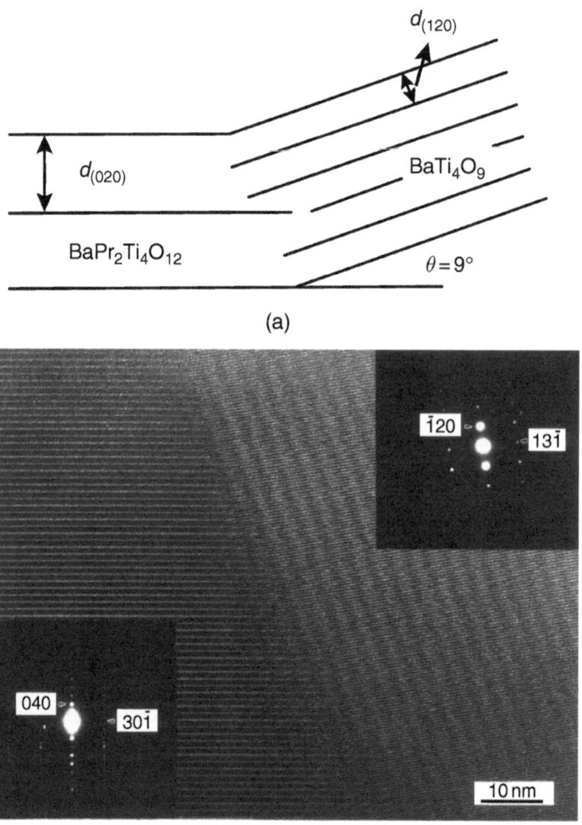

(b)

Figure 3.5 TEM photographs of (1–x)$BaTi_4O_9$–$x$$BaPr_2Ti_4O_{12}$ ceramic samples for $x = 0.36$ (after Ref. [53], courtesy Springer Science and Business Media).

One may expect that the interface of each phase consists of a disturbed array of atoms with a thickness of a few nanometers. But in contrast the lattices of each phase are well matched obliquely (about 9°). The dielectric properties of the mixture composition can be predicted using the following empirical relationships [63, 69]

$$\ln \varepsilon = V_1 \ln \varepsilon_1 + V_2 \ln \varepsilon_2$$
$$\tau_f = V_1 \tau_{f1} + V_2 \tau_{f2}$$
$$1/Q = V_1/Q_1 + V_2/Q_2$$

Figures 3.6–3.8 show the experimental results as a function of vol% and is in agreement with that calculated using the above equations. As the volume fraction of $BaPr_2Ti_4O_{12}$ increases the ε_r, τ_f increases and *Qf* decreases. A similar composite system consisting of $BaTi_4O_9$ and $BaEu_2Ti_4O_{12}$ was reported [70] in $0.64BaTi_4O_9$–$0.36BaEu_2Ti_4O_{12}$.

The ceramics prepared with WO_3, and ZnO–Nb_2O_5 or ZnO–Ta_2O_5 dopants show the best quality factors. It may be noted that the $BaTi_4O_9$ usually contains secondary phases and their amounts vary depending on the preparation conditions and initial raw materials. Pure $BaTi_4O_9$ does not have an optimum performance as a DR since its τ_f is relatively high. Hence the $BaTi_4O_9$ material is often modified by using additives [49, 53, 71, 72].

Several authors studied [50, 54, 55, 73–76] the effect of glass addition on the dielectric properties of $BaTi_4O_9$. The resultant performance of ceramics with glass additives strongly depends on the densification, microstructure and interaction between the glass and the ceramics. The liquid-phase sintering using glass additives is one of the most effective and least expensive method of reducing sintering temperature. Addition of zinc borate considerably lower the sintering temperature to a level useful for LTCC and are discussed in Chapter 12. Addition [55, 73, 76] of MCAS glass (magnesium calcium alumium silicate, MgO–CaO–Al_2O_3–SiO_2) in the ratio (5:19:26:50) to $BaTi_4O_9$ improved densification and lowered the sintering temperature. However, the density

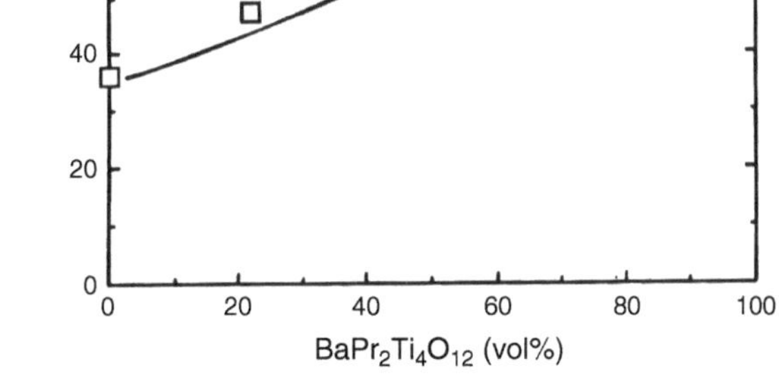

Figure 3.6 The plot of room temperature dielectric constant ε_r versus volume fraction of $BaPr_2Ti_4O_{12}$ in the $BaTi_4O_9$–$BaPr_2Ti_4O_{12}$ system. The solid curve represents values from the mixing relation (after Ref. [53]).

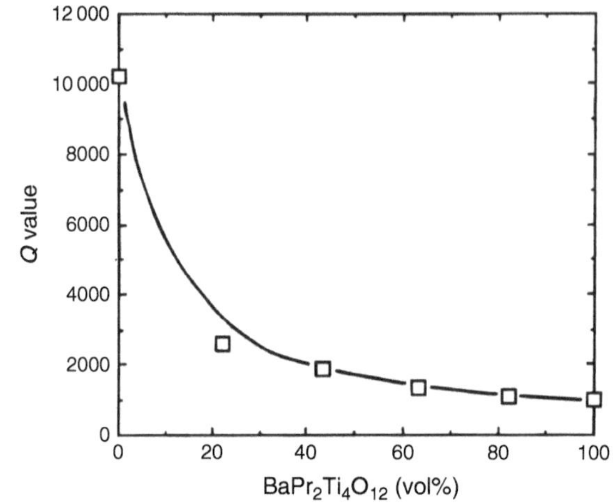

Figure 3.7 Q value versus volume fraction of $BaPr_2Ti_4O_{12}$ in the $BaTi_4O_9$–$BaPr_2Ti_4O_{12}$ system (after Ref. [53]).

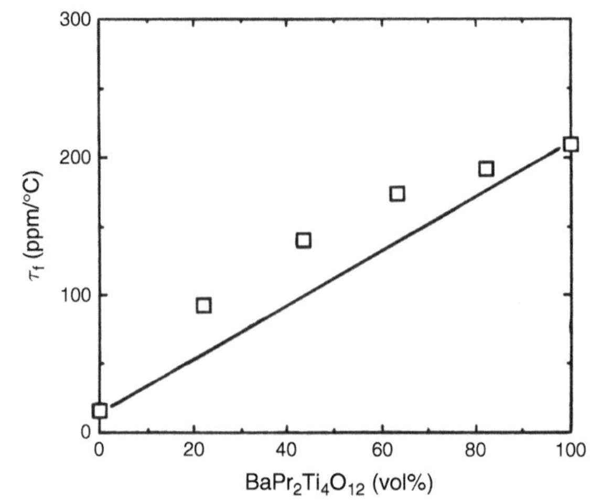

Figure 3.8 τ_f versus volume fraction of $BaPr_2Ti_4O_{12}$ in the $BaTi_4O_9$–$BaPr_2Ti_4O_{12}$ system (after Ref. [53]).

decreased with increasing amount of MCAS glass. X-ray diffraction study of 6 wt% MCAS-added $BaTi_4O_9$ and sintered at 1200°C showed the presence of $BaTi_4O_9$ and cordierite only. Addition of 1 wt% MCAS and sintering at 1300°C/4 h gave 98% of the theoretical density. MCAS glass has a low ε_r of about six and hence the glass addition decreased the permittivity. Figure 3.9 shows the variation of Qf with sintering temperature for different wt% of MCAS glass content. The highest quality factor was for 3 wt% glass-added $BaTi_4O_9$ and sintered at 1200°C. For 3 wt% addition of MCAS, the sintering temperature decreased to 1200°C from 1350°C and has $\varepsilon_r = 34.6$, $Qf = 42\,050$ GHz, and $\tau_f = 14.2$ ppm/°C.

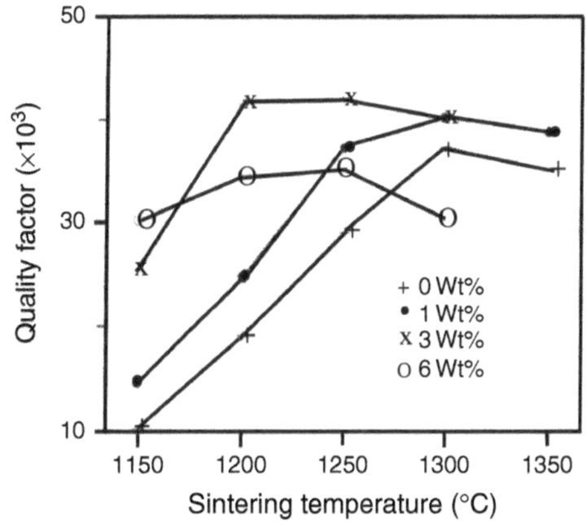

Figure 3.9 The quality factor (Qf) of $BaTi_4O_9$ ceramics as a function of sintering temperature and the amount of MCAS glass (after Ref. [55]).

Mhaisalkar et al. [77] investigated the properties of Mn-, Sn-, Zr-, Ca- and Sr-doped $BaTi_4O_9$ by analyzing its infrared reflection data. The reflectance data was converted to dielectric data using Kramers–Kronig analysis as described in Chapter 2. The dielectric parameters calculated from the reflectance data were found to be in reasonable agreement with those measured by microwave methods. The dielectric properties of $BaTi_4O_9$ sintered with different dopants and glass additions are given in Table 3.2.

3.3 $BaTi_5O_{11}$

Tillmanns was the first to report synthesis of $BaTi_5O_{11}$ [19] but single-phase material was not obtained. O'Bryan and Thomson [18] prepared $BaTi_5O_{11}$ as the major phase in a mixture of $BaTi_4O_9$ and rutile. It is difficult to obtain $BaTi_5O_{11}$ as a single-phase material by solid state reaction method [16, 78]. Kirby and Wechsler [14] reported that the compound formed as a result of non-equilibrium cooling from the liquid state. $BaTi_5O_{11}$ crystallizes from the liquid or can form as a reaction intermediate [18]. However, single-phase $BaTi_5O_{11}$ was prepared by wet chemical methods [33, 35, 79, 78–82]. Ritter et al. [83] and Javadpour and Eror [35] produced single-phase $BaTi_5O_{11}$ material by the controlled hydrolysis of ethoxide precursors and by using a liquid mixing technique. In their study, Ritter et al. [83] reported that when precipitates with $Ti/Ba = 5$ were heated between 700°C and 1110°C for extended periods (up to 2 weeks), the product consisted of only $BaTi_5O_{11}$. They concluded that production of $BaTi_5O_{11}$ is favored by an alkoxide mixing route. They also reported that at 1200°C, the $BaTi_5O_{11}$ decomposed into TiO_2, $Ba_2Ti_9O_{20}$ and/or $BaTi_4O_9$. Javadpour and Eror [35] observed the same stability range for $BaTi_5O_{11}$ and $BaTi_4O_9$. Lu et al. [82] prepared $BaTi_5O_{11}$ by first hydrolyzing titanium alkoxide and then mixing the resulting titania sol with a barium alkoxide–methanol solution. After drying, the xerogels of the precursors of barium titanates were sintered at temperatures from 700°C to 1200°C for 110 hours or longer. At 700°C $BaTi_5O_{11}$ was formed. A single-phase $BaTi_5O_{11}$ was formed on heating the powder at 1000°C/4 h. Figure 3.10 shows the XRD pattern of $BaTi_5O_{11}$ prepared by the sol–gel method as a function of temperature. When heated to 1200°C a mixture of $BaTi_5O_{11}$ and $Ba_2Ti_9O_{20}$ were formed. On prolonged heating at 1200°C $BaTi_5O_{11}$ decomposed [32, 35, 78, 82, 84] to $Ba_2Ti_9O_{20}$ and TiO_2 as shown in Figure 3.10. Figure 3.11 shows the Raman spectra of $BaTi_5O_{11}$ as a function of temperature corresponding to the XRD pattern. Choy et al. [33] used citrate route to prepare $BaTi_5O_{11}$ using $Ba(NO_3)_2$ and $TiCl_4$ solutions. Single-phase $BaTi_5O_{11}$ was formed on prolonged heating at 700°C but heating at higher temperatures decomposes into $BaTi_4O_9 + Ba_2Ti_9O_{20} + TiO_2$.

Fukui et al. [78] studied the effect of heating rate on sintering of alkoxide-derived $BaTi_5O_{11}$ powder. As the heating rate increased up to 30°C/min, density increased, and

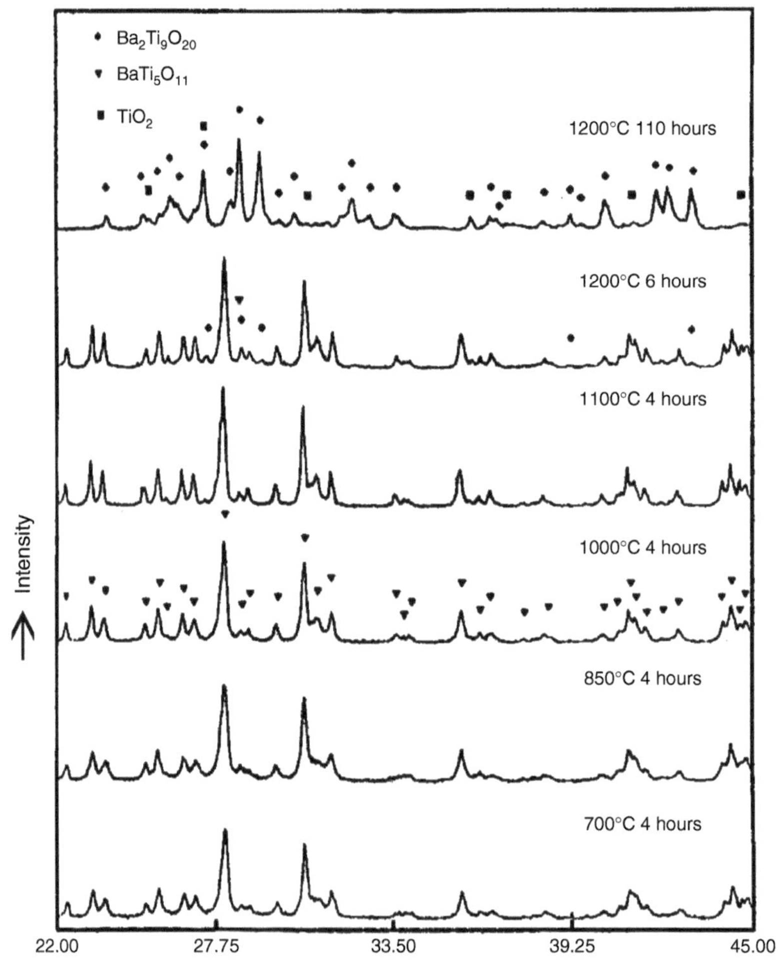

Figure 3.10 X-ray diffraction patterns of $BaTi_5O_{11}$ as a function of temperature (after Ref. [82]).

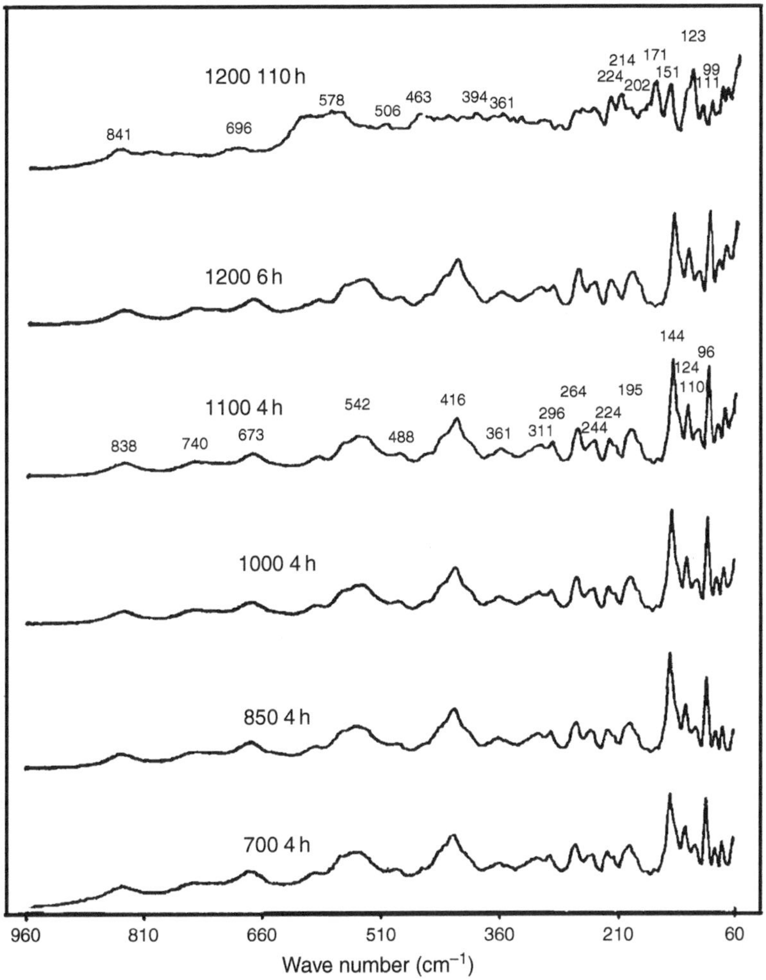

Figure 3.11 Raman spectra of $BaTi_5O_{11}$ as a function of temperature in the range 700–1200°C (after Ref. [82]).

the density decreased when the heating rate exceeded 30°C/min (Figure 3.12). Rapid heating is effective for producing high-density fine grain-size ceramics since fine particle had large specific surface area (SSA) maintained up to the sintering temperature with increase in heating rate, thus enhancing sinterability of ceramics. Maintaining high SSA, however, activates decomposition of $BaTi_5O_{11}$ phase. At a heating rate up to 30°C/min, the sinterability of $BaTi_5O_{11}$ thus increases, whereas at a heating rate over 30°C/min, density is reduced owing to a decomposition of $BaTi_5O_{11}$ into $Ba_2Ti_9O_{20}$ + rutile.

Tillmanns [19] grew small single crystals of $BaTi_5O_{11}$ by melting a BaO–$4TiO_2$ composition between 1400°C and 1500°C. The $BaTi_5O_{11}$ crystals grew intimately with rutile crystals. They have a monoclinic structure with *P2/n* space group. The structure consists of six close packed layers of barium and oxygen ions.

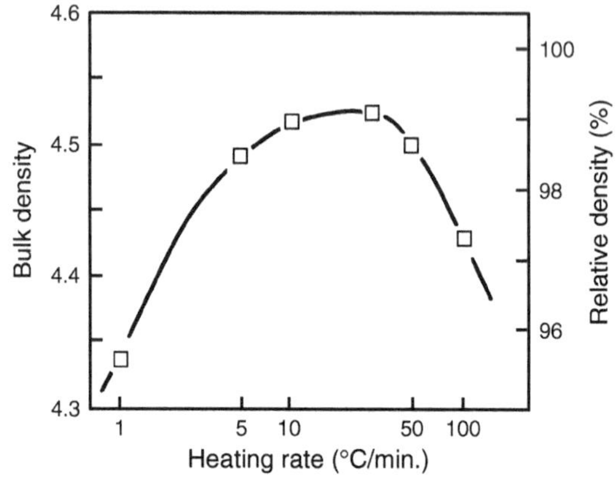

Figure 3.12 A dependence of bulk density on a heating rate of $BaTi_5O_{11}$ ceramics sintered at 1120°C for 48 hours (after Ref. [78]).

Table 3.3 Microwave dielectric properties of ceramics in $BaTi_5O_{11}$

	Sintering temperature (°C)	ε_r	Qf (GHz)	τ_f (ppm/°C)	Reference
$BaTi_5O_{11}$(Alkoxide + hotpressed)	1050/48 h	41	46 000	40	[81]
$Ba_{0.79}Sr_{0.21}Ti_5O_{11}$ hot pressed	1050/72 h	42	39 000	44	[85]
$BaTi_5O_{11}$ (Alkoxide)	1120/48 h	42	61 110	39	[78]

The sol-gel-prepared $BaTi_5O_{11}$ samples have $\varepsilon_r = 42$, $Qf > 60\,000$ GHz and $\tau_f = 39$/ppm/°C [78]. Hirano and Otsuka [81] obtained dense ceramics with a $Qf = 46\,000$ GHz by hot pressing at 1050°C for 48 hours under a pressure of 8.5 MPa. $(Ba_{1-x}Sr_x)Ti_5O_{11}$ and $BaTi_{1-x}Sn_xO_{11}$ were synthesized as single-phase materials by heating the fine particles prepared by the controlled hydrolysis of metal alkoxides [85] using barium, strontium-butoxide and titanium isopropoxide or titanium ethoxide and tin isoproxide. The substitution of Sn^{4+} for Ti^{4+} in $BaTi_5O_{11}$ increased the crystallization temperature and lowered the decomposition temperature. The $(Ba_{0.79}Sr_{0.21})Ti_5O_{11}$ had excellent properties with $Qf = 39\,000$ GHz. However, the τ_f remains unchanged even up to 20 mol% Sr substitution at the Ba site. The microwave dielectric properties of $BaTi_5O_{11}$ are given in Table 3.3.

3.4 $Ba_2Ti_9O_{20}$

The compound $Ba_2Ti_9O_{20}$ was first reported by Jonker and Kwestroo [3], who produced it by solid state reaction in the temperature range 1300–1400°C from binary compositions with >80 mol% TiO_2. They reported that single-phase $Ba_2Ti_9O_{20}$ could be obtained only by substituting a small amount of SnO_2 or ZrO_2 for TiO_2. However, O'Bryan et al. [15, 86] showed that single-phase $Ba_2Ti_9O_{20}$ can be prepared without SnO_2 or ZrO_2 addition. The $Ba_2Ti_9O_{20}$ undergoes a peritectoid reaction at 1420°C and decomposes into $BaTi_4O_9 + TiO_2$ [15, 16]. Thus it is difficult to obtain $Ba_2Ti_9O_{20}$ phase by cooling from the melt. The $BaTi_5O_{11}$ ceramics on heating below 1300°C yielded [16] $Ba_2Ti_9O_{20} + TiO_2$.

3.4.1 Preparation

The $Ba_2Ti_9O_{20}$ can be prepared by mixing $BaCO_3$ and TiO_2 in stoichiometric proportions and calcining at about 1200°C. The calcined material is again powdered, shaped and then sintered at about 1400°C/3 h [62, 87–89]. The preparation of dense and phase pure $Ba_2Ti_9O_{20}$ is generally difficult using solid state method due to the existence of thermodynamically stable $BaTi_4O_9$ compound in the vicinity of the desired composition. Generally the ceramics contain certain amounts of secondary phases consisting of $BaTi_4O_9$ or rutile or both. It is known from the phase diagram of BaO–TiO_2 system that $Ba_2Ti_9O_{20}$ decomposes into $BaTi_4O_9 + TiO_2$ above 1420°C. Small deviations from stoichiometry [28] or even chemical inhomogeneities of stoichiometric specimens are sufficient for these phases to originate also below 1420°C. O'Bryan et al. [28, 86] have mentioned that the calcination temperature has much influence on the amount of secondary phase. Hennings and Schnabel [87] reported that the presence of $BaTi_4O_9$ secondary phases can be avoided by calcining above 1170°C. Even in stioichiometric $Ba_2Ti_9O_{20}$ large amounts of $BaTi_4O_9$ secondary phases are formed below 1170°C due to thermodynamic reasons. $Ba_2Ti_9O_{20}$ has been difficult to synthesize as single phase without additives which form a solid solution. Tin ion dopants have been used in the form of SnO_2 and $BaSnO_3$ additives in several studies [26, 88–93] to stabilize $Ba_2Ti_9O_{20}$ and their microwave dielectric properties have been evaluated. The substitution of Sn^{4+} cations at Ti^{4+} cations reduced the τ_f without degrading ε_r considerably [26, 88–93]. Lin and Speyer [88, 94] prepared ZrO_2- and SnO_2-doped $Ba_2Ti_9O_{20}$ with minimum porosity by a rate-controlled-sintering process (RCS). The samples prepared by isostatic pressing and sintering at 1390°C gave a density of 99%. They studied the effect of solid solution additives, their concentration and thermal processing schedule on the microstructural evolution and microwave properties of $Ba_2Ti_9O_{20}$ [88]. It was found that a small amount of SnO_2 (0.82 mol%) resulted in a significantly greater concentration of $Ba_2Ti_9O_{20}$. Six hours of sintering at a temperature of 1390°C was adequate to sinter $Ba_2Ti_9O_{20}$ to a high density. Longer duration of sintering at 1390°C caused a density reduction. Addition of a small amount of Al_2O_3, Bi_2O_3, SnO_2, or ZrO_2 is also known to promote the formation of $Ba_2Ti_9O_{20}$ [3, 91, 95].

Several authors reported the preparation of $Ba_2Ti_9O_{20}$ by wet chemical methods [32, 82, 91, 96–102]. Wang and Chung [100] prepared single phase $Ba_2Ti_9O_{20}$ by ethyl diaminetetraacetic acid (EDTA) gel method using $Ba(NO_3)_2$ and $Ti(NO_3)_4$ raw materials. Lu et al. [82] prepared $Ba_2Ti_9O_{20}$ by first hydrolyzing titanium alkoxide and then mixing the resulting titania sol with barium alkoxide methanol solution. Heat treatment of the xerogels at temperatures in the range 700–1100°C produced $BaTi_5O_{11}$ whereas prolonged heating at 1200°C produced single-phase $Ba_2Ti_9O_{20}$. Figures 3.13 and 3.14 show the XRD pattern and Raman spectra of $Ba_2Ti_9O_{20}$ respectively as a function of temperature. The formation of different secondary phases at different temperatures is evident from the figures. Choy et al. [32] prepared $Ba_2Ti_9O_{20}$ by the citrate route. The crystallite size of $Ba_2Ti_9O_{20}$ obtained by heating

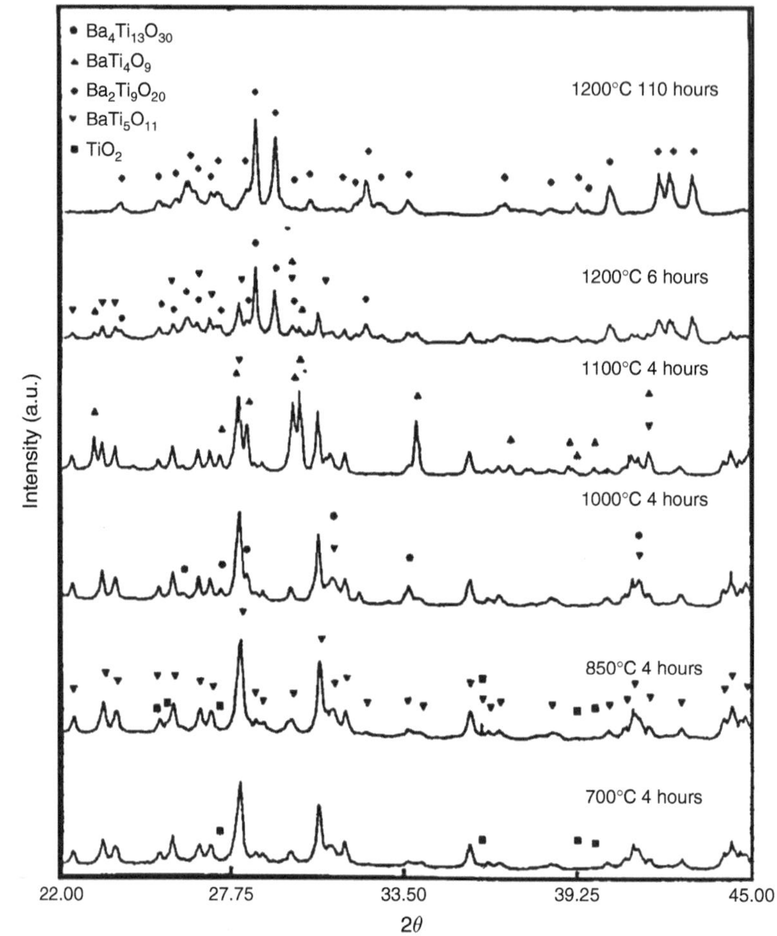

Figure 3.13 Raman spectra of $Ba_2Ti_9O_{20}$ as a function of temperature (after Ref [82]).

the citrate precursor at 1100°C for 1 hour was about 35 nm. The crystallization and sintering temperatures were significantly lowered due to the large reactive surface area of citrate-derived precursor powders. Wu and Wang [91] prepared $Ba_2Ti_9O_{20}$ by co-precipitation using $TiCl_4$ and $BaCl_2$ in 0.1 M $(NH_4)_2CO_3 + NH_4OH$. The co-precipitated powder at temperature <1100°C gave $BaTi_4O_9$ only. As the calcination temperature increased above 1100°C, $Ba_2Ti_9O_{20}$ phase was formed which on sintering at 1400°C/3 h resulted in a single-phase $Ba_2Ti_9O_{20}$. Chu et al. [101] prepared $Ba_2Ti_9O_{20}$ by co-precipitation using $BaCl_2{\cdot}2H_2O$ and $TiCl_4$ raw materials in the presence of NH_4OH. The powders calcined at 1000°C and samples sintered at 1300°C/4 h showed 98% density. Huang et al. [102] prepared $Ba_2Ti_9O_{20}$ nanopowders by hydrothermal method using $Ba(OH)_2{\cdot}8H_2O$, TiO_2 and de-ionized water in an autoclave at 130–200°C. Javadpour and Eror [35] prepared $BaTi_5O_{11}$, $BaTi_4O_9$ and $Ba_2Ti_9O_{20}$ by firing the material produced by reacting barium carbonate with tetraisopropyl titanate in an ethylene glycol-acid solution. Several authors reported [103–108] the effects of glass

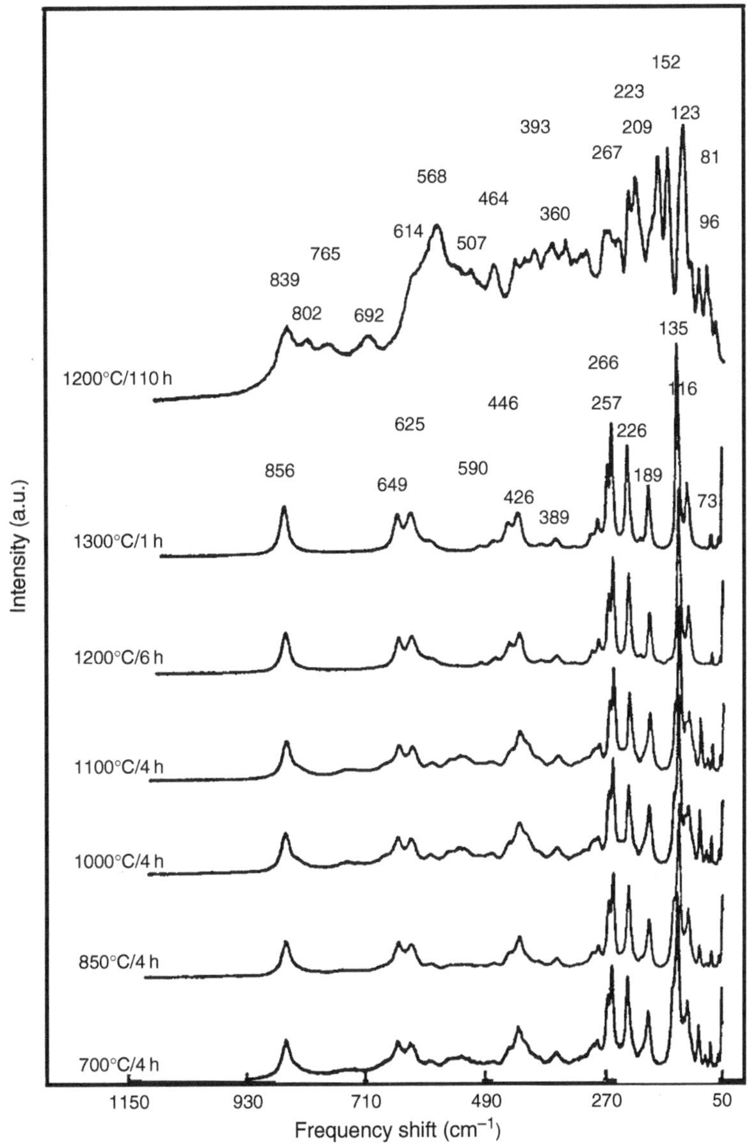

Figure 3.14 X-ray diffraction pattern of $Ba_2Ti_9O_{20}$ as a function of temperature (after Ref. [82]).

addition on the preparation condition and properties of $Ba_2Ti_9O_{20}$. Wu and Wang [91] reported that $Ba_2Ti_9O_{20}$ can be easily formed during sintering whenever calcined powders contain $BaTi_5O_{11}$. The higher the $BaTi_5O_{11}$ content, the greater the amount of $Ba_2Ti_9O_{20}$ formed. They [91] reported that the $Ba_2Ti_9O_{20}$ is the slowest one to form as compared to $BaTi_4O_9$ and $BaTi_5O_{11}$ and is due to the high activation energy for nucleation and higher interfacial energy. They also reported that $Ba_2Ti_9O_{20}$ forms more readily at low temperatures since it is able to form $BaTi_4O_9$ [3, 15, 16, 33] and $BaTi_5O_{11}$ [33, 35, 79, 82, 91, 109] through intermediate phases. However, $BaTi_5O_{11}$ is rarely observed during the solid state reaction [3, 14, 15, 91]. Ritter et al. reported [83] the formation of a mixture of $Ba_2Ti_9O_{20}$, $BaTi_4O_9$ and $BaTi_5O_{11}$ by chemical method. When heated above 1100°C, all these converted into $Ba_2Ti_9O_{20}$. They also reported that $Ba_2Ti_9O_{20}$ did not form until $BaTi_5O_{11}$ began to decompose. Javadpour and Eror found [35] that $BaTi_5O_{11}$ on heating at 1200°C gets converted to a mixture of $BaTi_5O_{11}$ and $BaTi_4O_9$ which on prolonged heating produced $Ba_2Ti_9O_{20}$ and TiO_2. Hence Cheng et al. suggested [76] the following reactions during heat treatment.

$$2BaTi_5O_{11} \rightarrow Ba_2Ti_9O_{20} + TiO_2$$
$$2BaTi_4O_9 + TiO_2 \rightarrow Ba_2Ti_9O_{20}$$
$$BaTi_4O_9 + BaTi_5O_{11} \rightarrow Ba_2Ti_9O_{20}$$

In order to eliminate secondary phases of $BaTi_4O_9$ and $BaTi_5O_{11}$ in $Ba_2Ti_9O_{20}$ a calcination temperature >1200°C and longer time >10 hours are needed.

3.4.2 Structure

O'Bryan et al. reported [45] from precision and Weissenberg XRD studies that the $Ba_2Ti_9O_{20}$ crystals are monoclinic. However, Tillmanns et al. [46] from a detailed X-ray diffraction study arrived at a triclinic cell with space group $P\bar{1}$ (see Table 3.1). Several authors reported [110–112] the occurrence of isolated defects in $Ba_2Ti_9O_{20}$ from high resolution electron microscopic studies. The common defect found in $Ba_2Ti_9O_{20}$ is the formation of polytype which results from periodic occurrence of stacking faults [113]. The polytype formation occurs by the systematic displacement of the Ba ions within the close-packed layers without any change in the overall anion stacking sequence. Each of the barium ion displacements result in a change in the titanium ion octahedral position above and below each closely packed layer. The samples prepared at lower temperatures are prone to be defective. The samples prepared at higher temperatures contained very few polytypic intergrowths. The $Ba_2Ti_9O_{20}$ is reported [111] to contain macroscopic twinning as well as microtwinning. In high resolution transmission electron microscopic study [111], a P1 polytype was observed to be intergrown with the parent $P\bar{1}$ phase. Yu et al. [115] observed diffused streaking in the electron diffraction pattern indicating the presence of planar defects.

The microstructures of slowly cooled $Ba_2Ti_9O_{20}$ samples are found to be heterogeneous. Interior and exterior sections of the same sample showed different phase distributions. A variety of phases and morphologies, which differed from the bulk, were observed on the surfaces of the slowly cooled surfaces. The physical stability and dielectric loss were also observed to depend strongly on slight changes in the initial Ba/Ti ratio [22]. Bryan and Thomson [22] studied the different phases present and their distribution in quenched and slowly cooled $Ba_2Ti_9O_{20}$. The microstructures of samples quenched were homogeneous whereas those slowly cooled from 1400°C were heterogeneous. Quenched samples having compositions with high TiO_2 contents showed well-dispersed grains of rutile whereas compositions with low mole fraction of TiO_2 (MFT) showed small grains of $BaTi_4O_9$. The single-phase composition depends on the quenching temperature. At 1400°C, composition with MFT 0.8168 is single phase, whereas at 1300°C and below the single-phase composition has an MFT ≥ 0.8180. Thus a decrease in temperature moves the single-phase region toward higher TiO_2 and a composition which is single-phase $Ba_2Ti_9O_{20}$ at 1400°C becomes $Ba_2Ti_9O_{20} + BaTi_4O_9$ at 1250°C when quenched. The heterogeneous phase distribution found in slowly cooled samples is attributed to the temperature dependence of the phase boundary. Quenching from 1400°C produces a uniform distribution of rutile within the barium-rich $Ba_2Ti_9O_{20}$ phase because there is insufficient time for dissolving the rutile. When the sample is cooled at 100°C/h some dissolution can occur and the microstructure shows a single-phase region near the sample surface. The core of this sample still has rutile in the $Ba_2Ti_9O_{20}$ matrix. For compositions which are single phase at 1400°C, equilibrium at lower temperatures (1250–1300°C) lies in the two-phase region $Ba_2Ti_9O_{20} + BaTi_4O_9$. Thus such samples contain the $BaTi_4O_9$ phase in the surface region. Since ceramics with a low microwave loss (high Q) must be fully oxidized, $Ba_2Ti_9O_{20}$ resonator must be cooled slowly or re-oxidized in the 1100–1200°C range. These heat treatments will produce secondary phase in the ceramic which is single phase at the sintering temperature of 1400°C. The high quality $Ba_2Ti_9O_{20}$ with low τ_f is usually found to have large surface grains of $BaTi_4O_9$.

3.4.3 Properties

The $Ba_2Ti_9O_{20}$ has a permittivity of 39, Qf of about 32 000 GHz and τ_f of 2 ppm/°C [86, 89, 116, 117]. Fang et al. [117] found that $Ba_2Ti_9O_{20}$ prepared by reacting $BaTi_4O_9$ with TiO_2 and sintered at 1390°C had $\varepsilon_r = 39$, $Qf = 42\,000$ GHz and $\tau_f = 5$ ppm/°C. The excellent properties of this may be due to the high sintered density of 99% and small grain size of about 2–3 μm. Plourde et al. [116] studied the microwave dielectric properties for TiO_2 mol% in the range 79–86 in the $BaO–TiO_2$ system. Figure 3.15 shows the variation of ε_r, τ_f, Q as a function of TiO_2 content measured at 4 GHz. From Figure 3.15, the composition with 81.8 mol% TiO_2 gives the best combination of DR properties $\varepsilon_r = 39.8$, $Qf = 32\,000$ GHz and $\tau_f = \pm 1$ ppm/°C which corresponds to single-phase $Ba_2Ti_9O_{20}$. O'Bryan et al. [28] chemically treated calcined and ball milled fine powder of $Ba_2Ti_9O_{20}$ with 1.5–15.7 mol% HNO_3 up to 16 hours. The ratio of the acid to solid powder was 200 ml to 100 g. The leached powder was washed dried and

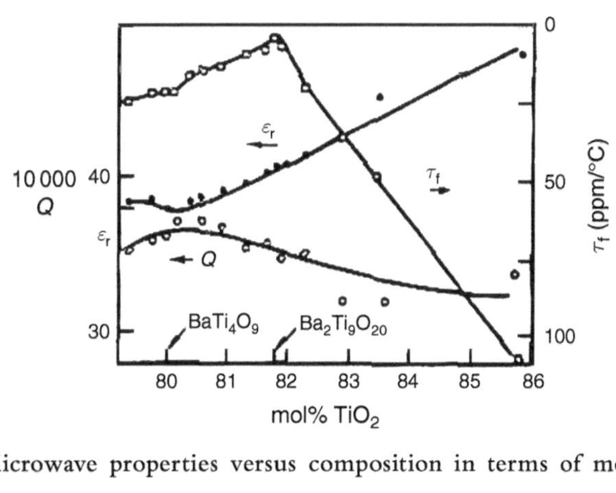

Figure 3.15 Microwave properties versus composition in terms of mol% TiO_2 (after Ref. [119]).

then DRs were made by sintering at 1350–1420°C. The DR made from leached powder had high quality factor as compared to the unleached ones. The microstructure showed the presence of small amount of rutile for compositions ≥81.6 mol% TiO_2. There was no change in the permittivity but the τ_f of the DR made from leached powder showed small changes. The $Ba_2Ti_9O_{20}$ (81.8 mol% TiO_2) phase is not tolerant to deviations in metal stoichiometry and deviations often resulted in the formation of secondary phases in the microstructure. The microstructure of the ceramic compositions with ≥82 mol% TiO_2 showed $BaTi_4O_9$ and $Ba_2Ti_9O_{20}$ phases. Thus the appearance of rutile in ceramics made from leached 81.6 mol% TiO_2 powders is evidence of the selective removal of Ba during the chemical treatment. They [28] found that the shift in τ_f is in agreement with Ba loss. O'Bryan et al. [28] reported that the low *Qf* for ceramics made from unleached powder is due to the temporary presence of low-melting Ba-rich compounds such as $Ba_4Ti_{13}O_{30}$ (MPt 1360°C), $Ba_6Ti_{17}O_{40}$ (MPt 1340°C) and $BaTi_2O_5$. The authors reported that the best *Q* is obtained for ceramics made from powders leached with 7.8 mol% HNO_3 for 3 hours with a particle size of 1 μm. Cerchez and Ciupeiu [118] also reported that chemical treatment of calcined powder using nitric acid improved the quality factor.

Varma et al. [120] studied the effect of the purity of raw materials on the properties and reported that the properties vary considerably with the purity of initial raw materials. It is difficult to obtain a single-phase $Ba_2Ti_9O_{20}$ and those prepared by conventional ceramic method often contain secondary phases of $BaTi_4O_9$ and traces of $BaTi_5O_{11}$. The dielectric properties vary depending on the amount of secondary phases. Usually single-phase $Ba_2Ti_9O_{20}$ is prepared by adding dopants. In such cases, the dielectric properties depend on the nature and amount of dopants added. Several authors studied [33, 62, 88, 89, 94, 119–124] the effects of dopant addition in $Ba_2Ti_9O_{20}$. The dielectric properties of the sintered ceramics containing different dopants are given in Table 3.4. Several authors studied [24, 49, 62, 125] the effect of Mn addition on the dielectric properties of $Ba_2Ti_9O_{20}$. Nomura et al. [62] reported that 1 mol% Mn-doped $Ba_2Ti_9O_{20}$ annealed at 1000°C in oxygen atmosphere for 24 hours showed $\varepsilon_r = 41$, $Qf = 46\,000$ GHz and τ_f of 3 ppm/°C. The Mn doping was done by dipping the calcined powder in a solution of manganese sulfate. The *Qf* increased with increase in Mn concentration and reached a maximum at 1 mol% and then gradually decreased as shown in Figure 3.16. The *Qf* increased with annealing in O_2. The Mn acts as an oxidizing agent. Mn behaves as a compensator in defect equilibrium probably helping to maintain Ti^{4+}. Conventional high temperature sintering route leads to compositional and structural fluctuations due to the reduction of Ti^{4+} to Ti^{3+}. This gives rise to degradation of the dielectric properties [22]. Pure $Ba_2Ti_9O_{20}$ has a gray color whereas with Mn addition changed the color from pale brown to dark brown. There was no appreciable change in ε_r and τ_f by Mn addition but *Qf* increased considerably [62]. Mn added after calcination gave the maximum *Qf*. Nomura et al. reported [126] the maximum *Qf* for 0.5 mol% Mn addition whereas Srivastava et al. observed [125] high *Qf* at about 3 mol%. In ZnO/Ta_2O_5-added samples the highest *Qf* was found for about 0.1 mol% Mn addition [49]. Annealing 6–12 hours at 1000–1275°C increases the *Qf* [62, 86, 127] as much as 50%. To be most effective the sintered samples after cooling below 600°C should be annealed at temperatures between 1000 and 1275°C for 6–12 hours. The cooling rates should be less than 200°C/h. It is found that addition of a small amount of ZrO_2, MnO_2, SnO_2 improves the quality factor [59, 86, 85, 88, 89, 93, 94]. SnO_2 doping increased the *Qf* up to 2.5 mol% whereas Zr doping increased it up to 1.64 mol% and further doping with Zr reduced the *Qf*. The solubility of Sn in $Ba_2Ti_9O_{20}$ was higher than that of Zr [88]. The quality factor decreased with increase of temperature (Figure 3.17). The $Ba_2Ti_9O_{20}$ with 1.64 mol%

Table 3.4 Microwave dielectric properties of ceramics in $BaTi_9O_{20}$

$Ba_2Ti_9O_{20}$ + Dopant	Sintering temperature (°C)	ε_r	Qf (GHz)	τ_f (ppm/°C)	Reference
$Ba_2Ti_9O_{20}$	1350/3 h	39	32 000	2	[86, 89, 116, 117]
Reacting $BaTi_4O_9$ with TiO_2	1390	39	42 000	5	[117]
$Ba_2Ti_9O_{20}$ + 2.5 mol%Nd_2O_3	1350/3 h	40	28 000	20	[119]
$Ba_2Ti_9O_{20}$ + 1 mol% Mn + Annealing at 1000°C/24h in O_2	1200/10 h	41	46 800	3	[62]
$Ba_2Ti_9O_{20}$ + 2 wt%ZrO_2	1390/6 h	40	33 000	–	[89]
$Ba_2Ti_9O_{20}$ + 6 wt% ZrO_2	1390/6 h	39	40 000		[89]
$Ba_2Ti_9O_{20}$ + 0.1 wt%MnO	1390/6 h	40	16 000	–	[89]
$Ba_2Ti_9O_{20}$ + 0.5 mol%MnO	1300/2 h	38	16 000	3	[62]
$Ba_2Ti_9O_{20}$ + ZnO + Nb_2O_5	1260/2 h	37	12 500	–	[121]
$Ba_2Ti_9O_{20}$ + 1.64 mol%ZrO_2	1390/6 h O_2	39.5	41 700	2.1	[88, 92]
$Ba_2Ti_9O_{20}$ + 1.64 mol%SnO_2	1390/6 h in O_2	39.3	38 400	–	[92]
$Ba_2Ti_9O_{20}$ + 1.4 wt%TiO_2	1390/2 h	42	31 000	–	[89]
$Ba_2Ti_9O_{20}$ + 2.46 mol%SnO_2	1390/6 h in O_2	38.8	38 700	1.4	[88]
$Ba_2Ti_9O_{20}$ + 1.64 mol%TiO_2	1390/6 h in O_2	40.2	6300	–	[92]
$Ba_2Ti_9O_{20}$ + 5 wt% (PbO–B_2O_3–SiO_2 glass)	1200	37.2	9800	9	[108]
$Ba_2Ti_9O_{20}$ + 10 wt% (PbO–B_2O_3–SiO_2 glass)	1100	34.8	4900	5	[108]
$Ba_2Ti_9O_{20}$ + BaO–B_2O_3–SiO_2 glass (ceramics: glass 1:1 vol ratio)	900/30 min	13.2	1150	–	[95]
$Ba_2Ti_9O_{20}$ (Hydrothermal)	130–200	25–55	1000–15 000		[102]
$Ba_2Ti_9O_{20}$ (Citrate route)	1300/10 h	37	57 000	–6	[32, 33]

(Continued)

Table 3.4 (Continued)

$Ba_2Ti_9O_{20}$ + Dopant	Sintering temperature (°C)	ε_r	*Qf* (GHz)	τ_f (ppm/°C)	Reference
$Ba_2Ti_9O_{20}$ (Microemulsion coprecipitation)	1250/4 h	39	28 000		[100, 101, 128]
$Ba_2Ti_9O_{20}$ (EDTA-gel method)	1275/3 h	38.3	33 000	6.5	[100]
$Ba_2Ti_9O_{20}$ (Chemically treated)	1410/6 h in O_2	40	36 000	−5	[28]
$0.84Ba_2Ti_9O_{20}$–$0.16BaTi_4O_9$	1340	38.2	36 000	–	[87]
$Ba_2Ti_9O_{20}$ + 5 wt% B_2O_3	1200	36.5	40 200	38	[129]
BaO–$4TiO_2$–$0.1WO_3$ + 5 wt% SiO_2	1200	36	4845		[50]
BaO–$4TiO_2$–$0.1WO_3$ + 5 wt% B_2O_3	1200	35	70 550		[50]
BaO–$4TiO_2$–$0.1WO_3$	1360	36	5200	0	[48, 50]
BaO–$4TiO_2$–$0.1WO_3$ + 5 wt% 5ZnO–$2B_2O_3$	1100	24	13 160		[50]
BaO–$4TiO_2$–$0.1WO_3$ + 5 wt% (ZnO–B_2O_3)	1100	29	6960		[50]
BaO–$4TiO_2$–$0.1WO_3$ + 5 wt% (ZnO–B_2O_3 + SiO_2)	1100	27	8400		[50]
BaO–$4TiO_2$–$0.1WO_3$ + 5 wt% (PbO–B_2O_3 + SiO_2)	1100	25	6600		[50]
BaO–$4TiO_2$–$0.1WO_3$ + 5 wt% (BaO–B_2O_3 + SiO_2)	1100	26	6100		[50]
BaO–$4TiO_2$–$0.1WO_3$ + 5wt% (Al_2O_3 + SiO_2)	1100	32	10 080		[50]
BaO–$4TiO_2$–$0.1WO_3$ + 5 wt% (PbO–Al_2O_3 + SiO_2)	1100	27	8540		[50]
ULF–280 + B_2O_3	940/2 h	28.3	10 800	−8.2	[103]
ULF–280 + 3ZnO–B_2O_3	940/2 h	27.3	8300	2.5	[104, 105]

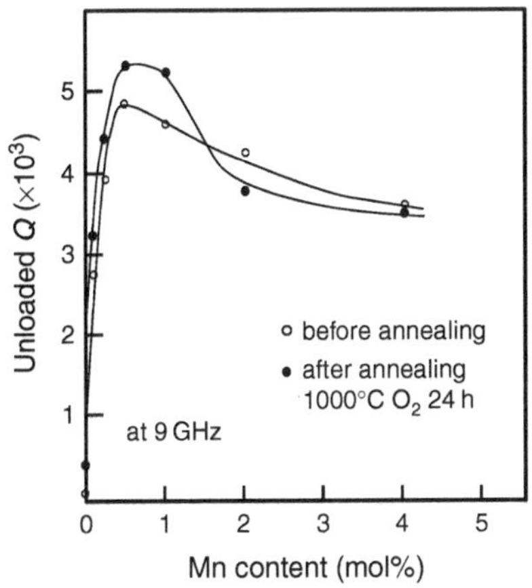

Figure 3.16 Unloaded Q value as a function of Mn concentration (after Ref. [62]).

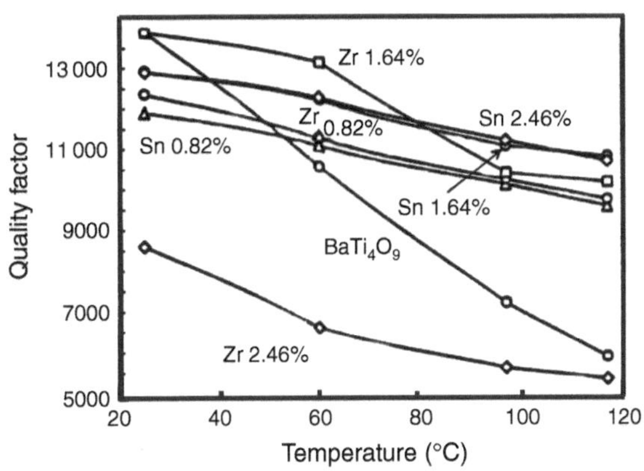

Figure 3.17 Influence of SnO_2 and ZrO_2 substitutions in $Ba_2Ti_9O_{20}$ on the quality factor at 3 GHz as a function of temperature. The sample was sintered at 139°C for 6 hours (after Ref. [88]).

of Sn showed a Qf value of 38 700 GHz whereas it was 41 700 GHz for 1.64 mol% of Zr doping both sintered at 1390°C for 6 hours.

Sreemoolanadhan et al. [130] and Chatterjee et al. [131] reported that Sr substitution at the Ba site increases the permittivity and τ_f. Sreemoolanadhan et al. [130] also reported that CeO_2 doping increases the density, ε_r and τ_f and lowers the quality factor. Addition of larger amount of CeO_2 degrades the dielectric properties whereas 0.5 wt% CeO_2 addition improves the properties. Sebastian and co-workers [130, 132] prepared $Ba_{2-x}Sr_xTi_9O_{20}$ and reported that the ε_r, τ_f and elastic constants increased with increase in x and undergoes a phase transition at $x = 1$. Jaakola et al. [119] reported that Nd addition increased ε_r and τ_f but Q is decreased. The effect is due to the formation of $BaNd_2Ti_5O_{14}$ which has $\varepsilon_r = 80$ and $\tau_f = 93$ ppm/°C. Samples annealed in oxygen increased Qf by 20% and hot-pressed samples have relatively higher Qf [86].

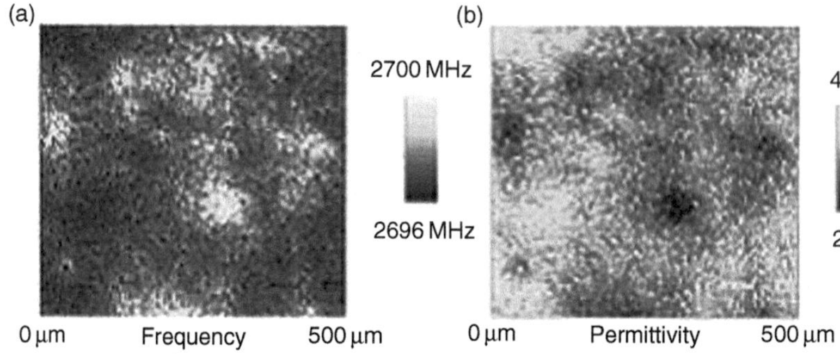

Figure 3.18 Frequency image (a) and dielectric constant (b) for $Ba_2Ti_9O_{20}$ from a sample calcined at 1000°C and sintered at 1340°C /4 h (after Ref. [133]).

Chen et al. [133] measured the dielectric image of $Ba_2Ti_9O_{20}$ using a scanning evanescent microwave probe (EMP) technique. Figure 3.18a shows the frequency image measured by EMP and the dielectric image derived for a sample prepared by calcining at 1000°C and sintering at 1300°C/4 h. Figure 3.18b shows that most of the regions examined possess high permittivity ($\varepsilon_r \sim 40$). The variation of ε_r over the sample is very limited. Only a few areas show either larger or lower ε_r. The average ε_r of the images is in good agreement with the results obtained by traditional cavity method. Different phases with different ε_r and shapes of grains can be seen directly by EMP.

Negas et al. [49] reported that the *Qf* of $Ba_2Ti_9O_{20}$ anomalously decrease by 10–25% irrespective of frequency and purity in the temperature from −60°C to 25°C. They suggested that this anomalous behavior may be due to a diffusion-less transition from triclinic space group $P\bar{1}$ to $P1$. This inversion of symmetry can produce a weak ferroelectric effect decreasing the quality factor. This is supported by the HRTEM observation of a P1 polytype intergrowth with the parent $P\bar{1}$ phase [111].

3.5 $BaTi_4O_9/Ba_2Ti_9O_{20}$ Composites

Addition of a small amount of $BaWO_4$ to $BaTi_4O_9$ considerably improved the microwave dielectric properties [48, 50]. X-ray diffraction study showed that the sintered ceramics is multiphased and consists of $BaTi_4O_9$, $Ba_2Ti_9O_{20}$ and $BaWO_4$. The composition BaO–4TiO$_2$–0.1WO$_3$ showed [50] excellent dielectric properties and is commercially known as N-35. It consists of 9.9 vol% $BaTi_4O_9$, 84.4 vol% $Ba_2Ti_9O_{20}$ and 5.7 vol% $BaWO_4$ with $\varepsilon_r = 36$, $Qf = 52\,000$ GHz and $\tau_f \sim 0$ [50]. Takada et al. [50] added 5–30 wt% glasses such as B_2O_3, SiO_2, 5ZnO–2B$_2$O$_3$ and commercial glasses to BaO–TiO$_2$–WO$_3$ ceramic (N-35). The density increased with sintering temperature and showed maximum density when sintered in the temperature range 1000–1200°C. As the sintering temperature increased, the TiO_2 content decreased and $Ba_2Ti_9O_{20}$ content increased for ceramics without glass addition. Addition of B_2O_3 lowered the sintering temperature and increased ε_r. The density and *Qf* increased with glass addition but larger amount of B_2O_3 decreased *Qf* and ε_r. The maximum density, ε_r and Q were found for samples containing 5 wt% B_2O_3 sintered at 1200°C. The sintered samples contain $BaTi_4O_9$, $Ba_2Ti_9O_{20}$, $BaWO_4$ and TiO_2. Addition of SiO_2 and 5ZnO–$2B_2O_3$ or commercial glass decreased the *Qf*.

Gormikov et al. [97] reported the formation of $Ba_3Ti_{12}Zn_7O_{34}$ in ZnO-added $BaTi_4O_9$. Later studies [47] revealed the formation of $BaTi_4Zn_2O_{11}$ in BaO–ZnO–TiO_2 system. Recently $BaTi_4O_9/Ba_2Ti_9O_{20}$ composites containing substantial amounts of ZnO/Ta_2O_5 have been reported [47] with *Qf* values 20–25% higher. Negas et al. [49] and Lee et al. [47, 52] reported excellent properties for $0.62BaTi_4O_9$–0.35ZnO–$0.3Ta_2O_5$ sintered at 1280°C. The powder samples were ball milled, calcined and sintered at 1220–1300°C up to 20 hours. X-ray diffraction analysis revealed that samples sintered at 1250–1300°C for 2 hours contained a mixture of four phases: $BaTi_4Zn_2O_{11}$, $BaTi_{4-x}Zn_{x/3}\,Ta_{2x/3}O_9$, $Ba_2Ti_{9-x}Zn_{x/3}Ta_{2x/3}O_{20}$ and a small amount of $Ba_3Ti_{4+5x}Ta_{4-4x}O_{21}$. Undoped samples had $Qf < 10\,000$ GHz whereas Mn doping increased *Qf* up to 50 000 GHz for $x = 0.1$. It was found that the samples calcined at 1000°C and sintered at 1280°C have negative τ_f and samples calcined at 1100°C and sintered at 1280°C have positive τ_f. Lee et al. [47, 52] believe that samples calcined at 1000°C have more $BaZn_2Ti_4O_{11}$ as compared to those calcined at 1100°C. The $BaZn_2Ti_4O_{11}$ has a negative τ_f [49]. Hence it is possible to slightly tune τ_f by varying the calcination temperature. Less expensive Nb can be used instead of Ta which slightly increased ε_r [49]. The $0.62BaTi_4O_9$–0.35ZnO–$0.3Nb_2O_5$ has $\varepsilon_r = 36.4$, $Qf = 48\,150$ GHz and $\tau_f = 0$ [49].

Many researchers have reported [58–61] that oxygen is lost from titanate systems during sintering in an oxygen-deficient atmosphere. This leads to formation of electrons, which result in the reduction of Ti^{4+} into Ti^{3+} [25]. Oxygen vacancies in undoped titanate systems are believed to occur due to (a) migration of host materials to the surface or to the grain boundary, (b) reduction of equilibrium with low oxygen activity in air at high temperature and (c) substitution of impurities in raw materials. When the $BaTi_4O_9$/$Ba_2Ti_9O_{20}$ composites were cooled to RT after sintering the samples, they show bluish cores and are characterized by a poor quality factor. This decrease of *Qf* is due to the reduction of Titanium (Ti^{4+} to Ti^{3+}) by equilibrium with low oxygen activity in an air atmosphere [62]. The $BaTi_4O_9$ and $Ba_2Ti_9O_{20}$ also showed such coring but can be eliminated by sintering in O_2 atmosphere or by annealing. But the new composites on annealing for reoxidation lead to the degradation of properties from intervening reactions. The Mn doping had a little effect on ε_r values. The undoped (no Mn) samples always showed the phenomena of coring or Ti reduction on sintering above 1220°C and showed low *Qf* values. A small amount of Mn doping considerably improve *Qf* values. The 0.1 wt% Mn-doped $0.615BaTi_4O_9 + 0.35ZnO + 0.035\ Ta_2O_5$ calcined at 1000°C and sintered at 1280°C/2 h showed a *Qf* of 50 700 GHz, τ_f of 5 ppm/°C and $\varepsilon_r = 35.8$. Addition of 0.02–0.1 wt% Mn yields uncored high *Qf* product. The addition of Mn helps to maintain Ti^{4+} probably by the reaction of

$$Mn^{3+} + Ti^{3+} \rightarrow Mn^{2+} + Ti^{4+}$$

It is also possible that Mn may substitute for Ta or Ti in $BaTi_{4-x}Zn_{x/3}Ta_{2x/3}O_9$ and $Ba_2Ti_{9-x}Zn_{x/3}Ta_{2x/3}O_{20}$ by one of the following reactions depending on the temperature

$$Mn^{2+} + 2Ta(Nb)^{5+} \rightarrow 3Ti^{4+}$$
$$Mn^{3+} + Ta(Nb)^{5+} \rightarrow 2Ti^{4+}$$
$$Mn^{4+} \rightarrow Ti^{4+}$$

The uniform distribution of Mn requires the use of solutions such as manganese nitrate. The Mn must be added to the raw materials and not to the calcined powder. The Mn precursor not only decomposes but catalyzes oxidation of the binder during initial stages of firing [49]. This may lead to rapid generation of gases which affect the properties.

Lee et al. [47, 52] made a detailed study of two compositions: (a) $0.62BaTi_4O_9 + 0.35ZnO + 0.03Ta_2O_5$ and (b) $0.615BaTi_4O_9 + 0.35ZnO + 0.035Ta_2O_5$ with 0, 0.05, 0.1, 0.3 wt% Mn. It was found that the *Qf* considerably increased by the addition of Mn as shown in Figure 3.19. The undoped (no Mn) ceramic samples were having a low *Qf* and contain bluish core. It is found that [52] the quality factor of (a) $0.62BaTi_4O_9 + 0.35ZnO + 0.03Ta_2O_5$ (b) $0.615BaTi_4O_9 + 0.35ZnO + 0.035\ Ta_2O_5$ sintered at 1280°C/2 h increase considerably by annealing in oxygen as shown in Figure 3.20. Improvement of *Qf* values by O_2 annealing indicates that oxygen vacancies exist in undoped titanate ceramics sintered in air. Figure 3.21 shows the effect of sintering time in air at 1250°C on *Qf* values for composition containing 0.1 wt% Mn. As the sintering time increased, the *Qf* value increases. This means Mn doping effectively removed almost all defects caused by sintering in air. It is possible to observe the presence of Ti^{3+} ions using Auger Electron Spectroscopy [134–136]. Lee et al. [52] directly observed the presence of Ti^{3+} ions in $BaTi_4O_9$–ZnO–Ta_2O_5 ceramics sintered in air by Scanning Auger Electron Spectroscopy. Ta^{5+} ions have a similar ionic radius to Ti^{4+} and therefore it can substitute for Ti^{4+}. Lee et al. have studied [52] the effect of Ta_2O_5 addition on the properties of $BaTi_4O_9$. Figure 3.22 shows the variation of *Q* as a function of mol% Ta_2O_5 in Ta_2O_5–$BaTi_4O_9$ ceramics. The *Qf* factor increased by the addition of Ta_2O_5 up to 1 mol% and further addition decreased the *Qf*. Undoped $0.62BaTi_4O_9 + 0.35ZnO + 0.03Ta_2O_5$ and $0.615BaTi_4O_9 + 0.035ZnO + 0.035\ Ta_2O_5$ sintered in air have low quality factors which may be due to electrons associated

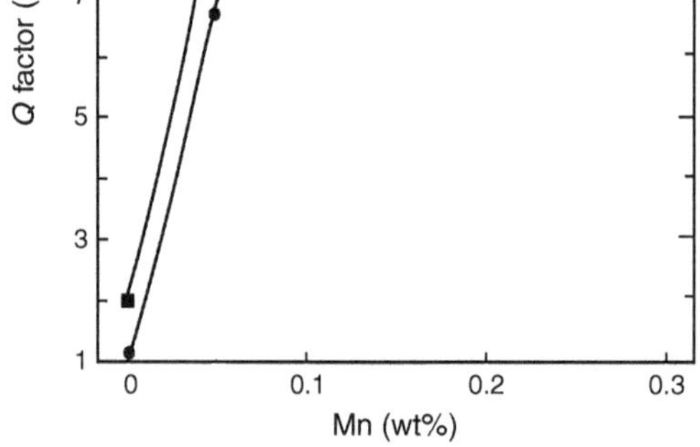

Figure 3.19 *Q*-factor variations of compositions a and b sintered at 1280°C/2 h as a function of Mn content (after Ref. [47]).

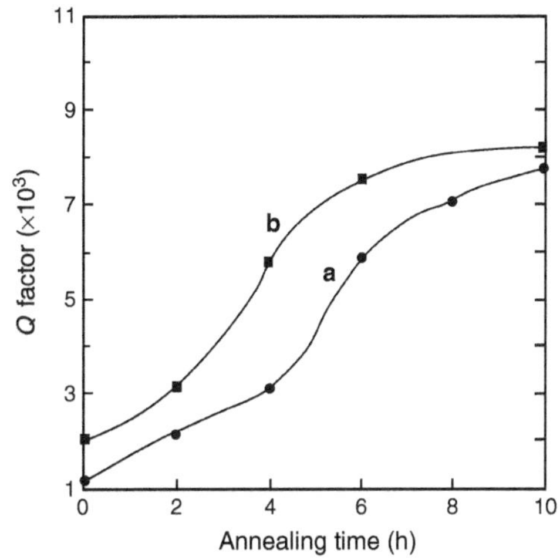

Figure 3.20 The variation in quality factors as a function of oxygen annealing time (after Ref. [52]).

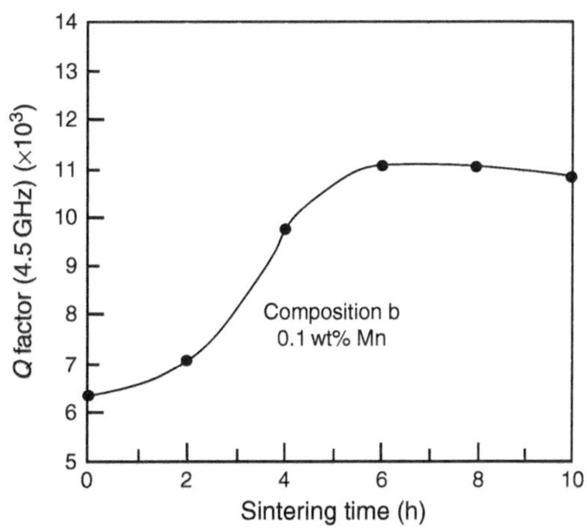

Figure 3.21 The effect of sintering time in air at 1250°C on the Q values for composition (b) (after Ref. [52]).

with intrinsic oxygen vacancies of titanates within the grains and Ta_2O_5 additives that exceed 1 mol% addition. These electrons are believed to reduce Ti^{4+} to Ti^{3+}. In Ta_2O_5-doped $BaTi_4O_9$ ceramics, the Q-factor increased up to 1 mol% of Ta_2O_5 addition that compensated existing oxygen vacancies. Further addition of Ta_2O_5 increased electron concentration resulting in the formation of Ti^{3+}. This is true for $0.62BaTi_4O_9 + 0.035ZnO + 0.035Ta_2O_5$. In the case of Mn-doped ceramics, Mn behaves as a compensator in defect equilibrium probably helping to maintain Ti^{4+} during cooling.

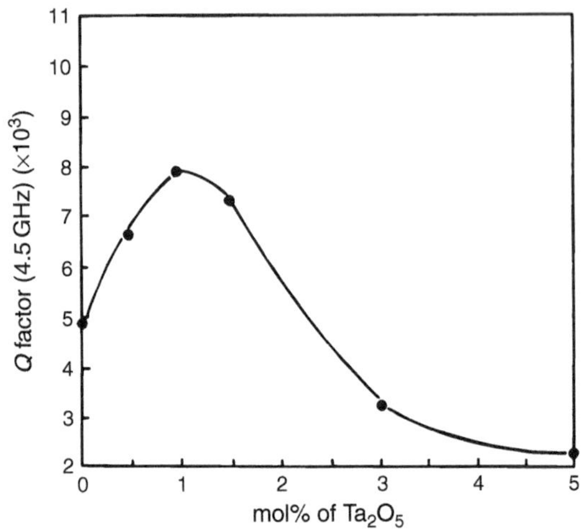

Figure 3.22 Q values as a function of Ta_2O_5 mol% in Ta_2O_5–$BaTi_4O_9$ ceramics (after Ref. [52]).

3.6 CONCLUSION

The BaO–TiO_2 system has three low loss dielectric compounds which are $BaTi_4O_9$, $BaTi_5O_{11}$ and $Ba_2Ti_9O_{20}$. It is difficult to produce these compounds in single-phase form by conventional solid state method since there are several compounds in the vicinity of the desired composition. The $BaTi_4O_9$ is prepared by calcining stoichiometric amount of TiO_2 and BaO at about 1100°C and sintering the shaped DR samples at about 1350°C. The $Ba_2Ti_9O_{20}$ is often found as a second phase in $BaTi_4O_9$ ceramics. The presence of $Ba_2Ti_9O_{20}$ in $BaTi_4O_9$ does not negatively affect the properties since both have nearly the same permittivity and quality factor with nearly a zero τ_f. $BaTi_4O_9$ has an orthorhombic symmetry with Pnmn space group. $BaTi_4O_9$ has $\varepsilon_r = 37$, $\tau_f \sim 15$ ppm/°C which can be tuned to zero by the addition of suitable dopants. It has a quality factor Qf up to 50 000 GHz depending on the preparation condition and dopant addition. Doping with oxides of Mn, Zr, W, and Sn improve quality factor. It is difficult to prepare $BaTi_5O_{11}$ as a single-phase compound by the conventional solid state ceramic route and is usually prepared by wet chemical methods. It has a monoclinic symmetry with $P2/n$ space group. The powder prepared by chemical methods on heating gives single-phase $BaTi_5O_{11}$ in the temperature range 700–1100°C. At temperatures above 1200°C, it decomposes into TiO_2, $Ba_2Ti_9O_{20}$ and/or $BaTi_4O_9$. $BaTi_5O_{11}$ has $\varepsilon_r = 42$, Qf up to 60 000 GHz and τ_f about +40 ppm/°C. $Ba_2Ti_9O_{20}$ is prepared by calcining the stoichiometric amount of TiO_2 and $BaCO_3$ at about 1200°C and sintering at about 1400°C. It usually contains a small amount of $BaTi_4O_9$ or TiO_2 or both as secondary phases. $Ba_2Ti_9O_{20}$ has $\varepsilon_r = 39$, Qf of about 32 000 GHz and $\tau_f \sim 2$ ppm/°C. Addition of dopants such as oxides of Mn, Zr, Sn and W improve quality factor. Samples prepared by chemical methods have Qf up to 57 000 GHz.

The composite 0.62$BaTi_4O_9$–0.35ZnO–0.03Ta_2O_5 + 0.33 wt% of Mn has $\varepsilon_r = 35.4$, $Qf = 48\,000$ GHz and $\tau_f = 0.5$ ppm/°C and 0.615$BaTi_4O_9$–0.35ZnO–0.035Ta_2O_5 + 0.33 wt% of Mn sintered at 1280°C has $\varepsilon_r = 35.8$ $Qf = 50\,800$ GHz and $\tau_f = 1.1$ ppm/°C.

The $BaO–4TiO_2–0.1WO_3$ sintered in oxygen at 1400°C has $\varepsilon_r = 35$ $Qf = 50\,400$ GHz and $\tau_f = -0.5$ ppm/°C. These materials are mixtures of $Ba_2Ti_9O_{20}$, $BaTi_4O_9$ and other phases and have excellent properties useful for practical applications. The $BaO–TiO_2$-based dielectric resonators sintered at high temperatures usually have oxygen vacancies associated with the reduction of Ti^{4+} to Ti^{3+}. Hence sintering in oxygen atmosphere or annealing improves the quality factor. Addition of a small amount of MnO (<1 wt%) which is an oxidizing agent also improves the quality factor. Addition of dopants such as ZrO_2, SnO_2, Ta_2O_5, WO_3 in small quantities also improves the quality factor. Zr, Sn, Ta, and W have ionic radii close to that of titanium and they substitute at the Ti site.

REFERENCES

[1] L. N. Bunting, G. A. Shelton, and A. S. Creamer. Properties of barium strontium titanate dielectrics. J. Am. Ceram. Soc. 30(1947)114–125.

[2] G. R. Shelton, A. S. Creamer, and E. N. Bunting. Properties of barium magnesium titanate dielectrics. J. Am. Ceram. Soc. 31(1948)205–212.

[3] G. H. Jonker and W. Kwestroo. Ternary systems $BaO–TiO_2–SnO_2$ and $BaO–TiO_2–ZrO_2$. J. Am. Ceram. Soc. 41(1958)390–394.

[4] J. Schwarzbach and L. Plocek. Contribution to the study of the system $BaO–TiO_2$ in the region of barium polytitanates. Silik. 11(1968)231–238.

[5] J. Naumann, J. Plotner, and K. Stellenberger. Dielectric properties of compositions in the barium oxide-titanium dioxide system. Hermsdorfer Tech. Mitt. 10(1970)947–950.

[6] D. J. Masse, R. A. Purcel, D. W. Readey, E. A. Maguire, and C. P. Hartwig. New low loss high k temperature compensated dielectric for microwave applications. Proc. IEEE. 59(1971)1628–1629.

[7] D. E. Rase and R. Roy. Phase equilibria in the system $BaO–TiO_2$. J. Am. Ceram. Soc. 38(1955)102–113.

[8] J. A. Bland. The crystal structure of barium orthotitanate Ba_2TiO_4. Acta Crystallogr. 14(1961)875–887.

[9] H. D. Megaw. Temperature changes in the crystal structure of barium titanium oxide. Proc. R. Soc. Ser. A 189(1947)261–283.

[10] R. D. Burbank, and H. T. Evans. The crystal structure of hexagonal barium titanate. Acta Crystallogr. 1(1948)330–336.

[11] F. W. Harrison. Unit cell and space group of $BaO{\cdot}4TiO_2$. Acta Crystallogr. 9(1956)198.

[12] K. Lukaszewicz. Crystal structure of barium tetratitanate $BaO–4TiO_2$. Rocz. Chem. 31(1957)1111–1122.

[13] E. K. Keler and N. B. Karpenko. Formation conditions of barium titanates. Russ. J. Inorg. Chem. 4(1959)511–516.

[14] K. W. Kirby and B. A. Wechsler. Phase relations in the barium titanate-titanium oxide system. J. Am. Ceram. Soc. 74(1991)1841–1847.

[15] H. M. O'Bryan Jr and J. Thomson Jr. Phase equilibria in the TiO_2-region of the system $BaOTiO_2$. J. Am. Ceram. Soc. 57(1974)522–526.

[16] T. Negas, R. S. Roth, H. S. Parker, and D. Minor. Subsolidus phase relations in the $BaTiO_3$–TiO_2 system. J. Solid State Chem. 9(1974)297–307.

[17] J. Guha. Subsolidus equilibria in the system $BaO–TiO_2–GeO_2$. J. Am. Ceram. Soc. 60(1977)246–249.

[18] H. M. O'Bryan and J. Thomson. Preparation of $BaTi_5O_{11}$ by solid state reaction. J. Am. Ceram. Soc. 58(1975)454–455.

[19] E. Tillmanns. Crystal structure of barium titanium oxide ($BaTi_5O_{11}$). Acta Crystallogr. B 25(1969)1444–1452.

[20] E. Tillmanns. Barium hexatitanate $BaTi_6O_{13}$. Cryst. Struct. Commun. 1(1972)1–4.
[21] W. O. Stratton. Phase diagram of the BaO–TiO_2 system. J. Chem. Phys. 19(1951)33–40.
[22] H. M. O'Bryan and J. Thomson, Jr. $Ba_2Ti_9O_{20}$ phase equilibria. J. Am. Ceram. Soc. 66(1983)66–68.
[23] H. M. O'Bryan and M. F. Yan. Second phase development in Ba-doped rutile. J. Am. Ceram. Soc. 65(1982)615–619.
[24] S. G. Mhaisalker, D. W. Readey, and S. A. Akbar. Microwave dielectric properties of doped $BaTi_4O_9$. J. Am. Ceram. Soc. 74(1991)1894–1898.
[25] S. G. Mhaisalker, D. W. Readey, and S. A. Akbar. Processing and characterization of $BaTi_4O_9$. J. Am. Ceram. Soc. 72(1989)2154–2158.
[26] W.-Y. Lin, R. A. Gerhardt, R. F. Speyer, and J. Y. Hsu. Thermal processing and properties of $BaTi_4O_9$ and $Ba_2Ti_9O_{20}$ dielectric resonators. J. Mater. Sci. 34(1999)3021–3025.
[27] M.-H. Weng, T. J. Liang, and C.-L. Huang. Lowering of sintering temperature and microwave dielectric properties of $BaTi_4O_9$ ceramics prepared by the polymeric precursor method. J. Eur. Ceram. Soc. 22(2002)1693–1698.
[28] H. M. O'Bryan, J. Thomson, and J. K. Plourde. Effects of chemical treatment on loss quality of microwave dielectric ceramics. Ber. Dt. Keram. Ges. 55(1978)348–351.
[29] C. Liou, C.-T. Wu, K.-H. Tseng, and T.-C. Chung. Synthesis of $BaTi_4O_9$ ceramics by reaction sintering process. Mater. Res. Bull. 40(2005)1483–1489.
[30] M. H. Weng, T.-J. Liang, and C.-L. Huang. Lowering of sintering temperature and microwave dielectric properties of $BaTi_4O_9$ ceramics prepared by the polymeric precursor method. J. Eur. Ceram. Soc. 22(2002)1693–1698.
[31] S. G. Lu, K. W. Kwok, H. L. W. Chen, and C. L. Choy. Structural and electrical properties of $BaTi_4O_9$ microwave ceramics incorporated with glass phase. Mater. Sci. Eng. B 99(2003)491–494.
[32] J. H. Choy, Y. S. Han, J. H. Sohn, and M. Itoh. Microwave characteristics of BaO–TiO_2 ceramics prepared via a citrate route. J. Am. Ceram. Soc. 78(1995)1169–1172.
[33] J.-H. Choy, Y.-S. Han, J.-T. Kim, and Y.-H. Kim. Citrate route to ultrafine barium polytitanates with microwave dielectric properties. J. Mater. Chem. 51(1995)57–63.
[34] J. H. Choy, Y.-S. Han, S.-H. Hwang, S.-H. Byeon, and G. Demazeau. Citrate route to Sn doped $BaTi_4O_9$ with microwave dielectric properties. J. Am. Ceram. Soc. 81(1998)3197–3204.
[35] J. Javadpour and N. G. Eror. Raman spectroscopy of higher titanate phases in the $BaTiO_3$–TiO_2 system. J. Am. Ceram. Soc. 71(1988)206–213.
[36] Y. Xu, Y. Yuan, P. Lu, and G. Huang. Effects of PH and citric acid contents on the synthesis of $BaTi_4O_9$ via polymeric precursor. Mater. Chem. Phys. 96(2006)427–432.
[37] R. Purohit, and A. Tyagi. Autoignition synthesis of nanocrystalline $BaTi_4O_9$ powder. J. Mater. Chem. 12(2002)312–316.
[38] W. Hoffmeaister, E. Tillmanns, and W. H. Buaer. Refinement of barium tetratitanate $BaTi_4O_9$ and hexa barium-17 titanate $Ba_6Ti_{17}O_{40}$. Acta Crystallogr. C 40(1984)1510–1512.
[39] I. Tanaka and H. Kojima. The growth of barium tetratitanate ($BaTi_4O_9$) by the float zone method. J. Cryst. Growth. 76(1986)311–316.
[40] I. Tanaka and H. Kojima. Dielectric properties of $BaTi_4O_9$ single crystals. J. Mater. Sci. 24(1989)959–962.
[41] I. Tanaka, J. Ishikawa, and H. Kojima. Crystal growth and dielectric properties of impurity doped barium titanate ($BaTi_4O_9$). Denki Kagaku Oyobi Kogyo Butsuri Kagaku 58(1990)503–506.
[42] D. H. Templeton and C. H. Dauben. Ploarised octahedral in barium tetratitanate. J. Chem. Phys. 32(1960)1515–1518.
[43] D. Wadsley. Alkali titanoniobates: Crystal structure of $KTiNbO_5$ and KTi_3NbO_9. Acta Crystallogr. 17(1964)623–628.
[44] R. W. G. Wyckoff. Crystal Structures, Vol. 1, 2nd Edition. Interscience, NY (1963).
[45] H. M. O'Bryan, W. H. Grodkiewicz, and J. L. Bernstein. Preparation and unit cell parameters of single crystals of $Ba_2Ti_9O_{20}$. J. Am. Ceram. Soc. 63(1980)309–310.
[46] E. Tillmanns, W. Hofmeister, and W. H. Baur. Crystal structure of the microwave dielectric resonator $Ba_2Ti_9O_{20}$. J. Am. Ceram. Soc. 66(1983)268–271.

[47] M. J. Lee, B. D. You, and D. S. King. Properties of Mn doped $BaTi_4O_9$–ZnO–Ta_2O_5 ceramics. Part 1. reaction sequence, microstructural analysis and microwave dielectric properties. J. Mater. Sci. Mater. El. 6(1995)165–172.
[48] S. Nishigaki, S. Yano, H. Kato, T. Hirai, and T. Nomura. BaO–TiO_2–WO_3 microwave ceramics and crystalline $BaWO_4$. J. Am. Ceram. Soc. 71(1988)C11–C17.
[49] T. Negas, G. Yeager, S. Bell, and N. Coats. $BaTi_4O_9$/$Ba_2Ti_9O_{20}$ based ceramics resurrected for modern microwave applications. Amer. Cer. Soc. Bull. 72(1993)80–89.
[50] T. Takada, S. F. Wang, S. Yoshikawa, S. J. Jang, and R. E. Newnham. Effect of glass additions on BaO–TiO_2–WO_3 microwave ceramics. J. Am. Ceram. Soc. 77(1994)1909–1916.
[51] K. Mashima, M. Shinohara, K. Kageyama, and H. Nagai. Influence of some oxide addition on dielectric properties of $BaTi_4O_9$. Huntai Oyobi Hummatsuyakin. 38(1991)735.
[52] M. J. Lee, C. Y. Kim, B. D. You, and D. S. King. Properties of Mn doped $BaTi_4O_9$–ZnO–Ta_2O_5 ceramics. Part II. Defect chemical relations. J. Mater. Sci. - Mater. Electron. 6(1995)173–177.
[53] K. Fukuda, R. Kitoh, and I. Awai. Microwave characteristics of mixed phases of $BaTi_4O_9$–$BaPr_2Ti_4O_{12}$ ceramics. J. Mater. Sci. 30(1995)1209–1216.
[54] D. W. Kim, D.-G. Lee, and K. S. Hong. Low temperature firing and microwave dielectric properties of $BaTi_4O_9$ with Zn–B–O glass system. Mater. Res. Bull. 36(2001)585–595.
[55] C.-F. Yang, Microwave characteristics of glass–$BaTi_4O_9$ ceramics. Jpn. J. Appl. Phys. 38(1999)3576–3579.
[56] G. H. Maher, C. E. Hutchins, and S. D. Ross. Preparation and characterization of ceramic fine powders produced by the emulsion process. J. Mater. Process. Tech. 56(1996)200–210.
[57] V. Ern and R. E. Newnham. Effect of WO_3 on dielectric properties of $BaTiO_3$ ceramics. J. Am. Ceram. Soc. 44(1961)199.
[58] Seuter. Philips Res. Rep. Suppl. 3(1974)210.
[59] R. J. Panlener and R. W. Blumenthal. Ti rich nonstoichiometric $BaTiO_3$-III. High temperature thermodynamic and X-ray diffraction measurements. J. Am. Ceram. Soc. 54(1971)610–614.
[60] Burn. Mn-doped polycrystalline $BaTiO_3$. J. Mater. Sci. 14(1979)2453–2458.
[61] M. Monneraye. The control of dielectric losses and temperature coefficient of permitivity by doping in some glass-ceramic materials. J. Glass Technol. 9(1968)70–77.
[62] S. Nomura, K. Tomoya, and K. Kaneta. Effect of Mn doping on the dielectric properties of $Ba_2Ti_9O_{20}$ ceramics at microwave frequency. Jpn. J. Appl. Phys. 22(1983)1125–1128.
[63] E. Paladino. Temperature compensated $MgTi_2O_5$–TiO_2 dielectrics. J. Am. Ceram. Soc. 54(1971)168–169.
[64] H. Sreemoolanadhan, J. Isaac, M. T. Sebastian, K. A. Jose, and P. Mohanan. Synthesis characterization and properties of $Ba_{1-x}Sr_x(Nd_{.5}Nb_{.5})O_3$ ceramics for application as dielectric resonator in microwave circuits. Ceram. Int. 21(1995)385–389.
[65] K. P. Surendran, M. R. Verma, P. Mohanan, and M. T. Sebastian. Microwave dielectric properties of $RE_{1-x}RE_x'TiNbO_6$ [RE = Pr, Nd, Sm, RE′ = Gd, Dy, Y] ceramics. J. Am. Ceram. Soc. 86(2003)1695–1699.
[66] K. P. Surendran, M. Kumar, P. Mohanan, and M. T. Sebastian. Tailoring the microwave dielectric properties of $GdTiNb_{1-x}Ta_xO_6$ and $Sm_{1-x}Y_xTiTaO_6$ ceramics. J. Eur. Ceram. Soc. 23(2003)2489–2495.
[67] P. V. Bijumon and M. T. Sebastian. Temperature stable microwave dielectric ceramics in the $Ca_5A_2Ti_{1-x}Zr_xO_{12}$ (A = Nb, Ta) system. J. Mater. Res. 19(2004)2922–2928.
[68] K. Fukuda, K. Shoda, A. Kunishinge, and I. Awai. Microstructures in mixed phases of $BaTi_4O_9$–$BaPr_2Ti_4O_{12}$ ceramics. J. Mater. Sci. Lett. 13(1994)1290–1292.
[69] H. Fukuda, R. Kitoh, and I. Awai. Microwave dielectric properties of TiO_2–Bi_2O_3 dielectric resonator. Jpn. J. Appl. Phys. 32(1993)4584–4588.
[70] Y. Song, F. Wang, and Z. Jiang. Effects of $BaTi_4O_9$ fibres of $0.64BaTi_4O_9 + 0.36BaEu_2Ti_4O_{12}$ composites. Microelectron. Eng. 66(2003)615–620.
[71] T. Negas, G. Yeager, S. Bell, R. Amren. Chemistry and properties of temperature compensated microwave dielectrics. NIST Spec. Publ. 804(1991)21.

[72] T. Negas and G. Yeager, US Patent 5262370 (1993).
[73] C.-F. Yang, S.-H. Lo, and C.-M. Cheng. Effect of glass addition on $BaTi_4O_9$ microwave ceramics. J. Mater. Sci. Lett. 17(1998)1029–1032.
[74] E. Guan, W. Chen, and L. Luo. Low firing and microwave dielectric properties of $BaTi_4O_9$ with B_2O_3–ZnO–La_2O_3–glass addition. Ceram. Int. 33(2007)1145–1148.
[75] L. Navis and R. L. Green. Dielectric properties of glasses at high frequencies and their relation to composition. J. Am. Ceram. Soc. 29(1946)267–276.
[76] C.-C. Cheng, C.-F. Yang, S.-H. Lo, and T.-Y. Tseng. Sintering of $BaTi_4O_9/Ba_2Ti_9O_{20}$ based ceramics by glass addition. J. Eur. Ceram. Soc. 20(2000)1061–1067.
[77] S. G. Mhaisalkar, D. W. Ready, S. A. Akbar, P. K. Dutta, and J. S. R. Rolhlin. Infrared reflectance spectra of doped $BaTi_4O_9$. J. Solid State Chem. 95(1991)275–282.
[78] T. Fukui, C. Sakurai, and M. Okayama. Effects of heating rate on sintering of alkoxide derived $BaTi_5O_{11}$ powder. J. Mater. Res. 7(1992)192–196.
[79] X. G. Tans, J. Z. Liu, K. W. Kwok, and H. L. W. Chan. Microwave dielectric properties of sol–gel derived $BaTi_4O_9$ and $Ba_2Ti_9O_{20}$:B_2O_3 ceramics. J. Electroceram. 14(2005)119–122.
[80] Y. Suwa, Y. Sugimoto, and S. Nalka. Jpn. Soc. Powder Metallurgy. 25(1978)164.
[81] S. Hirano and J. Otsuka. Sol-gel processing and properties of $BaTi_5O_{11}$ microwave dielectrics. Annu. Meet. Am. Ceram. Soc. Indianopolis, IN (1989).
[82] H.-C. Lu, L. E. Burkhart, and G. L. Schrader. Sol-gel process for the preparation of $Ba_2Ti_9O_{20}$ and $BaTi_5O_{11}$. J. Am. Ceram. Soc. 74(1991)968–972.
[83] J. J. Ritter, R. S. Roth., and J. E. Blendell. Alkoxide precursor synthesis and characterization of phases in the barium titanium oxide system. J. Am. Ceram. Soc. 69(1986)155–162.
[84] G. Pfaff. Sol–gel synthesis barium titanate powders of various compositions. Mater. Chem. 2(1992)591–594.
[85] S. Hirano, T. Hayashi, K. Kikuta, and J. Otsuka. Chemical processing and properties of $BaTi_5O_{11}$ microwave dielectrics. Ceram. Trans. 12(1990)733–740.
[86] H. M. O'Bryan, J. Thomson Jr, and J. K. Plourde. A new BaO–TiO_2 compound with temperature-stable high permitivity and low microwave loss. J. Am. Ceram. Soc. 57(1974)450–453.
[87] D. Hennings and P. Schnabel. Dielectric characterisation of $Ba_2Ti_9O_{20}$ type ceramics at microwave frequencies. Philips Res. J. 38(1983)295–311.
[88] W.-Y. Lin, R. F. Speyer, W. S. Hackenberger, and T. R. Shrout. Microwave properties of $Ba_2Ti_9O_{20}$ doped with zirconium and tin oxide. J. Am. Ceram. Soc. 82(1999)1207–1211.
[89] S. F. Wang, Y. F. Hsu, T.-H. Weng, C.-C. Chiang, J. P. Chu, and C.-Y. Huang. Effects of additives on the microstructure and dielectric properties of $Ba_2Ti_9O_{20}$ microwave ceramics. J. Mater. Res. 18(2003)1179–1187.
[90] I . Naositi and Akita. Dielektisches keramisches material. German Patent 3238858 (1983).
[91] J.-M. Wu and H-W. Wang. Factors affecting the formation of $Ba_2Ti_9O_{20}$. J. Am. Ceram. Soc. 71(1988)869–875.
[92] K. H. Yoon, J. B. Kim, and E. S. Kim. Effect of $BaSnO_3$ on the microwave dielectric properties of $Ba_2Ti_9O_{20}$. J. Mater. Res. 11(1996)1996–2001.
[93] H. M. O'Bryan, J. K. Plourde, and J. Thomson. US Patent. 4563 661 (1986).
[94] W.-Y. Lin and R. F. Speyer. Dielectric properties of microstructure controlled $Ba_2Ti_9O_{20}$ resonators. J. Am. Ceram. Soc. 82(1999)325–330.
[95] W. Huang, K.-S. Liu, L.-W. Chu, G.-H. Hsiue, and I.-N. Lin. Microwave dielectric properties of LTCC materials consisting of glass–$Ba_2Ti_9O_{20}$ composites. J. Eur. Ceram. Soc. 23(2003)2559–2563.
[96] I. Teoreanu, E. Andronescu, and A. Folea. Microwave processing of $Ba_2Ti_9O_{20}$ ceramic. Ceram. Int. 22(1996)305–307.
[97] Y. I. Gormikov, Z. Ya Makarova, A. G. Belous, L. G. Gavrilova, V. M. Paskov, and V. P. Chalyi. Effect of zinc oxide additions on the phase compositions and dielectric properties of barium tetratitanate. Sov. Prog. Chem. 50(1984)1243.
[98] N. Virkar, C. Chatterjee, and A. Paul. Synthesis and characterization of $Ba_2Ti_9O_{20}$. J. Mater. Sci. Lett. 7(1988)179–181.

[99] L. W. Chu, G.-H. Hsiue, Y.-J. Chiang, K.-S. Liu, and I.-N. Lin. Ultrafine $Ba_2Ti_9O_{20}$ microwave dielectric materials synthesized by chemical process. J. Eur. Ceram. Soc. 24(2004)1781–1785.
[100] H. W. Wang and M.-R. Chung. Formation of $Ba_2Ti_9O_{20}$ ceramics from EDTA gel derived powders. Mater. Chem. Phys. 77(2002)853–859.
[101] L.-W. Chu, G.-H. Hsiue, Y.-J. Chiang, K.-S. Liu, and I.-N. Lin. Ultrafine $Ba_2Ti_9O_{20}$ powders synthesized by inverse microemulsion processing and their microwave dielectric properties. J. Am. Ceram. Soc. 88(2005)3405–3411.
[102] J. R. Huang, Z. X. Xiong, C. Fang, and B. L. Feng. Hydrothermal synthesis of $Ba_2Ti_9O_{20}$ nanopowder for microwave ceramics. Mater. Sci. Eng. B 99(2003)226–229.
[103] Y.-C. Lee, W.-H. Lee, and F.-S. Shieu. Effects of B_2O_3 addition on the microwave dielectric properties of $Ba_2Ti_9O_{20}$ ceramics. Jpn. J. Appl. Phys. 42(2002)1311–1314.
[104] Y.-C. Lee, W.-H. Lee, and F.-S. Shieu. Microwave dielectric properties and microstructure of $Ba_2Ti_9O_{20}$ based ceramics with $3ZnO–B_2O_3$ addition. J. Eur. Ceram. Soc. 25(2005)3459–3468.
[105] Y.-C. Lee, W.-H. Lee, and F.-S. Shieu. Low temperature sintering and microwave dielectric properties of $Ba_2Ti_9O_{20}$ with $3ZnO–B_2O_3$ addition. Jpn. J. Appl. Phys. 41(2002)6049–6053.
[106] S.-F. Wang, C.-C. Chiang, and C.-H. Wang. Effects of B_2O_3 on the phase formation of $Ba_2Ti_9O_{20}$ ceramic. Mater. Chem. Phys. 79(2003)256–260.
[107] S.-F. Wang, C.-C. Chiang, C.-H. Wang, and J. P. Chu. Phase stability of B_2O_3 added $Ba_2Ti_9O_{20}$ ceramic processing effects. J. Mater. Res. 18(2003)201–207.
[108] M.-H. Weng and C.-L. Huang. Single phase $Ba_2Ti_9O_{20}$ microwave dielectric ceramics prepared by low temperature liquid phase sintering. Jpn. J. Appl. Phys. 39(2000)3528.
[109] W. Y. Lin and R. F. Speyer. Fabrication of undoped near monophase $Ba_2Ti_9O_{20}$ via rapid thermal processing. J. Mater. Res. 14(1999)1939–1943.
[110] G. Grzinic, L. A. Bursill, and D. J. Smith. The hollandite related structure of $Ba_2Ti_9O_{20}$. J. Solid State Chem. 47(1983)151–163.
[111] P. K. Davies and R. S. Roth. Defect intergrowths in barium polytitanates. J. Solid State Chem. 71(1987)490–502.
[112] A. Pring, D. A. Jefferson, and J. M. Thomas. Defect structures in $Ba_2Ti_9O_{20}$: Intergrowth with a Synroc hollandite phase $BaAl_2Ti_6O_{16}$. J. Solid State Chem. 55(1984)125–132.
[113] M . T. Sebastian and P. Krishna. Random Non Random and Periodic Faulting in Crystals. Gordon & Breach Science Publishers, Reading (1994)28–60.
[114] G. Pfaff. Synthesis and characterization of $BaTi_4O_9$. J. Mater. Sci. Lett. 10(1991)129–131.
[115] J. Yu, H. Zhao, J. Wang, and F. Xia. Effect of Al_2O_3 and Bi_2O_3 on the formation mechanism of Sn doped $Ba_2Ti_9O_{20}$. J. Am. Ceram. Soc. 77(1994)1052–1056.
[116] J. K. Plourde, D. F. Lim, H. M. O'Bryan, and J. Thomson Jr. $Ba_2Ti_9O_{20}$ as a microwave dielectric resonator. J. Amer. Ceram. Soc. 58(1975)418–420.
[117] R.-T. Fang, J.-J. Shiue, and S.-C. Liou. Formation mechanism and sintering behaviour of $Ba_2Ti_9O_{20}$. J. Eur. Ceram. Soc. 22(2002)79–85.
[118] L. Cerchez and A. Ciupeiu. Preparation and sintering of $Ba_2Ti_9O_{20}$ based ceramics and their properties for dielectric resonators. Ceram. Int. 26(2000)99–103.
[119] T. Jaakola, J. Mottonan, A. Uusimaki, R. Rautioaho, and S. Leppavuori. Preparation of Nd doped $Ba_2Ti_9O_{20}$ ceramics for use in microwave applications. Ceram. Int. 13(1987)151–157.
[120] M. R. Varma, H. Sreemoolanathan, and V. Chandrasekharan. Effect of different raw materials on the microwave dielectric properties of $Ba_2Ti_9O_{20}$ ceramics. J. Mater. Sci: Mater. Electron. 15(2004)345–349.
[121] S. Wu, G. Wang, Y. Zhao, and H. Su. $BaO-TiO_2$ microwave ceramics. J. Eur. Ceram. Soc. 23(2003)2565–2568.
[122] S.-F. Wang, T. C. K. Yang, C.-C. Chiang, and S. H. Y. Tsai. Effects of additives on the phase formation and microstructural evolution of $Ba_2Ti_9O_{20}$ microwave ceramic. Ceram. Int. 29(2003)77–81.

[123] W. Lin, R. Gerhardt, and R. Speyer. Thermal processing and properties of $BaTi_4O_9$ and $Ba_2Ti_9O_{20}$ dielectric resonators. J. Mater. Sci. 34(1999)3021–3025.

[124] K. Mashima, M. Shinohara, K. Kageyama, and H. Nagai. Effect of SnO_2 addition on the microstructure and dielectric property of $Ba_2Ti_9O_{20}$. Huntai Oyobi Hummatsuyakin. 38(1991)740.

[125] G. P. Srivastava, M. Srivastava, G. S. Tyagi, V. Mathew, and R. P. Tandom. Studies on Mn doped barium nanotitanate dielectric resonators. J. Mater. Sci. lett. 15(1996)1134–1136.

[126] S. Nomura, K. Kaneta, J. Kuwata, and K. Uchino. Piezoelectric ceramics of the $PbTiO_3$–$La(Me_{2/3}Nb_{1/3})O_3$ (MeMg,Zn) solid solid solution system. Ferroelectrics. 37(1981)587–590.

[127] H. M. O'Bryan. Identification of surface phases on $BaTiO_3$-TiO_2 ceramics. Am. Ceram. Soc. Bull. 66(1987)677–680.

[128] L.-W. Chu, G.-H. Hsiue, and I.-N. Lin. Microwave dielectric materials synthesized by inverse microemulsion process. Mater. Chem. Phys. 79(2003)252–255.

[129] C.-L. Huang, M.-H. Weng, C.-T. Lion, and C.-C. Wu. Low temperature sintering and microwave dielectric properties of $Ba_2Ti_9O_{20}$ ceramics using glass additions. Mater. Res. Bull. 35(2000)2445–2456.

[130] H. Sreemoolanadhan, J. Isaac, P. Koshy, M. T. Sebastian, K. A. Jose, and P. Mohanan. Preparation and microwave characterisation of $Ba_{2-x}Sr_xTi_9O_{20}$ ceramics. Br. Ceram. Trans. 94(1995)157–160.

[131] C. Chatterjee, A. N. Virkar, and A. Paul. Effects of SrO addition on densification and dielectric properties of barium nano titanate. J. Mater. Sci. Lett. 9(1990)1049–1051.

[132] N. Rodrigues, J. Philip, H. Sreemoolanathan, and M. T. Sebastian. Elastic properties of $Ba_2Ti_9O_{20}$ ceramics. Jpn. J. Appl. Phys. 35(1996)673–676.

[133] Y.-C. Chen, H.-F. Cheng, G. Wang, X.-D. Xiang, Y.-C. Chiang, K. S. Liu, and I.-N. Lin. Microwave dielectric imaging of $Ba_2Ti_9O_{20}$ materials with a scanning tip microwave near field microscope. J. Eur. Ceram. Soc. 23(2003)2671–2675.

[134] H. Madden. Chemical information from Auger Electron spectroscopy. J. Vac. Sci. Technol. 18(1981)677–689.

[135] V. Heinrich, G. Dresselhaus, and H. Zeuger. Surface defects and electronic structure of $SrTiO_3$ surfaces. Phys. Rev. B17(1978)4908–4921.

[136] S. B. Desu. Ceramic dielectric compositions, processing and properties. In: Ceramic Transactions. H. C. Ling and M. F. Yan (Eds), American Ceramic Society, Westerville, OH (1990), Vol. 8, p. 157.

[137] M. Cernea, E. Chirtop, D. Neacsu, I. Pasuk, and S. Iordanescu. Preparation of $BaTi_4O_9$ from oxalates. J. Am. Ceram. Soc. 85(2002)499–503.

[138] R. Natarajan, S. Ansari, D. Krishnan, B. Krishnan, S. Kunnambath, and U. Gopinath. Synthesis of polytitanates by carbonate-gel precipitation route. Mater Res. Bull. 40(2005)1408–1414.

[139] Z. Zhang and P. K. Gallaher. Combustion synthesis for $BaTi_4O_9$ and $Pb_xBa_{1-x}Ti_4O_9$. J. Mater. Res. 11(1996)162–168.

[140] F. Li, L.-Q. Weng, G.-Y. Xu, S.-H. Song, and J. Yu. Synthesis and characterization of microwave dielectric $BaTi_4O_9$ ceramics via EDTA-citrate gel process. Mater. Lett. 59(2005)2973–2976.

CHAPTER FOUR

Zirconium Tin Titanate

4.1 Introduction

Zirconium titanate solid solutions have a variety of applications ranging from wireless communications [1–3] to effective acid–base bifunctional catalysts [4] and as high temperature pigments in chemical industry [5]. ZrO_2- and TiO_2-based oxides have also found interesting applications in composites [6] and sensors [7]. $ZrTiO_4$ thin films have potential technological applications in optics [8]. In 1941, Rath [9] reported $ZrTiO_4$ as a useful temperature-stable dielectric ceramic. In the 1950's several laboratories investigated ZrO_2–TiO_2–SnO_2 solid solutions for applications where a low temperature coefficient of capacitance was required [10]. Wakino and co-workers [11–13] first reported $Zr_xTi_ySn_zO_4$ as a temperature-stable low loss dielectric material for applications in the microwave frequency region. Recently zirconium tin titanate has been reported [14, 15] as a suitable material to substitute for amorphous silicon dioxide gate dielectric. The use of silicon dioxide insulating thin layer is limited because the insulating properties break down at higher electric fields due to miniaturization of electronic devices. Zirconium tin titanate is reported [14, 15] to have high figure of merit (FOM). The composition $Zr_{0.8}Sn_{0.2}TiO_4$ gives the best microwave dielectric properties.

4.2 Preparation

4.2.1 Solid state method

The $Zr_{1-x}Sn_xTiO_4$ ceramic is usually prepared [16–18] by the conventional solid-state ceramic route by ball milling the metal oxides, drying and calcining at about 1100°C. The calcined powder is again ball milled and pressed into suitable shapes and then sintered at temperatures in the range 1400–1700°C depending on the composition. $Zr_{0.8}Sn_{0.2}TiO_4$ powders do not readily sinter by solid-state diffusion. It is difficult to fully densify $Zr_{0.8}Sn_{0.2}TiO_4$ without sintering aids if the powder is prepared by the mixed oxide route. Even sintering at high temperatures such as 1600°C was unable to give good densification. Hence ZnO is usually added as a sintering aid. The ZnO additive form a liquid phase at the grain boundaries during sintering thus helping densification through rapid transport of matter through liquid phase and considerably lower the sintering temperatures. Wang et al. [18] prepared $ZrTiO_4$ ceramics with additives such as ZnO, CuO and Y_2O_3 and reported that the microstructure and microwave dielectric properties are sensitive to the presence of additives and processing conditions. Yamamoto et al. [19] prepared single crystals of $ZrTiO_4$ by heating a mixture of $ZrTiO_4$, Li_2MoO_4 and

Dielectric Materials for Wireless Communications
ISBN-13: 978-0-08-045330-9

MoO_3 in the molar ratio 1:2:5 at 1300°C for 5 hours and slowly cooled to 800°C at a rate of 3°C/h and then quenched. Macias et al. [20] obtained crystalline $ZrTiO_4$ powders by heating reactive amorphous precursors. Several authors [21–23] studied the effect of high energy ball milling on the formation of zirconium titanate. Single phase $ZrTiO_4$ was formed on annealing the milled powders and the grain size increased with prolonged annealing. Houivet et al. [21] reported that $Zr_{0.8}Sn_{0.2}TiO_4$ prepared by milling in an Industrial Attrition mill and sintered at 1370°C/2 h in oxygen flow give very low porosity <0.1%.

4.2.2 Wet chemical methods

Wet chemical methods including sol–gel, co-precipitation, and hydrothermal synthesis were developed to obtain ultrafine, very homogeneous and high purity powders. In these ways it is possible to prepare highly reactive precursors and new synthetic compositions which allow one to lower the sintering temperatures and to control the microstructure.

Bianco et al. [24, 25] prepared $ZrTiO_4$ by a polymeric precursor route by dissolving $Ti(OBut)_4$, $ZrOCl_2{\cdot}8H_2O$ and citric acid in ethylene glycol. The solution was heated at 110°C until a viscous transparent gel was formed. Polymeric precursors were obtained by keeping the gels in an oven at 120°C for several days. The polymeric precursors were heat treated at 300°C and then calcined to obtain the ceramic powders. Crystalline powders of $ZrTiO_4$ were formed on calcining at 750°C. The calcined and ground-fine powder was shaped into pellets and sintered at 1600°C. The ceramics sintered up to 98% of the theoretical density. They also prepared [26] zirconium tin titanate thin films through an aqueous polymeric precursor route from titanium isoproxide, $ZrOCl_4{\cdot}8H_2O$ and $SnCl_4{\cdot}5H_2O$.

Hirano et al. [27] produced dense samples of (>96%) pure $Zr_{0.8}Sn_{0.2}TiO_4$ using a metal alkoxide route. The starting materials zirconium n-propoxide, titanium isopropoxide and tin isopropoxide were dissolved in absolute 1-propanol solvent. It was then refluxed and hydrolyzed. The resultant suspension was refluxed to get a precipitate. This method of controlled hydrolysis of metal alkoxides produced a fine (0.3 μm) monosized powders with chemical homogeneity. The particle size and morphology of the as-precipitated particles of $Zr_{0.8}Sn_{0.2}TiO_4$ was strongly dependent on the concentration of metal alkoxides, hydrolysis temperature and the pH of the solution. Dense $Zr_{0.8}Sn_{0.2}TiO_4$ ceramics up to 99% density were obtained by sintering the samples at 1600°C for 24 hours. Ho et al. [28] prepared single-phase $Zr_{0.8}Sn_{0.2}TiO_4$ nanopowders by the sol–gel method using high purity zirconium oxychloride, stannic chloride and tetrabutyl titanate starting raw materials. The gel formed on calcining at 700°C resulted in the crystallization of $Zr_{0.8}Sn_{0.2}TiO_4$ powders. 0.5 wt% ZnO was added to the nanopowders and on sintering at 1300°C for 3 hours gave a density of 97% with a grain size of about 3–9 μm. Khairulla and Phule [29] prepared $ZrTiO_4$ pre-ceramic gels by chemical polymerization of titanium alkoxide-acetate and zirconium acetylacetonate and also by titanium and zirconium alkoxides in the presence of acetic acid. The gelation of sols prepared from zirconium acetylacetonate took several days and the gelation can be expedited by heating at 70°C for 2 hours. However, sols prepared by modified zirconium *n*-propoxide gelated rather quickly. A single-phase $ZrTiO_4$ crystalline powder was formed by gel pyrolysis in the temperature range 200–500°C and heat treatment at 700°C.

Several authors reported [30–33] the preparation of $(Zr,Sn)TiO_4$ by co-precipitation method. Leoni et al. [30] and Zhang et al. [31] obtained $ZrTiO_4$ powders by hydrolysis

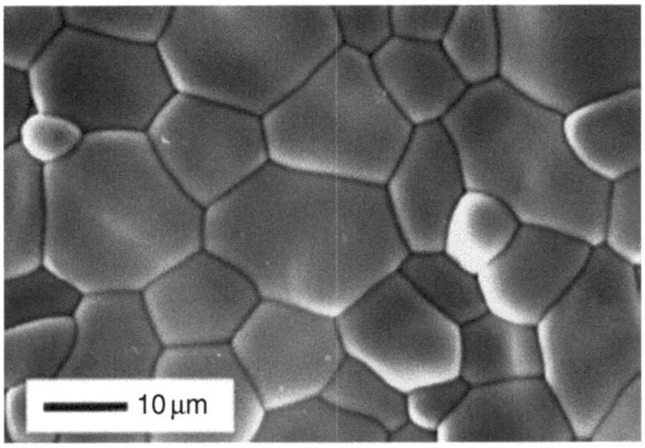

Figure 4.1 Microstructure of a $ZrTiO_4$ ceramic sintered at 1700°C (99.8% of the theoretical density). The samples were thermally etched at 1350°C for 15 min. (after Ref. [30]).

of an aqueous solution of $ZrOCl_2$ and $TiCl_4$ or titanium sulfate using ammonia. Single-phase $ZrTiO_4$ crystalline powders were obtained by calcining the powder at 1200°C for 4 hours. The shaped samples sintered at 1700°C/2 h gave a density up to 99.8%. Figure 4.1 shows the SEM picture of a typical $ZrTiO_4$-sintered ceramic with an average grain size of 11 µm. Han et al. [33] prepared low-temperature sinterable $Zr_{0.8}Sn_{0.2}TiO_4$ powders by adding 3 mol% $Zn(NO_3)_2$ to $Zr_{0.8}Sn_{0.2}TiO_4$ powder derived from co-precipitation of $ZrOCl_2$–$TiCl_4$ solution with ammonia in the presence of SnO_2. The samples sintered at 1250°C for 4 hours showed a density of 99%. Kudesia et al. [32] prepared $Zr_{0.8}Sn_{0.2}TiO_4$ by a co-precipitation technique using zirconyl nitrate, titanium (IV) isopropoxide and tin metal. Samples sintered at 1325°C/4 h with La_2O_3 and ZnO as sintering aid showed 94% density and increased up to 98% by sintering for 16 hours. The sintered density decreased for samples containing more than 0.5 wt% additives. The $Zr_{0.8}Sn_{0.2}TiO_4$ with 0.15 wt% of La_2O_3 and ZnO gave excellent dielectric properties.

The $Zr_{1-x}Sn_xTiO_4$ can also be prepared by hydrothermal methods [34, 35]. Xiong et al. [35] prepared $Zr_{0.85}Sn_{0.15}TiO_4$ nanopowders by a hydrothermal method. The raw materials $ZrOCl_2$, $ZrO(NO_3)_2$ and $SnCl_2$ in de-ionized water were stirred in an autoclave at 160–230°C. The resulting powders with particle size of 10–120 nm were washed, dried, pressed and sintered at 1300°C.

There are a number of publications [36–42] which report the efficacy of liquid phase sintering aids which were employed to increase densification and to improve the microwave dielectric properties of $Zr_{0.8}Sn_{0.2}TiO_4$ ceramics. Liquid-phase sintering aids such as $BaCuO_2 + CuO$, CuO and V_2O_5 were added to $Zr_{0.8}Sn_{0.2}TiO_4$ for the synthesis by the solid state reaction method. Copper oxide and vanadium pentoxide are flux formers and hence it was added to lower the sintering temperature of $Zr_{0.8}Sn_{0.2}TiO_4$ ceramics. Huang et al. [36–39] investigated the effects of using additives such as CuO, Bi_2O_3, V_2O_5 on $Zr_{0.8}Sn_{0.2}TiO_4$. The $Zr_{0.8}Sn_{0.2}TiO_4$ ceramics with 1 wt% ZnO and 1 wt% CuO could be sintered at 1220°C with 96.5% density [37, 39]. Jean and Lin [40] found addition of liquid-phase sintering aid such as $BaCuO_2 + CuO$ densified the ceramics at a low temperature of about 1000°C. Addition of more than 5 wt% CuO retarded the densification process with the formation of secondary phases. Glass addition is a common method for lowering the sintering temperature of ceramics and is the most effective and inexpensive technique [41]. Takada et al. [42] reported that addition of borate and silicate glasses in ZnO-added $Zr_{0.8}Sn_{0.2}TiO_4$ decreased the sintering temperature, permittivity and quality factor. The samples could be sintered in the temperature range 800–1200°C using SiO_2, B_2O_3, $5ZnO–2B_2O_3$ and commercially available glasses.

4.3 Crystal Structure and Phase Transformation

The $ZrTiO_4$ is orthorhombic with space group *Pbcn* [16, 43, 44] with lattice parameters $a = 4.806$ Å, $b = 5.447$ Å, $c = 5.032$ Å and two formula units in the cell with theoretical density of $5.15\,gm/cm^3$. It has been known for years that the lattice parameters of $ZrTiO_4$ quenched from higher temperatures are different from those furnace cooled [45, 46]. The difference has been attributed to an order–disorder phase transition [43, 47–50]. The phase transition in $ZrTiO_4$ is at a temperature between 1100 and 1200 °C [47, 51–54] and was first noticed by Coughanover et al. [45]. This transition is from α-PbO_2-type high-temperature phase with a random distribution of Zr and Ti ions to the low temperature phase with commensurate or incommensurate ordered phase of cations [54–56]. Figure 4.2(a) shows the X-ray diffraction patterns of the high temperature form recorded from a sample heated to 1500°C for 5 hours and then quenched and Figure 4.2(b) shows that of a low form cooled from 1200 to 800°C at the rate of 1.2°C/h and held for 5040 hours. The superstructure reflections corresponding to ordering is clear in Figure 4.2(b). The transition is accompanied by a slight decrease in cell volume [51]. It was found that *b* lattice parameter increases with increase in cooling rate whereas *a* and *c* remains nearly a constant when cooled in the temperature range 1200–1000°C. Figure 4.3 shows the variation of *b* lattice parameter as a function of temperature. In the disordered form of $ZrTiO_4$, the Zr and Ti ions are randomly distributed on octahedral sites in the α-PbO_2-type structure as shown in Figure 4.4a

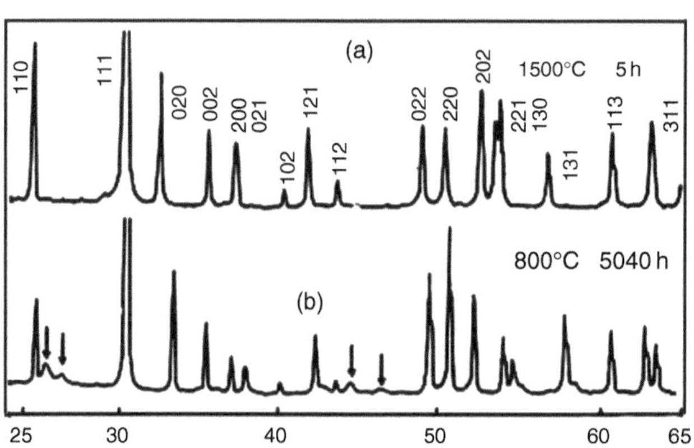

Figure 4.2 X-ray powder diffraction patterns (a) recorded from $ZrTiO_4$ heated to 1500°C for 5 hours and quenched (high temperature form) (b) low temperature form cooled from 1200 to 800°C at the rate of 1.2°C/h and held for 5040 hours. The arrows indicate superstructure reflections (after Ref. [57]).

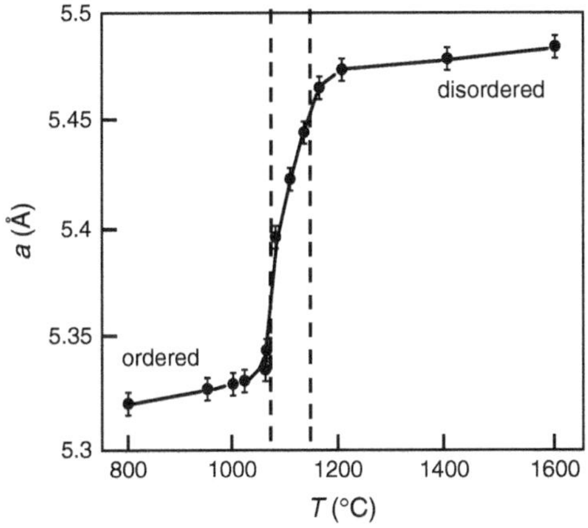

Figure 4.3 Variation of the *b*-lattice parameter of (Zr, Ti)O_4 as a function of temperature. (after Ref. [58]).

[43, 51, 59, 60]. The low temperature form has an ordered arrangement of atoms characterized by incommensurate superstructure diffraction along the a^* axis. In the low–temperature, ordered form the alteration of two distorted Zr layers and two Ti layers give rise to a doubled axis along a-direction [19, 56]. The ordering is accompanied by a decrease in *b*-axis length and an ordered arrangement of metal ions along the *a*-axis [43, 47, 50, 51, 59]. As a result of the order–disorder transition in $ZrTiO_4$, changes of *b*-lattice parameter, electrical conductivity and superlattice reflection pattern were experimentally observed [47, 48, 57]. The incommensurate structure has been analyzed as an antiphase domain structure [19, 61]. The Ti-rich samples near $ZrTi_2O_6$ (e.g., $Zr_5Ti_7O_{24}$) adopt a structure in which the larger Zr ions order onto every third layer along the *a*-axis (see Figure 4.4b) and distort the adjacent oxygen anions to adopt an approximately 8-fold coordinate site [62]. This commensurate phase has 'a' repeat exactly three times that of the disordered structure with a [..ZrTiTiZr..] repeat sequence. The ordered $ZrTiO_4$ end member has a [..ZrZrTiTi..] layer sequence and a commensurate 2× repeat along *a*-axis as shown in Figure 4.4c. The intermediate composition shows incommensurate diffraction effects and their structures consist of a random mixture of the ..ZrTiTiZr.. and ZrZrTiTi.. modules of the two end members. The kinetics of the transition being sluggish, a long range ordered phase of $ZrTiO_4$ can be prepared either by slow cooling of the samples or by annealing them at 1080 °C for a longer duration [47, 49]. The sluggish kinetics of the transformation are indicative of a structural change requiring large scale re-arrangements of the cations in the lattice. Ikawa et al. [57] found differences in the X-ray photoelectron spectra (XPS) for the high temperature form (quenched from 1500°C) and low temperature form (quenched from 1000 °C) of $ZrTiO_4$ and its solid solutions. The O 1s spectrum of the high temperature form consisted of two peaks of nearly equal intensities with a spacing of 2.2 eV while the low temperature form was almost a single peak. The high temperature form had an unassignable spectrum with a binding energy of 25 eV between the peaks of Zr 4P and O 2s.

Wolfram and Gobel investigated [16] the detailed phase diagram of the system ZrO_2–TiO_2–SnO_2 ($Zr_xTi_ySn_zO_4$) where $x+y+z=2$ and reported that single-phase ceramics existed over a limited range of composition. The single-phase region was formed of a

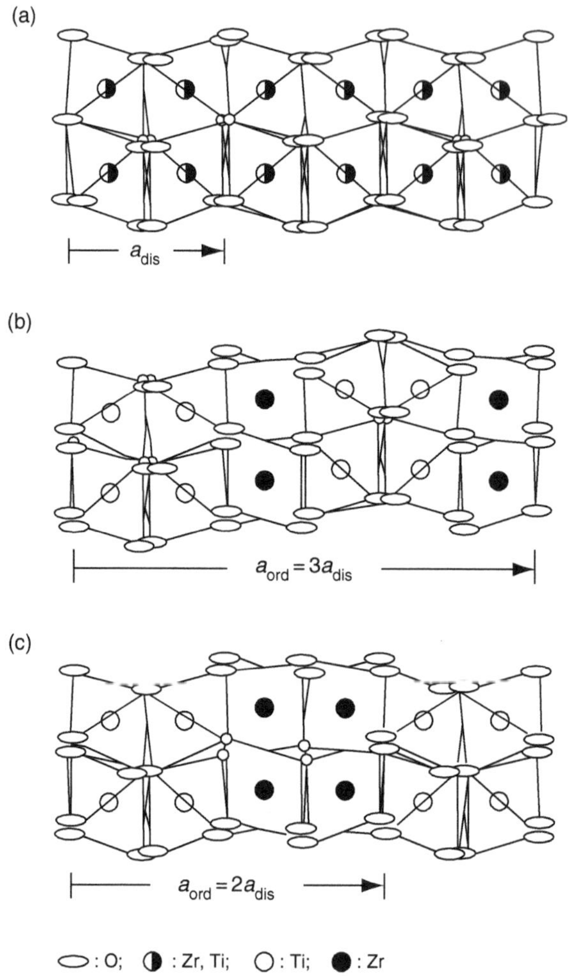

Figure 4.4 [010] projections of $ZrTiO_4$ structures (a) disordered $ZrTiO_4$ with α-PbO_2 structure (b) ordered structure for composition approaching $ZrTi_2O_6$ ($a_{ord} = 3a_{dis}$) (c) ordered structure of $ZrTiO_4$ ($a_{ord} = 2a_{dis}$) (after Ref. [59]).

solid solution with orthorhombic $ZrTiO_4$ structure having space group *Pbcn*. The substitution of Sn in $ZrTiO_4$ stabilizes the high-temperature phase with random arrangement of cations [33] and broadening the phase transition in $ZrTiO_4$ [63, 64]. Figure 4.5 shows the variation of specific heat capacities (Cp) of $Zr_{0.8}Sn_{0.2}TiO_4$ as a function of temperature. The addition of Sn to $ZrTiO_4$ broadens and smears the peak in specific heat capacities and is associated with the fact that Sn substitution for Zr in $ZrTiO_4$ inhibits the order–disorder transition. The DTA curves also show a similar behavior. The increase of x in $Zr_{1-x}Sn_xTiO_4$ makes the thermal anomalies of DTA broaden and weaken. In a similar way increasing Sn content decreases the dielectric anomalies (Figure 4.6). The thermal anomalies imply that the order–disorder transitions do not occur at a discrete temperature but in the temperature range 1116–1124°C. The intensity of the

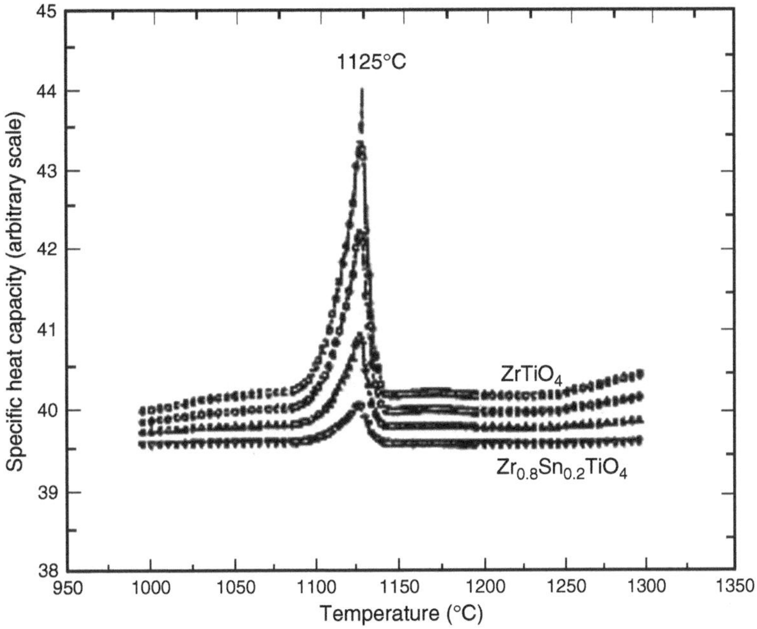

Figure 4.5 The variation of specific heat capacity of tin-modified $ZrTiO_4$. (after Ref. [63]).

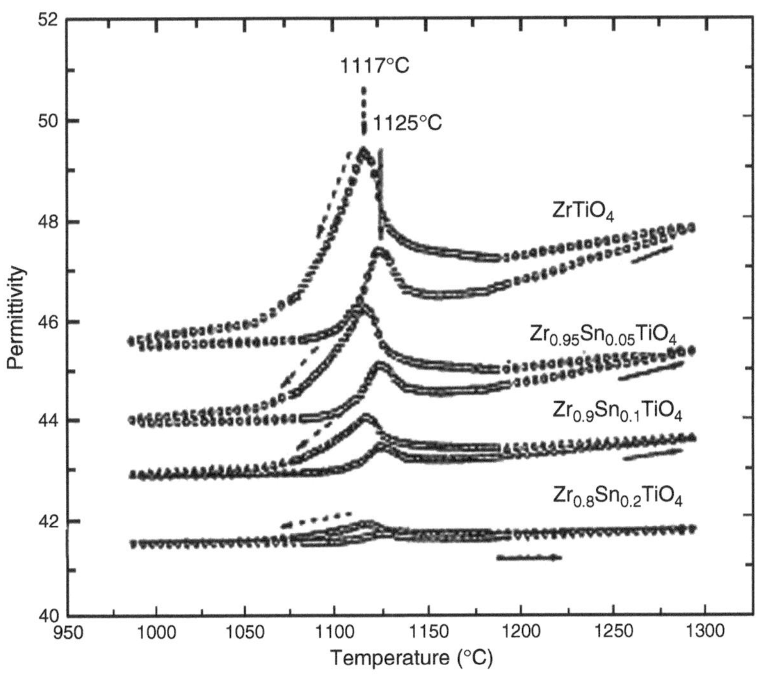

Figure 4.6 The variation of dielectric constant with temperature in $Zr_{1-x}Sn_xTiO_4$ for different values of x. (after Ref. [63]).

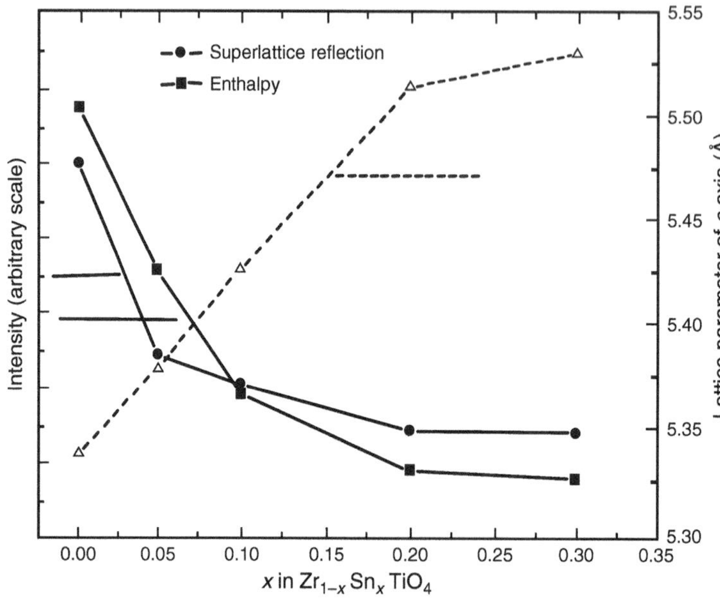

Figure 4.7 Variation of intensity of superlattice reflection, enthalpy and b-lattice parameter as a function of composition x in $Zr_{1-x}Sn_xTiO_4$ ceramics. The cooling rate in the temperature range 1100–1200°C was 1°C/h. (after Ref. [63]).

superlattice reflection and enthalpy decreased and the *b*-lattice parameter increased with increase in the Sn content and with increase in the cooling rate (see Figure 4.7). The ionic radius of Sn^{4+} (0.69 Å in octahedral coordination) is smaller than that of Zr^{4+} (0.84 Å in 8-fold coordination), whereas the addition of Sn to $ZrTiO_4$ increases the *b*-lattice parameter and unit cell volume [16, 43, 47]. Ikawa et al. [65] ascribed that the increase of *b*-axis lattice parameter to the fact that the addition of Sn to $ZrTiO_4$ weakens the crystal force. In addition, annealing for extended periods at high temperatures is known to induce order in $ZrTiO_4$ [47]. The normal to incommensurate phase transition at about 1120 °C is initiated by the short range positional ordering of Zr and Ti cations and has been experimentally observed by X-ray diffraction, selected area electron diffraction and Raman scattering methods [65–68]. The long-range cation ordering occurs only when the disordered $ZrTiO_4$ is cooled with a cooling rate less than 10 °C/1 h and the order parameter increases with decreasing cooling rate [68]. Christofferson et al. [66] and Azough and Freer [67] found that the electron diffraction patterns recorded from rapidly cooled and slow-cooled $Zr_{1-x}Sn_xTiO_4$ were different. The rapidly cooled samples do not show any sign of satellite reflections. However, the diffraction pattern of slow-cooled samples show well-developed incommensurate satellite peaks. The Sn substitution inhibits ordering but this effect is progressive and ordering proceeds to an observable degree in samples with $x < 0.2$. The size of the ordered domains was found [62] to decrease with increase in Sn content. Christofferson et al. [66] investigated the effects of Sn substitution on the cation ordering in $ZrTiO_4$ using HRTEM and reported that Sn substitution reduces the domain size of Zr-rich layer and Ti-rich layer along *a*-axis. The role of Sn is believed to be [66] to stabilize the interface between Zr-rich and Ti-rich domains on (100) which form during cation ordering transformation. Davies suggested [62] that stabilization of the Zr–Ti domain boundaries by Sn may reduce the dielectric loss of the ordered structures.

Several authors studied [68–73] the ordering in $ZrTiO_4$ using Raman scattering. The Factor group analysis of normal vibration modes of the high temperature phase of $ZrTiO_4$ predicts 18 distinctive Raman active optical phonon modes [60, 69]. These modes can be represented as

$$\Gamma_{RA} = 4A_g(xx, yy, zz) + 5B_{ig}(xy) + 4B_{2g}(xz) + 5B_{3g}(yz)$$

Figure 4.8 shows the Raman spectra of polycrystalline $ZrTiO_4$ prepared by cooling at different rates between 1250 and 1000°C and subsequently quenched at 1000°C [68, 70]. The Raman spectra showed splitting of the low frequency bands and decrease of line broadening with decreasing cooling rate. It is known that atomic positional ordering changes the local symmetry and gives rise to mode splitting [71, 72]. Thus the satellite peaks in the XRD and electron diffraction pattern and splitting of Raman bands are due to the long range cation ordering.

Kudesia et al. [74, 75] from a comprehensive structural study of $Zr_{0.8}Sn_{0.2}TiO_4$ using X-ray diffraction, Extended X-ray Absorption Fine Structure (EXAFS), Raman spectroscopy and neutron diffraction reported that the coordinating oxygen polyhedron around cations is severely deformed due to the large difference between the sizes of Zr^{4+}, Sn^{4+} and Ti^{4+}. They found that Zr ions possess eight coordinating oxygens, whereas Ti ions possess 6-fold co-ordination [75]. Raman spectral analysis of tin containing solid solution composition showed that the tin atoms prefer to be on Ti sites. This result is consistent

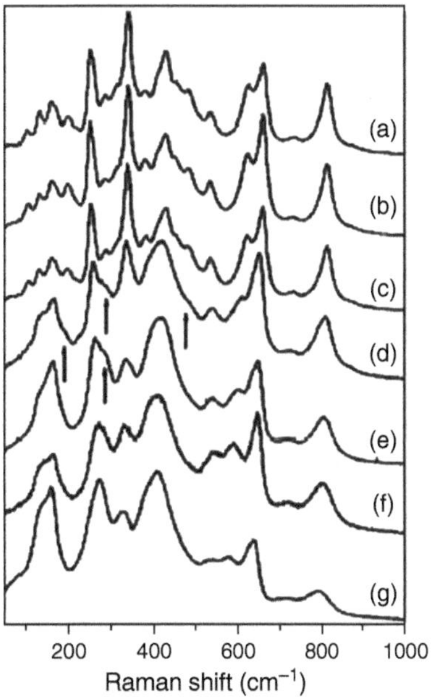

Figure 4.8 Room temperature Raman spectra of $ZrTiO_4$ ceramics prepared employing various cooling rates between 1000 and 1250°C (a) 1°C/h (b)5°C/h (c) 50°C/h (d) 100°C/h (e) 200°C/h (f) 300°C/h and (g) rapidly quenched. The arrows indicate peak splittings. (after Ref. [68]).

with tin having 6-fold coordination as Ti is always surrounded by six oxygens in both high temperature and low temperature structures. XPS studies indicated [74, 75] the existence of a small amount of Sn in the reduced state (Sn^{2+}). These reduced tin ions would create oxygen defects that undergo short range-ordering, creating a superlattice reflection. This ordering phenomenon exists only over small regions of about 80 Å.

4.4 Microwave Dielectric Properties

Several authors [18, 27, 28] reported microwave dielectric properties of $ZrTiO_4$ ceramics prepared by both solid-state ceramic and wet chemical methods. The $ZrTiO_4$ has $\varepsilon_r = 42$, $Qf = 28\,000$ GHz and $\tau_f = 58$ ppm/°C [2, 16, 17, 33, 56, 76–79]. Wang et al. [18] prepared $ZrTiO_4$ ceramics with additives such as ZnO, CuO and Y_2O_3 and found that the microstructure and microwave dielectric properties are sensitive to the presence of additives and processing conditions. Bianco et al. [24, 25] prepared Hf-doped $ZrTiO_4$ by thermal decomposition of polymeric precursors and reported a Qf of 29 700 GHz and $\varepsilon_r = 36$. Wolfram and Gobel [16] reported the microwave dielectric properties of $Zr_xTi_ySn_zO_4$ ($x + y + z = 2$) containing 1 wt% ZnO and 1.5 wt% La_2O_3 as sintering aids and sintered in the range 1300–1350°C for 4 hours. The ε_r decreased initially up to $z = 0.4$ and then it increased for $y = 1$. The quality factor increased with Sn content up to $z = 0.4$ and for $y = 1$ as shown in Figure 4.9. Figure 4.10 shows the variation of τ_f as a function of Sn content (z). The τ_f decreased for $y = 1$ with increasing amount of Sn (z) up to $z = 0.3$ and further increase in z considerably increased the τ_f. Compositions in the range $0.15 < z < 0.3$ and $1-z/2 < y < 1.05$ were found to be of particular interest for dielectric

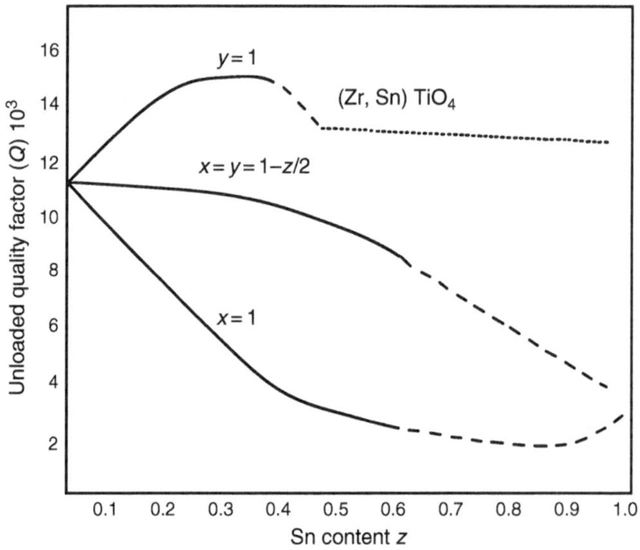

Figure 4.9 Variation of the unloaded quality factor with Sn in $Zr_xSn_zTi_yO_4$ ceramics. (after Ref. [16]).

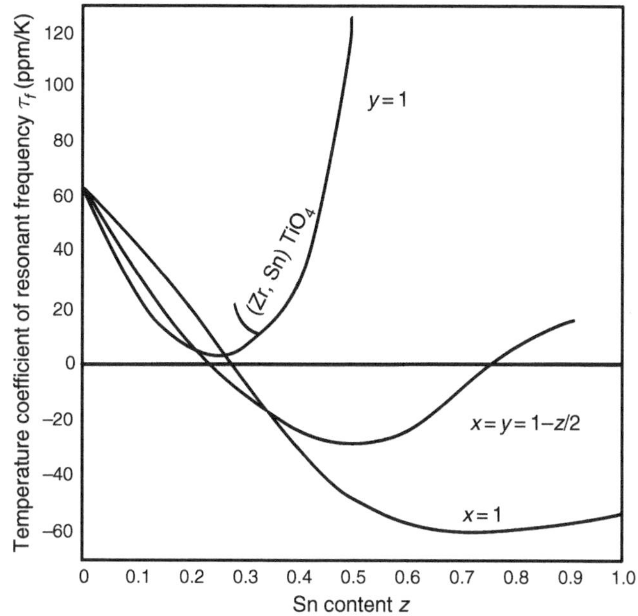

Figure 4.10 Variation of the temperature coefficient of resonance frequency τ_f with Sn content. (after Ref. [16]).

resonator application because they have high quality factors, low τ_f and a high ε_r of greater than 35. The relationship between composition of the solid solution and its dielectric properties is very complex. However, in general increasing the TiO_2 content at the expense of ZrO_2 or SnO_2 increases quality factor and ε_r, whereas increasing SnO_2 content at the expense of ZrO_2 increases quality factor with little effect on ε_r [16, 75]. The composition $Zr_{0.8}Sn_{0.2}TiO_4$, which has excellent microwave dielectric properties such as $\tau_f \sim 0$ ppm/°C, $\varepsilon_r = 38$, $Q = 7000$ at 7 GHz [1], is very much useful for practical applications. Heiao et al. [80] reported that use of anatase raw material improved quality factor as compared to rutile.

In $ZrTiO_4$ improvement in ordering increases the quality factor. Kudesia et al. [32] noted that although the *Qf* appears to increase with larger grains produced by longer sintering times, it is not the grain size itself that is controlling the *Qf*. In general disordered arrangement of cations can be regarded as a deviation from a perfect periodic lattice and thus would be expected to increase dielectric loss. It has been reported [16, 33, 29, 66, 81] that Sn substitution decreases the cation ordering but at the same time improves the quality factor. HREM studies have shown that ordering still occurs in Sn-substituted samples [82]. However, the size of the ordered domains was found to decrease as the Sn content was increased [82]. It is clear that the increase in *Qf* attributable to the replacement of Zr by Sn cannot be explained by cation ordering [82]. Christofferson et al. [66] proposed that segregation of Sn to the domain boundaries reduce their contribution to the dielectric loss of the ceramics.

$Zr_{0.8}Sn_{0.2}TiO_4$ ceramics is one of the most popular commercial dielectric materials for microwave devices [17, 83–85]. However, a major disadvantage of pure $Zr_{0.8}Sn_{0.2}TiO_4$ ceramics is their poor sinterability at <1600°C. Hence extensive research

[29, 30, 36, 42, 79, 81, 85–91] was done on adding minor components to $Zr_{0.8}Sn_{0.2}TiO_4$ system to lower the sintering temperature and to improve densification without causing significant deterioration of the required dielectric properties. Therefore sintering aids such as ZnO have been invariably added to achieve good densification at temperatures between 1200 and 1500 °C. Although addition of ZnO considerably improves densification at a relatively lower temperature [1, 32, 81, 85] of 1400 °C, the microwave quality factor is not improved. Mc Hale et al. [92] proposed a defect model in which a limited solubility of the sintering aid or dopant significantly affect final state of densification. Sintering aids used for $Zr_{0.8}Sn_{0.2}TiO_4$ ceramics are added as a combination of two or more oxides as given in Table 4.1. Huang et al. [88] reported that a small amount of ZnO (1 wt%) and WO_3 (up to 1 wt%) can significantly improve the density and dielectric properties of $Zr_{0.8}Sn_{0.2}TiO_4$ ceramics. WO_3 larger than 0.25 wt% reduces the Qf factor. Table 4.1

Table 4.1 Dielectric properties of $[Zr_{0.8}Sn_{0.2}]TiO_4$ with different additives/dopants

Sintering temperature °C	Dopants	ε_r	Qxf GHz	τ_f	Reference
1600/4 h	No dopant	38.9	51 500	0.7	[93]
1360/4 h	1.5 wt% ZnO + 0.5 wt% Fe_2O_3	38	2500		[1]
1360/4 h	1.5 wt% ZnO + 0.5 wt% NiO + 0.5 wt% Fe_2O_3	38	39 000	0	[1]
1360/6 h	1 wt% CdO + 0.5 wt% NiO + 0.5 wt% La_2O_3	35.5	42 750	−7	[80]
1300/3 h	1 wt% ZnO + 0.5 wt% Bi_2O_3	38.6	31 160	−1.8	[34]
1370/2 h	2 wt% La_2O_3, 1 wt% NiO	37	62 000	−9	[21, 94]
1370/20 h	La_2O_3, NiO + annealing	37.1	50 000		[95]
1400/4 h	1 wt% $BaCO_3$	39.5	47 500	0.3	[93]
1400/4 h	1 wt% $SrCO_3$	38.6	44 500	8.9	[93]
1400/4 h	1 wt% $CaCO_3$	39.2	46 500	0.1	[93]
1450/4 h	1 wt% $MgCO_3$	38.9	4000	−2.3	[93]
1380/4 h	0.5 wt% ZnO + 0.2 wt% NiO	38	58 800	−0.1	[17, 87]
1380/4 h	0.5 wt% ZnO + 0.2 wt% NiO + Fe_2O_3	38.1	5670	–	[87]
1350/4 h	1 wt% La_2O_3 + 2 wt% BaO	41	18 000	−3	[96, 97]
1300/3 h	1 wt% ZnO + 0.5 wt% Bi_2O_3	38.6	31 160	−1.8	[36]

Table 4.1 (Continued)

Sintering temperature °C	Dopants	ε_r	Qxf GHz	τ_f	Reference
1280/2 h	1 wt% ZnO + 2 wt% La_2O_3 + 0.2 wt% NiO	35.9	56 576	–	[91]
1220	1 wt% ZnO + 1 wt% CuO	38	50 000	3	[36–38]
1400/5 h	1 wt%ZnO + 1 mol%Nd_2O_3	38	27 000	–	[89]
1500/2 h	0.5 wt% B_2O_3 + 0.3 wt% La_2O_3 + 1 wt% $NiTa_2O_6$	39.3	52 000	1	[98]
1400	1.5 wt% ZnO + slow cooled	36	51 000	–	[67]
1100/4 h	5 wt% SiO_2	10	12 610	–	[42]
1100/4 h	5 wt% B_2O_3	9	12 360		[42]
1100/4 h	5 wt% 5ZnO–2B_2O_3	15	23 660		[42]
1100/4 h	5 wt% ZnO–B_2O_3	16	14 410		[42]
1100/4 h	5 wt% PbO–B_2O_3–SiO_2	17	11 760		[42]
1100/4 h	5 wt% ZnO–B_2O_3–SiO_2	20	17 220		[42]
1100/4 h	5 wt% BaO–B_2O_3–SiO_2	17	12 460		[42]
1100/4 h	5 wt% Al_2O_3–B_2O_3–SiO_2	16	4920		[42]
1100/4 h	5 wt% PbO–Al_2O_3–SiO_2	12	8100		[42]
1100/4 h	5 wt% Al_2O_3–SiO_2	15	8820		[42]
1400/5 h	1 wt% ZnO + 1 mol% Sb_2O_3	37.5	28 000	–	[95]
1400/5 h	1 wt% ZnO + 1 mol% Sb_2O_5	40.8	60 900		[99, 100]
1400/5 h	1 wt% ZnO + 0.5 wt% Nb_2O_5	38	47 600		[97, 100]
1450/5 h	1 wt% ZnO + 1 mol% Ta_2O_5	40.5	53 200		[100]
1400/5 h	1 wt% ZnO + 1 mol% Nb_2O_5	40	47 600		[100]
1400/2 h	2 mol% ZnO + 0.5 mol% Nb_2O_5	37	48 000	–4	[101]
1400/4 h	1 wt% ZnO + 0.5 wt% Ta_2O_5	38.5	53 200		[99, 100]

(Continued)

Table 4.1 (Continued)

Sintering temperature °C	Dopants	ε_r	Qxf GHz	τ_f	Reference
1380/4 h	0.5 wt% ZnO + 0.2 wt% NiO + Ta_2O_5	37.8	60 200	–	[87]
1340/3 h	1 wt% ZnO + 0.25 wt% WO_3	37.8	61 000	−3.9	[88]
1300/3 h	1 wt% ZnO + 1 wt% V_2O_5	37.2	51 200	−2.1	[36]
1300/5 h	1 mol% Sb_2O_5 + 0.35 wt% B_2O_3–Li_2O + slow cooled	37	62 000	–	[102]
1300/5 h	1 mol% WO_3 + 0.35 wt% B_2O_3–Li_2O + slow cooled	37	60 000	–	[102]
1000/2 h	2.5 wt% $BaCuO_2$ + CuO	36	35 000	–	[40]
–	2 wt% $MnCO_3$	35	51 000	26	[103]
–	ZnO + CuO + Bi_2O_3	39	35 000	−40	[103]
1250	Coprecipitation	40.9	49 000	−2	[30]
1260/1 h	Hydrothermal	40	20 000	–	[35]
1600/3 h	Alkoxide	40	50 000	3	[27]
1300/3 h	Sol–gel + 0.5wt% ZnO	38	55 000	0.5	[28]
1325/16 h	0.15 wt% (La_2O_3 + ZnO) Co-pption	37.6	53 750	−3	[32]
130/20 h	1 wt% NiO + 2 wt% La_2O_3 + industrial attrition milled	37.6	62 000	−9	[21, 94]
1350 hot forging	$Zr_{0.8}Sn_{0.2}TiO_4$	38	40 900		[66]

gives the dielectric properties of the $Zr_{0.8}Sn_{0.2}TiO_4$ ceramics prepared with different sintering aids and dopants. Fe_2O_3 decreases *Qf*, but Fe_2O_3 + NiO retains the high *Qf* [1]. Huang et al. [36] reported that addition of ZnO + V_2O_5 and sintered at 1300 °C retains the high quality factor of $Zr_{0.8}Sn_{0.2}TiO_4$, whereas Lee [104] found a decrease by V_2O_5 addition due to the formation of V_2O_5–TiO_2 rich secondary phase at the grain boundary.

Several authors studied [67, 95, 102] the effect of annealing the $Zr_{0.8}Sn_{0.2}TiO_4$ ceramics on the dielectric properties. It was found that slow-cooled ceramic has a much higher quality factor as compared to rapidly cooled ceramic. Azough and Freer [67] selected $Zr_{0.8}Sn_{0.2}TiO_4$ samples containing 1.5 wt% of ZnO and sintered at 1400°C/4 h and annealed at 700–1100°C range up to 1000 hours. Rapid cooling from the sintering temperature yielded a disordered structure having a low *Qf* value. The permittivity was nearly independent of the cooling rate but the quality factor increased considerably with slow cooling. Ahn et al. [102] studied the effect of cooling rate on grain size and dielectric properties of WO_3- and Sb_2O_5-doped $Zr_{0.8}Sn_{0.2}TiO_4$ containing 0.35 wt% B_2O_3–Li_2O sintering aid. The slow-cooled $Zr_{0.8}Sn_{0.2}TiO_4$ samples showed a *Qf* greater than 60 000 GHz. The increase in *Qf* with decrease of cooling rate is attributed to ordering of cations with decrease of cooling rate.

Tamura and co-workers investigated [1, 87, 105] the role of lattice defects such as oxygen vacancy in $Zr_{0.8}Sn_{0.2}TiO_4$. They studied four samples: (a) a benchmark sample prepared by usual solid state method; (b) sample annealed in an oxygen-free atmosphere. The annealing introduced about 1000 ppm oxygen vacancies; (c) sample prepared by doping 1.3 mol% Fe_2O_3 and (d) sample doped with 0.5 mol% Ta_2O_5. The $\tan\delta$ was higher for samples annealed in oxygen-free atmosphere. The samples doped with 1.3 mol% Fe_2O_3 had a much higher $\tan\delta$ whereas that doped with Ta_2O_5 had a $\tan\delta$ close to the bench mark as shown in Figure 4.11. The addition of Fe ions creates oxygen vacancies which led to an increase in $\tan\delta$. Addition of a small amount of Fe_2O_3 drastically reduces the *Qf* factor [1, 87, 105]. If the Fe_2O_3 was added along with a similar amount of NiO, the effect was greatly reduced [1]. The effect was attributed to the lower diffusion of Fe ions into the grains in the presence of Ni. The Ni ions gather at the grain boundaries, forming low melting phases such as (Zn,Ni)TiO_4 and (Zn,Ni)Fe_2O_4. Even though these structures have a very low *Qf*, they have less effect on the total *Qf* because they are present in low volume, whereas if the Fe is diffused into the grain, it affects the entire material volume. Fe ions in the grain cause loss because they occupy the Zr^{4+}, Ti^{4+} and Sn^{4+} sites as Fe^{3+}. This creates oxygen vacancies to maintain charge neutrality [87]. Figure 4.12 represents a defect model of Fe^{3+}- and Ta^{5+}-doped $Zr_{0.8}Sn_{0.2}TiO_4$, where M and O denote Ti^{4+} and oxygens. Substitution of Ta^{5+} leads to a positive charge and Fe^{3+} leads to a negative charge. The presence of oxygen vacancies gives positive charges. The presence of Fe ions and the associated oxygen vacancies degrade the quality factor. The high temperature sintering may lead to reduction of Ti^{4+} to Ti^{3+}.

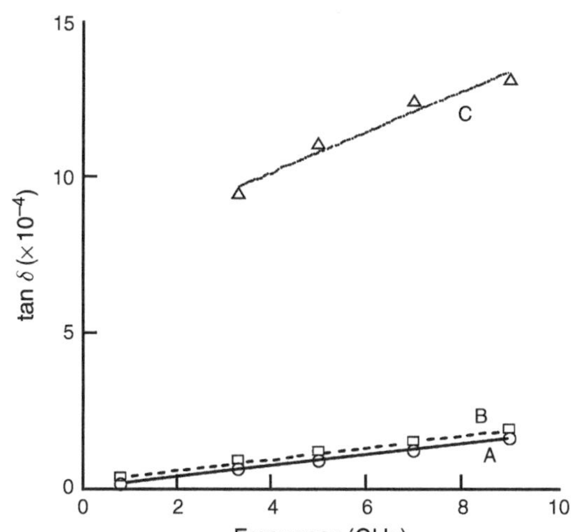

Figure 4.11 The frequency dependence of loss tangent in Fe- and Ta-doped zirconium tin titanate ceramic. A-Benchmark, B-0.5 mol% Ta_2O_5, C-1.3 mol% Fe_2O_3 (after Ref. [17, 105]).

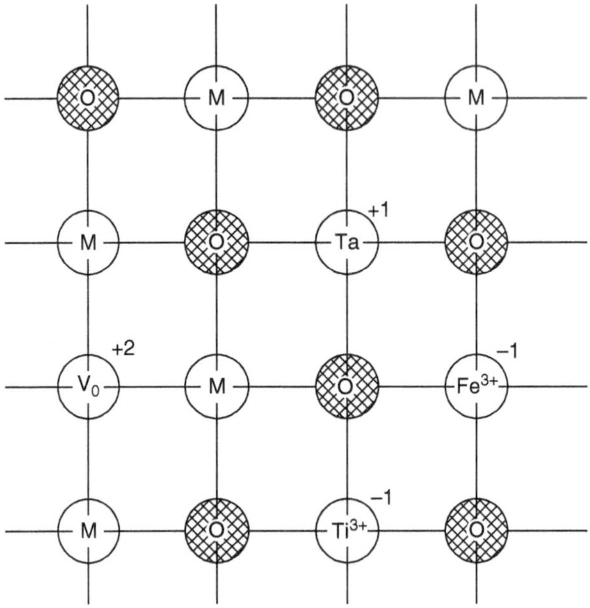

Figure 4.12 Simplified defect model for charge deviation of substitution of Ta^{5+} and Fe^{3+} for tetravalent cations and of oxygen vacancies. M and O denote cation and oxygen respectively. (after Ref. [105]).

In Ta-doped $Zr_{0.8}Sn_{0.2}TiO_4$, the Ta^{5+} may compensate for Ti^{3+} to maintain charge neutrality [87, 99, 100, 105]. This is supported by the fact that doping with Nb_2O_5, Sb_2O_5, V_2O_5 and WO_3 increases quality factor (see Table 4.1). Additives such as $ZnO + NiO + Ta_2O_5$, $ZnO + Ta_2O_5$, $ZnO + Nb_2O_5$, $ZnO + Sb_2O_5$, $Sb_2O_5 + B_2O_3$, $+ Li_2O$, $WO_3 + B_2O_3 + Li_2O$, $ZnO + NiO$ improve *Qf* [87, 88, 100–102]. Iddles et al. [81] reported that donor ions such as Nb^{5+} reduced the number of oxygen vacancies and decreased the dielectric loss, while acceptor ions such as La^{3+} produced electrons and increased the dielectric loss. The Nb^{5+}, Ta^{5+} and Sb^{5+} ions could substitute for Zr^{4+}, Sn^{4+}, Ti^{4+} ions in the $Zr_{0.8}Sn_{0.2}TiO_4$ lattice because they all have a similar atomic radius. Yoon et al. [99, 100] studied the microwave dielectric properties of $Zr_{0.8}Sn_{0.2}TiO_4$ as a function of the amount of additives such as Nb_2O_5, Ta_2O_5, Sb_2O_5. 1 wt% ZnO was also added as sintering aid and the samples were sintered at 1400 °C/5 h in oxygen or N_2 atmosphere. The maximum density was found for those samples sintered in oxygen atmosphere for 1 mol% Nb_2O_5. The highest ε_r was for Sb_2O_5 which also showed the highest *Qf*. Figure 4.13 shows the variation of *Qf* as a function of mol% of different additives. As the amount of additives increased up to 1 mol%, the unloaded *Qf* increased due to decrease of oxygen vacancies [81, 95] in the $Zr_{0.8}Sn_{0.2}TiO_4$ lattice and increase in density. The Q then decreased with further addition of dopants. It was found that Sb_2O_5-added samples show the highest unloaded *Qf*. The samples sintered in oxygen atmosphere showed the best properties. It is interesting to note that Sb_2O_5 addition increases the *Qf* factor and the addition of Sb_2O_3 decreases the *Qf* value [89].

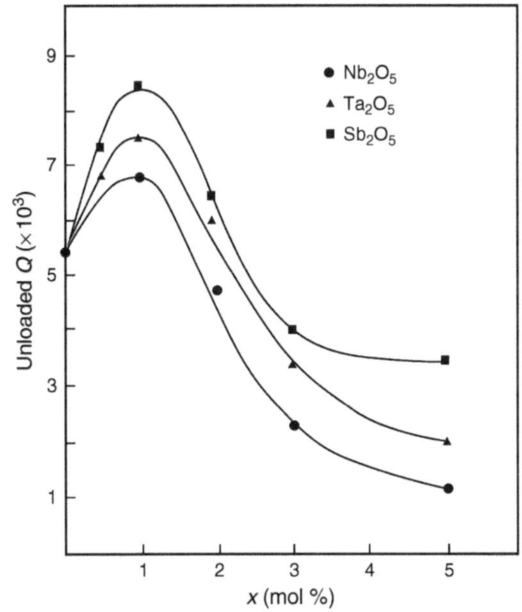

Figure 4.13 Variation of unloaded quality factor at 7 GHz with the amount of Nb_2O_5 (•) Ta_2O_5(Δ) and Sb_2O_5 (♦)for sintered $Zr_{0.8}Sn_{0.2}TiO_4$ (after Ref. [100]).

Huang et al. [36–39] reported that the addition of CuO, Bi_2O_3, V_2O_5 in $Zr_{0.8}Sn_{0.2}TiO_4$ reduced sintering temperature to 1220°C due to liquid-phase effects. The ε_r and τ_f are not significantly affected but the Qf is very much affected. Bi_2O_3 addition reduced the Qf considerably whereas 1 wt% CuO and V_2O_5 addition gave excellent quality factor. Huang et al. [37, 39] reported that $Zr_{0.8}Sn_{0.2}TiO_4$ ceramics with 1 wt% ZnO and 1 wt% CuO or V_2O_5 could be sintered at 1220°C with 96.5% density and showed ε_r of 38, Qf of 50 000 GHz and τ_f close to zero ppm/°C. Houivet et al. [21] reported that $Zr_{0.8}Sn_{0.2}TiO_4$ prepared by milling in an Industrial Attrition mill using zirconia grinding media gave very low porosity <0.1% and the samples had $\varepsilon_r = 37$, $Qf = 55\,000$ GHz. The Qf increased to 62 000 GHz when the grinding media was magnesia-stabilized zirconia balls [94].

Several authors reported [27–34, 106] that the quality factor of $Zr_{0.8}Sn_{0.2}TiO_4$ prepared by wet chemical methods are comparable or higher than those prepared by conventional solid state method. Hirano et al. [27] reported that $Zr_{0.8}Sn_{0.2}TiO_4$ prepared by a metal alkoxide route and sintered at 1600°C for 3 hours had $\varepsilon_r = 40$, $Qf = 50\,000$ GHz and $\tau_f = 3$ ppm/K at 10 GHz. The Qf was increased to 53 000 GHz by annealing the samples in O_2 at 1450°C for 15 hours. The lower Qf of the unannealed material was due to the presence of oxygen defects. This agrees with Wakino [106], who noted that large specimens of ZST exhibit a color change from surface to core (coring) that has been caused by oxygen vacancies. Hirano et al. [27] also quenched samples from 1600°C and found they were black in color. This is characteristic of ceramic sample containing reduced titanium.

A decrease in $\tan\delta$ on cooling to low temperatures is expected since lattice vibrations are suppressed at low temperatures. The temperature dependence of the

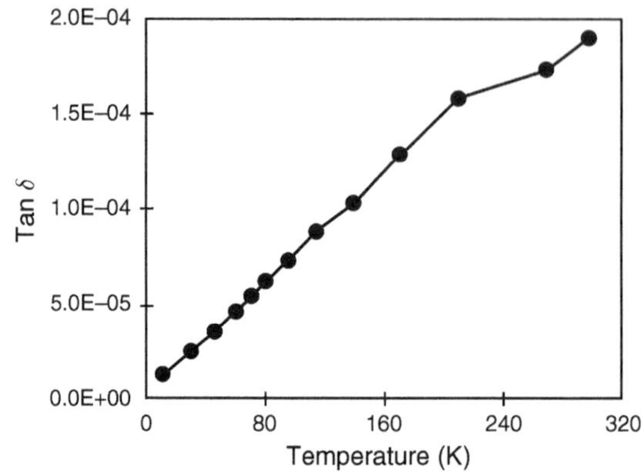

Figure 4.14 The temperature dependence of tan δ of ZST normalized to 10 GHz (after Ref. [107]).

tan δ of an undoped $Zr_{0.8}Sn_{0.2}TiO_4$ sample is shown in Figure 4.14 [107]. The tan δ decreases approximately linearly down to the lowest temperature measured (15 K). Tamura et al. [108] measured low temperature properties of $Zr_{0.8}Sn_{0.2}TiO_4$ prepared using raw materials of high and low purity. The τ_f of $Zr_{0.8}Sn_{0.2}TiO_4$ is not affected by the purities of the initial raw materials. The tan δ of high purity sample decreased with decrease in temperature whereas it increased for the low purity sample.

Several authors investigated the effect of suitable substitutions for Zr and Ti in $ZrTiO_4$ [95, 106, 109–111] and the results are given in Table 4.2. Okuyama et al. [112] prepared

Table 4.2 Dielectric properties of $[Zr_{0.8}Sn_{0.2}]TiO_4$ analogous materials

Material	Sintering temperature °C	ε_r	Qf GHz	τ_f	Reference
$Zr_{0.513}Hf_{0.487}TiO_4$	1600	41	20 400	13	[110]
$Zr_{0.648}Sn_{0.332}TiO_4 + La_2O_3 + NiO$	1370/20 h	37	41 500		[95]
$ZrO_2{-}MgO{-}Nb_2O_5{-}TiO_2$		43	52 000	0	[94]
$Zr_{0.91}Sn_{0.09}TiO_4$	1350 hotforging	36	30 141	30	[66]
$(Zr_{0.7})(ZnTa)_{0.3}TiO_4$	1300/3 h	43	40 200	1.1	[109]
$ZrTiO_4$	1600	42	28 000	58	[2, 16, 78, 79]
$ZrTiO_4$	1600 + 0.5 wt%Hf Polymeric precursor	36	29 700		[24]

novel ceramics in ZrO_2–TiO_2–MgO–Nb_2O_5 system. $\varepsilon_r = 42$–44, $Qf = 52\,000$ GHz $\tau_f = -4$ to +4 and density 4.9 g m/cm^3. The solid solution $(Zr_{1-x})(ZnTa)_xTiO_4$ ($0.2 \le x \le 0.6$) sintered into dense ceramics without any sintering aid at 1300 °C [109]. The ε_r is nearly the same for $x < 0.35$ and then increased with increasing amount of x. For $x = 0.3$, the $(Zr_{0.7})(ZnTa)_{0.3}TiO_4$ has $\varepsilon_r = 42.5$, $Qf = 40\,200$ GHz and $\tau_f = 1.1$ ppm/°C. The ionic size of $Zn^{2+} = 0.74$ Å with coordination number (CN) 6, $Ta^{5+} = 0.64$ Å with CN = 6 and $Zr^{4+} = 0.72$ Å with CN = 6. Therefore Zr^{4+} ion could be substituted by one-third of Zn^{2+} ions and two-thirds by Ta^{5+} in the $ZrTiO_4$ ceramics. Figure 4.15 shows the XRD pattern of $(Zr_{1-x})(ZnTa)_xTiO_4$ for different values of x. Rutile-type $Zn_{1/3}Ta_{2/3}$ TiO_4 secondary phase developed for $x > 0.35$. Figure 4.16 shows the variation of ε_r with $(Zn_{1/3}Ta_{2/3})O_2$ content (x). Figure 4.17 shows the variation of τ_f and Figure 4.18 the variation of Qf with x. The ε_r and τ_f showed a slight decrease initially and for $x > 0.35$ they rapidly increased. The Qf increased up to $x = 0.3$ and then decreased with increasing value of x.

Thin films of $ZrTiO_4$ and $Zr_{0.8}Sn_{0.2}TiO_4$ were prepared [111, 113–116] by different techniques such as sol–gel, dc or RF magnetron sputtering and have ε_r in the range 38–42 and tan $\delta = 10^{-2}$–10^{-4}. Kim et al. [116] studied microwave dielectric properties of $ZrTiO_4$ sputtered films and as the deposition temperature and RF power density were increased, the film crystallinality increased and dielectric loss decreased. The films deposited at 700 °C showed an ε_r of 38.

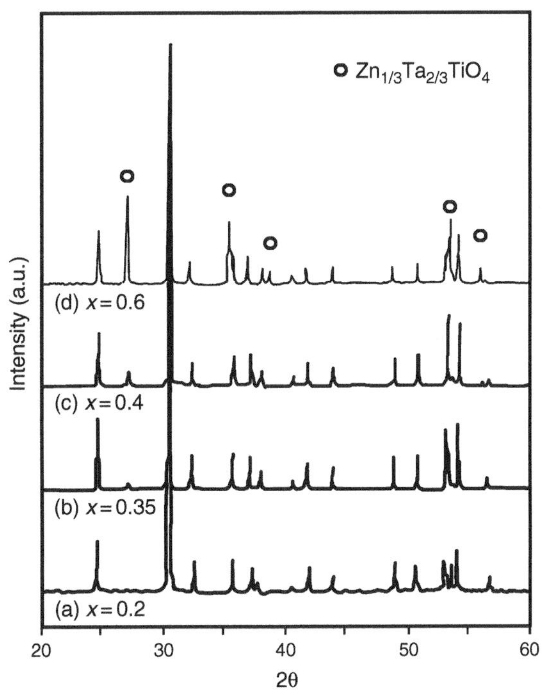

Figure 4.15 X-ray powder diffraction patterns of $Zr_{1-x}(Zn_{1/3}Ta_{2/3})_xTiO_4$ sintered at 1300°C (a) $x = 0.2$ (b) $x = 0.35$ (c) $x = 0.4$ (d) $x = 0.6$ (after Ref. [109]).

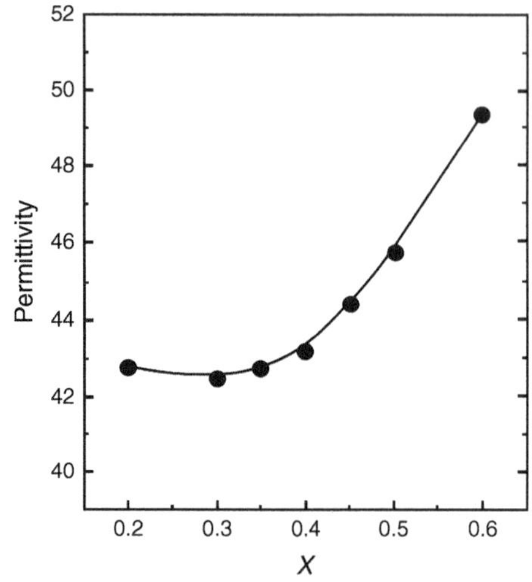

Figure 4.16 Variation of permittivity of $Zr_{1-x}(Zn_{1/3}Ta_{2/3})_xTiO_4$ ceramics with x (after Ref. [109]).

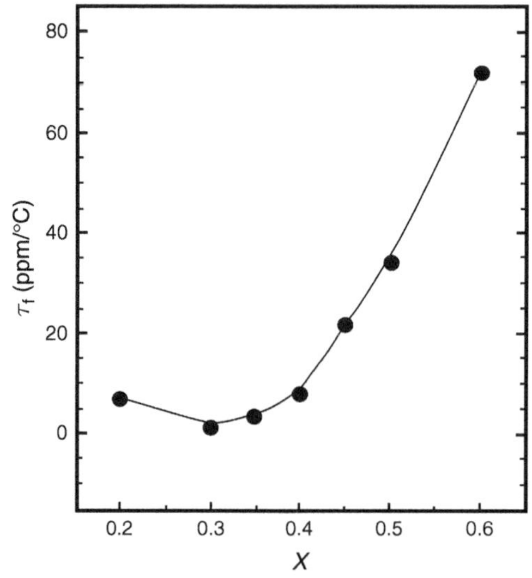

Figure 4.17 Variation of τ_f of $Zr_{1-x}(Zn_{1/3}Ta_{2/3})_xTiO_4$ ceramics with x (after Ref. [109]).

4.5 Conclusion

$ZrTiO_4$ is orthorhombic and undergoes an order–disorder transition. This transition is from α–PbO_2-type high-temperature structure with a random distribution of Zr and Ti ions to the low temperature structure with ordered arrangement of cations. Partial substitution of Zr by Sn inhibits the order–disorder transition. The

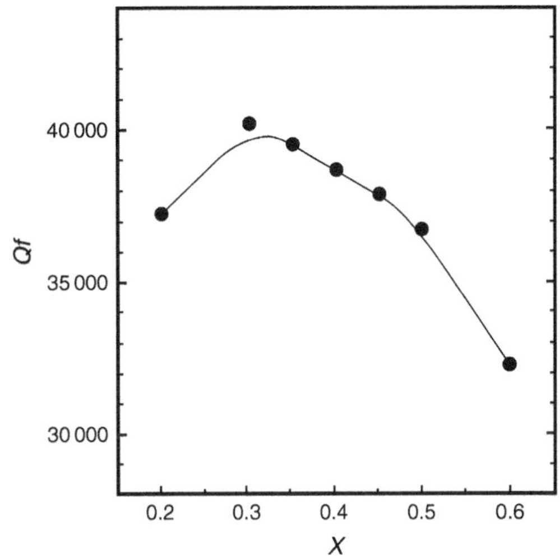

Figure 4.18 Variation of Qf of $Zr_{1-x}(Zn_{1/3}Ta_{2/3})_xTiO_4$ ceramics with x (after Ref. [109]).

transition to the incommensurate (IC) phase is initiated by the formation of short-range cation-ordered clusters in the disordered matrix. As the transformation proceeds, the growing ordered region will be surrounded with disordered boundary region. Finally the faulted boundary region becomes a sharp faulted boundary having one atomic layer that is characterized by a wrong-site occupation of Zr and Ti cations. Thus the incommensurately ordered structure can be viewed as a commensurately 1:1 cation-ordered structure with the interface modulation along the *a*-axis by faulted boundaries [56].

$ZrTiO_4$ has permittivity of 42, Qf of about 28 000 GHz and τ_f of +56 ppm/°C. Annealing $ZrTiO_4$ increases the order parameter and improves the dielectric quality factor. Substitution of 20 mol% Zr by Sn to form $Zr_{0.8}Sn_{0.2}TiO_4$ considerably improved the τ_f and the quality factor. The $Zr_{0.8}Sn_{0.2}TiO_4$ has a high sintering temperature of about 1600°C and is difficult to densify. Hence ZnO is usually added as a sintering aid. However, ZnO addition leads to the formation of lossy secondary phase which degrades the quality factor. Moreover, sintering at high temperature leads to reduction of Ti^{4+} to Ti^{3+} and creation of oxygen vacancies. The presence of oxygen vacancies also reduces the quality factor. Hence pentavalent metal oxides such as Ta_2O_5, Sb_2O_5, Nb_2O_5 or WO_3 are added in small quantities to compensate for the titanium reduction and oxygen vacancies. The pentavalent dopants substitute for Ti^{4+} and the remaining one + charge compensates for the Ti^{3+} making the crystal electrically neutral and thereby improving the quality factor. By careful preparation and using pentavalent doping, one can achieve a temperature-stable $Zr_{0.8}Sn_{0.2}TiO_4$ with a Qf factor of about 60 000 GHz. The ceramics prepared by wet chemical methods although expensive also give reasonably good quality factor. One of the advantages of $Zr_{0.8}Sn_{0.2}TiO_4$ is that by varying the Sn content, one can control its τ_f without drastically affecting the other properties. This is important for applications because a τ_f of precisely zero is not always required. A non-zero τ_f is often preferable to compensate for frequency variation due to the effect of temperature change on the resonator housing and dielectric support structure. Zirconium tin titanate is an important dielectric material with excellent properties useful for applications in wireless communication. The raw materials required for its commercial production are inexpensive.

References

[1] K. Wakino, K. Minai, and H. Tamura. Microwave characteristics of (Zr,Sn)TiO_4 and BaO–PbO–Nd_2O_3–TiO_2 dielectric resonator. J. Am. Ceram. Soc. 67(1984) 278–281.

[2] P. C. Osbond, R. W. Whatmore, and F. W. Ainger. The properties and microwave applications of zirconium titanate stannate ceramics. Br. Ceram. Proc. 36(1985) 167–178.

[3] A. J. Moulson and J. M. Herbet. Electroceramics. Chapman & Hall, London (1990).

[4] F. P. Daly, H. Ando, J. L. Schmodt, and E. A. Sturn. Hydrosulfurization over a TiO_2–ZrO_2 supported CoO–MoO_3 catalyst. J. Catal. 108(1987)401–408.

[5] F. Hund. $ZrTiO_4$–Miscphasenpigmente. Z. Anorg. Allg. Chem. 525(1985)221–229.

[6] F. J. Parker. Al_2TiO_5–$ZrTiO_5$–ZrO_2 composites: A new family of low thermal expansion ceramics. J. Am. Ceram. Soc. 73(1990)929–932.

[7] S. Yang and J. M. Wu. ZrO_2–TiO_2 ceramic humidity sensors. J. Mater. Sci. 26(1991) 631–636.

[8] D. A. Chang, P. Lin, and T. Tseng. Optical properties of $ZrTiO_4$ films grown by radio-frequency magnetron sputtering, J. Appl. Phys. 77(1995)4445–4451.

[9] W. Rath. Keramische sindermassen fur die Elektronik Fortschritter auf dem Greblet der keramischen isollerstoff fur die Electroteknik II. Keram. Radsch. 49(1941)137–139.

[10] Brit Patent No. 692468 (1952).

[11] K. Wakino, T. Nishikawa, S. Tamura, and Y. Ishikawa. Microwave band pass filters containing dielectric resonators with improved temperature stability and spurious response. Proc. IEEE MTT Symposium (New York) (1975)63–65.

[12] K. Wakino, T. Nishikawa, S. Tamura, and Y. Ishikawa. Miniaturised band pass filters using half wave dielectric resonators with improved spurious response. Proc IEEE MTT Symposium (1978)230–132.

[13] K. Wakino, T. Nishikawa, S. Tamura, and Y. Ishikawa. Proc IEEE MTT Symposium (1979)278.

[14] R. B. Van Dover, L. F. Schneemeyer, and R. M. Fleming. Discovery of a useful thin-film dielectric using a composition-spread approach. Nature. 392(1998)162–164.

[15] J. F. Eichler, O. Just, and W. S. Rees. The design and synthesis of heterometallic alkoxide-amides and their application in the MOCVD of zirconium-tin-titanate (ZTT) J. Mater. Chem. 14(2004)3139–3143.

[16] W. Wolfram and H. E. Gobel. Existence range, structural and dielectric properties of $Zr_xTi_ySn_zO_4$ ceramics ($x + y + z = 4$). Mater. Res. Bull. 16(1981)1455–1463.

[17] H. Tamura. Microwave loss quality of $(Zr_{0.8}Sn_{0.2})TiO_4$. Am. Ceram. Soc. Bull. 73(1994) 92–95.

[18] C. L. Wang, H. Y. Lee, F. Azough, and R. Freer. The microstructure and microwave dielectric properties of zirconium titanate ceramics in the solid solution system $ZrTiO_4$–$Zr_5Ti_7O_{24}$. J. Mater. Sci. 32(1997)1693–1701.

[19] A. Yamamoto, T. Yamada, H. Ikawa, O. Fukunga, K. Tanaka, and I. Marumo. Modulated structure of zirconium titanate. Acta Crystallogr. C 47(1991)1588–1591.

[20] M. Macias, P. J. Sanchez-Soto, and J. A. Navio. Kinetic study of crystallization in zirconium titanate from an amorphous reactive prepared precursor. J. Non-Cryst. Solids 147(1992) 262–265.

[21] D. Houivet, J. El. Fallah, and J. M. Haussonne. Effect of the grinding process on different characteristics of (Zr,Sn)TiO_4 microwave dielectric ceramics. Key Engineering Materials, Vols 132–136, Trans-Tech publications Inc, Zurich, Switzerland (1997) pp. 1199–1202.

[22] L. B. Kong, J. Ma, W. Zhu, and O. K. Tan. Phase formation and thermal stability of $(Zr_{1-x}Ti_x)O_2$ solid solution via a high energy ball milling process. J. Alloys Compd 335(2002) 290–296.

[23] M. Stubicar, V. Bermance, N. Stubicar, D. Dudrnovski, and D. Drumes. Microstructure evolution of an equimolar powder mixture of ZrO_2–TiO_2 during high energy ball milling and post annealing. J. Alloys Compd 316(2001)316–320.

[24] A. Bianco, G. Gusmano, R. Freer and P. Smith. Zirconium titanate microwave dielectrics prepared via polymeric precursor route, J. Eur. Ceram. Soc. 19(1999)959–963.

[25] A. Bianco, M. Paci, and R. Freer. Zirconium titanate: from polymeric precursors to bulk ceramics. J. Eur. Ceram. Soc. 18(1998)1235–1243.

[26] A. Bianco, M. Viticoli, G. Gusmano, M. Paci, G. Padeletti, and P. Scardi. Zirconium tin Titanate thin films via aqueous polymeric precursor route. Mater. Sci. Eng. C 15(2001) 211–213.

[27] S. Hirano, T. Hayashi, and A. Hattori. Chemical processing and microwave characteristics of (Zr, Sn)TiO_4 microwave dielectrics. J. Am. Ceram. Soc. 74(1991)1320–1324.

[28] Y. Ho, M.-H. Weng, B.-T. Dai, and S.-S. Wang. Nanopowder and microwave dielectric properties of sol–gel derived $Zr_{.8}Sn_{.2}TiO_4$ ceramics. Jpn. J. Appl. Phys. 44(2005)6152–6155.

[29] F. Khairulla and P. Phule. Chemical synthesis and structural evolution of zirconium titanate Mater. Sci. Eng. B. 12(1992)327–336.

[30] M. Leoni, M. Viviani, G. Battilana, A. M. Fiorella, and M. Viticoli. Aqueous synthesis and sintering of zirconium titanate powders for microwave components. J. Eur. Ceram. Soc. 21(2001)1739–1741.

[31] S. X. Zhang, J. B. Li, J. Cao, H. Z. Zhai, and B. Zhang. Preparation, microstructure and microwave dielectric properties of $Zr_xTi_{1-x}O_4$($x = 0.40$–0.60) ceramics. J. Eur. Ceram. Soc. 21(2001)2931–2936.

[32] R. Kudesia, A. E. McHale, and R. L. Snyder. Effect of La_2O_3/ZnO additives on microstructure and microwave dielectric properties of $Zr_{0.8}Sn_{0.2}TiO_4$ ceramics. J. Am. Ceram. Soc. 77(1994)3215–3220.

[33] K. R. Han, J.-W. Jang, S.-Y. Cho, D.-Y. Jeong, and K.-S. Hong. Preparation and dielectric properties of low temperature-sinterable $(Zr_{0.8}Sn_{0.2})TiO_4$ powder. J. Am. Ceram. Soc. 81(1998)1209–1214.

[34] D. Chen, X. Jiao, and M. Zhang. Hydrothermal synthesis and characterization of $(Zr_{1-x}Sn_x)TiO_4$ ($x = 0.05$–0.20) powders. Mater. Res. Bull. 35(2000)2101–2108.

[35] Z. X. Xiong. J. R. Huang, C. Fang, and Z. Y. Pan. Hydrothermal synthesis of (Zr,Sn)TiO_4 nano-powders for microwave ceramics. J. Eur. Ceram. Soc. 23(2003)2515–2518.

[36] C.-L. Huang, M.-H. Weng, and H.-L. Chen. Effects of additives on microstructures and microwave dielectric properties of (Zr, Sn) TiO_4 ceramics. Mater. Chem. Phys. 71(2001) 17–22.

[37] C.-L. Huang and M.-H. Weng. Liquid phase sintering of (Zr,Sn)TiO_4 microwave dielectric ceramics. Mater. Res. Bull. 35(2000)1881–1888.

[38] C.-L. Huang, M.-H. Weng, C. C. Wu, and C. C. Wei. Microwave dielectric properties and microstructures of V_2O_5 modified $Zr_{0.8}Sn_{0.2}TiO_4$ ceramics. Jpn. J. Appl. Phys. 40(2001) 698–702.

[39] C.-L. Huang and M.-H. Weng. Liquid phase sintering of (Zr,Sn)TiO_4 microwave dielectric ceramics. Mater. Res. Bull. 35(2001)1881–1888.

[40] J.-H. Jean and S.-C. Lin. Low fire processing of ZrO_2–SnO_2–TiO_2 ceramics. J. Am. Ceram. Soc. 83(2000)1417–1422.

[41] T. Takada, S. F. Wang, S. Yoshikawa, S. J. Jang, and R. E. Newnham. Effect of glass additions on BaO–TiO_2–WO_3 microwave ceramics, J. Am. Ceram. Soc. 77(1994) 1909–1916.

[42] T. Takada, S. F. Wang, S. Yoshikawa, S.-J. Jang, and R. E. Newnham. Effects of glass on (Zr, Sn)TiO_4 for microwave applications. J. Am. Ceram. Soc. 77(1994)2485–2488.

[43] R. E. Newnham. Crystal structure of $ZrTiO_4$. J. Am. Ceram. Soc. 50(1967)216.

[44] G. Blasse. Compounds with α-PbO_2 structure. J. Anorg. Allg. Chem. 345(1966)222–224.

[45] L. W. Coughanover, R. S. Roth, and V. A. De Prosse. Phase equilibria relations in the system lime–titania and zirconia–titania. J. Res. Natl. Bur. Stand. 52(1954)37–42.

[46] A. Cocco and G. Torriano. Reaction between the solid phase in the system ZrO_2–TiO_2. Ann. Chim. (Rome) 52(1965)153.

[47] A. E. Mc Hale and R. S. Roth. Investigation of the phase transition in $ZrTiO_4$ and $ZrTiO_4$–SnO_2 solid solutions. J. Am. Ceram. Soc. 66(1983)C-18–20.

[48] A. E. Mc Hale and R. S. Roth. Low-temperature phase relationships in the system ZrO_2–TiO_2. J. Am. Ceram. Soc. 69(1986)827–832.

[49] H. Ikawa, A. Iwai, K. Hiruta, H. Shimojima, K. Urabe, and S. Udagawa. Phase transformation and thermal expansion of zirconium and hafnium titanate and their solid solutions. J. Am. Ceram. Soc. 71(1988)120–127.
[50] Y. Park. Influence of order disorder transition on microwave characteristics of tin-modified zirconium titanate. J. Mater. Sci. Lett. 14(1995)873–875.
[51] P. Bordet, A. E. Mc Hale, A. Santaro, and R. S. Roth. Powder neutron diffraction study of $ZrTiO_4$, $Zr_5Ti_7O_{24}$, and $FeNb_2O_6$. J. Solid State Chem. 64(1986)30–46.
[52] Y. Park and K. M. Knowles. Glassy dynamics of the incommensurate–commensurate phase transition in $Zr_{0.98}Hf_{0.02}TiO_4$ ceramics. J. Appl. Phys. 85(1999)6434–6439.
[53] A. Yamamoto, T. Yamada, H. Ikada, O. Fukunaga, K. Tanaka, and F. Marumo. Modulated structure of zirconium titanate. Acta Crystallogr. C- 47(1991)1588–1591.
[54] H.Ikawa, H. Shimojima, K. Ukrabe, T. Yamada, and S. Udagawa. Polymorphism in $ZrTiO_4$. Science of Ceramics. D. Taylor (Ed.). Institute of Ceramics, Shelton, Uk (1988) pp. 509–514.
[55] H. Ikawa, S. Shimojima, K. Urabe, T. Yamada, and S. Udagawa. Science of Ceramics. D. Taylor (Ed.), Shelton, Stoke on Trent, UK (1988) Vol. 14, p. 769.
[56] R. Christerfferson and P. K. Davies. Structure of commensurate and incommensurate ordered phase in the system $ZrTiO_4$–$Zr_5Ti_7O_{24}$. J. Am. Ceram. Soc. 75(1992)563–569.
[57] H. Ikawa, T. Yamada, K. Kojima, and S. Matsumoto. X-ray photoelectron spectroscopy study of high- and low-temperature forms of zirconium titanate. J. Am. Ceram. Soc. 74(1991) 1459–1462.
[58] U. Toitzct and D. J. Hills. The ZrO_2–TiO_2 phase diagram. J. Mater. Sci. 40(2005)4571–4545.
[59] Y. Zhang and P. K. Davies. Stabilization of ordered zirconium titanates through the chemical substitution of Ti^{4+} by Al^{3+}/Ta^{5+}. J. Am. Ceram. Soc. 77(1994)743–748.
[60] F. Azough, R. Freer, and J. Petzelt. A Raman spectral characterization of ceramics in the system ZrO_2–TiO_2. J. Mater. Sci. 28(1993)2273–2276.
[61] T. Yamada, K. Urabe, H. Ikawa, and H. Shimojima. Modulated structure of low temperature form of zirconium titanate. J. Ceram. Soc. Jpn. 99(1991)380–383.
[62] P. K. Davies. Influence of structural defects on the dielectric properties of ceramic microwave resonators. In: Materials and Processes for Wireless Communications. T. Negas and H. Ling (Eds), Ceramic Transaction, American Ceramic Society, Westerville, OH (1995), Vol. 53, pp. 183–197.
[63] Y. Park and Y. Kim. Order-disorder transition of tin-modified zirconium titanate. Mater. Res. Bull. 31(1996)7–15.
[64] Y. Park, Y. Kim, and H. G. Kim. Structural-phase transition and electrical conductivity in tin-modified zirconium titanate. Solid State Ionics. 90(1996)245–249.
[65] H. Ikawa, H. Shimosa, K. Urabe, and O. Fukunaga. Thermal expansion of solid solutions in the $ZrTiO_4$–In_2O_3–M_2O_5($M = Sb$, Ta) system. J. Am. Ceram. Soc. 74(1991)1899–1904.
[66] R. Christofferson, P. K. Davies, X. Wei, and T. Negas. Effect of Sn substitution on cation ordering in $(Zr_{1-x} Sn_x) TiO_4$ microwave dielectric ceramics. J. Am. Ceram. Soc. 77(1994) 1441–1450.
[67] F. Azough and R. Freer. Microstructural development and microwave dielectric properties of $ZrTiO_4$ based ceramics. Proceedings of 7th IEEE International Symposium Oon Applied Ferroelectrics, Illinois (1990) 198–201.
[68] Y. K. Kim and H. M. Jang. Lattice contraction and cation ordering of $ZrTiO_4$ in the normal-to-incommensurate phase transition. J. Appl. Phys. 89(2001)6349–6355.
[69] M. A. Krebs and R. A. Condrate. Raman spectral characterization of various crystalline mixtures in the ZrO_2–TiO_2 and HfO_2–TiO_2 systems. J. Mater. Sci. Lett. 7(1988)1327–1330.
[70] Y. K. Kim and H. M. Jang. Polarization leakage and asymmetric Raman line broadening in microwave dielectric $ZrTiO_4$. J. Phys. Chem. Solids. 64(2003)1271–1278.
[71] Th. Held, P. Pfeiffer, and W. Kuhn. Influence of isotopic disorder on phonon frequencies and phonon linewidths of an anharmonic crystal. Phys. Rev. B. 55(1997)231–236.
[72] P. Li, W. Yang, P. Tan, H. Wen, and Z. Zhao. Raman forbidden mode and oxygen ordering in $Bi_2Sr_{2-x}La_xCuO_{6+y}$ single crystals annealed in oxygen. Phys. Rev. B 61(2000)11324–11327.
[73] Y. K. Kim and H. M. Jang. Raman line-shape analysis of nano-structural evolution in cation-ordered $ZrTiO_4$-based dielectrics. Solid State Commun. 127(2003)433–437.
[74] R. Kudesia, R. L. Snyder, A. E. Mc Hale, and R. A. Condrate. Proc. 95th Annual Meeting the Amer. Cer. Soc., Cincinnati, Westerville, OH, USA (1993).

[75] R. Kudesia, R. L. Snyder, R. A. Condrate, and A. E. McHale. Structural study of $Zr_{0.8}Zn_{0.2}TiO_4$. J. Phys. Chem. Solids. 54(1993)671.
[76] Y. Park and H. G. Kim. X-ray scattering study of incommensurate phase in $(Zr_{1-x}Sn_x)TiO_4$ ceramics. Mater. Lett. 15(1996)853.
[77] K. Wakino, M. Katsube, H. Tamura, N. Nishigaki, and Y. Ishikawa. Joint, Conv. Record of 4th Inst Electr. Eng. Japan. 285(1976)6.
[78] A. V. Zagordnyuk, L. V. Sadkovskaya, G. V. Shamrai, I. P. Kovaleskaya, R. L. Mugunov, and G. A. Teferin. Neodymium (3^+) – titaniumdioxide system. Zh. Neorg. Khim. 31(1986)2389–2389.
[79] F. Azough, R. Freer, C. L. Wang, and G. W. Lorimer. The relationship between the microstructure and microwave dielectric properties of zirconium titanate ceramics. J. Mater. Sci. 31(1996)2539–2549.
[80] Y. C. Heiao, L. Wu, and C. C. Wei. Microwave dielectric properties of $(ZrSn)TiO_4$ ceramic. Mater. Res. Bull. 23(1988)1687–1692.
[81] D. M. Iddles, A. J. Bell, and A. J. Moulson. Relationship between dopants, microstructure and the microwave dielectric properties of ZrO_2–TiO_2–SnO_2 ceramics, J. Mater. Sci. 27(1992) 6303–6310.
[82] P. K. Davies. Mater. Influence of internal interfaces on the dielectric properties of ceramic dielectric resonators. Res. Soc. Symp. Proc. 357(1994)351–361.
[83] W. Wersing. High frequency ceramic dielectrics and their applications for microwave components. In: Electroceramics, B. C. H. Steele (Ed.), Elsevier Applied Sciences, London and New York (1991). pp. 67–119.
[84] K. Wakino. Recent developments of dielectric resonator materials and filters in Japan. Ferroelectrics. 91(1989)69–86.
[85] F. Azough and R. Freer. Proc. The microstructure and low frequency dielectric properties of some zirconium titanate stannate (ZTS) ceramics. Proc. Br. Ceram. Soc. 42(1989)225–233.
[86] V. L. Arantes and D. P. F. de Souza. Microstructural development and microwave properties of ZnO doped tin titanate zirconate. Mater. Sci. Eng. A. 398(2005)220–226.
[87] N. Michiura, T. Tatekawa, Y. Higuchi, and H. Tamura. Role of donor and acceptor ions in the dielectric loss tangent of $(Zr_{0.8}Sn_{0.2})TiO_4$ dielectric resonator materials. J. Am. Ceram. Soc. 78(1995)793–796.
[88] C.-L. Huang, C.-S. Hsu, and R.-J. Lin. Improved high-Q microwave dielectric resonator using ZnO and WO_3-doped $Zr_{0.8}Sn_{0.2}TiO_4$ ceramics. Mater. Res. Bull. 36(2001)1985–1993.
[89] W. S. Kim, T. H. Kim, E. S. Kim, K. H. Yoon. Microwave dielectric properties and far infrared reflectivity spectra of the $(Zr_{0.8}Sn_{0.2})TiO_4$ ceramics with additives, Jpn. J. Appl. Phys. 37(1998)5367–5371.
[90] Y. Higuchi, N. Michiura, T. Tatekawa, H. Tamura. Effects of dopants and oxygen vacancies on the dielectric loss tangent of $(Zr_{0.8}Sn_{0.2})TiO_4$ dielectric ceramics. In: Materials and Processes for Wireless Communications, T. Negas and H. Ling (Eds), Ceramic Transaction, American Ceramic Society, Westerville, OH (1995), Vol. 53(1995) pp. 153–167.
[91] A. Ioachin, M. G. Banau, M. I. Toacsan, L. Nedelcu, D. Ghetu, H. V. Alexander, G. Toica, G. Annino, M. Cassettari, and M. Martnelli. Nickel doped $(Zr_{.8}Sn_{.2})TiO_4$ for microwave and millimeter wave applications. Mater. Sci. Eng. B 118(2005)205–209.
[92] E. Mc Hale, R. Kudesia, and A. H. Lee. Sintering additions as dopants in zirconium titanate microwave dielectric materials. Proceedings of the Symposium on Electroceramics and Solid State Ionics. H. L. Tuller and D. M. Smythe (Eds), The Electrochemical Society (1994)3–15.
[93] D.-J. Kim, J.-W. Hahn, G.-P. Han, S.-S. Lee, and T.-G. Choy. Effects of alkaline-earth-metal addition on the sinterability and microwave characteristics of $(Zr,Sn)TiO_4$ dielectrics. J. Am. Ceram. Soc. 83(2000)1010–1012.
[94] D. Houivet and J. El. Fallah, J. M. Haussone. Phases in La_2O_3 and NiO doped $(Zr, Sn)TiO_4$ microwave dielectric ceramics. J. Eur. Ceram. Soc. 19(1999)1095–1099.
[95] D. Houivet, J. El. Fallah, B. Lamagenere, and J. M. Haussonne. Effect of annealing on the microwave properties of $(Zr,Sn)TiO_4$ ceramics. J. Eur. Ceram. Soc. 21(2001)1727–1730.
[96] S. X. Zhang, J. B. Li, H. Z. Zhai, and J. H. Dai. Synthesis and characterization of La_2O_3/BaO-doped $(Zr_{0.8}Sn_{0.2})TiO_4$ microwave ceramics. Ceram. Int. 28(2002)407–411.

[97] S. X. Zhang, J. B. Li, H. Z. Zhai, and J. H. Dai. Low temperature sintering and dielectric properties of $(Zr_{0.8}Sn_{0.2})TiO_4$ microwave ceramics using La_2O_3/BaO additives. Mater. Chem. Phys. 77(2002)470–475.
[98] J.-R. Yoon, K.-Y. Kim, and H.-Y. Lee. Proceedings of 5th International Conference Properties and Applications of Dielectric Materials, Seoul, Korea, (1997) p. 960.
[99] K. H. Yoon and E. S. Kim. Dielectric characteristics of zirconium tin titanate ceramics at microwave frequencies. Mater. Res. Bull. 30(1995)813–820.
[100] K. H. Yoon, Y. S. Kim, and E. S. Kim. Microwave dielectric properties of $(Zr_{0.8}Sn_{0.2})TiO_4$ ceramics with pentavalent additives. J. Mater. Res. 10(1995)2085–2090.
[101] Y. H. Park, J. M. Ryu, M. Y. Shin, K. H. Ko, D. W. Kim, and K.-S. Hong. Effect of Nb_2O_5/ZnO addition on microwave properties of $(Zr_{0.8}Sn_{0.2})TiO_4$ ceramics. J. Am. Ceram. Soc. 84(2001)2542–2546.
[102] Y. S. Ahn, K. H. Yoon, and E. S. Kim. Effect of cooling rate on loss quality of $(Zr_{.8}Sn_{.2})TiO_4$ ceramics with additives. J. Eur. Ceram. Soc. 23(2003)2519–2523.
[103] H. G. Yu, C. X. Song, Z. H. Shen, and Z. X. Xiong. Dielectric properties of zirconium tin titanate with additives of ZnO, CuO, $MnCO_3$, and Bi_2O_3. Key Eng. Mater. 280–283(2005)91–94.
[104] K.-H. Lee. J. Microelectronics and Packaging Society 8(2001)27.
[105] H. Tamura Microwave dielectric losses caused by lattice defects. J. Eur. Ceram. Soc. 26(2006)1775–1780.
[106] K. Wakino. Ceramic Dielectrics, composition, Processing and Properties. In: Ceramic Transaction, H. C. Ling and M. F. Fan (Eds), American Ceramic Society, Westerville, OH (1990), Vol. 8, p. 305.
[107] N. Mc N. Alford, S. J. Penn, A. Templeton, X. Wang, and S. J. Webb. Materials and processing developments in microwave ceramics. Proc. 9th CIMTEC, World Ceramics Congress & Forum on New Materials, Florence, June 14–19 (1998).
[108] T. Tamura, H. Matsumoto, and K. Wakino. Low temperature properties of microwave dielectrics. Jpn. J. Appl. Phys. 28: Suppl. 28-2 (1989) 21–23.
[109] W. S. Kim, J. H. Kim, J. Han Kim, K. H. Hur, and J. Y. Lee. Microwave dielectric properties of the ZrO_2–ZnO–Ta_2O_5–TiO_2 systems. Mater. Chem. Phys. 79(2003)204–207.
[110] H. Ikawa, H. Narita, and O. Fukanaga. Microwave dielectric properties of solid solutions in the system $ZrTiO_4$–$HfTiO_4$. J. Jpn. Ceram. Soc. 98(1990)860–863.
[111] M. Viticoli, G. Padeletti, S. Kaciulis, G. M. Ingo, L. Pandolfia, and C. Zaldo. Structural and dielectric properties of $ZrTiO_4$ and $Zr_{0.8}Sn_{0.2}TiO_4$ deposited by pulsed laser deposition. Mater. Sci. Eng., B 118(2005)87–91.
[112] K. Okuyama, Y. Yokotani, S. Kawashima, K. Kugimiya, K. Nishimoto, T. Ishizaki, J. Katoh, and Y. Hakotani. A compact and low loss dielectric resonator for 800 MHz band mobile communication base stations. Proceedings of Asia Pacific Microwave Conference, The Institute of Electronics Information and Communication Engineers Vol. II (1994)751–754.
[113] T. P. Alexander, D. R. Ulhmann, G. Teowe, F. McCarthy, K. Mc Carthy, and T. J. Bukowsky. Dielectric characterization of sol–gel derived Sn doped $ZrTiO_4$ thin films. Integr. Ferroelectr. 17(1997)221.
[114] O. Nakagawara, Y. Toyota, M. Kobayashi, Y. Yoshino, Y. Katayama, K. Hitoshi, and T. Kawai. Electrical properties of $(Zr,Sn)TiO_4$ dielectric thin film prepared by pulsed laser deposition. J. Appl. Phys. 80(1996)388–392.; Mater. Res. Soc. Proc. 401(1996)97.
[115] E. S. Ramakrishnan, K. D. Cornett, G. H. Shapiro, and W. Y. Howung. Dielectric properties of radio frequency magnetron sputter deposited zirconium titanate-based thin films. J. Electrochem. Soc. 145(1998)358–362.
[116] Y. Kim, J. Oh, T.-G. Kim, and B. Park. Influence of the microstructures on the dielectric properties of $ZrTiO_4$ thin films at microwave frequency range. Jpn. J. Appl. Phys. 40(2001)4599–4603.

CHAPTER FIVE

Pseudo-Tungsten Bronze-Type Dielectric Materials

5.1 Introduction

In 1968 Bolton [1] made a systematic study of the TiO_2-rich region of $BaO–Nd_2O_3–TiO_2$ ternary system. Among various binary phases, he found two unknown ternary compounds, which had high relative permittivity and low temperature coefficient of relative permittivity. Mudroliubova et al. in 1977 [2] reported the barium rare earth titanate materials in a Russian patent. Later Kolar et al. [3–5] reported the dielectric properties of the ternary phases $BaNd_2Ti_3O_{10}$ ($BaO–Ln_2O_3–3TiO_2$; 1:1:3; Ln = lanthanide) and $BaNd_2Ti_5O_{14}$ ($BaO–Ln_2O_3–5TiO_2$; 1:1:5). During 1980–1981 Razgon et al. [6] and Gens et al. [7] reported the $BaNd_2Ti_4O_{12}$ ($BaO–Nd_2O_3–4TiO_2$; 1:1:4) composition. Figure 5.1 shows the phase diagram for the ternary $BaO–RE_2O_3–TiO_2$ system. Kolar et al. [4], Polyakov et al. [9] and Fukuda et al. [10] all assigned the 1:1:5 compound for $BaNd_2Ti_5O_{14}$ with orthorhombic lattice parameters practically coinciding with those of Gens et al. [7] for the 1:1:4 [$BaNd_2Ti_4O_{12}$], a stoichiometry first suggested by Razgon et al. [6]. In 1986, Jaakola et al. [11] reported that the true stoichiometry for the 1:1:4 and 1:1:5 phase must be closer to 15:19:72 ($Ba_{3.75}Nd_{9.5}Ti_{18}O_{54}$). Later several authors [11–13] found that the 1:1:5 composition is not single phase but contained $BaTi_4O_9$, TiO_2, $Ln_2Ti_2O_7$, $Ln_4Ti_9O_{24}$ etc. as secondary phases. Matveeva et al. [14] proposed the composition $Ba_{3.75}Pr_{9.5}Ti_{18}O_{54}$. However, many workers [10, 15–21] represented the composition as 1:1:5. Several authors [4, 11, 14, 22–50] investigated the crystal structure of $BaO–Ln_2O_3–TiO_2$ tungsten bronze-type dielectrics. Recently, Negas and Davies [28], Ohsato [8] and Ubic et al. [51] reviewed the structure and properties of these materials. Ohsato and co-workers at Nagoya Institute of Technology have done extensive work on the preparation, characterization, structure and properties of tungsten bronze-type materials. These materials have relatively high ε_r, low loss and are extensively used in cell phone industry.

5.2 Crystal Structure

Matveeva et al. [14] determined the crystal structure of $Ba_{3.75}Pr_{9.5}Ti_{18}O_{54}$ based on the fundamental unit cell by single crystal X-ray diffraction. This fundamental structure was later confirmed by several authors [28–31]. Varfolomeev et al. [35] explained $Ba_{3.75}Nd_{9.5}Ti_{18}O_{54}$ and $BaNd_2Ti_4O_{12}$ compounds as specific compositions of the solid solution phase $Ba_{6-3x}Nd_{8+2x}Ti_{18}O_{54}$, $0 \leq x \leq 0.75$. Ohsato et al. [36] later

Dielectric Materials for Wireless Communications
ISBN-13: 978-0-08-045330-9

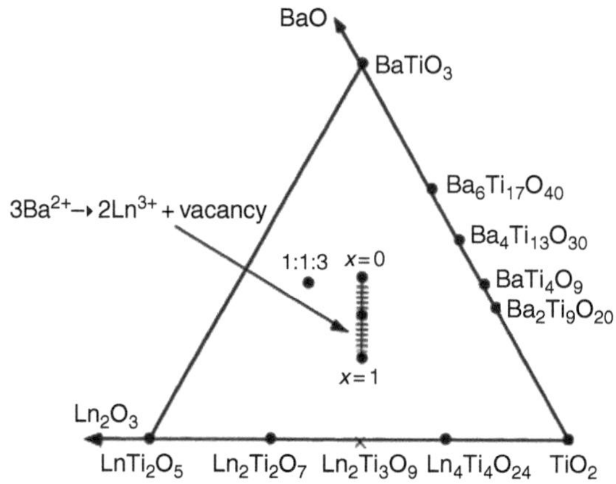

Figure 5.1 Phase diagram of the ternary system $BaO-Ln_2O_3$ TiO_2 (R = Ln = lanthanide) (after Ref. [3, 5, 6, 8]).

confirmed the formation of $Ba_{6-3x}Nd_{8+2x}Ti_{18}O_{54}$ solid solution phases. $BaNd_2$-Ti_4O_{12} (114) represents $x=0.5$ and $Ba_{3.75}Nd_{9.5}Ti_{18}O_{54}$ for $x=0.75$. The crystal structure of these materials essentially consists of a three-dimensional framework of corner sharing perovskite-like TiO_6 octahedra similar to tetragonal tungsten bronzes [14, 40, 52]. The $Ba_{3.75}Ln_{9.5}Ti_{18}O_{54}$ is orthorhombic with the general formula $A1_{10}A2_4Ti_{18}O_{54}$. The perovskite-like columns are joined at their corners to form alternating three- and five-membered rings of octahedra enclosing cavities which can accommodate cations. The crystal structure of tungsten bronze-type solid solutions is shown in Figure 5.2. Cations with large ionic radii are located within the three-dimensional framework of TiO_6 octahedra connected at all vertices. The atomic ratio of titanium to oxygen is 1:3 on the framework which is found in perovskite-type structure. There are three different cations with different ionic radii in the crystal structure which occupy different sites. The middle-sized R ions mainly occupy the A1 rhombic sites and the largest Ba ions mainly occupy pentagonal A2 sites. If the composition is Ba rich, small

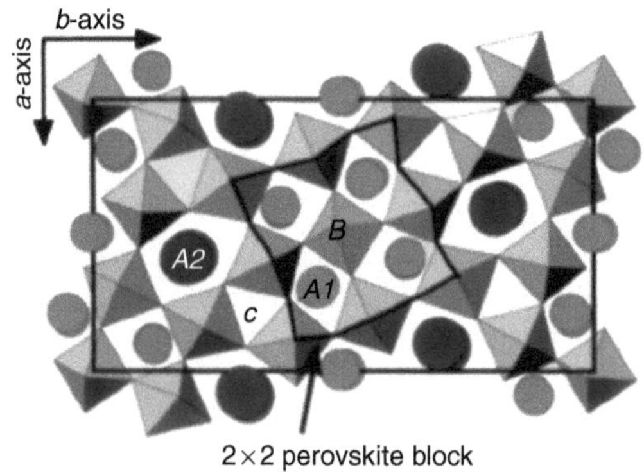

Figure 5.2 The tungsten bronze-type structure of $Ba_{6-3x}Ln_{8+2x}Ti_{18}O_{54}$ solid solutions. The R and Ba ions occupy the rhombic A1 and pentagonal A2 sites respectively. C is the trigonal site. (after Ref. [46]).

amount of Ba ions occupy also A1 sites. The smallest Ti ions alone occupy octahedral B sites. There are four trigonal C sites which are unoccupied and are the smallest sites [8]. There are 10 A1 sites and 4 A2 sites in the fundamental lattice ($Al_{10}A2_4Ti_{18}O_{54}$). The Ti cation is in an octahedron. The end composition $Ba_6Ln_8Ti_{18}O_{54}$ for $x = 0$ is represented as $[Ba_2Ln_8]A1[Ba_4]A_2Ti_{18}O_{54}$ in which the two largest Ba ions together with eight middle-sized R ions occupy A1 sites. The A2 sites are occupied only by four Ba ions. The structural formula of the solid solution is derived [8] from the end composition according to the substitution scheme

$$3Ba^{2+} \leftrightarrow 2Ln^{3+} + V_{A1} \text{ (where V is a vacancy)}$$

The chemical formula of the solid solution can be written as $Ba_{6-3x}Ln_{8+2x}Ti_{18}O_{54}$ and the structure as $[Ba_{6-3x}Ln_{8+2x}V_x]_{A1}\ [Ba_4]_{A2}Ti_{18}O_{54}$ in the range $0 \leq x \leq 2/3$

When all the Ba ions in A1 sites are substituted by Ln ions, $x = 2/3$, the resulting formula is $Ba_4Ln_{9.33}Ti_{18}O_{54}$. For $x \geq 2/3$, Ba ions in A2 sites become substituted by R ions. The decrease of Ba ions produces vacancies in A2 sites and may lead to unstable crystal structures as given by the limit of solid solubility located near $x = 0.7$. Substitution of $2Sm^{3+}$ for $3Ba^{2+}$ in the large cation sites causes considerable change in the crystal structure. One is the creation of vacancies in the large cation sites, because Sm^{3+} ions occupy only two-thirds of three Ba^{2+} sites to maintain charge neutrality. The amount of vacancy is proportional to x value. Moreover, the Ba^{2+} has much larger size than Sm^{3+}. The increase in the number of vacancies and the difference in ionic size lead to changes in properties and lattice parameters.

The structural formula can then be written as

$$[Ln_{9+1/3+2(x-2/3)}V_{2/3-2(x-2/3)}]_{A1}[Ba_{4-3(x-2/3)}V_{3(x-2/3)}]$$
$$A_2Ti_{18}O_{54} \quad \text{for the range } 2/3 \leq x \leq 1$$

According to this substitution of $2Ln^{3+}$ for $3Ba^{2+}$, the cell volume of the solid solution decreases as shown in Figure 5.3. The cell volume decreases with composition (x) up to about

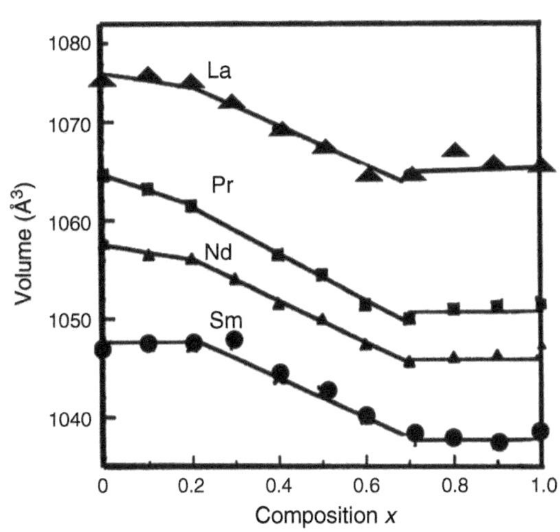

Figure 5.3 Variation of cell volume with composition in $Ba_{6-3x}Ln_{8+2x}Ti_{18}O_{54}$ (R = Sm, Nd, Pr, La) (after Ref. [44]).

$x \sim 0.7$ which is the solid solution limit. Ohsato et al. [36] reported that the solid solution range is $0 \leq x \leq 0.7$ for Nd and $0.3 \leq x \leq 0.7$ for Sm and Negas et al. [28] reported it as $0.2 < x < 0.7$ for Sm. For La-based tungsten bronze, the range is $0.07 < x < 0.77$ [53], and for Pr, $0.0 < x < 0.75$ [35]. The range of solid solution formation region narrows with decrease in the ionic radius of the rare earth [48]. Valant et al. reported [48] that for Eu and Gd, $Ba_{4.5}Gd_9Ti_{18}O_{54}$ or $Ba_{4.5}Eu_9Ti_{18}O_{54}$ [$x = 0.5$] only exist whereas Negas reported the range for Eu as $0.4 < x < 0.5$ [28]. The $Ba_{4.5}Gd_9Ti_{18}O_{54}$ decomposed [48] to $Gd_2Ti_2O_7$, $BaTi_2O_5$ and other polytitanates at 1340°C. Cruickshank et al. [39] have reported that the upper limit of solubility increased with increased processing temperature and reaches $x = 0.8$ at 1400°C.

Ohsato [8] studied in detail the crystallography of tungsten bronze type like $Ba_{6-3x}Ln_{8+2x}Ti_{18}O_{54}$ solid solutions [Ln = Sm]. The $Ba_{6-3x}Ln_{8+2x}Ti_{18}O_{54}$ ceramics have orthorhombic crystal lattice with *Pbam* space group. The space groups, which were reported by many researchers, were confirmed as Pbam for fundamental lattice and *Pbnm* for superlattice. The crystal structure is refined as tungsten bronze type-like structure with 2 × 2 perovskite blocks by X-ray crystal structure analysis. Matveeva et al. [14] and Kolar et al. [31] reported superstructure with twice the fundamental lattice spacing along *c*-axis. The presence of superstructure was confirmed by single-crystal X-ray diffraction and electron diffraction studies [33, 50]. The formation of the superstructure reflections was attributed to the occurrence of TiO_6 octahedral tilting [34]. Ohsato [8] found splitting in the oxygen ion position of TiO_6 octahedra on the Fourier map. The splitting is attributed to the tilting of octahedra as shown in Figure 5.4. The superstructure is caused by the tilting of octahedral strings along the c-axis which was deduced from splitting of oxygen ion position of the fundamental lattice. The structural data for the fundamental and superlattice is given in Table 5.1.

Powder diffraction pattern recorded from $Ba_{6-3x}Sm_{8+2x}Ti_{18}O_{54}$ using synchrotron radiation (wave length 1.2 Å) showed that the superstructure reflections are more intense

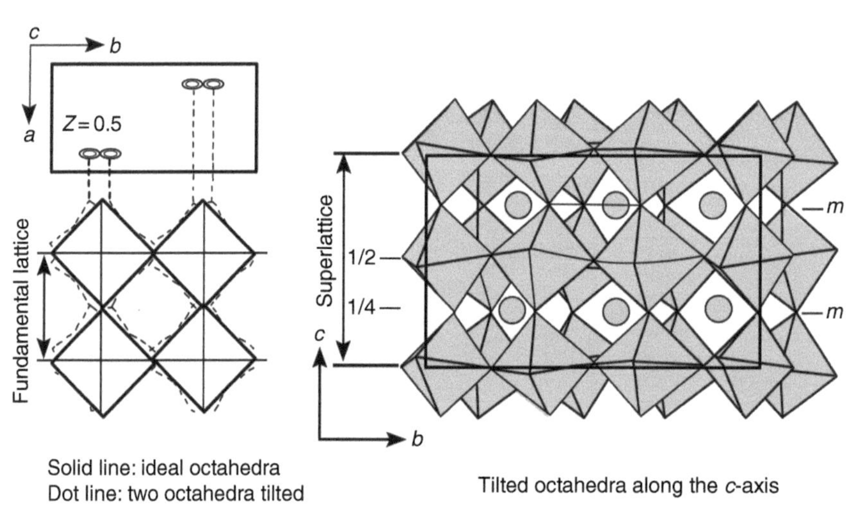

Figure 5.4 Tilting of TiO_6 octahedra deduced from splitting of oxygen ion on the fundamental structure (after Ref. [8]).

Table 5.1 Crystal data of fundamental lattice and superstructure (after Ref. [8])

	Fundamental lattice	Superstructure
Chemical formula	$Ba_{6-3x}Sm_{8+2x}Ti_{18}O_{54}$ (x = 0.7)	$Ba_{6-3x}Sm_{8+2x}Ti_{18}O_{54}$ (x = 0.7)
Crystal lattice type	Orthorhombic	Orthorhombic
Space group	*Pba2*(No.32) or *Pbam* (No.35)	$Pba2_1$(No.33) or *Pbnm* (No.62)
a (Å)	12.131(13)	12.131(13)
b (Å)	22.271(5)	22.271(5)
c (Å)	3.819(2)	7.639(5)
X-ray density	5.91	5.91
Z	1	2

Table 5.2 The observed relative intensities (%) of some of the superstructure reflections (after Ref. [8])

	Intensity	Intensity	Intensity	Intensity
Bragg reflection	x = 0.3	x = 0.5	x = 0.67	x = 0.71
461	2.8	3.7	3.9	3.8
481	1.8	2.5	2.8	2.4

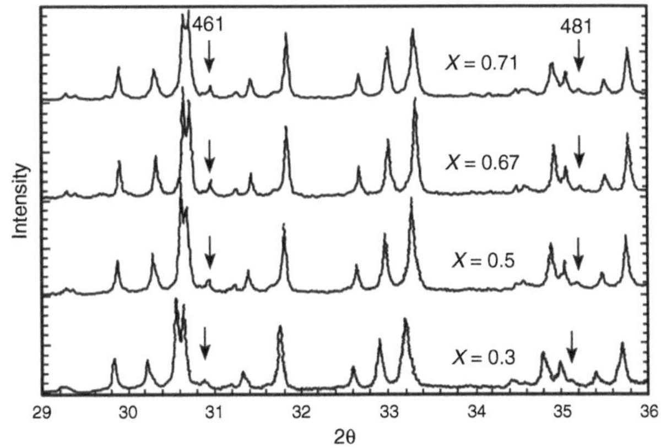

Figure 5.5 Powder diffraction pattern of $Ba_{6-3x}Sm_{8+2x}Ti_{18}O_{54}$ at $\lambda = 1.2$ Å. Dots observed profile, solid lines represent profiles calculated by program WPPF. Vertical arrows with indices point to superlattice reflections. (after Ref. [54]).

for $x = 2/3$ [54]. The observed intensities of 461 and 481 reflections normalized with the most intense 125 reflection are given in Table 5.2. Figure 5.5 shows the powder diffraction pattern of $Ba_{6-3x}Sm_{8+2x}Ti_{18}O_{54}$ for $x = 0.3$, 0.5, 0.67 and 0.71. The superstructure reflections are indicated by arrows.

5.3 PREPARATION OF $Ba_{6-3x}Ln_{8+2x}Ti_{18}O_{54}$

The $Ba_{6-3x}Ln_{8+2x}Ti_{18}O_{54}$ ceramics are prepared by calcining the stoichiometric amount of initial raw materials ($BaCO_3$, TiO_2 and Ln_2O_3) at about 1100°C and sintering at temperatures in the range 1350–1430°C. Figure 5.6 shows the microstructure of a typical $Ba_{6-3x}Nd_{8+2x}Ti_{18}O_{54}$ ($x = 1.5$) ceramic sintered at 1380°C/5 h. Several workers analyzed the sequence of phase formation during high temperature treatment of homogenized powder mixtures [17, 19, 56–58]. It is generally agreed that the first compounds to form at around 900°C are $BaTi_4O_9$, $Ba_4Ti_{13}O_{30}$ and $Ln_2Ti_2O_7$. At higher temperatures in the range 1050–1100°C these phases react among themselves and unreacted TiO_2 to form $Ba_{6-x}Ln_{8+2x}Ti_{18}O_{54}$. $BaTi_4O_9$, $Ba_2Ti_9O_{20}$, and TiO_2 are very often found as a secondary phase especially in $BaO–Ln_2O_3–4TiO_2$ (114) and $BaO–Ln_2O_3–5TiO_2$ (115) compositions.

Several authors [24, 59–64] prepared $Ba_{6-3x}Ln_{8+2x}Ti_{18}O_{54}$ by wet chemical methods. Takahashi [24] et al. prepared $BaLn_2Ti_4O_{12}$ [Ln = La, Nd, Sm] by a co-precipitation method. Valant et al. [62] prepared the rare earth tungsten bronze-type materials by a precipitation method using $NdCl_3$, $BaCl_2$ and ethanolic solution of $TiCl_4$. Xu et al. [59] prepared $Ba_{6-3x}Sm_{8+2x}Ti_{18}O_{54}$ ($x = 2/3$) ceramic powders by a sol–gel method using ethylene diaminetetra acetic acid as the complexing agent and butyl acetate, $BaCO_3$, Sm_2O_3 and nitric acid as the other starting chemicals. Hoffmann and Waser [60] prepared $Ba_{6-3x}Sm_{8+2x}Ti_{18}O_{54}$ [La, Ce, Nd, Sm] by sol–gel method using barium acetate, rare earth acetate, tetrabutyl orthotitanate and acetic acid starting chemicals. The samples were sintered at 1350°C/15 h and were also prepared by hot forging at 1250°C. Xu et al. [61] prepared $Ba_{6-3x}Nd_{8+2x}Ti_{18}O_{54}$ [$x = 0.75$] by the reaction method via citric acid precursor. The single-phase powders were obtained directly

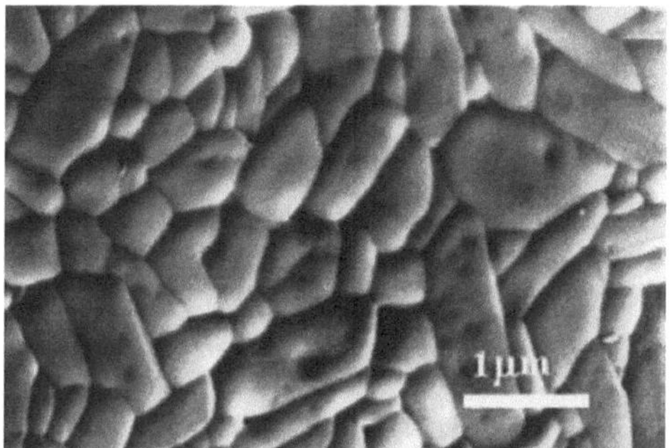

Figure 5.6 SEM image of $Ba_{6-3x}Nd_{8+2x}Ti_{18}O_{54}$ ($x = 1.5$) sintered at 1380°C/5 h (after Ref. [55]).

from the precursor at 900°C. Valant et al. reported [62] that samples made from powders leached in HNO_3 for 3 hours before sintering improve the quality factor. Nenesheva and Kartenko [63] prepared $A_xBa_{1-x}B_yLn_{2-y}Cn_zTi_{4-z}Zr_zO_{12}$ by chemical co-precipitation. Katayama et al. [64] prepared $BaNd_2Ti_4O_{12}$ powder at 1000–1200°C by the molten salt method using KCl. Ceramic samples obtained by chemical methods have about 100°C lower sintering than that by solid state method. Several authors [34, 39, 50, 65] prepared single crystals of $Ba_{6-3x}Ln_{8+2x}Ti_{18}O_{54}$. Ohsato [50] obtained single crystals of $Ba_{3.75}Ln_{9.5}Ti_{18}O_{54}$ (Ln = La, Nd, Sm) by melting $BaLn_2Ti_5O_{14}$ at 1520°C and slowly cooling to 1300°C at a rate of 12°C/h. Kolar et al. [4] and Valant et al. [65] grew single crystals of $BaLn_2Ti_5O_{14}$ [Ln = La, Pr, Nd, Sm, Gd] and $BaNd_2Ti_3O_{10}$ and studied their crystal structure.

5.4 Dielectric Properties

Several authors studied [3, 5, 10, 11, 13, 16, 19, 21, 29, 41, 57, 56, 66–84] the dielectric properties of $Ba_{6-3x}Ln_{8+2x}Ti_{18}O_{54}$ ceramics for different values of x and Ln = La, Pr, Sm, Nd, Eu, Gd and are given in Table 5.3. A study of Table 5.3 shows that the ceramics based on Eu, Sm and Nd have high quality factors (Qf) up to 14 000 GHz. The La-based ceramics have the highest relative permittivity. In general, the relative permittivity and τ_f decreases and the quality factor increases with decrease in ionic radii of the rare earth ion.

Ohsato et al. [8, 67, 68] made a detailed study of the dielectric properties of $Ba_{6-3x}Ln_{8+2x}Ti_{18}O_{54}$ solid solutions as a function of the composition x. The variation of ε_r, Qf and τ_f for the different rare earth ion is shown in Figure 5.7 for different values of x. The Qf varied non-linearly with x whereas ε_r and τ_f showed a near linear variation. The ε_r and τ_f decreased with x whereas Qf increased with x and reached a maxima at about $x \sim 0.7$ and then decreased. Negas et al. [28] observed similar variations for Sm- and Nd-based ceramics. Fukuda et al. [76] found a similar enhancement of Qf at $x = 2/3$ in $Ba_{6-3x}Pr_{8+2x}Ti_{18}O_{54}$. At $x = 2/3$, Ln and Ba ions occupy separately the rhombic sites (A1) and the pentagonal sites (A2) respectively and at this composition the ceramics showed a maxima in Qf. The microwave dielectric properties of $Ba_{6-3x}Ln_{8+2x}Ti_{18}O_{54}$ system are strongly dependent on the x value. As x increases beyond 0.7, ε_r, *Qf*, τ_f decreases. The compositions in the range $x = 0.5$–0.7 were found to have the best dielectric properties. The dielectric properties of the solid solutions are highly influenced by crystal structure, which is characteristic of these solid solutions. The octahedral B site containing Ti ions in perovskite blocks shrinks due to the substitution of smaller Sm^{3+} ion for larger Ba^{2+} ion to produce smaller ε_r and larger Qf. Ohsato et al. [67] established a relation between the microwave dielectric properties of $Ba_{6-3x}Sm_{8+2x}Ti_{18}O_{54}$ $(0.3 \leq x \leq 0.7)$ solid solution and the crystal structure. The permittivity decreased with increase in x. A Qf of about 10 000 GHz can be obtained in the composition range of $x = 0.5$–0.7 in the solid solution region. Substitution of $2Sm^{3+}$ for $3Ba^{2+}$ in the large cation sites results in the creation of vacancies and crystal distortion showing that the shrinkage of the TiO_6 octahedral site plays an important role for improvement of the microwave dielectric properties.

Ohsato et al. [8] attributed the increase in Qf at $x = 2/3$ to the ordering of Ln and Ba ions in the A1 and A2 sites respectively. The ordered distribution of the ions might reduce the internal strain and result in a non-linear variation of Qf. Ohsato et al. [8]

Table 5.3 Dielectric properties of $Ba_{63x}Ln_{8+2x}Ti_{18}O_{54}$ ceramics

Composition	Sintering temperature	ε_r	Qf GHz	τ_f ppm °C	Reference
$Ba_{2-x}Sm_{(4+2/3x)}Ti_9O_{26}$ x = 0	1360/4 h	77.5	11200	–3.4	[85]
$Ba_{2-x}Sm_{(4+2/3x)}Ti_9O_{26}$ x = 0.05	1360/4 h	78.1	11900	–2.1	[85]
$Ba_{2-x}Sm_{(4+2/3x)}Ti_9O_{26}$ x = 0.1	1360/4 h	77.8	12700	–1.2	[85]
$Ba_{2-x}Sm_{(4+2/3x)}Ti_9O_{26}$ x = 0.15	1360/4 h	76.1	12800	0.8	[85]
$Ba_{2-x}Sm_{(4+2/3x)}Ti_9O_{26}$ x = 0.2	1360/4 h	74.8	10900	2.4	[85]
$Ba_{2-x}Sm_{(4+2/3x)}Ti_9O_{26}$ x = 0.25	1360/4 h	71.5	10700	4.3	[85]
$Ba_{2-x}Sm_{(4+2/3x)}Ti_9O_{26}$ x = 0.3	1360/4 h	69.4	9700	5.8	[85]
$Ba_{6-3x}Sm_{8+2x}Ti_{18}O_{54}$ x = 0.75		78.67	8700	–	[50]
$Ba_{6-3x}Sm_{8+2x}Ti_{18}O_{54}$ x = 0.5		78.91	8400	–19.2	[50, 70]
$BaSm_2Ti_4O_{12}$ + 1 wt% CuO	1200/3 h	75.8	4910	–7.65	[86]
$Ba_{6-3x}Sm_{8+2x}Ti_{18}O_{54}$ x = 0.3	1450/2 h	85.77	429	–133.9	[69]
$Ba_{6-3x}Sm_{8+2x}Ti_{18}O_{54}$ x = 0.4	1450/2 h	83.49	1308	–42.3	[69]
$Ba_{6-3x}Sm_{8+2x}Ti_{18}O_{54}$ x = 0.5	1450/2 h	83.22	4384	–16.5	[69, 87]
$Ba_{6-3x}Sm_{8+2x}Ti_{18}O_{54}$ x = 0.6	1450/2 h	82.46	10050	–11.5	[69]

$Ba_{6-3x}Sm_{8+2x}Ti_{18}O_{54}$ x = 2/3	1450/2 h	80.96	10500	−11.3	[69, 87]
$Ba_{6-3x}Sm_{8+2x}Ti_{18}O_{54}$ x = 2/3 Sol–gel	1360/3 h	80.8	11330	–	[88]
$Ba_{6-3x}Sm_{8+2x}Ti_{18}O_{54}$ x = 0.7	1450/2 h	79	7500	−15	[67, 69]
$Ba_{6-3x}Sm_{8+2x}Ti_{18}O_{54}$ x = 0.8	1450/2 h	77.4	2000	4.1	[69]
$Ba_{6-3x}Sm_{8+2x}Ti_{18}O_{54}$ x = 0.9	1450/2 h	72.47	2100	25.1	[69]
$BaSm_2Ti_5O_{14}$		77	9300	12	[50, 56]
$BaO–Sm_2O_3–TiO_2$		74	12000	10	[10]
$[0.15(Ba_{0.93}Sr_{0.07})O–0.15Sm_2O_3–0.7TiO_2]$ + 1.5 wt% SnO_2 + 1 wt%CdO		80.7	16800	−4	[18]
$0.15(Ba_{0.93}Sr_{0.07})O–0.15(Sm_{0.4}La0.6)_2O_3–0.7TiO_2]$	1370	86.2	16700	95	[18]
$Ba_{6-3x}(Sm_{0.2}Nd_{0.8})_{8+2x}Ti_{18}O_{54}$ x = 2/3 + 1 wt% Bi_2O_3	1200/3 h	82.1	8530	17.3	[89]
$Ba_{6-3x}(Sm_{0.2}Nd_{0.8})_{8+2x}Ti_{18}O_{54}$ x = 2/3	1330/3 h	80.8	8100	35.6	[89]
$Ba_{6-3x}(Sm_{1-y}Nd_y)_{8+2x}Ti_{18}O_{54}$ x = 0.6 y = 0	1500/2 h	83	8950	−12.8	[71]
$Ba_{6-3x}(Sm_{1-y}Nd_y)_{8+2x}Ti_{18}O_{54}$ x = 0.6 y = 0.1	1500/2 h	83	8930	−5.8	[71]
$Ba_{6-3x}(Sm_{1-y}Nd_y)_{8+2x}Ti_{18}O_{54}$ x = 0.6 y = 0.3	1500/2 h	85	9160	8.6	[71]
$Ba_{6-3x}(Sm_{1-y}Nd_y)_{8+2x}Ti_{18}O_{54}$ x = 0.6 y = 0.5	1500/2 h	86	9170	24.8	[71]
$Ba_{6-3x}(Sm_{1-y}Nd_y)_{8+2x}Ti_{18}O_{54}$ x = 0.6 y = 0.7	1500/2 h	86	7920	42.4	[71]

(*Continued*)

Table 5.3 (Continued)

Composition	Sintering temperature	ε_r	Qf GHz	τ_f ppm °C	Reference
$Ba_{6-3x}(Sm_{1-y}Nd_y)_{8+2x}Ti_{18}O_{54}$ x = 2/3, y = 0.8	1340/12 h	85	9500	0.8	[72]
Ba_{6-3x} $(Sm_{1-y}Nd_y)_{8+2x}Ti_{18}O_{54}$ x = 0.6 y = 0.2	1500/2 h	84	9000	0	[71]
$Ba_{6-3x}(Sm_{1-y}Nd_y)_{8+2x}Ti_{18}O_{54}$ x = 0.6 y = 0.9	1500/2 h	88	8500	64.2	[71]
$Ba_{6-3x}(Sm_{1-y}Nd_y)_{8+2x}Ti_{18}O_{54}$ x = 0.6 y = 1.0	1500/2 h	88	8300	76	[71]
$Ba_{6-3x}(Sm_{1-y}Nd_y)_{8+2x}Ti_{18}O_{54}$ x = 0.5 y = 0	1400/10 h	85.5	7680	76	[90]
$Ba_{6-3x}(Sm_{1-y}Nd_y)_{8+2x}Ti_{18}O_{54}$ x = 0.5 y = 0.67	1400/10 h	86	7850	23.5	[90]
$Baa_{6-3x}(Sm_{1-y}Nd_y)_{8+2x}Ti_{18}O_{54}$ x = 2/3 y = 0.6	1400/10 h	83.4	10700	11	[90]
$Baa_{6-3x}(Sm_{1-y}Nd_y)_{8+2x}Ti_{18}O_{54}$ x = 2/3 y = 0.72	1400/10 h	82.7	10500	3.8	[90]
$Baa_{6-3x}(Sm_{1-y}Nd_y)_{8+2x}Ti_{18}O_{54}$ x = 2/3 y = 0.84	1400/10 h	81.7	10500	2.1	[90]
$Ba_{6-3x}(La_{1-y-z}Sm_yBi_z)_{8+2x}Ti_{18}O_{54}$ x = 2/3 y = 0.5 z = 0.0	1350/3 h	86.9	7365	83	[137]
$Ba_{6-3x}(La_{1-y-z}Sm_yBi_z)_{8+2x}Ti_{18}O_{54}$ x = 2/3 y = 0.5 z = 0.08	1320/3 h	95	3509	3	[137]
$Ba_{6-3x}(La_{1-y-z}Sm_yBi_z)_{8+2x}Ti_{18}O_{54}$ x = 2/3 y = 0.5 z = 0.12	1320/3 h	111.3	2470	–30	[137]
$Ba_{6-3x}(La_{1-y-z}Sm_yBi_z)_{8+2x}Ti_{18}O_{54}$ x = 2/3 y = 0.5 z = 0.2	1300/3 h	124.5	1430	–9	[137]
$Ba_{6-3x}(La_{1-y-z}Sm_yBi_z)_{8+2x}Ti_{18}O_{54}$ x = 2/3 y = 0.7 z = 0.04	1350/3 h	88.4	6690	1	[137]

$Ba_{6-3x}(La_{1-y-z}Sm_yBi_z)_{8+2x}Ti_{18}O_{54}$ x = 2/3 y = 0.7 z = 0.2	1275/3 h	117	1780	–36	[137]
$Ba_4Sm_{(28-y)/3}Li_yTi_{18}O_{54}$ y = 0	1460	79.6	9435	–12.1	[137]
$Ba_4Sm_{(28-y)/3}Li_yTi_{18}O_{54}$ y = 0.25	1400	82.1	5640	–2.1	[47]
$Ba_4Sm_{(28-y)/3}Li_yTi_{18}O_{54}$ y = 1	1400	85.4	5045	45.1	[47]
$Ba_4Sm_{(28-y)/3}Li_yTi_{18}O_{54}$ y = 2	1400	92.7	3720	88.8	[47]
$Ba_4Sm_{(28-y)/3}Li_yTi_{18}O_{54}$ y = 4	1350	94.8	1000	142	[47]
$Ba_4Sm_{(28-y)/3}Li_yTi_{18}O_{54}$ y = 8	1300	98.8	280	515	[47]
$BaO–(Sm_{1-x}La_x)_2O_3–5TiO_2$ x = 0.1	1350/4 h	90.7	8900	4.2	[73]
Ba_{6-3x} $(Sm_{1-y}Bi_y)_{8+2x}Ti_{18}O_{54}$ y = 0.05, x = 2/3	1420	82.3	8810	–16.5	[91]
Ba_{6-3x} $(Sm_{1-y}Bi_y)_{8+2x}Ti_{18}O_{54}$ y = 0.1 x = 2/3	1380	84.1	7840	–21	[91]
Ba_{6-3x} $(Sm_{1-y}Bi_y)_{8+2x}Ti_{18}O_{54}$ y = 0.15 x = 2/3	1360	88.9	6620	–19.9	[91]
Ba_{6-3x} $(Sm_{1-y}Bi_y)_{8+2x}Ti_{18}O_{54}$ y = 0.2 x = 2/3	1360	92.4	5680	–11.6	[91]
Ba_{6-3x} $(Sm_{1-y}Bi_y)_{8+2x}Ti_{18}O_{54}$ y = 0.3 x = 2/3	1320	103.3	2980	8.6	[91]
$Ba_{6-3x}(Sm_{1-y}La_y)_{8+2x}$ $Ti_{18}O_{54}$ x = 0.6, y = 0.1	1500	84	9050	1.6	[71]
$Ba_{6-3x}(Sm_{1-y}La_y)_{8+2x}$ $Ti_{18}O_{54}$ x = 0.6, y = 0.3	1500	88	8050	44.6	[71]
$Ba_{6-3x}(Sm_{1-y}La_y)_{8+2x}Ti_{18}O_{54}$ x = 0.6, y = 0.5	1500/2 h	93	1300	118	[71]
$Ba_{6-3x}(Sm_{1-y}La_y)_{8+2x}Ti_{18}O_{54}$ x = 2/3, y = 0 to 0.1		84	9000	4.2	[73]

(Continued)

Table 5.3 (Continued)

Composition	Sintering temperature	ε_r	Qf GHz	τ_f ppm °C	Reference
Ba_{6-3x} $Sm_{8+2x}Ti_{18}O_{54}$–0.1TiO_2 x = 2/3	1350/2 h	79.8	9880	–17.6	[87]
Ba_{6-3x} $Sm_{8+2x}Ti_{18}O_{54}$–0.1TiO_2 –0.3TiO_2 x = 2/3	1350/2 h	78.8	10750	–21.1	[87]
Ba_{6-3x} $Sm_{8+2x}Ti_{18}O_{54}$–0.1TiO_2 –0.5TiO_2 x = 2/3	1350/2 h	79	10000	–23.6	[87]
Ba_{6-3x} $Sm_{8+2x}Ti_{18}O_{54}$–0.1TiO_2 –0.9TiO_2 x = 2/3	1350/2 h	79.3	10220	–12.1	[87]
Ba_{6-3x} $Sm_{8+2x}Ti_{18}O_{54}$–0.1TiO_2 –1.4TiO_2 x = 2/3	1350/2 h	81.5	10400	–0.3	[87]
Ba_{6-3x} $Sm_{8+2x}Ti_{18}O_{54}$–0.1TiO_2 –1.9TiO_2 x = 2/3	1350/2 h	72	10300	7.2	[87]
Ba_{6-3x} $Sm_{8+2x}Ti_{18}O_{54}$ + 0.5 wt% B_2O_3 glass (x = 2/3)	1200	76.1	10500	–19.4	[92, 93]
$Ba(Nd_{0.8}Sm_{0.2})_2Ti_4O_{12}$ + 1 wt% B_2O_3	1020	43	5500		[94]
Ba_{6-3x} $Sm_{8+2x}Ti_{18}O_{54}$ + 0.5 wt% GeO_2 glass (x = 2/3)	1150	77.3	8900	–12.6	[92]
Ba_{6-3x} $Sm_{8+2x}Ti_{18}O_{54}$ + 0.5 wt% GeO_2 + 0.5 wt%B_2O_3, x = 2/3	1150	77.3	8900	–129	[92]
Ba_{6-3x} $Sm_{8+2x}Ti_{18}O_{54}$ (x = 2/3) + 0.5 wt%(22MgO–22Al_2O_3–56SiO_2)	1200	71	5890	–19	[93]
Ba_{6-3x} $Sm_{8+2x}Ti_{18}O_{54}$ (x = 2/3) + 0.5 wt%(71ZnO–29B_2O_2)	1200	73	4830	–14	[93]
Ba_{6-3x} $Sm_{8+2x}Ti_{18}O_{54}$ (x = 2/3) + 0.5 wt%40 B_2O_3–60SiO_2)	1200	73	7900	–16	[93]
$Ba_{6-3x}Sm_{8+2x}Ti_{18}O_{54}$ (x = 2/3) + 0.5 wt%(60ZnO–30BB_2O_3–10SiO_2)	1200	72	4530	–17	[93]
Ba_{6-3x} $Sm_{8+2x}Ti_{18}O_{54}$ (x = 2/3) + 0.5 wt%(40PgO–40B_2O_3–20SiO_2)	1200	75	6500	–17	[93]

$Ba_{6-3x}Sm_{8+2x}Ti_{18}O_{54}$ (x = 2/3) + 0.5 wt%(40MgO–40B_2O_3–20SiO_2)	1200	72	4450	–16	[93]
$Ba_{6-3x}Sm_{8+2x}Ti_{18}O_{54}$ (x = 2/3) + 0.5 wt%(50ZnO–50B_2O_3)	1220	74	5330	–17	[93]
$Ba_{6-3x}Sm_{8+2x}Ti_{18}O_{54}$ (x = 2/3) + 0.5 wt%(50Al_2O_3–50SiO_2)	1220	70	8500	–21	[93]
$Ba_{6-3x}Sm_{8+2x}Ti_{18}O_{54}$ (x = 2/3) + 0.5 wt%(44Al_2O_3–30B_2O_3_26SiO_2)	1220	70	8600	–12	[93]
$Ba_{6-3x}Sm_{8+2x}Ti_{18}O_{54}$ (x = 2/3) + 0.5 wt%(30BaO–60B_2O_3–10SiO_2)	1220	76	9050	–5	[93]
$Ba_{6-3x}Sm_{8+2x}Ti_{18}O_{54}$ (x = 2/3) + 0.5 wt%(30BaO–40B_2O_3–30SiO_2)	1220	76	9100	–7	[93]
$Ba_{6-3x}Sm_{8+2x}Ti_{18}O_{54}$ (x = 2/3) + 0.5 wt%(35Bi_2O_3–32ZnO–6SiO_2–27B_2O_3)	1200	71	8900	–10	[93]
$Ba_{6-3x}Sm_{8+2x}Ti_{18}O_{54}$ (x = 2/3) + 3 wt% Al_2O_3–B_2O_3–SiO_2	1175	64	8500	0	[93]
BaO–Nd_2O_3–4TiO_2 + 0.5 wt%Al_2O_3 + 8 wt% Bi_2O_3		88	8000	0	[95]
0.14(BaO–Nd_2O_3–4TiO_2) + 0.86(BaO–Al_2O_3–4TiO_2)		71	8200	0	[95]
BaO–Al_2O_3–4TiO_2		35	5000	–15	[95]
$Ba_{6-3x}(Sm_{1-y}La_y)_{8+2x}Ti_{18}O_{54}$ x = 0.6 y = 0.5	1500/2 h	93	1290	118	[71]
CaO:BaO:$Li_2O(Sm_{1-y}N_d)_2O_3$:TiO_2 (14:4:8:12:63) y = 0.17	1400/3 h	103	7200	2	[96]
CaO:BaO:$Li_2O(Sm_{1-y}N_d)_2O_3$:TiO_2 (14:4:8:12:63)y = 0.33	1400/3 h	106	6600	22	[96]
$Ba_{6-3x}Nd_{8+2x}Ti_{18}O_{54}$ x = 0.25		94	3000	126	[97]
$Ba_{6-3x}Nd_{8+2x}Ti_{18}O_{54}$ x = 0.5		84	7900	88	[97]
$Ba_{6-3x}Nd_{8+2x}Ti_{18}O_{54}$ x = 0.75		79	10400	65	[97]
BaO–Nd_2O_3–5TiO_2	1450/2 h	83	10500	70	[10]

(Continued)

Table 5.3 (Continued)

Composition	Sintering temperature	ε_r	Qf GHz	τ_f ppm °C	Reference
$Ba_{0.98}Sr_{0.02})Sm_2Ti_4O_{12}$	1375/6	73	7920	–6	[98]
Ba_{6-3x} $Nd_{8+2x}Ti_{18}O_{54}$–$x = 0.75$		86	10450	–	[50]
$BaO–Bi_2O_3–TiO_2–Nd_2O_3$		88	5500	8	[21]
$2Bi_2O_3–3TiO_2–24$ wt%$BaTiO_3–66$ wt%$Nd_2O_3–3TiO_2$	1310	88	5500	8	[21]
$BaNd_2Ti_5O_{14} + 25$ wt% $Nd_2O_3 + 0.5$ mol%PbO	1250/2 h	90	6000	–20	[99]
$BaO–Bi_2O_3–Nd_2O_3–TiO_2 + 0.4$ wt%$Mn(CH_3COO)_2 + WO_3$	1320	80	7000	0	[62]
$Ba_{6-3x}(Nd_{0.85}Bi_{0.15})_{8+2x}Ti_{18}O_{54}$ $x = 2/3$	1300/3 h	99.1	5290	–5.5	[100]
$Ba_{6-3x}[Nd_{1-y}Bi_y]_{8+2x}Ti_{18}O_{54}$ $x = 2/3,\ y = 0.05$	1420	82.3	8810	–16.5	[101, 91]
$Ba_{6-3x}[Nd_{1-y}Bi_y]_{8+2x}Ti_{18}O_{54}$ $x = 2/3,\ y = 0.1$	1380	84.1	7840	–21	[101, 91]
$Ba_{6-3x}[Nd_{1-y}Bi_y]_{8+2x}Ti_{18}O_{54}x = 2/3,\ y = 0.2$	1360	92.4	5680	–11.6	[101, 91]
$Ba_{6-3x}[Nd_{1-y}Bi_y]_{8+2x}Ti_{18}O_{54}x = 2/3, y = 0.3$	1320	103.3	2980	8.6	[101, 91]
$Ba_{6-3x}[Nd_{1-y}Bi_y]_{8+2x}Ti_{18}O_{54}$ $x = 2/3, y = 0.1$	1340	83	7220	20.2	[102]
$Ba_{6-3x}[Nd_{1-y}Bi_y]_{8+2x}Ti_{18}O_{54}$ $x = 2/3, y = 0.16$	1340	89.5	6050	17.1	[102]
$Ba_{6-3x}[Nd_{1-y}Bi_y]_{8+2x}Ti_{18}O_{54}$ $x = 2/3, y = 0.18$	1340	92.9	5010	15.9	[102]
$Ba_{6-3x}[Nd_{1-y}Bi_y]_{8+2x}Ti_{18}O_{54}$ $y = 0.2$ $x = 2/3$	1340	94.8	4390	17.8	[102]

$Ba_{6-3x}[Nd_{1-y}Bi_y]_{8+2x}Ti_{18}O_{54}$ x = 2/3, y = 0.25	1340	94.6	2800	28.9	[102]
$Ba_{6-3x}[Nd_{1-y}Bi_y]_{8+2x}Ti_{18}O_{54}$ x = 2/3, y = 0.0	1460	82.2	9765	62.2	[101]
$Ba_{6-3x}[Nd_{1-y}Bi_y]_{8+2x}Ti_{18}O_{54}$ x = 2/3, y = 0.05	1380	82.2	9765	62.2	[102, 101]
$Ba_{6-3x}[Nd_{1-y}Bi_y]_{8+2x}Ti_{18}O_{54}$ x = 2/3, y = 0.1	1360	90.71	7024	24.4	[102, 101]
$Ba_{6-3x}[Nd_{1-y}Bi_y]_{8+2x}Ti_{18}O_{54}$ x = 2/3, y = 0.15	1360	93.68	6350	17.3	[101, 102]
$Ba_{6-3x}[Nd_{1-y}Bi_y]_{8+2x}Ti_{18}O_{54}$ x = 2/3, y = 0.2	1340	98.03	4700	20.2	[102, 101]
$Ba_{6-3x}[Nd_{1-y}Bi_y]_{8+2x}Ti_{18}O_{54}$ x = 2/3, y = 0.3	1320	114.13	2700	43.8	[102, 101]
$BaO–(Nd_{0.7}Bi_{0.3})_2O_3–4TiO_2$	1275	115	2100	26	[103]
$BaO–(Nd_{0.9}Bi_{0.1})_2O_3–4TiO_2$	1300	93	5900	15	[103]
$Ba(Nd_{0.7}Sm_{0.12}Bi_{0.18})Ti_4O_{12}$		105	4110	–	[103]
$Ba_{6-3x—z}Sr_z[Nd_{1-y}Bi_y]_{8+2x}Ti_{18}O_{54}$ y = 0, z = 0.9 x = 0.5		86.7	7200	63	[28]
$Ba_{6-3x—z}Pb_z[Nd_{1-y}Bi_y]_{8+2x}Ti_{18}O_{54}$ y = 0.0, z = 1.0 x = 0.5		91	5500	21	[28]
$Ba_{6-3x-zSrz}[Nd_{1-y}Bi_y]_{8+2x}Ti_{18}O_{54}$ y = 0.5, x = 0.5		93.4	5700	40	[28]
$Ba_{6-3x-zSrz}[Nd_{1-y}Bi_y]_{8+2x}Ti_{18}O_{54}$ y = 1, x = 0.5		97	5500	22	[28]
$Ba_{6-3x—z}Pb_z[Nd_{1-y}Bi_y]_{8+2x}Ti_{18}O_{54}$ y = 1, z = 0.5, x = 0.5		99	5300	112	[28]
$Ba_{6-3x—z}Pb_z[Nd_{1-y}Bi_y]_{8+2x}Ti_{18}O_{54}$ y = 1, z = 1.0 x = 0.5		101	4000	–4	[28]
$Ba_{6-3x-}[Nd_{1-y}Bi_y]_{8+2x}Ti_{18}O_{54}$ y = 2, x = 0.5		112	3000	25	[28]

(*Continued*)

Table 5.3 (Continued)

Composition	Sintering temperature	ε_r	Qf GHz	τ_f ppm °C	Reference
$Ba_{6-3x}Nd_{8+2x}Ti_{18}O_{54}$ x = 0.5 + 10 wt%$Bi_4Ti_3O_{12}$	1300/3 h	94.9	5620	21.4	[105]
$Ba_{4.5}Nd_9Ti_{18}O_{54}$ + 15 mol% $Ba_{4.5}Gd_9Ti_{18}O_{54}$	1350/10 h	92	5000	0	[74]
$Ba_4(Nd_{28/3-y}Eu_y)Ti_{18}O_{54}$ y = 1	1480/2 h	82.6	10400	47.3	[77]
$Ba_4(Nd_{28/3-y}Dy_y)Ti_{18}O_{54}$ y = 1	1480/2 h	78.6	10040	33.8	[77]
$Ba_4(Nd_{28/3-y}Ho_y)Ti_{18}O_{54}$ y = 1	1480/2 h	79.3	9690	31.1	[77]
$Ba_4(Nd_{28/3-y}Er_y)Ti_{18}O_{54}$ y = 1	1480/2 h	79.5	8290	32.5	[77]
$Ba_4(Nd_{28/3-y}Yb_y)Ti_{18}O_{54}$ y = 1	1480/2 h	78.4	6785	53.1	[77]
$Ba(Nd_{0.82-z}Sm_zBi_{0.18})Ti_4O_{12}$ z = 0.7		89	6880		[104]
$BaO–(Nd_{0.8}Bi_{0.2})_2O_3–4TiO_2$ + 10 wt%$Li_2O–B_2O_3–SiO_2–Al_2O_3–CaO$	900	68	2200	55	[103]
$(BaO–Nd_2O_3–4TiO_2)$ + 2 mol%$Li_2O–Nd_2O_3–4TiO_2$	1475/4 h	86	5500	–	[78]
$(BaO–Nd_2O_3–4TiO_2)$ + 10 mol%$Li_2O–Nd_2O_3–4TiO_2$	1475/4 h	87	5500	–	[78]
$BaNd_{2(1-x)}Sm_{2x}Ti_5O_{14}$ x = 0.5–1		70–72	6400–8000	4 to −36	[75]
$BaNd_{2(1-x)}Sm_{2x}Ti_5O_{14}$ x = 0.6	1300	68.8	9180	22	[79]
$BaNd_{2(1-x)}Sm_{2x}Ti_5O_{14}$ x = 0.8	1300	63.2	8570	12	[79]

$BaNd_{2(1-x)}Sm_{2x}Ti_5O_{14}$ x = 0.5		71.1	6410	4	[79]
$BaNd_{2(1-x)}Sm_{2x}Ti_5O_{14}$ x = 0.6		71.8	8050	29	[79]
$BaNd_{2(1-x)}Sm_{2x}Ti_5O_{14}$ x = 0.7		66.5	7210	–23	[79]
24 wt%$BaTiO_3$–76 wt% Nd_2O_3–3TiO_2	1220	77	11000	123	[21]
$Ba_{4.5}Gd_9Ti_{18}O_{54}$	1340	76	3300	–36	[48, 74, 102]
BaO–$(Nd_{0.8}Bi_{0.2})_2O_3$–4TiO_2 + 10 wt%(Li_2O–B_2O_3–SiO_2–Al_2O_3–CaO) glass	900	68	2200	55	[106]
$Ba(Nd,Sm)_2Ti_4O_{12}$ + 20 wt%SiO_2–Al_2O_3 glass	1250	30	8700		[107]
BaO–La_2O_3–4.7TiO_2 + glass		64	4000	15	[108]
$BaNd_2Ti_4O_{12}$ + (B_2O_3–Bi_2O_3–SiO_2–ZnO) glass	900	67	6000	4	[109]
$BaNd_2Ti_5O_{14}$ + La_2O_3–B_2O_3–TiO_2	850	19.9	8200	77	[110]
$BaEu_2Ti_4O_{12}$			14000		[28]
BaO–La_2O_3–4.7TiO_2	1300/1 h	92	5000	40.3	[108]
$BaPr_2Ti_5O_{14}$	1450/2 h	81	9000	130	[10, 76]
0.92$Ba_{4.5}(Nd_{1-y}Bi_y)_9Ti_{18}O_{54}$–0.08$BaTi_4O_9$ y = 0.12		96	5590	27	[111]
0.92$Ba_{4.5}(Nd_{1-y}Bi_y)_9Ti_{18}O_{54}$–0.08$BaTi_4O_9$ y = 0.145		98	5500	17	[111]
0.92$Ba_{4.5}(Nd_{1-y}Bi_y)_9Ti_{18}O_{54}$–0.08$BaTi_4O_9$ y = 0.15		99	5400	15	[111]
$Ba_{6-3x}La_{8+2x}Ti_{18}O_{54}$ (x = 0.5) hotpressed	1200	110	2450	340	[60]

(Continued)

Table 5.3 (Continued)

Composition	Sintering temperature	ε_r	Qf GHz	τ_f ppm °C	Reference
$Ba_{6-3x}La_{8+2x}Ti_{18}O_{54}$ (x = 0.5)	1350/15 h	93	2340	329	[60]
$Ba_{6-3x}La_{8+2x}Ti_{18}O_{54}$ (x = 0.7) hotpressed	1200	102	2380	399	[60]
$Ba_{6-3x}La_{8+2x}Ti_{18}O_{54}$ (x = 0.7)	1350/15 h	96	2240	240	[60]
$Ba_{6-3x}Nd_{8+2x}Ti_{18}O_{54}$ (x = 0.3) hotpressed	1200	99	3600	110	[60]
$Ba_{6-3x}Nd_{8+2x}Ti_{18}O_{54}$ (x = 0.3)	1350/15 h	71	3500	75	[60]
$Ba_{6-3x}Nd_{8+2x}Ti_{18}O_{54}$ (x = 0.7) hotpressed	1200	88	4920	55	[60]
$Ba_{6-3x}Nd_{8+2x}Ti_{18}O_{54}$ (x = 0.7)	1350/15 h	73	4800	53	[60]
$Ba_{6-3x}Sm_{8+2x}Ti_{18}O_{54}$ (x = 0.3) hotpressed	1200	96	1440	–14	[60]
$Ba_{6-3x}Sm_{8+2x}Ti_{18}O_{54}$ (x = 0.5) hotpressed	1200	91	10800	–17	[60]
$Ba_{6-3x}Sm_{8+2x}Ti_{18}O_{54}$ (x = 0.5)	1350/15 h	82	10100	–14	[60]
$Ba_{6-3x}Sm_{8+2x}Ti_{18}O_{54}$ (x = 0.7) hotpressed	1200	84	9960	–14	[60]
$Ba_{6-3x}Sm_{8+2x}Ti_{18}O_{54}$ (x = 0.7)	1350/15 h	72	9880	–19	[60]
$(Ba_{1-\alpha}Sr_{\alpha})_{4.2}Sm_{9.2}Ti_{18}O_{54}$ $\alpha = 0.0$	1450/2 h	80.3	9500	–8.6	[112]
$(Ba_{1-\alpha}Sr_{\alpha})_{4.2}Sm_{9.2}Ti_{18}O_{54}$ $\alpha = 0.01$	1450/2 h	80	8890	–11.3	[112]
$(Ba_{1-\alpha}Sr_{\alpha})_{4.2}Sm_{9.2}Ti_{18}O_{54}$ $\alpha = 0.04$	1450/2 h	80.6	9590	–11.9	[112]

$(Ba_{1-\propto}Sr_{\propto})_{4.2}Sm_{9.2}Ti_{18}O_{54}$ $\propto = 0.06$	1450/2 h	80.2	10075	–7.4	[112]
$(Ba_{1-\propto}Sr_{\propto})_{4.2}Sm_{9.2}Ti_{18}O_{54}$ $\propto = 0.1$	1450/2 h	77	6680	–11.4	[112]
$(Ba_{1-\propto}Sr_{\propto})_{4.2}Sm_{9.2}Ti_{18}O_{54}$ $\propto = 0.2$	1450/2 h	82.3	2860	0.4	[112]
$Ba_2Sr_2Sm_2Ti_{4+x}Ta_{6-x}O_{30-x/2}$ $x = 2$	1340/3 h	114	151		[113]
$Ba_2Sr_2Sm_2Ti_{4+x}Ta_{6-x}O_{30-x/2}$ $x = 3$	1340/3 h	111	200		[113]
$(Ba_{1-z}Pb_z)_{6-x}Nd_{8+2/3x}Ti_{18}O_{54}$ $x = 2/3$ $z = 0.22$	1400/2 h	88	5500	0	[114]
$(Ba_{1-z}Pb_z)_{6-x}Nd_{8+2/3x}Ti_{18}O_{54}$ $x = 2/3$ $z = 0.4$	1400/2 h	87	4000	–32	[114]
$Ba_{6-3x}(Sm_{1-y}Nd_y)_{8+2x}(Ti_{1-z}Sn_z)_{18}O_{54}$ $y = 0$ $z = 0$	1360/3 h	81	9240	–11	[115, 116]
$Ba_{6-3x}(Sm_{1-y}Nd_y)_{8+2x}(Ti_{1-z}Sn_z)_{18}O_{54}$ $y = 0$ $z = 0.05$	1360/3 h	76	6260	2	[115, 116]
$Ba_{6-3x}(Sm_{1-y}Nd_y)_{8+2x}(Ti_{1-z}Sn_z)_{18}O_{54}$ $y = 0.1$ $z = 0.05$	1360/3 h	76	7200	–26	[115, 116]
$Ba_{6-3x}(Sm_{1-y}Nd_y)_{8+2x}(Ti_{1-z}Sn_z)_{18}O_{54}$ $y = 0.3$ $z = 0.05$	1360/12 h	77	8180	1	[115, 116]
$Ba_{6-3x}(Sm_{1-y}Nd_y)_{8+2x}(Ti_{1-z}Sn_z)_{18}O_{54}$ $y = 0.5$ $z = 0.05$	1360/12 h	80	10050	5	[115, 116]
$Ba_{6-3x}(Sm_{1-y}Nd_y)_{8+2x}(Ti_{1-z}Sn_z)_{18}O_{54}$ $y = 0.8$ $z = 0.05$	1360/12 h	80	10600	11	[115, 116]
$Ba_{6-3x}(Sm_{1-y}Nd_y)_{8+2x}(Ti_{1-z}Sn_z)_{18}O_{54}$ $x = 2/3$ $y = 0$ $z = 0$		81	9240	–10.6	[115]
$Ba_{6-3x}(Sm_{1-y}Nd_y)_{8+2x}(Ti_{1-z}Sn_z)_{18}O_{54}$ $x = 2/3$ $y = 0$ $z = 0.05$		76	6260	2	[115]
$Ba_{6-3x}(Sm_{1-y}Nd_y)_{8+2x}(Ti_{1-z}Sn_z)_{18}O_{54}$ $x = 2/3$ $y = 0$ $z = 0.1$		68	4020	20	[115]
$Ba_{6-3x}(Sm_{1-y}Nd_y)_{8+2x}(Ti_{1-z}Sn_z)_{18}O_{54}$ $x = 2/3$ $y = 0$ $z = 0.2$		59	610	80	[115]
$Ba_{6-3x}(Sm_{1-y}Nd_y)_{8+2x}(Ti_{1-z}Sn_z)_{18}O_{54}$ $x = 2/3$ $y = 0$ $z = 0.3$		57	160	140	[115]

(Continued)

Table 5.3 (Continued)

Composition	Sintering temperature	ε_r	Qf GHz	τ_f ppm °C	Reference
$Ba_{6-3x}(Sm_{1-y}Nd_y)_{8+2x}(Ti_{1-z}Sn_z)_{18}O_{54}$ x = 2/3 y = 0 z = 0.5		77	80	967	[115]
$(Ba_{0.8}Ca_{0.2})_{6-3x}Sm_{8+2x}Ti_{18}O_{54}$ x = 1.5	1350	82	10000	–20	[118]
$(Ba_{0.7}Pb_{0.3})_{6-3x}Sm_{8+2x}Ti_{18}O_{54}$ x = 0.75	1350	88	6000	20	[118]
$(Ba_{0.6}Pb_{0.4})_{6-3x}La_{8+2x}Ti_{18}O_{54}$ x = 1.5	1380	95	6000	200	[118]
$Ba_2Sr_2Sm_2Ti_{4+x}Ta_{6-x}O_{30-x/2}$ x = 2	1340/3 h	114	150	–	[113]
$Ba_2Sr_2Sm_2Ti_{4+x}Ta_{6-x}O_{30-x/2}$ x = 2.5	1340/3 h	114	140	–	[113]
$Ba_2Sr_2Sm_2Ti_{4+x}Ta_{6-x}O_{30-x/2}$ x = 3	1340/3 h	111	110	–	[113]
$(Ba_{1-x}Ca_x)O–Sm_2O_3–4.5TiO_2$ x = 0		78	10000	–15	[119]
$(Ba_{1-x}Ca_x)O–Sm_2O_3–4.5TiO_2$ x = 0.03		79	10500	–5	[119]
$(Ba_{1-x}Ca_x)O–Sm_2O_3–4.5TiO_2$ x = 0.05		81	9500	2	[119]
$(Ba_{1-x}Ca_x)O–Sm_2O_3–4.5TiO_2$ x = 0.1		84	9500	25	[119]
$(Ba_{4.2}Sm_{9.2})_{\propto}Ti_{18-y}Al_yO_{54}$ y = 0.2 $\propto$ = 1 + y/36 x = 0.6	1460	78	8200	–18	[120]
$(Ba_{4.2}Sm_{9.2})_{\propto}Ti_{18-y}Al_yO_{54}$ y = 0.4 $\propto$ = 1 + y/36 x = 0.6	1440	76.1	3800	–33	[120]
$(Ba_{4.2}Sm_{9.2})_{\propto}Ti_{18-y}Al_yO_{54}$ y = 0.8 $\propto$ = 1 + y/36 x = 0.6	1440	72.1	4600	–42	[120]
$(Ba_{4.2}Sm_{9.2})_{\propto}Ti_{18-y}Al_yO_{54}$ y = 1 $\propto$ = 1 + y/36 x = 0.6	1440	70.2	4350	–57	[120]

$(Ba_{4.2}Sm_{9.2})_{\propto}Ti_{18-y}Al_yO_{54}$ $y = 1.4$ $\propto = 1 + y/36$ $x = 0.6$	1440	67	1500	–90	[120]
$(Ba_{6-3x}Sm_{8+2x})_{\alpha}Ti_{18-y}Al_yO_{54}$ $y = 1.61$ $\alpha = 1 + y/36$ $x = 0.6$		80–63	8600–1050	–118	[120]
$(Ba_{1-\alpha}Sr_{\alpha})_{6-3x}$ $Sm_{8+2x}Ti_{18}O_{54}$ $\alpha = 0.06$, $x = 0.6$		80	10075	–7	[112]
$Ba_{0.98}Sr_{0.02}Sm_2Ti_4O_{12}$		75	7920	–6	[98]
$(Ba_{.875}Pb_{.125})_{3.75}Nd_{9.5}Ti_{18}O_{54}$ $x = 0.75$		72	6100	8	[97]
$BaO–PbO–Nd_2O_3–TiO_2$		88	5000	0	[15]
$(Ba_{1-y}Pb_y)_{6-x}Nd_{8+2/3x}Ti_{18}O_{54}$ $x = 2/3$ $y = 0$ to 0.6		83–89	4500–7500	80 to –20	[114]
$Sr(Bi_{1-x}Nd_x)_8Ti_7O_7$ $x = 0.05$		87	193		[121]
$Sr(Bi_{1-x}Nd_x)_8Ti_7O_7$ $x = 0.1$		97	745		[121]
$Sr(Bi_{1-x}Nd_x)_8Ti_7O_7$ $x = 0.3$		104	350		[121]
$Sr(Bi_{1-x}Nd_x)_8Ti_7O_7$ $x = 0.4$		108	2000		[121]
$Ba_3La_3Ti_5Ta_5O_{30}$	1425	126.6		100	[122]
$Ba_4La_2Ti_4Ta_6O_{30}$	1425	131.8		500	[122]
$Ba_3LaTi_3Ta_7O_{30}$	1425	146.3		600	[122]

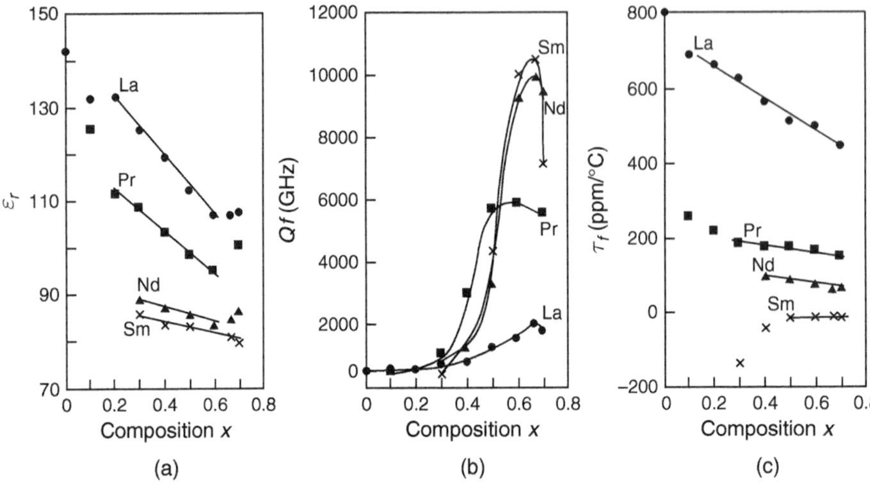

Figure 5.7 Microwave dielectric properties of $Ba_{6-3x}Ln_{8+2x}Ti_{18}O_{54}$ (R = La, Pr, Nd, and Sm) (after Ref. [8]).

evaluated the internal strain η for $x = 0.3$, 0.5, 2/3 and 0.7 and is given in Figure 5.8. The internal strain η was obtained [123] from the relation

$$\beta - 2\eta \tan \Theta,$$

where β is the full width at half maximum of the X-ray diffraction peaks assuming that the grain size of the ceramics is sufficiently large. The internal strain is the minimum for $x = 2/3$ in $Ba_{6-3x}Ln_{8+2x}Ti_{18}O_{54}$ solid solutions. This low internal strain originates from the distribution of cations in the rhombic and pentagonal sites in the tungsten bronze-type structure. This means that Ln ions are ordering in the rhombic A1 and Ba ions in the pentagonal A2 sites. This ordering leads to a decrease in the strain. As the x values decrease in the structural formula $[Ln_{8+2x}Ba_{2-3x}]A1[Ba_4]A2Ti_{18}O_{54}$ in the range $0 \leq x \leq 2/3$, Ba ions with larger ionic radii will occupy also part of the smaller rhombic sites. The occupation of Ba ions in A1 sites leads to internal strain around themselves and lowers

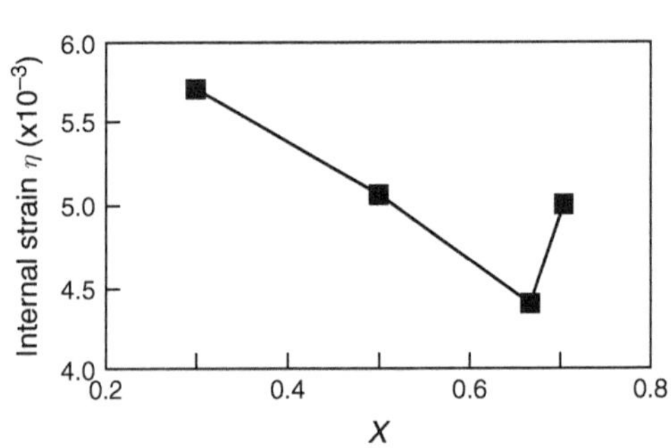

Figure 5.8 Variation of internal strain composition (after Ref. [8]).

the Qf value. Moreover, the vacancies generated in A1 sites by the substitution of 3Ba by 2R may lower the internal strain and improve the Qf. The Ba ions in A2 sites are substituted by Ln ions when the x value in the structural formula $[Ln_{9.33+2(x-2/3)}V_{0.66-(x-2/3)}A1[Ba_{4-3(x-2/3)}V_{3(x-2/3)}]A2Ti_{18}O_{54}$ increases to the range $2/3 \leq x \leq 0.7$. The decrease of number of Ba ions produce vacancies in A2 sites and may lead to unstable crystal structures as shown [8] by the limit of solid solubility located near $x = 0.7$ composition. Moreover, the decrease in the number of vacancies in the rhombic sites accompanied by the decrease of Ba ions in the pentagonal sites may lead to an additional internal strain. The Qf value of each Ln compound with $x = 2/3$ in the $Ba_{6-3x}Ln_{8+2x}Ti_{18}O_{54}$ solid solutions increased [8] according to a decrease in the rare earth ion size (lanthanoid contraction). The crystal structure with the largest difference in ionic size between Ba and Ln showed excellent quality factor since it has the lowest strain. The Ln ion Eu has the smallest size and has the highest quality factor among La, Pr, Sm, Nd, Dy and Eu. However, it may be noted that Yb which has the smallest atomic size exhibit very low Qf values since such small Ln ions also substitute at octahedral or triangular C sites. The Sm-, Nd- and Eu-based compounds have better Qf than the La compound. The crystal structure is maintained by the size difference of large cations such as Ba and Ln.

The relative permittivities are found to be affected by three factors (a) volume of the octahedron (b) tilting of the octahedron (c) polarizabilities of Ln and Ba ions. The relative permittivities of the solid solutions are proportional to the lattice parameters or cell volumes as shown in Figure 5.9a. As the volume of the unit cell increases, the ε_r and τ_f increased almost linearly. In the perovskite structure, the polarizability of Ti ions in the octahedra depends on the octahedral volume. Valant et al. [65] reported that the dielectric properties are affected by the tilting of TiO_6 octahedra. The tilting angle between the c-axis and the central axis of the octahedra is shown in Figure 5.10. The La-based compound has the highest polarizability and has the highest ε_r also. The polarizability of Ba is 6.4 and that of La is 6.03 [124]. When 3Ba is substituted by 2La (Ln) + V(vacancy), the total polarizability is reduced from 3×6.4 to 2×6.03. The τ_f also varies almost linearly with increase in cell volume as shown in Figure 5.9b.

Valant et al. [65] studied the τ_ε derived from Clausius–Mossotti equation as a function of ratio of mean radii (r_A/r_B) of A and B site ions. In accordance with the relationships described by the Clausius–Mossotti equation [125, 126], the permittivitty decreases with decreasing polarizability of the rare earth ions from La to Gd. The permitivitty also varies as a function of composition (x) within each rare earth solid solubility region, where it

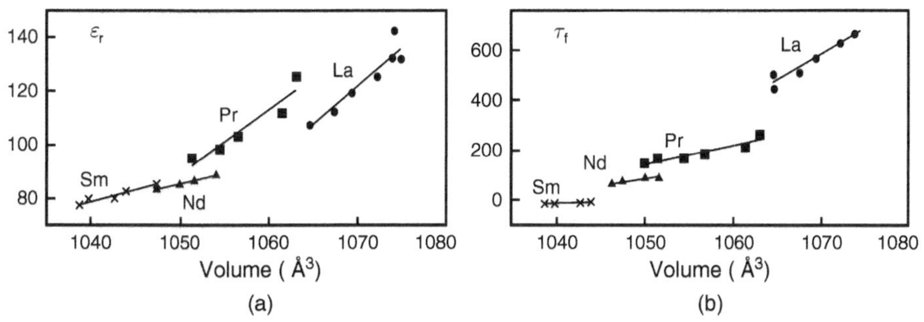

Figure 5.9 Variation of (a) ε_r and (b) the τ_f as a function of the unit cell volume (after Ref. [8]).

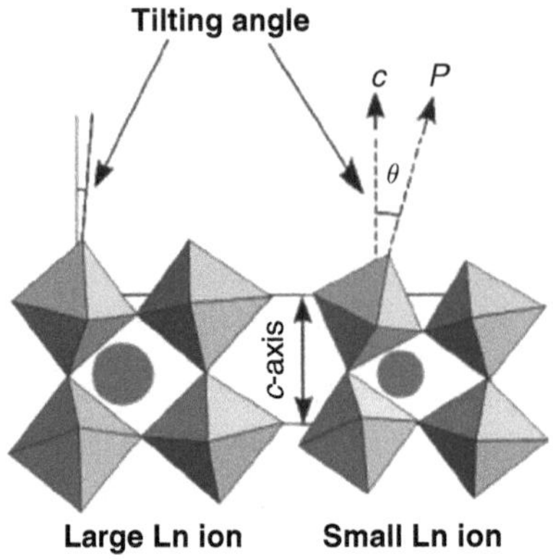

Figure 5.10 Octahedral tilting in $Ba_{6-3x}Ln_{8+2x}Ti_{18}O_{54}$ (after Ref. [8]).

decreases with increasing rare earth concentration (increasing x) as a result of decrease in the total polarizability. The τ_f is highly positive for La analogue but decreases with decreasing size of the rare earth ions and reaches negative values for Sm, Eu and Gd analogues. Inside each solid solubility region, τ_f decreases as x increases. The τ_ε study of $Ba_{6-3x}Ln_{8+2x}Ti_{18}O_{54}$ (Ln = La–Gd) solid solutions revealed a high degree of correlation with the ratio of mean radii (r_A/r_B) of A and B site ions [65]. Single crystal X-ray diffraction analysis of La, Sm and Pr analogues showed that r_A/r_B can be further correlated to tilting of the octahedra in the network of $Ba_{6-3x}Ln_{8+2x}Ti_{18}O_{54}$ crystal structure. Changes in the tilting of the octahedra induce changes in the crystal field and, subsequently, changes in τ_ε as well as dielectric loss. The correlation is distinct inside the solid solubility region, although deviations from the observed correlation occur, which can be explained by the limited flexibility of the octahedral network. Variations in dielectric properties are limited by the flexibility of crystal structure. Variations in the tilting of the octahedra resulting from changes in the composition are highly anisotropic, which explains the markedly high electrical anisotropy of $Ba_{6-3x}Ln_{8+2x}Ti_{18}O_{54}$ solid solution. Single crystal X-ray diffraction analysis revealed that the crystal structure of $Ba_{6-3x}Ln_{8+2x}Ti_{18}O_{54}$ adapts to variations in the composition by accommodating octahedral tilt angles. The increase in the average size of A site ions decreases the amplitude of tilting.

In adapting $Ba_{6-x}Ln_{8+2x}Ti_{18}O_{54}$ ceramics for microwave applications, the main tasks are to retain a high ε_r and Qf while tuning τ_f to zero. Two approaches are used. Improvement of τ_f can be achieved by additives and proper combination of rare earth elements in the solid solution. The effect of additives on the dielectric properties will be discussed in the next section. Ohsato et al. [77] investigated the effect of partial substitution of Eu, Dy, Ho, Er and Yb for Nd in $Ba_{6-3x}(Nd_{1-y}\text{–}Ln_y)_{8+2x}Ti_{18}O_{54}$ ceramics for $x = 2/3$ and $y = 0.0$ to 1. The lattice parameters along the a- and c-axes decreased linearly and those of b-axis increased linearly as the amount of substituted Ln ions increased. The sintered samples for each Ln ion series formed solid solutions in the whole range of y examined. The volume of the unit cell decreased linearly with composition y for all Ln ions except for Yb. All Ln ions occupy A1 sites, but Yb ions occupy different smaller sites such as B and C sites after occupying the A1 sites. The occupancy of the A1 sites by Ln ions was revealed by the decrease of the lattice parameters and volume values. The smaller the ionic radius is, the higher the rate of

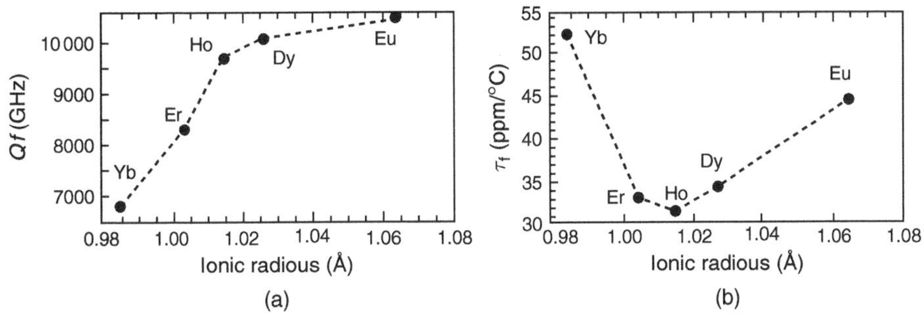

Figure 5.11 Variation of quality factor (Qf) and τ_f of $Ba_4(Nd_{28/3-y}R_y)Ti_{18}O_{54}$ for $y=1$ (after Ref. [77]).

decrease is. In the case of Yb which has the smallest ionic radius, the volume start increasing in the vicinity of $y=0.4$. This indicates that Yb ions occupy smaller sites such as trigonal and octahedral sites. The Qf improved up to $y=0.3$ for all Ln ions substituted for Nd. The τ_f values of all Ln compounds improved with y. Figure 5.11a and b shows the variation of Qf and τ_f with ionic radii of Ln for $y=1$ composition. The Eu series exhibited the best Qf for $y=1$ with $\varepsilon_r=83$, Qf = 10 500 GHz, $\tau_f=47.3$ ppm/°C. It may be noted that Negas and Davies [28] reported a high Qf value of 14 000 GHz for the Eu-based ceramics. In the case of Yb having the smallest ionic radius, the Qf values deteriorated at $y=1$ because the Yb ions substitute into octahedral or triangular C sites.

Several authors [71, 72, 75, 79, 127] succeeded in tuning the τ_f by partially substituting Sm by Nd. Ohsato et al. [71] studied the effect of substitution of La and Nd for Sm in $Ba_{6-3x}Sm_{8+2x}Ti_{18}O_{54}$ ($x=0.6$). The lattice parameters changed linearly in both Sm–La and Sm–Nd series indicating the formation of solid solution phases. Both La- and Nd-based ceramics have positive τ_f and the Sm based has a negative τ_f. Thus partial La and Nd substitution for Sm resulted in a temperature-stable ceramics. $Ba_{6-3x}(Sm_{0.8}Nd_{0.2})_{8+2x}Ti_{18}O_{54}$ ceramic showed [71, 72] $\varepsilon_r=84$, Qf of 9000 GHz and $\tau_f=0$ and $Ba_{6-3x}(Sm_{0.9}La_{0.1})_{8+2x}Ti_{18}O_{54}$ has $\varepsilon_r=84$, Qf = 9000 GHz and $\tau_f=1.6$ ppm/°C. Figure 5.12a shows the variation of ε_r, τ_f and Qf of Sm–Nd series and Figure 12b the Sm–La series. Ichinose and Amada [73] prepared $BaO–(Sm_{1-x}La_x)_2O_3–5TiO_2$ for $x=0$ to 1. The lattice parameters increased linearly with amount of La which is due to the larger size La ions. At $x=0.1$ the composition had $\varepsilon_r=91$, Qf = 8900 GHz and $\tau_f=4.2$ ppm/°C. La substitution increased ε_r and decreased Qf but changed negative τ_f to a positive τ_f. Valant et al. [48] prepared $Ba_{6-3x}Gd_{8+2/3x}Ti_{18}O_{54}$ for $x=0.5$ which corresponds to $Ba_{4.5}Gd_9Ti_{18}O_{54}$. The compound formed at 1230°C decomposed into $Gd_2Ti_2O_7$, $BaTi_2O_5$ etc. around 1350°C. $Ba_{4.5}Gd_9Ti_{18}O_{54}$ has $\varepsilon_r=76$, Qf = 2050 GHz, $\tau_f=-35$ ppm/°C. Addition of 15 mol% Bi_2O_3 showed $\varepsilon_r=92$, Qf = 5000 GHz and $\tau_f=0$. The Gd-based system has limited solid solubility range which is in agreement with the report of Kolar et al. [80] that only two ternary compounds $Ba_{4.5}Gd_9Ti_{18}O_{54}$ and $Ba_6Gd_2Ti_4O_{17}$ were formed. Huang and co-workers [81, 82] investigated $Ba_{2-x}Sm_{4+2x/3}Ti_{8+y}O_{24+2y}$ [$x=0.1$ $y=0$–2] prepared by sintering at 1310–1430°C/4 h. The ε_r values are in the range 63–85, Qf = 8500–13 000 GHz, $\tau_f=-12$ to $+17$ ppm/°C. The sintered ceramic was found to contain secondary phases of TiO_2 and $Ba_2Ti_9O_{20}$.

In order to study the effect of the rare earth ions on dielectric properties, Fukuda et al. [10] investigated infrared reflection spectra of $BaO–TiO_2–Pr_6O_{11}$, $BaO–TiO_2–Nd_2O_3$ and $BaO–TiO_2–Sm_2O_3$ ceramics from 50 to 4000 cm^{-1} at room temperature. The

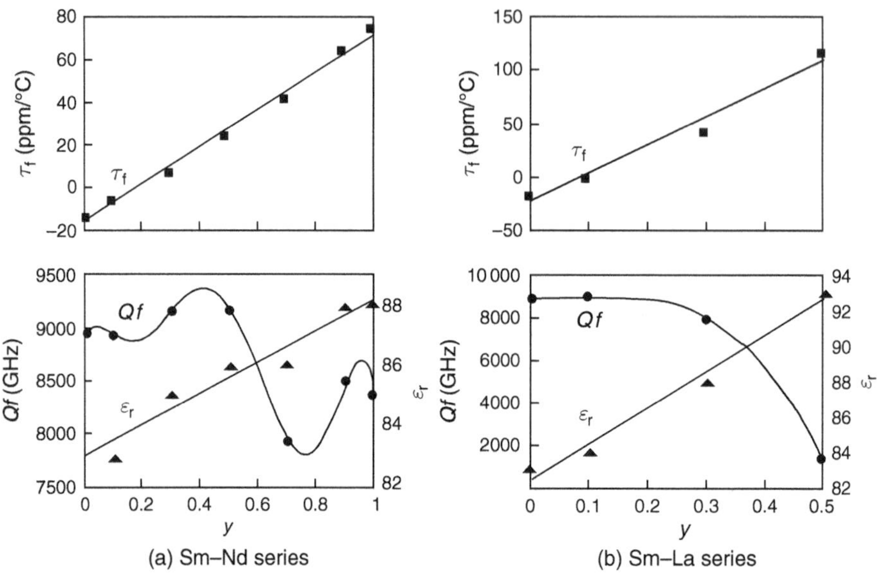

Figure 5.12 Variation of ε_r, Qf and τ_f of $Ba_4(Nd_{28/3-y}Ln_y)Ti_{18}O_{54}$ (Ln = Sm, La) (a) Sm—Nd (b) Sm—La (after Ref. [71]).

reflectance data were transformed to dielectric data using the model proposed by Servoin et al. [128] instead of the classical dispersion theory, because vibrational modes with different energy levels should not be equally damped. The Q value decreased in the order Sm, Nd, Pr. This is attributed to the fact that optic-mode frequencies which are in proportion to Q shift slightly to lower frequencies. The fourth-order dielectric constants

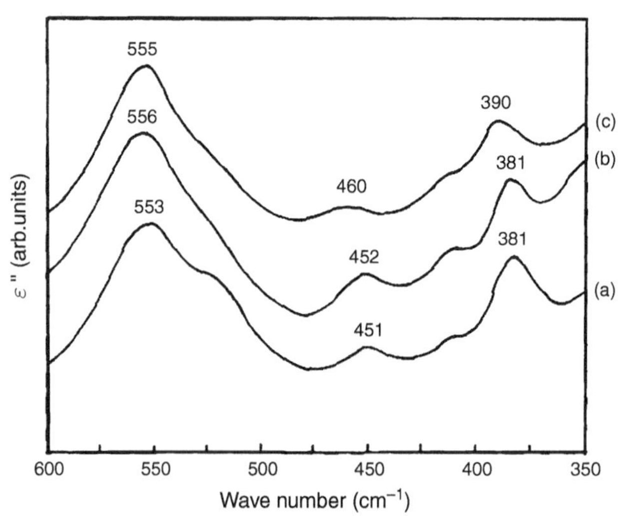

Figure 5.13 The effect of rare earth substitution on the infrared spectra (a) BaO—Pr_6O_{11} TiO_2 (b) BaO—Nd_2O_3—TiO_2 (c) BaO—Sm_2O_3—TiO_2 (after Ref. [10]).

which lead to anharmonicity of ceramics increased in the order of Sm, Nd and Pr. Figure 5.13 shows the effect of the rare earth substitution on the infrared active mode (ε'' versus wave number).

5.4.1 Effect of dopants

Several authors studied [66, 78, 87–130] the effect of additives on the sintering temperature, densification and microwave dielectric properties of $Ba_{6-3x}Ln_{8+2x}Ti_{18}O_{54}$ ceramics. The common among additives which improved the dielectric properties of $Ba_{6-3x}Ln_{8+2x}Ti_{18}O_{54}$ are PbO, Bi_2O_3 or $Bi_4Ti_3O_{12}$ and TiO_2. The $Ba_{6-3x}Sm_{8+2x}Ti_{18}O_{54}$ for $x=2/3$ has excellent properties: $\varepsilon_r=80.4$, Qf > 9000 GHz. However, the τ_f is −13 ppm/°C and hence TiO_2 which has a high positive τ_f was added to tailor the negative τ_f of $Ba_{6-3x}Sm_{8+2x}Ti_{18}O_{54}$ ceramics [87, 129]. The variations of permittivity, Qf and τ_f are shown in Figure 5.14 as a function of TiO_2 content. Addition of rutile considerably decreased the sintering temperature and improved the dielectric properties and could tune τ_f close to zero with ε_r of about 82 and Qf = 9000–12 000 GHz [87, 129]. XRD and Electron Probe Microanalysis (EPMA) studies showed the presence of $Ba_2Ti_9O_{20}$ and TiO_2 secondary phases in addition to $Ba_{6-3x}Sm_{8+2x}Ti_{18}O_{54}$.

The beneficial influence of Bi_2O_3 or Bi_2O_3–TiO_2 addition in improving densification and τ_f of $Ba_{6-3x}Ln_{8+2x}Ti_{18}O_{54}$ ceramics has been reported by several researchers [5, 21, 42, 101–105, 111, 131–135]. Wersing [131] found that the addition of Bi_2O_3 in BaO–Nd_2O_3–5TiO_2 decrease τ_f remarkably with an increase in ε_r and a decrease in Qf. The dielectric properties and sintering temperature of Bi_2O_3-added ceramics are given in Table 5.3. Recently Suvorov et al. [111] showed that zero τ_f can be achieved by the addition of 2.3 mol% Bi_2O_3 to the $Ba_{6-3x}Nd_{8+2x}Ti_{18}O_{54}$ ($x=0.5$) compound. This amount of Bi coincides with the solid solubility limit of Bi which substitutes for Nd in $Ba_{6-3x}Nd_{8+2x}Ti_{18}O_{54}$ ($x=0.5$). Further addition of Bi caused the formation of Bi-rich phase, accompanied with considerable reduction of Qf value and increase of τ_f. They also reported that the properties of Bi containing $Ba_{6-3x}Nd_{8+2x}Ti_{18}O_{54}$ ($x=0.5$) ceramics can be significantly improved by leaching off the unreacted secondary phases prior to sintering. This increased Qf to about 5800 GHz with $\varepsilon_r=88$ and $\tau_f \sim 0$ ppm/°C.

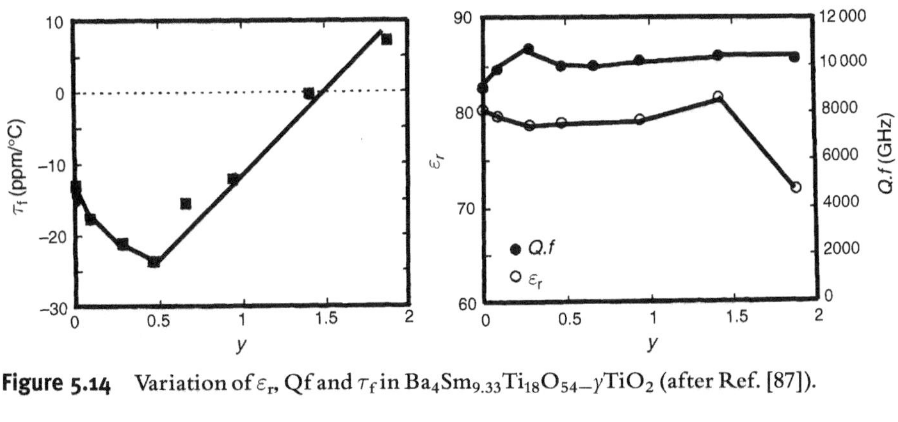

Figure 5.14 Variation of ε_r, Qf and τ_f in $Ba_4Sm_{9.33}Ti_{18}O_{54}$–$y$$TiO_2$ (after Ref. [87]).

Addition of small amount of Mn and WO_3 further improved Qf to about 7000 GHz. The Bi^{3+} substituted for Nd to form the solid solution $Ba_{4.5}(Nd_{1-y}Bi_y)_9Ti_{18}O_{54}$ and the solid solubility limit was determined [111] to be at $y = 0.15$ (2.5 mol%). The saturated phase can be represented as $Ba_{4.5}(Nd_{0.85}Bi_{0.15})Ti_{18}O_{54}$. Wu and Chen found [105] that the addition of $Bi_4Ti_3O_{12}$ led to the formation of $BaTi_4O_9$ secondary phase whereas Bi_2O_3 addition gave a single-phase material. EXAFS investigation of the local environment of Bi^{3+} ions incorporated in $Ba_{4.5}Nd_9Ti_{18}O_{54}$ showed [134, 135] that Bi^{3+} ions selectively substitute for the Nd^{3+} ions. The Bi^{3+} does not substitute for Nd^{3+} randomly on all possible sites but rather selectively enters one of the possible channels previously occupied by Nd^{3+}. Valant et al. investigated the solid solubility limit of Bi incorporation in $Ba_{6-3x}Ln_{8+2x}Ti_{18}O_{54}$ [for Ln = Nd, $x = 0.8$, 2) and $Ba_{4.5}Gd_9Ti_{18}O_{54}$ [102]. The microstructural investigations and microanalysis of a series of $Ba_{6-3x}Ln_{8+2x}Ti_{18}O_{54}$ (Ln = Nd, Gd) revealed that the solid solubility limit of Bi substitution of Ln depends on the composition of the $Ba_{6-3x}Ln_{8+2x}Ti_{18}O_{54}$ phase. For Nd, the solid solubility limit (y) in $Ba_{6-3x}(Nd_{1-y}Bi_y)_{8+2x}Ti_{18}O_{54}$ decreases with a decrease in x from $y = 0.16$ for $x = 2/3$ to $y = 0.1$ for $x = 0.8/3$. An even lower solid solubility limit $y = 0.06$ was found for the $Ba_{4.5}(Gd_{1-y}Bi_y)_9Ti_{18}O_{54}$ compound ($x = 0.5$). All Bi- substituted Nd compounds show high ε_r (83–99) and lower Qf than the parent compositions. By exceeding the solid solubility limit, abrupt changes in the dielectric properties were observed. The $Ba_{4.5}Gd_9Ti_{18}O_{54}$ is the least stable in the family of $Ba_{6-3x}Ln_{8+2x}Ti_{18}O_{54}$ compositions [48] and as such it is not able to accommodate higher concentrations of substituents.

Ohsato and co-workers investigated [46, 133, 101, 91] the effect of Bi substitution on the structure and properties of $Ba_{6-3x}Ln_{8+2x}Ti_{18}O_{54}$, for $x = 2/3$ which has the best quality factor. The lattice parameters of La and Nd showed a change in slope at $y = 0.05$. The b-lattice parameter of La decreased up to $y = 0.05$ and then increased with further increase in y. The decrease in b-lattice parameter for La and Nd indicates that the Ba ions located in the A2 pentagonal sites might be substituted by Bi ions since the Bi ion is smaller than Ba ion. Sakashita et al. [32, 136] derived the ionic radii for coordination number 12 as 1.61 and 1.45 Å respectively for Ba and Bi from the relationship between the coordination number and the effective ionic radii following Shannon [124]. The lattice parameters for La- and Nd-based compounds increase for $y > 0.05$ which indicate that the Bi ions are substituting for Ln ion located in the A1 site. The ionic radii of Ln ions (La, Nd Sm, Eu, Gd) for the coordination number 8 are smaller than that of Bi [124]. Hence the substitution of La, Nd, Sm, Eu, Gd at A1 site increase the lattice parameters. The microwave dielectric properties are very much influenced by the site occupancies. Figure 5.15 shows the variation of microwave dielectric properties of $Ba_{6-3x}(Ln_{1-y}Bi_y)_{8+2x}Ti_{18}O_{54}$ [$x = 2/3$] as a function of composition (y). The dielectric properties of La and Nd showed abrupt changes in the vicinity of $y = 0.05$ which is due to the change in the substitution site of Bi from pentagonal A2 site to A1 site. For La and Nd the ε_r decreased initially up to $y = 0.05$ and then increased with further increase in y. EPMA (back scattered electron probe micro analysis) indicated the formation of $Ba_2Ti_9O_{20}$ secondary phases for $y > 0.1$ due to the evaporation of Bi. The τ_f of La- and Nd-based ceramics decreased up to $y = 0.15$ and then continuously increased. In the case of Sm, Eu and Gd, the τ_f slightly increased with y. Zheng et al. [89] studied the effect of Bi addition in $Ba_{6-3x}(Sm_{0.2}Nd_{0.8})_{8+2x}Ti_{18}O_{54}$ [$x = 2/3$]. Bi_2O_3 addition lowered the sintering temperature, increased ε_r, and lowered τ_f and improved the Qf up to 1 wt% Bi_2O_3 addition. Addition of 1 wt% Bi_2O_3 and sintered at 1200°C/3 h gave a Qf of 8500 GHz and $\tau_f = -17$ ppm/°C, whereas 2 wt% Bi_2O_3 addition lowered the

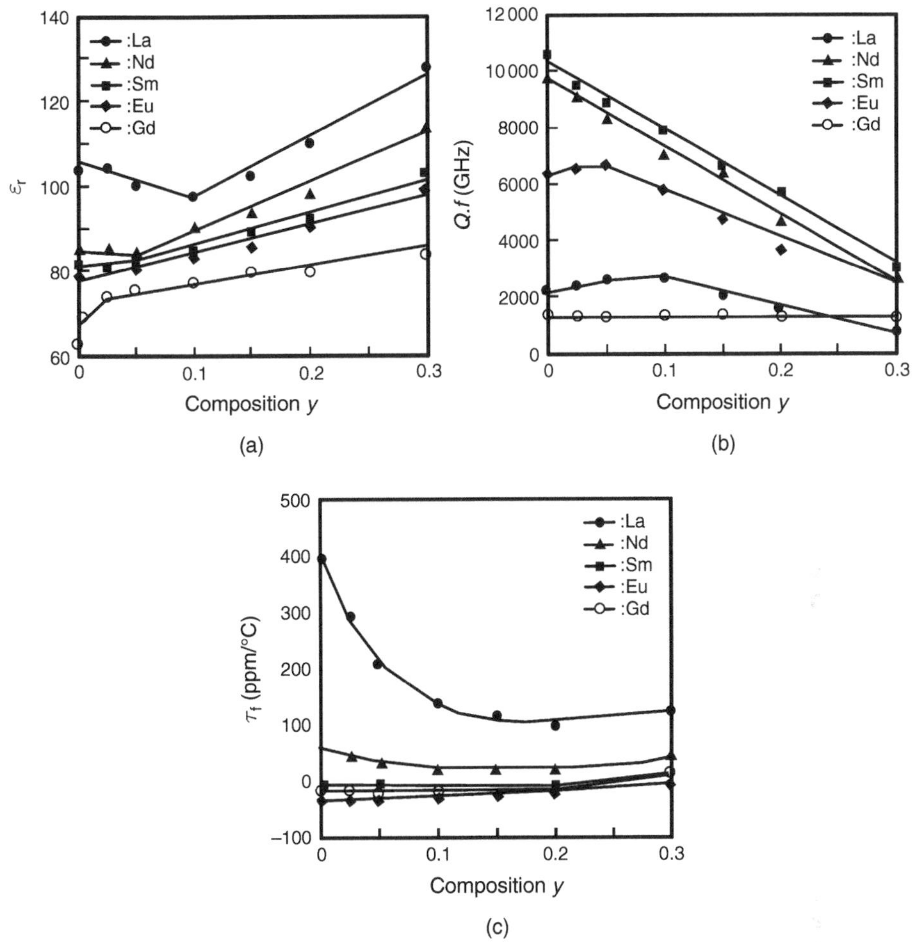

Figure 5.15 Variation of ε_r, Qf and τ_f in $Ba_{6-3x}(Ln_{1-y}Bi_y)_{8+2x}Ti_{18}O_{54}$ (after Ref. [133]).

sintering temperature to 1175°C/3 h and the ceramics had $\varepsilon_r = 83.5$, Qf = 7600 GHz and $\tau_f = 14$ ppm/°C. A secondary phase of $Ba_2Ti_9O_{20}$ formed when 2 wt% Bi_2O_3 was added. The *b*-axis lattice parameter and cell volume decreased with Bi_2O_3 addition indicating that Ba ions in the A2 pentagonal sites were substituted by Bi since Bi is smaller than Ba. Qin and Chen [137] reported that Sm/Bi co-substitution for La significantly improves the dielectric properties of $Ba_{6-3x}La_{8+2x}Ti_{18}O_{54}$ ($x = 2/3$). A single-phase solid solution was formed in $Ba_{6-3x}(La_{1-y-z}Sm_{1-y}Bi_z)_{8+2x}Ti_{18}O_{54}$ for $0 < z < 0.2$ for $y = 0.5$ and $0 < z < 0.16$ for $y = 0.7$. The ε_r increased and Qf decreased with increase in Bi content. $Bi_4Ti_3O_{12}$ secondary phase was formed when the Bi content exceeded this range. Figure 5.16 shows the variation of τ_f of $Ba_{6-3x}(La_{1-y-z}Sm_{1-y}Bi_z)_{8+2x}Ti_{18}O_{54}$ as a function of z for $y = 0.5$ and 0.7. The τ_f decreased up to the solid solubility limit.

Ohsato et al. studied [47] the effect of Li substitution for Sm in $Ba_{6-3x}Sm_{8+2x}Ti_{18}O_{54}$ [$x = 2/3$]. X-ray diffraction study showed that a single-phase $Ba_4Sm_{(28-y)/3}Li_yTi_{18}O_{54}$

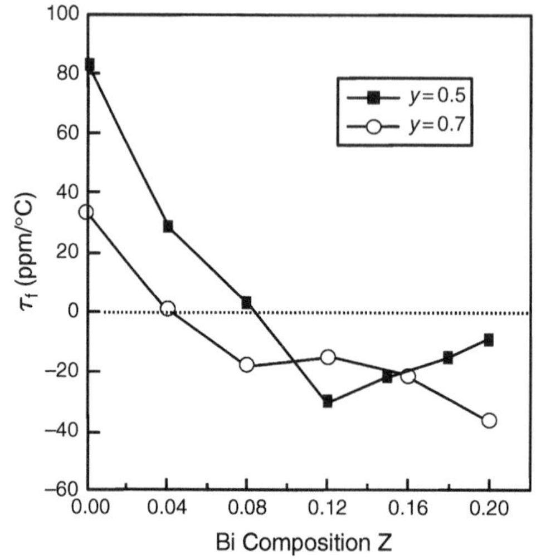

Figure 5.16 Variation of τ_f as a function of Bi^{3+} content in $Ba_{6-3x}(La_{1-y-z}Sm_yBi_z)_{8+2x}Ti_{18}O_{54}$ (after Ref. [137]).

formed for $y \leq 7$. The lattice parameters a and c decreased sharply until $y = 1$ and then increased linearly up to $y = 7$ as shown in Figure 5.17. In the range $0 \leq y \leq 1$ the lattice parameters decreased indicating that Li ions occupied only the A1 sites. In the region $1 \leq y \leq 7$, the a- and c-lattice parameters increased indicating that Li ions occupied

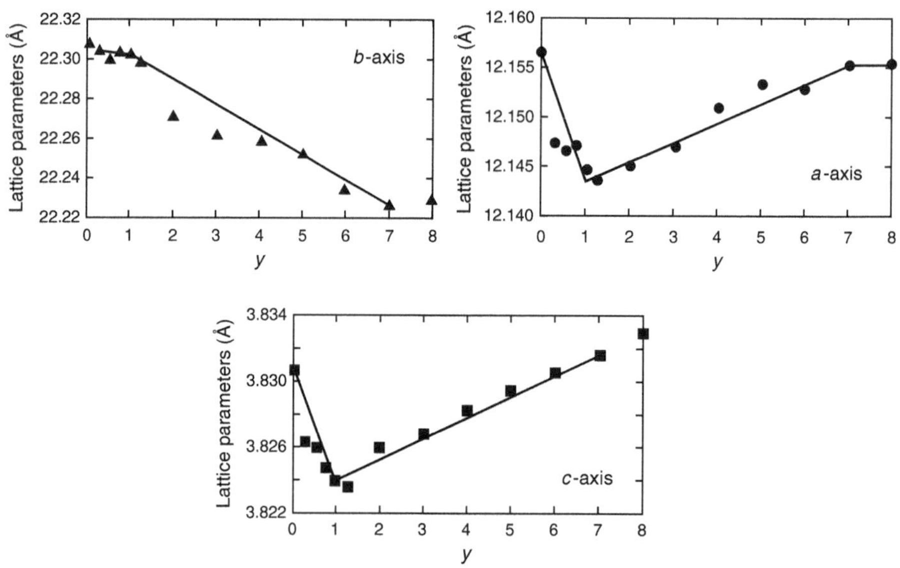

Figure 5.17 Variation of lattice parameters as a function of composition in $Ba_4Sm_{(28-y)/3}Li_yTi_{18}O_{54}$ solid solutions (after Ref. [47]).

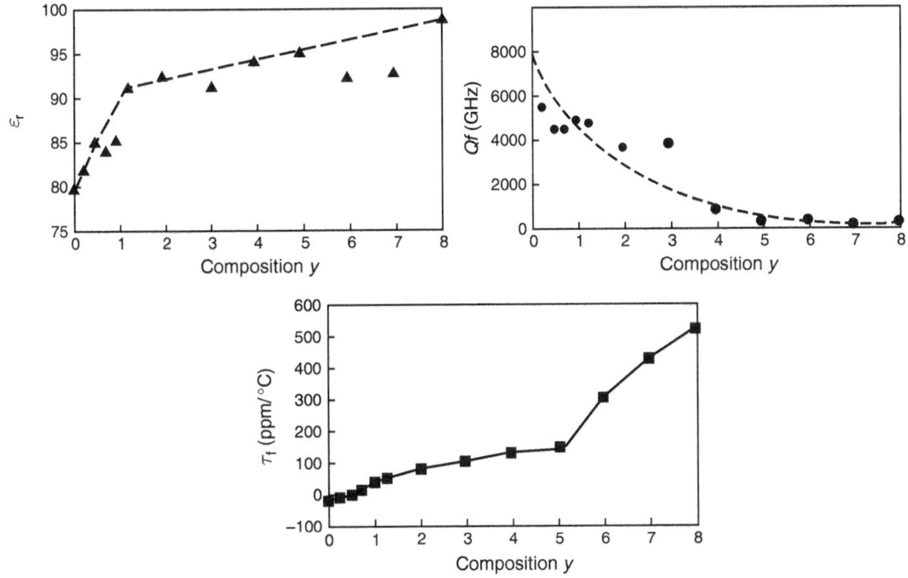

Figure 5.18 Variation of ε_r, Qf and τ_f as a function of composition in $Ba_4Sm_{(28-y)/3}Li_yTi_{18}O_{54}$ solid solutions (after Ref. [47]).

C sites. The site occupancy of Li changes at $y = 1$. One samarium ion can be substituted by three Li ions forming two vacancies. At $y = 1$, the vacancies of the A1 sites were fully occupied by Li ions. For $y > 1$, Li ions also occupy the C sites after filling the A1 sites. The dielectric properties of $Ba_4Sm_{(28-y)/3}Li_yTi_{18}O_{54}$ vary in relation to the variation in lattice parameters and are given in Figure 5.18 as a function of y. The ε_r increased with increase in y whereas the Qf decreased. In the vicinity of $y = 0.3$, $\varepsilon_r = 83$, Qf $= 5000$ GHz and $\tau_f \sim 0$. Xiong et al. [121] prepared $Sr(Bi_{1-x}Nd_x)_8Ti_7O_{27}$ for $x = 0.05$, 0.1, 0.2, 0.3, 0.4, 0.5 and reported ε_r values in the range 87–108 and Qf up to 2000 GHz. The major phase in the ceramic was $SrBi_8Ti_7O_{27}$ with aurivillus-type structure and a minor phase of $Bi_4Ti_3O_{12}$.

5.4.2 Substitution for Ba

Several authors [15, 18, 29, 63, 114, 118, 138] reported that addition of Pb improves the microwave dielectric properties of $Ba_{6-3x}Ln_{8+2x}Ti_{18}O_{54}$ ceramics. The Pb has a smaller atomic size and hence Pb substitution decreased the lattice parameters. The Pb substitution for Ba decreased τ_f. The high α_D (dielectric polarizability) of Pb increased ε_r. However, Ubic et al. [97, 139] reported a slight decrease in ε_r by Pb substitution. It was reported [114] that vaporization of Pb leads to the formation of secondary phases such as TiO_2. Belous et al. [118, 138] investigated solid solubility limits of $(Ba_{1-y}Pb_y)_{6-x}La_{8+2/3x}Ti_{18}O_{54}$, $(Ba_{1-y}Pb_y)_{6-x}Nd_{8+2/3x}Ti_{18}O_{54}$ and $(Ba_{1-y}Ca_y)_{6-x}Ln_{8+2/3x}Ti_{18}O_{54}$ for a wide range of y and x. The Pb^{2+} and Ca^{2+} ions, when partially substituting for Ba^{2+} ions, occupy first the A1 sites and then the pentagonal A2-sites. Dielectric properties of the $[Ba_{1-y}M_y)_{6-3x}Ln_{8+2x}Ti_{18}O_{54}$ [R = La, Nd, Sm M = Pb, Ca] materials strongly depend on the distribution of the Pb^{2+} and Ca^{2+} ions at different crystallographic sites. By partial isovalent substitution of Pb^{2+} and Ca^{2+} for Ba^{2+}, the τ_f of $[Ba_{1-y}M_y)_{6-3x}Ln_{8+2x}Ti_{18}O_{54}$ can be tuned to zero. Belous and Ovchar [118, 138] reported that $(Ba_{1-y}Pb_y)_{6-x}Nd_{8+2/3x}Ti_{18}O_{54}$ samples sintered at 1330–1380°C/2–3 h were single phase for the region $0<y<0.7$ for $x=0$ and $0<y<0.6$ for $x=0.25$ and $0<y<0.4$ for $x=0.5$. $Nd_4Ti_9O_{24}$ secondary phase was observed [114, 115, 118, 138] outside the single-phase region. It was found that the dielectric properties varied non-linearly with Pb concentration. Non-linear variation of dielectric characteristics against the Pb concentration was shown to be due to the location of Pb^{2+} ions in different crystallographic sites in the unit cell. It has been reported that Pb enters the Ba^{2+} sites [15, 29, 114, 115, 118]. Valant et al. [134, 135] studied actual sites of Pb incorporated in $Ba_{4.5}Nd_9Ti_{18}O_{54}$ using extended X-ray absorption fine structure (EXAFS). EXAFS studies on $Ba_{4.5-y}Pb_yNd_9Ti_{18}O_{54}$ revealed that Pb^{2+} does not substitute for Nd^{3+} or Ba^{2+} randomly on all possible sites. The Pb^{2+} selectively enters the A1 site previously shared by Ba^{2+} and Nd^{3+}. However, for larger concentration of Pb, also substitute for Ba in the A2 sites [138]. For $x=0$, this occurs at $y=0.33$ and for $x=0.25$ at $y=0.238$. From Figure 5.19 it was found that Q maxima occur at around $y=0.4$ for $x=0$ and 0.25. It was also found that *Qf* maximum occurs at a Pb concentration which corresponded to the complete substitution of Ba on the A1 sites. It is found that Pb substitution decreased τ_f up to $y=0.5$. The ε_r decreased up to $x=0.5$ and then increased for larger concentration of Pb.

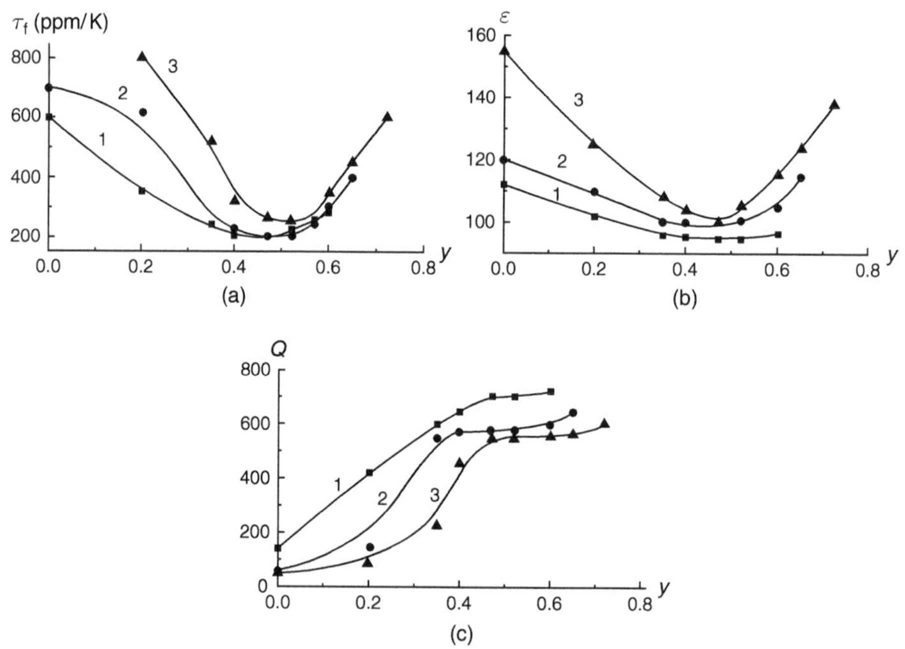

Figure 5.19 Variation of ε_r, Q and τ_f as a function of composition y in $(Ba_{1-y}Pb_y)_{6-x}La_{8+2x/3}Ti_{18}O_{54}$: (1) $x=1.5$ (2) $x=0.75$ (3) $x=0$ (after Ref. [118]).

The effect of Sr substitution for Ba in $Ba_{6-3x}Sm_{8+2x}Ti_{18}O_{54}$ has been investigated by several workers [16, 18, 112, 139–145]. Sun et al. [18] reported that it is possible to tune the τ_f from −13 to +30 ppm/°C by adjusting the Sr content from 0 to 25 mol% in the $0.15(Ba_{1-x}Sr_x)O$–$0.15Sm_2O_3$–$0.7TiO_2$. The τ_f became zero for $x = 0.07$ (7 mol% of Sr). Addition of SnO_2 as sintering aid lowered the sintering temperature whereas addition of CdO improved Q. Nishigaki et al. reported [16] the dielectric properties of 5 mol% Sr substitution for Ba in ceramics with molar composition $0.15BaO$–$0.15Sm_2O_3$–$0.7TiO_2$ which is equivalent to 1:1:4.7. The $Ba_{1-x}Sr_xSm_2Ti_{4.7}O_{14}$ with $x = 0.05$ sintered at 1350–1380°C/2 h showed excellent dielectric properties, i.e., $\varepsilon_r = 80$, Qf = 11 000 GHz and τ_f zero. The sintered ceramic showed the presence of small amounts of TiO_2 and $Ba_2Ti_9O_{20}$ secondary phases in accordance with the finding that 1:1:4.7 is not a single-phase compound. The presence of TiO_2 with high positive τ_f makes the 1:1:4.7 ceramics more positive. Imaeda et al. [112] reported the effects of substitution of Sr for Ba in $Ba_{6-3x}Sm_{8+2x}Ti_{18}O_{54}$ solid solutions in terms of the lattice parameters and microwave dielectric properties. Figure 5.20 shows the variation of lattice parameters with the composition α in $(Ba_{1-\alpha}Sr_\alpha)_{6-3x}Sm_{8+2x}Ti_{18}O_{54}$. For $x = 0.6$, the formula can be written as $Ba_{4.2}Sm_{9.2}Ti_{18}O_{54}$ or $(Sm_{9.2}Ba_{0.2})A1(Ba_4)A2Ti_{18}O_{54}$. The 0.2Ba ions in the A1 sites produce internal strain since the Ba ions are large in size for the A1 sites. When 0.2Ba ions are completely substituted by Sr ions, Qf value improved indicating that strain in the crystal structure has relieved. For $\alpha = 0.048$, 0.2Ba in $(Ba_{1-\alpha}Sr_\alpha)_{6-3x}Sm_{8+2x}Ti_{18}O_{54}$ is substituted by Sr.

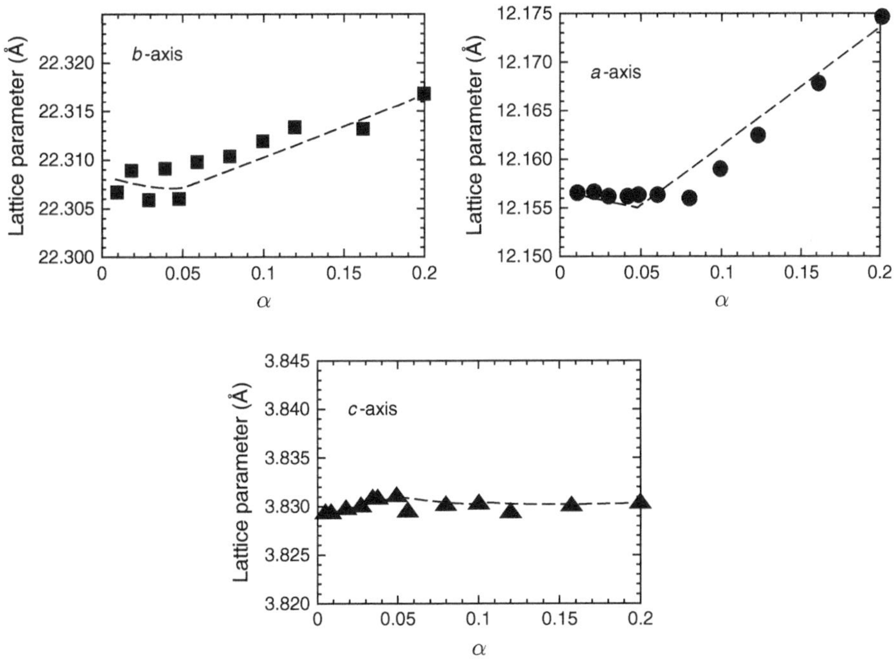

Figure 5.20 Variation of lattice parameters of $(Ba_{1-\alpha}Sr_\alpha)_{4.2}Sm_{9.2}Ti_{18}O_{54}$ solid solutions as a function of composition α (after Ref. [112]).

At $\alpha = 0.048$, the mode of substitution changes. For α in the range 0.0–0.048, *a*- and *b*-axis decreased linearly whereas for $\alpha = 0.048$–0.2, they increased linearly. These results show that the sites of substitution of Sr are different in the crystal structure. The decrease in the lattice parameters for the structure with the formula $(Sm_{9.2}Ba_{0.2-4.2\alpha}\ Sr_{4.2\alpha})(Ba_4)Ti_{18}O_{54}$ for compositions from $\alpha = 0.0$ to 0.048 means that the Ba ions located in the A1 sites in the perovskite blocks are substituted by the Sr ions, because Sr is smaller than Ba ions. For compositions $\alpha = 0.048$–0.2, the increase in the lattice parameters mean that the vacant sites located in perovskite blocks are preferentially occupied by the Sr ions as they substitute for Ba ions, thereby increasing the vacancies in the pentagonal columns. The structural formula can be represented as $(Sm_{9.2}Sr_{4.2\alpha})(Ba_{1-\alpha})_{4.2}Ti_{18}O_{54}$. If the Sr ions occupy the pentagonal A2 sites, the lattice parameters should decrease because of the difference in ionic radii. Figure 5.21 shows the variation of ε_r, Qf and τ_f of $(Ba_{1-\alpha}Sr_{\alpha})_{4.2}Sm_{9.2}Ti_{18}O_{54}$ solid solutions as a function of α. For $\alpha = 0.048$ for which all the Ba ions in the A1 sites are replaced by Sr ions showed the highest quality factor. The improvement in Qf is related to lowering of strain in the crystal structure due to Sr substitution. The composition in which Sr is substituted for 0.2Ba is $\alpha = 0.048$ in the $(Ba_{1-\alpha}Sr_{\alpha})_{6-3x}Sm_{8+2x}Ti_{18}O_{54}$ substitutional formula. The dielectric properties depend on the lattice parameters, the values of which change at

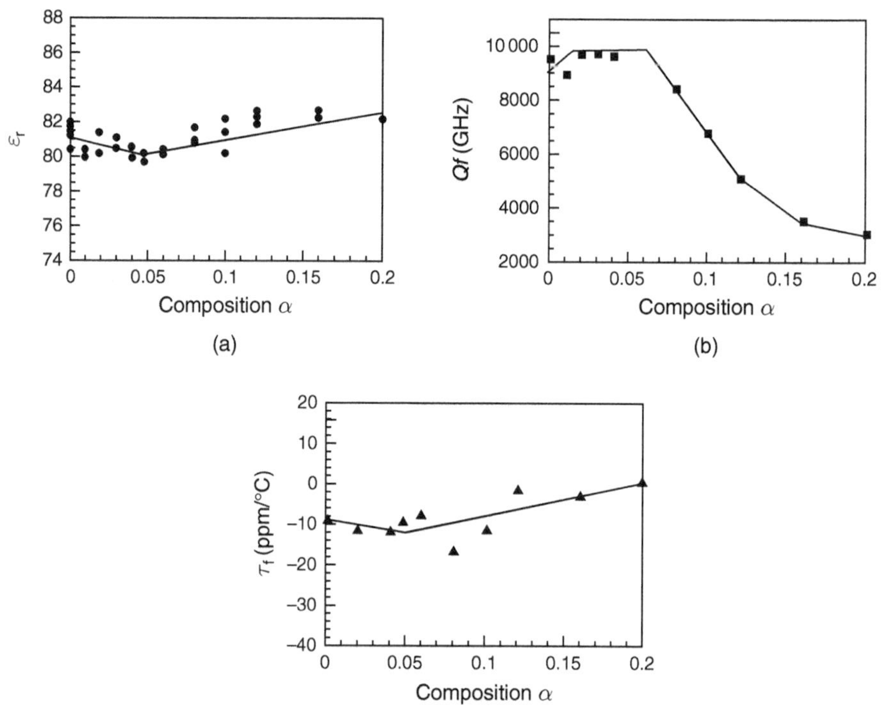

Figure 5.21 Variation of ε_r, Qf and τ_f in $(Ba_{1-\alpha}Sr_{\alpha})_{4.2}Sm_{9.2}Ti_{18}O_{54}$ solid solutions as a function of composition α (after Ref. [112]).

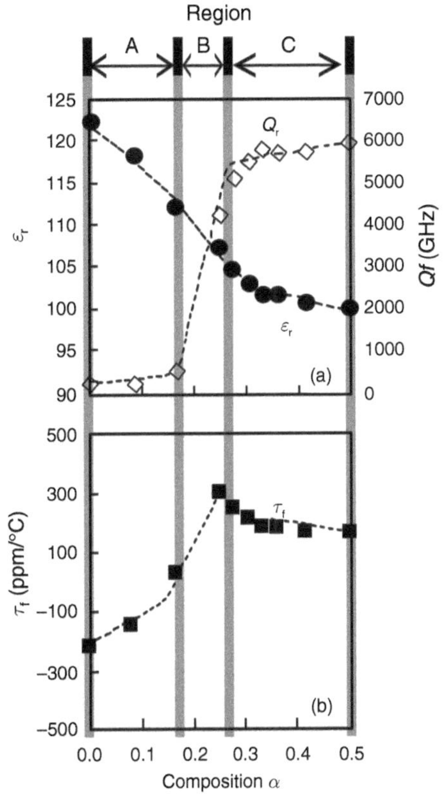

Figure 5.22 Variation of ε_r, Qf and τ_f in $(Ba_{1-\alpha}Sr_\alpha)_6Nd_8Ti_{18}O_{54}$ solid solutions as a function of composition (after Ref. [144]).

the composition $\alpha = 0.048$ due to change in the substitutional mode of Sr ions. In a similar way, Ohsato and co-workers [144, 145] improved the quality factor of $Ba_{6-3x}Nd_{8+2x}Ti_{18}O_{54}$ ($x = 0$) by partial substitution of Ba by smaller Sr ions. For $x = 0$, the structural formula can be written as $(Nd_8Ba_2)_{A1}(Ba_4)_{A2}Ti_{18}O_{54}$. The A2 sites are occupied by four Ba ions and A1 sites by eight Nd ions and two Ba ions. For $x = 0$, the structure has the highest internal strain and therefore exhibits low quality factor. Figure 5.22 shows the variation of ε_r, Qf and τ_f of $(Ba_{1-\alpha}Sr_\alpha)_6Nd_8Ti_{18}O_{54}$ as a function of composition (α). The Qf increased and ε_r decreased with increase in the value of α in the range $0 \leq \alpha \leq 0.5$. The Qf gradually increased with α in the range $0 \leq \alpha \leq 0.16$ and then abruptly increased in the range $0.16 \leq \alpha \leq 0.26$. The increase in Qf with α is small in the range $0.26 \leq \alpha \leq 0.5$. The ε_r and τ_f also showed a similar gradual change in the range $0 \leq \alpha \leq 0.16$ and abrupt change in the range $0.16 \leq \alpha \leq 0.26$. The ε_r varied in the range 122–103 and Qf in the range 200–5880 GHz and τ_f from −220 to +300 ppm/°C for α in the range 0–0.5. The substitution of smaller Sr for Ba ions in the A1 sites decreased the internal strain of $(Ba_{1-\alpha}Sr_\alpha)_6Nd_8Ti_{18}O_{54}$ and hence the quality factor increased as evidenced by Figure 5.22 in regions A and B. In the C region Sr substitution for Ba ions in A2 sites have not much effect in the improvement of Qf. The atomic size of Sr is too small for the A2 site to decrease the internal strain. XRD study showed [144] a linear decrease of lattice parameters indicating the formation of solid solution. However, $BaTiO_3$ secondary phase was observed for $\alpha = 0$ and $SrTiO_3$ for $\alpha = 0.5$ and $Ba_{1-x}Sr_xTiO_3$ for $0 \leq \alpha \leq 0.5$. The ferroelectric $BaTiO_3$ is very lossy as compared to the dielectric $SrTiO_3$ and is one of the reasons for low quality factor for low values of α. Nagatomo et al. [144] also investigated the effect of Y substitution for Nd on the dielectric properties of $(Ba_{1-\alpha}Sr_\alpha)_{6-x}(Nd_{1-y}Y_y)_{8+2x}Ti_{18}O_{54}$ ($x = 0$). It was found that Y substitution for Nd increased the Qf and decrease ε_r and τ_f in the range up to $y = 0.3$. The dielectric properties showed a sudden change at $y = 0.3$. For $y > 0.3$ the ε_r and τ_f increased and the Qf decreased. Presence of secondary phases of $Y_2Ti_2O_7$ and $Ba_4Ti_{18}O_{30}$ were detected for $y > 0.3$ and may be the reason for decrease in the quality factor. Nagamoto et al. reported that $y = 0.32$ is the limit of the solid solution formation in $Ba_4Sr_2(Nd_{1-y}Y_y)Ti_{18}O_{54}$.

The partial substitution of Ca for Ba in $Ba_{6-3x}Ln_{8+2x}Ti_{18}O_{54}$ considerably improved [118, 119, 138] τ_f with slight improvement in ε_r and Qf. The composition $(Ba_{0.95}Ca_{0.05})$–Sm_2O_3–$4.5TiO_2$ showed $\varepsilon_r = 81$, Qf = 11 000 GHz and $\tau_f = 2$ ppm/°C [119]. Ubic et al. [97, 139] found that the substitution of Ca and Sr for Ba in $Ba_{6-3x}Nd_{8+2x}Ti_{18}O_{54}$ severely degraded the Qf but increased ε_r and τ_f. $Nd_4Ti_9O_{24}$ was found as a secondary

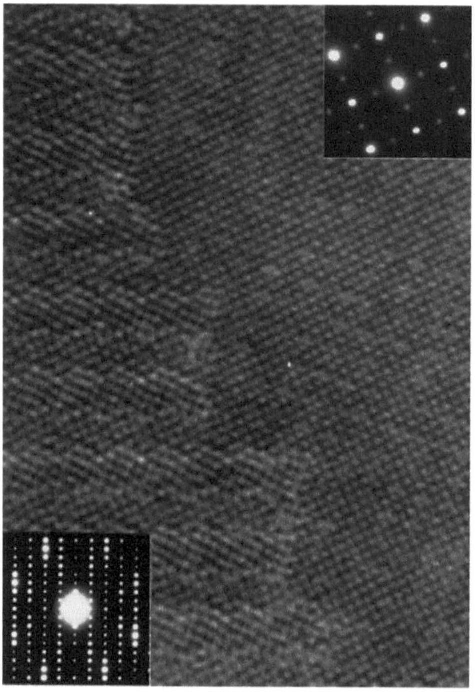

Figure 5.23 High resolution TEM image of the coherent boundary between $(Ba_{1-\alpha}Sr_\alpha)_{6-3x}Nd_{8+2x}Ti_{18}O_{54}$ and $NdTiO_3$ ceramics. The BNT is on the left and on the right is $NdTiO_3$. The zone axis is [001]. (after Ref. [140], Courtesy Materials Research Society).

phase in Sr-, Ca- and Pb-substituted ceramics. They also observed a third-phase $NdTiO_3$ in $(Ba_{0.5}Ca_{0.5})_{6-3x}Nd_{8+2x}Ti_{18}O_{54}$ and $(Ba_{0.5}Sr_{0.5})_{6-3x}Nd_{8+2x}Ti_{18}O_{54}$ in addition to $Nd_4Ti_9O_{24}$. Figure 5.23 shows a high resolution TEM image of the coherent boundary between Sr-doped $Ba_{6-3x}Nd_{8+2x}Ti_{18}O_{54}$ and $NdTiO_3$. The $NdTiO_3$ phase has a tilted perovskite structure and is stabilized by the presence of Ca^{2+} or Sr^{2+}. Ubic et al. [140] established an orientational relationship between $Ba_{6-3x}Nd_{8+2x}$ $Ti_{18}O_{54}$ and $NdTiO_3$.

The synthesis of Bi- and Pb-substituted $Ba_{6-3x}Ln_{8+2x}Ti_{18}O_{54}$ is related to the conditions necessary to maintain exact stoichiometry and reproducibility due to the high partial pressure of Bi_2O_3 and PbO at elevated temperatures [109]. Slight changes in stoichiometry can lead to dramatic changes in the microstructural development during sintering. Incomplete substitution of Pb for Ba and Bi for Nd due to the evaporation of PbO or Bi_2O_3 during sintering leads to the formation of multiphase ceramics. In such ceramics the resultant dielectric properties depend on the properties of each individual single phase.

5.4.3 Substitution for Ti

The microwave dielectric properties of $Ba_{6-3x}Ln_{8+2x}Ti_{18}O_{54}$ can also be tailored by suitable substitution at the B site. Attempts were made to substitute Al, Sn, Zr and Hf for Ti [115, 116, 120, 146]. Mizuta et al. [120] prepared $(Ba_{4.2}Sm_{9.2})_{\alpha}Ti_{18-y}Al_yO_{54}$ $(0 < y < 1.61)$ and $(\alpha = 1 + y/36)$ where the B site is substituted by Al and $x = 0.6$, i.e., $(Ba_{4.2}Sm_{9.2})_{\alpha}$ $Ti_{18-y}Al_yO_{54}$ $(0 < y < 1.61)$. The ε_r and Qf decreased and τ_f became more negative with Al substitution. The results indicate the possibility of tuning τ_f to zero in the La- and Nd-based ceramics by partial aluminium substitution at B sites. Azough et al. [95] reported that the addition of up to 1 wt% Al_2O_3 to the starting mixtures reduced the sintering temperatures by about 100°C with an increase in Qf. The substitution of small amount of Al to the Ti sites led to a decrease in ε_r and improvement in τ_f [95, 120]. Chen et al. [115, 116] made A and B site substitution in $Ba_{6-3x}Sm_{8+2x}$ $Ti_{18}O_{54}$ by preparing $Ba_{6-3x}Sm_{8+2x}(Ti_{1-z}Sn_z)_{18}O_{54}$ $(x = 2/3,\ z = 0.05, 0.1, 0.2, 0.3, 0.5, 0.8)$ and Ba_{6-3x} $(Sm_{1-y}Nd_y)_{8+2x}(Ti_{1-z}Sn_z)_{18}O_{54}$ $(x = 2/3,\ z = 0.05,\ y = 0.1, 0.3, 0.5, 0.8)$. The ε_r decreased up to $z = 0.3$ and then increased whereas the Qf value decreased. The τ_f increased with z in $Ba_{6-3x}Sm_{8+2x}(Ti_{1-z}Sn_z)_{18}O_{54}$. The $Ba_{6-3x}Sm_{8+2x}$ $(Ti_{1-z}Sn_z)_{18}O_{54}$ with $x = 2/3$, $z = 0.05$ showed $\varepsilon_r = 76$, Qf = 6260 GHz and $\tau_f = 2$ ppm/°C. For $z > 0.1$ secondary phases of $BaSm_2O_4$, $Sm_2Sn_2O_7$ appeared which degraded the dielectric properties. In $Ba_{6-3x}(Sm_{1-y}Nd_y)_{8+2x}(Ti_{1-z}Sn_z)_{18}O_{54}$ with $z = 0.05$, the ε_r, Qf and τ_f increased with increase in y as shown in Figures 5.24 and 5.25. The ceramic $Ba_{6-3x}(Sm_{1-y}Nd_y)_{8+2x}(Ti_{1-z}Sn_z)_{18}O_{54}$ $(x = 2/3)$ with $y = 0.8$, $z = 0.05$ sintered at 1360°C/12 h showed $\varepsilon_r = 80$, Qf = 10 600 GHz $\tau_f = +11$ ppm/°C. The ε_r and Qf decreased with increase in z. For $y = 0$, $z = 0.05$, nearly zero τ_f with $\varepsilon_r = 76$ and Qf = 6260 GHz. It was also found that prolonged sintering improves the quality factor. Several authors [63, 97, 146] studied the effect of Zr substitution for Ti in $Ba_{6-3x}R_{8+2x}Ti_{118}O_{54}$ ceramics. Zr substitution [97, 146] decreased ε_r, Qf and τ_f. $Nd_2Zr_2O_7$ secondary phase was found in Zr-substituted solid solutions [97]. The solid solution limit of Zr in $BaNd_2Ti_4O_{12}$ compound is limited to the range between 25 and 50 mol%. The density increased with Zr substitution with improvement in τ_ε but the ε_r and Qf decreased. Hf doping increased the dielectric loss and the samples did not resonate [97].

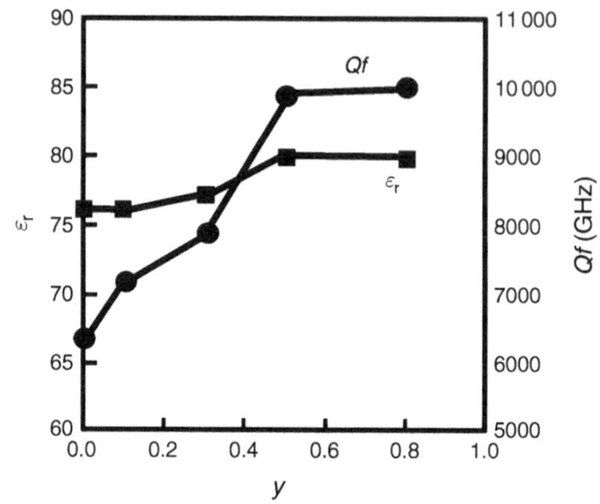

Figure 5.24 Variation of ε_r and Qf in $Ba_{6-3x}(Sm_{1-y}Nd_y)_{8+2x}(Ti_{1-z}Sn_z)_{18}O_{54}$ ($x=2/3$, $z=0.05$) as a function of composition y. (after Ref. [115]).

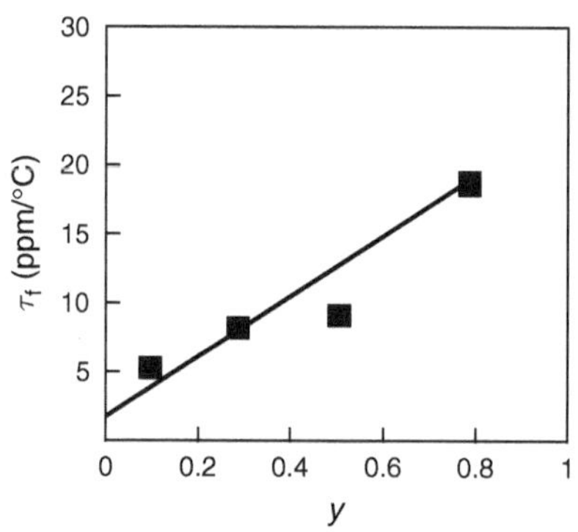

Figure 5.25 Variation of τ_f in $Ba_{6-3x}(Sm_{1-y}Nd_y)_{8+2x}(Ti_{1-z}Sn_z)_{18}O_{54}$ ($x=2/3$, $z=0.05$) as a function of composition y. (after Ref. [115]).

5.4.4 Texturing

The grains in the sintered $Ba_{6-3x}Ln_{8+2x}Ti_{18}O_{54}$ ceramics are *c*-axis-elongated due to the structural anisotropy. Large anisotropy in the crystal structure can lead to anisotropic electrical properties. Thus it is possible to tailor the electrical properties by making grain-oriented ceramics [147–150]. Negas and Davies [28] reported texturing in Bi-substituted $Ba_{6-3x}Nd_{8+2x}Ti_{18}O_{54}$ ceramics as evidenced by the presence of preferred alignment of grains in the X-ray diffraction pattern. Such ceramics show anisotropic dielectric properties and the Bi-substituted samples showed much larger anisotropy. Hoffman and Waser [60] prepared $Ba_{6-3x}Ln_{8+2x}Ti_{18}O_{54}$ with R = La, Ce, Nd, Sm by hot forging at a temperature in the range 1200–1250°C and at a pressure of about 34 MPa. The SEM and X-ray diffraction studies showed that the grains in the hot-forged ceramics are elongated in the *c*-direction. The hot-pressed samples have higher densities and higher ε_r but not much change in the quality factor. This difference is attributed to texturing (oriented grains) in the hot-pressed samples. The ε_r was found higher in the *c*-direction. Wada et al. [151–153] prepared grain-oriented $Ba_4Sm_{9.33}Ti_{18}O_{54}$ ceramics. The ε_r and Qf values were not much affected by the direction of grain orientation. In contrast high anisotropy in the τ_f was found between the directions parallel and perpendicular to the orientation direction. Wada et al. [151–153] prepared textured $Ba_{6-3x}Ln_{8+2x}Ti_{18}O_{54}$ [$x = 2/3$ and Ln = Sm] by a templated growth process [117, 154]. In the templated grain-growth processs, the oriented ceramics was made by aligning a small amount of anisotropic particles in a fine matrix powder followed by sintering. Wada et al. initially prepared elongated columnar $Ba_{6-3x}Sm_{8+2x}Ti_{18}O_{54}$ (BST) template particles by NaCl–KCl molten salt method. These template particles were then mixed with fine matrix powder of BST prepared by the conventional method. It was then made into slurry with polymer and then tape casted. After drying, the sheets were cut and stacked into a green block. The blocks were then pressed cold isostatically and then sintered at 1460°C/2 h. Two types of samples were prepared (a) sample with circular plane of disc and casting plane are parallel BST (ll) and (b) sample with circular plane of disc and casting plane are perpendicular BST (⊥) as shown schematically in Figure 5.26 SEM investigation revealed the presence of rod-like grains parallel to the casting direction in BST (ll) whereas only cross-sectional view observable in BST (⊥). The X-ray diffraction pattern of BST (*ll*) and BST(⊥) samples with 15% template concentration showed that in BST (*ll*) diffraction peaks of $\{hko\}$ became stronger and $\{00l\}$ became weaker. But in BST (⊥) the $\{00l\}$ became intense. The texture fraction was estimated using Lotergering's method [155].

The variation of orientation degree as a function of template concentration is shown in Figure 5.27. The orientation degree was increased up to 15% of template concentration. The degree of <*hko*> and <00*l*> orientation in BST (*ll*) and BST (⊥) approached approximately 0.89 and 0.66 respectively at the template concentration of 15%. As the template concentration increased, the ε_r and Qf slightly increased for BST (*ll*) whereas they slightly decreased for BST (⊥) as shown in Figure 5.28a, b. The τ_f of BST (*ll*)

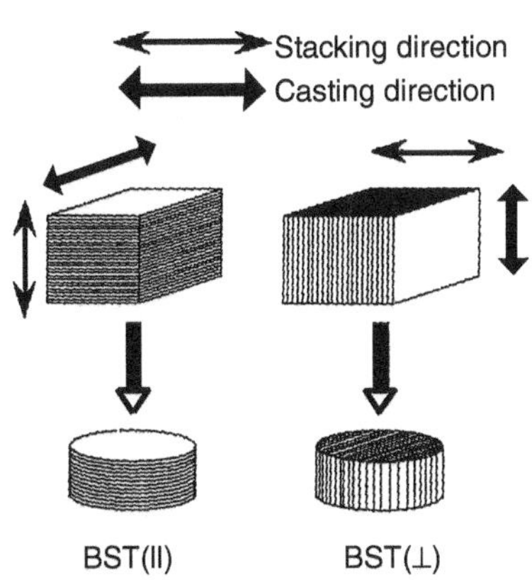

Figure 5.26 Schematic sketch of preparing textured samples whose circular plane was parallel to ($BST_{//}$) and perpendicular ($BST_{\perp}$) to the casting directions. (after Ref. [151], Courtesy Japanese Society of Applied Physics).

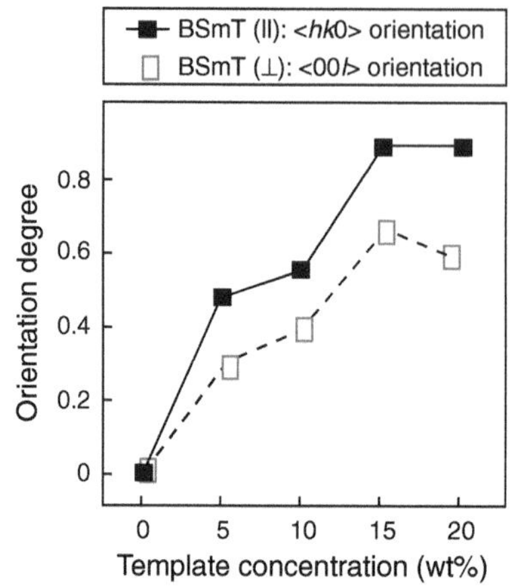

Figure 5.27 Variation of the degree of orientation as a function of template concentration. BSmT$_{//}$ <*hk*0> orientation, dotted line BSmT$_{\perp}$ <00*l*) orientation. (after Ref. [152]).

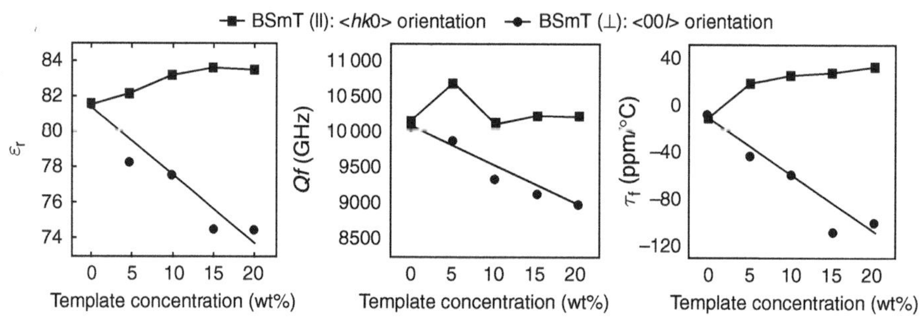

Figure 5.28 Variation of microwave dielectric properties ε_r, Qf and τ_f as a function of template concentration. (after Ref. [152]).

increased with template concentration whereas it became more negative for BST (⊥) as shown in Figure 5.28c. The results indicate that the presence of <*hk*0> orientation within the circular plane of disc leads to positive τ_f and <00*l*> orientation causes negative τ_f. Figure 5.29 shows the variation of τ_f with the degree of orientation. The large anisotropy between *ab*-axes and *c*-axis affects the change in τ_f. It is thus evident that one can obtain a desired τ_f value by controlling the grain orientation.

5.4.5 Effect of glass

Several authors studied [92, 106, 108–110, 130, 150, 156–159] the effect of addition of glasses on the sintering temperature and the dielectric properties of $Ba_{6-3x}Ln_{8+2x}Ti_{18}O_{54}$ ceramics. The glass addition lowered the sintering temperature and changed the dielectric properties. The grain size increased with the increase of dopant concentration. Lu and Huang [159] reported that the addition of less than 2 wt% B_2O_3 in $Ba_{6-3x}Nd_{8+2x}Ti_{18}O_{54}$ lowered the sintering temperature by about 200°C and improved densification whereas

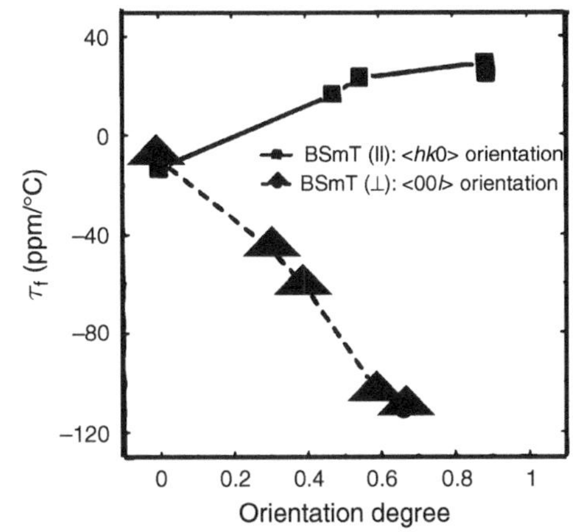

Figure 5.29 Variation of τ_f for BSmT (*ll*) and BSmT ($_\perp$) as a function of degree of orientation (after Ref. [152]).

addition of more than 5 wt% B_2O_3 considerably decreased ε_r and led to the formation of $Ba_2Ti_9O_{20}$ secondary phases. Ota et al. [92] reported the effect of addition of crystalline and glassy B_2O_3 and GeO_2 in $Ba_{6-3x}Sm_{8+2x}Ti_{18}O_{54}$. Addition of 0.5 wt% B_2O_3 lowered the sintering temperature from 1400 to about 1170°C with $\varepsilon_r = 76$, Qf = 10 500 GHz, and $\tau_f = -19.4$ ppm/°C and 0.5 wt% of GeO_2 reduced the sintering temperature to about 1150°C with $\varepsilon_r = 77$, Qf = 8900 GHz and $\tau_f = -19.7$ ppm/°C. However, addition of 0.5 wt% B_2O_3 + 0.5 wt% GeO_2 reduced the Qf to 5200 GHz. Addition of 0.5 wt% glassy B_2O_3 considerably reduced the Qf to about 4300 GHz and ε_r to 69.6 when sintered at 1200°C. Chang et al. [107, 156] reported that the addition of alumina-doped silica glass to $Ba_{6-3x}(Nd,Sm)_{8+2x}Ti_{18}O_{54}$ ceramic lowered the sintering temperature and permittivity and improved the Qf. Addition of 20 wt% alumina-doped silica and sintered at 1250°C gave $\varepsilon_r = 30$, Qf ~ 8700 GHz. Santha et al. [93] investigated the effect of several glass systems on the sintering temperature, microwave dielectric and elastic properties of $Ba_{6-3x}Sm_{8+2x}Ti_{18}O_{54}$ ceramics. It is found that addition of 3–4 wt% Al_2O_3–B_2O_3–SiO_2 and sintered at 1175°C resulted in a ceramic with $\varepsilon_r = 64$, Qf = 8500 GHz and zero τ_f. The Youngs modulus decreased with increasing amount of glass irrespective of the composition of the glass [93].

5.5 Phase Transition

Butko et al. [160] reported diffuse peaks in the plot of ε_r (T) with temperature and they attributed it to displacive transitions. Belous and Ovchar [119] reported that the maxima in permittivities of BaO–Sm_2O_3–$n TiO_2$ depend on the TiO_2 content and the temperature. The maxima in ε_r move toward a lower temperature with a decrease in the n value as shown in Figure 5.30. A small amount of Ca substitution for Ba led to a shift of permittivity maxima toward lower temperatures for $Ba_{1-x}Ca_x$–Sm_2O_3–4.5TiO_2 as shown in Figure 5.31. The lattice parameters, ε_r and Qf, change only slightly but the τ_ε

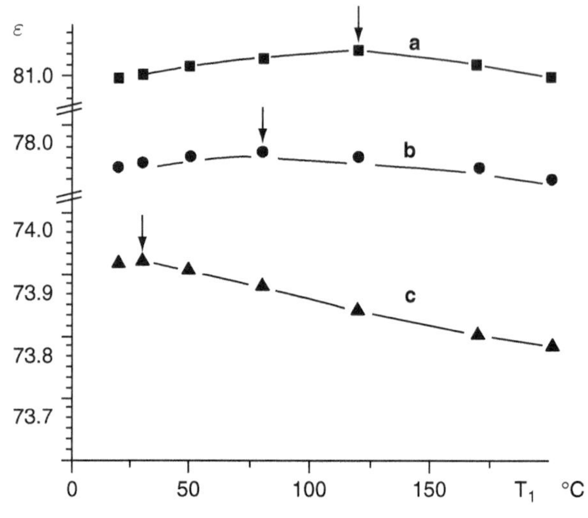

Figure 5.30 Variation of ε_r of $BaO{-}Sm_2O_3{-}nTiO_2$ as a function f temperature at 10 GHz (a) $n = 4.5$ (b) $n = 4.0$ (c) $n = 3$ (after Ref. [119]).

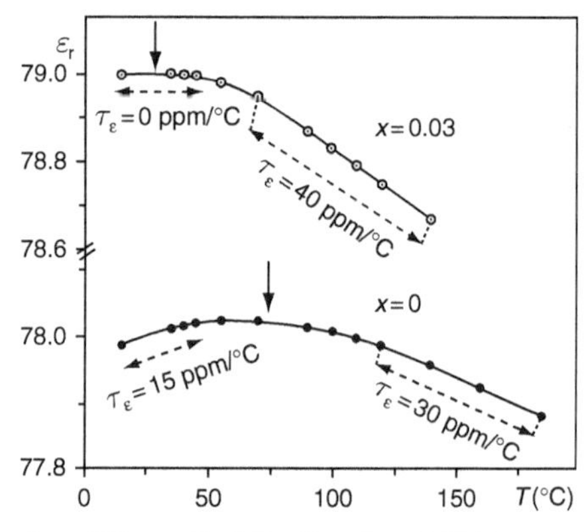

Figure 5.31 Variation of ε_r of $(Ba_{1-x}Ca_x)O{-}Sm_2O_3{-}4.5TiO_2$ as a function of composition x at 10 GHz (after Ref. [119]).

varies continuously from −15 to +25 ppm/°C for $x = 0$ to 0.1. Belous et al. [55, 161, 162] studied the temperature dependence of ε_r and tan δ for different values of x in $Ba_{6-x}Ln_{8+2x}Ti_{18}O_{54}$ (Ln = La, Nd, Sm, Gd). The studies showed the presence of diffuse anomalies in the temperature dependencies of ε_r and tan δ and are below room temperature for La and Nd and above room temperature for Sm- and Gd-based solid solutions. Figure 5.32 shows the temperature variations of ε_r and tan δ of $Ba_{4.5}Ln_9Ti_{18}O_{54}$ ($x = 0.5$) at 10 GHz. Mercurio et al. [163] and Poplakov et al. [164] reported that in $Ba_6Ln_8Ti_{18}O_{54}$ low frequency ε_r versus temperature curve shows dielectric peaks in the ε_r versus temperature plots indicating a phase transition. $Ba_6Ln_8Ti_{18}O_{54}$ (Ln = La, Pr, Nd, Sm) showed a strongly pronounced maxima in the ε_r with temperature. It was

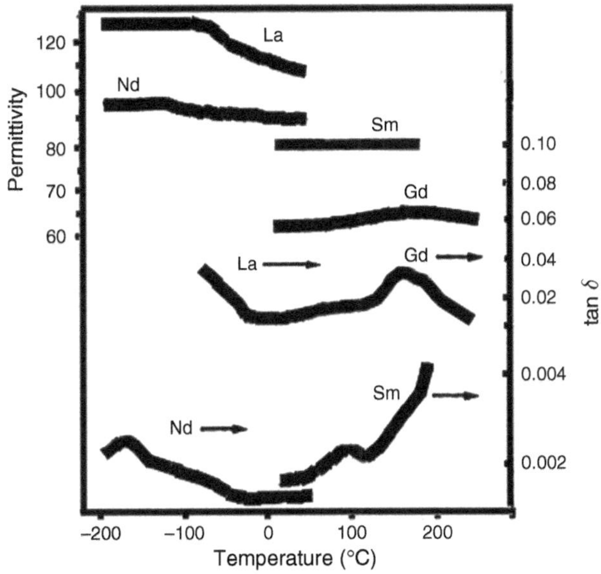

Figure 5.32 Variation of permittivity and loss tangent in $Ba_{6-x}Ln_{8+2x}Ti_{18}O_{54}$ (R = La, Nd, Sm, Gd) measured at 10 GHz (after Ref. [162]).

found [160, 162] that the maxima shift toward higher temperatures when the ionic radii of the rare earth element decreased. The diffuse maxima shift as a function of x. For example, in the Sm-based ceramic the diffuse maxima shifts toward lower temperatures with increase in x value. However, XRD and DSC studies did not indicate any evidence of a structural transition. Belous and Co-workers attributed the observed anomalies to the competition between harmonic and anharmonic contributions of the phonons of the $Ba_{6-3x}Ln_{8+2x}Ti_{18}O_{54}$ crystal lattice, which show an opposite temperature behavior of permittivity and tan δ [160, 162].

5.6 Conclusions

The low loss dielectric ceramics in the $BaO–Ln_2O_3–TiO_2$ ceramic system have the general formula $Ba_{6-3x}Ln_{8+2x}Ti_{18}O_{54}$. The samples are prepared usually by solid state method by calcining at about 1100°C and sintering at temperatures in the range 1350–1430°C. The $Ba_{6-3x}Ln_{8+2x}Ti_{18}O_{54}$ has an orthorhombic crystal structure with four Ba atoms occupying the A2 pentagonal sites and Ba and or Ln ions occupying the 10 A1 sites and Ti ions occupying the B sites. The Eu, Sm and Nd have high quality factors. Substitution of Al, Sn for Ti can tune τ_f but decreases ε_r, and τ_f. The Sm-based ceramic has a negative τ_f and Nd-based positive τ_f. Hence a solid solution phase between them results in a ceramic with excellent properties with zero τ_f and are useful for the mobile phone handset applications. Addition of glasses decrease the sintering temperature with a decrease in ε_r, Qf and τ_f. Addition of 0.5 wt% B_2O_3 to $Ba_{6-3x}Sm_{8+2x}Ti_{18}O_{54}$ lowers the sintering temperature to about 1170°C with $\varepsilon_r = 76$, Qf = 10 500 GHz and $\tau_f = -19$ ppm/°C. The dielectric properties, especially τ_f, can be tuned by partial substitution of Bi for Ba or Ln and Pb, Sr for Ba. The Bi substitution decreases the sintering temperature and considerably improves τ_f. However, the quality factor is degraded. Substitution of Pb for Ba increases ε_r and the τ_f and the τ_f can be tuned to zero in Sm-based compounds. The Pb-substituted ceramics shows a non-linear variation of dielectric properties and is attributed to the location of Pb^{2+} ions in crystallographic sites in the unit cell. For smaller concentration, Pb enters the A1 site and for larger concentration it also enters the A2 site. Substitution of Sr for Ba ions in the A site decreases the internal strain and improves the quality factor and τ_f in $Ba_{6-3x}Sm_{8+2x}Ti_{18}O_{54}$. The lowest internal strain is found for $x = 2/3$. Selection of structure without internal strain is important in getting high Qf materials. The internal strain is reduced by the ordering of cations. Ordering of Ba and Ln ions lowers internal strain and increases *Qf*. The highest order is for $x = 2/3$ and hence compositions with $x = 2/3$ show the highest Qf factor. Eu has the smallest ionic radii and smallest internal strain and hence it shows the highest quality factor. The solid solubility depends on the rare earth. La has the highest range of solid solubility, $0.07 \leq x \leq 0.77$, and Gd has the lowest, $x = 0.5$ only. The $Ba_{6-3x}Ln_{8+2x}Ti_{18}O_{54}$ solid solutions show excellent microwave dielectric properties useful for practical applications. The dielectric properties are dependent on the composition (x) and the rare earth element.

References

[1] R. L. Bolton. Temperature compensating ceramic capacitors in the system barium-rare earth oxide titania, Ph.D Thesis. The University of Illinois, (1968).

[2] L. P. Mudroliubova, B. A. Rotenberg, N. F. Kartenko, A. N. Borsch, V. G. Prohvatilova, V. G. Kostikov, P. Yu, and M. P. Ivanova. Barium Lanthanoid Tetratitanate, USSA Certificate of Recognition. 632176(23 February 1977).

[3] D. Kolar, Z. Stadler, S. Gaberscek, and D. Suvorov. Ceramic and dielectric properties of selected compositions in BaO-TiO_2-Nd_2O_3 system. Ber. Deut. Keram. Ges. 55(1978) 346–348.

[4] D. Kolar, S. Gaberscek, B. Volavsek, H. S. Parker, and R. S. Roth. Synthesis and crystal chemistry of $BaNd_2Ti_3O_{10}$, $BaNd_2Ti_5O_{14}$ and $Nd_4Ti_5O_{24}$. J. Solid State Chem. 38(1981)158–164.

[5] D. Kolar, S. Gaberscek, Z. Stadler, and D. Suvorov. High stability, low loss dielectrics in the system BaO–Nd_2O_3–TiO_2–Bi_2O_3. Ferroelectrics. 27(1988)269–272.

[6] E. S. Razgon, A. M. Gens, M. B. Varfolomeev, S. S. Korovin, and V. S. Kostomarov. The complex barium and lanthanum titanates. Russ. J. Inorg. Chem. 25(1980)945–947.

[7] M. Gens, M. B. Varfolomeeva, V. S. Kostomarov, and S. S. Korovin. Crystal-chemical and electrophysical properties of complex titanates of barium and the lanthanides. Russ. J. Inorg. Chem. 26(1981)896–898.

[8] H. Ohsato. Science of tungsten bronze type like $Ba_{6-3x}R_{8+2x}Ti_{18}O_{54}$ (R = rare earth) microwave dielectric solid solutions. J. Eur. Ceram. Soc. 21(2001)2703–2711.

[9] S. M. Polyakov, E. I. Gindin, V. G. Prokhvatilav, and S. V. Shtelmakh. Composition and structure of a barium lanthanide tetratitanate. Russ. J. Inorg. Chem. 28(1983)1686–1687.

[10] K. Fukuda, I. Fijui, R. Kitoh, Y. Cho, and I. Awai. Influence of rare earth ions on BaO–TiO_2–rare earth oxide ceramics for microwave applications. Jpn. J. Appl. Phys. 32(1993)1712–1715.

[11] T. Jaakola, A. Uusimaki, R. Rautioahu, and S. Leppauvuori. Matrix phase in ceramics with composition near BaO–Nd_2O_3–$5TiO_2$. J. Am. Cer Soc. 69(1986)C234–235.

[12] X. M. Chen, Y. Suzuki, and N. Sato. Microstructures and microwave dielectric characterization of ceramics with the composition $BaO–Nd_2O_3–5TiO_2$. J. Mater. Sci: Mater. Electron. 6(1995)10–16.
[13] E. A. Nenashev, N. M. Siderova, N. F. Kastenko, L. R. Mudrolubova, B. A. Rotenberg, and B. B. Gluskhevoa. Compounds of the system $BaLn_2O_4–TiO_2$ [Ln = La,Nd. Neorg. Mater. 21(1985)1524–1527.
[14] R. G. Matveeva, M. B. Varfolomeeva, and L. S. Il'yushenko. Refinement of the composition and crystal structure of $Ba_{3.75}Pr_{9.5}Ti_{18}O_{54}$. Russ. J. Inorg. Chem. 29(1984)17–19.
[15] K. Wakino, K. Minai, and H. Tamura. Microwave characteristics of (Zr, Sn)TiO_4 and $BaO–Nd_2O_3–TiO_2$ dielectric resonators. J Am. Ceram. Soc. 67(1984)278–281.
[16] S. Nishigaki, H. Kato, S. Yano, and R. Kamimura. Microwave dielectric properties of (BaSr)O–$Sm_2O_3–TiO_2$ ceramics. Bull. Am. Ceram. Soc. 66(1987)1405–1411.
[17] J.-M. Wu, M.-C. Chang, and P. C. Yao. Reaction sequence and effects of calcination and sintering on microwave properties of (Ba,Sr)O–$Sm_2O_3–TiO_2$ ceramics. J. Am. Ceram. Soc. 73(1990)1599–1605.
[18] J.-S. Sun, C.-C. Wei, and L. Wu. Dielectric properties of (Ba,Sr)O–$(Sm,La)_2O_3–TiO_2$ ceramics at microwave frequencies. J. Mater. Sci. 27(1992)5818–5822.
[19] X. M. Chen, Y. Suzuki, and N. Sato. Preparation, densification of $BaNd_2Ti_5O_{14}$ dielectric ceramics. Mater. Lett. 16(1993)75–78.
[20] H. Sremoolanathan, M. T. Sebastian, and P. Mohanan. Dielectric resonators in $BaO–Ln_2O_3–5TiO_2$ system (Ln = La, Pr, Nd, Sm). Br. Ceram. Trans. 95(1995)79–81.
[21] J. M. Durand and J. P. Boilot. Microwave dielectric characteristics of $BaO–Bi_2O_3–TiO_2–Nd_2O_3$ dielectric resonators. J. Mater. Sci. Lett. 6(1987)134–136.
[22] J. P. Guha. Synthesis and characterisation of Barium lanthanum titanates. J Am. Ceram. Soc. 74(1991)878–880.
[23] J. Takahashi, T. Ikegemi, and K. Kageyama. Occurrence of dielectric 1:1:4 compounds in the ternary system $BaO–Ln_2O_3–TiO_2$ (Ln = La, Nd, and Sm): 1- an improved coprecipitation method for preparing a single phase powder of ternary compound in $BaO–La_2O_3–TiO_2$ system. J. Am. Ceram. Soc. 74(1991)1868–1872.
[24] J. Takahashi, T. Ikegemi, and K. Kageyama. Occurrence of dielectric 1:1:4 compounds in the ternary system $BaO–Ln_2O_3–TiO_2$ (Ln = La, Nd, and Sm): II – Reexamination of formation of isostructural ternary compounds in identical systems. J. Am. Ceram. Soc. 74(1991)1873–1879.
[25] M. Valant, D. Suvorov, and D. Kolar. High permittivity microwave ceramics-inferences of $Ba_{4.5}Gd_9Ti_{18}O_{54}$ addition on the temperature stability of dielectric properties: in Proc Conf Electroceramics IV, Aachen, Germany. Eur. Ceram. Soc. 1(1994)69–72.
[26] M. Valant, D. Suvorov, and D. Kolar. In Proc Conf Fourth Euroceramics, Riccione, Italy. Eur. Ceram. Soc. 5(1995)211–218.
[27] S. Kawashima, M. Nishida, I. Ueda, and H. Ouchi. Dielectric properties at microwave frequencies of the ceramics in $BaO–Sm_2O_3–TiO_2$ system. 87th Ann Meeting Amer Cer Soc Cincinnati. (1985) (paper No. 15-E-85).
[28] T. Negas and P. K. Davies. Influence of chemistry and processing on the electrical properties of $Ba_{6-3x}Ln_{8+2x}Ti_{18}O_{54}$ solid solutions. Materials and Processes for Wireless Communication. T. Negas and H. Ling (Eds), American Ceramic Society, Westerville, Oh (1995), pp.179–196.
[29] R. S. Roth, F. Beach, A. Antorov, P. K. Davies, J. L. Soubeyronx, and M. Zucchi. Structure of the stoichiometric solid solutions $Ba_2RE_4(Ba_x+RE_{2/3-2/3x})Ti_9O_{27}$ (RE=Nd,Sm). Abstract 07.9–9.14th IUCr congress Aug 12–20, 1987 Perth, Australia. Act Cryst. A43(1987)C138–C139.
[30] H. Ohsato, M. Mizuta, T. Okuda, and S. Nishigaki. Crystal structure and microwave dielectric properties of tungsten bronze type $Ba_{6-3x}R_8+2xTi_{18}O_{54}$(R=La,Pr, Nd, and Sm) solid solutions. Proc. XVII Conf. App. Crystallogr., World Scientific Publishers Singapore (1998) pp. 440–447.
[31] D. Kolar, S. Gabecek, and D. Suvorov. Structural and dielectric properties of perovskite like rare earth titanates. 3rd Euroceramics. P. Duran and J. F. Fernnandez (Eds), Vol. 2, Faenza Editrice, Iberica. S. L. (1993) pp. 229–234.

[32] H. Ohsato, Y. Futamata, H. Sakashita, N. Araki, K. Kakimoto, and S. Nishigaki. Configuration and coordination number of cation polyhedra of tungsten bronze type like Ba_{6-3x} $Sm_{8+2x}Ti_{18}O_{54}$ solid solutions. J. Eur. Ceram. Soc. 23(2003)2529–2534.
[33] F. Azough, P. Setasuwon, and R. Freer. The structure and microwave dielectric properties of ceramics based on $Ba_{3.75}Nd_{9.5}Ti_{18}O_{54}$, in Materials and Processes for Wireless Communication. Am. Ceram. Soc. Westerville, OH, Cerem. Trans. 53(1995)215.
[34] C. J. Rawn, D. P. Birnie, III, M. A. Bruck, J. H. Enemark, and R. S. Roth. Structural investigation of $Ba_{6-3x}Ln_{8+2x}Ti_{18}O_{54}$ ($x = 0.27$, Ln = Sm) by single crystal X-ray diffraction in space group *Pnma*(No.62). J. Mater. Res. 13(1998)187–196.
[35] M. B. Varfolomeeva, A. S. Miranov, V. S. Kostamarov, and L. A. Golubchova. The synthesis and homogeneity ranges of the phases $Ba_{6-3x}Ln_{8+2x}Ti_{18}O_{54}$. Russ. J. Inorg. Chem. 33(1998)607.
[36] H. Ohsato, T. Ohhashi, S. Nishigaki, T. Okuda, K. Sumiya, and S. Suzuki. Formation of solid solutions of new tungsten bronze type microwave dielectric compounds $Ba_{6-3x}R_{8+2x}$ $Ti_{18}O_{54}$ (R = Nd,Sm $0 \leq x \leq 1$). Jpn. J. Appl. Phys. 32(1993)4323–4326.
[37] A. Oslon and R. S. Roth. Crystal structure determination of $BaNd_2Ti_3O_{10}$ using high resolution electron microscopy. J. Solid State Chem. 60(1985)437–357.
[38] C. C. Tang, M. A. Roberts, F. Azough, and R. Freer. Structural studies of $Ba_{6-3x}Nd_{8+2x}$ $Ti_{18}O_{54}$ microwave dielectric ceramics at $T = 10$–195K. Nucl. Instru. Methods Phys. 199 (2003)64–66.
[39] K. M. Cruikshank, X. Jing, G. Wood, E. E. Lachowski, D. B. Cruiksank, and A. R. West. Barium neodymium titnate electroceramics: Phase equilibria studies of $Ba_{6-3x}Nd_{8+2x}$ $Ti_{18}O_{54}$ solid solution. J. Am. Ceram. Soc. 79(1996)1605–1610.
[40] P. B. Jamieson, S. C. Abrahams, and J. L. Bernstein. Ferroelectric tungsten bronze type crystal structures. I. Barium strontium niobate $Ba_{0.27}Sr_{0.75}Nb_{2}o_{5.78}$ J. Chem. Phys. 48(1968)5048–5057.
[41] H. Ohsato, T. Ohhashi, K. Sumiya, S. Suzuki, and T. Okuda. Lattice parameters of bronze type $3BaO–R_2O_3–9Tio_2$ (R = Sm,Nd) solid solutions for microwave dielectric ceramic. Advances In X- Ray Analysis. 37(1994)79–85.
[42] J. S. Kim, C. Cheon, T.-R. Park, and H.-S. Shim. Dielectric properties and crystal structure of $Ba_{6-3x}(NdM)_{8+2x}Ti_{18}O_{54}$ (M = La,Bi,Y) microwave ceramics. J. Mater. Sci. 35(2000)1487.
[43] F. Azough, P. E. Champness, and R. Freer. Determination of the space group of ceramic $BaO.Pr_2O_3.4TiO_2$ by electron diffraction. J. Appl. Crystallogr. 28(1995)577.
[44] H. Ohsato, M. Imaeda, A. Komura, and T. Okuda. Non-linear microwave quality factor change based on the site occupancy of cations on the tungsten bronze typ $Ba_{6-3x}R_{8+2x}Ti_{18}O_{54}$ (R = rare earth) solid solutions. Dielectric Ceramic Materials. Ceram. Trans. Vol. 100, pp. 41–50.
[45] H. Ohsato and M. Imaeda. The quality factor of the microwave dielectric materials based on the crystal structure-as an example: $Ba_{6-3x}R_{8+2x}Ti_{18}O_{54}$ (R = rare earth) solid solutions. Mater. Chem. Phys. 79(2003)208–212.
[46] T. Okawa, M. Imaeda, and H. Ohsato. Site occupancy of Bi ions and microwave dielectric properties in $Ba_{6-3x}Nd_{8+2x}Ti_{18}O_{54}$ solid solutions. Mater. Sci. Engn., B. 88(2002)58–61.
[47] H. Ohsato, A. Komura, and S. Nishigaki. Occupation in the trigonal columns of tungsten bronze like $Ba_{6-3x}Sm_{8+2x}Ti_{18}O_{54}$ solid solutions. J. Ceram. Soc. Jpn. 112:Suppl. 1(2004)1618–1621.
[48] M. Valant, D. Suvorov, and D. Kolar. X-ray investigations and determination of the dielectric properties of the compound $Ba_{4.5}Gd_9Ti_{18}O_{54}$. Jpn. J. Appl. Phys. 35(1996)144–150.
[49] H. Ohsato, Y. Futamata, H. Sakashita, N. Araki, K. Kakimoto, and S. Nishigaki. Configuration and coordination number of cation polyhedra of tungsten bronze type like Ba_{6-3x} $R_{8+2x}Ti_{18}O_{54}$ solid solutions. J. Eur. Ceram. Soc. 23(2003)2529–2533.
[50] H. Ohsato, S. Nishigaki, and T. Okuda. Superlattice and dielectric properties of $BaO–R_2O_3–TiO_2$ (R = La, Nd, Sm) microwave dielectric compounds. Jpn. J. Appl. Phys. 31(1992) 3136–3138.
[51] R. Ubic, I. M. reaney, and W. E. Lee. Microwave dielectric solid solution phase in system $BaO–Ln_2O_3–TiO_2$ (Ln = lanthanide cation). Int. Mater. Rev. 43(1998)205–219.
[52] E. C. Subharao, G. Shirane, and F. Jona. X-ray dielectric and optical study of ferroelectric lead meta tantalite and related compounds. Acta Crystallogr. 13(1960)226–231.
[53] S. Skapin, D. Kolar, D. Suvorov, and Z. Samardzija. Phase equilibria in the $BaTiO_3$–La_2TiO_5–TiO_2 system. J. Mater. Res. 13(1998)1327–1334.

[54] H. Okudera, H. Nakamura, H. Toraya, and H. Ohsato. Tungsten bronze type solid solutions Ba_{6-3x} $Sm_{8+2x}Ti_{18}O_{54}$ ($x = 0.3$, 0.5, 0.67, 0.71) with superstructure. J. Solid State Chem. 142(1999)336–343.
[55] A. Belous, O. Ovchar, M. Valant, and D. Suvorov. Abnormal behaviour of the dielectric parameters of $Ba_{6-x}Ln_{8+2x/3}Ti_{18}O_{54}$ (Ln = La–Gd) solid solutions. J. Appl. Phys. 92(2002)3917–3922.
[56] P. Laffezz, G. Desgardin, and B. Raveau. Influence of calcinations, sintering and composition upon microwave properties of the $Ba_{6-x}Sm_{8+2x/3}Ti_{18}O_{54}$ type oxide. J. Mater. Sci. 27(1992)5229–5238.
[57] D. Suvorov, M. Valant, and D. Kolar. The role of dopants in tailoring the microwave properties of $Ba_{6-3x}R_{8+2x}Ti_{18}O_{54}$ (La = La–Gd) ceramics. J. Mater. Sci. 32(1997)6483–6485.
[58] N. M. Sidorova, E. A. Nenasheva, N. F. Kartenko, L. P. Mudrolubova, and V. B. Gluskhova. Neorg. Mater. 27(1991)1714–1717 (in Russian).
[59] Y. Xu, G. Huang, and Y. He. Sol-gel preparation of $Ba_{6-3x}Sm_{8+2x}Ti_{18}O_{54}$ microwave dielectric ceramics. Ceram. Int. 31(2005)21–25.
[60] C. Hoffmann and R. Waser. Hotforging of $Ba_{6-3x}Re_{8+2x}Ti_{18}O_{54}$ ceramics (RE = La, Ce, Nd, Sm). Ferroelectrics. 201(1997)127–135.
[61] Y. Xu, X. M. Chen, and Y. J. Wu. Synthesis of $Ba_{6-3x}Nd_{8+2x}Ti_{18}O_{54}$ ceramic powders via citric acid precursor. J. Mater. Sci. Mater. Electr. 11(2000)633–636.
[62] M. Valant, D. Suvorov, and D. Kolar. Microwave ceramics of high dielectric constant. 3rd Euroceramics. P. Duram and J. F. Fernandez, (Eds), Faenza Editrice Iberice S. L. (Spain) Vol. 2, (1993) pp. 235–240.
[63] E. A. Nenesheva and N. F. Kartenko. High dielectric constant microwave ceramics. J. Eur. Ceram. Soc. 21(2001)2697–2701.
[64] K. Katayama, Y. Azuma, and Y. Takahashi. molten salt synthesis of single phase $BaNd_2$-Ti_4O_{12} powder. Mater. Sci. 34(1999)301–305.
[65] M. Valant, D. Suvorov, and C. J. Rawn. Intrinsic reasons for variations in dielectric properties of $Ba_{6-3x}R_{8+2x}Ti_{18}O_{54}$ (R = La-Gd) solid solution. Jpn. J. Appl Phys. 30(1999)2820–2826.
[66] A. Yamada, Y. Utsumi, and H. Watarai. Effect of Mn addition on dielectric properties and microstructure of $BaO–Nd_2O_3–TiO_2$ ceramics. Jpn. J Appl. Phys. 30(1991)2350–2353.
[67] H. Ohsato, T. Ohhashi, H. Kato, S. Nishigaki, and T. Okuda. Microwave dielectric properties and structure of the $Ba_{6-3x}Sm_{8+2x}Ti_{18}O_{54}$ solid solutions. Jpn. J. Appl. Phys. 34(1995) 187–191.
[68] H. Ohsato, M. Mizuta, T. Ikoma, Y. Onogi, S. Nishigaki, and T. Okuda. Microwave dielectric properties of tungsten bronze type $Ba_{6-3x}R_{8+2x}Ti_{18}O_{54}$ (R = La,Pr, Nd, and Sm) solid solutions. J. Ceram. Soc. Jpn. 106(1998)178–182.
[69] H. Ohsato, M. Imaeda, Y. Takagi, A. Komura, and T. Ohuda. Microwave quality factor improved by ordering of Ba and rare earth on the tungsten bronze type $Ba_{6-3x}R_{8+2x}Ti_{18}O_{54}$ (R = La, Nd, Sm) solid solutions. Proc. Symp. Applications of Ferroelectrics XI, Switzerland, August. (1998) pp. 509–512.
[70] P.-S. Cheng, C.-F. Yang, Y.-C. Chen, and W.-C. Tzou. The reaction sequence and dielectric properties of $BaSm_2Ti_4O_{12}$ ceramics. Ceram. Int. 26(2000)877.
[71] H. Ohsato, H. Kato, M. Mizuta, S. Nishigaki, and T. Okuda. Microwave dielectric properties of the $Ba_{6-3x}(Sm_{1-y}R_y)_{8+2x}Ti_{18}O_{54}$ (R = Nd and La) solid solutions with zero temperature coefficient of the resonant frequency. Jpn. J. Appl. Phys. 34(1995)5413–5417.
[72] Y. Li and X. M. Chen. Effects of sintering conditions on microstructures and microwave dielectric properties of $Ba_{6-3x}(Sm_{1-y}Nd_y)_{8+2x}Ti_{18}O_{54}$ ceramics ($x = 2/3$). J. Eur. Ceram. Soc. 22(2002)715–719.
[73] N. Ichinose and H. Amada. Preparation and microwave dielectric properties of the $BaO{\cdot}(Sm_{1-x}La_x)_2O_3{\cdot}5TiO_2$ ceramic system. J. Eur. Ceram. Soc. 21(2001)2751–2753.
[74] M. Valant, D. Suvorov, and D. Kolar. High permitivitty microwave ceramics-The influence of $Ba_{4.5}Gd_9Ti_{18}O_{54}$ addition on the temperature stability of dielectric properties. Proc. Conf. Euroceramics, Aachen. Ed. Waser (1994) pp. 146–147.
[75] J. P. Mercurio, M. Manir, and B. Frit. Microwave dielectric properties of $BaNd_{(2-x)}Sm_{2x}$ Ti_5O_{14} ceramics. Mater. Lett. 8(1989)112–114.

[76] K. Fukuda, R. Kitoh, and I. Awai. Microwave characteristics of $BaPr_2Ti_4O_{12}$ and $BaPr_2Ti_5O_{14}$ ceramics. J. Mater. Res. 10(1995)312–319.
[77] H. Ohsato, J. Sugino, A. Komura, S. Nishigaki, and T. Okuda. Microwave dielectric properties of $Ba_4(Nd_{28/3-y}R_y)Ti_{18}O_{54}$ (R = Eu, Dy, Ho, Er, and Yb) solid solutions. Jpn. J. Appl. Phys. 38(1999)5625–5628.
[78] A. Silva, F. Azough, R. Freer, and C. Leach. Microwave dielectric ceramics in the system $BaO–Li_2O–Nd_2O_3–TiO_2$. J. Eur. Ceram. Soc. 20(2000)2727.
[79] V. Satheesh, P. Murigavel, V. R. K. Murthy, and B. Viswanathan. Synthesis and role of Nd and Sm on the microwave dielectric properties of $BaNd_{2(1-x)}Sm_{2x}Ti_5O_{14}$ dielectric resonator. Mater. Sci. Eng., B. 48(1997)202–204.
[80] D. Kolar, S. D. Skapin, and D. Suvorov. Phase equilibria in the system $BaO–TiO_2–Gd_2O_3$. Acta. Chim. Slov. 46(1999)193–202.
[81] C. L. Huang and Y.-C. Chen. Microwave dielectric properties and microstructure of $Ba_{2-x}Sm_{4+2x/3}Ti_{8+y}O_{24+2y}$ ceramics. Mater. Sci. Eng., A. 345(2003)106–112.
[82] Y.-C. Chen and C.-L. Huang. Microwave dielectric properties of $Ba_{2-x}Sm_{4+2/3x}Ti_9O_{26}$ with zero temperature coefficient. Mater. Sci. Eng., A. 334(2003)250–256.
[83] H. Sreemoolanathan, M. T. Sebastian, and P. Mohanan. Dielectric ceramics in the $BaO–Ln_2O_3–5TiO_2$ composition. Ferroelectrics. 189(1996)43–46.
[84] R. Ratheesh, H. Sreemoolanathan, M. T. Sebastian, and P. Mohanan. Preparation characterization and dielectric properties of ceramics in the $BaO–Nd_2O_3–TiO_2$ system. Ferroelectrics. 211(1998)1–8.
[85] Y.-C. Chen and C.-L. Huang. Microwave dielectric properties of $Ba_{2-x}Sm_{4+2/3x}Ti_9O_{26}$ ceramics with zero temperature coefficient. Mater. Sci. Eng., A334(2002)250–256.
[86] M.-W. Zuo, W. Li, J.-L. Shi, and Q. Zeng. Influence of CuO addition to $BaSm_2Ti_4O_{12}$ microwave ceramics on sintering behaviour and dielectric properties. Mater. Res. Bull. 41(2006)1127–1132.
[87] H. Ohsato, A. Komura, Y. Takagi, S. Nishigaki, and T. Okuda. Microwave dielectric properties and sintering of $Ba_{6-3x}R_{8+2x}Ti_{18}O_{54}$ (R = Sm, $x = 2/3$) solid solutions with added rutile. Jpn. J Appl. Phys. 37(1998)5357–5359.
[88] Y. Xu, G. Huang, and Y. He. Sol-gel preparation of $Ba_{6-3x}Sm_{8+2x}Ti_{18}O_{54}$ microwave dielectric ceramics. Ceram. Int. 31(2005)21–25.
[89] Y. Zheng, X. Zhao, W. Lei, and S. Wang. Effects of Bi_2O_3 addition on the microstructures and microwave dielectric characteristics of $Ba_{6-3x}(Sm_{0.2}Nd_{0.8})_{8+2x}Ti_{18}O_{54}$ ($x = 2/3$) ceramics. Mater. Lett. 60(2005)459–463.
[90] A. Webhoffer and A. Feltz. Microwave dielectric properties of ceramics of the system $Ba_{6-3x}(Sm_yNd_{1-y})_{8+2x}Ti_{18}O_{54}$. J. Mater. Sci. Lett. 18(1999)719–721.
[91] T. Okawa, M. Imaeda, and H. Ohsato. Microwave dielectric properties of $Ba_4(Sm_{1-y}Bi_y)_{9+1/3}Ti_{18}O_{54}$ solid solutions. Key Engn. Mater. 214–215(2002)37–42.
[92] Y. Ota, K.-I. Kakimoto, H. Ohsato, and T. Okawa. Low temperature sintering of $Ba_{6-3x}Sm_{8+2x}Ti_{18}O_{54}$ ceramics by B_2O_3 and GeO_2 addition. J. Eur. Ceram. Soc. 24(2004) 1755–1760.
[93] N. Santha, M. T. Sebastian, V. George, and J. Philip. Microwave dielectric and elastic properties of glass added $Ba_{6-3x}R_{8+2x}Ti_{18}O_{54}$ ($x = 2/3$). Mater. Chem. Phys. 100(2006) 423–429.
[94] L.-C. Chang and B.-S. Chiou. Effect of B_2O_3 nano coating on the sintering behaviour and electrical microwave properties of $Ba(Nd_{2-x}Sm_x)Ti_4O_{12}$ ceramics. J. Electroceram. 13(2004) 829–837.
[95] F. Azough, T. Lowe, and R. Freer. Control of microwave dielectric properties in the system $BaO{\cdot}Nd_2O_3{\cdot}4TiO_2–BaO{\cdot}Al_2O_3{\cdot}4TiO_2$. J. Electroceram. 15(2005)183–192.
[96] L. Huang, J. T. Tsai, and J.-L. Li. Microwave dielectric properties of $(1-x)CaO–xBaO–Li_2O–(1-y)Sm_2O_3–yNd_2O_3–TiO_2$ ceramics system. J. Mater. Sci. 35(2000)4901–4905.
[97] R. Ubic, I. M. Reaney, W. E. Lee, J. Samuels, and E. Evangelins. Properties of the microwave dielectric phase $Ba_{6-3x}Nd_{8+2x}Ti_{18}O_{54}$. Ferroelectrics. 228(1999)271–282.
[98] C.-F. Yang. Microwave properties of $(Ba_{1-x}Sr_x)Sm_2Ti_4O_{12}$ ceramics and its application as dielectric resonator antenna. Jpn. J. Appl. Phys. 39(2000)1430–1434.

[99] A. Ioachim, M. I. Toacsan, M. G. Banciu, L. Nedeku, H. Alexandru, C. Berbecaru, D. G. Hetur, and G. Stoica. BNT ceramics synthesis and characterisation. Mater. Sci. Eng. B. 109(2004)183–187.

[100] Y. J. Wu and X. M. Chen. Structures and microwave dielectric properties of $Ba_{6-3x}(Nd,Bi_y)_{8+2x}Ti_{18}O_{54}$ ($x = 2/3$) solid solution. J. Mater. Res. 16(2001)1734–1738.

[101] T. Okawa, M. Imaeda, and H. Ohsato. Microwave dielectric properties of Bi added $Ba_4Nd_{9+1/3}Ti_{18}O_{54}$ solid solutions. Jpn. J. Appl. Phys. 39(2000)5645–5649.

[102] M. Valant and D. Suvorov. Incorporation of bismuth into $Ba_{6-3x}R_{8+2x}Ti_{18}O_{54}$ (R = Nd, Gd). J. Mater. Sci. 36(2001)2991–2997.

[103] I.-S. Cho, D.-W. Kim, J.-R. Kim, and K. S. Hong. Low temperature sintering and microwave dielectric properties of $BaO–(Nd_{1-x}Bi_x)_2O_3–4TiO_2$ by the glass additions. Ceram. Int. 30(2004)1181–1185.

[104] Y. J. Wu and X. M. Chen. Bismuth/Samarium cosubstitutied $Ba_{6-3x}Nd_{8+2x}Ti_{18}O_{54}$ microwave dielectric ceramics. J. Am. Ceram. Soc. 83(2000)1837–1839.

[105] Y. J. Wu and X. M. Chen. Modified $Ba_{6-3x}Nd_{8+2x}Ti_{18}O_{54}$ microwave dielectric ceramics. J. Eur. Ceram. Soc. 19(1999)1123–1126.

[106] I.-S. Cho, D.-W. Kim, J.-R. Kim, and K. S. Hong. Low temperature sintering and microwave dielectric properties of $BaO \cdot (Nd_{1-x}Bi_x)_2O_3 \cdot 4TiO_2$ by the glass additions. Ceram. Int. 30(2004)1181–1185.

[107] L.-E. Chang, B.-S. Chiou, and W.-H. Lee. Effect of glass additions on the sintering behaviour and electrical microwave properties of $BaO–Nd_2O_3–Sm_2O_3–TiO_2$ ceramics. J. Mater. Sci.: Mater. Electron. 15(2004)153–158.

[108] C. C. Lee and P. Lin. Effect of glass addition on microwave properties of $BaO–La_2O_3–4.7TiO_2$. Jpn. J. Appl. Phys. 37(1998)6048–6054.

[109] O. Dernovsek, A. Naeini, G. Preu, W. Wersing, M. Eberstein, and W. A. Schiller. LTCC glass ceramic composites for microwave application. J. Eur. Ceram. Soc. 21(2001) 1693–1697.

[110] B.-H. Jung, S.-J. Hwang, and H.-S. Kim. Glass-ceramic for low temperature co-fired dielectric ceramic materials based on $La_2O_3–B_2O_3–TiO_2$ glass with BNT ceramics. J. Eur. Ceram. Soc. 25(2005)3187–3193.

[111] D. Suvorov, M. Valant, and D. Kola. Chemistry and microwave properties of Bi_2O_3 doped $Ba_{4.5}Nd_9Ti_{18}O_{54}$ based ceramics. In: Materials and Processes for Wireless Communication. T. Negas and H. Ling (Eds), American Ceramic Society, Westerville, OH, Ceramic Transaction Vol. 53 (1995) pp. 197–206.

[112] M. Imaeda, K. Ito, M. Mizuta, H. Ohsato, S. Nishigaki, and T. Okuda. Microwave dielectric properties of $Ba_{6-3x}Sm_{8+2x}Ti_{18}O_{54}$ solid solutions with Sr substituted for Ba. Jpn. J Appl. Phys. 36(1997)6012–6015.

[113] J. Li, X. M. Chen, and Y. J. Wu. Strontium/titanoum cosubstituted $Ba_4Sm_2Ti_4Ta_6O_{30}$ dielectric ceramics. J. Eur. Ceram. Soc. 22(2002)87–91.

[114] M. Podlinik, M. Valant, and D. Suvorov. The influence of isovalent doping on the microwave dielectric properties of $Ba_{6-3x}Nd_{8+2x}Ti_{18}O_{54}$ based ceramics. Key Engn. Mater. 132–136(1997)1211–1214.

[115] X. M. Chen, Y. Li. A- and B-Site cosubstituted $Ba_{6-3x}Sm_{8+2x}Ti_{18}O_{54}$ microwave dielectric ceramics. J. Am. Ceram. Soc. 85(2002)579–584.

[116] S. Y. Wu, Y. Li, and X. M. Chen. Raman spectra of Nd/Sn cosubstituted $Ba_{6-3x}Sm_{8+2x}Ti_{18}O_{54}$ microwave dielectric ceramics. J. Appl. Phys. 96(2004)5683–5686.

[117] C. Duran, S. Troiler Mc Kinsty, and G. L. Messing. Fabrication and electrical properties of textured $Sr_{.53}Ba_{.47}Nb_2O_6$ ceramics by templated grain growth. J. Am. Ceram. Soc. 83(2000)2203.

[118] A. G. Belous, O. Ovchar, M. Valant, D. Suvorov, and D. Kolar. The effect of partial isovalent substitution in the A sublattice on MW properties of materials based on $Ba_{6-x}Ln_{8+2x/3}Ti_{18}O_{54}$ solid solutions. J. Eur. Ceram. Soc. 21(2001)2723–2730.

[119] A. G. Belous and O. V. Ovchar. MW dielectrics with perovskite like structure based on Sm containing systems. J. Eur. Ceram. Soc. 19(1999)1119–1122.

[120] M. Mizuta, K. Uenoyama, H. Ohsato, S. Nishigaki, and T. Okuda. Formation of tungsten bronze type $(Ba_{6-3x}Sm_{8+2x})Ti_{18-y}Al_yO_{54}$ ($\alpha = 1 + y/36$) solid solutions and microwave dielectric properties. Jpn. J Appl. Phys. 35(1996)5065–5068.

[121] Z. X. Xiong, C. Fang, J. R. Huang, H. Qiu, and M. Lin. Preparation and properties of $Sr(Bi_{1-x}Nd_x)_8Ti_7O_{27}$ ceramics for wireless technologies. Mater. Sci. Eng., B. 99(2003) 138–142.

[122] X. M. Chen, Y. H. Sun, and X. H. Zheng. High permittivity and low loss dielectric ceramics in the $BaO–La_2O_3–TiO_2–Ta_2O_5$ system. J. Eur. Ceram. Soc. 23(2003)1571–1575.

[123] A. R. Stokes and A. J. C. Wilson. The diffraction of X-rays by distorted crystal aggregate-1. Proc. Phys. Soc. 1(1944)174–182.

[124] R. D. Shannon. Dielectric polarizibilties of ions in oxides and fluorides. J. Appl. Phys. 73(1993)348–366.

[125] H. Frolichs. Theory of Dielectrics. Clarendon Press, Oxford (1958).

[126] J. Bosman and E. E. Havinga. Temperature dependence of dielectric constants of cubic ionic compounds. Phys. Rev. 129(1963)1593.

[127] D. Suvorov, M. Valant, and D. Kolar. The role of dopants in tailoring the microwave properties of $Ba_{6-3x}R_{8+2x}Ti_{18}O_{54}$ (R = La–Gd) ceramics. J. Mater. Sci. 32(1997) 6483–6485.

[128] J. L. Servoin, F. Gervais, A. M. Quillett, and Y. Luspin. Infrared and Raman responses in ferroelectric perovskite crystals: Apparent inconsistencies. Phys. Rev. B21(1980) 2038–2041.

[129] T. Negas, G. Yeager, S. Bell, and R. Amren. Chemistry and properties of temperature compensated microwave dielectrics. In: Chemistry of Electronic Ceramic Material. P. K. Davies and R.S. Roth (Eds), Vol. 804, NIST Spec Publication (1991) p. 27.

[130] C.-C. Cheng, T. E. Hsieh, and I.-N. Lin. Effects of composition on low temperature sinterable Ba–Nd–Sm–Ti–O microwave dielectric materials. J. Eur. Ceram. Soc. 24(2004) 1787–1790.

[131] W. Wersing. Microwave ceramics for resonators and filters. Curr. Opin. Solid State Mater. Sci. 1(1996)715–731.

[132] P. L. Mudrulobva, K. E. Lisker, B. A. Rotenberg, T. F. Limari, and N. Borsch. Ceramic materials based on $BaLn_2Ti_4O_{12}$ compound for high permittivity Capacitors. Tech. Ser. Radio Detajili. Radiokomp. 46(1982)3–8.

[133] T. Okawa, M. Imaeda, H. Ohsato, and A. Harada. Site occupancy of Bi ions and microwave dielectric properties of Bi doped $Ba_{6-3x}R_{8+2x}Ti_{18}O_{54}$ (R = rare earth, $x = 2/3$) solid solutions. Mater. Chem. Phys. 79(2003)199–203.

[134] M. Valant, I. Arcon, D. Suvorov, A. Kodre, T. Negas, and R. Frahim. Extended X-ray absorption fine structure study of incorporation of Bi and Pb atoms into the crystal structure of $Ba_{4.5}Nd_9Ti_{18}O_{54}$. J. Mater. Res. 12(1997)799–804.

[135] M. Valant, D. Suvorov, and I. Arson. EXAFs investigations of Bi and Pb doped $Ba_{6-3x}Nd_{8+2x}Ti1_9O_{54}$. Electroceramics V. Book 2(1996)73.

[136] H. Sakashita, H. Ohsato, A. Komura, and T. Okuda Configuration of A1 and A2 sites on the microwave dielectric properties of $Ba_{6-3x}Sm_{8+2x}Ti_{18}O_{54}$ ($x = 2/3$) solid solutions. Proc. 18th IUCR Congress and General Assembly, Glasgow, (1999) P. 495.

[137] N. Qin and X. M. Chen. Effects of Sm/Bi cosubstitution on microstructures and microwave dielectric characteristics of $Ba_{6-3x}La_{8+2x}Ti_{18}O_{54}$. Mater. Sci. Eng., B. 111(2004)90–94.

[138] G. Belous and O. V. Ovchar. Pb-induced temperature stabilization of high dielectric constant in barium neodymium titanates. Ferroelectrics. 238(2000)9–16.

[139] R. Ubic, I. M. Reaney, W. E. Lee, J. Samuels, and E. Evangelinos. $Ba_{6-3x}Nd_{8+2x}Ti_{18}O_{54}$ microwave dielectric resonators. Ferroelectrics. 223(1999)293–300.

[140] R. Ubic, I. M. Reaney, W. E. Lee, and Perovskite. $NdTiO_3$ in Sr- and Ca doped $BaO–Nd_2O_3–TiO_2$ microwave dielectric ceramics. J. Mater. Res. 14(1999)1576–1580.

[141] J. Takahashi, K. Kageyama, and K. Kodaira. Microwave dielectric properties of lanthanide titanate ceramics. Jpn. J. Appl. Phys. 32(1993)4327–4331.

[142] C. C. Lee and P. Lin. Microwave dielectric properties and microstructures of $(Ba_{1-x}Pb_x)O–La_2O_3–4{\cdot}TiO_2$ ceramics. Jpn. J Appl. Phys. 37(1998)878–884.

[143] D. Mercurio, M. Abou-Salama, and J.-P. Mercurio. Investigations of the tungsten bronze type $(Ba_{1-\alpha}Sr_\alpha)_{6-x}La_{8+2x/3}Ti_{18}O_{54}$ $(0 \leq x \leq 1)$ solid solutions. J. Eur. Ceram. Soc. 21(2001)2713–2716.
[144] T. Nagamoto, T. Otagiri, M. Suzuki, and H. Ohsato. Microwave dielectric properties and crystal structure of the tungsten bronze type $(Ba_{1-\alpha}Sr_\alpha)_6(Nd_{1-\beta}Y_\beta)_8Ti_{18}O_{54}$ solid solutions. J. Eur. Ceram. Soc. 26(2006)1865–1898.
[145] M. Suzuki, H. Ohsato, T. Nagomoto, K.-Kakimoto, and T. Otagiri. Crystal structure and microwave dielectric properties of the tungsten bronze-type like $(Ba_{1-\alpha}Sr_\alpha)_6(Nd_{1-\beta}Y_\beta)_8Ti_{18}O_{54}$ solid solutions. J. Eur. Ceram. Soc. 26(2006)2035–2038.
[146] Y. J. Wu and X. M. Chen. Dielectric ceramics of $Ba_{6-3x}Nd_{8+2x}(Zr,Ti)_{18}O_{54}$. Ferroelectrics. 233(1999)271–277.
[147] T. Takenaka and K. Sakata. Grain orientation and electrical properties of hotforged $Bi_4Ti_3O_{12}$ ceramics. Jpn. J. Appl. Phys. 19(1980)31–39.
[148] K. Nakgata and K. Okazaki. One directional grain oriented lead metaniobate ceramics. Jpn. J. Appl. Phys. 24(1985)812–814.
[149] T. Takeuchi, T. Tani, and Y. Sait. Piezoelectric properties of bismuth layer-structured ferroelectric ceramics with a preferred orientation processed by the reactive templated grain growth method. Jpn. J. Appl. Phys. 38(1999)5553–5556.
[150] E. M. Sabolsky, S. Troiler, Mc Knisty, and G. L. Messing. Dielectric and piezoelectric properties of <001> fibre textured $0.675\,Pb(Mg_{1/3}\ Mg_{2/3})O_3$–$0.325PbTiO_3$ ceramics. J. Appl. Phys. 93(2003)4072–4080.
[151] K. Wada, K.-I. Kakimoto, and H. Ohsato. Grain orientation control and microwave dielectric properties of $Ba_4Sm_{9.33}Ti_{18}O_{54}$ ceramics. Jpn. J. Appl. Phys. 42(2003)6149.
[152] K. Wada, K. Kakimoto, and H. Ohsato. Anisotropic microwave dielectric properties of textured $Ba_4Sm_{9.33}Ti_{18}O_{54}$ ceramics. Key Engn. Mater. 269(2004)207–210.
[153] K. Wada, K. Kakimoto, and H. Ohsato. Microstructure and microwave dielectric properties of $Ba_4Sm_{9.33}Ti_{18}O_{54}$ ceramics containing columnar crystals. J. Eur. Ceram. Soc. 23(2003)2535–2539.
[154] T. Tani. Crystalline oriented piezoelectric bulk ceramics with a perovskite type structure. J. Korean Phys. Soc. 32(1998)S1217–S1221.
[155] F. K. Lotergering. Topotactical reactions with ferromagnetic oxides having hexagonal crystal structures-1. J. Inorg. Nucl. Chem. 9(1959)113–123.
[156] L.-C. Chang and B.-S. Chiou. Electrical behaviour of BaO–Nd_2O_3–Sm_2O_3–TiO_2 with glass/oxide additives analysed by impedance spectroscopy. J. Electroceram. 15(2005)75–81.
[157] C.-C. Cheng, T.-E. Hsieh, and I.-N. Lin. Microwave dielectric properties of glass ceramics composites for low temperature cofirable ceramics. J. Eur. Ceram. Soc. 23(2003)2553–2558.
[158] C.-C. Cheng, T.-E. Hsieh, and I.-N. Lin. The effect of composition on Ba–Nd–Sm–Ti–O microwave dielectric materials for LTCC application. Mater. Chem. Phys. 79(2003) 119–123.
[159] C.-H. Lu and Y.-H. Huang. Densification and dielectric properties of barium neodymium titanate oxide ceramics. Mater. Sci. Eng. B. 98(2003)33–37.
[160] V. I. Butko, A. G. Belous, and Y. A. Nenasheva. Microwave dielectric properties of barium lanthanide tetratitanates. Fizyka Tvyordogo Tela. 26(1980)2951–2956.
[161] A. G. Belous, O. V. Ovchar, M. Valant, and D. Suvorov. Anomalies in the temperature dependence of the microwave dielectric properties of $Ba_{6-3x}Sm_{8+2x}Ti_{18}O_{54}$. Appl. Phys Lett. 77(2000)1707–1709.
[162] G. Belous, O. V. Ovchar, and D. O. Mischuk. Temperature trends of the permittivity in complex oxides of rare earth elements with perovskite type structure. Condens. Matter Phys. 6(2003)251–259.
[163] J. P. Mercurio, M. Manier, and B. Frit. Dielectric properties of ceramics within the BaO–Ln_2O_3–TiO_2 system. Ferroelectrics. 127(1992)35–40.
[164] Yu. Poplakov, B.A. Rotenberg, V. N. Borisev, E. A. Nenesheva, V. I. Bitko, and V. M. Paskov. Dielectric properties of $BaLa_2Ti_4O_{12}$ compounds at low temperature. Sovt. Phys. Solid State. 28(1986)882–883.

CHAPTER SIX

ABO_3 Type Perovskites

6.1 Introduction

The perovskite family of materials is of considerable technological importance, particularly with regard to physical and electrical properties such as pyro and piezo electricity, linear and non-linear electro-optic effects, dielectric and superconducting properties. Many of these properties are gross effects varying considerably from one perovskite to another, yet the differences in the crystal structures are hardly apparent. Compounds having the ideal perovskite structure can be represented by the general formula ABX_3 (X = O,F) where A is a large metal cation close-packed in layers with oxygen ions and B is a smaller metal ion situated in an octahedrally coordinated hole between the close-packed layers. Since ABF_3 materials in general have relatively high dielectric loss factor, we restrict rest of this chapter to ABO_3 type materials only. The perovskite structure is one of the most extensively studied structures in materials science. German chemist and mineralogist Gustav Rose discovered the mineral $CaTiO_3$ in 1839. Rose named $CaTiO_3$ after Lev Alexeievitch Perovsky, a Russian military official and dignitary [1]. The atomic arrangement for ABO_3 perovskite structure was first found for the mineral perovskite $CaTiO_3$. It was earlier thought that the $CaTiO_3$ unit cell could be represented by Ca (A) ions at the corners of a cube with Ti (B) ions at the body center and oxygen ions at the center of the faces as shown in Figure 6.1. This simple cubic structure is called perovskite, even though, $CaTiO_3$ was later determined to be of orthorhombic [2] symmetry at room temperature. The ideal perovskite structure having cubic symmetry with one ABO_3 formula per unit cell is shown in Figure 6.1. Many perovskite materials have symmetry different from cubic at room temperature, but transform to cubic symmetry at high temperatures [2–7]. The low temperature (low symmetry) structure is called hettotype and the high temperature cubic structure is called aristotype [2]. Perovskites having the ideal structure adopt the cubic space group $Pm\bar{3}m$. The stability of the ABO_3 perovskite structure is mainly derived from the electrostatic energy achieved by arranging the B cations in corner shared octahedra [5, 6]. Thus the first pre-requisite for a stable ABO_3 "perovskite" is the existence of a stable polar, octahedral BO_3 skeletal sub-array. Given the BO_3 skeletal sub-array, additional stabilization is achieved by accommodating a large cation within this skeleton. Galasso [3] has described the structure and properties of several perovskite type compounds in detail. Excellent monographs and reviews have also appeared in the literature about the structure and properties of perovskites [3, 5–8]. The perovskite structure is viable to wide departures in compositions from the ideal formula ABO_3. These variations can be achieved by isomorphous substitutions at A or B sites, and cationic and anionic deficiencies. These can be represented as $A_{1-x}A'_xBO_3$, $AB_{1-x}B'_xO_3$, A cationic vacancies like $A_x\square_{1-x}BO_3$, oxygen deficient systems and cationic ordering as in $A(B'B'')O_3$.

Dielectric Materials for Wireless Communications
ISBN-13: 978-0-08-045330-9

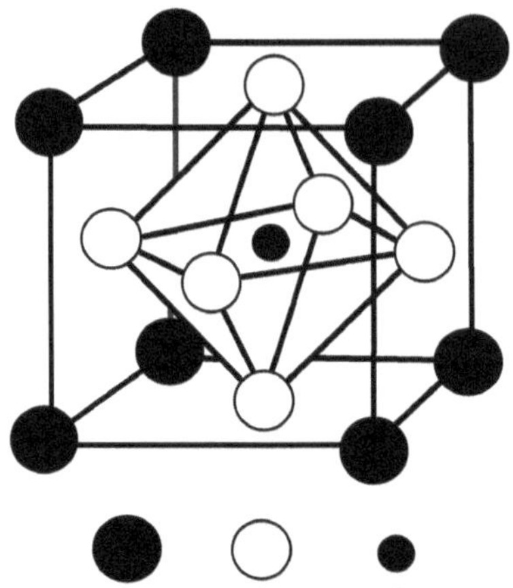

Figure 6.1 Ideal perovskite structure.

Frequently, the composition changes are accompanied by structural changes both of which have profound effects on the electrical and magnetic properties of the material. Non-ferroelectric perovskite ceramics with high permittivity, low τ_f and low loss represent one of the most important material families for applications in wireless communication.

One interesting feature of the perovskites is that they can accommodate more than one aliovalent elements for both A and B sites and such compositions which can be typically represented as $A(B'_xB''_y)O_3$ and termed as complex perovskites. The B′ and B″ are two elements in different oxidation states and $x+y=1$. Roy [4] has studied the possible multiple substitution in perovskites

The complex perovskite type compounds $A(B'_xB''_y)O_3$ can be divided into

1. $A(B'_{2/3}B''_{1/3})O_3$ which contain twice as much lower valence state element as higher valence state element.
2. $A(B'_{1/3}B''_{2/3})O_3$ which contain twice as much of the higher valence state element as the lower valence state.
3. $A(B'_{1/2}B''_{1/2})O_3$, contain the B elements in equal amounts.
4. Oxygen deficient phases $A(B'_xB''_y)O_{3-\delta}$.
5. Hexagonal perovskites of the type $A_nB_{n-1}O_{3n}$ and related compounds.

In this chapter, we discuss the preparation, characterization and microwave dielectric properties of ABO_3 and $Ca(Li_{1/3}A_{2/3})O_{3-\delta}$ [A = Nb,Ta] perovskites. The other complex perovskites will be discussed in the following chapters.

6.2 Tolerance Factor (t) and Perovskite Cell Parameter (a_p)

The tolerance factor "t" was first suggested by Megaw [2] to determine the stability of the perovskite phase for a given set of anions and cations. The tolerance factor and perovskite cell parameter are two important parameters related to the symmetry of perovskites that significantly affect the dielectric properties. In the perovskite structure, the A cations are coordinated with 12 oxygen ions and the B cations with six oxygen ions. The A cation is normally found to be somewhat larger than the B cation. The oxygen anions are coordinated by two B site cations and four A cations. Thus from Figure 6.1 it can be seen that in order to have contact between the A, B and O ions, $R_A + R_O$ should be equal to $\sqrt{2}(R_B + R_O)$; where R_A, R_B and R_O are the ionic radii of the constituent ions.

Goldschmidt [9] defined the tolerance factor (t) to account for the limits of the size of the cations to form a perovskite structure as

$$t = \frac{R_A + R_O}{\sqrt{2}(R_B + R_O)} \tag{6.1}$$

For complex perovskites of the type $A(B'_{1/2}B''_{1/2})O_3$, the above equation can be modified as

$$t = \frac{R_A + R_O}{\sqrt{2\left[\frac{(R_B' + R_B'')}{2} + R_O\right]}} \tag{6.2}$$

In general when the value of t is close to 1, the perovskite phase will be formed. If t is very far from 1, then the perovskite will not be formed as demonstrated by $MgTiO_3$, $t = 0.81$ which adopts the illmenite structure. Both $BaZrO_3$ and $SrZrO_3$ exhibit the perovskite structure. At room temperature $BaZrO_3$ is cubic with $t = 1.01$ and $SrZrO_3$ is orthorhombic with $t = 0.94$. Identical trends can be readily found for other simple perovskites such as $SrTiO_3$ with $t = 1.00$ is cubic, $CaTiO_3$ with $t = 0.97$ is orthorhombic. The symmetry is lower for the material which has a lower tolerance factor. If $t > 1$, the A cations have large size and the B cations are small so that B ions have larger room to move. For $t < 1$, the B cations have large size. In fact, t is related to the packing of ions in the perovskite cell. When $t = 1$, a perfect cubic cell is formed. When t deviates from 1, the perovskite cell gets deformed and the symmetry is lowered.

It was suggested [10–12] that the lowering of symmetry in perovskites is often due to tilting of octahedra. When the ionic radii of A site ions are too small to occupy fully the available volume at a critical value of t, the octahedra tilts. Hence at a given temperature, the octahedra rotates in order to reduce the size of cubo-octahedral interstices of the oxygen sublattice. Setter and co-workers reported [13, 14] that the tilting depends on tolerance factor, temperature and composition. The tilting may be in-phase or anti-phase depending upon the minimum energy configuration at that temperature. Figure 6.2 shows the un-tilted, in-phase tilted and anti-phase tilted octahedral framework. Glazer gave a description of an octahedral frame, tilted through small angles around the principal axis, which occur in perovskite compounds [11, 12]. The tolerance factor indicates the extent to which the A site species fill the cubo-octahedral interstices created by the B site octahedra.

The side of the cube or lattice parameter of the ideal perovskite is given by

$$a_p = \frac{R_A + R_O}{\sqrt{2}} + (R_B + R_O) \tag{6.3}$$

The lattice parameter a_p calculated using the ionic radii are found to be slightly larger than the experimentally observed values.

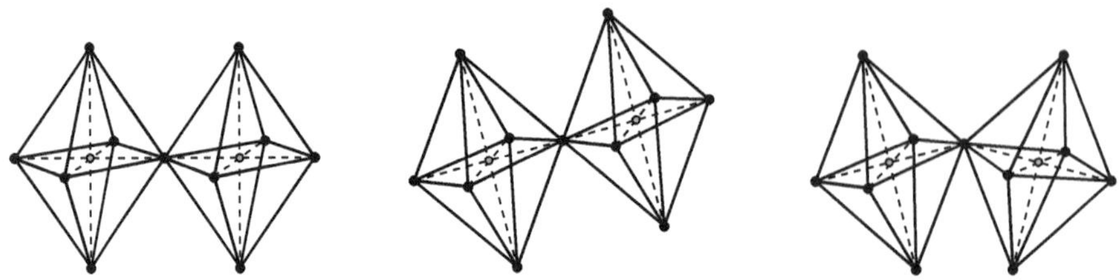

Figure 6.2 Tilting of octahedral framework: (a) un-tilted, (b) in-phase tilted and (c) anti-phase tilted (after Ref. [11]).

More recently Jiang et al. [15] reported an empirical equation establishing a relationship between lattice parameter, ionic radii and tolerance factor of ABX_3 cubic perovskites. However, they used the ionic radii of A, B and X ions corresponding to the coordination number six. Moreira and Dias [16] corrected this and obtained the empirical expression for obtaining the lattice parameter using coordination numbers twelve for A, six for B and two for X ions respectively as

$$a_{pred} = 2\beta(R_B + R_X) + \gamma t - \delta \tag{6.4}$$

where $\beta = 0.9109$, $\gamma = 1.1359$ and $\delta = 0.7785$.
Ubic obtained [17] another empirical relation to predict the lattice constant.

$$a_{pred} = 0.6742 + 0.49533(R_A + R_X) + 1.2856(R_B + R_X) \tag{6.5}$$

Several low loss ABO_3 ceramic materials have been reported in the last decade for applications as dielectric resonators, substrates etc. in the microwave frequency region. The dielectric properties of the ABO_3 type microwave ceramics are given in Table 6.1.

6.3 $ATiO_3$ (A = Ba, Sr, Ca)

Since the discovery of ferroelectric properties of $BaTiO_3$ by Von Hippel, extensive work was done on $BaTiO_3$ and other ABO_3 type compounds. These studies have led to the discovery of many new ferroelectric and piezoelectric materials. Barium titanate has the ideal cubic perovskite structure above the Curie point (~130°C) and on cooling below this temperature the oxygen and titanium ions shift to form tetragonal crystal symmetry. The tetragonal distortion in the structure of $BaTiO_3$ results in the formation of dipoles and the material becomes ferroelectric. Feteira et al. [18, 19] prepared $BaTi_{1-2y}Ga_yNb_yO_3$ ($0 \leq y \leq 0.35$) by sintering at 1300°C. The room temperature tetragonal crystal structure changed to cubic with increasing y. For $y \leq 0.15$ the ceramics exhibit relaxor type behavior having a high $\varepsilon_r > 2000$ with high dielectric loss. For $y = 0.15$ the ε_r peak at about 350 at 200 K and exhibit weak relaxor behavior. For $y > 0.15$ it is paralectric and has a low ε_r of about 50–100 and low losses and they resonate in the microwave frequency region with a Qf of about 2500–3700 GHz. The $BaTi_{0.92}Ga_{0.08}O_{2.96}$ has a hexagonal symmetry with space group $P6_3/mmc$ similar to 6H $BaTiO_3$ [18]. The 6H $BaTiO_3$ is prepared by rapid quenching [103] in air or by processing under low oxygen

Table 6.1 Microwave dielectric properties of ABO_3 perovskites

Material	Sintering temperature °C	ε_r	*Qf* (GHz)	τ_f (ppm/°C)	Reference
$BaTi_{1-2y}Ga_yNb_yO_3$ ($y=0.25$)	1500	86	3000	–	[18]
$BaTi_{1-2y}Ga_yNb_yO_3$ ($y=0.3$)	1500	66	3700		[18]
$BaTi_{1-2y}Ga_yNb_yO_3$ ($y=0.35$)	1500	47	2500		[18]
$BaTi_{0.92}Ga_{0.08}O_{2.96}$ in oxygen	1450	74	7800		[19]
$BaTi_{0.92}Ga_{0.08}O_{2.96}$	1450	72	5500		[19]
$BaTi_{0.95}Mn_{.05}O_{3-\delta}$	1450/2 h	71	7700		[20]
$BaTi_{0.95}Fe_{0.05}O_{3-\delta}$	1450/2h	82	4800		[20]
$BaTi_{0.95}Co_{0.05}O_{3-\delta}$	1450/2 h	74	1300		[20]
$BaTi_{0.95}Ni_{0.05}O_{3-\delta}$	1450/2 h	56	2400		[20]
$La_{0.4}Ba_{0.6}Ti_{0.6}Y_{0.4}O_3$	1600/4 h	57	750	12	[21]
$La_{0.4}Ba_{0.6}Ti_{0.6}Yb_{0.4}O_3$	1600/4 h	65	4500	1	[21]
$CaTiO_3$	1400	160	7000	850	[22–24]
$Ca(Zr_{0.2}Ti_{0.8})O_3$		153	4400		[25]

(Continued)

Table 6.1 (Continued)

Material	Sintering temperature °C	ε_r	Qf (GHz)	τ_f (ppm/°C)	Reference
$Ca(Zr_{0.4}Ti_{0.6})O_3$		118	6400		[25]
$Ca(Zr_{0.6}Ti_{0.4})O_3$		78	7800		[25]
$Ca(Zr_{0.8}Ti_{0.2})O_3$		49	10800		[25]
$Ca[Ti_{1-x}(Mg_{1/3}Nb_{2/3})_x]O_3$ ($x = 0.7$)	1450	42	29400	–13	[26]
$Ca[Ti_{1-x}(Mg_{1/3}Nb_{2/3})_x]O_3$ ($x = 0.65$)	1450	44	28300	–2	[26]
$Ca[Ti_{1-x}(Mg_{1/3}Nb_{2/3})_x]O_3$ ($x = 0.6$)	1450	47	25600	8	[26]
$Ca[Ti_{1-x}(Mg_{1/3}Nb_{2/3})_x]O_3$ ($x = 0.5$)	1450	54	22900	39	[26]
$Ca[Ti_{1-x}(Mg_{1/3}Nb_{2/3})_x]O_3$ ($x = 0.4$)	1450	63	12200	93	[26]
$Ca(Mg_{1/3}Ta_{2/3})_{0.6}Ti_{0.4}O_3$		60	36900	–10	[27]
$Ca_{0.85}Nd_{0.1}[Ti_{1-x}(Mg_{1/3}Nb_{2/3})_x]O_3$ ($x = 0.5$)	1400	54	7600	0.8	[28]
$Ca_{0.4}(Li_{1/2}Nd_{1/2})_{0.6}TiO_3$		113	5000	8	[27]
$CaTi_{1-x}(Al_{1/2}Nb_{1/2})_xO_3$ ($x = 0.5$)	1500/5 h	48	26100	–4	[29]
$CaTi_{1-x}(Al_{1/2}Nb_{1/2})_xO_3$ ($x = 0.5$) + 1 wt% Li_3NbO_4	1300/5 h	48	32100	–2	[29]
$CaTi_{1-x}(Al_{1/2}Nb_{1/2})_xO_3$ ($x = 0.7$) + 1 wt% Li_3NbO_4	1300/5 h	36	38900	–57	[29]
$CaTi_{1-x}(Al_{1/2}Nb_{1/2})_xO_3$ ($x = 0.3$) + 1 wt% Li_3NbO_4	1300/5h	69	21500	145	[29]

$CaTi_{1-x}(Al_{1/2}Nb_{1/2})_xO_3$ ($x=0.4$) + 1 wt% Li_3NbO_4	1300/5 h	57	28000	53	[29]
$CaTi_{1-x}(Al_{1/2}Nb_{1/2})_xO_3$ ($x=0.6$)+1 wt% Li_3NbO_4	1300/5 h	41	36100	–36	[29]
$CaTi_{0.7}(Al_{1/2}Ta_{1/2})_{0.3}O_3$	1500/15	65	20000	113	[30]
$CaTi_{0.6}(Al_{1/2}Ta_{1/2})_{0.4}O_3$	1500/15	52	13200	37	[30]
$CaTi_{0.54}(Al_{1/2}Ta_{1/2})_{0.46}O_3$	1500/15	47	27300	0	[30]
$CaTi_{0.5}(Al_{1/2}Ta_{1/2})_{0.5}O_3$	1500/15	41	26100	–20	[30]
$CaZrO_3$		27	20800		[25]
$SrZrO_3$		30	13600	–67	[25, 31]
$BaZrO_3$		35	8800		[25]
$LaGaO_3$		27	97000	–80	[22, 32]
$SrTiO_3$		300	3300	1650	[23, 24]
$Sr_xCa_{1-x}TiO_3$ ($x=0.5$)		236	4100	1230	[23]
$Sr_{n+1}Ti_nO_{3n+1}$ ($n=1$)		37	8000	137	[23]
$Sr_{n+1}Ti_nO_{3n+1}$ ($n=2$)		58	18600	317	[23]

(Continued)

Table 6.1 (Continued)

Material	Sintering temperature °C	ε_r	*Qf* (GHz)	τ_f (ppm/°C)	Reference
$Sr_{n+1}Ti_nO_{3n+1}$ ($n=3$)		76	12500	576	[23]
$Sr_{n+1}Ti_nO_{3n+1}$ ($n=4$)		100	4000	801	[23]
$(Sr_xCa_{1-x})_3Ti_2O_7$ (x = 0.8)		58	2500	359	[23]
$(Sr_xCa_{1-x})_3Ti_2O_7$ (x = 0.2)		60	2600	232	[23]
$(Sr_xCa_{1-x})_3Ti_2O_7$ (x = 0.0)		46	2600	50	[23]
$Sr_xCa_{1-x}TiO_3$ (x = 0.4)		218	7200	1164	[23]
$Sr_xCa_{1-x}TiO_3$ (x = 0.2)		181	3900	991	[23]
$Sr_xCa_{1-x}TiO_3$ (x = 0.8)		145	8200	990	[23]
$Sr_xCa_{1-x}TiO_3$ (x = 0.1)		170	8300	930	[23]
$Sr_{1.6}Ca_{0.4}TiO_4$	1600	39	8100	195	[33]
$Sr_{2.4}Ca_{0.6}Ti_2O_7$	1600	58	25000	359	[33]
$Ba_{0.2}Sr_{0.8}TiO_3$	1450	363	2400	–	[34]
$Ba_{0.4}Sr_{0.6}TiO_3$	1450	672	1600	–	[34]
$Ba_{0.6}Sr_{0.4}TiO_3$	1450	838	300	–	[34]
$Pb_{0.75}Ca_{0.15}ZrO_3$	1250	167	1000	111	[35]

$Pb_{0.63}Ca_{0.37}ZrO_3$	1450	110	3000	13	[35]
$Pb_{0.6}Ca_{0.4}ZrO_3$	1450	94	3600	–10	[35]
$Pb_{0.5}Ca_{0.5}ZrO_3$	1500	74	4000	–17	[35]
$(Pb_{0.5}Ca_{0.5})(Zr_{0.95}Ti_{0.05})O_3$	1350	72	4100	2	[36]
$Ba_{0.2}Sr_{0.71}(Zr_{0.951}Ti_{0.039}Ta_{0.01})O_3$		38	1700	0	[31]
$Ba_{0.29}Sr_{.71}(Zr_{.973}Ti_{.027})O_3$		38	2000	–40	[31]
$0.8CaTiO_3$–$0.2(Li_{1/2}Nd_{1/2})TiO_3$		134	13800	200	[37]
$0.2CaTiO_3$–$0.68(Li_{1/2}Nd_{1/2})TiO_3$–$0.12La_{1/3}Nd_{1/3}TiO_3$	1350	133	1450	–17	[37]
$0.3CaTiO_3$–$0.4(Li_{1/2}Nd_{1/2})TiO_3$–$0.3La_{1/3}Nd_{1/3}TiO_3$	1350	110	1400	–93	[37]
$0.3CaTiO_3$–$0.4(Li_{1/2}Nd_{1/2})TiO_3$–$0.3Dy_{1/3}Nd_{1/3}TiO_3$	1350	98	5100	0	[37]
$CaTiO_3$–$0.7Li_{1/2}Sm_{1/2}TiO_3$	1300/3 h	114	3700	12	[38]
$(Sm_{1/2}Li_{1/2})TiO_3$		52	1000	–260	[39, 40, 41]
$(Nd_{1/2}Li_{1/2})TiO_3$		80	700	–310	[37, 40]
$(La_{1/2}Na_{1/2})TiO_3$		122	9800	480	[39, 44]

(Continued)

Table 6.1 (Continued)

Material	Sintering temperature °C	ε_r	Qf (GHz)	τ_f (ppm/°C)	Reference
$(Nd_{1/2}Na_{1/2})TiO_3$		98	2700	190	[39]
$Sm_{1/2}Na_{1/2}TiO_3$	1300	80	13000		[41]
$0.55Ca_{0.61}Nd_{0.39}TiO_3$–$0.45Li_{1/2}Nd_{1/2}TiO_3$	1400	101	5300	13	[42]
$0.4(Na_{1/2}Sm_{1/2})TiO_3$–$0.6(Li_{1/2}Nd_{1/2})TiO_3$		115	3800	15	[43]
$0.3(Na_{1/2}La_{1/2})TiO_3$–$0.7(Li_{1/2}Sm_{1/2})TiO_3$	1300	117	2300	–19	[44]
$0.3Sm_{1/2}Li_{1/2}TiO_3$–$0.7Sm_{1/2}Na_{1/2}TiO_3$	1300	95	1000	–240	[41]
$0.5Sm_{1/2}Li_{1/2}TiO_3$–$0.5Sm_{1/2}Na_{1/2}TiO_3$	1300	90	1500	–140	[41]
$0.6Sm_{1/2}Li_{1/2}TiO_3$–$0.4Sm_{1/2}Na_{1/2}TiO_3$	1300	75	2000	60	[41]
$0.42(La_{1/2}Na_{1/2})TiO_3$–$0.58Ca(Fe_{1/2}Nb_{1/2})O_3$	1300/10	59	14000	0	[45]
$0.97La_{2/3}TiO_3$–$0.03NiTiO_3$	1350	70	17000	18	[46]
$0.96La_{2/3}TiO_3$–$0.04LaAlO_3$	1325	72	24000	123	[47]
$0.9La_{2/3}TiO_3$–$0.1LaAlO_3$	1350	63	26100	82	[47]
$0.8La_{2/3}TiO_3$–$0.2LaAlO_3$	1400	54	29000	35	[47]
$0.7La_{2/3}TiO_3$–$0.3LaAlO_3$	1425	45	33300	7	[47]
$0.6La_{2/3}TiO_3$–$0.4LaAlO_3$	1400	40	50800	–15	[47]

$0.5La_{2/3}TiO_3$—$0.5LaAlO_3$	1425	34	45000	–23	[47]
$0.7Ca_{2/5}Sm_{2/5}TiO_3$–$0.3Li_{1/2}Nd_{1/2}TiO_3$	1200/3 h	99	6200	9	[48, 49]
$0.5Ca_{2/5}Sm_{2/5}TiO_3$–$0.5Li_{1/2}Nd_{1/2}TiO_3$		106	3700		[49]
$0.4Ca_{2/5}Sm_{2/5}TiO_3$–$0.6Li_{1/2}Nd_{1/2}TiO_3$		110	3400		[49]
$0.3Ca_{2/5}Sm_{2/5}TiO_3$–$0.7Li_{1/2}Nd_{1/2}TiO_3$		106	3100		[49]
$0.67Ca_{2/5}Sm_{2/5}TiO_3$–$0.33Li_{1/2}Sm_{1/2}TiO_3$	1300/3 h	96	7200	0	[50]
$Ca_{2/5}Sm_{2/5}TiO_3$–$Li_{1/2}Nd_{1/2}TiO_3$–$0.6TiO_2$	1300/3 h	107	3300	0	[51]
$Ca_{2/5}Sm_{2/5}TiO_3$–$Li_{1/2}Sm_{1/2}TiO_2$–$0.8TiO_2$	1300/3 h	92.5	4900	9	[51]
$(Ca_{0.275}Sm_{0.4}Li_{0.25})TiO_3 + 0.5$ wt%B_2O_3–Li_2O	1200/3 h	99	5900	–4	[52]
CaO–SrO–Li_2O–(1–x)Sm_2O_3–xLa_2O_3–TiO_2 ($x = 0.17$)		118	4100	15	[53]
CaO–SrO–Li_2O–(1–x)Sm_2O_3–xPr_2O_3–TiO_2 ($x = 0.17$)		114	4800	14	[53]
CaO–SrO–Li_2O–(1–x)Sm_2O_3–xNd_2O_3–TiO_2 ($x = 0.17$)		113	4900	14	[53]
CaO–SrO–Li_2O–(1–x)Sm_2O_3–xSm_2O_3–TiO_2 ($x = 0.17$)		108	5500	15	[53]
CaO–SrO–Li_2O–(1–x)Sm_2O_3–xDy_2O_3–TiO_2 ($x = 0.17$)		100	5900	30	[53]

(Continued)

Table 6.1 (Continued)

Material	Sintering temperature °C	ε_r	*Qf* (GHz)	τ_f (ppm/°C)	Reference
$CaO–SrO–Li_2O–(1–x)Sm_2O_3–xYb_2O_3–TiO_2$ ($x = 0.17$)		96	2700	36	[53]
$La_{3/4}Mg_{2/4}Nb_{1/4}W_{1/4}O_3$	1400/4 h	25	16400	–56	[54]
$La_{3/4}Mg_{2/4}Ta_{1/4}W_{1/4}O_3$	1350/4 h	24	13600	–46	[54]
$LaAlO_3$+0.25 wt% CuO	1460	21	48000	–80	[55]
$NdAlO_3$+0.25 wt%V_2O_5	1400	22	64000	–30	[56]]
$NdAlO_3$+0.25 wt% CuO	1410	22	63000	–30	[57]
$NdAlO_3$	1650	22.3	58000	–33	[58]
$LaAlO_3$	1650	23	68000	–44	[58]
$PrAlO_3$	1650	23	51000	–25	[58]
$ErAlO_3$	1650	16	44000	–40	[58]
$DyAlO_3$	1650	18	38000	–34	[58]
$GdAlO_3$	1650	18	11000	–54	[58]
$SmAlO_3$	1650	20	65000	–74	[58]
$YAlO_3$	1650	16	68000	–59	[58]
$Ca_{0.7}Nd_{0.3}Ti_{0.7}Al_{0.3}O_3$	1600	44	33000	0	[59]

$0.5LaAlO_3–0.5SrTiO_3 + 0.25$ wt% B_2O_3	1430	35	43200	–11	[60]
$Ca_{0.6}La_{0.2667}TiO_3–0.5Li_{1/2}Nd_{1/2}TiO_3$	1400/4 h	105	7000	4	[61]
$0.73CaTiO_3–0.27NdAlO_3$	1450/10	45	31000	–15	[62]
$0.71CaTiO_3–0.29NdAlO_3$	1450/10	45	38400	6	[62]
$0.7CaTiO_3–0.3NdAlO_3$	1450/10	44	34800	–2	[62]
$0.67CaTiO_3–0.33NdAlO_3$	1450/10	42	43000	45	[62]
$Ca_{0.7}Ti_{0.7}La_{0.3}Al_{0.3}O_3 + 0.25$ wt%Al_2O_3	1450/12	46	38300	12	[63]
$Ca_{1-x}Y_xTi_{1-x}Al_xO_3$ ($x = 0.3$)	1650/6h	38	14200	–14	[64]
$0.6CaTiO_3–0.4NdAlO_3$	1450/10	37	40700	114	[62]
$0.7CaTiO_3–0.3NdAlO_3$ slow cooled	1450/10	45	44000	3	[62, 65]
$0.75CaTiO_3–0.25SmAlO_3$	1450/12	51	31000	31	[62]
$0.7CaTiO_3–0.3SmAlO_3$	1450/12	45	42000	1	[62]
$0.65CaTiO_3–0.35SmAlO_3$	1450/12	41	42000	–18	[62]
$0.66CaTiO_3–0.34LaAlO_3$	1450/12	44	30000	–3	[62]

(Continued)

Table 6.1 (Continued)

Material	Sintering temperature °C	ε_r	Qf (GHz)	τ_f (ppm/°C)	Reference
$0.96CaTiO_3$–$0.04La_{2/3}TiO_3$		90	27000	190	[66]
$0.9CaTiO_3$–$0.1Pb(Fe_{1/2}Nb_{1/2})O_3$	1200/3h	164	6200	580	[67]
$0.4CaTiO_3$–$0.6Li_{0.5}Nd_{0.5}TiO_3$	1225	126	2600	127	[68]
$0.4CaTiO_3$–$0.6Li_{0.5}Nd_{0.5}TiO_3 + 10$ wt% $Bi_2Ti_2O_7$	1175	150	2400	70	[68]
$0.2CaTiO_3$–$0.8Li_{1/2}Nd_{1/2}TiO_3 + 5$ wt% $Bi_2Ti_2O_7$	1325	130	2400	20	[69]
$0.2CaTiO_3$–$0.8Sr(Mg_{1/3}Nb_{2/3})O_3$	1600	45	9000	0	[70]
$0.7CaTiO_3$–$0.3(La_{0.5}Nd_{0.5})AlO_3$	1540	41.5	37000	4	[71]
$0.7CaTiO_3$–$0.3LaGaO_3$	1540	49	29000	22	[62, 71]
$0.7CaTiO_3$–$0.3Nd(Ga_{0.5}Al_{0.5})O_3$	1540	45	38000	12	[71]
$0.7CaTiO_3$–$0.3La(Ga_{0.5}Al_{0.5})O_3$	1540	43	40000	13	[71]
$0.7CaTiO_3$–$0.3(La_{0.5}Nd_{0.5})(Ga_{0.5}Al_{0.5})O_3$	1540	45	43000	9	[71]
$0.64CaTiO_3$–$0.36LaGaO_3$	1540	46.5	48000	–3	[71, 72]
$0.66CaTiO_3$–$0.34LaGaO_3$	1540	48	46000	4	[71]
$0.66CaTiO_3$–$0.34(La_{0.5}Nd_{0.5})$ GaO_3	1540	49	43000	0	[71]
$0.65CaTiO_3$–$0.35LaGaO_3$	1600	47	40000	0	[22]

$0.7CaTiO_3$–$0.3NdGaO_3$	1450/12	49	32000	35	[62]
$0.6CaTiO_3$–$0.4NdGaO_3$	1450/12	44	30000	–18	[62]
$0.65CaTiO_3$–$0.35NdGaO_3$	1450	45	46000	–2	[62, 73]
$0.6CaTiO_3$–$0.4LaGaO_3$	1450/12	45	34000	–20	[62]
$0.7CaTiO_3$–$0.3SmGaO_3$	1450/12	51	18000	41	[62]
$0.6CaTiO_3$–$0.4SmGaO_3$	1450/12	42	35000	–11	[62]
$0.8Ca_{0.85}Nd_{0.1}TiO_3$–$0.2SmAlO_3$	1400	58	14000	13	[74]
$0.46LaAlO_3$–$0.54SrTiO_3 + 0.25$ wt%B_2O_3	1460	35	38000	–1	[75]
$LaAlO_3 + 15$ mol%$Sr_2Nb_2O_7$	1600/3 h	23	20500	5	[76]
$0.65CaTiO_3$–$0.35LaAlO_3$	1650	37	47000	5	[77]
$AgNb_{0.52}Ta_{0.48}O_3$	1200	415	700		[78]
$AgTa_{0.57}Nb_{0.43}O_3$	1200	380	800		[79]
$Ag(Nb_{2/4}Ta_{2/4})O_3 + 1$ wt%CuO	900	386	380		[80]
$Ag(Nb_{3/4}Ta_{1/4})O_3 + 1$ wt% CuO	925	487	200		[80]

(*Continued*)

Table 6.1 (Continued)

Material	Sintering temperature °C	ε_r	Qf (GHz)	τ_f (ppm/°C)	Reference
$Ag(Nb_{1/3}Ta_{2/3})O_3 + 1$ wt% CuO	875	271	800		[80]
$(Ca_{1-x}Nd_{2x/3})TiO_3$ ($x = 0.3$) + 25 vol% $3ZnO–2B_2O_3$	900	60	3700	62	[81]
$(1-x)LaCa_{0.5}Zr_{0.5}O_3–xCaTiO_3$ ($x = 1/3$)		18	16000	–75	[82]
$(1-x)LaCa_{0.5}Zr_{0.5}O_3–xCaTiO_3$ ($x = 1/2$)		26	13500	–67	[82]
$Ca_{1-x}Nd_{2x/3}TiO_3$ (x–0.2)	1400	119	4100	433	[42]
$Ca_{1-x}Nd_{2x/3}TiO_3$ (x–0.3)	1400	107	6600	316	[42]
$Ca_{1-x}Nd_{2x/3}TiO_3$ (x–0.42)	1400	93	6900	228	[42]
$Ca_{1-x}Sm_{2x/3}TiO_3$ ($x = 0$)	1450/3 h	170	7000		[83]
$Ca_{1-x}Sm_{2x/3}TiO_3$ ($x = 0.2$)	1450/3 h	119	12300		[83]
$Ca_{1-x}Sm_{2x/3}TiO_3$ ($x = 0.4$)	1450/3 h	101	14000		[83]
$Ca_{1-x}Sm_{2x/3}TiO_3$ ($x = 0.6$)	1450/3 h	95	14900		[83]
$(Ca_{0.85}Nd_{0.1})[(Mg_{1/3}Nb_{2/3})_xTi_{1-x}]O_3$ ($x = 0.5$)	1400/3 h	54	7600	0	[84]
$Ca[(Li_{0.33}Nb_{0.67})_{0.9}Ti_{0.1}]\ O_{3-\delta} + 10$ wt% LiF	900	25	19300	–15	[85]
$Ca[(Li_{0.33}Nb_{0.67})_{0.9}Ti_{0.1}]\ O_{3-\delta} + 20$ wt% LiF	840	21	20400	–18	[85]
$Ca[(Li_{1/3}Nb_{2/3})_{0.8}Ti_{0.2}]\ O_{3-\delta} + 12$ wt% $B_2O_3–ZnO–SiO_2–PbO$	900/3 h	40	12500	–8	[86]

$(Sr_{1-x}Ca_x)[(Li_{1/4}Nb_{3/4})_{1-y}Ti_y]O_3$		42	31500	0	[87]
$Ca(Li_{1/3}Nb_{2/3})_{1-x}Ti_x]O_{3-\delta}$ $(x=0.1)+1$ wt% B_2O_3	940	50	6500	–7	[88]
$Ca(Li_{1/3}Nb_{2/3})O_{3-\delta}$	1150	30	40000	–21	[89]
$Ca(Li_{1/3}Nb_{2/3})O_{3-\delta}+4$ wt% B_2O_3	1000	31	31000	–18	[90]
$Ca(Li_{1/3}Ta_{2/3})O_{3-\delta}$	1200	24	42000	–40	[91]
$Ca[(Li_{1/3}Nb_{2/3})_{0.9}\ Zr_{0.1}]O_{3-\delta}$	1150	30	36300	–5	[89]
$Ca[(Li_{1/3}Nb_{2/3})_{1-x}Ti_x])\ O_{3-\delta}+5$ wt% Bi_2O_3 $(x=0.05)$	900/3 h	20	6500	–4	[92]
$Ca[(Li_{1/3}Nb_{2/3})_{1-x}Ti_x])\ O_{3-\delta}+5$ wt% Bi_2O_3 $(x=0.2)$	900/3 h	35	11000	13	[92]
$Ca[(Li_{1/3}Ta_{2/3})_{1-x}Ti_x]\ O_{3-\delta}+3$ wt% B_2O_3 $(x=0.05)$	1050/4 h	26	22000	–97	[91]
$Ca[(Li_{1/3}Ta_{2/3})_{1-x}Ti_x]\ O_{3-\delta}+3$ wt% B_2O_3 $(x=0.1)$	1000/4 h	28	9800	–	[91]
$Ca[(Li_{1/3}Ta_{2/3})_{1-x}Ti_x]\ O_{3-\delta}+3$ wt% B_2O_3 $(x=0.15)$	1050/4 h	29	20700	–57	[91]
$Ca[(Li_{1/3}Ta_{2/3})_{1-x}Ti_x]\ O_{3-\delta}+3$ wt% B_2O_3 $(x=0.2)$	1050/4 h	28	12900	–15	[91]
$Ca[(Li_{1/3}Ta_{2/3})_{1-x}Ti_x]O_{3-\delta}+3$ wt% B_2O_3 $(x=0.2)$	1100	30	22800	–	[91]
$Ca[(Li_{1/3}Ta_{2/3})_{1-x}Ti_x]\ O_{3-\delta}+3$ wt% B_2O_3 $(x=0.3)$	1050/4 h	35	22800	–4	[91]
$Ca[(Li_{1/3}Ta_{2/3})_{1-x}Ti_x]O_{3-\delta}+3$ wt% B_2O_3 $(x=0.5)$	1150	48	21000	–	[91]
$Ca[(Li_{1/3}Ta_{2/3})_{1-x}Ti_x]O_{3-\delta}+3$ wt% B_2O_3 $(x=0.5)$	1050	45	12300	75	[91]

(Continued)

Table 6.1 (Continued)

Material	Sintering temperature °C	ε_r	*Qf* (GHz)	τ_f (ppm/°C)	Reference
$Ca[(Li_{1/3}Nb_{2/3})_{1-x}Sn_x]\ O_{3-\delta}\ (x=0)$	1150/3 h	30	40000	–21	[93]
$Ca[(Li_{1/3}Nb_{2/3})_{1-x}Sn_x]\ O_{3-\delta}\ (x=0.15)$	1150/3 h	25	49100	–25	[93]
$Ca[(Li_{1/3}Nb_{2/3})_{1-x}Sn_x]\ O_{3-\delta}\ (x=0.3)$	1150/3 h	23	46300	–39	[93]
$Ca[(Li_{1/3}Nb_{2/3})_{1-x}Ti_x]\ O_{3-\delta}\ (x=0.1)$	1150/3 h	35	27200	–2	[93]
$Ca[(Li_{1/3}Nb_{2/3})_{1-x}Ti_x]\ O_{3-\delta}\ (x=0.15)$	1150/3 h	39	26100	0	[93]
$Ca[(Li_{1/3}Nb_{2/3})_{1-x}Ti_x]\ O_{3-\delta}\ (x=0.3)$	1150/3 h	45	22500	20	[93]
$Ca[(Li_{1/3}Nb_{2/3})_{1-x}Ti_x]\ O_{3-\delta}\ (x=0.5)$	1150/3 h	55	18600	83	[93]
$Ca[(Li_{1/3}Nb_{2/3})_{0.84}Ti_{0.16}]O_{3-\delta} + 2\ wt\%\ LiF + 3\ wt\%\ B_2O_3$	900/2 h	34	17400	–5	[94]
$(1-x)Sr(Li_{1/4}Nb_{3/4})O_3{-}xSr(Li_{2/5}W_{3/5})O_3(x=0.385)$	1450	30	21200	–33	[95]
$(1-x)Ca(Li_{1/4}Nb_{3/4})O_3{-}xCa(Li_{2/5}W_{3/5})O_3(x=0.238)$	1150	30	22700	–33	[95]
$(1-x)Ca(Li_{1/4}Nb_{3/4})O_3{-}xCa(Li_{2/5}W_{3/5})O_3(x=0.333)$	1150	29	15700	–35	[95]
$0.677Ba(Li_{1/4}Nb_{3/4})O_3{-}0.333Ba(Li_{2/5}W_{3/5})O_3$	1470/5 h	31	19000	18	[96]
$(Pb_{1-x}Ca_x)(Li_{1/4}Nb_{3/4})O_3\ (x=0.25)$		93	2100	630	[97]
$(Pb_{1-x}Ca_x)(Li_{1/4}Nb_{3/4})O_3\ (x=0.5)$		78	2000	460	[97]
$(Pb_{1-x}Ca_x)(Na_{1/4}Nb_{3/4})O_3\ (x=0.25)$		79	400	550	[97]

$(Pb_{1-x}Ca_x)(Li_{1/4}Nb_{3/4})O_3$ ($x = 0.5$)		72	1500	230	[97]
$0.7Ca(Li_{1/4}Nb_{3/4})O_3$–$0.3CaTiO_3$	1300	44	12000	–9	[98]
$[Ca_{0.6}(Li_{0.5}Nd_{0.5})_{0.4}]_{0.45}Zn_{0.55}TiO_3 + 2$ wt% $0.33ZnO$–$0.67H_3BO_3$	875	42	10300	195	[99]
$Sr_2Ce_2Ti_5O_{16}$ ($Sr_{1-3x/2}Ce_xTiO_3$ $x = 0.4$)	1300/2 h	113	8000	306	[100, 101]
$Sr_3Ce_2Ti_6O_{19}$ ($Sr_{1-3x/2}Ce_xTiO_3$ $x = 0.333$)	1350/2 h	123	10000	392	[101]
$Sr_4Ce_2Ti_7O_{22}$ ($Sr_{1-3x/2}Ce_xTiO_3$ $x = 0.286$)	1350/2 h	136	10800	428	[101]
$Sr_5Ce_2Ti_8O_{25}$ ($Sr_{1-3x/2}Ce_xTiO_3$ $x = 0.25$)	1375/2 h	143	11000	478	[101]
$Sr_6Ce_2Ti_9O_{28}$ ($Sr_{1-3x/2}Ce_xTiO_3$ $x = 0.222$)	1375/2 h	150	9600	497	[101]
$Sr_7Ce_2Ti_{10}O_{31}$ ($Sr_{1-3x/2}Ce_xTiO_3$ $x = 0.2$)	1375/2 h	157	9300	544	[101]
$Sr_8Ce_2Ti_{11}O_{34}$ ($Sr_{1-3x/2}Ce_xTiO_3$ $x = 0.182$)	1375/2 h	167	8000	601	[101]
$Sr_9Ce_2Ti_{12}O_{37}$ ($Sr_{1-3x/2}Ce_xTiO_3$ $x = 0.167$)	1375/2 h	173	3000	637	[101]
$Sr_{10}Ce_2Ti_{13}O_{40}$ ($Sr_{13x/2}Ce_xTiO_3$ $x = 0.154$)	1400/2 h	179	8000	724	[101]
$Sr_{11}Ce_2Ti_{14}O_{43}$ ($Sr_{1-3x/2}Ce_xTiO_3$ $x = 0.154$)	1400/2 h	185	6000	789	[101]
$PbZrO_3$–CeO_2		140	2500	–1080	[102]

partial pressure at high temperatures [104]. Feteira et al. [18] found that 6H $BaTiO_3$ can be stabilized by partial substitution of Ga for Ti. The $BaTi_{0.92}Ga_{0.8}O_{2.96}$ sintered at 1450°C in oxygen showed ε_r of 74, $Qf \sim$ 8000 GHz. However, it has a high τ_f of 550 ppm/°C. Keith et al. [20] studied dielectric properties of dopant stabilized 6H $BaTiO_3$. i.e., $Ba(Ti_{1-x}M_x)O_3$ M = Mg, Al, Al, Cr, Fe, Ge, Zn, Ni and In. The dielectric properties of these B site substituted samples are given in the Table 6.1. The Mn substituted samples sintered at 1400°C have the highest quality factor. The $CaTiO_3$ has a high ε_r of about 160 with good quality factor of about 7000 GHz but its τ_f is too high (+850 ppm/°C) for practical applications [22, 24]. The $SrTiO_3$ has a very high ε_r of about 300 and Qf of about 3000 GHz and a very large τ_f of about 1650 ppm/°C [23, 24].

6.4 $Ag(Nb_{1-x}Ta_x)O_3$

Recently $Ag(Nb_{1-x}Ta_x)O_3$ was reported [78–80, 105, 106] as a high permittivity low loss dielectrics for microwave applications. The $Ag(Nb_{1/2}Ta_{1/2})O_3$ is monoclinic with space group $P2/m$ [105]. Single crystals of $AgNbO_3$ and $AgTaO_3$ were grown by the flux and slow cooling methods [107, 108]. The $Ag(Nb_{1-x}Ta_x)O_3$ ceramics are prepared by initially mixing Nb_2O_5 and Ta_2O_5 and pre-reacting at about 1200°C. This will give a homogeneous solid solution of $(Nb_{1-x}Ta_x)_2O_5$ [109]. It is then ground and mixed well with appropriate amount of Ag_2O and calcined at about 1100°C. The ground powder is compacted and then sintered at temperatures in the range 1200–1300°C. As the amount of Ta content increases, the sintering temperature also increases. It was reported that with increasing sintering temperature, increasing amount of black silver precipitates in the ceramic [105, 109]. One of the problem with $Ag(Nb,Ta)O_3$ dielectric is its decomposition during sintering which leads to degradation of properties [105] and can be controlled by processing in an oxygen atmosphere [78, 105]. The decomposition of $Ag(Nb,Ta)O_3$ is due to volatilization of silver [78, 105, 109]. In order to avoid decomposition, Valant et al. kept the pressed samples in a corundum tube or crucible during sintering [78, 105]. The other technique to prevent decomposition at high temperatures is to lower the sintering temperature. The sintering temperature can be lowered using fine powder obtained by wet chemical methods such as alkoxide and citrate route [110, 111], use of low melting additives such as CuO [112] or glasses [113]. Li et al. [113] reported that addition of 2 wt% glass lowered the sintering temperature to about 960°C and prevented the silver decomposition problem. However, they do not reveal what glass they have used. Guo et al. [114] found that addition of 2.5 wt% Sb_2O_5 lowers the dielectric loss.

The unusually high ε_r, fairly low dielectric loss and low sintering temperature makes this material important for passive dielectric components. However, the temperature dependence of permittivity is relatively high. Recently Valant and co-workers [115–117] deliberately generated an inhomegeneity on the B sites of $Ag(Nb,Ta)O_3$ by controlling the powder morphology and carefully selecting the firing conditions. The different $Ag(Nb,Ta)O_3$ phases present in such ceramics exhibited different temperature dependencies of the permittivity, which produced a compensated overall temperature dependence of such ceramics. Valant et al. [115] suppressed the high temperature dependence of permittivity in $Ag(Nb,Ta)O_3$ based ceramics by preparing a mixture of $Ag(Nb_{0.65}Ta_{0.35})O_3$ and $Ag(Nb_{0.35}Ta_{0.65})O_3$ which have monotonically increasing and decreasing permittivity variation with temperature. By mixing a mixture of 45 wt% of $Ag(Nb_{0.65}Ta_{0.35})O_3$ and 55 wt% $Ag(Nb_{0.35}Ta_{0.65})O_3$

and controlling its microstructure by using coarse grains and preventing the reaction to form $Ag(Nb_{0.5}Ta_{0.5})O_3$, they could achieve a ε_r of 430, $Qf = 700$ GHz and temperature dependence of permittivity only 40 ppm/°C. The microwave dielectric properties of $Ag(Nb_{1-x}Ta_x)O_3$ for various values of x and additives are given in Table 6.1. Porokhonsky et al. [118] studied the dielectric properties of $AgNbO_3$ and $Ag_{1-x}Li_xNbO_3$ in the millimeter wavelength region. Substitution of 5% of Ag with Li induces orthorhombic to rhombohedral phase transition. The temperature characteristics of the ε_r significantly vary with Li substitution and the material becomes a useful ferroelectric ceramic [107, 119–121] having surprisingly high piezoelectric properties [122]. Hence it is a potential lead-free piezoelectric ceramic although costwise it is not attractive. Recently it was shown that the compositions $Ag(Nb_{0.8}Ta_{0.2})O_3$ and $Ag(Nb_{0.9}Ta_{0.1})O_3$ are useful for possible applications in tunable microwave devices such as phase shifters, tunable filters, varactors [123–126]. You and Koh [127] prepared highly polar-axis oriented $Ag(Ta_{0.38}Nb_{0.62})O_3$ films on $LaAlO_3$ single crystals by pulsed laser ablation. The $Ag(Ta_{0.38}Nb_{0.62})O_3/LaAlO_3$ thin film capacitors exhibited superior overall performance–low loss tangent with ε_r of 224. The $Au/Cr/Ag(Ta,Nb)O_3/LaAlO_3$ interdigital capacitors show low loss tangent of about 0.0033 at 1 MHz and a high figure of merit (K factor) greater than 48 at 40 V. The $Ag(Nb,Ta)O_3$ is also reported as a photocatalyst with ability to split water into oxygen and hydrogen [128, 129]. Silver niobate is a weak ferroelectric and antiferroelectric undergoing a rich sequence of phase transitions from simple perovskite cubic above 853°K to antiferroelectric M_3 phase below 620°K, disordered antiferroelectric M_2 below 530°K and weak ferroelectric M_1 phase below 340°K. This sequence of displacive phase transitions occur due to tilting of the oxygen octahedra [130–137] and these phase transitions considerably influence the dielectric properties. This material is useful for several microwave electronic components such as band pass filters, high capacity NPO multilayer capacitors. Addition of 1 wt% CuO to $Ag(Nb_{1/2}Ta_{1/2})O_3$ lowers sintering temperature to 900°C [112] and hence a promising material for LTCC technology. However, it was shown later that Cu ions incorporate into the $Ag(Nb,Ta)O_3$ crystal structure [138].

6.5 $Ca(Li_{1/3}Nb_{2/3})O_{3-\delta}$

The $Ca(Li_{1/3}B_{2/3})O_{3-\delta}$ (B = Nb,Ta) is a nonstoichiometric perovskite and is multiphase when prepared by the conventional ceramic route [93, 98]. However, substitution of Sn^{4+} and Ti^{4+} for $(Li_{1/3}Nb_{2/3})^{3.67+}$ by forming $Ca[(Li_{1/3}Nb_{2/3})_{1-x}M_x]O_{3-\delta}$ (M = Ti, Sn) gives single-phase perovskites in the vicinity of $x = 0.2$ [88, 91–93]. The $Ca[(Li_{1/3}Nb_{2/3})_{1-x}M_x]O_{3-\delta}$ (M = Ti, Sn) dielectric resonators were prepared by calcining the ball milled raw powders at 800–900°C/2 h and sintering at about 1150°C [93]. Choi et al. kept the samples in a closed pt box to prevent the escape of Li [93]. The stability of the perovskite increased with increasing substitution of Ti^{4+} and Sn^{4+} up to $x = 0.3$ and secondary phase appeared for $x > 0.3$. The bulk density increased with x for Sn and decreased for Ti. The XRD pattern could be indexed based on orthorhombic $CaTiO_3$. The ε_r decreased with increase in x for Sn and increased for Ti as shown in Figure 6.3. Figure 6.4 shows the variation of Qf with x. With Sn, the Qf increased initially up to $x = 0.15$ and then decreased whereas Ti substitution always decreased Qf. Sn substitution significantly improved quality factor and for $x = 0.15$ showed a Qf of 52 000 GHz. For Sn, the τ_f slightly improved for $x = 0.1$ and then increased towards the negative side whereas Ti substitution considerably improved the τ_f as shown in Figure 6.5. A zero τ_f was reached for $x = 0.2$ in $Ca[(Li_{1/3}Nb_{2/3})_{1-x}Ti_x]O_{3-\delta}$ with $\varepsilon_r = 39$,

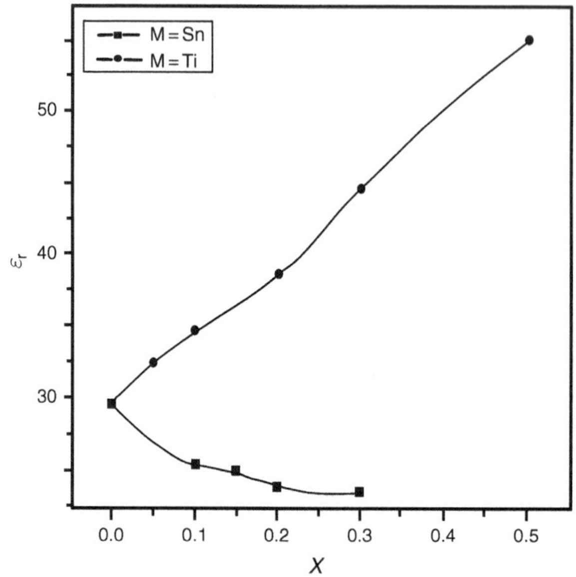

Figure 6.3 Variation in ε_r of $Ca[(Li_{1/3}Nb_{2/3})_{1-x}M_x]O_{3-\delta}$(M = Sn, Ti) ceramics sintered at 1150°C as a function of composition (after Ref. [93]).

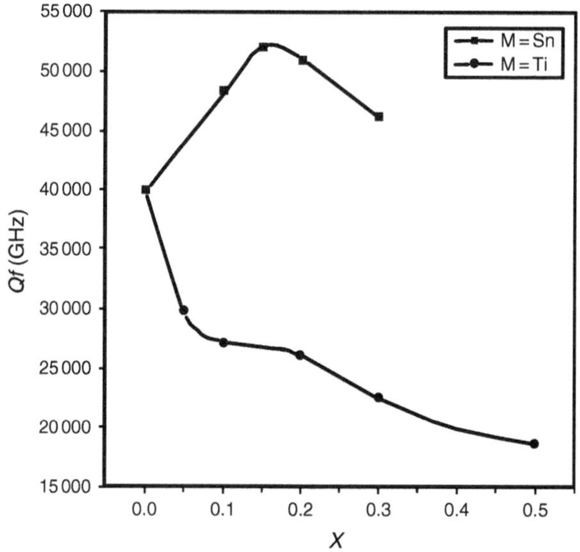

Figure 6.4 Variation in *Qf* of $Ca[(Li_{1/3}Nb_{2/3})_{1-x}Mx]O_{3-\delta}$(M = Sn, Ti) ceramics sintered at 1150°C as a function of composition (after Ref. [93]).

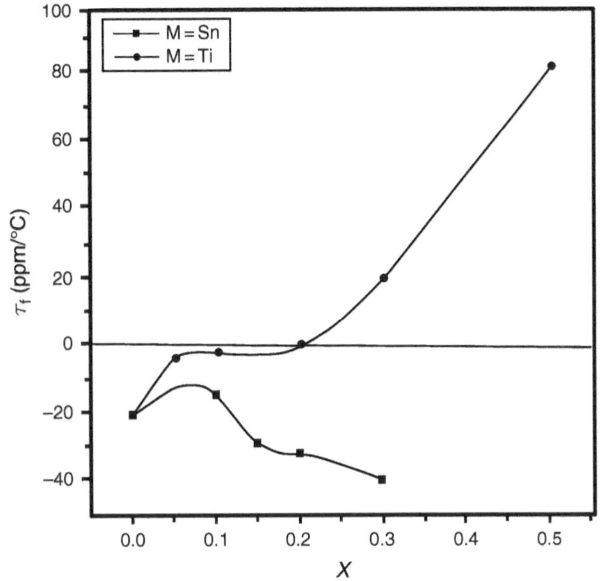

Figure 6.5 Variation in τ_f of $Ca[(Li_{1/3}Nb_{2/3})_{1-x}M_x]O_{3-\delta}$ (M = Sn, Ti) ceramics sintered at 1150°C as a function of composition (after Ref. [93]).

Qf = 26 000 GHz when sintered at 1150°C. The dielectric properties of $Ca[(Li_{1/3}B_{2/3})_{1-x}M_x]O_{3-\delta}$ (B = Nb,Ta, M = Ti, Sn) prepared under different conditions with various additives are given in Table 6.1.

Several authors [85, 86, 88, 90, 92, 94] tried to reduce the escape of volatile lithium by lowering the sintering temperature by the addition of low melting additives such as B_2O_3, Bi_2O_3, glass frit or LiF. Choi et al. [89] substituted $(Li_{1/3}Nb_{2/3})$ by Zr^{4+} up to 30 mol% to form $Ca[(Li_{1/3}Nb_{2/3})_{1-x}Zr_x]O_{3-\delta}$. Secondary phase was formed for $x = 0.1$ and they disappeared with increasing Zr content. The ε_r increased with Zr content and Qf and τ_f initially increased and then decreased with x. The ceramics with $x = 0.1$ and sintered at 1150°C showed $\varepsilon_r = 30$, $Qf = 36\ 300$ GHz and $\tau_f = -5$ ppm/°C. In order to tune the τ_f and to lower the sintering temperature, Liu et al. [91] substituted $(Li_{1/3}Ta_{2/3})$ by Ti and added B_2O_3 in $Ca(Li_{1/3}Ta_{2/3})O_{3-\delta}$. The Ti substituted $Ca(Li,Ta)O_{3-\delta}$ showed a single-phase ordered perovskite structure as shown in Figure 6.6. The substitution of smaller Ti^{4+} for $(Li_{1/3}Ta_{2/3})$ decreased the cell volume. X-ray diffraction study showed that substitution of Ti for (Li,Ta) stabilized the orthorhombic perovskite phase but decreased the 1:2 ordering. The ordering peak disappeared with Ti substitution. As x increased to 0.5, ε_r increased linearly from 24 to 48 and τ_f decreased and became positive. At $x = 0.3$, $Ca[(Li_{1/3}Nb_{2/3})_{0.7}Ti_{0.3}]O_{3-\delta}$ + 3 wt% B_2O_3 sintered at 1050°C/4 h showed $\varepsilon_r = 35$, $Qf = 22\ 800$ GHz and $\tau_f = -4$ ppm/°C. Yoon et al. [98] studied the effect of $CaTiO_3$ addition in $Ca(Li_{1/4}Nb_{3/4})O_{3-\delta}$. The orthorhombic $Ca(Li_{1/4}Nb_{3/4})O_{3-\delta}$ ceramics sintered at 1275°C has $\varepsilon_r = 26$, $Qf = 13\ 000$ GHz and $\tau_f = -49$ ppm/°C. In order to tune τ_f, they added $CaTiO_3$ which formed a solid solution with $Ca(Li_{1/4}Nb_{3/4})O_{3-\delta}$. The $0.7Ca(Li_{1/4}Nb_{3/4})O_{3-\delta}$–$0.3CaTiO_3$ sintered at 1300°C showed $\varepsilon_r = 44$, $Qf = 12\ 000$ GHz and $\tau_f = -9$ ppm/°C. These properties are different from nonstoichiometric $Ca[(Li_{1/3}Nb_{2/3})_{1-x}Ti_x]O_{3-\delta}$ where the zero τ_f is for $x = 0.2$ [93].

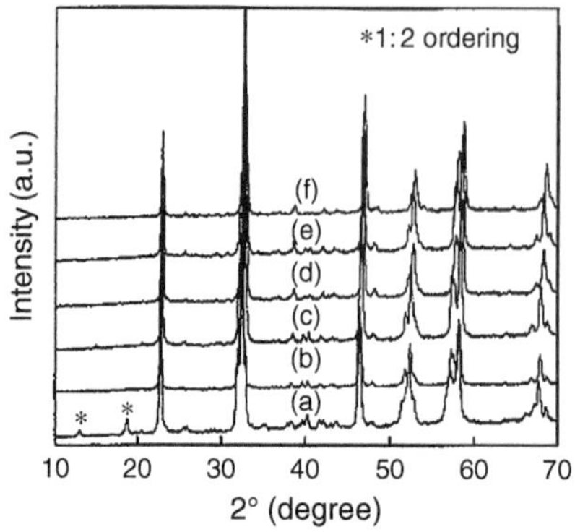

Figure 6.6 X-ray diffraction pattern of $Ca[(Li_{1/3}Ta_{2/3})_{1-x}Ti_x]O_{3-\delta}$ ceramics doped with 3 wt% B_2O_3 and sintered at 1050°C/4 h: (a) $x = 0.05$, (b) 0.1, (c) 0.15, (d) 0.2, (e) 0.3 and (f) 0.5 (after Ref. [91]).

6.6 $CaO-Ln_2O_3-TiO_2-Li_2O$ System

In 1989 Yokohama et al. [139] reported that $La_{2/3}TiO_3$ has high ε_r of about 130 at 1 MHz with relatively high dielectric loss. Pure $La_{2/3}TiO_3$ is an unstable perovskite due to the high amount of vacancies in the A cationic sublattice. Hence it is difficult to prepare single-phase lanthanum titanate ($La_{2/3}TiO_3$). It can be produced either in the presence of some alkaline earth or under an oxygen pressure lower than that of air [140]. Abe and Uchino [141] reported the preparation of $La_{2/3}TiO_{3-\delta}$ in a reduced atmosphere in the presence of a mixture of CO_2 and H_2. The resultant product was perovskite having one-third of lanthanide sites (A sites) vacant and was slightly oxygen deficient. The structure is found to depend on oxygen stoichiometry. For a small δ, the structure was orthorhombic double perovskite. They showed the existence of an ordering in the cationic sublattice of the perovskite between the La^{3+} cations and the vacancies, leading to the substructure of the lattice that is doubled along the *c*-direction. The $La_{2/3}TiO_3$ can be stabilized by substitution of (+2) or (+1) ions on to the A sites [142–144]. Several dopants such as Ca, Sr, Pb, Sc, Cr, Al and Nb have been used to stabilize the $La_{2/3}TiO_3$ phase with useful dielectric properties [145–147].

Recently it has been reported that $LaAlO_3$ addition to $La_{2/3}TiO_3$ could stabilize the structure and decrease the ε_r and the loss factor [47, 147–149]. Suvorov et al. [47] reported the formation of $La_{2/3}TiO_3$–$LaAlO_3$ (up to 50 mol%) solid solution when sintered at temperatures in the range 1325–1400°C. Figure 6.7 shows the variation of ε_r, τ_f and Qf with $LaAlO_3$ content. The complete stabilization of $La_{2/3}TiO_3$ occurred when 4 mol% of $LaAlO_3$ was added. Addition of $LaAlO_3$ up to 4 mol% increased the τ_f and ε_r and further addition decreased ε_r and τ_f (Figure 6.7). In $La_{2/3}TiO_3$ addition of 30 mol% of $LaAlO_3$ gives ε_r of 45, $Qf = 33\ 000$ and $\tau_f = +7$ ppm/°C. The dielectric properties of the samples sintered in air and in oxygen are nearly the same. Houivet et al. [46] reported that formation of $(1-x)La_{2/3}TiO_3$–$xNiTiO_3$ solid solution ceramics by sintering in an oxygen flow in the temperature range 1340–1380°C resulting in excellent properties for $x = 0.03$ with $\varepsilon_r = 70$, $Qf = 17\ 000$ GHz and $\tau_f = 18$ ppm/°C. The dielectric properties deteriorated when $x > 0.03$ due to formation of secondary phases of $La_2Ti_2O_7$ and $La_4Ti_9O_{24}$.

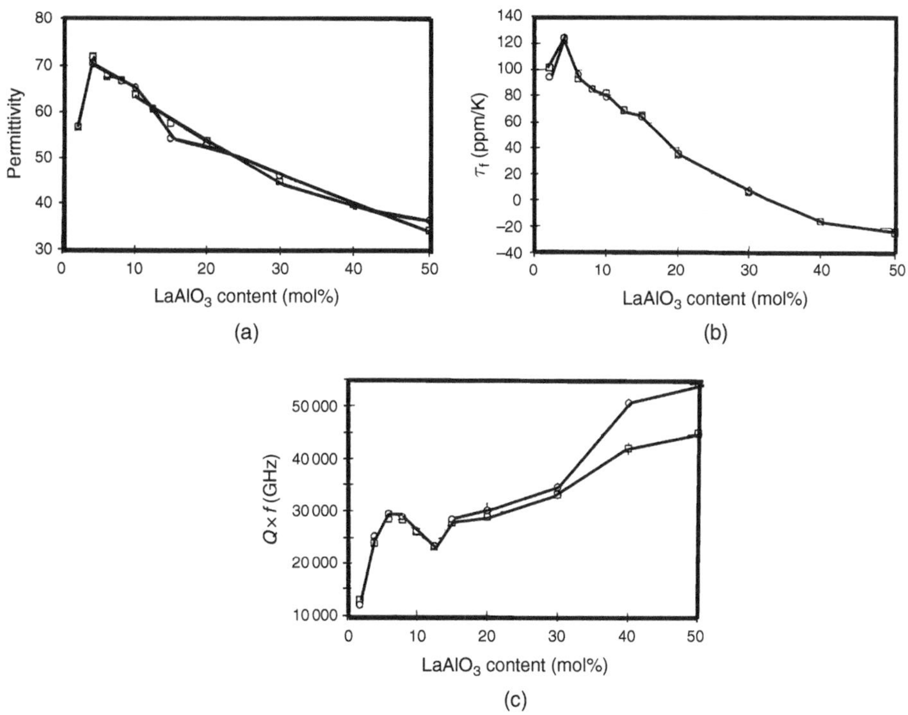

Figure 6.7 Variation of (a) ε_r, (b) τ_f and (c) Qf of $La_{2/3}TiO_3$–$LaAlO_3$ as a function of $LaAlO_3$ content sintered in air and oxygen (after Ref. [47]).

It has been reported [42, 66, 83, 150–152] that substitution of Ca^{2+} in $CaTiO_3$ by trivalent La, Nd or Sm ions form Ca^{2+} ion vacancies influencing the dielectric properties. The ceramics can be represented as $Ca_{1-x}La_{2x/3}\,\square_{x/3}TiO_3$ where □ is an A site vacancy. The number of vacancies increases with x and the formation of an ordered arrangement of the cations and the vacancies resulted in the doubling of the perovskite unit cell. These materials showed ε_r of about 100 and $Qf > 10\,000$ GHz. Kim et al. [66] prepared $(1-x)$ $CaTiO_3$–$xLa_{2/3}TiO_3$ by sintering at 1400°C/12 h. The structure changed from pseudocubic perovskite to tetragonal double perovskite at $x = 0.7$ and to orthorhombic double perovskite at $x = 0.9$ as shown in Figure 6.8. The superstructure reflections were detected for the composition $x = 0.8$ and $x = 0.96$. The sintered samples were annealed in oxygen gas flow at 1000°C/48 h to prevent Ti^{4+} reduction. Figure 6.9 shows the variation of ε_r, τ_f and Q with x. The ε_r and τ_f decreased with increase in La content. The Qf increased rapidly at first and then increased steadily and almost linearly for $x > 0.3$. The ceramic with $x = 0.96$ has $\varepsilon_r = 90$, $Q = 27\,000$ GHz and $\tau_f = 190$ ppm/°C. The τ_f varies from +450 ppm/°C to +180 ppm/°C. The analogous $Ca_{1-x}Nd_{2x/3}TiO_3$ was also reported with interesting dielectric properties [42, 150, 151]. Yoshi [153] first reported the crystal structure of $Nd_{2/3}TiO_3$ as orthorhombic with space group *Pmmm* and later it was reported [154–156] as isostructural with $La_{2/3}TiO_3$ with orthorhombic *Cmmm* space group. The $Nd_{2/3}TiO_3$ is unstable and slight oxygen deficiency [157] or low level doping with $NdTiO_3$ [156] or $NdAlO_3$ [158] stabilizes the structure. The $Ca_{1-x}Nd_{2x/3}TiO_3$ has a sintering temperature of 1400°C/12 h with $\varepsilon_r \sim 100$–180 and $Qf \sim 1000$–9000 GHz [42, 150].

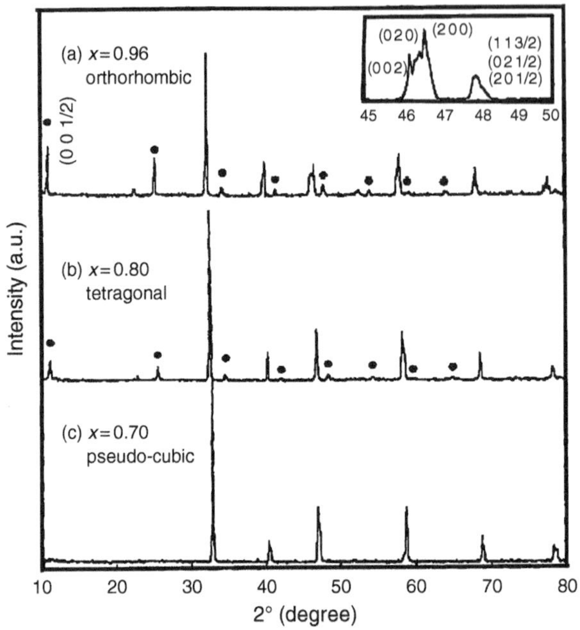

Figure 6.8 XRD patterns of $(1-x)CaTiO_3-xLa_{2/3}TiO_3$: (a) $x=0.7$ pseudocubic, (b) $x=0.8$ tetragonal and (c) 0.96 orthorhombic (after Ref. [66]). The superstructure reflections are marked by filled circles.

Yoon and co-workers [83, 152] studied the effect of Sm^{3+} substitution in $CaTiO_3$ by preparing $Ca_{1-x}Sm_{2x/3}TiO_3$ for $0 \leq x \leq 0.8$. The $Ca_{1-x}Sm_{2x/3}TiO_3$ ($x=0$–0.8) was calcined at 1250°C/3 h and sintered at 1450°C/3 h followed by annealing at 1200°C/24 h in an oxygen atmosphere to prevent Ti^{4+} reduction. A single perovskite phase with the $CaTiO_3$ type structure was obtained for $x=0$–0.6 as evidenced by the X-ray diffraction. The Sm^{3+} substitution decreased the tolerance factor from 0.966 to 0.822 affecting the stability of the perovskite phase. The relative density and Qf increased up to $x=0.6$ and then decreased because of the formation of secondary phases (Figure 6.10). The symmetry changed from orthorhombic ($x=0$) to tetragonal at $x=0.6$. Substitution of Sm^{3+} for Ca introduced A site vacancies and increased the quality factor. For $x>0.6$ secondary phases of $Sm_2Ti_2O_7$ appeared which lowered the quality factor. The permittivity gradually decreased with increasing Sm substitution (Figure 6.11). The far infrared reflectivity data showed that Sm substitution at the Ca^{2+} site affected the modes contributing to the dielectric loss and dielectric function and resulted in a decrease of dielectric loss and permittivity. The dielectric loss of these ceramics calculated from the reflectivity data was in agreement with those measured by conventional microwave methods.

The high positive τ_f of $Ca_{1-x}Ln_{2/x}TiO_3$ can be tailored [48–51, 61] by adding $Li_{1/2}Ln_{1/2}TiO_3$. Yoon et al. [50] studied $Ca_{1-x}Sm_{2/x}TiO_3$–$Li_{1/2}Ln_{1/2}TiO_3$ (Ln = Sm,Nd) as a function of the amount of Sm substitution ($x=0.0$–1.00). The $Li_{1/2}Sm_{1/2}TiO_3$ is orthorhombic with $\tau_f = -260$ ppm/°C and $Li_{1/2}Nd_{1/2}TiO_3$ is cubic with $\tau_f = -310$ ppm/°C [39]. Hence Yoon et al. [50] tailored the high τ_f of $Ca_{1-x}Ln_{2x/3}TiO_3$ (Ln = Sm,Nd) by forming solid solutions with $Li_{1/2}Ln_{1/2}TiO_3$ (Ln = Nd, Sm) samples by sintering at 1300°C/3 h. The ε_r and τ_f decreased with increase in x. The τ_f decreased

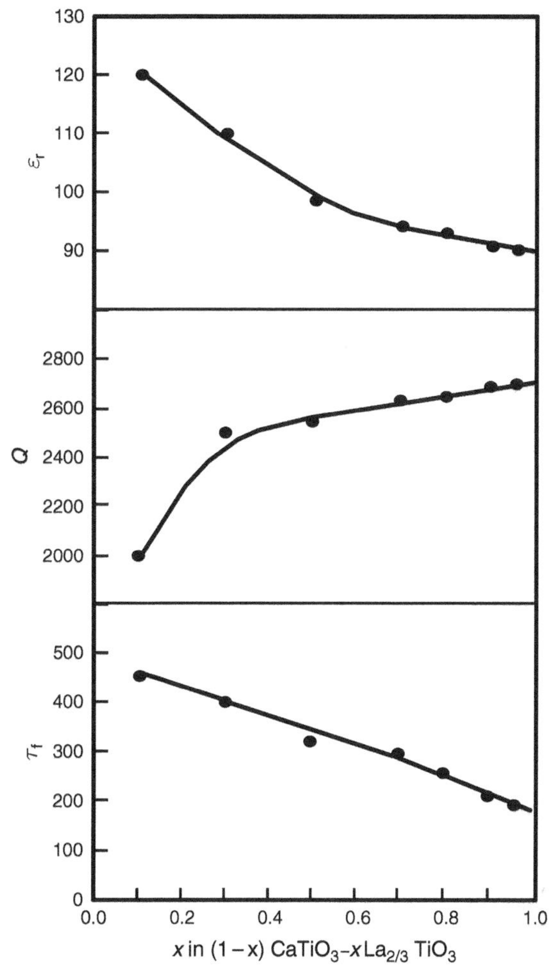

Figure 6.9 Variation of ε_r, Qf and τ_f of $(1-x)CaTiO_3-xLa_{2/3}TiO_3$ as a function of x (after Ref. [66]).

linearly and became negative with increase in x as shown in Figure 6.12. Figure 6.13 shows the variation of Qf with composition (x). The Qf of $Ca_{1-x}Sm_{2x/3}TiO_3$–$Li_{1/2}Nd_{1/2}TiO_3$ increased with x for the entire range. However, the Qf increased for $Ca_{1-x}Sm_{2x/3}TiO_3$–$Li_{1/2}Sm_{1/2}TiO_3$ up to $x=0.6$ which is the solid solution limit and then decreased. The τ_f of $Ca_{1-x}Nd_{2x/3}TiO_3$ can also be tailored by substituting $(Mg_{1/3}\ Nb_{2/3})^{4+}$ for Ti^{4+} [26]. The $(Ca_{0.85}Nd_{0.1})[(Mg_{1/3}Nb_{2/3})_xTi_{1-x}]O_3$ formed a single-phase orthorhombic perovskite solid solution when sintered at 1400°C/3 h with $\varepsilon_r=54$, $Qf=7600$ GHz and $\tau_f=1$ ppm/°C for $x=0.5$.

Several authors [37, 50, 53, 69, 159, 160] studied $CaTiO_3-Li_xLn_{1-x}TiO_3$ (Ln = Sm,Nd) ceramics because of its interesting dielectric properties. Both the end members in the system are distorted perovskites with high ε_r and opposite τ_f values. Takahashi et al. studied [53] CaO–Li_2O–$(1-x)Sm_2O_3$–xLn_2O_3–TiO_2 and reported the effect of

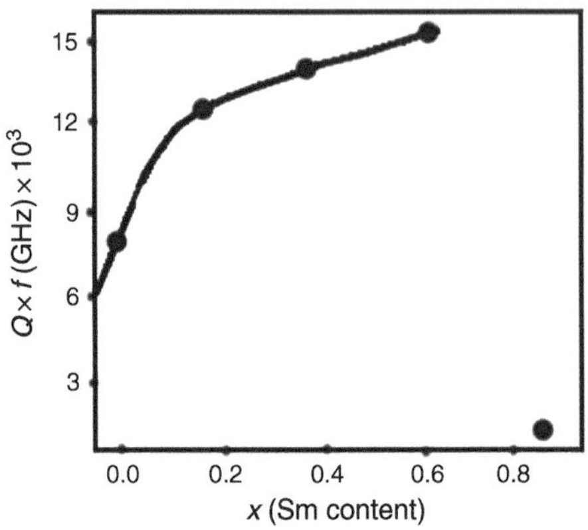

Figure 6.10 Variation of Qf of $Ca_{1-x}Sm_{2x/3}TiO_3$ as a function of x sintered at 1450°C/3 h. The samples were annealed in oxygen at 1200°C/24 h (after Ref. [83]).

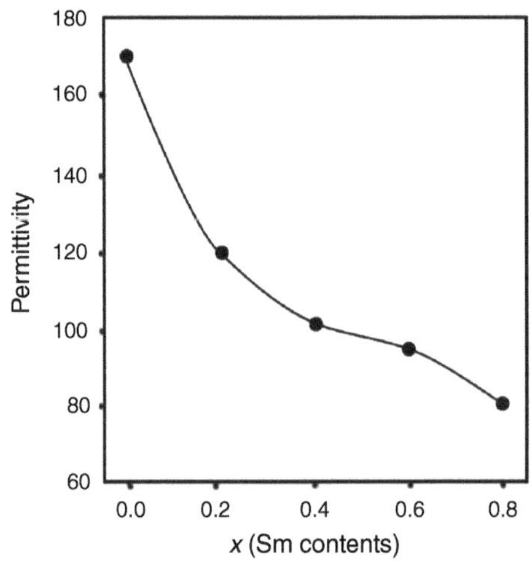

Figure 6.11 Variation of ε_r of $Ca_{1-x}Sm_{2x/3}TiO_3$ as a function of x sintered at 1450°C/3 h. The samples were annealed in oxygen at 1200°C/24 h (after Ref. [83]).

replacing Sm with other lanthanides. The ε_r increases linearly with increase in ionic radius as shown in Figure 6.14 whereas the Qf decreases linearly with ionic radius except in the case of Yb. On varying x to 1, the ε_r increased in the case of La, Nd, Pr whereas it decreased for Dy. When x increased secondary phases also appeared which leads to sudden decrease in quality factor. Lowe et al. [69, 159] reported that addition of Bi_2O_3 is beneficial in improving the properties of $CaTiO_3$–$Li_{1/2}Nd_{1/2}Ti_{1/2}O_3$ based ceramics. The $0.2CaTiO_3$–$0.8Li_{1/2}Nd_{1/2}TiO_3$ + 5 wt% $Bi_2Ti_2O_7$ showed $\varepsilon_r = 130$, $Qf = 2400$ GHz and $\tau_f = 20$ ppm/°C. Zhao and co-workers [26, 27] substituted $(Li_{1/2}Nd_{1/2})^{2+}$ in to the A site and $(Mg_{1/3}Ta_{2/3})^{4+}$ at the B site of $CaTiO_3$. The substitutions improved their dielectric properties. They prepared $Ca_{1-x}(Li_{1/2}Nd_{1/2})_xTiO_3$ ($x = 0.1$, 0.3, 0.6, 0.7, 0.9) and $Ca[(Mg_{1/3}Ta_{2/3})_yTi_{1-y}]O_3$ ($y = 0.2$, 0.4, 0.6, 0.8) by sintering them in the temperature

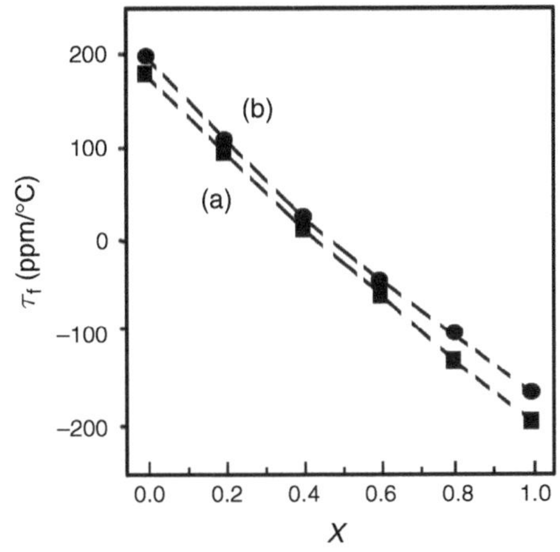

Figure 6.12 Variation of τ_f as a function of x in $Ca_{1-x}Sm_{2x/3}TiO_3-Li_{1/2}Ln_{1/2}TiO_3$ sintered at 1300°C/3 h: (a) Ln = Sm and (b) Nd (after Ref. [50]).

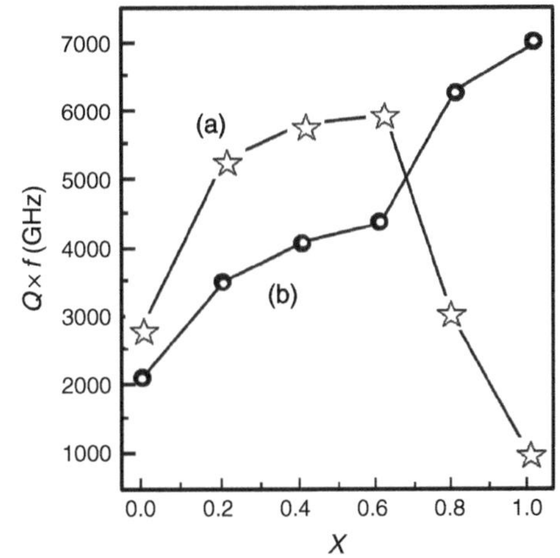

Figure 6.13 Variation of Qf as a function of x in $Ca_{1-x}Sm_{2x/3}TiO_3-Li_{1/2}Ln_{1/2}TiO_3$ sintered at 1300°C/3 h: (a)Ln = Sm and (b) Nd (after Ref. [50]).

range 1300–1400°C. X-ray diffraction study revealed the formation of solid solutions for both systems without the formation of any secondary phase. The ε_r, Qf and τ_f decreased with increase of x. The composition $Ca_{0.4}(Li_{1/2}Nd_{1/2})_{0.6}TiO_3$ has $\varepsilon_r = 113$, $Qf = 5000$ GHz and $\tau_f = 8$ ppm/°C. In the case of $(Mg_{1/3}Ta_{2/3})^{4+}$ substitution for Ti^{4+}, the ε_r and τ_f decreased and Qf increased with increase in y. The composition $Ca(Mg_{1/3}Ta_{2/3})_{0.6}Ti_{0.4}O_3$ has $\varepsilon_r = 60$, $Qf = 36\,900$ GHz and $\tau_f = -10$ ppm/°C. Kucheiko et al. [30] and Levin et al. [161] used the negative τ_f of $Ca(Al_{1/2}Ta_{1/2})O_3$ and $Ca(Al_{1/2}Nb_{1/2})O_3$ to lower the τ_f of $CaTiO_3$. They prepared $CaTiO_3-Ca(Al_{1/2}B_{1/2})O_3$ [$CaTi_{1-x}(Al_{1/2}B_{1/2})_xO_3$, B = Ta, Nb: $x = 0.3-0.5$] which has a perovskite structure. The partial substitution of Ti^{4+} by

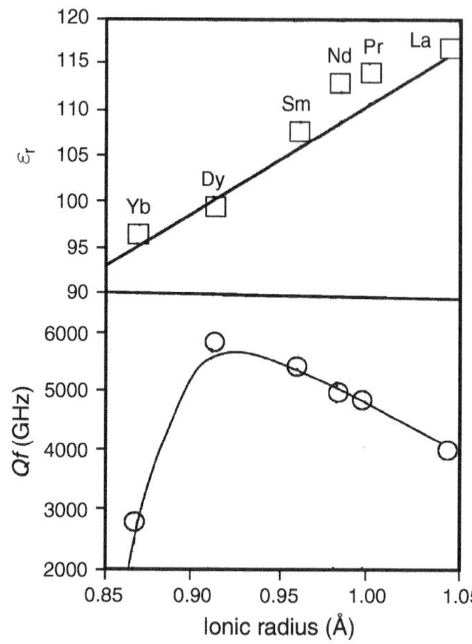

Figure 6.14 Variation of ε_r and Qf with Ln ionic radii in $CaO–SrO–Li_2O–Sm_2O_3–Ln_2O_3–TiO_2$ system (after Ref. [53]).

Al^{+}/Ta^{5+} improved the quality factor. The τ_f decreased and reached close to zero for $x \sim 0.5$ and then became negative with further increase in x. Addition of 1 wt% of Li_3NbO_4 lowers the sintering temperature from 1500°C to 1300°C with an improvement in quality factor.

6.7 $LnAlO_3$

It was reported [162, 163] that $LaAlO_3$ has a very low dielectric loss and is a suitable material as substrate for YBCO superconductor. Cho et al. [58] prepared rare earth aluminate ($LnAlO_3$ Ln = Dy, Er, Gd, La, Nd, Pr, Sm, Y) by calcining at 1400°C/2 h and sintering at 1650°C. The $LnAlO_3$ with Ln = La, Pr and Nd have a rhombohedral symmetry and those with Ln = Y, Er, Ho, Dy, Nd and Sm have an orthorhombic symmetry. Figure 6.15 shows the relative permittivities of $LnAlO_3$ as a function of their tolerance factor (t). The permittivity increases with increase in t. The microwave dielectric properties of $LnAlO_3$ are given in Table 6.1. Huang and co-workers [55, 56] lowered the sintering temperature of $LnAlO_3$ by the addition of CuO and V_2O_5. The $LaAlO_3$ sintered with 0.25 wt% CuO at 1460°C showed $\varepsilon_r = 20.7$, $Qf = 48\,000$ GHz and $\tau_f = -80$ ppm/°C. But addition of more than 1 wt% V_2O_5 or 0.5 wt% CuO leads to formation of secondary phases of $NdAl_{11}O_{18}$ and $Nd_4Al_2O_9$ which considerably degraded the quality factor. Several authors studied the microwave dielectric properties of $LaAlO_3$ single crystals [163–166] and reported that the quality factor of single crystals of $LaAlO_3$ are higher by an order of magnitude as compared to $LaAlO_3$ ceramics.

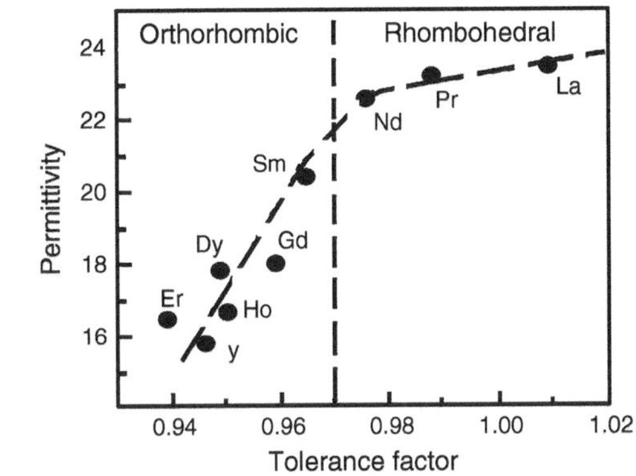

Figure 6.15 Variation of ε_r as a function of tolerance factor t in $LnAlO_3$ (after Ref. [58]).

The high negative τ_f of $LaAlO_3$ and $NdAlO_3$ can be compensated [59, 60, 62, 64, 74, 76, 156, 167–174] by the addition of TiO_2 or solid solution formation with $SrTiO_3$ and $CaTiO_3$. Cho et al. [167] prepared $(1-x)LaAlO_3-xSrTiO_3$ ($x=0$, 0.2, 0.4, 0.6, 0.8) by sintering at 1550–1650°C/2 h. X-ray diffraction pattern as shown in Figure 6.16 indicates that $LaAlO_3$ and $SrTiO_3$ combine to form a solid solution. Figure 6.17 shows the variation of ε_r, Qf and τ_f as a function of composition x. The variations in the dielectric properties are not linear. X-ray diffraction showed that $LaAlO_3$ and $SrTiO_3$ form a complete solid solution but the permittivity and τ_f values exhibited non-monotonic mixture-like behavior. The dielectric properties were dominated by the properties of $LaAlO_3$ end member up to ($x=0.6$) 60 mol% of $SrTiO_3$. The addition of TiO_2 to $LaAlO_3$ also showed [157] a non-monotonic behavior as shown in Figure 6.18 with

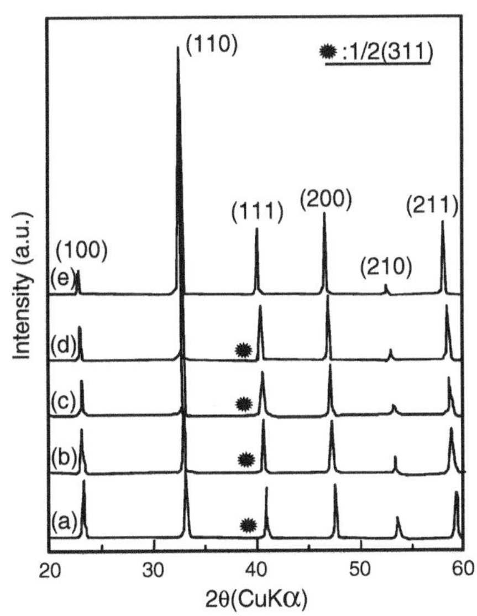

Figure 6.16 X-ray diffraction pattern of $(1-x)LaAlO_3-xSrTiO_3$ ceramics: (a) $x=0$, (b) $x=0.2$, (c) $x=0.4$, (d) $x=0.6$ and (e) $x=0.8$ (after Ref. [167]).

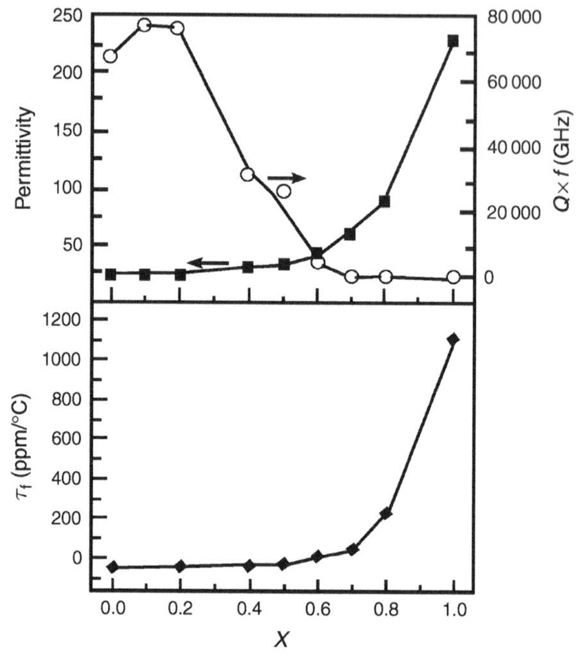

Figure 6.17 Variation of the microwave dielectric properties of (1–x)LaAlO$_3$–xSrTiO$_3$ ceramics (after Ref. [167]).

abrupt change at $x = 0.6$. The $\varepsilon_r = 37$, $Qf = 37\,000$ GHz and $\tau_f = -1$ ppm/°C for $x = 0.5$. The X-ray diffraction study of samples shows presence of secondary phases such as $LaAl_{11}O_{18}$, $LaTi_2Al_9O_{17}$, $La_4Ti_9O_{24}$ and rutile. It was reported that addition of B_2O_3 [60, 75] in $LaAlO_3$–$SrTiO_3$ lowers the sintering temperature and improves the microwave dielectric properties. The ε_r and Qf decreased with increasing B_2O_3 content. 0.5$LaAlO_3$–0.5$SrTiO_3$ with 0.25 wt% B_2O_3 and sintered at 1430°C had $\varepsilon_r = 34.5$, $Qf = 43\,200$ GHz and $\tau_f = -11$ ppm/°C.

The high negative τ_f of $LaAlO_3$ has also been compensated by forming a solid solution with $CaTiO_3$ [62, 77, 169]. Moon et al. [77] prepared $(1-x)CaTiO_3$–$xLaAlO_3$ (CTLA) by sintering at 1500–1650°C/3 h. As x increases the following phase transformation occurred orthorhombic ($x \leq 0.4$) to pseudocubic ($x = 0.5$) to rhombohedral ($x \geq 0.6$). Khalyavin et al. [172] reported a composition induced structural transition in $(1-x)CaTiO_3$–$xLaAlO_3$ ceramics. They reported that with increasing amount of $LaAlO_3$ the octahedra tilts and the space group changes from *Pnma* to *Inma* between $x = 0.4$ and 0.5 and then from *Inma* to *R3c* at x between 0.5 and 0.6. Moon et al. [77] lowered sintering temperature of CTLA from 1650°C to 1450°C by the addition of Bi_2O_3 with Al_2O_3 or NiO but this reduced the quality factor. Ravi et al. [170] reported that addition of a small amount of Al_2O_3 and slow cooling improves the quality factor. The $Ca_{0.7}Ti_{0.7}La_{0.3}Al_{0.3}O_3$ has $\varepsilon_r = 46$, $Qf = 38\,300$ GHz and $\tau_f = 12$ ppm/°C. They found that addition of a small amount (0.25 wt%) of Al_2O_3 and sintering at 1500°C followed by a slow cooling rate of 5°C/h increased the grain size and density. Figure 6.19 represents a typical microstructure of 0.25 wt% Al_2O_3 added CTLA. X-ray diffraction

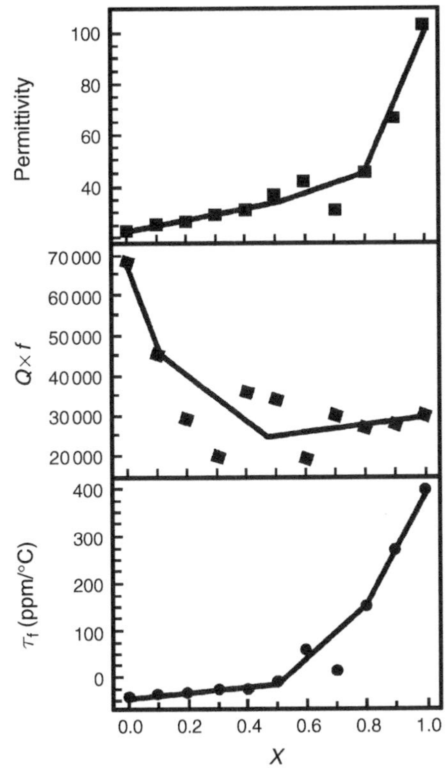

Figure 6.18 Variation of the microwave dielectric properties of $(1-x)LaAlO_3-xTiO_2$ ceramics as a function of composition x (after Ref. [157]).

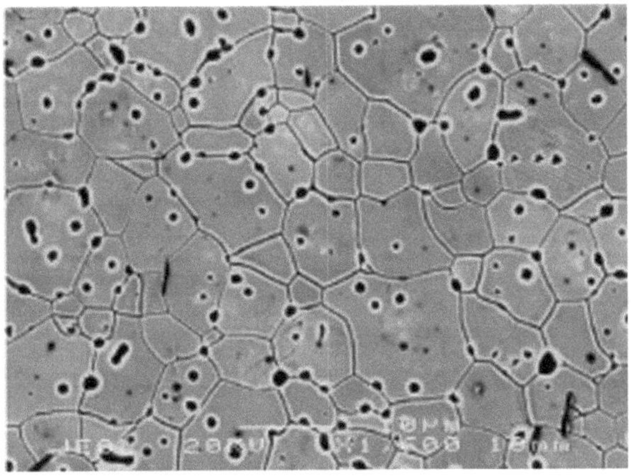

Figure 6.19 Surface morphology of $CaTiO_3-LaAlO_3 + 0.25$ wt% Al_2O_3 ceramics (after Ref. [170]).

(a) (b) (c)

Figure 6.20 Selected are diffraction patterns recorded from different domains: (a) [1$\bar{1}$0] zone axis SAD of a single domain, (b) [001] zone axis SAD of twinned domain and (c) multidomain diffraction pattern (after Ref. [170]).

study showed that $Ca_{0.7}Ti_{0.7}La_{0.3}Al_{0.3}O_3 + 0.25$ wt% of Al_2O_3 is orthorhombic with *Pbnm* space group. They found [170] that ceramics contain (112) and (110) twins and antiphase domain boundaries which are formed due to the high temperature ordered to the low temperature disordered phase transition. Figure 6.20 shows the selected area diffraction pattern recorded from different domains showing (a) untwinned domains, (b) twinned domains and (c) multiple domains.

The high negative τ_f of $NdAlO_3$ (−33 ppm/°C) can also be compensated by forming a solid solution with $CaTiO_3$. The (1−x)$CaTiO_3$–$x$$NdAlO_3$ [CTNA] is prepared [59, 62, 71, 173] by mixing $CaCO_3$, TiO_3, Al_2O_3, Nd_2O_3 and calcining at about 1350°C followed by sintering at 1450–1600°C. X-ray diffraction study showed that $CaTiO_3$ and $NdAlO_3$ form a solid solution across the whole compositional range ($0 < x < 1$). Because of the similar ionic radii between Ca^{2+} (1.34 Å, coordination No. 12) and Nd^{3+} (1.27 Å, coordination No. 6) and between Ti^{4+} (0.605 Å) coordination No. 6) and Al^{3+} (0.535 Å, coordination No. 6) [174] the most probable mechanism for solid solution formation is the substitution of Nd on the Ca(A) site and Ti^{4+} on the B (Al)-site in the perovskite structure. The solid solution can be represented as $Ca_{1-x}Nd_xTi_{1-x}Al_xO_3$. Figure 6.21 shows the variation of ε_r, Qf and τ_f as a function of x. As x increases from 0 to 0.3, τ_f decreases from 800 ppm/°C to about 0 ppm/°C. Partial substitution of Ca^{2+} and Ti^{4+} by Nd^{3+} and Al^{3+} causes a significant decrease in the high positive τ_f of $CaTiO_3$. The ε_r and τ_f initially decrease sharply with x and the quality factor increase for $x < 0.3$. As x increases further, the decrease in ε_r and τ_f become slow and reaches the value of $NdAlO_3$. The $Ca_{0.7}Nd_{0.3}Ti_{0.7}Al_{0.3}O_3$ shows temperature-stable resonant frequency with $Qf = 33\,000$ GHz, $\varepsilon_r = 44$ and $\tau_f = 0$ ppm/°C [65]. Suvorov et al. [62, 173] reported from SEM, TEM, EDAX study that the heating conditions during sintering and subsequent cooling strongly affect the microstructural development of CTNA. Defects such as dislocations, twin or antiphase boundaries degrade the quality factor. Jancar et al. [174] reported that $CaAl_{12}O_{19}$ is formed as an intermediate phase during preparation which degrades the dielectric properties in $0.7CaTiO_3$–$0.3NdAlO_3$. In order to minimize the amount of $CaAl_{12}O_{19}$, prolonged calcinations and intermediate homogenizations at temperatures above 1300°C are required. Kipoech et al. [168] determined the crystal structure of $Ca_{0.7}Nd_{0.3}Ti_{0.7}Al_{0.3}O_3$ using synchrotron X-ray diffraction. They proposed an othorhombic perovskite lattice with space group *Pbnm* and tilted oxygen octahedra surrounding the Al/Ti atoms. The

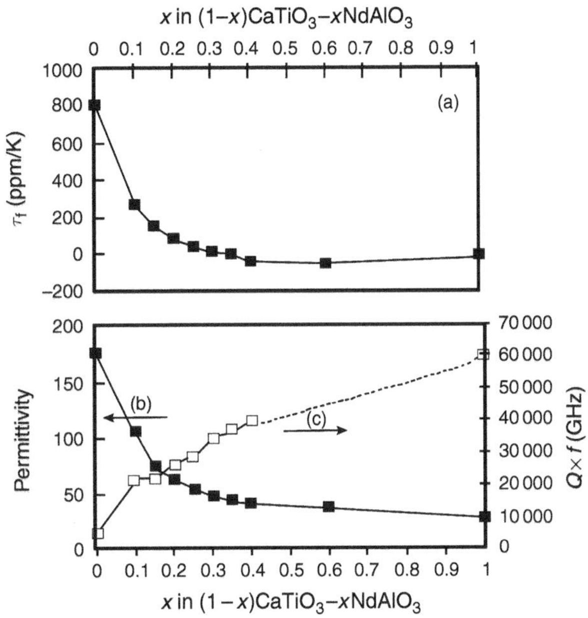

Figure 6.21 Variation of ε_r, Qf and τ_f of (1–x)$CaTiO_3$–$x$$NdAlO_3$ as a function of x (after Ref. [65]).

oxygen octahedra tilted both in phase and antiphase. Figure 6.22 shows the structure of $CaTiO_3$–$NdAlO_3$ solid solution viewed along the c-axis. Zheng et al. [175] reported from Raman spectroscopic study that Al^{3+} and Ti^{4+} are distributed in a non-random way on the B site.

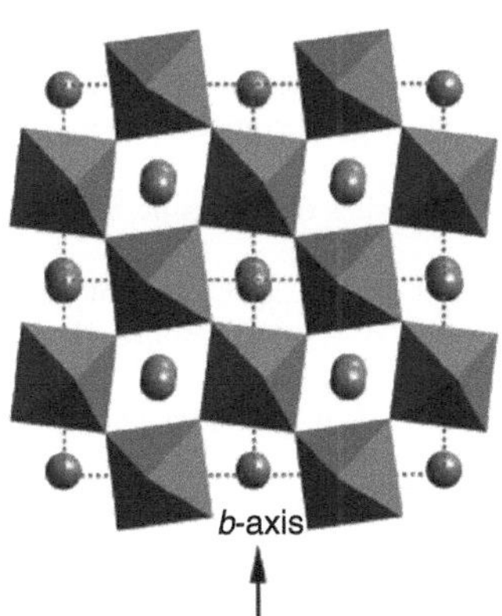

Figure 6.22 The structure of $Ca_{0.7}Nd_{0.3}Al_{0.3}Ti_{0.7}O_3$ viewed along the c-axis. The octahedral represents Al/TiO_6 molecules and the spheres either Ca or Nd cations (after Ref. [168]).

6.8 Conclusions

The ABO_3 perovskite materials represents one of the most important family of materials used in the electronic industry. The $CaTiO_3$ has a high relative permittivity (ε_r) and low loss but its τ_f is very high, at +850 ppm/°C. It is possible to obtain temperature-stable ceramics by adding $CaTiO_3$ with negative τ_f materials such as $NdAlO_3$, $LaGaO_3$ and $LaAlO_3$. The $0.7CaTiO_3$–$0.3NdAlO_3$, $0.6CaTiO_3$–$0.4LaGaO_3$ and $0.66CaTiO_3$–$0.64LaAlO_3$ are three of the temperature-stable materials with high ε_r and high quality factor >35 000 GHz useful for applications in base station devices. The $Ag(Nb_{1-x}Ta_x)O_3$ has a high dielectric constant ε_r of about 400 and a quality factor of about 700 GHz. The $Ag(Nb_{0.65}Ta_{0.35})O_3$ and $Ag(Nb_{0.35}Ta_{0.35})O_3$ have opposite temperature dependence of ε_r. By preparing a mixture of 45 wt% $Ag(Nb_{0.65}Ta_{0.35})O_3$ and 55 wt% $Ag(Nb_{0.35}Ta_{0.65})O_3$ and controlling the microstructure by using coarse grains it is possible to get nearly temperature-stable ceramics. Partial substitution of Ag by Li leads to a useful ferroelectric material with excellent piezoelectric properties. The $Ag(Nb_{1-x}Ta_x)O_3$ are also useful materials for tunable microwave devices.

References

[1] A. Navrotsky and D. J. Weidner. Perovskite: A Structure of Great Interest to Geophysics and Materials Science. A. Navrotsky and D. J. Weidener (Eds), American Geophysics Union, Washington, DC (1989) Vol. 45, p. xi.
[2] H. D. Megaw. Crystal structure of double oxides of the perovskite type. Proc. Phys. Soc. 58(1946)133–152.
[3] F. S. Galasso. Structure Properties and Preparation of Perovskite Type Compounds. Pergamon Press, Oxford (1969) pp. 3–41.
[4] R. Roy. Multiple ion substitution in perovskite lattice. J. Am. Ceram. Soc. 37(1954)581–588.
[5] J. B. Goodenough and J. M. Longo.Crystallographic and Magnetic Properties of Perovskite and Perovskite Related Compounds. Landolt-Bornstein New Series group III, Springer Verlag, New York (1970) Vol. 4a, p. 126.
[6] M. T. Anderson, K. B. Greenwood, G. A. Taylor, and K. R. Poepplmeir. B cation arrangements in double perovskites. Prog. Solid State Chem. 22(1993)197–233.
[7] F. S. Galasso. Perovskites and High Temperature Superconductors. Gordon & Breach Science Publishers (1990) pp. 3–55.
[8] F. S. Galasso. Structure and Properties of Inorganic Solids. Pergamon Press, Oxford (1970) pp. 163–204.
[9] V. M. Goldschmidt. Skrifer Norsk Videnskapes-Akadem Oslo.I. Mater. Nat. K1. No.8 (1926).
[10] H. D. Megaw. Crystal Structures: A Working Approach. WB Saunders Company, London (1973) p. 293.
[11] A. M. Glazer. The classification of tilted octahedral perovskites. Acta Crystallogr. B 28(1972) 3384–3392.
[12] A. M. Glazer. Simple ways of determining perovskite structures. Acta Crystallogr. A 31(1975) 756–762.
[13] E. L. Colla, I. M. Reaney, and N. Setter. Effect of structural changes in complex perovskite on the temperature coefficient of relative permittivity. J. Appl. Phys. 74(1993)3414–3425.
[14] I. M. Reaney, E. L. Colla, and N. Setter. Dielectric and structural characteristics of Ba and Sr based complex perovskites as a function of tolerance factor. Jpn. J. Appl. Phys. 33(1994) 3984–3990.
[15] L. Q. Jiang, J. K. Guo, H. B. Liu, M. Zhu, X. Zhou, P. Wu, and C. H. Li. Prediction of lattice constant in cubic perovskites. J. Phys. Chem. Sol. 67(2006)1531–1536.

[16] R. L. Moreira and A. Dias. Comment on prediction of lattice constant in cubic perovskites. J. Phys. Chem. Solids. 68(2007)1617–1622.
[17] R. Ubic. Revised method for the prediction of lattice constants in cubic and psuedocubic perovskites. J. Am. Ceram. Soc. 90(2007)3326–3330.
[18] A. Feteira, D. C. Sinclair, I. M. Reaney, and M. T. Lanagan. Structure property relationship of $BaTi_{1-2y}Ga_yNb_yO_3$ ($0 \leq y \leq 0.3$) ceramics. J. Am. Ceram. Soc. 88(2005)3055–3062.
[19] A. Feteira, K. Sarma, N. Mc, N. Alford, I. M. Reaney, and D. C. Sinclair. Microwave dielectric properties of Ga doped hexagonal barium titanate ceramics. J. Am. Ceram. Soc. 86(2003)511–513.
[20] G. M. Keith, M. J. Pampling, K. Sarma, N. Mc, N. Alford, and D. C. Sinclair. Synthesis and characterization of doped 6H $BaTiO_3$ ceramics. J. Eur. Ceram. Soc. 24(2004)1721–1724.
[21] A. Feteira, R. Elsebrock, A. Dias, R. L. Moreira, M. T. Lanagan, and D. C. Sinclair. Synthesis and characterisation of $La_{0.4}Ba_{0.6}Ti_{0.6}RE_{0.4}O_3$ [RE = Y,Yb] ceramics. J. Eur. Ceram. Soc. 26(2006)1947–1951.
[22] H. Zheng, G. D. C. de Gyorgyfalva, and I. M. Reaney. Microstructure and microwave properties of $CaTiO_3$–$LaGaO_3$ solid solutions. J. Mater. Sci. 40(2005)5207–5214.
[23] P. L. Wise, I. M. Reaney, W. E. Lee, T. J. Price, D. M. Iddles, and D. S. Cannell. Structure-microwave property relations in $(Sr_xCa_{1-x})_{n+1}Ti_nO_{3n+1}$. J. Eur. Ceram. Soc. 21(2001) 1723–1726.
[24] Y. Konishi. Novel dielectric wave guide components – microwave applications of new ceramic materials. Proc. IEEE 79(1991)726–740.
[25] V. Sivasubramanyan, V. R. K. Murthy, and B. Viswanathan. Microwave dielectric properties of certain simple alkaline earth perovskite compounds as a function of tolerance factor. Jpn. J. Appl. Phys. 36(1997)194–197.
[26] F. Zhao, X. Yue, Y. Zhang, Z. Gui, and L. Li. Microstructure and microwave dielectric properties of $Ca[Ti_{1-x}(Mg_{1/3}Nb_{2/3})_x]O_3$ ceramics. J. Eur. Ceram. Soc. 25(2005)3347–3352.
[27] T. Liu, X.-Z. Zhao, and W. Chen. A/B site modified $CaTiO_3$ dielectric ceramics for microwave applications. J. Am. Ceram. Soc. 89(2006)1153–1155.
[28] E. S. Kim, S. H. Kim, and K. H. Yoon. Microwave dielectric properties of $[Ca_{0.85}Nd_{0.1}][Ti_{1-x}(Mg_{0.33}Nb_{0.67})_x]O_3$ ceramics. J. Ceram. Soc. Jpn. 112–1(2004)S1641–S1644.
[29] S. Kucheiko, D. H. Yeo, J.-W. Choi, S.-J. Yoon, and H.-J. Kim. Microwave dielectric properties of $CaTiO_3$–$CaAl_{1/2}Nb_{1/2}O_3$ ceramics doped with Li_3NbO_4. J. Am. Ceram. Soc. 85(2002)1327–1329.
[30] S. Kucheiko, J.-W. Choi, H.-J. Kim, and H.-J. Jung. Microwave dielectric properties of $CaTiO_3$–$Ca(Al_{1/2}Ta_{1/2})O_3$ ceramics. J. Am. Ceram. Soc. 79(1996)2739–2743.
[31] J. Joseph, T. M. Vimala, K. C. J. Raju, and V. R. K. Murthy. Structural investigations on $(Ba_xSr_{1-x})(Zr_yTi_{1-y+\delta}Ta_\delta)O_3$ dielectric resonator used for microwave applications. Jpn. J. Appl. Phys. 35(1996)179–181.
[32] K. Wakino. Recent development of dielectric resonator materials and filters in Japan. Ferroelectrics 91(1989)69–86.
[33] P. L. Wise, I. M. Reaney, Y. Ubic, and W. E. Lee. Periodicity and its relation to microwave dielectric properties in the series $Sr_{0.8}Ca_{0.2}TiO_3$, $Sr_{1.6}Ca_{0.4}TiO_4$ and $Sr_{2.4}Ca_{0.6}Ti_2O_7$. Inst of Physics Conf. Ser. No 161. Section 4, paper presented at Electron microscopy and Analysis group, Sheffield (1999)161–164.
[34] L. Wu, Y.-C. Chen, L. J. Chen, Y.-P. Chou, and Y.-J. Tsai. Preparation and microwave characterisation of $Ba_{1-x}Sr_xTiO_3$ ceramics. Jpn. J. Appl. Phys. 38(1990)5612–5615.
[35] J. Kato.Material produces small resonators with high dielectric constant. JEE Sept (1991)114–118.
[36] C.-L. Huang, and M.-H. Weng. The effect of PbO loss on microwave dielectric properties of $(Pb,Ca)(Zr,Ti)O_3$ ceramics. Mater. Res. Bull. 36(2001)683–691.
[37] J.-S. Kim, C. I. Cheon, H.-J. Kang, C.-H. Lee, K.-Y. Kim, S. Nahm, and J.-D. Byun. Crystal structure and microwave dielectric properties of $CaTiO_3$–$(Li_{1/2}Nd_{1/2})TiO_3$–$(Ln_{1/3}Nd_{1/3})TiO_3$(Ln = La, Dy) ceramics. Jpn. J. Appl. Phys. 38(1999)5633–5637.
[38] E. S. Kim and K. H. Yoon. Microwave dielectric properties of $(1-x)CaTiO_3$-$xLi_{1/2}Sm_{1/2}TiO_3$ ceramics. J. Eur. Ceram. Soc. 23(2003)2397–2401.

[39] H. Takahashi, Y. Baba, and K. Ezaki. Dielectric characteristics of $A^{1+}{}_{1/2}A^{3+}{}_{1/2}TiO_3$ ceramics at microwave frequencies. Jpn J. Appl. Phys. 30(1991)2339–2342.

[40] K. Ezaki, Y. Baba, H. Takahashi, and K. Shibata. Microwave dielectric properties of $CaO–Li_2O–Sm_2O_3–TiO_2$ ceramics. Jpn. J. Appl. Phys. 32(1993)4319–4322.

[41] A. G. Belous and O. V. Ovchar. Temperature compensated microwave dielectrics based on lithium containing titanates. J. Eur. Ceram. Soc. 23(2003)2525–2528.

[42] H.-L. Chen and C.-L. Huang. Microwave dielectric properties and microstructures of $Ca_{1-x}Nd_{2x/3}TiO_3–Li_{1/2}Nd_{1/2}TiO_3$ ceramics. Jpn. J. Appl. Phys. 41(2002)5650–5653.

[43] N. Ichinose and N. Chida. Microwave dielectric properties of $(Li_{1/2}Nd_{1/2})TiO_3–(Na_{1/2}Ln_{1/2})TiO_3$ (Ln = La, Nd, Sm) ceramic system. Applications of Ferroelectrics 1998. Proc. IEEE Int. Symp. (1998)513–514.

[44] N. Ichinose and K. Mutoh. Microwave dielectric properties in the $(1-x)(Na_{1/2}La_{1/2})TiO_3–x(Li_{1/2}Sm_{1/2})TiO_3$ ceramic system. J. Eur. Ceram. Soc. 23(2003)2455–2459.

[45] H.-J. Kim, S. Kucheiko, S.-J. Yoon, and H.-J. Jung. Microwave dielectrics in the $(La_{1/2}Na_{1/2})TiO_3–Ca(Fe_{1/2}Nb_{1/2})O_3$ system. J. Am. Ceram. Soc. 80(1997)1316–1318.

[46] D. Houivet, J. El. Fallah, J. Bernad, F. Roulland, and J. M. Haussonne.Microwave properties and microstuctures of $La_{2/3}TiO_3$ stabilised with NiO. J. Eur. Ceram. Soc. 21(2001) 1715–1718.

[47] D. Suvorov, M. Valant, S. Skapin, and D. Kolar. Microwave dielectric properties of ceramics with compositions along the $La_{2/3}TiO_3–LaAlO_3$ tie-line. J. Mater. Sci. 33(1998)85–89.

[48] G. Huang, D. Zhou, J. Yu, Z. Zheng, and S. Gong. Synthesis of $(1-x)Ca_{2/5}Sm_{2/5}TiO_3–xLi_{1/2}Nd_{1/2}TiO_3$ ceramic powder via ethylenediaminetetracacetioc acid precursor. Mater. Res. Bull. 40(2005)13–19.

[49] W. S. Kim, K. H. Yoon, and E. S. Kim. Microwave dielectric characteristics of the $Ca_{2/5}Sm_{2/5}TiO_3–Li_{1/2}Nd_{1/2}TiO_3$ ceramics. Jpn. J. Appl. Phys. 39(2000)5650–5653.

[50] K. H. Yoon, Y. H. Chang, W. S. Kim, J. B. Kim, and E. S. Kim. Dielectric properties of $Ca_{1-x}Sm_{2x/3}TiO_3–Li_{1/2}Ln_{1/2}TiO_3$ ceramics. Jpn. J. Appl. Phys. 35(1996)5145–5149.

[51] W. S. Kim, E. S. Kim, and K. H. Yoon. Effect of excess TiO_2 on the microwave dielectric properties of the $Ca_{2/5}Sm_{2/5}TiO_3–Li_{1/2}Ln_{1/2}TiO_3$ (Ln = Sm,Nd) ceramics. Ferroelectrics. 223(1999)277–284.

[52] K. H. Yoon, M. S. Park, J. Y. Cho, and E. S. Kim. Effect of $B_2O_3–Li_2O$ on microwave dielectric properties of $Ca_{0.275}Sm_{0.4}Li_{0.25})TiO_3$ ceramics. J. Eur. Ceram. Soc. 23(2003)2423 –2427.

[53] H. Takahashi, Y. Baba, K. Ezaki, and K. Shibata. Microwave dielectric properties and crystal structure of $CaO–Li_2O–(1-x)Sm_2O_3–xLn_2O_3–TiO_2$ (Ln = lanthanide) ceramic system. Jpn. J. Appl. Phys. 35(1996)5069–5073.

[54] R. Rejini and M. T. Sebastian. Microwave dielectric properties of $La_6Mg_4B_2O_{24}$ [B = Ta,Nb] ceramics. J. Am. Ceram. Soc. 90(2007)2472–2475.

[55] C.-S. Hsu and C.-L. Huang. Effect of CuO additive on sintering and microwave dielectric behaviour of $LaAlO_3$ ceramics. Mater. Res. Bull. 36(2001)1939–1947.

[56] C.-L. Huang and Y.-C. Chen. Influence of V_2O_5 additions to $NdAlO_3$ ceramics on sintering temperature and microwave dielectric properties. J. Eur. Ceram. Soc. 23(2003)167–173.

[57] C.-L. Huang, and Y.-C. Chen. Liquid phase sintering and microwave dielectric properties of $NdAlO_3$ ceramics. Jpn. J. Appl. Phys. 41(2002)1459–1464.

[58] S.-Y. Cho, I.-T. Kim, and K. S. Hong. Microwave dielectric properties and applications of rare earth aluminates. J. Mater. Res. 14(1999)114–119.

[59] B. Jancar, D. Suvrorov, and M. Valant. Microwave dielectric properties of $CaTiO_3–NdAlO_3$ ceramics. J. Mater. Sci. Lett. 20(2001)71–72.

[60] C.-L. Huang and C.-S. Hsu. Improved high Q value of $0.5LaAlO_3–0.5SrTiO_3$ microwave dielectric ceramics at low sintering temperatures. Mater. Res. Bull. 36(2001)2677–2687.

[61] C.-L. Huang, J.-T. Tsai, and Y.-B. Chen. Dielectric properties of $(1-y)Ca_{1-x}La_{2x/3}TiO_{3-y}(Li_{1/2}Nd_{1/2})TiO_3$ ceramic system at microwave frequency. Mater. Res. Bull. 36 (2001)547–558.

[62] D. Suvorov, M. Valant, B. Jancar, and S. D. Skapin. $CaTiO_3$-based ceramics: Microstructural development and dielectric-properties. Acta Chimi. Slov. 48(2001) 87–99.

[63] G. A. Rai, F. Azough, and R. Freer. Effect of Al_2O_3 on the structure and microwave dielectric properties of $Ca_{0.7}Ti_{0.7}La_{0.3}Al_{0.3}O_3$. J. Eur. Ceram. Soc. 27(2007) 2855–2860.

[64] A. Feteira, D. C. Sinclar, and M. T. Lanagan. Structure and microwave dielectric properties of $Ca_{1-x}Y_xTi_{1-x}Al_xO_3$ (CYTA) ceramics. J. Mater. Res. 20(2005)2391–2399.

[65] B. Jancar, D. Suvorov, M. Valant, and D. Drazic. Characterisation of $CaTiO_3$–$NdAlO_3$ dielectric ceramics. J. Eur. Ceram. Soc. 3(2003)1391–1400.

[66] I.-S. Kim, W. H. Jung, Y. Inaguma, T. Nakamura, and M. Itoh.Dielectric properties of A site deficient perovskite type lanthanum-calcium-titanium oxide solid solution system $(1-x)La_{2/3}TiO_3$–$x$$CaTiO_3$ $(0.1 \leq x \leq 0.96)$. Mater. Res. Bull. 30(1995)307–316.

[67] X. Hu, X. M. Chen, and S. Y. Wu. Preparation properties and characterisation of $CaTiO_3$ modified $Pb(Fe_{1/2}Nb_{1/2})O_3$ dielectrics. J. Eur. Ceram. Soc. 23(2003)1919–1924.

[68] T. Lowe, F. Azough, and R. Freer. The microwave dielectric properties of $CaTiO_3$–$Li_{0.5}Nd_{0.5}TiO_3$ ceramics prepared with $Bi_2Ti_2O_7$ additive. J. Eur. Ceram. Soc. 23(2003)2429–2433.

[69] T. Lowe, F. Azough, and R. Freer. Control of $0.2CaTiO_3$–$0.8Li_{0.5}Nd_{0.5}TiO_3$ microwave dielectric ceramics by the addition of $Bi_2Ti_2O_7$. J. Electroceram. 16(2006)257–261.

[70] H. Bagshaw, D. Iddles, R. Quimby, and I. M. Reaney. Structure property relations in $xCaTiO_3$–$(1-x)Sr(Mg_{1/3}Nb_{2/3})O_3$ based microwave dielectrics. J. Eur. Ceram. Soc. 23(2003)2435–2441.

[71] E. A. Nenesheva, L. P. Mudroliuba, and N. F. Kartenko. Microwave dielectric properties of ceramics based on $CaTiO_3$–$LnMO_3$ system (Ln = La, Nd; M = Al, Ga). J. Eur. Ceram. Soc. 23(2003)2443–2448.

[72] H. Zheng, G. D. C. Csette de Gyorgyfalva, and I. M. Reaney. Microstructure and microwave properties of $CaTiO_3$–$LaGaO_3$ solid solutions. J. Mater. Sci. 40(2005)5207–5214.

[73] M.-H. Kim, S. Nahm, C.-H. Choi, H.-J. Lee, and H.-M. Park. Dielectric properties of $(1-x)NdGaO_3$–$xCaTiO_3$ solid solution at microwave frequencies. Jpn. J. Appl. Phys. 41(2002)717–721.

[74] E. S. Kim, B. G. Chun, and D. H. Kang. Effects of structural characteristics on microwave dielectric properties of $(1-x)Ca_{0.85}Nd_{0.1}TiO_3$–$xLnAlO_3$ (Ln = Sm, Er, and Dy) ceramics. J. Eur. Ceram. Soc. 27(2007)3005–3010.

[75] C.-L. Huang and K.-H. Chiang. Dielectric properties of B_2O_3 doped $(1-x)LaAlO_3$–$xSrTiO_3$ ceramic system at microwave frequency range. Mater. Res. Bull. 37(2002)1941–1948.

[76] Y. Q. Liu and X. M. Chen. Dielectric and mechanical characteristics of lanthanum aluminate ceramics with strontium niobate addition. J. Eur. Ceram. Soc. 24(2004)1999–2004.

[77] J. H. Moon, H. M. Jung, H. S. Park, J. Y. Shin, and H. S. Kim. Sintering behaviour and microwave dielectric properties of (Ca,La)(Ti,Al)O_3 ceramics. Jpn. J. Appl. Phys. 38(1999) 6821–6827.

[78] M. Valant and D. Suvorov. New high permittivity $AgNb_{1-x}Ta_xO_3$ microwave ceramics. Part II: Dielectric characteristics. J. Am. Ceram. Soc. 82(1999)88–93.

[79] J. Petzel, S. Kamba, E. Buixaders, V. Boutin, Z. Zikmund, A. Kanica, V. Koukal, J. Pokormy, J. Polivka, V. Pashkov, G. Komandin, and A. Volkov. Infrared and microwave dielectric response of disordered and antiferroelectric Ag(Ta,Nb)O_3 system. Ferroelectrics 223(1999)235–246.

[80] H. T. Kim, T. Shrout, C. Rindall, and M. Lanagan. Low temperature sintering and dielectric properties of Ag(Nb,Ta)O_3 composite ceramics. J. Am. Ceram. Soc. 85(2002)2738–2744.

[81] C.-H. Wei and J.-H. Jean. Low fire processing $(Ca_{1-x}Nd_{2x/3})TiO_3$ microwave ceramics. J. Am. Ceram. Soc. 86(2003)93–98.

[82] I. Levin, T. A. Vanderah, and R. Coutts. Phase equlibria and dielectric properties in perovskite like $(1-x)LaCa_{0.5}Zr_{0.5}O_3$–$xATiO_3$ (A = Ca,Sr) ceramics. J. Mater. Res. 17 (2002)1729–1734.

[83] W. S. Kim, E. S. Kim, and K. H. Yoon. Effect of Sm^{3+} substitution on dielectric properties of $Ca_{1-x}Sm_{2x/3}TiO_3$ ceramics at microwave frequencies. J. Am. Ceram. Soc. 82(1999) 2111–2115.

[84] E. S. Kim, S. H. Kim, S. G. Kang, and K. H. Yoon. Microwave dielectric properties of $(Ca_{0.85}Nd_{0.1})[Ti_{1-x}(Mg_{0.33}Nb_{0.67})_x]O_3$ ceramics. J. Ceram. Soc. Jpn. Supl. 112–1(2004) S1641–S1644.
[85] J. X. Tong, Q. L. Zhang, H. Yang, and J. L. Zou. Low temperature firing and microwave dielectric properties of $Ca[(Li_{0.33}Nb_{0.67})_{0.9}Ti_{0.1}]O_{3-\delta}$ ceramics with LiF addition. Mater. Lett. 59(2005)3252–3255.
[86] J.-Y. Ha, J.-W. Choi, C.-Y. Kang, S.-J. Yoon, D. J. Choi, and H. J. Kim. Low temperature sintering of $Ca[(Li_{1/3}Nb_{2/3})_{1-x}Ti_x]O_{3-\delta}$ microwave dielectric ceramics with glass frit. Jpn. J. Appl. Phys. 44(2005)1322–1325.
[87] N. Ichinose, T. Mizutani, H. Hiraki, and H. Ookuma. Microwave materials in the system $(Sr_{1-x}Ca_x)[Li_{1/4}Nb_{3/4}]_{1-y}Ti_y]O_3$. Ceramurga Int. 3(1977)100–102.
[88] P. Liu, E. S. Kim, S. G. Kang, and H. S. Jang. Microwave dielectric properties of $Ca[(Li_{1/3}Nb_{2/3})_{1-x}Ti_{3x})O_{3-\delta}$ ceramics with B_2O_3. Mater. Chem. Phys. 79(2003)270–272.
[89] J. W. Choi, J.-Y. Ha, S.-J. Yoon, H.-J. Kim, and K. H. Yoon. Microwave dielectric characteristics of $[Ca(Li_{1/3}Nb_{2/3})_{1-x}Zr_x]O_{3-\delta}$ ceramics. Jpn. J. Appl. Phys. 43(2004)223–225.
[90] P. Liu, E. S. Kim, and K. H. Yoon. Low temperature sintering and microwave dielectric properties of $Ca(Li_{1/3}Nb_{2/3})O_{3-\delta}$ ceramics. Jpn. J. Appl. Phys. 40(2001)5769–5773.
[91] P. Liu, H. Ogawa, E. S. Kim, and A. Kan. Low temperature sintering and microwave dielectric properties of $Ca(Li_{1/3}Ta_{2/3})O_{3-\delta}$–$xCaTiO_3$ ceramics. J. Eur. Ceram. Soc. 23(2003)2417–2421.
[92] J.-Y. Ha, J.-W. Choi, S.-J. Yoon, D. J. Choi, K. H. Yoon, and H.-J. Kim. Microwave dielectric properties of Bi_2O_3 doped $Ca[(Li_{1/3}Nb_{2/3})_{1-x}Ti_x]O_{3-\delta}$ ceramics. J. Eur. Ceram. Soc. 23(2005)2413–2416.
[93] J. W. Choi, C.-Y. Kang, S.-J. Yoon, H.-J. Kim, H.-J. Jung, and K. H. Yoon. Microwave dielectric properties of $Ca[(Li_{1/3}Nb_{2/3})_{1-x}M_x]O_{3-\delta}$ [M = Sn, Ti] ceramics. J. Mater. Res. 14(1999)3567–3570.
[94] J.-X. Tong, Q.-L. Zhang, H. Yang, and J.-l. Zou. Low temperature firing and microwave dielectric properties of $Ca[Li_{1/3}Nb_{2/3}]_{0.84}Ti_{0.16}]O_{3-\delta}$ ceramics for LTCC applications. J. Am. Ceram. Soc. 90(2007)845–849.
[95] H. Wu and P. K. Davies. Ordered perovskites in the $A^{2+}(Li_{1/4}Nb_{3/4})O_3$–$A^{2+}(Li_{2/5}W_{3/5})O_3$ (A^{2+} = Sr, Ca) systems. J. Solid State Chem. 177(2004)4305–4315.
[96] H. Wu and P. K. Davies. Non-stoichiometric 1:2 ordered perovskites in the $Ba(Li_{1/4}Nb_{3/4})O_3$–$Ba(Li_{2/5}W_{3/5})O_3$ system. J. Solid State Chem. 177(2004)3469–3478.
[97] Kato, H. Kagata, and K. Nishimoto. Dielectric properties of $(Pb,Ca)(Me,Nb)O_3$ at micrwave frequencies. Jpn. J. Appl. Phys. 31(1992)3144–3147.
[98] S. O. Yoon, D. M. Kim, S. H. Shim, J. K. Park, and K. S. Hong. Microwave dielectric properties of $Ca[Li_{1/4}Nb_{3/4}]O_3$–$CaTiO_3$ ceramic systems. J. Eur. Ceram. Soc. 26(2006) 2023–2026.
[99] E. S. Kim, B. S. Chu, J. D. Kim, and K. H. Yoon. Low temperature sintering and microwave dielectric properties of $[Ca_{0.6}(Li_{0.5}Nd_{0.5})_{0.4}]_{0.45}TiO_3$ ceramics. Mater. Sci. Eng., B 99(2003) 243–246.
[100] G. Subodh and M. T. Sebastian. Microwave dielectric properties fo $Sr_2Ce_2Ti_5O_{16}$ ceramics. Mater. Sci. Eng., B 136(2007)50–56.
[101] G. Subodh, J. James, M. T. Sebastian, R. M. Paniago, A. Dias, and R. L. Moreira. Structure and microwave dielectric properties of $Sr_{2+n}Ce_2Ti_{5+n}O_{15+3n}$ ($n \leq 10$) homologous series. Chem. Mater. 19(2007)4077–4082.
[102] K. Tatsuki, K. Murano, T. Kawamura H. Kato, S. Yano, and S. Nishigaki, Jpn. J. Appl. Phys. 26, Suppl. 26-2(1987)80.
[103] K. W. Kirby and W. A. Wechesler. Phase relations in barium titanate- titanium oxide system. J. Am. Ceram. Soc. 74(1991)1841–1847.
[104] D.-C. Sinclair, J. M. S. Skakle, F. D. Morroson, R. I. Smith, and T. P. Beales. Structure and electrical properties of oxygen efficient hexagonal $BaTiO_3$. J. Mater. Chem. 9(1999)1327–1331.
[105] M. Valant and D. Suvorov. A new high permittivity $AgNb_{1-x}Ta_xO_3$ microwave ceramics. Part-1. Crystal structures and phase composition relations. J. Am. Ceram. Soc. 82(1999)81–87.

[106] A. Kanaria. $AgNb_{1-x}Ta_xO_3$ solid solutions – dielectric properties and phase transitions. Phase Transitions 31(1983)131–140.
[107] M. Lukasszewski, A. Kania, and A. Ratuszna. Flux growth of single crystals of $AgNbO_3$ and $AgTaO_3$. J. Cryst. Growth. 48(1980)493–495.
[108] A. Saito, S. Uraki, H. Kakemoto, T. Tsurumi, and S. Wada. Growth of lithium doped silver niobate single crystals and their piezoelectric properties. Mater. Sci. Eng. B 120(2005) 166–169.
[109] M. Valant, A.-K. Axelsson, N. Alford. Review of $Ag(Nb,Ta)O_3$ as a functional material. J. Eur. Ceram. Soc. 27(2007)2549–2560.
[110] M. B. Telli. Chemical solution deposition of silver tantalate niobate $Ag(Ta_xNb_{1-x})O_3$ thin films. Doctoral Thesis. Pennsylvannia State University. USA (2005).
[111] M. Xiao, X. M. Wu, and K. X. Song. $Ag_{1-x}Na_x(Nb_{1-y}Ta_y)O_3$ dielectric ceramic prepared through citrate precursor method. Key Eng. Mater. 280–283(2005)15–17.
[112] H. T. Kim, T. Shrout, C. Rindall and M. Lanagan. Low temperature sintering and dielectric properties of $Ag(Nb.Ta)O_3$ composite ceramics. J. Am. Ceram. Soc. 85(2002)2738–2744.
[113] L. Li, D. Wang, and X. Wu. Temperature stable and high k dielectric ceramics based on $Ag(Nb_{1-x}Ta_x)O_3$. J. Eur. Ceram. Soc. 26(2006)1991–1993.
[114] X. Guo, M. Xiao, N. Ding, and Z. Zhang. The effect of Sb_2O_5 additions on the dielectric properties of $Ag(Nb_{0.8}Nb_{0.2})O_3$ ceramics. Mater. Lett. 60(2006)3651–3654.
[115] M. Valant, D. Suvorov, C. Hoffmann, and H. Sommaria. $Ag(Nb,Ta)O_3$ based ceramics with suppressed temperature dependence of permittivity. J. Eur. Ceram. Soc. 21(2001)2647–2651.
[116] M. Valant, D. Suvorov, C. Hoffmann, and H. Sommaria. Method for producing a ceramic silver niobium tantalate body. US patent 6,843,956 (2002), German Patent 10042349 (2001).
[117] C. Hoffmann, H. Sommaria, D. Suvorov and M. Valant. Metalized ceramics. US Patent 6956001 (2005).
[118] V. Porokhonsky, V. Bovtun, S. Kamba, E. Buixaderas, J. Petzelt, A. Kania, S. Miga, and Y. Yakimenko. Microwave dielectric properties of the $Ag_{1-x}Li_xNbO_3$ ($x = 0$–0.06) ceramics. Ferroelectrics 238(2000)134–138.
[119] Y. Sakabe, Y. Ogiso, and N. Wada. Ferroelectric properties of $(AgLi)(Nb,Ta)O_3$ ceramics. Jpn. J. Appl. Phys. 40(2001)5675–5678.
[120] A. Kanaria. Dielectric properties of $Ag_{1-x}A_xNbO_3$ [A = K, Na, Li] and $AgNb_{1-x}Ta_xO_3$ solid solutions in the vicinity of diffuse phase transitions. J. Phys. D. 34(2001)1447–1455.
[121] A. Kanaria and S. Miga. Preparation and dielectric properties of $Ag_{1-x}Li_xNbO_3$ (ALN) solid solutions ceramics. Mater. Sci. Eng. B 86(2001)128–133.
[122] T. Takada, Y. Takahashi, N. Wada, and Y. Sakabe. Effects of substitution of Na, K ions for Ag ion in $(Ag,Li)NbO_3$ ceramics. Jpn. J. Appl. Phys. 42(2003)6023–6026.
[123] F. Zimmermann, W. Menesklou, and E. Ivers-Tiffee. Investigation of $Ag(Ta,Nb)O_3$ as tuneable microwave dielectrics. J. Eur. Ceram. Soc. 24(2004)1811–1814.
[124] F. Zimmermann, W. Menesklou, and E. Ivers-Tiffee. Electrical properties of silver tantalite-niobate thick films. Integr. Ferroelectrics 50(2002)181–188.
[125] J. H. Koh and A. Grishim. Electrically tuneable $Ag(Ta,Nb)O_3$ thin film structures on oxide substrates. Integr. Ferroelectrics 39(2001)1281–1288.
[126] J. H. Koh and A. Grishin. $Ag(Ta,Nb)O_3$ thin film low loss variable interdigital capacitors. Appl. Phys. Lett. 79(2001)2234–2236.
[127] H.-W. You and J.-H. Koh. $Ag(Ta,Nb)O_3$ thin film interdigital capacitors for microwave applications. Microelectron. J. 38(2007)222–226.
[128] H. Kato, H. Kobayashi, and A. Kudo. Role of Ag+ in the band structures and photocatalytic properties of $AgMO_3$ (M = Ta,Nb) with the persovskite structure. J. Phys. Chem. B 106 (2002)12441.
[129] A. Kudo. Development of photocatalyst materials for water splitting. Int. J. Hydrogen Energy 31(2006)197–202.
[130] M. H. Francombe and B. Lewis. Structural and electrical properties of silver niobate and silver tantalite. Acta Crystallogr. 11(1958)175–178.

[131] A. Kanaria and A. Ratusza. Phase transitions in $AgTaO_3$ single crystals. Phase Transitions 2(1981)7–14.
[132] M. Lukaszewski, M. Pawelczyk, and A. Kanaria. On the phase transition in silver niobate $AgNbO_3$. Phase Transitions 3(1983)247–258.
[133] M. Pawelczyk.Phase transitions in $AgTa_xNb_{1-x}O_3$ solid solutions. Phase Transitions 8(1999)273–292.
[134] G. E. Kugel, M. D. Fontana, M. Hafid, K. Roleder, A. Kanaria, and M. Pawelczyk. A Raman study of silver tantalite ($AgTaO_3$) and its structural phase transition sequence. J. Phys. C. Solid State Phys. 20(1987)1217–1230.
[135] A. Ratuszna, J. Pawluk, and A. Kanaria. Temperature evolution of the crystal structure of $AgNbO_3$. Phase Transitions 76(2003)611–620.
[136] A. Kanaria. An additional phase Transition in silver niobate. Ferroelectrics 205(1998)19–28.
[137] P. Sciau, A. Kanaria, B. Dkhil, E. Suard, and A. Ratuszna. Structural investigation of $AgNbO_3$ phase using X-ray and neutron diffraction. J. Phys.: Condens. Matter. 16(2004)2795–27810.
[138] S. Duguey, R. Leburgeois, C. Grattepain, J. M. Haintz, and J. P. Ganne. Study of copper substitution in $Ag(Nb_xTa_{1-x})O_3$. J. Eur. Ceram. Soc. 27(2007)1171–1175.
[139] M. Yokohama, T. Ota, I. Yamai, and J. Takahashi. Flux growth of perovskite type $La_{2/3}TiO_{3-x}$ crystals. J. Cryst. Growth 96(1989)490–496.
[140] J. B. Mc Chesney and H. A. Sauer. The system La_2O_3–TiO_2 phase equilibria and electrical properties. J. Am. Ceram. Soc. 45(1962)416–422.
[141] M. Abe and K. Uchino. X-ray study of the deficient perovskite $La_{2/3}TiO_3$. Mater. Res. Bull. 10(1974)147–156.
[142] A. I. Leonov, M. M. Piyutte, J. K. Keler, and Izvest. Akdem Nauk SSSR. Ceram. Khim. 5(1966)787.
[143] T. Y. Tien and F. A. Hummel. Solid solutions in the system $SrTiO_3$–(La_2O_3·$3TiO_2$). Trans. Br. Ceram. Soc. 66(1967)233–245.
[144] A. G. Belous, L. G. Gavrilova, S. V. Polyaneckaya, Z. J. A. Makarova, and V. P. Chalij. Ukr. Khim. Zh. 50(1984)460.
[145] E. A. Nenasheva, B. A. Rotenberg, E. I Gindin, and V. G. Prokhvatilov. High frequency capacitor dielectrics based on lanthanum and divalent metal titanates. Akadem. Nauk. SSSR Neorg. Mater. 15(1980) 1489–1491.
[146] Y. J. Zhong, F. Azough, and R. Freer. The effect of $La_2Ti_3O_9$ second phase on the microstructure and dielectric properties of $La_2Ti_2O_7$ ceramics. J. Eur. Ceram. Soc. 15(1995)255–263.
[147] T. Negas. G. Yeager, S. Bell, and R. Armen. Proceedings of International Conference on the Chemistry of Electronic Ceramic Materials. P. K. Davies and R. S. Roth (Eds) (NIST SP 804) NIST, Gaithersberg, MD (1991) pp. 21–33.
[148] J. H. Moon, H. S. Park, K. T. Lee, J. H. Choi, D. H. Yeo, and H. J. Kim. Microwave dielectric properties of the $(1-x)La_{2/3}TiO_3$–$LnAlO_3$ system. Jpn J. Appl. Phys. 36(1997)6814–6817.
[149] H. Yoshioka. Unusual dielectric behaviour La–Ti–Al–Oceramics with perovskite structure. Jpn. J. Appl. Phys. 33(1994)L945–L948.
[150] M. Yoshida, N. Hara, T. Takada, and A. Seki. Structure and dielectric properties of $(Ca_{1-x}Nd_{2x/3})TiO_3$. Jpn. J. Appl. Phys. 36(1997)6818–6823.
[151] C.-H. Wei and J. H. Jean. Low fire processing $Ca_{1-x}Nd_{2x/3}TiO_3$ microwave ceramics. J. Am. Ceram. Soc. 86(2003)93–98.
[152] K. H. Yoon, W. S. Kim, and E. S. Kim. Dependence of octahedral bond valence on microwave dielectric properties of $Ca_{1-x}Sm_{2x/3}TiO_3$ ceramics. Mater. Sci. Eng. B 99(2003)112–115.
[153] K. Yoshii. Synthesis and magnetic properties of $Ln_{2/3}TiO_3$ (Ln = Pr and Nd). J. Solid State Chem. 149(2000)354–359.
[154] C. J. Howard and Z. Zheng. Structure of perovskites with layered ordering of A site cations. Acta Crystallogr. B 60(2004)249–251.

[155] C. J. Howard and Z. Zheng. Structure of perovskites with layered ordering of A site cations. Erratum Acta Crystallogr. B 60(2004)763.

[156] A. S. Sefat, G. Anow, M.-Y. Wu, G. A. Bottom, and J. E. Greedan. High resolution EELS study of the vacancy doped metal /insulator system $Nd_{1-x}TiO_3$ $x=0$–0.33. J. Solid State Chem. 178(2005)1008–101.

[157] D. G.`Lim, B. H. Kim, T.-G. Kim, and H. J. Jung. Microwave dielectric properties of $(1-x)LaAlO_3-xTiO_2$ system. Mater. Res. Bull. 34(1999)1577–1582.

[158] H. J. Lee, J. O. Son, S. H. Oh, Y. K. Cho, and S. Nahm. Microstructural changes in $(1-x)Nd_{2/3}TiO_3-xNdAlO_3$ system. Jpn. J. Appl. Phys. 43(2004)7587–7591.

[159] T. Lowe, F. Azough and R. Freer. The microwave dielectric properties and structure of $xCaTiO_3-(1-x)Li_{0.5}Nd_{0.5}TiO_3$ ceramics. J. Korean Ceram. Soc. 40(2003)328–332.

[160] E. S. Kim, B. S. Chun, D. W. Yoo, and K. H. Yoon. Microwave dielectric properties of $(1-x)(Ca_{0.7}Nd_{0.2})TiO_3-x(Li_{0.5}Nd_{0.5})TiO_3$ ceramics. Mater. Sci. Eng. B 99(2003)247–251.

[161] I. Levin, J. Y. Chan, J. E. Maslar, T. A. Vanderah and S. Bell. Phase transitions and microwave dielectric propertiesin the perovskite like $Ca(Al_{0.5}Nb_{0.5})O_3-CaTiO_3$ system. J. Appl. Phys. 90(2001)904.

[162] J. M. Philips, M. P. Siegel, R. B. Dover, T. H. Tiefel, J. H. Marshall, C. D. Brandle, G. Berstresser. A. J. Straus, R. E. Fahey, S. Sengupta, A. Cassanho, and H. P. Jenssen. Comparison of $Ba_2YCu_3O_{7-d}$ thin films grown on various perovskite substrate by co-evaporation. J. Mater. Res. 7(1992)2650–2657.

[163] T. Shimada, K. Kakimoto and H. Ohsato. Microwave dielectric properties of lanthanum aluminate ceramics and single crystals. J. Eur. Ceram. Soc. 25(2005)2901–2905.

[164] C. Zuccaro, W. Winter, N. Klein, and K. Urban. Microwave absorption in single crystals of lanthanum aluminate. J. Appl. Phys. 82(1997)5695–5704.

[165] N. Mc, N. Alford, J. Breeze, X. Wang, S. J. Penn, S. Dalla, and S. Webb. Dielectric loss of oxide single crystals and polycrystalline analogues from 10 to 320 K. J. Eur. Ceram. Soc. 21(2001)2605–2611.

[166] D. Mateika, H. Kohler, H. Landern, and E. V. Ikel. Mixed perovskite substrate for high Tc superconductors. J. Cryst. Growth 109(1991)447–456.

[167] S.-Y. Cho, K. S. Hong, and K. H. Ko. Mixture like behaviour in the microwave dielectric properties of the $(1-x)LaAlO_3-xSrTiO_3$ system. Mater. Res. Bull. 34(1999)511–516.

[168] E. R. Kipoech, F. Azough, R. Freer, C. Leach, S. P. Thompson, and C. C. Tang. Structural study of $Ca_{0.7}Nd_{0.3}Ti_{0.7}Al_{0.3}O_3$ dilectric ceramics using synchrotron X-ray absorption. J. Eur. Ceram. Soc. 23(2003)2677–2682.

[169] E. A. Nenesheva, L. P. Mudroliubova, and N. F. Kartenko. Dielectric properties of ceramics based on $CaTiO_3-LnMO_3$ system (Ln = La, Nd, M = Al, Ga). J. Eur. Ceram. Soc. 23(2003) 2443–2448.

[170] G. A. Ravi, F. Azough and R. Freer. Effect of Al_2O_3 on the structure and microwave dielectric properties of $Ca_{0.7}Ti_{0.7}La_{0.3}Al_{0.3}O_3$. J. Eur. Ceram. Soc. 27(2007)2855–2860.

[171] P. Sun, T. Nakamura, Y. J. Shan, Y. Inaguma, M. Itoh, and T. Kitamura. Dielectric behaviour of $(1-x)LaAlO_3-xSrTiO_3$ solid solution system at microwave frequencies. Jpn. J. Appl. Phys. 37(1998)5625–5629.

[172] D. Khalyavin, A. N. Salak, A. M. R. Senos, P. Q. Mantas, V. M. Ferreira. Structure sequence in the $CaTiO_3-LaAlO_3$ microwave ceramics-Revised. J. Am. Ceram. Soc. 89(2006)1721–1723.

[173] D. Suvorov, G. Drazic, M. Valant and B. Jancar. Microstructural characterization of $CaTiO_3-NdAlO_3$ based ceramics. Korean J. Crystallogr. 11(2000)195–199.

[174] B. Jancar, M. Valant, and D. Suvorov. Solid state reactions occurring during the synthesis of $CaTiO_3-NdAlO_3$ perovskite solid solutions. Chem. Mater. 16(2004)1075–1082.

[175] H. Zheng, G. D. C. Csete de Gyorgyfalva, R. Quimby, H. Bagshaw, R. Ubic, I. M. Reaney, and J. Yarwood. Raman spectroscopy of B site order-disorder in $CaTiO_3$ microwave ceramics. J. Eur. Ceram. Soc. 23(2003)2653–2659.

CHAPTER SEVEN

$A(B'_{1/2}B''_{1/2})O_3$ [A = A^{2+} OR A^{3+}; B′ = B^{2+},B^{3+}; B″ = B^{4+},B^{5+},B^{6+}] COMPLEX PEROVSKITES

7.1 INTRODUCTION

One of the largest group of complex perovskite-type materials is of the general formula $A(B'_{1/2}B''_{1/2})O_3$ [A = A^{2+} or A^{3+}; B′ = B^{2+}, B^{3+}; B″ = B^{4+}, B^{5+} and B^{6+}], where the structure usually has an ordered arrangement of B′ and B″ atoms. Perovskites with 1:1 B-cation ordering, also known as double perovskites, are derived from the simple perovskite structure by substituting a mixture of two cations (B′ and B″) on the octahedral B site. Octahedral B-site cation ordering occurs when the stoichiometry is $A(B'_{1/2}B''_{1/2})O_3$ with a large difference in atomic size and/or charge between the B′ and B″ cations [1–3]. The preparation conditions and polarization of certain cations can also influence cation ordering. Ordered perovskites containing a highly charged main group cation such as Sb^{5+} are more ordered than transition metal (e.g. Nb^{5+} or Ta^{5+}) cations [4]. Figure 7.1 shows the structure of an ordered $A(B'_{1/2}B''_{1/2})O_3$ with the corner sharing of $B'O_6$ and $B''O_6$ octahedra. A cations are typically large, low oxidation state ions such as Ba^{2+}, Ca^{2+}, Sr^{2+}, Pb^{2+}, La^{3+}. The ordering of the octahedral site B cations doubles the unit cell of the simple undistorted ABO_3 perovskite, changing the space group symmetry from $Pm\bar{3}m$ to $Fm\bar{3}m$. The change in the space group leads to doubling of the primitive cubic lattice parameter (a_p). The perovskites are prone to a variety of distortion mechanisms that lower the symmetry of the cubic aristotype structure. Generally, double perovskites with large A cations such as Ba^{2+} exhibit cubic symmetry and those with smaller A cations such as Ca^{2+} exhibit lower symmetries such as monoclinic or orthorhombic. A large number of $A(B'_{1/2}B''_{1/2})O_3$ complex perovskites were reported in the 1960's [6–12]. The X-ray diffraction peaks with *hkl* values that are all even are independent of cation ordering and are termed "subcell reflections". Reflections with all odd miller indices are indicative of cation ordering and are referred to as the superstructure reflections. In a completely disordered structure, the supercell reflections are absent due to the random distribution of B cations on the same crystallographic site. The subcell reflections are still present but with reduced indices by a factor of two. The supercell diffraction peak positions, peak shape and intensity depend on the type and extent of ordering. The peak shift and peak broadening of supercell reflections also indicate antiphase boundary (APB) defects. APBs are defects in the ordered structure due to the mismatch of the cations $B'O_6$ and $B''O_6$ ordered regions from one another (e.g. B′-B″-B′-B″-**B′-B′**-B″-B′-B″... in one dimension). The two ordered domains become 180° out of phase with respect to the B″/B″ cation distribution before and after the occurrence of an antiphase boundary region.

Galasso has listed a large number of compounds belonging to the $A(B'_{1/2}B''_{1/2})O_3$ complex perovskite compounds in his books [13–15] with lattice parameters, densities

Dielectric Materials for Wireless Communications

ISBN-13: 978-0-08-045330-9

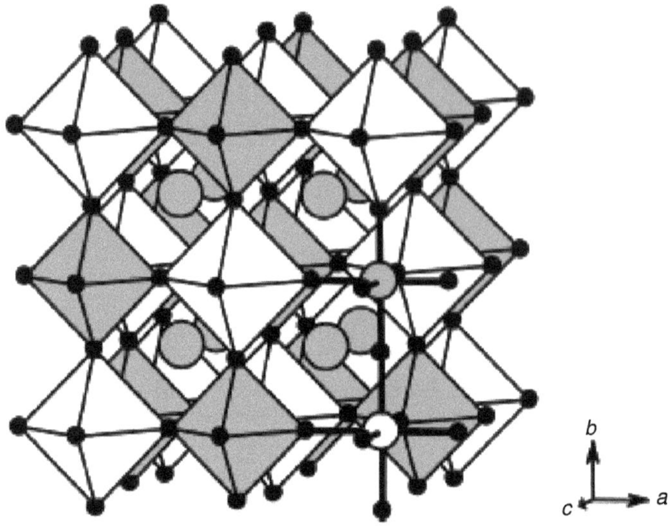

Figure 7.1 Conventional view of the cubic aristoytpe double perovskite (space group $Fm\bar{3}m$) showing $B'O_6$ and $B''O_6$ octahedra with the A cations residing in the cavities formed by the octahedral network. $B'O_6$–white octahedral/sphere. $B''O_6$–grey octahedral/sphere (after Ref. [5], Courtesy IUCR).

and crystal symmetries. In 1960, Agranovskaya [16] outlined the dielectric properties of $A(B'_{1/2}B''_{1/2})O_3$. Takata and Kageyama [17] were the first to investigate the microwave dielectric properties of $A(B'_{1/2}B''_{1/2})O_3$-type perovskites [A = Ba, Sr, Ca: B′ = rare earths and B″ = Nb or Ta]. Since then several authors [18–30] have investigated the microwave dielectric properties of this family of materials. The sintering temperature, tolerance factors and the microwave dielectric properties of $A(B'_{1/2}B''_{1/2})O_3$ materials are given in Table 7.1.

7.2 $Ba(B'_{1/2}Nb_{1/2})O_3$ Ceramics

$Ba(B'_{1/2}Nb_{1/2})O_3$ [B′ = La, Pr, Nd, Sm, Eu, Gd, Tb, Dy, Ho, Y, Yb, and In] ceramics are prepared by ball milling the stoichiometric amounts of oxide or carbonate raw materials. The ball milled raw materials are then calcined at temperatures in the range 1200–1400°C [17, 18, 22, 84–86] followed by sintering the shaped samples in the temperature range 1550–1650°C. Figure 7.2 shows the X-ray diffraction patterns of Ba $(B'_{1/2}Nb_{1/2})O_3$ ceramics. The X-ray diffraction patterns of some of the compounds showed splitting of the main reflections, indicating a non-cubic symmetry. The amount of splitting of the main reflections increased with a decrease of the tolerance factor. Galasso reported [13–15] that these compounds are face-centered cubic with $(NH_4)_3FeF_6$ structure. However, later, several authors [1, 9, 18–21, 85–89] reported that the room temperature symmetry of these compounds may be different from cubic. These studies showed that the room temperature symmetry of $Ba(B'_{1/2}B''_{1/2})O_3$ can be cubic, tetragonal, orthorhombic or monoclinic depending on the tolerance factor. It was suggested that the deviation from the cubic symmetry is due to the tilting of octahedra.

Table 7.1 Dielectric properties of $A(B'_{1/2}B''_{1/2})O_3$ ceramics

Material	Tolerance factor (*t*)	Sintering temperature (°C)	ε_r	*Qf* (GHz)	τ_f (ppm/°C)	Reference
$Pb_{0.6}Ca_{0.4}(Fe_{1/2}Nb_{1/2})O_3$	0.979	–	154	1700	135	[31]
$Pb_{0.5}Ca_{0.5}(Fe_{1/2}Nb_{1/2})O_3$	0.975	–	104	4000	26	[31]
$Pb_{0.45}Ca_{0.55}(Fe_{1/2}Nb_{1/2})O_3$	0.972	1100	93	6000	2	[31, 32]
$Pb_{0.2}Ca_{0.8}(Fe_{1/2}Nb_{1/2})O_3$	0.959	–	53	10 000	–69	[31]
$Pb_{0.75}Ca_{0.25}(Al_{1/2}Nb_{1/2})O_3$	1.021	–	35	1080	133	[31]
$Pb_{0.5}Ca_{0.5}(Al_{1/2}Nb_{1/2})O_3$	1.008	–	30	1500	–23	[31]
$Pb_{0.75}Ca_{0.25}(Cr_{1/2}Nb_{1/2})O_3$	1.073	–	48	3600	8	[31]
$Pb_{0.5}Ca_{0.5}(Cr_{/2}Nb_{1/2})O_3$	0.982	–	43	3800	293	[31]
$Pb_{1-x}Ca_x[(Fe_{1/2}Nb_{1/2})_{1-y}Sn_y]O_3$ ($x=0.55$, $y=0.05$)	0.989	1150/3 h	86	6300	2	[33]
$Pb_{1-x}Ca_x[(Fe_{1/2}Nb_{1/2})_{1-y}Sn_y]O_3$ ($x=0.6$, $y=0.05$)	0.983	1150/3 h	81	4830	3	[33]
$Pb_{1-x}Ca_x[(Fe_{1/2}Nb_{1/2})_{1-y}Sn_y]O_3$ ($x=0.55$, $y=0.1$)	0.964	1150/3 h	85	8600	0	[33]
$[(Pb_{0.5}Ca_{0.5})_{0.95}La_{0.05}](Fe_{1/2}Nb_{1/2})O_3 + 1$ wt% PbO–B_2O_3–V_2O_5 glass	–	1050/3 h	101	5400	6	[34]
$(Pb_{0.45}Ca_{0.55})[(Fe_{1/2}Nb_{1/2})_{0.9}Sn_{0.1}]O_3 + 0.2$ wt% CuO + 0.4 wt% Bi_2O_3	–	1000/3 h	86	4300	8	[35]
$[(Pb_{0.5}Ca_{0.5})_{0.92}La_{0.08}](Fe_{1/2}Nb_{1/2})O_3 + MnO_2 + Bi_2O_3$	–	1050/4 h	91	4800	19	[36]

(Continued)

Table 7.1 (Continued)

Material	Tolerance factor (t)	Sintering temperature (°C)	ε_r	Qf (GHz)	τ_f (ppm/°C)	Reference
$(Pb_{0.48}Ca_{0.52})(Fe_{1/2}Nb_{1/2})O_3 + 2.2$ mol% CeO_2	–	1190	94	6800	2	[37]
$(Pb_{0.5}Ca_{0.5})(Fe_{1/2}Ta_{1/2})O_3$	0.975	1250	84	6700	–	[38]
$(Pb_{0.4}Ca_{0.6})(Fe_{1/2}Ta_{1/2})O_3$	0.969	1050/3 h	62	9000	–15	[39]
$[(Pb_{0.5}Ca_{0.5})_{0.98}Nd_{0.02}](Fe_{1/2}Nb_{1/2})O_3$	0.974		110	5800	17	[40]
$[(Pb_{0.5}Ca_{0.5})_{0.95}La_{0.05}][(Fe_{1/2}Nb_{1/2})_{0.9}Ti_{0.1}]O_3$	0.975	1150	117	5000	17	[41]
$[(Pb_{0.5}Ca_{0.5})_{0.94}(La_{0.5}Nd_{0.5})0_{.06}](Fe_{1/2}Nb_{1/2})O_3$	0.972	1150	100	5800	0	[42]
$(Pb_{1-x}Ca_x)[(Fe_{1/2}Nb_{1/2})_{1-y}Zr_y]O_3$ ($y = 0.05$, $x = 0.55$)	0.974	1200	87	8500	–10	[43]
$(Pb_{1-x}Ca_x)[(Fe_{1/2}Nb_{1/2})_{1-y}Zr_y]O_3$ ($y = 0.1$, $x = 0.55$)	0.978	1200	85	8600	–1	[43]
$[(Pb_{0.7}Ca_{0.3})_{1/2}La_{1/2}](Mg_{1/2}Nb_{1/2})O_3$	0.953	1350/3 h	50	86 000	0	[44]
$(Pb_{0.4}Ca_{0.6})[(Mg_{1/2}Nb_{1/2})O_3Sn_x]$ ($x = 0.01$)	0.952	1350	65	7100	136	[45]
$[(Pb_{0.5}Ca_{0.5})_{0.95}Nd_{0.05}]$ $(Fe_{1/2}Nb_{1/2})O_3$	0.972		100	5800	0	[46]
$(Pb_{0.45}Ca_{0.55})[Fe_{0.5}(Nb_{0.96}Ta_{0.04})_{1/2}]O_3$	0.972	1150/5 h	82	7700	–5	[47]
$(1-x)PbZrO_3-xCa(Fe_{1/2}Nb_{1/2})O_3$ ($x = 0.2$)	–	1350/3 h	336	300	386	[48]
$(1-x)PbZrO_3-xCa(Fe_{1/2}Nb_{1/2})O_3$ ($x = 0.4$)	–	1350/3 h	141	1800	120	[48]
$(1-x)PbZrO_3-xCa(Fe_{1/2}Nb_{1/2})O_3$ ($x = 0.6$)	–	1350/3 h	85	3000	40	[48]
$(1-x)PbZrO_3-xCa(Fe_{1/2}Nb_{1/2})O_3$ ($x = 0.8$)	–	1350/3 h	55	500	–52	[48]

$Ba(Mg_{1/2}W_{1/2})O_3$	1.033	1550/6 h	17	57 000	–34	[49, 50]
$x$$Ba(Mg_{1/2}W_{1/2})O_3$–$(1–x)BaTiO_3$ ($x = 0.92$)	–	1500/6 h	20	37 000	–19	[49]
$x$$Ba(Mg_{1/2}W_{1/2})O_3$–$(1–x)BaTiO_3$ ($x = 0.72$)	0–	1500/6 h	12.3	11 000	–5	[49]
$x$$Ba(Mg_{1/2}W_{1/2})O_3$–$(1–x)BaTiO_3$ ($x = 2/3$)	–	1500/6 h	13	35 000	–6	[49]
$x$$Ba(Mg_{1/2}W_{1/2})O_3$–$(1–x)BaTiO_3$ ($x = 0.6$)	–	1500/6 h	17	15 000	12	[49]
$x$$Ba(Mg_{1/2}W_{1/2})O_3$–$(1–x)BaTiO_3$ ($x = 0.5$)	–	1500/6 h	31	8200	125	[49]
$BaO + 0.34MgO + 0.32WO_3 + 0.34TiO_2$	–	1500/12 h	14.5	107 000	–8	[49]
$[Pb_{(1-3x)/2}La_x]$ $(Mg_{1/2}W_{1/2})O_3$ ($x = 0.56$)	–	–	29	18 100	–6	[51]
$0.42(La_{1/2}Na_{1/2})TiO_3$–$0.58Ca(Fe_{1/2}Nb_{1/2})O_3$	–	1300/10 h	59	14 000	0	[52]
$Ba(Co_{1/2}W_{1/2})O_3$	1.063	1390	19	21 000	–55	[50, 53]
$Ba(Ni_{1/2}W_{1/2})O_3$	1.06	1450	18	52 000	–45	[50, 53]
$Ba(Zn_{1/2}W_{1/2})O_3$	1.03	1330	29	36 000	–31	[50, 53]
$Sr(Co_{1/2}W_{1/2})O_3$	0.996	1450	21	14 000	–73	[53]
$Sr(Ni_{1/2}W_{1/2})O_3$	1.03	1570	18	56 000	–50	[53]
$Sr(Zn_{1/2}W_{1/2})O_3$	0.97	1360	27.5	51 000	–45	[53]

(Continued)

Table 7.1 (Continued)

Material	Tolerance factor (*t*)	Sintering temperature (°C)	ε_r	*Qf* (GHz)	τ_f (ppm/°C)	Reference
$La(Mg_{1/2}Ti_{1/2})O_3$	0.94	1650/2 h	29	114 000	–81	[54–57]
$La(Mg_{1/2}Ti_{1/2})O_3$ sol–gel	0.94	–	27	74 500	-	[58]
$La(Mg_{1/2}Ti_{1/2})O_3 + 1$ wt% CuO	–	1500	30	33 800	–68	[59]
$0.5BaTiO_3$–$0.5La(Mg_{1/2}Ti_{1/2})O_3$	–	1550	60.9	9600	–2	[55]
$0.5La(Mg_{1/2}Ti_{1/2})O_3$–$0.5CaTiO_3$	–	1600	43	28 000	–13	[56, 60]
$Dy(Mg_{1/2}Ti_{1/2})O_3$	–	1650/2 h	23	36 800	–6	[54]
$0.9Nd(Mg_{1/2}Ti_{1/2})O_3$–$0.1CaTiO_3$	–	1500	42	43 000	–10	[61]
$Nd(Mg_{1/2}Ti_{1/2})O_3$	0.915	1650/2 h	26	60 000	–72	[54, 62]
$Pr(Mg_{1/2}Ti_{1/2})O_3$	–	1650/2 h	28	27 800	–17	[54]
$Sm(Mg_{1/2}Ti_{1/2})O_3$	0.905	1650/2 h	25	65 500	–26	[54]
$Y(Mg_{1/2}Ti_{1/2})O_3$	–	1650/2 h	22	33 700	–46	[54]
$La(Zn_{1/2}Ti_{1/2})O_3$	0.942	1650/2 h	34	59 000	–52	[63]
$La(Zn_{1/2}Ti_{1/2})O_3$ sol–gel	0.942	1350	30	60 000	–71	[64]
$0.9La(Mg_{1/2}Ti_{1/2})O_3$–$0.1La_{2/3}TiO_3$	–	–	28	56 000	–66	[65]
$0.8La(Mg_{1/2}Ti_{1/2})O_3$–$0.2La_{2/3}TiO_3$	–	–	31	43 000	–54	[65]

$0.55La(Mg_{1/2}Ti_{1/2})O_3$–$0.45La_{2/3}TiO_3$	–	–	42	4500	–30	[65]
$0.5La(Mg_{1/2}Ti_{1/2})O_3$–$0.5La_{2/3}TiO_3$	–	–	38	300	23	[65]
$La(Mg_{1/2}Sn_{1/2})O_3$	0.927	1600	20	63 000	–78	[66]
$0.5La(Zn_{1/2}Ti_{1/2})O_3$–$0.5CaTiO_3$	–	1550/3 h	50	3800	0	[67]
$Sm(Zn_{1/2}Ti_{1/2})O_3$	0.9	1310/2 h	31	37 000	–19	[68]
$Nd(Zn_{1/2}Ti_{1/2})O_3$	0.911	1330/4 h	31.6	170 000	–42	[69]
$0.5Nd(Zn_{1/2}Ti_{1/2})O_3$–$0.5CaTiO_3$	–	1300/4 h	45	56 000	0	[70]
$La(Co_{1/2}Ti_{1/2})O_3 + 0.25$ wt%CuO	–	1380	30	64 000	–56	[71]
$Sm(Co_{1/2}Ti_{1/2})O_3$	0.923	1360/4 h	26	76 000	–16	[72]
$La(Co_{1/2}Ti_{1/2})O_3$	0.965	1550	25	67 000	–42	[73, 74]
$La(Co_{1/2}Ti_{1/2})O_3 + 0.25$ wt%B_2O_3	–	1350/6 h	30	64 600	–48	[75]
$Nd(Co_{1/2}Ti_{1/2})O_3 + 0.75$ wt%B_2O_3	–	1320/4 h	27	153 000	0	[76]
$0.52Nd(Co_{1/2}Ti_{1/2})O_3$–$0.48CaTiO_3$	–	1550/6 h	43	4000		[77]
$0.48La(Co_{1/2}Ti_{1/2})O_3$–$0.52CaTiO_3$	–	1550/6 h	45	5000		[77]
$Ba(La_{1/2}Nb_{1/2})O_3$	0.956	1600	45	5700	7	[22]

(Continued)

Table 7.1 (Continued)

Material	Tolerance factor (*t*)	Sintering temperature (°C)	ε_r	*Qf* (GHz)	τ_f (ppm/°C)	Reference
$Ba(Pr_{1/2}Nb_{1/2})O_3$	0.966	1600	45	28 500	–22	[22]
$Ba(Nd_{1/2}Nb_{1/2})O_3$	0.967	1600	44	11 700	10	[22]
$Ba(Sm_{1/2}Nb_{1/2})O_3$	0.972	1600	43	18 400	9	[22]
$Ba(Eu_{1/2}Nb_{1/2})O_3$	0.975	1600	40	40 400	7	[22]
$Ba(Gd_{1/2}Nb_{1/2})O_3$	0.977	1600	40	5700	5	[22]
$Ba(Tb_{1/2}Nb_{1/2})O_3$	0.98	1600	39	52 400	–2	[22]
$Ba(Dy_{1/2}Nb_{1/2})O_3$	0.983	1600	39	20 600	–3.6	[22]
$Ba(Ho_{1/2}Nb_{1/2})O_3$	0.985	1600	38	21 600	–11	[22]
$Ba(Y_{1/2}Nb_{1/2})O_3$	0.986	1600	37	49 600	15	[22]
$Ba(Yb_{1/2}Nb_{1/2})O_3$	0.99	1600	36	38 100	2	[22]
$Ba(In_{1/2}Nb_{1/2})O_3$	1.01	1600	39	30 700	17	[22]
$0.95Ba(Yb_{1/2}Nb_{1/2})O_3–0.05Ca(Y_{1/2}Nb_{1/2})O_3$	0.98	1600	34	47 500	1	[29]
$Ba_{0.95}Sr_{0.05}(Y_{1/2}Ta_{1/2})O_3$	0.978	1600	33	47 300	0	[26]
$Ba_{0.85}Sr_{0.15}(Y_{1/2}Ta_{1/2})O_3$	0.98	1600	31	32 000	0	[26]
$Ba(La_{1/2}Ta_{1/2})O_3$	0.956	1625/4 h	37	20 950	–36	[23]
$Ba(Nd_{1/2}Ta_{1/2})O_3$	0.967	1625/4 h	39	12 000	–4	[23]

$Ba(Sm_{1/2}Ta_{1/2})O_3$	0.972	1625/4 h	38	15 000	–10	[23]
$Ba(Eu_{1/2}Ta_{1/2})O_3$	0.975	1625/4 h	37	41 200	–16	[23]
$Ba(Gd_{1/2}Ta_{1/2})O_3$	0.977	1625/4 h	36	3200	–18	[23]
$Ba(Tb_{1/2}Ta_{1/2})O_3$	0.981	1625/4 h	36	31 900	–38	[23]
$Ba(Dy_{1/2}Ta_{1/2})O_3$	0.983	1625/4 h	34	20 600	–48	[23]
$Ba(Ho_{1/2}Ta_{1/2})O_3$	0.98	1625/4 h	34	24 000	130	[23]
$Ba(Y_{1/2}Ta_{1/2})O_3$	0.986	1625/4 h	33	50 200	120	[23]
$Ba(Yb_{1/2}Ta_{1/2})O_3$	0.98	1625/4 h	32	35 900	112	[23]
$Sr(La_{1/2}Nb_{1/2})O_3$	0.898	1575/4 h	37	4000	–20	[27]
$Sr(Pr_{1/2}Nb_{1/2})O_3$	0.907	1575/4 h	38	3300	–34	[27]
$Sr(Nd_{1/2}Nb_{1/2})O_3$	0.908	1575/4 h	37	20 100	–40	[27]
$Sr(Sm_{1/2}Nb_{1/2})O_3$	0.913	1575/4 h	36	32 300	–47	[27]
$Sr(Eu_{1/2}Nb_{1/2})O_3$	0.916	1575/4 h	35	44 000	–52	[27]
$Sr(Gd_{1/2}Nb_{1/2})O_3$	0.917	1575/4 h	34	8800	–56	[27]
$Sr(Tb_{1/2}Nb_{1/2})O_3$	0.92	1575/4 fh	34	36 300	–61	[27]

(Continued)

Table 7.1 (Continued)

Material	Tolerance factor (t)	Sintering temperature (°C)	ε_r	Qf (GHz)	τ_f (ppm/°C)	Reference
$Sr(Dy_{1/2}Nb_{1/2})O_3$	0.923	1575/4 h	33	32 700	–63	[27]
$Sr(Ho_{1/2}Nb_{1/2})O_3$	0.925	1600/4 h	32	20 400	–65	[27]
$Sr(Y_{1/2}Nb_{1/2})O_3$	0.925	1600/4 h	32	38 800	–66	[27]
$Sr(Er_{1/2}Nb_{1/2})O_3$	0.927	1575/4 h	32	36 100	–67	[27]
$Sr(Yb_{1/2}Nb_{1/2})O_3$	0.932	1600/4 h	31	26 600	–73	[27]
$Sr(In_{1/2}Nb_{1/2})O_3$	0.947	1600/4 h	26	32 700	–62	[27]
$Sr(Al_{1/2}Nb_{1/2})O_3$ (O_2)	1.01	1550	19	16 000	–5	[78]
Sr_4AlNbO_8	–	1525	22	3700	-	[78]
$Sr(Fe_{1/2}Nb_{1/2})O_3$	1.007	1450/4 h	45	4800	–24	[27]
$Sr(Cr_{1/2}Nb_{1/2})O_3$	0.991	1600/4 h	35	6400	–80	[27]
$Sr(Al_{1/2}Nb_{1/2})O_3$	1.01	1600/24 h	31	10 800	–27	[27, 79]
$Sr(Ga_{1/2}Ta_{1/2})O_3$	0.99	1350/3 h	27	91 000	–50	[80, 81]
$Sr(La_{1/2}Ta_{1/2})O_3$	0.898	1600/4 h	31	4500	–42	[26]
$Sr(Pr_{1/2}Ta_{1/2})O_3$	0.907	1600/4 h	32	8400	–50	[26]
$Sr(Nd_{1/2}Ta_{1/2})O_3$	0.909	1600/4 h	32	38 500	–55	[26]
$Sr(Sm_{1/2}Ta_{1/2})O_3$	0.913	1600/4 h	31	45 200	–61	[26]

$Sr(Eu_{1/2}Ta_{1/2})O_3$	0.916	1600/4 h	30	45 500	–43	[26]
$Sr(Gd_{1/2}Ta_{1/2})O_3$	0.918	1600/4 h	30	4000	–66	[26]
$Sr(Tb_{1/2}Ta_{1/2})O_3$	0.921	1600/4 h	29	34 200	–70	[26]
$Sr(Dy_{1/2}Ta_{1/2})O_3$	0.923	1600/4 h	28	34 200	–73	[26]
$Sr(Ho_{1/2}Ta_{1/2})O_3$	0.925	1600/4 h	28	38 800	–75	[26]
$Sr(Y_{1/2}Ta_{1/2})O_3$	0.926	1600/4 h	28	54 300	–77	[26]
$Sr(Er_{1/2}Ta_{1/2})O_3$	0.928	1600/4 h	27	22 100	–77	[26]
$Sr(Yb_{1/2}Ta_{1/2})O_3$	0.932	1600/4 h	26	32 300	–79	[26]
$Sr(Al_{1/2}Ta_{1/2})O_3$	1.01		12	6500	-	[81]
$Sr(Sm_{1/2}Ta_{1/2})O_3 + 0.2$ wt% TiO_2	–	1600	32	46 400	–46	[26]
$Sr(Sm_{1/2}Ta_{1/2})O_3 + 3$ wt% TiO_2	–	1600	42	8800	3	[26]
$Ca(La_{1/2}Nb_{1/2})O_3$	0.866	1550/4 h	23	31 000	–43	[29]
$Ca(Pr_{1/2}Nb_{1/2})O_3$	0.875	1550/4 h	24	31 500	–39	[29]
$Ca(Nd_{1/2}Nb_{1/2})O_3$	0.879	1550/4 h	25	31 800	–37	[29]
$Ca(Sm_{1/2}Nb_{1/2})O_3$	0.888	1550/4 h	25	33 200	–34	[29]

(Continued)

Table 7.1 (Continued)

Material	Tolerance factor (*t*)	Sintering temperature (°C)	ε_r	*Qf* (GHz)	τ_f (ppm/°C)	Reference
$Ca(Eu_{1/2}Nb_{1/2})O_3$	0.88	1550/4 h	25	35 800	–30	[29]
$Ca(Gd_{1/2}Nb_{1/2})O_3$	0.885	1550/4 h	26	11 000	–26	[29]
$Ca(Tb_{1/2}Nb_{1/2})O_3$	0.888	1550/4 h	27	34 600	–13	[29]
$Ca(Dy_{1/2}Nb_{1/2})O_3$	0.890	1550/4 h	32	32 500	5	[29]
$Ca(Ho_{1/2}Nb_{1/2})O_3$	0.892	1550/4 h	23	32 000	3	[29]
$Ca(Y_{1/2}Nb_{1/2})O_3$	0.893	1550/4 h	31	35 000	–13	[29]
$Ca(Er_{1/2}Nb_{1/2})O_3$	0.894	1550/4 h	32	31 800	–18	[29]
$Ca(Yb_{1/2}Nb_{1/2})O_3$	0.899	1550/4 h	30	32 500	–25	[29]
$Ca(In_{1/2}Nb_{1/2})O_3$	0.91	1550/4 h	30	37 900	–33	[29]
$Ca(Al_{1/2}Nb_{1/2})O_3$	0.975	–	25	7500	–87	[82]
$Ca(Fe_{1/2}Nb_{1/2})O_3$	0.947	1500/6 h	40	20 000	–76	[48, 82]
$Ca(La_{1/2}Ta_{1/2})O_3$	0.867	1600/4 h	23	20 600	–32	[83]
$Ca(Pr_{1/2}Ta_{1/2})O_3$	0.875	1600/4 h	24	22 200	–31	[83]
$Ca(Nd_{1/2}Ta_{1/2})O_3$	0.879	1600/4 h	24	22 400	–30	[83]
$Ca(Sm_{1/2}Ta_{1/2})O_3$	0.881	1600/4 h	26	25 000	–25	[83]
$Ca(Eu_{1/2}Ta_{1/2})O_3$	0.883	1600/4 h	27	23 600	–22	[83]

$Ca(Gd_{1/2}Ta_{1/2})O_3$	0.885	1600/4 h	27	26 000	–16	[83]
$Ca(Tb_{1/2}Ta_{1/2})O_3$	0.888	1600/4 h	28	28 400	–10	[83]
$Ca(Dy_{1/2}Ta_{1/2})O_3$	0.890	1600/4 h	30	26 500	–6	[83]
$Ca(Ho_{1/2}Ta_{1/2})O_3$	0.893	1600/4 h	28	23 700	–8	[83]
$Ca(Y_{1/2}Ta_{1/2})O_3$	0.893	1600/4 h	27	42 300	–9	[83]
$Ca(Er_{1/2}Ta_{1/2})O_3$	0.894	1600/4 h	26	29 600	–12	[83]
$Ca(Yb_{1/2}Ta_{1/2})O_3$	0.900	1600/4 h	26	59 200	–21	[83]
$Ca(In_{1/2}Ta_{1/2})O_3$	0.914	1600/4 h	24	16 700	–35	[83]
$0.6Ca(Yb_{1/2}Ta_{1/2})O_3$–$0.4\ Ba(Yb_{1/2}Ta_{1/2})O_3$	–	1600/4 h	28	48 000	2	[83]
$0.6Ca(Y_{1/2}Ta_{1/2})O_3$–$0.4\ Ba(Y_{1/2}Ta_{1/2})O_3$	–	1600/4 h	27	42 000	–1	[83]
$Ca(Yb_{1/2}Ta_{1/2})O_3 + 4$ mol% $CaTiO_3$	–	1600/4 h	28	41 000	–2	[83]
$Ca(Al_{1/2}Ta_{1/2})O_3$	0.975	–	20	8500	–90	[82]
$Ca(Fe_{1/2}Ta_{1/2})O_3$	0.947	–	32	20 000	–61	[82]
$Ca(Ga_{1/2}Ta_{1/2})O_3$	0.936	1500/2 h	25	80 000	–81	[80]
$Ca_{0.5}Ba_{0.5}(Sc_{1/2}Nb_{1/2})O_3$	–	–	47	28 000		[30, 84]

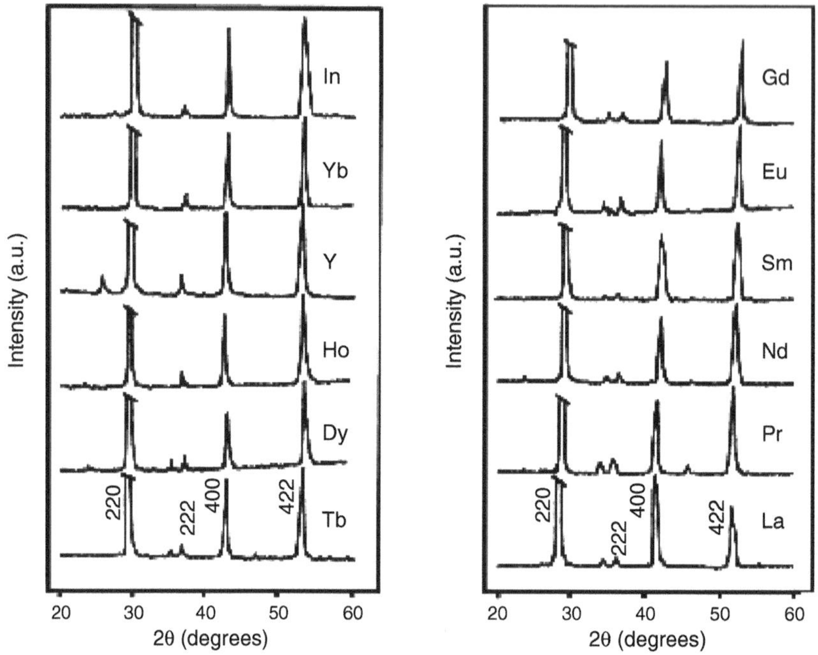

Figure 7.2 X-ray diffraction patterns of $Ba(B'_{1/2}Nb_{1/2})O_3$ ceramics. (after Ref. [22]).

The symmetry depends on the size of the rare earth ion and the resulting tolerance factor. It has been suggested that the difference in crystal symmetry is due to the small tilting of the octahedra, antiphase or inphase, which causes a splitting in some of the X-ray diffraction peaks, and the determination of the correct crystal structure is difficult [1, 22, 90, 91]. The oxygen having only eight electrons scatters very weakly and finding oxygen positions in the presence of heavier atoms such as Sr or Ta can be very difficult. X-ray diffraction is a powerful tool to observe cation order when the difference in the number of electrons between the two B cations are large. When B′ and B″ have similar number of electrons such as Y^{3+} and Nb^{5+} it is difficult to distinguish between them by X-ray powder diffraction. A detailed description about the tilting of octahedra and their effect on the symmetry of perovskites was reported by several authors [2, 3, 91–100].

Howard et al. [99] made a theoretical group analysis and discussed how to recognize the presence and type of octahedral tilting distortion from the analysis of reflection splitting and systematic absences. It was shown [18–20] that for compounds with $t < 0.985$, the symmetry is reduced from cubic due to antiphase or inphase tilting of octahedra. Hence the splitting observed in the X-ray diffraction pattern in Figure 7.2 is due to lowering of symmetry. It is difficult to establish the correct symmetry and structure of these compounds from XRD because the scattering power of the oxygen sublattice is low and the tilt angle is small [1, 18–20]. Several authors [1, 18–20] reported that complex perovskite compounds with non-cubic symmetry at room temperature are transformed to cubic at higher temperatures.

Fu and Ijdo reported [90] that $Ba(B'_{1/2}Nb_{1/2})O_3$ exhibits the sequence $I2/m$ monoclinic ($a^-a^-c^\circ$ - tilt system)–$I4/m$ tetragonal ($a^\circ a^\circ c^-$) to $Fm\bar{3}m$ cubic symmetry with no octahedral tilting. The symmetry increases as the ionic radius of the lanthanide decreases and is consistent with the increase of the tolerance factor. An increase in the tolerance factor indicates that the volume of the BO_6 octahedron is better matched to the size of the AO_{12} polyhedron, reducing the need for the octahedral tilting to accommodate the A-site cation. Since octahedral tilting is responsible for lowering the symmetry from cubic, the symmetry tends to increase as the B-type cation gets smaller. Fu and Ijdo [90] calculated the average tilt angles and found that the tilt angles increase with increasing ionic radii of the lanthanide ion. In other words, the tilt angles decrease with increasing tolerance factor as shown in Figure 7.3. Henmi et al. [87] reported that all members of $Ba(B'_{1/2}Nb_{1/2})O_3$ are monoclinic with $P2_1/n$ space group, whereas Fu and Ijdo [90] suggested that they may adopt a tetragonal symmetry $I4/m$ [B′ = Eu, Gd, Tb and Dy, monoclinic $I2/m$ [B′ = La, Pr, Nd, and Sm] and $Fm\bar{3}m$ [B′ = Y, Ho]. Dias et al. [101] investigated the crystal structure and phonon modes of $Ba(B'_{1/2}Nb_{1/2})O_3$ ceramics by Raman spectroscopy. They determined the vibrational bands of the ceramics with different chemical substitution at the B′ site (La, Nd, Sm, Gd, Tb, Y) and investigated the crystal structure on the basis of group theory analysis. Dias et al. [101], from a detailed study of $Ba(B'_{1/2}Nb_{1/2})O_3$ ceramics using X-ray diffraction and Raman spectroscopy, reported that $Ba(B'_{1/2}Nb_{1/2})O_3$ with B′ = La, Nd, and Sm are orthorhombic with *Pbnm* space group with 24 Raman active modes. $Ba(B'_{1/2}Nb_{1/2})O_3$ with B′ = Gd, Tb, Y belong to the tetragonal symmetry with space group $I4/m$ with nine Raman active modes. Figure 7.4 shows the Raman spectra of $Ba(B'_{1/2}Nb_{1/2})O_3$ with B′ = La, Gd, and Y. It is difficult to assign the correct symmetry and space group by X-ray diffraction since the X-ray atomic scattering factor of oxygen is small and the tilt angle is also small. Neutron diffraction is much more sensitive to the positions of the oxygen anions and is better suited to distinguish between two monoclinic space groups $P2_1/n$ or $I2/m$. Recent synchrotron X-ray and neutron diffraction study [90, 91] showed that $Ba(La_{1/2}Nb_{1/2})O_3$ and

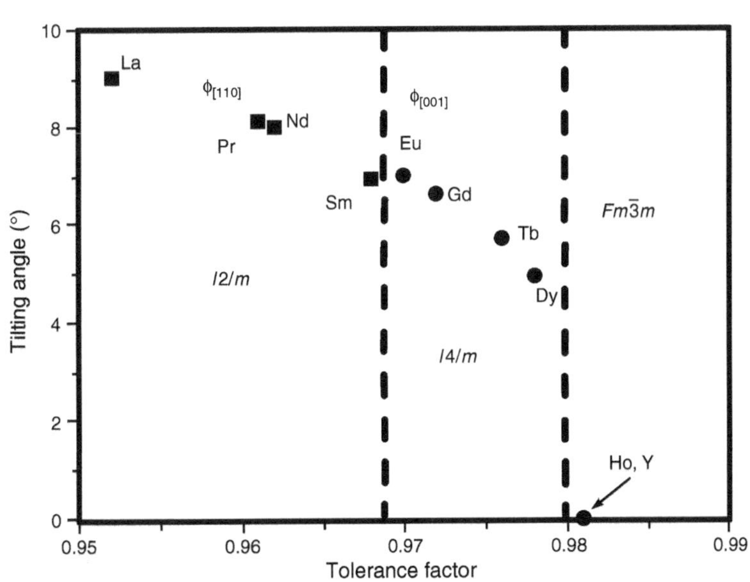

Figure 7.3 Calculated tilt angles in $Ba(B'_{1/2}Nb_{1/2})O_3$ ceramics {B′ = La, Pr, Nd, Sm, Eu, Gd, Tb, Dy, Ho, Y} as a function of tolerance factor. (after Ref. [90]).

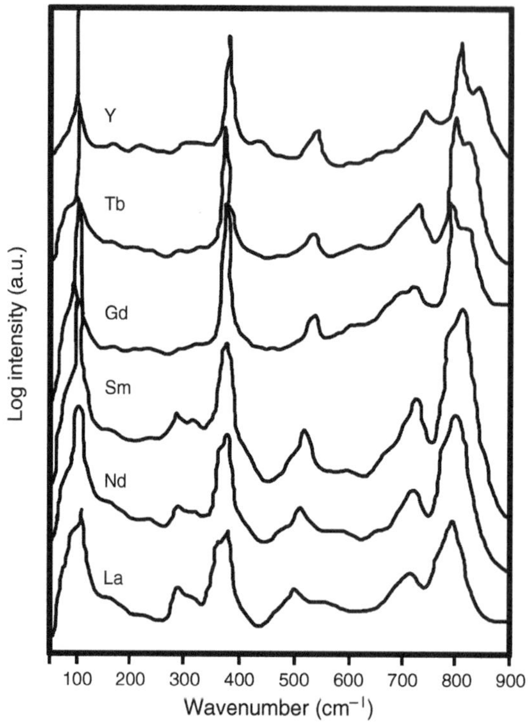

Figure 7.4 Raman spectra of $Ba(B'_{1/2}Nb_{1/2})O_3$ ceramics. Intensity in log scale (after Ref. [101]).

$Ba(Pr_{1/2}Nb_{1/2})O_3$, $Ba(Nd_{1/2}Nb_{1/2})O_3$ adopt a monoclinic symmetry with $I2/m$ space group and not $P2_1/n$ space group proposed by Henmi et al. [87]. $Ba(B'_{1/2}Nb_{1/2})O_3$ B' = Sm, Eu, Gd, Tb and Dy are tetragonal with $I4/m$ symmetry. $Ba(B'_{1/2}Nb_{1/2})O_3$ [B' = Ho, Y, Yb] are cubic with $Fm\bar{3}m$ space group. Saines et al. [91] found that $Ba(Nd_{1/2}Nb_{1/2})O_3$ undergoes a monoclinic $I2/m$ to tetragonal $I4/m$ phase transition on heating and both the phases co-exist in the temperature range 235–355°C. Above 425°C the $Ba(Nd_{1/2}Nb_{1/2})O_3$ transforms to the cubic $Fm\bar{3}m$ symmetry. Saines et al. [91] also found that $Ba(Sm_{1/2}Nb_{1/2})O_3$ undergoes a first-order $I2/m$ to $I4/m$ transition near room temperature which on heating to about 300°C transforms to the cubic symmetry.

The microwave dielectric properties of $Ba(B'_{1/2}Nb_{1/2})O_3$ ceramics are given in Table 7.1. The different $Ba(B'_{1/2}Nb_{1/2})O_3$ ceramics have ε_r in the range 36–45. Figure 7.5 shows the variation of ε_r and B' ionic polarizability of $Ba(B'_{1/2}Nb_{1/2})O_3$ ceramics with B' ionic radii. The ε_r and ionic polarizability of B' ions increase with increase in B' ionic radii. The variation of ε_r with tolerance factor t is shown in Figure 7.6 which is in agreement with the report of Reaney et al. [20] for complex perovskites. The inphase, antiphase and untilted regions are marked in Figures 7.5 and Figure 7.6. The ε_r of $Ba(B'_{1/2}Nb_{1/2})O_3$ ceramics decrease with increase in t as shown in Figure 7.6. Since t is related to packing of ions in the perovskite cell, when t deviates from one, the perovskite cell gets distorted and the symmetry is lowered from cubic. This deviation from cubic symmetry results in additional polarization and is reflected in the permittivity. The larger the deviation from cubic symmetry, the larger is the ε_r (see Table 7.1) for a particular family of complex perovskites. The ε_r increases with increase in dielectric polarizability [22, 102–104]. $Ba(B'_{1/2}Nb_{1/2})O_3$ have high quality

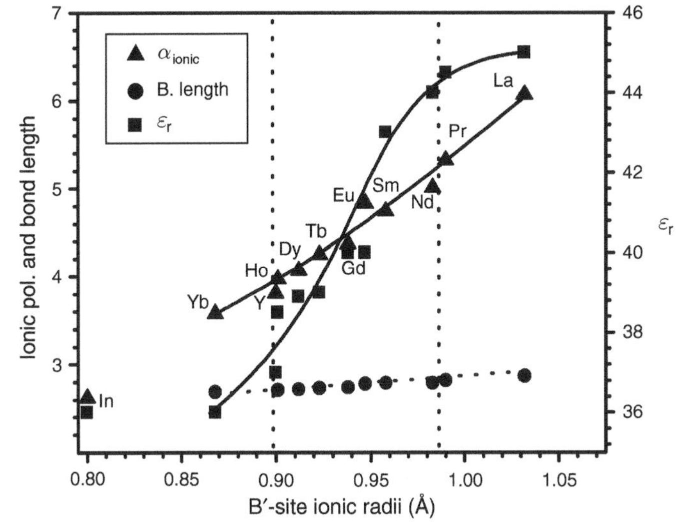

Figure 7.5 Variation of permittivity, ionic polarizability and bond length with ionic radii of B′ ions. Dotted lines separate the untilted (U), antiphase tilted (A) and inphase tilted (I) regions given by Reaney et al. [20]. In and Y are non-lanthanides (after Ref. [22]).

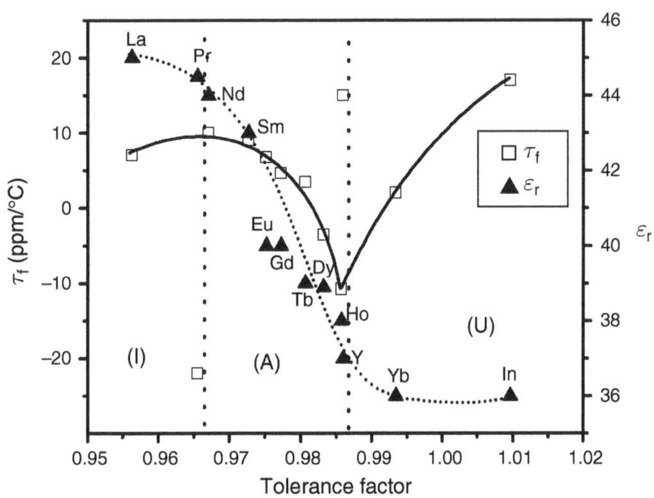

Figure 7.6 Permittivity and temperature coefficient of resonant frequency are related to the tolerance factor. Dotted lines separate the untilted (U), antiphase. (after Ref. [22]).

factor up to about 52 000 GHz [17, 22]. $Ba(B'_{1/2}Nb_{1/2})O_3$ have relatively low τ_f values in the range from −22 to + 17 ppm/°C [17, 22]. The τ_f values vary non-linearly with the tolerance factor as shown in Figure 7.6. Khalam et al. [22] calculated the bond valence of $Ba(B'_{1/2}Nb_{1/2})O_3$ ceramics using the bond parameters of Brown and Altermaut [105].

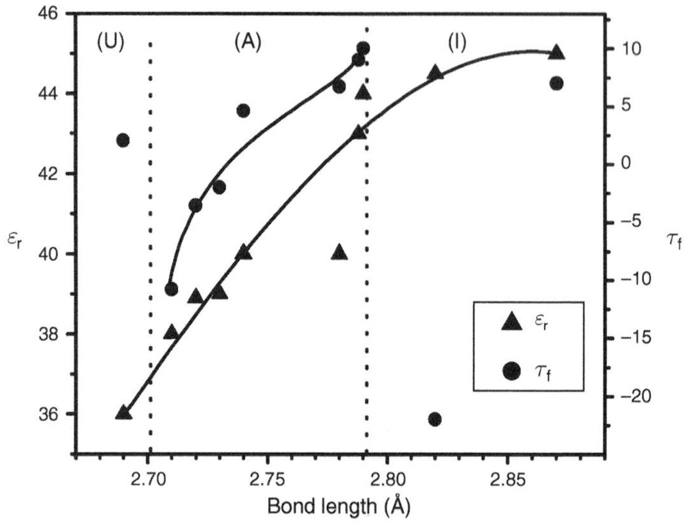

Figure 7.7 ε_r and τ_f of $Ba(B'_{1/2}Nb_{1/2})O_3$ ceramics are related to bond length of rare earth ions. (U), (A) and (I) are untilted, antiphase tilted and inphase tilted, respectively (after Ref. [22]).

Figure7.7 shows the variation of ε_r and τ_f of $Ba(B'_{1/2}Nb_{1/2})O_3$ ceramics with bond length. The ε_r increases with increase in B′-O bond length and bond valence. Koshy and co-workers [86, 106] reported that $Ba(B'_{1/2}Nb_{1/2})O_3$ ceramics are useful as substrates for YBCO since there is no chemical reaction between them.

7.3 $Ba(B'_{1/2}Ta_{1/2})O_3$

Several people reported [8, 13–15, 17, 18, 23, 25, 89, 107] the preparation and characterization of $Ba(Ln_{1/2}Ta_{1/2})O_3$ complex perovskites. The ceramics are prepared by calcining stoichiometric amount of mixed raw materials at temperatures in the range 1200–1400°C and sintering at temperatures in the range 1550–1650°C [17, 23]. Galasso et al. [8] obtained single crystals of $Ba(Ln_{1/2}Ta_{1/2})O_3$ [Ln = La, Gd, Y, Sc, Lu] up to 0.5 mm in size by the flux method using BaF_2. Several authors [2, 17, 18, 19, 23, 107–113] investigated the crystal structure of $Ba(Ln_{1/2}Ta_{1/2})O_3$ and reported to have cubic, tetragonal, orthorhombic and monoclinic symmetries depending on the size of rare earth ions. Figure 7.8 shows the X-ray diffraction pattern recorded from $Ba(B'_{1/2}Ta_{1/2})O_3$ ceramics. $Ba(Y_{1/2}Ta_{1/2})O_3$ was reported to be [8, 13, 18, 23, 108, 109] cubic belonging to the space group $Fm\bar{3}m$ (Z = 4) with a low-temperature phase transition (PT) to a tetragonal ($I4/m$, Z = 4) symmetry at around 253 K [18, 108, 110]. The same phase transition sequence with temperature was observed at high temperatures for Gd and Nd compounds, so that these materials would have the tetragonal structure at room temperature, confirming the proposition of Galasso et al. [8, 13] for Gd. On the other hand, Doi and Hinatsu [109] recently reported that $Ba(B'_{1/2}Ta_{1/2})O_3$ would be cubic for B′ = Y, tetragonal for B′ = Lu–Dy and monoclinic $P2_1/n$ symmetries for B′ = Tb–La. However, Khalam and co-workers [23] and Gregoria et al. [108] from X-ray diffraction and Raman spectroscopic studies reported that the

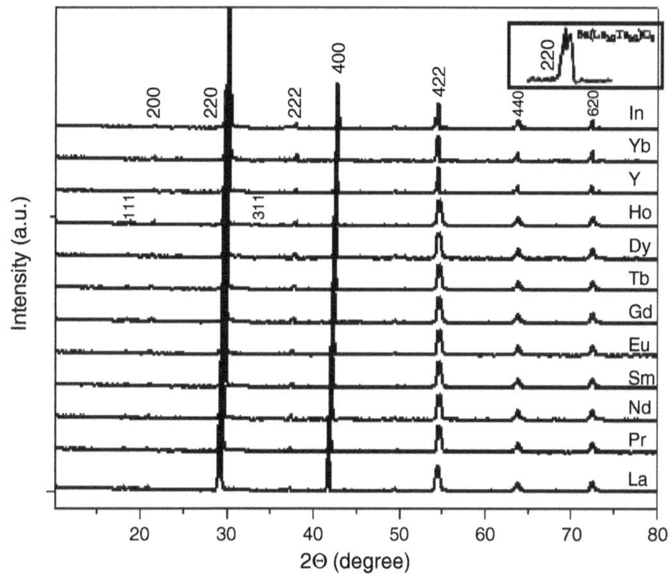

Figure 7.8 The X-ray diffraction patterns of $Ba(B'_{1/2}Ta_{1/2})O_3$ [B' = La, Pr, Nd, Sm, Eu, Gd, Tb, Dy, Ho, Y, Yb and In] ceramics (after Ref. [23]).

$Ba(B'_{1/2}Ta_{1/2})O_3$ ceramics based on Yb, Y and Ho are cubic with four intense Raman bands, ceramics based on Dy, Tb, Gd and Eu are suggested to be tetragonal showing nine Raman active bands. The materials with B' = La, Nd, Sm are orthorhombic with space group *Pbnm* and exhibited 16 observed bands of the 24 predicted ones. Figure 7.9 (a,b,c) shows the Raman spectra of these three groups of materials. $Ba(In_{1/2}Ta_{1/2})O_3$ belongs to

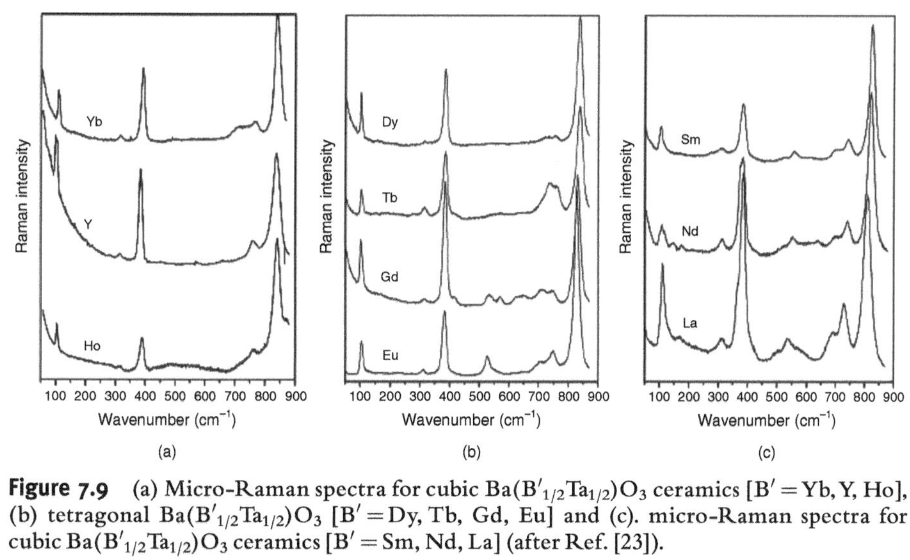

Figure 7.9 (a) Micro-Raman spectra for cubic $Ba(B'_{1/2}Ta_{1/2})O_3$ ceramics [B' = Yb, Y, Ho], (b) tetragonal $Ba(B'_{1/2}Ta_{1/2})O_3$ [B' = Dy, Tb, Gd, Eu] and (c). micro-Raman spectra for cubic $Ba(B'_{1/2}Ta_{1/2})O_3$ ceramics [B' = Sm, Nd, La] (after Ref. [23]).

the tetragonal *P4/mnc* space group [114]. There are controversies about the correct crystal symmetries of $Ba(B'_{1/2}Ta_{1/2})O_3$ [13, 23, 107–109, 113] and further detailed study is needed to establish the correct symmetries of this group of materials.

The microwave dielectric properties of the $Ba(Ln_{1/2}Ta_{1/2})O_3$ ceramics are given in Table 7.1. The permittivity varies in the range 32–39 and *Qf* up to 50 000 GHz. The permittivity linearly increases with increase in the B′ ionic radii except in the case of lanthanum. Khalam [115] reported that the addition of a small amount (0.5 wt%) of MgO, ZnO or CuO significantly improved the quality factor of $Ba(Sm_{1/2}Ta_{1/2})O_3$ ceramics. The τ_f changes drastically as a function of the tolerance factor as shown in Figure 7.10. The τ_f of $Ba(B'_{1/2}Ta_{1/2})O_3$ varies from a high positive value of 120 ppm/°C to a negative value of −48 ppm/°C depending on the rare earth ion and the tolerance factor. The variation of ε_r and τ_f with bond length of B′-site ions is shown in Figure 7.11 . The ε_r increases linearly with increase in bond length except for La. Both ε_r and τ_f vary with bond valence also in a way similar with that of B′–O bond length. Bond length and bond valence are related to ionic radius and tolerance factor (*t*). The tilting of oxygen octahedra depends on *t* and the dielectric properties of $Ba(B'_{1/2}Ta_{1/2})O_3$ ceramics are closely related to ionic radius, tolerance factor and bond valence.

Zurmuhlen et al. [18, 19] and Gregoria et al. [108] reported that $Ba(Y_{1/2}Ta_{1/2})O_3$, $Ba(Gd_{1/2}Ta_{1/2})O_3$, $Ba(Y_{1/2}Nb_{1/2})O_3$ and $Ba(Gd_{1/2}Nb_{1/2})O_3$ exhibit a high temperature phase $Fm\bar{3}m$ and a low temperature tetragonal *I4/m* phase. These materials undergo a ferroelastic second-order transition characterized by an antiphase tilt of the oxygen octahedra around the <001> axis. Neutron diffraction study showed that the oxygen octahedra rotation angle is the order parameter for the phase transition. Neutron scattering, TEM and DSC studies indicated that the improper ferroelastic second-order transition in $Ba(Y_{1/2}Ta_{1/2})O_3$ occurs at the critical temperature 253 K characterized by an antiphase tilting of the oxygen octahedra around the *c*-axis. Due to

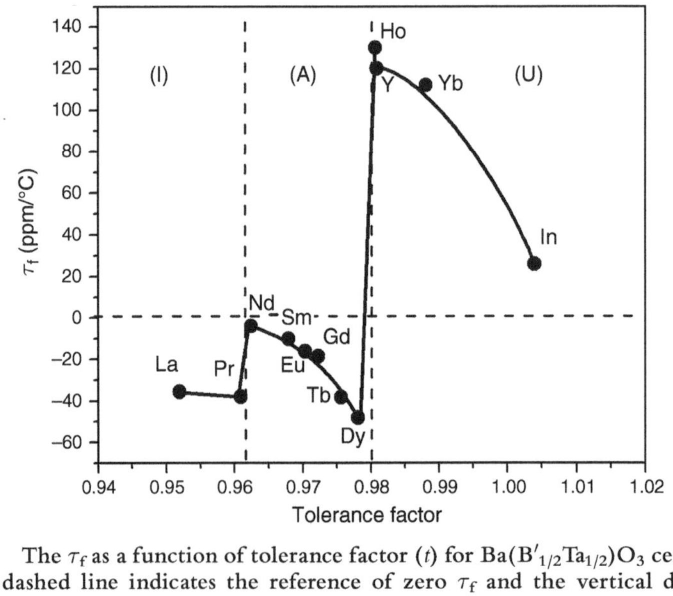

Figure 7.10 The τ_f as a function of tolerance factor (*t*) for $Ba(B'_{1/2}Ta_{1/2})O_3$ ceramics. The horizontal dashed line indicates the reference of zero τ_f and the vertical dashed lines indicate approximately the critical values of t according to Reaney et al. [20] (after Ref. [23]).

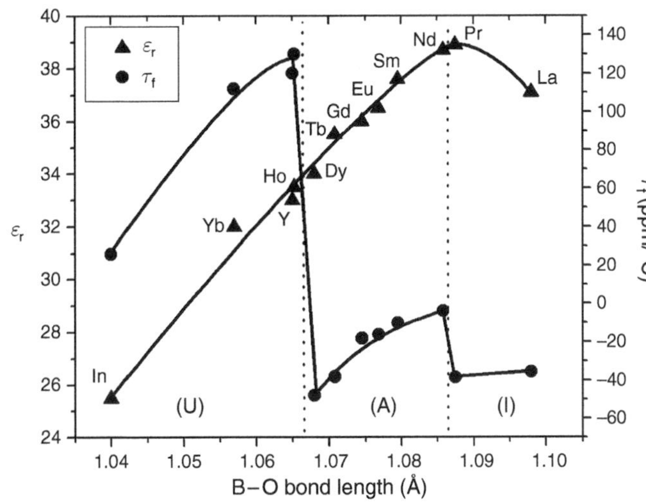

Figure 7.11 Variation of ε_r and τ_f of $Ba(B'_{1/2}Ta_{1/2})O_3$ ceramics versus bond length (B′-O). (I), (A) and (U) are the inphase, anti phase and untilted transition phases, respectively. (after Ref. [23]).

small change in specific heat associated with the transition, large amount of samples (216 mg) and high heating rate (40 K/min) were required to have a sufficiently large signal.

7.4 $Sr(B'_{1/2}Nb_{1/2})O_3$

$Sr(B'_{1/2}Nb_{1/2})O_3$ ceramics were prepared [17, 27] by calcining the stoichiometric amounts of the ball milled raw materials at about 1250°C and sintering the pellets at about 1600°C/4 h. Khalam and Sebastian [27] reported that it is difficult to densify $Sr(B'_{1/2}Nb_{1/2})O_3$ [B′ = La, Pr, Nd, Sm, Eu, Gd, Tb, Dy, Ho, Y, Er, Yb and In] ceramics. Hence they added 0.5 wt% CeO_2 to the calcined powder to improve the percentage densification of the ceramics to about 97% on sintering at temperatures in the range 1575–1600°C/4 h. Figure 7.12 shows a typical scanning electron micrograph of 0.5 wt% CeO_2-added $Sr(B'_{1/2}Nb_{1/2})O_3$ ceramic showing well-packed grains of the size of about 5–10 μm. Since the melting point of CeO_2 is about 2000°C, no liquid phase sintering is expected due to the addition of the CeO_2 sintering aid. X-ray diffraction patterns of $Sr(B'_{1/2}Nb_{1/2})O_3$ ceramics with B′ = La, Pr, Nd, Sm, Eu, Gd, Tb, Dy, Ho, Y, Er, Yb and In are shown in Figure 7.13. The X-ray diffraction patterns of $Sr(B'_{1/2}Nb_{1/2})O_3$ ceramics with B′ = La, Pr and Nd are in agreement for the cubic symmetry. In the $Sr(B'_{1/2}Nb_{1/2})O_3$ ceramics the B-site ions have a charge difference of two and a size difference between 0.16 and 0.39 Å. Thus both the charge and size differences of B′-site and Nb ions lead to B-site cation ordering. In the X-ray diffraction pattern of $Sr(B'_{1/2}Nb_{1/2})O_3$ ceramics (Figure 7.13), the intensities of such superstructure reflections indicating ordering decrease gradually from In to Nd in the series. It has been reported [28] that strontium-based perovskites undergo a structural transition from cubic

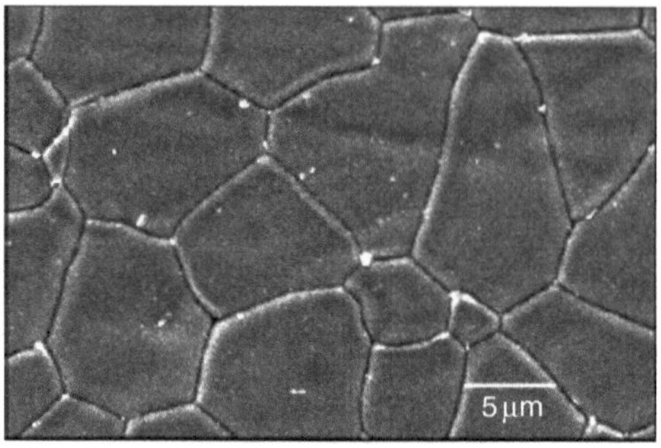

Figure 7.12 Surface morphology of 0.5 wt.% CeO_2-added $Sr(Sm_{1/2}Nb_{1/2})O_3$ ceramic (after Ref. [27]).

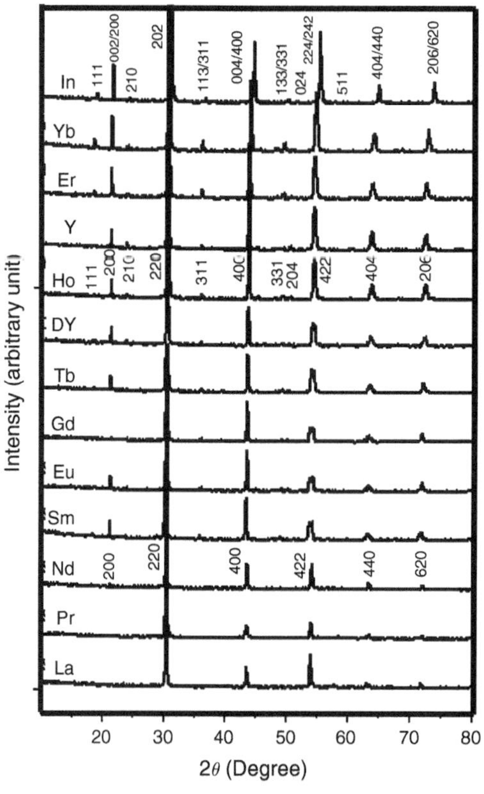

Figure 7.13 X-ray diffraction pattern recorded from $Sr(B'_{1/2}Nb_{1/2})O_3$ ceramics. (after Ref. [27]).

to a structure with a lower symmetry with decreasing ionic size of the B′-site element. The XRD patterns of $Sr(B'_{1/2}Nb_{1/2})O_3$ ceramics with B′ = Sm, Eu, Gd, Tb, Dy, Ho, Y, Er, Yb and In show splitting of some of the reflections. The X-ray diffraction patterns of $Sr(B'_{1/2}Nb_{1/2})O_3$ ceramics with B′ = Sm, Eu, Gd, Tb, Dy and Ho are comparable to that of

$Sr(Sm_{1/2}Ta_{1/2})O_3$ [116] which have a tetragonal symmetry with space group *I4/m*. The X-ray diffraction patterns of $Sr(Y_{1/2}Nb_{1/2})O_3$, $Sr(Er_{1/2}Nb_{1/2})O_3$, $Sr(Yb_{1/2}Nb_{1/2})O_3$ and $Sr(In_{1/2}Nb_{1/2})O_3$ are similar to the orthorhombic symmetry reported [117] for $Sr(Ni_{0.5}W_{0.5})O_3$ with space group *Pnma*. However, Howard et al. reported [118] a monoclinic symmetry with space group $P2_1/n$ for $Sr(Y_{1/2}Nb_{1/2})O_3$. The unit cell volumes of $Sr(B'_{1/2}Nb_{1/2})O_3$ ceramics increase and tolerance factors decrease with increase in the B′-site ionic radii. There is disagreement about the correct crystal symmetries of $Sr(B'_{1/2}Nb_{1/2})O_3$ ceramics and monoclinic, tetragonal, cubic or orthorhombic symmetries have been proposed [27, 28, 116, 119, 120]. The tolerance factors of $Sr(B'_{1/2}Nb_{1/2})O_3$ ceramics are calculated using Shannon's ionic radii [121] and are given in Table 7.1. $Sr(B'_{1/2}Nb_{1/2})O_3$ ceramics have a permittivity in the range 26–45. The ε_r increases linearly with increase in B′-site ionic radii with the exception of La (Figure 7.14). $Sr(B'_{1/2}Nb_{1/2})O_3$ ceramics have quality factors (*Qf*) in the range 3250–44 000 GHz (see Table 7.1). $Sr(B'_{1/2}Nb_{1/2})O_3$ ceramics with B′ = La, Pr, Nd and Gd have poor ordering and show relatively poor quality factors as compared to other ceramics in the system. The ceramics with lower symmetries show B-site ordering as evidenced by superstructure reflections in the X-ray diffraction pattern (Figure 7.13) and they show relatively high quality factors. $Sr(Eu_{1/2}Nb_{1/2})O_3$ has the highest quality factor of 44 000 GHz. The τ_f of $Sr(B'_{1/2}Nb_{1/2})O_3$ ceramics varies from −20 to −73 ppm/°C (see Table 7.1). The τ_f increases in the negative side with decrease in ionic radii of B′-site ions. The $|\tau_f|$ increases linearly with increase in the tolerance factor with the exception of $Sr(In_{1/2}Nb_{1/2})O_3$ as shown in Figure 7.15. It should be noted that $Sr(In_{1/2}Nb_{1/2})O_3$ does not belong to the lanthanide family. A similar variation of temperature coefficient of permittivity versus tolerance factor was reported by Reaney et al. [20, 122] for Ba- and Sr-based complex perovskites. Ratheesh et al. [28] from Raman spectroscopic study found that $Sr(B'_{1/2}Ta_{1/2})O_3$ compounds have a higher degree of order as compared to their niobium analogues. They also found a correlation between the tolerance factor *t* and the A_{1g} mode as shown in Figure 7.16. As the tolerance factor decreases from 1.018 to 0.903, the A_{1g} mode shows a decrease in the wave number from 849 to 762 cm^{-1}. This shift of the A_{1g} mode is

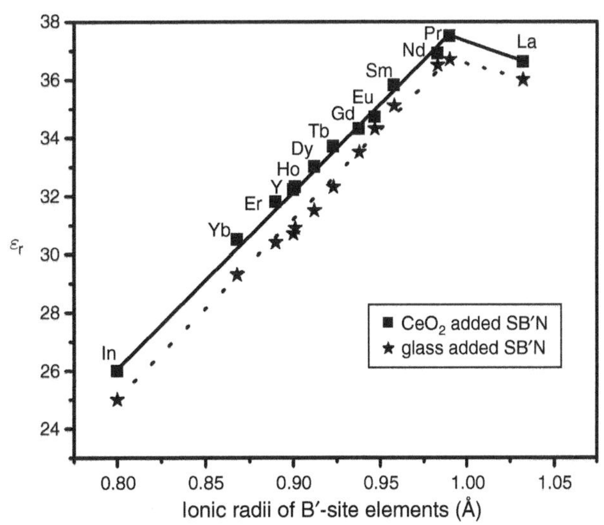

Figure 7.14 The variation in the ε_r of CeO_2-added and B_2O_3-added $Sr(B'_{1/2}Nb_{1/2})O_3$ ceramics versus B′-site ionic radii (after Ref. [27]).

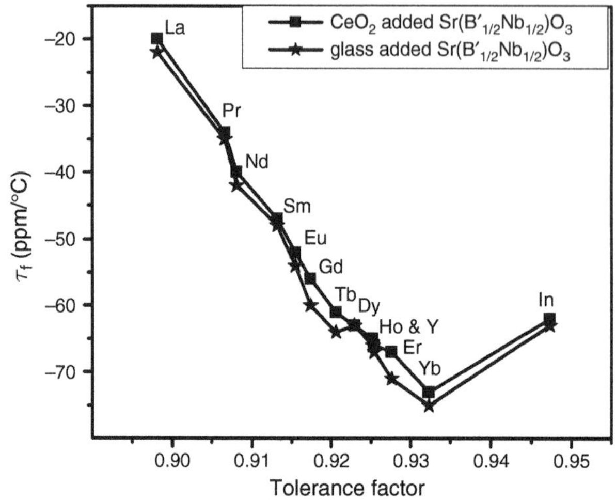

Figure 7.15 The variation in the τ_f of CeO_2- and B_2O_3-added $Sr(B'_{1/2}Nb_{1/2})O_3$ ceramics versus tolerance factor (after Ref. [27]).

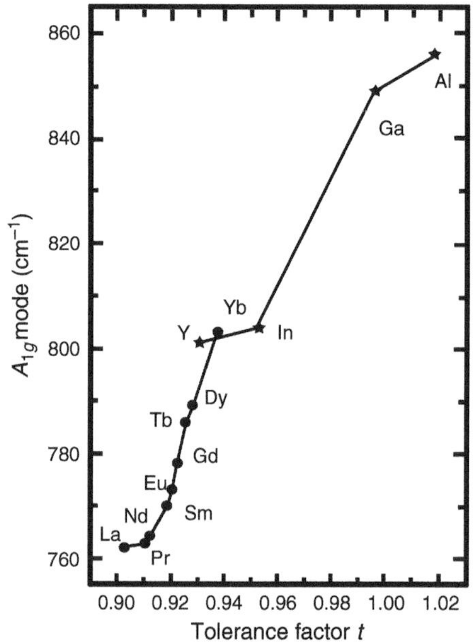

Figure 7.16 Frequency of the A_{1g} mode tolerance factor t for rare earths (solid squares), In, Ga and Al (stars) occupying a B′ site. The solid lines are only guides for the eye (after Ref. [28]).

attributed to the influence of B′ ions since all the compounds have the same octahedral coordination with Nb.

$Sr(B'_{1/2}Nb_{1/2})O_3$ [B′ = La, Pr, Nd, Sm, Eu, Gd, Tb, Dy, Ho, Y, Er, Yb, In, Al and Cr] have a relatively very high sintering temperature (≈1600°C). Hence, in order to

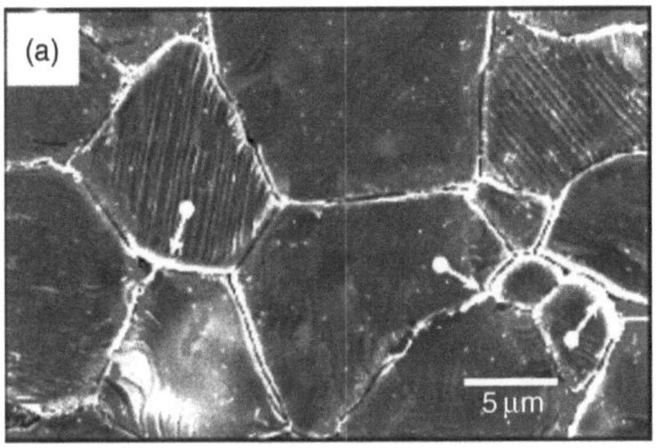

Figure 7.17 Scanning electron micrographs of the 0.2 wt% B_2O_3-added $Sr(Sm_{1/2}Nb_{1/2})O_3$ ceramic sample (after Ref. [27]).

lower the sintering temperature of $Sr(B'_{1/2}Nb_{1/2})O_3$ ceramics, Khalam [115] added B_2O_3 glass. It was found that addition of 0.2 wt% B_2O_3 glass lowered the sintering temperature to about 1350°C without any appreciable change in the microwave dielectric properties. Addition of more than 0.2 wt% B_2O_3 glass to $Sr(B'_{1/2}Nb_{1/2})O_3$, considerably degraded the dielectric properties although the sintering temperature is further lowered. The addition of 0.2 wt% B_2O_3 to the calcined powders of $Sr(B'_{1/2}Nb_{1/2})O_3$ [B' = La, Pr, Nd, Sm, Eu, Gd, Tb, Dy, Ho, Y, Er, Yb, In, Al, Fe and Cr] has promoted the densification of the ceramics to 95–98.5%. Figure 7.17 shows the SEM recorded from a 0.2 wt% B_2O_3-added $Sr(Sm_{1/2}Nb_{1/2})O_3$ sample. The grains appeared as well packed with average size of 5–10 µm. The 0. 2 wt% B_2O_3 added ceramics have a slightly lower ε_r and τ_f as shown in Figures 7.14 and Figure 7.15.

7.4.1 Tailoring of τ_f in $Sr(B'_{1/2}Nb_{1/2})O_3$ ceramics

The dielectric properties of ceramics especially τ_f can be tuned [27, 81, 123] by suitable substitutions at A and/or B sites of $A(B'_{1/2}B''_{1/2})O_3$ ceramics. Several authors [21, 27, 85] tailored the properties of $Sr(B'_{1/2}Nb_{1/2})O_3$ by suitable substitution at A or B sites. Khalam and Sebastian [27] prepared $(Sr_{1-x}Ba_x)(Y_{1/2}Nb_{1/2})O_3$ compositions for different values of x in order to tune the τ_f. $(Sr_{1-x}Ba_x)(Y_{1/2}Nb_{1/2})O_3$ ceramic undergoes orthorhombic–tetragonal–cubic phase transitions with the substitution of larger Ba^{2+} at the A site. $Ba(Y_{1/2}Nb_{1/2})O_3$ ceramic has $\varepsilon_r = 37$, $Qf = 49\,600$ GHz and $\tau_f = +15$ ppm/°C, whereas $Sr(Y_{1/2}Nb_{1/2})O_3$ ceramic has $\varepsilon_r = 32$, $Qf = 38\,900$ GHz and $\tau_f = -66$ ppm/°C. The tolerance factors of $Ba(Y_{1/2}Nb_{1/2})O_3$ and $Sr(Y_{1/2}Nb_{1/2})O_3$ are 0.981 and 0.925 respectively. The ε_r and Qf increased with the substitution of bigger Ba for the smaller Sr ions in the A site. The τ_f of $(Sr_{1-x}Ba_x)(Y_{1/2}Nb_{1/2})O_3$ ceramic changed from −66 to +15 ppm/°C for x between 0 and 1. The variation of τ_f with tolerance factor (t) is shown in Figure 7.18. The figure shows two abrupt changes in τ_f at $t = 0.961$ and 0.98 which are due to phase transitions. The minimum value of τ_f with the variation of t is due to the lowest energy state attained by the perovskite due to the octahedral tilting [124]. The τ_f of the composition changed from a negative to a positive value by the substitution of Sr by Ba in the A site. The solid solution $(Sr_{1-x}Ba_x)(Y_{1/2}Nb_{1/2})O_3$ with $x = 0.65$ has zero τ_f with $\varepsilon_r = 34$ and $Qf = 45\,600$ GHz.

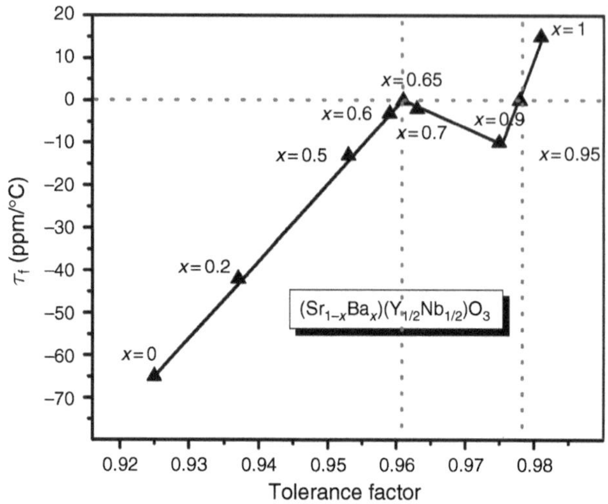

Figure 7.18 The variation of τ_f versus tolerance factor of the solid solution $(Sr_{1-x}Ba_x)(Y_{1/2}Nb_{1/2})O_3$ (after Ref. [27]).

7.5 $Sr(B'_{1/2}Ta_{1/2})O_3$

$Sr(B'_{0.5}Ta_{0.5})O_3$ [B' = La, Pr, Nd, Sm, Eu, Gd, Tb, Dy, Ho, Y, Er and Yb] ceramics are prepared from stoichiometric mixtures of high purity $SrCO_3$, rare earth oxides and Ta_2O_5 by the solid-state ceramic route [17, 26]. The mixed powders are calcined at 1250°C/4 h. $Sr(B'_{0.5}Ta_{0.5})O_3$ ceramics with B' = La, Pr, Nd, Sm, Eu, Gd, Tb, Dy, Er are sintered into dense form at 1600°C/4 h. It was found that addition of a small amount of (0.25 wt%) Nb_2O_5 improved the densification of $Sr(B'_{0.5}Ta_{0.5})O_3$ ceramics. It was reported that Nb_2O_5 addition improve the sinterability of $Ba(B'_{1/2}Ta_{1/2})O_3$ ceramics [23] and $MgTiO_3$ ceramics [125]. Figure 7.19 shows the X-ray diffraction patterns of sintered and powdered $Sr(B'_{0.5}Ta_{0.5})O_3$ ceramics. X-ray diffraction patterns of the sintered ceramics showed superstructure reflections which are marked by ★ in Figure 7.19. The doubling of perovskite unit cell in the $Sr(B'_{0.5}Ta_{0.5})O_3$ structure indicates the ordering of B-site cations. $Sr(Y_{0.5}Ta_{0.5})O_3$ and $Sr(La_{0.5}Ta_{0.5})O_3$ have rhombohedral symmetry with space group $R3m$ [26, 110]. However, more recently Howard et al. [118] reported that $Sr(Y_{0.5}Ta_{0.5})O_3$ is monoclinic with $P2_1/n$ space group. $Sr(B'_{1/2}Ta_{1/2})O_3$ have rhombohedral, monoclinic or tetragonal structure [26, 110, 116, 120] depending on the ionic radii or tolerance factor. Several authors [81, 126, 127] reported $Sr(Ga_{1/2}Ta_{1/2})O_3$ as cubic with $Fm\bar{3}m$ space group. However, later works showed [94, 128] that $Sr(Ga_{1/2}Ta_{1/2})O_3$ is not cubic but is tetragonal with space group $I4/m$. $Sr(B'_{0.5}Ta_{0.5})O_3$ ceramics were sintered up to 98% of the theoretical density with average grain size 5–15 μm. The dielectric properties of $Sr(B'_{0.5}Ta_{0.5})O_3$ ceramics with the addition of 0.5 wt% of Nb_2O_5 are given in Table 7.1. Khalam and Sebastian [26] reported an inverse relation between ε_r and tolerance factor (t) of the ceramics. The ε_r decreases as the tolerance factor increases. The tolerance factors of $Sr(B'_{0.5}Ta_{0.5})O_3$ ceramics are calculated using Shannon's ionic radii [121]. The B-site ionic size difference ($r_{B'}-r_{Ta}$) increases as B'-site elements vary from Yb to La. The variation of ε_r of

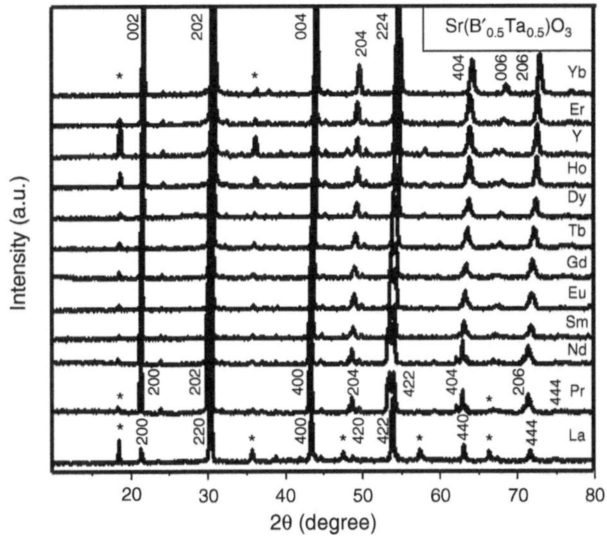

Figure 7.19 X-ray powder diffraction pattern of sintered and powdered $Sr(B'_{0.5}Ta_{0.5})O_3$ [B' = La, Pr, Nd, Sm, Eu, Gd, Tb, Dy, Ho, Y, Er and Yb] ceramics. ⋆ represents the superlattice reflections (after Ref. [26]).

$Sr(B'_{0.5}Ta_{0.5})O_3$ ceramics with the size of B'-site elements is shown in Figure 7.20. $Sr(B'_{0.5}Ta_{0.5})O_3$ [B' = La, Pr and Gd] ceramics show relatively poor quality factors. $Sr(Y_{0.5}Ta_{0.5})O_3$) has a high quality factor of 54 300 GHz and $Sr(Ga_{1/2}Ta_{1/2})O_3$ shows the highest quality factor of 91 000 GHz [81]. The temperature coefficients of resonant

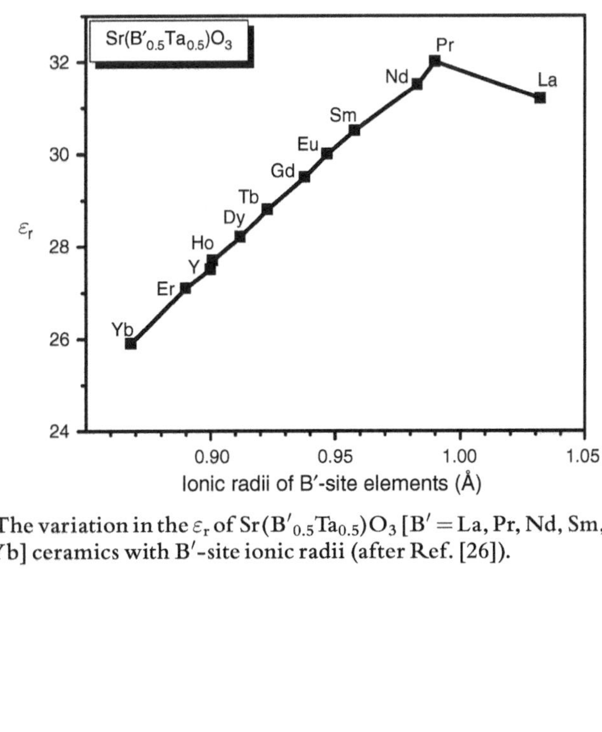

Figure 7.20 The variation in the ε_r of $Sr(B'_{0.5}Ta_{0.5})O_3$ [B' = La, Pr, Nd, Sm, Eu, Gd, Tb, Dy, Ho, Y, Er and Yb] ceramics with B'-site ionic radii (after Ref. [26]).

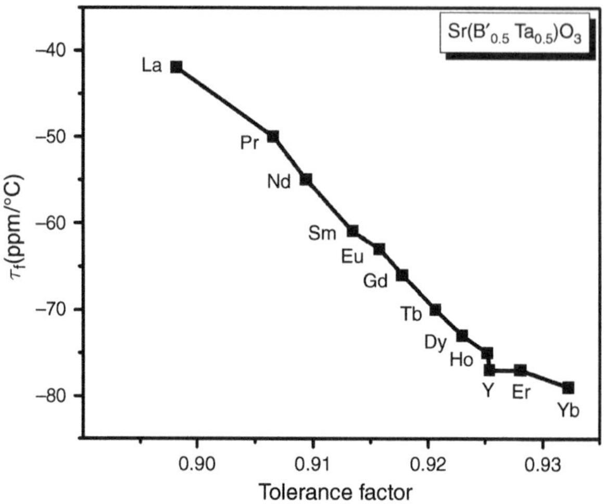

Figure 7.21 The variation in the τ_f of $Sr(B'_{0.5}Ta_{0.5})O_3$ [B' = La, Pr, Nd, Sm, Eu, Gd, Tb, Dy, Ho, Y, Er and Yb] ceramics with tolerance factor (after Ref. [26]).

frequency (τ_f) of $Sr(B'_{0.5}Ta_{0.5})O_3$ ceramics is inversely related to the B′-site ionic radii (Table 7.1). The variation of τ_f versus tolerance factor is plotted in Figure 7.21. The negative value of τ_f increases with increase in t. The τ_f of $Sr(B'_{0.5}Ta_{0.5})O_3$ ceramics varies from −79 to −42 ppm/°C.

7.5.1 Effect of non-stoichiometry on the dielectric properties of $Sr(B'_{0.5}Ta_{0.5})O_3$ ceramics

Khalam and Sebastian [26] reported that a slight non-stoichiometry of A- or B-site ions has considerable influence on the microwave dielectric properties of $Sr(B'_{0.5}Ta_{0.5})O_3$ ceramics. They prepared $Sr_{1+x}(Eu_{0.5}Ta_{0.5})O_3$, $Sr(Eu_{0.5+y}Ta_{0.5})O_3$ and $Sr(Eu_{0.5}Ta_{0.5+z})O_3$ with $x = y = z = -0.02$–0.1 by conventional solid-state ceramic route. $Sr(Eu_{0.5}Ta_{0.5})O_3$ has $\varepsilon_r = 30$, $Q_u \times f = 45\ 500$ GHz and $\tau_f = -63$ ppm/°C. Figure 7.22a shows the variations in the ε_r of $Sr_{1+x}(Eu_{0.5}Ta_{0.5})O_3$, $Sr(Eu_{0.5+y}Ta_{0.5})O_3$ and $Sr(Eu_{0.5}Ta_{0.5+z})O_3$ for $x = y = z = -0.02$–0.1. The relative permittivities of $Sr_{1+x}(Eu_{0.5}Ta_{0.5})O_3$ and $Sr(Eu_{0.5}Ta_{0.5+z})O_3$ have decreased for both positive and negative values of x and z respectively. The ε_r of $Sr(Eu_{0.5+y}Ta_{0.5})O_3$ has decreased with positive value of y (excess of Eu) and increased for Eu deficiency as shown in Figure 7.22a. With the deficiency of Eu in $Sr(Eu_{0.5+y}Ta_{0.5})O_3$, the density decreased, which is attributed to the formation of $Sr_4Ta_2O_9$ secondary phase. $Sr_4Ta_2O_9$ has a high relative permittivity of 38 and low quality factor (5600 GHz) as compared to that of $Sr(Eu_{0.5}Ta_{0.5})O_3$ ceramic. The permittivity of $Sr(Eu_{0.5+y}Ta_{0.5})O_3$ has increased for negative values of y (deficiency) due to the presence of $Sr_4Ta_2O_9$ secondary phase which has a higher ε_r. Figure 7.22b shows the variations in the normalized quality factors of $Sr_{1+x}(Eu_{0.5}Ta_{0.5})O_3$, $Sr(Eu_{0.5+y}Ta_{0.5})O_3$ and $Sr(Eu_{0.5}Ta_{0.5+z})O_3$ for $x = y = z = -0.02$–0.1. The Q-factor of $Sr_{1+x}(Eu_{0.5}Ta_{0.5})O_3$ has increased when x becomes negative (Sr deficient). It reached a maximum (59 500 GHz) at $x = -0.005$ and then decreased. The Qf has decreased for positive value of x (excess Sr). A similar increase in the Qf value in $Ba_{1-x}(Mg_{1/3}Ta_{2/3})O_3$

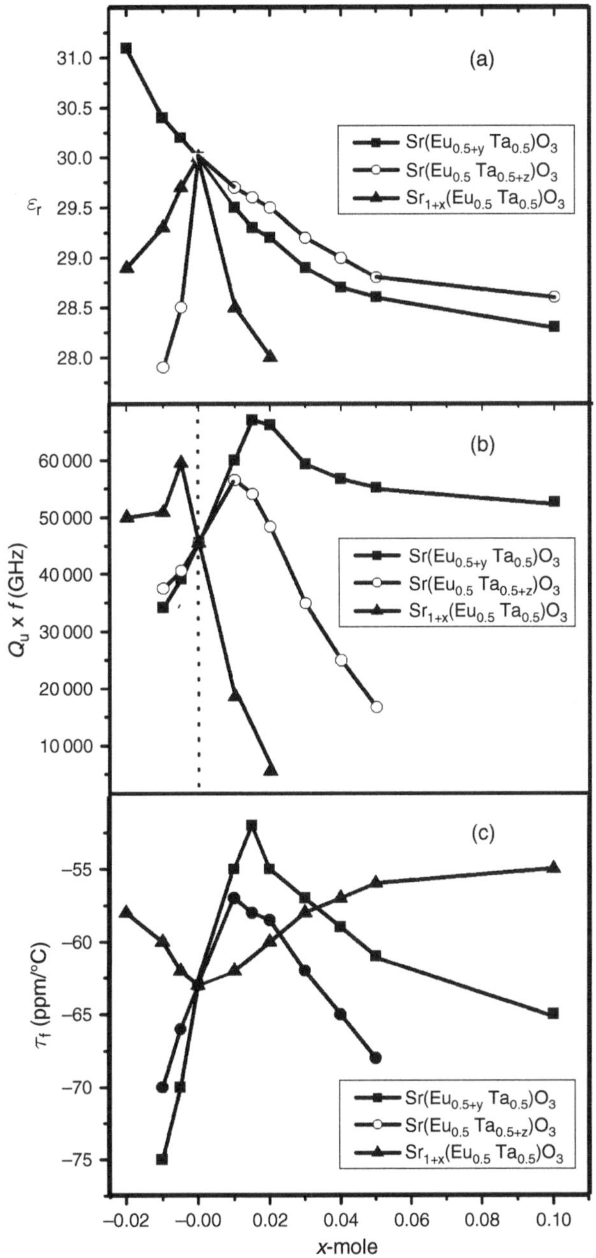

Figure 7.22 The variation in the dielectric properties of $Sr_{1+x}(Eu_{0.5}Ta_{0.5})O_3$, $Sr(Eu_{0.5+y}Ta_{0.5})O_3$ and $Sr(Eu_{0.5}Ta_{0.5+z})O_3$ with $x = y = z = -0.02$–0.1 (after Ref. [26]).

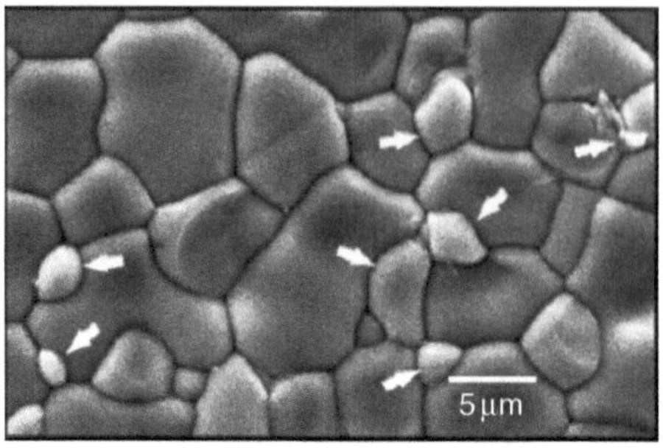

Figure 7.23 The surface morphology of $Sr(Eu_{0.5+y}Ta_{0.5})O_3$ ceramic with $y = -0.02$. The $Sr_4Ta_2O_9$ phase is represented by white arrows (after Ref. [26]).

was reported [129] with Ba deficiency. The *Q*-factor of $Sr(Eu_{0.5+y}Ta_{0.5})O_3$ and $Sr(Eu_{0.5}Ta_{0.5+z})O_3$ have increased for positive values of y and z (excess Eu and Ta). These materials show maximum *Qf* values (67 000 and 56 500 GHz) for $y = 0.015$ and $z = 0.01$ respectively as shown in Figure 7.22b. A deficiency of Sr ion or an excess amount of smaller Eu and Ta ions has increased the *Q*-factor of $Sr(Eu_{0.5}Ta_{0.5})O_3$. In fact the quality factor has improved with a slight non-stoichiometry and large deviation from stoichiometry and associated point defects increase the dielectric loss factor. The microstructure of Eu-deficient $Sr(Eu_{0.5}Ta_{0.5})O_3$ revealed the presence of $Sr_4Ta_2O_9$ secondary phase as depicted in Figure 7.23.

The stoichiometric $Sr(Eu_{0.5}Ta_{0.5})O_3$ has a τ_f value of −63 ppm/°C. The τ_f of $Sr_{1+x}(Eu_{0.5}Ta_{0.5})O_3$ ceramics became less negative for both positive and negative values of x as shown in Figure 7.22c. The τ_f of $Sr_{1+x}(Eu_{0.5}Ta_{0.5})O_3$ has changed to −55 ppm/°C as x varied from 0 to 0.1 and it has increased to −57 ppm/°C when x varied to −0.02. $Sr(Eu_{0.5+y}Ta_{0.5})O_3$ and $Sr(Eu_{0.5}Ta_{0.5+z})O_3$ have lower values of τ_f (−52 and −57 ppm/°C) for $y = 0.015$ and $z = 0.01$ respectively. The $|\tau_f|$ of both compositions have found to increase for all other values of y and z as shown in Figure 7.22c. $Sr(Eu_{0.5+y}Ta_{0.5})O_3$ and $Sr(Eu_{0.5}Ta_{0.5+z})O_3$ showed lower values of $|\tau_f|$ where they showed maximum *Q*-values.

7.5.2 Effect of A- and B-site substitutions

It is possible to tune the τ_f's of $Sr(B'_{0.5}Ta_{0.5})O_3$ ceramics by making a solid solution with $Ba(Y_{0.5}Ta_{0.5})O_3$ and $Ba(Yb_{0.5}Ta_{0.5})O_3$ which are having positive τ_f's. In order to tailor the properties, Khalam and Sebastian [26] made solid solutions of $(Sr_{1-x}Ba_x)(Y_{0.5}Ta_{0.5})O_3$ and $(Sr_{1-x}Ba_x)(Yb_{0.5}Ta_{0.5})O_3$. The X-ray diffraction patterns of $Sr(Y_{0.5}Ta_{0.5})O_3$ and $Sr(Yb_{0.5}$-$Ta_{0.5})O_3$ show rhombohedral and orthorhombic symmetries respectively [26], while $Ba(Y_{0.5}$-$Ta_{0.5})O_3$ and $Ba(Yb_{0.5}Ta_{0.5})O_3$ were reported as cubic [23]. X-ray diffraction study [26] showed that a phase transition occurs from rhombohedral to cubic in $(Sr_{1-x}Ba_x)(Y_{0.5}Ta_{0.5})O_3$ and orthorhombic to cubic in $(Sr_{1-x}Ba_x)(Yb_{0.5}Ta_{0.5})O_3$ ceramics as x increases from 0.8 to 1.0. Similar phase transitions have also been reported by Fuji et al. [110] for the solid solutions $(Sr_{1-x}Ba_x)(Y_{0.5}Ta_{0.5})O_3$ and $(Sr_{1-y}Ca_y)(Y_{0.5}Ta_{0.5})O_3$ ceramics. The substitution of large-sized Ba for the relatively small Sr ion in the A site has increased the unit cell volume and led to transformation to the cubic symmetry. Figure 7.24 shows the variations in the dielectric properties of $(Sr_{1-x}Ba_x)(Y_{0.5}Ta_{0.5})O_3$ and $(Sr_{1-x}Ba_x)(Yb_{0.5}$-$Ta_{0.5})O_3$ ceramics with $x = 0$–1. The ε_r of $(Sr_{1-x}Ba_x)(Y_{0.5}Ta_{0.5})O_3$ and $(Sr_{1-x}Ba_x)(Yb_{0.5}Ta_{0.5})O_3$ increased while *Qf* decreased

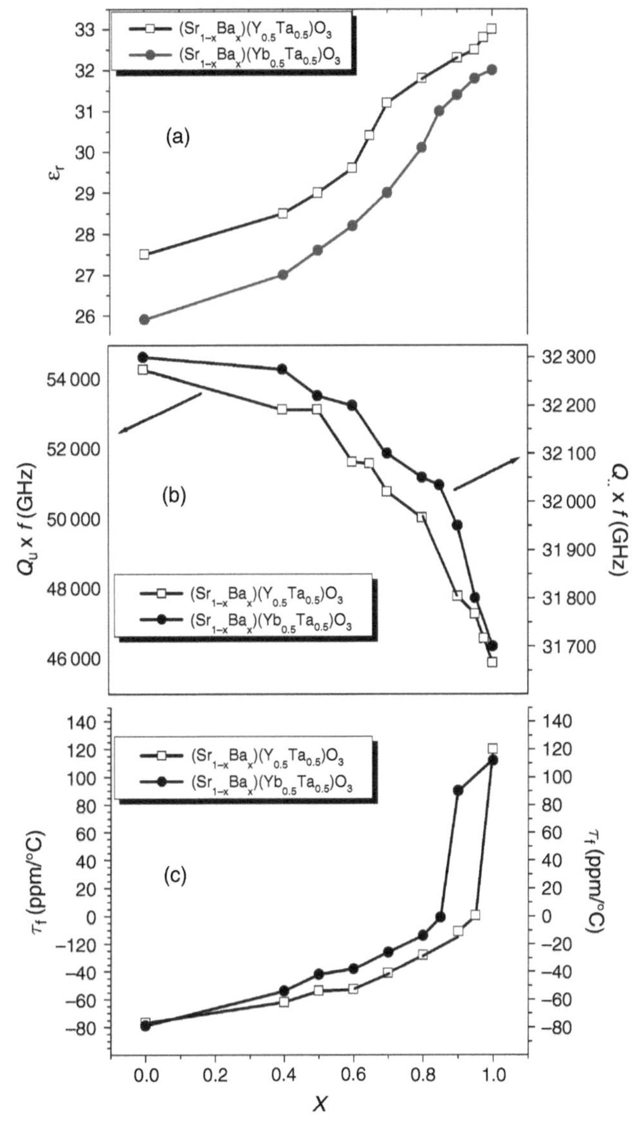

Figure 7.24 The variations in the dielectric properties of $(Sr_{1-x}Ba_x)(Y_{0.5}Ta_{0.5})O_3$ and $(Sr_{1-x}Ba_x)(Yb_{0.5}Ta_{0.5})O_3$ ceramics with the variation of x (after Ref. [26]).

with the increase of x as shown in Figure 7.24a and b. Figure 7.24c shows the variation of τ_f of $(Sr_{1-x}Ba_x)(Y_{0.5}Ta_{0.5})O_3$ and $(Sr_{1-x}Ba_x)(Yb_{0.5}Ta_{0.5})O_3$ ceramics with $x=0$–1. The τ_f of $(Sr_{1-x}Ba_x)(Y_{0.5}Ta_{0.5})O_3$ decreased linearly from −77 to −11 ppm/° up to $x=0.9$ and then abruptly changed to 120 ppm/°C at $x=1$. This abrupt change is attributed to the phase transition of the ceramic from rhombohedral to cubic. It is found that $(Sr_{1-x}Ba_x)(Y_{0.5}Ta_{0.5})O_3$ has nearly zero τ_f (−0.4 ppm/°C) at $x=0.95$. Similar variations are observed in the case of $(Sr_{1-x}Ba_x)(Yb_{0.5}Ta_{0.5})O_3$ ceramics also, for x between 0 and 1. $(Sr_{1-x}Ba_x)(Yb_{0.5}Ta_{0.5})O_3$ with

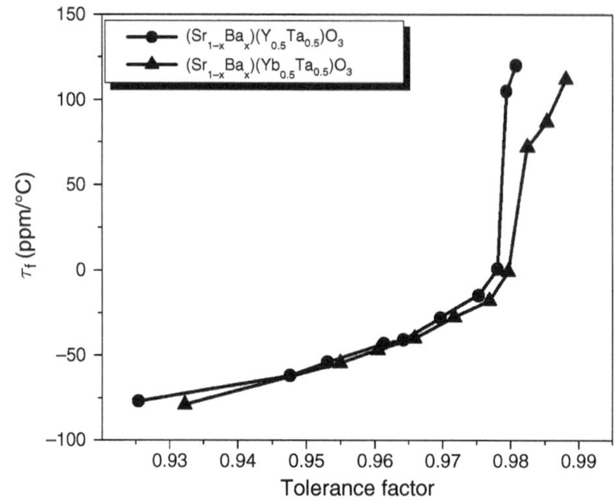

Figure 7.25 The variations of τ_f of $(Sr_{1-x}Ba_x)(Y_{0.5}Ta_{0.5})O_3$ and $(Sr_{1-x}Ba_x)(Yb_{0.5}Ta_{0.5})O_3$ ceramics versus tolerance factor corresponding to the x values (after Ref. [26]).

$x = 0.85$ has $\tau_f = -0.8$ ppm/°C. The tolerance factors of $(Sr_{0.05}Ba_{0.95})(Y_{0.5}Ta_{0.5})O_3$ and $(Sr_{0.15}Ba_{0.85})(Yb_{0.5}Ta_{0.5})O_3$ are 0.978 and 0.98 respectively. It should be noted that the τ_f of both $(Sr_{1-x}Ba_x)(Y_{0.5}Ta_{0.5})O_3$ and $(Sr_{1-x}Ba_x)(Yb_{0.5}Ta_{0.5})O_3$ have changed from a negative value to a positive value at $t \approx 0.98$ as shown in Figure 7.25 in agreement with the report of Reaney et al [20], [122]. $(Sr_{0.05}Ba_{0.95})(Y_{0.5}$-$Ta_{0.5})O_3$ and $(Sr_{0.15}Ba_{0.85})(Yb_{0.5}Ta_{0.5})O_3$ compositions have high permittivities ($\varepsilon_r = 33$ and 31) and high Q-factors ($Qf = 47$ 300 and 32 050 GHz) as shown in Figure 7.24a and b.

7.5.3 Effect of rutile addition

Addition of materials with positive τ_f like rutile can also compensate the negative τ_f of $Sr(B'_{0.5}Ta_{0.5})O_3$ ceramics, provided they do not chemically react. Hence, Khalam and Sebastian [26] added a small amount of TiO_2 to $Sr(B'_{0.5}Ta_{0.5})O_3$ ceramic to tune the τ_f. The sintering temperature of $Sr(B'_{0.5}Ta_{0.5})O_3$ decreased by 50°C for every 1 wt% addition of TiO_2. X-ray diffraction studies showed that TiO_2 did not react with $Sr(B'_{1/2}Ta_{1/2})O_3$ but remain as a mixture of $Sr(B'_{0.5}Ta_{0.5})O_3$ and TiO_2. The addition of rutile increased the ε_r of $Sr(Sm_{0.5}Ta_{0.5})O_3$ ceramic as shown in Table 7.1. Addition of 0.2 wt% TiO_2 improved the densification of the ceramics with an increase in the Q-factor (46 400 GHz) of $Sr(B'_{0.5}Ta_{0.5})O_3$ ceramic. More than 0.2 wt% TiO_2 addition to $Sr(Sm_{0.5}Ta_{0.5})O_3$ decreased the density and Q-factor. Tsunooka et al. [130] reported a similar variation of sinterability and dielectric properties for the forsterite ($2MgO{\cdot}SiO_2$) ceramics with less than 5 wt% TiO_2 additions. The τ_f of the $Sr(B'_{0.5}Ta_{0.5})O_3$ ceramic improved with TiO_2 addition. The variations of τ_f of $Sr(B'_{0.5}Ta_{0.5})O_3 + x$ wt% of TiO_2 [$x = 0$–5] are plotted in Figure 7.26. It can be seen that the plot of τ_f versus wt% of TiO_2 crossed zero τ_f at different levels of TiO_2 addition for different rare earth based compounds. As the ionic radii of B′-site elements decrease, the $Sr(B'_{0.5}Ta_{0.5})O_3$ ceramics attain near zero τ_f with higher concentration of TiO_2 as shown in Figure 7.26. $Sr(B'_{0.5}Ta_{0.5})O_3 + x$ wt% TiO_2 ceramics have attained nearly zero τ_f for $B' = La$, Pr

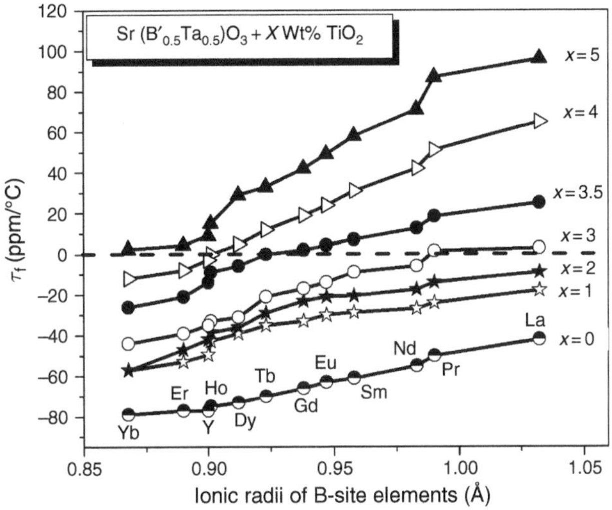

Figure 7.26 The variation of τ_f versus B′ ionic radii of $Sr(B'_{1/2}Ta_{1/2})O_3$ [B′ = La, Pr, Nd, Sm, Su, Gd, Tb, Dy, Ho, Y, Er and Yb] ceramics with the addition of 0–5 wt% TiO_2. The zero τ_f is indicated by the dashed horizontal line (after Ref. [26]).

and Nd at $x = 3$, for B′ = Sm, Eu, Gd and Tb at $x = 3.5$, for B′ = Dy, Ho and Y at $x = 4$ and for B′ = Er and Yb at $x = 5$. The materials with near zero τ_f possess high ε_r also, but with relatively lower values of *Qf*. A Similar behavior was reported [131] for $M^{2+}Nb_2O_6$ [M^{2+} = Zn, Mg, Ca and Co] ceramics.

7.6 $Ca(B'_{1/2}Nb_{1/2})O_3$

$Ca(B'_{1/2}Nb_{1/2})O_3$ [B′ = La, Pr, Nd, Sm, Eu, Gd, Tb, Dy, Ho, Y, Er, Yb and In] samples can be prepared by the conventional solid state method by calcining the mixed raw materials at 1200°C/4 h and then sintering at 1500°C/4 h [29]. Several people studied [10, 30, 82, 85, 132, 133] the structure and symmetry of $Ca(B'_{1/2}B''_{1/2})O_3$ and reported them as monoclinic, triclinic, orthorhombic or rhombohedral depending on the B′ atomic size and tolerance factor. Galasso [14] reported monoclinic symmetry for $Ca(B'_{1/2}Nb_{1/2})O_3$ [B′ = La, Pr, Nd, Sm, Gd, Tb, Dy, Ho, Y, Er, Yb] ceramics. Trunov et al. reported [132] monoclinic cell for $Ca(B'_{1/2}Nb_{1/2})O_3$ with B′ = Pr–Tb and B′ = La–Gd have a symmetry closer to orthorhombic. Filipev and Fesenko [10] claimed a structural boundary (monoclinic–triclinic) between B′ = Tb and Dy in $Ca(B'_{1/2}B''_{1/2})O_3$ [B′ = La, Pr, Nd, Sm, Eu, Gd, Tb, Dy, Ho, Y, Er, Yb; B″ = Nb, Ta] perovskites. B′ = Dy–In based materials were reported [10] as monoclinic while B′ = La–Tb based materials as triclinic. The X-ray diffraction patterns of $Ca(B'_{1/2}Nb_{1/2})O_3$ [B′ = La, Pr, Nd, Sm, Eu, Gd, Tb, Dy, Ho, Y, Er, Yb and In] ceramics are shown in Figure 7.27 Khalam and Sebastian [29] reported that $Ca(B'_{1/2}Nb_{1/2})O_3$ [B′ = La, Pr, Nd, Sm, Eu, Gd and Tb] are monoclinic, and those with B′ = Dy, Ho, Y, Er, Yb and In are orthorhombic. The X-ray diffraction patterns (Figure 7.27) of $Ca(B'_{1/2}Nb_{1/2})O_3$

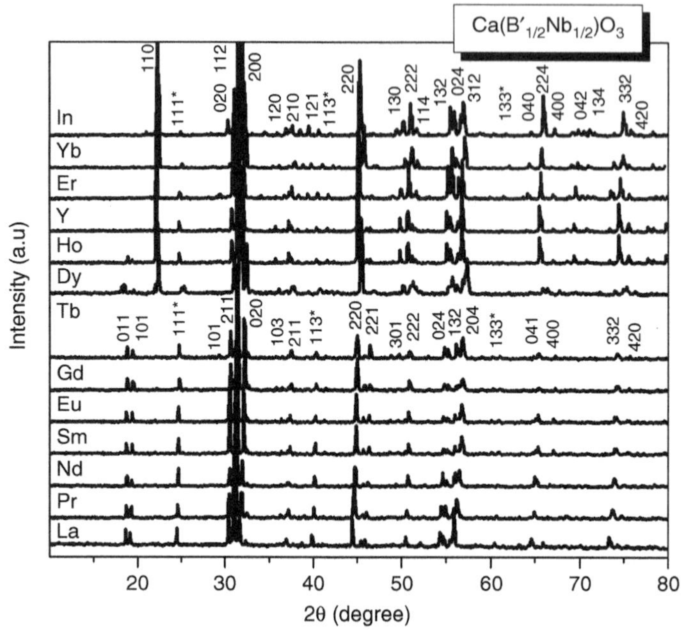

Figure 7.27 X-ray diffraction patterns of $Ca(B'_{1/2}Nb_{1/2})O_3$ [B' = La, Pr, Nd, Sm, Eu, Gd, Tb, Dy, Ho, Y, Er, Yb and In] ceramics. * represents the super structure reflections (after Ref. [29]).

ceramics clearly show two different symmetries with a structural boundary between $Ca(Tb_{1/2}Nb_{1/2})O_3$ and $Ca(Dy_{1/2}Nb_{1/2})O_3$ similar to the report of Filipev and Fesenko [10].

Superstructure reflections are observed in the X-ray diffraction patterns of all $Ca(B'_{1/2}Nb_{1/2})O_3$ ceramics indicating an ordered structure. B-site ions have a size difference of 0.16–0.392 Å and a charge difference of 2. It was reported [10] that Ba, Sr and Ca compounds of $A(B'_{1/2}B''_{1/2})O_3$ type occur with an ordered arrangement of B' and B'' cations when.

$$K=\frac{|R_{B'}-R_{B''}|}{R_{B'}}\geq 0.09 \tag{7.1}$$

where $R_{B'}$ and $R_{B''}$ are the ionic radii of B' and B'' cations. For $Ca(B'_{1/2}Nb_{1/2})O_3$ ceramics, this value is $0.25\leq K\leq 0.6125$ indicating an ordered B-site and is evidenced by the superstructure reflections shown in Figure 7.27. The intensities of superstructure reflections in $Ca(B'_{1/2}Nb_{1/2})O_3$ ceramics increase with increase in size difference ($R_{B'}-R_{Nb}$) between the B-site ions (see Figure 7.27).

$Ca(B'_{1/2}Nb_{1/2})O_3$ [B' = La, Pr, Nd, Sm, Eu, Gd, Tb, Dy, Ho, Y, Er, Yb, In] ceramics are densified to about 95–98% of their theoretical densities. Figure 7.28 shows a typical microstructures recorded from Ho and Sm belonging to orthorhombic and monoclinic symmetries respectively. In $Ca(B'_{1/2}Nb_{1/2})O_3$ ceramics, Ca and Nb ions are common and any variation in the dielectric properties is attributed to the variation of B'-site ions. Figure 7.29 shows the ε_r of $Ca(B'_{1/2}Nb_{1/2})O_3$ ceramics versus B'-site ionic

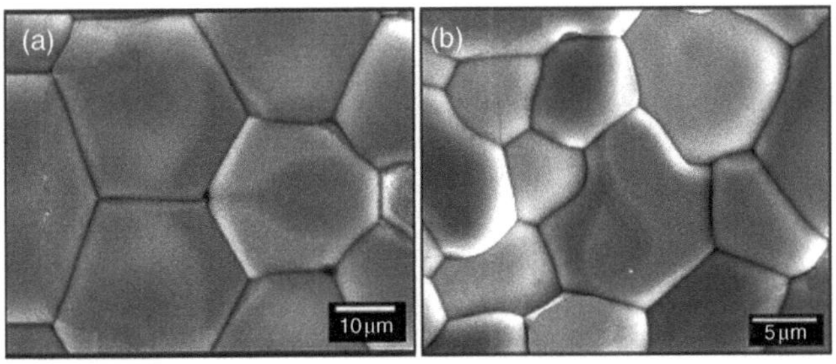

Figure 7.28 The surface morphology of (a) $Ca(Ho_{1/2}Nb_{1/2})O_3$ and (b) $Ca(Sm_{1/2}Nb_{1/2})$ O_3ceramics (after Ref. [29]).

radii. The ε_r of these ceramics increases linearly up to B′ ionic radius of 0.91 Å and then it abruptly decreases beyond $R_{B'}$ of 0.92 Å. This change in ε_r is related to the orthorhombic to monoclinic structural transformation of $Ca(B'_{1/2}Nb_{1/2})O_3$ ceramics at $R_{B'} = 0.91$–0.92 Å. $Ca(B'_{1/2}Nb_{1/2})O_3$ ceramics except $Ca(B'_{1/2}Nb_{1/2})O_3$ have normalized quality factor in the range 31 000–38 000 GHz as given in Table 7.1. However, a low quality factor ($Q_u \times f = 11\ 000$ GHz) was obtained for $Ca(Gd_{1/2}Nb_{1/2})O_3$ ceramic.

The temperature coefficient of resonant frequency (τ_f) of $Ca(B'_{1/2}Nb_{1/2})O_3$ ceramics are plotted versus ionic radii of B′ -site elements as shown in Figure 7.29. The $\tau_{f'}$ s have a similar variation as that of ε_r versus B′-site ionic radii. The τ_fs of $Ca(B'_{1/2}Nb_{1/2})O_3$ ceramics with B′ = Dy–In vary linearly from −33 to 5 ppm/°C while that of B′ = Tb–La vary from −13 to −43 ppm/°C with B′ ionic radii (see Figure 7.29). Corresponding to the curved portion in Figure 7.29 τ_f reversal is observed due to the structural transformation. The structural change at the boundary between $Ca(Tb_{1/2}Nb_{1/2})O_3$ and $Ca(Dy_{1/2}Nb_{1/2})$ O_3 is clearly evidenced in the X-ray diffraction patterns (Figure 7.27) also. In general Ca-based perovskites were reported [17, 29, 85] with negative τ_f. However, Khalam and Sebastian observed [29] small positive τ_f values (5 and 3 ppm/°C) for $Ca(Dy_{1/2}Nb_{1/2})O_3$

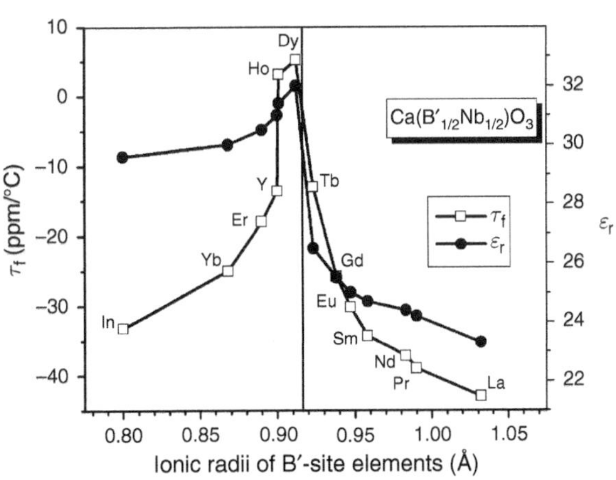

Figure 7.29 The permittivity and temperature coefficient of resonant frequency of $Ca(B'_{1/2}Nb_{1/2})O_3$ ceramics versus ionic radii of B′-site elements (after Ref. [29]).

and $Ca(Ho_{1/2}Nb_{1/2})O_3$, which have tolerance factors 0.89 and 0.892 respectively. The sudden change in ε_r and τ_f of $Ca(B'_{1/2}Nb_{1/2})O_3$ ceramics at $t \approx 0.89$ is attributed to the phase transition from monoclinic to orthorhombic.

7.6.1 Tailoring the properties of $Ca(B'_{1/2}Nb_{1/2})O_3$ by addition of TiO_2 and $CaTiO_3$

The negative τ_f $Ca(B'_{1/2}Nb_{1/2})O_3$ [B′ = La, Pr, Nd, Sm, Eu, Gd, Tb, Y, Er, Yb and In] ceramics can be tuned by the addition of rutile and $CaTiO_3$ [29]. It was found that addition of about 1 mol% rutile considerably improved densification and thereby the quality factor as shown in Figure 7.30. TiO_2 did not react with $Ca(B'_{1/2}Nb_{1/2})O_3$ but remained as a mixture of rutile and $Ca(B'_{1/2}Nb_{1/2})O_3$. Addition of more than 1 mol% of rutile improved τ_f and ε_r but degraded the quality factor. A similar improvement in dielectric properties was reported for $Sr(B'_{1/2}Ta_{1/2})O_3$ [26], $Ba(Zn_{1/3}Ta_{2/3})O_3$ [134] and $2MgO{\cdot}SiO_2$ ceramics with the addition of TiO_2 [130]. Figure 7.30 shows the variation of relative density, ε_r, *Qf*. The $|\tau_f|$ of $Ca(Eu_{1/2}Nb_{1/2})O_3$ ceramic decreases with TiO_2

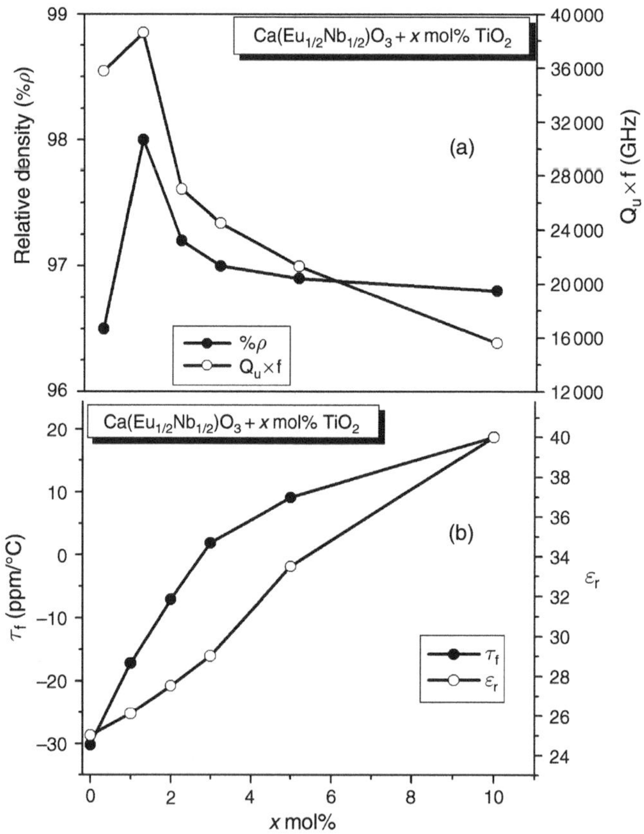

Figure 7.30 (a) The variation of relative density and *Qf* versus x and (b) the variation of ε_r and τ_f versus x of $Ca(Eu_{1/2}Nb_{1/2})O_3 + x$ mol% TiO_2 ceramics (after Ref. [29]).

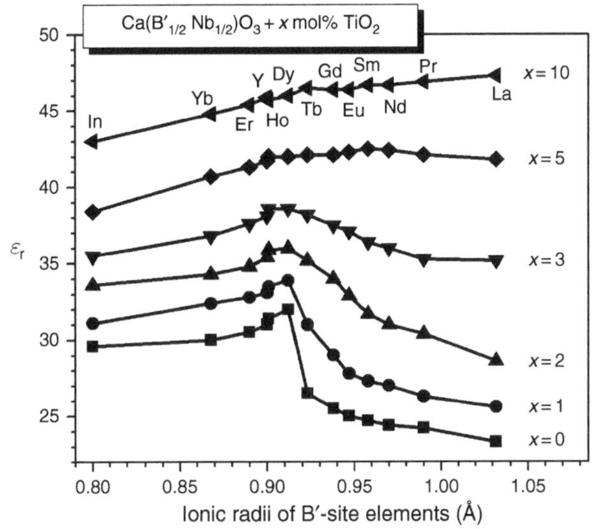

Figure 7.31 The variation of ε_r of $Ca(B'_{1/2}Nb_{1/2})O_3$ ceramics with the addition of 0–10 mol% TiO_2 (after Ref. [29]).

addition. Addition of more than 1 mol% TiO_2 decreases the quality factor of $Ca(Eu_{1/2}Nb_{1/2})O_3$ ceramic (see Figure 7.30a) with the appearance of TiO_2 phase in the X-ray diffraction pattern. The variation of ε_r with B′ ionic radii of $Ca(B'_{1/2}Nb_{1/2})O_3 + x$ mol% TiO_2 ceramics, where $1 \leq x \leq 10$, is shown in Figure 7.31. The increase in ε_r and shift of τ_f toward the positive value are due to the higher ε_r and high positive τ_f of the rutile. Figure 7.32 shows the variations of τ_f versus tolerance factor of $Ca(B'_{1/2}Nb_{1/2})O_3 + x$ mol% TiO_2 ceramics, where $1 \leq x \leq 10$. Almost all $Ca(B'_{1/2}Nb_{1/2})O_3$ ceramics crossed zero τ_f for $x \leq 3$ with a corresponding increase in their ε_r. The variation of ε_r versus ionic radius and τ_f versus tolerance factor of $Ca(B'_{1/2}Nb_{1/2})O_3 + \leq 2$ mol% TiO_2 ceramics shows a similar trend as shown in Figures 7.31 and Figure 7.32. When $x \geq 5$, the ε_r and τ_f of $Ca(B'_{1/2}Nb_{1/2})O_3$ ceramics with B′ ionic radii and tolerance factor tend to have a linear variation (Figures 7.31 and Figure 7.32).

Khalam and Sebastian [29] also tailored the τ_f of $Ca(B'_{1/2}Nb_{1/2})O_3$ by adding 1–10 mol% of $CaTiO_3$ which has a positive τ_f. The orthorhombic $CaTiO_3$ has [135, 136] $\varepsilon_r = 170$, $Qf \sim 7000$ GHz and $\tau_f = 850$ ppm/°C. Figure 7.33 shows the surface morphology of $CaTiO_3$-added $Ca(Eu_{1/2}Nb_{1/2})O_3$ revealing their mixture behavior with grain sizes nearly 2 and 10 μm respectively. The EDAX analysis showed that the smaller grains of size 2 μm are that of $CaTiO_3$. It was found [29] that the intensity of superstructure reflections of $Ca(Eu_{1/2}Nb_{1/2})O_3$ decreases with $CaTiO_3$ addition. The ε_r and τ_f of $Ca(B'_{1/2}Nb_{1/2})O_3$ ceramics increased with increasing amount of $CaTiO_3$. The quality factor of $Ca(B'_{1/2}Nb_{1/2})O_3$ ceramics linearly decreased (from 35 750 to 28 400 GHz) with the addition of 1–10 mol% $CaTiO_3$. The ε_r of $Ca(B'_{1/2}Nb_{1/2})O_3$ ceramics increased whereas the τ_f became less negative and then increased to positive value with the addition of $CaTiO_3$ up to 10 mol%. The increase in ε_r and τ_f is due to the presence of $CaTiO_3$ phase which has high ε_r and positive τ_f. Figure 7.34 shows a gradual variation of τ_f with tolerance factor and $CaTiO_3$ content in the $Ca(B'_{1/2}Nb_{1/2})O_3$

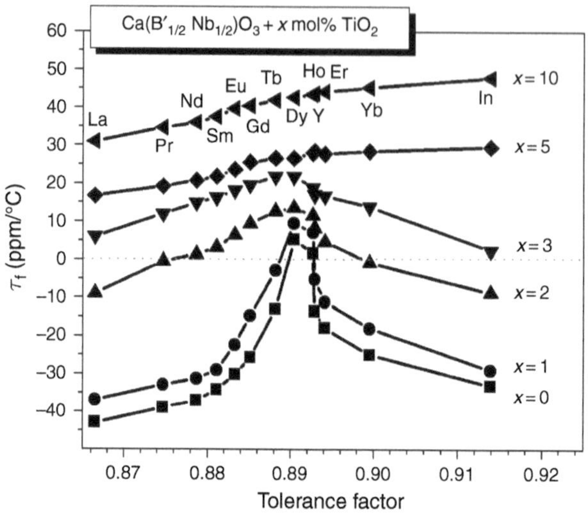

Figure 7.32 The variation of τ_f of $Ca(B'_{1/2}Nb_{1/2})O_3$ ceramics with the addition of 0–10 mol% TiO_2. The dotted horizontal line represents zero τ_f (after Ref. [29]).

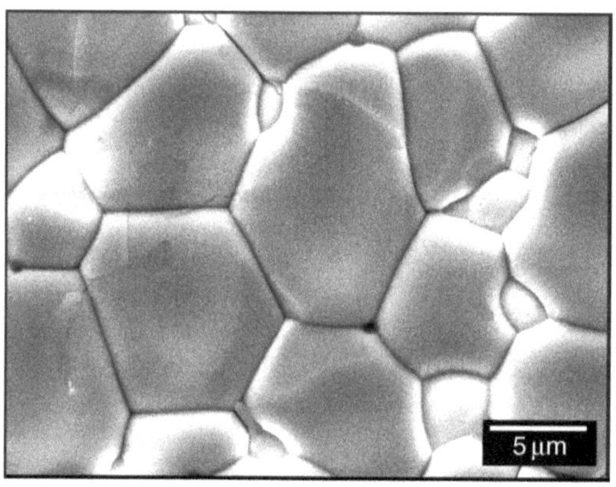

Figure 7.33 The microstructure of the $Ca(Eu_{1/2}Nb_{1/2})O_3$–$CaTiO_3$ ceramics mixture of $Ca(Eu_{1/2}Nb_{1/2})O_3$ (large grains) and $CaTiO_3$ (small grains) when 3 mol% $CaTiO_3$ is added to $Ca(Eu_{1/2}Nb_{1/2})O_3$ and sintered at 1550°C (after Ref. [29]).

ceramics. With the increase of x in $Ca(B'_{1/2}Nb_{1/2})O_3 + x$ mol% $CaTiO_3$ ceramics, the peak heights of τ_f curves decreased. The peak disappeared for $\geq$ 10 mol% $CaTiO_3$.

7.6.2 Effect of A- and B-site substitution on the structure and dielectric properties

The properties of $Ca(B'_{1/2}Nb_{1/2})O_3$ can be tailored by suitable A- and B-site substitutions. $Ba(Y_{1/2}Nb_{1/2})O_3$ is cubic with a positive τ_f with high quality factor whereas $Ca(Y_{1/2}Nb_{1/2})O_3$ is orthorhombic with a negative τ_f. Hence it is possible to tune τ_f to zero by the

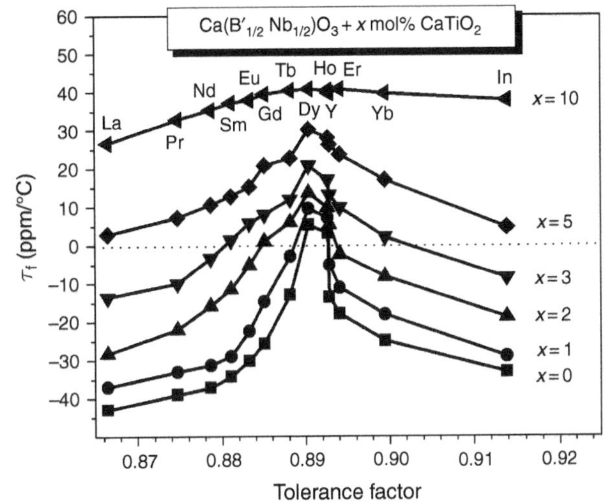

Figure 7.34 The variation of τ_f of $Ca(B'_{1/2}Nb_{1/2})O_3$ ceramics with the addition of 0–10 mol% $CaTiO_3$ (after Ref. [29]).

formation of a solid solution of $(Ca_{1-x}Ba_x)(Y_{1/2}Nb_{1/2})O_3$ [29]. However, X-ray diffraction study showed that $Ba(Y_{1/2}Nb_{1/2})O_3$ and $Ca(Y_{1/2}Nb_{1/2})O_3$ do not form a solid solution for the entire range and a mixture of these two compounds are formed in the range x = 0.4–0.9. The difficulty to accommodate the large Ba^{2+} ion on the Ca site leads to a mixture of $Ba(Y_{1/2}Nb_{1/2})O_3$ and $Ca(Y_{1/2}Nb_{1/2})O_3$. X-ray diffraction study showed that the orthorhombic peaks of $Ca(Y_{1/2}Nb_{1/2})O_3$ are maintained up to $x = 0.3$ and the diffraction peaks of cubic $Ba(Y_{1/2}Nb_{1/2})O_3$ start appearing for $x \geq 0.4$ in addition to the $Ca(Y_{1/2}Nb_{1/2})O_3$ peaks. The orthorhombic $Ca(Y_{1/2}Nb_{1/2})O_3$ phase disappeared for $x \geq 0.9$. Figure 7.35 shows the variation in the dielectric properties of $Ca(Y_{1/2}Nb_{1/2})O_3$ ceramics with the addition of $Ba(Y_{1/2}Nb_{1/2})O_3$. The *Qf* of $(1-x)Ca(Y_{1/2}Nb_{1/2})O_3$–$xBa(Y_{1/2}Nb_{1/2})O_3$ compound decreased from 35 000 to 26 400 GHz for $0 \leq x \leq 0.3$ and then increased gradually to 49 600 GHz for $0.4 \leq x \leq 1$. The variation in the ε_r of $(1-x)Ca(Y_{1/2}Nb_{1/2})O_3 - xBa(Y_{1/2}Nb_{1/2})O_3$ was also found to have a similar trend as that of *Qf*. The ε_r of $(1-x)Ca(Y_{1/2}Nb_{1/2})O_3$–$xBa(Y_{1/2}Nb_{1/2})O_3$ decreased to 27 with $x = 0.3$ and then linearly increased to 37 for $0.4 \leq x \leq 1$. The τ_f of the compositions varied from −14 to −17 ppm/°C for $x = 0$–0.3 and then changed to −7 ppm/°C at $x = 0.9$. When x varied from 0.9 to 1, the τ_f abruptly increased to a positive value (15 ppm/°C) with $0.9 < x \leq 1$. The τ_f of the composition could be tuned to near zero value (1 ppm/°C) for $x = 0.95$. The tolerance factor of $0.05Ca(Y_{1/2}Nb_{1/2})O_3$–$0.95Ba(Y_{1/2}Nb_{1/2})O_3$ is 0.98, which is close to the antiphase tilted–untilted phase boundary of octahedral distortion reported by Reaney et al. [20, 122].

$Ca(Y_{1/2}Nb_{1/2})O_3$ is orthorhombic with space group *Pbnm* [10] while $Sr(Y_{1/2}Nb_{1/2})O_3$ is orthorhombic with space group *Pnma* [120]. Khalam and Sebastian [29] prepared a solid solution of $Ca_{1-x}Ba_x(Y_{1/2}Nb_{1/2})O_3$. The X-ray diffraction patterns of some selected compositions of $(Ca_{1-x}Sr_x)(Y_{1/2}Nb_{1/2})O_3$ are shown in Figure 7.36. $(Ca_{1-x}Sr_x)(Y_{1/2}Nb_{1/2})O_3$ has maintained the space group (orthorhombic) *Pbnm* for values of x up to 0.7 whereas *Pnma* has been found with $x \geq 0.8$. This indicates that by the substitution of $\geq$ 80% of Ca^{2+} ($r_{Ca} = 1.34$ Å) ion with comparatively bigger Sr^{2+}

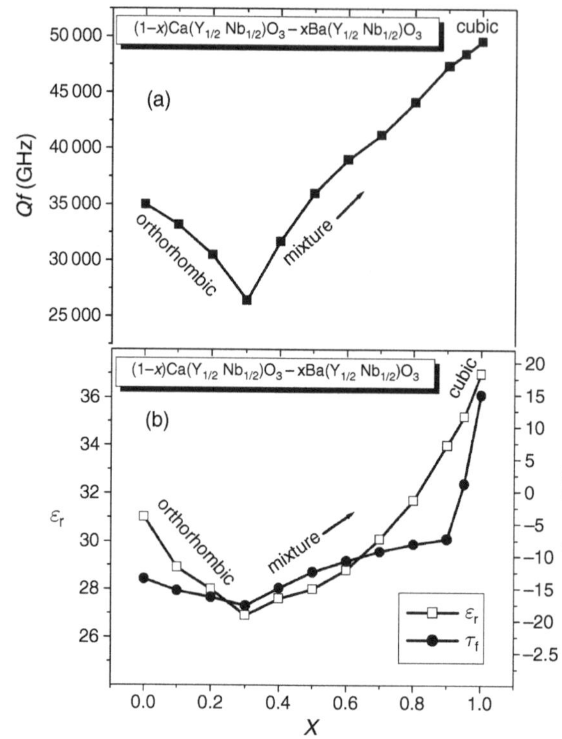

Figure 7.35 (a) The variation of $Q_u \times f$ versus x and (b) the variation of ε_r and τ_f versus x of the solid solution $(1-x)Ca(Y_{1/2}Nb_{1/2})O_3-xBa(Y_{1/2}Nb_{1/2})O_3$ceramics (after Ref. [29]).

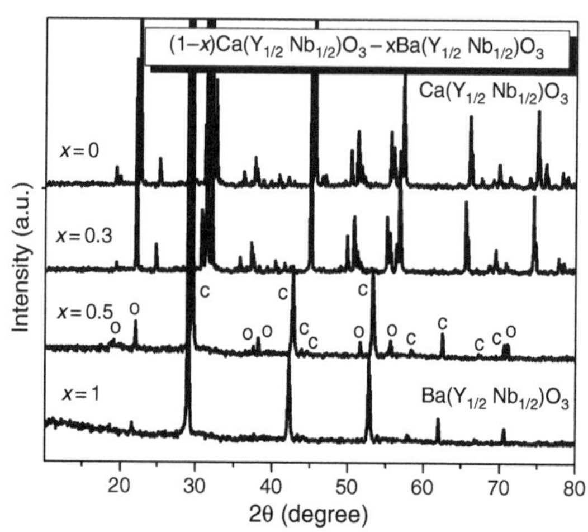

Figure 7.36 The XRD patterns of some selected compositions of $(1-x)Ca(Y_{1/2}Nb_{1/2})O_3-x$ $Ba(Y_{1/2}Nb_{1/2})O_3$ ceramics. ^{C} represents the cubic peaks of $Ba(Y_{1/2}Nb_{1/2})$ and ^{O} represents the orthorhombic peaks of $Ca(Y_{1/2}Nb_{1/2})$ ceramics (after Ref. [29]).

(r_{Sr} = 1.44 Å) ion leads to a symmetry change. Shimada reported [123] a cubic to orthorhombic phase transformation in $Sr(Ga_{1/2}Ta_{1/2})O_3$ when the A site Sr^{2+} was substituted by Ca^{2+} ions. The variations in the dielectric properties of $Ca(Y_{1/2}Nb_{1/2})O_3$ by the substitution of Sr for Ca ions are shown in Figure 7.37. The *Qf*, ε_r and τ_f show a sudden change at $x = 0.7$ indicating a structural transformation. The quality factor of $Ca(Y_{1/2}Nb_{1/2})O_3$ (35 000 GHz) linearly decreased to 23 500 GHz when 70% of Ca was substituted by Sr and then it showed an abrupt increase to 38 850 GHz with $0.7 \leq x \leq 1$ as shown in Figure 7.37a. The variation of ε_r and τ_f of $(Ca_{1-x}Sr_x)(Y_{1/2}Nb_{1/2})O_3$ are as shown in Figure 7.37b. The ε_r of $(Ca_{1-x}Sr_x)(Y_{1/2}Nb_{1/2})O_3$ decrease linearly from 31 to 26 up to $x = 0.7$ whereas the variation is abrupt from 26 to 32 for $0.7 \leq x \leq 1$. Similarly, the τ_f of $(Ca_{1-x}Sr_x)(Y_{1/2}Nb_{1/2})O_3$ changed to more negative value from −14 to −35 ppm/°C up to $x = 0.7$ and then varied abruptly to −66 ppm/°C for $0.7 \leq x \leq 1$. These sudden changes in ε_r, *Qf* and τ_f at $x \approx 0.7$ was due to the transformation of orthorhombic $Ca(Y_{1/2}Nb_{1/2})O_3$ with space group *Pbnm* to that of $Sr(Y_{1/2}Nb_{1/2})O_3$ with space group *Pnma*.

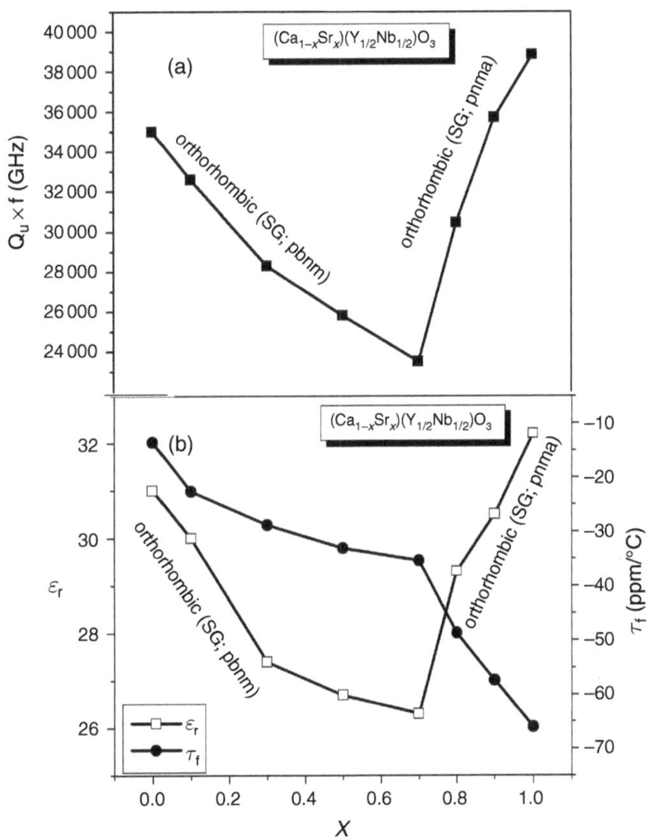

Figure 7.37 The variation of (a) $Q \times f$ versus x and (b) ε_r and τ_f versus x of the solid solution $Ca_{1-x}Sr_x(Y_{1/2}Nb_{1/2})O_3$ (after Ref. [29]).

7.7 $Ca(B'_{1/2}Ta_{1/2})O_3$ [B′ = Lanthanides, Y and In] System

$Ca(B'_{1/2}Ta_{1/2})O_3$ [B′ = La, Pr, Nd, Sm, Eu, Gd, Tb, Dy, Ho, Y, Er, Yb and In] ceramics have a relatively high sintering temperature of about 1550–1600°C [17, 83]. Figure 7.38 shows the microstructure of two representative samples of $Ca(La_{1/2}Ta_{1/2})O_3$ and $Ca(Yb_{1/2}Ta_{1/2})O_3$ ceramics. Large uniform grains of about 20 μm were observed for ceramics with smaller B′ ionic radii, whereas relatively smaller non-uniform duplex grains were observed when the ionic radii of the B′ were increased to $Ca(La_{1/2}Ta_{1/2})O_3$. Figure 7.39 shows the X-ray diffraction patterns of $Ca(B'_{1/2}Ta_{1/2})O_3$ [B′ = La, Pr, Nd, Sm, Eu, Gd, Tb, Dy, Ho, Y, Er, Yb and In] ceramics. The ceramics [B′ = La, Pr, Nd, Sm, Eu, Gd, Tb, Dy, Ho, Y, Er, Yb and In] have a monoclinic symmetry with space group $P2_1/n$ [10, 13, 80, 82, 110, 133, 137]. $Ca(B'_{1/2}Ta_{1/2})O_3$ ceramics have an ordered structure as evidenced by the presence of odd, odd, odd superlattice reflections in the X-ray diffraction patterns. Filipev and Fesenko [10] have reported the possibility of intersubstitution between the Ca and B′ lanthanides because of their comparable ionic sizes. The Ca^{2+} and La^{3+} have ionic radii of 1.34 and 1.36 Å for the 12-fold co-ordination respectively [138]. The maximum substitution of La^{3+} on the A site can be represented by $La_{1/2}Ca_{1/2}(Ca_{1/2}Ta_{1/2})O_3$. Khalam et al. obtained the XRD patterns of $Ca(B'_{1/2}Ta_{1/2})O_3$ ceramics by substituting different amounts of B′ cations at the Ca site and Ca at the B′ site using the powder cell software. Figure 7.40 shows the theoretical variation of the diffracted intensity of the 011 and 101 reflections for various amounts of intersubtitutions. It is clear from Figure 7.40 that the intensities of 011 and 101 reflections increase with an increase in the intersubstitution between Ca and B′ elements. On comparing with the experimental profile in Figure 7.39, it is evident that the intersubstitution is weak for La and that it increases with a decrease in the B′ site ionic size (La to Yb). The 011 and 101 reflections are relatively weak for La, Pr and Nd but intense for Sm, Eu, Gd, Tb, Dy, Ho, Y, Er, Yb and In. The relative intensity of the B-site ordering peaks 011 and 101 depends on the difference in the scattering factors of B′ and B″ ions. When the B′ cation has the largest difference in the atomic number compared with B″ (Ta), the strongest ordering peaks will result. Because Ta has the atomic number 70 and that of La is 57, the largest difference and strongest 011 and 101 reflections are expected.

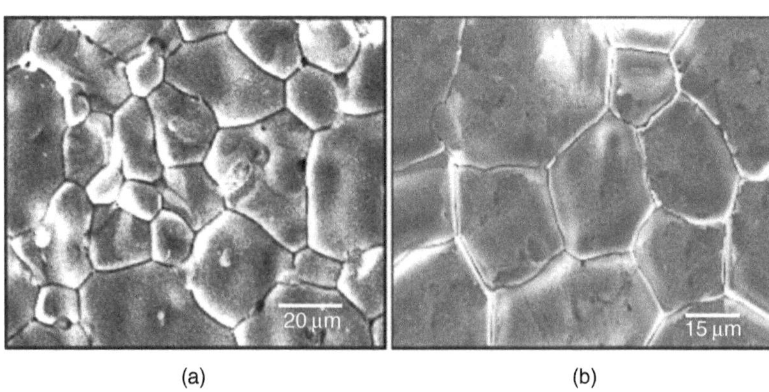

Figure 7.38 The microstructures of (a) $Ca(La_{1/2}Ta_{1/2})O_3$ and (b) $Ca(Yb_{1/2}Ta_{1/2})O_3$ ceramics (after Ref. [83]).

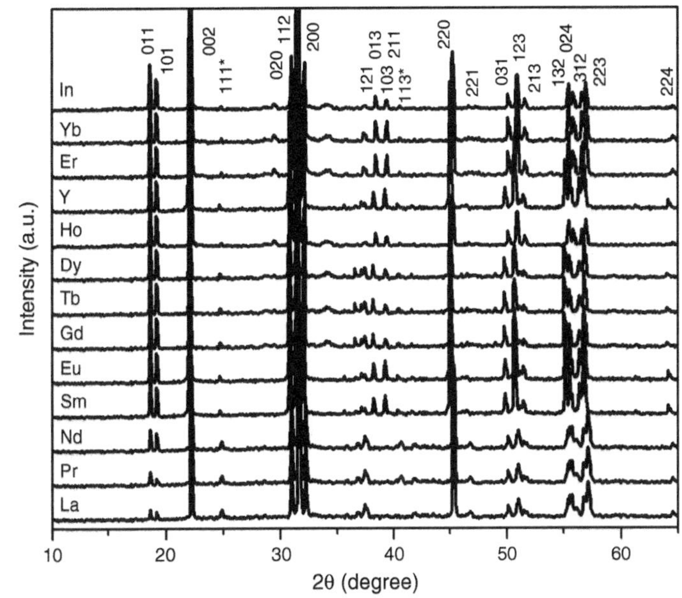

Figure 7.39 XRD patterns of Ca $(B'_{1/2}Ta_{1/2})O_3$ [B' = La, Pr, Nd, Sm, Eu, Gd, Tb, Dy, Ho, Y, Er, Yb and In]. * represents superstructure reflections (after Ref. [83]).

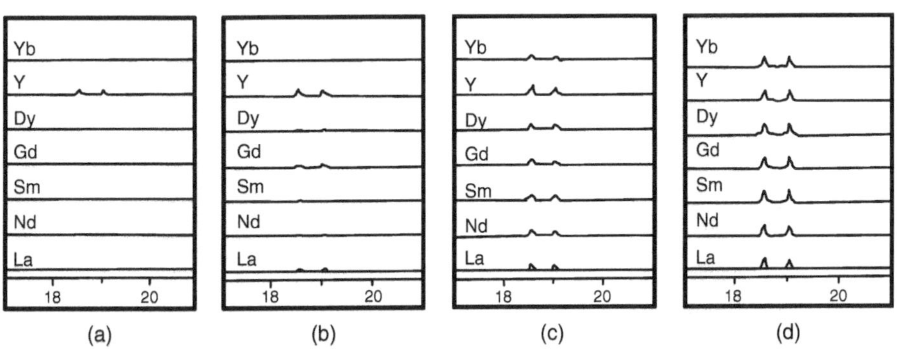

Figure 7.40 Theoretical variation of the diffracted intensity for 011 and 101 reflections of $Ca(B'_{1/2}Ta_{1/2})O_3$ [Ln = La, Nd, Sm, Gd, Dy, Y, Yb] ceramics for various amounts intersubstitutions (a) $Ca(B'_{1/2}Ta_{1/2})O_3$, (b) $Ca_{0.9}\ B'_{0.1}(Ca_{0.1}B'_{0.4}Ta_{0.5})O_3$ and (c) $Ca_{0.7}\ B'_{0.3}(Ca_{0.3}B'_{0.2}Ta_{0.5})O_3$ (d) $Ca_{0.5}B'_{0.5}(Ca_{0.5}Ta_{0.5})O$ (after Ref. [83]).

However, the strongest reflections were found for Sm–Yb, Y and In, indicating strong intersubstitution. Hence, e.g., the Yb-based ceramics can be represented by $(Ca_{1/2}Yb_{1/2})(Ca_{1/2}Ta_{1/2})O_3$ for the extreme case of intersubstitution. Thus the largest difference in atomic number between Ta (70) and B' Ca (20) and the strongest 011 and 101 reflections occur. In a similar manner the Sm-, Eu-, Gd-, Tb-, Dy- and In-based ceramics can be represented. Although the atomic size of La matches more with that of A-site Ca, the intersubstitution is weak. The lanthanides Sm-Yb showed a considerable amount of intersubstitution, although the matching of the atomic size with Ca (A site) is weak as compared with La.

The tolerance factors and dielectric properties of $Ca(B'_{1/2}Ta_{1/2})O_3$ [$B'=La$, Pr, Nd, Sm, Eu, Gd, Tb, Dy, Ho, Y, Er, Yb and In] ceramics are given in Table 7.1. The *Qf*s of $Ca(B'_{1/2}Ta_{1/2})O_3$ ceramics are in the range 20 600–59 200 GHz, with maximum value for $Ca(Yb_{1/2}Ta_{1/2})O_3$. The ε_r of $Ca(B'_{1/2}Ta_{1/2})O_3$ ceramics increased from 24 to 30 with B′-site ionic radii up to $R_{Dy}=0.912$ Å and then decreased from 28 to 23 up to $R_{La}=1.032$ Å. The variation of τ_f of $Ca(B'_{1/2}Ta_{1/2})O_3$ is similar to the variation in ε_r. The decrease in ε_r of $Ca(B'_{1/2}Ta_{1/2})O_3$ ceramics with large-sized B′-cations ($R_{B'} \geq 0.923$ Å) may be due to the intersubstitution of comparable size Ca^{2+} and B'^{3+} ions and the related structural changes [83]. It may be noted that for the extreme case of intersubstitution, half of the Ca ions will be in 12-fold coordination and the other half in 6-fold coordination and the B′ cations at the 12-fold coordination. Thus due to intersubstitution, larger Ca ions may occupy B′ site and smaller lanthanide ion may occupy the A site. This decreases the tolerance factor, leading to octahedral tilting and possible phase changes affecting the dielectric properties. It may be noted that phase transitions in complex perovskites brought about by tilting of the $B'O_6$ octahedra are difficult to discern by X-ray diffraction because the tilt angle is small and the atomic scattering factor of the oxygen sublattice is small [20, 139].

Khalam and Sebastian [83] tuned the negative τ_f of $Ca(B'_{1/2}Ta_{1/2})O_3$ by the addition of $CaTiO_3$ and TiO_2 which formed mixture phase and also by adding $Ba(B'_{1/2}Ta_{1/2})O_3$ in a way similar to the case $Ca(B'_{1/2}Nb_{1/2})O_3$ described in Section 7.6. The ceramic $0.6Ca(Yb_{1/2}Ta_{1/2})O_3$–$0.4Ba(Yb_{1/2}Ta_{1/2})O_3$ has $\varepsilon_r = 27.5$, $Qf = 48\,000$ GHz and $\tau_f = 1.8$ ppm/°C.

7.8 $(Pb_{1-x}Ca_x)(Fe_{1/2}B''_{1/2})O_3$ [B′ = Nb, Ta]

The lead-based perovskite materials such as $(Pb,Ca)ZrO_3$, $(Pb,Ca)HfO_3$ and $(Pb,Ca)(Fe,Nb)O_3$ have high relative permittivity and small temperature coefficient of permittivities [140, 141]. These materials called as relaxors are well studied as multilayer capacitors. Recently it has been found that these types of materials possess excellent properties in the microwave frequency region. Kato and co-workers reported [31, 142, 143] $Pb_{1-x}Ca_x(Fe_{1/2}Nb_{1/2})O_3$ as single-phase perovskites with high ε_r and high *Qf* for the first time. As the amount of Ca substitution increases to 0.55, the ε_r and τ_f decreased and the *Qf* increased. At $x=0.55$, the ceramic has $\varepsilon_r=91$, $Qf=4950$ GHz, $\tau_f=2$ ppm/°C. $(Pb,Ca)(M_{1/2}Nb_{1/2})O_3$ [M = Al, Cr] are not perovskite and have low quality factors. Yang et al. [42] reported that partial substitution of La^{3+} and Nd^{3+} at the A site of $(Pb_{0.5}Ca_{0.5})(Fe_{1/2}Nb_{1/2})O_3$ improves the microwave dielectric properties. $[(Pb_{0.5}Ca_{0.5})_{1-x}(La_{0.5}Nd_{0.5})_x](Fe_{1/2}Nb_{1/2})O_{3+\delta}$ with $x=0.06$ and sintered at 1150°C showed $\varepsilon_r=100$, $Qf=5820$ GHz and $\tau_f=0$ ppm/°C. For $x>0.04$ a pyrochlore secondary phase was formed degrading the dielectric properties. The ε_r decreased with increasing amount of $(La, Nd)^{3+}$. Yang et al. [40, 42] suggested that substitution of a small amount of $(La,Nd)^{3+}$ for $(Pb,Ca)^{2+}$ could eliminate oxygen vacancies and improve the quality factor. However, substitution of Ti^{4+} for $(Fe_{1/2},Nb_{1/2})$ increased the ε_r and τ_f [41]. $[(Pb_{0.5}Ca_{0.5})_{0.95}La_{0.05}][(Fe_{1/2}Nb_{1/2})_{1-y}Ti_y]O_{3+\delta}$ [PCLFNT] with $y=0.1$ showed

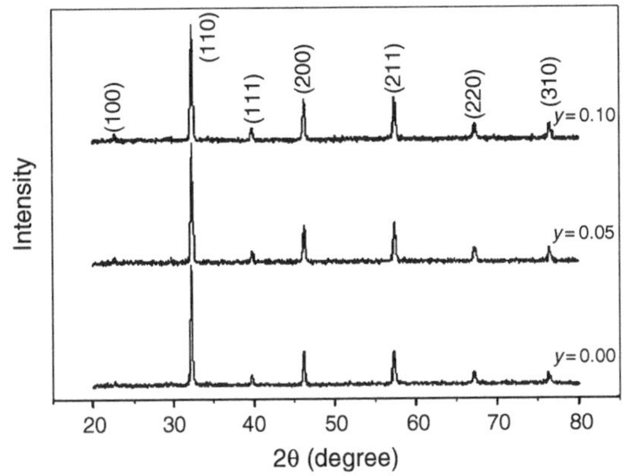

Figure 7.41 X-ray diffraction patterns of $[(Pb_{0.5}Ca_{0.5})_{0.95}La_{0.05}](Fe_{1/2}Nb_{1/2})O_{1-y}Ti_y)O_{3+}$ for various values of y (after Ref. [41]).

$\varepsilon_r = 117$, $Qf = 4950$ GHz and $\tau_f = 17$ ppm/°C. Yang et al. [41] reported that substitution of Ti increased the ε_r although the ionic polarizability of Ti is less than that of $(Fe,Nb)^{4+}$. This is attributed to the increase in rattling of B-site ions by the smaller Ti^{4+} substitution. Figure 7.41 shows the X-ray diffraction of a typical PCLFNT for various amounts of Ti showing the single-phase perovskite structure. Kucheiko et al. [33] reported that substitution of Fe^{3+}/Nb^{5+} by Sn^{4+} at the B site of the perovskite $(Pb,Ca)(Fe,Nb)O_3$ considerably improves the quality factor without appreciable changes in the τ_f and ε_r. For $[Pb_{1-x}Ca_x][(Fe_{1/2}Nb_{1/2})_{1-y}Sn_y]O_3$ with $x = 0.55$, $y = 0.1$ has $Qf = 8600$ GHz, $\tau_f = 0$ ppm/°C and $\varepsilon_r = 85$ when sintered at 1150°C. Choi et al. obtained [45] a single-phase $(Pb_{0.4}Ca_{0.6})[(Mg_{1/3}Nb_{2/3})_{1-x}Sn_x]O_3$ with tetragonal perovskite structure by sintering at 1280°C. As the concentration of Sn increased the Qf increased but ε_r and τ_f decreased. For $x = 0.1$, it has $\varepsilon_r = 52$, $Qf = 8200$ GHz and $\tau_f = -3$ ppm/°C. Several authors improved [33–37, 40–43, 46–48, 144, 145] the properties of $[Pb_{1-x}Ca_x][(Fe_{1/2}Ta_{1/2})$ [PCFT] and $[Pb_{1-x}Ca_x][(Fe_{1/2}Nb_{1/2})$ [PCFN] by suitable substitutions at A or B sites and by the addition of dopants. Xiang et al. [37] investigated the effect of CeO_2 addition in $(Pb_{0.48}Ca_{0.52})(Fe_{1/2}Nb_{1/2})O_3$. The cerium entered the A site by the addition of CeO_2 upto 1.5 mol% and when more than 1.5 mol% is added, the cerium entered the B site. It was found that the addition of 2.2 mol% CeO_2 increased the Qf to 6800 GHz with $\varepsilon_r = 94$ and $\tau_f = 4$ ppm/°C.

The partial substitution of La for Pb in $Pb(Mg_{1/2}Nb_{1/2})O_3$ improves [146–148] ordering and the degree of ordering becomes unity for the composition $(Pb_{1/2}La_{1/2})(Mg_{1/2}Nb_{1/2})O_3$ [PLMN] [149]. However, the PLMN has a high τ_f [150]. Hence, Liu and Wu [44] introduced smaller Ca for Pb by forming $Pb_{1-x}Ca_x(Mg_{1/2}Nb_{1/2})O_3$ and investigated the effect of Ca substitution on the structure and microwave dielectric properties. They prepared $[(Pb_{1-x}Ca_x)_{1/2}La_{1/2}][Mg_{1/2}Nb_{1/2}]O_3$ with $x = 0.01$–0.5 by sintering at 1350°C. X-ray diffraction study showed that all the materials have $A(B'_{1/2}B''_{1/2})O_3$ type perovskite structure. However, the space group changed from $Fm\bar{3}m$ to $Pa3$ by the substitution of 10 mol% Ca. As the Ca content is increased to 50 mol% the symmetry changed to $R\bar{3}$. The ε_r decreased from 80 to 44 and Qf increased from 50 000 to 90 000 GHz for $x = 0.4$ and τ_f changed from 120 to −30 ppm/°C. $[(Pb_{1-x}Ca_x)_{1/2}La_{1/2}][Mg_{1/2}Nb_{1/2}]O=$ for $x = 0.3$ sintered at 1350°C/3 h showed $\varepsilon_r = 50$, $Qf = 86\,000$ GHz and $\tau_f = 0$ ppm/°C.

Several people [34–36, 39, 151] lowered the sintering temperature of PCFN and PCFT by adding low melting dopants and glasses or by lowering the particle size of the starting precursor powders. Yoon et al. [39] used nanometer-sized powders of $(Pb_{0.4}Ca_{0.6})(Fe_{1/2}Ta_{1/2})O_3$ obtained by the high energy mechanochemical milling method to lower the sintering temperature. By mechanochemical process for 60 hours and without calcination, a single-phase complex perovskite $Pb_{0.4}Ca_{0.6}(Fe_{1/2}Ta_{1/2})O_3$ was formed at 100°C. The sample has a lower sintering temperature of 1050°C/3 h and showed $\varepsilon_r = 62$, $Qf = 9000$ GHz and $\tau_f = -15$ ppm/°C.

7.9 $Ln(A_{1/2}Ti_{1/2})O_3$ [Ln = LANTHANIDE, A = Zn, Mg, Co]

The compound $La(Mg_{1/2}Ti_{1/2})O_3$ was first reported by Roy in 1954 [152]. Later several members of the group $Ln(Mg_{1/2}Ti_{1/2})O_3$ [Ln = Lanthanide] were reported [54, 153–158]. $La(Mg_{1/2}Ti_{1/2})O_3$ [LMT] can be prepared by calcining the oxide and/or carbonate raw materials at about 1200°C and sintering at about 1650°C [54–57]. Seabra et al. prepared LMT by the citrate chemical method [159, 160]. Among these several compounds, $La(Mg_{1/2}Ti_{1/2})O_3$ received considerable attention because of its high quality factor. German and Kovba [153] and Harshe et al. [161] reported that $La(Mg_{1/2}Ti_{1/2})O_3$ is cubic. Negas et al. [162] and Meden and Ceh [163] and Godzhiev et al. [155] found it to be orthorhombic. More recently it was reported to be monoclinic with space group $P2_1/n$ [154, 159]. However, it becomes orthorhombic on doping with small amounts of $SrTiO_3$ [164] or $BaTiO_3$ [165] or $La_{2/3}TiO_3$ [160]. The high negative τ_f of $La(Mg_{1/2}Ti_{1/2})O_3$ [LMT] has been tailored by the addition of $SrTiO_3$, $BaTiO_3$, $La_{2/3}TiO_3$, or $CaTiO_3$ which are having positive τ_f [159, 160, 164–170]. Recent X-ray and electron diffraction studies have shown [159] that LMT and NMT have 1:1 ordered B-site ions with monoclinic $P2_1/n$ space group having a-a-c+ tilt system. Porotnikov et al. [156] reported from Raman and infrared spectra studies that $Ln(Mg_{1/2}Ti_{1/2})O_3$ [Ln = La, Pr, Nd, Sm, Eu, Gd, Tb, Ho] have a high degree of order. Addition of $CaTiO_3$ or $SrTiO_3$ destroys the B-site ordering and reduce the microwave quality factor. It was reported [54, 158] that $Ln(Mg_{1/2}Ti_{1/2})O_3$ with Ln = Nd, Sm, Eu, Gd, are orthorhombic. However, it was reported later that $Ln(Mg_{1/2}Ti_{1/2})O_3$ [Ln = Nd, Tb, Dy, Ho, Y, Er, Yb] are monoclinic [154, 157]. Seabra et al. reported $Nd(Mg_{1/2}Ti_{1/2})O_3$ with a very high quality factor of 151 000 GHz [171]. The microwave dielectric properties of the various $Ln(Mg_{1/2}Ti_{1/2}O_3$ are given in Table 7.1. The microwave dielectric properties of LMT reported by different research groups are very much different: ε_r in the range 25–34, Qf in the range 40 000–114 000 GHz and τ_f of −70 to −100 ppm/°C [54, 59, 159, 160, 162, 171]. Salak et al. [160] investigated the structure and microwave dielectric properties of $(1-x)La(Mg_{1/2}Ti_{1/2})O_3-xLa_{2/3}TiO_3$ for $x \leq 0.5$ prepared by the citrate-based chemical route. The solid solution has a perovskite orthorhombic symmetry for $x = 0.1-0.3$. As the $La_{2/3}TiO_3$ content increased the Mg/Ti ordering decreased and the structure becomes pseudocubic for $x = 0.5$. The addition of $La_{2/3}TiO_3$ increased ε_r and decreased Qf and the τ_f becomes less negative and then becomes zero for $x = 0.5$. The air-sintered samples have a low Qf due to Ti^{4+} reduction to Ti^{3+} and the Qf improved on annealing. Seabra et al. [159] prepared $(1-x)La(Mg_{1/2}Ti_{1/2})O_3-xCaTiO_3$ $(0 < x < 1)$ by chemical route based on Pechini method. X-ray diffraction study showed the formation of a solid solution in the entire range. The ordered structure with $P2_1/n$

space group become disordered and transformed to the *Pbnm* space group for $x \geq 0.3$. As the $CaTiO_3$ content increased, the ε_r increased and *Qf* decreased and τ_f became less negative and ultimately became a high positive value. The relatively high sintering temperature of $La(Mg_{1/2}Ti_{1/2})O_3$ of about 1650°C can be lowered [172, 173] by the addition of a small amount of CuO or B_2O_3. However, large amounts of the additives degrade the microwave quality factor [172, 173]. Harshe et al. [161] and Cho et al. [54] proposed $La(Mg_{1/2}Ti_{1/2})O_3$ as a suitable substrate for YBCO superconductors with its low dielectric loss, matching lattice parameters and thermal expansion.

$La(Zn_{1/2}Ti_{1/2})O_3$ [LZT] was first reported by Ramadas [174]. Several authors [64, 155, 175] investigated the crystal structure of $Ln(Zn_{1/2}Ti_{1/2})O_3$ [Ln = La, Nd, Pr, Sm, Gd, Tb, Dy, Ho] and reported them as orthorhombic. More recently Ubic and co-workers [176, 177] made a detailed study on the structure of LZT. The Reitveld refinement of the neutron diffraction data showed $La(Zn_{1/2}Ti_{1/2})O_3$ to be monoclinic with space group $P2_1/n$. The microwave dielectric properties of the different $Ln(Zn_{1/2}Ti_{1/2})O_3$ are given Table 7.1. Cho et al. [63] investigated the effect of ZnO evaporation on the microwave dielectric properties of $La(Zn_{1/2}Ti_{1/2})O_3$ and found that samples sintered in ZnO atmosphere has a lower Q. Yeo et al. [67] tailored the properties of $La(Zn_{1/2}Ti_{1/2})O_3$ by forming a solid solution of LZT with $CaTiO_3$ which has a positive τ_f. The composition $0.5CaTiO_3$–0.5LZT sintered at 1550°C/3 h show $\varepsilon_r = 50$, $Qf = 38\,000$ GHz and $\tau_f = 0$ ppm/°C. Kucheiko et al. [64] prepared LZT by the sol–gel method. The sinterability of the nanopowder helped to sinter the samples at 1350°C with $\varepsilon_r = 30$, $Qf = 60\,000$ GHz and $\tau_f = -71$ ppm/°C. However, sintering at high temperatures led to evaporation of ZnO and decomposition of $La(Zn_{1/2}Ti_{1/2})O_3$ to La_2TiO_5 and ZnO.

Huang and Tseng [178] reported $La(Co_{1/2}Ti_{1/2})O_3$ which belongs to the space group $P2_1/n$ as a high Q dielectric material with $\varepsilon_r = 30$, $Qf = 67\,000$ GHz and $\tau_f = -64$ ppm/°C when sintered at 1440°C. Tseng and Huang [71, 75] added small amount of CuO or B_2O_3 to lower the sintering temperature to about 1350°C without significant degradation in the microwave dielectric properties. Cairns et al. [73] prepared single-phase $La(Co_{1/2}Ti_{1/2})O_3$ by the solid state method by sintering at 1550°C. XRD and TEM study showed that $La(Co_{1/2}Ti_{1/2})O_3$ has either orthorohombic *Pbnm* with disordered Co/Ti ions or $P2_1/n$ (ordered) space group symmetry consistent with $a^-a^-c^+$ tilt system. Rodriguez et al. [179] studied the crystal structure of $La(Co_{1/2}Ti_{1/2})O_3$ at room temperature. They reported that $La(Co_{1/2}Ti_{1/2})O_3$ is partially ordered having a monoclinic space group $P2_1/n$. It has an $a^-a^-c^+$ tilt system [71, 179]. Cairns et al. [77] prepared solid solutions of $La(Co_{1/2}Ti_{1/2})O_3$–$CaTiO_3$ and $Nd(Co_{1/2}Ti_{1/2})O_3$–$CaTiO_3$ to tune the high negative τ_f close to zero. Song et al. [72] prepared $Sm(Co_{1/2}Ti_{1/2})O_3$ by calcining at 1100°C and sintering at 1360°C/4 h. The sintered sample has an orthorhombic crystal structure with $\varepsilon_r = 26$; $Qf = 76\,000$ GHz and $\tau_f = -16$ ppm/°C. $Ln(A_{1/2}Ti_{1/2})O_3$ (A = Mg, Zn, Co) ceramics have negative τ_f and can be tailored by adding $BaTiO_3$, $SrTiO_3$, $CaTiO_3$ or $La_{2/3}TiO_3$. However, addition of these τ_f compensators decrease the quality factor.

7.10 Conclusions

$A(B'_{1/2}B''_{1/2})O_3$ is the largest group of low loss complex perovskite compounds. $A(B'_{1/2}B''_{1/2})O_3$ [A = Ba, Sr, Ca and B′ = lanthanide, B″ = Nb,Ta] have a relatively high sintering temperatures of 1600–1650°C. They have poor sinterability and sintering additives are usually needed to improve their densification. This group of compounds generally has an ordered arrangement of B-site ions. This ordering causes doubling of the simple perovskite unit cell and the compounds are called double perovskites. This group of materials is prone to a variety of distortions which lower the crystal symmetry. The space group of $A(B'_{1/2}B''_{1/2})O_3$ [A = Ba, Ca, Sr] depends on the tolerance factor or the size of A- and B-site cations. Depending on the size of A and B′ cations, the B-O_6 octahedra undergoes tilting. The tilt angle is small and the scattering factor of the oxygen sublattice being small, it is difficult to find the correct space group symmetry of these group of materials by conventional X-ray diffraction techniques. The splitting of X-ray reflections will be small and difficult to observe. There are controversies about the correct space group symmetry of these compounds. Several authors used X-ray diffraction, neutron diffraction and Raman spectroscopy to investigate the structure and space group. However, space group symmetry reported by the different groups of many of the compounds are not in agreement with each other. It may be noted that presence of a small amount of impurities also affect the space group symmetry. In $Ca(B'_{1/2}Ta_{1/2})O_3$ the ionic size of 12-fold coordination B′ and Ca are comparable and X-ray diffraction study revealed intersubstitution. There are several compounds with excellent permittivity and quality factor but have poor temperature stability. The properties of materials with negative τ_f have been tailored by the addition of compounds such as TiO_2, $CaTiO_3$, $SrTiO_3$ or other complex perovskites having a positive τ_f. These materials either form a solid solution with the parent material or form mixture phases. By adjusting the amount of the additives the τ_f can be tuned close to zero. Some of the temperature-stable and useful materials in this family of compounds are $Ba(Tb_{1/2}Nb_{1/2})O_3$ with $\varepsilon_r=39$, $Qf=52\ 400$ GHz; $0.95Ba(Yb_{1/2}Nb_{1/2})O_3$–$0.05Ca(Y_{1/2}Nb_{1/2})O_3$ with $\varepsilon_r=34$, $Qf=47\ 500$ GHz; $Ba_{0.95}Sr_{0.05}(Y_{1/2}Ta_{1/2})O_3$ with $\varepsilon_r=33$, $Qf=47\ 500$ GHz; $0.6Ca(Yb_{1/2}Ta_{1/2})O_3$–$0.4Ba(Yb_{1/2}Ta_{1/2})O_3$ with $\varepsilon_r=28$, $Qf=48\ 000$ GHz; $La(Co_{1/2}Ti_{1/2})O_3$–$0.5CaTiO_3$ with $\varepsilon_r=30$, $Qf=56\ 000$ GHz. These materials have high ε_r and high quality factors, and are potential materials for use in mobile phone base station applications. A small amount of non-stoichiometry, such as deficiency of A-site Sr or slight excess of B-site Eu or Ta, is found to improve the quality factor in $Sr(Eu_{1/2}Ta_{1/2})O_3$.

REFERENCES

[1] M. T. Anderson, K. B. Greenwood, G. A. Taylor, and K. K. Poeplmeir. B cation arrangements in double perovskites. Prog. Solid State Chem. 22(1993)197–233.

[2] P. K. Davies. Cation ordering in complex oxides. Curr. Opin. Solid State Mater. Sci. 4(1999)467–471.

[3] R. H. Mitchell. Perovskites. Modern and Ancient. Thunder Bay, MI Almaz Press (2002).

[4] E. J. Cussen, J. F. Vente, P. D. Battle, and T. C. Gibb. Neutron diffraction study of the influence of structural disorder on the magnetic disorder of Sr_2FeMO_6 [M=Nb,Ta]. J. Mater. Chem. 7(1997)459–463.

[5] P.W. Barnes, M. W. Lufasso, and P. M. Woodward. Structure determination of $A_2M^{3+}TaO_6$ and $A_2M^{3+}NbO_6$ ordered perovskites: octahedral tilting and psuedosymmetry. Acta Crystallogr. B 62(2006)384–396.

[6] F. Galasso and W. Darby. Ordering of the octahedrally coordinated cation position in the perovskite structure. J. Phys. Chem. 66(1962)131–132.

[7] L. Brixner. Preparation and crystallographic study of some new rare earth compounds. J. Inorg. Nucl. Chem. 15(1960)352–355.

[8] F. Galasso, G. Layden, and D. Flichbagh. The $Ba(B'_{0.5}Ta_{0.5})O_3$ ordered perovskite type compounds possible new laser host materials. J. Chem Phys. 44(1966)2703–2707.

[9] V. S. Filipev and E. G. Fesenko. Unit cells of some compounds of $A_2B'B''O_6$ types. Sov. Phys. Crystallogr. 6(1962)616–618.

[10] V. S. Filipev and E. G. Fesenko. Preparation and structure of complex perovskites of type $Ca_2B'B''O_6$. Sov. Phys. Crystallogr. 10(1965)243–247.

[11] F. J. Fresia, L. Katz, and R. Ward. Cation substitution in perovskite like phases. J. Am. Chem. Soc. 81(1959)4783–4785.

[12] E. G. Steward and H. P. Rooksby. Psuedocubic alkaline earth tungstate and molybdnates of the R_3MX_6 types. Acta Crystallogr. 4(1951)503–507.

[13] F. S. Galasso. Structure and properties of Perovskite compounds. Pergammon Press, Headington Hill, Oxford (1969).

[14] F.S. Galasso. Structure and properties of inorganic solids. Pergamon Press, Headington Hill Oxford (1970).

[15] F. S. Galasso. Perovskite and high Tc superconductors. Gordon & Breach Science Publishers (1990).

[16] A. I. Agranovskaya. Physical-chemical investigation of formation of complex ferroelectrics with the perovskite structure. Bull. Acad. Sci. USSR Phys. Ser. 24(1960)1271.

[17] M. Takata and K. Kageyama. Microwave characteristics of $A(B^{3+}{}_{1/2}B^{5+}{}_{1/2})O_3$. J. Am. Ceram. Soc. 72(1989)1955–1959.

[18] R. Zuhrmuhlen, J. Petzelt, S. Kamba, G. Kozlov, B. Volkov, B. Gorshunov, D. C. Dube, A. Tagentsev, and N. Setter. Dielectric spectroscopy of $Ba(B'_{/2}B''_{1/2})O_3$ complex perovskite ceramics and microwave dielectric properties. Correlation between ionic parameters and microwave dielectric properties. J. Appl. Phys. 77(1995)5341–5350.

[19] R. Zurmuhlen, E. L. Colla, D. C. Dube, J. Petzelt, I. M. Reaney, A. Bell, and N. Setter. Structure of $Ba(Y^{3+}{}_{1/2}Ta^{5+}{}_{1/2})O_3$ and its dielectric properties in the range 10^2–10^{14} Hz, 20–600 K. J. Appl. Phys. 76(1994)5864–5873.

[20] M. Reaney, E. L. Colla, and N. Setter. Dielectric and structural characteristics of Ba and Sr based complex perovskites as a function of tolerance factor. Jpn. J. Appl. Phys. 33(1994)3984–3990.

[21] H. Sreemoolanathan, J. Isaac, M. T. Sebastian, K. A. Jose, and P. Mohanan. Synthesis, characterization and properties of $Ba_{1-x}Sr_x(Nd_{1/2}Nb_{1/2})O_3$ ceramics for applications as dielectric resonators in microwave circuits. Ceram. Int. 21(1995)385–389.

[22] L. A. Khalam, H. Sreemoolanathan, R. Ratheesh, P. Mohanan, and M. T. Sebastian. Preparation characterisation and microwave dielectric properties of $Ba(B'_{1/2}B''_{1/2})O_3$ [B′=La, Pr, Nd, sm, Eu, Gd, Tb, Dy, Ho, Y, Yb, In] ceramics. Mater. Sci. Eng. B 107(2004)264–270.

[23] R. L. Moreira, L. A. Khalam, M. T. Sebastian, and A. Dias. Raman spectroscopic investigations on the crystal structure and phonon modes of $Ba(Re_{1/2}Ta_{1/2})O_3$ microwave ceramics. J. Eur. Ceram. Soc. 27(2007)2803–2810.

[24] H. Sreemoolanathan, R. Ratheesh, M. T. Sebastian, and P. Mohanan. $Ba(Yb_{1/2}Nb_{1/2})O_3$: a new ceramic microwave dielectric resonator. Mater. Lett. 33(1997)161–165.

[25] S. K. Korchagina and Y. A. Shevchek. Low frequency and microwave dielectric properties of Ba_2LnTaO_6 [Ln=La, Pr, Sm, Dy, Ce, Gd, Nd, Tm, Tb] ceramics. Inorg. Mater. 42(2006)66–69.

[26] L. A. Khalam and M T Sebastian. Effect of cation substitution and non-stoichiometry on the microwave dielectric properties of $Sr(B'_{0.5}Ta_{0.5})O_3$ [B′=lanthanide] perovskites J. Am. Ceram. Soc. 89(2006)3689–3695.

[27] L. A. Khalam and M. T. Sebastian. Microwave dielectric properties of $Sr(B'_{1/2}Nb_{1/2})O_3$ [B′=La, Pr, Nd, Sm, Eu, Gd, Tb, Dy, Ho, Y, Er, Yb and In] ceramics. Int. J. Appl. Ceram. Tech. 3(2006)364–374.

[28] R. Ratheesh, M. Wohlecke, B. Berge, Th. Wahnbrink, H. Haeuseler, E. Ruhl, R. Blachnik, P. Balan, N. Santha, and M. T. Sebastian. Raman study of the ordering in $Sr(B'_{0.5}Nb_{0.5})O_3$ compounds. J. Appl. Phys. 88(2000)2813–2818.

[29] L. A. Khalam and M. T. Sebastian. Low loss dielectrics in the $Ca(B'_{1/2}B''_{1/2})O_3$ [B′=lanthanides, y] system. J. Am. Ceram. Soc. 90(2007)1467–1474.

[30] H. Ikawa, M. Ohara, T. Iida, M. Takayama, and M. Takemoto. Products and microwave dielectric properties of ceramics with nominal composition $Ba_{1-x}Ca_x(Sc_{1/2}Nb_{1/2})O_3$. J. Ceram. Soc. Jpn. Suppl. 112–1. Special issue (2005)S1614–S1617.
[31] J. Kato, H. Kagata, and N. Nishimoto. Dielectric properties of $(Pb,Ca)(Me,Nb)O_3$ at microwave frequency. Jpn. J. Appl. Phys. 31(1992)3144–3147.
[32] E. S. Kim, J. S. Jeon, and K. H. Yoon. Effect of sintering method on the microwave dielectric properties of $(Pb_{0.45}Ca_{0.55})(Fe_{0.5}Nb_{0.5})O_3$ ceramics. J. Eur. Ceram. Soc. 23(2003)2583–2587.
[33] S. Kucheiko, J. W. Choi, H-J. Kim, S. J. Kim, and H. J. Jung. Microwave characteristics of $(Pb,Ca)(Fe,Nb,Sn)O_3$ dielectric materials. J. Am. Ceram. Soc. 80(1997)2937–2940.
[34] Q. H. Yang, E. S. Kim, Y. J. Kim, and P. K. Kim. Effect of $PbO–B_2O_3–V_2O_5$ glass on the microwave dielectric properties of (Pb, Ca, La)$(Fe,Nb)O_3$ ceramics. Mater. Chem. Phys. 79(2003)239–242.
[35] J.-Y. Ha, J-W. Choi, H-J. Kim, S-J. Yoon, and K. H. Yoon. The effect of sintered aid $CuO–Bi_2O_3$ on the microwave dielectric properties of $(Pb_{0.45}Ca_{0.55})[(Fe_{0.5}Nb_{0.5})_{0.9}Sn_{0.1}]O_3$ ceramics. Mater. Chem. Phys. 79(2003)261–265.
[36] H. M. Zhe, Z.D. Xiang, G. H. Huang, W. Hao, Z. Tianjing, and Z. Bin. Influence of Bi_2O_3 and MnO_2 doping on the microwave properties of $[(Pb,Ca,La)][Fe,Nb]O_3$ dielectric ceramics. Mater. Sci. Eng., B. 117(2005)199–204.
[37] Z. D. Xiang, H. M. Zhe, J. S. Lin, and H. Jing. Microwave properties of CeO_2 modified $(Pb,Ca)(Fe,Nb)O_3$ dielectric ceramics. Mater. Lett. 58(2004)2684–2687.
[38] E. S. Kim, H. S. Park, and K. H. Yoon. Porosity dependence of microwave dielectric properties of complex perovskite $(Pb_{0.5}Ca_{0.5})(Fe_{0.5}Ta_{0.5})O_3$ ceramics. Mater. Chem. Phys. 79(2003)213–217.
[39] K. H. Yoon, H. S. Park, J. Y. Cho, and E. S. Kim. Microwave dielectric properties of $(Pb_{0.4}Ca_{0.6})(Fe_{0.5}Ta_{0.5})O_3$ ceramics by mechanochemical processing. J. Eur. Ceram. Soc. 23(2003)2579–2582.
[40] O. H. Yang, E. S. Kim, and J. Xu. Effect of Nd^{3+} on the microwave dielectric properties of $(Pb_{0.5}Ca_{0.5})(Fe_{0.5}Nb_{0.5})O_3$ceramics. Mater. Sci. Eng. B 113(2004)224–227.
[41] Q. Yang, E. S. Kim, J. Xu, and Z. Meng. Microwave dielectric properties of $(Pb,Ca,La)(Fe,Nb)O_{3+d}$ ceramics substituted by Ti in B site. Mater. Sci. Eng. B 99(2003)332–335.
[42] Q. Yang, E.S. Kim, J. Xu, and Z. Meng. Effect of La^{3+} and Nd^{3+} on the microwave dielectric properties of $(Pb_{0.5}Ca_{0.5})(Fe_{0.5}Nb_{0.5})O_3$ ceramics. Mater. Sci. Eng. B 99(2004)259–261.
[43] M. Hu, D. Zhou, D. Zhang, W. Lu, B. Li, J. Huang, and S. Gong. Microwave dielectric properties of $(Pb,Ca)(Fe,Nb,Zr)O_3$ ceramics. Mater. Sci. Eng. B 99(2003)403–407.
[44] C.-L. Liu and J.-B. Wu. Effects of calcium substitutions on the structure and microwave dielectric characteristics of $[(Pb_{1-x}Ca_x)_{1/2}La_{1/2}](Mg_{1/2}Nb_{1/2})O_3$ ceramics. J. Am. Ceram. Soc. 84(2001)1291–1295.
[45] J.-W. Choi, J.-Y. Ha, C.-Y. Kang, S.-J. Yoon, H.-J. Kim, and K. H. Yoon. Microwave dielectric properties of $(Pb,Ca)(Mg,Nb,Sn)O_3$ ceramics. Jpn. J. Appl. Phys. 39(2000)5923–5926.
[46] Q.H. Yang, E.S. Kim, and J. Xu. Effect of Nd^{3+} on the microwave dielectric properties of $(Pb_{0.5}Ca_{0.5})(Fe_{0.5}Nb_{0.5})O_3$ ceramics. Mater. Sci. Eng. B 113(2004)224–227.
[47] K. H. Yoon, E. S. Kim, and J-S. Jeon. Understanding the microwave dielectric properties of $(Pb_{0.45}Ca_{0.55})[(Fe_{0.5}(Nb_{1-x}Ta_x)_{0.5})O_3$ ceramics. J. Eur. Ceram. Soc. 23(2003)2391–2396.
[48] X. Hu and X. M. Chen. Microstructure and dielectric properties of $Ca(Fe_{1/2}Nb_{1/2})O_3$ modified by $PbZrO_3$ ceramics. Jpn. J. Appl. Phys. 43(2004)215–218.
[49] H. Takahashi, K. Ayusewa, and N. Sakamoto. Microwave dielectric properties of $Ba(Mg_{1/2}W_{1/2})O_3–BaTiO_3$ ceramics. Jpn. J. Appl. Phys. 36(1997)5597–5599.
[50] D. D. Khalyvin, J. Han, A. M. R. Senas, and P. Q. Mantas. Synthesis and dielectric properties of tungsten based complex perovskites. J. Mater. Res. 18(2003)2600–2606.
[51] J. Jiang, G. H. Bo, and W. Xiaowu. Microwave dielectric properties of $(Pb_{1-3x/2}La_x)(Mg_{1/2}W_{1/2})O_3$. Mater. Res. Bull. 39(2004)2127–2135.
[52] H.-J. Kim, S. Kucheiko, S-J. Yoon, and H.-J. Jung. Microwave dielectric properties in the $(La_{1/2}Na_{1/2})TiO_3–Ca(Fe_{1/2}Nb_{1/2})O_3$ system. J. Am. Ceram. Soc. 80(1997)1316–1318.

[53] F. Zhao, Z. Yue, Z. Gui, and L. Li. Preparation characterization and microwave dielectric properties of A_2BWO_6 [A=Sr, Ba, B=Co, Ni, Zn] double perovskite ceramics. Jpn. J. Appl. Phys. 44(2005)8070.
[54] S.-Y. Cho, C.-H. Kim, D.-W. Kim, and K. S. Hong. Dielectric properties of $Ln(Mg_{1/2}Ti_{1/2})O_4$ as substrates for high Tc superconductor thin films. J. Mater. Res. 14(1999)2484–2487.
[55] M. P. Seabra, M. Avdeev, V. M. Ferreira, R. C. Pullar, Mc N. Alford, and I. M. Reaney. Structure property relations in xBaTiO$_3$–(1–x)La(Mg$_{1/2}$Ti$_{1/2}$)O$_3$ solid solutions. J. Am. Ceram. Soc. 87(2004)584–590.
[56] M. P. Seabra, M. Avdeev, V. M. Ferreira, R. C. Pullar, and Mc N. Alford. Structure and microwave dielectric properties of $La(Mg_{0.5}Ti_{0.5})O_3$–$CaTiO_3$ system. J. Eur. Ceram. Soc. 23(2003)2403–2408.
[57] D.-Y. Lee, S.-J. Yeon, J.-H. Yeo, S. Nahm, J.-H. Park, and K.C. Whang. Crystal structure and microwave properties of $La(Mg_{1/2}Ti_{1/2})O_3$ ceramics. J. Mater. Sci. Lett. 19(2000)131–134.
[58] M. P. Seabra and V. M. Ferreira. Synthesis and characterisation of microwave $La(Mg_{1/2}Ti_{1/2})O_3$ ceramics. Key Eng. Mater. 206–213(2001)1501–1504.
[59] Y.-B. Chen, C.-L. Huang, and C.-W. Lo. Microwave dielectric properties and microstructures of $La(Mg_{1/2}Ti_{1/2})O_3$ with CuO doped. Mater. Sci. Eng. B 128(2006)98–102.
[60] C.-L. Huang, Y.-B. Chen, and C.-W. Lo. Microwave dielectric properties and microstructures of $0.5La(Mg_{1/2}Ti_{1/2})O_3$–$CaTiO_3$ ceramics with B_2O_3 addition. Jpn. J. Appl. Phys. 44(2005)6706–6708.
[61] C.-L. Huang, Y.-B. Chen, and S.-H. Lin. New dielectric material system $Nd(Mg_{1/2}Ti_{1/2})$-O_3–$CaTiO_3$ at microwave frequency. Solid State Electron. 49(2005)1921–1924.
[62] J. B. Kim, K. H. Yoon, and Y. S. Cho. Compositional dependence of microwave dielectrics in (1–x)(Na$_{1/2}$Nd$_{1/2}$)TiO$_3$–xNd(Mg$_{1/2}$Ti$_{1/2}$)O$_3$ ceramics. J. Am. Ceram. Soc. 86(2003)1159–1161.
[63] S.-Y. Cho, M.-K. Seo, K. S. Hong, S. J. Park, and I.-T. Kim. Influence of ZnO evaporation on the microwave dielectric properties of $La(Zn_{1/2}Ti_{1/2})O_3$. Mater. Res. Bull. 32(1997)725–735.
[64] S. Kucheiko, H.-J. Kim, D.-H. Yeo, and H.-J. Jung. Microwave dielectric properties of $LaZn_{0.5}Ti_{0.5}O_3$ ceramics prepared by sol-gel process. Jpn. J. Appl. Phys. 35(1966)668–675.
[65] T. A. Vanderah, V. L. Miller, I. Levin, S. M. Bell, and T. Negas. Phase relations, crystal chemistry and dielectric properties in sections of La_2O_3–CaO–MgO–TiO_2 system. J. Solid State Chem. 177(2004)2023–2038.
[66] G. S. Babu, V. Subramanian, and V. R. K. Murthy. Structure determination and microwave dielectric properties of $La(MgSn)_{0.5}O_3$ ceramics. J. Eur. Ceram. Soc. 27(2007)2973–2976.
[67] D. H. Yeo, H. J. Kim, H. H. Moon, S. J. Yoon, and H. J. Kim. Dielectric properties of (1–x)CaTiO$_3$–xLa(Zn$_{1/2}$Ti$_{1/2}$)O$_3$ ceramics at microwave frequencies. Jpn. J. Appl. Phys. 35(1996)663–667.
[68] C.-H. Hsu, H.-T. Song, C.-C. Yu, C.-L. Huang, and M.-T. Ku. The effect of sintering temperature and time on microwave properties of $Sm(Zn_{1/2}Ti_{1/2})O_3$ ceramic resonators. Ceram. Int. 33(2007)951–955.
[69] C.-F. Tseng, C.-L. Huang, N.-R. Yang, and C.-H. Hsu. Microwave dielectric properties of $Nd(Zn_{1/2}Ti_{1/2})O_3$ ceramics at microwave frequencies. J. Am. Ceram. Soc. 89(2006) 1465–1470.
[70] C.-F. Tseng, C.-L. Huang, N.-R. Yang, and C.-H. Hsu. Dielectric characterization of xNd(Zn$_{1/2}$Ti$_{1/2}$)O$_3$–(1–x)CaTiO$_3$ ceramics. Mater. Lett. 61(2007)4054–4057.
[71] C.-F. Tseng and C.-L. Huang. Microwave dielectric properties of CuO doped $La(Co_{1/2}Ti_{1/2})O_3$ ceramics. Jpn. J. Appl. Phys. 45(2006)826–830.
[72] H.-T. Song, C.-S. Hsu, M.-T. Kuo, and C.-L. Huang. Microwave characteristics of $Sm(Co_{1/2}Ti_{1/2})O_3$ dielectric resonators. Mater. Lett. 58(2004)2829–2833.
[73] D. L. Cairns, I. M. Reaney, H. Zheng, D. Iddles, and T. Price. Synthesis and characterization of $La(Co_{1/2}Ti_{1/2})O_3$. J. Eur. Ceram. Soc. 25(2005)433–439.
[74] C.-L. Huang and J.-F. Tseng. Dielectric characteristics of $La(Co_{1/2}Ti_{1/2})O_3$ ceramics at microwave frequencies. Mater. Lett. 58(2004)3732–3736.

[75] C.-F. Tseng and C.-L. Huang. Effect of B_2O_3 addition to $La(Co_{1/2}Ti_{1/2})O_3$ ceramics on sintering behaviour and microwave dielectric properties. Jpn. J. Appl. Phys. 45(2006) 8766–8768.

[76] C.-F. Tseng and C.-L. Huang. Improved microwave dielectric properties of B_2O_3 doped $Nd(Co_{1/2}Ti_{1/2})O_3$ ceramics with near temperature coefficient of resonant frequency. Mater. Res. Bull. 42(2006)9–16.

[77] D. L. Cairns, I. M. Reaney, N. Otten, D. Iddles, and T. Price. Structural determination and microwave properties of xRe$(Co_{1/2}Ti_{1/2})O_3$–$(1-x)CaTiO_3$[Re=La.Nd] solid solutions. J. Eur. Ceram. Soc. 26(2006)875–882.

[78] J. Y. Chen, I. Levin, T. A. Vanderah, K. G. Geyer, and R. S. Roth. Subsolidus phase relations and dielectric properties in the SrO–Al_2O_3–Nb_2O_5 systems. Int. J. Inorg. Mater. 2(2000)107–114.

[79] R. Guo, S. Bhalla, J. Sheen, F. W. Ainger, S. Edei, E. C. Subbarao, and L. E. Cross. Strontium aluminum tantalum oxide and strontium aluminum niobium oxide as potential substrates for highTc thin films. J. Mater. Res. 10(1995)18–25.

[80] K. Kageyama. Microwave dielectric properties of CaO–Ga_2O_3–Ta_2O_5 ceramics. Ferroelectrics. 109(1990)173–178.

[81] J. Takahashi, K. Kageyama, T. Fuji, Y. Yamada, and K. Kodaira. Formation and microwave dielectric properties of $Sr(Ga_{0.5}Ta_{0.5})O_3$ based complex perovskites. J. Mater. Sci. – Mater. Electron. 8(1997)79–84.

[82] H. Kagata and J. Kato. Dielectric properties of Ca based complex perovskite at microwave frequencies. Jpn. J. Appl. Phys. 33(1994)5463–5465.

[83] L. A. Khalam and M. T. Sebastian. Temperature stable and low loss microwave dielectrics in the $Ca(B'_{1/2}Ta_{1/2})O_3$ [B' = lanthanides, Y, In] system. J. Am. Ceram. Soc. 90(2007)2476–2486.

[84] T. Fuji, M. Takemoto, M. Takayama, and H. Ikawa. Reinvestigation of extraordinary products and microwave dielectric properties of ceramics with nominal composition $(Ba_{1-x}Ca_x)(Sc_{1/2}Nb_{1/2})O_3$. MMA (2004), Inuyama, Japan.

[85] H. Ikawa and M. Takamoto. Products and microwave dielectric properties of ceramics with nominal compositions $(Ba_{1-x}Ca_x)(B'_{1/2}B''_{1/2})O_3$ [$B'=Y^{3+}$, Nd^{3+}, Gd^{3+}, $B''=Nb^{5+}$, Ta^{5+}]. Mater. Chem. Phys. 79(2003)222–225.

[86] K. V. Paulose, J. Koshy, and A. D. Damodaran. YBa_2NbO_6: Synthesis, properties and compatibility with $YBa_2Cu_3O_{7-d}$. Physica C. 193(1992)273–276.

[87] K. Henmi, Y. Hinatsu, and N. M. Masaki. Crystal structures and magnetic properties of ordered perovskites Ba_2LnNbO_6 (Ln = lanthanide elements). J. Solid State Chem. 148(1999)353–360.

[88] V. Ting, Y. Liu, R. L. Withers, and E. Krausz. An electron diffraction and bond valence sum study of the space group symmetries and structures of the photocatalytic 1:1 ordered A_2InNbO_6 double perovskites ($A = Ca^{2+}$, Sr^{2+}, Ba^{2+}) J. Solid State Chem. 177(2004)979–986.

[89] A. A. Evdokimov and N. F. Menshenina. Unit cell parameters of Ba_2MEO_6[E = Nb,Ta]. Russ. J. Inorg. Chem. 27(1982)1208–1209.

[90] W. T. Fu and D. J. W. Ijdo. New insight into the symmetry and the structure of the double perovskites Ba_2LnNbO_6 (Ln = lanthanides and Y). J. Solid State Chem. 179(2006)1022–1028.

[91] P. J. Saines, B. J. Kennedy, and M. M. Elcombe. Structural phase transition and crystal chemistry of the series $Ba_2LnB'O_6$ [Ln = lanthanide, $B' = Nb^{5+}$, Sb^{5+}]. J. Solid State Chem. 180(2007)401–409.

[92] A. M. Glazer. The classification of tilted octahedral in perovskites. Acta Crystallogr. B 28(1975)3384–3392.

[93] A. M. Glazer. Simple ways of determining perovskite structures. Acta Crystallogr. A 31(1975)756–762.

[94] K. S. Alexandrov and S. V. Misyul. Sovt. Phys. Crystallogr. 26(1981)612–618.

[95] O. Bock and U. Muller. Symmetriverwarnstadten bei variaten des perowskit tryps. Acta Crystallogr. B 58(2002)594–606.

[96] P. M. Woodward. Octahedral tilting in perovskites.I. Geometrical considerations. Acta Crystallogr. B 53(1997)32–43.
[97] D. I. Woodward and I. M. Reaney. Electron diffraction of tilted perovskites. Acta Crystallogr. B 61(2005)387–399.
[98] M. W. Lufasso, P. W. Barns, and P. M. Woodward. Structure prediction of ordered and disordered multiple octahedral cation perovskites using SPUDS. Acta Crystallogr. B 62(2006)397–410.
[99] C. J. Howard, B. J. Kennedy, and P. M. Woodward. Ordered double perovskites – a group theoretical analysis. Acta Crystallogr. B 59(2005)463–471.
[100] P. W. Barnes. Ph. D. Dissertation. Exploring structural changes and distortions in quarternary perovskites and defect pyrochlores using powder diffraction techniques. Ohio State University, Columbus (2003).
[101] A. Dias, L. A. Khalam, M T Sebastian, C. W. A. Paschol, and R. L. Moreira. Chemical substitution in $Ba(RE_{1/2}Nb_{1/2})O_3$(Re = La, Nd, Sm, Gd, Tb, and Y) microwave ceramics and its influence on the crystal structure and phonon modes. Chem. Mater. 18(2006)214–220.
[102] H. S. Park and K. H. Yoon. Effect of bond valence on microwave dielectric properties complex perovskite ceramics. Mater. Chem. Phys. 79(2003)181.
[103] E. S. Kim, B. S. Chun, D. W. Yoo, and K. Y. Yoon. Microwave dielectric properties of $(1-x)$ $(Ca_{0.7}Nd_{0.2})TiO_3$–$xLi_{0.5}Nd_{0.5}TiO_3$ ceramics. Mater. Sci. Eng. B 99(2003)247–251.
[104] E. S. Kim, Y. H. Kim, J. H. Chae, D. W. Kim, and K. Y. Yoon. Dielectric properties of $(Pb_{0.2}Ca_{0.8})(Ca_{1/3}Nb_{2/3})O_3$ ceramics at microwave frequency. Mater. Chem. Phys. 79(2003)230–232.
[105] I. D. Brown and D. Altermaut. Bond valence parameters obtained from a systematic analysis of inorganic crystal structure database. Acta Crystallogr. B. 41(1985)244–247.
[106] J. Koshy, J. Kurian, J. K. Thomas, Y. P. Yadava, and A. D. Damodaran. Rare earth barium niobates. A new class of potential substrates for $VBa_2Cu_3O_{7-d}$ superconductor. Jpn. J. Appl. Phys. 33(1994)117–121.
[107] N. F. Fedorov, O. V. Melnikov, and Yu. N. Smironov. Some properties of complex perovskites of the M_2TaNdO_6 type. Russ. J. Inorg. Chem. 22(1977)1719–1720.
[108] I. Gregoria, J. Petzelt, J. Pokorny, V. Vorlicek, Z. Zikmund, Z. Zurmuhlen, and N. Setter. Raman spectroscopy of zone center improper ferroelastic transition in ordered $Ba(Y_{1/2}Ta_{1/2})$ O_3 complex perovskite ceramic. Solid State Commun. 94(1995)899–903.
[109] Y. Doi and Y. Hinatsu. Magnetic properties of ordered perovskites Ba_2LnTaO_6 [Ln = lanthanides]. J. Phys.: Condens. Matter. 13(2001)4191–4202.
[110] T. Fuji, J. Takahashi, and S. Shimada. Phase transformation and temperature dependence of relative permittivity of A site substituted $Sr(Y_{0.5}Ta_{0.5})O_3$ perovskites. J. Electroceram. 3(1999)387–397.
[111] S. K. Korchigina, N.V. Goluko, Yu. A. Shebvchuk, and V. V. Gagulin. Dilatometry and dielectric properties of Ba_2RTaO_6 [R = La, Pr, Sm, Dy] ceramics. Inorg. Mater. 40(2004)1088.
[112] N. Taira and Y. Hinatsu. Magnetic susceptibility of Ba_2YbTaO_6 with the ordered perovskite structure and electron paramagnetic resonance of Yb^{3+}doped in Ba_2LuTaO_6. J. Solid State Chem. 150(2000)31–35.
[113] P. J. Saines, J. R. Spencer, B. J. Kennedy, and M. Avdeev. Structures and crystal chemistry of the double perovskites $Ba_2LnB'O_6$ (Ln = lanthanide, $B' = Nb^{5+}$ and Ta^{5+}): Part I. investigation of Ba_2LnTaO_6 using synchrotron X-ray and neutron powder diffraction. J. Solid State Chem. 180(2007)2991–3000.
[114] A. Dias, L. A. Khalam, M. T. Sebastian, and R. L. Moreira. Raman spectroscopic investigation of Ba_2InTaO_6 and Sr_2InTaO_6 perovskites. J. Solid State Chem. 180(2007)2143–2148.
[115] L. A. Khalam. $A(B'_{1/2}B''_{1/2})O_3$ [A = Ba,Sr, Ca, Mg; B' = rare earths B'' = Nb Ta] microwave ceramics for wireless communications. PhD Thesis. Kerala University (2007).
[116] J. James, M. Sankar, S. Senthil Kumar, and K. V. O. Nair. Preparation and properties of Ba_{2-x} $SrSmTaO_6$: A group of new perovskite materials. Mater. Chem. Phys. 83(2004)328–333.

[117] T. Fuji, H. Sato, S. Ito, and G. L. Messing. Crystal structure and dielectric properties of complex perovskite (Sr,Ba,Ca)($Ni_{0.5}W_{0.5}$)O_3. Solid State Ionics 172(2004)425–429.
[118] C. J. Howard, P. W. Barnes, B. J. Kennedy, and P. M. Woodward. Structures of the ordered double perovskites Sr_2YTaO_6 and Sr_2YNbO_6. Acta Crystallogr. B 61(2005) 258–262.
[119] V. P. Sirotinkin, V. A. Efremov, and V. K. Trunov. The crystal structures of the compounds Sr_2SmNbO_6 and Sr_2TmNbO_6. Russ. J. Inorg. Chem. 30(1985)981–983.
[120] V. K. Trunov, V. P. Sirotkin, and A. A. Evdokimov. An x-ray diffraction study of the compounds Sr_2LnEO_6 [E = Nb, Ta]. Russ. J. Inorg. Chem. 28(1983)349–350.
[121] R. D. Shannon. Revised effective ionic radii and systematic study of inter atomic distances in halides and chalcogenides. Acta Crystallogr. A 32(1976)751–757.
[122] I. M. Reaney and R. Ubic. Dielectric and structural characteristics of perovskites and related materials as a function of tolerance factor. Ferroelectrics. 228(1999)23–38.
[123] T. Shimada. Far infrared reflections $Sr_{1-x}Ca_x(Ga_{1/2}Ta_{1/2})O_3$ ceramics. J. Eur. Ceram. Soc. 21(2001)2617–2620.
[124] R. H. Mitchell. Perovskites – Modern and Ancient. Aemaz Press, Thunder Bay, ON. Canada (2002).
[125] V. M. Ferreira and J. L. Baptista. Role of niobium in magnesium titanate microwave ceramics. J. Am. Ceram. Soc. 79(1996)1697–1698.
[126] C. D. Brandle and V. J. Fratelle. Preparation of perovskite oxides for high Tc superconductor substrates. J. Mater. Res. 5(1990)2160–2164.
[127] P. Woodward, R. D. Hoffmann, and A. W. Sleight. Order–disorder in $A_2M^{3+}M^{5+}O_6$ perovskite. J. Mater. Res. 9(1994)2118–2127.
[128] P. M. Woodward. PhD Thesis. Oregon State University. Corvallis. Oregon, USA (1997).
[129] K. P. Surendran, M. T. Sebastian, P. Mohanan, R. L. Moreira, and A. Dias. Effect of nonstoichiometry on the structure and microwave dielectric properties of Ba($Mg_{0.33}$-$Ta_{0.67}$)O_3. Chem. Mater. 17(2005)142–151.
[130] T. Tsunooka, M. Androu, Y. Higashida, H. Sugiura, and H. Ohsato. Effects of TiO_2 on sinterability and dielectric properties of high Q forsterite. Ceramics. J. Eur. Ceram. Soc. 23(2003)2573–2578.
[131] R. C. Pullar, K. Okeme, N. Mc Neil Alford. Temperature compensated niobate microwave ceramics with the columbite structure $M^{2+}Nb_2O_6$. J. Eur. Ceram. Soc. 23(2003)2479–2483.
[132] V. K. Trunov, L. I. Konstatinova, Yu. A. Velikodnaji, A. A. Evdokinov, and A. M. Frotov. The compounds of Ca_2MNbO_6. Russ. J. Inorg. Chem. 26(1981)1738–1741.
[133] V. B. Vyazov, A. M. Frolov, Yu. M. Golovin, and V. V. Fonichev. The effect of inversion of the cations on the vibrational spectra of Ca_2LnEO_6 [E = Nb, or Ta]. Russ. J. Inorg. Chem. 30(1985)1576–1578.
[134] M. H. Kim, S. Nahm, W. S. Lee, M. J. Yoo, N. G. Gang, and H J. Lee. Structural variation and microwave dielectric properties of TiO_2 doped $Ba(Zn_{1/3}Ta_{2/3})O_3$ ceramics. J. Eur. Ceram. Soc. 24(2004)3547–3552.
[135] J. S. Kim, C. Chen, H. J. Kang, C. H. Lee, K.-Y. Kim, S. Nahm, and J. D. Byun. Crystal structure and microwave dielectric properties of $CaTiO_3$–$(Li_{1/2}Nd_{1/2})TiO_3$–$(Ln_{1/3}Nd_{1/3})$ TiO_3 (Ln = La,Y) ceramics. Jpn. J. Appl. Phys. 38(1999)5633–5637.
[136] H. Zheng, G. D. C. Gyrorgtalova, and I. M. Reaney. Microstructure and microwave properties of $CaTiO_3$–$LaGaO_3$ solid solutions. J. Mater. Sci. 40(2005)5207–5214.
[137] V. K. Trunov, L. I. Konstatinova, and A. A. Evdokimov. An X-ray diffraction study of the compounds Ca_2LnTaO_6. Russ. J. Inorg. Chem. 28(1983)807–810.
[138] R. D. Shannon. Revised effective ionic radii and systematic studies of interatomic distances in halides and chalcogenides. Acta Crystallogr. A 32(1976)751–767.
[139] M. T. Anderson, K. B. Greenwood, G. A. Taylor, and K. B. Poeplemeir. B-cation arrangements in double perovskites. Prog. Solid State Chem. 22(1993)197–233.
[140] J. Kato. Material produces small resonators with high dielectric constant. JEE (Japan) Sept (1991)114–118.

[141] J. Kato, H. Kagata, and K. Nishimoto. Dielectric properties of lead-alkaline earth zirconate at microwave frequencies. Jpn. J. Appl. Phys. 30(1991)2343–2346.

[142] H. Kagata, J. Kato, and K. Nishimoto. Rap. Techn. Meet. Dielectric and electrical insulation. IEE Japan, Tsukuba 1992. DEI-92–32 (in Japanese).

[143] H. Kagata J. Kato, K. Nishimoto, and T. Inoue. Dielectric properties of Pb based perovskite substituted by Ti for B site at microwave frequencies. Jpn. J. Appl. Phys. 32(1993)4332–4334.

[144] Y. S. Cho, K. H. Yoon, B. D. Lee, H. R. Lee, and E. S. Kim. Understanding microwave dielectric properties of Pb based perovskite ceramics via bond valence. Ceram. Int. 30(2004)2247–2250.

[145] M. Nakano, K. Suzuki, T. Miura, and M. Kobayashi. Low temperature fireable dielectric material $Pb(Fe_{1/2}W_{1/2})O_3$–$(PbCa)(Fe_{1/2}Nb_{1/2})O_3$ for microwave use. Jpn. J. Appl. Phys. 32(1993)4314–4318.

[146] J. Chen, H. M. Chen, and M. P. Harmer. Ordering and dielectric properties of undoped and La/Na doped $(Pb(Mg_{1/2}Nb_{2/3})O_3$ ceramics. J. Am. Ceram. Soc. 72(1989)593–598.

[147] N. Kim, W. Huebner, S. J. Jang, and T. R. Shrout. Dielectric and piezoelectric properties of lanthanum modified lead magnesium niobate and lead titaanate ceramics. Ferroelectrics. 93(1989)341–349.

[148] N. de Nathan, E. Hausson, P. Gaucher, and A. Morel. Modification of the B site order of $Pb(Mg_{1/3}Nb_{2/3})O_3$ ceramics by thermal annealing or by La doping. Mater. Res. Bull. 25(1990)427–434.

[149] L. J. Lin and T. B. Wu. Ordering behaviour of lead magnesium niobate ceramics with A site substitution. J. Am. Ceram. Soc. 73(1990)1253–1254.

[150] L. J. Lin. Structure and dielectric properties of $Pb(Mg_{1/3}Nb_{2/3})O_3$–$La(Mg_{2/3}Nb_{1/3})O_3$–$PbTiO_3$. PhD Thesis. National Hua University, Taiwan (1991).

[151] Y. Quihong, K. E. Soo, X. Jun. J. Chin. Ceram. Soc. 31(2002)554–558.

[152] R. Roy. Multiple ion substitution in the perovskite lattice. J. Am. Ceram. Soc. 37(1954)581–588.

[153] M. German and L. M. Kovba. New ternary oxides with a perovskite structure. Russ. J. Inorg. Chem. 28(1983)586–587.

[154] W. A. Gron, F. P. F. Berkel, and D.J. Ijdo. Dineodymium magnesium titanate (IV). A rietveld refinement of neutron powder diffraction data. Acta Crystallogr. C 42(1986)1472–1475.

[155] O. V. Godzhiev, N. V. Porotnikov, and K. I. Petrov. A physicochemical study of compounds based on zinc, titanium and lanthanide oxides. Russ. J. Inorg. Chem. 32(1987)1676–1677.

[156] N. V. Porotnikov, M. German, and L. M. Kovba. The vibrational spectra of complex oxides Ln_2MgTiO_6. Russ. J. Inorg. Chem. 29(1984)823–828.

[157] M. German and L. M. Kovba. New ternary oxides with a perovskite structure. Russ. J. Inorg. Chem. 28(1983)586–587.

[158] N. Santha, I. N. Jawahar, P. Mohanan, and M. T. Sebastian. Microwave dielectric properties of $(1-x)CaTiO_{3-x}Sm(Mg_{1/2}Ti_{1/2})O_3$ $[0.1 \leq x \leq 1]$ ceramics. Mater. Lett. 54(2002)318–322.

[159] M. P. Seabra, M. Avdeev, V. M. Ferreira, R. C. Pullar, N. Mc, and N Alford. Structure and microwave dielectric properties of $La(Mg_{0.5}Ti_{0.5})O_3$–$CaTiO_3$ system. J. Eur. Ceram. Soc. 23(2003)2403–2408.

[160] A. N. Salak, M. P. Seabra, and V. M. Ferreira. $La(Mg_{1/2}Ti_{1/2})O_3$–$La_{2/3}TiO_3$ microwave dielectric ceramics. J. Eur. Ceram. Soc. 23(2003)2409–2412.

[161] G. Harshe, A. S. Bhalla, and L. E. Cross. Synthesis and dielectric properties of cubic perovskite $La(Mg_{1/2}Ti_{1/2})O_3$. Mater. Lett. 18(1994)173–175.

[162] T. Negas, G. Yeager, S. Bell, and R. Amren. Chemistry and properties of temperature compensated microwave dielectrics. NIST special publications 804(1991)21–34.

[163] A. Meden and M. Ceh. Structure determination and Rietveld refinement of $La(Mg_{0.5}Ti_{0.5})O_3$. Mater. Sci. Forum. 278–282(1998)773–778.

[164] M. Avdeev, M. P. Seabra, and V. M. Ferreira. Structure evolution in $La(Mg_{0.5}Ti_{0.5})TiO_3$–$SrTiO_3$ system. Mater. Res. Bull. 37(2002)1459–1468.

[165] M. Avdeev, M. P. Seabra, and V. M. Ferreira. Crystal structure of dielectric ceramics in the $La(Mg_{0.5}Ti_{0.5})O_3$–$BaTiO_3$ system. J. Mater. Res. 17(2002)1112–1117.
[166] A. N. Salak, D. D. Khalyavin, P. Q. Mantas, A. M. R. Senos, and V M. Ferreira. Structure dependent microwave dielectric properties of $(1-x)La(Mg_{1/2}Ti_{1/2})O_3$–$xLa_{2/3}TiO_3$ ceramics. J. Appl. Phys. 98(2005)034101. (online page no.)
[167] M. P. Seabra, A. N. Salak, V. M. Ferreira, J. L. Riterio, and L. G. Viera. Dielectric properties of $(1-x)La(Mg_{1/2}Ti_{1/2})O_3$–$xSrTiO_3$ ceramics. J. Eur. Ceram. Soc. 24(2004) 2995–3002.
[168] A. N. Salak, M. P. Seabra, V. M. Ferreira, J. L. Riterio, and L. G. Viera. Dielectric characterisation of the $(1-x)La(Mg_{1/2}Ti_{1/2})TiO_3$–$BaTiO_3$ microwave ceramics. J. Phys. D. 37(2004)914–920.
[169] C.-L. Huang and Y.-B. Chen. Microwave properties of B_2O_3 doped $Nd(Mg_{1/2}Ti_{1/2})O_3$–$CaTiO_3$ dielectric resonators at microwave frequency. Mater. Lett. 60(2005)198–202.
[170] M. P. Seabra, M. Avdeev, V. M. Ferreira, R. C. Pullar, Mc N. Alford, and I. M. Reaney. Structure property relations in $xBaTiO_3$–$(1-x)La(Mg_{1/2}Ti_{1/2})O_3$ solid solutions. J. Am. Ceram. Soc. 87(2004)584–590.
[171] M. P. Seabra, A. N. Salak, M. Avdeev, and V. M. Ferreira. Structure and dielectric characterisation of the $La(Mg_{1/2}Ti_{1/2})O_3$–$Nd(Mg_{1/2}Ti_{1/2})O_3$ system. J. Phys.: Condens. Matter. 15(2003)4229–4238.
[172] Y.-B. Chen, C.-L. Huang, and C.-W. Lo. Microwave dielectric properties and microstructures of $La(Mg_{1/2}Ti_{1/2})O_3$ with CuO doped. Mater. Sci. Eng. B 128(2006)98–102.
[173] C.-L Huang, Y.-B. Chen, and C.-W. Lo. Microwave dielectric properties and microstructures of $0.5La(Mg_{1/2}Ti_{1/2})O_3$–$0.5CaTiO_3$ ceramics with B_2O_3 addition. Jpn. J. Appl. Phys. 44(2005)6706–6708.
[174] N. Ramadas, J. Gopalakrishnn, and M.V.C Sastri. Preparation and characterisation of La_2TiMO_6[Co,Ni,Cu,Zn] perovskites. J. Inorg. Nucl. Chem.40(1978)1453–1454.
[175] N. V. Porotnikov, O. V. Godzhieva, and L. N. Margolin. An investigation of the structure of quaternary oxides with composition Ln_2ZnTiO_6 by vibrational spectroscopy. Russ. J. Inorg. Chem. 32(1987)793–795.
[176] R. Ubic, Y. Hu, K. Khamoushi, and I. Abrahams. Structure and properties of $La(Zn_{1/2}Ti_{1/2})O_3$. J. Eur. Ceram. Soc. 26(2006)1787–1790.
[177] R. Ubic, Y. Hu, and I. Abrahams. Neutron and electron diffraction studies of $La(Zn_{1/2}Ti_{1/2})O_3$ perovskite. Acta Crystallogr. B 62(2006)521–529.
[178] C.-L. Huang and J. F. Tseng. Dielectric characteristics of $La(Co_{1/2}Ti_{1/2})O_3$ ceramics at microwave frequencies. Mater. Lett. 58(2004)3732–3736.
[179] E. Rodriguez, M. L. Lopez, J. Campa, M. L. Viera, and L.C. Pico. Crystal and magnetic structure of perovskite La_2MTiO_6 [M = Co,Ni]. J. Mater. Chem. 12(2002)2798–2802.

CHAPTER EIGHT

$A(B'_{1/3}B''_{2/3})O_3$ Complex Perovskites

8.1 Introduction

The $A(B'_{1/3}B''_{2/3})O_3$ perovskites are the most widely studied family of materials in microwave ceramics. A broad range of chemical substitutions such as Ba, Sr, Ca, at A site; Mg, Zn, Ni, Co, Sr, Ca, Mn, Cd at B′ site and Nb and Ta at B″ site are possible and the substitutions enable the tailoring of the dielectric properties. $Ba(Mg_{1/3}Ta_{2/3})O_3$ (BMT), $Ba(Zn_{1/3}Ta_{2/3})O_3$ (BZT) and $Ba[(Zn,Co)_{1/3}Nb_{2/3}]O_3$ (BZCN) are the most widely studied materials in this family and are commercially produced for applications in wireless communication. A large number of compounds with niobium or tantalum as the B″ ions in the $A(B'_{1/3}B''_{2/3})O_3$ complex perovskite were reported by Roy [1] and Galasso and co-workers [2]. These oxides have twice the B″ ions as the B′ ions. It was believed that $A(B'_{1/3}B''_{2/3})O_3$ has a cubic perovskite cell with three layers of BaO_3, one layer of B^{2+} and two layers of B^{5+} in the unit cell. Galasso and co-workers observed extra weak reflections on the X-ray diffraction patterns of some of these complex perovskites. Later, Galasso et al. found [3] that one of the compound $Ba(Sr_{1/3}Ta_{2/3})O_3$ has an ordered structure in which Sr and Ta occupy ordered positions which account for the extra reflections. Galasso reported [4–6] that the ordered $Ba(Sr_{1/3}Ta_{2/3})O_3$ has a hexagonal unit cell whose *c*-axis is equivalent to the <111> direction of the disordered cubic perovskite cell. It was found [4–6] that many of the $A(B'_{1/3}B''_{2/3})O_3$ compounds show ordered perovskite structure. It was also reported [7] that 1:2 ordering in the B site for compounds of general formula $A(B'_{1/3}B''_{2/3})O_3$ is a strong function of radius mismatch between B′ and B″ ions. Galasso and Pyle observed [7] that the ordering increased with increasing size difference between the B^{2+} and the B^{5+} ions. It was found [4] that the ordering increased when the samples were annealed and was attributed to the existence of small ordered domains which grew on annealing at high temperatures. The crystal structure, lattice parameters and cell volumes of several $A(B'_{1/3}B''_{2/3})O_3$ complex perovskite materials are listed by Galasso in his books [4–6]. In 1977, Kawashima and co-workers reported [8] for the first time useful microwave dielectric resonator materials in the $Ba(B'_{1/3}B''_{2/3})O_3$ system. They reported $Ba(Zn_{1/3}Nb_{2/3})O_3$ (BZN) with $\varepsilon_r = 41$, $Q = 5600$, $\tau_f = +28$ ppm/°C and $Ba(Zn_{1/3}Ta_{2/3})O_3$ [BZT] with $\varepsilon_r = 30$, $\tau_f = 0 \pm 0.5$ ppm/°C and $Q = 6500$. Since then several papers appeared in the literature reporting microwave dielectric properties of several $A(B'_{1/3}B''_{2/3})O_3$ complex perovskites and are given in Table 8.1 The tolerance factors as calculated using Shannon's ionic radii [9] are also given in Table 8.1.

Dielectric Materials for Wireless Communications
ISBN-13: 978-0-08-045330-9

Table 8.1 Microwave dielectric properties of $A(B'_{1/3}B''_{2/3})O_3$ ceramics

Composition	Tolerance factor	Sintering temperature (°C)	ε_r	Qf (GHz)	τ_f (ppm/°C)	Reference
$Ba(Mg_{1/3-x}Nb_{2/3})O_{3-\delta}$ ($x = 0.02$)	1.037	1450	32	96 000	30	[10]
$Ba(Ni_{1/3}Nb_{2/3})O_3$	1.035	1400	31	48 000	−18	[11]
$Ba(Ni_{1/3}Ta_{2/3})O_3$	1.035	1500	23	49 700	−18	[12]
$Ba(Mg_{1/3}Ta_{2/3})O_3$	1.029	1640	24	430 000	8	[13]
$Ba(Mg_{1/3}Nb_{2/3})O_3$	1.029	1350	31	46 000	18	[14]
$Ba(Zn_{1/3}Nb_{2/3})O_3$	1.027	1390	41	87 000	30	[15, 16]
$Ba(Zn_{1/3}Ta_{2/3})O_3$	1.027	1350/120 h	28	168 000	0.5	[17]
$Ba(Co_{1/3}Nb_{2/3})O_3$	1.026	1400	32	78 000	−12	[18,19]
$Ba(Co_{1/3}Ta_{2/3})O_3$	1.026	1500	25	71 400	−16	[12]
$Ba(Mn_{1/3}Ta_{2/3})O_3$	1.012	1600/air	27	15 500	45	[20]
$Ba(Mn_{1/3}Ta_{2/3})O_3$	1.012	1600/N_2	27	104 000	45	[20]
$Ba(Cd_{1/3}Ta_{2/3})O_3 + B_2O_3$	0.993	1350	32	50 000	80	[21]
$Ba(Cd_{1/3}Ta_{2/3})O_3$ + 2 wt% ZnO		1550	33	37 500	80	[22]
$Ba(Ca_{1/3}Ta_{2/3})O_3$	0.986	1500	30	27 300	145	[12]
$Sr(Ni_{1/3}Ta_{2/3})O_3$	0.977	1500	23	21 000	−57	[12]
$Sr(Mg_{1/3}Nb_{2/3})O_3$	0.972	1500	33	32 000	−14	[15, 23]
$Sr(Mg_{1/3}Ta_{2/3})O_3$	0.972	1500	22	5600	−50	[12]
$Sr(Co_{1/3}Ta_{2/3})O_3$	0.97	1500	23	17 500	−71	[12]
$Sr(Zn_{1/3}Nb_{2/3})O_3$	0.968	1500	40	44 000	−39	[15, 24]
$Sr(Zn_{1/3}Ta_{2/3})O_3$	0.968	1500	28	21 700	−54	[12]
$Ca(Ni_{1/3}Nb_{2/3})O_3$	0.963	#	26	11 000	−78	[25]
$Ca(Ni_{1/3}Ta_{2/3})O_3$	0.963	#	22	21 000	−80	[25]

Table 8.1 (Continued)

Composition	Tolerance factor	Sintering temperature (°C)	ε_r	Qf (GHz)	τ_f (ppm/°C)	Reference
$Ca(Cu_{1/3}Ta_{2/3})O_3$	0.96	#	23	5500	#	[25]
$Ca(Cu_{1/3}Nb_{2/3})O_3$	0.96	#	27	3300	#	[25]
$Ca(Mg_{1/3}Ta_{2/3})O_3$	0.958	#	21	78 000	–61	[25]
$Ca(Mg_{1/3}Nb_{2/3})O_3$	0.958	#	28	58 000	–48	[25, 26]
$Ca(Zn_{1/3}Nb_{2/3})O_3$	0.955	#	35	16 000	–43	[25]
$Ca(Zn_{1/3}Ta_{2/3})O_3$	0.955	#	25	25 000	–66	[25]
$Ca(Co_{1/3}Ta_{2/3})O_3$	0.954	#	23	12 000	–65	[25]
$Ca(Co_{1/3}Nb_{2/3})O_3$	0.954	#	29	6200	–65	[25]
$Sr(Ca_{1/3}Ta_{2/3})O_3$	0.93	1500	22	27 300	–91	[12]
$Ca(Ca_{1/3}Ta_{2/3})O_3$	0.92	#	22	22 000	–41	[25]
$Ca(Ca_{1/3}Nb_{2/3})O_3$	0.892	#	28	17 000	–22	[25]
$(Pb_{0.5}Ca_{0.5})(Mg_{1/3}Nb_{2/3})O_3$	0.963	#	86	4600	34	[27]
$(Pb_{0.4}Ca_{0.6})(Mg_{1/3}Nb_{2/3})O_3$	0.958	#	73	4100	3.7	[27]
$(Pb_{0.25}Ca_{0.75})(Mg_{1/3}Nb_{2/3})O_3$	0.950	#	46	8700	–34	[27]
$(Pb_{0.4}Ca_{0.6})(Mg_{1/3}Nb_{2/3})_{1-x}Sn_xO_3$ $x = 0.0$	#	1280	66	6900	3	[28]
$(Pb_{0.4}Ca_{0.6})(Mg_{1/3}Nb_{2/3})_{1-x}Sn_xO_3$ $x = 0.01$	#	1280	65	7100	0.0	[28]
$(Pb_{0.4}Ca_{0.6})(Mg_{1/3}Nb_{2/3})_{1-x}Sn_xO_3(x = 0.03)$	#	1280	63	7500	–4	[28]
$(Pb_{0.4}Ca_{0.6})(Mg_{1/3}Nb_{2/3})_{1-x}Sn_xO_3$ $x = 0.05$	#	1280	57	8100	–4	[28]

(Continued)

Table 8.1 (Continued)

Composition	Tolerance factor	Sintering temperature (°C)	ε_r	Qf (GHz)	τ_f (ppm/°C)	Reference
$(Pb_{0.4}Ca_{0.6})(Mg_{1/3}Nb_{2/3})_{1-x}Sn_xO_3$ $x = 0.1$	#	1280	52	8200	−4	[28]
$(Pb_{0.25}Ca_{0.75})(Mg_{1/3}Nb_{2/3})_{0.75}Ti_{0.25}O_3$	#	#	60	11 000	0	[29]
$(Pb_{0.6}Ni_{0.4})(Mg_{1/3}Nb_{2/3})O_3$	#	#	94	3800	130	[27]
$(Pb_{0.5}Ni_{0.5})(Mg_{1/3}Nb_{2/3})O_3$	#	#	73	4900	52	[27]
$(Pb_{0.4}Ni_{0.6})(Mg_{1/3}Nb_{2/3})O_3$	#	#	59	7100	6.2	[27]
$(Pb_{0.5}Co_{0.5})(Mg_{1/3}Nb_{2/3})O_3$	#	#	75	1400	16	[27]
$(Pb_{0.2}Ca_{0.8})(Ca_{1/3}Nb_{2/3})O_3$	0.91	1350	36	12 500	−27	[30]
$0.99Ba(Co_{1/3}Nb_{2/3})O_3$–$0.01Ba(Y_{1/2}Nb_{1/2})O_3$	#	1380	34	38 690	#	[31]
$0.9Ba(Co_{1/3}Nb_{2/3})O_3$–$0.1Ba(Y_{1/2}Nb_{1/2})O_3$	#	1380	37	25 560	#	[31]

Data not available in the respective literature.

8.2 $Ba(Zn_{1/3}Ta_{2/3})O_3$ [BZT]

8.2.1 Preparation

BZT is usually prepared by the conventional solid state method by calcining the mixed stoichiometric amounts of ZnO, $BaCO_3$ and Ta_2O_5 raw materials at temperatures in the range 1100–1200°C and sintering at temperatures in the range 1500–1550°C [8, 17, 32–38]. Figure 8.1 shows the variation of density and grain size as a function of sintering temperature. The density increased with sintering temperature and reaches a

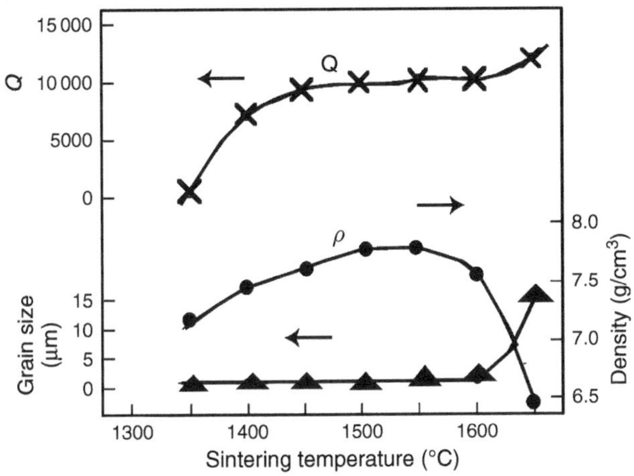

Figure 8.1 Variation of Q and grain size of BZT with sintering temperature (after Ref. [17]).

maximum at about 1550°C and further increase in sintering temperature decreased density. Abnormal grain growth occurred on sintering above 1600°C and the density decreased sharply. The sinterability and the dielectric properties are very much influenced by the preparation conditions, stoichiometry, and origin and purity of the raw materials [38, 39]. Sintering at temperatures above 1500°C or prolonged heat treatment leads to volatilization of ZnO [38–40]. The escape of ZnO led to poor densification near the surface of the samples. However, homogeneous densification was found when BZT was muffled in ZnO powder [37–39, 41, 42]. Davies et al. [37, 41] avoided ZnO loss during heat treatment by a double pre-calcination procedure. Zn loss usually results in poor densification or a low density white skin on the surface. EDAX spectra taken from the surface of sintered samples showed that the as-sintered surface is depleted of Zn concentration and enriched in Ta concentration [39]. X-ray diffraction study of the as-sintered surface showed the presence of new phases. The depletion of ZnO at the surface of BZT during heat treatment at high temperatures leads to formation of zinc-deficient phases such as $Ba_8ZnTa_6O_{24}$, $Ba_3Ta_2O_8$, and $BaTa_2O_6$ [38–40, 43–46]. The Zn loss during the processing (calcination, sintering, annealing) plays an important role in controlling the properties and also the crystal structure. The weight loss increased with increasing sintering temperature, slowly up to 1500°C and rapidly above 1500°C. Kawashima et al. [17] prepared BZT by hot pressing at 1400°C which showed a density close to the theoretical density of 7.92 gm/cm^3. It has been reported [44–46] that attrition milling produces a very fine and monomodal powder that allows good densification at a relatively lower temperature of about 1450°C. Nomura and co-workers [47] reported that it is difficult to densify BMT and BZT without Mn addition. Figure 8.2 shows the variation of bulk density of BZT with Mn concentration. Mn addition increased bulk density and the maximum density was found for 2 mol% Mn addition. Several authors tried to lower the sintering temperature of BZT by the addition of low melting additives such as CuO, B_2O_3, LiF, Li_2CO_3, $BaTi_4O_9$ etc, [45, 46, 48–50]. Roulland and Marinel [48] reported that addition of 10 mol% of B_2O_3 and 5 mol% LiF lower sintering temperature to about 1150°C. The sintering temperature can be further lowered to 870°C by the addition of $B_2O_3 + CuO$ to BZT [50]. The formation of

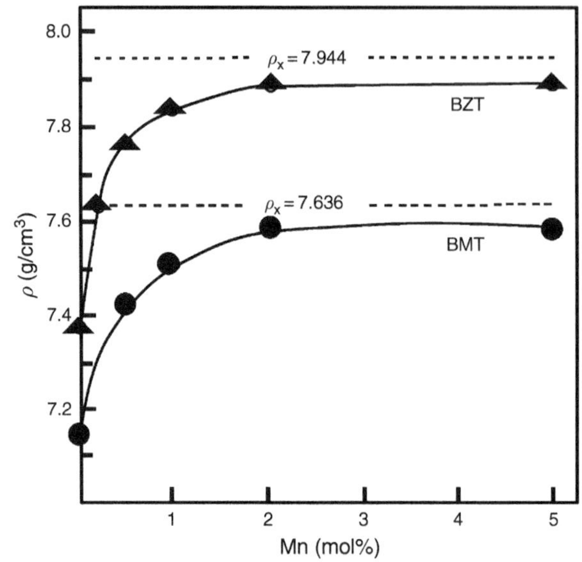

Figure 8.2 Variation of density of BZT and BMT as a function of mol% of Mn addition (after Ref. [47]).

$BaCu(B_2O_5)$ phase during sintering as a liquid phase helped in the densification of BZT at relatively lower temperatures.

Renoult et al. [51] prepared BZT by the alkoxide route which involves the reaction of bimetallic alkoxide $Ta_2Zn(OEt)_{12}$ with $Ba(OH)_2 \cdot 8H_2O$. Single-phase BZT was formed on calcining the precipitated powder at 650 C. Varma et al. [52] prepared BZT nanopowder by the decomposition of a citrate precursor gel. The sinterability of BZT ceramics made from nanopowder was very poor and on sintering at high temperatures caused depletion of ZnO which led to the complete conversion of BZT to $BaTa_2O_6$. However, Varma et al. [52] could succeed in sintering BZT ceramics from nanopowder by muffling in BZT powder obtained by solid state method. Mc Laren et al. [43] prepared BZT by hydrothermal methods. However, the BZT so formed were found to be Ba and Zn deficient. Thirumal and Ganguli [53] reported preparation of BZT by molten salt method using NaCl–KCl flux at a temperature of about 900°C. Koga et al. [54] found that slight deviation from stoichiometry leads to formation of secondary phases.

8.2.2 Crystal structure and ordering

The ordering behavior of the B cations in the $A(B'_{1/3}B''_{2/3})O_3$ type complex perovskites has a crucial role on the physical and electrical properties of these compounds [55]. Jacobson et al. studied [56] the long range ordering in BZT by X-ray diffraction method and refined the structure using neutron diffraction data. The study revealed the existence of hexagonal ordered and cubic disordered structures depending on the preparation conditions. The disordered structure is cubic and has Zn and Ta ions at the B sites arranged in a random way. Table 8.2 gives the lattice parameters, space group and

Table 8.2 Crystallographic data of important $A(B_{1/3}'B''_{2/3})O_3$ materials [4, 6, 59, 60]

	BMN	BMT	BZT	BZN	BNN	BNT	SZN
a (Å)	5.7754	5.77385	5.7812	5.78 207	5.75 496	5.75 513	5.66
c (Å)	7.08 762	7.09 376	7.0823	7.09 731	7.06 695	7.07 480	6.95
c/a	1.2272	1.2286	1.2251	1.2275	1.2280	1.2293	1.2279
Space group (ordered)	$P\bar{3}m1$	$P\bar{3}m1$	$P\bar{3}m1$	$P\bar{3}m1$	$P\bar{3}m1$	$P\bar{3}m1$	$P\bar{3}m1$
Disordered	$Pm3m$	$Pm3m$	$Pm3m$	$Pm3m$	$Pm3m$	$Pm3m$	$Pm3m$
Theoretical density (g/cm^3)	6.236	7.636	7.92	6.511	6.554	8.017	5.667

density of BZT and the related compounds. When BZT phases are first prepared, they crystallize in an apparently disordered cubic perovskite structure and on annealing become ordered hexagonal structure with Zn–Ta–Ta repeat sequence along [111] direction of the parent cubic cell. This type of ordering is usually called 1:2 ordering. BZT is normally annealed at temperatures in the range 1300–1400°C for various lengths of time up to 100 hours for getting an ordered structure. Figure 8.3 (a) and (b) shows the

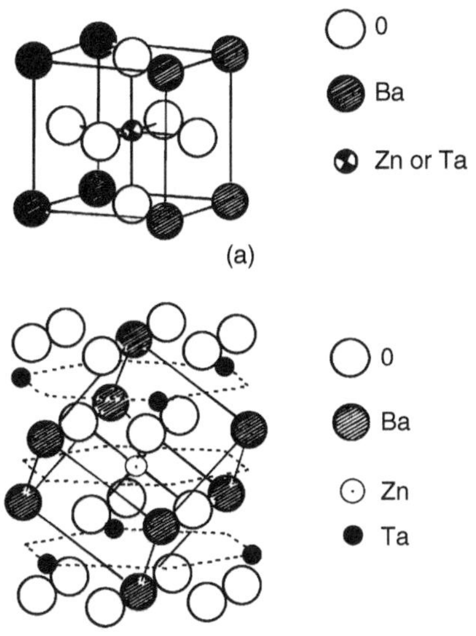

Figure 8.3 Crystal structure of $Ba(Zn_{1/3}Ta_{2/3})O_3$ (A) Disordered and (B) Ordered (after Ref. [33]).

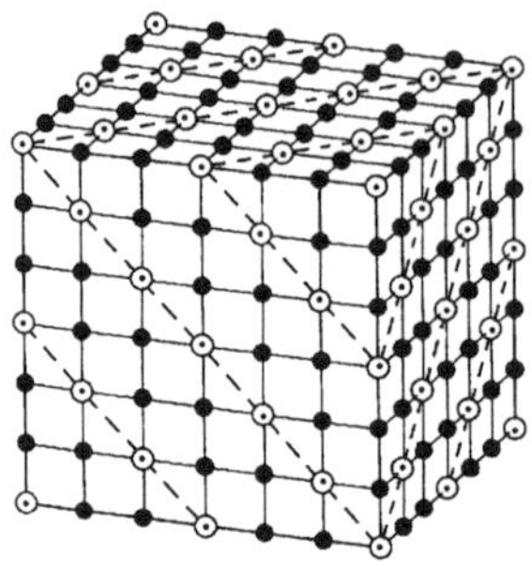

Figure 8.4 Zn and Ta stacking sequence in the ordered structure in BZT (after Ref. [33]).

cubic disordered and hexagonal ordered BZT structures respectively. The large and small solid circles represent Ba and Ta ions respectively. The Zn and Ta ions in the ordered structure follow the stacking sequence as shown in Figure 8.4. The variation in the ordering of the B-site cations with annealing time and temperature was studied using the powder X-ray diffraction method [17, 39]. The degree of order as gauged by the intensity and sharpness of the superstructure reflection lines in the X-ray patterns increased with prolonged annealing, and as a completely ordered structure is approached, the unit cell undergoes a small hexagonal distortion. The lattice distortion arises from a small expansion of the parent cubic cell in a direction normal to the ordered (111) planes. The phase transformation is second order. Sagala and Nambu [33] calculated the lattice energy of ordered and disordered BZT and reported that the lattice energy of ordered BZT is 2.15 eV which is lower than that of the disordered structure. They also [36] theoretically calculated the loss tangent at microwave frequencies for BZT with respect to the degree of B-site structural order. Starting from the equations of ion motion, dielectric loss was expressed in terms of the pair correlation functions corresponding to the ordering of Zn and Ta ions on B sites. It was found [36] that the loss tangent decreased (10^{-3} to 10^{-6}) with increasing B-site order. This is consistent with the fact that the degree of ordering increases with annealing. Figure 8.5 shows the predicted dielectric loss tangents due to disorder of the distribution of mass and charge for the frequencies at 10 and 20 GHz. Madelung energy calculations of several compounds of

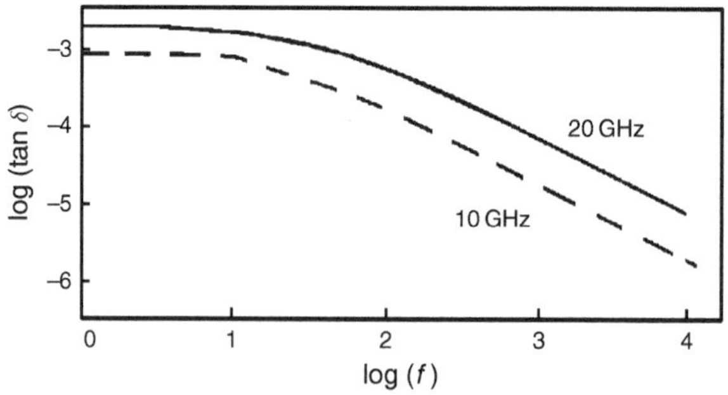

Figure. 8.5 Predicted values of dielectric loss factor due to disorder of the distribution of mass and charge for frequencies 10 and 20 GHz (after Ref. [36]).

the type $A^{2+}(B^{2+}_{1/3}B^{5+}_{2/3})O_3$ showed that the electrostatic ordering energy increases with the increasing square of the difference in the charge of B-site ions [35]. This suggests that electrostatic interaction plays a major role in the long range order of B-site ions. It is also known [7, 17, 56] from experimental work that a large difference in the radius of B-site ions also tends to lead to long range order, which implies size effects. Bellaiche and Vanderbit [57], Bellaiche et al. [58] investigated the effect of electrostatic interactions on ground-state ordering and proposed a simple model which includes the long range coulomb interactions between ions. In this model energy is taken to be proportional to the electrostatic energy of an ideal system of ionic charges. Bellaiche et al. showed that this model provides a systematic understanding of the complicated ordering behavior of complex perovskite alloys, thereby providing strong evidence that coulomb interaction between ions is the dominant factor in determining such ordering. They suggested that the main driving mechanism responsible for long range order (LRO) occurring in heterovalent perovskites is simply the electrostatic interaction between different species.

The ordering in BZT produces superstructure reflections and deviation of c/a ratio from the value of $1.2247(\sqrt{3/2})$ for an undistorted unit cell [4, 7, 17, 39, 60, 61]. The samples sintered at temperatures above 1350°C show ordering. The ordering increases with sintering temperature, sintering time and annealing. A fully ordered hexagonal structure has $c/a > 1.2247$. The lattice distortion due to Zn–Ta ordering causes splitting of (422) and (226) reflections. Figure 8.6 shows splitting of the (226) and (422) reflections with sintering time at 1350°C. The (422) and (226) reflections split with increase in sintering time. The principal signatures of the cation ordering used to evaluate the relationship of the structure to quality factor are the appearance and relative changes in the intensity of the superstructure peaks originating from the chemical order and splitting of reflections associated with the deviation of the c/a ratio of 1.2247. The degree of order parameter (S) is defined as

$$S = \left(\frac{I_{100}/I_{(110,012)\text{obs}}}{I_{100}/I_{(110,012)\text{cal}}} \right)^{1/2} \tag{8.1}$$

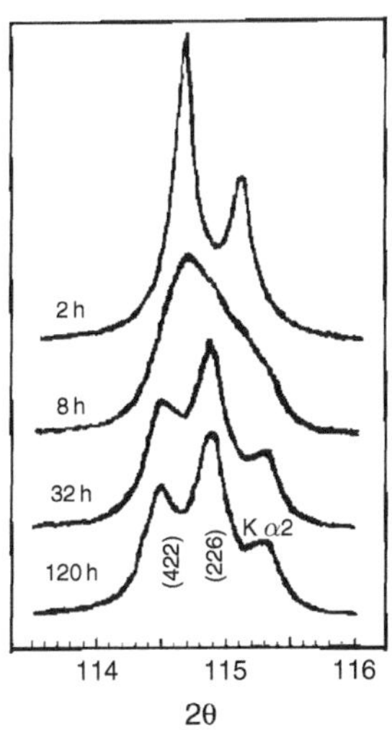

Figure. 8.6 Splitting of 422 and 226 reflections in BZTdue to ordering (after Ref. [17]).

$[I_{100}/I_{(110,012)}]_{obs}$ is the ratio of the observed intensity of the 100 superstructure reflection to that of the 110 and 012 reflections from the sub-cell. The $[I_{100}/I_{(110,012)}]_{cal}$ is the corresponding intensity ratio calculated for a fully ordered structure [56].

Annealing studies [17, 39] showed that disordered BZT become ordered on heating at about 1400°C for various time lengths up to 120 hours. It was found that the diffracted peak intensity of the superstructure reflections arising from the ordering of B-site ions increases with increasing lengths of the annealing time. Desu and Bryan found [39] that the ordering is dependent on the processing condition. Figure 8.7 shows the variation of 100 superstructure reflection intensity with sintering time at 1400°C in oxygen for bulk BZT. The intensity of the 100 reflection increased with increase in sintering time, showing a resultant increase in Zn–Ta ordering, and reached a saturation value at about 60 hours of sintering. The *c/a* ratio of the bulk sample also increased steadily with sintering time. Reaney and co-workers [34, 62] studied the order–disorder transition in BZT using XRD and TEM. Figure 8.8 shows the X-ray diffraction pattern of as-sintered samples annealed and quenched from different temperatures. The X-ray diffraction patterns recorded from samples annealed and quenched from 1600°C showed maximum ordering. The samples quenched from 1625°C did not show superlattice reflections indicating that order–disorder transition occur between these two temperatures. They found an order of magnitude increase in the size of the ordered domains (100–400 nm) in annealed from that observed in as-sintered samples (20–40 nm). It was found that the order–disorder phenomenon is reversible [34, 63]. Electron diffraction patterns recorded from samples quenched from 1600 and 1625°C as shown in Figure 8.9 reveal the presence of 1:2 B-site cation ordering in samples quenched from 1600°C. Figure 8.9a shows the superlattice reflections at $h \pm 1/3$, $k \pm 1/3$ and $l \pm 1/3$ positions. The samples quenched from 1625°C did not show

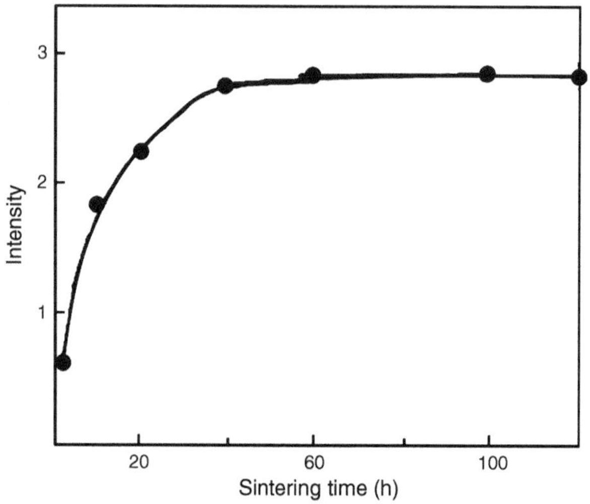

Figure. 8.7 Variation of the intensity of (100) superstructure reflection of BZT with sintering time at 1400°C (after Ref. [39]).

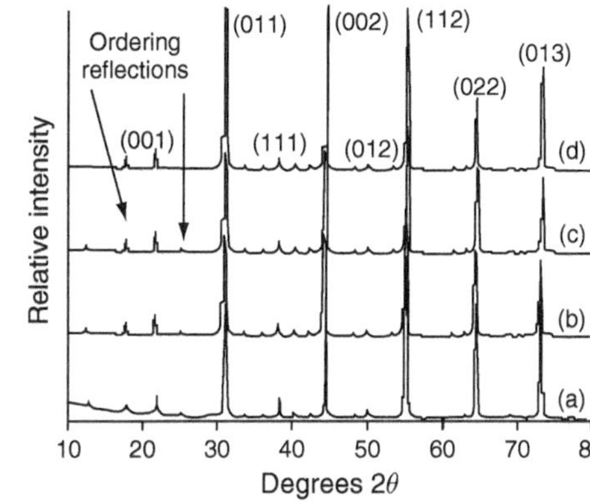

Figure 8.8 X-ray diffraction pattern recorded from BZT (a) sintered at 1475°C, annealed and quenched from (b) 1575°C, (c) 1600°C and (d) 1625°C (after Ref. [34]).

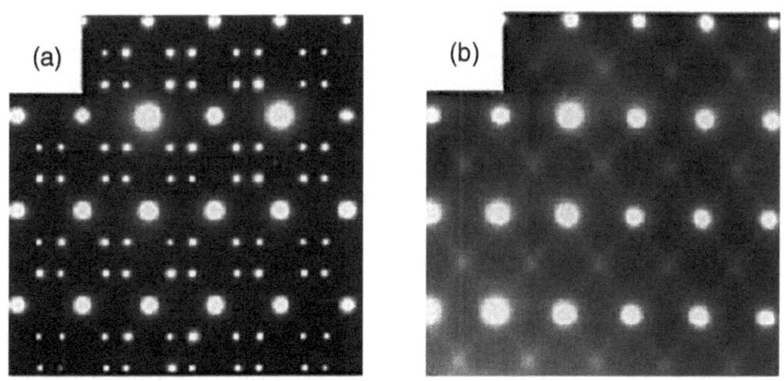

Figure 8.9 Electron diffraction patterns recorded from BZTwith the beam perpendicular to a pseudocubic {110} direction obtained from the center of a grain in a sample quenched from (a) 1600°C and (b) 1625°C (after Ref. [34]).

the 1:2 ordered superlattice reflections. However, they show diffuse spots corresponding to 1:1 ordering. Qazi et al. [34] believed that this is an intermediate metastable structure frozen in during quenching. Barber et al. [64] observed 1:1 B-site ordering in the early stages of ordering in BZT. Lee et al. [65] reported addition of a small amount of B_2O_3 improved ordering. The order–disorder transition in pure BZT has been found to be between 1600 and 1650°C [62, 66] whereas in 0.95BZT–0.05$Sr(Ga_{1/2}Ta_{1/2})O_3$, the order–disorder transition occurs at about 1500°C [40].

Galasso [4] described the evolution of ordered and disordered phases in terms of the nucleation and growth of small ordered domains with increasing annealing time and temperature. Rossiensky and co-workers [67–69] proposed a mechanism in which a fully ordered BZT phase grows at the expense of a slightly Zn-deficient and partially ordered phase. The formation of Zn vacancies from the loss of ZnO was reported [67, 68] to enhance the cation diffusion and domain growth, and a higher extent of cation order in free BZT powders as compared to a pellet. High resolution neutron and X-ray scattering studies revealed two BZT phases with slightly different cell parameters and *c/a* ratios [67, 68]. Refinement of the occupancies of these phases led one phase stoichiometric and the other containing Zn vacancies. Because of the significant line broadening associated with small domain sizes, accurate measurements of the true structural state of these systems (the degree of order within the domains and the domain size) using X-ray or neutron methods require careful analysis of the peak widths and intensities using profile refinement methods. Recent structure refinements of the closely related $Sr(Al_{1/2}Ta_{1/2})O_3$ systems have confirmed the validity of the domain nucleation/growth mechanism [70]. In situ powder diffraction study using synchrotron radiation and neutron diffraction revealed [69] the development of B-cation ordering with time. Usually the overall degree of ordering of the samples is estimated by X-ray powder diffraction method. Electron microscopic technique provides evidence for local ordering which is more microscopic than XRD. It is also possible to study ordering in $A(B'_{1/3}\ B''_{2/3})O_3$ type complex perovskite compounds based on vibrational spectroscopy which is highly sensitive to the short range ordering. It has already been reported that far infrared spectroscopy and Raman spectroscopy can give information on ordering [71–74].

8.2.3 Dielectric Properties

Studies of the structure and electrical properties of the BZT revealed that the degree of ordering of the Zn^{2+} and Ta^{5+} ions has a pronounced effect on the dielectric loss at microwave frequencies. By inducing long range cation order through long time high temperature (≥1350°C) annealing, the *Qf* values of BZT increased from about 5000 to > 150000 GHz [17, 75–77]. The most significant improvements in the *Qf* value with annealing occur [77] when the order parameter $S > 0.75$. These also coincide with the appearance and increase in the lattice distortion associated with the long range chemical order [77]. Although it was suggested [39] that the improvements in *Qf* could be related to the volatilization of Zn, several subsequent studies of BZT, in which volatilization was minimized or eliminated, have clearly demonstrated that changes in the losses arise from alterations in the intrinsic degree of order [37, 79]. The density of BZT decreased and grain size and *Qf* increased on sintering above 1550°C [17, 40]. Figure 8.1 shows the variation of grain size and quality factor with sintering temperature. Hot-pressed samples had lower *Qf* as compared to those prepared by solid-state route [17]. Kim et al. [42] reported that the addition of 0.5–1.5 mol% $BaWO_4$ improved the sinterability of BZT and increased *Qf* to 160 000–200 000 GHz when sintered at 1570–1580°C for 3 hours in air. Further increase of $BaWO_4$ content lowered the *Qf* values. The XRD analysis showed that $Ba_7Ta_6O_{22}$ phase was a major extra phase in all the air-sintered specimens, while in ZnO-muffled specimens the formation of $Ba_7Ta_6O_{22}$ was suppressed. However, BZT sintered with ZnO muffling showed a very low *Qf* and the *Qf* decreased with increasing sintering time [38, 39, 42] regardless of the degree of long range order. The microwave dielectric properties of BZT prepared under different processing conditions and doping are given in Table 8.3.

It has been reported [40, 79] that addition of a small amount of $Sr(Ga_{1/2}Ta_{1/2})O_3$ to BZT improve the dielectric properties and shorten the time required to get the best properties. There is a close relationship between domain and grain growth in BZT–SGT system as cation order proceeds, although the ordered domain size is smaller than that in

Table 8.3 Microwave dielectric properties of BZT with dopants

Composition/dopant	Sintering temperature (°C/h)	ε_r	Qf (GHz)	τ_f (ppm/°C)	Reference
BZT	1350/2 h, 1350/120 h	28	168 000	1	[17]
BZT	1500/60 h	#	140 000	#	[38]
BZT	1570/3 h, 1450/10 h in O_2	#	114 000	#	[44]
BZT ZnO muffling	1500/6 h	#	30 000	#	[38]
BZT ZnO muffling	1500/60 h	#	12 000	#	[38, 42]
BZT + 1 mol% Mn	1550	30	145 000	0	[15]
$Ba(Zr_{0.05}Zn_{0.32}Ta_{0.63})O_3$	1500/4 h	30	148 000	8	[61, 78]
BZT + 5 mol% $BaZrO_3$	–	#	98 000	#	[40]
0.95BZT–0.05$Sr(Ga_{1/2}Ta_{1/2})O_3$	1550/2 h, 1450/24 h	29	162 000	#	[40, 79]
0.95BZT–0.05$(Sr_{0.25}Ba_{0.75})(Ga_{1/2}Ta_{1/2})O_3$	1550/2 h, 1450/24 h	#	210 000	#	[79]
0.95BZT–0.05$Sr(Ga_{1/2}Ta_{1/2})O_3$	1525/2 h, 1275/24 h	29	156 000	0	[40]
0.95BZT–0.05$BaZrO_3$ + 1 wt%CuO	1430/4 h	30	93 000	3	[49]
$Ba_3(Zr_{0.0375}Zn_{0.79}Ni_{0.1975}Ta_{1.975})O_9$	1510/24 h	28	114 480	–5	[37, 80]
$Ba_3(Zr_{0.0645}Zn_{0.816}Ni_{0.1625}Ta_{1.957})O_9$	1510/24 h	29	126 860	–2	[37, 80]
$Ba_3(Zr_{0.09}Zn_{0.845}Ni_{0.125}Ta_{1.94})O_9$	1510/24 h	29	100 180	1	[37, 80]
$Ba_3(Zr_{0.1275}Zn_{0.885}Ni_{0.0725}Ta_{1.915})O_9$	1510/24 h	30	119 090	2	[37, 80]
$Ba_3[Zr_{0.0645}Ni_{0.1625}Zn_{0.816}Ta_{1..957}]O_9$	1520/48 h	28	136 770	–3	[80]
$Ba_3[Zr_{0.09}$ $Ni_{0.125}Zn_{0.845}Ta_{1..94}]O_9$	1520/48 h	30	138 710	–1	[80]
BZT + 1 mol% $BaWO_4$	1580	#	200 000	14	[42]
BZT + 5 mol% B_2O_3 + 10 mol% CuO	875/12 h	26	11 000	0	[50]

(*Continued*)

Table 8.3 (Continued)

Composition/dopant	Sintering temperature (°C/h)	ε_r	Qf (GHz)	τ_f (ppm/°C)	Reference
$Ba_{0.8}Sr_{0.2}(Zn_{1/3}Ta_{2/3})_{0.94}Ti_{0.03}O_3$	1400	42	82 000	−13	[81]
$Ba_{0.75}Sr_{0.25}(Zn_{1/3}Ta_{2/3})_{0.94}Ti_{0.03}O_3$	1400	40	65 000	−2	[81]
BZT + 0.75 mol% Al_2O_3	1580/10 h	#	140 000	#	[82]
BZT + 1.5 mol% SnO_2	1580/10 h	#	142 000	#	[83]
BZT + 1 mol% TiO_2	1580/10 h	#	135 000	#	[84]
BZT + 1 mol% $CaTiO_3$	1450	30	100 000	0	[85]
BZT + 2 mol% ZrO_2	1550/10 h	#	164 000	0	[86]
BZT + 1 mol% Ga_2O_3	1550/10 h	#	161 000	−2.5	[87]
BZT + 1 mol% Cr_2O_3	1525/6 h, 1350/5 h	28	125 500	−1.6	[88]
BZT + 1 mol% CeO_2	1525/6 h, 1350/5 h	27	123 000	14	[88]
BZT + 0.5 mol% In_2O_3	1525/6 h,	25	105 600	9.6	[88]
BZT + 0.5 mol% Sb_2O_5	1525/6 h, 1350/5 h	28	103 200	4.4	[88]
BZT + 0.3 mol% Ta_2O_5	1620/10 h	29	152 000	#	[89]

Data not available.

the pure BZT [40]. Kageyama and Takahashi prepared [75, 79] $xBa(Zn_{1/3},Ta_{2/3})O_3$–$(1-x)$ $[Sr_yBa_{1-y}(Ga_{1/2}Ta_{1/2})O_3]$ ($x=0$, 0.1–1, $y=0$, 0.25–1) by sintering at temperatures in the range 1500–1550°C followed by annealing at 1450°C for 24 hours. The surface layers were ground off to remove Zn-depleted phases. In the X-ray diffraction pattern, the peaks corresponding to the hexagonal superstructure lines disappeared completely by the addition of 5 mol% of $Sr(Ga_{1/2}Ta_{1/2})O_3$ (0.95 BZT–0.05SGT). However, compositions with $x < 0.95$, e.g., $x = 0.925$, on annealing at 1450°C/24 h showed superstructure lines. This suggests that the temperature of transition from order to disorder decreased with the addition of SGT to BZT. The compositions in the range $x \le 0.4$ showed a 1:1 ordering of B-site ions in the cubic structure. The maximum Qf value ($Q \times f =$ 162 000 GHz) was obtained for 0.95BZT–0.05SGT by sintering at 1550°C/2 h followed by 24 hours annealing at 1450°C. The solid solution phases showed a non-linear variation of τ_f with x. In order to further tune the properties, Sr was substituted by Ba in SGT complex perovskites. A maximum Q value of (Qf =210 000 GHz) was attained for the composition $0.95BZT–0.05[(Sr_{0.25}Ba_{0.75})(Ga_{1/2}Ta_{1/2})]O_3$.

It is well established that the 1:2 ordering in BZT improves the quality factor. However, it was reported that addition of Ga_2O_3, SnO_2 and ZrO_2 decrease the 1:2 ordering and improve the quality factor and gradually transform to 1:1ordered cubic structure [37, 87, 90, 91]. It was found [87] that addition of Ga_2O_3 suppressed the formation of $Ba_5Ta_4O_{15}$ which was found as a secondary phase in most of the BZT specimens. Addition of Ga_2O_3 significantly improved the quality factor and the maximum Qf of 161 000 GHz was found for BZT containing 1 mol% Ga_2O_3. Microstructural investigation indicated a lowering of porosity followed by an increase in grain size as revealed by SEM micrographs shown in Figure 8.10. Kim et al. [91] reported that the Qf increased to about 145 000 GHz by the addition of about 1.4 mol% SnO_2. In the case of Ga, Sn and Zr doping, the Zn vacancies are replaced by Zr/Ga/Sn and alter the 1:2 ordering to the 1:1 ordering. The mechanism of improvement in the quality factor with a decrease in order parameter and the 1:2 to 1:1 transformation is described in the next section. Lee et al. found [92, 93] that lanthanum (La) substitution at the A site in BZT decreased the 1:2 ordering and at higher concentrations transformed to 1:1 ordered phase. The ordered 1:2 domains of $Ba_{1-x}La_x[Zn_{(1+x)/3}Ta_{(2-x)/3}]O_3$ first shrink in size with x, then transform to 1:1 ordered domains and the size of the new domains increase with x again.

Addition of Ta_2O_5, TiO_2, and Al_2O_3 considerably improved the quality factor and retain the 1:2 ordering without changing to 1:1 ordering [82, 84, 89]. For all dopants, grain growth occurred. The increase in Qf value for small amount of doping is attributed to increase in density and grain growth. However, addition of large amount (>2 mol%) of these dopants decrease density and quality factor. In all cases, $Ba_5Ta_4O_{15}$ was formed as a secondary phase. The peak intensity of $Ba_5Ta_4O_{15}$ secondary phase did not decrease and the intensity of the 1:2 ordering peak did not change with increasing dopant (Ta_2O_5, Al_2O_3, or TiO_2) content. X-ray diffraction and electron diffraction studies showed a 1:2 ordered structure and the absence of 1:1 ordering. The Ti (0.605 Å), Al (0.67 Å) or Ta (0.64 Å) are smaller than Zn (0.74 Å) with a difference in charge and hence not likely to

Figure 8.10 SEM image of BZT doped with 1.5 mol% Ga_2O_3. The sample was calcined at 1200°C and sintered at 1550°C (after Ref. [87], Courtesy Japanese Society of Applied Physics).

enter Zn site and disturb the 1:2 ordering and they may remain in the grain boundaries. The relative density increased up to 1 mol% TiO_2 addition. The Qf value increased from 80 000 up to 135 000 GHz for 1 mol% TiO_2 added BZT and sintered at 1580°C/10 h. Increasing the sintering temperature to 1630°C, decreased density and Qf factor. A small amount of TiO_2, Al_2O_3 or Ta_2O_5 increased the density and large amount of these dopants led to abnormal grain growth with a decrease in density and quality factor. Takada et al. reported [85] that addition of a small amount of $CaTiO_3$ up to 1 mol% gave $\varepsilon_r = 30$ with $\tau_f = 0$ ppm/°C and $Qf > 100\,000$ GHz. The Qf increased to about 125 000 GHz with $\varepsilon_r \sim 28$ for 0.1 mol% $CaTiO_3$ when sintered at 1550°C. The ε_r and τ_f increased with increasing amount of $CaTiO_3$, but Qf decreased for larger amount of $CaTiO_3$. Zhao et al. [94] reported that addition of $Ba(Cd_{1/3}Nb_{2/3})O_3$ to BZT improved sinterability and quality factor.

Varma et al. [88] made a detailed investigation on the effect of dopant addition in BZT. They added several dopants of varying valencies, ionic size and concentrations and studied the variations in densification, and microwave dielectric properties. It was found that the quality factor increased when the ionic radii of the dopant is close to that of B-site ions (Zn or Ta). Figure 8.11 shows the variation of the Qf in BZT as a function of the ionic radii of the dopants. The Qf increased when the ionic radii of the dopant is close to that of Zn (0.74 Å) or to that of Ta (0.64 Å). An amount of 0.5 mol% of Mg, Ni, Cr, In, Ga, Sn, Zr, Ce, Mn, and Sb improved the quality factor. When the amount of dopant was increased to 1 mol%, the Qf was found to increase only for Cr, Ga, Zr, Ce, and Sn. The highest Qf was found for doping with Zr, Cr and Ce. In the doped samples the quality factor is very much improved although the order parameter is decreased. Since these dopants having ionic radii close to that of the B-site ions improve Qf, it implies that these dopants are substituting for Zn or Ta in BZT. By introducing long range cation order by prolonged annealing at temperatures above 1350°C and by adding a small amount of suitable dopants, the Qf of BZT can be increased up to 200 000 GHz. The highest Qf reported for BZT ~210 000 GHz was with the addition of $Ba_{0.75}Sr_{0.25}(Ga_{1/2}Ta_{1/2})O_3$ and $BaWO_4$.

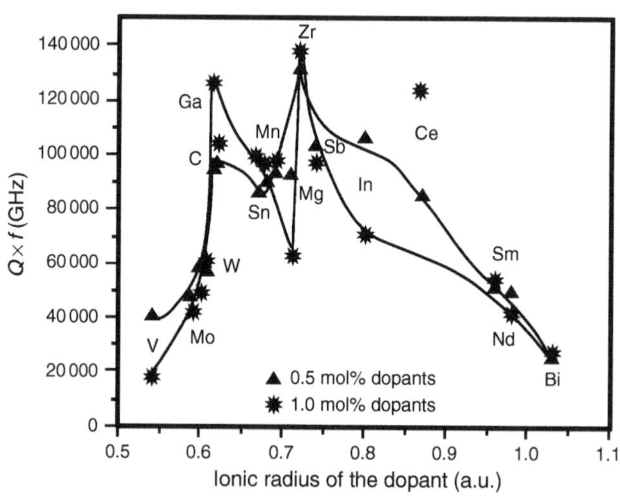

Figure 8.11 Variation of Qf with ionic radii of dopants for 1 and 0.5 mol% dopant addition in BZT (after Ref. [88]).

8.2.4 Effect of $BaZrO_3$ addition in BZT

Pure BZT initially crystallizes in an apparently disordered cubic structure and on prolonged annealing it becomes an ordered hexagonal structure. It has been reported that this ordering process is brought about by the nucleation and growth of small domains with increasing annealing time and temperature [4]. It was also reported that annealing increases the Qf of BZT ceramics and the improvement in quality factor was explained [37, 39, 90] on the basis of the increase in the degree of 1:2 cation ordering. However, long period sintering and annealing times at high temperatures are required to obtain BZT ceramics with a high Qf value [37, 39]. The conventionally prepared BZT ceramics usually contain $Ba_5Ta_4O_{15}$ as a secondary phase. Tamura et al. [12] in 1984 found that the addition of $BaZrO_3$ and $SrTiO_3$ to BZT improved sinterability and crystallization. Addition of a small amount of $BaZrO_3$ decreased the ordering in BZT and improved the Qf factor. When about 4 mol% $BaZrO_3$ was added, the superstructure reflections disappeared. The $SrTiO_3$ addition in a similar way increased ε_r but decreased Qf. X-ray diffraction and far infrared spectroscopic studies showed that addition of ZrO_2 decreased ordering [32]. Wakino et al. analyzed [78] the far infrared spectra of the system by the classical dispersion theory and found that the dielectric loss as obtained by the spectroscopic method for BZT was 0.495×10^{-4} whereas it was 0.191×10^{-4} for $Ba(Zr,Zn,Ta)O_3$. For almost all resonant modes the $Ba(Zr,Zn,Ta)O_3$ had smaller damping constants than BZT in spite of the fact that it contains Zr ions as impurity in the lattice. Tamura et al. reported [12] a reduction in the annealing times required to reach the high Qf state in $BaZrO_3$ added BZT. They examined the microwave characteristics and cell geometries of BZT–$x$$BaZrO_3$ and BZT–$x$$SrTiO_3$ solid solutions. In the case of $BaZrO_3$ substitution, the high Qf value was maintained for additions up to approximately 4 mol% and then, gradually degraded at higher concentrations. In the $SrTiO_3$ system, the Qfs are deteriorated rapidly for all concentrations. It was found that both substitutions accelerated crystallization of the perovskite structure and the low $BaZrO_3$-added samples yielded high Qf ceramics after very short annealing times (4 hours at 1500°C). In this high Qf −4 mol% $BaZrO_3$-doped samples, the X-ray diffraction reflections corresponding to the ordered structure disappeared. The low concentration $SrTiO_3$-added BZT had a similar effect on the structural order though the Qf was decreased. The high Qf values reported by Tamura et al. [12] for the BZT–$x$$BaZrO_3$ system have been reproduced by several groups [32, 37, 41, 61, 90, 78, 86, 95] and $BaZrO_3$ additions is currently used to produce commercial resonators. The Qf value initially increased and then gradually decreased with x whereas the ε_r and τ_f increased gradually with x. The $Ba(Ni_{1/3}Ta_{2/3})O_3$ has a negative τ_f of 17 ppm/°C [12]. Hence partial substitution of Ni for Zn tuned the τ_f of BZT to zero. Addition of a small amount of $BaZrO_3$ further improved the Qf factor. Addition of a small amount of $BaZrO_3$ typically less than 5 mol% dramatically reduce the annealing times <10 h required to reach the high Q state. Therefore commercial high Qf perovskite dielectrics are formulated from a mixture containing 89 mol% BZT, 7.5 mol% $Ba(Ni_{1/3}Ta_{2/3})O_3$ and 3.5 mol% $BaZrO_3$ [12, 37, 80]. This material has $\varepsilon_r = 29$, $Qf >$ 120 000 GHz and τ_f close to zero [12, 37, 80]. Addition of ZrO_2 to BZT also behaved in a manner similar to $BaZrO_3$ addition. Yang et al. [86, 90] studied the effect of ZrO_2 addition on the chemical order, microstructure and microwave dielectric properties of BZT. Addition of a small amount of ZrO_2 disturbed the 1:2 cation ordering. The average size of the BZT grains significantly increased with the addition of ZrO_2. The relative density increased with the addition of a small amount of ZrO_2 but it decreased when the

ZrO_2 content was increased to >1.5 mol%. The Qf value significantly improved with the addition of ZrO_2 and the maximum Qf of 164 000 GHz was found when 2 mol% ZrO_2 was added and sintered at 1550°C for 10 hours. $Ba_5Ta_4O_{15}$ secondary phase was found in all samples and their concentration decreased with increase in ZrO_2 content. However, when more than 1 mol% ZrO_2 was added to BZT, $BaTa_2O_6$ was formed as a secondary phase [86, 90]. The 1:2 ordered hexagonal structure changed to 1:1 ordered cubic with the addition of ZrO_2 [12, 86, 90]. The Zr^{4+} ions may enter the B site of BZT because of the ionic size. For ZrO_2-added ceramics, the 1:2 ordering was disturbed with the addition of ZrO_2, thus the improvement of Qf value is not related to 1:2 ordering. Huang et al. [49] reported that addition of 1 wt% of CuO as a sintering aid in 0.95BZT–0.05 $BaZrO_3$ effectively lowered the sintering temperature to about 1430°C. The sample sintered at 1430°C/4 h with 1 wt% CuO gave $\varepsilon_r = 29.6$, Qf = 93 000 GHz and $\tau_f = 3.1$ ppm/°C. X-ray diffraction study showed that the sample has a disordered cubic crystal structure (*Pm*3*m*).

The origin of the improved quality factor, explanation for the decrease in annealing time to reach the high Qf state and the observation of an apparently disordered structure with very small levels of $BaZrO_3$ or ZrO_2 remained unanswered until the recent work of Davies and co-workers [37, 41]. The formation of high Qf ceramic with random cation arrangement contradicts the experimental and theoretical studies of the undoped barium zinc tantalite system. It was reported [12, 32, 37, 61, 78] that addition of $BaZrO_3$ or ZrO_2 decrease the degree of 1:2 ordering with an improvement in quality factor. As the ZrO_2 or $BaZrO_3$ content increased, the superstructure reflections corresponding to 1:2 ordering disappeared and the ceramic transformed to a 1:1 ordered structure as evidenced by the appearance of superstructure reflections corresponding to 1:1 ordering in the X-ray diffraction and electron diffraction patterns. Davies et al. [37, 41] made a detailed study of $BaZrO_3$-added BZT using X-ray diffraction and TEM and suggested that the low Qf values reported for disordered or partially disordered BZT samples results from losses associated with domain boundaries. During annealing, the domains with hexagonal structure are nucleated and grow in size with increase in order parameter and that the improved losses of fully ordered ceramics originate from the growth and eventual formation of single domain grains. The formation of domain structure implies that disordered or partially ordered samples contain high levels of unstable domain boundaries. Two types of boundaries are likely to be produced during the ordering of the B-site cation [37, 41]: (a) twin boundary separating domains in which the nucleation and growth of the ordered structure has propagated along different <111> orientations, (b) an anti-phase boundary (APB) associated with each set of ordering orientations. From energetic and structural point of view, these boundaries are highly unfavorable since they produce a region with local cation disorder and considerable associated elastic strain. It was found [37] that annealing time required to reach the high Q state decreased with increasing amount of $BaZrO_3$. Figure 8.12 shows the X-ray diffraction pattern of $Ba[(ZnNi)_{1/3}Ta_{2/3}]O_3$ solid solutions with 1.25, 2.15, and 3 mol% $BaZrO_3$. The sample with 1.25 mol% $BaZrO_3$ has an ordered hexagonal structure with order parameter $S = 0.99$ whereas the sample with 2.15 mol% has $S = 0.75$ and the superstructure peaks were broadened indicating a decrease in ordering. The cation ordering disappeared for samples containing 3 and 5 mol% $BaZrO_3$ and the X-ray diffraction patterns showed an ideal cubic perovskite cell as shown in Figure 8.12c. Figure 8.13a shows the electron diffraction pattern of $Ba[(ZnNi)_{1/3}Ta_{2/3}]O_3$ solid solution with 1.25 mol% $BaZrO_3$ recorded along the [110] direction of the pseudocubic perovskite cell. The superstructure

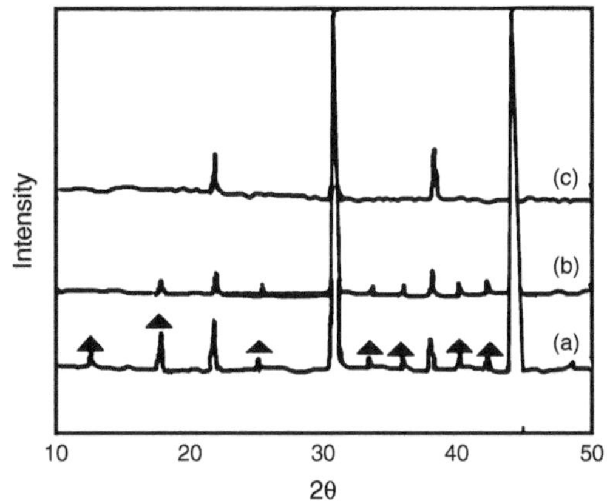

Figure 8.12 X-ray powder diffraction patterns of BZT solid solutions with $BaZrO_3$ (a) 1.25 mol% (b) 2.15 mol% (c) 3 mol% $BaZrO_3$ (after Ref. [37]).

reflections at $h \pm 1/3$, $k \pm 1/3$, $l \pm 1/3$ are due to 1:2 chemical ordering and correspond to the tripling of the unit cell along one <111> direction. The lattice image of $Ba[(ZnNi)_{1/3}Ta_{2/3}]O_3$ containing 1.25 mol% $BaZrO_3$ revealed that the structure is ordered with twin and antiphase boundaries. A typical lattice image of BZT with 1.25 mol% $BaZrO_3$ is shown in Figure 8.14. The size of the ordered domains was found to be more than 200 nm [37]. The electron diffraction pattern of $Ba[(ZnNi]_{1/3}Ta_{2/3}]O_3$ samples containing 2.15 mol% $BaZrO_3$ showed the presence of $h \pm 1/3$, $k \pm 1/3$, $l \pm 1/3$ superstructure reflections in both the allowed, <111>reflections as shown in Figure 8.13b. This indicates that the diffracting volume contained twinned ordered domain structure. This is further supported by the presence of ordered twinned domains in the lattice image as shown in Figure 8.15. The size of the ordered domain decreased to about 40 nm. As the amount of $BaZrO_3$ increased, the domain size further decreased to about 2.5 nm for samples containing 3–5 mol% $BaZrO_3$. The electron diffraction showed only weak diffuse spots at $h \pm 1/3$, $k \pm 1/3$, $l \pm 1/3$ but strong superstructure reflections appeared at $h \pm 1/2$, $k \pm 1/2$, $l \pm 1/2$ indicating 1:1 ordering. The electron diffraction pattern recorded at different temperatures confirmed that the reflections at $h \pm 1/2$, $k \pm 1/2$, $l \pm 1/2$ are due to chemical order and not due to octahedral tilting which is temperature sensitive [95]. The increase in Zr content lowers cation ordering and reduces the size of the 1:2 ordered domain and the annealing time required to reach the high Qf state. In pure BZT lowering of cation order decrease the quality factor. In Zr-doped BZT, cation ordering is decreased and contains a very high volume of domain boundaries but retained the high Qf. Davies et al. [37] proposed that the stabilization of boundaries due to segregation of Zr to the boundary regions is responsible for retention of the high Qf. They also suggested that the lowering of the losses at these regions is directly related to the reduction in the degree of strain at the interface which arises from the preferential segregation of the Zr ions to the boundary regions. The Zr substitution in BZT destabilizes the 1:2 cation ordering and promotes a cubic 1:1 ordered structure with a doubled perovskite repeat. Chai and Davies reported [96] that the homogeneity range of the 1:1 phase extends from $x = 0.04$ to approximately $x = 0.25$; substitutions beyond this range stabilize a disordered perovskite.

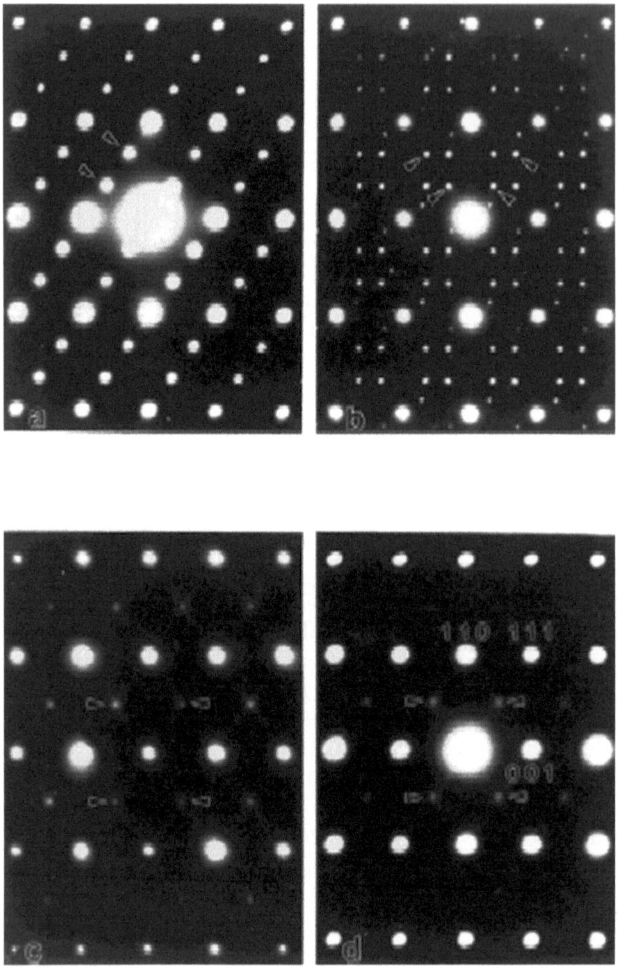

Figure 8.13 Electron diffraction patterns of BZT doped with $BaZrO_3$ recorded along [110] (a) 1.2 mol%, (b) 2.15 mol%, (c) 3.0 mol% and (d) 4.25 mol% $BaZrO_3$ (after Ref. [37], Courtesy, Wiley-Blackwell Publishing Ltd).

Davies et al. [37] proposed a random layer model in which an ordering arises from an alteration of a Ta layer with a second layer that contains a random distribution of Zn, Zr and remaining Ta ions which can be represented as $Ba\{[Zn,Ta,Zr]_{1/2}Ta_{1/2}\}O_3$.

Figure 8.16 shows a simplified model [97] of the arrangements of B-site ions in $Ba\{[Zn,Ta,Zr]_{1/2}Ta_{1/2}\}O_3$. The combination of Zr, Zn and Ta ions in the B′ plane has the average charge 3^+ and the crystal transformed to a 1:1 ordered structure. In this model, the B″ positions are assumed to be occupied exclusively by Ta^{5+} and B′ sites by a random distribution of Zn, Zr and the remaining Ta cations. The validity of the model where the ordered solid solution can be represented by $Ba\{[Zn_{(2-y)/3}Ta_{(1-2y)/3}Zr_y]_{1/2}Ta_{1/2}\}O_3$ was confirmed by Rietveld refinement made using data collected with a synchrotron X-ray source [37].

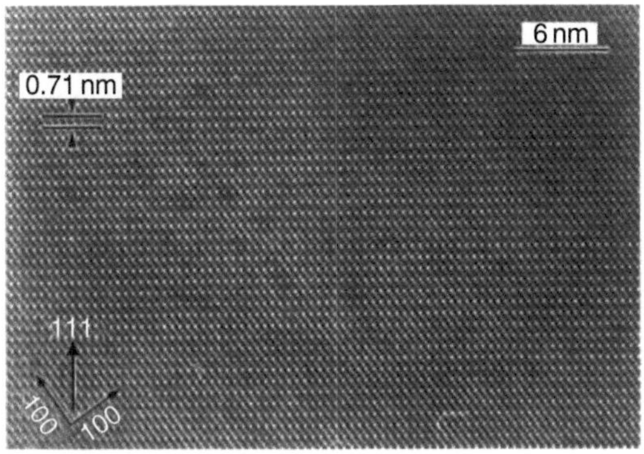

Figure 8.14 Lattice image recorded along [110] from BZT doped with 1.25 mol% $BaZrO_3$ (after Ref. [37], Courtesy, Wiley-Blackwell Publishing Ltd).

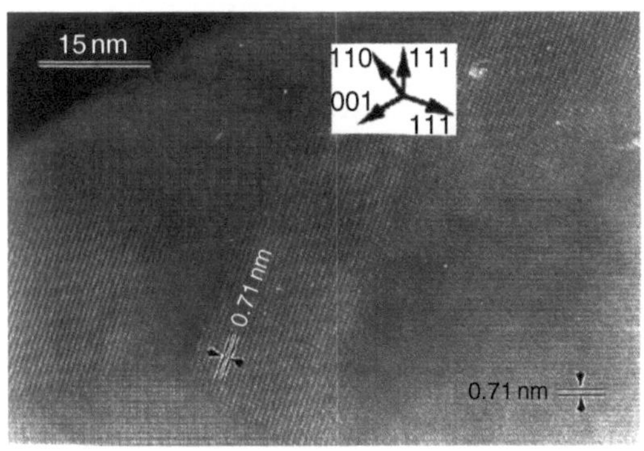

Figure 8.15 [110] lattice image recorded from BZT sample doped with 2.15 mol% $BaZrO_3$ (after Ref. [37], Courtesy, Wiley-Blackwell Publishing Ltd).

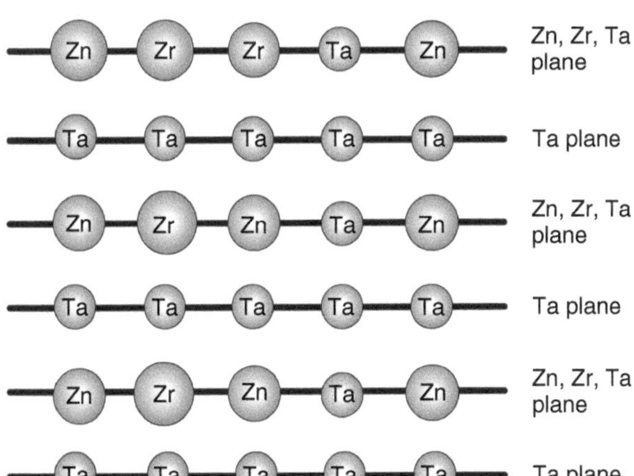

Figure 8.16 Simplified model of doubling ordering of B-site ions in $Ba(Zr,Zn,Ta)O_3$ crystal (after Ref. [97]).

Wakino and co-workers in 1986 reported [78, 98] the lattice vibration of ordered $Ba(B'_{1/3}B''_{2/3})O_3$ for the first time. As the dielectric characteristics at microwave frequencies mainly depend on the ionic polarization brought about by lattice vibration, information concerning lattice vibration is important for understanding dielectric loss in a crystal. Wakino et al. studied lattice vibrations in BZT using far infra red reflection and Raman spectra. The cubic perovskites which belong to the space group O^1_h has no Raman active mode but has three infra red active modes. The BZT with a hexagonal superstructure ($P\bar{3}m1$), however, has 9 Raman active modes and 16 infrared active modes.

$$\Gamma = 4A_{1g}[A, B'', O] + 5E_g[A, B,'' O] \tag{8.2}$$

However, the 1:1 ordered perovskite with space group $Fm\bar{3}m$ symmetry allows four strong Raman lines [63, 98]:

$$\Gamma = A_{1g}[O] + E_g[O] + 2F_{2g}[A + O] \tag{8.3}$$

where the square brackets show the ions involved in a particular vibration mode.

Figure 8.17 and Figure 8.18 show the Raman and infrared reflection spectra of BZT ceramics. Dispersion parameters were calculated from Figure 8.17 and Figure 8.18 by the method described in Chapter 2. The ε_r calculated from the dispersion parameter was 30.3, which is in agreement with the measured value of 29.1. The quality factor calculated was much greater than that measured by microwave methods.

Webb et al. [99] investigated Ga-doped BZT using Raman spectroscopy and reported that the frequency of the strongest Raman modes increased and the modes became narrower with increase in the quality factor.

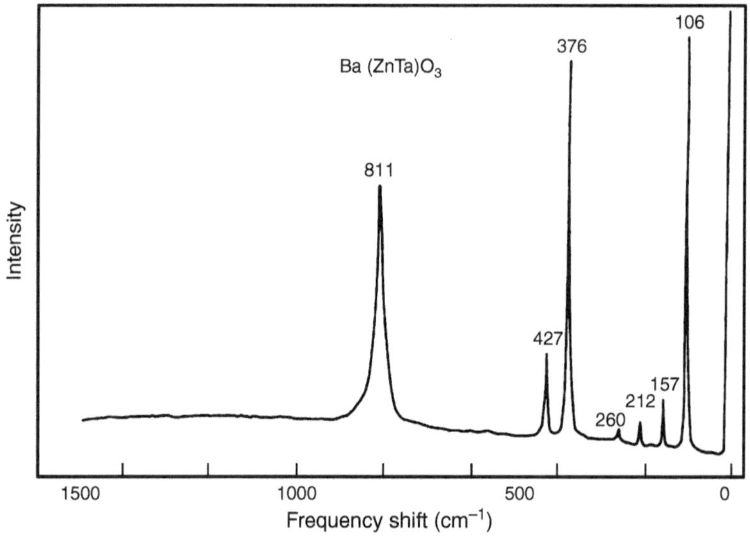

Figure 8.17 Raman spectra of $Ba(Zn_{1/3},Ta_{2/3})O_3$ ceramic (after Ref. [98]).

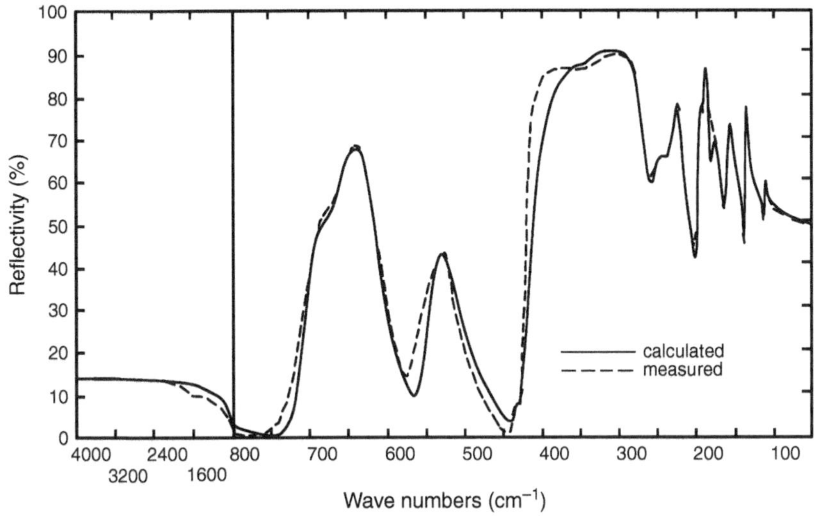

Figure 8.18 Measured and calculated far-infra red reflectivity of $Ba(Zn_{1/3}Ta_{2/3})O_3$ ceramic (after Ref. [98]).

8.3 $Ba(Mg_{1/3}Ta_{2/3})O_3$ (BMT)

8.3.1 Preparation

In 1982, Nomura and co-workers [47] reported $Ba(Mg_{1/3}Ta_{2/3})O_3$ (BMT) as another member of the $A^{2+}(B'_{1/3}B''_{2/3})O_3$ family of high Q dielectric materials. BMT with its melting point of about 3000°C is considered as one of the most refractory oxides [100]. Several people [13, 15, 101] have prepared BMT ceramics by the solid state method. The BMT is usually prepared by calcining the oxides or carbonates at temperatures in the range 1250–1350°C for 4–8 hours. The samples when sintered in the temperature range 1600–1650°C give the best density and dielectric properties. The as-received MgO is usually hydrated and carbonated from the air (contains $MgCO_3$ or $Mg(OH)_2$). Hence in order to maintain the correct stoichiometry, the MgO has to be calcined at high temperatures of about 1000°C before weighing. The phase evolution and reaction sequence of $Ba(Mg_{1/3}Ta_{2/3})O_3$ using powder diffraction pattern is plotted as a function of temperature in Figure 8.19. The X-ray diffraction patterns suggest that the formation of $Ba(Mg_{1/3}Ta_{2/3})O_3$ is starting above 800°C. But even at temperatures above 900°C trace amounts of Ta_2O_5 and $BaTa_2O_6$ are present which will be completely eliminated only on a heat treatment above 1100°C. Longer calcination durations at higher temperatures above 1350°C are detrimental to densification and dielectric properties [102–106]. This is due to the formation of secondary phases such as $Ba_5Ta_4O_{15}$, $BaTa_2O_6$, $Ba_4Ta_2O_9$, $Ba_7Ta_6O_{22}$, $Mg_4Ta_2O_9$. Calcination conditions are important [103] parameters which influence the structure, secondary phase formation, and dielectric resonator properties. Secondary phases play an important role in controlling the microstructure, density, grain growth, and microwave dielectric properties of BMT. The origin and purity of the initial raw materials have considerable influence on the sinterability and sintering temperature, sintering duration, ordering, effect of annealing on

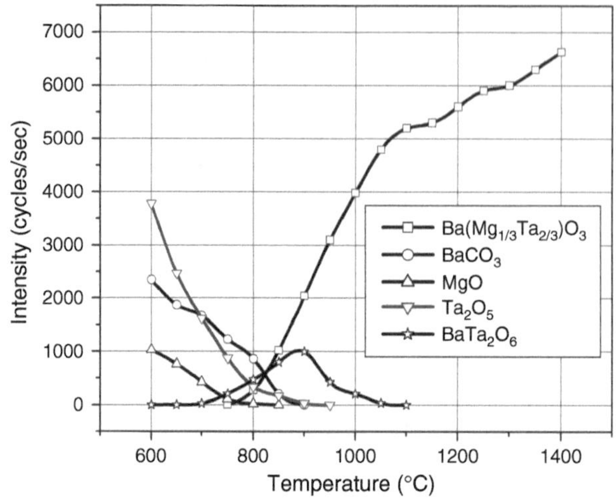

Figure 8.19 Variation of most intense (*hkl*) reflection in the reaction sequence of $Ba(Mg_{1/3}Ta_{2/3})O_3$ as a function of temperature (after Ref. [102]).

ordering and dielectric properties. It may be noted that there may be small variations in calcination, sintering temperature and durations with the purity and origin of the initial raw materials. Ichinose and Shimada [106] studied the relation between the dielectric properties and the grain size in BMT. The samples with different grain size were obtained by controlling the sintering time of BMT at 1500°C. Figure 8.20 shows the microstructure of BMT sintered for 4 and 50 hours at 1500°C. A significant difference in grain size was observed for samples sintered for the two different durations. The grain size increased with sintering time. The average grain sizes were about 5, 8 and 14 μm for BMT sintered for 100, 150 and 300 hours at 1500°C. Sintering for long periods produce secondary phases such as $Ba_5Ta_4O_{15}$, $BaTa_2O_6$, $Ba_7Ta_6O_{22}$ which degrade the dielectric properties [104]. EDX study of samples sintered for long periods showed [104] less Mg

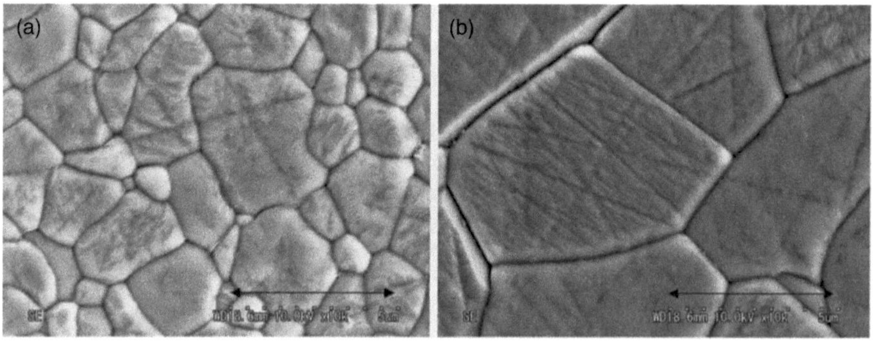

Figure 8.20 SEM image of BMT sintered at 1873 K (a) for 4 hours and (b) 50 hours (after Ref. [106]).

content at the surface than inside. This indicates that Mg evaporation caused the presence of secondary phases such as $BaTa_2O_6$.

Lan et al. [107] reported that preparation of 0.5 mol% $BaZrO_3$ separately and adding to BMT could suppress the formation of $Ba_5Ta_4O_{15}$ with pronounced improvement in the *Qf* factor. Li [108] studied the effect of sintering processes on the microstructures and reported that large grains contain wavy polar domain boundaries, dislocations and even dislocation networks which are detrimental to microwave dielectric properties. Nomura and co-workers [15, 47] reported that it is difficult to densify BMT without Mn addition. Mn addition increased bulk density and the maximum density was found for 2 mol% Mn addition as shown in Figure 8.2. Matsumoto et al. [76] utilized a fast firing technique to improve sinterability of BMT without adding dopants. It has been reported [102, 108] that the use of $(MgCO_3)_4 \cdot Mg(OH)_2 \cdot 5H_2O$ rather than MgO or $MgCO_3$ can suppress the presence of detrimental secondary phases. Chen and Wu reported [109] that pure complex perovskite phase of BMT can be obtained by calcining at about 1000°C by using $(MgCO_3)_4 \cdot Mg(OH)_2 \cdot 5H_2O$ which is much less than that by using MgO (about 1300°C). This lower formation temperature is due to the higher reactivity between $(MgCO_3)_4 \cdot Mg(OH)_2 \cdot 5H_2O$, $BaCO_3$ and Ta_2O_5 powders.

Several authors succeeded in improving the sinterability and in lowering the sintering temperature by the addition of low melting additives [110–115], glasses [113, 116, 117], wet chemical methods [51, 118–127] and by slight non-stoichiometry [104, 110, 128, 129]. Addition of a small amount of glasses substantially lower the sintering temperature to about 1300–1350°C without much degradation of dielectric properties. It has been reported [104, 110, 128, 129] that a slight deficiency of Ba or Mg in BMT can improve the density, ordering and dielectric properties. The vacancies created by the deficiency of Ba or Mg improve material transport and improve densities. However, large deficiencies decrease the density and degrade properties. It may be noted that excess Ba or Mg always degrade density and dielectric properties [129]. It was found [129] that reducing the barium content in BMT substantially improved the densification and ordering. A two-step process (Columbite route) consisting of first preparing $MgTa_2O_6$ and then reacting $MgTa_2O_6$ with $BaCO_3$ markedly [107, 130–133] reduce the sintering temperature and soaking time required to densify BMT and to achieve a high *Qf* state. The two-step columbite route improved the sintering behavior with higher density at a relatively lower sintering temperature of about 1550°C/4 h. TGA/DTA and X-ray diffraction analysis indicated that the columbite route suppressed the formation of secondary phase $Ba_5Ta_4O_{15}$ which deteriorates the *Qf* value. The BMT prepared by the two-step process has better densification than the one-step process. Figure 8.21 a and b shows the SEM images of BMT prepared by the conventional one-step and columbite route and sintered at 1650°C/20 h respectively. The SEM picture of sample prepared by one-step method shows the presence of $Ba_5Ta_4O_{15}$ secondary phase.

Several authors reported [51, 102, 118–128, 134–138] the preparation of BMT by wet chemical methods. Scholler and Wersing in 1991 were the first to report [118, 121] the preparation of BMT by sol–gel method. They prepared the gel by dissolving barium acetate and magnesium acetate in boiling acetic acid and then cooled to room temperature. Stoichiometric amount of tantalum ethoxide was added and mixed well by shaking. The solution was then hydrolyzed to form the gel by adding a mixture of water and 2-methoxy ethanol. The xerogel thus formed was dried and ball milled and then heated to 380°C to remove major portion of organic components. The BMT powder thus obtained was calcined and used for preparing DR samples. Katayama et al. reported [125, 126] the formation of BMT powders by the hydrolysis of barium magnesium and

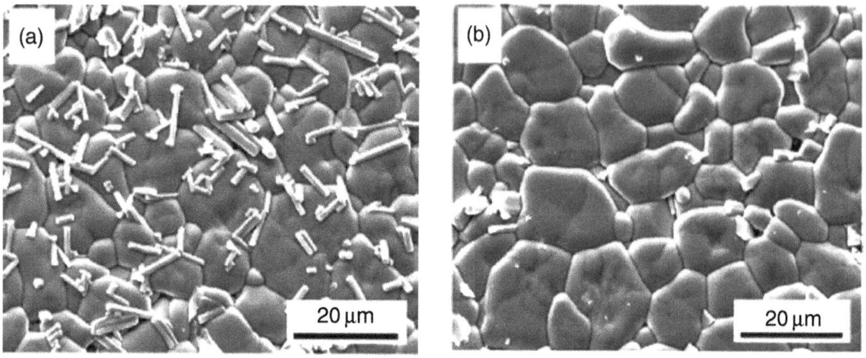

Figure 8.21 SEM image of BMT prepared by (a) one-step method and (b) two-step method (after Ref. [131]).

tantalum ethoxides. FTIR and Raman spectroscopic studies of the Ba–Mg–Ta ethoxide precursor revealed that Ba, Mg, Ta ethoxides were bonded with each other via alkoxy bridging to form a trimetallic alkoxide. The hydrolysis of the trimetallic alkoxide with large amount of water under refluxing resulted in crystalline BMT with a cubic perovskite structure at about 100°C. The BMT powder calcined at 1000°C provided a sintered body with a relative density of 94–98% at a low firing temperature of 1400°C. Renoult et al. [51, 122] and Ravichandran et al. [123] prepared BMT by the reaction of bimetallic alkoxide of Mg and Ta with hydrated barium hydroxide or metallic barium. BMT powder obtained by calcining at 600–650°C showed a cubic perovskite structure with a particle size of about 0.1 μm. The samples sintered at 1400°C showed an ordered trigonal structure with 98% density.

Several reports were published [51, 102, 105, 119, 120, 135–138] on the preparation of BMT fine powders by co-precipitation using inorganic salts using $Ba(NO_3)_2$, $Mg(NO_3)_2$ and $TaCl_5$ starting materials. The semi-coprecipitation method using oxine by Kakegawa et al. [105] was the first attempt to prepare BMT nanopowders. Maclaren and Ponton [135] obtained BMT by hydrothermal method at moderately low temperatures of 160–350°C/2 h. X-ray diffraction study showed formation of BMT at temperatures as low as 200°C. The achievement of desired stoichiometry and homogeneity was found to be difficult. Tsai et al. [139] prepared highly ordered BMT by hot isostatic pressing (HIP) at about 1300°C under a pressure of 1500 kg/cm^2 in Ar atmosphere. Galasso and Pinto [140] grew BMT single crystals up to 2 mm in size by flux method using BF_2 fluxes (B = Mg, Zn, Ni,Ca). More recently Guo et al. [100] successfully grew BMT single crystal fibers by laser heated pedestal growth technique (LPHG). The twin-free BMT single crystal fibers were 300–1000 μm diameter and 1–2 cm in length. The molten zone temperature was about 3000°C. The tip of the fiber (at the beginning of the growth) was found to be Mg deficient. The Mg-deficient region is found to contain $Ba_5Ta_4O_{15}$. The single crystals were heavily disordered with a cubic symmetry.

8.3.2 Crystal structure and ordering

BMT shows the phenomenon of order–disorder similar to that in BZT. When Mg and Ta cations are disordered over the B site, the crystal structure of BMT is a pseudo-cubic perovskite with *Pm*3*m* (O_{1h}) space group symmetry. The ordered compound adopt a hexagonal symmetry with D^3_{3d} space group due to distortion along the [111] direction of the cubic cell. The ordered BMT has a structure similar to $Ba(Sr_{1/3},Ta_{2/3})O_3$ based on $P\bar{3}m1$ space group. When fully ordered, the Mg and Ta ions form planes parallel to the close-packed layers with a three-layer repeat sequence (two layers of Ta^{5+} ions and one layer of Mg^{2+} ions). Ordering on the octahedral sites induces the layer sequence –Mg–Ta–Ta–Mg– along the [111] direction of the pseudo cubic cell. This long range ordering will expand the unit cell along the [111] direction and contract it along the direction normal to [111], which leads to the splitting of (422) and (226) reflections [39, 122]. The B-site ordering increases the c/a value and is slightly higher than $\sqrt{3/2} = 1.2247$. This lattice distortion (c/a ratio changes) is observed by the increased splitting of (422) and (226) reflections with sintering time at 1400°C. The X-ray diffraction study using Rietveld technique of calcined powder showed [141–143] a disordered structure. The ceramics sintered at 1600°C showed B-site ordering. Superstructure reflections are hardly visible on non-sintered samples. When the preparation temperature was increased from 1300 to 1600°C the crystal symmetry changed from cubic *Pm*3*m* to hexagonal $P\bar{3}m1$.

The ratio of intensities for a completely ordered structure $(I_{100}/I_{110,102})_{ordered}$ given in Equation (8.1) is calculated to be 8.3% putting all the atoms in approximate ideal positions in BMT crystal lattice [100]. The recent structural analysis carried out by Janaswamy et al. [142] and Lufasso [59] suggests that a more accurate value of $(I_{100}/I_{110,102})_{theor}$ is equal to 8.7%. The value of *S* varies from zero for a disordered to one for a completely ordered structure. Janasamy et al. [142] concluded that in addition to charge imbalance and size difference in the cations, the temperature of synthesis has a pronounced effect on the ordering of the cations in BMT. The degree of ordering depends [17, 56, 143–147] on the sample preparation conditions and can be improved by annealing the ceramics at high temperatures in the range 1300–1600°C or by introducing small amounts of certain dopants. Figure 8.22 shows the effect of annealing at 1500°C on the order parameter. The order parameter increased with annealing time. TEM and SEM

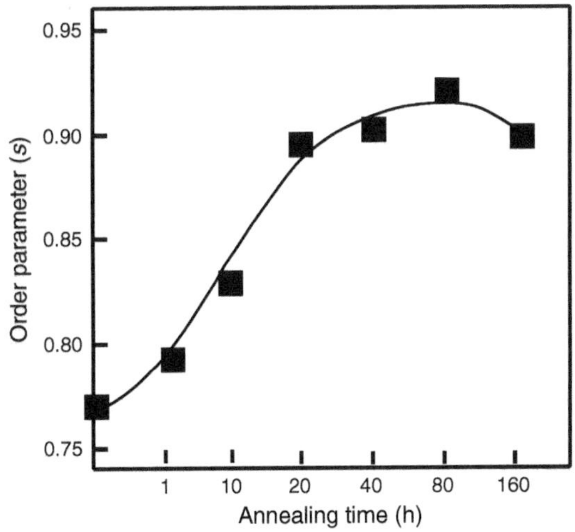

Figure 8.22 Variation of order parameter as a function of annealing time in BMT (after Ref. [146]).

studies showed presence of secondary phases at the grain boundaries in the prolonged sintered or annealed samples. Sugiyama et al. [148] reported from X-ray diffraction and TEM studies that during sintering ordered domains with hexagonal structure are nucleated and grow to the equivalent four <111> directions in the cubic disordered matrix. Impingement of the ordered domains results in the formation of sub-boundaries in the parent grain. The sub-boundaries are always in a highly strained condition and dislocations are often introduced at the interface. They suggested that the dielectric loss at the domain boundary is high. It was found [149] that disordered BMT consists of short range ordered nanodomains with diffuse streaking along all <111> directions. Annealing yields 1:2 ordered nuclei which grow on further annealing. Figure 8.23 shows the X-ray diffraction pattern of BMT after various heat treatments. The samples were annealed and quenched after each heat-treatment. The intensity of the 1:2 superstructure reflection peaks increased with annealing at 1600°C. Lei et al. [149] reported that the microstructure of the partially ordered crystals contain nanosized domains. These nanodomains serve as nuclei for the growth of large domains during annealing.

Chai et al. studied [150] the effect of $BaZrO_3$ addition on the cation ordering in BMT using TEM, XRD and neutron diffraction techniques. A transformation from 1:2 to a

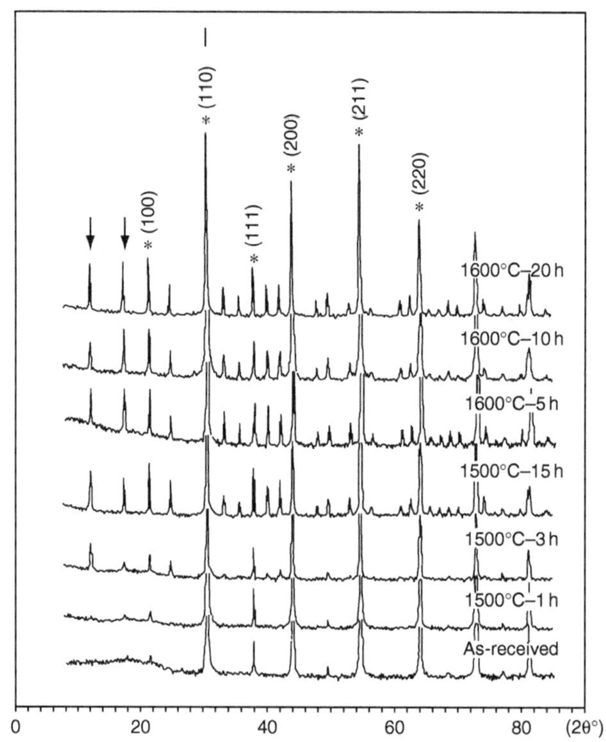

Figure. 8.23 X-ray diffraction pattern of BMT after various heat treatments. The intensity of 1:2 superstructure reflections increased with annealing time (after Ref. [149], Courtesy-Taylor and Francis).

1:1 ordered cubic ($Fm\bar{3}m$) phase occurs for substitution levels between 10—and 25 mol% $BaZrO_3$. The occupancies of the two cation positions, refined using Rietveld method, were found to be consistent with the random site model in which B″ is occupied by Ta and B′ by a random distribution of the remaining cations (Mg, Ta, Zr). Compared to BZT–$BaZrO_3$ system, the kinetics of equilibrium in (1–x)BMT–$x$$BaZrO_3$ were quite slow and a stable phase assemblage was only achieved after heating at 1600°C for 40–60 hours. The XRD patterns indicate that the changes in the cation order induced by the substitution of $BaZrO_3$ into BMT are quite similar to those of BZT–$BaZrO_3$ system. The range of stability of the trigonal 1:2 ordered structure of BMT is very narrow. Samples with $x = 0.05$ comprised of a two phase mixture of the 1:2 end member and a cubic 1:1 ordered structure. Chai et al. also studied [151] the effect of different tetravalent Ce, Sn and Ti octahedral site substitutions on the B-site cation ordering in BMT to understand the factors that control the cation ordering transformation. They found a 1:1 ordered phase for 7–25 mol% of $BaCeO_3$ and 1:2 + 1:1 ordered phases for 3–5 mol% $BaCeO_3$. No evidence of 1:2 to 1:1 transformation was found with $BaTiO_3$ and $BaSnO_3$ additions in BMT. Sagala and Kayasu investigated [144] the dielectric function of the ordered BMT ceramics by the infrared reflectance spectra taken over the 50–4000 cm^{-1} range. The reflectance spectrum was analyzed on the basis of the four parameter semi-quantitative model assuming 16 infrared active vibrational modes allowed for the related D^3_{3d} space group. The lowest frequency optical mode was found at 60 cm^{-1} which can be assumed to involve the motions of the heavy TaO_6 octahedra.

Burton has studied ordering [152, 153] and disordering in $A(B_{1/3}'B''_{2/3})O_3$ perovskites in a phenomenological way as well as the effect of substitutions at A- and B-site cations on disordering temperature using first principles. Takahashi et al. studied [154–156] the order–disorder phenomena of B-site cations in the various barium-based $Ba(B_{1/3}B''_{2/3})O_3$ complex perovskites using first principles, cluster expansion technique and Monte Carlo simulations. The calculations confirmed the experimentally observed hexagonal superstructure with space group superstructure [$P\bar{3}m1(D^3_{3d})$] as the ground state. The order–disorder transition temperature is estimated from the energy difference between a hexagonal 1:2 ordered structure and an approximated disordered state. Takahashi's calculations indicate that many $Ba(B'_{1/3}B''_{2/3})O_3$ [B′ = Co,Mg, Mn, Ni, Zn, B″ = Ta, Nb] compounds in equilibrium should be fully ordered in the 1:2 B site structure with space group $P\bar{3}m1(D^3_{3d})$. Their calculations disagree with experimental results for $Ba(Mn_{1/3}Nb_{2/3})O_3$ but is consistent with experimental observations on other $Ba(B'_{1/3}B''_{2/3})O_3$ compositions. The authors found that the greater the stability of the ordered 1:2 structure with respect to the disordered phase, the higher the experimentally measured microwave Qf factor. The charge density distribution and electron density of states (DOS) for the 1:2 ordered structure indicate that Ta and O atoms posses some degree of covalency and some overlap between the O-2p orbitals and the Ta-5d orbitals. Takahashi also investigated the electrostatic stability of BMT using Madelung energy calculations for different B site arrangements. These results show the importance of 1:2 ordering of Mg and Ta cations along the [111] direction of the simple perovskites in determining electrostatic stability of BMT. Liu et al. [157] investigated the atomic ordering in complex perovskite alloys by the cluster variation method. It has been reported [99, 158–161] that Raman scattering studies are very useful in evaluating the order of complex perovskites materials. The ordered ceramics show narrow and intense Raman lines. The vibrational spectra of complex perovskite-type compounds are functions of both disordered and ordered regions, allowing the appearance of specific Raman scattering. Thus in real systems, a combination of ordered and disordered regions makes the analysis of Raman data difficult [162]. The effects of ordering on the Raman spectra of BMT ceramics have been studied by a number of researchers [124, 159, 163].

8.3.3 Properties

It was Nomura et al. [47] in 1982 who first reported the microwave dielectric properties of BMT which was doped with Mn. Since then several authors investigated the microwave dielectric properties of BMT [13, 47, 106, 131, 164–169]. The BMT has ε_r of about 24–25, Qf up to 430 000 GHz and τ_f close to zero ppm/°C. The BMT showed excellent dielectric properties when calcined in the range 1250–1350°C and sintered at temperatures in the range 1600–1650°C. Kolodiazhyni et al. [165, 167] reported from EPR studies that $Ba(B_{1/3}'B_{2/3}'')O_3$ contain a substantial amount of paramagnetic point defects which contribute to extrinsic loss at microwave frequencies degrading the Qf factor. It has been shown [8, 33, 76, 143, 170] that the Qf depends on calcination and sintering temperature and their durations, annealing and degree of long range order (LRO) in terms of B-site ions. The as-prepared samples will be usually partially ordered. They are normally annealed in the temperature range 1400–1500°C to get an ordered high Q ceramic. Sintering for a long duration and slow cooling improve the ordering and thereby the quality factor. Figure 8.24 shows the variation of unloaded quality factor and ε_r as a function of annealing time at 1500°C [146]. Annealing improved the quality factor considerably and also slightly increased the permittivity. Thus in pure BMT high Qf is related to a high degree of ordering. Ichinose and Shimada [106] studied the relation between the dielectric properties and the grain size in BMT. The samples with different grain size were obtained by controlling the sintering time of BMT at 1500°C. A significant difference in grain size was observed for samples sintered for different durations. Figure 8.25 shows the variations of the quality factor (Qf) as a function of the grain size. The Qf gradually increased with the grain size and reached a

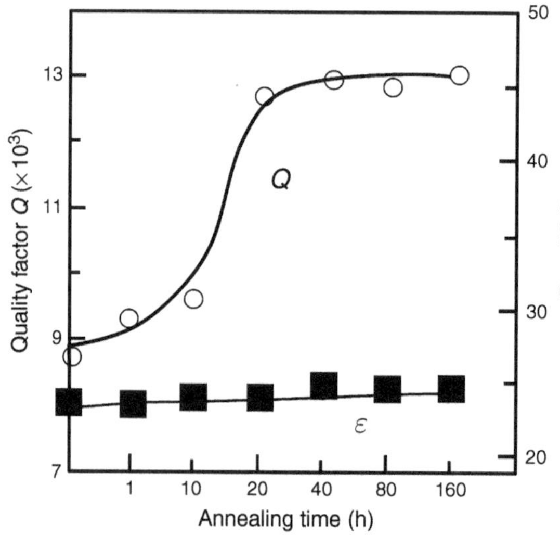

Figure 8.24 Variation of Qf of BMT as a function of annealing time at 1500°C (after Ref. [146]).

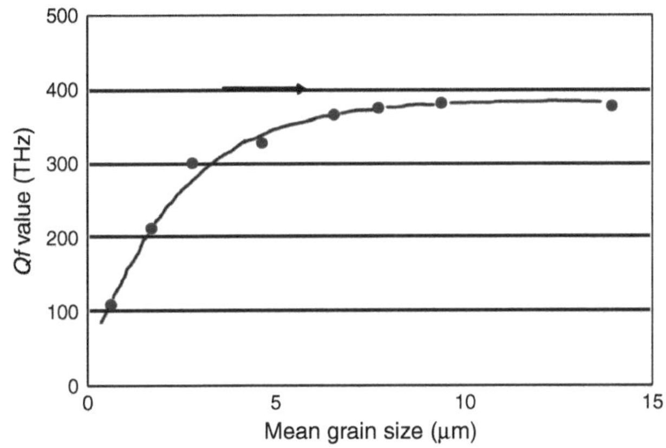

Figure 8.25 Variation of quality factor Qf as a function of mean grain size (after Ref. [106]).

maximum of 400 000 GHz for the average grain size of about 9 μm. The Qf saturates for grain size larger than 9 μm and becomes nearly independent of grain size. The microwave dielectric properties of BMT prepared under different processing conditions and additives are given in Table 8.4

The BMT samples prepared [182, 183] by sintering at high temperatures usually contain secondary phases such as $BaTa_2O_6$, $Ba_5Ta_4O_{15}$ which degrade dielectric properties. The BMT prepared by wet chemical methods have poor sinterability, less ordered with $c/a < 1.2247$, and have a relatively low quality factor [51, 120–122, 126]. Several authors reported [131–133, 139] that the BMT samples prepared by a two-step (columbite) process can improve Qf. X-ray diffraction and TGA/DTA analysis indicated that the columbite route suppressed the formation of $Ba_5Ta_4O_{15}$ and thus improved the quality factor. Tsai et al. [139] prepared BMT by hot isostatic pressing (HIP) and two-step process (columbite). The hot-pressed samples had lower Qf as compared to those by two-step process. Cheng et al. [183] investigated the BMT prepared by one-step and two-step methods by Evanescent Microwave Probe (EMP) method. Figure 8.26a and b shows the mappings of permittivity derived from EMP measurements synthesized by one-step and two-step methods. The mappings showed that the ε_r for one-step processed sample vary from 22.3 to 38.3. This variation over a large range is attributed to the presence of secondary phases. The ε_r of samples synthesized by two-step method varied only between 27 and 32. However, these range of values are very different from that is expected for BMT. Ratheesh et al. measured [164] the Qf of the BMT samples by the Whispering Gallery Mode (WGM) method and reported a Qf of 325 000 GHz for the $TM_{4,1,0}$ mode at a frequency of 13.25 GHz.

Shimada [169] investigated the variation of Qf value of BMT ceramics sintered at 1600°C for different durations of 4, 50, and 250 hours by far infrared reflectivity measurements. The Qf considerably increased with increasing sintering time. Figure 8.27 shows the infrared reflectivity spectra of three BMT samples sintered at 1600°C for 4, 50, and 250 hours respectively. The spectra of the sample sintered at 250 hours showed an increase in reflectivity at 250 cm^{-1} and the shape of the spectrum changed and became more steep. Shimada [169] found a variation in the damping constants of third, fourth and fifth modes with sintering time. The increase in the Qf factor is attributed to variations in the damping constants. However, he could not explain the reason why the three

Table 8.4 Microwave dielectric properties and sintering temperature of BMT with different dopants

Material	Dopant	Sintering temperature (°C)	ε_r	Qf (GHz)	τ_f (ppm/°C)	Reference
BMT	–	1640/20 h	24	430 000	5	[13]
BMT	–	–	25	360 000	–	[76]
BMT	–	1600	24	300 000	0	[167, 168]
BMT	–	1600/50 h	25	400 000	2	[106, 169]
BMT	Two step	1580	23	275 000	–0	[131]
BMT	0.03 mol% Ni	1650	–	84 000		[171]
BMT	0.05 mol% $BaWO_4$	1650/2 h	24	157 500		[172]
BMT	0.5 mol% $Ba(Mg_{1/2}W_{1/2})O_3$	1600/20 h	25	400 000		[173]
BMT	Nd_2O_3	1550	26	52 300	59	[174]
BMT	1 wt% MCAS glass	1450	23	6500		[116]
BMT	6 wt% MCAS glass	1350	16	9610	12	[113]
BMT	0.25 wt% V_2O_5	1500/3 h	24	149 000	7	[114]
BMT	1 wt% Mn	1600	25	175 000	3	[15, 47]
BMT	1 mol% $BaSnO_3$	1640/20 h	24	330 000	1	[13]

BMT	3 wt% NaF	1340/3 h	26	50 400	11	[175]
BMT	SnO_2	–	25	200 000		[176]
BMT	1 wt% B_2O_3	1325/4 h	24	124 700	–1	[117]
BMT	0.2 wt% $ZnO–B_2O_3$	1410/4 h	24	136 500		[117]
BMT	1 wt% ($5ZnO–2B_2O_3$)	1350/4 h	22	141 800		[117]
BMT	0.2 wt% ($ZnO-B_2O_3–SiO_2$)	1360/4 h	26	152 800	–2	[117]
BMT	1 wt% $0.2BaO–0.8WO_3$	1400/6 h	25	159 000	7	[110]
BMT	0.2 mol% $MnCO_3$	1625/4 h	26	162 800	5	[115]
BMT	0.5 mol% ZnO	1625/4 h	26	124 500	12	[115]
BMT	0.5 mol% ZrO_2	1625/4 h	26	112 500	0	[115]
BMT	0.5 mol% SnO_2	1625/4 h	24	122 500	–2	[115]
BMT	0.1 mol% Sb_2O_3	1625/4 h	25	172 500	10	[115]
BMT	0.5 mol% WO_3	1625/4 h	25	144 500	15	[115]
0.7BMT—0.3BCN	–	1575/5 h	28	235 450	–1	[177]
$0.85BMT–0.15BaSnO_3$	–	1640/20 h	24	330 000	–1	[13]
$Ba(Mg_{0.3183}Ta_{0.67})O_3$	–	1650/4 h	25	120 500	3	[129]

(Continued)

Table 8.4 (Continued)

Material	Dopant	Sintering temperature (°C)	ε_r	Qf (GHz)	τ_f (ppm/°C)	Reference
$Ba_{0.9925}(Mg_{1/3}Ta_{2/3})O_3$	–	1600/4 h	25	15 280	1	[129]
$Ba[(Mg_{0.4}Zn_{0.6})Ta_{2/3}]O_3$	–	1600/4 h	27	109 900	4	[178]
$0.7BMT–0.3Ba(Co_{1/3}Nb_{2/3})O_3$	–	1530/5 h	27	172 700	–1	[177]
$Ba[Mg_{1/3}(Nb_{1/4}Ta_{3/4})]O_3$	–	1650/9 h	26	140 660	5	[179]
$Ba[(Co_{0.125}Mg_{0.875})_{1/3}Ta_{2/3}]O_3$	–	1600/54 h	23	103 100		[126]
0.5BMT–0.5BZT	0.1 wt% SiO_2	1500/6 h	27	126 000	2	[180]
$Ba(Mg_{1/3}Ta_{(2-2x)/3}W_{x/3}Ti_{x/3})O_3$ (x 0.1)		1550/4 h	22	90 000	–3	[181]
BMT	Sol–gel	1400/5 h	24	50 000		[121, 122]
BMT	Sol–gel	1400/5 h	24	45 000	–	[51]
BMT	Inverse microemulsion	1500/4 h	26	65 200	–	[120]

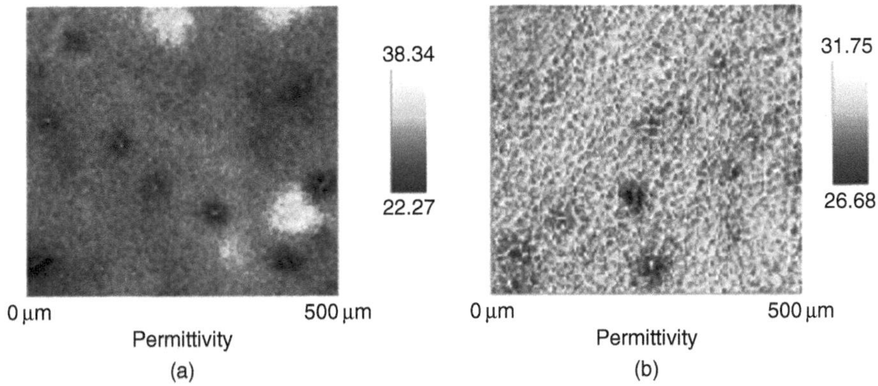

Figure 8.26 Evanescent Microprobe of BMT prepared by one-step and two-step methods (after Ref. [183]).

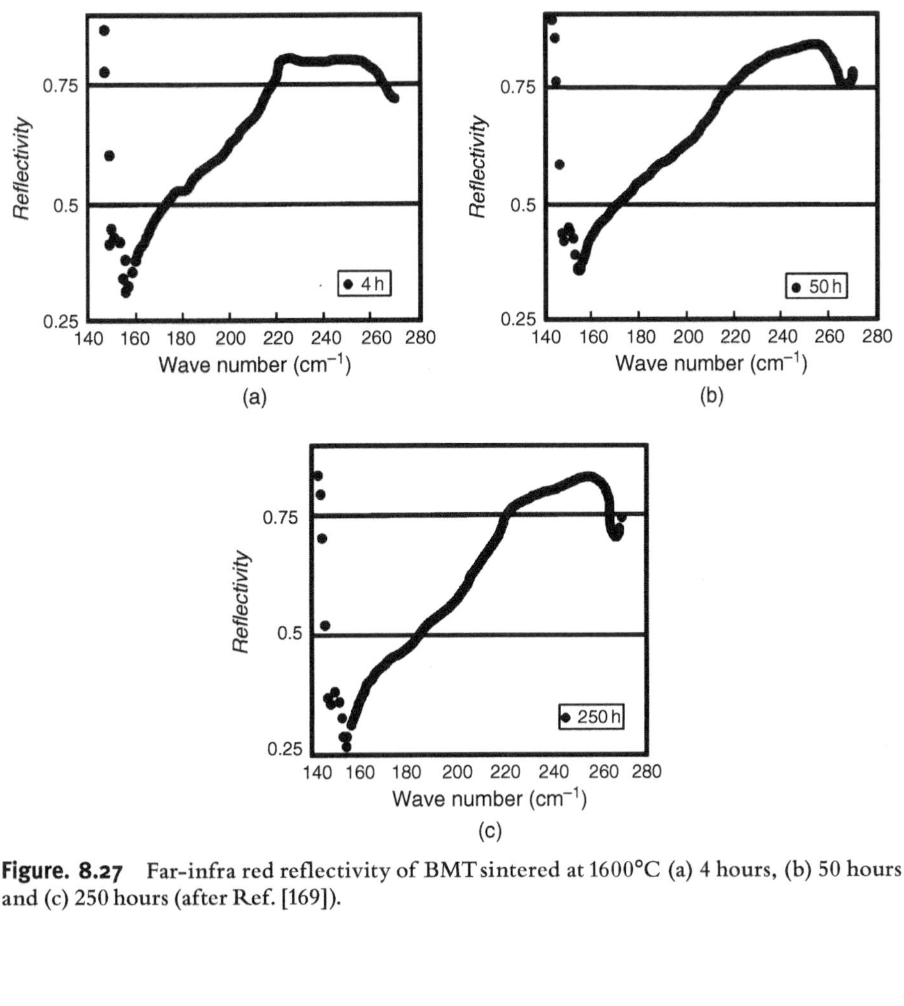

Figure. 8.27 Far-infra red reflectivity of BMT sintered at 1600°C (a) 4 hours, (b) 50 hours and (c) 250 hours (after Ref. [169]).

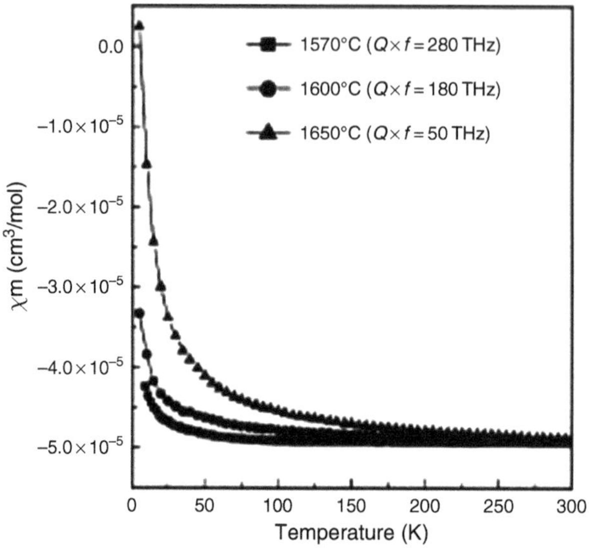

Figure 8.28 Effect of processing temperature on the molar magnetic susceptibility of BMT (after Ref. [165]).

damping constants varied with sintering time. It was inferred that the variations was due to variations in the crystal chemistry of BMT.

Kolodiazhyni et al. [165] reported that the presence of lattice defects in BMT leads to deviation from the temperature-independent diamagnetic behavior with a non-zero magnetic moment. The BMT is diamagnetic. Figure 8.28 shows the variation of molar magnetic susceptibility (χ_m) with temperature for BMT sintered at different temperatures and having different Qf factor. At sufficiently high temperatures the χ_m is about $^-50 \times 10^{-6}$ cm^3/mol. At low temperatures the χ_m is temperature dependent and is attributed to the presence of paramagnetic lattice defects. At low temperatures the amount of lattice defects increases exponentially with processing temperature and χ_m also increases with increasing processing temperatures at lower temperatures (Figure 8.28). The measured molar magnetic susceptibility of BMT was found to be higher than the calculated value of -78.3×10^{-6} cm^3/mol [184]. The results of Kolodiazhyni et al. [165] is in agreement with the first principle calculations of Takahashi et al. [154] that a purely ionic picture of BMT is not valid and some covalent bonding generate additional paramagnetic contribution to the χ_m.

8.3.4 Effect of dopants

The B-site cation ordering has a large influence on the dielectric loss at microwave frequencies in BMT. The most efficient way of reducing the dielectric loss at microwave frequency range is by the formation of the stoichiometric 1:2 long range ordered clusters. However, due to poor sinterability, sintering at elevated temperatures (~1650°C) and a subsequent long time annealing are required for the fabrication of high quality factor BMT [128, 185]. In order to overcome this problem, various additives with their melting points lower than BMT were investigated and their effect on dielectric properties were examined [15, 76, 115]. Nomura and co-workers reported [47] that Mn addition improve density and quality factor and about 1 mol% of Mn addition gives the highest quality factor. Since the work of Nomura et al. [47], several authors reported [16, 110, 127, 130, 133, 149, 165, 172, 173, 181, 183] improvement in the properties of BMT using different preparation conditions and additives. It was reported [110, 172, 186] that addition of $BaWO_4$ and the subsequent sintering at temperatures above 1430°C increased the unloaded quality factor of BMT. The order parameter S, density and c/a increased with $BaWO_4$ addition up to about 0.05 mol% beyond which there was no change. The Qf increased up to 0.05 mol% addition of $BaWO_4$ beyond which it decreased. $BaWO_4$ has [187] $Q = 2250$, $\varepsilon_r = 8$ and $\tau_f = -33$ ppm/°C. Hence addition of large amount of $BaWO_4$ decreased the quality factor, ε_r and τ_f. Yoon et al. attributed [110, 172] the increase in Qf to the enhancement of the B-site cation ordering and the elimination of oxygen vacancies by the substitution of W^{6+} ions for Ta^{5+} ions in the B-site sublattice of the perovskites. Substitution of W^{6+} increases the charge difference between B′ and B″, increases the ionic radii difference [Mg = 0.72 Å, Ta = 0.64 Å, W = 0.58 Å]. Thus the ordering is promoted by $BaWO_4$ addition.

Several authors reported [107, 115, 132] improvement in the quality factor of BMT by the addition of ZrO_2 or $BaZrO_3$. Lan reported [107] that the effect of ZrO_2 addition on the dielectric properties in BMT very much depend on the processing routes. Addition of $BaCO_3 + ZrO_2$ to the initial raw materials of BMT ($BaCO_3$, MgO and Ta_2O_5) degrades the quality factor whereas addition of $BaZrO_3$ to the initial BMT raw materials improves the Qf factor. The formation of secondary phases depends on the processing route which influenced the dielectric properties. Ra and Phule [132] reported that addition of $BaZrO_3$ lowered the order parameter but the Qf remains nearly the same. These authors suggested that the Qf does not depend on the order parameter. Hence they concluded that increase in Qf on long periods of annealing is limited to the changes in the concentration of point defects such as oxygen vacancies that are introduced as a result of non-stoichiometry, impurities in raw materials and processing. It was reported [13, 146, 188] that addition of $BaSnO_3$ decreased quality factor except for very small amount of $BaSnO_3$ (0.5 mol%). X-ray, electron diffraction and Raman study showed a decrease in ordering with $BaSnO_3$ addition and the crystal symmetry changed from the hexagonal superstructure to the cubic perovskite structure. Kim et al. [171] and Shimada [189] reported Ni substitution for Mg decreased density, degree of ordering, permittivity and Qf but increased τ_f.

Surendran et al. [115] made a detailed investigation of the effect of dopants of different valencies and ionic radii in BMT. Figures 8.29 and Figure 8.30 show the variation of the Qf and τ_f of BMT doped with various amounts of different dopants. It is found that small amount of dopants such as Sb_2O_5, MnO, ZrO_2, WO_3 and ZnO improve the quality factors. Large amount of all dopants degrade the microwave dielectric properties due to the formation of secondary phases. Figure 8.31 shows the variation of the Qf of BMT with 0.1 mol % dopant as a function of the dopant ionic radii. It is found that when the ionic radii of the dopants are between 0.6 and 0.7 Å the quality factor increases. The weighted average ionic radii of the B-site ions ($Mg_{1/2}Ta_{2/3}$) is 0.653 Å. This means that the Qf increases when the dopant ionic radii is close to the average ionic radii of B-site ions. However, cations such as Ga (IR = 0.64 Å), Nb (IR = 0.64 Å) and Fe (IR = 0.645 Å) are not favorable for improving the quality factor. It was reported [190] that about 10 mol% La substitution for Ba in BMT resulted in a 1:2 to 1:1 transition in ordering type. The $[Ba_{1-x}La_x]$ $[Mg_{0.3391+x})Ta_{0.67[1-.0.5x]}O_3$ sintered at 1650°C for 2 hours showed an fcc cell with 1:1 ordering for $x = 0.1$. For x in the range 0.02–0.008, both

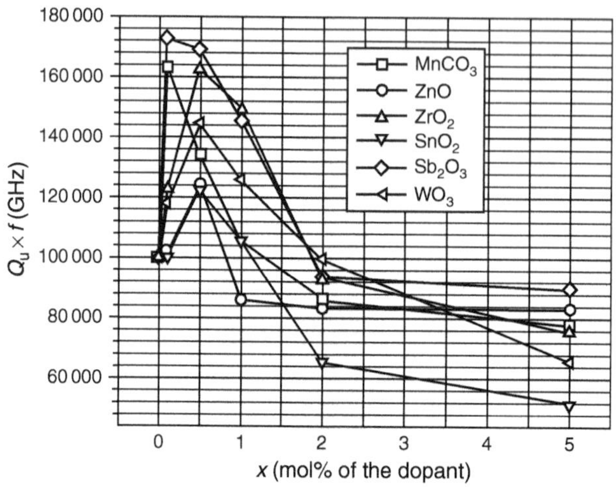

Figure 8.29 Variation of quality factor of BMT with concentration of dopant (after Ref. [115]).

1:2 and 1:1 ordering coexist as evidenced by X-ray diffraction study. The ε_r and τ_f increased and Qf decreased with x. The ε_r, τ_f and Qf variation with x show a discontinuity at x between 0.02 and 0.04. This discontinuity appeared at the composition at which 1:1 order starts appearing. The authors suggested that the co-existence of two ordering types originates from the overlap of the stability region of each ordering type.

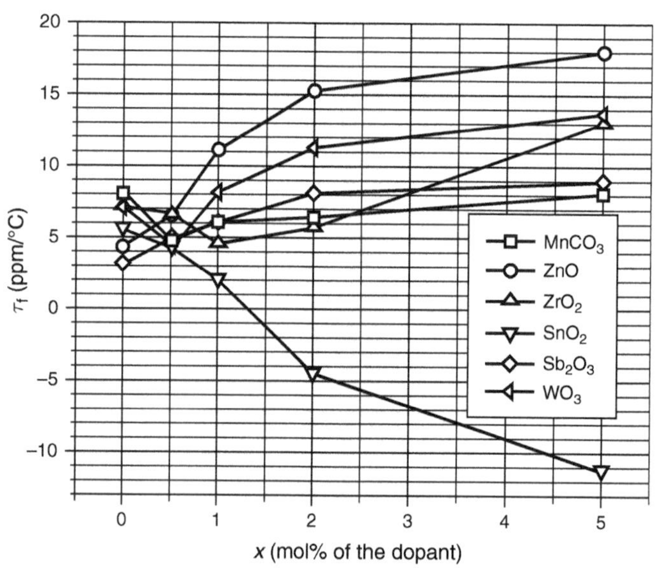

Figure 8.30 Variation of τ_f of BMT with dopant concentration (after Ref. [115]).

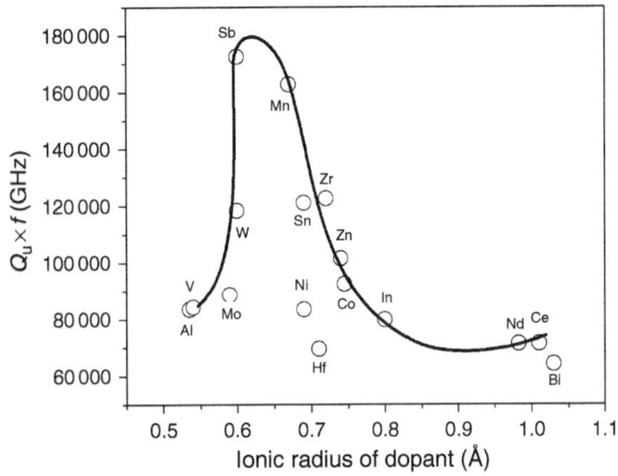

Figure 8.31 Plot of quality factor Qf with ionic radius of the dopants in BMT (after Ref. [115]).

Several investigators reported [106, 178, 180, 191] microwave dielectric properties of BMT–BZT solid solution. Ouchi et al. reported [180] that addition of small amounts (0.1 wt%) of SiO_2, MnO_2, and Bi_2O_3 in BMT–BZT solid solution improved the density to about 98% and also improved ε_r, Qf and τ_f. Large amount of these additives lowered the density and dielectric properties. The variation of ε_r with x is not linear and the non-linear variations are attributed to octahedral tiltings. The BMT and BZT form a solid solution in the entire compositional range [106, 178, 191] and increasing BZT content increases ε_r with a decrease in Qf. It was found that prolonged sintering produced $BaTa_2O_6$ secondary phase which decreased the quality factor [106]. The properties of 1:2 ordered complex perovskites can be tailored by (a) replacing A-site ion with another alkaline earth metal or replacing complex cation at B site $[(B^{2+}{}_{1/3}B^{5+}{}_{2/3})]^{4+}$ with a tetravalent metal. Surendran et al. [178] prepared $Ba[(Mg_{1/3}Ta_{2/3})_{1-x}Ti_x]O_3$ for different values of x. X-ray diffraction study showed the appearance of $BaTiO_3$ peaks for $x > 0.4$. The Ti substitution at B site decreases the quality factor and increases ε_r and τ_f [151, 178, 192, 193]. Chai and Davies [151] reported that in BMT–$BaTiO_3$ system, a complete solid solution is limited to about 10 mol% of $BaTiO_3$ in BMT matrix. The dielectric loss increased considerably for $x > 0.5$ and they do not resonate. Low frequency measurements showed properties typical of ferroelectric materials for $x > 0.5$.

The $(Ti_{1/3}W_{1/3})$ substitution for $Ta_{2/3}$ in $Ba(Mg_{1/3}Ta_{2/3})O_3$ decreased ε_r and Qf. The τ_f decreased and became negative [181]. The $Ba(Mg_{1/3}W_{1/3}Ti_{1/3})O_3$ showed $\varepsilon_r = 15.4$, $\tau_f = -25$ and $Qf = 35\ 400$ GHz. The $Ba[Mg_{1/3}Ta_{(2-2x)/3}W_{x/3}Ti_{x/3}]O_3$ ($x = 0.1$) has $\varepsilon_r = 22$, $Qf = 90\ 000$ GHz and $\tau_f \sim -3$ ppm/°C. Higuchi and Tamura [194] reported zero porosity $Ba(Sn,MgTa)O_3$ with $\varepsilon_r = 24$, and $\tan\delta = 1.7 \times 10^{-4}$ at 60 GHz. By improving the density, the material became optically translucent. At optical frequency, ionic polarization is negligible and only electronic polarization is dominant and contributes to refractive index. Figure 8.32 shows the transmittance of $Ba(Sn,Mg,Ta)O_3$ translucent substrate of thickness 0.6 mm. It has a high refractive index of 2.074 and

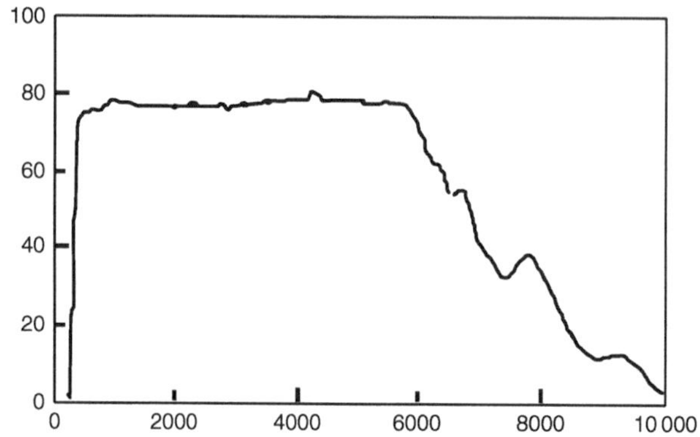

Figure 8.32 Transmittance spectrum of $Ba(Mg,Sn,Ta)O_3$ translucent substrate (after Ref.[194]).

no birefringence. Such characteristics which are not found in conventional glasses may be useful for miniaturizing optical elements.

8.3.5 Effect of glass addition

The BMT has a relatively poor sinterability with a relatively high sintering temperature. The formation of secondary phases such as $Ba_5Ta_4O_{15}$, $Ba_4Ta_2O_9$, $Ba_7Ta_6O_{12}$ during synthesis also affect the sinterability. Cheng et al. reported [113, 116] that addition of a small amount of $MgO–CaO–Al_2O_3–SiO_2$ (MCAS) glass improved sinterability and lowered sintering temperature with a decrease in the amount of secondary phases. Recently, Surendran et al. investigated [117] the effect of several glasses on the sinterability, sintering temperature, ordering and dielectric properties of BMT. They added B_2O_3, SiO_2, $B_2O_3–SiO_2$, $ZnO–B_2O_3$, $5ZnO–2B_2O_3$, $Al_2O_3–SiO_2$, $Na_2O–2B_2O_3$, $10H_2O$, $BaO–B_2O_3–SiO_2$, $MgO–B_2O_3–SiO_2$, $PbO–B_2O_3–SiO_2$, $ZnO–B_2O_3–SiO_2$, and $2MgO–Al_2O_3–5SiO_2$. Addition of all these glasses lowered the sintering temperature considerably. However, only the addition of small amounts of B_2O_3, $ZnO–B_2O_3$, $5ZnO–2B_2O_3$, $ZnO–B_2O_3–SiO_2$ improved the densification and microwave dielectric properties. Addition of about 0.2 wt% of these glasses lowered sintering temperature to about 1350°C without appreciable degradation in the dielectric properties and suppressed the formation of $Ba_5Ta_4O_{15}$, $BaTa_2O_6$, $Ba_4Ta_2O_9$ and $Ba_7Ta_6O_{22}$ secondary phases and improved the ordering parameter. Other glasses reacted with BMT to form secondary phases and hindered densification. It may be noted that addition of more than 2 wt% of all glasses produced glass-based secondary phases with a decrease in the quality factor. Addition of 1 wt% B_2O_3 increased the order parameter, density and Qf and they reached a maximum on sintering at 1325°C and then decreased with increasing sintering temperature. The permittivity in general decreased with addition of glass although there is a slight increase for very small amount of glass which is due to improved densification. Figure 8.33 shows the variation of quality factor with addition of glass in different wt%. The quality factor of BMT increased for a small amount of B_2O_3. The addition of 1 wt% of B_2O_3 and $5ZnO–2B_2O_3$ and 0.2 wt% $ZnO–B_2O_3$ and $ZnO–B_2O_3–SiO_2$ gave the best quality factors.

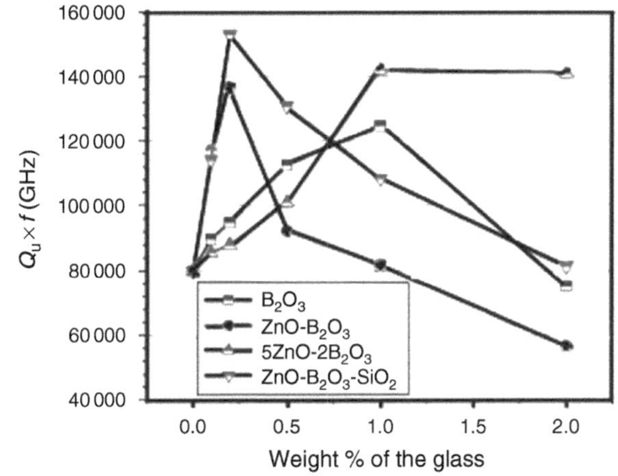

Figure 8.33 Variation of quality factor of BMT with the concentration of glass additives (after Ref. [117]).

8.3.6 Non-stoichiometry

Lu and Tsai reported [195] that reducing the Ba content in BMT improved densification and enhanced ordering whereas excess Ba content leads to the decrease in density and disordering. Surendran et al. [129, 163] investigated the effect of slight A- and B-site cation non-stoichiometry in BMT sintered at 1600°C/4 h and slow cooled. It was found that a slight deficiency of Ba or Mg increases the density and order parameter as shown in Figure 8.34 and Figure 8.35. Excess amount of Ba or Mg decreases the order parameter and density. The $Ba(Mg_{0.33-x}Ta_{0.67})O_3$ for $x = 0.015$ and $Ba_{1-x}(Mg_{0.33}Ta_{0.67})O_3$ for $x = 0.0075$ both showed about 98% densification. Deficiency of more than 1.2 mol% of Mg or Ba decreased the density and order parameter, and X-ray diffraction study showed

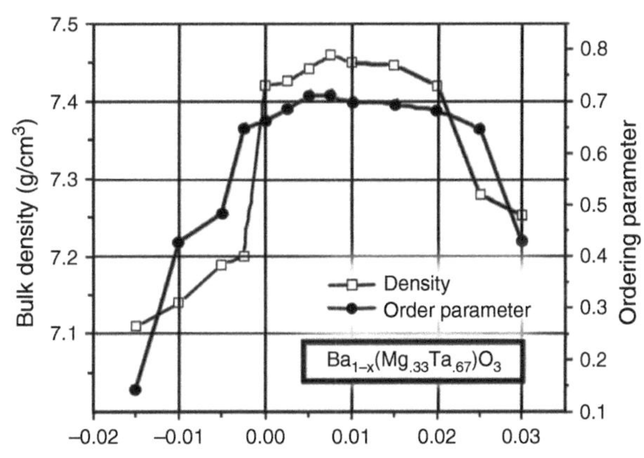

Figure 8.34 Variation of the bulk density and order parameter with x in $Ba_{1-x}(Mg_{1/3}Ta_{2/3})O_3$ ceramics (after Ref. [129]).

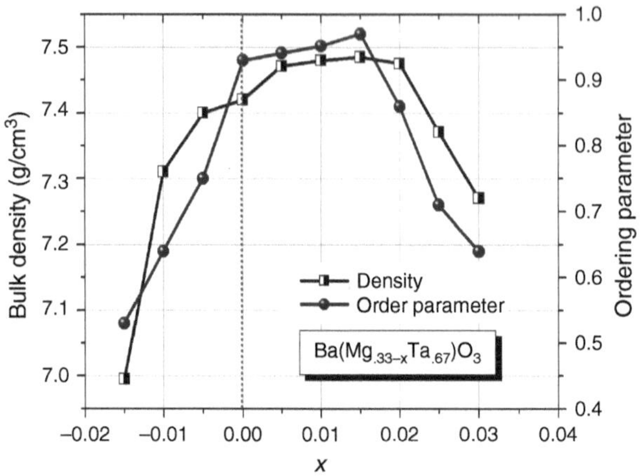

Figure 8.35 Variation of the bulk density and order parameter with x in $Ba(Mg_{0.33-x}Ta_{0.67})O_3$ ceramics (after Ref. [129]).

the presence of $BaTa_2O_6$ and $Ba_5Ta_4O_{15}$ in Mg-deficient BMT and $MgTa_2O_6$ in Ba-deficient BMT. Figure 8.36 shows the X-ray diffraction patterns of $Ba(Mg_{0.33-x}Ta_{0.67})O_3$ for different values of x showing the (422) and (226) reflections. The splittings of (422) and (226) X-ray reflections due to the lattice distortion (ordering) of $Ba(Mg_{0.33-x}Ta_{0.67})O_3$ for different values of x are evident from the figures. The splitting is pronounced for $x = 0.015$ in $Ba(Mg_{0.33-x}Ta_{0.67})O_3$ which showed the highest density and order parameter. In a similar way the maximum profile splitting was observed for $x = 0.0075$ in $Ba_{1-x}(Mg_{0.33}Ta_{0.67})O_3$. Raman spectroscopic study also showed an increase in order with a small amount of Mg and Ba deficiency. Figure 8.37 shows the variation of ε_r and τ_f with x in $Ba(Mg_{0.33-x}Ta_{0.67})O_3$. The ε_r steadily increased with increase in Mg deficiency up to $x = 0.015$ beyond which it decreased. The τ_f decreased with Mg deficiency up to $x = 0.015$ and further deficiency increased τ_f sharply. Figure 8.38 shows the variation of ε_r and τ_f of Ba-deficient BMT as a function of x. The ε_r increased and τ_f decreased with Ba deficiency up to $x = 0.0075$ beyond which ε_r decreased and τ_f increased. The variation of Qf with Ba and Mg stoichiometry is shown in Figure 8.39. The maximum Qf in $Ba_{1-x}(Mg_{0.33}Ta_{0.67})O_3$ was found for $x = 0.0075$ for which maximum density, order parameter, ε_r and lowest τ_f were found. In the case of $Ba(Mg_{0.33-x}Ta_{0.67})O_3$, the maximum Q f was found for x = 0.015 for which the highest density, ε_r, order parameter and lowest τ_f were observed. Excess amount of Ba and Mg always degrade the dielectric properties. The presence of a very small amount of vacancies facilitates material transport improving the densification process and thereby the dielectric properties.

8.3.7 Dielectric properties at low temperatures

Several authors [115, 141, 194, 196–198] have studied the low temperature dielectric properties of 1:2 ordered perovskites. As discussed in Chapter 2, the classical dispersion theory predicts that tan δ is proportional to frequency in the microwave frequency region [78]. Figure 8.40 shows the dependence of tan δ on temperature. As expected

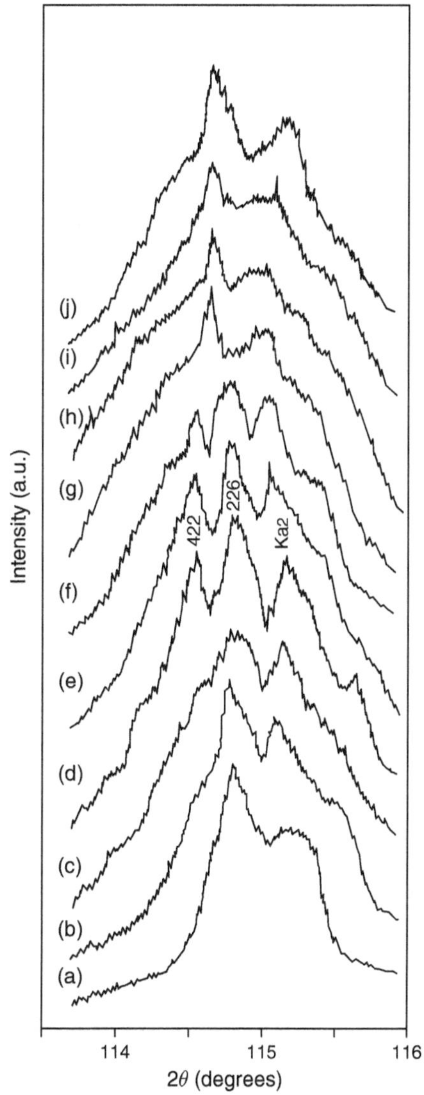

Figure 8.36 X-ray diffraction patterns of 422 and 226 reflections for different values of x in $Ba(Mg_{0.33-x}Ta_{0.67})O_3$ for x (a) $x=0.03$ (b) $x=0.025$ (c) $x=0.02$ (d) $x=0.015$ (e) $x=0.01$ (f) $x=0.005$ (g) $x=0.0$ (h) $x=-0.005$ (i) $x=-0.010$ (j) $x=-0.015$ (after Ref. [129]).

from the theory, the tan δ decreased for BMT on cooling. In the case of $Ba(ZnZrTa)O_3$ the tan δ decreased on cooling up to 100 K and further cooling slightly increased the tan δ. The tan δ of $Ba(Sn,Mg,Ta)O_3$ decreased and that of $Ba(Zr,Zn,Ta)O_3$ increased at low temperatures. Similar experiments on $(Zr,Sn)TiO_4$ with low and high purity showed that the loss increased at low temperatures for the low purity resonators [194]. This increase in tan δ for certain materials at cryogenic temperatures can be due to paramagnetic defects, oxygen vacancies or impurities. It was found [198] that the third

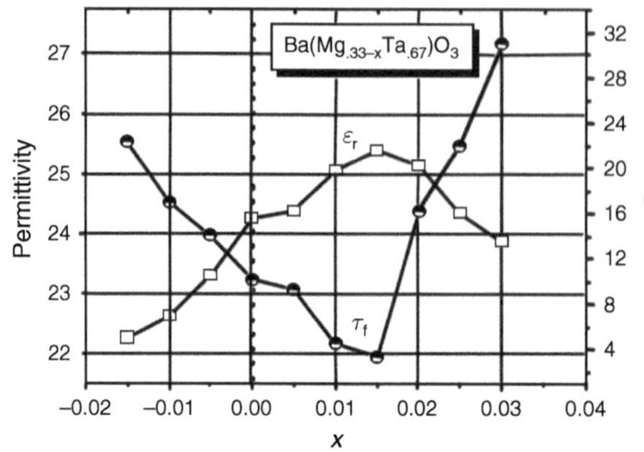

Figure 8.37 Variation of permittivity and τ_f as a function of x in $BaMg_{0.33-x}Ta_{0.67}O_3$ (after Ref. [129]).

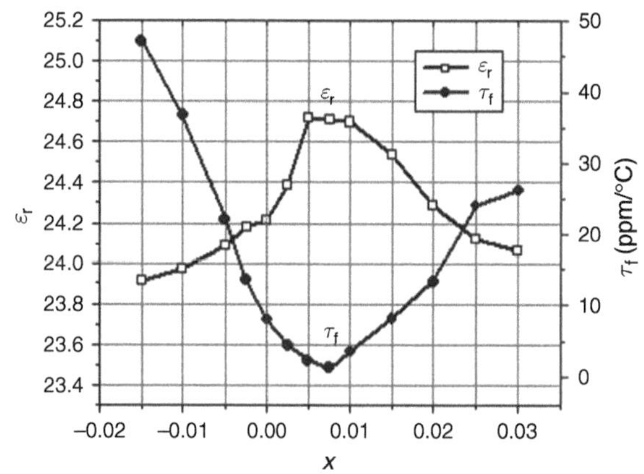

Figure 8.38 Variation of permittivity and τ_f as a function of x in $Ba_{1-x}Mg_{1/3}Ta_{2/3}O_3$ (after Ref. [129]).

harmonic distortion levels have a strong correlation with dielectric loss tangents. This phenomenon was explained by considering the anharmonic terms in the potential energy. Vincent et al. [141] also observed an increase in loss on cooling. Loss increased rather fast below 150 K. It is evident from Figure 8.40 that $Ba(Mg,Sn,Ta)O_3$ is a promising candidate for low temperature microwave dielectric applications as it has extremely high Qf value of about 1 000 000 GHz at 70 K [194, 196].

8.4 $BaSr(Mg_{1/3}Ta_{2/3})O_3$

Several authors reported [199–207] that the dielectric properties of $Ba_{1-x}Sr_x(B'_{1/3}B''_{2/3})O_3$ compounds exhibit a characteristic change with Sr content. The ε_r and τ_f increased linearly and reached a maximum at $x = 0.55$ and then decreased linearly with increasing Sr content.

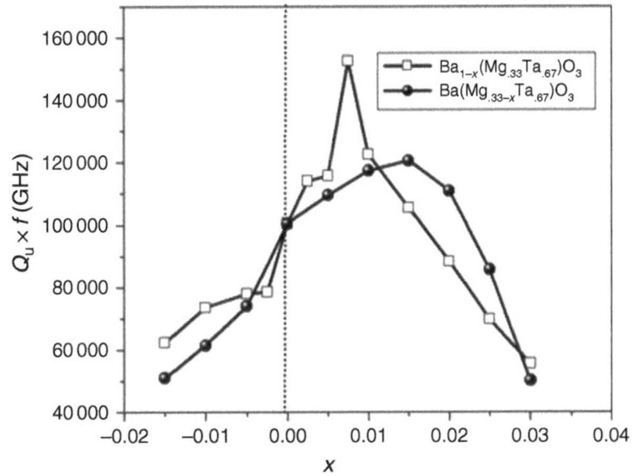

Figure 8.39 Variation of quality factor as a function of x in $Ba_{1-x}(Mg_{1/3}Ta_{2/3})O_3$ and $Ba(Mg_{0.33-x}Ta_{0.67})O_3$ (after Ref. [129]).

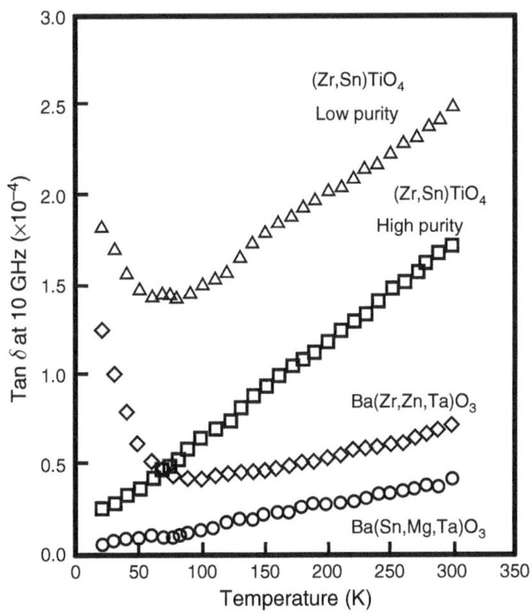

Figure 8.40 Temperature dependence of tan δ for some DR materials (after Ref. [194]).

Figure 8.41 shows the variation of ε_r, τ_ε as a function of Sr concentration. The τ_ε decreased with x up to $x = 0.5$ and then increased with further increase in Sr concentration. Nagai et al. [200] reported that $Ba_{1-x}Sr_x(Mg_{1/3}Ta_{2/3})O_3$ forms a solid solution with a linear variation in the lattice parameters. In spite of a linear change of lattice constants in the

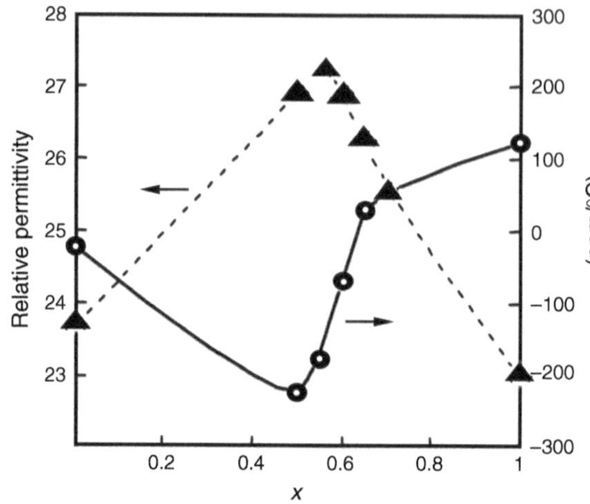

Figure 8.41 Variation of the permittivity and τ_ε of $Ba_{1-x}Sr_x(Mg_{1/3}Ta_{2/3})O_3$ ceramics as a function of x (after Ref. [204]).

solid solution, the τ_f and τ_ε change abruptly at $x = 0.5$ [202]. X-ray diffraction, electron diffraction, dielectric properties, infrared and Raman spectroscopic studies [199–206] showed that the solid solution undergoes a reversible structural change. Figure 8.42 shows the Raman spectrum of $Ba_{1-x}Sr_x(Mg_{1/3}Ta_{2/3})O_3$ for different values of x. With increasing Sr content two of the peaks shifted to lower frequencies and one peak shifted to a higher frequency. For $x = 0.65$ and 0.7 three new additional peaks marked by arrows

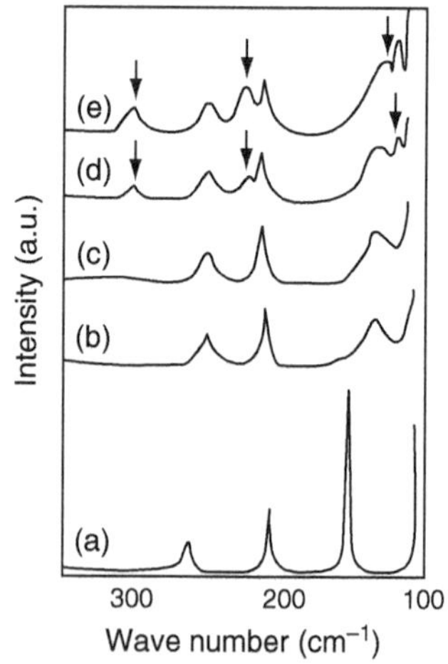

Figure 8.42 The Raman spectrum of $Ba_{1-x}Sr_x(Mg_{1/3}Ta_{2/3})O_3$ ceramics as a function of x. (a) $x = 0$ (b) $x = 0.5$ (c) $x = 0.6$ (d) $x = 0.65$ (e) $x = 0.7$ (after Ref. [201]).

appeared. These three additional peaks are due to the formation of a low symmetry phase. On heating to above 80°C the additional peaks in the Raman spectra disappeared and on cooling it reappeared. Sugiyama and Nagai [202] studied the temperature dependence of structure in $Ba_{1-x}Sr_x(Mg_{1/3}Ta_{2/3})O_3$ by TEM and Raman methods. The study revealed the existence of a new low temperature phase in the strontium-rich composition and it transformed to an ordered hexagonal perovskite with increasing temperature. The compounds with negative τ_f are the low temperature phase at room temperature. The appearance of new peaks is attributed to a lower symmetry phase which arose from a phase transformation due to a tilt of the oxygen octahedra. Nagai et al. extended [205] Glazer's oxygen octahedral tilting model for structural changes in cubic perovskites. Based on Glazers model, electron diffraction and X-ray diffraction, they derived a monoclinic unit cell with space group *C2h6* with an angle >90° from antiphase tilting of oxygen octahedra. It has been suggested [208–210] that the apparent lowering of symmetry in perovskites often results from tilting of the octahedra. The tilting arises because the ionic radii of the site species are too small to occupy fully the available volume (at a critical value of the tolerance factor). Therefore, at a given temperature, the octahedra rotates in order to reduce the size of the cubo-octahedral interstices of the oxygen sublattice. It has been reported that the onset of tilt transitions [211, 212] is the major factor which influences the behavior of the τ_ε as a function of temperature and composition. The τ_ε is an important parameter which determines the properties of dielectric resonators. Ba- and Sr-based complex perovskite [213] show positive and negative values of τ_ε respectively at room temperature. Reaney et al. made [212] a detailed study on the effect of tolerance factor on structural transitions. Figure 8.43 gives a plot of τ_ε versus tolerance factor in complex perovskites. The τ_ε is close to zero when $t \sim 1.05$, and decreases to a minimum at $t = 0.985$. It rises sharply to become positive and continue to increase slowly as tolerance factor decreases.

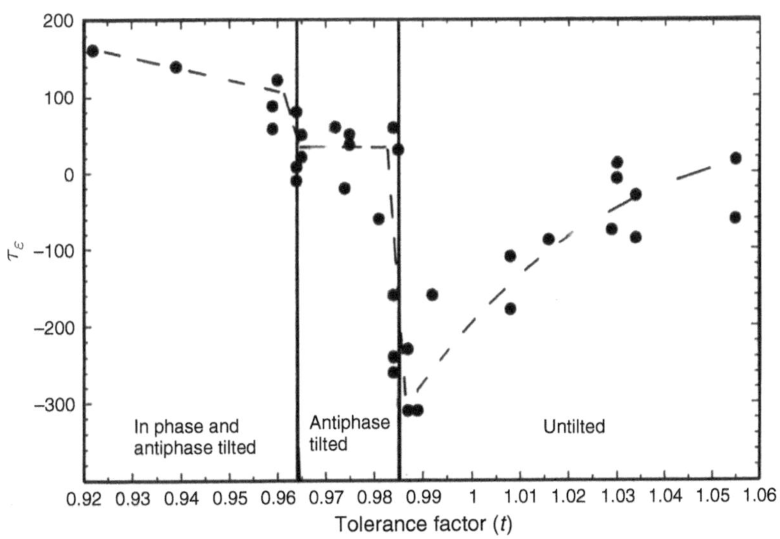

Figure 8.43 The change of τ_ε with tolerance factor in complex perovskites (after Ref. [212]).

8.5 $Ba(Zn_{1/3}Nb_{2/3})O_3$ (BZN)

8.5.1 Preparation

In 1977, Kawashima et al. reported [8] $Ba(Zn_{1/3}Nb_{2/3})O_3$ (BZN) as a low loss microwave dielectric material. Since then several authors investigated [15, 55, 211–224] the microwave dielectric properties of BZN. The BZN is prepared by the conventional solid-state ceramic route by ball milling the stoichiometric amounts of raw materials ($BaCO_3$, ZnO, Nb_2O_5) and calcining in the temperature range 1100–1200°C followed by sintering in the temperature range 1400–1500°C. It has been reported [15, 48, 218, 223–226] that the sintering temperature of BZN can be lowered by the addition of additives such as CuO, Sb_2O_3, B_2O_3, $B_2O_3 + LiF$, $Ba_3W_2O_9$. Roulland and coworkers reported [48, 218, 225] that BZN can be sintered at about 1000°C by the addition of 10 mol% B_2O_3 + 5 mol% LiF. A slight non-stoichiometry at A site (~1%) can further lower sintering temperature to about 900°C [218] enabling silver co-sintering applications. Secondary phases of $Ba_5Nb_4O_{15}$, $BaNb_2O_6$, etc. are often found in the sintered BZN ceramics. Liou et al. [227] prepared $Ba_xSr_{1-x}(Zn_{1/3}Nb_{2/3})O_3$ by a reaction sintering process with 3 wt% CuO without the calcination process at 1450°C. However, secondary phases of $ZnNb_2O_6$ and $(Cu_2Zn)Nb_2O_8$ were found in the sintered ceramics. Liang et al. [228] prepared BZN powder by a spray pyrolysis technique and the ceramic sintered at 1250°C showed presence of $Ba_5Nb_4O_{15}$ secondary phase. The powders obtained by the wet chemical methods have small particle size and hence ZnO can escape easily which led to the formation of $Ba_5Nb_4O_{15}$ and $BaNb_2O_6$ secondary phases. Kolodiazhyni from a study using ESR, positron annihilation spectroscopy and dielectric spectroscopy reported [165] that $Ba(B'_{1/3}B''_{2/3})O_3$ ceramics contain substantial amount of lattice vacancy defects and some of them contain unpaired electron spin. The BZN when prepared at high temperatures has a disordered cubic structure with lattice parameter $a = 4.09$ Å [213]. BZN undergoes a transition from a 1:2 ordered structure to a disordered B site arrangement at about 1375°C [55, 215–218, 229]. The lower thermal stability means that BZN needs to be annealed at temperatures lower than 1375°C to increase the ordering and thereby increase the quality factor.

8.5.2 Dielectric properties

The BZN has ε_r of 40, Qf of about 80 000 GHz, and τ_f of about 30 ppm/°C [8, 15, 222, 230]. Noh et al. [231] sintered BZN over a range of temperatures and concluded that the grain size and the density were more important in controlling the quality factor. They reported that volatility of ZnO and the formation of secondary phases degrade the quality factor [231]. However, Wu and Davies [215] found that BZN when sintered in ZnO-rich environment severely degraded the quality factor. Chemical analysis showed that the degradation of the properties were due to the uptake of ZnO to form non-stoichiometric BZN from the muffling agent. This is supported by the fact that addition of ZnO to BZN considerably reduces the quality factor [215]. Slow cooling of the ceramic after sintering through the order–disorder region greatly improve Qf [224]. The high Qf is obtained by annealing the sintered samples below the order–disorder transition temperature. Addition of $BaSnO_3$, $BaZrO_3$, V_2O_5, $Ba(Co_{1/3}Nb_{2/3})O_3$, $Ba(Ga_{1/2}Ta_{1/2})O_3$, CeO_2, $Ba_3W_2O_9$ improves the quality factor of BZN [223, 224, 226, 232–236]. Some of the additives increase the order–disorder transition temperature and others lower it. When the transition temperature is lowered, then the annealing temperature will also lower. In such cases prolonged annealing is needed to get a high Qf material since at lower annealing temperature the ordering process is sluggish. Huang et al. reported [232] that in BZN–$BaZrO_3$, the Qf value, density, and ε_r increased with sintering temperature and reached a maximum on sintering at 1400°C and then decreased. The samples sintered at 1450°C/2 h showed $\varepsilon_r = 42$, $Qf = 96\,000$ GHz and $\tau_f = 27$ ppm/°C. X-ray diffraction study showed that the BZN–$BaZrO_3$ exhibited a disordered cubic structure (*Pm3m*). The dielectric properties of BZN prepared under different conditions and with different additives are given in Table 8.5. Varma and Sebastian [222] added several dopants of varying valencies and ionic radii in different mol% to BZN and reported that dopants with ionic radii close to that of Zn or Nb improve the quality factor of BZN. Figure 8.44 shows the variation of quality factor of BZN as a function of the dopant ionic radii.

BZN has a relatively high τ_f of about 30 ppm/°C which limits its use in practical applications. Hence several attempts were made to lower the τ_f of BZN by adding dopants such as B_2O_3, $Sb_2O_3 + B_2O_3$, and by forming solid solution with $Ba(Mg_{1/3}Nb_{2/3})O_3$ (BMN), $Ba(Co_{1/3}Nb_{2/3})O_3$ (BCN), $BaSnO_3$, $Ba(Ni_{1/3}Nb_{2/3})O_3$ (BNN), $Ca(Zn_{1/3}Nb_{2/3})O_3$ (CZN), $Sr(Zn_{1/3}Nb_{2/3})O_3$ (SZN), and $Ba(Ga_{1/2}Ta_{1/2})O_3$ (BGT) [16, 219–221, 224, 226, 230, 233–236, 240, 242]. The BCN, BNN, SZN, and CZN are having negative τ_fs. Hence it is possible to compensate for the large positive τ_f of BZN by making a solid solution of BZN with BCN, SZN, BNN, and BGT. The BZN has a disordered cubic structure when prepared above 1375°C. $Sr(Zn_{1/3}Nb_{2/3})O_3$ (SZN) is a hexagonally ordered perovskites with $a = 5.66$ Å and $c = 6.95$ Å [60, 213] with $\varepsilon_r = 40$, $Qf = 20\,000$ GHz and $\tau_f = -38$ ppm/°C. Several authors tailored [95, 211–213, 243] the τ_f of BZN by forming a solid solution with SZN and obtained a nearly temperature compensated ceramics although the quality factor was reduced. The ε_r varied non-linearly with x with the maximum of 46 at $x = 0.6$ (0.6BZn–0.4SZN). A τ_f value close to zero was obtained for 0.3BZN–0.7SZN with $\varepsilon_r = 40$ with Qf of 30 500 GHz. It was reported [95, 211, 213, 243] that in $Ba_xSr_{1-x}(ZnNb)O_3$ (BSZN) the substitution of Sr for Ba monotonously change the τ_f, τ_c and ε_r and an anomaly existed at about $x = 0.5$. Reaney and co-workers found [95, 211, 212] a strong correlation between tolerance factor and τ_ε and the BZN–SZN solid solution undergoes a structural transition involving octahedral tilting. The octahedral tilt transitions lead to doubling of the unit cell and appearance of superstructure reflections [209, 210].

Several authors tailored [223, 224, 233–237, 242, 244] the dielectric properties of $(1-x)Ba(Co_{1/3}Nb_{2/3})O_3$–$xBa(Zn_{1/3}Nb_{2/3})O_3$. The ε_r and τ_f increases with x (BZN content). It was found [219, 235, 237, 242] that for $0.7Ba(Co_{1/3},Nb_{2/3})O_3$–$0.3Ba(Zn_{1/3}, Nb_{2/3})O_3$, the resonant frequency is nearly independent of temperature. This composition has $\varepsilon_r = 35$, with a $Qf \sim 97\,000$ GHz. However, Scott et al. have observed [244] the zero τ_f composition to be $0.4Ba(Co_{1/3}Zn_{2/3})O_3$–$0.6Ba(Zn_{1/3}Nb_{2/3})O_3$ with $\varepsilon_r = 36$ and $Qf = 86\,000$ GHz. The Qf varies with sintering temperature and has the maximum Qf when sintered at 1400°C [237]. Annealing the samples improved the ordering and thereby increased the quality factor [216]. Azough et al. [235] reported that addition of a small amount of CeO_2 (0.5 wt%) considerably improve densification of $Ba[(Co_{0.7}Zn_{0.3})_{1/3}Nb_{2/3}]O_3$ and thereby the Qf. BCN undergoes a transition from 1:2 ordered to a disordered structure at about 1425°C [245]. Hence the BCN–BZN can form an ordered solid solution by prolonged annealing at 1300–1400°C.

Addition of a small amount of $Ba(Ga_{1/2}Ta_{1/2})O_3$ (BGT) considerably improve the microwave dielectric properties of BZCN [234, 238, 246]. The $0.9Ba[(Zn_{0.6}Co_{0.4})_{1/3}Nb_{2/3}]O_3$–$0.1Ba(Ga_{1/2}Ta_{1/2})O_3$ showed $\varepsilon_r = 35$, $Qf = 97\,600$ GHz [234]. The sintered

Table 8.5 Microwave dielectric properties of BZN prepared under different conditions and additives

Composition	Dopant	Sintering temperature	ε_r	Qf (GHz)	τ_f (ppm/°C)	Reference
BZN	–	1390	40	87 000	30	[15, 16]
BZN	5 mol% B_2O_3	900	32	3500	20	[16]
BZN	5 mol% B_2O_3 + 5 mol% CuO	875	36	19 000	21	[16]
BZN	1 mol% WO_3	1450	38	95 150	39	[222]
BZN	1 mol% SnO_2	1450	37	83 200	29	[222]
BZN	1 mol%ZrO_2	1450	40	77 800	26	[222]
BZN	0.5 mol% Al_2O_3	1450	40	77 400	29	[222]
BZN	Annealed in N_2	1500	41	90 000	4	[15]
0.95BZN–0.05$BaZrO_3$		1450/2 h	42	96 000	27	[232]
0.95BZN–0.05$BaZrO_3$	1 wt% CuO	1360/2 h	40	70 000	17	[221]
$Ba(Zn_{1/3}Nb_{2/3})_{0.9}Zr_{0.1}O_3$		1400	38	61 000	15	[81]
$Ba(Sn_{0.226}Zn_{0.258}Nb_{0.516})O_3$		1500	32	970 000	12	[226]
0.95BZN–0.05$Ba(Ga_{1/2}Ta_{1/2})O_3$			38	102 955	19	[234]

$0.9BZN–0.1\ Ba(Ga_{1/2}Ta_{1/2})O_3$			37	93 500	15	[234]
$xBa(Zn_{1/3}Nb_{2/3})O_3–(1–x)Ba(Mg_{1/3}Nb_{2/3})O_3$ (x 0.25)		1500	34	76 700	–4	[225]
0.3BZN–0.7SZN			40	30 500	0	[213]
0.7BCN–0.3BZN		1400/20 h	35	97 000	0	[223, 237]
0.7BCN–0.3BZN		1400/6 h	35	80 000	0	[223, 237]
$Ba[(Zn_{0.8}Co_{0.2})_{1/3}Nb_{2/3}]O_3$		1410	40	50 135	18	[223]
$Ba[(Zn_{0.6}Co_{0.4})_{1/3}Nb_{2/3}]O_3$		1400	38	55 000	14.3	[223]
$Ba[(Zn_{0.4}Co_{0.6})_{1/3}Nb_{2/3}]O_3$		1400	36	54 400	0	[223]
$Ba[(Zn_{0.2}Co_{0.8})_{1/3}Nb_{2/3}]O_3$		1400	34	7000	–10	[223]
$Ba[(Zn_{0.3}Co_{0.7})_{1/3}Nb_{2/3}O_3$	0.025 wt% V_2O_5	1450/4 h	35	85 000	0	[233]
$Ba([(Co_{0.7}Zn_{0.3})_{1/3}Nb_{2/3}]O_3$	0.4 wt% CeO_2	1450	35	84 000	0	[235]
0.6BZN–0.4BCN		1400	36	86 000	0	[244]
$0.99Ba[Zn_{0.3}Co_{0.7}]_{1/3}Nb_{2/3}O_3–0.01Ba_3W_2O_9$		1380	34	82 000	0	[219]
$0.9Ba[(Zn_{0.6}Co_{0.4})_{1/3}Nb_{2/3}]O_3–0.1Ba(Ga_{1/2}Ta_{1/2})O_3$			35	97 600	0	[234, 238]
0.35BNN–0.65BZN		1450/4 h	35	25 000	1	[239]

(Continued)

Table 8.5 (Continued)

Composition	Dopant	Sintering temperature	ε_r	Qf (GHz)	τ_f (ppm/°C)	Reference
$Ba[(Ni_{0.6}Zn_{0.4})_{1/3}Nb_{2/3}]O_3$	0.5 mol% B_2O_3	1340	34	42 100	–8	[220]
$Ba[(Ni_{0.6}Zn_{0.4})_{1/3}Nb_{2/3}]O_3$	1 mo% B_2O_3	1300	33	39 700	–4	[220]
$Ba[(Ni_{0.6}Zn_{0.4})_{1/3}Nb_{2/3}]O_3$	2 mol% B_2O_3	1300	33	32 500	–11	[220]
$Ba[(Ni_{0.6}Zn_{0.4})_{1/3}Nb_{2/3}]O_3$	1 mol% B_2O_3 + 0.5 mol% Sb_2O_3	1300	35	4300	0	[220]
$Ba[(Ni_{0.6}Zn_{0.4})_{1/3}Nb_{2/3}]O_3$	Pechini method	1520/2 h	36	57 440	5	[238]
$Ca_{0.9}Ba_{0.1}(Zn_{1/3}Nb_{2/3})O_3$			36	16 170	–12	[240]
$Ca(Zn_{1/3}Nb_{2/3})O_3$			35	16 000	–43	[24]
$(1-x)Ba_3(ZnNb_2)O_9-xBa_3W_2O_9$ (x 0.007)		1380	39	118 000	21	[219]
$Sr(Zn_{1/3}Nb_{2/3})O_3$		100/1 h	40	20 000	–38	[213]
$Ba_{1-x}La_x[Zn_{(1-x)/3}Nb_{(2-x)/3}]O_3$ (x 0.0)		1350/4 h	40	112 280	19	[241]
$Ba_{1-x}La_x[Zn_{(1-x)/3}Nb_{(2-x)/3}]O_3$ (x 0.05)		1350/4 h	43	46 530	35	[241]
$Ba_{1-x}La_x[Zn_{(1-x)/3}Nb_{(2-x)/3}]O_3$ (x 0.1)		1350/4 h	44	2850	45	[241]
$Ba_{1-x}La_x[Zn_{(1-x)/3}Nb_{(2-x)/3}]O_3$ (x 0.3)		1350/4 h	45	1990	8	[241]

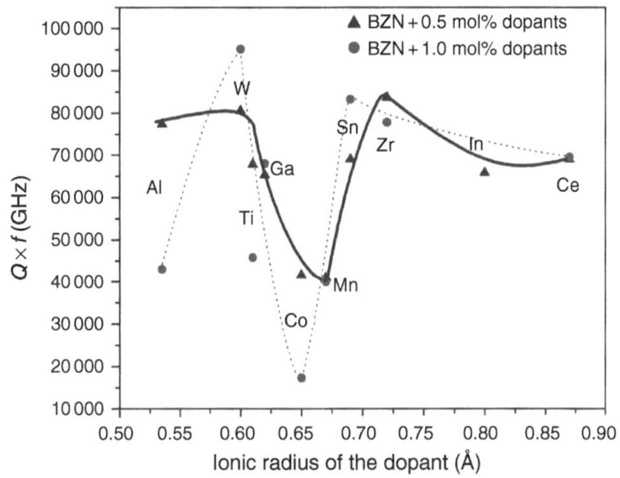

Figure 8.44 Variation of the quality factor Qf as a function of the dopant ionic radii in BZN (after Ref. [222]).

ceramics showed the presence of secondary phase of barium niobate which may be due to escape of volatile ZnO. Reaney et al. sintered [246] $0.9Ba[(Zn_{0.6}Co_{0.4})_{1/3}Nb_{2/3}]O_3$–$0.1Ba(Ga_{1/2}Ta_{1/2})O_3$ at 1350°C/8 h and then annealed and quenched at different temperatures 1100, 1200, 1300 and 1400°C. Figure 8.45 shows the SEM images of

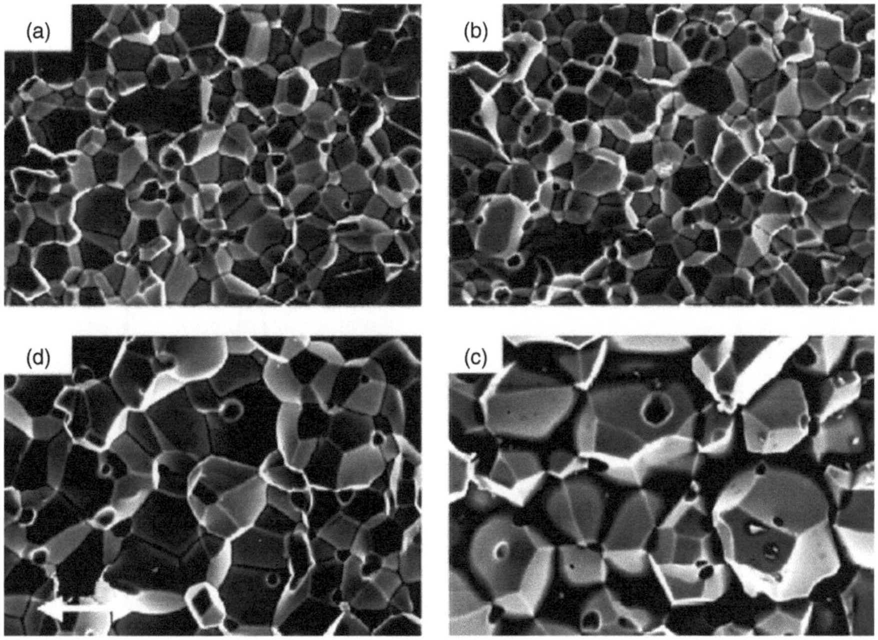

Figure 8.45 SEM micrographs recorded from BCZN–BGT samples annealed and quenched from (a) 1100, (b) 1200, (c) 1300 and (d) 1400°C (after Ref. [246]).

BCZN–BGT samples annealed and quenched from different temperatures. The grain size increased from 10 to 20 μm with increase in annealing temperatures. X-ray diffraction study showed that the samples annealed and quenched from different temperatures are cubic (*Pm3m*) with no evidence of ordering. However, Raman spectra and TEM study revealed order–disorder transition at about 1200°C. Figure 8.46 shows the Raman spectra of the samples annealed and quenched at different temperatures. The F_2g band which was present in samples annealed and quenched at 1100°C became weak in the samples annealed and quenched at 1200°C indicating a decrease in the long range ordering. It has been reported earlier that F_2g band is due to long range ordering and A_1g band can be due to long range order (LRO) or short range order (SRO) [160, 247]. The samples annealed and quenched from 1250 to 1350°C did not show the F_2g bands but the A_1g band is still present although it is weak. This indicates that SRO is still present in samples annealed and quenched above 1200°C although long range ordering is absent. Electron diffraction study showed the presence of discrete superstructure reflections at ($h \pm 1/3$, $k \pm 1/3$, $l \pm 1/3$) indicating 1:2 ordering.

Several authors [220, 236, 239, 240, 248] tuned the τ_f of BZN by forming solid solutions or mixture phases with $Ca(Mg_{1/3}Nb_{2/3})O_3$ (CMN), $Ca(Zn_{1/3}Nb_{2/3})O_3$(CZN), $Ba(Ni_{1/3}Nb_{2/3})O_3$ (BNN) etc. Li and Chen [249] tuned the positive τ_f of BZN by stacking $Ca(Mg_{1/3}Nb_{2/3})O_3$ which has a negative τ_f. The resultant ε_r and τ_f depended on the volume fraction of $Ca(Mg_{1/3}Nb_{2/3})O_3$. The kinetics of cation ordering was slower in BZN–BNN, BZN–BCN, BNN, BNT. Hence long-term sintering, slow cooling or prolonged low temperature annealing are needed to enhance the cation ordering and to get reasonably good quality factor [223, 224, 242, 245, 250, 251].

Wu and Davies [219] reported that substitution of a small amount of $Ba_3W_2O_9$ in BZN accelerates the kinetics of cation ordering, increases the stability of order and improves the sinterability resulting the highest Qf of 118 500 GHz. The (1–x)BZN–$x$$Ba_3W_2O_9$ ceramics were sintered at 1370–1400°C/4–6 h and then annealed at 1300°C/6–24 h to develop ordering. Addition of more than 2 mol% $Ba_3W_2O_9$ led to formation of secondary phases, mainly $BaWO_4$. The substitution of W for Nb in BZN is

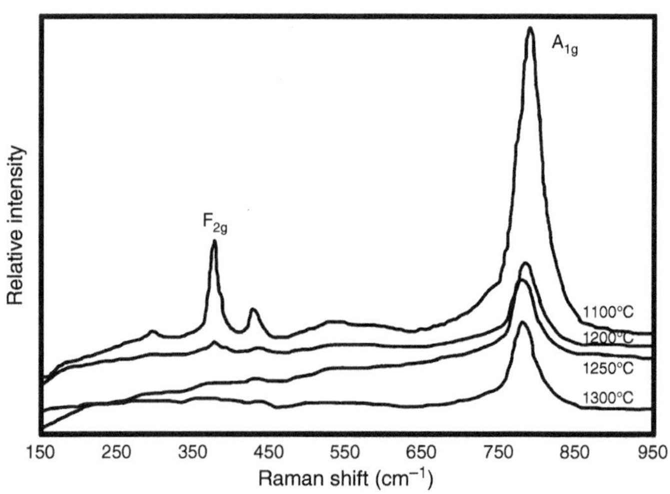

Figure 8.46 Raman spectrum of BCZN–BGTceramic (after Ref. [246]).

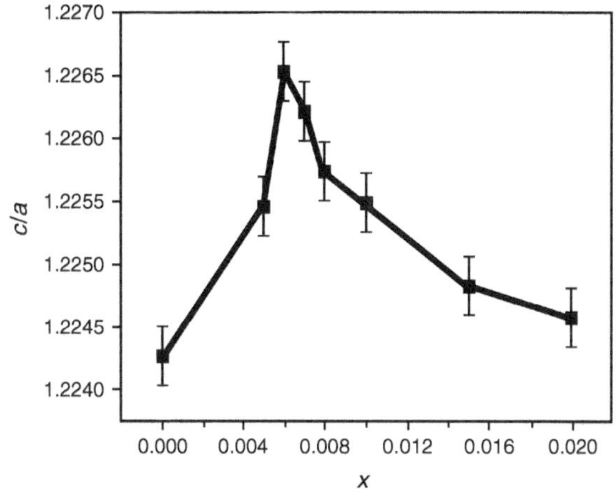

Figure 8.47 Variation of c/a ratio with x in (1–x)BZN–xBW after annealing for 24 hours ($x = 0$) and 12 hours ($x > 0$) at 1300°C (after Ref. [219]).

charge compensated by the introduction of vacancies on the Zn sites. The introduction of cation (Zn) vacancies enhanced the stability of the 1:2 B-site ordered structure. $Ba(Zn_{1-x}\Box_x)_{1/3}(Nb_{1-x}W_x)_{2/3}O_3$ underwent an order–disorder transition at 1410°C which is higher than that of pure BZN. The kinetics of the ordering process in W containing BZN are much faster than pure BZN. It was found that the intensity of the (100) ordered reflection is higher than that of pure BZN and increased at a faster rate on annealing as compared to pure BZN. Figure 8.47 shows the variation of c/a with composition x. The maximum c/a value was observed for $x = 0.006$ which is indicative of long range ordering. Lattice imaging study using HRTEM showed the presence of superlattice fringes along both of the allowed <111> directions. Figure 8.48 shows the typical microstructures of (1–x)BZN–$x$$Ba_3W_2O_9$ for two different values of $x = 0$ and 0.005

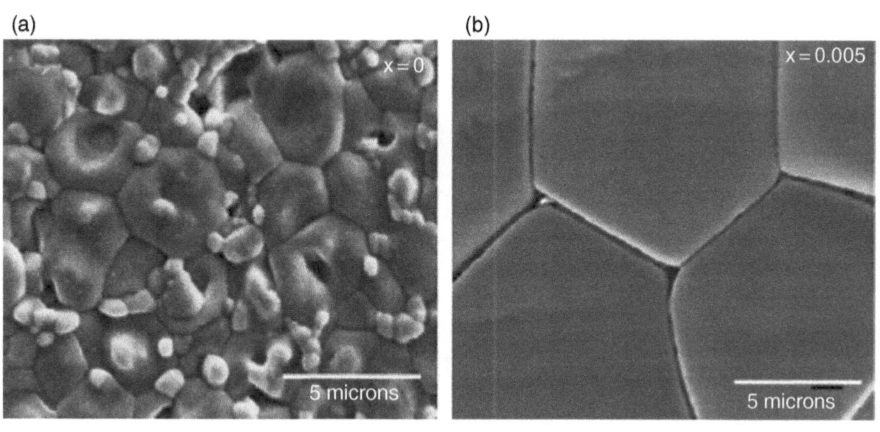

Figure 8.48 SEM micrograph of (1–x)BZN–xBW after sintering at 1390°C/4 h (a) $x = 0.0$ and (b) $x = 0.005$ (after Ref. [219], Courtesy, Wiley-Blackwell Publishing Ltd).

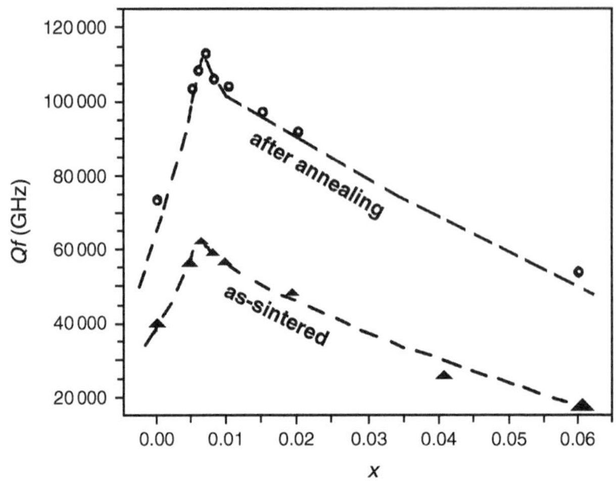

Figure 8.49 Variation of the Qf with annealing time at 1300°C in (1–x)BZN—xBW for different values of x (after Ref. [219]).

sintered at 1390°C/4 h. It can be clearly seen that the grain size increased with $Ba_3W_2O_9$ addition. Figure 8.49 shows the variation of Qf as a function of composition x. The highest Qf was reported for $x=0.007$. There is no appreciable variation in ε_r and τ_f with BW addition. The Qf also increased with annealing time at 1300°C as shown in Figure 8.50. The Figures 8.47 and Figure 8.50 indicate that addition of BW and annealing improve ordering and Qf. In both pure BZN and BW-added ceramics the order parameter and Qf increased with annealing. Addition of $Ba_3W_2O_9$ lowers the sintering temperature by about 50°C through liquid phase sintering. In BZN–$Ba_3W_2O_9$

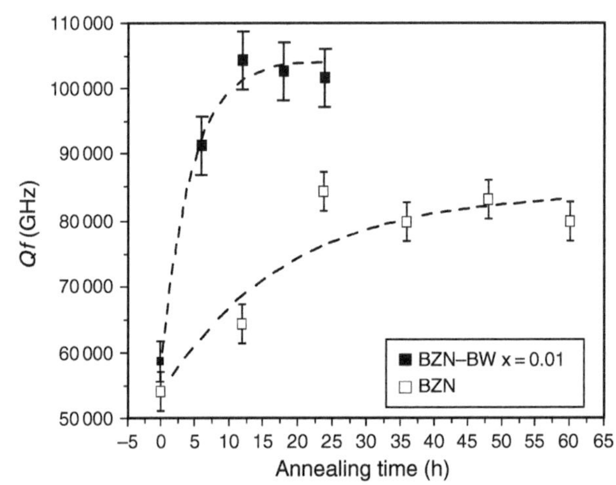

Figure 8.50 Variation of Qf of (1–x)BZN–xBW for $x=0.01$ as a function of annealing time at 1300°C (after Ref. [219]).

initially small ordered domains nucleate and these isolated ordered clusters or nanodomains grow in size until the domains impinge at domain boundaries. After impingement of the domains, the subsequent coarsening occurs through the disappearance and reorientation of the domain boundaries. The growth of the domains is accompanied by a continuous expansion of the *c*-axis. Single-phase (1–*x*)BZN–*x*$Ba_3W_2O_9$ ceramics are formed in the limited range $0 \leq x \leq 0.02$. They form a B-site vacancy containing ordered structure with a stoichiometry $Ba(Zn_{1-x}\square_x)_{1/3}(Nb_{1-x}W_x)_{2/3}O_3$. $Ba_3W_2O_9$ enhanced the relative stability of the 1:2 ordered structure and increased the order–disorder temperature by 35°C. Wu and Davies reported that the Zn vacancies enhance the kinetics of the ordering and coarsening process.

It has been reported [219, 250, 252] that small deviations in stoichiometry considerably influence the order parameter, density and microwave dielectric properties. Wu and Davies [250] made a detailed study of excess and deficiency of ZnO, BaO, and Nb_2O_5 in BZN. It was found that slight deficiency of (<1 mol%) BaO and ZnO and slight excess of Nb_2O_5 improved order parameter and microwave quality factor. Figure 8.51 shows the variation of *Qf* of BZN with excess and deficiency of ZnO, BaO, and Nb_2O_5 for the as-sintered (sintered at 1420°C/4 h) and samples annealed at 1330°C/24 h. There is no

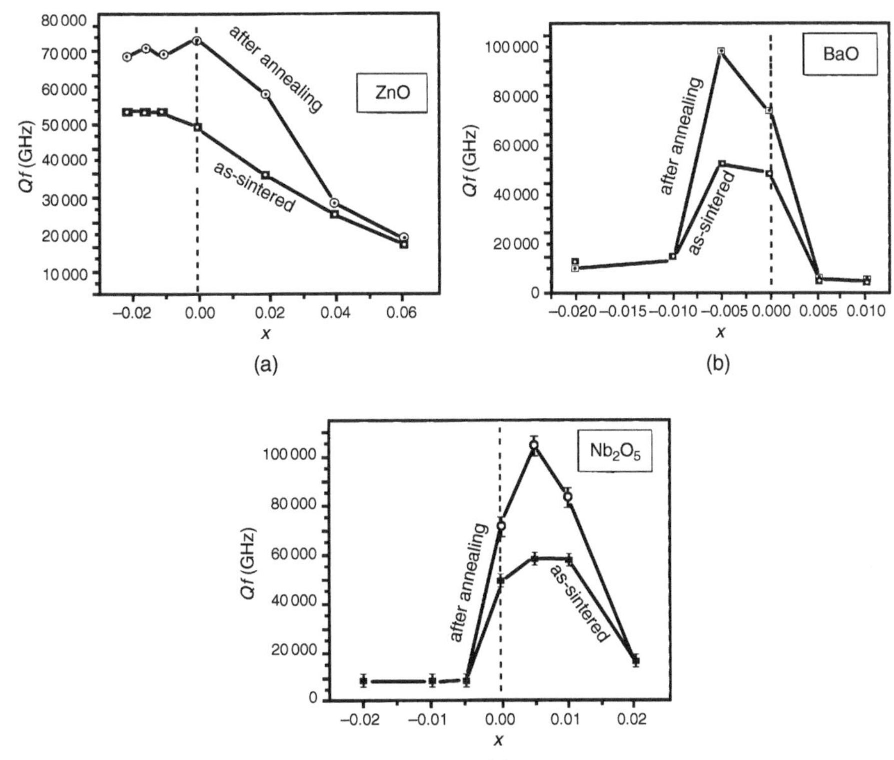

Figure 8.51 Variation of *Qf* in BZN-with excess and deficiency of (a) ZnO, (b) BaO and (c) Nb_2O_5. As-sintered samples are indicated by solid squares and annealed ceramics by open squares (after Ref. [250]).

significant effect of ZnO deficiency on the Q factor but excess amount of ZnO reduced the quality factor considerably. Excess BaO considerably lowered the quality factor whereas BaO deficiency considerably improved the quality factor. The BaO excess samples were having a disordered structure whereas the BaO-deficient ceramics were ordered. In contrast to BaO- and ZnO-deficient ceramics, Nb_2O_5 deficiency lowered the quality factor whereas excess Nb_2O_5 considerably improved the quality factor. The highest *Qf* factor was for 0.5 mol% excess Nb_2O_5. This ceramic composition was very dense and had a well-ordered structure. In the case of $Ba(Zn_{1/3}Nb_{2/3})O_3$–$Ba_3W_2O_9$ the *Qf* decreased considerably for both BaO-excess and -deficient ceramics. However, $Ba(Zn_{1/3}Nb_{2/3})O_3$–$Ba_3W_2O_9$ with ZnO deficiency showed considerable improvement in *Qf* factor. The highest *Qf* was 105 000 GHz found for ZnO deficiency of less than 1 mol%. More than 1 mol% excess or deficiency of BaO, ZnO or Nb_2O_5 leads to the formation of secondary phases such as $ZnNb_2O_6$, $Ba_5Nb_4O_{15}$, $BaNb_2O_6$, $Ba_4Nb_2O_9$, Ba_3ZnO_4 which degrade *Qf* values. A gradual decrease in ordering was found in big samples from the surface to the interior. The ordering gradient is related with the increased constraint on the growth of the 1:2 ordered structure within the interior of large samples and can be improved by prolonged annealing. Wu and Davies reported that [215] BZN pellets muffled in ZnO powder showed a considerable decrease in *Qf*. Detailed analysis showed that BZN absorbed ZnO to form non-stoichiometric solid solutions with reduced cation order and Q.

8.6 $Ba(Ni_{1/3}Nb_{2/3})O_3$

Several authors investigated [11, 59, 251, 253–258] $Ba(Ni_{1/3}Nb_{2/3})O_3$ (BNN) and its solid solution phases. The BNN undergoes an disorder to order transition from cubic perovskite with $Pm3m$ to hexagonal $P\bar{3}m1$ at about 1500°C [11, 251, 255, 257] which is accompanied by the formation of a liquid phase at the grain boundary at or above the transition temperature. The amount of secondary phase increased when sintered above 1500°C. Park et al. [257] reported remnants of this liquid phase which remains as an intergranular secondary phase, can dissolve back into the primary material during annealing. The liquid-phase sintered samples had a lower quality factor. The as-prepared samples are disordered and the degree of ordering increased with annealing at 1300°C. TEM study showed that the increase in long range order parameter with annealing is due to the growth of the small ordered region. X-ray diffraction study showed a decrease [251] in the intensity of the superstructure reflections on increasing the sintering temperature and the superstructure reflections disappear on sintering at 1550°C. A large decrease in bulk density and a large increase in grain size observed when sintered above 1500°C. The samples sintered at 1450°C had the maximum density and quality factor [11, 251]. The relative permittivity and τ_f show an increase on sintering at 1500°C and is followed by a decrease in relative density.

8.7 $Ba(Co_{1/3}Nb_{2/3})O_3$

Several authors investigated [18, 19, 31, 223, 259] the microwave dielectric properties of $Ba(Co_{1/2}Nb_{1/2})O_3$ (BCN). Melodetsky and Davies studied [31] the effect of different solid solution formation and heat-treatment on the cation ordering and dielectric properties of $Ba(Co_{1/3}Nb_{2/3})O_3$. Pure BCN undergoes an order–disorder transition at 1400°C. The substitution of $BaZrO_3$ destabilizes the 1:2 order and a 1:1 ordered phase is formed for 10–20 mol% $BaZrO_3$. The order–disorder transition temperatures for the 1:1 BCN–$BaZrO_3$ phases are quite low (<1300°C) and lead to lower degrees of order and lower *Qf* values. The substitution of $Ba(Y_{1/2}Nb_{1/2})O_3$ also induces a transition to 1:1 ordering, but in this case the stability of the order is significantly higher and the samples remain ordered to at least 1550°C. The high degree of order in the BYN-based system is accompanied by a higher *Qf* value compared to their $BaZrO_3$ counterparts. However, none of the samples reach the *Qf* values of the ordered BCN end member. Ahn et al. [18, 223] reported that BCN has a 1:2 ordered hexagonal structure when sintered below 1400°C and showed the highest *Qf* of 78 000 GHz with $\varepsilon_r = 32$ and $\tau_f = -12$ ppm/°C. Sintering above 1400°C led to evaporation of CoO and the formation of Ba- and Nb-rich liquid phases which led to grain growth and the degradation of microwave dielectric properties. It was found [19] that Co deficiency in BCN degraded the quality factor. In Co-deficient compositions, cobalt-deficient secondary phases such as $Ba_8CoNb_6O_{24}$ and $Ba_5Nb_4O_{15}$ were developed.

8.8 $Ba(Mg_{1/3}Nb_{2/3})O_3$

Several people prepared BMN by both chemical and solid state methods [10, 260–264]. The samples sintered at high temperatures were disordered with a cubic *Pm3m* structure. The samples prepared by alkoxide–hydroxide route [260] and hydrothermal methods [261–263] when sintered at 1350°C showed a density of 99%. The order parameter increased with sintering temperature and became larger than 0.9 when sintered at 1350°C. The as-prepared disordered BMN becomes ordered on annealing at about 1300°C. Tian et al. [264] investigated the effect of addition of $BaWO_4$ on the microstructure, phase formation, 1:2 ordering and microwave dielectric properties of BMN. Secondary phases of $Ba_5Nb_4O_{15}$ and $BaWO_4$ are found in all $BaWO_4$-added samples sintered at 1500°C/4 h. The density, *c*/*a* ratio and order parameter increased with $BaWO_4$ addition. The increase in order parameter was attributed to the larger difference of ionic radii and charge valences between Mg^{2+} and W^{6+} than those between Mg^{2+} and Nb^{5+}. The 1:2 long range order is also promoted by the niobate–oxygen vacancies when the Nb^{5+} was substituted by W^{6+} ion. The melting point of $BaWO_4$ is 1430°C and addition of $BaWO_4$ leads to liquid-phase sintering. The increase in *Qf* in $BaWO_4$-added BMN is attributed to increase in density and order parameter. Addition of more than 3 mol% $BaWO_4$ degraded the *Qf* factor. Addition of 3 mol% $BaWO_4$ gave the highest $Qf = 82\,300$ GHz with $\varepsilon_r = 31$, $\tau_f = 32$ ppm/°C. However, the sintered ceramics contain $BaWO_4$ and $Ba_5Nb_4O_{15}$ secondary phases. Paik et al. [10] reported that the sinterability and microwave dielectric properties of BMN are improved by a slight deficiency of Mg. The order parameter and density and *Qf* increased with slight Mg deficiency and reached their maximum for $x = 0.02$ in $Ba(Mg_{1/3-x}Nb_{2/3})O_{3-\delta}$. This composition showed the highest quality factor of 96 000 GHz, with $\varepsilon_r = 32$ and $\tau_f = 30$ ppm/°C when sintered at 1450°C/2 h. Further increase in *x* decreased the order parameter, density and *Qf*. Kim et al. [265] reported that according to sintering conditions, the BMN ceramics show local variations in the structure which in turn leads to variations in the microwave dielectric properties. The BMN has a relatively high τ_f which precludes its use in practical applications. The $La(Mg_{2/3}Nb_{1/3})$ O_3 [LMN] has a negative τ_f and is being used to adjust the positive τ_f of BMN [266–268]. Akbas and Davies [266] have found that a complete range of perovskite solid solutions are possible in $Ba_{1-x}La_x(Mg_{(1+x)/3}Nb_{(2-x)/3})O_3$ [BLMN].

8.9 Conclusion

The quality factor of complex perovskite microwave materials is very sensitive to a number of different processing and structural variables. The losses can vary from sample to sample even when they have same nominal composition due to small differences in the intrinsic crystal structure, microstructure, density and impurity concentration, ordering and stoichiometry. It was found that lowest losses are observed when the two different B cations adopt a long range 1:2 ordering, i.e., B′,B″B″ repeat along the [111] direction of the cubic perovskite subcell. Clear correlations between the dielectric loss and degree of cation order and/or the size of the resultant ordered domains have been established in BZT, BMT, BMN and BCN. The ordering makes changes in the relative intensity of the superstructure peaks, deviation of the *c/a* ratio, and the size of the ordered domains. The ideal undistorted cell has $c/a=\sqrt{3/2}=1.2247$. Studies of the kinetics and mechanism of the ordering have established that the intensity of the superstructure reflections, the domain size, and the associated lattice distortion (*c/a*) in BZT-type perovskites increase with increasing annealing or sintering and coincide with improvements in *Qf*. The structural order and presence of secondary phase considerably affect the quality factor. Hence suppression of secondary phase, increasing the chemical order are needed to get a low loss material. Several types of planar defects such as APBs that separate regions where the order nucleates on the same set of {111} planes but is out of phase and orientation (twin) boundaries separating regions that nucleate on different sets of {111} planes may be formed during the ordering phenomena. The final ceramics often consists of domains of the various orientational and translational variants and their size depends very much on the heat treatment and bulk solid state chemistry [62].

The factors influencing *Qf* values of BZT/BZN have been considered to be long range ordering of cations, ZnO evaporation, point defects and stabilization of microdomain boundaries [5, 7, 14, 20]. The LRO of B-site cations is the most frequently asserted parameter since the high *Qf* values could be obtained after an extended high temperature annealing step. Conversely, Tamura et al. and Davies et al. showed [12, 41] that *Qf* values of BZT could be enhanced even with the reduction of LRO by $BaZrO_3$ doping. Although, the studies of pure $Ba(A_{1/3}T_{2/3})O_3$ (A = Mg, Zn, Ni) perovskites seem to yield a consistent picture of the relationship between the cation ordering, the associated structural distortions and dielectric loss, the behavior of $BaZrO_3$-doped systems is not straight forward. The substitution of $BaZrO_3$ into low loss BZT microwave ceramics induces a series of complex changes in the ordering of the cations. At very low levels of substitution $\leq$2.15 mol% the solid solution retain the ordered 1:2 structure of BZT end member but are comprised of ordered domains whose size decrease as the Zr concentration is increased. In the undoped BZT, the high losses of partially ordered ceramics are attributed to the formation of the elastically strained ordering induced domain boundaries. Quality factor can be considerably improved by adding a small amount of (3 mol%) $BaZrO_3$ [12] and the partial segregation of Zr was suggested to stabilize the domain boundaries [37] and decrease the annealing time required to get a low loss state.

Reaney and co-workers [211, 212] related the temperature coefficient of permittivity to the tolerance factor, which indirectly influences the microscopic polarizability of the ions through tilting of octahedra. Depending upon the minimum energy configuration at a given temperature, the octahedra in one plane can tilt in either inphase or antiphase with respect to the other plane about a pseudo-cubic axis. The magnitude of tilting increases with decrease in the tolerance factors. The BMT, BZT, BZCN ceramics belonging to the $A(B'_{1/3}B''_{2/3})O_3$ family have high *Qf*, high ε_r and low τ_f and are currently used in commercial applications. BZT is prepared by calcining the raw materials at temperatures in the range 1100–1200°C and sintering at 1500–1550°C. By introducing long range order by prolonged annealing at 1350–1400°C and by adding a small amount of suitable dopants, the *Qf* of BZT can be increased up to about 200 000 GHz with ε_r of 29. The BMT is prepared by calcining the raw materials at temperatures in the range 1200–1400°C and sintering in the temperature range 1600–1650°C. The BMT shows ε_r of about 25 and *Qf* up to 430 000 GHz. The BZCN prepared by calcining at 1100–1200°C and sintered at temperatures in the range 1400–1500°C showed a ε_r of about 35 and *Qf* up to 100 000 GHz.

Excess amount of Zn degrades the quality factor in BZT and BZN. A slight deficiency of Ba or Mg is found to improve ordering and quality factor in BMT. In BZN, deficiency of Ba or Zn or a small excess amount of Nb_2O_5 improve *Qf*. Recently it has been reported that Zn-deficient BZT contains a ternary hexagonal perovskite as a secondary phase which has a high *Qf* and is not expected to degrade Q of BZT [269, 270]. Wu and Davies [219] reported that the formation of small concentration (<1 mol%) of zinc vacancies on the B sites, introduced through the substitution of $Ba_3W_2O_9$ into BZN, enhanced the relative stability of the 1:2 order, accelerate the kinetics of the ordering and domain-coarsening process, and allow the *c*/*a* lattice distortion to approach its maximum value. The $Ba_3W_2O_9$ addition also lowered the sintering temperature of BZN to 1380°C and the dense ceramics have *Qf* values up to 118 500 GHz and ε_r of 39. It is found that addition of a small amount, typically 0.2–2 wt%, of dopants with ionic radii close to that of the B-site ions considerably improves the quality factor in BMT, BZT and BZN ceramics.

REFERENCES

[1] R. Roy. Multiple ion substitution in the perovskite lattice. J. Am. Ceram. Soc. 27(1954) 581–588.
[2] F. S. Galasso, L. Katz, and R. Ward. Substitution in the octahedrally coordinated cation positions in compounds of perovskite type. J. Amer. Chem. Soc. 81(1959)820–823.
[3] F. S. Galasso, J. R. Barrante, and L. Katz. Alkaline earth tantalum oxygen phases including the crystal structure of an ordered perovskite compound $Ba_3SrTa_2O_9$. J. Amer. Chem. Soc. 83(1961)2830–2832.
[4] F. S. Galasso. Structure and properties of perovskite compounds. Pergamon Press, Headington Hill, Oxford (1969).
[5] F. S. Galasso. Structure and properties of inorganic solids. Pergamon Press, Headington Hill, Oxford (1970).

[6] F. S. Galasso. Perovskite and high Tc superconductors. Gordon & Breach Science Publishers Reading, UK (1990).
[7] F. Galasso and J. Pyle. Ordering in compounds of the $A(B'_{0.33}Ta_{0.67})O_3$ type. Inorg. Chem. 23(1963)482–484.
[8] S. Kawashima, M. Nishida, I. Ueda, H. Ouchi, and S. Hayakawa. Dielectric properties of $Ba(Zn_{1/3}Nb_{2/3})O_3$–$Ba(Zn_{1/3}Ta_{2/3})O_3$ ceramics. Proc. 1st Meeting Ferroelectric Materials & Their Applications. O. Omoto and A. Kumada, (Eds), Keihin Printing Co., Ltd, Kyoto, Tokyo (1977) pp. 293–296.
[9] R. D. Shannon. Revised effective ionic radius and systematic studies of interatomic distances in halides and chalcogenides. Acta Crystallogr. A 32(1976)751–767.
[10] J. H. Paik, S. Nahm, J.-D. Byulin, M. H. Kim, and H. J. Lee. The effect of Mg deficiency on the microwave dielectric properties of $Ba(Mg_{1/3}Nb_{2/3})O_3$. J. Mater. Sci. Lett. 17(1998)1777–1780.
[11] I.-T. Kim, Y. H. Kim, and S. J. Chung. Ordering and microwave dielectric properties of $Ba(Ni_{1/3}Nb_{2/3})O_3$ ceramics. J. Mater. Res. 12(1997)518–525.
[12] H. Tamura, T. Konoike, Y. Sakabe, and K. Wakino. Improved high Q dielectric resonators with complex perovskite structure. J. Am. Ceram. Soc. 67(1984)C59–C61.
[13] H. Matsumoto, H. Tamura, and K. Wakino. $Ba(MgTa)O_3$–$BaSnO_3$ high Q dielectric resonator. Jpn. J. Appl. Phys. 30(1991)2347–2349.
[14] Y.-W. Kim, J.-H. Park, and J.-G. Park. Local cationic ordering behaviour in $Ba(Mg_{1/3}Nb_{2/3})O_3$ ceramics. J. Eur. Ceram. Soc. 24(2004)1775–1779.
[15] S. Nomura. Ceramics for microwave dielectric resonator. Ferroelectrics 49(1983)61–70.
[16] Z. Yue, F. Zhao, Y. Zhang, Z. Gui, and L. Li. Microwave dielectric properties of $Ba[(Zn_{1-x}Co_x)_{1/3}Nb_{2/3}]O_3$ ceramics. Mater. Lett. 58(2004)1830–1834.
[17] S. Kawashima, M. Nishida, I. Ueda, and H. Ouchi. $Ba(Zn_{1/3}Ta_{2/3})O_3$ ceramics with low dielectric loss at microwave frequencies. J. Am. Ceram. Soc. 66(1983)421–423.
[18] C.-W. Ahn, S. Nahm, Y.-S. Lim, W. Choi, H.-M. Park, and H.-J. Lee. Microstructure and microwave dielectric properties of $Ba(Co_{1/3}Nb_{2/3})O_3$ ceramics. Jpn. J. Appl. Phys. 41(2002) 5277–5280.
[19] F. Azough, C. Leach, and R. Freer. Effect of non-stoichiometry on the structure and microwave dielectric properties of $Ba(Co_{1/3}Nb_{2/3})O_3$ ceramics. J. Eur. Ceram. Soc. 26(2006) 2877–2884.
[20] S. Nomura and K. Kaneta. $Ba(Mn_{1/3}Ta_{2/3})O_3$ ceramic with ultra low loss at microwave frequency. Jpn. J. Appl. Phys. 33(1984)507–508.
[21] S. Liu, J. Sun, R. Taylor, D. J. Smith, and N. Newmann. Microstructure and dielectric properties of $Ba(Cd_{1/3}Ta_{2/3})O_3$ microwave ceramics synthesized with boron oxide sintering aid. J. Mater. Res. 19(2004)3526–3533.
[22] S. Liu, R. Taylor, N. S. Petrovic, L. Budd, M.V. Schilfgaarde, and N. Newmann. Experimental and theoretical investigation of the structural, chemical, electronic and high frequency dielectric properties of barium cadmium tantalate based ceramics. J. Appl. Phys. 97(2005) 014105.
[23] H. Badshaw, D. Iddles, R. Quimby, and I. M. Reaney. Structure property relations in $xCaTiO_3$–$(1-x)Sr(Mg_{1/3}Nb_{2/3})O_3$ based microwave dielectrics. J. Eur. Ceram. Soc. 23(2003)2435–2441.
[24] K. Fukuda, R. Kitoh, and I. Awai. Far infrared reflection spectra of dielectric ceramics for microwave applications. J. Am. Ceram. Soc. 77(1994)149–154.
[25] H. Kagata and J. Kato. Dielectric properties of Ca based complex perovskite at microwave frequencies. Jpn. J. Appl. Phys. 33(1994)5463–5465.
[26] C.-L. Huang, R.-Y. Yang, and M.-H. Weng. Dielectric properties of $CaTiO_3$–$Ca(Mg_{1/3}Nb_{2/3})O_3$ ceramic system at microwave frequency. Jpn. J. Appl. Phys. 39(2000)6608–6611.
[27] J. Kato, H. Kagata, and K. Nishimoto. Dielectric properties of $(Pb,Ca)(Me,Nb)O_3$ at microwave frequencies. Jpn. J. Appl. Phys. 31(1992)3144–3147.
[28] J. W. Choi, J.-Y. Ha, C.-Y. Kang, S. J. Yoon, H.-J. Kim, and K. H. Yoon. Microwave dielectric properties of $(Pb,Ca)(Mg,Nb,Sn)O_3$ ceramics. Jpn. J. Appl. Phys. 39(2000) 5923–5926.

[29] H. Kagata, J. Kato, K. Nishimoto, and T. Inoue. Dielectric properties of Pb based perovskite substituted by Ti for B site at microwave frequencies. Jpn. J. Appl. Phys. 32(1993)4332–4334.

[30] E. S. Kim, Y. H. Kim, J. H. Chae, D. W. Kim, and K. H. Yoon. Dielectric properties of $[Pb_{0.2}Ca_{0.8}]$ $(Ca_{1/3}Nb_{2/3})O_3$ ceramics at microwave frequencies. Mater. Chem. Phys. 79(2003)230–232.

[31] I. Melodetsky and P. K. Davies. Effect of $Ba(Y_{1/2}Nb_{1/2})O_3$ and $BaZrO_3$ on the cation order and properties of $Ba(Co_{1/3}Nb_{2/3})O_3$ microwave ceramics. J. Eur. Ceram. Soc. 21(2001) 2587–2591.

[32] K. Tochi, S. Nakamura, T. Ohgaku, and N. Takeuchi. Lattice vibration modes of $(1-x)Ba(Zn_{1/3}Ta_{2/3})O_3-xBaZrO_3$. J. Mater. Sci. Lett. 8(1989)1144–1146.

[33] D. A. Sagala and S. Nambu. Lattice energy calculation for ordered and disordered $Ba(Zn_{1/2}Ta_{2/3})O_3$. J. Phys. Soc. Jpn. 61(1992)1791–1797.

[34] I. Qazi, I. M. Reaney, and W. E. Lee. Order–disorder phase transition in $Ba(Zn_{1/3}Ta_{2/3})O_3$. J. Eur. Ceram. Soc. 21(2001)2613–2616.

[35] R. D. Rosenstein and R. Schor. Superlattice madelung energy of idealized ordered perovskites. J. Chem. Phys. 38(1963)1789–1790.

[36] D. A. Sagala and S. Nambu. Microscopic calculation of dielectric loss at microwave frequencies for complex perovskite $Ba(Zn_{1/3}Ta_{2/3})O_3$. J. Am. Ceram. Soc. 75(1992)2573–2575.

[37] P. K. Davies, J. Tong, and T. Negas. Effect of ordering induced domain boundaries of low loss $Ba(Zn_{1/3}Ta_{2/3})O_3-BaZrO_3$ perovskite microwave dielectrics. J. Am. Ceram. Soc. 80(1997)1727–1740.

[38] S. Kawashima. Influence of ZnO evaporation on microwave dielectric loss and sinterability of $Ba(Zn_{1/3}Ta_{2/3})O_3$ ceramics. Amer. Ceram. Soc. Bull. 72(1993)120–126.

[39] S. B. Desu and H. M. O Bryan. Microwave loss quality of $Ba(Zn_{1/3}Ta_{2/3})O_3$ ceramics. J. Am. Ceram. Soc. 68(1985)546–551.

[40] I. M. Reaney, P. L. Wise, I. Qazi, C. A. Miller, T. J. Price, D. S. Cannell, D. M. Idles, M. J. Rosienky, S. M. Moussa, M. Bieringer, L. D. Noailles, and R. M. Ibberson. Ordering and quality factor in $0.95Ba(Zn_{1/3}Ta_{2/3})O_3-0.05Sr(Ga_{1/2}Ta_{1/2})O_3$ production resonators. J. Eur. Ceram. Soc. 23(2003)3021–3034.

[41] P. K. Davies. Influence of structural defects on the dielectric properties of ceramic microwave resonators. In Materials and processes for wireless communication. T. Negas and H. Ling. (Eds.), Ceramic. Transaction, Amercan Ceramic Society, Westerville, USA (1995), Vol. 53, pp. 183–197.

[42] J. S. Kim, J.-W. Kim, C. I. Cheon, Y.-S. Kim, S. Nahm, and J. D. Byun. Effect of chemical element doping and sintering atmosphere on the microwave dielectric properties of barium zinc tantalates. J. Eur. Ceram. Soc. 21(2001)2599–2604.

[43] I. Mc Laren, S. Wannkukorale, and C. B. Ponton. Hydrothermal synthesis and characterization of perovskite $Ba(Zn_{1/3}Ta_{2/3})O_3$. J. Mater. Chem. 9(1999)2663–2670.

[44] V. Tolmer and G. Desgardin. Low temperature sintering and influence of the process on the dielectric properties of $Ba(Zn_{1/3}Ta_{2/3})O_3$. J. Am. Ceram. Soc. 80(1997)1981–1991.

[45] F. Roulland, R. Terras, and S. Marinel. Influence of both milling conditions and lithium salt addition on the sinterability of $Ba(Zn_{1/3}Ta_{2/3})O_3$. Mater. Sci. Eng. B 104(2003)156–162.

[46] W.-C. Tzou, C.-C. Chan, P.-S. Cheng, and C.-F. Yang. Effect of $BaTi_4O_9$ on the sintering and microwave dielectric characteristics of $Ba(Zn_{1/3}Ta_{2/3})O_3$ ceramics. J. Mater. Sci. 40(2005)4711–4714.

[47] S. Nomura, K. Toyoma, and K. Kaneta. $Ba(Mg_{1/3}Ta_{2/3})O_3$ ceramics with temperature stable high dielectric constant and low microwave loss. Jpn. J. Appl. Phys. 21(1982)L624–L626.

[48] F. Roulland and S. Marinel. Effects of glass phase additives and stoichiometry on the $Ba(Zn_{1/3}X_{2/3})O_3$ (X=Ta or Nb) sinterability and dielectric properties. J. Electroceram. 14(2005) 239–246.

[49] C.-L. Huang, R.-J. Lin, and J.-F. Tzeng. Dielectric properties of copper oxide doped $0.95Ba(Zn_{1/3}Ta_{2/3})O_3-0.05BaZrO_3$ ceramics at microwave frequency. Mater. Chem. Phys. 97(2006)256–260.

[50] M.-H. Kim, S. Nahm, W.-S. Lee, M.-J. Yoo, N.-K. Kang, and H.-T. Kim. Effect of B_2O_3 and CuO on the sintering temperature and microwave dielectric properties of $Ba(Zn_{1/3}Ta_{2/3})O_3$ ceramics. Jpn. J. Appl. Phys. 44(2005)3091–3094.
[51] O. Renoult, J. P. Boilot, F. Chaput, R. Papiernik, L.G. Hubert-Pfalzgraf, and M. Lejeune. Sol–gel powders for microwave dielectric resonators. Ceramics today-Tomorrow ceramics. P. Vincezini. (Ed.), Elsevier Publishers B.V. (1991)1991–1997.
[52] M. R. Varma, S. Biju, and M. T. Sebastian. Preparation of phase pure $Ba(Zn_{1/3}Ta_{2/3})O_3$ nano powders for microwave dielectric resonator application. J. Eur. Ceram. Soc. 26(2006) 1903–1908.
[53] M. Thirumal and A. K. Ganguli. Studies on dielectric oxide materials containing niobium and tantalum. Progr. Cryst. Growth Mater. Charact. 44(2002)147–154.
[54] E. Koga, Y. Yamagishi, H. Moriake, K. Kakimoto, and H. Ohsato. Large Q factor variation with dense, highly ordered $Ba(Zn_{1/3}Ta_{2/3})O_3$ system. J. Eur. Ceram. Soc. 26(2006)1961–1964.
[55] B.-K. Kim, H. Hamaguchi, I.-T. Kim, and K.S. Hong. Probing of 1:2 ordering in $Ba(Mg_{1/3}Nb_{2/3})O_3$ and $Ba(Zn_{1/3}Nb_{2/3})O_3$ ceramics by XRD and Raman spectroscopy. J. Am. Ceram. Soc. 78(1995)3117–3120.
[56] B. M. Jacobson, B. E. Collins, and F. Fender. A powder neutron and X-ray diffraction determination of the structure of $Ba_3Ta_2ZnO_9$: An investigation of perovskite phases in the system Ba–Ta–Zn–O and the preparation of $Ba_2TaCdO_{5.5}$ and $Ba_2CeInO_{5.5}$. Acta Crystallogr. B 32(1976)1083–1087.
[57] L. Bellaichead and D. Vanderbit. Electrostatic model of atomic ordering in complex perovskite alloys. Phys. Rev. Lett. 81(1998)1318–1321.
[58] L. Bellaiche, J. Padilla, and D. Van Derbit. Heterovalent and A atom effects in $A(B'B'')O_3$ perovskite alloys. Phys. Rev. B 59(1999)1834–1839.
[59] M. Lufasso. Crystal structures, modeling and dielectric property relationship of 2:1 ordered $Ba_3MM'_2O_9$ (M=Mg,Zn, Ni; M′=Nb,Ta) perovskites. Chem. Mater. 16(2004)2148–2156.
[60] F. S. Galasso and J. Pyle. Preparation and study of ordering in $A(B_{0.33}Nb_{0.67})O_3$ perovskite type. J. Phys. Chem. 67(1963)1561–1562.
[61] K. Wakino, T. Nishigaki, S. Tamura, and H. Tamura. An X-band GaAs FET oscillator using a dielectric resonator. Proc. 37th Annual Freq Contl. Symp. Sterling Va (1984).
[62] I. M. Reaney, I. Qazi, and W. E. Lee. Order–disorder behaviour in $Ba(Zn_{1/3}Ta_{2/3})O_3$. J. Appl. Phys. 88(2000)6708–6714.
[63] G. A. Smolensky, I. G. Siny, R. V. Pisarev, and E. G. Kuzrninov. Raman Scattering in ordered and disordered perovskite type crystals. Ferroelectrics. 12(1976)135–136.
[64] D. J Barber, K. M. Moulding, J. Zhou, and M. Li. Structural order in $Ba(Zn_{1/3}Ta_{2/3})O_3$, $Ba(Zn_{1/3}Nb_{2/3})O_3$ $Ba(Mg_{1/3}Ta_{2/3})O_3$ microwave dielectric ceramics. J. Mater. Sci. 32(1997)1531–1544.
[65] C. J. Lee, G. Pezzotti, S. H. Kang, D. J. Kim, and K. S. Hong. Quantitative analysis of lattice distortion in $Ba(Zn_{1/3}Ta_{2/3})O_3$ microwave dielectric ceramics with added B_2O_3 using Raman spectroscopy. J. Eur. Ceram. Soc. 26(2006)1385–1391.
[66] I. M. Reaney, E Colla, and N. Setter. Dielectric and structural characterisation of Ba and Sr based complex perovskites as a function of tolerance factor. Jpn. J. Appl. Phys. 33(1994) 3984–3990.
[67] M. Bieringer, S. M. Moussa, L. D. Noailles, A. Burrows, C. J. Kiely, M. J. Rossiensky, and R. M. Ibberson. Cation ordering, domain growth and zinc loss in the microwave dielectric oxide $Ba_3ZnTa_2O_{9-\delta}$. Chem. Mater. 15(2003)586–597.
[68] S. Moussa, R.M. Ibberson, M. Bieringer, A. N. Fitsch, and M.J. Rossiensky. In situ measurements of cation order and domain growth in an electroceramic. Chem. Mater. 15(2003)2527–33.
[69] R. M. Ibberson, S. M. Moussa, M. J. Rossiensky, A. N. Fitch, D. Iddles, and T. Price. In situ neutron and X-ray diffraction study of cation ordering and domain growth in the dielectric ceramic $Ba_3ZnTa_2O_9$–Sr_2GaTaO_6. J. Am. Ceram. Soc. 89(2006)1827–1833.

[70] P. Woodward, R. D. Hoffman, and A. W. Sleight. Order–disorder in $A^{2+}M^{3+}M^{5+}$ perovskites. J. Mater. Res. 9(1994)2118–2127.
[71] I. M. Reaney, J. Petzelt, V.V. Voisckhovska, E. Chu, and N. Setter. B site order and infra red reflectivity in $A(B'B'')O_3$ complex perovskite ceramics. J. Appl. Phys. 76(1994)2086–2092.
[72] N. Setter and I. Laulicht. The observation of B site ordering by Raman scattering in A(B′B ″) O_3. Appl. Spectroscop. 41(1987)526–528.
[73] E. Husson, I. Abello, and A. Morell. Short range order in $PbMg_{1/3}Nb_{2/3}O_3$ ceramics by Raman spectroscopy. Mater. Res. Bull. 25(1990)539–548.
[74] I. G. Siny and T.A. Smirnova. Preceding paraphrases in diffuse transition. Ferroelectrics. 90(1989)191–194.
[75] K. Kageyama and J. Takahashi. Microstructures and microwave dielectric properties of $(Ba,Sr)(Ga_{1/2}Ta_{1/2})O_3$ (in Japanese). Jpn. Soc. Powder Metall. 38(1991)745–748.
[76] K. Matsumoto, T. Hiuga, K. Takada, and H. Ichimura. $Ba(Mg_{1/3}Ta_{2/3})O_3$ ceramics with ultralow loss at microwave frequencies. Proc. 6th IEEE Intl. Symp. On Applications of Ferroelectrics IEEE (NY) (1986), p. 118.
[77] I. T. Kim, T. S.-Oh, and Y.-H. Kim. Lattice distortion of $Ba(Zn_{1/3}Ta_{2/3})O_3$ with ordering of B site cations. J. Mater. Sci. Lett. 12(1993)182.
[78] K. Wakino, M. Murata, and H. Tamura. Far infrared reflection spectra of Ba(Zn,Ta) O_3–$BaZrO_3$ dielectric resonator material. J. Am. Ceram. Soc. 69(1986)34–37.
[79] K. Kageyama. Crystal structure and microwave dielectric properties of $Ba(Zn_{1/3}Ta_{2/3})$ O_3–$(Sr,Ba)(Ga_{1/2}Ta_{1/2})O_3$ ceramics. J. Am. Ceram. Soc. 75(1992)1767–1771.
[80] G. Rong, N. Newman, B. Shaw, and D. Cronin. Role of Ni and Zr doping on the electrical optical magnetic and structural properties of barium zinc tantalate ceramics. J. Mater. Res. 14(1999)4011–4019.
[81] V. D. Parvanova, and S. K. Andreev. Microwave properties of the systems $Ba_{1-x}Sr_x(Zn_{1/3}Ta_{2/3})_{0.94}Ti_{0.06}O_3$ and $Ba(Zn_{1/3}Nb_{2/3})_{1-x}Zr_xO_3$. J. Mater. Sci.: Mater Electron. 16(2005)249–252.
[82] M.-H. Kim, S. Nahm, W.-S. Lee, M.-J. Yoo, J.-C. Park, and H.-J. Lee. Effect of microstructure on microwave dielectric properties of Al_2O_3 added $Ba(Zn_{1/3}Ta_{2/3})O_3$ ceramics. Jpn. J. Appl. Phys. 43(2004)1438–1441.
[83] H.-K. Kim, B.-J. Kim, S. Nahm, W.-S. Lee, M.-J. Yoo, N.-K. Kang, H.-J. Lee, Y.-S. Kim, and S.-Y. Ryon. Microstructure and microwave dielectric properties of SnO_2-added $Ba(Zn_{1/3}Ta_{2/3})O_3$ ceramics. Jpn. J. Appl. Phys. 43(2004)4259–4262.
[84] M.-H. Kim, S. Nahm, W.-S Lee, M.-J. Yoo, N.-G. Gang, and H.-J. Lee. Structural variation and microwave dielectric properties of TiO_2-added $Ba(Zn_{1/3}Ta_{2/3})O_3$. J. Eur. Ceram. Soc. 24 (2004)3547–3552.
[85] T. Takada, K. Kageyama, M. Yonemura, N. Hara, and S. Tako. Microwave dielectric properties of mixed phase ceramics, $Ba(Zn_{1/3}Ta_{2/3})O_3$–$x$$CaTiO_3$ and $x$$MgTiO_3$–$y$$CaTiO_3$–$z$ $(Nd_2O_3$–$wTiO_2)$. J. Mater. Sci.: Mater. Elctron. 14(2005)205–214.
[86] J.-I. Yang, S. Nahm, C.-H. Choi, H.-J. Lee, and H.-M. Park. Microstructure and microwave dielectric properties of $Ba(Zn_{1/3}Ta_{2/3})O_3$ with ZrO_2 addition. J. Am. Ceram. Soc. 85(2002)165–168.
[87] J.-I. Yang, S. Nahm, C.-H. Choi, H.-J. Lee, J.-C. Kim, and H.-M. Park. Effect of Ga_2O_3 on the microstructure and microwave dielectric properties of $Ba(Zn_{1/3}Ta_{2/3})O_3$ ceramics. Jpn. J. Appl. Phys. 41(2002)702–706.
[88] M. R. Varma, R. Reghunathan, and M. T. Sebastian. Effect of dopants on the microwave dielectric properties of $Ba(Zn_{1/3}Ta_{2/3})O_3$ ceramics. Jpn. J. Appl. Phys. 44(2005)298–303.
[89] Y.-H. Jeong, M.-H. Kim, S. Nahm, W.-S. Lee, M.-J. Yoo, N.-K. Kang, and H.-J. Lee. Effect of Ta_2O_5 on microstructure and microwave dielectric properties of $Ba(Zn_{1/3}Ta_{2/3})O_3$ ceramics. Jpn. J. Appl. Phys. 44(2005)956–960.
[90] J. I. Yang, S. Nahm, S.-J. Yoon, H.-M. Park, and H-J. Lee. Structural variation and microwave dielectric properties of ZrO_2 added $Ba(Zn_{1/3}Ta_{2/3})O_3$ ceramics. Jpn. J. Appl. Phys. 43(2004)211–214.

[91] M.-H Kim, B.-J. Kim, S. Nahm, W.-S Lee, M.-J Yoo, N.-K. Kang, H.-J. Lee, Y.-S. Kim, and S.-Y. Ryon. Microstructure and microwave dielectric properties of SnO_2 added $Ba(Zn_{1/3}Ta_{2/3})O_3$ ceramics. Jpn. J. Appl. Phys. 43(2004)4259–4262.

[92] C.-C. Lee, C.-C. Chou, and D.-S. Tsai. Effect of La/K A site substitution on the ordering of $Ba(Zn_{1/3}Ta_{2/3})O_3$. J. Am. Ceram. Soc. 80(1997)2885–2890.

[93] C.-C. Lee, C.-C. Chou, and D.-S. Tsai. Variation in the ordering of $Ba(Zn_{1/3}Ta_{2/3})O_3$ with A site substitutions. Ferroelecrics. 206–207(1998)293–305.

[94] Z. L. Zhao, S. L. Lai, and A. S. Liu. $Ba(Zn_{1/3}Ta_{2/3})O_3$–$Ba(Cd_{1/3}Nb_{2/3})O_3$ dielectric ceramics with low microwave loss. Electron. Lett. 26(1990)1605–1606.

[95] E. L. Colla, I. M. Reaney, and N. Setter. Effect of structural changes in complex perovskites on the temperature coefficient of the relative permittivity. J. Appl. Phys. 74(1993)3414–3425.

[96] L. Chai and P. K. Davies. Formation and structural characterization of 1:1 ordered perovskites in the $Ba(Zn_{1/3}Ta_{2/3})O_3$–$BaZrO_3$ system. J. Am. Ceram. Soc. 80(1997)3193–3198.

[97] H. Tamura. Microwave dielectric losses caused by lattice defects. J. Eur. Ceram. Soc. 26(2006)1775–1780.

[98] H. Tamura, D. A. Sagala, and K. Wakino. Lattice vibrations of $Ba(Zn_{1/3}Ta_{2/3})O_3$ crystal with ordered perovskite structure. Jpn. J. Appl. Phys. 25(1986)787–791.

[99] S. J. Webb, J. Breeze, R. I. Scott, D. S. Cannell, D. M. Iddles, and N. Mc N. Alford. Raman spectroscopic study of gallium doped $Ba(Zn_{1/3}Ta_{2/3})O_3$. J. Am. Ceram. Soc. 85(2002)1753–1756.

[100] R. Guo, A. S. Bhalla, and L. E. Cross. $Ba(Mg_{1/3}Ta_{2/3})O_3$ single crystal fibre grown by laser heated pedestal growth technique. J. Appl. Phys. 75(1994)4704–4708.

[101] M. Sugiyama and T. Nagai. Anomaly of dielectric constant $Ba_{1-x}Sr_x(Mg_{1/3}Ta_{2/3})O_3$ solid solution and its relation to structural change. Jpn. J. Appl. Phys. 32(1993)4360.

[102] K. P. Surendran. Investigations on low loss dielectric ceramic materials for wireless communication. PhD Thesis, Kerala University (2005).

[103] Y. Fang, A. Hu, S. Ouyang, and J. J. Oh. The effect of calcination on the microwave dielectric properties of $Ba(Mg_{1/3}Ta_{2/3})O_3$. J. Eur. Ceram. Soc. 21(2001)2745–2750.

[104] H.-J. Youn, K.-Y. Kim, and H. Kim. Microstructural characteristics of $Ba(Mg_{1/3}Ta_{2/3})$ O_3 ceramics and its related microwave dielectric properties. Jpn. J. Appl. Phys 35(1996) 3947–3953.

[105] K. Kakegawa, T. Wakabayashi, and Y. Sasaki. Preparation of $Ba(Mg_{1/3}Ta_{2/3})O_3$ using oxine. J. Am. Ceram. Soc. 69(1986)C82–C84.

[106] N. Ichinose and T. Shimada. Effect of grain size and secondary phase on microwave dielectric properties of $Ba(Mg_{1/3}Ta_{2/3})O_3$ and $Ba([Mg,Zn]_{1/3}Ta_{2/3})O_3$ systems. J. Eur. Ceram. Soc. 26(2006)1775–1760.

[107] W. A. Lan, M.-H. Liang, C.-T. Hu, K.-S. Liu, and I.-N Lin. Influence of Zr doping on the microstructure and microwave dielectric properties of $Ba(Mg_{1/3}Ta_{2/3})O_3$ materials. Mater. Chem. Phys. 79(2003)266–269.

[108] M. Li. Synthesis of $Ba(Mg_{1/3}Ta_{2/3})O_3$ with complex perovskite type structure. Ferroelectrics 195(1997)87–91.

[109] X. M. Chen and Y. J. Wu. A low temperature approach to synthesize pure complex perovskite $Ba(Mg_{1/3}Ta_{2/3})O_3$ powders. Mater. Lett 26(1996)237–239.

[110] K. H. Yoon, J.J. Yoon, W.S. Kim, J. B. Kim, and S. Kim. Effect of BaO/WO_3 addition on the microwave loss quality of $Ba(Mg_{1/3}Ta_{2/3})O_3$. Jpn. J. Appl. Phys. 38(1999) 5616–5620.

[111] X. M. Chen and Y. J. Wu. Effects of NaF upon sintering temperature of $Ba(Mg_{1/3}Ta_{2/3})O_3$ dielectric ceramics. J. Mater. Sci.: Mater. Electron. 7(1996)427–431.

[112] X. M. Chen and Y. J. Wu. Dielectric characteristics of $Ba(Mg_{1/3}Ta_{2/3})O_3$ ceramics sintered at low temperatures. J. Mater. Sci.: Mater. Electron. 7(1996)369–371.

[113] C.-M. Cheng, Y.-T. Hsieh, and C.-F. Yang. The sintering and microwave dielectric characteristics of MgO–CaO–Al_2O_3–SiO_2 glass fluxed $Ba(Mg_{1/3}Ta_{2/3})O_3$ ceramics. Mater. Lett. 57(2003)1471–1476.

[114] C.-L. Huang, K.-H. Chiang, and S.-C. Chuang. Influence of V_2O_5 additions to $Ba(Mg_{1/3}Ta_{2/3})O_3$ ceramics on sintering behaviour and microwave dielectric properties. Mater. Res. Bull. 39(2004)629–636.

[115] K. P. Surendran, M. T. Sebastian, P. Mohanan, and M. V. Jacob. The effects of dopants on the microwave dielectric properties of $Ba(Mg_{0.33}Ta_{0.67})O_3$ ceramics. J. Appl. Phys. 98(2005)094114.

[116] C.-C. Cheng, Y.-T. Hsieh, and C.-F. Yang. Effects of glass doping for the sinterability of $Ba(Mg_{1/3}Ta_{2/3})O_3$ ceramics. Ceram. Int. 28(2002)255–260.

[117] K. P. Surendran, P. Mohanan, and M. T. Sebastian. Effect of glass additives on the microwave dielectric properties of $Ba(Mg_{1/3}Ta_{2/3})O_3$. J. Solid State Chem. 177(2004) 4031–4046.

[118] M. Scholler and W. Wersing. Sol-gel synthesis of advanced microwave ceramics and their dielectric properties. In: European Mater. Res. Soc. Monographs Euro-gel 91. Vilminot I.S., R. Noss and H. Schimdt. Elsevier Science Publishers, B.V. (1992) pp. 255–263.

[119] M.-H. Liang, S.-Y. Wu, C.-T. Hu, and I.-N. Lin. Enhancing the sinterability of $Ba(Mg_{1/3}Ta_{2/3})O_3$ dielectrics by using chemically derived powders. Mater. Chem. Phys. 79(2003)276–278.

[120] Y. C. Lee, M. H. Liang, C.T. Hu, and I. N. Lin. Microwave dielectric properties of $Ba(Mg_{1/3}Ta_2/3)O_3$ materials synthesized by inverse microemulsion process. J. Eur. Ceram. Soc. 21(2001)2755–2758.

[121] M. Scholler and W. Wersing. Sol–gel processing of advanced dielectric ceramics for microwave applications. Mater. Res. Soc. Symp. Proc. 155(1989)45–52.

[122] O. Renoult, J. P. Boilot, F. Chaput, R. Pappienik, L. G. Hubert-Pfalzgraf, and M. Lejeune. Sol–gel processing and microwave characterization of $Ba(Mg_{1/3}Ta_{2/3})O_3$ dielectrics. J. Am. Ceram. Soc. 75(1992)3337–3340.

[123] D. Ravichandran, B. Jim, R. Roy, and A. S. Bhalla. Raman spectroscopy of sol-gel derived $Ba(Mg_{1/3}Ta_{2/3})O_3$ perovskites. Mater. Lett. 25(1995)257–259.

[124] J. Zhou, Q.-X Su, K. M. Moulding, and D. J. Barber. Sol–gel derived $Ba(Mg_{1/3}Ta_{2/3})O_3$ thin films: preparation and structure. J. Mater. Res. 12(1997)596–599.

[125] S. Katayama and M. Sekine. Low temperature synthesis of $Ba(Mg_{1/3}Ta_{2/3})O_3$ perovskite by the metal alkoxide method. J. Mater. Chem. 2(1998)889–890.

[126] S. Katayama, I. Yoshinaga, N. Yamada, and T. Nagai. Low temperature synthesis of $Ba(Mg_{1/3}Ta_{2/3})O_3$ ceramics from Ba–Mg–Ta alkoxide precursor. J. Am. Ceram. Soc. 79(1996)2059–2064.

[127] C.-H. Lu and C.-C. Tsai. A new sol-gel route using inorganic salts for synthesizing barium magnesium tantalate ceramics. Mater. Lett. 31(1997)271–274.

[128] C.-H. Lu and C. C. Tsai. Reaction kinetics, sintering characteristics and ordering behaviour of microwave dielectrics: barium magnesium tantalates. J. Mater. Res. 11(1996)1219–1227.

[129] K. P. Surendran, M. T. Sebastian, P. Mohanan, R. L. Moreira, and A. Dias. Effect of non-stoichiometry on the structure and microwave dielectric properties of $Ba(Mg_{0.33}Ta_{0.67})O_3$. Chem. Mater. 17(2005)141–151.

[130] M.-H. Liang, C.-T. Hu, H.-F. Cheng, I.-N. Lin, and J. Steeds. Effects of sintering process on microstructure characteristics of $Ba(Mg_{1/3}Ta_{2/3})O_3$ ceramics and their microwave dielectric properties. J. Eur. Ceram. Soc. 21(2001)2759–2763.

[131] T. V. Kolodiazhnyi, A. Petrice, G. P. Johari, and A. G. Belous. Effect of preparation conditions on cation ordering and dielectric properties of $Ba(Mg_{1/3}Ta_{2/3})O_3$ ceramics. J. Eur. Ceram. Soc. 22(2002)2013–2014.

[132] S. H. Ra and P. P. Phule. Processing and microwave dielectric properties of barium magnesium tantalates ceramics for high quality personnel communication service filters. J. Mater. Res. 14(1999)4259–4265.

[133] M. H. Liang, C.-T. Hu, C.-G. Chiou, Y.-N. Tsai, and I.-N. Lin. Effect of $Ba_5Ta_4O_{15}$ incorporation on sintering behaviour and microwave dielectric properties of $Ba(Mg_{1/3}Ta_{2/3})$ O_3 materials. Jpn. J. Appl. Phys. 38(1999)5621–5624.

[134] C.-H. Lu and C.-C. Tsai. Homologous precipitation synthesis and sintering behaviour of microwave dielectric $Ba(Mg_{1/3}Ta_{2/3})O_3$. Mater. Sci. Eng. B 55(1998)95–101.
[135] I. McLaren and C. B. Ponton. Low temperature hydrothermal synthesis of $Ba(Mg_{1/3}Ta_{2/3})O_3$ sol–gel derived powders. J. Mater. Sci. 33(1998)17–22.
[136] R. Abothu, S. Komarneni, D.-S. Paik, and H. G. Jentsch. Nanocomposite versus sol-gel processing of barium magnesium tantalate. J. Electroceram. 31(1999)65–71.
[137] L. Xu, F. Lian, J. Yang, and, J. M.F. Ferreira. A novel wet chemical process to synthesize $Ba(Mg_{1/3}Ta_{2/3})O_3$ nanopowders. Ceram. Int. 38(2002)542–552.
[138] W. Wang, S. Wu, P. Sun, and Y. Zhang. Solution liquid phase coating processing and microwave characteristics of $Ba(Mg_{1/3}Ta_{2/3})O_3$ dielectrics. Mater. Sci. Eng. B 116(2005)91–94.
[139] T. R. Tsai, M.-H. Liang, C. T. Hu, C. C. Chi, and I.-N. Lin. Terahertz spectroscopic technique for characterizing the microwave dielectric properties of $Ba(Mg_{1/3}Ta_{2/3})O_3$ materials. J. Eur. Ceram. Soc. 21(2001)2787–2790.
[140] F. S. Galasso and J. Pinto. Growth of single crystals of $Ba(B'_{0.33}Ta_{0.67})O_3$ perovskite type compounds. Nature. 207(1965)70–72.
[141] H. Vincent, C. Perrier, P. l'heritier, and M. Labeyrie. Crystallographic study by Rietveld method of barium magnesium tantalate oxide based ceramics for use as dielectric resonator. Mater. Res. Bull. 28(1993)951–958.
[142] S. Janaswamy, G. S. Murthy, E. D. Dias, and V. R. K. Murthy. Structural analysis of $Ba(Mg_{1/3}Ta_{2/3})O_3$ ceramics. Mater. Lett. 55(2002)414–419.
[143] M. Furuya. Microwave dielectric properties and phonon characteristics of $Ba(Mg_{1/3}Ta_{2/3})O_3$–$A(Mg_{1/2}W_{1/2})O_3$ (ABa, Ca, Sr) ceramics. J. Kor. Phys. Soc. 32(1998)S353–S357.
[144] D. A. Sagala and S. Kayasu. Infrared reflection of $Ba(Mg_{1/3}Ta_{2/3})O_3$ ceramics. J. Am. Ceram. Soc. 76(1993)2433–2436.
[145] Y.-K. Im, K.-M. Lee, and H. M. Jang. B site cation ordering and associated defect processes in WO_3 doped $Ba(Mg_{1/3}Ta_{2/3})O_3$. J. Kor. Phys. Soc. 32(1988)S292–S295.
[146] E.S. Kim and K. H. Yoon. Microwave dielectric properties of complex perovskite $Ba(Mg_{1/3}Ta_{2/3})O_3$. Ferroelectrics. 133(1992)187–192.
[147] K. Tochi, S. Nakamura, T. Ohgaku, and N. Takeuchi. Lattice vibration modes of $(1-x)Ba(Zn_{1/3}Ta_{2/3})O_3$–$xBaZrO_3$. J. Mater. Sci. Lett. 8(1989)1144–1146.
[148] M. Sugiyama, T. Inuzuka, and H. Kubo. Effects of processing on microstructure and dielectric properties in $Ba(Mg_{1/3}Ta_{2/3})O_3$ ceramics. Ceram. Trans. 15(1990)153–166.
[149] C. H. Lei, S. Amelinckx, and G. Van Tendeloo. Disordered $Ba(Mg_{1/3}Ta_{2/3})O_3$ and its ordering transition. Phil. Mag. A 82(2002)2321–2332.
[150] L. Chai, M. A. Akbas, P. K. Davies, and J. B. Praise. Cation ordering transformations in $Ba(Mg_{1/3}Ta_{2/3})O_3$–$BaZrO_3$ perovskite solid solutions. Mater. Res. Bull. 32(1997) 1261–1269.
[151] L. Chai and P. K. Davies. Effect of M^{4+} [Ce,Sn,Ti] B-site substitutions on the cation ordering in $Ba(Mg_{1/3}Ta_{2/3})O_3$. Mater. Res. Bull. 33(1998)1283–1292.
[152] B. P. Burton. Empirical cluster expansion models of cation order–diosorder in $A(B'_{1/3}B''_{2/3})O_3$ perovskites. Phys. Rev. B 59(1999)6087–6091.
[153] B. P. Burton. Why $Pb(BB'')O_3$ perovskite disorder at a lower temperature than $Ba(B,B')O_3$ perovskites. Phys. Rev. B 60(1999)12542–12545.
[154] T. Takahashi, E. J. Wu, A. Van Der Ven, and G. Ceder. First principle investigation of B site ordering in $Ba(Mg_xTa_{1-x})O_3$ microwave dielectrics with complex perovskite structure. Jpn. J. Appl. Phys. 39(2000)1241–1248.
[155] T. Takahashi. First principles investigation of the phase stability for $Ba(B'_{1/3}B''_{2/3})O_3$ microwave dielectrics with the complex perovskite structure. Jpn. J. Appl. Phys. 39(2000)5637–5641.
[156] T. Takahashi, E. J. Wu, and G. Ceder. First principle estimate of the order-disorder transition in $Ba(Zn_xNb_{1-x})O_3$ ceramics by XRD and Raman spectroscopy. J. Mater. Res. 15(2000)2061–2064.

[157] Z.-R. Liu, J.-S. Liu, B.-L. Gu, and X.-W. Zhang. Cluster variation method studies on the atomic ordering in complex perovskite alloys. Phys. Rev. B 61(2000)11918–11921.
[158] A. Dias, V. S. T. Ciminelli, F. M. Matinaga, and R. L. Moreira. Raman scattering and X-ray diffraction investigations on hydrothermal barium magnesium niobate ceramics. J. Eur. Ceram. Soc. 21(2001)2739–2744.
[159] G. Siny, R. S. Katiyar, and A. S. Bhalla. Cation arrangement in the complex perovskite and vibrational spectra. J. Raman Spectrosc. 29(1998)385–390.
[160] G. Siny, R. Tao, R. S. Katiyar, R. Guo, and A. S. Bhalla. Raman spectroscopy of Mg-Ta-order-disorder in $Ba(Mg_{1/3}Ta_{2/3})O_3$. J. Phys. Chem. Solids. 99(1998)181.
[161] I.-N. Lin, C.-T. Chia, H.-L. Liu, Y.-C. Chen, and H.-F. Cheng. High frequency dielectric properties of $Ba(Mg_{1/3}Ta_{2/3})O_3$ complex perovskite ceramics. J. Eur. Ceram. Soc. 23(2003)2633–2637.
[162] R. L. Moreira, F. M. Matinaga, and A. Dias. Raman spectroscopic evaluation of the long range order in $Ba(B'_{1/3}B''_{2/3})O_3$ ceramics. Appl. Phys. Lett. 78(2001)428–430.
[163] K. P. Surendran and M. T. Sebastian. Tailoring the microwave dielectric properties of $Ba(Mg_{1/3}Ta_{2/3})O_3$ ceramics. J. Eur. Ceram. Soc. 26(2006)1791–1800.
[164] R. Ratheesh, M. T. Sebastian, P. Mohanan, M. E. Tobar, J. Hartett, and D. G. Blair. Whispering Gallery mode microwave characterization of $Ba(Mg_{1/3}Ta_{2/3})O_3$ dielectric resonators. J. Phys. D 32(1999) 2821–2826.
[165] T. Kolodiazhnyi, G. Annino, A. Younker, P. Malysz, P. Maschen, and H. Haneda. Probing point defects in $Ba(B'_{1/3}B''_{2/3})O_3$ by ESR, PAS and dielectric spectroscopy. J. Eur. Ceram. Soc. 26(2006)1921–1924.
[166] M. Sugiyama, T. Inuzuka, and H. Kubo. Ceram. Transaction. K. M. Nair, R. Pohanka, and R.C. Buchanam, (Eds), American Ceramic Society, Westerville, USA (1990), Vol. 15, pp. 153–166.
[167] T. Kolodiazhnyi, A. Petric, A. Belous, O. Vyuno, and O. Yannchevskij. Synthesis and dielectric properties of barium tantalates and niobates with complex perovskite structure. J. Mater. Res. 17(2002)3182–3189.
[168] Murata Electronic North America, RF and Microwave Products Catalog (1996).
[169] T. Shimada, Dielectric loss and damping constant of lattice vibration in $Ba(Mg_{1/3}Ta_{2/3})O_3$. J. Eur. Ceram. Soc. 23(2003)2647–2651.
[170] P. K. Davies. Cation ordering in complex oxides. Curr. Opin. Solid State Mater. Sci. 4(1999)467–471.
[171] E. S. Kim and K. H. Kim. Effect of nickel on microwave dielectric properties in $Ba(Mg_{1/3}Ta_{2/3})O_3$. J. Mater. Sci. 29(1994)830–834.
[172] K. H. Yoon, D. P. Kim, and E. S. Kim. Effect of $BaWO_4$ on the microwave dielectric properties of $Ba(Mg_{1/3}Ta_{2/3})O_3$. J. Am. Ceram. Soc.77(1994)1062–1066.
[173] F. Furuya and A. Ouchi. Microwave dielectric properties of $Ba(Mg_{1/3}Ta_{2/3})O_3$-$A(Mg_{1/2}W_{1/2})O_3$ (A=Ba,Sr,Ca). Jpn. J. Appl. Phys. 33(1994)5482–5487.
[174] X. Hu, X. M. Chen, Y. J. Wu. Nd substituted $Ba(Mg_{1/3}Ta_{2/3})O_3$ microwave dielectric ceramics. Mater. Lett. 54(2002)279–283.
[175] Y. J. Wu and X. M. Chen. Microwave dielectric characteristics of $Ba(Mg_{1/3}Ta_{2/3})O_3$ ceramics sintered at low temperatures. Mater. Sci. Eng. B 100(2003)244–247.
[176] H. Tamura, D. A. Sagala, M. Murata, and K. Wakino. High Q dielectric resonator material for millimeter wave frequency. 3rd US–Japan seminar on Dielectric & Piezoelectric Ceramics (1986)53–54.
[177] M. K. Lee, J.-H. Chung, K. W. Ryu, and Y. H. Lee. Microwave dielectric properties of $Ba(Mg,Ta)O_3$–$Ba(Co,Nb)O_3$ ceramics. Proc. Intl. Symp. Insulating Materials, Toyahashi, Japan, (1998) pp. 139–142.
[178] K. P. Surendran and M. T. Sebastian. Low loss dielectrics in $Ba[(Mg_{1/3}Ta_{2/3})_{1-x}Ti_x]O_3$ and $Ba[(Mg_{1-x}Zn_x)_{1/3}Ta_{2/3}]O_3$ systems. J. Mater. Res. 20(2005)2919–2928.
[179] C.-F. Lin, H.-H. Lu, T.-I. Chang, and J.-L. Huang. Microstructural characteristics and microwave dielectric properties of $Ba[(Mg_{1/3}(Nb_{x/4}Ta_{(4-x)/4})_{2/3}O_3$ ceramics. J. Alloys Compd. 407(2006)318–325.

[180] H. Ouchi, M. Okajima, and H. Ito. Effect of sintering additives on dielectric properties of $Ba(Zn_{1/3}Ta_{2/3})O_3$–$Ba(Mg_{1/3}Ta_{2/3})O_3$ ceramics. Electroceram–V. Book 2(1996)1–4.
[181] K. P. Surendran, P. Mohanan, and M.T. Sebastian. Low loss dielectric $Ba[(Mg_{1/3}Ta_{2/3})_{1-x}(Ti_{x/3}W_{x/3})]O_3$ systems. Mater. Res. Bull. 41(2006)784–790.
[182] Y. Fang, A. Hu, Y. Gu, and Y.-J. Oh. Synthesis of $Ba(Mg_{1/3}Ta_{2/3})O_3$ microwave dielectrics by solid state processing. J. Eur. Ceram. Soc. 23(2003)2497–2502.
[183] H.-F. Cheng, Y.-C. Chen, G. Wang, X.-D. Xiang, G.-Y. Chen, K.-S. Liu, and I.-N. Lin. Study of secondary phases in $Ba(Mg_{1/3}Ta_{2/3})O_3$ materials by microwave field microscopy. J. Eur. Ceram. Soc. 23(2003)2667–2670.
[184] R. W. Selwood. Magnetochemistry. Interscience Publishers, New York (1956).
[185] X. M. Chen, Y. Suzuki, and N. Sato. Sinterability improvement of $Ba(Mg_{1/3}Ta_{2/3})O_3$ dielectric ceramics. J. Mater. Sci. 5(1994)244–247.
[186] Y. K. Kim, K.-M. Lee, and H. M. Jang. 1:2 long range ordering and defect mechanism of WO_3 doped perovskite $Ba(Mg_{1/3}Ta_{2/3})O_3$. J. Mater. Sci. 35(2000)4885–4893.
[187] S. Nishigaki, S. Yano, H. Kato, T. Hirai, and T. Nomura. BaO–TiO_2–WO_3 microwave ceramics and crystalline $BaWO_4$. J. Am. Ceram. Soc. 71(1988)C11–C16.
[188] L.-C. Tien, C.-C. Chou, and D.-S. Tsai. Microstructure of $Ba(Mg_{1/3}Ta_{2/3})O_3$–$BaSnO_3$ microwave dielectrics. Ceram. Int. 26(2000)57–62.
[189] T. Shimada. Effect of Ni substitution on the dielectric properties and lattice vibration of $Ba(Mg_{1/3}Ta_{2/3})O_3$. J. Eur. Ceram. Soc. 26(2006)1781–1785.
[190] H.-K. Yoon, K. S. Hong, and H. Kim. Co-existence of 1:2 and 1:1 long range ordering types in La modified $Ba(Mg_{0.33}Ta_{0.67})O_3$ ceramics. J. Mater. Res. 12(1997)589–592.
[191] T. Shimada. Far infrared reflection and microwave properties of $Ba(Mg_{1-x}Zn_x)_{1/3}Ta_{2/3})O_3$ ceramics. J. Eur Ceram. Soc. 24(2004)1799–1803.
[192] E. Andorenscu, A. Folea, A. Rahaianu. Proc. 4th Intl Conf. Electroceramics and Applications. Electroceramics-IV. Dielectric ceramics based on $1\text{-}xBaTiO_3$–$xBa(Mg_{1/3}Ta_{2/3})O_3$. (Ed.) R. Waser, Verlag der Augustinus Buchhhandl, Aachen, Germany. (1994)148.
[193] H. Choi, S. Nahm, and Y.W. Song. Effect of TiO_2 and SnO_2 on the microwave dielectric properties of $Ba(Mg_{1/3}Ta_{2/3})O_3$ ceramics. J. Kor. Phys. Soc. 35(1999)S410.
[194] Y. Higuchi and H. Tamura. Recent progress on the dielectric properties of dielectric resonator materials with their applications from microwave to optical frequencies. J. Eur. Ceram. Soc. 23(2003)2683–2688.
[195] C.-H. Lu and C.-C. Tsai. Reaction kinetics, sintering characteristics, and ordering behaviour of microwave dielectrics: Barium magnesium tantalates. J. Mater. Res. 11(1996) 1219–1227.
[196] H. Tamura, H. Matsumoto, and K. Wakino. Low temperature properties of microwave materials. Jpn. J. Appl. Phys. Suppl. 28/2(1989)21–23.
[197] S. Penn and Mc N. Alford. High dielectric constant low loss dielectric resonator materials. Final Report of EPSRC grant GR/K70649 SBU. (2000).
[198] K. Wakino and H. Tamura. Microwave dielectric materials. Ceram. Trans. 15(1990) 305–314.
[199] S. Nomura, K. Toyoma, and K. Kaneta. Proc. 4th Meeting ferroelectric materials and their applications. Kyoto, Jpn. J. Appl. Phys. 22. suppl 22–2(1983)83.
[200] T. Nagai, T. Inuzuka, and M. Sugiyama. Contribution of dielectric constant to change in temperature coefficient of resonant frequency in $Ba_{1-x}Sr_x(Mg_{1/3}Ta_{2/3})O_3$ compounds. Jpn. J. Appl. Phys. 31(1992)3132–3135.
[201] T. Nagai, and M. Sugiyama.Contributions of phase stability to dielectric constant of solid solutions $(Ba_{1-x}Sr_x)(Mg_{1/3}Ta_{2/3})O_3$. Advanced Materials 93, 1/B. Trans. Mater. Res. Soc. Jpn. Vol. 14B, (1994) Elsevier Science B.V. pp. 1735–1738.
[202] M. Sugiyama and T. Nagai. Anomaly in dielectric constant of $Ba_{1-x}Sr_x(Mg_{1/3}Ta_{2/3})O_3$ solid solution and its relation to structural change. Jpn. J. Appl. Phys. 32(1993)4360–4363.
[203] O. Steiner, E. L. Colla, I. M. Reaney, and N. Setter. Dielectric behaviour as a function of temperature and structure of complex perovskites $Ba_xSr_{1-x}(Mg_{0.33}Ta_{0.67})O_3$ and

$Ba_xSr_{1-x}(In_{0.5}Nb_{0.5})O_3$. In: P. Duran and J.F. Fernandez Third Euroceramics, Faenza Editrice Iberia, Spain. Vol. 2 (1993)223–228.,.

[204] T. Nagai, M. Sugiyama, M. Sando, and K. Niihara. Anomaly in the infra red active phonon modes and its relationship with dielectric constant $Ba_{1-x}Sr_x(Mg_{1/3}Ta_{2/3})O_3$ compound. Jpn. J. Appl. Phys. 35(1996)5163–5167.

[205] T. Nagai, M. Sugiyama, M. Sando, and K. Niihara. Structural changes in $Ba(Sr_{1/3}Ta_{2/3})O_3$ type perovskite compounds upon tilting of oxygen octahedra. Jpn. J. Appl. Phys. 36(1997)1146–1153.

[206] T. Nagai, M. Sugiyama, M. Sando, and K. Niihara. Controlled temperature coefficient of dielectric constant through phase transformation in (BaSr) $(Mg_{1/3}Ta_{2/3})O_3$. Mater. Res. Soc. Symp. Proc. 453(1997)467–472.

[207] K. Matsumoto and T. Hiuga. Erekutoronika seramikasu 19(1988) March. p 34 (in Japanese).

[208] H. D. Megaw. Crystal Structure: A Working Approach, WB Sanders Company London, (1978) p. 293.

[209] A. M. Glazer. The classification of tilted octahedra in perovskites. Acta Crystallogr. B 28(1972)3384–3392.

[210] A. M. Glazer. Simple ways of determining perovskite structures. Acta Crystallogr. A 31(1975)756–762.

[211] E. L. Colla, I.M. Reaney, and N. Setter. The temperature coefficients of the relative permittivity of complex perovskite and its relation to structural transformations. Ferroelectrics. 133(1992)217–222.

[212] I. M. Reaney, E. L. Colla, and N. Setter. Dielectric and structural characteristics of Ba and Sr based complex perovskites as a function of tolerance factor. Jpn. J. Appl. Phys. 33(1994)3984–3990.

[213] M. Onoda, J. Kuwata, K. Kaneta, K. Toyoma, and S. Nomura. $Ba(Zn_{1/3}Nb_{2/3})$ O_3–$Sr(Zn_{1/3}Nb_{2/3})O_3$ solid solution ceramics with temperature stable high dielectric constant and microwave loss. Jpn. J. Appl. Phys. 21(1982)1707–1710.

[214] M. A. Akbas and P. K. Davies. Cation ordering transformations in the $Ba(Zn_{1/3}Nb_{2/3})O_3$–$La(Zn_{2/3}Nb_{1/3})O_3$ system. J. Am. Ceram. Soc. 81(1998)1061–1064.

[215] H. Wu and P K. Davies. Influence of non-stoichiometry on the structure and properties of $Ba(Zn_{1/3}Nb_{2/3})O_3$ microwave dielectrics. III. Effect of the muffling environment. J. Am. Ceram. Soc. 89(2006)2264–2270.

[216] K. Endo, K. Fujimoto, and K. Murakawa. Dielectric properties of ceramics in $Ba(Co_{1/3}Nb_{2/3})O_3$–$Ba(Zn_{1/3}Nb_{2/3})O_3$ solid solutions. J. Am. Ceram. Soc. 70(1987) C215–C218.

[217] J. S. Kim, J.-J. Yu, J.-C. You, and C.-C. Wei. Microwave dielectric properties of $Ba(Sn_xZn_{(1-x)/3}Nb_{2/3})O_3$ (x=0–0.32) ceramics.J. Mater. Sci. 28(1993)2163–2168.

[218] F. Roulland and S. Marinel. $Ba(Zn_{1/3}Nb_{2/3})O_3$ sintering temperature lowering for silver co-sintering applications. Ceram. Int. 32(2006)377–383.

[219] H. Wu and P. K. Davies. Influence of non-stoichiometry on the structure and properties of $Ba(Zn_{1/3}Nb_{2/3})O_3$ microwave dielectrics. I. Substitution of $Ba_3W_2O_9$. J. Am. Ceram. Soc. 89(2006)2239–2249.

[220] J.-H. Lee, Y. I. Jang, H. J. Youn, J.W. Jang, and B. K. Kim. Low firing and microwave dielectric properties of $Ba[(Ni_{0.6}Zn_{0.4})_{0.33}Nb_{0.64}]O_3$ ceramics doped with Sb_2O_5 and B_2O_3. J. Mater. Sci. 34(1999)625–628.

[221] J.-L. Wang, C.-H. Hsu, C.-L. Huang, and R.-J. Lin. Effect of CuO additives on sintering and microwave dielectric properties of $0.95Ba(Zn_{1/3}Nb_{2/3})O_3$–$0.05BaZrO_3$ ceramics. Jpn. J. Appl. Phys. 44(2005)8039–8042.

[222] M. R. Varma and M. T. Sebastian. Effects of dopant addition on the microwave dielectric properties of $Ba(Zn_{1/3}Nb_{2/3})O_3$ ceramics. J. Eur. Ceram. Soc. 27(2007) 2827–2833.

[223] C.-W. Ahn, H.-J. Jang, S. Nahm, H.-M. Park, and J.-J. Lee. Effects of microstructure on the microwave dielectric properties of $Ba(Co_{1/3}Nb_{2/3})O_3$ and $(1-x)Ba(Zn_{1/3}Nb_{2/3})O_3$–$xBa(Co_{1/3}Nb_{2/3})O_3$ ceramics. J. Eur. Ceram Soc. 23(2003)2473–2478.

[224] R. I. Scott, M. Thomas, and C. Hapson. Development of low cost, high performance $Ba(Zn_{1/3}Nb_{2/3})O_3$ based materials for microwave resonator applications. J. Eur. Ceram. Soc. 23(2003)2467–2471.

[225] A. Veres, S. Marinel, and F. Roulland. Dielectric properties of $Ba(Mg,Zn)_{1/3}Nb_{2/3})O_3$ and effect of B_2O_3 and LiF addition. J. Eur. Ceram. Soc. 25(2005)2759–2762.

[226] J.-S. Sun, J.-J. Yu, J.-C. You, and C. C. Wei. Microwave dielectric properties of $Ba(Sn_xZn_{(1-x)/3}Nb_{(1-x)2/3})O_3$ ($0 \leq x \leq 0.$ 32) ceramics. J. Mater. Sci. 28(1995)2163–2168.

[227] Y.-C. Liou, J.-H. Chen, H.-W. Wang, and C.-Y. Liu. Synthesis of $Ba_xSr_{1-x}(Zn_{1/3}Nb_{2/3})O_3$ ceramics by reaction sintering process and microstructure. Mater. Res. Bull. 41(2006)455–460.

[228] M. H. Liang, C.-T. Hu, H.-Y. Chang, and I.-N. Lin. $Ba(Zn_{1/3}Nb_{2/3})O_3$ ceramics synthesized by spray pyrolysis techniques. Ferroelectrics. 231(1999)243–248.

[229] K. S. Hong, I.-T. Kim, and C.-D. Kim. Order-disorder phase transformation in complex perovskite compounds $Ba(Ni_{1/3}Nb_{2/3})O_3$ and $Ba(Zn_{1/3}Nb_{2/3})O_3$. J. Am. Ceram. Soc. 79(1996)3218–3224.

[230] M.-H. Kim, Y.-H. Jeong, S. Nahm, H.-T. Kim, and H.-J. Lee. Effect of B_2O_3 and CuO additives on the sintering temperature and microwave dielectric properties of $Ba(Zn_{1/3}Nb_{2/3})O_3$ ceramics. J. Eur. Ceram. Soc. 26(2006)2139–2142.

[231] S.-Y. Noh, M. J. Yoo, S. Nahm, C. H. Choi, H. M. Park, and H. J. Lee. Effect of structural changes on the microwave dielectric properties of $Ba(Zn_{1/3}Nb_{2/3})O_3$ ceramics. Jpn. J. Appl. Phys. 41(2002)2978–2981.

[232] C.-L. Huang, C.-S. Hsu, and S.-J. Lu. Dielectric properties of $0.95Ba(Zn_{1/3}Nb_{2/3})O_3$–$0.05BaZrO_3$ ceramics at microwave frequency. Mater. Lett. 57(2003)3602–3605.

[233] F. Azough, L. Leach, and R. Freer. Effect of V_2O_5 on the sintering behaviour, cation order and properties of $Ba_3Co_{0.7}Zn_{0.3}Nb_2O_9$ ceramics. J. Eur. Ceram. Soc. 25(2005)2839–2841.

[234] H. Hughes, D. M. Iddles, and I. M. Reaney. Niobate based dielectrics suitable for third generation mobile phone base stations. Appl. Phys. Lett. 79(2001)2952–2954.

[235] F. Azough, C. Leach, and R. Freer. Effect of CeO_2 on the sintering behaviour cation order and properties of $Ba_3Co_{0.7}Zn_{0.3}Nb_2O_9$ ceramics. J. Eur. Ceram. Soc. 26(2006) 1883–1887.

[236] H. Wu and P. K. Davies. Influence of non-stoichiometry on the structure and properties of $Ba(Zn_{1/3}Nb_{2/3})O_3$ microwave dielectrics. IV. Tuning τ_f and the part size dependence of *Qf*. J. Am. Ceram. Soc. 89(2006)2271–2278.

[237] C.-W. Ahn, S. Nahm, S.-J. Yoon, H.-M. Park, and H.-J. Lee. Microstructure and microwave dielectric properties of $(1-x)Ba(Co_{1/3}Nb_{2/3})O_3$–$xBa(Zn_{1/3}Nb_{2/3})O_3$ ceramics. Jpn. J. Appl. Phys. 42(2003)6964–6968.

[238] S. Kamba, H. Hughes, D. Noujini, S. Santha, R. C. Pullar, P. Samoukhim, J. Petzelt, F. Freer, N. Mc N. Alford, and D. Iddles. Relationship between microwave and lattice vibration properties in $Ba(Zn_{1/3}Nb_{2/3})O_3$ based microwave dielectric ceramics. J. Phys. D 37(2004)1980–1986.

[239] M. Barwick, F. Azough, and R. Freer. Structure and dielectric properties of perovskite ceramics in the system $Ba(Ni_{1/3}Nb_{2/3})O_3$–$Ba(Zn_{1/3}Nb_{2/3})O_3$. J. Eur. Ceram. Soc. 26(2006)1767–1773.

[240] D. Liu, X. M. Chen, X. Hu. Microwave dielectric ceramics in $Ca_{1-x}Ba_x(Zn_{1/3}Nb_{2/3})O_3$ system. Mater. Lett. 61(2007)1166–1169.

[241] C.-T. Lee, Y.-C. Lin, C.-Y. Huang, C.-Y. Su, and C.-L. Hu, Structural and dielectric characteristics of barium lanthanum zinc niobate. J. Am. Ceram. Soc. 89(2007) 3662–3668.

[242] H. Ikawa, M. Touma, T. Shirakami, and O. Fukunaga. Effects of A site ion substitution on paraelectric properties of perovskite compounds. Electroceramics IV, Aachen. In: Waser, I.S. Hoffmann, D. Bonnenberg, C. Hoffmann. (Eds), (1994) pp. 41–46

[243] J. S. Kim, J. H. Lee, Y.-S. Lim, J. W. Jang, and I.-T. Kim. Revisit to the anomaly in dielectric properties of $Ba_{1-x}Sr_x$ $(Zn_{1/3}Nb_{2/3})O_3$ solid solution systems. Jpn. J. Appl. Phys. 36(1997)5558–5581.

[244] R. J. Scott, M. Thomas, and C. Hampson. Development of low cost, high performance $Ba(Zn_{1/3}Nb_{2/3})O_3$ based materials for microwave resonator applications. J. Eur. Ceram. Soc. 23(2003)2467–2471.

[245] I. Molodetsky and P. K Davies. Effect of $Ba(Y_{1/2}Nb_{1/2})O_3$ and $BaZrO_3$ on the cation order and properties of $Ba(Co_{1/3}Nb_{2/3})O_3$ microwave ceramics. J. Eur. Ceram. Soc. 21(2001)2587–2591.

[246] M. Reaney, Y. Iqbal, H. Zheng, A. Feteira, H. Highes, D. Iddles, D. Miur, and T. Price. Order-disorder behaviour in $0.9Ba[(Zn_{0.6}Co_{0.4})_{1/3}Nb_{2/3}]O_3$–$0.1Ba(Ga_{1/2}Ta_{1/2})O_3$. J. Eur. Ceram. Soc. 25(2005)1183–1188.

[247] G. Burns and B. A. Scott. Index of refraction in dirty displacive ferroelectrics. Solid State Commun. 13(1973)423–426.

[248] S.-Y. Cho, H.-J. Youn, K.-S. Hong, I.-T. Kim, and Y.-H. Kim. A new microwave dielectric ceramics based on the solid solutions systems between $Ba(Ni_{1/3}Nb_{2/3})O_3$ and $Ba(Zn_{1/3}Nb_{2/3})O_3$. J. Mater. Res. 12(1997)1558–1562.

[249] L. Li and X. M Chen. Effects of stacking scheme on microwave dielectric characteristics of $Ca(Mg_{1/3}Nb_{2/3})O_3$–$Ba(Zn_{1/3}Nb_{2/3})O_3$ layered dielectric resonators. J. Am. Ceram. Soc. 89(2006)1514–2521.

[250] H. Wu and P. K. Davies. Influence of non-stoichiometry on the structure and properties of $Ba(Zn_{1/3}Nb_{2/3})O_3$ microwave dielectrics—II. Compositional variations in pure BZN. J. Am. Ceram. Soc. 89(2006)2250–2263.

[251] I.-T. Kim, Y.-H. Kim, and S.-J. Chung. Order–disorder transitions and microwave dielectric properties of $Ba(Ni_{1/3}Nb_{2/3})O_3$ ceramics. Jpn. J. Appl. Phys. 34(1995)4096–4103.

[252] I. T. Kim, K. S. Hong, and S. J. Yoon. Effect of non-stoichiometry and chemical inhomogeneity on the order–disorder phase formation in the complex perovskite compounds $Ba(Ni_{1/3}Nb_{2/3})O_3$ and $Ba(Zn_{1/3}Nb_{2/3})O_3$. J. Mater. Sci. 30(1995)514–521.

[253] H. Banno, F. Mizuno, T. Takeuchi T. Tsunooka, and K. Ohya. Dielectric properties of $Sr(Ni_{1/3}Nb_{2/3})O_3$–$Ba(Ni_{1/3}Nb_{2/3})O_3$ ceramics at microwave frequencies. Jpn. J. Appl. Phys. 24(1985)87–89.

[254] S.-Y. Cho, J. Youn, K. S. Kong, I. T. Kim, and Y. H. Kim. A new microwave dielectric ceramic based on the solid solution system between $Ba(Ni_{1/3}Nb_{2/3})O_3$ and $Ba(Zn_{1/3}Nb_{2/3})$ O_3. J. Mater. Res. 12(1997)1558–1562.

[255] K. S. Hong, I. T Kim, and C. D. Kim. Order-disorder phase formation in the complex perovskite compounds $Ba(Ni_{1/3}Nb_{2/3})O_3$ and $Ba(Zn_{1/3}Nb_{2/3})O_3$. J. Am. Ceram. Soc. 79(1996)3218–3224.

[256] I.-T. Kim, K. S. Hong, and S. J. Yoon. Effects of non-stoichiometry and chemical inhomogeneity on the order-disorder phase formation in the complex perovskite compound $Ba(Ni_{1/3}Nb_{2/3})O_3$ and $Ba(Zn_{1/3}Nb_{2/3})O_3$. J. Mater. Sci. 30(995)514–521.

[257] J.-K. Park, D.-Y. Kim, and D. Y. Yoon. Discontinuous dissolution of a liquid phase in $Ba(Ni_{1/3}Nb_{2/3})O_3$ ceramics. J. Am. Ceram. Soc. 84(2001)218–220.

[258] S.-Y. Cho, H. J. Youn, and K. S. Hong. A new microwave dielectric ceramic based on the solid solution system between $Ba(Ni_{1/3}Nb_{2/3})O_3$ and $Ba(Zn_{1/3}Nb_{2/3})O_3$. J. Mater. Res. 12(1997)1558–1562.

[259] A. Veres, S. Marinel, and M. Pollet. Physical properties of $Ba(Mg_{1/3}Nb_{2/3})O_3$-$Ba(Co_{1/3}$ $Nb_{2/3})O_3$ solid solutions. J. Phys. Chem. Solids. 67(2006)822–827.

[260] T. Fukui, C. Sakari, and M. Okayama. Preparation of $Ba(Mg_{1/3}Nb_{2/3})O_3$ ceramics as microwave dielectrics through alkoxide hydroxide route. J. Mater. Res. 7(1992)1883–1887.

[261] R. L. Moreira, F. M. Fatinaga, and A. Dias. Raman spectroscopic evaluation of the long range order in $Ba(B'_{1/3}B''_{2/3})O_3$ ceramics. Appl. Phys. Lett. 78(2001)428–430.

[262] A. Dias, V. S. T. Ciminelli, F. M. Matinaga, and R. L. Moreira. Raman scattering and X-ray diffraction investigations on hydrothermal barium magnesium niobate ceramics. J. Eur. Ceram. Soc. 21(2001)2739–2744.

[263] A. Dias, C. W. A. Paschoal, and R. L. Moreira. Infrared spectroscopic investigations in ordered barium magnesium niobate ceramics. J. Am. Ceram. Soc. 86(2003)1985–1987.

[264] Z.-Q. Tian, H.-X Liu, H.-T. Yu, and S.-X Ouyang. Effect of $BaWO_4$ on microstructure and microwave dielectric properties of $Ba(Mg_{1/3}Nb_{2/3})O_3$. Mater. Chem. Phys. 86(2004)228–232.
[265] Y.-W. Kim, J.-H. Park, and J.-G. Park. Local ordering behaviour in $Ba(Mg_{1/3}Nb_{2/3})O_3$ ceramics. J. Eur. Ceram. Soc. 23(2003)1775–1779.
[266] M. A. Akbas and P. K. Davies. Structural and dielectric properties of $Ba(Mg_{1/3}Nb_{2/3})O_3$–La$(Mg_{2/3}Nb_{1/3})O_3$ system. J. Am. Ceram. Soc. 81(1998)2205–2208.
[267] H. M. Park, H. J. Lee, Y. K. Cho, and S. Nahm. Crystal structure of $Ba_{1-x}La_x(Mg_{(1+x)/3}Nb_{(2-x)/3})O_3$ with $0.9 \leq x \leq 1.0$. J. Mater. Res. 18(2003)1003–1010.
[268] H. J. Lee, H. M. Park, Y. K. Cho, H. Rya, J. H. Park, S. Nahm, and J.-D. Byan. Dielectric and structural characteristics in barium lanthanum niobate. J. Am. Ceram. Soc. 83(2000) 937–942.
[269] P. K. Davies, A. Borisevich, and M. Thirumal. Communicating with wireless perovskites. Cation order and zinc volatilization. J. Eur. Ceram. Soc. 23(2003)2461–2466.
[270] S. M. Moussa, J. B. Claridge, M. J. Rossiensky, and S. Clarke. $Ba_8ZnTa_6O_{24}$. A high Q microwave dielectric from a potentially diverse homologous series. Appl. Phys. Lett. 82(2003)4537–4539.

CHAPTER NINE

CATION-DEFICIENT PEROVSKITES

9.1 INTRODUCTION

The ABO_3 perovskite structure is flexible and can incorporate a considerable amount of vacancies. A large number of anion- or cation-deficient perovskites exist and both A-site and B-site vacant perovskites have been reported in the literature [1–5]. The vacancies are often ordered forming superstructures with interesting dielectric properties. The B-site vacancies are more often found and these structures, with the general formula $A_nB_{n-1}O_{3n}$, have blocks of (n–1) corner sharing octahedra separated by vacant octahedral layers. These oxides which form a series of trigonal structures are observed [6] in various systems with $A = Ba^{2+}$, Sr^{2+}, Ca^{2+}, La^{3+} and $B = Mg^{2+}$, Al^{3+}, Ti^{3+}, Ti^{4+}, Nb^{5+} and Ta^{5+}. The cubic ABO_3 perovskite can be described as stacking of close packed AO_3 layers in the ABCABC... stacking sequence. The B cations occupy 1/4 of the available octahedral cavities in such a way that the BO_6 octahedra forms a three-dimensional network of corner sharing octahedra. The cation-deficient $A_nB_{n-1}O_{3n}$ perovskite is different from the cubic close packing sequence since they have mixed cubic and hexagonal AO_3 stacking sequence. This packing is usually denoted by h or c depending on whether neighboring layers are alike or different. The letters h and c denote the hexagonal and cubic close-packed layers [7] respectively. The existence of such mixed hexagonal–cubic sequences results in the appearance of BO_6 face sharing octahedra. Trolliard et al. [8] subdivided the hexagonal perovskites into two classes (a) shifted perovskites and (b) twinned perovskites, depending on the distribution of B cations. Figure 9.1 represents the (hhc...c)-type sequences and can be described as (n–1) corner sharing octahedra with a vacant site at the centre of the face sharing trimers. The (hhc...c)-type sequence is called "shifted perovskite" because of the periodic shift of $1/3\langle 10\bar{1}0\rangle_H$ in the stacking of corner sharing octahedral blocks. In the case of (hc...c)-type sequence, the B-site vacancies does not result in fully vacant octahedral layer, instead a partial occupation of face sharing dimers showing twin plane boundaries and are called twinned perovskites [8].

9.2 $A_4B_3O_{12}$ CERAMICS

Several authors [9–14] reported low loss dielectric materials in the cation-deficient $A_4B_3O_{12}$ perovskite family such as $Ba_3LaNb_3O_{12}$, $Ba_2La_2TiNb_2O_{12}$ and $BaLa_3Ti_2NbO_{12}$. The Ba and La ions occupy the A sites with coordination number 12 and Nb and Ti ions occupy the B sites with coordination number 6. The $Ba_3LaNb_3O_{12}$ has high negative τ_f and substitution of Ti at the Nb site lowered the τ_f close to zero value. The $Ba_3LaNb_3O_{12}$ has a 12R-type hexagonal perovskite structure with $(hhcc)_3$ stacking sequence. The microwave dielectric properties of $A_4B_3O_{12}$-type cation-deficient perovskites are given in Table 9.1.

Dielectric Materials for Wireless Communications
ISBN-13: 978-0-08-045330-9

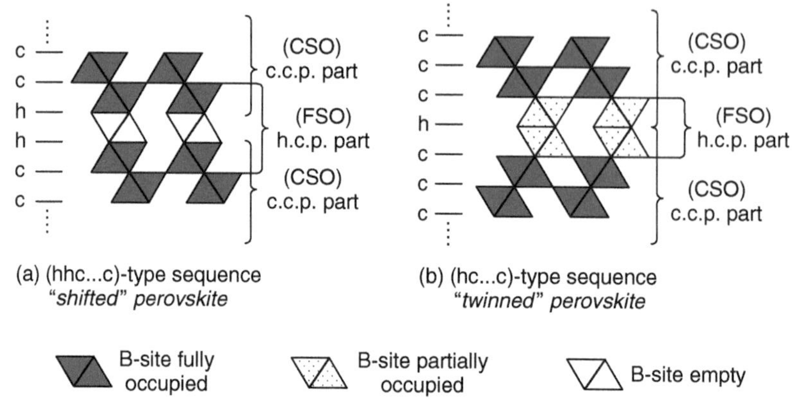

Figure 9.1 Schematic representation of the shifted and twinned hexagonal perovskite (after Ref. [6]).

9.3 $A_5B_4O_{15}$

Galasso and co-workers were the first to report [2, 3] the existence of B-site vacant $Ba_5Nb_4O_{15}$, $Sr_5Ta_4O_{15}$ and $Ba_5Ta_4O_{15}$ compounds. The crystal structure of a series of cation-deficient perovskites of the type $A_5B_4O_{15}$ (A = Ba, Sr, B = Nb.Ta) compounds have been studied extensively [2, 3, 54, 55]. These compounds have hexagonal symmetry and crystallize in the space group with one formula per unit cell. The $Ba_5Nb_4O_{15}$ have five-layer close packing of oxygen and barium ions [2, 3, 54, 55]. The $A_5B_4O_{15}$-type materials are called cation-deficient perovskites if written in the form ABO_3: the $A_5B_4O_{15}$ reduced to $AB_{0.8}O_3$. There is a vacancy of 0.2B cation per 1A cation to accommodate the charge neutrality. Thus there are 1B cation vacancies per 5A cation. The $Ba_5Nb_4O_{15}$ is a hexagonal polytype, i.e., the (5H) member of a series of polytypes characterized by Hutchinson [56] containing 4-, 5-, 6-, 8-, 10- and 12-layered (4H, 5H, 6H, 8H, 10H, 12H) polytype species. Whiston and Smith [57] reported $Sr_5Nb_4O_{15}$ as hexagonal whereas Weiden et al. reported it as monoclinic [58]. Later, it was shown [15, 59, 60] to be hexagonal from Raman Scattering, IR and single crystal X-ray diffraction studies. Lee et al. [59] and Ratheesh et al. [15] prepared $Ba_{5-x}Sr_xNb_4O_{15}$ and investigated its crystal structure using XRD and Raman and reported that the symmetry changed from $P\bar{3}m1$ to $P\bar{3}c1$ with increasing Sr substitution. Kasper [61] reported tri-pseudo-brookite super structure for $Mg_5Nb_4O_{15}$ and $Mg_5Ta_4O_{15}$. Abbatista et al. [62] reported the existence of $Mg_5Nb_4O_{15}$ in the MgO–Nb_2O_5-system. Pagola et al. [63] studied the structure of both $Mg_5Nb_4O_{15}$ and $Mg_5Ta_4O_{15}$ compounds with the help of neutron diffraction and reported that the compounds are isostructural with pseudo-brookite Fe_2TiO_5. The $Mg_5Nb_4O_{15}$ and $Mg_5Ta_4O_{15}$ crystallize in the orthorhombic symmetry with space group $D^{17}{}_{2h}$-$Cmcm$ and $Z = 4$.

Sreemoolanathan et al. [16] were the first to report the microwave dielectric properties of cation-deficient hexagonal $Ba_5Nb_4O_{15}$ perovskite which they prepared by sintering at 1420°C for 2 hours. Since then several authors [17, 18, 64] reported the microwave dielectric properties of different $A_5B_4O_{15}$ ceramics. The microwave dielectric properties of the doped and undoped $Ba_5Nb_4O_{15}$ and other cation-deficient

Table 9.1 Microwave dielectric properties of cation-deficient perovskites

Material	Sintering temperature (°C)	ε_r	*Qf* (GHz)	τ_f	References
$La_{5/3}MgTaO_6$	1500/6 h	22.5	5000	–80	[1]
$Ba_3LaNb_3O_{12}$	1500	43.5	9000	–100	[10], [11]
$Ba_3LaTa_3O_{12}$	1500	39.4	26 800	–49	[14]
$Ba_xLa_4Ti_{3+x}O_{12+3x}$ (x = 0.2)	–	42	86 000	–17	[9]
$Ba_xLa_4Ti_{3+x}O_{12+3x}$ (x = 0.4)	–	45	60 000	–15	[9]
$Ba_xLa_4Ti_{3+x}O_{12+3x}$ (x = 0.6)	–	45	50 000	–13	[9]
$Ba_2La_2TiNb_2O_{12}$	1440	44.2	31 600	–5	[11], [13]
$BaLa_3Ti_2NbO_{12}$	1460	42.4	33 600	6	[11], [12]
$Ba_5Nb_4O_{15}$	1400	40	53 000	78	[15], [16], [17]
$Ba_4SrNb_4O_{15}$	–	48	14 600	140	[18]
$Ba_3Sr_2Nb_4O_{15}$	–	50	16 500	232	[18]
$Ba_2Sr_3Nb_4O_{15}$	–	51	21 200	117	[18]
$BaSr_4Nb_4O_{15}$	–	45	23 300	82	[18]
$Sr_5Nb_4O_{15}$	–	40	19 400	55	[18]
$Ba_5NbTa_3O_{15}$	1500	31.7	21 500	16	[19]
$Ba_5Nb_2Ta_2O_{15}$	1475	34.2	10 500	22	[19]

(*Continued*)

Table 9.1 (Continued)

Material	Sintering temperature (°C)	ε_r	*Qf* (GHz)	τ_f	References
$Ba_5Nb_3TaO_{15}$	1435	37.2	4500	35	[19]
$Ba_4SrTa_4O_{15}$	1575	32.9	9000	8	[18], [19]
$Ba_3Sr_2Ta_4O_{15}$	1575	34.3	4000	–15	[19]
$Ba_2Sr_3Ta_4O_{15}$	1575	35.2	2400	–25	[19]
$BaSr_4Ta_4O_{15}$	1500	31.7	2800	–60	[19]
$Ba_5Ta_4O_{15}$	1550/40 h	30	31 600	12	[18], [19], [14]
$Sr_5Ta_4O_{15}$	1510	43.5	2400	–	[19]
$Ba_5Nb_4O_{15}$+3 wt% B_2O_3	925	39	18 700	0	[20]
0.84$Ba_5Nb_4O_{15}$–0.16 $BaNb_2O_6$ +0.3 wt% B_2O_3	900	42	28 000	0	[21]
0.17$Ba_5Nb_4O_{15}$–0.83 $BaNb_2O_6$	1200/2 h	35.2	59 300	0	[22]
$Ba_2La_3Ti_3NbO_{15}$	1460	42.8	21 700	–8	[23]
$Ba_4LaTiNb_3O_{15}$	1450/4 h	52	15 650	93	[24]
$Ba_4LaTiTa_3O_{15}$	1520	42.3	28 800	33	[25]
$Ba_4LaSnNb_3O_{15}$	1480/4 h	39	14 900	–29	[24]
$BaLa_4Ti_4O_{15}$	1600/2 h	46	46 000	–11	[26], [27]
$CaLa_4Ti_4O_{15}$	1550/24 h	41.1	50 200	–25.5	[26], [28], [29]

$Mg_5Nb_4O_{15}$	1450	11	37 400	54	[18]
$Mg_5Ta_4O_{15}$	1560	11	18 100	54	[18]
$SrLa_4Ti_4O_{15}$	1530/48 h	43.7	46 200	–8.4	[27]
$La_5CrTi_3O_{15}$	1650	34.8	34 000	–35	[30]
$La_4PrCrTi_3O_{15}$	1575	34.6	23 700	–22	[30]
$La_4NdCrTi_3O_{15}$	1650	35.6	19 400	–34	[30]
$La_4SmCrTi_3O_{15}$	1650	34.5	17 300	–38	[30]
$Ca(La_{1-x}Nd_x)_4Ti_4O_{15}$ (x = 0.125)	1550	43.4	32 900	–13	[29]
$Ca(La_{1-x}Nd_x)_4Ti_4O_{15}$ (x = 0.25)	1550	43.1	29 800	–9	[29]
$Ca(La_{1-x}Nd_x)_4Ti_4O_{15}$ (x = 0.5)	1525	42.3	15 200	–6	[29]
$Ba_4NdTiNb_3O_{15}$	1430/3 h	38.1	18 700	12	[31]
$Ba_3Nd_2Ti_2Nb_2O_{15}$	1450/3 h	46.8	19 500	28	[31]
$Ba_2La_3Ti_3TaO_{15}$	1520	45.4	26 800	–1	[32]
$Ba_3La_2Ti_2Ta_2O_{15}$	1540	45.1	31 000	–13	[33]

(Continued)

Table 9.1 (Continued)

Material	Sintering temperature (°C)	ε_r	*Qf* (GHz)	τ_f	References
$Ba_3La_2Ti_2NbTaO_{15}$	1500	46.5	27 100	–4	[33]
$Ba_3La_2Ti_2Nb_2O_{15}$	1460	49.8	22 000	6.9	[33]
$Ba_{1-x}Sr_xLa_4Ti_4O_{15}$ (x = 0.2)	1450	46.8	24 500	–7.5	[34]
$Ba_{1-x}Sr_xLa_4Ti_4O_{15}$ (x = 0.4)	1450	45.7	44 200	–5.5	[34]
$Ba_{1-x}Sr_xLa_4Ti_4O_{15}$ (x = 0.6)	1550	45.4	47 500	–7.3	[34]
$Ba_{1-x}Sr_xLa_4Ti_4O_{15}$ (x = 0.8)	1600	46.1	52 800	–3	[34]
$Ba_{1-x}Ca_x$ $La_4Ti_4O_{15}$ (x = 0.2)	1575	47.7	47 100	–7.9	[34]
$Ba_{1-x}Ca_x$ $La_4Ti_4O_{15}$ (x = 0.6)	1575	48.9	41 200	–6.8	[34]
$La_5Zn_{0.5}Ti_{3.5}O_{15}$	1550/30 h	37.7	23 000	–38.6	[35]
$La_5GaTi_3O_{15}$	1600/30 h	36.2	30 300	–54.5	[35]
$La_5AlTi_3O_{15}$	1600/30 h	33	28 600	–39.2	[35]
$Ba_{1.2}La_4Ti_{4.2}O_{15.6}$		44.9	48 200	–17	[36]
$Ba_3La_3Ti_4NbO_{18}$	1480/4 h	47.4	17 700	35	[37], [38]
$Ba_4La_2Ti_3Nb_2O_{18}$	1450/6 h	55.1	21 300	61	[39]
$Ba_5LaTi_2Nb_3O_{18}$	1420/6 h	57.3	18 500	135	[39]

$Ba_2La_4Ti_5O_{18}$		46	31 800	–36.4	[17]
$Sr_6Nb_4TiO_{18}$	1625/2 h	46.2	6700	26	[40]
$Sr_6Nb_4ZrO_{18}$+2 wt% Bi_2O_3-B_2O_3 glass	1625/2 h	36.4	21 000	–8	[40]
$Sr_6Ta_4TiO_{18}$+3 wt% Bi_2O_3-B_2O_3 glass	1625/2 h	34.8	5600	–19	[40]
$Sr_6Ta_4ZrO_{18}$+3 wt% Bi_2O_3-B_2O_3 glass	1625/2 h	28.4	9100	–39	[40]
$Sr_2La_4Ti_5O_{18}$	1625/2 h	48	27 350	20	[40]
$Sr_2La_4Ti_5O_{18}$+0.3 wt% Bi_2O_3-B_2O_3 glass	1625/2 h	48.7	23 000	22	[40]
$La_6MgTi_4O_{18}$	1625/2 h	40.2	35 000	–39	[40]
$La_6ZnTi_4O_{18}$	1600/4 h	40.8	21 900	–37	[40]
$Ca_2La_4Ti_5O_{18}$		49.3	20 100	6	[41]
$Ba_6Nb_4TiO_{18}$	1425/2 h	45	12 000	18	[42], [40]
$Ba_6Nb_4ZrO_{18}$	1625/2 h	35.9	52 000	25	[40]
$Ba_6Ta_4TiO_{18}$	1550/2 h	29.3	27 500	45	[40]
$Ba_6Ta_4ZrO_{18}$+2 wt% Bi_2O_3-B_2O_3 glass	1625/2 h	30.1	41 000	5	[40]
$Ba_5SrNb_4TiO_{18}$	1450/4 h	50.8	7000	83	[40]
$Ba_5SrNb_4ZrO_{18}$	1600/4 h	39.5	36 000	68	[40]

(Continued)

Table 9.1 (Continued)

Material	Sintering temperature (°C)	ε_r	*Qf* (GHz)	τ_f	References
$Ba_5SrTa_4TiO_{18}$	1550/4 h	34.9	33 000	65	[40]
$Ba_5SrTa_4ZrO_{18}$ +2 wt% Bi_2O_3-B_2O_3 glass	1525/4 h	30	18 500	37	[40]
$Ba_6Nb_4TiO_{18}$		44.9	12 000	33	[43]
$Ba_4Nd_2Ti_3Nb_2O_{18}$	1450/8 h	44.6	13 100	18	[44]
$Ba_3La_4Ti_6O_{21}$		63	9100	198	[36]
$Ba_4La_4Ti_7O_{24}$		82.2	500	317	[36]
$Ba_8ZnTa_6O_{24}$	1350	30.5	62 300	36	[45], [46]
$Ba_8Ta_{4+0.8x}Ti_{3-x}O_{24}$ ($x = 0 = 0$)	1400/40 h	40	12 900	–	[47]
$Ba_8Ta_{4+0.8x}Ti_{3-x}O_{24}$ ($x = 0 = 0.4$)	1400/40 h	36	12 000	–	[47]
$Ba_8Ta_{4+0.8x}Ti_{3-x}O_{24}$ ($x = 0 = 0.8$)	1400/40 h	44	9700	–	[47]
$Ba_8Ta_6NiO_{24}$		27.5	81 800	33	[48]
$Ba_8Nb_4Ti_3O_{24}$	1450	44.1	22 000	115	[49]
$La_6Mg_4Ta_2W_2O_{24}$	1350/4 h	25.2	13 600	–45.7	[4]
$La_6Mg_4Ta_2W_2O_{24}$	1400/4 h	25.8	16 400	–56	[4]
$Ba_8Ta_6(Ni_{1-x}Zn_x)O_{24}$ ($x = 0.25$)		27.9	80 600	34	[48]

$Ba_8Ta_6(Ni_{1-x}Zn_x)O_{24}$ ($x=0.5$)		27.4	83 800	36	[48]
$Ba_8Ta_6(Ni_{1-x}Zn_x)O_{24}$ ($x=0.75$)		27.7	91 700	37	[48]
$Ba_8Ta_6(Ni_{1-x}Zn_x)O_{24}$ ($x=1$)		29	85 000	40	[48]
$Ba_8Ta_6(Ni_{1-x}Mg_x)O_{24}$ ($x=0.25$)		27.9	81 500	32	[48]
$Ba_8Ta_6(Ni_{1-x}Mg_x)O_{24}$ ($x=0.5$)		26.6	86 800	31	[48]
$Ba_8Ta_6(Ni_{1-x}Mg_x)O_{24}$ ($x=0.75$)		24.3	93 100	26	[48]
$Ba_8Ta_6(Ni_{1-x}Mg_x)O_{24}$ ($x=1$)		26.6	80 900	18	[48]
$Ba_3Ti_5Nb_6O_{28}$+5 wt%B_2O_3	900/2 h	15.9	14 000	−13	[44]
$Ba_3Ti_5Nb_6O_{28}$+5 wt% CuO	900/2 h	26.6	14 100	21	[44]
$Ba_3Ti_5Nb_6O_{28}$+3 wt%B_2O_3+1 wt% CuO	900/2 h	38.6	29 800	5	[44]
$Ba_3Ti_5Nb_6O_{28}$+1 wt%B_2O_3+3 wt% CuO	900/2 h	40.3	32 500	9	[44]
$Ba_{10}Ta_{8-0.8x}Ti_xO_{30}$ ($x=1.2$)	1400/40 h	35	25 700	64	[47]
$Ba_{10}Ta_{8-0.8x}Ti_xO_{30}$ ($x=0.9$)	1400/40 h	35	23 600	69	[47]
$Ba_{10}Ta_{8-0.8x}Ti_xO_{30}$ ($x=0.6$)	1400/40 h	34	30 800	57	[47]
$Ba_{10}Mg_{0.25}Ta_{7.9}O_{30}$	1600/12 h	28.2	33 500	29	[50]
$Ba_{10}Co_{0.25}Ta_{7.9}O_{30}$	1600/24 h	29.7	36 700	29	[50]

(*Continued*)

Table 9.1 (Continued)

Material	Sintering temperature (°C)	ε_r	*Qf* (GHz)	τ_f	References
$Ba_{11}TiNb_8O_{33}$	1400	42.3	27 000	47	[51], [42]
$Ba_{16}Nb_{12}TiO_{48}$		42.9	29 000	25	[42]
$Ba_{21}Nb_{16}TiO_{63}$		42.7	19 000	25	[42]
$La_{2/3}(Mg_{1/2}W_{1/2})O_3$	1325	24	32 500	–43	[52], [53]
$La_{(2-x)/3}Na_x(Mg_{1/2}W_{1/2})O_3$ (x = 0.1)	1250	23	19 700	–34	[52]
$La_{(2-x)/3}Na_x(Mg_{1/2}W_{1/2})O_3$ (x = 0.2)	1250	23	16 700	–27	[52]
$La_{(2-x)/3}Na_x(Mg_{1/2}W_{1/2})O_3$ (x = 0.3)	1250	23	11 500	–45	[52]
$La_{(2-x)/3}Na_x(Mg_{1/2}W_{1/2})O_3$ (x = 0.4)	1250	22	5500	–47	[52]
$La_{(2-x)/3}Na_x(Mg_{1/2}W_{1/2})O_3$ (x = 0.5)	1250	21	5700	–47	[52]
$La_{2/3}(Mg_{1/2}W_{1/2})O_3$+2 mol% TiO_2	1325	23.6	14 800	–10	[53]

materials are given in Table 9.1. Zhao et al. [65] prepared $Ba_5Nb_4O_{15}$ hydrothermally at 240 °C for 3 hours. Liou et al. [66] prepared $Ba_5Nb_4O_{15}$ and $Sr_5Nb_4O_{15}$ by a reaction sintering process without calcination. Addition of 1 wt% CuO lowered the reaction sintering temperature. The $Ba_5Nb_4O_{15}$ has ε_r of 40, *Qf* about 53 000 GHz and τ_f of +78 ppm/°C [17, 18, 67]. Lee et al. [59] reported the effect of Sr substitution for Ba on the microwave dielectric properties of $Ba_5Nb_4O_{15}$. Figure 9.2 shows the variation of microwave dielectric properties of $Ba_{5-x}Sr_xNb_4O_{15}$ [59]. The *Qf* decreased, ε_r and τ_f increased with Sr substitution up to $x = 0.5$ and further increase in x lowered ε_r and τ_f. This sudden change in the dielectric properties is due to the change in crystal symmetry at $x = 0.5$. Kim et al. [20] lowered the sintering temperature of $Ba_5Nb_4O_{15}$ from 1400 to 925 °C by the addition of 3 wt% B_2O_3. The B_2O_3 reacted with BaO forming BaB_2O_4 which has a low melting point. The reaction of BaO with B_2O_3 caused the formation of niobium-rich $BaNb_2O_6$. The hexagonal $BaNb_2O_6$ has $\tau_f = -800$ ppm/°C which effectively lowered the high positive τ_f of $Ba_5Nb_4O_{15}$ to zero. Kim et al. [22] reported that $0.17Ba_5Nb_4O_{15}$–$0.83BaNb_2O_6$ mixture has a zero τ_f with Qf = 59 500 GHz and $\varepsilon_r = 38.2$. Zhao et al. [68] attempted to substitute vanadium for Nb in $Ba_5Nb_4O_{15}$. However, vanadium did not substitute for Nb and a multiphase ceramic consisting of

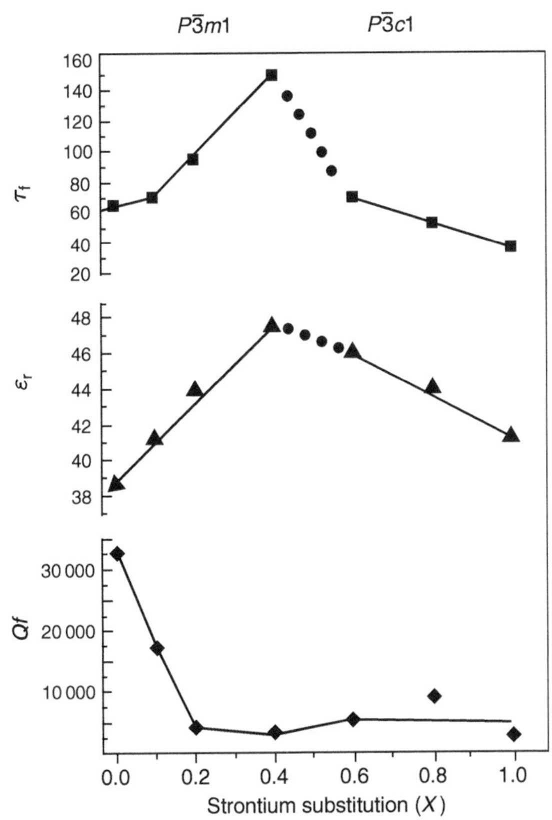

Figure 9.2 The variation of *Qf*, ε_r and τ_f in $Ba_{5-x}Sr_xNb_4O_{15}$ as a function of x (after Ref. [59]).

$BaNb_2O_6$ and $Ba_3(VO_4)_2$ were formed. Kamba et al. [69] investigated the high frequency dielectric properties of $A_5B_4O_{15}$ ceramics using a combination of far infrared reflection and time-resolved terahertz transmission spectroscopy. It was found that the microwave permittivity of the $A_5B_4O_{15}$ ceramics is determined by the polar phonon contributions and that linear extrapolation of the submillimeter dielectric loss ε'' down to the microwave region is in agreement with the microwave data. Jawahar et al. [19] reported the microwave dielectric properties of $Ba_5Nb_xTa_{4-x}O_{15}$, $Ba_{5-x}Sr_xTa_4O_{15}$ and $Sr_5Nb_xTa_{4-x}O_{15}$. The substitution of Sr for Ba in $Ba_5Ta_4O_{15}$ increased the ε_r, and decreased Qf and the τ_f decreased from +12 ppm/°C and became a high negative value of −60 ppm/°C for $BaSr_4Ta_4O_{15}$. Figures 9.3 and 9.4 show the variation of ε_r and τ_f as a function of x for $Ba_{5-x}Sr_xTa_4O_{15}$ and $Ba_5Nb_xTa_{4-x}O_{15}$ respectively. The Ta substitution for Nb in $Ba_5Nb_4O_{15}$ decreased ε_r and τ_f. Jawahar et al. [18] made a detailed study of several $A_5B_4O_{15}$ compounds. They reported that $Zn_5Nb_3O_{15}$ was not formed but a mixture of $Zn_3Nb_2O_8$ and $ZnNb_2O_6$ was formed with $\varepsilon_r \sim 21$, $Qf = 88\,000$ GHz and τ_f of −73 ppm/°C. It was found [18] that the use of uncalcined MgO to prepare

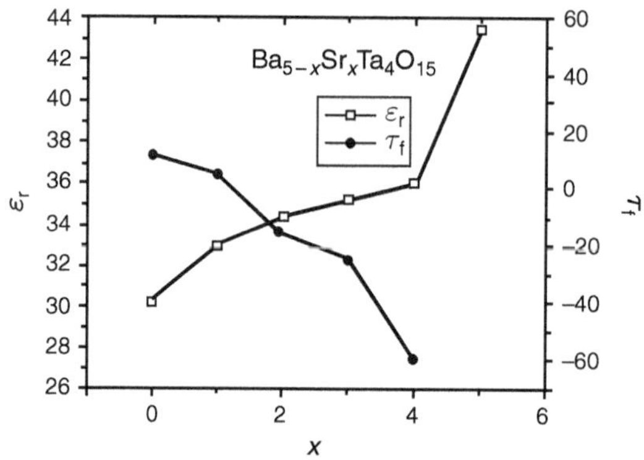

Figure 9.3 Variation of ε_r and τ_f in $Ba_{5-x}Sr_xTa_4O_{15}$ as a function of x (after Ref. [19]).

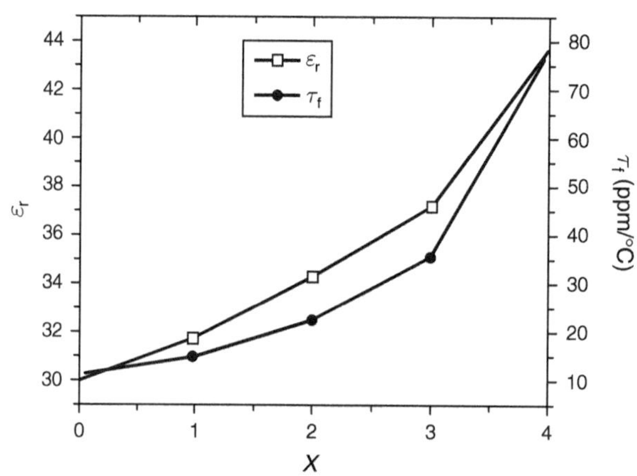

Figure 9.4 Variation of ε_r and τ_f in $Ba_5Nb_xTa_{4-x}O_{15}$ as a function of x (after Ref. [19]).

$Mg_5Nb_4O_{15}$ and $Mg_5Ta_4O_{15}$ led to the formation of secondary phase $MgTa_2O_6$ or $MgNb_2O_6$. This led to the lowering of *Qf* and increased ε_r since $MgNb_2O_6$ and $MgTa_2O_6$ have higher ε_r than $Mg_5Nb_4O_{15}$ and $Mg_5Ta_4O_{15}$. The use of MgO calcined at 1000 °C resulted in phase pure $Mg_5Nb_4O_{15}$ and $Mg_5Ta_4O_{15}$ with both having ε_r of 11. It was found that a slight deficiency of Mg leads to the formation of $MgNb_2O_6$ or $MgTa_2O_6$ secondary phases which considerably influence the dielectric properties.

Vineis et al. reported [17] a few hexagonal perovskite ceramics of the general formula $Ba_nLa_4Ti_{3+n}O_{12+3n}$ with n = 1 and 2 with useful microwave dielectric properties. The crystal structure of $Ba_nLa_4Ti_{3+n}O_{12+3n}$ was investigated by several authors [28, 70, 71]. The $BaLa_4Ti_4O_{15}$ belongs to the crystal system with $P\bar{3}c1$ space group [71]. Okawa et al. [26, 72] investigated the phase composition and microwave dielectric properties of $Ba_nLa_4Ti_{3+n}O_{12+3n}$ homologous series. It was found that as *n* increased, the ε_r and τ_f also increased and the quality factor *Qf* decreased as shown in Figure 9.5. The $BaLa_4Ti_4O_{15}$ ($n = 1$) ceramic has the highest quality factor of 47 000 GHz with $\varepsilon_r = 46$ and $\tau_f = -11$ ppm/°C. Jawahar et al. [41] and Tohdo et al. [27] reported the microwave dielectric properties of $(A,La)_nTi_{n-1}O_{3n}$ (A = Ca, Sr, $n = 5,6$) ceramics, i.e., $CaLa_4Ti_4O_{15}$, $SrLa_4Ti_4O_{15}$, $BaLa_4Ti_4O_{15}$, and they crystallize in the hexagonal perovskite structure [28, 70, 71, 73–76]. These compounds belong to the cation-deficient hexagonal perovskite family $A_nB_{n-1}O_{3n}$ ($n = 5$), where Ba, Ca, La occupy the A sites. The crystal lattice of the cation-deficient perovskite related phases $A_nB_{n-1}O_{3n}$ ($n = 5$) can be derived from the basic perovskite structure by the periodic introduction of intrinsic stacking faults in the cubic close packing of the AO_3-mixed layers with hexagonal symmetry. Alternatively the structure may be defined as consisting of identical perovskite like blocks of n corner sharing octahedral where successive blocks are shifted by 1/3<10$\bar{1}$0> vector. One-*n*th of the octahedral holes are kept vacant in such a way that B cations are omitted from the face sharing octahedral holes to form the cation-deficient structure. The lattice parameters and the atom positions of the homologous phases

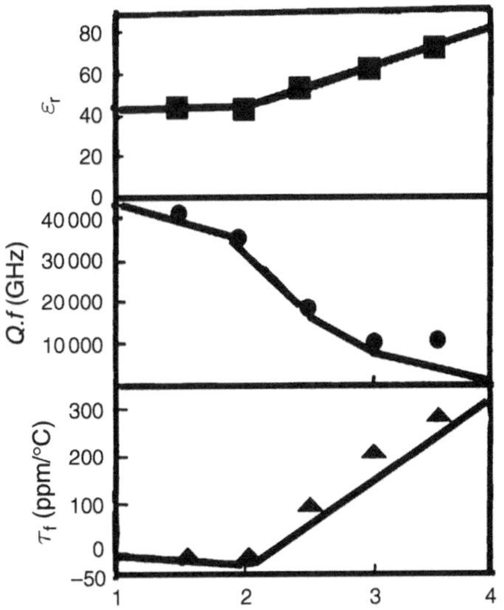

Figure 9.5 Variation of ε_r *Qf* and τ_f in $Ba_nLa_4Ti_{3+n}O_{12+3n}$ (after Ref. [26]).

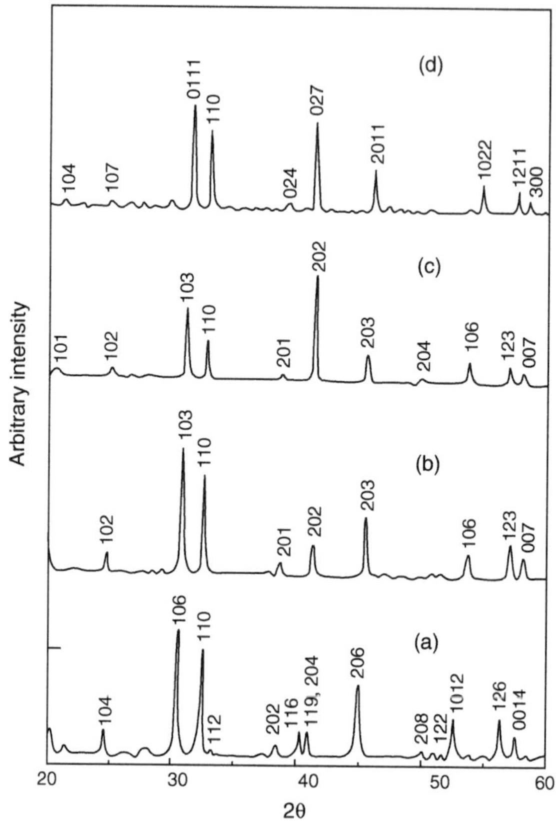

Figure 9.6 X-Ray diffraction patterns of (a) $BaLa_4Ti_4O_{15}$ (b) $SrLa_4Ti_4O_{15}$ (c) $CaLa_4Ti_4O_{15}$ and (d) $Ca_2La_4Ti_5O_{18}$ ceramics (after Ref. [41]).

$BaLa_4Ti_4O_{15}$ and $Ba_2La_4Ti_5O_{18}$ were precisely determined by Harre et al. [70, 75] based on single crystal X-ray diffraction and later confirmed by neutron diffraction studies [76]. Figure 9.6 shows the X-ray diffraction patterns of sintered and powdered ceramics of $BaLa_4Ti_4O_{15}$, $SrLa_4Ti_4O_{15}$, $CaLa_4Ti_4O_{15}$ and $Ca_2La_4Ti_5O_{18}$. Tohdo et al. grew [27] $CaLa_4Ti_4O_{15}$ single crystals by the flux method and studied its crystal structure. The $CaLa_4Ti_4O_{15}$ belongs to the trigonal system with $P\bar{3}c1$ (165) space group. Among these materials, the $CaLa_4Ti_4O_{15}$ shows the highest quality factor of 50 200 GHz with ε_r of 41 and $\tau_f = -26$ ppm/°C. The $BaLa_4Ti_4O_{15}$ ceramics showed the highest ε_r because of the highest polarizability of Ba, and $CaLa_4Ti_4O_{15}$ showed the highest Qf due to the smallest internal strain because of the comparable ionic radii with La ions [28]. Ohsato and co-workers [77, 78] prepared textured or highly grain-oriented $BaLa_4Ti_4O_{15}$ ceramics to tailor the dielectric properties. They added plate-like $BaLa_4Ti_4O_{15}$ particles prepared by molten salt method to $BaLa_4Ti_4O_{15}$ powders. It was then pressed and sintered. The sintered ceramics with template particles showed anisotropy in the microstructure and crystalline phase. As a result of anisotropy, improved microwave dielectric properties of $\varepsilon_r = 53$, $Qf = 41\,000$ GHz, and $\tau_f = -1$ ppm/°C was observed for cylindrical specimen with a [001] grain orientation in the circular plane as compared to those of a randomly

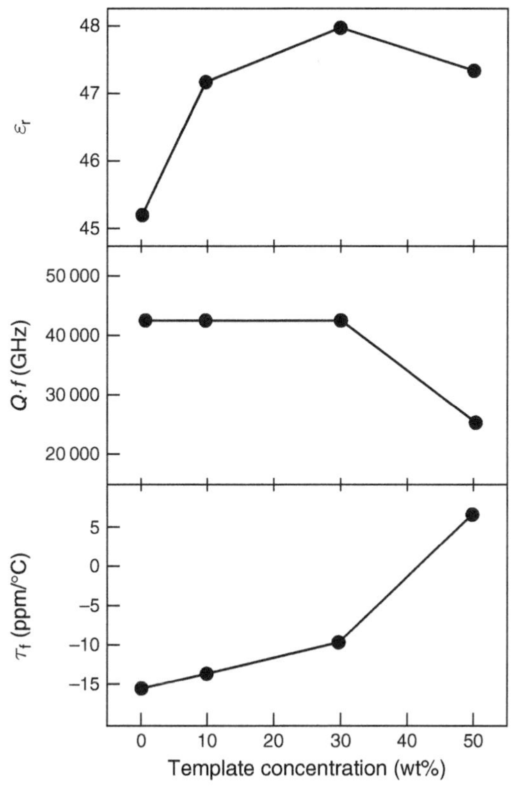

Figure 9.7 Variation of ε_r, *Qf* and τ_f in $Ba_4La_4Ti_5O_{15}$ as a function of plate-like crystals in the initial mixture (after Ref. [77]).

oriented ceramics. It was found that the microwave dielectric properties vary with template concentration as shown in Figure 9.7. Thus it is possible to tailor the dielectric properties by controlling the template concentration. As the template concentration increased, the τ_f decreased and became close to zero for a template concentration of about 45 wt%. Cho et al. [51] reported that in cation-deficient hexagonal perovskites $Ba_{5+n}Ti_nNb_4O_{15+n}$ ($0.3 < n < 1.2$) the ε_r increased and *Qf* decreased and τ_f became less negative with increasing value of *n*. Yue and co-workers prepared [29, 34] $Ca[La_{1-x}Nd_x]_4Ti_4O_{15}$ for $x = 0$–0.5 by sintering at 1500 °C. The XRD study showed that for $x > 0.25$ a secondary phase of $Ca(La_{1-x}Nd_x)_4Ti_5O_{17}$ with orthorhombic perovskite structure was formed. The amount of $Ca(La_{1-x}Nd_x)_4Ti_5O_{17}$ secondary phase increased with increase in *x*. As Nd content (*x*) increased, ε_r and *Qf* decreased and τ_f became less negative as shown in Figure 9.8.

Fang et al. [23, 24, 31–33] and Liou and Liu [79] reported several A and B site co-substituted $A_5B_4O_{15}$ ceramics. The $Ba_3La_2Ti_2Nb_2O_{15}$ and $Ba_2La_3Ti_3NbO_{15}$ [23] are two cation-deficient hexagonal perovskites in the BaO–La_2O_3–Nb_2O_5 system with both having a *Qf* of about 20 000 GHz, high ε_r and both have a low τ_f close to zero. The tantalum analogue $Ba_2La_3Ti_3TaO_{15}$ has a very low τ_f of 1 ppm/°C with $\varepsilon_r \sim 43$ and *Qf* of about 27 000 GHz [32]. The substitution of Ta for Nb in $Ba_3La_2Ti_2Nb_2O_{15}$ increased

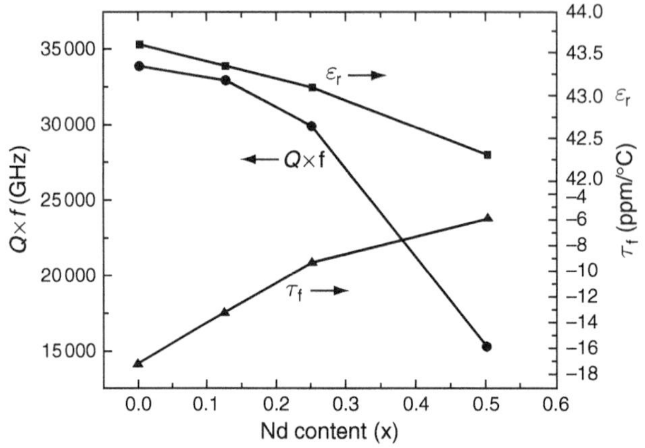

Figure 9.8 Variation of microwave dielectric properties of $Ca(La_{1-x}Nd_x)_4Ti_4O_{15}$ as a function of x (after Ref. [29]).

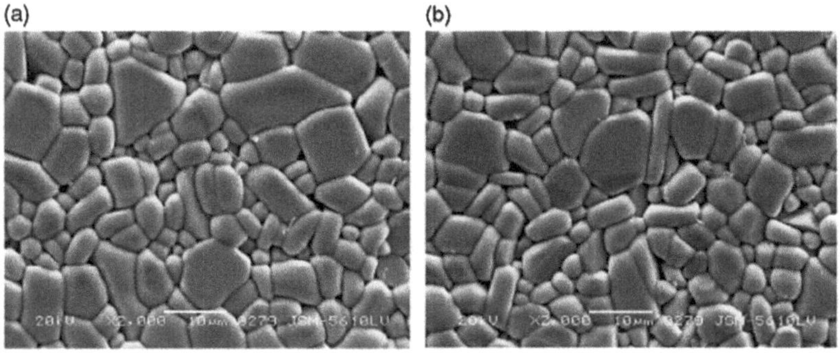

Figure 9.9 Microstructure of (a) $Ba_3La_2Ti_2Nb_2O_{15}$ (b) $Ba_3La_2Ti_2Ta_2O_{15}$ (after Ref. [33]).

Qf and improved τ_f but slightly decreased ε_r [33]. Figure 9.9 shows the microstructures of $Ba_3La_2Ti_2Nb_2O_{15}$ and $Ba_3La_2Ti_2Ta_2O_{15}$ with grains up to 15 μm size. Liou and Liu [79] prepared $Ba_3La_2Ti_2Nb_2O_{15}$ by a reaction sintering process at 1470°C for 2 hours without the calcination step. However, this process lowered the quality factor. Fang et al. [31] reported two more cation-deficient hexagonal perovskites $Ba_4NdTiNb_3O_{15}$ and $Ba_3Nd_2Ti_2Nb_2O_{15}$ having $Qf > 18\,000$ GHz and positive τ_f. Fang et al. [23] also reported that $Ba_4LaTiNb_3O_{15}$ and $Ba_4LaSnNb_3O_{15}$ have hexagonal crystal symmetry similar to $Ba_5Nb_4O_{15}$, where Ba and La occupy the A site and Nb and Sn/Ti occupy the B site. The Ti-based ceramic has a high positive τ_f of 93 ppm/°C and Sn-based a negative τ_f of 29 ppm/°C. Kuang et al. [35] reported the microwave dielectric properties of $A_nB_{n-1}O_{3n}$ [A = La, B = Zn, Ga, Al] where n = 5. The $La_5Zn_{0.5}Ti_{3.5}O_{15}$, La_5Ga Ti_3O_{15} and $La_5AlTi_3O_{15}$ have a 10H structure with an AO_3 stacking sequence of $(hhccc)_2$ with vacant octahedral sites between two hexagonal layers. The Zn, Ga, and Al preferably occupy the octahedral sites between two cubic layers than between the cubic and hexagonal layers [35] and these materials have negative τ_fs. The $La_5CrTi_3O_{15}$

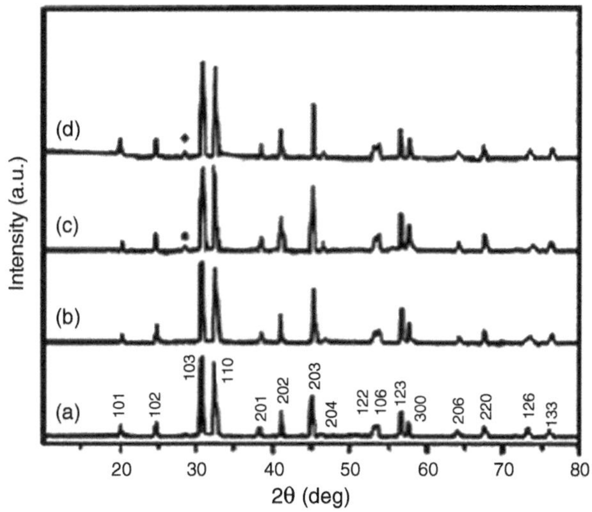

Figure 9.10 XRD Patterns of (a) $La_5CrTi_3O_{15}$ (b) $La_4PrCrTi_3O_{15}$ (c) $La_4NdCrTi_3O_{15}$ and (d) $La_4SmCrTi_3O_{15}$ ceramics (after Ref. [30]).

is known to crystallize in the $A_5B_4O_{15}$-type cation-deficient hexagonal structure with La in the A site with coordination number 12 and Cr and Ti in the B site with coordination number 6 [30]. Rejini and Sebastian [30] prepared $La_5CrTi_3O_{15}$ and $La_4MCrTi_3O_{15}$ [M = Pr, Nd, Sm] low loss dielectric ceramics. The X-ray diffraction patterns of $La_4PrCrTi_3O_{15}$, $La_4NdCrTi_3O_{15}$ and $La_4SmCrTi_3O_{15}$ are similar to that of $La_5CrTi_3O_{15}$ as shown in Figure 9.10. However, trace amount of orthorhombic $La_4MCrTi_4O_{17}$ marked by ★ in Figure 9.10 appeared for Nd- and Sm-based ceramics. The $La_5CrTi_3O_{15}$ sintered at 1650°C/2 h showed ε_r of 34.8, $Qf = 34\,000$ GHz and $\tau_f = -35$ ppm/°C. Partial substitution of La by Pr, Nd, Sm decreased the quality factor although ε_r and τ_f are not significantly affected.

9.4 $A_6B_5O_{18}$

Several authors investigated [37–40, 42, 44] the microwave dielectric properties of $A_6B_5O_{18}$ [A = Ba, Sr, La, Nd; B = Nb, Ta, Zr, Ti, Mg, Zn] type cation-deficient dielectric ceramics. The microwave dielectric properties of this group of materials are given in Table 9.1. Fang and co-workers reported [37, 38] the microwave dielectric properties of cation-deficient hexagonal perovskite $Ba_3La_3Ti_4NbO_{18}$. The samples sintered at 1480 °C/6 h showed ε_r of 47.4, Qf of 17 800 GHz and $\tau_f = 5.2$ ppm/°C. The $Ba_5LaTi_2Nb_3O_{18}$ and $Ba_4La_2Ti_3Nb_2O_{18}$ belong to the space group $R\bar{3}m$ and have a high ε_r of about 55 with high quality factor [39]. Fang and co-workers [44] also reported the microwave dielectric properties of Nd analogue $Ba_4Nd_2Ti_3Nb_2O_{18}$ which is isostructural with $Ba_4La_2Ti_3Nb_2O_{18}$. Trolliard et al. [6] reported the existence of various ordered intergrowth compounds with the nominal compositions between $Ba_5Nb_4O_{15}$ (5*H*) and $Ba_6Nb_4TiO_{18}$ (6*R*) such as $Ba_{11}Nb_8TiO_{33}$ ($Ba_5Nb_4O_{15}$–$Ba_6Nb_4TiO_{18}$:

$5H^1$,$6R^1$), $Ba_{16}Nb_{12}TiO_{48}$ ($2Ba_5Nb_4O_{15}$–1 $Ba_6Nb_4TiO_{18}$: $5H^2$,$6R^1$) and $Ba_{21}Nb_{16}TiO_{63}$ (3 $Ba_5Nb_4O_{15}$ –1 $Ba_6Nb_4TiO_{18}$: $5H^3$,$6R^1$). Zhao et al. [42] studied the microwave dielectric properties of $Ba_5Nb_4O_{15}$, $Ba_6Nb_4TiO_{18}$ and their intergrowth compounds such as $Ba_{11}Nb_8TiO_{33}$, $Ba_{16}Nb_{12}TiO_{48}$ and $Ba_{21}Nb_{16}TiO_{63}$. These compounds have ε_r of about 42 and $Qf > 19\ 000$ GHz. Santha and Sebastian [40] investigated the crystal structure and microwave dielectric properties of several cation-deficient $A_6B_5O_{18}$ [A = Ba, Sr, La; B = Nb,Ta,Zr, Ti, Mg, Zn] ceramics. The $Sr_6TiNb_4O_{18}$ is isostructural with $Ba_6TiNb_4O_{18}$ [80]. Abakumov et al. [43] prepared $A_nB_{n-1}O_{3n}$ homologues based on $Ba_5Ta_4O_{15}$ and $AZrO_3$ (A = Ba,Sr) and investigated their crystal structure. They could not obtain a single-phase $Ba_6Ta_4ZrO_{18}$ but only a mixture of $Ba_6Ta_4ZrO_{18}$ and $Ba_7Ta_4Zr_2O_{21}$. Several authors [81–83] reported the formation of several phases in La_2O_3–MgO–TiO_2 system. Santha and Sebastian [40] reported the microwave dielectric properties of different $A_nB_{n-1}O_{3n}$ compounds such as $Sr_2La_4Ti_5O_{18}$, $La_6MgTi_4O_{18}$, $La_6ZnTi_4O_{18}$, $Ba_6Nb_4TiO_{18}$, $Ba_6Nb_4ZrO_{18}$, $Ba_6Ta_4ZrO_{18}$, $Ba_5SrNb_4TiO_{18}$, $Ba_5SrNb_4ZrO_{18}$, $Ba_5SrTa_4TiO_{18}$, $Ba_5SrTa_4ZrO_{18}$, $Sr_6Nb_4ZrO_{18}$, $Sr_6Ta_4TiO_{18}$ and $Sr_6Ta_4ZrO_{18}$. The microwave dielectric properties of these ceramics are given in Table 9.1.

9.5 $A_8B_7O_{24}$

In $Ba(Zn_{1/3}Ta_{2/3})O_3$ [BZT], the Zn escapes when sintered at high temperatures of about 1550 °C. In Zn-deficient BZT, $Ba_8ZnTa_6O_{24}$ is formed as a secondary phase [84]. Several authors [45, 46, 85] showed that hexagonal $Ba_8ZnTa_6O_{24}$ can be sintered to high density with excellent dielectric properties when sintered at 1350 °C/24 h. It has a hexagonal perovskite structure with $\varepsilon_r = 30.5$, $Qf = 62\ 300$ GHz and $\tau_f = 36$ ppm/°C. Figure 9.11 shows the crystal structure of $Ba_8ZnTa_6O_{24}$ [85]. A closely related $Ba_8NiTa_6O_{24}$ perovskite has also been reported [86]. Kawaguchi et al. [48] studied the microwave dielectric properties of $Ba_8(Ni_{1-x}Zn_x)Ta_6O_{24}$ and $Ba_8(Ni_{1-x}Mg_x)Ta_6O_{24}$. The samples sintered into dense ceramics at 1450–1650 °C. The $Ba_8(Ni_{1-x}Zn_x)Ta_6O_{24}$ solid solution shows a single-phase composition in the entire range of $x = 0$ to 1, whereas $Ba_8(Ni_{1-x}Zn_x)Ta_6O_{24}$ forms a solid solution up to $x = 0.75$. For $x = 0$, $Ba_8(Ni_{1-x}Zn_x)Ta_6O_{24}$ showed secondary phases of BMT, $Ba_5Ta_4O_{15}$ and an unknown phase. Figure 9.12 shows the variation of microwave dielectric properties as a function of composition x for both Zn- and Mg-based ceramics. The Qf of both Mg- and Zn-based materials increased with x, reached a maximum of about 90 000 GHz at $x = 0.75$ and then gradually decreased. The τ_f of Zn-based ceramic slightly increased with x whereas those based on Mg slightly decreased with x. However, the τ_fs need to be further lowered for commercial application. Kawaguchi et al. also confirmed [48] the existence of superstructures which is associated with the complex ordering of Ti, Nb atoms and vacancies on the B sites as suggested earlier [86, 87]. Rawal et al. [49] reported the microwave dielectric properties of $Ba_8Nb_4Ti_3O_{24}$ which has a twinned 8H hexagonal perovskite with $(hccc)_2$ stacking sequence [7, 88]. As compared with B-site cation vacant perovskites, the A-site vacant low loss perovskites are relatively less. The crystal structure and cation vacancy ordering in the perovskite system with A-site vacancies have drawn much interest due to their attractive properties such as ionic conductivity, dielectric behavior and magnetic properties [5]. Khalyavin et al. [89] reported the crystal structure of A-site deficient compound $La_6Mg_4Ta_2W_2O_{24}$. It has a monoclinic

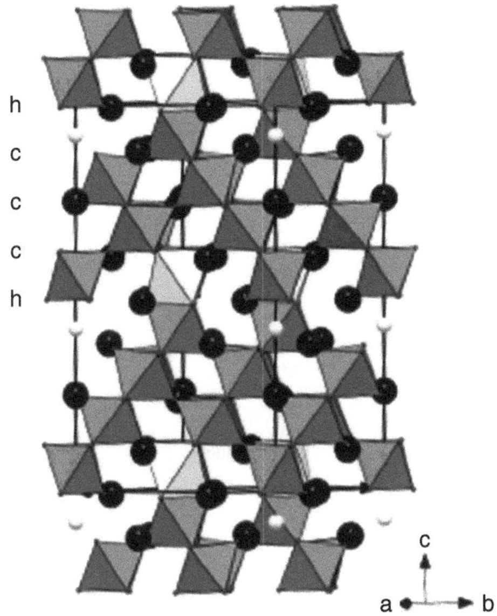

Figure 9.11 Crystal structure of $Ba_8ZnTa_6O_{24}$. The gray octahedrals are TaO_6, white are ZnO_6, large black circles are barium cations and open circles are the cation sites. (after Ref. [85], Courtesy American Institute of Physics).

symmetry with *I2/a* space group. More recently Rejini and Sebastian [4] prepared $La_6Mg_4B_2W_2O_{24}$ [B = Ta and Nb] ceramics and reported their microwave dielectric properties. $La_6Mg_4BW_2O_{24}$ [B = Ta and Nb] can be written in the perovskite $A_{1-\delta}BO_3$ form as $La_{3/4}Mg_{2/4}B_{1/4}W_{1/4}O_3$ where La is in the A site and the rest of the cations occupy the B site. The coordination number of A-site ions is 12 and that of B-site ions is 6. Figure 9.13 shows the X-ray diffraction pattern of $La_6Mg_4BW_2O_{24}$ [B = Ta and Nb] ceramics. The $La_6Mg_4Ta_2W_2O_{24}$ had a maximum densification of 95% when sintered at 1350°C with a Qf = 13 600 GHz, $\varepsilon_r = 25.2$ and $\tau_f = -45.7$ ppm/°C. The $La_6Mg_4Nb_2W_2O_{24}$ sintered at 1400 °C had 98% densification with Qf = 16 400 GHz, $\varepsilon_r = 25.8$ and $\tau_f = -56.0$ ppm/°C. Khalyavin et al. [1] reported A-site deficient perovskite $La_{5/3}MgTaO_6$ with $\varepsilon_r = 22.5$, Qf = 5000 and $\tau_f = -80$ ppm/°C. X-ray and neutron diffraction study revealed a monoclinic *I2/m* space group. The structure has about 17% A-site vacancies. X-ray and neutron diffraction study did not show vacancy ordering but electron diffraction study revealed superstructure reflections indicating local vacancy ordering [1].

9.6 $La_{2/3}(Mg_{1/2}W_{1/2})O_3$

The cubic $A(B_{1/2}W_{1/2})O_3$ [A = Ba, Sr, Ca; B = Co, Ni, Zn] ceramics have excellent microwave dielectric properties with ε_r in the range 13–30, Qf up to 56 000 GHz and τ_f in the range −73 to −31 ppm/°C [see Chapter 7]. Recently, complex perovskite

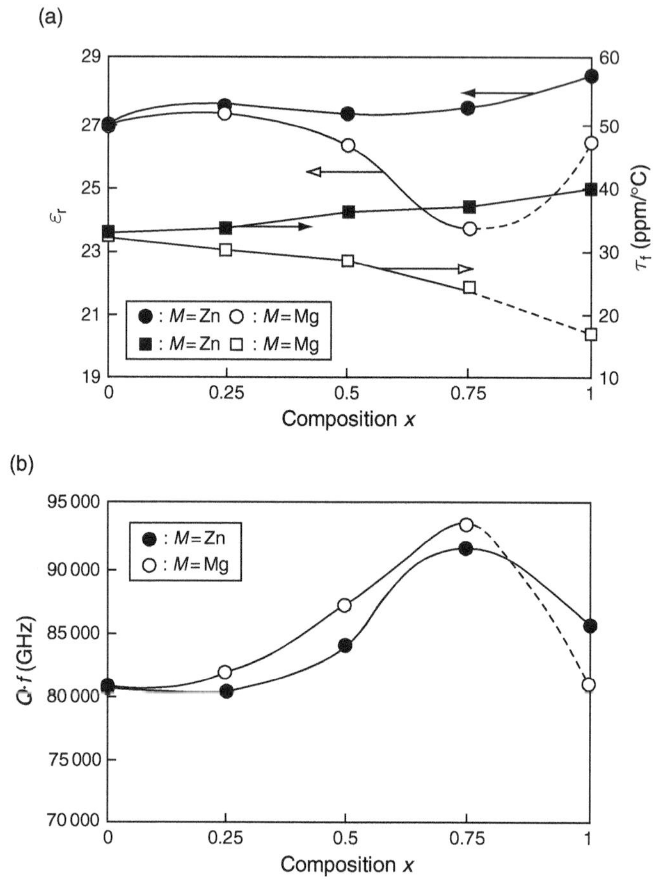

Figure 9.12 Variation of microwave dielectric properties in $Ba_8Ta_6(Ni_{1-x}M_x)O_{24}$ (M = Zn and Mg) as a function of x (after Ref. [48]).

$La_{2/3}(Mg_{1/2}W_{1/2})O_3$ with both A- and B-site cation ordering was reported [52, 53, 90] to have good dielectric properties $\varepsilon_r = 24$, $Qf = 32\,500$ GHz and $\tau_f = -43$ ppm/°C. Its high Qf is attributed [90] to the A-site vacancy ordering. Bian et al. [53] added TiO_2 to tailor the high negative τ_f of $La_{2/3}(Mg_{1/2}W_{1/2})O_3$. Addition of 2 mol% TiO_2 lowered τ_f to less than −10 ppm/°C without significantly affecting the ε_r. However, the Qf decreased from 32 500 to 14 800 GHz when sintered at 1325° C due to the formation of secondary phase of $La_{2/3}TiO_3$ in the sintered ceramic. Bian et al. [52] investigated the effect of coupled A-site substitution of Na, La on the cation order and microwave dielectric properties of $La_{(2-x)/3}Na_x(Mg_{1/2}W_{1/2})O_3$ $(0 \leq x \leq 0.5)$. The ordered $La_{2/3}(Mg_{1/2}W_{1/2})O_3$ structure consists of one layer of vacancies alternating with a second layer that contain apparently a random distribution of 1/3La and 2/3 vacancies [91]. Figure 9.14 shows the X-ray diffraction pattern of $La_{(2-x)/3}Na_x(Mg_{1/2}W_{1/2})O_3$. As the Na substitution increases, the superstructure reflections (marked by asterisk) became weak indicating a decrease in the A-site cation ordering. The reflections marked by filled

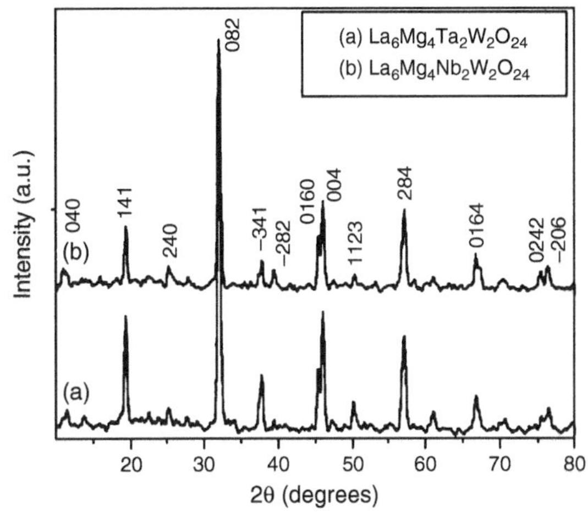

Figure 9.13 XRD Patterns of $La_6Mg_4Ta_2W_2O_{24}$ and $La_6Mg_4Nb_2W_2O_{24}$ ceramics (after Ref. [4]).

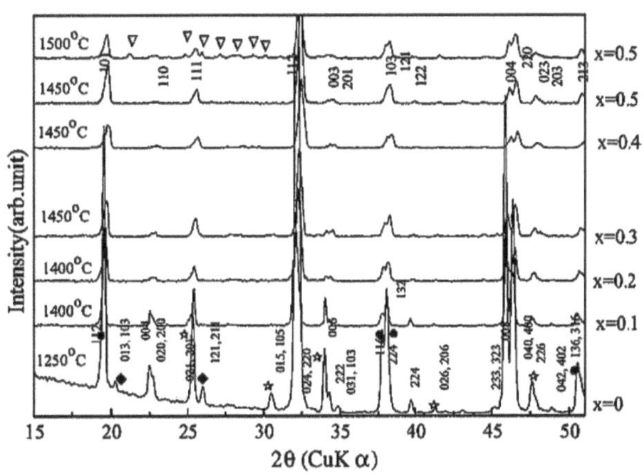

Figure 9.14 X-Ray diffraction patterns of $La_{(2-x)/3}Na_x(Mg_{1/2}W_{1/2})O_3$ (after Ref. [52]).

diamond which are caused by the antiparallel displacement of B-site cations along the c-axis also became weak. The reflections marked by filled circles are caused by the B-site ordering. The $La_{(2-x)/3}Na_x(Mg_{1/2}W_{1/2})O_3$ has an orthorhombic symmetry [52] for $x \leq 0.3$. The orthorhombic phase transformed to monoclinic for $0.3 \leq x \leq 0.5$. The ε_r decreased slightly and the *Qf* decreased considerably with increase in Na substitution. The τ_f decreased from −43 ppm/°C and changed to positive values for $x > 0.3$. Bian et al. [90] prepared $(Pb_{1-3x/2}La_x)(Mg_{1/2}W_{1/2})O_3$ solid solution which has a cubic perovskite structure with random distribution of A-site vacancies for $0 < x < 0.5$. For $0.5 < x < 2/3$, it has a tetragonal or orthorhombic symmetry with the ordering of A-site vacancies. As x increased, the ε_r and τ_f decreased. The composition $x = 0.56$ has $\varepsilon_r = 29$, $Qf = 18\,000$ GHz, $\tau_f = -6$ ppm/°C.

9.7 CONCLUSIONS

A large number of low loss A- or B-site vacant perovskite materials exist. The vacancies often get ordered which improves the microwave dielectric properties. This group of materials have ε_r in the range 11–83 and *Qf* up to 93 000 GHz. The ceramics $0.17Ba_5Nb_4O_{15}$–$0.83BaNb_2O_6$, $Ba_5La_4Ti_4O_{15}$, $Ba_{0.4}Sr_{0.6}Ti_4O_{15}$, $Ba_8Ta_6NiO_{24}$ and $Ba_6Ta_4ZrO_{18}$ are having excellent microwave dielectric properties.

REFERENCES

[1] D. D. Khalyavin, A. M. R. Senos, P. Q. Mantas, D. N. Agyriou, I. Tarraso Gomes, L. G. Viera, and J. L. Riebeiro.Structure and dielectric characterization of a new A-site deficient $La_{5/3}MgTaO_6$ perovskite. J. Solid. State. Chem. 180(2007)40–47.
[2] F. Galasso and L. Katz. Preparation and structure of $Ba_5Ta_4O_{15}$. Acta Crystallogr. 14(1961)647–650.
[3] F. S. Galasso. Structure and Properties of Inorganic Solids. Pergamon Press, NY 1970.
[4] R. Rejini and M. T. Sebastian. Microwave Dielectric Properties of $La_6Mg_4B_2W_2O_{24}$[B = Ta and Nb] Ceramics. J. Am. Ceram. Soc. 90(2007)2472–2475.
[5] V. Vashook, L. Vasylchko, M. Kuzneccov, P. Otchik, J. Zosel, and U. Guth. A site deficient perovskite type compounds in the ternary $CaTiO_3$-$LaCrO_3$–$La_{2/3}TiO_3$ system. J. Alloys Compd. 419(2006)271–280.
[6] G. Trolliard, N. Tenezee, Ph. Boullay, M. Manier, and D. Mercurio. HRTEM study of cation deficient perovskite related $A_nB_{n-1}O_{3n}$ ($n \geq 4\delta$) microphases in $Ba_5Nb_4O_{15}$–$BaTiO_3$ system. J. Solid. State. Chem. 173(2003)91–100.
[7] M. T. Sebastian and P. Krishna. Random, non-random and periodic faulting in crystals. Gordon & Breach Science Publishers, Reading, UK (1994).
[8] G. Trolliard, N. Tenezee, Ph. Boullay, and D. Mercurio. TEM study of cation deficient-perovskite related $A_nB_{n-1}O_{3n}$ compounds: the twin shift option. J. Solid. State. Chem. 177(2004)1188–1196.
[9] H. Yamada, T. Okawa, Y. Tohdo, and H. Ohsato. Microwave dielectric properties of $BaLa_xTi_{3+x}O_{12+3x}$ (x=0.0 –1)ceramics. J. Eur. Ceram. Soc. 26(2006)2059–2062.
[10] R. Rawal, A. Feteira, N. C. Hyatt, D. C. Sinclair. , K. Sarma, and N. Mc N. Alford. Microwave dielectric properties of hexagonal 12R-$Ba_3LaNb_3O_{12}$ ceramics. J. Am. Ceram. Soc. 89(2006)332–335.
[11] L. Fang, S-S. Meng, H. Zhang , and B-I. Wu. Preparation and characterization of new microwave dielectric ceramics $Ba_{3-n}La_{1+n}Nb_{3-n}O_{12}$. Trans. Nonferrous. Metals Soc. China. 16(2006)S521–523.
[12] L. Fang, H-F. Liu, H. Zhang, and Z-Q. Liu. Preparation and microwave characterisation of a new $A_4B_3O_{12}$-type cation deficient perovskite $BaLa_4Ti_2NbO_{12}$. J. Mater. Sci. 41(2006)1281–1284.
[13] L. Fang, S-S. Meng, H. Zhang, and Z-Q. Liu, $Ba_2La_2TiNb_2O_{12}$: A new microwave dielectric of $A_4B_3O_{12}$ type cation deficient perovskites. Mater. Lett. 60(2006)1147–1150.
[14] L. Fang, Y. Wu, C-Z He, P. Chen, and H. Zhang. $Ba_3LaTa_3O_{12}$: A new microwave dielectric of $A_4B_3O_{12}$-type cation deficient perovskites. Mater. Lett. 60(2006)1147–1150.
[15] R. Ratheesh, H. Sreemoolanathan, and M. T. Sebastian. Vibrational analysis of $Ba_{5-x}Sr_xNb_4O_{15}$ microwave dielectric ceramic resonators. J. Solid. State. Chem. 131(1997)2–8.
[16] H. Sreemoolanathan, M. T. Sebastian, and P. Mohanan. High permittivity and low loss ceramics in the BaO–SrO–Nb_2O_5-system. Mater. Res. Bull. 30(1995)653–658.
[17] C. Vineis, P. K. Davies, T. Negas, and S. Bell. Microwave dielectric properties of hexagonal perovskites. Mater. Res. Bull. 31(1996)431–437.

[18] I. N. Jawahar, P. Mohanan, and M. T. Sebastian. $A_5B_4O_{15}$ [A = Ba, Sr, Ca, Zn:, B = Nb,Ta] microwave dielectricceramics. Mater. Lett. 57(2003)4043–4048.

[19] I. N. Jawahar, M. T. Sebastian, and P. Mohanan. Microwave dielectric properties of $Ba_{5-x}Sr_xTa_4O_{15}$, $Ba_5Nb_xTa_{4-x}TaO_{15}$ and $Sr_5Nb_xTa_{4-x}O_{15}$ ceramics. Mater. Sci. Eng. B. 106(2004)207–212.

[20] D-W. Kim, J. R. Kim, S-H. Yoon, K. S. Hong, and C. K. Kim. Microwave dielectric properties of low fired $Ba_5Nb_4O_{15}$. J. Am. Ceram. Soc. 85(2002)2759–2762.

[21] D-W. Kim, K. S. Hong, C-S. Yoon, and C. K. Kim. Low temperature sintering and microwave dielectric properties of $Ba_5Nb_4O_{15}$–$BaNb_2O_6$ for LTCC applications. J. Eur. Ceram. Soc. 23(2003)2597–2601.

[22] D-W. Kim, H-J. Youn, K. S. Hong, and C. K. Kim. Microwave dielectric properties of $(1-x)Ba_5Nb_4O_{15}$–$xBaNb_2O_6$ mixtures. Jpn. J. Appl. Phys. 41(2002)1812–3816.

[23] L. Fang, H. Zhang, X. K. Hong, F. C. Meng, J. F. Yang, and T. H. Huang. Preparation and characterization of two new dielectric ceramics $Ba_3La_2Ti_2Nb_2O_{15}$ and $Ba_2La_3Ti_3NbO_{15}$. Mater. Lett. 58(2004)3884–3887.

[24] L. Fang, H. Zhang, T. H. Huang, R. Z. Yuan, and R. Dronskowski. Preparation, structure and dielectric properties of $Ba_4LaMNb_3O_{15}$ (M = Ti, Sn) ceramics. Mater. Res. Bull. 39(2004)1649–1654.

[25] L. Fang, G. Zhang, C. Hu, and H. Zhang. Preparation and microwave dielectric properties of a new $A_5B_4O_{15}$ type cation deficient perovskites: $Ba_4LaTiTaO_{15}$. Mater. Lett. 62(2008)670–672.

[26] T. Okawa, K. Kiuchi, H. Okabe, and H. Ohsato. Microwave dielectric properties of $Ba_nLa_4Ti_{3+n}O_{12+3n}$ homologous series. Jpn. J. Appl. Phys. 40(2001)5779–5782.

[27] Y. Tohdo, K. Kakimoto, H. Ohsato, H. Yamada, and T. Okawa. Microwave dielectric properties and crystal structureof homologous compounds $ALa_4Ti_4O_{15}$ (A = Ba, Ca, Sr) for base station applications. J. Eur. Ceram. Soc. 26(2006)2039–2043.

[28] Y. Tohodo, T. Okawa, H. Okabe, K. Kakimoto, and H. Ohsato. Microwave dielectric homologous materials $ALa_4Ti_4O_{15}$ (A = Ba, Ca, Sr) with high Q, high dielectric constant for base station. Key Eng. Mater. 269(2004)203–206.

[29] Z. Yue, F. Zhao, Z. Gui, and L. Li. Phases, microstructure and microwave dielectric properties of hexagonal perovskites $Ca(La_{1-x}Nd_x)_4Ti_4O_{15}$ ceramics. J. Alloys Compd. 395(2005)126–131.

[30] R. Rejini and M. T. Sebastian. Microwave dielectric properties of $La_5CrTi_3O_{15}$ and $La_4MCrTi_3O_{15}$ (M = Pr, Nd, and Sm) ceramics. Intl. J. Appl. Cer. Technol. 3(2006)230–235.

[31] L. Fang, C. L. Diao, H. Zhang, R. Z. Yuan, R. Dronskowski, and H. X. Liu. Preparation and characterization of two new dielectric ceramics $Ba_4NdTiNb_3O_{15}$ and Ba_3Nd_2 $Ti_2Nb_5O_{15}$. J. Mater. Sci.: Mater. Electron. 15(2004)803–805.

[32] L. Fang, Q. Yu, H. Zhang, C. Z. Hu, and B. Wu. Microwave dielectric properties of a new $A_5B_4O_{15}$ type cation deficient perovskite $Ba_2La_3Ti_3TaO_{15}$. J. Am. Ceram. Soc. 90(2007)1626–1628.

[33] L. Fang, Q. Yu, H. Zhang, and C-Z. Hu. Microwave dielectric properties of $Ba_3La_2Ti_2Nb_{2-x}$ Ta_xO_{15} ceramics. Mater. Lett. 61(2007)3093–3095.

[34] J. Pei, Z. Yue, F. Zhao, Z. Gui, and L. Li. Microwave dielectric ceramics of hexagonal $(Ba_{1-x}A_x)$ $La_4Ti_4O_{15}$ (A = Sr,Ca) for base station applications. J. Alloys Compd. (2008) in press.

[35] X. Kuang, M. M. B. Allix, J. B. Claridge, H. J. Niu, M. J. Rossiensky, K. M. Ibberson, and D. M. Iddles. Crystal structure, microwave dielectric properties and ac conductivity of B cation deficient hexagonal perovskites $La_5M_xTi_{4-x}O_{15}$ (x = 0.5, 1: M = Zn, Mg, Ga, Al). J. Mater. Chem. 16(2006)1038–1045.

[36] H. Ohsato, A. Harada, T. Okawa, and H. Okabe. Microwave dielectric composition and method for producing the same. Eur. Pat. EP142207A1.

[37] L. Fang, H. Zhang, R. Z. Yuan, and H. X. Liu. $Ba_3La_3Ti_4NbO_{18}$: A new microwave dielectric of $A_6B_5O_{18}$ type cation deficient perovskites. Mater. Lett. 59(2005)535–538.

[38] H. Zhang, L. Fang, R. Elsebrock, and R. Z. Yuan. Crystal structure and microwave dielectric properties of a new $A_6B_5O_{18}$ type cation deficient perovskite $Ba_3La_3Ti_4NbO_{18}$. Mater. Chem. Phys. 93(2005)450–454.

[39] H. Zhang, L. Fang, R. Dronskowski, P. Mueller, and R. Z. Yuan. Some $A_6B_5O_{18}$ cation deficient perovskites in the $BaO–La_2O_3–TiO_2$-Nb_2O_5 system. J. Solid. State. Chem. 177(2004)4007–4012.
[40] N. I. Santha and M. T. Sebastian. Microwave dielectric properties of $A_6B_5O_{18}$ type perovskites. J. Am. Ceram. Soc. 90(2007)496–501.
[41] I. N. Jawahar, N. I. Santha, and M. T. Sebastian. Microwave dielectric properties of MO-La_2O_3–TiO_2 (M = Ca, Sr, Ba) ceramics. J. Mater. Res. 17(2002)3084–3089.
[42] F. Zhao, Z. Yue, J. Pei, H. Zhuans, Z. Gui, and L. Li. Improvement on the temperature coefficient of resonant frequency of hexagonal perovskites through intergrowth structures. Appl. Phys. Lett. 89(2006)202901.
[43] A. M. Abakumov, R. V. Shpanchenko, and E. V. Antipov. Synthesis and structural study of hexagonal perovskites in the $Ba_5Ta_4O_{15}$–$MZrO_3$ (M = Ba, Sr) system. J. Solid State Chem. 141(1998)492–499.
[44] H. Zhang, L. Fang, R. Elsebrocke, R. Z. Yuan, and H. X. Liu. Characterization of materials and microwave dielectric properties of a new $A_6B_5O_{18}$ type cation deficient perovskite-$Ba_4Nd_2Ti_3Nb_2O_{18}$. J. Mater. Sci. 40(2005)4427–4429.
[45] M. Thirumal and P. K. Davies. $Ba_8ZnTa_6O_{24}$: A new high Q dielectric perovskite. J. Am. Ceram. Soc. 88(2005)2126–2128.
[46] P. K. Davies, A. Borisevich, and M. Thirumal. Communicating with wireless perovskites. Cation order and zinc volatilization. J. Eur. Ceram. Soc. 23(2003)2461–2466.
[47] A. N. Baranov and Y-J. Oh. Microwave frequency dielectric properties of hexagonal perovskites in the $Ba_5Ta_4O_{15}$–$BaTiO_3$ system. J. Eur. Ceram. Soc. 25(2005)3451–3457.
[48] S. Kawaguchi, H. Ogawa, A. Kan, and S. Ishihara. Microwave dielectric properties of $Ba_8Ta_6(Ni_{1-x}M_x)O_{24}$ (M = Zn and Mg) ceramics. J. Eur. Ceram. Soc. 26(2006)2045–2049.
[49] R. Rawal, A. Feteira, N. C. Hyatt, A. R. West, D. C. Sinclair, K. Sarma, and N. McN. Alford.Dielectric properties of the twinned 8H-hexagonal perovskite $Ba_8Nb_4Ti_3O_{24}$. J. Am. Ceram. Soc. 89(2006)336–339.
[50] P. Mallison, J. B. Claridge, D. Iddles, T. Price, R. M. Ibberson, M. Allix, and M. J. Rossiensky. New10 layer hexagonal perovskites. Relationship between cation and vacancy ordering and microwave dielectric loss. Chem. Mater. 18(2006)6227–6238.
[51] I-S. Cho, J-R. Kim, D. W. Kim, and K. S.Hong. Microwave dielectric properties and far infra red spectroscopic analysis of $Ba_{5+n}Ti_nNb_4O_{15+3n}$ ($0.3 < n < 1.2$) ceramics. J. Eur. Ceram. Soc. 2: 7(2007)3081–3086.
[52] J. J. Bian, K. Y. Yan, and J. Ji. Structure and microwave dielectric properties of $La_{(2-x)/3}$ $Na_x(Mg_{1/2}W_{1/2})O_3$. J. Eur. Ceram. Soc. 26(2006)1957–1960.
[53] J. J. Bian, K. Yan, and H. B. Gao. Effect of TiO_2 addition on the microwave dielectric properties of $La_{1/2}(Mg_{1/2}W_{1/2})O_3$. Mater. Chem. Phys. 96(2006)349–352.
[54] L. M. Kovba, L. N. Lykova, M. V. Paromova, L. M. Lopato, and A. V. Shevchenko. Russ. J. Inorg. Chem. 22(1997)1544.
[55] J. Shannon and L. Katz. A refinement of the structure of barium tantalum oxide $Ba_5Ta_4O_{15}$. Acta Crystallogr. B. 26(1970)102–105.
[56] J. L. Hutchinson. Electron microscopy of perovskite related structures. Chem. Scripta. 14(1978–79)181.
[57] C. D. Whiston and A. J. Smith. Double oxide containing niobium and or tantalum. II. Systems involving strontium or barium. Acta Crystallogr. 23(1967)82–84.
[58] M. Weiden, A. Grauel, J. Norwig, S. Horn, and F. Steglich. Crystalline structure of the strontium niobates $Sr_4Nb_2O_9$ and $Sr_5Nb_4O_{15}$. J. Alloys Compd. 218(1995)13–16.
[59] C-T. Lee, C-C. Ou, Y-C. Lim, C-Y. Huang, and C-Y. Su. Structure and microwave dielectric property relations in $(Ba_{1-x}Sr_x)_5Nb_4O_{15}$ systems. J. Eur. Ceram. Soc. 27(2007)2273–2280.
[60] N. Tenezee, D. Mercurio, G. Trolliard, and J. C. C. Mesjard. Reinvestigation of the crystal structure of pentastrontium tetraniobate $Sr_5Nb_4O_{15}$. Z. Krist. 215(2000)11–12.
[61] H. Kasper. Die tripseudobrookitephasen $Mg_5Nb_4O_{15}$ and $Mg_5Ta_4O_{15}$ ein nuen strukturtypus undie Lichtabsorption Von $Co^{2+}Ni^{2+}$ und Cu^{2+} in pseudobrookite und tripseudobrookitegitter. Z. Anorg. Chem. 354(1967)208–224.

[62] F. Abbattista, P. Rolondo, and G. Boroni Grass. Magnesium oxide-niobium pentoxide systems.Ann. Chim. 60(1970)426–435.
[63] S. Pagola, R. E. Carbonio, M.T. Fernandez-diaz, and J. A. Alonso. Crystal structure refinement of $Mg_5Nb_4O_{15}$ and $Mg_5Ta_4O_{15}$ by Rietveld analysis of neutron diffraction data. J. Solid State Chem. 137(1998)359–364.
[64] W-H. Jung, J-H. Sohn, Y. Inaguma, and M. Itoh. $Ba_5Nb_4O_{15}$ ceramics with temperature stable high dielectric constant and low microwave loss. Korean J. Ceram. 2(1996)111–113.
[65] H. Zhao, S. Feng, W. Xu, Y. Sho, Y. Mao, and X. Zhu. A rapid chemical route to niobates hydrothermal synthesis and transport properties of ultrafine $Ba_5Nb_4O_{15}$. J. Mater. Chem. 10(2000)965–968.
[66] Y-C. Liou, W-H. Shiu, and C-Y. Shih. Microwave ceramics $Ba_5Nb_4O_{15}$ and $Sr_5Nb_4O_{15}$ prepared by a reaction sintering process. Mater. Sci. Eng. B 31(2006)142–146.
[67] R. Ratheesh, M. T. Sebastian, P. Mohanan, H. Hartnett, M. E. Tobar, R. Woode, and D. G. Balir. Microwave characterization of $BaCe_2Ti_5O_{15}$ and $Ba_5Nb_4O_{15}$ ceramic dielectric resonators using whispering Gallery Mode. Mater. Lett. 45(2000)279–285.
[68] F. Zhao, H. Zhuang, Z. Yue, J. Pei, Z. Gui, and L. Li. Phase relations and microwave dielectric properties of vanadium modified $Ba_5Nb_4O_{15}$ ceramics. Mater. Lett. 61(2006)3466–3468.
[69] S. Kamba, J. Petzelt, E. Buixaderas, D. Haubrich, P. Vanek, P. Kuzel, I. N. Jawahar, M. T. Sebastian, and P. Mohanan. High frequency dielectric properties of $A_5B_4O_{15}$ microwave ceramics. J. Appl. Phys. 89(2001)3900–3906.
[70] N. Harre, D. Mercurio, G. Trolliard, and B. Frit. Crystal structure of $BaLa_4Ti_5O_{15}$ member $n = 5$ of the homologous series $(Ba,La)_nTi_{n-1}O_{3n}$ homologous series cation deficient perovskite related compounds. Mater. Res. Bull. 33(1998)1537–1548.
[71] V. A. Saltikova, O. V. Melinokova, N. V. Leonova, and N. F. Federov. The $La_4Ti_3O_{12}$–$BaTiO_3$ system. Russ. J. Inorg. Chem. 30(1985)105–107.
[72] T. Okawa, K. Kiuchi, H. Okabe, and H. Ohsato. Microwave dielectric properties of $Ba_nLa_4Ti_{3+n}O_{12+n}$ homologous compounds and substitution of trivalent cations for La. Ferroelectrics. 272(2002)345–350.
[73] V. A. Saltikova, O. V. Melinokova, and N. F. Federov. Phase formation in the $La_4Ti_3O_{12}$–$MTiO_3$ systems [M = Ca, Sr, Ba]. Russ. J. Inorg. Chem. 34(d1989)758–759.
[74] M. German and L. M. Kovba. The structure of the hexagonal phase $A_nB_{n-1}O_{3n}$. Russ. J. Inorg.Chem. 30(1985)176–180.
[75] N. Harre, D. Mercurio, G. Trolliard, and B. Frit. Crystal structure of $Ba_2La_4Ti_5O_{18}$, member $n = 6$ of the homologous series $(Ba,La)_nTi_{n-1}O_{3n}$ of cation deficient perovskite related compounds. Eur. J. Solid State Inorg. Chem. 35(1998)77–90.
[76] N. Teneze, D. Mercurio, G. Trolliard, and B. Frit. Cation deficient perovskite related compounds$(Ba,La)_nTi_{n-1}O_{3n}$; n = 4,5 and 6, Rietveld refinement from neutron diffraction data. Mater. Res. Bull. 35(2000)1603–1614.
[77] Y. Fukami, K. Wada, K. Kakimoto, and H. Ohsato. Microstructure and microwave dielectric properties of $BaLa_4Ti_4O_{15}$ceramics with template particles. J. Eur. Ceram. Soc. 26(2006)2055–2058.
[78] K. Wada, Y. Fukami, K-Kakimoto, and H. Ohsato. Microwave dielectric properties of textured $BaLa_4Ti_4O_{15}$ ceramics with layered perovskite structure. Jpn. J. Appl. Phys. 44(2005)7094–7097.
[79] Y-C. Liou and C-Y. Liu. A simple and effective process for $Ba_3La_2Ti_2Nb_2O_{15}$ microwave-ceramics. Mater. Sci. Eng. A 448(2007)351–355.
[80] A. R. Drews, W. Wong-Ng, T. A. Vanderah, and R. S. Roth. Preparation and crystal structure of $Sr_6TiNb_4O_{18}$. J. Alloys Compd. 255(1997)243–247.
[81] T. A. Vanderah, V. L. Miller, I. Levin, S. M. Bell, and T. Negas. Phase relations, crystal-chemistry and dielectric properties in sections of the La_2O_3–CaO–MgO–TiO_2 system. J. Solid State Chem. 177(2004)2023–2038.
[82] M. German and L. M. Kovba. Hexagonal perovskite phases in La_2O_3–TiO_2–MO system. (M = Mg,Ca, Sr, Ba). Russ. J. Inorg. Chem. 28(1983)2377–2379.

[83] M. German and L. M. Koba. The structure of the hexagonal phase $A_nB_{n-1}O_{3n}$. Russ. J.org. Chem. 30(1985)317–322.

[84] V. Tolmer and G. Desgardin. Low temperature sintering and influence of the process on the dielectric properties of $Ba(Zn_{1/3}Ta_{2/3})O_3$. J. Am. Ceram. Soc. 85(2002)1753–1756.

[85] S. M. Moussa, J. B. Claridge, M. J. Rossiensky, S. Clarke, R. M. Ibberson, T. Price, D. M. Iddles, and D. C. Sinclair. $Ba_8ZnTa_6O_{24}$: a high Q microwave dielectric from a potentially diverse homologous series. Appl. Phys Lett. 82(2003)4537–4540.

[86] A. M. Abakumov, G. Van Tendeloo, A. A. Schaglov, R. V. Shpanchenkov, and E. V. Antipov. The crystal structure of $Ba_8Ta_6NiO_{24}$: cation ordering in hexagonal perovskites. J. Solid State Chem. 125(1996)102–107.

[87] R. V. Shanchenko, I. N. Nistor, G. Van Tendeloo, J. Van Landyut, S. Amelinckx, A. M. Abakumov, E. V. Antipov, and L. M. Kovba. Structural studies on new ternary oxides $Ba_8Ta_4Ti_3O_{24}$ and $Ba_{10}Ta_{7.04}Ti_{1.2}O_{30}$ J. Solid State Chem. 114(1995)560–574.

[88] T. Teneze, Ph. Boullay, V. Petricek, G. Trolliard, and D. Mercurio. Structural study of thecation ordering in the ternary oxide $Ba_8Ti_3Nb_4O_{24}$. Solid State Sci. 4(2002)1129–1136.

[89] D. D. Khalyavin, A. B. Lopes, A. M. R. Senos, and P. Q. Mantas, Crystal structure of $La_6Mg_4Ta_2W_2O_{24}$ oxide: A representative of a novel $A_{3n}B'_{2n}B''_{2n}O_{12n}$ homologous series with $n=2$, Chem. Mater. 18(2006)3843–3849.

[90] J. J. Bian, H. B. Gao, X. W. Wang. Microwave dielectric properties of $[Pb_{(1-x)/3}La_x](Mg_{1/2}W_{1/2})O_3$. Mater. Res. Bull. 39(2004)2127–2155.

[91] Y. Torii and T. Sekiya. Crystallographic and dielectric properties of the $Pb_2(MgW)O_6$–$La_{1.33}(MgW)O_6$ solid solutions. Mater. Res. Bull. 16(1981)1153–1158.

CHAPTER TEN

$Ca(Ca_{1/4}B_{2/4}Ti_{1/4})O_3$ (B = Nb, Ta) Complex Perovskites

10.1 Introduction

The A- and B-site cations in simple perovskite structures can be substituted by a combination of other cations to form complex perovskites with the general formula $A(BB')O_3$, $(AA'')BO_3$ and $(AA')(BB')O_3$. A large number of permutations employing various cations in perovskite structure exists [1] because of the potential to tolerate a wide range of elements of varying size and charge. An unusual composition with the complex perovskite structure was reported by Cava and co-workers [2–5] in the $Ca(Ca_{1/4}B_{2/4}Ti_{1/4})O_3$ (B = Nb, Ta) system, where mixing of three different cations occurs on the B site. This material is actually a composition derived [6] from $CaTiO_3$–$Ca_4B_2O_9$ (B = Nb, Ta) polymorphs. High temperature phases of both $Ca(Ca_{1/4}Nb_{2/4}Ti_{1/4})O_3$ and $Ca(Ca_{1/4}\ Ta_{2/4}Ti_{1/4})O_3$ ceramics are disordered, whereas the low temperature phase of tantalum-based material exhibits 1:2 type ordering and a mixing of 1:1 + 1:3 ordering for the niobium analogue. Consequent to the complexity in ordering, the materials showed interesting dielectric properties when measured at 1 MHz [2–5]. The $CaTiO_3$–$Ca_4Nb_2O_9$ polymorphs has drawn much attention because of the opposite signs of τ_f for $CaTiO_3$ ($\varepsilon_r = 162$, $Qf = 13\,000$ GHz and $\tau_f = +850$ ppm/°C [7] and $Ca_4Nb_2O_9$ with $\varepsilon_r = 28$, $Qf = 17\,000$ GHz and $\tau_f = -22$ ppm/°C [8]), which indicates the possibility of tuning the τ_f to zero. The $CaTiO_3$ has orthorhombic symmetry with *Pnma* space group at room temperature. In 1997, Hervieu et al. reported [9] the existence of a high temperature orthorhombic and low temperature monoclinic phase for $Ca_4Nb_2O_9$. They proposed that both the $Ca_4Nb_2O_9$ polymorphs are derivatives of the perovskite structure with one-fourth of the Ca ions occupying the B site, when represented in $Ca(Ca_{1/3}Nb_{2/3})O_3$ form. Recently Levin et al. [10] made a detailed investigation on the octahedral tilting and cation ordering in $Ca(Ca_{1/3}Nb_{2/3})O_3$ polymorphs. They suggested that four distinct perovskite-related polymorphs exist in $Ca(Ca_{1/3}Nb_{2/3})O_3$ which was identified with structures that combine octahedral tilting and different ordering of Ca/Nb ions on the B site. The polymorphs include two high temperature phases which exhibit disordering or 1:1 ordering and two low temperature phases which have 1:2 ordering of Ca/Nb ions in the octahedral site.

10.2 Structure and Properties of $Ca_5B_2TiO_{12}$ [B = Nb, Ta]

Recently Bendersky et al. studied [6] the phase equilibrium and microstructure of $x$$CaTiO_3$–$(1-x)$$Ca_4Nb_2O_9$ ceramics and found that all the phases participating in equilibrium are solid solutions of the binary end members. Since both the end

Dielectric Materials for Wireless Communications
ISBN-13: 978-0-08-045330-9

members have perovskite-based structures, it is expected that the solid solution phases form as $Ca(CaNbTi)O_3$. For $x = 0.5$, the composition can be represented in the usual complex perovskite form as $Ca(Ca_{1/4}Nb_{2/4}Ti_{1/4})O_3$ ($Ca_5Nb_2TiO_{12}$) with A site occupied by Ca and B site by Ca, Nb and Ti in 1:2:1 proportion. They found that the X-ray diffraction patterns of compositions in $xCaTiO_3$–$(1-x)Ca_4Nb_2O_9$ with $0 < x < 1$ were indexable by an orthorhombic lattice with parameters close to that of $CaTiO_3$. This result was confirmed by transmission electron microscopic (TEM) studies by observing different [110]-type selected area electron diffraction (SAED) patterns, and also supported the *Pnma* symmetry. In 2001, Bendersky et al. [5] investigated the structure and microstructure of $Ca_5Nb_2TiO_{12}$ and $Ca_5Ta_2TiO_{12}$ ceramics using TEM techniques. They found different types of ordering between (111) planes, namely 1:1, 1:2 and 1:3, as well as distortions by octahedral tilting. Both the compounds in the as-sintered conditions have microdomain structure but with a different type of ordering, 1:3 for $Ca_5Nb_2TiO_{12}$ and 1:2 for $Ca_5Ta_2TiO_{12}$. Moreover both the ceramics undergo a tilting phase transition from the disordered *Pm*3*m* to distorted *Pnma* (with the $a^-b^+a^-$ tilt of octahedra) structures at 1500°C for $Ca_5Nb_2TiO_{12}$ and 1550°C for $Ca_5Ta_2TiO_{12}$ ceramics respectively. They reported that the structures of $Ca_5Nb_2TiO_{12}$ and $Ca_5Ta_2TiO_{12}$ ceramics sintered at temperatures below 1450°C were different; the disordered structure was found for $Ca_5Ta_2TiO_{12}$, and a 1:1 ordered structure was found for $Ca_5Nb_2TiO_{12}$. Because of kinetic reasons, the 1:1 ordering was only weakly developed during continuous cooling for $Ca_5Nb_2TiO_{12}$ sintered at temperatures above 1500°C. Figure 10.1 shows the X-Ray diffraction patterns of $Ca_5Nb_2TiO_{12}$ and $Ca_5Ta_2TiO_{12}$ powdered ceramics. The patterns are similar for both materials with a slight shift in the position of peaks. Both these materials are having [2] orthorhombic crystal symmetry. The lattice parameters for $Ca_5Nb_2TiO_{12}$ are $a = 5.510(4)$ Å, $b = 7.907(9)$ Å and $c = 5.688(0)$ Å. The theoretical density of $Ca_5Nb_2TiO_{12}$ is

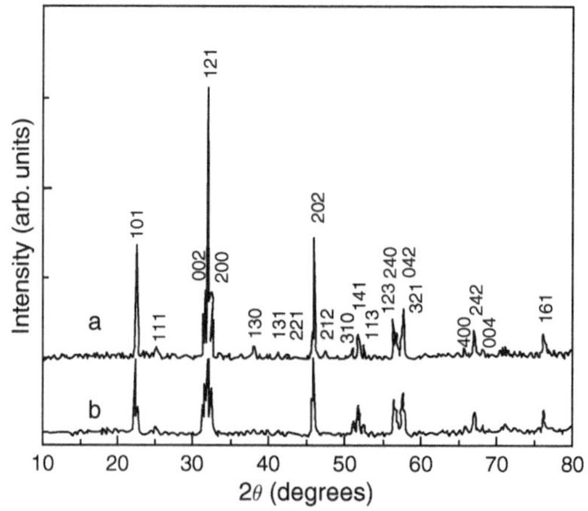

Figure 10.1 X-ray diffraction patterns of (a) $Ca_5Nb_2TiO_{12}$ (b) $Ca_5Ta_2TiO_{12}$ ceramics (after Ref. [15]).

4.19 g/cm^3. The $Ca_5Ta_2TiO_{12}$ has $a = 5.502(2)$ Å, $b = 7.893(1)$ Å and $c = 5.668(5)$ Å with theoretical density 5.41 g/cm^3. The tolerance factor (t) calculated using equation.

$$t = \frac{R_{Ca} + R_O}{\sqrt{2}\{[R_{Ca}/4 + R_{Nb/Ta}/2 + R_{Ti}/4] + R_O\}} \qquad (10.1)$$

where R denotes the radii of corresponding cations reported [11] by Shannon. The value was found to be the same for both materials and is equal to 0.8823, which is much less than that for an ideal cubic structure ($t = 1$). The sintered samples when kept in boiling water for 2 hours did not show any change in density, dielectric properties or in XRD pattern, indicating excellent chemical and thermal stability of the ceramics [12]. Figure 10.2 shows the microstructure of thermally etched samples which show uniformly distributed grains of relatively large size up to 10 µm. Cava et al. reported [2, 3] the 1 MHz dielectric properties of $Ca_5Nb_2TiO_{12}$ and $Ca_5Ta_2TiO_{12}$ ceramics. They studied the dielectric properties at different processing temperatures of the specimens and reported that in the vicinity of ambient temperature, the relative permittivities are approximately 35 and 23 for Nb- and Ta-based ceramics respectively, and dielectric losses of the order of 0.0002 and temperature variation of relative permittivity, $\tau_\varepsilon < 5$ ppm/°C. Bijumon et al. were the first to investigate [12–14] the microwave dielectric properties of $Ca_5Nb_2TiO_{12}$ and $Ca_5Ta_2TiO_{12}$. They optimized the calcination temperature, sintering temperature and their durations of $Ca_5Nb_2TiO_{12}$ and $Ca_5Ta_2TiO_{12}$ materials to get the best density and dielectric properties. The best density and dielectric properties of $Ca_5Nb_2TiO_{12}$ ceramics are found for powders calcined at 1350°C followed by sintering at 1550°C/4 hours. In the case of $Ca_5Ta_2TiO_{12}$ ceramics the calcination temperature was the same as that of $Ca_5Nb_2TiO_{12}$, but the sintering temperature was 1625°C/4 h. The $Ca_5Nb_2TiO_{12}$ has $\varepsilon_r = 48$, $Qf > 26\,000$ GHz and $\tau_f = 40$ ppm/°C, whereas $Ca_5Ta_2TiO_{12}$ has $\varepsilon_r = 38$, $Qf > 33\,000$ GHz and $\tau_f = 10$ ppm/°C. In both materials no significant improvement in density or dielectric properties were observed on annealing. The ionic radii [11] and charge are the same for both Nb and Ta ions and hence the $Ca_5Nb_{2-x}Ta_xTiO_{12}$ forms a complete range of solid solution for all values of x with properties changing linearly with x. The $Ca_5Nb_{2-x}Ta_xTiO_{12}$ $[0 \le x \le 2]$ ceramics show intermediate dielectric properties between the end members $Ca_5Nb_2TiO_{12}$ and $Ca_5Ta_2TiO_{12}$ [15]. The relative permittivity and τ_f decreased and the density and Qf increased with increasing value of x as shown Figures 10.3 and Figure 10.4.

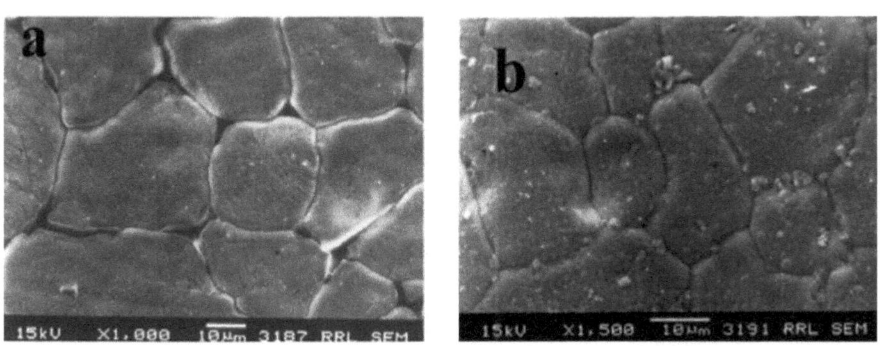

Figure 10.2 Typical SEM photographs of (a) $Ca_5Nb_2TiO_{12}$ (b) $Ca_5Ta_2TiO_{12}$ (after Ref. [15]).

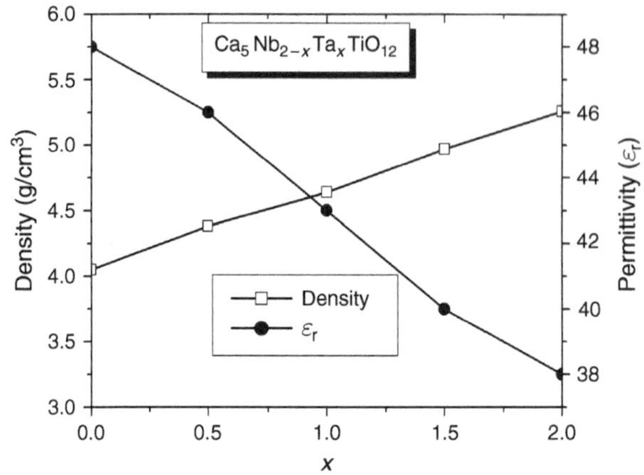

Figure 10.3 Variation of density and ε_r of $Ca_5Nb_{2-x}Ta_xTiO_{12}$ ceramics with x (after Ref. [15]).

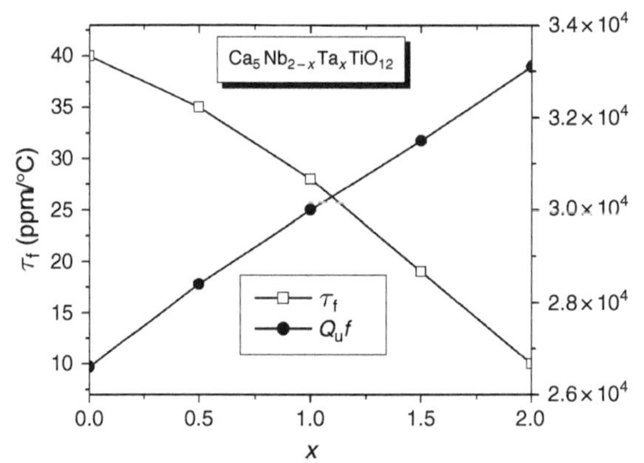

Figure 10.4 Variation of τ_f and Qf of $Ca_5Nb_{2-x}Ta_xTiO_{12}$ ceramics as a function of x (after Ref. [15]).

10.3 Effect of Dopant Addition in $Ca_5B_2TiO_{12}$ (B=Nb, Ta) Ceramics

Bijumon and Sebastian studied [16] the effect of several divalent, trivalent, tetravalent and pentavalant dopant additions in both $Ca_5Nb_2TiO_{12}$ and $Ca_5Ta_2TiO_{12}$. It was found that the addition of 0.5–1 mol% MgO, ZnO, NiO, CuO, Cr_2O_3, SnO_2, Sb_2O_5 and Co_3O_4 in both $Ca_5Nb_2TiO_{12}$ and $Ca_5Ta_2TiO_{12}$ improved the quality factor with a slight decrease in ε_r. The τ_f has improved except for CuO addition. The quality factor of $Ca_5Ta_2TiO_{12}$ reached a maximum of more than 40 000 and 38 000 GHz respectively with 0.5 mol% doping of MgO and CuO. Doping of more than 1 mol% dopants was detrimental to the quality factor of $Ca_5B_2TiO_{12}$ (B=Nb, Ta) ceramics as they form additional phases. $Ca_5Ta_2TiO_{12}$ ceramics doped with 1 mol% each of Cr_2O_3 and In_2O_3

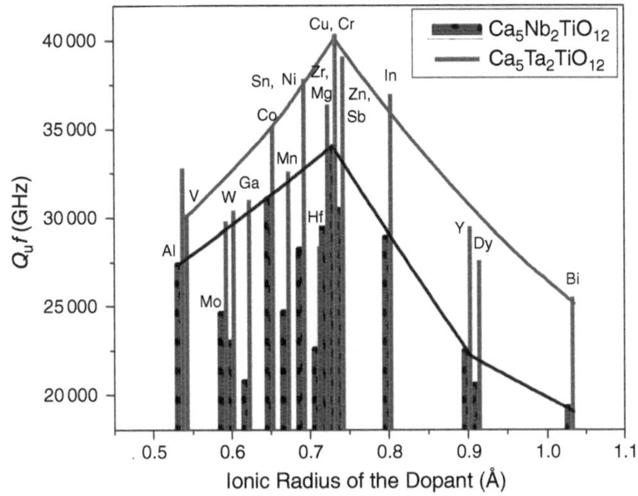

Figure 10.5 Plot of quality factor–frequency product as a function of ionic radius of dopants in 1 Mol% doped $Ca_5Nb_2TiO_{12}$ and $Ca_5Ta_2TiO_{12}$ ceramics (after Ref. [16]).

have Qf = 40 500 and 37 000 GHz respectively. The highest quality factor was found for doping with 1 mol% CuO, MgO and Cr_2O_3. Bijumon and Sebastian [16] found that the quality factor improved when the ionic radius of the dopant is close to the weighted average ionic radius of the B -site ions. Figure 10.5 shows the variation of quality factor of $Ca_5Nb_2TiO_{12}$ and $Ca_5Ta_2TiO_{12}$ ceramics doped with 1 mol% of different dopants as a function of ionic radii of the dopants. The average B-site ionic radius of both $Ca_5Nb_2TiO_{12}$ [$Ca(Ca_{1/4}Nb_{2/4}Ti_{1/4})O_3$] and $Ca_5Ta_2TiO_{12}$ [$Ca(Ca_{1/4}Ta_{2/4}Ti_{1/4})O_3$] are the same and is equal to 0.7213 Å. The ionic radius of the dopants for the coordination number of 6 is taken since the investigated dopants can possibly get substituted in the B site of $Ca_5B_2TiO_{12}$ [B = Nb, Ta] complex perovskites. In general when the ionic radii of the dopants are between 0.65 and 0.75 Å (i.e., close to the average ionic radii of the B-site ion in $Ca_5B_2TiO_{12}$ [B = Nb, Ta] ceramics), the quality factors reached highest values. However, Mn^{2+}, Hf^{4+} and Zr^{4+} doping (0–2 mol%) has lowered the Qf even though ionic radii of the dopants are comparable to that of the average B-site ionic radius of the parent materials. A similar observation was recently reported [17, 18] in $Ba(Zn_{1/3}Ta_{2/3})O_3$ and $Ba(Mg_{1/3}Ta_{2/3})O_3$ ceramics and the Qf improved when the ionic radius of the dopant was close to the average B-site ionic radius of the complex perovskite material and are discussed in Chapter 8.

10.4 Effect of Glass Addition

Bijumon and Sebastian [19, 20] made a detailed study on the effect of addition of several glasses on the densification and microwave dielectric properties of $Ca_5B_2TiO_{12}$ [B = Nb, Ta] ceramics. The crystal structure of $Ca_5B_2TiO_{12}$ (B = Nb, Ta) ceramics were unaffected by the addition of small amount of primary, binary and ternary glasses.

However, glass-based additional phases appeared in the XRD patterns for higher concentration of all glasses. Boron oxide-based glasses were found to be more effective in lowering the sintering temperature although the microwave dielectric properties were degraded. 2 wt% additions of boron-based glasses lowered the firing temperature of $Ca_5Nb_2TiO_{12}$ ceramics even down to 1320°C from 1550°C whereas the sintering temperature of $Ca_5Ta_2TiO_{12}$ was brought down to 1450°C from 1625°C. It was observed that a small amount of SiO_2, $MgO–B_2O_3–SiO_2$, $Al_2O_3–SiO_2$ and $Al_2O_3–B_2O_3–SiO_2$ and $2MgO–Al_2O_3–5SiO_2$ increased the density and improved the microwave dielectric properties of $Ca_5B_2TiO_{12}$ (B = Nb, Ta) ceramics. $Ca_5B_2TiO_{12}$ (B = Nb, Ta) ceramics mixed with a small amount of Al_2O_3- and SiO_2-based glass compositions exhibited good microwave dielectric properties. The improvement of microwave dielectric properties were more pronounced with ternary glasses than with primary and binary glasses. Marginal increase of 2% density, 14% quality factor and 4% permittivity were attained when $Ca_5Nb_2TiO_{12}$ ceramics were sintered with small amount of SiO_2, $Al_2O_3–SiO_2$, $Al_2O_3–B_2O_3–SiO_2$, $MgO–B_2O_3–SiO_2$ and $2MgO–Al_2O_3–5SiO_2$. Addition of 0.1 wt% of $Al_2O_3–SiO_2$, $MgO–B_2O_3–SiO_2$ or $2MgO–Al_2O_3–5SiO_2$ to $Ca_5Ta_2TiO_{12}$, produced an enhancement of 4% in ε_r and 22% in Qf values with a decrease in τ_f value. For 0.2 wt% of $Al_2O_3–B_2O_3–SiO_2$ the microwave dielectric properties of $Ca_5Ta_2TiO_{12}$ ceramics were $\varepsilon_r = 38$, $Qf = 38\,000$ GHz and $\tau_f = 8$ ppm/°C, whereas for 0.1 wt% addition of $2MgO–Al_2O_3–5SiO_2$ $\varepsilon_r = 38$, $Qf = 40\,000$ GHz and $\tau_f = 5$ ppm/°C. Addition of B_2O_3, $Al_2O_3–B_2O_3–SiO_2$, $MgO–B_2O_3–SiO_2$ and $2MgO–Al_2O_3–5SiO_2$ glasses to $Ca_5Ta_2TiO_{12}$ in 1–2 wt% shifted the τ_f of the ceramics from positive to negative values, forming temperature-stable compositions. Alumina-based glasses were more effective in improving the temperature variation of resonant frequency.

10.5 Effect of Cationic Substitutions at A and B Sites of $Ca_5B_2TiO_{12}$ Ceramics (B = Nb, Ta)

Bijumon and Sebastian [12, 21–24] studied the effect of both A- and B-site substitutions on the microwave dielectric properties $Ca_5Nb_2TiO_{12}$ and $Ca_5Ta_2TiO_{12}$. It is well known [25] that in a perovskite compound bigger cation will occupy the A-site of the perovskite structure and hence the Ba- and Sr-substituted $Ca_{5-x}A_xNb_2TiO_{12}$ (A = Ba, Sr) ceramics can be represented as $Ca_{3/4}A_{1/4}(Ca_{1/4}Nb_{2/4}Ti_{1/4})O_3$, $Ca_{2/4}A_{2/4}(Ca_{1/4}Nb_{2/4}Ti_{1/4})O_3$, $Ca_{1/4}A_{3/4}$ $(Ca_{1/4}Nb_{2/4}Ti_{1/4})O_3$, and $A(Ca_{1/4}Nb_{2/4}Ti_{1/4})O_3$ [A = Ba, Sr] for $x = 1$, 2, 3 and 4, respectively. X-ray diffraction and spectroscopic study showed that a single-phase compound was formed only for $x = 4$; i.e., $Ba(Ca_{1/4}Nb_{2/4}Ti_{1/4})O_3$ and $Ba(Ca_{1/4}Ta_{2/4}Ti_{1/4})O_3$. With the substitution of one Ba^{2+} ion for Ca^{2+}, a multiphase ceramic consisting of $Ca_4Nb_2O_9$ or $Ca_4Ta_2O_9$ and $BaTiO_3$ was formed instead of single-phase $Ca_4Ba(Nb/Ta)_2TiO_{12}$. For $x = 2$ and 3, a multiphase ceramic consisting of mainly $Ba(Ca_{1/3}Nb_{2/3})O_3$ or $Ba(Ca_{1/3}Ta_{2/3})O_3$ and trace amount of $Ca_4Nb_2O_9$ or $Ca_4Ta_2O_9$ and $BaTiO_3$ phases were formed. For $x = 4$, phase pure $Ba(Ca_{1/4}[Nb/Ta]_{2/4}$ $Ti_{1/4})O_3$ was formed with cubic symmetry and for $x = 5$, a mixture of $Ba_4[Nb/Ta]_2O_9–BaTiO_3$ and $BaNb_2O_6$ were formed. Sr substitution for Ca also showed a similar effect as that of Ba where $SrTiO_3$ is formed instead of $BaTiO_3$. The Ba and Sr substitutions for Ca decreased the quality factor and increased ε_r and τ_f.

Bijumon et al. [23] also investigated the effect of substitution of Mg, Zn, Ni and Co for Ca on the microwave dielectric properties of $Ca_5Nb_2TiO_{12}$ and $Ca_5Ta_2TiO_{12}$ ceramics. They prepared solid solution phases of $Ca_{5-x}Mg_xNb_2TiO_{12}$ $Ca_{5-x}Zn_xNb_2TiO_{12}$, $Ca_{5-x}Ni_xNb_2TiO_{12}$, $Ca_{5-x}Co_xNb_2TiO_{12}$. The $Ca_{5-x}A'_xNb_2TiO_{12}$ (A′ = Mg, Zn, Ni and Co) ceramics form solid solutions only for x up to 1 and beyond this limit they form a mixture of different phases. However, the microwave dielectric properties of the ceramics were improved for $0 \leq x \leq 1$. Within the solid solution range, Mg, Zn, Ni and Co substitution in $Ca_5Nb_2TiO_{12}$ ceramics resulted in the enhancement of quality factor, decrease in ε_r and improvement in τ_f. Figures 10.6 and 10.7 show the variation of the microwave dielectric properties of $Ca_{5-x}Zn_xNb_2TiO_{12}$ and $Ca_{5-x}Zn_xTa_2TiO_{12}$ ceramics respectively as a function of x. The Mg-, Ni- and Co-based ceramics showed a similar behavior, although the range of properties

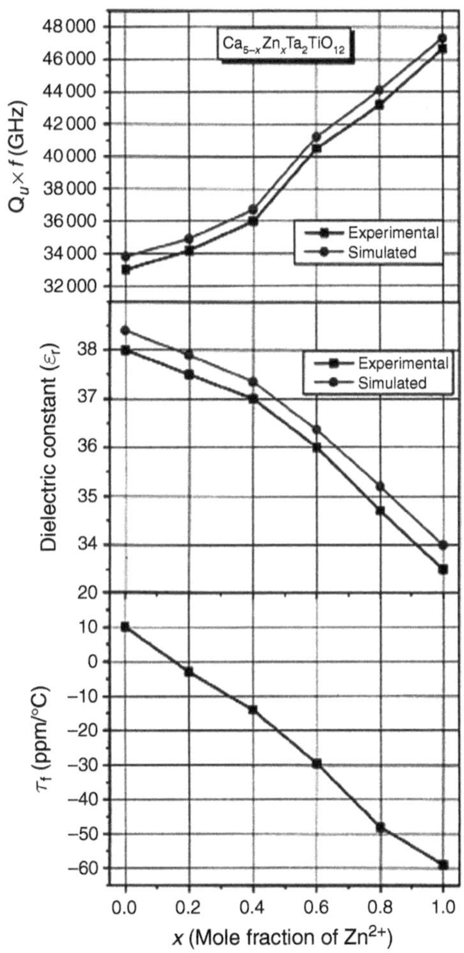

Figure 10.6 Experimental and simulated microwave dielectric properties of $Ca_{5-x}Zn_xNb_2TiO_{12}$ ceramics as a function of x (after Ref. [23]).

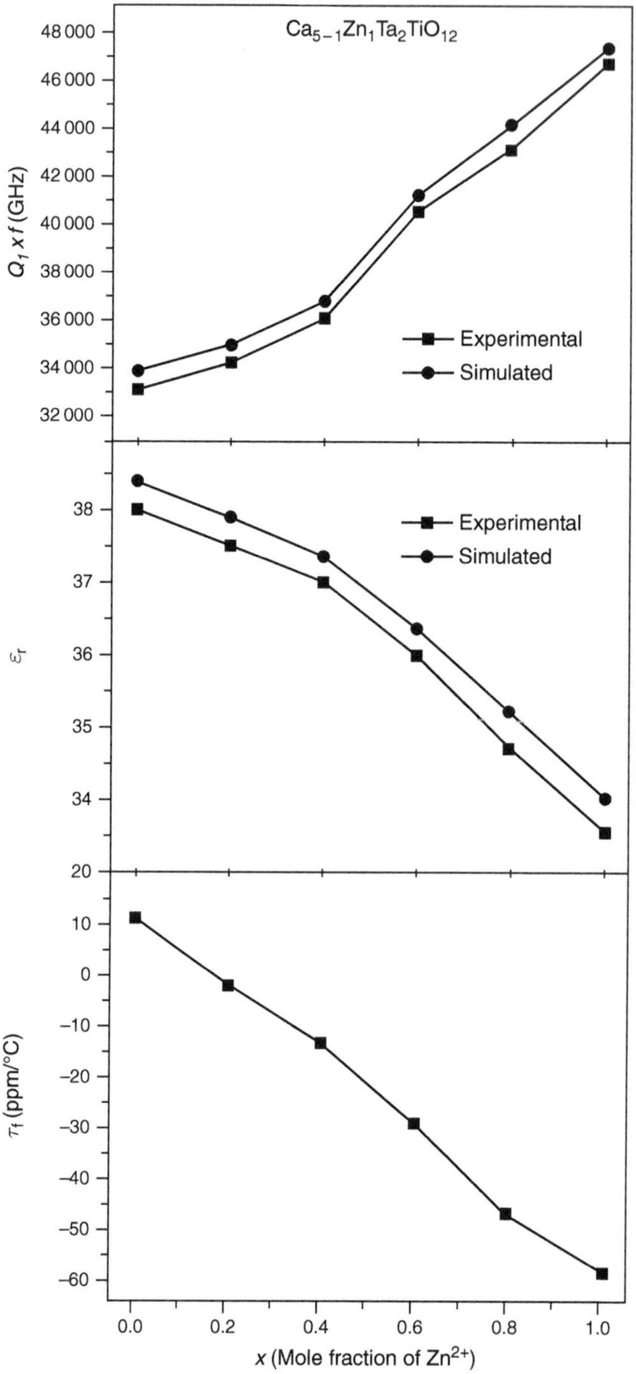

Figure 10.7 Experimental and simulated microwave dielectric properties of $Ca_{5-x}Zn_xTa_2TiO_{12}$ ceramics as a function of x (after Ref. [23]).

varied slightly. At certain values of x, the compositions showed a near zero temperature coefficient of resonant frequency. They are $Ca_{4.35}Mg_{0.65}Nb_2TiO_{12}$ with $\varepsilon_r = 41$, $Qf = 33\,000$ GHz; $Ca_{4.36}Zn_{0.64}Nb_2TiO_{12}$, with $\varepsilon_r = 43$, $Qf = 29\,000$ GHz; $Ca_{4.38}Ni_{0.62}Nb_2TiO_{12}$ with $\varepsilon_r = 42$, $Qf = 28\,200$ GHz and $Ca_{4.18}Co_{0.82}Nb_2TiO_{12}$ with $\varepsilon_r = 37$, $Qf = 30\,000$ GHz. For $2 \leq x \leq 5$, all the compositions formed mixture phases with low ε_r and high negative τ_f. However, compositions with $x = 5$, like $5MgO–Nb_2O_5–TiO_2$ sintered at 1325°C, has $\varepsilon_r = 15$, $Qf = 59\,000$ GHz and $\tau_f = -77$ ppm/°C, whereas $5CoO–Nb_2O_5–TiO_2$ sintered at 1010°C has $\varepsilon_r = 9$, $Qf = 41\,000$ GHz and $\tau_f = -59$ ppm/°C. These low loss, low ε_r composite materials may find applications as substrates for MICs. Bijumon et al. also computed [23] the resonant frequency, unloaded quality factor and ε_r of the ceramic resonators excited in the fundamental transverse-electric mode by means of Three Dimensional (3D) Transmission Line Matrix Method. The simulated values of microwave dielectric properties were found to be in excellent agreement with experimental results with a tolerance of about 2.5% in Qf and 1% in ε_r. The ability of 3D Transmission Line Matrix method to compute the resonant frequency and dielectric properties of a shielded cylindrical ceramic resonator was established by simulating their transmission mode resonance spectrum. The microwave dielectric properties calculated from the simulated resonance spectrum showed excellent agreement with experimental results as shown in Figures 10.6 and 10.7.

Cava et al. reported [4] that partial substitution of Zr for Ti had significant effect on the τ_ε values of $Ca_5Nb_2TiO_{12}$ ceramics. The low temperature coefficient of relative permittivity was at the borderline between ordered and disordered states of the Ca, Nb, and Ti ions in the B sites of the perovskite lattice. The pure Zr analogue $Ca_5Nb_2ZrO_{12}$, had a positive τ_ε value at all firing temperatures, as compared to the negative τ_ε values for $Ca_5Nb_2TiO_{12}$. Bijumon and Sebastian [21, 22] investigated the effect of Zr and Hf substitutions for Ti on the microwave dielectric properties in $Ca_5Nb_2TiO_{12}$ and $Ca_5Ta_2TiO_{12}$ ceramics. XRD study of $Ca_5Nb_2Ti_{1-x}Zr_xO_{12}$ and $Ca_5Ta_2Ti_{1-x}Zr_xO_{12}$ indicated the formation of a solid solution phase [21] as shown in Figure 10.8 for $Ca_5Nb_2Ti_{1-x}Zr_xO_{12}$. The density and cell volume increased linearly with Zr content. Figure 10.9 shows the variation of bond valence and electronegativity with Zn content of $Ca_5Nb_2Ti_{1-x}Zr_xO_{12}$. The bond valence increased linearly with increase in x. The larger ionic radius of Zr compared to Ti leads to increased bond length. With increase in Zr content, the electronegativity decreased, which indicates that the bonding strength between oxygen and the B-site ion is weakened. The ε_r decreased with increase in bond valence as shown in Figure 10.10. The ε_r decreased with increasing amount of Zr substitution and it varied from 51 to 28 in $Ca_5Nb_2Ti_{1-x}Zr_xO_{12}$ and from 40 to 23 in $Ca_5Ta_2Ti_{1-x}C_xO_{12}$. Figures 10.11 and 10.12 show the variation of τ_f and Qf with Zr content. The Qf and τ_f decrease with increasing amount of Zr substitution. In $Ca_5Nb_2Ti_{1-x}Zr_xO_{12}$ for $x = 0.8$, the ceramic has $\varepsilon_r = 34$, $Qf = 24\,000$ GHz and $\tau_f = 0$ ppm/°C, and in $Ca_5Ta_2Ti_{1-x}Zr_xO_{12}$ it has $\varepsilon_r = 36$, $Qf = 28\,000$ GHz and $\tau_f = 0$ ppm/°C for $x = 0.3$. $Ca_5Nb_2Ti_{1-x}Hf_xO_{12}$ and $Ca_5Ta_2Ti_{1-x}Hf_xO_{12}$ ceramics with $x > 0.6$ have poor sinterability and addition of a small amount of B_2O_3 improved the densification [22]. X-ray diffraction study showed the formation of a complete solid solution in the whole range in both $Ca_5Nb_2Ti_{1-x}Hf_xO_{12}$ and $Ca_5Ta_2Ti_{1-x}Hf_xO_{12}$ ceramics. The ε_r, Qf and τ_f decreased with increasing amount of Hf substitution. The τ_f decreased from a positive value, crossed zero and became negative as x increased. The composition $Ca_5Nb_2Ti_{0.4}Hf_{0.6}O_{12}$ has $\varepsilon_r = 32$, $Qf = 22\,000$ GHz and $\tau_f = 0$ ppm/°C and $Ca_5Ta_2Ti_{0.6}Hf_{0.4}O_{12}$ has $\varepsilon_r = 34$, $Qf = 26\,000$ GHz with $\tau_f = 0$ ppm/°C. The microwave dielectric properties of $Ca(Ca_{1/2}B_2Ti)O_3$ [B = Nb, Ta] type materials are given in Table 10.1.

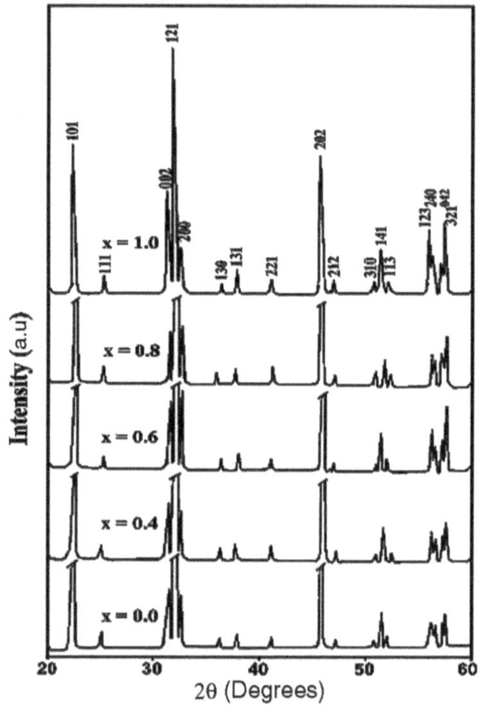

Figure 10.8 XRD Patterns of $Ca_5Nb_2Ti_{1-x}Zr_xO_{12}$ ceramics (after Ref. [21]).

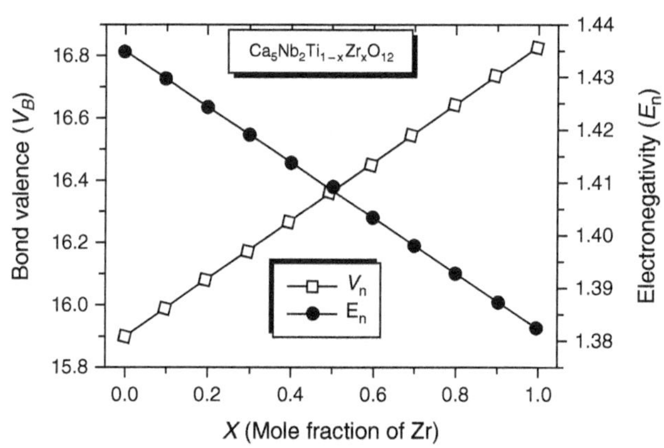

Figure 10.9 Variation of bond valence and electronegativity as a function of x in $Ca_5Nb_2Ti_{1-x}Zr_xO_{12}$ (after Ref. [21]).

Bijumon et al. [26] fabricated broadband dielectric resonator loaded microstrip patch antennas by two separate methods. Dielectric resonators (DRs) were placed over the patch and as an alternate method where they were positioned over the feedline as shown in Figure 10.13. In both cases, the position of DR on the patch surface/feedline as well as

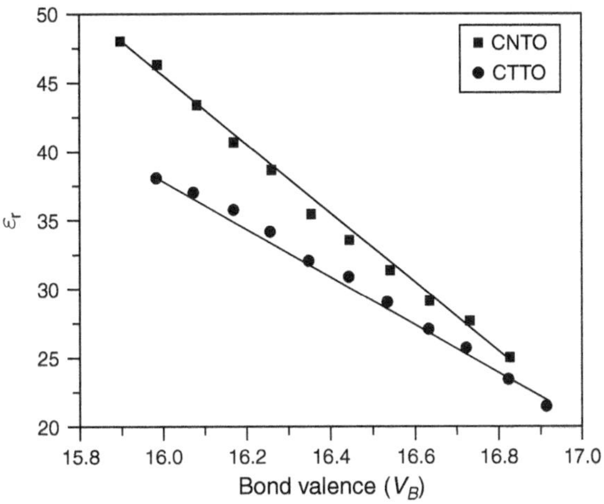

Figure 10.10 Variation of ε_r with bond valence in $Ca_5Nb_2Ti_{1-x}Zr_xO_{12}$ and $Ca_5Ta_2Ti_{1-x}Zr_xO_{12}$ ceramics (after Ref. [21]).

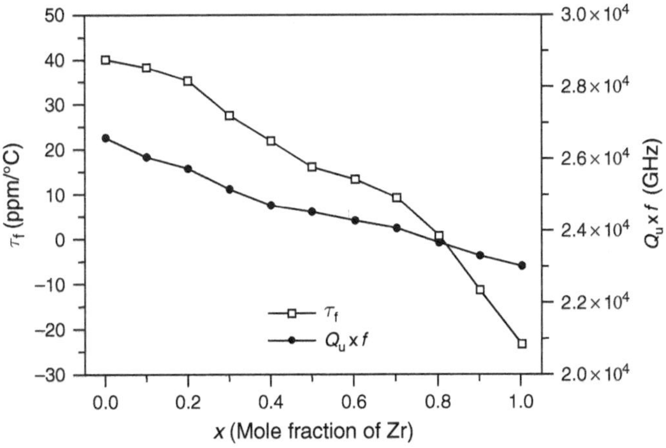

Figure 10.11 Variation of τ_f and Qf with x in $Ca_5Nb_2Ti_{1-x}Zr_xO_{12}$. (after Ref. [21]).

the value of relative permittivity and resonant frequency for maximum percentage bandwidth is optimized. Cylindrical dielectric resonators with ε_r varying in the range 9–92 were used for the study. The $Ca_5Nb_2TiO_{12}$ with $\varepsilon_r = 48$ was found to be the best suited for maximum enhancement of impedance bandwidth of the antenna as shown in Figure 10.14. Furthermore, it was found that the maximum coupling of the electromagnetic energy between the feed and the DR occurs when the resonant frequency of the DR matches with that of the patch antenna. A bandwidth of more than 10% is achieved by loading a dielectric resonator of $\varepsilon_r = 48$ over the patch and about 14% by

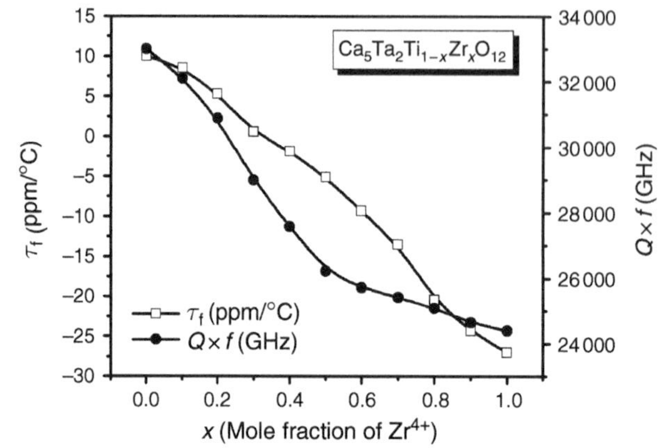

Figure 10.12 Variation of τ_f and Qf with x in $Ca_5Ta_2Ti_{1-x}Zr_xO_{12}$. (after Ref. [21]).

loading the same DR on the feedline. DR loading produced microstrip patch antennas with 5–7-fold enhancement in their impedance bandwidth. The studies also revealed that these methods will not adversely affect other properties of the antenna especially its gain and radiation efficiency.

10.6 Conclusions

The $Ca_5Nb_2TiO_{12}$ and $Ca_5Ta_2TiO_{12}$ ceramics can be prepared by the conventional solid-state ceramic route by sintering at 1550 and 1625°C. These complex perovskite materials with three types of cations in the B site have orthorhombic symmetry and belong to the *Pnma* space group. The $Ca_5Nb_2TiO_{12}$ has $\varepsilon_r = 48$, $Qf > 26\,000$ GHz and $\tau_f = +40$ ppm/°C, whereas $Ca_5Ta_2TiO_{12}$ has $\varepsilon_r = 38$, $Qf > 33\,000$ GHz and $\tau_f = 10$ ppm/°C. Addition of divalent dopants such as MgO, NiO, ZnO and CuO, trivalent Cr_2O_3, In_2O_3 and pentavalent Sb_2O_5 were found to improve the microwave dielectric properties of $Ca_5Nb_2TiO_{12}$ and $Ca_5Ta_2TiO_{12}$ ceramics. It was found that dopants with ionic radii comparable to that of the average B-site ionic radius improve the microwave dielectric properties of $Ca_5Nb_2TiO_{12}$ and $Ca_5Ta_2TiO_{12}$ ceramics. Addition of a small amount of SiO_2, $MgO–B_2O_3–SiO_2$, $Al_2O_3–SiO_2$ and $Al_2O_3–B_2O_3–SiO_2$ and $2MgO–Al_2O_3–5SiO_2$ increased the density and improved the microwave dielectric properties of $Ca_5B_2TiO_{12}$ (B = Nb, Ta) ceramics. The improvement of microwave dielectric properties were more pronounced with ternary glasses than that with primary and binary glasses. Partial substitution of Mg, Ni, Co and Zn for Ca increases Qf, decreases ε_r and improves τ_f. Mg, Zn, Ni and Co substitute only up to $x = 1$ and for $x > 1$ a multiphase ceramic is formed. It is found that dielectric resonators with ε_r close to 48 increase the bandwidth of microstrip antennas with excellent radiation characteristics. A bandwidth of more than 10% is achieved by loading a dielectric resonator of $\varepsilon_r = 48$ over the patch and about 14% by loading the same DR on the feedline.

Table 10.1 Microwave dielectric properties of $A(A_{1/4}B_{2/4}C_{1/4})O_3$ dielectric ceramics

Material	Sintering temperature (°C)	ε_r	Qf (GHz)	Frequency	τ_f (ppm/°C)	References
$Ca_5Nb_2TiO_{12}$	1550	48	26 000	3.683	40	[12, 13]
$Ca_{4.35}Mg_{0.65}Nb_2TiO_{12}$	1550	41	33 000	4.106	≈0	[23]
$5MgO–Nb_2O_5–TiO_2$	1325	15	59 000	6.801	−77	[23]
$Ca_{4.36}Zn_{0.64}Nb_2TiO_{12}$	1550	43	29 000	4.004	≈0	[23]
$Ca_{4.38}Ni_{0.62}Nb_2TiO_{12}$	1550	42	28 200	4.074	≈0	[23]
$Ca_{4.18}Co_{0.82}Nb_2TiO_{12}$	1550	37	30 000	4.308	≈0	[23]
$5CoO–Nb_2O_5–TiO_2$	1010	9.0	41 000	7.459	−59	[23]
$Ca_5Nb_2Ti_{0.2}Zr_{0.8}O_{12}$	1670	34	24 000	4.413	≈0	[21]
$Ca_5Nb_2Ti_{0.4}Hf_{0.6}O_{12}$	1675	32	22 000	4.458	≈0	[22]
$Ca_5Nb_2TiO_{12}$ + 0.1 wt% 2 + $MgO–Al_2O_3–5SiO_2$	1520	50	30 000	–	38	[19]
$Ca_5Ta_2TiO_{12}$	1625	38	33 000	4.253	+10	[13]

(*Continued*)

Table 10.1 (Continued)

Material	Sintering temperature (°C)	ε_r	Qf (GHz)	Frequency	τ_f (ppm/°C)	References
$Ca_5NbTaTiO_{12}$	1580	43	30 000	–	28	[24]
$Ca_{4.82}Mg_{0.18}Ta_2TiO_{12}$	1625	37	36 000	4.356	≈0	[23]
$5MgO–Ta_2O_5–TiO_2$	1325	18	114 000	6.611	–56	[23]
$Ca_{4.85}Zn_{0.15}Ta_2TiO_{12}$	1625	37	35 000	4.154	≈0	[23]
$Ca_{4.75}Ni_{0.25}Ta_2TiO_{12}$	1625	35	34 000	4.496	≈0	[23]
$Ca_{4.88}Co_{0.12}Ta_2TiO_{12}$	1625	36	35 000	4.488	≈0	[23]
$5CoO–Ta_2O_5–TiO_2$	1150	14	48 000	6.528	–43	[23]
$Ca_5Ta_2Ti_{0.7}Zr_{0.3}O_{12}$	1650	36	28 000	4.413	≈0	[21]
$Ca_5Ta_2Ti_{0.6}Hf_{0.4}O_{12}$	1675	34	26 000	4.357	≈0	[22]
$Ca_5Ta_2TiO_{12}$ + 0.2 wt% $Al_2O_3–B_2O_3–SiO_2$	1550	38	38 000		8	[20]
$Ca_5Ta_2TiO_{12}$ + 0.1 wt% $2MgO–Al_2O_3–5SiO_2$	1550	38	40 000		5	[20]

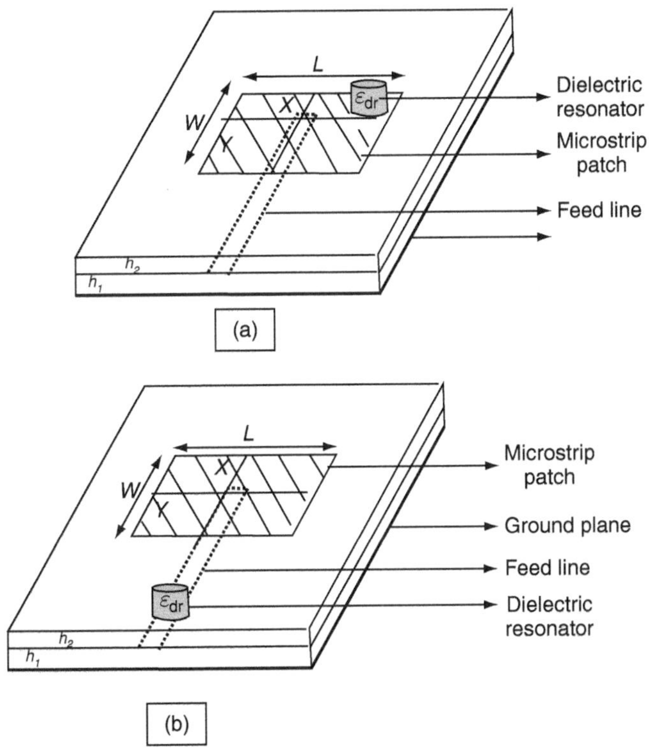

Figure 10.13 Geometry of DR-loaded microstrip patch antenna (a) DR over the patch (b) DR on the feedline (after Ref. [26]).

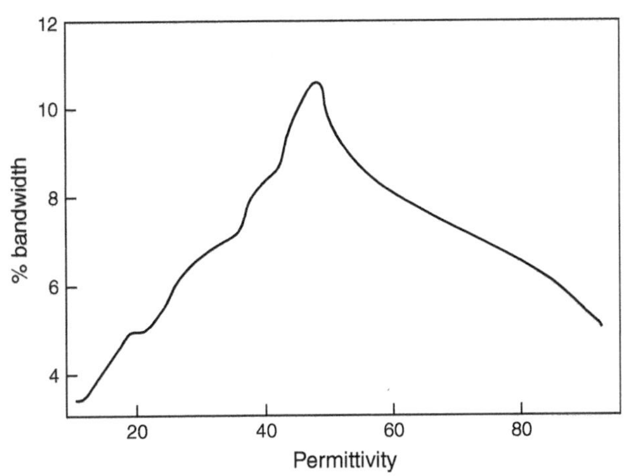

Figure 10.14 Variation of percentage bandwidth of the DR-loaded antenna as a function of permittivity of the DR (after Ref. [26]).

REFERENCES

[1] R. Roy. Multiple ion substitution in the perovskite lattice. J. Am. Ceram. Soc. 37(1954) 581–588.
[2] R. J. Cava, J. J. Krajewski, and R. S. Roth. $Ca_5Nb_2TiO_{12}$ and $Ca_5Ta_2O_{12}$: low temperature coefficient low loss dielectric materials. Mater. Res. Bull. 34(1999)355–362.
[3] R. J. Cava. Dielectric materials for applications in microwave communications. J. Mater. Chem. 11(2001)54–62.
[4] R. J. Cava, J. J. Krajewski, and R. S. Roth. Stabilization of the low temperature coefficient of dielectric constant of $Ca_5Nb_2TiO_{12}$ by Zr doping. Mater. Res. Bull. 34(1999)1817–1824.
[5] L. A. Bendersky, J. J. Krajewski, and R. J. Cava. Dielectric properties and microstructure of $Ca_5Nb_2TiO_{12}$ and $Ca_5Ta_2O_{12}$. J. Eur. Ceram. Soc. 21(2001)2653–2658.
[6] L. A. Bendersky, I. Levin, R. S. Roth, and A. J. Shapiro. $Ca_4Nb_2O_9$–$CaTiO_3$: phase equilibria and microstructures. J. Solid State Chem. 160(2001)257–271.
[7] P. L. Wise, I. M. Reaney, W. E. Lee, T. J. Price, D. M. Iddles, and D. S. Cannell. Structure-microwave property relations in $(Sr_xCa_{1-x})_{n+1}Ti_nO_{3n+1}$. J. Eur. Ceram. Soc. 21(2001)1723–1726.
[8] H. Kagata and J. Kato. Dielectric properties of Ca based complex perovskite at microwave frequencies. Jpn. J. Appl. Phys. 33(1994) 5463–5465.
[9] M. Hervieu, F. Studer, and B. Raveu. Oxydes de type perovskite du systeme Ca–Nb–O. J. Solid State Chem. 22(1977)273–289.
[10] I. Levin, L. A. Bendersky, J. P. Cline, R. S. Roth, and T. A. Vanderah. Octahedral tilting and cation ordering in perovskite like $Ca_4Nb_2O_9$: $Ca(Ca_{1/3}Nb_{2/3})O_3$ polymorphs. J. Solid State Chem. 150(2000)43–61.
[11] R. D. Shannon. Revised effective ionic radii and systematic studies of interatomic distances in halides and chalcogenides. Acta Cryst. A 32(1976)751–767.
[12] P. V. Bijumon, Ph D. Thesis. Novel low loss $A(A_{1/4}B_{2/4}C_{1/4})O_3$ dielectrics and their applications in broadband antennas. Cochin University of Science & Technology. Cochin (2006).
[13] P. V. Bijumon, P. Mohanan, and M. T. Sebastian. Synthesis, characterization and properties of $Ca_5A_2TiO_{12}$ (A = Nb, Ta) ceramic dielectric materials for applications in microwave communication systems. Jpn. J. Appl. Phys. 41(2002)3834–3835.
[14] A. Dias, P. V. Bijumon, M. T. Sebastian, and R. L. Moreira. Vibrational spectroscopy and microwave dielectric properties of $Ca_{5-x}Ba_xNb_2TiO_{12}$ and $Ca_{5-x}Ba_xTa_2TiO_{12}$ ceramics. J. Appl. Phys. 98(2005)084105.
[15] P. V. Bijumon, P. Mohanan, and M. T. Sebastian. High dielectric constant low loss microwave dielectric ceramics in the $Ca_5Nb_{2-x}Ta_xTiO_{12}$ system. Mater. Lett. 57(2003)1380–1384.
[16] P. V. Bijumon and M. T. Sebastian. Doped $Ca(Ca_{1/4}A_{2/4}Ti_{1/4})O_3$ (A = Nb, Ta) dielectrics for microwave telecommunication applications. Int. J. Appl. Ceram. Technol. 4(2007)60–74.
[17] K. P. Surendran, M. T. Sebastian, P. Mohanan, and M. V. Jacob. The effect of dopants on the microwave dielectric properties of $Ba(Mg_{0.33}Ta_{0.67})O_3$ ceramics. J. Appl. Phys. 98(2005) 094114.
[18] M. R. Varma and M. T. Sebastian. Effect of dopants on microwave dielectric properties of $Ba(Zn_{1/3}Ta_{2/3})O_3$ ceramics. Jpn. J. Appl. Phys. 44(2005)298–303.
[19] P. V. Bijumon and M. T. Sebastian. Influence of glass additives on the microwave dielectric properties of $Ca_5Nb_2TiO_{12}$ ceramics. Mater. Sci. Eng. B 123(2005)31–40.
[20] P. V. Bijumon and M. T. Sebastian. Tailoring the microwave dielectric properties of $Ca_5Ta_2TiO_{12}$ ceramics through glass addition. J. Am. Ceram. Soc. 88(2005)3433–3439.
[21] P. V. Bijumon and M. T. Sebastian. Temperature stable microwave dielectric ceramics in the $Ca_5A_2Ti_{1-x}Zr_xO_{12}$ (A = Nb, Ta) system. J. Mater. Res. 19(2004)2922–2928.
[22] P. V. Bijumon and M. T. Sebastian. Microwave dielectric properties of temperature stable $Ca_5A_2Ti_{1-x}Hf_xO_{12}$ (A = Nb, Ta) ceramics. J. Electroceram. 16(2006)239–245.
[23] P. V. Bijumon, M. T. Sebastian, and P. Mohanan. Experimental investigations and three dimensional transmission line matrix simulation of $Ca_{5-x}A_xB_2TiO_{12}$ (A = Mg, Zn, Ni and Co: B = Nb and Ta) ceramic resonators. J. Appl. Phys. 98(2005)124105.

[24] P. V. Bijumon, M. T. Sebastian, A. Dias, R. L. Moreira, and P. Mohanan. Low loss $Ca_{5-x}Sr_x A_2TiO_{12}$ [A = Nb, Ta] ceramics. Microwave dielectric properties and vibrational spectroscopic analysis. J. Appl. Phys. 97(2005)104108.
[25] F. S. D. Gallaso. Perovskite and High Tc Superconductors. Gordon & Breach Science Publishers, Reading UK. (1990).
[26] P. V. Bijumon, S. K. Menon, M. T. Sebastian, and P. Mohanan. Enhanced bandwidth microstrip patch antennas loaded with high permittivity dielectric resonators. Microwave and Opt. Technol. Lett. 35(2002)327–330.

CHAPTER ELEVEN

ALUMINA, TITANIA, CERIA, SILICATE, TUNGSTATE AND OTHER MATERIALS

11.1 ALUMINA

Alumina has a high melting point of about 2050°C with ε_r of about 10, relatively high thermal conductivity and low dielectric loss and is a well-known ceramic packaging material. Most of the properties of the high-purity alumina such as thermal expansion, thermal conductivity, elastic modulus, melting point, Poisson's ratio and permittivity fall within a narrow range for different samples from a wide range of sources. However, the dielectric loss is highly variable from sample to sample varying over several orders of magnitude. Powder purity is an important factor in the production of low-loss alumina [1–4]. In polycrystalline alumina, the quality factor is dramatically decreased [5, 6] by the presence of a very small amount of alkali ions and metallic impurities such as K, Na and Fe but the presence of a very small amount of TiO_2 considerably improved the quality factor [1]. It was found [1] that addition of 0.5 wt% TiO_2 lower the sintering temperature to about 1500°C with considerable increase in the quality factor up to 50 300 at 9 GHz which is close to that of a single-crystal sapphire. Figure 11.1 shows the variation of Q as a function of wt% of TiO_2 addition. Addition of more than 0.5 mol% TiO_2 considerably lowers the quality factor. The *Qf* increased with increase in density in polycrystalline alumina [1]. The presence of porosity has a tremendous influence on dielectric loss. Alford and coworkers [5, 6] studied the effect of porosity and grain size on the microwave (MW) dielectric properties of sintered alumina. The purity of the starting powder, the porosity and the grain size have been varied and the influence of these variations on ε_r and tan δ was investigated. The samples were sintered at 1000–1600°C for up to 30 hours to vary the density and thus porosity. Figure 11.2 shows the effect of porosity on the loss factor in polycrystalline alumina [5, 6]. As the pore volume increased, the permittivity decreased as expected. The dielectric loss was found to depend strongly on the pore volume with only a small degree of porosity having a very marked effect on the loss factor. A variation in the grain size did not affect the permittivity in nearly dense samples and caused an increase in the dielectric loss when the grain size in the alumina exceeded approximately 3–4 μm. Very low loss of about 2.4×10^{-5} has been observed in polycrystalline alumina with grain sizes less than about 3 μm. The loss increased with increase in grain size as shown in Figure 11.3. The loss decreased on cooling and the decrease was rather fast as the perfection of the sample improved [5]. The polycrystalline ceramics always exhibit higher losses than single crystals but in high-quality ceramics the difference can be small. Polycrystalline alumina with extremely low MW dielectric loss is reported to have properties analogues to theoretical ensemble of randomly oriented single-crystal sapphire grains [5]. Bragisnsky and Ilschenko [7] have shown that below 50 K the loss is strongly dependent on the level of

Dielectric Materials for Wireless Communications
ISBN-13: 978-0-08-045330-9

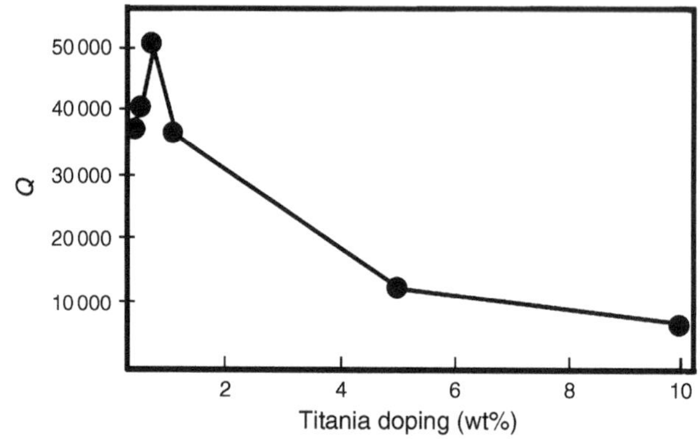

Figure 11.1 Variation of Q factor of alumina as a function of wt% TiO_2 doping (after Ref. [1]).

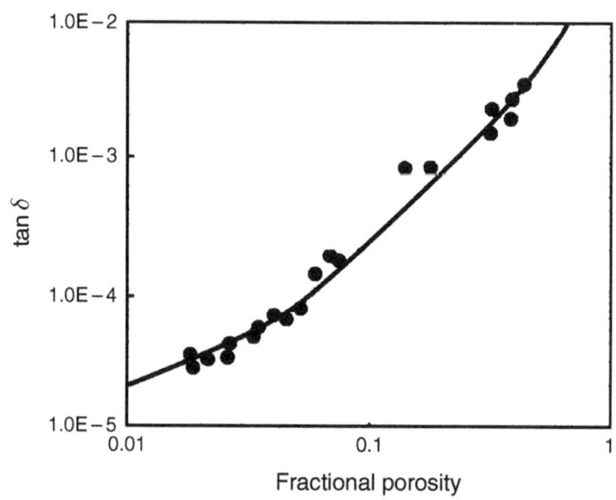

Figure 11.2 Variation of tan δ with fractional porosity (after Ref. [5]).

crystal perfection as determined by the rate of crystal growth. Ikeda and Miura [8] studied the tan δ of single crystals of alumina and reported that alumina without disorientation had a low tan δ in agreement with the theory of Gurevich and Tagantsev [9]. The crystal disorientation caused a deleterious influence on dielectric loss. By avoiding deleterious impurities and by careful control of microstructure, Breeze et al. [10] showed that grain boundaries in alumina have only limited influence on the dielectric loss. They reported that high-purity, fine grained polycrystalline alumina possess very low dielectric loss of about 8.33×10^{-6} at 10 GHz at 300°K. This is the lowest loss factor reported for a ceramic material at room temperature. It is difficult to densify pure alumina without sintering aids. It is therefore important that these sintering aids do not adversely affect the loss. Alford and Penn [1] showed that the dielectric loss of sintered materials can approach to that measured in single crystals. Several authors [1, 11, 12] investigated the loss factor of alumina of different purity levels and concluded that

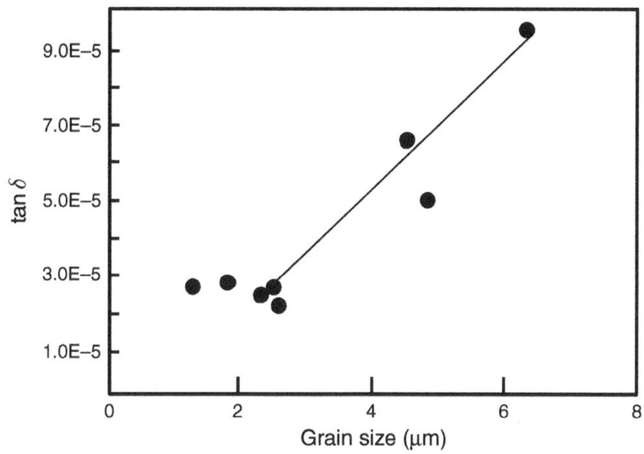

Figure 11.3 Influence of grain size on tan δ in alumina (after Ref. [6]).

impure alumina always gives a poor loss. High purity, proper doping, correct processing and thus good microstructure are required for low dielectric loss. The Q factor of alumina varies from manufacturer to manufacturer. The results indicate that purity alone is a poor indicator of the resulting dielectric loss tangent. It is found that purity influence the quality factor but purity alone is not a guarantee for high Q [1, 11, 12]. Several authors reported [1, 2, 4, 6, 13–18] a low dielectric loss in alumina at MW frequency range. Figure 11.4 shows the variation of ε_r, Qf and τ_f as a function of sintering temperature. The ε_r and τ_f are not much affected by the variation in the sintering temperature, whereas the Qf increases with increase in sintering temperature. Ohsato et al. [13] attributed the high purity of the alumina raw material as the origin of the high Qf value. Huang et al. [19, 20] found that use of TiO_2 additive can considerably lower sintering temperature with excellent quality factor. The sample containing 8 wt% of nano-TiO_2 sintered at 1350°C for 4 hours had $\varepsilon_r = 10.8$ with $Qf = 338\,000$ GHz and $\tau_f = 1.3$ ppm/°C.

Several authors lowered the sintering temperature of alumina by adding low melting glasses and are discussed in Chapter 12. Molla et al. [3] studied the effect of moisture on tan δ in alumina. They measured the MW dielectric loss in highly porous alumina in both the dry and moist atmospheres. Figure 11.5 shows the variation of tan δ with time in 42% porous alumina after introduction of dry nitrogen gas in to the resonant measuring cavity. The tan δ measured at 15 GHz decreased with time on introducing dry gas. This means that moisture content in the cavity increased the tan δ. Molla et al. [3] attributed the ionic conductivity on the water surfaces to be the reason for the origin of high dielectric losses in porous alumina. Alford et al. [21] investigated the effect of different binders on the dielectric properties of alumina ceramic. They observed significant differences in the quality factor of alumina using different binders. They found that polyethylene glycol (PEG) 3350 and PEG 20 000 give the best quality factors. The green body strength increased with increase in the binder concentration. However, the porosity left by the binders and the impurities present in the binders can affect the quality factor considerably [21].

Several people studied [22–29] the quality factor of sapphire using WGM. WGM mode resonators are very attractive to design high Q and high spectral purity MW

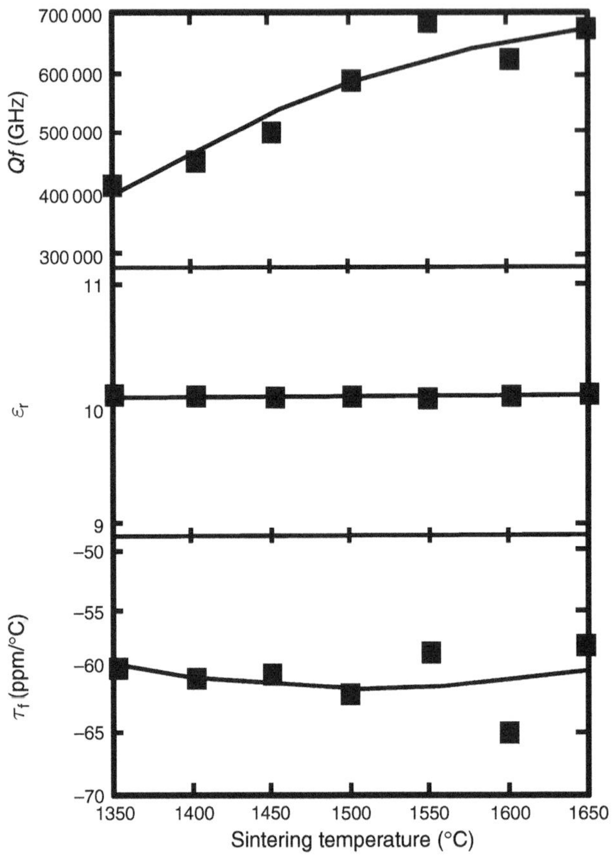

Figure 11.4 Variation of ε, Qf, τ_f as a function of sintering temperature in high-purity Al_2O_3 (after Ref. [13]).

oscillators. It has been reported that use of cooled sapphire DRs resulted in a stable ultrahigh-frequency control elements. Resonator operation of high-order modes has resulted in attainment of unloaded Q's in the range 10^7–10^9 at X band in liquid nitrogen to helium temperature range [23]. The loss tangent of sapphire decreases significantly with temperature [12] from 10^{-5} at room temperature to about 10^{-7} at 77°K. At liquid helium temperatures (4.2 K), the Q factors $>10^9$ have been reported [5, 25–30]. The helium-cooled systems are larger in size and expensive. However, liquid-nitrogen-cooled systems are smaller and less expensive to maintain. The Q of liquid-nitrogen-cooled resonators (77 K) are low as compared to 4.2 K but is still significantly higher than the Q factor measured at room temperature [6, 30, 31]. Sapphire WGM mode resonator shows a Q as high as 200 000 at ambient temperature and higher than 10 million at 77 K [13]. Woode et al. [32] reported a sapphire-loaded cavity resonator with $Q = 6 \times 10^7$ at 8.95 GHz and at 77 K. Tobar et al. [30] reported a $Q = 2 \times 10^5$ at 300 K and a $Q = 5 \times 10^7$ at 77 K. Luiten et al. [33] reported a $Q = 8.3 \times 10^9$ at 12.7 GHz at 1.55 K. This is the highest quality factor reported for a material. Driscoll et al. [24] reported an

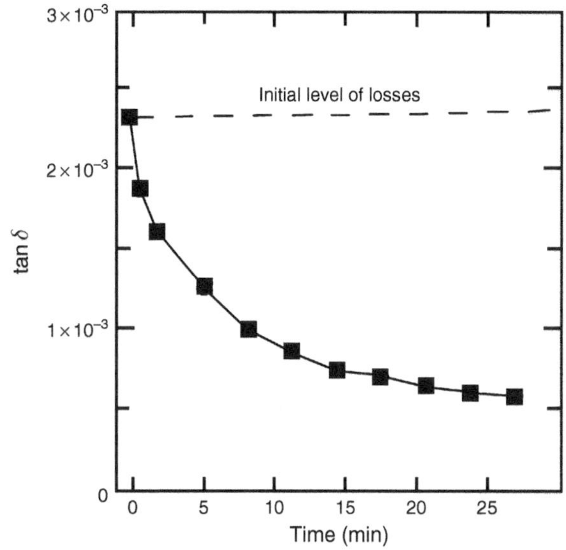

Figure 11.5 Variation of tan δ with time in 42% porous alumina after introduction of dry N_2 gas (after Ref. [3]).

unloaded quality factor of 5×10^6 [24] for sapphire DRs operating on a low-order TE (transverse electric) mode at 77°K employing high-temperature superconducting films.

The low dielectric loss, suitable ε_r and relatively high thermal conductivity makes alumina a suitable material for electronic packaging applications. The thermal conductivity of alumina is very high at room temperature (30 W/m.K) resulting in better dissipation of heat in high-power filters. However the τ_f of alumina is too large. Alumina has a very high *Qf* value of about one million [1, 2, 10, 13–18] with $\varepsilon_r = 9.8$ and $\tau_f = -60$ ppm/°C at room temperature and TiO_2 has a high *Qf* of 48 000 GHz, $\varepsilon_r = 100$ and a positive τ_f of 450 ppm/°C [34]. Hence the possibility of tailoring the τ_f of alumina with the use of rutile as a τ_f compensator. Alford and coworkers [34, 35] achieved temperature compensation in alumina by coating a film of TiO_2 over the surface of alumina disk. The composite resonators obtained by firing at 1400°C showed temperature compensation depending on the volume fraction of TiO_2. The result was a dense layer of TiO_2 on a dense alumina disc. The sintered TiO_2 films have thickness in the range 0–20 μm corresponding to volume fractions of 0.003–0.03. Figure 11.6 shows the variation of τ_f with volume fraction of TiO_2. A volume fraction of 0.015 yielded a $\tau_f = 0$ with $Q = 30\,000$. Figure 11.7 shows the variation of Q with volume fraction of TiO_2. Tzou et al. [36] obtained temperature compensation in Al_2O_3–TiO_2 composite with glass additives. Although they obtained a $\tau_f = 0$, the *Qf* was very much degraded due to the formation of Al_2TiO_5 secondary phase. Ohsato and coworkers [14, 15] prepared $0.9Al_2O_3$–$0.1TiO_2$ by sintering at different temperatures in the range up to 1550°C. They found the formation of secondary phase of Al_2TiO_5 on sintering at high temperatures. On annealing at about 1000°C, the Al_2TiO_5 formed at high temperatures decomposed into Al_2O_3 and TiO_2 compensating the negative τ_f of alumina. Figure 11.8 shows the X-ray diffraction (XRD) pattern of $0.9Al_2O_3$–$0.1TiO_2$ sintered at 1350°C and annealed at different temperatures. It is evident from the figure that annealing decomposed Al_2TiO_5 into alumina and rutile.

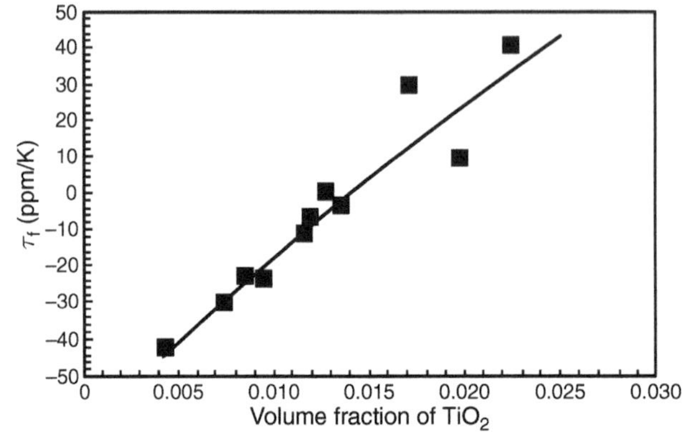

Figure 11.6 Variation of τ_f with volume fraction of TiO_2 (after Ref. [34]).

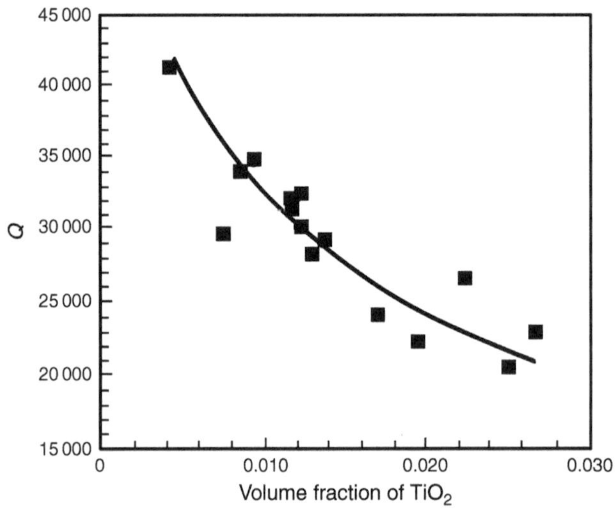

Figure 11.7 Variation of Q factor of composite dielectric resonator with TiO_2 volume fraction (after Ref. [34]).

The sample annealed at 1000°C for 2 hours had $Qf = 117\,000$ GHz with $\varepsilon_r = 12.4$ and $\tau_f = 1.5$ pm/°C. Figure 11.9 shows the variation of τ_f for as-sintered and postannealed $0.9Al_2O_3$–$0.1TiO_2$ as a function of sintering temperature. Kono et al. [37] reported that the addition of 0.4 mol% MnO in $0.9Al_2O_3$–$0.1TiO_2$ lowered the sintering temperature to 1300°C and thereby suppress the formation of Al_2TiO_5. Tobar and coworkers showed that it is possible to compensate the frequency–temperature dependence of a WGM sapphire resonator by making a sapphire–rutile composite [28, 38]. The composite consists of a sapphire resonator with rutile rings at the end faces of the sapphire resonator. Hartnett et al. [38] reported frequency–temperature compensated sapphire–rutile resonator with Q factors more than 10^7. The high negative τ_f of sapphire was compensated by the use of

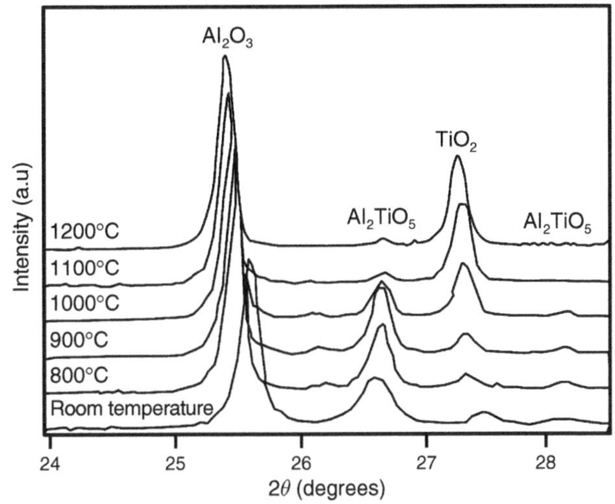

Figure 11.8 XRD pattern of $0.9Al_2O_3$–$0.1TiO_2$ (after Ref. [15]).

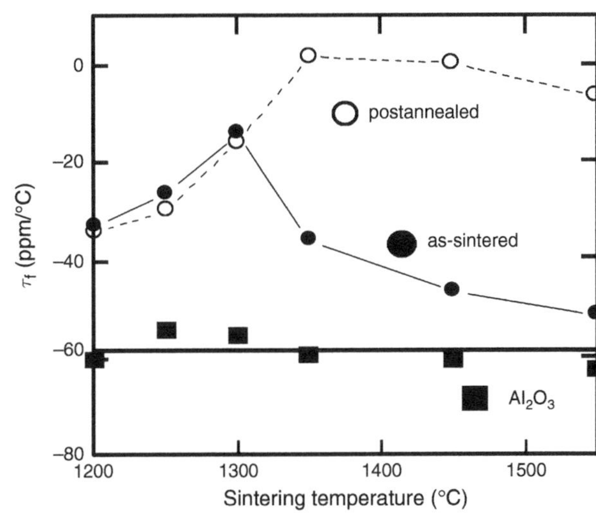

Figure 11.9 Variation of τ_f of $0.9Al_2O_3$–$0.1TiO_2$ ceramics with sintering temperature (after Ref. [15]).

two thin rings of rutile at the top and bottom of the sapphire resonator. The frequency–temperature dependence was annulled when a specific balance of electric energy in the rutile and sapphire was reached at 56 K in a WGE_{900} mode at 13.1 GHz with a Q factor of 30 million. Thus, it is possible to annul the frequency–temperature dependence of sapphire DRs using a dielectric of the opposite frequency–temperature dependence. Tobar et al. [30] using thin $SrTiO_3$ wafers, the τ_f dependence was annulled at around 100°K with Q factors varying in the range 20 000–50 000. Residual paramagnetic impurities in a sapphire resonator can provide [39] a means of temperature compensation so that the effect of temperature fluctuations on the frequency is compensated. High-purity monocrystalline sapphire at cryogenic temperatures is an excellent dielectric for high-stability resonator applications such as ultralow noise, ultrastable MW oscillators, owing to its mechanical rigidity, low thermal expansion and very low dielectric loss [25, 40–42].

11.2 TITANIA

The titanium dioxide crystallizes in three forms: rutile, brookite and anatase. Brookite and anatase converts irreversibly to rutile in the temperature range 700–920°C. Rutile is tetragonal, and in certain oxidation states, is an oxide semiconductor. If sintering occurs in an air atmosphere or under low oxygen partial pressure, a slight reduction of TiO_2 occurs (Ti^{4+} to Ti^{3+}). Sintering temperature up to 1500°C are required to attain dense samples of rutile, although lower sintering temperatures are possible in much finer powders [43]. Okaya and Barash in 1962 were the first to experiment a DR and they did it using a rutile single crystal [44]. In 1968 Cohen [45] for the first time experimentally determined the MW dielectric properties of rutile resonator and reported that it has $\varepsilon_r = 100$, $Q = 10\,000$ at 3.45 GHz and $\tau_f = +400$ ppm/°C. Egerton and Thomson [46] reported dense rutile ceramics having low loss at MW frequency range prepared by hot pressing. The dense rutile with quality factor $Q > 10\,000$ at the MW frequency range was obtained by hot pressing the rutile powder in a graphite die. High-purity zirconia was used to isolate the TiO_2 from direct contact with graphite. They reported that the heat treatment of the samples at elevated temperatures in oxygen atmosphere was needed to restore oxygen stoichiometry and to get the high quality factor of the ceramic. Addition of trivalent cations like Fe, Cr, Co or Al to rutile gives better results. These additives ease the critical nature of the reoxidation cycle.

Templeton et al. [47] prepared TiO_2 using rutile raw powder as the starting material. The raw powders were mixed with dopants, ball-milled and sintered to almost full density at 1500°C. Figure 11.10 shows the plot of Q of undoped sintered TiO_2 as a

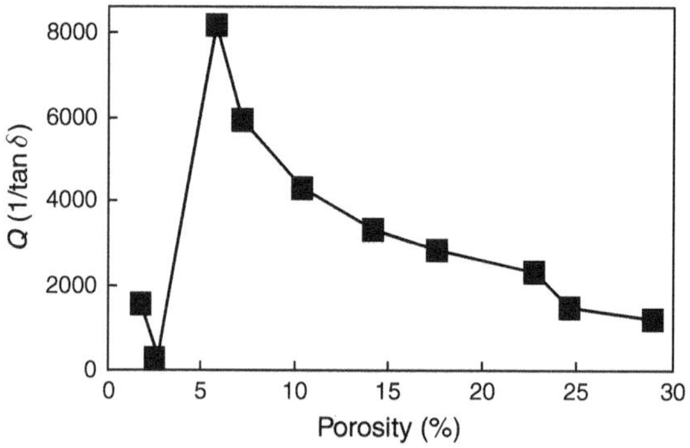

Figure 11.10 Plot of Q versus porosity in TiO_2 (after Ref. [47]).

function of porosity. The Q value steadily increased from a porosity of 30% to a value of about 6%. However, at a porosity <6% the Q value rapidly decreased. This was attributed [47] to oxygen deficiency in the dense or almost fully dense material. During sintering at high temperatures of about 1500°C the material achieves >97% of the theoretical density and there is no interconnected porosity. At these temperatures, rutile is slightly oxygen deficient [48] and the diffusion coefficient [49] even at 1400°C is too low to allow oxygenation of the thick pellets. The dense TiO_2 appears dark tan in color. However, on polishing the sample, a well-defined dark core region was observed indicating the presence of a reduced Ti species (Ti^{3+}). Such coring due to Ti^{4+} reduction has also been reported in (Zr, Sn)TiO_4 and $Ba_2Ti_9O_{20}$/$BaTi_4O_9$ [50, 51]. It was reported [50, 51] from EPMA studies that the reduction was below 0.2% and this limited reduction was enough to make severe degradation of the quality factor. Templeton et al. also doped TiO_2 with ions in the valency range +1 to +5 [47]. They found that divalent and trivalent dopants with ionic radius in the range 0.5–0.95 Å improved the quality factor of TiO_2. It was found that doping with 0.05 mol% of Fe^{3+}, Al^{3+}, Zn^{2+}, Cu^{2+}, Mn^{2+}, Y^{3+} and Mg^{2+} increased the quality factor. The maximum quality factor was found for Fe^{3+}- and Zn^{2+}- doped samples. No evidence of coring was observed and the color of these doped ceramics was similar to that for the undoped TiO_2. Reduction of the Ti^{4+} ion was prevented by a favorite compensation mechanism when TiO_2 is doped with these ions. Figure 11.11 plots the tan δ versus temperature for undoped, 0.05 mol% alumina-doped TiO_2 and single-crystal rutile. In the undoped TiO_2, the loss is fairly constant and high until a temperature of ~100 K is reached. On cooling below 100 K, the tan δ decreased sharply which was attributed to possible freeze out of charge carriers in the conduction band [47]. In the alumina-doped and rutile single crystal, the loss was significantly reduced throughout the temperature range. For undoped TiO_2, the Q value, relative to pore volume displayed unusual behavior of achieving a maximum at 5–7% porosity, which was coincident with the onset of interconnected porosity (<5–7%). For <5% porosity, the Q value was very poor (<2000), due to oxygen deficiency. To overcome this problem, doping with divalent or trivalent ions with ionic radii in the range 0.5–0.95 Å can enhance the Q value considerably, to ~17 000 at a frequency of 3 GHz and temperature 300 K.

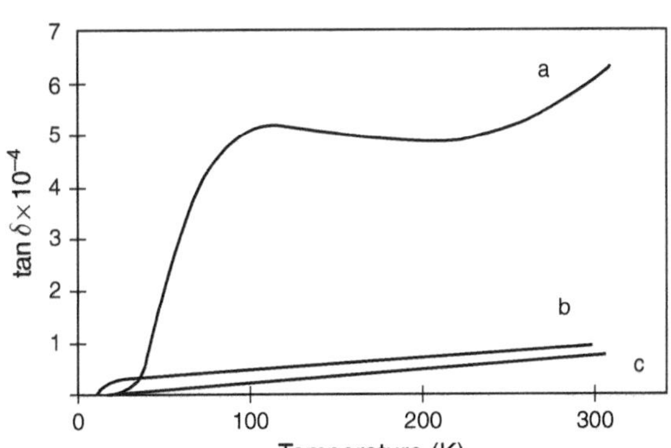

Figure 11.11 Variation of tan δ versus temeperature for (a) undoped TiO_2, (b) 0.5 mol% Al_2O_3-doped TiO_2 and (c) single-crystal rutile measured at 3 GHz (after Ref. [47]).

In TiO_2-doped with divalent and trivalent ions in the preferred size range 0.5–0.95 Å, the tan δ value decreased smoothly from 6×10^{-5} at a temperature of 300 K to a value of 5×10^{-6} ($Q \sim 200\,000$) at 15 K which is comparable to that of single crystals. Kim et al. [52, 53] reported that addition of 1 wt% CuO considerably reduced the sintering temperature. The TiO_2 + 1 wt% CuO sintered at 950°C had $\varepsilon_r = 100$ and $Qf = 16\,400$ GHz. Addition of more than 2 wt% CuO lowered the Qf with a decrease in τ_f. XRD study showed a mixture of CuO and TiO_2 with no secondary phases. The CuO has $\varepsilon_r = 11.2$, $Q = 70$ at 10 MHz and $\tau_f = -110$ ppm/°C. Thus the decrease in ε_r, Q and τ_f in the CuO-doped TiO_2 ceramics. Several authors [54, 55] studied the effect of addition of glass in TiO_2–ceramic composites. Glass–ceramic composites containing TiO_2 and modified borosilicate glasses were prepared and their sintering behavior, phase evolution, interface reactions and MW properties were investigated in detail for low-temperature cofired ceramics (LTCC) and are discussed in Chapter 12.

Klein et al. [56] investigated the MW dielectric properties of single-crystal rutile. Figure 11.12 shows the Q and resonant frequency versus temperature in high-temperature superconductor (HTS)-shielded rutile resonator. High-purity rutile single crystals exhibit extraordinary low losses at MW frequencies and cryogenic temperatures. In combination with HTS films, miniaturized resonators with Q's in the 10^5–10^6 range were obtained. Tobar et al. [57] used WGM to measure the Q of rutile single crystals at low temperatures. The rutile had $Q = 10^4$ at 300 K, 10^5 at 80 K and 10^7 at 10 K. The rutile shows anisotropy both in permittivity and loss factor as shown in Figures 11.13 and 11.14. The $\varepsilon_{\parallel}$ and $\varepsilon_{\perp}$ and tan $\varepsilon_{\parallel}$ and tan $\varepsilon_{\perp}$ are the relative permittivities and the loss factors parallel and perpendicular to the anisotropic c-axis. Tobar et al. [57, 58] reported a high Q sapphire–rutile frequency compensated DR s with a Q factor of the order of a few million below 90 K.

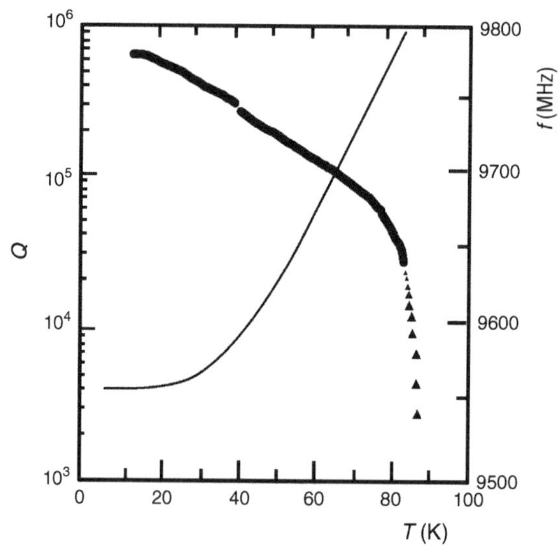

Figure 11.12 Variation of Q and resonant frequency in megahertz with temperature with HTS-shielded rutile resonator (after [56]).

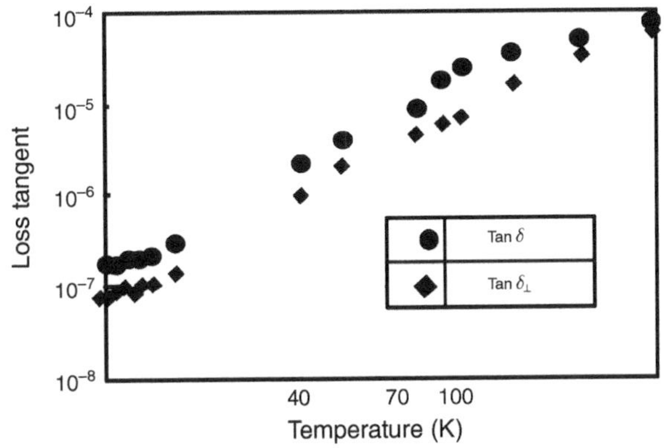

Figure 11.13 Variation of loss tangent component with temperature showing anisotropy (after Ref. [57]).

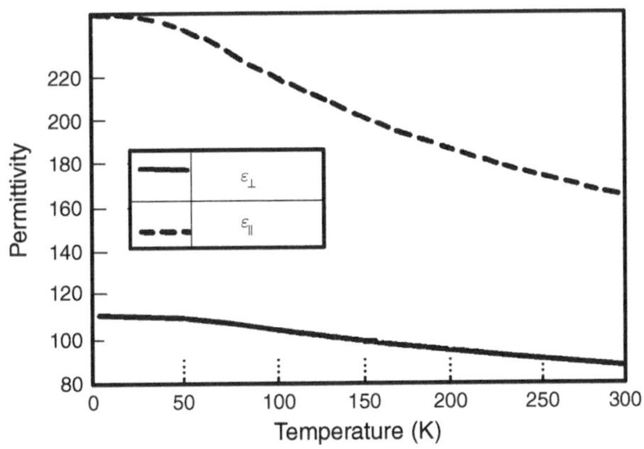

Figure 11.14 Variation of permittivity with temperature showing anisotropy in TiO_2 (after Ref. [57]).

11.3 CeO_2

Cerium oxide (CeO_2) with a cubic fluorite structure and space group *Fm*3*m* is an attractive insulating material because of its excellent chemical stability, high relative permittivity and close lattice match with silicon. It has found applications as capacitors and buffer layers of superconducting materials [59]. Ceria, either in its pure form or doped with alien cations (Ca^{2+}, Mg^{2+}, Sc^{2+}, Y^{3+}, Zr^{4+}, etc.) is used in a number of applications including gas sensors, electrode materials for solid oxide fuel cells, oxygen pumps, amperometric oxygen monitors and catalytic supports for automobile exhaust system [60]. Studies on the electrical conductivity of ceria with respect to different dopants and dopant concentrations have shown that yttria is most soluble in the ceria lattice with excellent ionic conductivity [61]. To elucidate the role of dopants on the

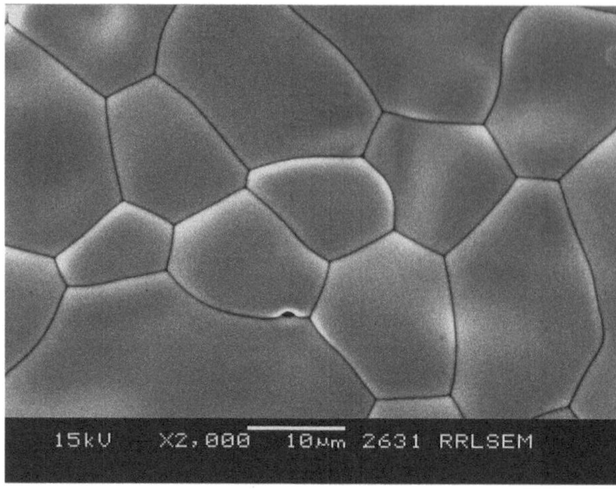

Figure 11.15 Microstructure of thermally etched CeO_2 sintered at 1650°C (courtesy P. S. Anjana).

morphology and chemical homogeneity of synthesized powder and on the sintering and grain growth of the compacted powders, Rahaman and Zhou [62] have chosen CeO_2 as a model system as it has the advantage of having a relatively simple cubic fluorite structure, high solid solubility for many cations and does not undergo any crystallographic transformation during the normal range of heating. Recently, Santha et al. [63] reported that CeO_2 sintered at 1650°C for 2 hours show about 94% of the theoretical density with $\varepsilon_r = 24$, Qf of about 65 000 GHz and $\tau_f = -53$ ppm/°C. The large negative τ_f of ceria precludes its immediate practical use. Figure 11.15 shows the microstructure of a thermally etched ceria sintered at 1650°C showing large grains. Addition of a small amount of Ca^{2+} (0.5–2 mol%) increased the density and the quality factor. Figure 11.16 shows the variation of quality factor ($Q \times f$ in GHz) ε_r and τ_f as a function of Ca content. Addition of a small amount of Ca (1 mol%) increased the Qf to about 120 000 GHz and is attributed to the improved densification. However, the addition of larger amounts of Ca^{2+} decreased the Qf as is evident from Figure 11.16. Doping with 1 mol% of rare-earth ions such as Gd^{3+} and Sm^{3+} also improved the Qf. Addition of TiO_2 improved the τ_f, but even a small amount of TiO_2 reduced the quality factor considerably [64, 65]. Sebastian et al. [65] tuned the τ_f of ceria by preparing a solid solution of CeO_2–$CaTiO_3$ and CeO_2–Sm_2O_3. The addition of $CaTiO_3$ and samarium oxide led to a zero temperature coefficient of the resonant frequency. The permittivity increased and the Qf decreased with $CaTiO_3$ addition and the τ_f changed from a negative to a positive value. Figure 11.17 shows the variation of τ_f of CeO_2–$CaTiO_3$ as a function of $CaTiO_3$ content. Complete temperature stability is achieved by the addition of 5.7 mol% of $CaTiO_3$ to CeO_2. Sm_2O_3 addition of up to 50 mol% increased Qf, decreased ε_r and improved τ_f. The MW dielectric properties of pure and doped ceria are given in Table 11.1.

Figure 11.18 shows the variation of Qf with temperature for pure CeO_2 and samples doped with 1 mol% $CaCO_3$ and TiO_2. At cryogenic temperatures, the MW dielectric properties of pure CeO_2 are improved. The Qf increased as the temperature was lowered from room temperature, to give a very high Qf value of about 600 000 GHz (at 5.58 GHz) at 30 K. Temperature dependence of the quality factor is nonmonotonous

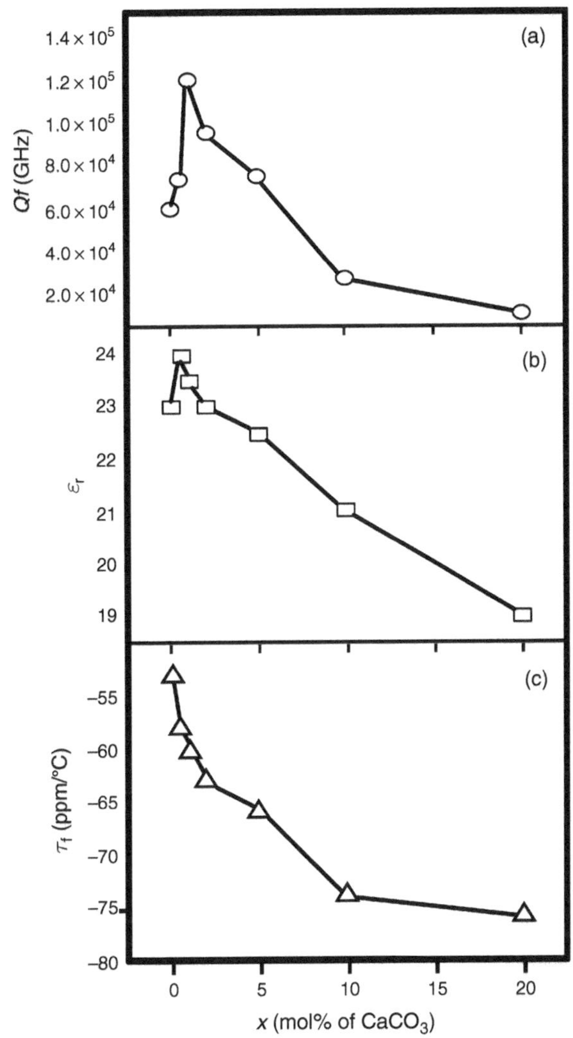

Figure 11.16 Variation of microwave dielectric properties of CeO_2 with $CaCO_3$ content: (a) variations of *Qf*, (b) variation of ε_r and (c) variation of τ_f (after Ref. [63]).

with a minimum near 100°K. It gives evidence about weak extrinsic relaxation whose frequency softens on cooling and reaching a frequency of 5.58 GHz near 100°K. Although higher at room temperature, the *Qf* of the 1 mol% Ca-doped ceria was lower at 30°K, with a value of 460 000 GHz. The TiO_2-doped CeO_2 begins with a lower *Qf* than pure CeO_2, and shows a much smaller increase than the other samples even at the lowest temperatures. None of the samples exhibited much variation in ε_r with temperature, with a decrease of typically around 2% when cooled from 300 to 20 K. The τ_f values calculated from the cryogenic resonant cavity method between 250 and 300 K show good agreements with those measured using the Hakki and Coleman method as shown in Figure 11.16. The τ_f value of pure CeO_2 was −47.4 ppm/°C

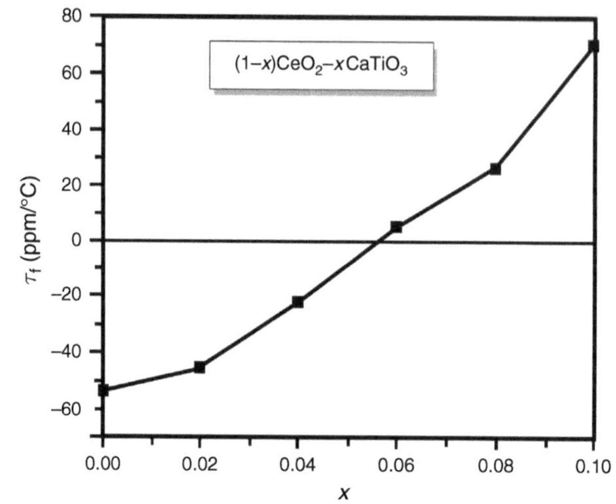

Figure 11.17 Variation of τ_f of $(1-x)CeO_2$–$xCaTiO_3$ as a function of $CaTiO_3$ content (after Ref. [65]).

Table 11.1 Microwave dielectric properties of pure and doped ceria

	Sintering temperature (°C)	ε_r	*Qf* (GHz)	τ_f (ppm/°C)	Reference
CeO_2	1675 for 2 hours	24	65 000	−55	[63]
CeO_2 + 1 wt% $CaCO_3$	1675 for 2 hours	24	120 000	−60	[63]
$CeO_2 + 0.06CaTiO_3$	1650 for 2 hours	29	25 000	0	[65]
CeO_2 + 1 mol% Sm_2O_3	1650 for 2 hours	24	90 000	−50	[65]

when measured over the temperature range 250–300 K, although it should be noted that the relationship is not linear, and that τ_f approaches zero around 30 K. A similar pattern is observed with 1 mol% Ca-doped ceria, and although τ_f was found to be higher than that of the pure ceria near room temperature (−57.7 ppm/°C), it is also nonlinear and approaches zero at 30 K.

IR reflectivity spectra of pure and doped CeO_2 are shown in Figure 11.19. The reflectivity spectra $R(\omega)$ were fitted [66] together with complex dielectric data obtained from time-domain terahertz transmission measurements (see the points in Figures 11.20 and 11.21) using Equations (2.41) and (2.48). The fit allows the direct determination of the real (ε') and imaginary (ε'') parts of the permittivity in the IR range and its extrapolation to the MW range. The results of the fits (together with experimental terahertz data) are shown in Figures 11.20 and 11.21. Although only one IR active

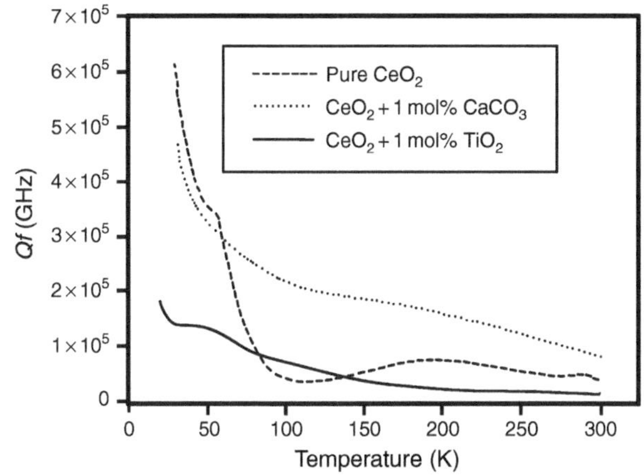

Figure 11.18 Variation of *Qf* of pure CeO_2, 1 mol% Ca doped and 1 mol% TiO_2 doped as a function of temperature (measured at 5.5, 5.4 and 6.2 GHz, respectively) (after Ref. [63]).

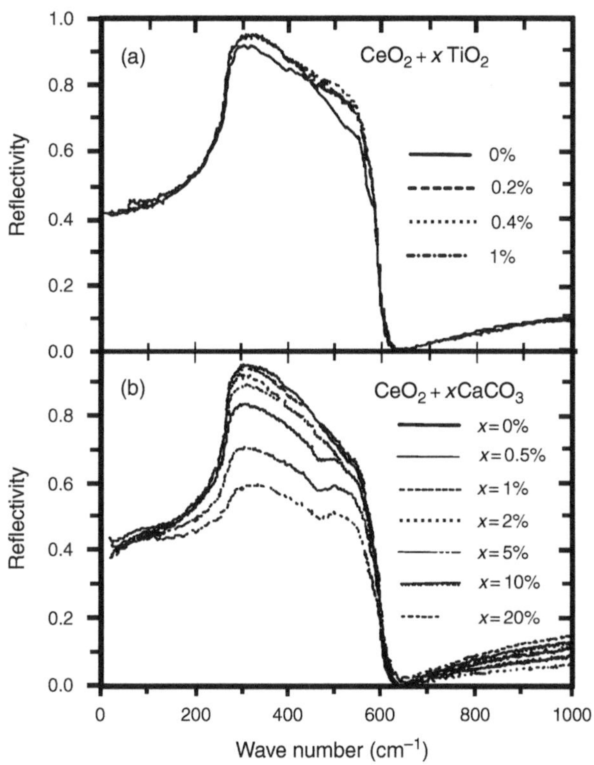

Figure 11.19 IR reflectvities of pure and doped ceria (after Ref. [63]).

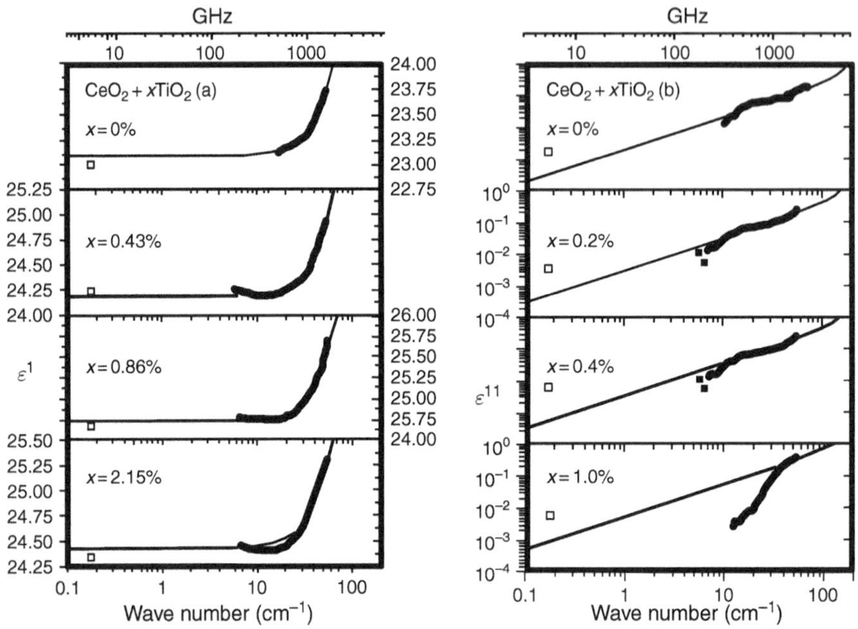

Figure 11.20 (a) Real and (b) imaginary parts of permittivity of $CeO_2 + xCaCO_3$ (time-domain transmission) measurement at gigahertz and terahertz (after Ref. [63]).

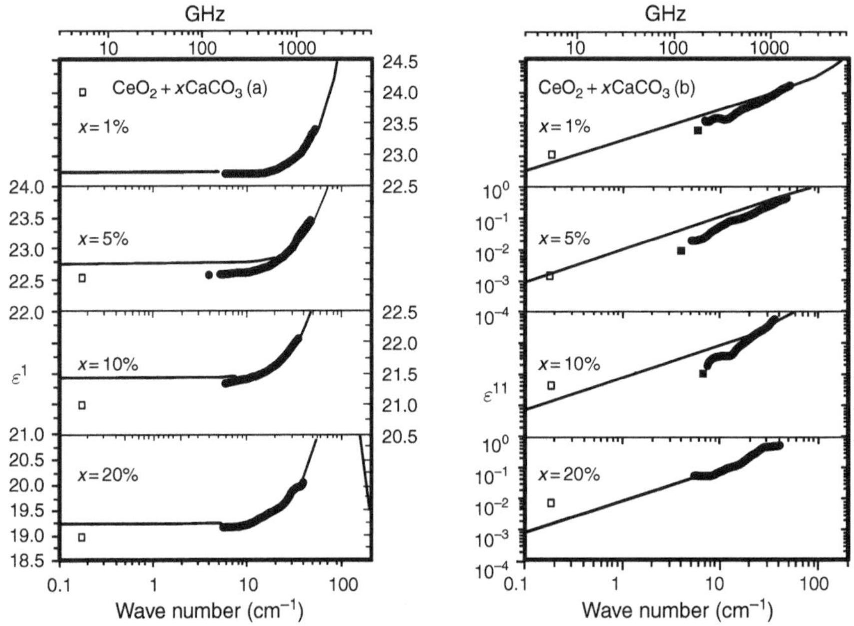

Figure 11.21 (a) Real and (b) imaginary parts of permittivity of $CeO_2 + xTiO_2$ (time-domain transmission measurement at gigahertz and terhertz (after Ref. [63]).

phonon of F_{1u} symmetry is allowed in the cubic $Fm3m$ structure of CeO_2, two oscillators were needed for the fits of each reflectivity. The second one has multiphonon (anharmonic) origin and has two orders of magnitude lower intensity (i.e., dielectric strength $\Delta\varepsilon$) than the F_{1u} mode. The second (multiphonon) mode is responsible for the decrease of reflectivity in the range 400–600 cm^{-1} (see Figure 11.19a). It is rather surprising that small TiO_2 doping causes increase of reflectivity in the range 450–600 cm^{-1}, in other words, small TiO_2 doping has an influence on the decrease of multiphonon absorption, i.e., it surprisingly decreases anharmonicity of the lattice. This effect is not seen by CaO-doped samples, where reflectivity decreases with doping due to the increase of F_{1u} mode damping (i.e., higher anharmonicity). The changes of reflectivity with doping are smaller in Figure 11.19a than in Figure 11.19b, because a 10 times higher doping in the latter case was studied. In pure CeO_2, the frequency of ω_1 and $\Delta\varepsilon_1$ of the F_{1u} polar phonon mode are 283 cm^{-1} and 17, respectively. With TiO_2 doping, ω_1 softens (its value decreases) and $\Delta\varepsilon_1$ increases, so $\omega_1 = 273\ cm^{-1}$ and $\Delta\varepsilon_1 = 18.8$ in $CeO_2 + 2.15$ mol% TiO_2. This explains the experimentally observed increase of the MW permittivity (from 23.0 to 24.9) with TiO_2 doping. In $CeO_2 + x$ mol% $CaCO_3$, ω_1 increases and $\Delta\varepsilon_1$ decreases with $CaCO_3$ doping which explains the decrease of the MW permittivity with doping. Doping above 10 mol% of $CaCO_3$ causes such a distinct breaking of cubic symmetry that a new heavily damped mode is activated in the spectra near 130 cm^{-1}. The high damping of all modes is responsible for high submillimeter dielectric loss in heavily doped samples and lower Q factor in the MW range.

Figures 11.20 and 11.21 show that the permittivity ε' is frequency independent below the phonon frequency. There is no distinct dielectric dispersion in the MW range and that the intrinsic dielectric losses ε'' extrapolated from the IR and the terahertz range down to the MW range are always lower than the experimental MW values. Their differences express extrinsic losses. One can see higher sensitivity of the losses on defects. The permittivity is negative close to the phonon frequency, because anomalous dispersion occurs where the electromagnetic wave cannot propagate through the sample due to the large absorption (see the maximum of losses in the same frequency range).

11.4 SILICATES

Materials with low ε_r and low loss are required as substrates, and dielectric waveguides for applications in the MW and millimeter wave devices. Recently silicate-based ceramics are reported to be excellent materials for substrate applications and millimeter wave communications. The silicates have predominantly covalent bonding which restricts the rattling of the atoms [67, 68]. This leads to a low ε_r and low dielectric loss. Recently, several low-loss silicate-based dielectric ceramics were reported [69–82] and are listed in Table 11.2 with properties. Cordierite ($2MgO–2Al_2O_3–5SiO_2$) is a very useful material with low dielectric loss and low ε_r [49]. It is difficult to sinter cordierite into dense ceramics and often glass is added to make it dense. However, the addition of glasses increase the dielectric loss factor. Okamura and Kishina [70] succeeded in obtaining a dense cordierite by adding rare-earth oxides. The dense $2MgO–2Al_2O_3–5SiO_2$ containing rare-earth oxide (Yb_2O_3) ceramics were obtained by sintering at temperatures in the range 1345–1420°C. XRD showed the presence of secondary phases of $Yb_2Si_2O_7$ and

Table 11.2 Microwave dielectric properties of silicates

Material	Sintering temperature (°C)	ε_r	Qf (GHz)	τ_f (ppm/°C)	Reference
Mg_2SiO_4 (forsterite)	1450	6.8	270 000	–67	[67, 68, 73]
Mg_2SiO_4 + 24 wt% TiO_2	1200	11	82 000	0	[68]
Zn_2SiO_4 (willemite)	1340	6.5	219 000	–63	[74]
Zn_2SiO_4 + 11 wt% TiO_2	1250	9.3	113 000	1	[74]
$Zn_{1.8}SiO_{3.8}$	1300	6.6	147 000	–22	[69]
$CaAl_2Si_2O_8$ (anorthite)	1500	7.4	12 000	–130	[76]
$NaAlSi_3O_8$ (albite)	1025	5.5	11 200	–5	[76]
$CaAl_2Si_2O_8$ + 5 wt% TiO_2	900	8	22 500	–50	[77]
$Ca_{0.99}Mg_{0.01}SiO_3$	1290	6.5	62 400	–43	[80]
$Ba_2MgSi_2O_7$	1350	7.2	30 000	–62	[79]
$Ba_2ZnSi_2O_7$	1350	7.5	48 000	–74	[79]
$Na_{0.8}Ca_{0.2}Al_{1.2}Si_{2.8}O_8$		6	17 600	0	[81]
$ZnO–0.6SiO_2$ + 5 wt% Li_2CO_3 + 4 wt% Bi_2O_3	910	6.7	33 000	–70	[75]
$Mg_2Al_4Si_5O_{18}$	1440	6.2	40 000	–25	[71]
$(Mg_{1-x}Ni_x)_2Al_4Si_5O_{18}$ ($x = 0.1$)	1440	6.2	99 100	–30	[71]
$Mg_2Al_4Si_5O_{18}$+7 wt% Yb_2O_3	1420	4.9	112 500		[70]
$Ba_{0.95}Sr_{0.05}Al_2Si_2O_8$	1500 for 40 hours	7.3	77 700	–22	[78]
$Ca_3SnSi_2O_9$	1400	8.4	54 800	–65	[82]
$Ca_3ZrSi_2O_9$	1400	10.6	93 300	–77	[82]

spinel $MgAl_2O_4$ in addition to cordierite. The room temperature ε_r and tan δ were determined to be 4.9 and 1.6×10^{-4} at 18 GHz. Okamura and Kishina also investigated the transmittance of millimeter wave through the cordierite dielectric waveguide. A transmission loss of <1 dB was found for a cordierite strip of 80 mm × 2.25 mm × 1 mm in the frequency range 55–65 GHz which is comparable to that of the commonly used polytetrafluroethylene (PTFE). Several authors developed cordierite glasses containing B_2O_3 and P_2O_5 for application in low-temperature sintered ceramic packaging [71, 83–85]. More recently, Terada et al. [72] reported that partial substitution of Mg by Ni improve the quality factor of cordierite. About 10 mol% Ni substitution in cordierite increased the Qf to 99 000 GHz.

Ohsato and coworkers [67, 68, 73] reported that forsterite (Mg_2SiO_4) as a high Q material for millimeter wave communication as well as a suitable substrate for the MICs. Mg_2SiO_4 sintered at 1450°C has $\varepsilon_r = 6.8$, $Qf = 270\,000$ GHz and $\tau_f = -67$ ppm/°C. The Qf of forsterite very much depends on the purity of the raw materials [68]. The high negative τ_f of forsterite has been tuned to zero by the addition of 24 wt% of TiO_2 with $\varepsilon_r = 11$ and $Qf = 82\,000$ GHz when sintered at 1200°C. Figure 11.22 shows the variation of Qf and τ_f as a function of TiO_2 content. Addition of TiO_2 lowers the sintering temperature. When fosterite is sintered with TiO_2 at high temperatures above 1200°C, the TiO_2 reacts with Mg to form $MgTi_2O_5$ which lowers the quality factor considerably [67]. Addition of a small amount of lithium borisilcate (LBS) glass lowered the sintering temperature and improved [86] the sinterability as evidenced by the SEM microstructures shown in Figure 11.23. The LBS glass addition increased the grain size and lowered the permittivity and τ_f. Addition of 15 wt% LBS sintering temperature to about 950°C [86]. Figure 11.23c shows that the LBS glass melted and segregated along the grain boundaries. The willemite (Zn_2SiO_4) is yet another low-loss silicate useful as a suitable substrate and millimeter wave communication [74]. The willemite ceramics sintered up to 97% density at 1340°C had $Qf = 219\,000$ GHz, $\varepsilon_r = 6.5$ and $\tau_f = -63$ ppm/°C. Guo et al. tailored the high negative τ_f of willemite by the addition of TiO_2. The willemite–rutile mixture ceramics have excellent dielectric properties. It was found that addition of 11 wt% TiO_2 and sintered at 1250/°C resulted in a temperature-stable

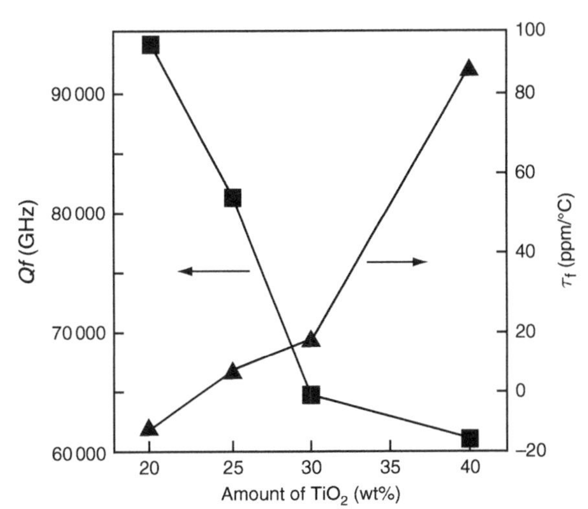

Figure 11.22 Variation of Qf and τ_f with rutile content in fosterite ceramics (after Ref. [68]).

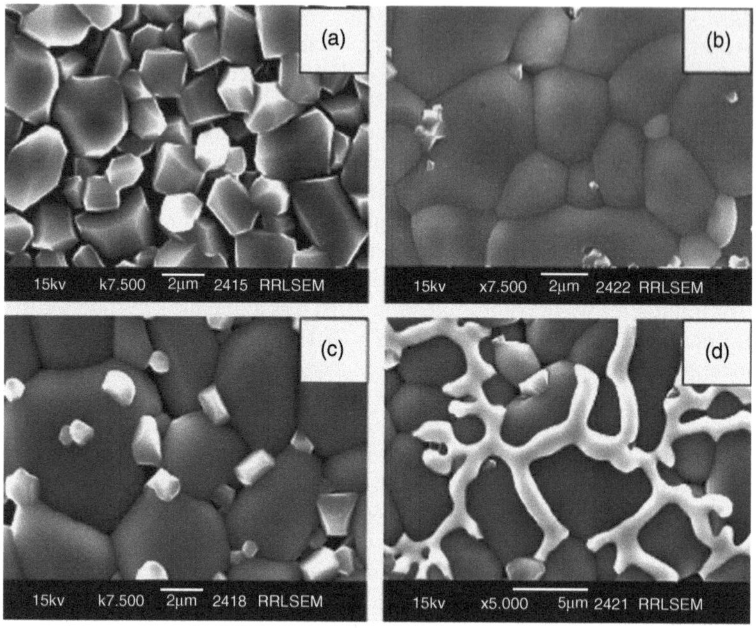

Figure 11.23 SEM micrographs of (a) pure Mg_2SiO_4 sintered at 1500°C, (b) Mg_2SiO_4+1 wt% LBS glass sintered at 1350°C, (c) Mg_2SiO_4+5 wt% LBS glass sintered at 1300°C and (d) Mg_2SiO_4 + 15 wt% LBS glass sintered at 950°C (after Ref. [86]).

ceramic with $\varepsilon_r=9.3$, $Qf=113\,000$ GHz and $\tau_f=1$ ppm/°C [74]. Zou et al. reported that ZnO–$xSiO_2$ is difficult to densify even at high temperatures above 1380°C [75]. However, the addition of 5 wt% Li_2CO_3 and 4 wt% Bi_2O_3 lowered the sintering temperature to 900 C. For $x < 0.5$, Zn_2SiO_4 was predominant and for $x > 0.5$, $Bi_4Si_3O_{12}$ and SiO_2 secondary phases appeared. ZnO–$0.6SiO_2$ ceramics sintered at 910°C for 2 hours with Li_2CO_3 and Bi_2O_3 had $\varepsilon_r=6.7$, $Qf=33\,000$ GHz and $\tau_f=-70$ ppm/°C.

There are several silicate-based ceramics and glasses with low losses and they sinter at relatively low temperatures and are discussed in Chapter 12 on LTCC.

11.5 Spinel

Compounds with the general formula AM_2O_4 (A = Mg, Zn, Co, Ni, Cu; M = Al, Ga, Fe) are referred to as spinels [87]. The aluminate spinel materials were reported to be ideal candidates for serving as radiation-resistant materials in nuclear bombardment experiments and as a support for Pt and Pt–Sn catalysts because of their high thermal stability and low acidity. Zinc aluminate spinels are also finding applications in industrial ceramics [88]. Recently, Surendran et al. [89] reported $ZnAl_2O_4$ as a useful material for the MW substrate applications. $ZnAl_2O_4$

has ε_r of 8.5 with $Qf \sim$ 60 000 GHz and a τ_f of −79 ppm/°C when sintered at 1425/°C followed by annealing at 1000°C. To compensate the high negative τ_f, they prepared $ZnAl_2O_4$–$x$$TiO_2$ composite ceramics. XRD study showed the formation of a mixture of spinel and rutile as shown in Figure 11.24. It was found that as x increased, ε_r also increased and τ_f decreased and became close to zero at about $x \sim 0.17$ and a further increase in x increased the τ_f to the positive side. Figures 11.25 and 11.26 show the variations in ε_r and τ_f with x. The Qf initially increased up to $x = 0.17$ and then gradually decreased as shown in Figure 11.27. The mixture ceramic for $x = 0.17$ had $\varepsilon_r = 12.67$ with a high Qf of about 100 000 GHz and a τ_f close to zero. However, Lei et al. [91]

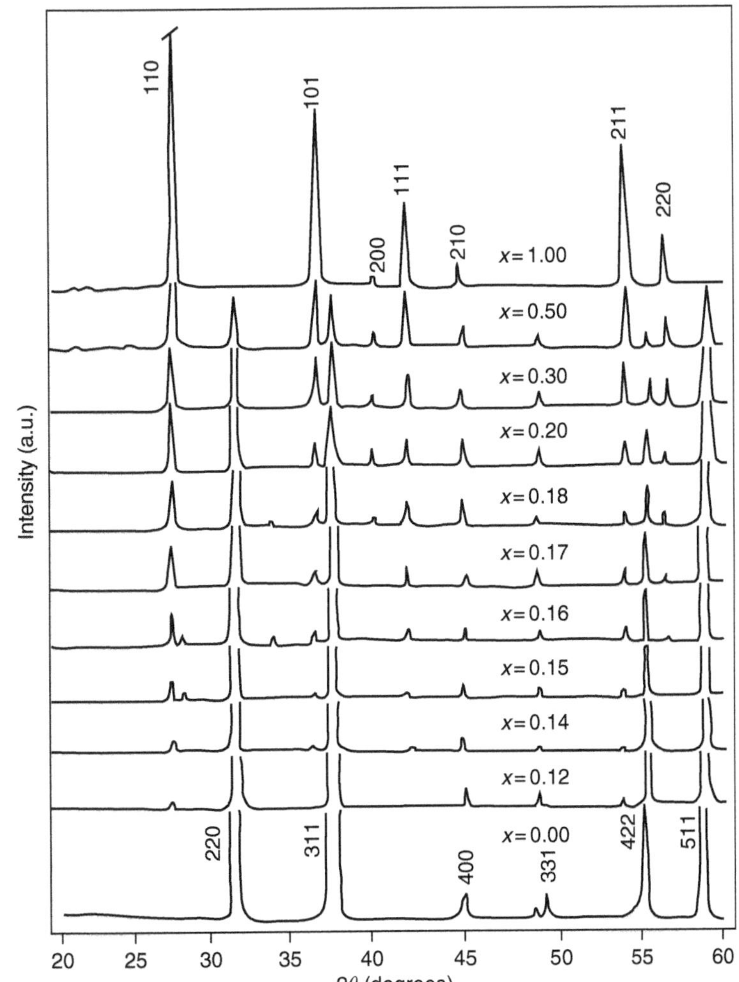

Figure 11.24 X-ray diffraction patterns of $(1 - x)ZnAl_2O_4$–$xTiO_2$ for different values of x (after Ref. [89]).

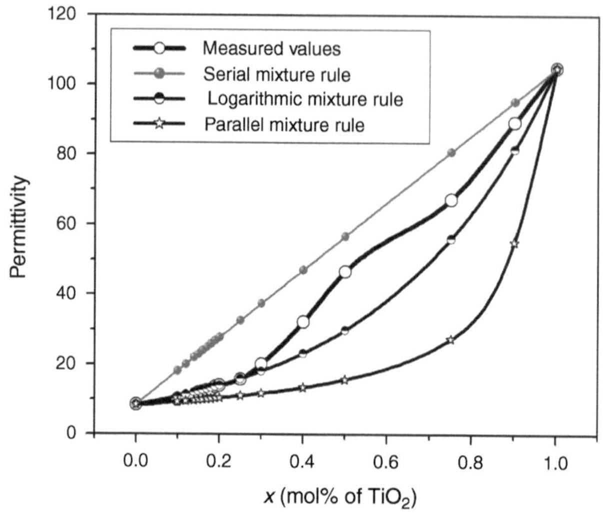

Figure 11.25 Variation of ε_r of $(1-x)ZnAl_2O_4$–$xTiO_2$ as a function of x (after [89]).

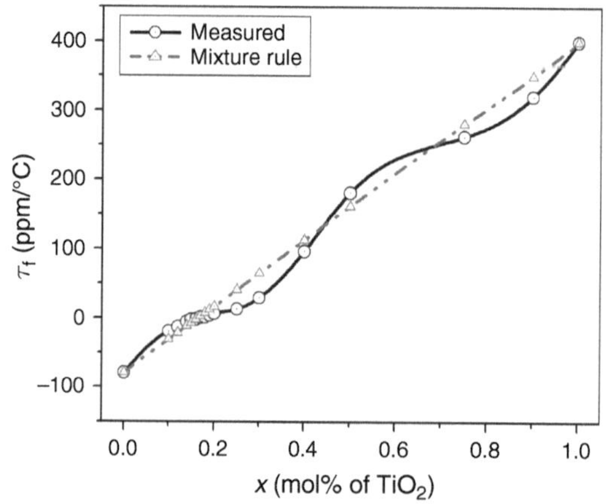

Figure 11.26 Variation of τ_f of $(1-x)ZnAl_2O_4$–$xTiO_2$ as a function of x (after Ref. [89]).

found the $\tau_f = 0$ composition close to $x = 0.21$. $ZnAl_2O_4$–$0.17TiO_2$ composite has a thermal expansion coefficient of 6.3 ppm/°C which is comparable to that of silicon [92]. All these properties are suitable for millimeter wave communication and as MW substrates. Magnesium aluminate is a potential refractory material with excellent high-temperature mechanical, thermal and chemical properties. Recently, Surendran et al. [90] reported that $MgAl_2O_4$ sintered at 1460°C has $\varepsilon_r = 8.75$, $Qf \sim 69\,000$ GHz and τ_f–75 ppm/°C in the MW frequency range. They added rutile to lower the τ_f. The addition of rutile increased ε_r and improved the τ_f. Figure 11.28 shows the variation

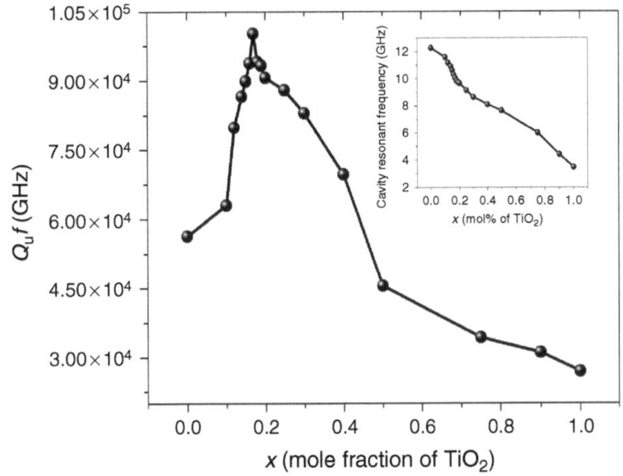

Figure 11.27 Variation of Qf of $(1-x)ZnAl_2O_4$–$xTiO_2$ as a function of x (after Ref. [89]).

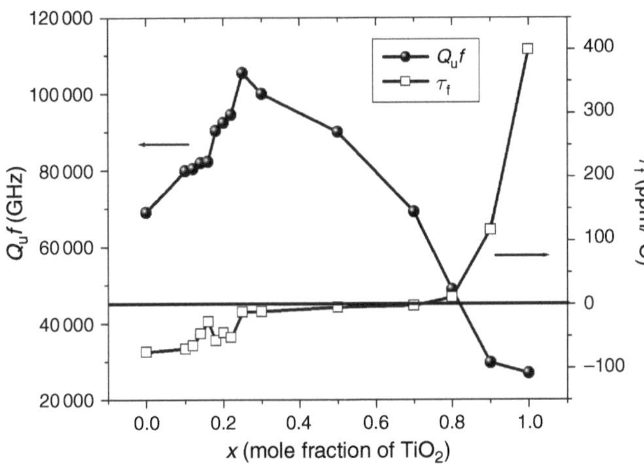

Figure 11.28 Variation of Qf of $(1-x)MgAl_2O_4$–$xTiO_2$ as a function of x (after Ref. [90]).

of Qf and τ_f with rutile content (x). The Qf initially increased up to $x=0.25$ and then decreased sharply. The mixture ceramics with $x=0.25$ had $Qf=105\,400$ GHz with $\varepsilon_r=11$ and τ_f close to zero. XRD study of $MgAl_2O_4$–TiO_2 showed the formation of a small amount of $MgTiO_3$ and $MgTi_2O_4$ secondary phases in rutile-added $MgAl_2O_4$ ceramics. Zheng et al. [93] reported the formation of a single-phase solid solution $(Mg_{1-x}Zn_x)Al_2O_4$ and the Qf increased with increase in x.

11.6 TUNGSTATES

Woframite and scheelite [AWO_4; A = Ba, Sr, Ca, Ni, Mg, Co] have been investigated as cathode, photoluminescent (blue and green light emission) and as scintillation detector materials [94, 95]. Nishigaki et al. [96] in 1988 reported the MW dielectric properties of $BaWO_4$. The mineral $CaWO_4$ (scheelite) is well known [97] for its optical properties and used in scintillation counters, lasers and optical fibers. The monoclinic $ZnWO_4$ mineral is known as sanmartinite. Yoon et al. [98] reported that the structure changed from wolframite to scheelite with increasing size of the A cations. Depending on the size of the A cations, tungstates can crystallize in either the scheelite or the wolframite structure [98, 99]. Scheelite structure with tungsten in the tetrahedral coordination is formed when the A cation has ionic radius >0.77 Å (Ca, Ba, Pb, Sr). For smaller A cations <0.77 Å (Mn, Co, Ni, Zn, Mg), the woframite structure is formed with tungsten in octahedral coordination [100]. Several tungstate compounds crystallize in the perovskite structure [101] and are discussed in the Chapter 7 on $A(B_{1/2}B''_{1/2})O_3$. Recently, Pullar et al. [102] reported the MW dielectric properties of $MgWO_4$, $ZnWO_4$, $NiWO_4$ and $CoWO_4$. XRD study showed that they have wolframite-like monoclinic *P2/c* structure. The MW dielectric properties of tungstate-based dielectric materials are given in Table 11.3.

Bijumon et al. [103] investigated the MW dielectric properties of the $Ln(Ti_{1/2}W_{1/2})O_4$ system. The $Ln(Ti_{1/2}W_{1/2})O_4$ with Ln = Pr, Nd, Sm, Gd, Tb and Dy crystallize in the tetragonal scheelite structures and $Y(Ti_{1/2}W_{1/2})O_4$ in the monoclinic fergusonite structure [106]. Brixner [106] reported that $Y(Ti_{1/2}W_{1/2})O_4$ undergoes a fergusonite to scheelite reversible transition at moderate temperatures. Bijumon et al. found that the $RE(Ti_{1/2}W_{1/2})O_4$ (Pr, Nd, Sm, Gd, Tb, Dy and Y) ceramics have a scheelite structure. The ceramics based on Ce, La, In and Al were multiphase.

11.7 AB_2O_6 (A = Zn, Co, Ni, Sr, Ca, Mg, B = Nb, Ta)

The ANb_2O_6 are mostly isostructural with orthorhombic mineral columbite ($ZnNb_2O_6$) with space group *Pnca* [107]. However, $SrNb_2O_6$ and $BaNb_2O_6$ are exceptions because of the large size of the M^{2+} ions [108]. ATa_2O_6 (A = Pb, Ni, Fe, Co, Zn, Cu) are known as functional inorganic materials with a wide application in gas sensors [109]. Maeda et al. [107] in 1987 was the first to report the MW dielectric properties of some of the AB_2O_6 compounds. In 1997 Lee et al. [110, 111] made a detailed study on several AB_2O_6 compounds. The ANb_2O_6 compounds have a sintering temperature in the range 1100–1400°C and the ATa_2O_6 in the range 1300–1600°C. In general, ANb_2O_6 compounds with columbite structure have negative τ_f. The niobates have higher *Qf* than the tantalates whereas the tantalates have higher ε_r. The MW dielectric properties of the different AB_2O_6 compounds are given in Table 11.4. The compounds based on Mg and Zn have excellent quality factors. In general, the compounds with a columbite structure have a better quality factor. Figure 11.29 shows the variation of permittivity with electronegativity in AB_2O_6 ceramics [111]. The ε_r increased with increase in electronegativity. This is expected as the relative permittivity at MW frequency range is mainly due to contributions from ionic polarizability. Kim et al. [126] reported that hexagonal $BaNb_2O_6$ is the low-temperature phase and can transform to the orthorhombic phase above 1150°C. The hexagonal to orthorhombic transition is

Table 11.3 Microwave dielectric properties of tungstate-based resonator materials

Material	Sintering temperature (°C)	ε_r	*Qf* (GHz)	τ_f (ppm/°C)	Reference
$ZnWO_4$	1200 for 2 hours	17.6	65 000	–60	[98, 102]
$NiWO_4$	1200 for 2 hours	13.3	24 900	–	[102]
$CoWO_4$	1200 for 2 hours	10.7	38 600	–	[102]
$MgWO_4$	1050	13.5	69 000	–58	[98, 102]
$MnWO_4$	1100	14.8	32 000	–64	[98]
$SrWO_4$	1150	8.1	56 000	–55	[98]
$BaWO_4$	1150	8.1	57 500	–78	[98]
$CaWO_4$	1200	8.7	75 000	–54	[98, 99]
$CaWO_4 + 1$ wt% $MnSO_4$	1050	6.6	129 500	–56	[99]
$CaWO_4 + 0.5$ wt% B_2O_3	1050	6.1	38 100	–47	[99]
$CaWO_4 + 0.5$ wt% $Bi_2O_3 + 9$ wt% H_3BO_3	850 for 30 minutes	8.7	70 200	–15	[104]
$0.7CaWO_4–0.3LaNbO_4$	1150	13.3	50 000	–8.7	[99]
$LiYW_2O_8$	900	14.8	9500	64	[99]
$LiYbW_2O_8$	900	19.7	8700	45	[99]
$ZnMnW_2O_8$	950	13.7	10 700	–17	[99]

(Continued)

Table 11.3 (Continued)

Material	Sintering temperature (°C)	ε_r	Qf (GHz)	τ_f (ppm/°C)	Reference
$BaWO_4 + 0.5$ wt% B_2O_3	950	8.2	32 700	–85	[99]
$(Mg_{1/2}Ca_{1/2})WO_4 + 1$ wt% Li_2WO_4	–	9.9	30 200	–63	[99]
$0.4LiY_2O_8$–$0.6BaWO_4$	900	11.7	19 700	14	[99]
$0.6LiY_2O_8$–$0.4BaWO_4 + 0.5$ wt% B_2O_3	930	10.2	24 300	–21	[99]
$Pr(Ti_{1/2}W_{1/2})O_4$	1300	20.2	6900	–20	[103]
$Nd(Ti_{1/2}W_{1/2})O_4$	1285	21.3	10 600	–22	[103]
$Sm(Ti_{1/2}W_{1/2})O_4$	1300	21.5	7100	–14	[103]
$Gd(Ti_{1/2}W_{1/2})O_4$	1375	22.1	5000	–10	[103]
$Tb(Ti_{1/2}W_{1/2})O_4$	1375	19.0	5900	–6	[103]
$Dy(Ti_{1/2}W_{1/2})O_4$	1425	19.8	6000	–5	[103]
$Y(Ti_{1/2}W_{1/2})O_4$	1425	19.3	6200	–19	[103]
$Nd(W_{0.5}Ti_{1.5})O_6$	1350 for 10 hours	42	26 000	9	[100]
$Sm(W_{0.5}Ti_{1.5})O_6$	1350 for 10 hours	39.4	35 400	–0.7	[100]
$Eu(W_{0.5}Ti_{1.5})O_6$	1375 for 10 hours	38.6	30 500	2.8	[100]
$Gd(W_{0.5}Ti_{1.5})O_6$	1375 for 10 hours	37.9	2600	–6.7	[100]
$Dy(W_{0.5}Ti_{1.5})O_6$	1450 for 10 hours	35.6	20 200	9.8	[100]

Table 11.4 Microwave dielectric properties of AB_2O_6 (A = Ba, Ca, Sr, Mg, Co, Ni, Cu, Mn; B = Nb, Ta) ceramics

Material	Sintering temperature (°C)	ε_r	*Qf* (GHz)	τ_f (ppm/°C)	Reference
$MgNb_2O_6$	1300	21.4	93 800	–70	[111]
$MgNb_2O_6 + 0.25$ wt% B_2O_3	1260	21.5	115 800	–48	[112]
$CaNb_2O_6$	1350	17.3	49 600	–53	[113–115]
$CuNb_2O_6$	1000	17.1	7100	–45	[114, 115]
$MgNb_2O_6 + 2$ wt% CuO	1170	19.9	110 000	–44	[116]
$CoNb_2O_6$	1150	20.5	41 700	–66	[114]
$NiNb_2O_6$	1150	22.6	40 100	–38	[111]
$ZnNb_2O_6$	1200	21.5	84 500	–76	[114, 115]
$ZnNb_2O_6 + 5$ wt% CuO	925	22.1	59 500	–65	[117]
78 wt% $ZnNb_2O_6$ + 22 wt% TiO_2	1200	29.6	27 700	21.7	[118]
78 wt% $ZnNb_2O_6$ + 22 wt% TiO_2	1150	22.5	15 000	–14.8	[118]
$0.5ZnNb_2O_6$–$0.5TiO_2$	1250 for 2 hours	34.3	42 500	–52	[119]
$0.58ZnNb_2O_6$–$0.42TiO_2$	1250 for 2 hours	45	6000	0	[119]
0.25ZnO–$0.5TiO_2$–$0.25Nb_2O_5$	1100	58	13 000	–9	[120]

(Continued)

Table 11.4 (Continued)

Material	Sintering temperature (°C)	ε_r	*Qf* (GHz)	τ_f (ppm/°C)	Reference
$0.83BaNb_2O_6$–$0.17Ba_5Nb_4O_{15}$ (hexagonal)	1250 for 2 hours	35	59 300	0	[121]
$0.42ZnNb_2O_6$–$0.58TiO_2$ + 10 wt% CuO	925	37	17 000	–7	[122]
$0.9ZnNb_2O_6$–$0.1(ZnO$–$V_2O_5)$	950	21.4	29 500	–	[123]
$0.9ZnNb_2O_6$–$0.1(2ZnO$–$V_2O_5)$	950	24.3	72 800	–	[123]
$ZnNb_2O_6$ + 1.5 wt% (CuO–V_2O_5–Bi_2O_3)	870 for 2 hours	32.7	67 100	–33	[124]
$ZnNb_2O_6$ + 10 wt%V_2O_5	900 for 2 hours	22.1	10 300	–83	[125]
$Zn(Nb_{0.94}V_{0.06})_2O_6$	875 for 2 hours	23.9	65 000	–73	[125]
$0.16BaNb_2O_6$–$0.84Ba_5Nb_4O_{15}$ + 0.3 wt% B_2O_3 + 0.3 wt% V_2O_5	900	43	19 500	0	[121]
$0.25ZnO$–$5TiO_2$–$0.25Nb_2O_5$ + 2 wt% $FeVO_4$	900	44	16 300	–10	[120]
$0.17Ba_5Nb_4O_{15}$–$0.83BaNb_2O_6$	1300	25.2	59 300	0	[121]
$SrNb_2O_6$	1300	20.1	16 900	–	[114, 115]
$BaNb_2O_6$ (orthorhombic)	1300	30	43 000	–45	[126]
$BaNb_2O_6$ (hexagonal)	1050	42	4000	–800	[126]
$MnTa_2O_6$	1350	20.3	16 500	–44	[111]

$MgTa_2O_6$	1550	30.3	59 600	30	[111]
$CaTa_2O_6$	1600	21.2	11 600	1	[111]
$CoTa_2O_6$	1500	29	2300	23	[111]
$NiTa_2O_6$	1600	25	31 000	35	[111]
$ZnTa_2O_6$	1400	30.3	87 580	9.5	[127]
$ZnTa_2O_6 + 0.5$ wt% CuO	1230	34.6	65 500	5	[128]
92 wt% $CoNb_2O_6$ + 8 wt% $CaTiO_3$	1150	22.8	29 000	−12	[118]
94 wt% $CoNb_2O_6$ + 6 wt% TiO_2	1150	29.6	20 300	4	[118]
90 wt% $CoNb_2O_6$ + 10 wt% $CaTiO_3$	1150	25.2	21 700	2	[118]
$Zn_{0.5}Mg_{0.5}Nb_2O_6$	1150	22	33 100	−29	[129]
$Zn_{1+x}Nb_2O_{6+x}$ ($x = 0.01$)		23.8	120 000	−73	[130]
$MgTa_2O_6 + 0.5$ wt% CuO	1400	28	58 000	18	[131]
$MgTa_{1.5}Nb_{0.5}O_6$	1450	27.9	33 100	−0.7	[132, 133]
$MgNb_2O_6 + 0.5$ wt% Fe_2O_3	1140	20.5	70 000	−49	[134]
$0.76ZrTi_2O_6$–$0.24ZnNb_2O_6$	1260	47	34 000	0	[135]
$MgTa_{1.3}Nb_{0.7}O_6$	1450	26.2	53 100	−4.1	[133]

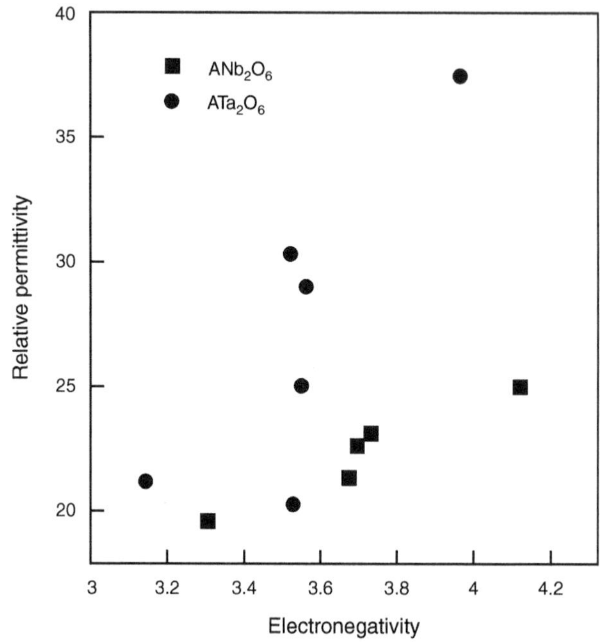

Figure 11.29 Variation of ε_r of ANb_2O_6 and ATa_2O_6 compounds as a function of electronegativity (after Ref. [111]).

reversible. $CuNb_2O_6$ also crystallizes into two polymorphic forms: the orthorhombic columbite which forms above 900°C and a monoclinic form present between 700°C and 900°C [136].

Several attempts were made [120–125, 137–141] to lower the sintering temperature of AB_2O_6 ceramics suitable for LTCC applications by the addition of CuO, CuO–Bi_2O_3–V_2O_5, V_2O_5, $BiVO_5$, CaF_2, etc. Several authors studied [116, 117, 122, 124, 128, 131] the effect of CuO additions on the sinterability of AB_2O_6. Addition of CuO to $ZnNb_2O_6$ decreased the sintering temperature but decreased the *Qf* and ε_r and made the τ_f more negative. The presence of a Cu-rich intergranular phase in $ZnNb_2O_6$ was observed which indicated the presence of a liquid phase $(ZnCu_2)Nb_2O_8$ formation during sintering. $(ZnCu_2)Nb_2O_8$ has $\varepsilon_r = 16.7$ and a *Qf* about 41 000 GHz and a τ_f of −77 ppm/°C [117]. In the case of $MgNb_2O_6$ with 1–4 wt% CuO, the *Qf* increased to 110 000 GHz [116]. Pullar et al. [141] reported that the addition of 1 wt% V_2O_5, CeO_2 and 2 wt% WO_3, improve the τ_f of $A^{2+}Nb_2O_6$ but in general the *Qf* is lowered. Belous et al. [130] reported that a slight excess of ZnO in $ZnNb_2O_6$ considerably improves the quality factor.

Zhang et al. [142] prepared solid solution phases between $ZnNb_2O_6$ and $ZnTa_2O_6$ $[(Zn(Nb_{1-x}Ta_x)_2O_6]$. As the Ta content increased, the sintering temperature and ε_r increased and the *Qf* decreased. Two types of solid solutions were formed based on $ZnNb_2O_6$ and $ZnTa_2O_6$, respectively. The τ_f became zero for $x = 0.65$. In the case of $(Zn_{1-x}Mg_x)Nb_2O_6$, as the sintering temperature increased, the *Qf*, τ_f and ε_r all decreased with increasing x [143]. ANb_2O_6 has a negative τ_f and hence several authors added TiO_2

to tune the high τ_f [118, 119, 122]. In the $(1-x)ZnNb_2O_6$–$xTiO_2$ system, the Ti^{4+} cation can substitute into the columbite structure which may cause cation disorder [146, 147]. Baumgarte et al. [146, 148] reported an α-PbO_2-related ixiolite structure for the $ZnNb_2O_6$–TiO_2 system. Four phase regions were observed with an increasing amount of TiO_2: (a) the columbite structure up to about 40 mol% TiO_2, (b) for 50 and 54 mol% of TiO_2, an ixiolite ($ZnTiNb_2O_8$) structure was formed, (c) For 57–60 mol% TiO_2, a narrow region consisting of ixiolite and rutile structure was observed and (d) for >70 mol%, rutile solid solution. Kim et al. [144] prepared $(1-x)ZnNb_2O_6$–$xTiO_2$ ceramics using both the anatase and the rutile forms of TiO_2. At a composition $x = 0.58$, a mixture region of ixiolite ($ZnTiNb_2O_6$) and rutile with $\tau_f = 0$ was observed. The Qf of $0.42ZnNb_2O_6$–$0.58TiO_2$ prepared from anatase and rutile was 6000 and 29 000 GHz, respectively, although ε_r and τ_f were comparable. The origin of the difference in Qf of both the samples was investigated by measuring electrical conductivity and by analysis of the anatase–rutile PT [144]. The anatase-derived sample had higher electrical conductivity which was related to the reduction of T^{4+}. The conductivity in the TiO_2 system originates from the Ti^{3+} and oxygen vacancies which were directly related to electron concentration associated with oxygen vacancy [149, 150]. The amount of Ti^{3+} and oxygen vacancies were higher in the anatase-derived $0.42ZnNb_2O_6$–$0.58TiO_2$. Therefore, the low Qf in this sample is interpreted to be due to oxygen vacancies caused by enhanced Ti^{4+} reduction. Templeton [47] suggested that there is only a very limited reduction of the TiO_2, and is sufficient to cause a severe deterioration in the dielectric loss. It was [122, 145] reported that the addition of CuO into $ZnNb_2O_6$–TiO_2 lowers the sintering temperature to below 900°C in comparison to 1200°C for samples without CuO. The quality factor decreased with increasing amounts of CuO due to the increase of secondary phases such as $(Cu_{0.85}Zn_{0.15})Nb_2O_6$ and $CuNb_2O_6$ in $(1-x)ZnNb_2O_6$–$xTiO_2$ ($x = 0.58$). The samples ($x = 0.58$) with 10 wt% CuO showed good MW properties with $Qf = 17\,000$ GHz, $\varepsilon_r = 37$ and $\tau_f = -7$ ppm/°C. Pullar et al. [118] doped ANb_2O_6 (A = Zn, Mg, Ca, Co) ceramics with $CaTiO_3$ and TiO_2 to tune the τ_f to zero. Figure 11.30 shows the variation of τ_f with wt% addition of TiO_2.

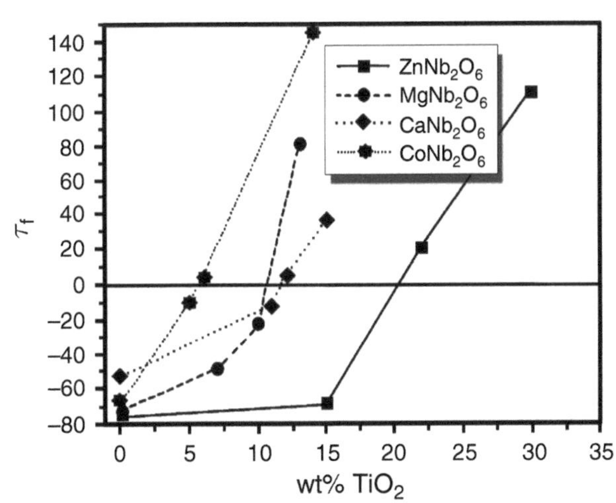

Figure 11.30 Variation of τ_f of ANb_2O_6 with wt% addition of TiO_2 (after Ref. [118]).

In all the cases, the τ_f reached zero with TiO_2 addition. Addition of TiO_2 and $CaTiO_3$ increases the permittivity but drastically reduces the quality factor.

$BaNb_2O_6$ has two polymorphs – orthorhombic and hexagonal structures [126, 151, 152]. $Ba_5Nb_4O_{15}$ has a positive τ_f of 78 ppm/°C [153]. $Ba_5Nb_4O_{15}$–$BaNb_2O_6$ sintered at 1250°C for 2 hours showed [121] a mixture of hexagonal $Ba_5Nb_4O_{15}$ and $BaNb_2O_6$ with orthorhombic structure. The ε_r and τ_f decrease with $BaNb_2O_6$ content, whereas the Qf does not show much change. The $0.17Ba_5Nb_4O_{15}$–$0.83BaNb_2O_6$ had $\tau_f=0$ with $Qf=59\,300$ GHz and $\varepsilon_r=35.2$. But the addition of 0.3 wt% V_2O_5 + 0.3 wt% B_2O_3 to this lowered the sintering temperature to 900°C and this led to the formation of hexagonal $BaNb_2O_6$. The 0.3 wt% B_2O_3 and 0.3 wt% V_2O_5 added to $(1-x)BaNb_4O_{15}$–$xBaNb_2O_6$ when sintered at 900°C showed a mixture of hexagonal $Ba_5Nb_4O_{15}$ and $BaNb_2O_6$. The composition $0.16BaNb_2O_6$–$0.84Ba_5Nb_4O_{15}$ had $\tau_f=0$ with $Qf=19\,500$ GHz and $\varepsilon_r=42$.

11.8 $A_4M_2O_9$ (M = Mg, Mn, Fe, Co; A = Ta, Nb)

The $A_4M_2O_9$ (A = Mg, Mn, Fe, Co; M = Nb, Ta) has been reported [154–156] to form compounds with an ordered corundum-type structure. It is found that a small amount of $MgNb_2O_6$ and or MgO as secondary phases may occur during the formation of $Mg_4Nb_2O_9$ [157, 158]. The MW dielectric properties of the corundum-type materials are given in Table 11.5. $Mg_4Nb_2O_9$ sintered at 1300°C for 10 hours had $Qf=217\,400$ GHz with $\varepsilon_r=12.9$ and $\tau_f=-70$ ppm/°C [159, 160]. Recently, Ogawa et al. [157, 158] reported the MW dielectric properties of $Mg_4(Nb_{2-x}Ta_x)O_9$ solid solutions. The samples were prepared by sintering in the temperature range 1350–1400°C for 10 hours followed by annealing at 850°C. Figure 11.31 (a and b) shows the variation of ε_r and Qf of $Mg_4(Nb_{2-x}Ta_x)O_9$ with composition x. The ε_r decreased and Qf increased with increase in x. Substitution of Ta for Nb slightly decreased the ε_r and is attributed to the covalent interaction of Ta–O bonding. It was found [157] that Ta–O bonds in TaO_6 octahedron become more covalent than Nb–O bonds in the NbO_6 octahedron. This leads to a decrease in the ionicity of the Ta^{5+} ion. Hence the ε_r decreases and Qf increases with increase in Ta content.

The $Mg_4Nb_2O_9$ powder prepared by a coprecipitation process lowered the sintering temperature to 1100°C and has $Qf=150\,000$ GHz [161]. It was reported that a small amount of V substitution for Nb is effective in further reducing the sintering temperature without much degradation in the dielectric properties [159, 162]. $Mg_4Nb_{2-x}V_xO_9$ with $x=0.0625$ and sintered at 1025°C has $Qf=160\,250$ GHz with $\varepsilon_r=11.6$. The V substitutes for Nb only up to $x=0.125$. It was found [159] that the limit of solid solution formation in $Mg_4(Nb_{2-x}V_x)O_9$ is lower than $x=0.125$. Kan et al. [162] reported a $Qf=200\,000$ GHz with $\varepsilon_r=12$ in $Mg_4(TaNb_{1-x}V_x)O_9$ for $x=0.025$. Kan et al. [158, 160] also found that Co substitution for Mg decreases Qf and τ_f, and increases ε_r. It was found [163, 166] that addition of 3 wt% LiF to $Mg_4Nb_2O_9$ lowers the sintering temperature to 850°C with a Qf of about 103 600 GHz. However, the τ_f is relatively large at –70 ppm/°C. Hence Yokoi et al. [163] added $CaTiO_3$ to lower the τ_f. The addition of $CaTiO_3$ increased ε_r and lowered Qf and improved τ_f. Figure 11.32 shows the variation of τ_f as a function of the $CaTiO_3$ content. The sample containing 6 wt% $CaTiO_3$ and sintered at 950°C had $\tau_f=-3$ ppm/°C with $Qf=23\,050$ GHz and $\varepsilon_r=15.7$.

Table 11.5 Microwave dielectric properties of $A_4B_2O_9$ (A = Mg,Co; B = Nb, Ta)

Material	Sintering temperature (°C)	ε_r	*Qf* (GHz)	τ_f (ppm/°C)	Reference
$Mg_4Nb_2O_9$	1300 for 10 hours	12.9	217 400	–70	[159, 160]
$Mg_4Nb_2O_9$ (precipitation)	950	8.5	150 000	–	[161]
$Mg_4(Nb_{2-x}V_x)O_9$ ($x = 0.0625$)	1025	11.6	160 250	–75	[159]
$Mg_4(TaNb_{1-x}V_x)O_9$ ($x = 0.025$)	1200	12	200 000	–73	[162]
$Mg_4Ta_2O_9$	1450	11.5	347 000	–70	[157, 160]
$Mg_3CoNb_2O_9$	1150	12.3	34 560	–64	[158]
$Mg_2Co_2Nb_2O_9$	1150	13.2	14 300	–51	[158]
$MgCo_3Nb_2O_9$	1100	15.2	2000	–27	[158]
$Co_4Nb_2O_9$	1100	16.4	5000	–11	[158], 160]
Mg_4NbTaO_9	1400	11.8	281 600	–66	[158]
$Mg_4Nb_2O_9$ + 3 wt% LiF	850/10 h	12.6	103 600	–70	[161]
$Mg_4Nb_2O_9$ + 3 wt% LiF + 6 wt% $CaTiO_3$	920/10 h	15.7	23 000	–3	[163]
$Mn_4Nb_2O_9$	–	16	50 000	–	[164]
Mg_4NbSbO_9	1450/10 h	10	280 000	–70	[165]

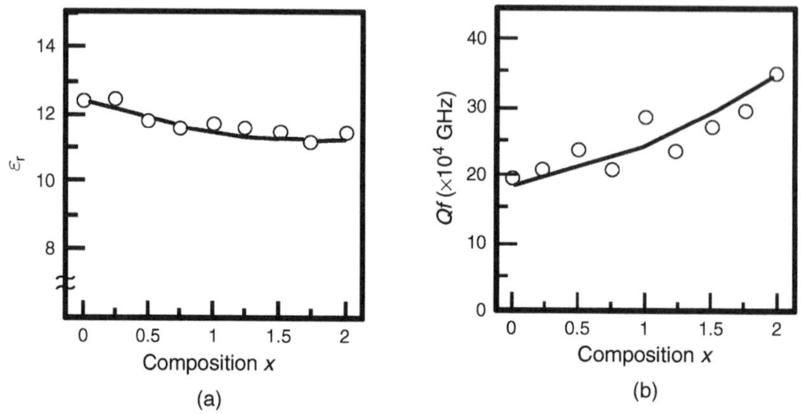

Figure 11.31 Variation of microwave dielectric properties of $Mg_4(Nb_{2-x}Ta_x)O_9$ solid solution as a function of composition x (after Ref. [157]).

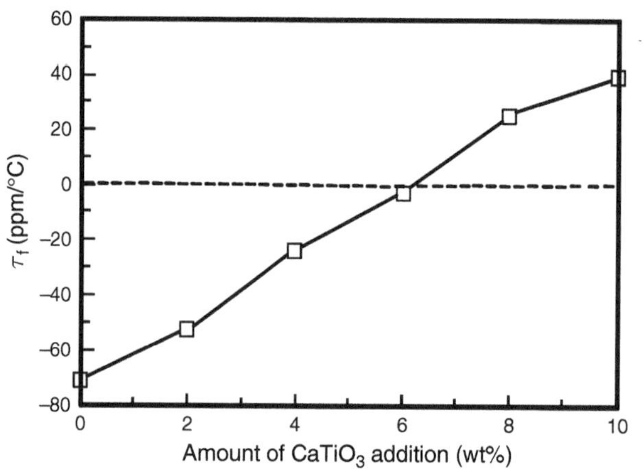

Figure 11.32 Variation of τ_f of $Mg_4Nb_2O_9 + 3$ wt% LiF as a function of wt% addition of $CaTiO_3$ sintered at 950°C (after Ref. [163]).

Mn substitution for Mg increased ε_r and decreased Qf [164] and is attributed to the decrease of the covalency of the Nb–O bond by Mn substitution. Ogawa et al. [165] reported formation of a solid solution in $Mg_4(Nb_{2-x}Sb_x)O_9$ for $x = 0$ to 1. For $x > 1$, diffraction peak corresponding to MgO and $Mg_7Sb_2O_{12}$ secondary phases appeared on the XRD pattern. In the composition range $x = 0$–1, the Qf of $(Mg_4Nb_{2-x}Sb_x)O_9$ increased from 196 000 to 280 000 GHz as shown in Figure 11.33. and the ε_r decreased to 10 whereas there was no significant change in the τ_f.

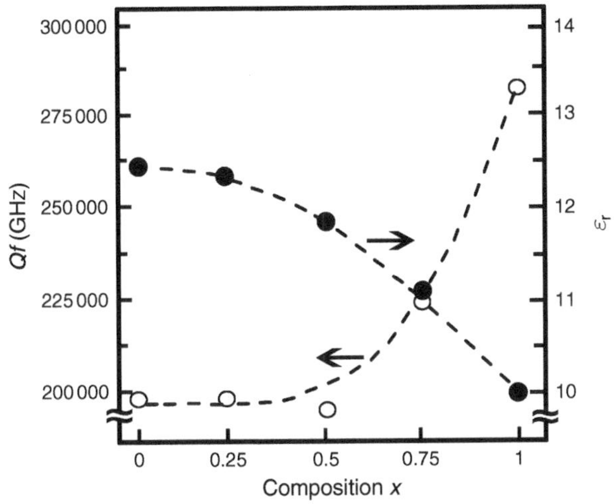

Figure 11.33 Variation of ε_r and *Qf* of $Mg_4(Nb_{2-x}Sb_x)O_9$ solid solution as a function of composition *x* (after Ref. [165]).

11.9 Ln_2BaAO_5 (Ln = Rare Earth; A = Cu, Zn, Mg)

$Ln_2BaCu_{1-x}Zn_xO_5$ crystallize in an orthorhombic structure with the *Pnma* space group when the ionic radius of Ln is smaller than that of Sm [167–169]. When the ionic radius of the Ln is larger than Sm, a tetragonal phase is formed [170, 171]. The crystal structure of Ln_2BaCuO_5 compounds were investigated by Salinez-Sanchez et al. [172] and Michel and Raveau [169]. Y_2BaCuO_5 is the insulator phase (green phase) of the high-temperature superconductor, $YBa_2Cu_3O_{6.5+x}$. The Nd_2BaCuO_5 has *P4/mbm* and Nd_2BaZnO_5 has *I4/mcm* symmetries. Roth et al. [167] reported the equilibrium phase diagram of the ternary system BaO–1/2Y_2O_3–CuO. It was found that Y_2BaCuO_5 decomposes into Y_2O_3 and a liquid phase at about 1280°C [167]. Substitution of Zn for Cu enhanced the decomposition temperature of Y_2BaCuO_5 [173]. $Y_2Ba(Cu_{1-x}Zn_x)O_5$ forms a solid solution in the complete range of *x*. Watanabe et al. [173] in 1998 for the first time reported the MW dielectric properties of $Y_2Ba(Cu_{1-x}Zn_x)O_5$ for $x = 0$–1. Since then several authors investigated the MW dielectric properties of Ln_2BaAO_5-type ceramics [174–189] and the dielectric properties are given in Table 11.6. This family of materials are usually sintered in the temperature range 1250–1350°C. Zn substitution slightly increased the ε_r and τ_f in Y_2BaCuO_5. The covalency of the Ln–O bond in Ln_2BaZnO_5 compounds was smaller than that of Ln_2BaCuO_5 [191]. It was also found that the covalency of the Ln–O bond in the Ln_2BaCuO_5 and Ln_2BaZnO_5 compounds decreases with increasing ionic radii of the Ln ions. The ε_r of Ln_2BaZnO_5 and Ln_2BaCuO_5 compounds increases with increasing ionic radii of the Ln ions as shown in Figure 11.34. The ε_r of Ln_2BaZnO_5 was higher than that of Ln_2BaCuO_5 although the ionic polarizability of Zn is smaller than that of Cu. Hence, Tsuji et al. [190] attributed the lower ε_r of Cu-based compounds to the covalency of the Ln–O bonds. Yoshida et al. [188] reported that the density and *Qf* increased considerably with increasing the sintering duration. The Y_2BaZnO_5 sintered at 1300°C for 2 hours shows 90% density

Table 11.6 Microwave dielectric properties of Ln_2BaAO_5 (A = Zn, Cu)

	ε_r	Qf (GHz)	τ_f (ppm/°C)	Reference
Y_2BaCuO_5	9.4	3800	–35	[173, 186]
Y_2BaCuO_5 (CIP)	8.3	53 300	–39.5	[186]
$Y_2BaCu_{0.5}Zn_{0.5}O_5$	14.2	110 600	–41.5	[173]
Y_2BaZnO_5	15.4	189 000	–41.5	[173, 187]
$Y_2Ba_{0.7}Sr_{0.3}CuO_5$	12.9	2963	+1.6	[175]
$Y_2Ba_{0.7}Sr_{0.3}Cu_{0.5}Zn_{0.5}O_5$	16.5	17 650	–1.6	[175]
$Y_2Ba_{0.7}Sr_{0.3}Cu_{0.15}Zn_{0.85}O_5$	16.5	23 600	–17.5	[175]
$Y_2Ba_{0.7}Sr_{0.3}ZnO_5$	16.7	4920	–35.3	[175]
$Y_2BaCu_{0.2}Zn_{0.8}O_5$	13.8	87 200	–16.3	[175]
$Y_2BaCu_{0.75}Zn_{0.25}O_5$	14	56 200	–39	[173]
$Y_2BaCu_{0.25}Zn_{0.75}O_5$	15.2	70 000	–41	[173]
$NdYBaZn_{0.45}Cu_{0.55}O_5$	17.1	100 300	–30	[189]
$Y_2BaCu_{0.2}Ni_{0.8}O_5$	13.8	87 200	–16.3	[184]
$Y_2BaCu_{0.6}Ni_{0.4}O_5$	9.7	36 000	–26.5	[184]
$Y_2BaCu_{0.4}Ni_{0.6}O_5$	13.1	45 200	–20.3	[184]
$Y_2BaCu_{0.8}Mg_{0.2}O_5$	9.5	42 300	–38	[186]
$Y_2BaCu_{0.8}Mg_{0.2}O_5$ (CIP)	9.8	49 200	–39.5	[186]
Nd_2BaZnO_5	22.6	12 450	4.6	[176, 189]
$Nd_2Ba_{0.5}Sr_{0.5}ZnO_5$	25.5	6120	25.5	[176]
$Nd_2Ba_{0.5}Ca_{0.5}ZnO_5$	26.4	6200	24	[176]
$Nd_2Ba(Zn_{1-x}Cu_x)O_5$ ($x = 0.15$)	22.1	7700	2	[189]
$Nd_2Ba(Zn_{1-x}Cu_x)O_5$ ($x = 0.2$)	20.7	11 600	–1.6	[189]
$Nd_2Ba(Zn_{1-x}Cu_x)O_5$ ($x = 0.3$)	20.8	19 800	–3.1	[189]
$Nd_2Ba(Zn_{1-x}Cu_x)O_5$ (x 0.5)	16.2	36 500	–13.2	[189]
$Nd_2Ba(Zn_{1-x}Cu_x)O_5$ ($x = 0.55$)	18.8	44 100	–19.9	[189]

Table 11.6 (Continued)

	ε_r	Qf (GHz)	τ_f (ppm/°C)	Reference
Nd_2BaCuO_5	17.6	2200	–18.4	[189]
Sm_2BaCuO_5	16.5	53 200	–5.2	[183]
$Sm_2BaCu_{0.5}Zn_{0.5}O_5$	18	65 700	–6.4	[173, 187]
Sm_2BaZnO_5	19.5	35 500	–6.4	[174, 187]
$Sm_2BaCu_{0.25}Zn_{0.75}O_5$	16.9	42 200	–4.6	[174]
$Sm_2BaCu_{0.99}Co_{0.01}O_5$	16.8	90 700	–9.2	[183]
$Sm_2Ba_{0.9}Sr_{0.1}ZnO_5$	23	8500	36.0	[185]
$Sm_2Ba_{0.5}Sr_{0.5}ZnO_5$	25.3	10 000	29.6	[185]
$Sm_2Ba_{0.15}Sr_{0.85}ZnO_5$	24.4	12 100	2.6	[185]
Sm_2SrZnO_5	24.1	19 300	–97	[185]
$Yb_2Ba(Cu_{0.5}Zn_{0.5})O_5$	14.2	20 600	–47.5	[177]
Yb_2BaCuO_5	7.9	7300	–44.4	[177]
$Yb_2Ba(Cu_{0.25}Zn_{0.75})O_5$	14.9	52 800	–44.9	[177]
Yb_2BaZnO_5	12.3	27 000	–59.9	[177]
$Yb_2Ba(Cu_{0.5}Ni_{0.5})O_5$	8.5	13 300	–46.4	[177]
$Yb_2Ba(Cu_{0.25}Ni_{0.75})O_5$	12.6	50 000	–40.9	[177]
Yb_2BaZnO_5	9.1	44 600	–37.5	[177]
$YTmBaCuO_5$	12.7	17 800	–27.3	[178]
Tm_2BaCuO_5	12.8	14 400	–14.8	[178]
$YErBaCuO_5$	13.3	16 000	–34.2	[178]
Er_2BaCuO_5	13.5	12 500	–26.1	[178]
$YDyBaCuO_5$	14	42 600	–22.1	[178]
Dy_2BaCuO_5	14.9	31 600	–6.4	[178]
$YGdBaCuO_5$	14.0	14 300	–35.2	[178]
Gd_2BaCuO_5	16	3300	–27.7	[178]

(Continued)

Table 11.6 (Continued)

	ε_r	*Qf* (GHz)	τ_f (ppm/°C)	Reference
$YSmBaCuO_5$	12.6	25 100	–29.9	[178]
$YSmBaZnO_5$	15.8	63 200	–44.5	[190]
$Y_{1.5}Sm_{0.5}BaZnO_5$	16	120 000	–32	[190]
$YHoBaCuO_5$	13.9	12 000	–29.8	[178]
Ho_2BaCuO_5	15.3	9360	–19.3	[178]
La_2BaZnO_5	20.3	17 800	–0.9	[180]
$NdLaBaZnO_5$	20.3	7900	–5	[179]
Eu_2BaCuO_5	17.1	9800	–25.4	[181]
$Eu_2BaCu_{0.5}Zn_{0.5}O_5$	17.9	49 800	–29.5	[181]
$Eu_2BaCu_{0.25}Zn_{0.75}O_5$	17.2	57 900	–28.8	[181]
Eu_2BaZnO_5	18.1	23 300	–25.4	[181]

CIP – cold isostatically pressed.

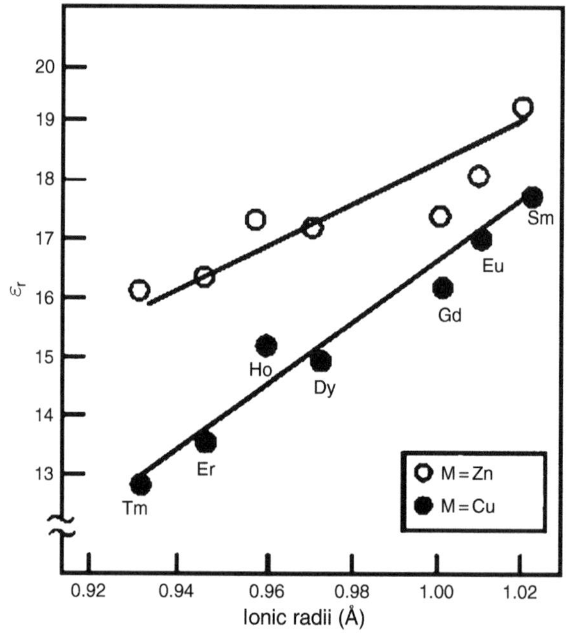

Figure 11.34 Variation of ε_r with ionic radii Ln_2BaMO_5 (M = Zn, Cu) (after Ref. [190]).

with $Qf = 50\,000$ GHz and increased to 97% with $Qf = 189\,000$ GHz on increasing the sintering duration to 50 hours at 1300°C. Zn substitution for Cu improved the Qf but not the τ_f. Hence, Kan et al. [175] tried Sr substitution at the Ba site with a view to improve the τ_f. They prepared $Y_2Ba_{0.7}Sr_{0.3}(Cu_{1-y}Zn_y)O_5$ and XRD study showed that a single-phase solid solution was formed in the range $y = 0$–0.85. Secondary phases appeared for $y > 0.85$ indicating the solid solubility of the ceramic is limited to $y < 0.85$. The Qf of $Y_2Ba_{0.7}Sr_{0.3}(Cu_{1-y}Zn_y)O_5$ solid solution increases smoothly with increasing value of y up to 0.85 which was the limit of the solid solution formation. The Qf decreased rapidly for $y > 0.85$. Thus the combined Sr for Ba and Zn for Cu substitutions improved Qf and τ_f. Kan et al. [174] substituted Y for Sm in the $Sm_2Ba(Cu_{1-x}Zn_x)O_5$ ($x = 0$–1) and a single-phase solid solution was formed for the entire range of x with orthorhombic crystal symmetry having the *Pnma* space group [174]. The Qf increased with x and reached a maxima at $x = 0.5$ and then decreased. Zn substitution improved the τ_f values in Sm_2BaCuO_5. A detailed analysis of the XRD pattern using Rietveld method indicated that the improvement in the quality factor is due to the lowering of strains in the Sm_2O_{11} polyhedron based on the ordering of Cu and Zn in the MO_5 (M = Cu, Zn) pyramid [174, 175, 178]. The substitution of Sr or Ca for Ba in the Nd-based compounds such as Nd_2BaZnO_5, decreased the quality factor and degraded the τ_f values and is attributed to instability of the crystal structure due to the partial substitution of Ba by Sr and Ca [176]. In $Sm_2(Ba_{1-x}Sr_x)ZnO_5$, the orthorhombic phase ($x = 0$) transformed to tetragonal for $x \geq 0$ [185]. As x increased ε_r, Qf and τ_f significantly changed due to the change in the crystal structure. Kawaguchi et al. [187] studied the effect of Sr substitution for Ba on the MW dielectric properties and the crystal structure of $Sm_2Ba(Cu_{0.5}Zn_{0.5})O_5$ ceramics. XRD study showed that the limit of solid solution formation was approximately $x = 0.4$ in $Sm_2Ba_{1-x}Sr_x(Cu_{0.5}Zn_{0.5})O_5$ ceramics. Figure 11.35 shows the variation of τ_f with x and

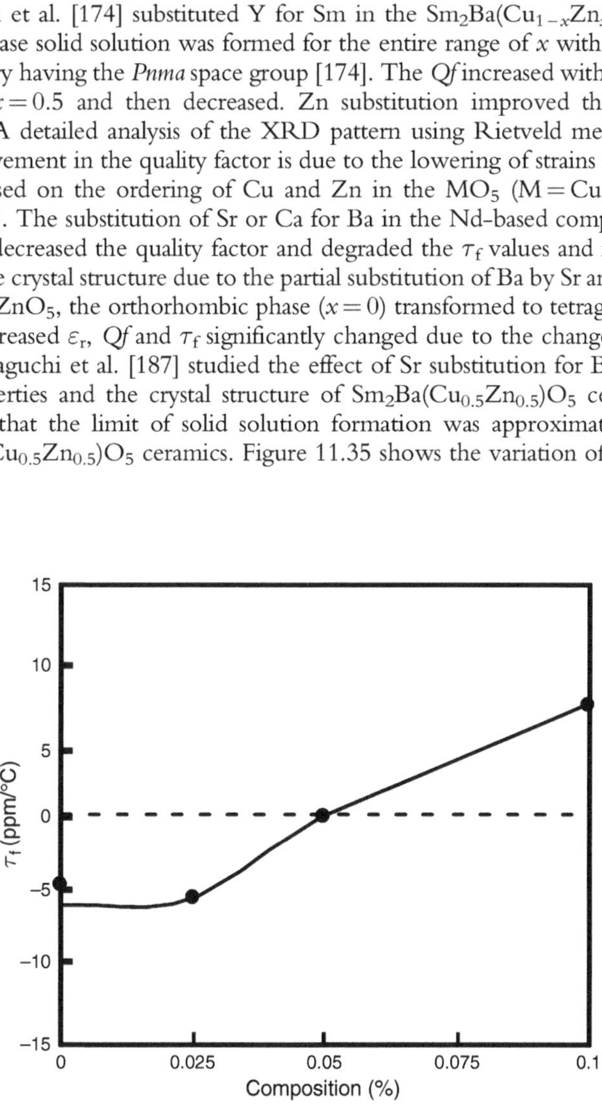

Figure 11.35 Variation of τ_f of $Sm_2(Ba_{1-x}Sr_x)(Cu_{0.5}Zn_{0.5})O_5$ ceramics as a function of sintering temperature (after Ref. [187]).

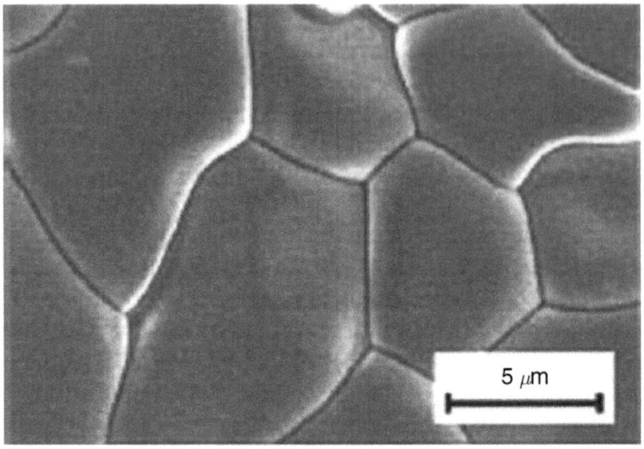

Figure 11.36 Microstructure of $Sm_2(Ba_{0.95}Sr_{0.05})(Cu_{0.5}Zn_{0.5})O_5$ ceramics sintered at 1225°C for 2 hours (after Ref. [187]).

the τ_f become close to zero for $x = 0.05$. Figure 11.36 shows the microstructure of a sample with $x = 0.05$ sintered at 1250°C for 2 hours.

The effect of Sm, Pr and La substitutions for Nd in $(Nd_{2-x}R_x)BaZnO_5$ was studied by Mori and coworkers [179, 180]. The limits of the solid solution formation with Pr and Sm substitution for Nd are up to $x = 1.5$, whereas La substitution for Nd show single phase over the whole composition range. Several authors studied [177, 182, 184] the effect of the partial substitution of Ni for Cu in Yb_2BaCuO_5 and Y_2BaCuO_5. The $Yb_2Ba(Cu_{1-x}Ni_x)O_5$ and $Y_2Ba(Cu_{1-x}Zn_x)O_5$ solid solutions were single phase over the whole composition range, whereas the limit of $Yb_2Ba(Cu_{1-x}Zn_x)O_5$ and $Y_2Ba(Cu_{1-x}Ni_x)O_5$ solid solution was $x = 0.75$ [177, 182, 184]. The formation of the solid solutions were closely related with the difference of ionic radii between the Yb^{3+} and the M^{2+} ions (Zn and Ni). Figure 11.37 (a and b) shows the variation of ε_r, Qf and τ_f of the Zn- and

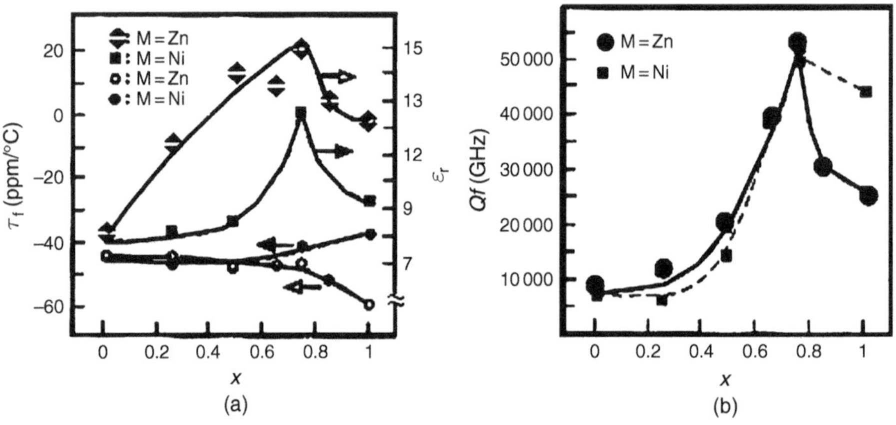

Figure 11.37 Variation of microwave dielectric properties of $Yb_2Ba(Cu_{1-x}M_x)O_5$ (M = Zn, Ni) solid solutions as a function of composition x (after Ref. [177]).

Ni-based ceramics. The properties vary suddenly at $x = 0.75$. In the case of Zn, it was due to the presence of secondary phases (above the solid solution formation limit) and in the case of Ni-based ceramics the porosity increased for $x > 0.75$. The ε_r varies between 7.9 and 14.9 and the *Qf* from 7300 to 52 800 GHz as x varies from 0 to 0.75. The maximum *Qf* was for $x = 0.75$ (see Table 11.6). The *Qf* factor was very much improved by Zn and Ni substitution for Cu in these compounds.

11.10 $LnTiAO_6$ (A = Nb, Ta)

Many researchers investigated the structural properties of ceramics of the type AB_2O_6 [192–194]. Among these, $A^{3+}B^{4+}C^{5+}O_6$ constitutes a special group whose single-phase occurrence was first substantiated by Kazantsev et al. in 1974 [194]. They established the optimum conditions of formation of double tantalates of rare-earth elements with titanium based on the formula $LnTiTaO_6$, where Ln is a lanthanide. In the same investigation, it was shown that rare-earth titanium tantalate compounds with rare-earth atomic number in the range 57–66 have orthorhombic aeschynite crystal symmetry, whereas compounds with rare-earth atomic number of 67–71 have orthorhombic euxenite symmetry. The principal difference between Ln^{3+} in these structures is that in aeschynites they lie in closely connected chains, whereas in euxenites the Ln^{3+} ions lie on densely packed parallel planes [195]. Later, Holcombe [196, 197] studied the crystal structure of ternary oxides such as $AlTiTaO_6$ and $YTiTaO_6$ as these compounds possess a unique low thermal expansion coefficient and high melting point. The single crystals of stoichiometric $LnTiNbO_6$ compounds are used as ideal gain media for miniature solid-state lasers [198] because of their exciting optical properties [195]. Single crystals of $LnTiNbO_6$ have been grown [192, 199, 200] by hydrothermal and Czochralski methods. Maeda et al. [107] suggested the possibility of using tantalates and niobates related to TiO_2 such as MTi(Ta, Nb)O_6 (M = Al, Y and Dy) for MW frequency applications. Recently, Sebastian and coworkers [201–205] made extensive studies on $LnTiAO_6$ (A = Nb, Ta) ceramics for MW applications. The Ce-, Pr-, Nd- and Sm-based ceramics have similar XRD patterns and have orthorhombic $CaTa_2O_6$-type structure with the space group *Pnma* (*D2h*16) [198, 199]. The XRD patterns of Gd-, Tb-, Dy- and Y-based ceramics are similar, and they have an orthorhombic columbite structure with the space group *Pbcn* (*D2h*14) [198]. The dielectric properties of $LnTiNbO_6$ are given in Table 11.7.

The structure and dielectric properties of $LnTiNbO_6$ depend on the ionic radii of the rare-earth ion. The ceramics with Ln = Ce, Pr, Nd and Sm show permittivity in the range 46–54. These high-permittivity materials show positive coefficient of thermal variation of resonant frequency (τ_f). The ceramics with Ln = Gd, Tb, Dy, Y and Yb have negative τ_f with permittivity in the range 19–22. The $EuTiNbO_6$ compound has a permittivity and τ_f between those of aeschynite and euxenite compounds. $EuTiNbO_6$ has a high quality factor and low-temperature variation of resonant frequency. Addition of a small amount of ZnO improves τ_f but with a decrease in the quality factor [209].

The members of aeschynite group have positive τ_f with a high permittivity, whereas euxenites have negative τ_f with a relatively lower permittivity. Surendran et al. [203] prepared $Pr_{1-x}Gd_xTiNbO_6$, $Nd_{1-x}Dy_xTiNbO_6$ and $Sm_{1-x}Y_xTiNbO_6$ solid solution phases and investigated the range of solid solution formation between the materials belonging to the aeschynite and euxenite groups. The variation in the MW dielectric properties, density and structure of these systems was investigated as a function of

Table 11.7 Microwave dielectric properties of $LnTiAO_6$ (A = Nb, Ta)

Material	Sintering temperature (°C)	ε_r	Qf (GHz)	τ_f (ppm/°C)	Reference
$CeTiNbO_6$	1360	54	6500	67	[202]
$PrTiNbO_6$	1370	53	12 300	56	[202]
$NdTiNbO_6$	1370	52	4500	46	[202]
$SmTiNbO_6$	1360	45	18 000	50	[202]
$EuTiNbO_6$	1370	32	17 200	5	[202]
$GdTiNbO_6$	1385	20	9000	–52	[202]
$TbTiNbO_6$	1385	21	15 700	–45	[202]
$DyTiNbO_6$	1385	22	19 100	–42	[202]
$YTiNbO_6$	1400	19	8800	–45	[202]
$YbTiNbO_6$	1400	22	11 000	–63	[202]
$LaTiTaO_6$	1530	24.4	45 300	39	[205]
$CeTiTaO_6$	1540	46.0	33 300	41	[205]
$PrTiTaO_6$	1500	45.8	32 300	33	[205]
$NdTiTaO_6$	1550	43.1	26 400	30	[205]
$SmTiTaO_6$	1500	41.8	24 500	24	[205, 206]
$EuTiTaO_6$	1525	41.3	59 500	19	[205]
$GdTiTaO_6$	1540	37.9	12 900	11	[205]
$TbTiTaO_6$	1525	36.8	32 300	10	[205]
$DyTiTaO_6$	1500	34.6	40 100	7	[205]
$HoTiTaO_6$	1550	23.1	46 900	–8	[205]
$YTiTaO_6$	1625	22.1	51 400	–20	[205]
$ErTiTaO_6$	1560	20.6	85 500	–29	[205]
$YbTiTaO_6$	1560	19.3	31 800	–41	[205]
In_2O_3–TiO_2–Ta_2O_5	1525	24.3	15 400	39	[205]

Table 11.7 (Continued)

Material	Sintering temperature (°C)	ε_r	Qf (GHz)	τ_f (ppm/°C)	Reference
Al_2O_3–TiO_2–Ta_2O_5	1575	28.1	10 000	20	[205]
$SmTiNb_{1/2}Ta_{1/2}O_6$	1600	39.3	19 600	33	[205]
$SmTiTa_{0.7}Zr_{0.3}O_6$	1650	31.1	37 500	−2	[206]
$Ce(Zr_{1/3}Ti_{2/3})O_6$	1600	33.4	15 800	14	[207]
$Pr(Zr_{1/3}Ti_{2/3})O_6$	1600	33.3	16 200	13.5	[207]
$Nd(Zr_{1/3}Ti_{2/3})O_6$	1600	31.4	15 800	5.5	[207]
$Eu(Zr_{1/3}Ti_{2/3})O_6$	1600	30.4	11 000	−4	[207]
$Ce_{0.25}Dy_{0.75}(Nb_{0.5}Ta_{0.5})TiO_6$	1500/10 h	30.9	23 700	0	[208]

composition (x). In $Pr_{1-x}Gd_xTiNbO_6$ ceramics the XRD patterns are similar to that of aeschynite structure for $x < 0.8$, and it is similar to that of euxenites when $x > 0.9$. The structural phase transition (PT) from *Pnma* to *Pbcn* symmetry occurs between $x = 0.8$ and 0.9. Both the euxenite and the aeschynite phases coexist near the transition region. Similar XRD patterns are found in the other two solid solutions and their transition points are different. In $Nd_{1-x}Dy_xTiNbO_6$ system the crystal structure is comparable to aeschynites for $x < 0.4$. The structural transformation occurs between $x = 0.4$ and 0.5. For $x > 0.5$, the euxenite structure prevails. In $Sm_{1-x}Y_xTiNbO_6$ solid solution, similar structural transition occurs between $x = 0.2$ and 0.3. It is found that the ceramics have poor sinterability near the transition region which is a two-phase region. The variation in permittivity in $Pr_{1-x}Gd_xTiNbO_6$, $Nd_{1-x}Dy_xTiNbO_6$ and $Sm_{1-x}Y_xTiNbO_6$ with x are shown in Figure 11.38. The permittivity of $PrTiNbO_6$ is 53 and that of $GdTiNbO_6$ is 20. The solid solution phases of $Pr_{1-x}Gd_xTiNbO_6$ is expected to have a permittivity between 53 and 20, depending on the value of x. With the substitution of Gd^{3+} ion on the Pr^{3+} site, the permittivity decreased steadily from 53 to 42 for $x < 0.8$, where the solid solution phases crystallize in the orthorhombic aeschynite symmetry. Again, for $x > 0.9$, the permittivity decreased linearly from 22 to 20 where the phase is euxenite. The permittivity decreased abruptly for values of x between 0.8 and 0.9. The ceramics have poor quality factor near the transition region which is due to poor sinterability. Similar variations in ε_r was observed for the $Nd_{1-x}Dy_xTiNbO_6$ and $Sm_{1-x}Y_xTiNbO_6$ solid solutions. The variation of the τ_f for the three solid solution systems is plotted in Figure 11.39. The variation of τ_f with composition (x) is similar to the variation of the permittivity with x (see Figure 11.38). In the case of $Pr_{1-x}Gd_xTiNbO_6$ the value of τ_f for $x = 0.85$ is −2.4 ppm/°C The average ionic radius of rare-earth ion in $Ln_{1-x}Ln'_xTiNbO_6$ (Ln = Pr, Nd, Sm; Ln′ = Gd, Dy, Y) is calculated using the data

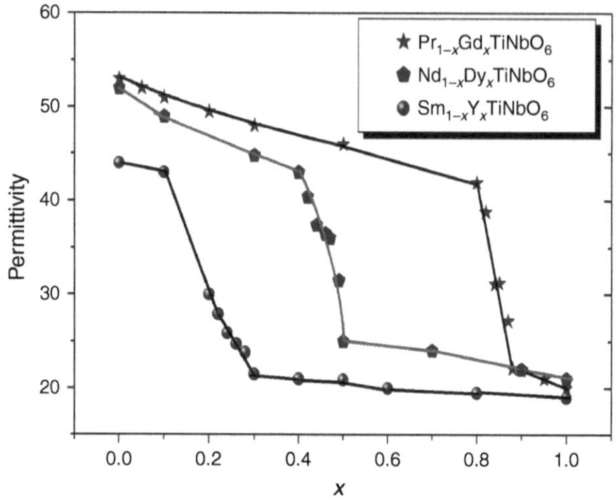

Figure 11.38 Variation of ε_r as a function of composition x in $Pr_{1-x}Gd_xTiNbO_6$ $Nd_{1-x}Dy_xTiNbO_6$, $Sm_{1-x}Y_xTiNbO_6$ (after Ref. [203]).

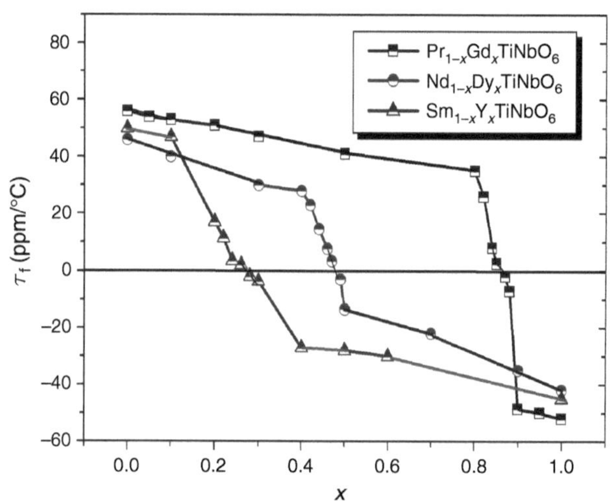

Figure 11.39 Variation of τ_f as a function of composition x in $Pr_{1-x}Gd_xTiNbO_6$, $Nd_{1-x}Dy_xTiNbO_6$, $Sm_{1-x}Y_xTiNbO_6$ (after Ref. [203]).

reported by Shannon [210]. Figures 11.40 and 11.41 show the variation of permittivity and τ_f with the average ionic radius (IR) of the rare-earth ions. The permittivity and τ_f values show a sharp and abrupt change when the average rare-earth ionic radius is 0.945 Å. This apparently indicates that the aeschynite to euxenite transition occurs when the average ionic radius of the rare-earth ions in $Ln_{1-x}Ln'_xTiNbO_6$ is 0.945 Å. The results show that $LnTiNbO_6$ compounds crystallize in the euxenite form for IR < 0.945 Å and in the aeschynite form when IR > 0.945 Å. Moreover, it

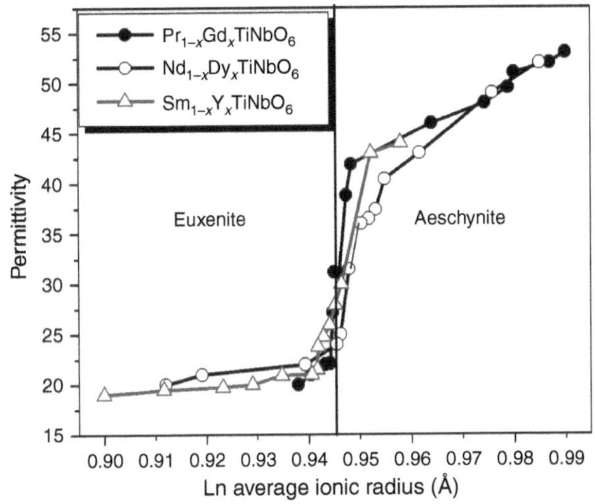

Figure 11.40 Variation of ε_r with average ionic radius of rare-earth ion in $Ln_{1-x}Ln'_xTiNbO_6$ (after Ref. [203]).

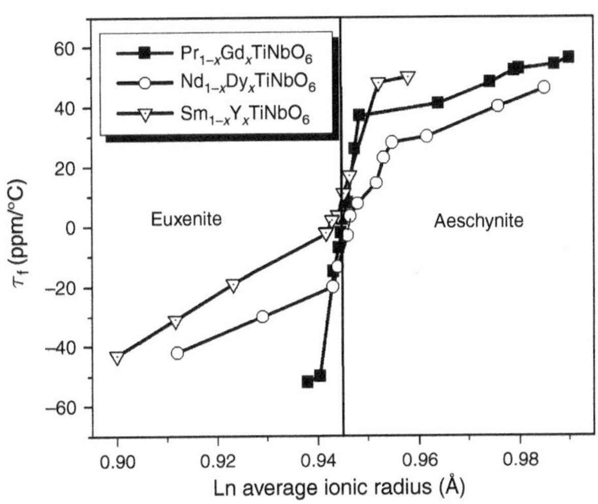

Figure 11.41 Variation of τ_f with average ionic radius of rare-earth ion in $Ln_{1-x}Ln'_xTiNbO_6$ (after Ref. [203]).

is found from Figure 11.41 that the sign of τ_f strongly depends on the average ionic radius of the rare earths. For IR > 0.945 Å, the material will have positive τ_f and for IR < 0.945 Å the τ_f will be negative. The results indicate that one can obtain a nearly zero τ_f material by tuning the average rare-earth ionic radius to be ~0.945 Å in the $LnTiNbO_6$ compounds. This is in agreement with the fact that $EuTiNbO_6$ (Eu IR = 0.947 Å) has a very low τ_f of –5 ppm/°C [201].

The tantalite analogues have higher quality factor as compared to the niobates [205]. The Ce-, Pr-, Nd-, Sm-, Eu-, Gd-, Tb- and Dy-based tantalates belong to the aeschynite family. The Ho-, Er- and Yb-based titanium tantalates belong to the euxenite family. XRD and SEM study revealed that $LaTiTaO_6$ and $InTiTaO_6$ are multiphase. The permittivities of aeschynite-type $LnTiTaO_6$ (Ln = Ce, Pr, Nd, Sm, Eu, Gd, Tb and Dy) are relatively high, varying from 46.9 to 34.6, while those of the euxenite-type ceramics, such as $HoTiTaO_6$, $YTiTaO_6$, $ErTiTaO_6$ and $YbTiTaO_6$, are low, varying from 23 to 21. These results agreed with those of $LnTiNbO_6$ in which aeschynites ($CeTiNbO_6$, $PrTiNbO_6$, $NdTiNbO_6$, $SmTiTaO_6$ and $EuTiTaO_6$) have high permittivities, whereas euxenites ($GdTiNbO_6$, $TbTiNbO_6$, $DyTiNbO_6$, $YTiNbO_6$ and $YbTiNbO_6$) have relatively lower permittivities. The τ_f was plotted against the ionic radius of the rare earths in $LnTiTaO_6$ (see Figure 11.42). It is evident that the rare-earth titanium tantalates belonging to aeschynite symmetry have positive τ_f and those belonging to euxenite symmetry have negative τ_f. This result is similar to the niobium analogues. It is interesting to note that the ionic radii of ceramics in the $LnTiTaO_6$ system with low τ_f ($DyTiTaO_6$ and $HoTiTaO_6$) are between 0.9 and 0.92 Å. The results of the τ_f measurements in $LnTiTaO_6$ indicate that the aeschynite–euxenite PT boundary lies between $DyTiTaO_6$ and $HoTiTaO_6$. In other words, one can tailor the value of τ_f to a minimum by making a solid solution between aeschynites and euxenites to bring down the average ionic radii of the rare earths to between 0.9 and 0.92 Å, similar to the rare-earth titanium niobates [202]. The boundary of the aeschynite to euxenite morphotropic change lies between $DyTiTaO_6$ and $HoTiTaO_6$, which is evidenced by the results on ε_r and τ_f measurements (see Figures 11.42 and 11.43). Most of the tantalate DRs have high *Qf* factors as compared to their niobium analogues. It is observed that low-loss ceramics like $ErTiTaO_6$ ($\varepsilon_r = 20.6$, $Qf = 85\,500$), $EuTiTaO_6$ ($\sim\varepsilon_r = 41.3$, $Qf = 59\,500$ GHz) and $YTiTaO_6$ ($\varepsilon_r = 22.1$, $Qf = 51\,400$ GHz) are potential candidates for DR applications. The material $LaTiTaO_6$ proved to be a low-loss ceramic with a high quality factor ($Qf \sim 46\,300$) in spite of it being of multiphase nature. Surendran et al. [211] also tuned the τ_f of $LnTiTaO_6$ ceramics by making solid solution phases between euxenites and aeschynites which are having τ_f of opposite sign.

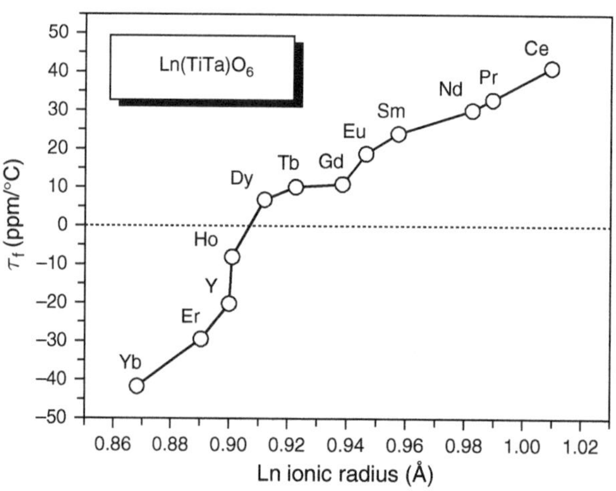

Figure 11.42 Variation of τ_f with ionic radius of the rare earth in $Ln(TiTa)O_6$ (after Ref. [205]).

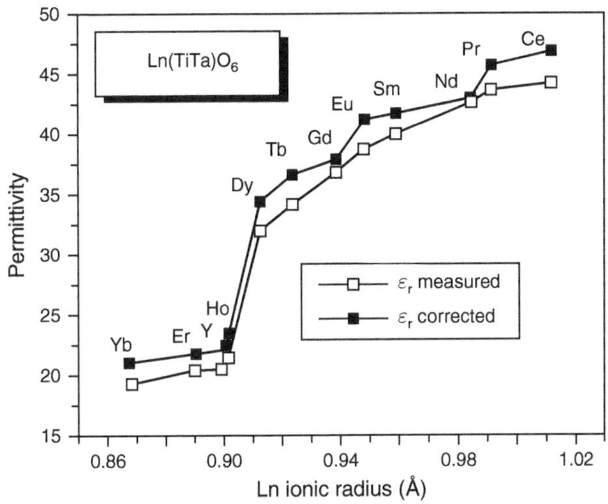

Figure 11.43 Variation of ε_r with ionic radius of the rare earth in $Ln(TiTa)O_6$ (after Ref. [205]).

11.11 $MgTiO_3$

$MgTiO_3$ has an ilmenite-type structure with $\varepsilon_r = 17$, $Qf = 160\,000$ GHz and $\tau_f = -50$ ppm/°C [50, 212]. Ferreira et al. [212–214] reported that $MgTiO_3$ dielectrics prepared by chemical method have better quality factor (166 400 GHz). It was found [212–216] that addition of La_2O_3, Cr_2O_3 or Fe_2O_3 lowered the Qf value although it improved densification. In the La-doped samples, secondary phases of $La_2Ti_2O_7$ was found which degraded quality factor. $MgTiO_3$ when sintered at 1400°C contain $MgTi_2O_5$ secondary phase [217]. Ichinose et al. [217] reported that the addition of 5 mol% B_2O_3 to $MgTiO_3$ lowers the sintering temperature and suppressed the formation of $MgTi_2O_5$. Yoo et al. [218] reported that the quality factor depends very much on the cooling rate. The slow cooled samples have lower amount of strain and thereby exhibit higher Qf. $MgTiO_3$ sintered at 1350°C and cooled at the rate of 1°/min had a high Qf of 220 000 GHz, whereas the sample cooled at a rate of 30°/min had $Qf = 170\,000$ GHz and quenched sample 150 000 GHz, respectively. To compensate for the negative τ_f of $MgTiO_3$, Wakino [50] prepared a composite ceramic of $MgTiO_3$–$CaTiO_3$ (MCT). It was found that the composition $0.95MgTiO_3$–$0.05CaTiO_3$ ceramics has $\tau_f = 0$. They do not form a solid solution because of the large difference in the ionic sizes of Mg^{2+} and Ca^{2+} and the difference in the crystallographic structure. The $0.95MgTiO_3$–$0.05CaTiO_3$ mixture ceramics have $\tau_f = 0$ with $\varepsilon_r \sim 21$ and $Qf \sim 56\,000$ GHz. Several authors [217–228] investigated the effect of small amounts of dopants and glass additives on the sintering temperature and MW dielectric properties of MCT. Ichinose et al. [217] reported that addition of 5 mol% B_2O_5 to MCT and fired at 1200°C showed a maximum $Qf = 86\,000$ GHz with $\varepsilon_r = 19.6$ and $\tau_f = -3$ ppm/°C. Addition of 5 mol% V_2O_5 in MCT lowers the sintering temperature to about 1000°C but the sintered ceramic was multiphase consisting of $MgTiO_3$, $MgTi_2O_5$ and $Ca_5Mg_4V_6O_{24}$ [217] with a degradation in the dielectric properties. Addition of low melting glasses lowered the sintering temperature to <950°C suitable for LTCC and are described in Chapter 12 on LTCC. Addition of 0.2°wt% Bi_2O_3 [221, 229] in $0.94MgTiO_3$–$0.06CaTiO_3$ lowered sintering temperature to about 1250°C resulting a τ_f of −2.9 ppm/°C with $Qf \sim 53\,000$ GHz and $\varepsilon_r \sim 22.6$. Further addition of Bi_2O_3 leads to formation of $Bi_2Ti_2O_7$ which lowered the *Qf* value. Cho et al. [226] reported that $(1-x)MgTiO_3$–$xSrTiO_3$ ($x = 0.036$) sintered at 1270°C for 2 hours showed temperature-stable ceramic with $Qf = 71\,000$ GHz. Hence they propose $(1-x)MgTiO_3$–$xSrTiO_3$ (MST) superior to MCT as MCT has a higher sintering temperature of 1400°C and the properties of MST and MCT are comparable.

11.12 ZnO–TiO_2 SYSTEM

Zinc titanate is a low sintering dielectric which can be sintered below 1100°C without sintering aids, or at 900°C by the addition of a small amount of glass [230–233]. Three compounds, i.e., $ZnTi_2O_4$ (cubic), $ZnTiO_3$ (hexagonal) and $Zn_2Ti_3O_8$ (cubic), are known to exist in the ZnO–TiO_2 system [234–242]. Zn_2TiO_4 can be easily prepared by conventional solid-state reaction between ZnO and TiO_2. The preparation of pure $ZnTiO_3$ is difficult because it decomposes to form Zn_2TiO_4 and rutile at about 945°C [239]. The presence of rutile increases the τ_f of the ceramic. Kim et al. [232, 233] succeeded in developing a $\tau_f = 0$ dielectric based on ZnO–TiO_2 by the substitution of Zn by Ba, Ca and Sr. They prepared $Zn_{1-x}(BaCa,Sr)_x$–TiO_2 ($x = 0$–0.09) by calcining at 1000°C and sintering at 1100–1300°C for 4 hours. The XRD of $Zn_{1-x}(Ba,Ca,Sr)_xTiO_3$ sintered at 1150°C for 4 hours showed the presence of secondary phases of rutile, $SrTiO_3$, $CaTiO_3$, $Sr_2Zn_4Ti_{15}O_{36}$, $Ca_2Zn_4Ti_{15}O_{36}$, $Ba_3Zn_7Ti_{12}O_{34}$ and Zn_2TiO_4. Figure 11.44 shows the MW dielectric properties of $(Zn_{1-x}M_x)TiO_3$ (M = Ba,Ca,Sr) sintered at 1150°C for 4 hours. In Ba-based ZnO–TiO_2, the ε_r decreased and the *Qf* increased with Ba content up to 7 mol% due to a decrease in the rutile phase. The τ_f changed from positive to negative value as the Ba content reached 7 mol%. The ε_r of Ca- and Sr-based ceramics also decreased as the doping increased to 3 mol% beyond which it increased due to formation of $CaTiO_3$ and $SrTiO_3$. The Q factor decreased and τ_f increased beyond 3 mol%. A $\tau_f = 0$ can be achieved with the substitution of 3–5 mol% of Ba, Ca or Sr in $ZnTiO_3$. Chaouchi et al. [243, 244] reported that the addition of 5 wt% ZnO–B_2O_3, ZnO–B_2O_3–SiO_2 or B_2O_3 + LiF lower the sintering temperature of $ZnTiO_3$ to about 900°C. The samples sintered at 900°C are single-phase $ZnTiO_3$. Liu et al. [245] reported that the addition of V_2O_5–B_2O_3 lower sintering temperature to about 875°C and avoid the problem of decomposition to Zn_2TiO_4. Recently, Golovchanski et al. [231] reported that $ZnTiO_3$ prepared by the sol–gel process and sintered at 930°C has $\varepsilon_r = 19$, $Qf \sim 30\,000$ GHz and $\tau_f = -55$ ppm/°C. Several authors tailored the dielectric properties of $ZnTiO_3$ by the addition of TiO_2, $MgTiO_3$, $CaTiO_3$ and $CoTiO_3$ [230, 246–249].

Kim et al. [230] tailored the negative τ_f of $ZnTiO_3$ by preparing $(1-x)ZnTiO_3$–$xTiO_2$ for $x = 0$–0.5 by the mixed oxide route. The ε_r, *Qf* and τ_f values increased with increase in TiO_2 content. The $\tau_f = 0$ value was obtained for $x = 0.35$. XRD study showed the presence of secondary phases of $Zn_2Ti_3O_8$, Zn_2TiO_4 and rutile when sintered at temperatures above 925°C. Addition of B_2O_3 glass decreased the sintering temperature to about 875°C and significantly enhanced the sinterability and density.

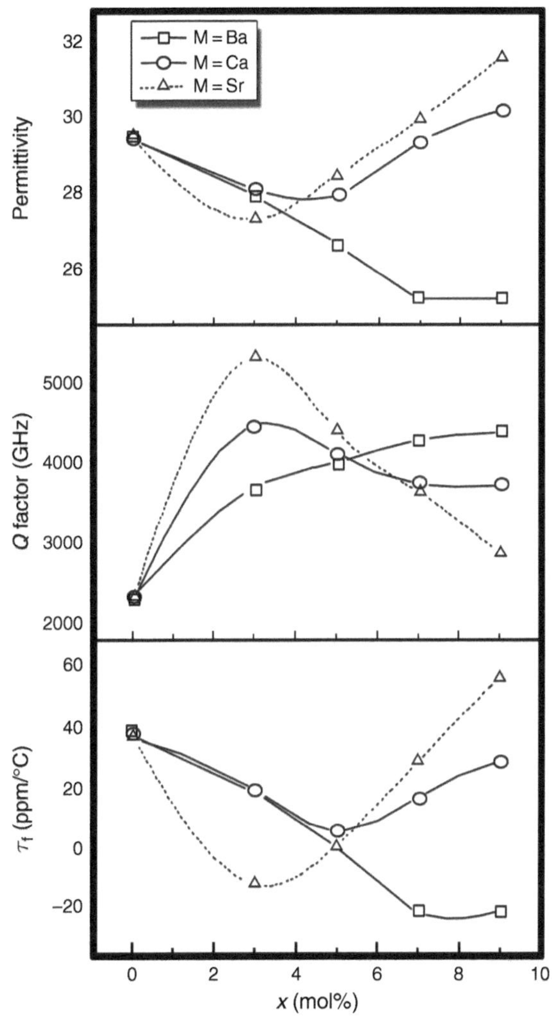

Figure 11.44 Variation of microwave dielectric properties of $(Zn_{1-x}M_x)O$–TiO_2 sintered at 1150°C as a function of x (after [232]).

Addition of about 1 wt% B_2O_3 gave the maximum Q factor of $Qf = 56\,000$ GHz, $\varepsilon_r = 30$ and $\tau_f = 10$ ppm/°C. Kim et al. [248] prepared $Zn_{1-x}Co_xTiO_3$ by sintering at 1150°C. The XRD pattern shows the presence of cubic spinel $(Zn,Co)_2TiO_4$ and rutile for $x = 0$–0.5. For $x = 0.5$–0.7, the spinel, rutile and ilmenite phases coexisted. For $x > 0.7$, a single-phase ilmenite structure was found. Figure 11.45 shows the variation of dielectric properties as a function of x. The ε_r initially increased up to $x = 0.5$ and then decreased. The ε_r, Qf and τ_f show changes depending on the compositional range. The α-region contained spinel + rutile, the β-region spinel+rutile+ilmenite and the γ-region contained ilmenite. The ε_r, Qf and τ_f were influenced by the relative amounts of the different phases. The microwave dielectric properties of ceramics in the ZnO–TiO_2 system are given in Table 11.8.

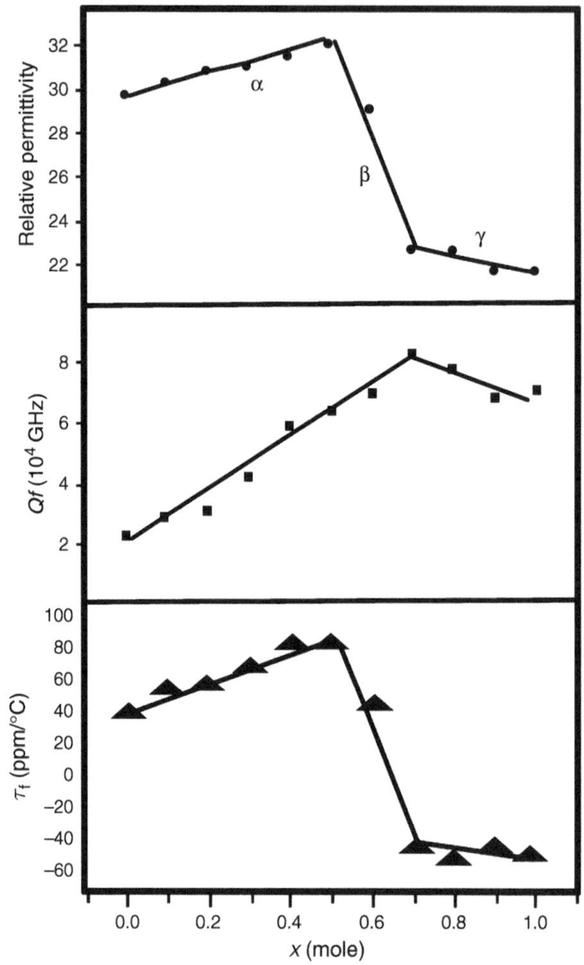

Figure 11.45 Variation of microwave dielectric properties of $Zn_{1-x}Co_xTiO_3$ sintered at 1150°C for 4 hours (after Ref. [245]).

11.13 Conclusions

Alumina is an important low-loss electronic packaging material. The dielectric quality factor very much depends on the purity and density of the sintered ceramics. The sintered dense alumina shows a very high quality factor (Qf) of about 1 million at room temperature. This is the highest quality factor reported for a dielectric ceramic at room temperature. However, alumina has a high negative τ_f of 60 ppm/°C. The negative τ_f of alumina has been compensated by the addition of a small amount of TiO_2 which has got a high positive τ_f. However, care must be taken to suppress the formation of Al_2TiO_5 which lowers the quality factor. The $0.9Al_2O_3$–$0.1TiO_2$ composite had $Qf = 117$ 000 GHz with $\varepsilon_r = 12.4$ and $\tau_f = 1.5$ ppm/°C. The single-crystal alumina (sapphire)

Table 11.8 Microwave dielectric properties of pure and doped $MgTiO_3$ and $ZnTiO_3$

Material	Sintering temperature (°C)	ε_r	Qf (GHz)	τ_f (ppm/°C)	Reference
$MgTiO_3$		17	166 400	–50	[50, 212]
$MgTiO_3$ + 1 mol% Nb_2O_5	1350	17.7	175 000	–	[220]
$0.95MgTiO_3$–$0.05CaTiO_3$	1400	20	56 000	0	[221]
$MgTiO_3$ (slow cooled 1°/min)	1350	16.5	220 000	–55	[219]
$(1 - x)MgTiO_3$–$xSrTiO_3$ ($x = 0.036$)	1270 for 2 hours	20.8	71 000	–1.3	[226]
$0.5MgTiO_3$–$0.5CaTiO_3$–$0.25(Nd_2O_3$–$TiO_2)$	1400	47.6	30 000	7.9	[227]
$0.95MgTiO_3$–$0.05CaTiO_3$ + 2 wt% B_2O_3	1200	21.2	62 000	4	[221]
$(Mg_{0.95}Ca_{0.05})TiO_3$ + 5 mol%B_2O_3	1200	19.6	86 000	–3	[217]
$(Mg_{0.95}Ca_{0.05})TiO_3$ + 3 mol%B_2O_3	1100	16.2	62 000	50.2	[217]
$0.95MgTiO_3$–$0.05CaTiO_3$ + 0.25 wt% CuO	1275	20	51 000	–8.3	[250]
$0.94MgTiO_3$–$0.06CaTiO_3$ + 0.25 wt% CuO	1275	20	48 000	–3	[250]

(Continued)

Table 11.8 (Continued)

Material	Sintering temperature (°C)	ε_r	Qf (GHz)	τ_f (ppm/°C)	Reference
$(Mg_{0.95}Ca_{0.05})TiO_3 + BaO–B_2O_3–SiO_2$ (50:50 wt%)	900	13.2	10 000	–	[251]
$(Mg_{0.93}Ca_{0.07})TiO_3$	1350	22.15	68 550	5.6	[224]
$0.95(Mg_{0.95}Co_{0.05})TiO_3–0.05CaTiO_3$	1275 for 4 hours	20.3	107 000	–22.8	[229]
$0.93(Mg_{0.95}Co_{0.05})TiO_3–0.07CaTiO_3$	1275 for 4 hours	21.6	92 000	–1.8	[229]
$(Mg_{0.95}Co_{0.05})TiO_3$	1275/4 hours	14.3	128 000	–51	[230, 229]
$0.94MgTiO_3–0.06CaTiO_3 + 0.2$ mol% Bi_2O_3	1250	22.6	53 000	–2.9	[228]
$0.95MgTiO_3–0.05CaTiO_3 + 0.2$ mol% Bi_2O_3	1250	21	55 600	–12.5	[228]
$0.7MgTiO_3–0.3MgTa_2O_6$	1460 for 3 hours	23	81 000	–2	[223]
$0.9MgTiO_3–0.1BaTiO_3$	1325	32.7	32 700	–85	[225]
$0.93MgTiO_3–0.07CaTiO_3$ (spark plasma sintering)	1150 for 10 minutes	23	7000	–	[252]
$Zn_{0.95}Mg_{0.05}TiO_3 + 0.25TiO_2 + 5$ wt% $Bi_2O_3 + 1$ wt% $2ZnO–B_2O_3$	920	24.6	4000	–14	[246]
$Zn_{0.6}Mg_{0.4}TiO_3 + 5$ wt% $B_2O_3–SiO_2–ZnO–K_2O$	1100	18	29 400	–	[253]

$0.91(Mg_{0.7}Zn_{0.3})TiO_3$–$0.09CaTiO_3$	1310 for 3 hours	22.5	86 000	3	[247]
$0.93(Mg_{0.6}Zn_{0.4})_{0.95}Co_{0.05}TiO_3$–$0.07CaTiO_3$	1200	23	79 500	1.4	[254]
$Mg_{0.95}Zn_{0.05}TiO_3$	1300/4 h	17.1	264 000	–40.3	[255]
$(Zn_{0.9}Mg_{0.1})TiO_3 + 4$ wt% Bi_2O_3	1000/4 h	25	70 000	–10	[222]
$0.85(Mg_{0.95}Zn_{0.05})TiO_3$–$0.15Ca_{0.6}La_{0.8/3}TiO_3$	1320 for 4 hours	26	86 000	0.5	[256]
$0.96(Mg_{0.95}Zn_{0.5})TiO_3$–$0.04SrTiO_3$	1300 for 4 hours	20.9	135 000	0	[257]
$ZnTiO_3 + 5$ wt% B_2O_3–SiO_2	850	22.2	52 500	–	[258]
$0.85(Mg_{0.95}Zn_{0.05})TiO_3$–$0.15Ca_{0.61}Nd_{0.26}TiO_3$	1300	24.3	112 000	–10.1	[259]
$ZnTiO_3$–$0.25TiO_2 + 1$ wt% B_2O_3	875	30	56 000	10	[230]
$(Mg_{0.95}Ni_{0.05})TiO_3$	–	17.2	18 000	–45	[222]

showed a very high quality factor of 8.3×10^9 at 12 GHz at 1.55 K. This is the highest quality factor reported for a dielectric material in the literature. The Zn_2SiO_4 (willemite), Mg_2SiO_4 (forsterite), Mg_4NbTaO_9, $Mg_4Nb_2O_9$, Y_2BaZnO_5 and $MgTiO_3$ have very high quality factor >200 000 GHz at room temperature. The τ_f of these materials have been tuned by suitable additives or by solid solution formation. The use of additives or τ_f compensators considerably lower the quality factor.

References

[1] N. Mc N. Alford and S.J. Penn. Sintered alumina with low dielectric loss. J. Appl. Phys. 80(1996)5895–5898.

[2] J. Molla, R. Heidenger, and A. Ibarro. Alumina ceramics for heating systems. J. Nucl. Mater. 212/215(1994)1029–1034.

[3] J. Molla, M. Gonzalez, R. Vila, and A. Ibarro. Effect of humidity on microwave dielectric losses of porous alumina. J. Appl. Phys. 85(1999)1727–1730.

[4] A. Ibarro, R. R. Heidenger, and J. Molla. New potentials for high mechanical strength grades of polycrystalline alumina for EC wave windows. J. Nucl. Mater. 191/194(1992)530–534.

[5] N. McN Alford, J. Breeze, X. Wang, S. J. Penn, S. Dalla, S. J. Webb, N. Ljepojevic, and X. Aupi. Dielectric loss of oxide single crystals and polycrystalline analogues from 10 to 320 K. J. Eur. Ceram. Soc. 21(2001)2605–2611.

[6] S. J. Penn, N. McN. Alford, A. Templeton, X. Wang, M. Xu, M. Reace, and K. Schrapel. Effect of porosity and grain size on the microwave dielectric properties of sintered alumina. J. Am. Ceram. Soc. 80(1997)1885–1888.

[7] V. B. Bragisnsky and V. S. Ilschenko. Experimental observation of fundamental microwave absorption in high quality dielectric crystals. Phys. Lett. A 120(1987)300–305.

[8] M. Ikeda and T. Miura. Measurement of complex dielectric constant above room temperature and its application to determination of intrinsic dielectric loss. Jpn. J. Appl. Phys. 37(1998) 5353–5356.

[9] V. I. Gurevich and A. K. Tagantsev. Intrinsic dielectric loss in crystals. Adv. Phys. 40(1991)719–767.

[10] J. Breeze, X. Aupi, and N. Mc N. Alford. Ultralow loss polycrystalline alumina. Appl. Phys. Lett. 81(2002)5021–5023.

[11] R. A. Woode, E. N. Ivanov, M. E. Tobar, and D. G. Blair. Measurement of dielectric loss tangent of alumina at microwave frequencies and room temperature. Electron. Lett. 30(1994)2120–2122.

[12] Giordano, Y. Kersale, O. Di. Monaco M. Chaubet. Progress report on the development of microwave spectral references at the LPMO. Eur. Phys. J. Appl. Phys. 8(1999)269–274.

[13] H. Ohsato, T. Tsunooka, M. Ando, Y. Ohishi, Y. Miyauchi, and K. Kakimoto. Millimeter–wave dielectric ceramics of alumina and fosterite with high quality factor and low dielectric constant. J. Korean Ceram. Soc. 40(2003)350–353.

[14] Y. Miyauchi, Y. Ohishi, S. Miyake, and H. Ohsato. Improvement of the dielectric properties of rutile-doped Al_2O_3 ceramics by annealing treatment. J. Eur. Ceram. Soc. 26(2006)2093–2096.

[15] Y. Ohishi, Y. Miyauchi, H. Ohsato, and K. Kakimoto. Controlled temperature coefficient of resonant frequency of Al_2O_3-TiO_2 ceramics by annealing treatment. Jpn. J. Appl. Phys. 43(2004)L749–L751.

[16] R. A. Woode, E. V. Ivanov, M. E. Tobar, and D. G. Blair. Cryogenic sapphire microwave resonator-oscillator with exceptional stability. Electron. Lett. 30(1994)417–419.

[17] J. Molla, A. Ibbara, J. Marginede, J. M. Zamaro, and A. Hernandez. Dielectric property measurement system at cryogenic temperature and microwave frequency. IEEE Trans. Instrum. Meas. 42(1993)817–821.

[18] Y. Kobayashi. Microwave measurement of dielectric properties of low loss materials by the dielectric rod resonator method. IEEE Trans. Microwave Theory and Technology. MTT 33(1985)586–592.
[19] C.-L. Huang, J.-J. Huang, and C.-Y. Huang. Sintering behaviour and microwave dielectric properties of nano-alpha alumina. Mater. Lett. 59(2005)3746–3749.
[20] C.-L. Huang, J.-J. Wang, and C.-Y. Huang. Microwave dielectric properties of sintered alumina using nano scaled powders of alumina. and TiO_2. J. Am. Ceram. Soc. 90(2007)1487–1493.
[21] N. McN Alford, X. Wang, S. J. Penn, M. Poole, and A. Jones. Effect of ceramic binders on microwave dielectric loss of alumina. Br. Ceram. Trans. 99(2000)212–214.
[22] D. G. Santiago, R. T. Wang, and G. J. Dick. Improved performance of a temperature compensated LN_2 cooled sapphire oscillator. IEEE Int. Freq. Contl. Symp. (1995)397–400.
[23] J. Dick and J. Saunders. Measurement and analysis of a microwave oscillator stabilized by a sapphire dielectric ring resonator for ultralow noise. Proc. 43th Ann. Freq. Contl. Symp. (1989)107–114.
[24] M. M. Driscoll, J. T Haynes, R. A. Jelen, K. W. Weinert, J. R. Gavaler, J. Tavacchio, G. R. Wagner, K. A. Zaki, X. P. Liang. Cooled, ultrahigh sapphire dielectric resonators for low noise microwave signal generation. IEEE Trans. Ultrason., Ferroelectr. Frequency Control 39(1992)405–411.
[25] G. J. Dick and J. Saunders. Measurement and analysis of a microwave oscillator stabilized by a sapphire dielectric ring resonator for ultra low noise IEEE Trans. Ultrason., Ferroelectr. Frequency Control 37(1990)339–346.
[26] A. Mann, A. N. Luiten, D. G. Blair, and M. J. Buckingham. Ultrastable cryogenic sapphire dielectric microwave resonators. Proc. IEEE 46th AFC. Hershey, PA (1992) pp. 167–171.
[27] M. E. Tobar, J. Krupka, E. N. Ivanova, and R. A. Woode. Dielectric frequency –temperature compensated microwave whispering gallery mode resonators. J. Phys. D. 30(1997)2770–2775.
[28] M. E. Tobar, J. Krupka, J. G. Hartnett, E. N. Ivanov, and R. A. Woode. High Q sapphire rutile frequency compensated microwave dielectric resonators. IEEE Trans. Microwave Theory Techniques MTT 45(1998)830–835.
[29] A. N. Luiten, A. G. Mann, and D. G. Blair. Ultra high Q factor cryogenic sapphire resonator. Electron. Lett. 29(1993)879.
[30] M. E. Tobar, A. G. Giles, S. J. Edwards, and J. H. Searles. High Q TE stabilized sapphire microwave resonators for low noise applications. IEEE Trans. Ultrson. Ferroelectr. Frequency Control 41(1994)391–396.
[31] M. M. Driscoll and R. W. Weinert. Spectral performance of sapphire dielectric resonator controlled oscillators operating in the 80 K to 275 K temperature range. Proc. IEEE Frequency Control Symposium (1995)401–412.
[32] R. A. Woode, M. E. Tobar, E. N. Ivanov, and D. G. Blair. An ultralow noise microwave oscillator based on a high Q liquid nitrogen cooled sapphire resonator. IEEE Trans. Microwave Theory Techniques MTT 43(1996)936–941.
[33] A. N. Luiten, A. G. Mann, and D. G. Blair. Ultra high Q factor cryogenic sapphire resonator. Electron. Lett. 29(1993)879–881.
[34] J. Breeze, S. J Penn, M. Poole, and N Mc N Alford. Layered Al_2O_3–TiO_2 composite dielectric resonators. Electron. Lett. 36(2000)883–884.
[35] N. McN. Alford, J. Breeze, S. J. Penn, and M. Poole. Layered Al_2O_3–TiO_2 composite dielectric resonators with tuneable temperature coefficient for microwave applications. IEEE Proc. Sci. Meas. Technol. 147(2000)269–273.
[36] W.-C. Tzou, Y.-C. Chen, S.L. Cheng, and C.-F. Yang. Microwave dielectric characteristics of glass added $(1-x)Al_2O_3$–$xTiO_2$ ceramics. Jpn. J. Appl. Phys. 41(2002)7422–7425.
[37] M. Kono, H. Takagi, T. Tatekawa, and H. Tamura. High Q dielectric resonator material with low dielectric constant for millimeter wave applications. J. Eur. Ceram. Soc. 26(2006)1909–1912.
[38] J. Hartnett, M. E. Tobar, E. N. Ivanov, and D. G. Cros. High frequency temperature compensated sapphire/rutile resonator. Electron. Lett. 36(2000)726–727.

[39] A. G. Mann, A. J Giles, D.G Blair, and M. J. Buckingham. Ultrastable cryogenic dielectric microwave resonators. Mode frequency temperature compensation by residual paramagnetic impurities. J. Phys. D 5(1992)1105–1109.
[40] A. J. Giles, A.G Mann, S.K Jones, and D.G Blair. A very high stability sapphire loaded superconducting cavity oscillator. Physica B 165(1990)145–146.
[41] G. J. Dick and R. T. Wang. Ultrastable performance of the superconducting maser. IEEE Trans. Instrum. Meas. 40(1991)174–177.
[42] S. K. Jones, D. G. Blair, and M. J. Buckingham. Effect of paramagnetic impurities on frequency of sapphire loaded superconducting cavity resonators. Electron. Lett. 24(1988)346–347.
[43] M. F. Yan and W.W. Rhodes. Low temperature sintering of TiO_2. Mater. Sci. Eng. 61(1983) 59–66.
[44] A. Okaya and L.F. Barash. The dielectric microwave resonator. Proc. IRE 50(1962)2081–2092.
[45] S. B. Cohen. Microwave band pass filters containing high Q dielectric resonators IEEE Microwave Theory and Techniques MTT- 16(1968)218–227.
[46] C. Egerton and J. Thomson. Ceram. Int. 50 (1971)924.
[47] A. Templeton, X. Wang, S. J. Webb, L. F. Cohen, and N. Mc N. Alford. Microwave dielectric loss of titanium oxide. J. Am. Ceram. Soc. 83(2000)95–148.
[48] F. A. Grant. Properties of rutile. Rev. Mod. Phys. 31(1959)646–674.
[49] W. D. Kingery, H.K. Bowen, and D.R. Uhlman. Introduction to Ceramics, 2nd Edition. Wiley, NY (1976).
[50] K. Wakino. Recent developments of dielectric resonator materials and filters (in Japanese). Ferroelectrics 91(1989)69–86.
[51] T. Negas, G. Yeager, S. Bell, N. Coats, and I. Minis $BaTi_2O_9/Ba_2Ti_9O_{20}$ based ceramics resurrected for modern microwave applications. Am. Ceram. Soc. Bull. 72(1983)80–89.
[52] D.-W. Kim, B. Park, J.-H. Chung, and K. S. Hong. Mixture behaviour and dielectric properties in the low fired TiO_2–CuO system. Jpn. J. Appl. Phys. 39(2000)2696–2700.
[53] D.-W. Kim, T.G. Kim, and K. S. Hong. Low firing of CuO doped anatase. Mater. Res. Bull. 34(1997)771–781.
[54] S.-H. Yoon, D.-W Kim, S.-Y. Cho, and K. S. Hong. Phase analysis and microwave dielectric properties of LTCC TiO_2 with glass system. J. Eur. Ceram. Soc. 23(2003)2549–2552.
[55] C. L. Lo, J.-G. Duh, B.-S Chiou, and W.-H Lee. Low temperature sintering and microwave dielectric properties of anorthite based glass ceramics. J. Am. Ceram. Soc. 85(2002)2230–2235.
[56] N. Klein C. Zuccaro, U. Dahe, H. Schulz, N. Tellmann, R. Kutzer, A. G. Zaitsev, and R. Wordenweber. Dielectric properties of rutile and its use in high temperature superconducting resonators. J. Appl. Phys. 78(1995)6683–6686.
[57] M. E. Tobar, J. Krupka, E. N. Ivanov, and R. A. Woode. Anisotropic complex permittivity measurement of monocrystalline rutile. J. Appl. Phys. 83(1998)1604–1609.
[58] M. E. Tobar, J. Krupka, J. G. Hartnett, and R. A. Woode. High Q sapphire rutile frequency temperature compensated microwave dielectric resonators. IEEE Trans Ultrason., Ferroelectr. and Frequency Control 45(1998)830–835.
[59] T. Nakazawa, T. Inoue, M. Satoh, and Y. Yamamoto. Electrical characteristics of metal/cerium dioxide/silicon structure. Jpn. J. Appl. Phys. 34(1995)548–553.
[60] G. Li, T. Ikegami, J. H. Lee, and T. Mori. Characterization and sintering of nanocrystalline CeO_2 powder synthesized by a mimic alkoxide method. Acta Mater. 49(2001)419–426.
[61] C. Tian and S. W. Chan. Ionic conductivities, sintering temperatures and microstructure of bulk ceramic CeO_2 doped with Y_2O_3. Solid State Ionics 134(2000)89–102.
[62] M. N. Rahaman and Y. C. Zhou. Effect of solid solution additives on the sintering of ultrafine CeO_2 powders. J. Eur. Ceram. Soc. 240–243(1995)939–950.
[63] N. Santha, M. T. Sebastian, P. Mohanan, N. Mc N. Alford, K. Sarma, R. C. Pullar, S. Kamba, A. Pashkin, P. Samukhlinia, and J. Petzelt. Effect of doping on the dielectric properties of cerium oxide in the microwave and far infrared frequency range. J. Am. Ceram. Soc. 87(2004)1233–1237.

[64] D. H. Kim, S. K. Lim, and C. An Microwave dielectric properties of $xTiO_2$–(1–x)CeO_2 ceramics. Mater. Lett. 52(2002)240–243.
[65] M. T. Sebastian, N. Santha, P. V. Bijumon, A.-K. Axelsson, and N. Mc N. Alford. Microwave dielectric properties of (1–x)CeO_2–x$CaTiO_3$ and (1–x)CeO_2–xSm_2O_3 ceramics. J. Eur. Ceram. Soc. 24(2004)2583–2589.
[66] J. Petzelt, S. Kamba, G. V. Kozlov, and A. A. Volokov. Dielectric properties of microwave ceramics investigated by infrared and submillimeter spectroscopy. Ferroelectrics 176(1996)145.
[67] T. Tsunooka, H. Sugiyma, K. Kakimoto, H. Ohsato, and H. Ogawa. Zero temperature coefficient τ_f and sinterability of forsterite ceramics by rutile addition. J. Ceram. Soc. Jpn. 112(2004)S1637–S1640.
[68] H. Ohsato, T. Tsunooka, A. Kan, Y. Ohishi, Y. Miyauchi, T. Tohda, T. Okawa, K. Kakimoto, and H. Ogawa. Microwave–millimetrewave dielectric materials. Key Eng. Mater. 269(2004)195–198.
[69] N.-H. Nguyen, J.-B. Lim, S. Nahm, J.-H. Paik, and J.-H. Kim. Effect of Zn/Si ratio on the microstructural and microwave dielectric properties of Zn_2SiO_4 ceramics. J. Am. Ceram. Soc. 90(2007)3127–3130.
[70] T. Okamura and T. Kishina. Dielectric properties of rare earth added cordierite at microwave and millimeterwave frequencies. Jpn. J. Appl. Phys. 37(1998)5364–5366.
[71] J.-M. Wu and H. L. Huang. Effect of crystallization on microwave dielectric properties of stoichiometric cordierite glasses containing B_2O_3 and P_2O_5 glasses. J. Mater. Res. 15(2000) 222–227.
[72] M. Terada, K. Kawamura, I. Kagomiya, K. Kakimoto, H. and Ohsato. Effects of Ni substitution on the microwave dielectric properties of cordierite. J. Eur. Ceram. Soc. 27(2007)3045–3048.
[73] T. Tsunooka, M. Androu, Y. Higashida, H. Sugiura, and H. Ohsato. Effects of TiO_2 on sinterability and dielectric properties of hgh Q forsterite ceramics. J. Eur. Ceram. Soc. 23(2003)2573–2578.
[74] Y. Guo, H. Ohsato, and K. Kakimoto. Characterization and dielectric behaviour of willemite and TiO_2- doped willemite ceramics at millimetrewave frequency. J. Eur. Ceram. Soc. 26(2006)1827–1830.
[75] J.-L. Zou, Q.-L. Zhang, H. Yang, and H.-P. Sun. A new system of low temperature sintering ZnO-SiO_2 dielectric ceramics. Jpn. J. Appl. Phys. 45(2006)4143–4145.
[76] M. M. Krzmanc, M. Valant, and D. Suvorov. A structural and dielectric characterization of Na_xCa_{1-x} $Al_{2-x}Si_{2+x}O_8$ ($x=0$, 1) ceramics. J. Eur. Ceram. Soc. 25(2005)2835–2838.
[77] C.-L. Lo, J.-G. Duh, B.-S. Chiou, and W.-H. Lee. Low temperature sintering and microwave dielectric properties of anorthite based glass. J. Am. Ceram. Soc. 85(2002)2230–2235.
[78] M.M. Krzmanc, M. Valant, and D. Suvorov. The synthesis and microwave dielectric properties of Sr_x Ba_{1x} $Al_2Si_2O_8$ and $Ca_yBa_{1-y}Al_2Si_2O_8$ ceramics. J. Eur. Ceram. Soc. 27(2007)1181–1185.
[79] Yokoi, H. Ogawa, A. Kan, and Y. Nakamura. Relationship between crystal structure and microwave dielectric properties of melilite type ceramics. J. Eur. Cer. Soc. 27(2007)2989–2993.
[80] H. Sun, Q. Zhang, H. Yang, and J. Zou. $(Ca_{1-x}Mg_x)SiO_3$: a low permittivity microwave dielectric system. Mater. Sci. Eng., B 138(2007)46–50.
[81] M. M. Krzmanc, M. Valant, B. Jancar, and D. Suvorov. Subsolidus synthesis and microwave dielectric characterization of plagioclase feldspars. J. Am. Ceram. Soc. 88(2005)2472–2479.
[82] E. Tanaka, H. Ogawa, K. Hibino, A. Yokoi, H. Ohsato, and A. Kan. Effect of Sn substitution for Zr on microwave dielectric properties of $Ca_3(Zr_{1-x}Sn_x)Si_2O_9$ solid solution Microwave Materials and their Applications Inuyama, Japan Poster. No.104(2004).
[83] S. Knickenecker, A. Kumar, and L. Herron. Cordierite glass-ceramic for multilayer ceramic packaging. Am. Ceram. Soc. Bull. 72(1993)90–95.
[84] E. A. Geiss, C. F. Guerci, G. F. Walker, and S. H. Wen. Isothermal sintering of spherodized cordierite-type glass powders. J. Am. Ceram. Soc. 68(1985)C328–C329.

[85] E. A. Geiss, J. P. Flecther, and L. W. Herron. Isothermal sintering of cordierite-type glass powders. J. Am. Ceram. Soc. 67(1984)549–552.
[86] T. S. Sasikala, M. N. Suma, P. Mohanan, C. Pavithran, and M. T. Sebastian. Forsterite based ceramic-glass composites for substrate applications and millimeter wave communications. J. Alloys Compd. (2008) in press.
[87] H. M. Buschbaum. Crystal chemistry of AM_2O_4 oxometallates. J. Alloys Compd. 349 (2003)49–104.
[88] A. Escardino, J. L. Amoros, A. Gozalbo, M. J. Orts, and A. Moreno. Gahnite devitrificationin ceramic frits: Mechanism and process kinetics. J. Am. Ceram. Soc. 83(2000)2938–2944.
[89] K. P. Surendran, N. Santha, P. Mohanan, and M. T. Sebastian. Temperature stable low loss ceramic dielectrics $(1-x)ZnAl_2O_4-xTiO_2$ system for microwave substrate applications. Eur. Phys. J. B. 41(2004)301–306.
[90] K. P. Surendran, P. V. Bijumon, P. Mohanan, and M.T. Sebastian. $(1-x)$ $MgAl_2O_4-xTiO_2$ dielectrics for microwave and millimetrewave applications. Appl. Phys. A 81(2005)823–826.
[91] W. Lei, W.-Z. Lu, J.-H. Zhu, and X.-H. Wang. Microwave dielectric properties of $ZnAl_2O_4$–$xTiO_2$ spinel based composites. Mater. Lett. 61(2007)4066–4069.
[92] K. P. Surendran, M. T. Sebastian, M. V. Manjusha, and J. Philip. A low loss dielectric substrate in $ZnAl_2O_4$-TiO_2 for microelectronic applications. J. Appl. Phys. 98(2005)044101.
[93] C. M. Zheng, S.-Y. Wu, X.M. Chen, and K. X. Song. Modification of $MgAl_2O_4$ microwave dielectric ceramics by Zn substitution. J. Am. Ceram. Soc. 90(2007)1483–1486.
[94] Z. Brynar, R. Grasser, Z. Potucek, A. Scharmann, and P. Bohacek. Real and near infrared luminescence of undoped $CaWO_4$. J. Lumin. 72–74(1997)643–645.
[95] L. Pankratov, D. Grigorjev, D. Millers, S. Chernov, and A. J. Voloshinozaki. Luminescence of some excited state absorption in tungstates. J. Lumin. 94–95(2001)427–432.
[96] S. Nishigaki, S. Yano, H. Kato, T. Hirai, and T. Nonomura. BaO-TiO_2-WO_3microwave ceramics and crystalline $BaWO_4$. J. Am. Ceram. Soc. 71(1988)C11–C17.
[97] A. Kuzmin and J. Purans. Local atomic and electronic structure of tungsten ions in AWO_4 crystals of Scheelite and wolframite types. Radiant. Meas. 33(2001)583–586.
[98] S. H. Yoon, D.-W. Kim, S.-Y. Cho, and K. S. Hong. Investigations of the relations between structure and microwave dielectric properties of divalent metal tungstates compounds. J. Eur. Ceram. Soc. 26(2006)2051–2054.
[99] M. H. Kim, J. C. Lee, C. I. Cheon, and H.-J. Kang. Crystal structures and low temperature cofiring ceramic property of (1–x)(Li,Re)W_2O_8–x$BaWO_4$ ceramics. (Re = Y, Yb) Jpn. J. Appl. Phys. 45(2006)7397–7400.
[100] A. Kan, H. Ogawa, T. Oishi, A. Yokoi, and H. Ohsato. Crystal structure and microwave dielectric properties of aeschenite type $R(W_{0.5}Ti_{1.5})O_6$ (R = Nd, Sm, Eu, Gd, Dy) ceramics. Jpn. J. Appl. Phys. 44(2005)7103–7106.
[101] F. Zhao, Z. Yue, Z. Gui, and L. Li. Preparation characterization and microwave dielectric properties of A_2BWO_6 [A = Sr, Ba, B = Co, Ni, Zn] double perovskite ceramics. Jpn. J. Appl. Phys. 44(2005)8066–8070.
[102] R. C. Pullar, S. Farrah, and N. McN. Alford $MgWO_4$, $ZnWO_4$, $NiWO_4$, and $CoWO_4$, microwave dielectric ceramics. J. Eur. Ceram. Soc. 27(2007)1059–1063.
[103] P. V. Bijumon, S. Solomon, M. T. Sebastian, and P. Mohanan. A new group of microwave dielectrics in the $RE(Ti_{0.5}W_{0.5})O_4$ [RE = Pr, Nd, Sm, Gd, Tb, Dy, and Y] system. J. Mater. Sci. – Mater. Electron. 14(2005)5–8.
[104] E. S. Kim, S. H. Kim, and B. I. Lee. Low temperature sintering and microwave dielectric properties of $CaWO_4$ ceramics for LTCC applications. J. Eur. Ceram. Soc. 26(2006)2101–2104.
[105] E. S. Kim, and S. H. Kim, Effects of structural characteristics on microwave dielectric properties of $(1-x)CaWO_4$–$xLaNbO_4$ ceramics. J. Electroceram. 17(2006)471–477.
[106] L. H. Brixner, Preparation and properties of some novel rare earth compounds of the type $LnTi_{0.5}M_0O_5$ and $LnTi_{0.5}W_{0.5}O_4$. J. Inorg Chem. 3(1964)600–602.
[107] M. Maeda, T. Yamamura, and T. Ikeda. Dielectric characteristics of several complex oxide ceramics at microwave frequencies. Jpn. J. Appl. Phys. Suppl. 26–2(1987)76–79.

[108] Y. Repelin, E. Hausson, and H. Brusso, Etude par spectroscopies d'absorption i.r. de diffusion Raman des composes $A^{II}B_2O_6$ de structure de type Blocs 1 × 2-1 Etude niobate de barium $BaNb_2O_6$. Spectrochem. Acta 34(1979)937–948.

[109] H. Obayashi, Y. Sakurai, and T. Gejo. Perovskite oxides as ethanol sensor. J. Solid State Chem. 17(1976)299–303.

[110] H.-J. Lee, K.-S Hong, S.-J. Kim, and I.-T. Kim. Dielectric properties of MNb_2O_6 compounds (where M = Ca, Mn, Co, Ni, or Zn). Mater. Res. Bull. 32(1997)847–855.

[111] H.-J. Lee, I.-T. Kim, and K.S. Hong. Dielectric properties of AB_2O_6 compounds at microwave frequencies (A = Ca, Mn, Co, Ni, Zn, and B = Nb, Ta). Jpn. J. Appl. Phys. 36(1997)L1318-L1320.

[112] C.-L. Huang, and K.-H. Chiang. Improved high Q microwave dielectric materials using B_2O_3 doped $MgNb_2O_6$ ceramics. Mater. Sci. Eng., A 474(2008)76–80.

[113] Y.-C. Liou, M.-H. Weng, and C.-Y. Shiue, $CaNb_2O_6$ ceramics prepared by a reaction sintering process. Mater. Sci. Eng., B 133(2006)14–19.

[114] R.C. Pullar, Mc N Alford, and J. D. Breeze. Microwave dielectric properties of columbite-structure niobate ceramics $M^{2+}Nb_2O_6$. Key. Eng. Mater. 224–226(2002)1–4.

[115] R.C. Pullar, J. D. Breeze, and N. Mc N. Alford. Characterzation and microwave dielectric properties of $M^{2+}Nb_2O_6$ ceramics. J. Am. Ceram. Soc. 88(2005)2466–2471.

[116] C.-S. Hsu, C.-L. Huang, J.-F. Tseng, and C.-Y Huang. Improved high Q microwave dielectric resonator using CuO doped $MgNb_2O_6$ ceramics. Mater. Res. Bull. 38(2003) 1091–1099.

[117] D.-W Kim, K. H. Ko, and K. S. Hong. Influence of copper(II) oxide additions to zinc niobate microwave ceramics on sintering temperature and dielectric properties. J. Am. Ceram. Soc. 84(2001)1286-1290.

[118] R. C. Pullar, K. Okemen, and N. Mc N Alford. Temperature compensated niobate microwave ceramics with columbite structure $M^{2+}Nb_2O_6$. J. Eur. Ceram. Soc. 23(2003)2479–2483.

[119] D.-W. Kim, D.-Y. Kim, and K.S. Hong. Phase relations and microwave dielectric properties of $ZnNb_2O_6$–TiO_2. J. Mater. Res. 15(2000)1331–1335.

[120] D.-H. Kim, S.-K. Lim, and C.-An. Proc. International Conference on Solid dielectrics Eindhoven, The Netherlands (2001) p. 195.

[121] D.-W. Kim, H. J. Youn, K. S. Hong, and C. K. Kim. Microwave dielectric properties of $(1-x)Ba_5Nb_4O_{15}$-$xBaNb_2O_6$ mixtures. Jpn. J. Appl. Phys. 41(2002)3812–3816.

[122] D.-W. Kim, H.B. Hong, and K. S. Hong. Structural transition and microwave dielectric properties of $ZnNb_2O_6$-TiO_2 sintered at low temperatures. Jpn. J. Appl. Phys. 41(2002)1465–1469.

[123] J. Wang, Z. Yue, Z. Gui, and L. Li. Low temperature sintered $ZnNb_2O_6$ microwave dielectric ceramics doped with ZnO–V_2O_5 additives. J. Mater. Sci. 40(2005)6581–6583.

[124] Y.-C. Zhang, Z. X. Yue, L. Gui, and L. T. Li. Microwave dielectric properties of CuO-V_2O_5-Bi_2O_3 doped $ZnNb_2O_6$ ceramics with low sintering temperature. J. Electroceram. 14(2005) 67–74.

[125] S.-H. Wee, D.-W. Kim, S.-I. Yoo, and K. S. Hong. Low temperature sintering of V_2O_5 added and substituted $ZnNb_2O_6$ microwave ceramics. Jpn. J. Appl. Phys. 43(2004)3511–11515.

[126] D.-W Kim, H. B. Hong, C. K. Kim, and D. J. Kim. The reversible phase transition and dielectric properties of $BaNb_2O_6$ polymorphs. Jpn. J. Appl. Phys. 41(2002)6045–6048.

[127] A. Kan, H. Ogawa, and H. Ohsato. Influence of microstructure on microwave dielectric properties of $ZnTa_2O_6$ ceramics with low dielectric loss. J. Alloys Compd. 337(2002) 303–308.

[128] C.-L. Huang and K.-H. Chiang. Characterization and dielectric behaviour of CuO doped $ZnTa_2O_6$ ceramics at microwave frequency. Mater. Res. Bull. 39(2004)1701–1708.

[129] Y.-C. Zhang, Z. X. Yue, Z. L. Gui, and L. T. Li. Microwave dielectric properties of $(Zn_{1-x}Mg_x)Nb_2O_6$ ceramics. Mater. Lett. 57(2003)4531–4534.

[130] A. G. Belous, O. V. Ovchar, A. V. Kramerenkov, B. Jancar, J. Bezjak, and D. Suvorov. Synthesis and microwave dielectric properties of $Zn_{1+x}Nb_2O_{6+x}$. Inorg. Mater. 43(2007)277–280.
[131] C.-L. Huang, K.-H. Chiang, and C.-Y. Huang. Microwave dielectric properties and microstructures of $MgTa_2O_6$ ceramics with CuO addition. Mater. Chem. Phys. 90(2005)373–377.
[132] C.-F. Yang, C.-C. Chan, C.-M. Cheng, and Y.-C. Chen. The sintering and microwave dielectric characteristics of $MgTa_{1.5}Nb_{0.5}O_6$ ceramics. J. Eur. Ceram. Soc. 25(2005) 2849–2852.
[133] C.-M. Cheng, Y.-C. Chen, C.-F. Yang, and C.-C. Chen. Sintering and compositional effects on the microwave dielectric characteristics of $Mg(Ta_{1-x}Nb_x)_2O_6$ ceramics with $0.2 \leq x \leq 0.35$. J. Electroceram. 18(2007)155–160.
[134] C.-H. Hsu, C.-F. Tseng, and C.-L. Huang. Microwave dielectric properties of $MgNb_2O_6$ ceramics with Fe_2O_3 additions. Jpn. J. Appl. Phys. 44(2005)8043–8047.
[135] P. R. Khoei, F. Azough, and R. Freer. The influence of $ZnNb_2O_6$ on the microwave dielectric properties of $ZrTi_2O_6$ ceramics. J. Am. Ceram. Soc. 89(2006)216–223.
[136] E. Wahlstorm and B.O. Marinder. Phase analysis studies in the copper–niobium–oxygen system. Inorg. Nucl Chem. Lett. 13(1977)559–564.
[137] Y. C. Zhang, Z. X. Yue, Z. L. Gui, and L. T. Li. Effect of CaF_2 addition on the microstructure and microwave dielectric properties of $ZnNb_2O_6$ ceramics. Ceram. Int. 29(2003)555–559.
[138] Y. C. Zhang, Z. X. Yue, Z. L. Gui, and L. T. Li. Low temperature sintered $ZnNb_2O_6$ microwave dielectric ceramics doped with $CuO–Bi_2O_3–V_2O_5$ additions. J. Mater. Sci. Lett. 22(2003) 593–597.
[139] S. H. Wee, D. W. Kim, and S. I. Yoo. Microwave dielectric properties of low fired $ZnNb_2O_6$ceramics with $BiVO_4$ addition. J. Am. Ceram. Soc. 87(2004)871–874.
[140] S.-H. Wee, D.-W. Kim, S-I. Yoo, and K.S. Hong. Low temperature sintering of V_2O_5 added and substituted $ZnNb_2O_6$ microwave ceramics. Jpn. J. Appl. Phys. 43(2004)3511–3515.
[141] R. C. Pullar, C. Vaughan, and N. Mc N Alford. The effects of sintering aids upon dielectric microwave properties of columbite $M^{2+}Nb_2O_6$. J. Phys. D: Appl. Phys. 37(2004)348–352.
[142] Y. C. Zhang, Z. X. Yue, X. Qi, B. Li, Z. L. Gui, and L.T. Li. Microwave dielectric properties of $Zn(Nb_{1-x}\ Ta_x)_2O_6$ ceramics. Mater. Lett. 58(2004)1392–1395.
[143] Y. C. Zhang, Z. X Yue, Z. L. Gui, and L. T. Li. Microwave dielectric properties of $(Zn_{1-x}Mg_x)\ Nb_2O_6$ ceramics. Mater. Lett. 57(2003)4531–4534.
[144] D.-W. Kim, K. H. Ko, D.-K. Kwon, and K. S. Hong. Origin of microwave dielectric loss in $ZnNb_2O_6–TiO_2$. J. Am. Ceram. Soc. 85(2002)1169–1172.
[145] H. J. Lee, H. J. Yoon, K. S. Hong, and B. K. Kim. US patent 5,756,412 (1998).
[146] A. Baumgarte and R. Blachnik. New $M^{2+}M^{4+}Nb_2O_8$ phases. J. Alloys Compd. 215(1994) 117–120.
[147] E. H. Nickel, J. F. Rouland, and R. C. Mc Adam. Ixiolite-a columbite structure. Am. Mineral. 48(1963)961–979.
[148] A. Baumgarte and R. Blachnik. Phase relations in the system titanium dioxide diniobium-zinc hexoxide. Mater. Res. Bull. 27(1992)1287–1294.
[149] R.G. Breckenridge and W.R. Hsaler. Electrical properties of titanium dioxide semiconductors. Phys. Rev. 91(1953)793–802.
[150] R. R. Hasigati, K. Minami, and H. Yonemitsu. Electrical resistivity and defect energy levels in reduced titanium dioxide at low temperatures. J. Phys. Soc. Jpn. 16(1961)2223–2226.
[151] M. H. Francombe. The relation between structure and ferroelectricity in lead barium and barium strontium niobate. Acta Crystallogr. 13(1960)131–140.
[152] O. Yamaguchi, K. Mastsui, and K. Shimiza. Crystallization of hexagonal $BaNb_2O_6$. J. Am. Soc. 68(1985)C173–C175.
[153] R. Ratheesh, M. T. Sebastian, P. Mohanan, M. E. Tobar, J. Hartnett, R. Woode, and D. G. Blair. Microwave characterization of $BaCe_2Ti_5O_{15}$ and $Ba_5Nb_4O_{15}$ ceramic dielectric resonators using whispering gallery mode method. Mater. Lett. 45(2000)279–285.

[154] E. F. Bertaut, L. Corliss, F. Forrat, R. Aleonard, and R. Raurthenet. Etude de niobates et tantalates de metaux de transition bivalents. J. Phys. Chem. Sol. 21(1961)234–251.
[155] N. Kumada, K. Taki, and N. Kinomura. Single crystal structure refinement of a magnesium niobium oxide: $Mg_4Nb_2O_9$. Mater. Res. Bull. 35(2000)1017–1021.
[156] W. Wong-Ng, H. F. Mc Murdie, B. Paretzkin, C. R. Hubbard, A. L. Drago, and J. M. Stewart. Standard X-ray diffraction pattern of sixteen ceramic phases. Powder Diffr. 2(1987)191.
[157] H. Ogawa, A. Kan, S. Ishihara, and Y. Higashida. Crystal structure of corundom type $Mg_4(Nb_{2-x}Ta_x)O_9$ microwave dielectric ceramics with low dielectric loss. J. Eur. Ceram. Soc. 23(2003)2485–2488.
[158] A. Kan, H. Ogawa, and H. Ohsato. Relationship between bond strength and microwave dielectric properties of corundom type $(Mg_{4-x}Co_x)Nb_2O_9$ and $Mg_4(Nb_{2-y}Ta_y)O_9$ solid solutions. J. Ceram. Soc. Jpn. Supl. 112–1(2004)S1622–S1626.
[159] A. Kan, H. Ogawa, A. Yokoi, and H. Ohsato. Low temperature sintering and microstructure of $Mg_4(Nb_{2-x}V_x)O_9$. Microwave dielectric ceramic by V substitution for Nb. Jpn. J. Appl. Phys. 42(2003)6154–6157.
[160] A. Kan, H. Ogawa, A. Yokoi, and Y. Nakamura. Crystal structural refinement of corundom structured $A_4M_2O_9$ (A = Co, and Mg, M = Nb and Ta) microwave dielectric ceramics by high temperature X-ray powder diffraction. J. Eur. Ceram. Soc. 27(2007)2977–2981.
[161] A. Kan and H. Ogawa. Low temperature synthesis and microwave dielectric properties of $Mg_4Nb_2O_9$ ceramics synthesized by a precipitation method. J. Alloys Compd. 364 (2004)247–249.
[162] A. Kan, H. Ogawa, and A. Yokoi. Sintering temperature dependence of microwave dielectric properties in $Mg_4(TaNb_{1-x}V_x)O_9$ compounds. Mater. Res. Bull. 41(2006)1178–1184.
[163] A. Yokoi, H. Ogawa, A. Kan, H. Ohsato, and Y. Higashida. Microwave dielectric properties of $Mg_4Nb_2O_9$-3.0 wt% LiF ceramics prepared with $CaTiO_3$ additions. J. Eur. Ceram. Soc. 25(2005)2871–2875.
[164] A. Kan, H. Ogawa, and A. Yokoi. Electric state and bonding of $(Mg_{4-x}Mn_x)Nb_2O_9$ microwave dielectric ceramics. J. Mater. Res. 20(2005)2456–2461.
[165] H. Ogawa, H. Taketani, A. Kan, A. Fujita, and G. Zouganelis. Evaluation of electronic state of $Mg_4(Nb_{2-x}Sb_x)O_9$ microwave dielectric ceramics by first principal calculation method. J. Eur. Ceram Soc. 25(2005)2859–2863.
[166] A. Yokoi, H. Ogawa, A. Kan, H. Ohsato, and Y. Higashida. Use of LiF to achieve a low temperature cofired ceramics (LTCC) with low dielectric loss. J. Ceram. Soc. Jpn. 112 Suppl(2004)S1633–S1636.
[167] R. S. Roth, C. J. Rawn, F. Beedia, J. D. Whitler, and J. O. Anderson. Ceramic Superconductors-II. Am. Ceram. Soc. 1988, p. 13.
[168] C. Michel and B. Raveau. Ln_2BaZnO_5 and $Ln_2BaZn_{1-x}Cu_xO_5$: a series of zinc oxide with zinc in pyramidal coordination. J. Solid State Chem. 49(1983)150–156.
[169] C. Michel and B. Raveau. Les oxides A_2BaCuO_5 (A=Y, Sm, Eu, Gd, Dy, Ho, Er, Yb). J. Solid State Chem. 43(1982)73–86.
[170] D. Dube, B. Champagne, and Y. Le. Page X-ray characterisation of Nd_2BaCuO_5. Mater. Lett. 8(1989)283–285.
[171] M. Tabi, J. Aride, J. Darriet, and A. Moquine. Structure cristallite de l'oxyde Nd_2BaZnO_5. J. Solid State Chem. 86(1996)233–237.
[172] A. Salinez-Sanchez, J. L. Gracia-Munoz, J. Rodriguez-Carvajal, R. Saez-Puche, and J. L. Martinez. Structural characterization of R_2BaCuO_5 (R = Y, La, Yb, Tmm, Er, Ho, Dy, Gd, Eu, and Sm) oxides by X-ray and neutron diffraction. J. Solid State Chem. 100(1992)201–211.
[173] M. Watanabe, H. Ogawa, H. Ohsato, and C. Humphreys. Microwave dielectric properties of $Y_2Ba(Cu_{1-x}Zn_x)O_5$ solid solutions. Jpn J. Appl. Phys. 37(1998)5360–5363.
[174] A. Kan, H. Ogawa, M. Watanabe, S. Hatanaka, and H. Ohsato. Microwave dielectric properties of $Sm_2Ba(Cu_{1-x}Zn_x)O_5$ solid solutions. Jpn. J. Appl. Phys. 38(1999)5629–5632.

[175] A. Kan, H. Ogawa, H. Ohsato, and S. Ishihara. Microwave dielectric properties and crystal structure of $Y_2(Ba_{1-x}Sr_x)(Cu_{1-y}Zn_y)O_5$ solid solutions synthesized by a solid state reaction method. Jpn. J. Appl. Phys. 39(2000)5654–5657.
[176] H. Ogawa, A. Kan, S. Yokoi, H. Ohsato, and S. Ishihara. Mictrowave dielectric properties and crystal structure of $Nd_2(Ba_{1-x}M_x)ZnO_5$ solid solutions. J. Eur. Ceram. Soc. 21(2001)1731–1734.
[177] A. Kan, H. Ogawa, H. Ohsato, and J. Sugishita. Influence of M [M = Zn, Ni] substitution for Cu on microwave dielectric characteristics of $Yb_2Ba(Cu_{1-x}M_x)O_5$ solid solutions. Jpn. J. Appl. Phys. 40(2001)5774–5778.
[178] A. Kan, H. Ogawa, H. Ohsato, and S. Ishihara. Effects of variations in crystal structure on microwave dielectric properties of Y_2BaCuO_5 system. J. Eur. Ceram. Soc. 21(2001)2593–2598.
[179] A. Kan, H. Ogawa, K. Mori, and J. Sugishita. Effects of La substitution for Nd in $Nd_{2-x}La_xBaZnO_5$ solid solutions on microwave dielectric properties and crystal structure. Mater. Res. Bull. 37(2002)1509–1518.
[180] A. Mori, H. Ogawa, A. Kan, H. Ohsato, and S. Ishihara. Crystal structural characterization of Nd_2BaZnO_5 type microwave dielectric ceramic with rare earth substitution for Nd. Mater. Chem. Phys. 79(2003)273–275.
[181] A. Kan, H. Ogawa, and H. Ohsato. Role of Zn substitution for Cu on the microwave dielectric properties and crystal strutcure of $Eu_2Ba(Cu_{1-x}Zn_x)O_5$ solid solutions. Physica B 322(2002)403–407.
[182] A. Kan, H. Ogawa, and H. Ohsato. Microwave dielectric properties of $R_2Ba(Cu_{1-x}M_x)O_5$ [R = Y and Yb, M = Zn and Ni] solid solutions. Mater. Chem. Phys. 79(2003)184–186.
[183] K. Mori, H. Ogawa, A. Kan, and H. Ohsato. Dielectric property microstructure relations in CoO doped $(Y_{2-x}Sm_x)BaCuO_5$ ceramics. J. Eur. Ceram. Soc. 23(2003)2603–2606.
[184] H.-C. Chen, M.-H. Weng, C.-I. Hung, M.-P. Houn, and Y.-H. Wang. Effect of Ni substitution for Cu on the dielectric properties of $Y_2Ba(Cu_{1-x}Ni_x)O_5$ solid solutions. Physica B 349(2004)304–309.
[185] K. Mori, H. Ogawa, A. Kan, and H. Ohsato. Relationships between Sr substitution for Ba and dielectric characteristics in Sm_2BaZnO_5 ceramics. J. Eur. Ceram. Soc. 24(2004) 1745–1748.
[186] K. Mori, H. Ogawa, A. Kan, and H. Ohsato. Microwave dielectric-microstructure relationships in $Y_2Ba(Cu_{1-x}Mg_x)O_5$ solid solutions. J. Eur. Ceram Soc. 24(2004)1749–1753.
[187] S. Kawaguchi, H. Ogawa, A. Kan, and E. Tanaka. Effect of Sr substitution for Ba on the microwave dielectric properties of $Sm_2(Ba_{2-x}Sr_x)(Cu_{0.5}Zn_{0.5})O_5$ ceramics. J. Eur. Ceram. Soc. 25(2005)2853–2858.
[188] A. Yoshida, H. Ogawa, A. Kan, and T. Kondo. Sintering time dependence of microwave dielectric properties and crystal structure in Y_2BaZnO_5 ceramics. J. Eur. Ceram. Soc. 25(2005)2897–2900.
[189] H. Ogawa, A. Kan, H. Taketani, and H. Ohsato. Phase transition and microwave dielectric properties of $Nd_2Ba(Zn_{1-x}Cu_x)O_5$ (x = 0 to 0.55) by Cu substitution for Zn. J. Ceram. Soc. Jpn. Suppl. 112–1(2004)S1627–S1632.
[190] Y. Tsuji, A. Kan, H. Ogawa, and S. Ishihara. Relationship between microwave dielectric properties and chemical bonding in R_2BaMO_5 [R = rare earth, M = Cu and Zn] coumpounds. J. Eur. Ceram. Soc. 25(2005)2883–2887.
[191] A. Kan, H. Ogawa, and H. Ohsato. Microwave dielectric characteristics of Y_2BaZnO_5 ceramics with Sm substitution for Y. Jpn. J. Appl. Phys. 41(2002)7226–7229.
[192] I. Komkov. Polymorphism of the compound $ReTiNbO_6$ type. Dokl. Acad. Nauk. SSSR 148(1963)1182–1183.
[193] G. Blasse. New compounds of the AB_2O_6 type. J. Inorg. Nucl. Chem. 28(1966)1122–1124.
[194] V.V. Kazantsev E.I. Krylov A.K. Borisov A.I. Chupin. Russian J. Inorg. Chem. 19(1974) 506–507.
[195] X. Qi, T. P. J. Han, H. G. Gallagher, B. Henderson, R. Illingworth, and I. S. Ruddock. Optical spectroscopy of $PrTiNbO_6$ $NdTiNbO_6$ and $ErTiNbO_6$. J. Phys. Condens. Matter. 8(1996)4837–4845.

[196] C. E. Holcombe, M. K. Morrow, D. D. Smith, and D. A. Carpenter. Survey Study of Low Expending, High Melting, Mixed Oxides, Y-1913 (Union Carbide Corporation, Nuclear Division, Oak Ridge, TN, *1974)*.

[197] C. E. Holcombe. Ternary tantalite compositions. J. Mater. Sci. Lett. 14(1974)2255–2260.

[198] X. Qi, R. Illingworth, H. G. Gallagher, T. P. J. Han, and B. Henderson. Potential laser gain media with stoichiometric formula $RETiNbO_6$. J. Cryst. Growth. 160(1996)111–118.

[199] X. Qi, H. G. Gallagher, T. P. J. Han, and B. Henderson. Modified Czochralski growth and characterization of $RETiTaO_6$ crystals. J. Cryst. Growth. 180(1997)73–80.

[200] H. Seifert and B. Beck. The metamiet minerals of the euxenite group. Fortschr. Mineral. 39(1961)36–39.

[201] M. T. Sebastian, S. Solomon, R. Ratheesh, J. George, and P. Mohanan. Preparation and microwave properties of $ReTiNbO_6$ [Re = Ce, Pr, Nd, Sm, Eu, Gd, Tb, Dy, Y and Yb] dielectric ceramics. J. Am. Ceram. Soc. 84(2001)1487–1489.

[202] S. Solomon, M. Kumar, K. P. Surendran, M. T. Sebastian, and P. Mohanan. Synthesis characterization and properties of $Re_{1-x}RE'_xTiNbO_6$ dielectric ceramics. Mater. Chem. Phys. 67(2001)291–293.

[203] K. P. surendran, P. Mohanan, and M. T. Sebastian. Microwave dielectric properties of $RE_{1-x}RE'_x$ $TiNbO_6$ [RE = Pr, Nd, Sm, RE' = Gd, Dy, Y] ceramics. J. Am. Ceram. Soc. 86(2003)1695–1699.

[204] M. T. Sebastian, R. Ratheesh, H. Sreemoolanathan, S. Solomon, and P. Mohanan. Samarium titanium niobate(SmT, NbO_6): A new microwave dielectric ceramic. Mater. Res Bull. 32(1997)1279–1284.

[205] K. P. Surendran, S. Solomon, M. R. Varma, P. Mohanan, and M. T. Sebastian. Microwave dielectric properties of $RETiTaO_6$ [RE = La, Ce, Pr, Nd, Sm, Eu, Gd, Tb, Dy, Ho, Y, Yb, nAl, and In] ceramics. J. Mater. Res. 17(2002)2561–2566.

[206] T. Oishi, A. Kan, H. Ohsato, and H. Ogawa. Crystal structure-microwave dielectric property relations in $Sm(Nb_{1-x}Ta_x)(Ti_{1-y}Zr_y)O_6$ ceramics. J. Eur. Ceram. Soc. 26(2006)2075–2079.

[207] S. Solomon, H. P. Kumar, L. Jacob, J. K. Thomas, and M. R. Varma. $Ln(Zr_{1/3}Ti_{2/3})TaO_6$ (Ln = Ce, Pr, Nd, and Eu) A novel group of microwave ceramics. J. Alloys Compd. (2008) in press.

[208] T. Oishi, H. Ogawa, A. Kan, and H. Ohsato. Synthesis and microwave dielectric properties of $(Ce_{1-y}Dy_y)(Nb_{1-x}Ta_x)TiO_6$ ceramics. J. Eur. Ceram. Soc. 25(2005)2889–2895.

[209] S. Solomon, J. J. Joseph, H. Padmakumar, J. K. Thomas. Effect of ZnO doping on the microwave dielectric properties of $LnTiNbO_6$ (Ln = Sm, Dy) ceramics. Mater. Lett. 60(2006)2814–2818.

[210] R. D. Shannon. Revised effective ionic radii and systematic studies of interatomic distances in halides and chalcogenides. Acta Crystallogr. A 32(1976)751–767.

[211] K. P. Surendran, P. Mohanan, and M. T. Sebastian. Tailoring the microwave dielectric properties of $GdTiNb_{1-x}Ta_xO_6$ and $Sm_{1-x}Y_xTiTaO_6$ ceramics. J. Eur. Ceram. Soc. 23(2003)2489–2495.

[212] V. M. Ferreira, F. Azough, R. Freer, and J. L. Baptista. The effect of Cr and La on $MgTiO_3$ and $MgTiO_3$-$CaTiO_3$ microwave dielectric ceramics. J. Mater. Res. 12(1997)3293–3299.

[213] V. M. Ferreira, F. Azough, J. L. Baptista, and R. Freer. Magnesium titanate microwave dielectric ceramics. Ferroelectrics 133(1992)127–132.

[214] V. M. Ferreira, S. Kamba, J. Petzelt, and J. L. Baptista. Dielectric Spectroscopy of $MgTiO_3$–based ceramics in the 10^9–10^{14} Hz region. J. Mater. Sci. 28(1993)5894–5900.

[215] V. M. Ferreira and J. L. Baptista. Preparation and microwave dielectric properties of pure and doped magnesium titanate ceramics. Mater. Res. Bull. 29(1994)1017–1023.

[216] V. M. Ferreira, J. L. Baptista, J. Petzelt, G. A. Komandin, and V. V. Voitsekhovskii. Loss spectra of pure and La doped $MgTiO_3$ microwave ceramics. J. Mater. Res. 10(1995)2301–2305.

[217] N. Ichinose and H. Yamamoto. Effect of additives on microwave dielectric properties in low temperature firing of $(Mg,Ca)TiO_3$ based ceramics. Ferroelectrics 201(1997)255–262.

[218] S. H. Yoo, K. H. Yoon, J. W. Choi, and S. J. Yoon. Effect of thermal strain on the quality factor of microwave $MgTiO_3$ ceramics. J. Ceram. Soc. Jpn. Suppl. 112–2(2004)S1606–S1609.

[219] M. L. Hsieh, L.-S. Chen, S.-M. Wang, C.-H. Sun, M.-H. Weng, M.-P. Houng, and S.-L. Fu. Low temperature sintering of microwave dielectrics $(Zn,Mg)TiO_3$. Jpn. J. Appl. Phys. 44(2005)5045-5048.

[220] V. M. Ferreira, J. L. Baptista. The role of niobium in magnesium titanate microwave dielectric ceramics. J. Am. Ceram. Soc. 79(1996)1697–1698.

[221] C.-L. Huang and M.-H. Weng. Improved high Q values of $MgTiO_3$-$CaTiO_3$ microwave dielectric ceramics at low sintering temperatures. Mater. Res. Bull. 36(2001)2741–2750.

[222] J. H. Sohn, Y. Inaguma, S. K. Yoon, M. Itoh, T. Nakamura, S. J. Yoon, and H. J. Kim. Microwave dielectric characteristics of ilmenite type titanates with high Q values. Jpn. J. Appl. Phys. 33(1994)5466–5470.

[223] C.-L. Huang and K.-H. Chiang. Structures and dielectric properties of a new dielectric material system $xMgTiO_3$-$(1-x)MgTa_2O_6$ at microwave frequency. J. Alloys Compd. 431(2007)326–330.

[224] J. Y. Cho, K. H. Yoon. E. S. Kim. Effect of stress on microwave dielectric properties of layered $Mg_{0.93}Ca_{0.07}TiO_3$-$(Ca_{0.3}Li_{0.114}Sm_{0.42})TiO_3$ ceramics. Mater. Chem. Phys. 79(2003)286–288.

[225] E. S. Choi, Y. H. Lee, and S. G. Bae. Microwave dielectric properties of $MgTiO_3$-$BaTiO_3$ ceramics. Proc. Int. Symp. Electrical Insulating Materials ISEIM-2001. pp. 99–102.

[226] W. W. Cho, K. Kakimoto, and H. Ohsato. High Q microwave dielectric $SrTiO_3$-doped $MgTiO_3$ materials with near zero temperature coefficient of resonant frequency. Jpn. J. Appl. Phys. 43(2004)6221–6224.

[227] T. Takada, K. Kageyama, M. Yonemura, N. Hara, and S. Takao. Microwave dielectric properties of mixed phase ceramics $Ba(Zn_{1/3}Ta_{2/3})O_3$–$xCaTiO_3$ and $xMgTiO_3$-$yCaTiO_3$-z (Nd_2O_3,TiO_2). J. Mater. Sci. – Mater. Electron. 14(2003)205–214.

[228] C.-L. Huang and C.-L. Pan. Low temperature sintering and microwave dielectric properties of $(1-x)MgTiO_3$–$xCaTiO_3$ ceramics using bismuth addition. Jpn. J. Appl. Phys. 41(2002) 707–711.

[229] C.-L. Huang, C.-L. Pan, and J.-F. Hsu. Dielectric properties of $(1-x)(Mg_{0.95}Co_{0.05})TiO_3$–$xCaTiO_3$ ceramic system at microwave frequency. Mater. Res. Bull. 37(2002)2483–2490.

[230] H. T. Kim, S. H. Kim, S. Nahm, J. D. Byun, and Y. Kim. Low temperature sintering and microwave dielectric properties of zinc metatitanate-rutile mixtures using boron. J. Am. Ceram. Soc. 82(1999)3043–3048.

[231] A. Golovchanski, H.T. Kim, and Y. H. Kim. Zinc titanates dielectric ceramics prepared by sol-gel process. J. Korean Phys. Soc. 32(1998)S346–S348.

[232] H. T. Kim, J. D. Byun, and Y. Kim. Microstructure and microwave dielectric properties of modified zinc titanates (I). Mater. Res. Bull. 33(1998)963–973.

[233] H. T. Kim, J. D. Byun, and Y. Kim. Microstructure and microwave dielectric properties of modified zinc titanates (II). Mater. Res. Bull. 33(1998)975–986.

[234] O. Yamaguchi, M. Morimi, H. Kawabata, and K. Shimuzu. Formation and transformation of $ZnTiO_3$. J. Am. Ceram. Soc. 70(1987)C97–C98.

[235] S.S. Cole and W. K. Nelson. The system zinc oxide-titanium dioxide: Zinc orthotitanate and solid solutions with titanium dioxide. J. Phys. Chem. 42(1938)245–251.

[236] F. H. Dulin and D. E. Rase. Phase equlibria in the system ZnO-TiO_2. J. Am. Ceram. Soc. 43(1960)125–131.

[237] S. F. Bartram and R. A. Stepetys. Compund formation and crystal structure in the system ZnO-TiO_2. J. Am. Ceram. Soc. 44(1961)493–499.

[238] U. Steinike and B. Wallis. Formation and structure of Ti-Zn oxides. Cryst. Res. Technol. 32 (1997)187–193.

[239] M. Sugiura and K. Ikeda. Studies on dielectrics of the TiO_2-ZnO system. J. Jpn. Ceram. Assoc. 55(1947)62–66.

[240] H.T. Kim, Y. H. Kim, M. Valant, and D. Suvrorov. Titanium incorporation in Zn_2TiO_4 spinel ceramics. J. Am. Ceram. Soc. 84(2001)1081–1086.

[241] K. Haga, T. Ishi, J. Mashiyama, and T. Ikeda. Dielectric properties of two phase mixture ceramics composed of rutile and its compounds. Jpn. J. Appl. Phys. 31(1992)3156–3159.

[242] Y. M. Poplakov. Microwave high dielectric constant ceramics. Electroceramics-V. (1996)51–60.

[243] A. Chaouchi, M. Sd'Astorg, S. Marinel, and M. Aliout. $ZnTiO_3$ ceramic sintered at low temperature with glass phase addition for LTCC applications. Mater. Chem. Phys. 103(2007)106–116.

[244] A. Chaouchi, M. Aliout, Sd'Astorg, S. Marinel, S. d'Astorg, and H. Bourahla. Effects of additives on the sintering temperature and dielectric properties of $ZnTiO_3$ based ceramic. Ceram. Int. 33(2007)245–248.

[245] X. Liu, F. Gao, L. Zhao, and C. Tian. Low temperature sintering and phase transition of zinc titanate ceramics with V_2O_5 and B_2O_3 addition. J. Alloys Compd. 436(2007)285–289.

[246] Y.-C. Lee and W.-H. Lee. Effect of glass addition on microwave dielectric properties of $Zn_{0.95}Mg_{0.05}TiO_3 + 0.25TiO_2$. Jpn. J. Appl. Phys. 44(2005)1838–1843.

[247] H. Su and W. Wu. Studies on the $(Mg,Zn)TiO_3$-$CaTiO_3$ microwave dielectric ceramics. Mater. Lett. 59(2005)2337–2341.

[248] H. T. Kim and M.T. Lanagan. Structure and microwave dielectric properties of $(Zn_{1-x}Co_x)TiO_3$ ceramics. J. Am. Ceram. Soc. 86(2003)1874–1878.

[249] H. T. Kim, S. Nahm, J. D. Byun, and Y. Kim. Low fired $(Zn,Mg)TiO_3$ microwave dielectrics. J. Am. Ceram. Soc. 82(1999)3476–3480.

[250] C.-L. Huang, C.-L. Pan, and S.-J. Shim. Liquid phase sintering of $MgTiO_3$-$CaTiO_3$ microwave dielectric ceramics. Mater. Chem. Phys. 78(2003)111–115.

[251] C.-S. Shen, C.-C. Chou, W.-J. Shih, K.-s. Liu, C.-S. Chen, and I.-N. Lin. Microwave dielectric properties of glass-ceramic composites for low temperature cofireable ceramics. Mater. Chem. Phys. 79(2003)124–128.

[252] S.H. Shim, B. G. Choi, J. S. H. Kim, and K. B. Shim. Microwave characteriticss of $MgTiO_3$- $CaTiO_3$ dielectric ceramics fabricated using spark plasma sintering. Jpn. J. Appl. Phys. 44(2005)5073–5075.

[253] Y.-R. Wang, S.-F. Wang, and Y.-M. Lin. Low temperature sintering of $(Zn_{1-x}Mg_x)TiO_3$ microwave dielectrics. Ceram. Int. 31(2005)905–909.

[254] H. J. Cha, D.H. Kang, and Y.S. Cho. Optimised microwave dielectric properties of Co- and Ca- substituted $Mg_{0.6}Zn_{0.4}TiO_3$. Mater. Res. Bull. 42(2007)265–273.

[255] C.-L. Huang and S.-S. Liu. Characterization of extremely low loss dielectrics $(Mg_{0.95}Zn_{0.05})TiO_3$ at microwave frequency. Jpn. J. Appl. Phys. 46(2007)283–285.

[256] J.-J. Wang, C.-L. Huang, and P.-H. Li. Microwave dielectric properties of $(1-x)$ $Mg_{0.95}Zn_{0.05}TiO_3$–$xCa_{0.6}La_{0.8/3}TiO_3$ ceramic system. Jpn. J. Appl. Phys. 45(2006) 6352–6356.

[257] C.-L. Huang, J.-J. Wang, and Y.-P. Chang. Dielectric properties of low loss $(1-x)$ $Mg_{0.95}Zn_{0.05}TiO_3$–$xSrTiO_3$ ceramics system at microwave frequency range. J. Am. Ceram. Soc. 90(2007)858–862.

[258] Y.-L. Chai, Y.-S. Chang, Y.-J. Hsiao, and Y.-C. Lian. Effects of borosilicate glass addition on the structure and dielectric properties of $ZnTiO_3$ ceramics. Mater. Res. Bull. 43(2008) 257–263.

[259] C.-L. Huang, C.-F. Tasi, Y.-B. Chen, and Y.-C. Cheng. New material system of $(Mg_{0.95}Zn_{0.05})TiO_3$–$Ca_{0.61}Nd_{0.26}TiO_3$ at microwave frequency. J. Alloys Compd. 452(2008)337–340.

CHAPTER TWELVE

Low Temperature Cofired Ceramics

12.1 Introduction

In the past the microwave devices have been traditionally machined from metal, and co-axial RF connections were made with connectors generally leading to expensive heavy and bulky packages [1, 2]. Moreover, electronic circuits for the automotive industry, entertainment electronics and telecommunications have to handle today a steady increasing amount of functions occupying as minimal space as possible. In the development of complex miniaturized circuits, flexible glass ceramic tapes, known as low temperature cofired ceramic (LTCC) tapes, play a decisive role as a base material. Recently, the LTCC has become crucial in the development of various modules and substrates [3–5]. In this technology, several thin layers of low-permittivity ceramic composites and conductors are combined, and the resulting multilayered LTCC modules that are generally used in the form of a 3D wiring circuit board today. The LTCC enables a versatile mix of passive microwave components like microstrips, striplines, antennas, filters, resonators, capacitors, inductors, phase shifters and dividers, making possible a whole matrix of design that are not practical on regular alumina or any soft substrates. Furthermore, these integrated components are interconnected with 3D stripline circuitry [3, 6, 7]. Among the various components that could be realized in LTCC packages, the resonators and internal capacitors are important in terms of the latest technology. The internal capacitors are required to realize decoupling capacitors monolithically in LTCC packages, and the resonators are needed for filters of quarter wavelengths on the LTCC layer. The appropriate relative permittivity range for the resonators and the internal capacitors is 20–100 [8, 9].

The most important parameter for the LTCC technology is the low sintering temperature, and it enables the advantageous utilization for today's packaging concepts in microelectronic and microwave modules. Since the LTCC tapes can be sintered at low temperatures (<950°C), the embedded microwave components and transmission lines can be fabricated using highly conductive and inexpensive metals such as silver, gold or copper with low conductor loss and low electrical resistance at high-frequencies. This is an advantage over other ceramic technologies. Recently, several novel LTCCs with low dielectric loss (tan δ) and moderate and temperature compensated relative permittivity have been developed. If low thermal expansion (close to that of silicon), high strength and high thermal conductivity of the LTCCs are also taken into account, this technology can be seen very desirable even against polymer materials, especially for low loss, high-frequency circuits required for high-speed data communications.

Dielectric Materials for Wireless Communications
ISBN-13: 978-0-08-045330-9

The low-permittivity materials enable fabrication of devices with high signal speed. The high-permittivity, low dielectric loss and temperature stability of the electrical properties enable fabrication of microwave filters, with convenient size and impedance matching, low insertion loss, steep cut-off of the performance curve and operational stability against ambient change [10].

The LTCC tapes are prepared by making a slurry of a mixture of glass and ceramic powder in binders and organic solvents. It is cast under "doctor blades" to obtain a certain desired tape thickness usually in the range 50–300 μm, which are mechanically punched or laser-aided micromachined to form holes or vias. After via filling with a conductive paste, stencil and screen printing processes are used to generate conductive patterns. Ceramic sheets are then stacked, laminated and sintered followed by post-firing metallization, electrical testing and final assembly. For details of tape casting and LTCC, the reader is referred to Refs [11, 12]. The fine line technique and positioning accuracy of different LTCC layers are very important in the LTCC process. This is especially important as the operational frequency increases, because more vias are needed to provide isolation between the different elements and because the dimensions of discrete elements decrease at higher frequencies. Several methods especially suitable for LTCC tapes enabling line widths <100 μm have been developed [13, 14]. The positioning accuracy needed depends on the size of the discrete structures in the module. Usually, the stacking of different layers is done mechanically using pins in punched registration holes. The achieved accuracy is about 60 μm, which is not enough for high-frequency applications. In such cases, more accurate optical positioning method is used [15]. The presence of large conductor areas, such as grounds, is not desirable in LTCC structures, since it may cause component warping during the cofiring process [16]. This forces designers to use meshed ground planes with metallization coverage below 50%, causing further decrease of the Q factor of resonators and degraded isolation properties.

12.2 Materials Selection and Requirements

The LTCC technology for high frequency applications has several advantageous features but its development is still in the early stages. The main problems relate to the rigorous demands placed on the materials requirement. In general, it is believed that the main difficulties in the development of new LTCC materials are not related to their dielectric properties but to their sintering behavior, thermomechanical properties, chemical compatibility, shrinkage, production cost and the range of variation of each parameter. The conductivity losses in the circuitry and other losses occurring in the glassy phase of the glass-ceramic composite limit the loss factor of the LTCC. The LTCC materials should also have a good thermal conductivity, good mechanical properties and should not react with the conductive materials used. Figure 12.1 shows the variation of the microwave dielectric properties of some common commercial LTCC and FR4.

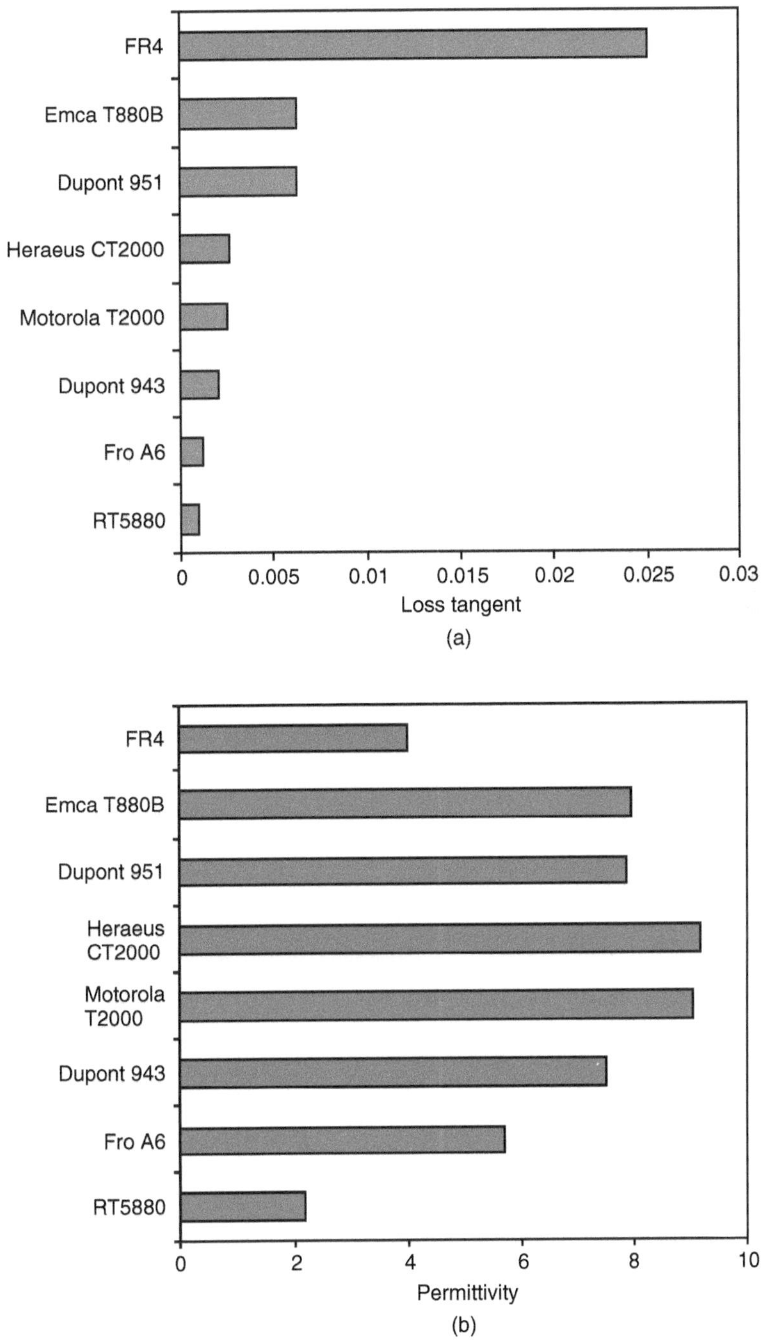

Figure 12.1 Microwave dielectric properties of common LTCC materials and FR-4: (a) loss tangent and (b) permittivity (coutesy Mohan Jacob).

12.3 The Important Characteristics Required for the Glass-Ceramic Composites

12.3.1 Low densification temperature

The densification or sintering temperature of the LTCC should be less than 950°C since the common electrode material, Ag, melts at 961°C. In the case of Cu- or Au-based electrodes, the sintering temperature should be less than 1050°C. The LTCC materials are cofired with an inner electrode structure to produce a module and, as a consequence, the sintering temperature must be lower than the melting point of the electrode. In addition, a chemical compatibility between the LTCC material and the electrode material must exist. Silver is an usual choice for the electrode, which means the sintering temperature is commonly adjusted to about 900°C. Copper with a melting point of about 1083°C is also another alternative electrode material but is less commonly used. Gold-or gold–palladium-based electrode materials can also be considered for LTCC although it is not attractive costwise. In the glass-ceramic composites, the main phase is a dielectric material having high sintering temperature. However, addition of a glass phase to the dielectric lowers the sintering temperature to a suitable level depending on the amount and type of the glass composition. In these composites, the main phase is the crystalline phase which makes a significant contribution to the dielectric properties. The glassy phase lowers relative permittivity and increases the dielectric loss. The dissolution of the ceramic particles in the glassy phase has an important influence on the viscosity of the melt during the firing. Sintering occurs through the viscous flow mechanism in which liquefaction of glass has a dominant role. Thus the selection of glass materials is very important and the glasses influence the mechanical and dielectric properties of the composites [11]. It should be noted that any densification or crystallization of the composite at lower temperatures, such as below 800°C, is undesirable as this can prevent the evaporation of the organics and solvents used in conductive pastes, binders and plasticizers causing residual carbon traces in the microstructure [3, 17]. Any residual carbon that may form during binder decomposition if left in the LTCC would adversely affect the dielectric properties as shown in Figure 12.2. It is estimated that the residual carbon content should be below 300 ppm to get the desirable

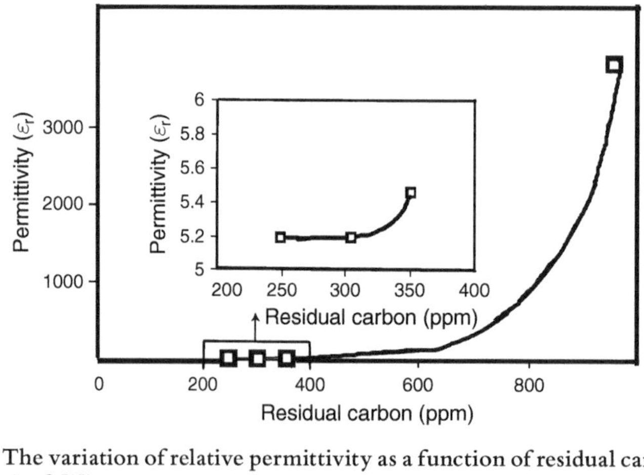

Figure 12.2 The variation of relative permittivity as a function of residual carbon in glass-ceramic (after Ref. [3]).

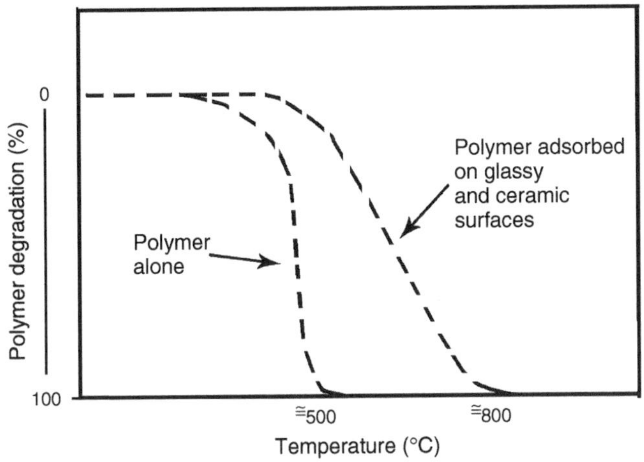

Figure 12.3 Polymer degradation in neutral atmosphere (after Ref. [3]).

dielectric properties [3]. The removal of last traces of carbon from the green sheet extends up to 800°C as shown in Figure 12.3. This means that the densification of the ceramic should start above this temperature [3]. Thus complete carbon removal makes the selection of the glass ceramic and identification of a suitable composition very challenging. Thus e.g., borosilicate based glass-ceramics are not often used since borosilicate glasses have a shallow viscosity–temperature relationship and they exhibit a softening point of about 750°C.

12.3.2 Permittivity in the range 4—100

The dielectric properties of a particular LTCC material determine its practical use. Low relative permittivity materials with $\varepsilon_r = 4$–12 are used for substrate layers while high permittivity materials are used as mainly for capacitor layers. The signal propagation is one of the most important aspects in electronic packaging. This is a direct function of the relative permittivity. In the case of ceramic packages, the relative permittivity of the ceramic over and within the metal lines is deposited or embedded governs the propagation delay, t_d, which is given by [18]

$$t_d = l\sqrt{\varepsilon_r}/c$$

where l is the line length, ε_r the relative permittivity of the substrate and c the speed of light. Thus substrates with low relative permittivity are required to increase the speed of the signal [3]. Figure 12.4 shows the propagation delay as a function of the relative permittivity of the ceramic materials.

The relative permittivities of commercial LTCCs, which are usually measured at low frequencies, are in the range 3–10. The permittivities of alumina and FR-4 also fall in this range. One great advantage of the LTCC technology is that it can provide compositions with much higher permittivity ($\varepsilon_r > 20$) enabling small embedded capacitors, inductors, filters, and antennas [19, 20]. In this case one must, however, pay more attention to the component fabrication, since smaller dimensions mean higher processing accuracy, especially in the alignment of different layers. Also to enable $50\,\Omega$ lines in high

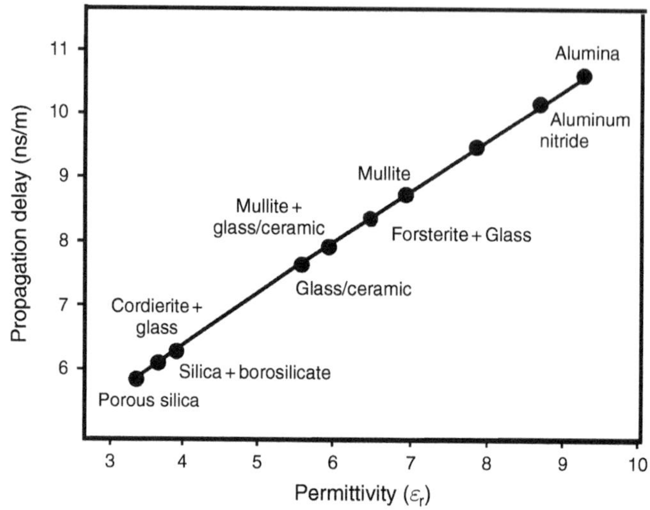

Figure 12.4 Variation of propagation delay as a function of relative permittivity of ceramic materials (after Ref. [3]).

permittivity substrate, narrower conductor lines and their spaces than possible with the commonly used screen printing process are needed. However, one advantage is that the permittivities of the commercial LTCCs are very stable, and batch-to-batch variations are generally less than 2%. Furthermore, the frequency dependence of the relative permittivity is commonly very low. For example, ε_r of DuPont 943 changes by about 1.6% in the frequency range 1–12 GHz [21], whereas the relative permittivity of FR-4 changes more than 10% in the frequency range from 1 KHz to 1 GHz [22].

12.3.3 Quality factor (Qf) > 1000 GHz

Recent material development has produced new generation LTCC systems (e.g., Dupont 943) with low insertion loss values easily satisfing the normal electrical specifications of high-frequency applications [23–25]. As shown in Figure 12.1, the dielectric loss values (loss tangent) of the commercial LTCCs even at low frequencies are competitive with the loss values of alumina and much lower than the values typical for FR-4. Actually in the case of the LTCC components, the main losses in the frequency range 4–44 GHz are conductor losses although their role also increases with the increase in frequency. The dielectric loss value of common LTCC materials, as expressed with the Q value ($Q \sim 1/\tan\delta$) multiplied by the measurement frequency, usually exceed the value 1000.

12.3.4 Temperature stability of dielectric properties

The temperature variation of the electrical properties is very important for practical application and not much attention has been paid on LTCC materials. The coefficient of temperature variation of the permittivity (τ_ε) or resonant frequency (τ_f) value of 10 ppm/°C causes a 0.11% shift in the resonant frequency (5.7 MHz at 5.2 GHz) within the temperature range −30°C to +80°C. Large τ_f values are especially problematic, because temperature compensation requires additional mechanical structures or electrical circuits [26, 27]. Novel LTCC materials with zero τ_ε values are now available [19, 28] (e.g., Heraues CT2000 with $\tau_\varepsilon < 10$ ppm/°C). In spite of this, the design engineer should be aware of the fact that the device structure itself may affect its temperature stability. The τ_f values of novel LTCCs are much better than those for FR-4 (+80 ppm/°C).

12.3.5 High thermal conductivity

The thermal conductivity of LTCC should be high enough to remove the heat generated by the device during operation. It is therefore necessary to maintain the temperature below 100°C for efficient and reliable operation of the device. Hence the removal of heat from the chip is a very important function of the package. The heat removal has become even more critical in recent years because of the ever-growing need to fabricate high density and high power devices that can operate at high speed. Advancement in technology and the continuing trends toward miniaturization of devices in the future will place even more stringent requirements on heat dissipation characteristics of the packaging LTCC. Considerable attention should also need to be paid to the thermal design as higher packaging density and operating frequencies increase the power density. One disadvantage of LTCC is its low thermal conductivity usually of the order of 2–5 W/mK, although, it is 10 times more than that for organic laminates. In high power applications, such as microwave amplifier packages, the low thermal conductivity of LTCCs may considerably affect the reliability of the device performance. A common method to improve thermal dissipation is to use a heat spreader, but a more advantageous alternative provided by LTCC technology is to place metallic via arrays under high power components [29].

12.3.6 Thermal expansion

The thermomechanical properties of the LTCC are an important material aspect. The thermal expansion match with silicon is important since systems based on silicon chips with a high device density mounted on LTCC substrates, and a thermal expansion mismatch would give rise to failure of area solder connections between the chip and the substrate. These affect the reliability of the designed components. The coefficient of thermal expansion (CTE) should thus be chosen such that it can match with the value of the mounting board and chip. This means that if the LTCC module is mounted on silicon, CTE should be about 4 ppm/°C while on alumina it should be 7–9 ppm/°C and on PCB 12–20 ppm/°C [3, 17].

12.3.7 Chemical compatibility with electrode material

The LTCC should not react with the electrode material. The formation of additional phases in the ceramic should be minimized since the reaction of the composites with the conducting electrode, degrade the performance of the microwave modules. A critical issue in manufacturing LTCC microelectronics is the precise and reproducible control of shrinkage on sintering. Antonio et al. [30] successfully applied master sintering curve theory as a tool to predict and control the LTCC sintering. The system itself is, however, complicated since in the screen printing of the conductive patterns, instead of pure metals, pastes containing conductive particles in glassy or fritless additives, are used. Thus when developing LTCC materials, one has to take into account reactions not only with the conductive material like silver but also with other additives of the conductor paste.

12.4 Commercial LTCC Materials

Table 12.1 gives a list of commercially available LTCCs with their dielectric properties. Most of these materials have relative permittivity less than 10. The lowest loss tangent values within these commercial LTCCs at microwave frequencies are close to 0.0002. However, solid comparisons is difficult to be made for a designer or a researcher. One reason is that measurements are done at different frequencies using different techniques. Especially since the dielectric loss depends on frequency, one has to perform accurate pre-measurements at desired frequency before selecting the material for a product or before comparing the composition. It may be noted that the temperature variation of dielectric properties (τ_f or τ_ε) is not commonly reported. Thus standardization of the measurement frequencies could provide more advantage and productive utilization of the LTCC technology.

12.5 Glass-Ceramic Composites

Most conventional electroceramics do not meet the basic requirements with regard to sinterability for LTCC technology since they have relatively high sintering temperature. In general, there are several methods to lower the sintering temperature of ceramics such as use of a glass with low melting temperature, addition of oxides as sintering aids, chemical processing, and using starting materials with small particle size. The chemical processing method or the use of small particles are less commonly used methods. The use of glasses, however, is explained here in more detail since it is found to be an effective way to decrease the firing temperature and also being the most commonly used method. Table 12.2 lists most commonly used glasses in the LTCCs with physical properties. In practice, there are two approaches to exploit glasses to obtain ceramic compositions sinterable below 1000°C.

The first approach is via the glass-ceramic route, which starts with a fully glassy system that devitrifies almost completely during the sintering process. The divitrification of glass greatly increases the viscosity of the system during firing, thereby improving the resistance to distortion [46]. The starting materials used in the approach are pure glasses, such as cordierite glass [46] that densifies first, followed by crystallization. The physical properties of the resulting composition are controlled by the degree of crystallization, which can be enhanced by the addition of a small amount of crystalline phase that act as a nucleating agent. During sintering, the glass recrystallizes to low loss phases and produces a low dielectric loss ceramic body [47]. Thus during the heat treatment when the glass is transformed into a glass-ceramic material, not only complete densification, but also sufficient crystallization must be achieved. Otherwise, high porosity or low degree of crystallinity can result in relatively poor mechanical properties. It may be noted that addition of sintering aid or nucleating agent would also lead to [41] difference in

Table 12.1 Dielectric properties of commercial LTCC

LTCC supplier	Composition/product name	ε_r	Q or tan δ (f)
Asahi Glass	35 wt%Al_2O_3 + 25 wt% forsterite + 40 wt% BSG glass	7.4	
Kyocera G 55	G 55 (BSG + SiO_2 + Al_2O_3 + cordierite)	5	800 (10 GHz)
Kyocera	GL 660	9.5	300 (10 GHz)
Kyocera	JHB62 (Pb-borosilicate glass + Al_2O_3 + SiO_2)	7.9	0.0002 (2 GHz)
Kyocera	JIB62	18.7	0.00025 (2 GHz)
Kyocera	AAB62	9.4	0.0005 (3.2 GHz)
Kyocera	GL 560	6	0.0017
Kyocera	GL530	4.9	0.0006 (2 GHz)
Murata	BAS (Celsian) (BaO–Al_2O_3–SiO_2)	6.1	300 (5 GHZ)
Murata	$CaZrO_3$ + glass	25	700 (5 GHz)
Murata	BaO–B_2O_3–Al_2O_3–CaO–SiO_2	6.1	0.0007
NEC glass	MLS-25M (Al_2O_3–B_2O_3–SiO_2)	4.7	300 (2.4 GHz)
NEC	MLS-41 (Nd_2O_3–TiO_2–SiO_2)	19	500 (2.4 GHz)
NEC	MLS-1000 (PbO–Al_2O_3–SiO_2)	8	500 (2.4 GHz)
NEC	MLS-61	8.1	150 (2.4 GHz)

(Continued)

Table 12.1 (Continued)

LTCC supplier	Composition/product name	ε_r	Q or tan δ (f)
NEC Vacuum Glass	GCS78 (PbO–BSG glass + Al_2O_3)	7.8	>300 (1 MHz)
NEC Vacuum glass	GCS 71	7.1	>300 (1 MHz)
NEC vacuum glass	GCS 60	6.0	>300 (1 MHz)
Sumitomo Metal	LFC ($CaO–Al_2O_3–SiO_2–B_2O_3 + Al_2O_3$)	7.7	–
NTK	GC-11	7.9	200 (3 GHz)
NTK	NOC-F1	6.3	0.0028 (40 GHz)
NTK	GC-21	7.6	0.003 (3 GHz)
Matsushita	MKE-100 (PbO–glass + Al_2O_3)	7.8	500 (1 MHz)
Niko	NL-Ag II	7.8	>300 (1 MHz)
Niko	NL-Ag III	7.1	>300 (1 MHz)
Maruwa	HA-995	9.7	–
Dupont	951 ($Al_2O_3 + CaZrO_3$ + glass)	7.8	300 (3 GHz)
Dupont	943	7.8	500 (40 GHz)
Ferro	A6M	5.9	500 (3 GHz)
Ferro	A6-B	6.5	0.005
Electro-Science Lab	41020-70C	7–8	200 (1 MHz)

Electro-Science Lab	41110	4.2	0.0037 (3 GHz)
Heraeus	CT700	7.5–7.9	450 (1 MHz)
Heraeus	CT2000	9.1	1000 (450 MHz)
EPCOS	K8	7.8	0.001
EMCA	T8800	7.2	0.002
Motorola	T2000	9.1	0.003
Amkor	GCS50	5.0	0.001 (10 GHz)
Amkor	GCS 71	7.1	0.005 (10 GHz)
Amkor	GCS60	6.0	0.001 (10 GHz)
Amkor	GCS 44	4.4	0.001 (10 GHz)
Amkor	GCS2000	18.0	0.006 (10 GHz)
Samsung	TCL-6A	6.3	
Samsung	TCL-70	6.8	
Taiyoyudan	Al_2O_3–CaO–SiO_2–ZrO_2–MgO–B_2O_3	6.7	0.001
Taiyoyudan	Al_2O_3–SiO_2–ZrO_2–MgO	7.3	0.002

(*Continued*)

Table 12.1 (Continued)

LTCC supplier	Composition/product name	ε_r	Q or tan δ (f)
Tektronix	MgO–CaO-silicate + Al_2O_3	5.8	0.0016
Toshiba	$BaSnB_2O_6$	8.5	
Toshiba	BaO–SnO_2–TiO_2–B_2O_3	7–13	0.0005–0.0008
Noritake	Al_2O_3-fosterite + glass	7.4	
Shoei	$BaZr(BO_3)_2$	7.0	0.001
Alcoa	Borosilicate glass + SiO_2 + dopants	3.9–4.2	<0.003
NGK	ZnO–MgO–Al_2O_3–SiO_2 (cordierite)	5.0	
IBM	Cordierite crystallised glass	5.0	
Corning	Crystallisable glass + crystalline cordierite	5.2	
Hitachi	(BaO–Al_2O_3–BSG) + Al_2O_3 + $ZrSiO_4$	7.0	
Hitachi	Pb-alumino-borosilicate +Al_2O_3, $CaZrO_3$	9.0–12	0.001–0.003
Fujitsu	Al_2O_3-BSG (50:50 wt%)	7.8	
Fujitsu	Borate glass + Al_2O_3, SiO_2	4.9	

Table 12.2 Common LTCC glasses and their physical properties

Glass	Density $g cm^{-3}$	T_s°C	T°cryt (°C)	T_g°C	ε_r	tan δ (%) at 1 MHz	Reference
$ZnO:B_2O_3$ (50:50)	3.65	582	718				[31]
$BaO:ZnO:B_2O_3$ (10:45:45)	3.85	552	622		6.9	0.009	[31, 32]
$BaO:ZnO:B_2O_3$ (15: 42.5: 42.5)	3.92	544	615				[31]
$ZnO–B_2O_3$ (71:29)	2.19	567			4.21	0.003	[33]
$BaO:ZnO:B_2O_3$ (20:40:40)	4.03	536	604				[31]
$BaO:ZnO:B_2O_3$ (30:35:35)	4.39	496	568				[31]
$BaO:ZnO:B_2O_3$ (40:30:30)	4.42	480	532				[31]
$La_2O_3–B_2O_3–ZnO$ (1:2:0.5)	–	660	750	640			[34]
$La_2O_3–B_2O_3–ZnO$ (1:3:0.5)	–	640	750	610			[34]
$La_2O_3–B_2O_3$ (26:74)	–	700	–	680			
$MgO–B_2O_3–SiO_2$ (42:45:13)		613					[35]
$SrO–B_2O_3–SiO_2$ (32.85:52.09:15.05)		583					[35]
$BaO–B_2O_3–SiO_2$ (42:45:13)		560					[35]
$CaO–B_2O_3–SiO_2$ (42:45:13)		613					[35]

(*Continued*)

Table 12.2 (Continued)

Glass	Density $g cm^{-3}$	T_s°C	T°cryt (°C)	T_g°C	ε_r	tan δ (%) at 1 MHz	Reference
$CaO–B_2O_3–SiO_2$ (P_2O_5 doped)	2.51		820		6.51	0.0018	[36]
$ZnO–B_2O_3–SiO_2$ (Ferro EG2730)	3.9	615	715				–
$SiO_2–B_2O_3–Al_2O_3$ (Asahi K801)	–	640					
$SiO_2–BaO–Al_2O_3$ (Asahi K807)	–	725					
$SiO_2–BaO–Al_2O_3$ (Asahi LS-5)	–	553					
$SiO_2–B_2O_3–CaO$ (Asahi BS-7)	–	789					
$SiO_2–MgO–Al_2O_3$ (Asahi FF201)	–	820					
$B_2O_3–Li_2O$ (Asahi K801)	–	450					
$La_2O_3–B_2O_3–TiO_2$ (20:60:20)		722	835				[37]
Li_2O(40–50):B_2O_3(32–40):SiO_2(12–30)		437–524		420–503	6.4–7.7	0.42–0.58	[38]
CaO(25–40): B_2O_3(20–31):SiO_2(30–40)		637–684		570–636	8.0–8.5	0.2–0.3	[38]
SiO_2(55–60):B_2O_3(20–22):Al_2O_3(2–4): AE(5–15)		732–770		642–685	4.8–5.8	0.15–0.22	[38]
SiO_2(65–72):B_2O_3(20–24):Al_2O_3(0–2): AE(2–8)		735–780		674–722	4.4–4.8	0.12–0.18	[38]
$Li_2O–B_2O_3–SiO_2$ (51.3:36.53:12.1)	2.38	422		403	7.21	0.004	[39]
$Li_2O–B_2O_3–SiO_2$ (35.4:31.66:33.2)	2.34	513		488	6.44	0.0036	[39]

Li_2O–B_2O_3–SiO_2 (56.92:37.59:5.49)	2.4	433	410	7.58	0.0045	[39]
Li_2O–B_2O_3–SiO_2 (50.:40.24:9.76)	22.4	398	379	8.15	0.0057	[39]
Li_2O–B_2O_3–SiO_2 CaO–Al_2O_3 (28:27:30:5:10)	2.36	484	456	8.12	0.0025	[39]
Li_2O–B_2O_3–SiO_2 CaO–Al_2O_3(25:30:33:5:7)	2.42	484	470	8.12	0.0023	[39]
Li_2O–SiO_2 CaO–Al_2O_3(28:27:27:8:10)	2.45	470	450	8.31	0.0027	[39]
Li_2O–B_2O_3–SiO_2 CaO–Al_2O_3 (52.45:31.06:11.99:2.25:)	2.31	389	373	8.76	0.0042	[39]
Li_2O–B_2O_3–SiO_2 CaO–Al_2O_3 (44.3:29.71:16.99:4:5)	2.32	427	409	8.52	0.0037	[39]
Li_2O–B_2O_3–SiO_2 CaO–Al_2O_3 (36.15:28.35:22:6:7.5)	2.38	464	444	8.42	0.0036	[39]
CaO–SiO_2–B_2O_3 (30–50:10–20:35–45 wt%) + 0.5 wt% P_2O_5 + 0.5 wt% ZnO	1.74			6.5	0.002	[40]
BaO–B_2O_3–SiO_2 (30:20:50)		717		7.28	0.008	[32]
BaO–B_2O_3–SiO_2 (30:40:30)		677		7.31	0.0057	[32]
BaO–B_2O_3–SiO_2 (30:60:10)		627		7.31	0.004	[32]
BaO–B_2O_3–SiO_2 (50:20:30)		595		9.61	0.01	[32]
BaO–B_2O_3–SiO_2 (50:30:20)		586		9.52	0.01	[32]

(Continued)

Table 12.2 (Continued)

Glass	Density gcm^{-3}	T_s°C	T°cryt (°C)	T_g°C	ε_r	tan δ (%) at 1 MHz	Reference
$BaO–B_2O_3–SiO_2$ (50:40:10)		577			9.15	0.01	[32]
$ZnO–B_2O_3–SiO_2$ (50:30:20)		614			7.08	0.0095	[32]
$ZnO–B_2O_3–SiO_2$ (50:40:10)		611			6.91	0.0095	[32]
$ZnO–B_2O_3–SiO_2$ (60:20:20)		604			7.51	0.009	[32]
$ZnO–B_2O_3–SiO_2$ (60:30:10)		581			7.56	0.011	[32]
$PbO–B_2O_3–SiO_2$ (30:60:10)		492			9.06	0.011	[32]
$PbO–B_2O_3–SiO_2$ (40:20:40)		442			12.11	0.01	[32]
$PbO–B_2O_3–SiO_2$ (40:40:20)		449			12.74	0.009	[32]
$PbO–B_2O_3–SiO_2$ (50:40:10)		408			13.78	0.012	[32]
$PbO–B_2O_3–SiO_2$ (60:20:20)		348			15.32	0.018	[32]
$PbO–B_2O_3–SiO_2$ (70:20:10)		312			19.57	0.02	[32]
$La_2O_3–B_2O_3–ZnO$(2:4:1)		660	750	640			[32]
$La_2O_3–B_2O_3–ZnO$(2:6:1)		640	750	610			[34]
$La_2O_3–B_2O_3–TiO_2$ (20:40:60)		722	778		12.6		[34]
$B_2O_3–SiO_2$ (70:25 trace amounts of Li_2O,K_2O, Na_2O)		700–750			4.1		[41]

B_2O_3–La_2O_3–MgO–TiO_2 (60:12:18:10)			847	661			[42]
BaO–ZnO–B_2O_3–SiO_2		575	600				[43]
B_2O_3–Bi_2O_3–SiO_2–ZnO (27:35:6:32 mol%)		430			21	0.01	[44]
La_2O_3–B_2O_3–TiO_2 (23:35:42 mol%) + ZrO–BaO–SrO <3 mol%		725			16		[44]
MgO–B_2O_3–SiO_2 (42:45:13 vol%)	2.32			613	6.64	0.00045	[35]
CaO–B_2O_3–SiO_2 (42:45:13 vol%)	2.77			613	7.47	0.00042	[35]
SrO–B_2O_3–SiO_2 (32.85:52.09:15.05 vol%)	2.29			653	7.12	0.00027	[35]
BaO–B_2O_3–SiO_2 (42:45:13 vol%)	2.71			623	7.63	0.00025	[35]
CaO–B_2O_3–SiO_2 (69.7:16.2:14.1)		830			7.3	0.0041	[45]
CaO–B_2O_3–SiO_2 (38.3:31.5:30.2)		705			7.3	0.0045	[45]
CaO–B_2O_3–SiO_2 (29.3:9.3:61.4)		700			3.9	0.0055	[45]
CaO–B_2O_3–SiO_2 (19.8:30.9:49.3)		680			4.1	0.0048	[45]
CaO–B_2O_3–SiO_2 (10.5:22.2:67.3)		710			4.1	0.0038	[45]
CaO–B_2O_3–SiO_2 (50.1:7.3:42.6)		710			7.9	0.0045	[45]

T_s°C = softening temperature; $T°_{cryst}$ C = crystallization temperature; T_g°C = glass transition temperature.

properties, such as the relative permittivity (ε_r), quality factor (Qf), CTE and temperature coefficient of the relative permittivity (τ_ε). Hence, optimization of glass-ceramic composition and further understanding of the related property differences with crystallinity is vital. A good example is Ferro A6 –M containing $CaO–SiO_2–B_2O_3$ glass. During firing, the crystallites of wollastonite ($CaSiO_3$) are formed, and some residual borosilicate glass is also present in the sintered product. This type of dielectric tape is suitable for 20–30 GHz applications such as in military and aerospace applications where very low losses are required.

In the second approach (glass + ceramic), the starting material consists of a low softening point glass and a crystalline ceramic [3, 46, 48, 49]. The densification of glass + ceramic has been described by a three-stage liquid phase sintering by particle re-arrangement, dissolution, precipitation and solid state sintering [50–59]. Depending on the type and amount of glass added, the densification of glass + ceramic can be further classified as non-reactive, partially reactive and completely reactive systems, depending on the reactivity between glass and ceramics [60, 61]. Poor dissolution of ceramic filler in glass during the sintering is observed for non-reactive systems, such as borosilicate glass (BSG) + cordierite, in which the densification is mainly achieved by particle rearrangement [55, 62]. The required amount of glass content to achieve densification decreases with increasing the particle size ratio between ceramic filler and glass [56]. In some cases, with a mixture of low melting glass and ceramic filler, the role of the glass is not only to serve as a bonding agent to hold the ceramic particles but is also to react with filler ceramic at the sintering temperature to form high-Q crystalline phases. In this type of reactive system, the microstructure, phases, and final properties are controlled by the sintering conditions such as heating rate, sintering temperature, and soaking time. For partially reactive systems such as BSG + alumina [57], the dissolution of ceramic filler in glass is localized and limited, and no particle growth and shape accommodation are observed. The required BSG content to achieve densification is close to that of non-reactive system such as BSG + cordierite [55, 56]. A lower and slower densification results when larger BSG or ceramic filler particle is used for a completely reactive system [58]. Jean and Lin [57, 59] studied the densification kinetics of non-reactive (BSG + cordierite) [63, 64], partially reactive (BSG + alumina; BSG + TiO_2) [63, 64], and reactive (BSG + high silica glass) [63] glass-ceramic systems. In the case of alumina–glass composites, very small amount of alumina dissolve in the glass at the sintering temperature. The small amount of alumina is enough to suppress crystallization of the glass or in some cases promote crystallization of the glass thus playing an important role in controlling the properties [12]. In the case of borosilicate glass when heat treated, cristobalite, which have large thermal expansion, is precipitated. This makes it difficult to control the thermal expansion of this kind of LTCC and may retard the densification process [63, 64]. However, precipitation of cristobalite can be suppressed and a composite with a matrix of amorphous glass can be obtained [65]. Jean et al. [41] reported that addition of a small amount of Ga_2O_3 in a mixture of BSG and silica can completely prevent formation of cristobalite. In addition, Nishigaki et al. [66] reported that precipitation of anorthite in alumina-$CaO–Al_2O_3–SiO_2–B_2O_3$ glass improved the mechanical strength of the composites.

The selection of glass materials is very important in the sintering process, since the liquefaction of glass takes a dominant role in the viscous flow mechanism among constituents. When sintering glass-ceramic composites, the liquefaction of the glass is the key mechanism, where the glass penetrates the 3D mesh structure formed by the ceramic particles, facilitating the wetting of each ceramic particle surface with glass melt.

Thus to improve the sintered density of glass-ceramic composition, it is necessary to control the softening point of the glass material, as well as its volume and powder particle size to increase its fluidity [67, 68]. It may be noted that the ceramic has the effect of hindering the flow of the glass and hence use of ceramic with large particle size lower the specific surface area and is beneficial in improving the sintered density. Glass fluidity, crystallization, foaming and reactions are all important in low-temperature firing. Additionally, the glass consists of different oxides, which all affect its properties [12]. Commercial LTCC tapes are mainly low-ε_r glass-ceramic compositions having typically four or five phases present and none of which should react with the electrode. SiO_2 and B_2O_3 glasses form commonly the network structures of glass. The SiO_2, for example, has a high melting point and high viscosity. Thus when the SiO_2 content is high, the glass has a high transition temperature, low thermal expansion and better chemical stability. Thus addition of B_2O_3 to quartz (SiO_2) glass lowers the viscosity. Na_2O, PbO, K_2O, Li_2O, CaO, MgO, BaO are modifier oxides. Na_2O lowers the softening point considerably but it increases the thermal expansion coefficient and degrades the stability. However, the addition of Na_2O and Li_2O modifiers increases the ionic conductivity and the Li_2O crystallizes readily. Furthermore, Al_2O_3 can form AlO_4 tetrahedrons and can connect to the network structure and has effect of controlling crystallization.

The glass basically only softens and wets the ceramic powder during sintering, providing a dense hermetic structure. At the same time it allows the dielectric to conform to the setter on which it is fired, bringing an extremely flat finished part. The volume fraction of the added glass determines the sintering characteristics, and the crystalline filler is a major determinant of electrical properties. It increases viscosity during sintering and thereby minimizes distortion. The crystalline phase also increases the mechanical strength of the final LTCC. Thus the properties of the final glass + ceramic are controlled by the ratio of glass to ceramic and the individual properties of the mixtures. Typically, the sintered properties in this system are designed to match the CTE of standard alumina ceramics and to have the relative permittivity of 6–9. The approach of glass + ceramic has been widely used apparently because simple and ease in controlling densification behavior. However, the composition can produce very complicated phase formation. Typical systems with this approach include the borosilicate glass (BSG) + alumina by Fujitsu [69] and the lead borosilicate glass + alumina by Dupont [70]. Table 12.3 presents the crystalline phases commonly considered for glass-ceramic systems.

Fujitsu and NEC have developed LTCC systems utilizing the above mentioned mixed phases' method requiring their reaction during the sintering process. Fujitsu developed two systems: Cordierite–borosilicate glass and alumina–borosilicate glass systems. Cordierite is interesting because it has got a low relative permittivity and its thermal expansion coefficient is very close to that of silicon [71]. In the first composition as against to internally nucleating a crystalline ceramic phase within the glass matrix, the glass was admixed into the crystalline composition to form a glass–ceramic mixture. The glass and cordierite system fabricated showed a high thermal expansion coefficient of the order of 17 ppm/°C due to the formation of cristobalite. However, the alumina–glass system did not show large difference in thermal expansion since cristobalite were not produced in the composite system. NEC developed a LTCC system consisting of alumina and 45 wt% lead borosilicate glass [72]. The use of lead-based glass increased the relative permittivity to 7.8, which is more than that of cordierite based, but less than alumina. The samples sintered at 900°C gave a thermal expansion coefficient of 4.2 ppm/°C, which is close to silicon. Crystallization occurs on heat treatment at 900°C as a result of the

Table 12.3 Crystalline phases considered for glass-ceramic composites [3]

Composition	Crystal phase	Coefft thermal expansion
$Li_2O–Al_2O_3–SiO_2$	Beta eucryptite	−10.0
$Li_2O–2SiO_2$	Lithium disilicide	11.00
$Li_2O–Al_2O_3–4SiO_2$	Beta spodumene	0.9
$Al_2O_3–TiO_2$	Aluminium titanate	0.5
$2MgO–2Al_2O_3–5SiO_2$	Cordierite	1.0
$BaO–Al_2O_3–2SiO_2$	Celsian	2.7
$CaO–Al_2O_3–2SiO_2$	Anorthite	4.5
$MgO–SiO_2$	Clinoenstatite	7.8
$MgO–TiO_2$	Magnesium titanate	7.9
$2MgO–SiO_2$	Forsterite	9.4
$CaO–SiO_2$	Wollastonite	9.4
SiO_2	Quartz	11.2

reaction between the glass and the ceramic particles, resulting in the formation of cristobalite. The formation of crystalline quartz also improved the mechanical strength.

As a conclusion, although the use of glass or addition of sintering aids is found an effective method to develop LTCC systems, the reduction of the sintering temperature of an original dielectric ceramic is usually accompanied by an abrupt degradation of the dielectric properties due to the formation of secondary phases. Only in a few cases could the sintering temperature be lowered without degradation of the dielectric properties due to the enhancement of the density of ceramics or elimination of oxygen vacancies [73]. The effectiveness of sintering aids depends on several factors, such as sintering temperature, viscosity, solubility and glass wettability [74]. The main requirement for liquid phase sintering is that the liquid phase should wet the grains of the ceramics. Generally, chemical reaction between sintering aids and ceramics can provide the best wetting condition [75]. However, chemical reaction often results in the formation of secondary phases. X-ray diffraction study of the glass materials heat treated at 800°C showed [35] that $MgO–B_2O_3–SiO_2$ is the most susceptible for crystallization, with $BaO–B_2O_3–SiO_2$ representing the least. For further analysis Kemethmuller et al. [76] proposed a method based on Rietveld refinement for X-ray analysis to determine quantitatively the amount of crystalline phases and also the amount of remaining amorphous phase of glass-ceramic composites. Recently, Ebberstein et al. [77] developed a software for calculating the permittivity and dielectric loss according to the effective medium theory and effective field theory. Finally, the instability source in development of LTCCs, glass fabrication, must also be controlled carefully. In almost all cases, initial glass preparation is needed which involves mixing raw materials to yield the chosen glass composition, melting the mixture between 900°C and 1500°C, quenching and pulverization. This high-temperature step involves volatilization of constituents such as Bi_2O_3, B_2O_3, and PbO, which can lead to undesirable variations in the final composition.

12.6 MICROWAVE DIELECTRIC PROPERTIES OF GLASSES

The microwave dielectric properties of ceramics with glass additives depend strongly on the densification, microstructure and interactions between glass and ceramics. Single components with low melting point are often used as sintering aids but they easily form complex compound with matrix ceramics degrading the dielectric properties [78]. However, complex additives and multicomponent glasses are much effective to reduce the sintering temperature of ceramics with good microwave dielectric properties [79]. Generally, low softening temperature glass materials were mixed with the ceramic materials to reduce the firing temperature. However, network formers contained in the glass materials may absorb the microwave power profoundly in high-frequency region, degrading the quality factor for the material [80]. At least three types of losses for glasses have been distinguished [81]; resonance type vibrational losses at very high frequency, migration losses caused by the movement of mobile ions (mainly alkali ions) and deformation losses by defect or deformation of the basic silicon oxide network. Resonant type vibrational losses are particularly important in the microwave region. Among the glasses, silica glass has the lowest tan δ value in the microwave region [82, 83]. Tan δ in fused quartz is less than 0.001 in the frequency range from 1×10^2 to 2.5×10^{10} Hz [84]. Although the loss level is useful, silica is not an effective flux to lower sintering temperature for microwave dielectrics if used alone. To lower the melting point, the rigid bonds in SiO_2 must be broken by modifiers such as alkali ions, but the use of alkali ions considerably increase the loss factor [84, 85]. The tan δ of silica-based binary glasses such as borosilicate (B_2O_3–SiO_2) is about 0.001 at 3 GHz [86]. The tan δ of ternary glasses based [84, 86] on borosilicates such as low potash lithium borosilicate at 3 GHz is about 0.0012, aluminium borosilicate about 0.002 (3 GHz), soda borosilicate about 0.004 at 3 GHz. Aluminum silicates such as cordierite (MgO–Al_2O_3–SiO_2) and celsian (BaO–Al_2O_3–SiO_2) also show low loss factors in the microwave frequency region [86, 87]. Wu and Huang [32] investigated the microwave dielectric properties of several low melting ZnO–B_2O_3–SiO_2, BaO–B_2O_3–SiO_2 and PbO–B_2O_3–SiO_2 glasses. The zinc- and barium-based glasses have low ε_r values in the range 7.0–9.5, whereas the Pb-based glasses have ε_r up to 19.5 depending on the composition (Table 12.2). The glasses, in general, have a negative τ_f with Qf up to 3400 GHz. The Pb-based glasses have a relatively low Qf. Zhu et al. [40] reported that CaO–SiO_2–B_2O_3 with 0.5 wt% P_2O_5 and 0.5 wt% ZnO sintered at 820°C showed ε_r of 6.5 and tan δ = 0.002 at 30 MHz. Mandai et al. [88–90] reported that BaO–CaO–Al_2O_3–B_2O_3–SiO_2 glass had $\varepsilon_r = 6.1$, $Qf = 400$ GHz at 10 GHz, BaO–SrO–SiO_2–ZrO_2 with $\varepsilon_r = 12$ and $Qf = 1000$ GHz; CaO–ZrO_2 based glass with $\varepsilon_r = 25$ and $Qf = 3500$ GHz. Wakino [83] reported that MgO–Al_2O_3–B_2O_3–SiO_2–TiO_2 has $\varepsilon_r = 6.1$ and $Qf = 4200$ GHz.

Takada et al. [34, 91] investigated the microwave dielectric properties of rare earth borates. $LaBO_3$ showed the highest quality factor (53 000 GHz) and highest ε_r of 12.5, although the sintering temperature was about 1200°C. The ε_r increased with increase in rare earth ionic size. It was found that the Qf improved when a mixture of amorphous melt and crystalline mixtures of La_2O_3–B_2O_3 was fired. The best dielectric properties, $Qf = 72\,000$ GHz and $\varepsilon_r = 8$ at 13 GHz, with low firing temperatures of 900°C were found when the mixing ratio of La_2O_3–$2B_2O_3$–$0.5ZnO$ was 20% crystalline melt and 80% amorphous melt. Kagata et al. [92] reported MgO–Sm_2O_3–Al_2O_3 ($MgSmAl_{11}O_{19}$) and Al_2O_3–SiO_2–B_2O_3 in equal amount, and sintering at 920°C resulted in LTCC with $\varepsilon_r = 7.8$, $Qf = 10\,000$ GHz and $\tau_f = +6$ ppm/°C. Chang and Jean [93] studied the crystallization kinetics and mechanism of low ε_r, low temperature, cofireable CaO–B_2O_3–SiO_2 glass ceramics using commercially available A-6 tapes. The samples fully densified during sintering at 850°C. Several crystalline phases of calcium silicate (Wollastonite $CaSiO_3$, $Ca_3Si_2O_7$, Ca_2SiO_4) and CaB_2O_4 formed during the firing. Crystalline phase of $CaSiO_3$ and CaB_2O_4 were the stable phases with $CaSiO_3$ as the major phase and other phases disappeared on firing at higher temperatures. The relative permittivity decreased and the CTE remains almost constant with increase in crystalline phase or with increased sintering duration at 850°C. The samples had a relative permittivity $\varepsilon_r = \sim 6$ and $\tan\delta < 0.2\%$ at MHz. In addition, Kobayashi and Kato [94] reported the low temperature preparation of anorthite ceramics, and Lo et al. [95] reported the effects of crystallinity in anorthite on the dielectric properties. They added 5–10 wt% TiO_2 as the nucleating agent for crystallization. At a constant heating rate of 5°C/min, the anorthite glass crystallized at 950°C. The LTCC dielectric had low $\varepsilon_r = 8$ and $Qf = 22\,500$ GHz for the samples containing 5 wt% TiO_2. Addition of 5 wt% B_2O_3 glass increased ε_r to 10.5 and lowered the quality factor. Kim et al. [43] studied the effect of addition of different types of fillers (TiO_2, ZrO_2, Al_2O_3, MgO, and cordierite) to BaO–ZnO–B_2O_3–SiO_2 [BZBS] (10:40:40:10) glass on CTE, optical reflectance and relative permittivity. All the fillers partially dissolved into the glass at 575°C. The BZBS glass has $\varepsilon_r = 9.9$. Lim et al. [96] studied the effect of BaO content on the crystallization, sintering behavior and properties of BaO–B_2O_3–SiO_2 glass. Since BaO acted as a network modifier, the glass transition temperature and crystallization temperature decreased as the BaO content increased. Crystallization occurred when more than 50 mol% BaO was added.

Wu and Huang [97] investigated the effect of crystallization on microwave dielectric properties of cordierite glasses containing B_2O_3 and P_2O_5. Figure 12.5 shows the relative permittivity measured in the microwave frequency region of about 15–17 GHz for (a) 90/5/5 (90 wt% cordierite, 5 wt% B_2O_3, 5 wt% P_2O_5), glasses heated at 860°C, 890°C, 920°C, and 950°C for various times. Cordierite glasses containing 5 wt% B_2O_3 and 5 wt% P_2O_5 (90/5/5) can be sintered to dense ceramic at temperatures as low as 860°C in 30 minutes. In the 90/5/5 glass the μ-cordierite crystallizes at temperatures lower than 860°C and the α-cordierite is dominant at 950°C. Increasing the amount of B_2O_3/P_2O_5 improves the μ-cordierite crystallization and prevents transformation to the α-form. Figure 12.6 shows the Qf measured at 15–17 GHz for 90/5/5 glasses heated at 860°C, 890°C, 920°C, and 950°C for various times. They reported that α-cordierite has a better Q factor than the glassy phase and μ-cordierite phase. Qf is about 6000 GHz for α-cordierite, 1200 GHz for glassy phase, and about 3000 GHz for μ-cordierite. $\tau_f = -55$ ppm/°C for μ-cordierite and that about -15 ppm/°C for α-cordierite. Increasing the amount of B_2O_3/P_2O_5 prevents the cystallization of the μ-cordierite and the transformation to α-cordierite.

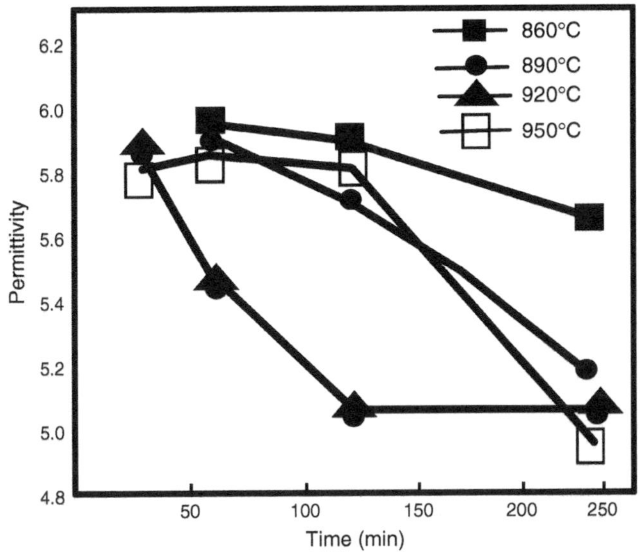

Figure 12.5 Variation of ε_r of 90 wt% cordierite/5 wt%P_2O_5/5 wt% B_2O_3 composite as a function of sintering time (after Ref. [97]).

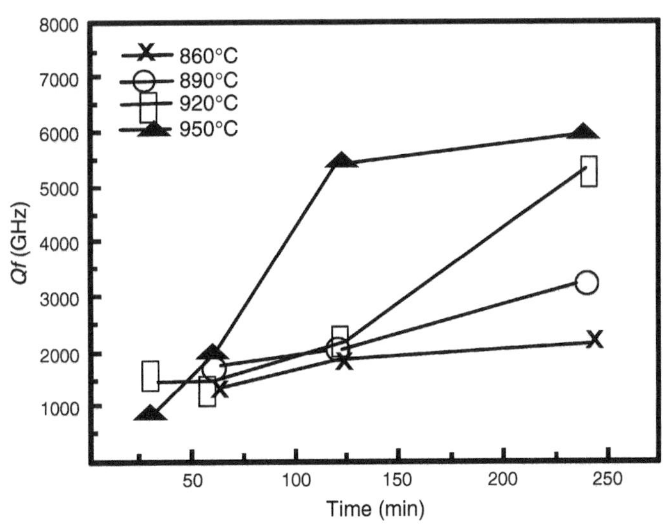

Figure 12.6 Variation of *Qf* 90 wt% cordierite/5 wt%P_2O_5/5 wt% B_2O_3 composite as a function of sintering time (after Ref. [97]).

12.7 LTCC Materials and Their Properties

Takada et al. [98] reported in 1994 for the first time the effect of glass additions on the microwave properties of dielectric ceramics. Since then several different types of low melting glasses and sintering aids have been added to several low loss dielectric ceramic and sintered at temperatures less than 1050°C. The sintering temperature, dielectric properties and references of the LTCC compositions are given in Appendix 2. The data are arranged in the order of increasing dielectric constant. It is possible to further lower the sintering temperature of these ceramics by adding suitable amount of low melting glasses and additives. The following sections discuss the important compositions researched and their microwave properties.

12.7.1 Alumina

Alumina, although has a high sintering temperature of about 1600°C and $\tau_f = -60$ ppm/°C, is widely used as a substrate material at high frequencies. It has the highest quality factor of about one million at room temperature among all dielectric ceramics [99]. Several authors investigated [100–106] the effect of addition of different glasses on the sintering behavior and microwave dielectric properties of alumina. Jei et al. [100] and Chen et al. [103] reported that densification and crystallization temperatures of the La_2O_3–Al_2O_3–B_2O_3 glass-alumina ceramics take place at about 800°C, and that they are compatible with Ag with no obvious interfacial diffusion of Ag. Seo et al. [102] conducted a detailed study of Al_2O_3 filled with different amounts of La_2O_3–B_2O_3 glass by firing at different temperatures up to 1150°C. They found that the formation of secondary phases $LaAl_{2.03}(B_4O_{10})O_{0.54}$ and $LaBO_3$ during firing resulted in the improvement of the microwave dielectric properties. The dielectric properties were sensitive to the filler content and the firing temperature. The firing temperature is a dominant factor in controlling the dielectric properties due to the formation of crystalline phases at certain temperatures. A composite containing 30 wt% alumina and 70 wt% glass and fired at 850°C showed excellent properties with $\varepsilon_r = 8.3$ and $Qf = 7600$ GHz at 17.2 GHz. However, the same composition sintered at 950°C showed $\varepsilon_r = 8.4$ with $Qf = 12\ 400$ GHz. A decrease or increase in firing temperatures can thus affect the dielectric properties due to changes in the crystalline phases with temperature. Jo et al. [105] reported excellent microwave dielectric properties in 40 wt% Al_2O_3 filled $20AO$–$20La_2O_3$–$60B_2O_3$ glass (A = Ca, Zn, Mg), which was sintered at 850–950°C. The Zn-based glass sintered at 850°C showed $\varepsilon_r = 8.3$ with $Qf = 18\,600$ GHz and increasing the sintering temperature to 950°C increased the Qf to 20 300 GHz. Lim et al. [101] investigated glass-ceramic mixtures consisting of cordierite-based crystallizable glass (MgO–Al_2O_3–SiO_2–GeO_2) with Al_2O_3 as a filler. The addition of zinc borate as a sintering additive lowered the sintering temperature to 900°C with promising microwave properties such as $Qf = 5590$ GHz and $\varepsilon_r = 5.92$ at 8.4 GHz.

Dai et al. [107] from Motorola developed a low loss and near zero τ_f LTCC ceramic (T2000) based on alumina. The zero τ_f was achieved by compensating the large negative τ_f of alumina-glass composites with TiO_2 which has a positive τ_f. X-ray diffraction study revealed that a portion of the TiO_2, in the starting formulation dissolved in the glass and extensive formation of crystalline titanium compounds was found through a nucleation and growth process. They believe that the dissolution of TiO_2 in the glass and subsequent formation of titanium-based

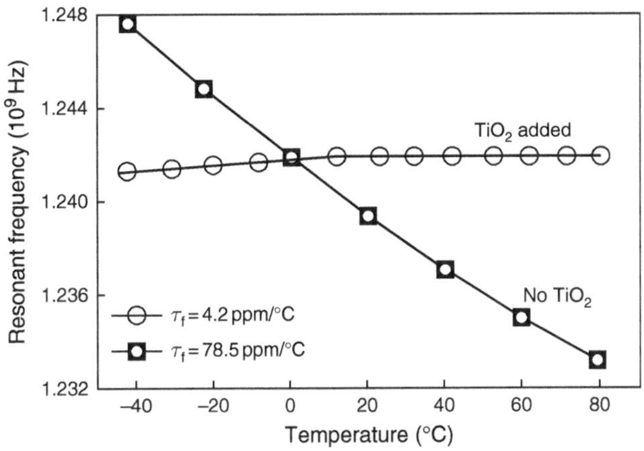

Figure 12.7 Variation of resonant frequency of a stripline T200 as a function of temperature with and without TiO_2 (after Ref. [107]).

compounds resulted in a relatively small amount of TiO_2 required to achieve a near zero τ_f in the sintered product. During sintering, Al_2O_3 reacted with the glass and formed anorthite-type crystalline phases $MSi_2Al_2O_8$ (M = Ca, Sr, or Ba). The consumption of SiO_2, CaO, SrO and BaO to form anorthite greatly decreased the volume of glass in the final structure and results in high quality factor. The dissolution of Al_2O_3 into the glass and subsequent formation of crystalline phases via a diffusion process was observed [108, 109]. The T2000 dielectric has $\varepsilon_r = 9.1$, and $Qf = 2500$ GHz. Figure 12.7 shows the change in the resonant frequency of a stripline as a function of temperature prepared by T2000 dielectric formulations with and without TiO_2. The τ_f of T2000 can be adjusted over a wide range depending on the amount of TiO_2 in the formulation. The dielectric has $\tau_f = -80$ ppm/°C in the absence of TiO_2. Extensive study using TEM and EDS analysis showed the presence of a significant amount of titanium in the crystalline phase as well as in the embedded glass matrix. Dai and co-workers suggested that during sintering, part of the TiO_2 dissolves into the glass matrix and acts as nucleation agent for the subsequent growth of the titanium-rich crystalline phases. This process makes the τ_f modification more efficient than a simple mixing of TiO_2 into the composition. The compensating mechanism of TiO_2 for τ_f can be significantly different from that of a TiO_2 addition to a non-reactive LTCC dielectric. Dai and co-workers [107–109] proposed that the τ_f of T2000 is adjusted by a combination of residual TiO_2, titanium in the glass matrix and the titanium-rich crystalline phases. The dissolved titanium and subsequently formed titanium compounds probably have much higher positive τ_f value than TiO_2, contributing to more adjustment of τ_f. In the case of LTCC dielectrics in which TiO_2 remains unreacted, nearly double the amount of TiO_2 is needed to achieve zero τ_f. Seo et al. [110] investigated the effects of particle size of alumina filler for the LTCC system. The densification temperature and crystallization strongly depended on the filler particle size. It was found that alumina particle size of 0.5–0.3 μm is ideal for best physical and microwave dielectric properties for the LTCC system.

12.7.2 TiO_2-based LTCC

TiO_2 is a high ε_r dielectric ceramic that has been extensively studied for LTCC applications. Several approaches including monosized [111] or nanosized [112] TiO_2 have been attempted to densify TiO_2 ceramics at temperatures lower than 1000°C. Kim et al. [113], in 1999, reported that the addition of CuO to anatase significantly lower the sintering temperature to about 900°C. It was reported [114] that low-fired CuO-doped TiO_2 is a mixture of CuO and rutile. Low-fired TiO_2 doped with 2 wt% CuO sintered at 900°C/2 h showed ε_r=98, Qf= 14 000 GHz and a high τ_f= 374 ppm/°C.

Several authors studied [115–120] the effect of borosilicate glass addition in TiO_2. Yano et al. [115] reported the modification of τ_f by adding TiO_2 in glass-ceramic in which the reaction of TiO_2 with glass was purposely minimized by coarsening the TiO_2 particles via heat treatment. They reported that 15 wt% TiO_2 was necessary to decrease the original τ_f of an undoped LTCC dielectric from −50 ppm/°C to almost 0 ppm/°C. Yoon et al. [116] prepared glass ceramic composites containing TiO_2 and modified borosilicate glasses and studied their sintering behavior, phase evolution and microwave properties. The TiO_2 + zinc borosilicate (ZBSG) sintered at 900°C showed ε_r = 74, Qf= 8000 GHz and τ_f= 340 ppm/°C. Figure 12.8 shows the microwave dielectric properties of TiO_2 sintered with BSG and ZBSG as a function of sintering temperature. It was found that BSG did not react with TiO_2 but ZBSG reacted with TiO_2 forming the secondary phase of Zn_2SiO_4. This indicates that the wetting behavior of BSG and ZBSG

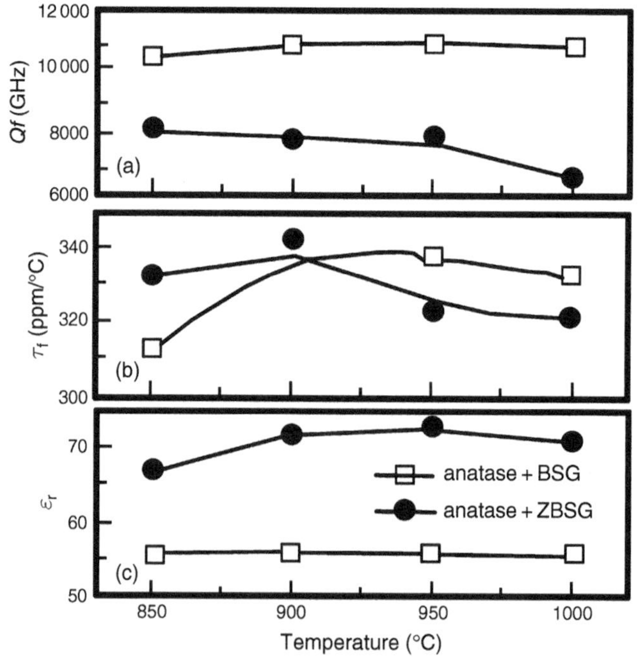

Figure 12.8 Microwave dielectric properties of low temperature sintered TiO_2 with BSG and ZBSG as a function of Sintering temperature: (a) quality factor; (b) temperature coefficient of resonant frequency; (c) relative permittivity (after Ref. [116]).

glasses is very different. The softening point of BSG is around 700°C and ZBSG at around 600°C. Borosilicate glasses (BSG) are the most commonly used glass materials in glass-ceramic composites for microelectronic packaging. It was reported [117] that TiO_2–BSG is a non-reactive system. Jean and Lin [117] used both rutile and anatase starting phases and reported that anatase has better wetting properties with borosilicate glass and hence resulted in greater densification than rutile. It was found that the densification was controlled by viscous flow of BSG. The CTE and volume fraction of rutile (V_R) increased with increasing firing time [117]. The CTE increased with increasing rutile content since rutile has a higher CTE than anatase [119].

The anorthite (CaO–Al_2O_3–$2SiO_2$: $CaAl_2Si_2O_8$) predominant glass ceramic system has been extensively investigated and regarded as a potential material for LTCC applications [95, 121–126]. Anorthite is an important material for LTCC substrates due to its lower coefficient of thermal expansion and lower relative permittivity than alumina [122]. It was reported [125, 126] that the sintering temperature can be lowered below 900°C in anorthite glass-ceramic by reducing the particle size of glass powders to the submicrometer scale. Lo et al. [124] reported that anorthite crystals show preferential nucleation at specific sites with rutile crystals precipitated from the glassy matrix. Rutile plays the role of nucleating agent to reduce the crystallization temperature lower than 900°C for anorthite-based glass-ceramics. Lo et al. [95] also investigated the sintering behavior of TiO_2 nucleated anorthite-based glass-ceramics. They have selected three different compositions: CAS–T10:CaO–Al_2O_3–SiO_2–TiO_2 (18.16:32.94:38.9:10), CAS-TB (calcium aluminium silicate): (17.24:31.3:36.96:10:5B_2O_3), and CAS-T5: (19.17, 34.77:41.06:5). The anorthite glass densified and crystallized on sintering at about 950°C. The glass ceramic showed a thermal expansion coefficient in the range 4–5 ppm/°C, which is very close to silicon. The addition of TiO_2 as a nucleating agent increased the relative permittivity. The temperature dependence of the permittivity was only slightly influenced by the TiO_2 addition. Figure 12.9 shows the variation of the quality factor with firing

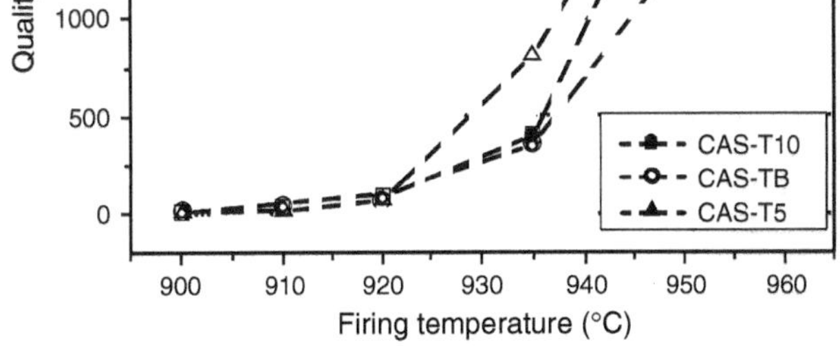

Figure 12.9 Correlation of the dielectric quality factor and firing temperature measured at 10 GHz (after Ref. [95]).

temperature. The samples showed a quality factor of about 2500 at 10 GHz. The B_2O_3 added samples showed a relatively lower *Q*.

12.7.3 Li_2O—M_2O_5—TiO_2 system (M = Nb, Ta)

Villafuerte-Castrejon et al. [127], in 1987, reported the formation of a series of solid solution phases $Li_{1+x-y}Nb_{1-x-3y}Ti_{x+4y}O_3$ ($0.05 \leq x \leq 0.3$, $0 \leq y \leq 0.182$) in the Li_2O–Nb_2O_5–TiO_2 system. Later, Borisevich and Davies [128–130] reported that $Li_{1+x-y}M_{1-x-3y}Ti_{x+4y}O_3$ (M = Nb, Ta; $x = 0.1$, $y = 0.05$–0.175) solid solutions are potential materials for LTCC applications with a quality factor up to 10 500 GHz when sintered at 1100°C. The structure of this system consists of intergrowths of $LiNbO_3$-type slabs with a length that depends on x and y values. The sintering temperature of $LiNb_{0.6}Ti_{0.5}O_3$ can be lowered below 900°C by the addition of V_2O_5 or Li_2O–V_2O_5 [131–133]. Zeng et al. [133–137] reported several low loss dielectric compositions in Li_2O–Nb_2O_5–TiO_2 system. The monoclinic $Li_{2.081}Ti_{0.676}Nb_{0.243}O_3$ sintered at 1100°C showed [134] $\varepsilon_r = 20$, $Qf = 50\,000$ GHz and $\tau_f = 13$ ppm/°C. Addition of 1.5 wt% B_2O_3 lowered the sintering temperature to 880°C with $\varepsilon_r = 20.9$, $Qf = 34\ 100$ GHz and $\tau_f = 8.3$ ppm/°C. Zeng et al. [135] also reported that addition of 1 wt% B_2O_3 to $5Li_2O$–Nb_2O_5–$5TiO_2$ and sintered at 900°C has $\varepsilon_r = 41.3$, Qf=9320 GHz, whereas $5Li_2O$–$0.583Nb_2O_5$–$3.248TiO_2$ + 1 wt% V_2O_5 sintered at 920°C has ε_r=21.5, Qf=33 000 GHz and τ_f=6.1 ppm/°C [136]. Although some of the Li based ceramics have excellent dielectric properties, the volatile Li has a deleterious effect on dielectric properties in Li-based ceramics. The deficiency of Li leads to decrease in density and lattice defects. This problem can be minimized by muffling the pellets in its own powders.

12.7.4 Bismuth based compounds

12.7.4.1 $BiAO_4$ (A = Nb, Ta)

Bismuth oxide is polymorphic with α-, β-, γ-, δ-forms. The δ-form is oxygen-deficient having a fluorite structure and is a high temperature solid state oxygen ion conductor [138]. The δ-form is stable in a narrow temperature range 729–825°C and the stability range can be brought down by suitable doping with Nb_2O_5, Ta_2O_5 or with rare earth oxides [139–145]. The δ-Bi_2O_3 has 25% oxygen sites vacant and substitution of Nb^{5+} for Bi^{3+} leads to a reduction in oxygen vacancies [146–150]. The low-temperature α-Bi_2O_3 transforms to face-centered cubic δ-Bi_2O_3 at 729°C [151, 152]. The δ-Bi_2O_3-based solid solutions have been intensely studied [139–145] as promising materials for fuel cells, sensors, catalysts, and so on. It was found that Nb_2O_5 is one of the best cation for stabilizing the δ-Bi_2O_3. The $(1-x)Bi_2O_3$–xNb_2O_5, $0.1 \leq x \leq 0.26$, solid solution shows one of the highest ionic conductivities among all Bi_2O_3-based fluorites [143–145]. The ionic conductivity of the solid solution increases with a decrease in the concentration of the substitution. A survey of literature shows that the phase relations in the Bi_2O_3-rich part of Bi_2O_3–Nb_2O_5 system differ significantly [153–159]. Bismuth-based dielectric ceramics are known as low-fired materials and have been studied for use as low-temperature coefficient ceramic capacitors [160]. Valant et al. [146, 153, 161] reported Bi_2O_3–Nb_2O_5 as high-permittivity glass-free dielectrics for LTCC applications. Valant and Suvorov [161] found an unusual scattering of dielectric properties as a function of preparation temperature for $x = 0.25$ and was attributed to the formation of a tetragonal fluorite phase [154]. Valant and co-workers [146, 161] reported that in

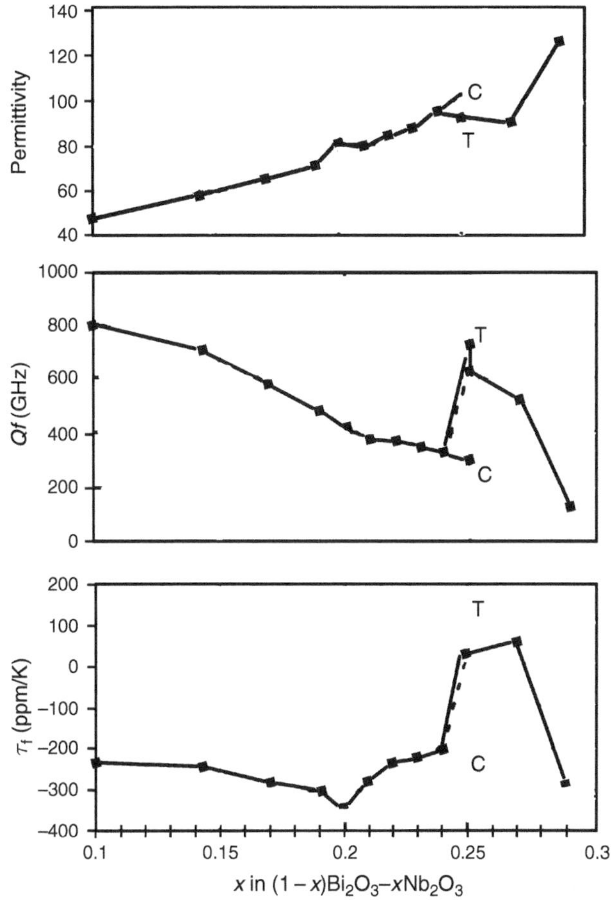

Figure 12.10 Microwave dielectric properties of the δ-BN ceramics sintered at 900°C (after Ref. [161]).

$(1-x)Bi_2O_3$–xNb_2O_5 ceramics, the ε_r and tan δ increases with increasing amount of Nb_2O_5. τ_f is negative for $x \leq 0.24$, and for $x = 0.25$ they found two structural modifications, tetragonal and cubic, depending on the preparation temperature with very different dielectric properties. At $x = 0.25$ the ceramics showed a positive τ_f and a discontinuity in the ε_r and Qf as a function of composition when sintered at 900°C as shown in Figure 12.10. The cubic phase has $\varepsilon_r = 100$, $Qf = 300$ GHz, and $\tau_f = -200$ ppm/°C, the tetragonal phase has $\varepsilon_r = 90$, $Qf = 750$ GHz and $\tau_f = 100$ ppm/°C.

Kagata et al. [19], in 1992, reported microwave dielectric properties of $BiNbO_4$ for the first time. They reported that $BiNbO_4$ ceramics is difficult to densify without the use of additives. Since then several authors [162–174] investigated the structure and the microwave dielectric properties of $BiNbO_4$ ceramics. $BiNbO_4$ has α-$SbTaO_4$-type low-temperature orthorhombic phase and transforms to high-temperature triclinic phase at 1020°C. The $BiNbO_4$ has ε_r in the range 40–45, Qf up to 30 000 GHz depending on the composition, additives or substituents. The τ_f can be tuned to zero by suitable A and B site substitution and additives. Addition of a small amount of CuO or V_2O_5 lowered the sintering temperature with improvement in quality factor [19, 172–174], and addition of a mixture of CuO and V_2O_5 further lowered the sintering temperature to about 900°C and could tune the τ_f values [170, 172]. Addition of 0.1 wt% CuO and 0.4 wt%V_2O_5 gave Qf = 21 000 GHz, τ_f = +6.8 ppm/°C and τ_f can be tuned by adjusting the CuO–V_2O_5 composition [170, 172]. Several authors studied [163, 167, 175–177] the effect of rare earth substitution for Bi on the structure, phase transition and microwave dielectric properties in $BiNbO_4$. The orthorhombic to triclinic phase transition temperature can be altered by the addition of rare earth oxide [163, 175]. The ε_r Qf factor and τ_f depend on the amount of rare earth oxide substitution, phase transition, sintering temperature and densification [163, 167, 175–178]. Fe and Sb substitutions in $BiNbO_4$ decreased the Qf value and lowered the phase transition temperature [162, 176, 178, 179]. In $BiNb_{1-x}Sb_xO_4$, the phase transition temperature lowered to 900°C [162]. Lee et al. [164] prepared $BiNbO_4$–$ZnNb_2O_6$ mixture phases: the Qf increased with addition of a small amount of $ZnNb_2O_6$ when sintered at 900°C, although the ε_r decreased. The highest Qf was obtained for $0.9BiNbO_4$–$0.1ZnNb_2O_6$. Addition of a small amount of CuV_2O_6 lowered sintering temperature to 850°C with ε_r = 41, Qf = 28 000 GHz and τ_f = 4 ppm/°C. However, co-sintering the ceramics with Ag electrode led to secondary phase formation with a degradation in dielectric properties [180]. It has been reported [181, 182] that V_2O_5 is reactive with silver and hence not suitable for LTCC. It was found that sintering in N_2 atmosphere lower the quality factor due to formation of oxygen vacancies [178, 179, 183, 184]. The $BiTaO_4$ has an orthorhombic $SbTaO_4$-type crystal structure at low temperatures and transforms to the triclinic phase at 1100°C [168], and addition of CuO lowers the transition temperature. The $BiTaO_4$ has ε_r = 43–44, Qf up to 12 000 GHz and a τ_f = −40 ppm/°C when sintered at 940°C. Substitution of Ta for Nb increased the sintering temperature [166, 185, 186]. The quality factor reached a maximum of 21 000 GHz for $BiNb_{0.4}Ta_{0.6}O_4$ when sintered at 940°C with τ_f = −30 ppm/°C. The τ_f value decreased with increasing Ta content and became negative and for $x = 0.12$ (12 mol% Ta) the τ_f = −0.2 ppm/°C.

12.7.4.2 Bi_2O_3–TiO_2

The rutile has a high positive τ_f about 400 ppm/°C and $Bi_2Ti_4O_{11}$ has ε_r of 51 and negative τ_f of −530 ppm/°C. Hence, Fukuda et al. [187] and Axelsson et al. [188, 189] investigated the ceramics based on TiO_2–Bi_2O_3 system to get a temperature-compensated ceramic. The composition $0.88TiO_2$–$0.12Bi_2O_3$ prepared by sintering at 1200°C showed excellent properties with ε_r = 83, Qf = 9300 GHz and τ_f = 3 ppm/°C. The $-\tau_f$ value of $Bi_2Ti_4O_{11}$ has been compensated by the $+\tau_f$ of TiO_2. The structural and microstructural analysis identified two separate phases: TiO_2 (rutile) and $Bi_2Ti_4O_{11}$. The separate grains of titania and bismuth titanate are distributed in the ceramic matrix as shown in Figure 12.11. The dielectric properties of this system proved to be related to the volume fraction of each component using mixture rules. Axelsson and Alford [189] found that nanosized (Degussa P25) TiO_2 powders resulted in excellent dielectric properties. The Bi_2O_3–$11.3TiO_2$ sintered at 1150°C showed ε_r = 83, Qf = 9500 GHz and τ_f = 0 ppm/°C. Addition of 0.112 wt% CuO lowered sintering temperature to 915°C with ε_r = 80, Qf = 8900 GHz and τ_f = 0 ppm/°C. This composite ceramic has the highest quality factor among the high ε_r LTCC ceramics. However, its compatibility with common electrode materials is not yet investigated.

Figure 12.11 Microstructure of BiO_2–TiO_2 ceramic (after Ref. [188]).

12.7.4.3 Bi_2O_3—ZnO—Nb_2O_5

Among the bismuth-based ceramics, Bi_2O_3–ZnO–Nb_2O_5 (BZN) ternary oxides received considerable attention [190–193]. The Chinese engineers in the 1970s developed [194] Bi_2O_3–ZnO–Nb_2O_5–(BZN)-based pyrochlore ceramics as low-temperature sinterable multilayer capacitors. There are two main phases in Bi_2O_3–ZnO–Nb_2O_5 system [160, 195]: a cubic pyrochlore phase $Bi_{1.5}ZnNb_{1.5}O_7$ with ε_r about 150 and $\tau_\varepsilon \sim -400$ ppm/°C and a monoclinic zirconolite like pyrochlore phase $Bi_2Zn_{2/3}Nb_{4/3}O_7$ with $\varepsilon_r \sim 80$ and $\tau_\varepsilon \sim +150$ ppm/°C. In BZN, Zn randomly substitutes for Bi on the A site and Nb on the B site of the pyrochlore structure [196, 197]. The BZN pyrochlore can be represented as $A_2B_2O_6O'$ where O′ is the seventh oxygen bonded only to the A site [197]. The A cations and O′ ions are randomly displaced from their ideal positions in the cubic pyrochlore with six possible positions for the A site and 12 for the O' site [194, 197]. Dielectric measurements of cubic BZN showed a broad relaxation, which weakened and broadened on cooling [198, 199]. This relaxation is attributed to hopping of atoms in A and O' sites of pyrochlore structure among several potential minima. No dielectric relaxation was observed in monoclinic BZN [200] which corresponds to an ordered structure.

Several people studied the Bi_2O_3–ZnO–Nb_2O_5 system as a useful low loss dielectric material for multilayer capacitor and LTCC applications [195, 198, 201–206]. The sintering temperature has been lowered and the properties tailored using additives such as $BaCO_3$–CuO, MoO_3–CuO, V_2O_5 [182, 199, 203–212]. Figure 12.12 shows the microstructure of a dense $Bi_{1.5}ZnNb_{1.5}O_7$ ceramics prepared by metalloorganic decomposition method and sintered at 950°C [213]. Du and Yao [214] investigated the effect of Ti substitution in bismuth zinc niobates by preparing $Bi_{1.5}Zn_{1-x/3}Ti_xNb_{1.5-2x/3}O_7$ ($0 \leq x \leq 1.5$). The Ti substitution linearly decreased the lattice parameter and the solid solution retained the cubic pyrochlore structure. The Ti substitution increased ε_r from 150 to 210 and τ_ε varied from −490 to −1290 ppm/°C with increase in x from 0 to 1.5 and the dielectric loss remained at a low value of about 1×10^{-4}. Wang et al. [215] studied the effect of substitution of Zn, Ca, Cd and Sr on the A site in $Bi_2Zn_{2/3}Nb_{4/3}O_7$. It was found that the structure and permittivity of $(Bi_{1.92}M_{0.08})(Zn_{0.64}Nb_{1.36})O_7$ (M = Ca, Zn, Cd, Sr) remain almost the same as in $Bi_2Zn_{2/3}Nb_{4/3}O_7$. Wang et al. [215] also studied the dielectric properties in the

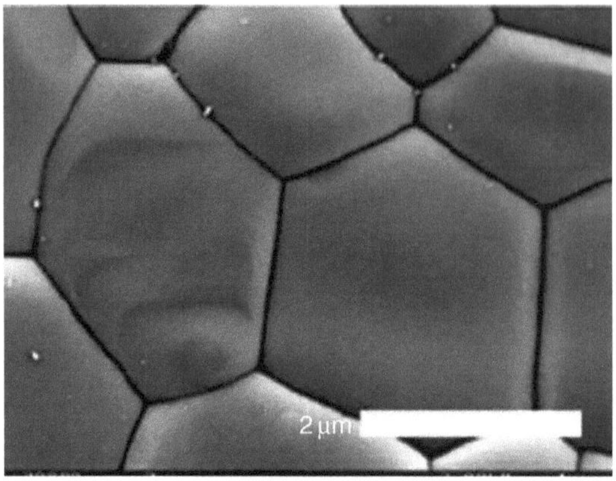

Figure 12.12 Microstructure of $Bi_{1.5}ZnNb_{1.5}O_7$ prepared by metallo-organic decomposition and sintered at 950°C (after Ref. [213]).

0.1–100 THz range. They found from a broad spectral range that the best microwave dielectric properties are for the Ca-substituted BZN due to reduced extrinsic contributions to dielectric losses. Nenesheva and Kartenko [212] investigated the effect of $ZnNb_2O_6$ addition on the microwave dielectric properties of $Bi_2ZnNb_2O_9$. The $Bi_2ZnNb_2O_9$ ($Bi_{2/3}$ $Zn_{1/3}Nb_{2/3}O_6$) has a cubic pyrochlore structure. The samples sintered at temperatures in the range 1000–1140°C showed a mixture of $Bi_2ZnNb_2O_9$ pyrochlore and columbite $ZnNb_2O_6$ phases. The $Bi_2ZnNb_2O_9$ with 80 wt% $ZnNb_2O_6$ and sintered at 1000°C had $\varepsilon_r = 30$, $Qf = 5000$ GHz and $\tau_\varepsilon = 0$. Addition of 3 wt% of glass based on PbO–Bi_2O_3–B_2O_3–ZnO–TiO_2 lowered sintering temperature further to 900°C without appreciable change in the dielectric properties.

Cubic bismuth zinc niobate pyrochlore ($Bi_{1.5}ZnNb_{1.5}O_7$–α BZN) in bulk [217–220] or thin films [221] show relatively low dielectric losses and a high ε_r. These properties make this compound an attractive candidate for capacitor and high-frequency filter applications in multilayered structures cofired with metal electrodes. The BZN thin films show large electric field dependence of permittivity [222–226]. The tunability of BZN with low loss is attractive for tunable microwave devices. The electric field-induced tunability of permittivity is mostly found in ferroelectrics close to the para-electric–ferroelectric transition temperature [227]. However, BZN is cubic pyrochlore and not ferroelectric and exhibits a high tunability of permittivity as shown in Figure 12.13. A tunability in ε_r up to 55 % was reported [223] under an applied bias field of 24 MV/cm. The origin of the large tunability of permittivity with electric field is not yet clearly understood. Tagantsev et al. [226] reported that permittivity tunability of $Bi_{1.5}ZnNb_{1.5}O_7$ very weakly depends on the temperature above the onset of dielectric relaxation. Thus BZN is advantageous over ferroelectric materials for temperature-stable tunable devices [226]. At temperatures below the onset of the dielectric relaxation (`150 K at 1 MHz) larger electric fields are required to achieve good tunabilities.

Cann et al. [228] were the first to report the dielectric properties of the tantalite analogue $Bi_2(Zn_{1/3}Ta_{2/3})_2O_7$ ceramics with $\varepsilon_r = 67$ and tan δ ~0.0001 at 1 MHz. Youn et al. [229] reported two compositions: cubic $(Bi_{3/2}Zn_{1/2})(Zn_{1/2}Ta_{3/2})O_7$ with $\varepsilon_r = 64$ and $Bi_2(Zn_{1/3}Ta_{2/3})_2O_7$ with $\varepsilon_r = \sim 60$ in the microwave frequency region. Hong et al. [230] reported the microwave dielectric properties of $Bi_2(Zn_{1/3}Ta_{2/3})_2O_7$ polymorphs. Addition of 0.5 wt% B_2O_3 lower the sintering temperature from

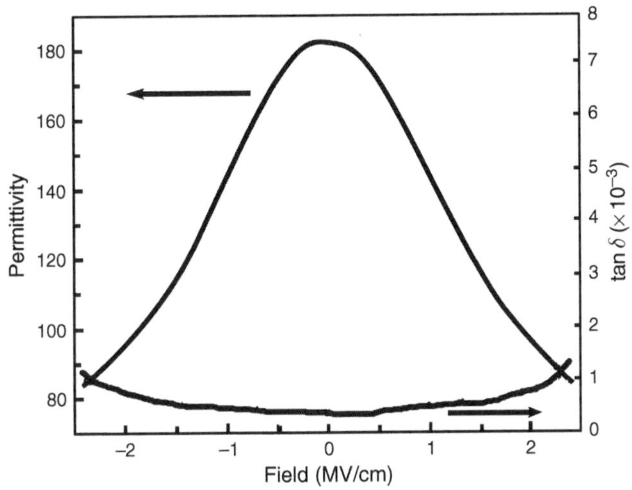

Figure 12.13 Dependence of relative permittivity and dielectric loss in BZN films deposited on Al_2O_3/Pt with applied bias fieled at 1 MHz (after Ref. [223]).

~1000 to ~850°C with low-temperature monoclinic symmetry with $\varepsilon_r = \sim 63$, $Qf = 3500$ GHz and $\tau_f = -14$ ppm/°C. Shen et al. [231] reported that addition of CuO + V_2O_5 lowers sintering temperature of $Bi_2Zn_{2/3}Ta_{4/3}O_7$ ceramics. The structure of CuO-doped sample remains as monoclinic zirconolite but >0.3 wt% V_2O_5 doping changed to fluorite phase. Addition of 0.05 wt% of CuO + 0.05 wt% V_2O_5 and sintering at 930°C resulted in $\varepsilon_r = 63$ and $Qf = 6800$ GHz.

Youn et al. [192] formulated high relative permittivity bismuth pyrochlore dielectric composition with low dielectric loss at microwave frequencies. These compositions closely matched [232, 233] the shrinkage profiles with certain commercially available LTCC systems, allowing for the cofiring of the two material systems. Plugs made of the high ε_r bismuth-based compounds were integrated into a low-ε_r LTCC matrix and cofired to produce a miniaturized, high performing 2.5 GHz microstrip band pass filter with excellent characteristics. By utilizing the 3-D capabilities of LTCC technology, Baker et al. [232, 233] used a series of loop inductors and parallel plate capacitors within a structure to create a unique miniaturized patch antenna with antenna size reduction up to 93% with good bandwidth and gain characteristics. Mixed dielectric structures combine the attributes of high and low relative permittivity materials. Low ε_r substrates are useful for impedance matching transmission lines and for minimizing cross talk. The incorporation of high ε_r bismuth pyrochlore capacitors in low ε_r LTCC substrates was found [233, 234] to decrease band pass filter dimensions with improvement in the overall filter characteristics.

12.7.4.4 $Bi_{12}MO_{20-\delta}$

The stoichiometric sillenites are compounds with the general formula $Bi_{12}MO_{20-\delta}$, where M represents a tetravalent ion or a combination of ions which gives an average charge of 4^+ [235]. Several different aliovalent substitution of M cation allows the formation non-stoichiometric $Bi_{12}MO_{20-\delta}$ sillenites with excess (negative δ) or deficiency (positive δ) in the oxygen sublattice. The charge balance is maintained by a decrease or increase in the average charge of the tetra-co-ordinated M ions, which occupy B sites of the sillenite structure. The A sites are fully occupied by Bi ions and sillenites crystallize in the I23 space group with a body-centered cubic cell. Valant and Suvorov [235] prepared sillenite compounds $Bi_{12}MO_{20-\delta}$ (M = Si, Ge, Ti, Pb, Mn, $B_{1/2}\,P_{1/2}$) by calcining at 650–800°C/20–50 h and sintering at 680–850°C. In $Bi_{12}PbO_{19}$, Pb is in 2^+ oxidation state and to maintain charge neutrality, a deficiency of oxygen sublattice occurs ($\delta = 1$) [236]. The Mn analogue is also not fully stoichiometric because of the partial oxidation of Mn^{4+}, the Mn sillenites exhibits a slight excess in the oxygen sublattice. The samples sintered to about 97% density at 680–850°C. This combined with the fact that they do not react with silver makes these materials compatible with LTCC technology. $Bi_{12}SiO_{20}$ with sintering temperature of 850°C, $Qf = 8100$ GHz, $\tau_f = -20$ ppm/°C and $\varepsilon_r = 37.6$ is the best material for LTCC among the sillenites.

12.7.5 TeO_2 type

Most tellurium-based oxide materials can be synthesized and sintered at temperatures below 900°C, which makes them potential candidate for use in LTCC technology. For ceramic processing, the most important oxidation states of Te are Te^{4+} and Te^{6+}, where Te^{4+} has a lone electron pair. Pure TeO_2 never oxidizes in air to form 6^+ oxidation state [237]. Udovic et al. [238] reported that TeO_2 has a poor sinterability and the ceramics sintered at 640°C/15 h with 20% porosity has $\varepsilon_r = 19.3$, $Qf = 30\,000$ GHz and $\tau_f = -119$ ppm/°C. Udovic et al. [239] prepared several bismuth tellurite compounds such as $Bi_2Te_2O_8$, Bi_2TeO_6, $Bi_6Te_2O_{15}$ and $7Bi_2O_3$–$2TeO_2$ in oxygen atmospheres with low sintering temperatures down to 650°C. They have ε_r in the range 33–56 and Qf up to 41 000 GHz and negative τ_f values. However, $Bi_2Te_2O_8$ and Bi_2TeO_6 compounds are found to react with silver whereas bismuth-rich compounds $Bi_6Te_2O_{15}$ and $7Bi_2O_3$–$2TeO_2$ do not react with silver.

Many authors [238–242] have reported the synthesis of $TiTe_3O_8$ with a cubic unit cell from oxides at 700°C. Maeda et al. [242] reported that it is difficult to prepare a dense $TiTe_3O_8$ ceramic. Udovic et al. [238] prepared single-phase dense $TiTe_3O_8$ by muffling in TeO_2 powder and the samples showed ε_r of 50, $Qf = 30\,600$ GHz, but the τ_f was high at +133 ppm/°C. Figure 12.14 shows a typical TEM image of a $TiTe_3O_8$

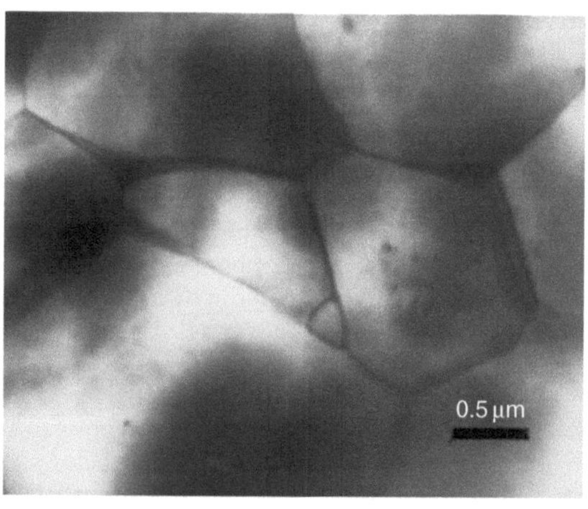

Figure 12.14 Typical TEM image of a $TiTe_3O_8$ ceramic sintered at 720°C/5 h and cooled at 100°C/min. Micrograph reveals no main structural defects (after Ref. [238]).

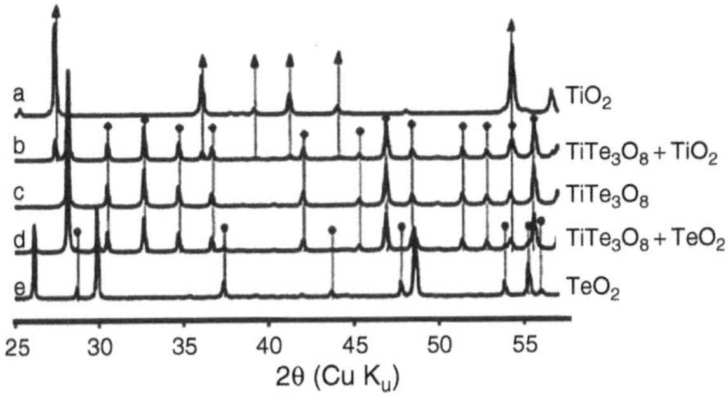

Figure 12.15 XRD patterns of various samples on the TiO_2–TeO_2 tie-line (after Ref. [238]).

ceramic sintered at 720°C/5 h. The low porosity led to considerable improvement in properties as compared to those of Maeda et al. [242]. To lower τ_f, they prepared $TiTe_3O_8$–TeO_2. Figure 12.15 shows the XRD pattern of various compounds in the TiO_2–TeO_2 system. It is found that $TiTe_3O_8$ and TeO_2 have negligible solid solubility. The $TiTe_3O_8$–TeO_2 sintered into dense ceramics at 670°C. Figure 12.16 shows the variation of the microwave dielectric properties of $TiTe_3O_8$ as a function of TeO_2 content. The composition $0.15TiTe_3O_8$–$0.85TeO_2$ showed a zero τ_f with ε_r=30 and Qf = 22 000 GHz. Valant and Suvorov [243] reported that $CaTeO_3$ and $CaTe_2O_5$ sinter at low temperatures (less than 900°C) with excellent quality factor. Kwon et al. [244] reported $BaTe_4O_9$ as a high-Q LTCC material. The $BaTe_4O_9$ sintered into dense ceramic at 550°C with ε_r = 17.5 and Qf = 54 700 GHz and τ_f = −90 ppm/°C. The $BaTe_4O_9$ ceramic reacts with silver. However, $BaTe_4O_9$ was found to be chemically compatible and successfully cofired with aluminum electrode maintaining good electrical

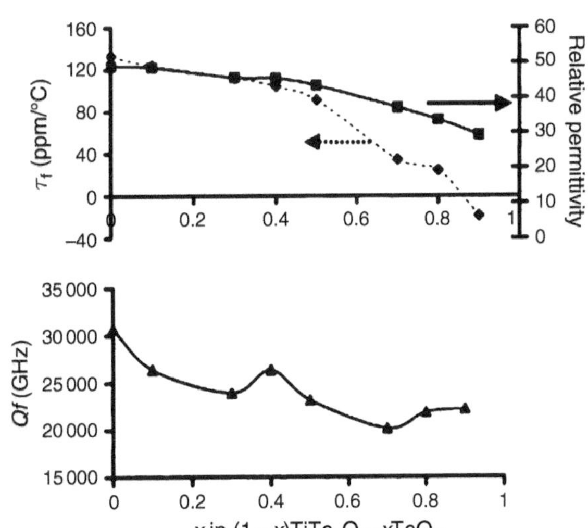

Figure 12.16 Microwave dielectric properties of composition from the $TiTe_3O_8$–TeO_2 tie-line (after Ref. [238]).

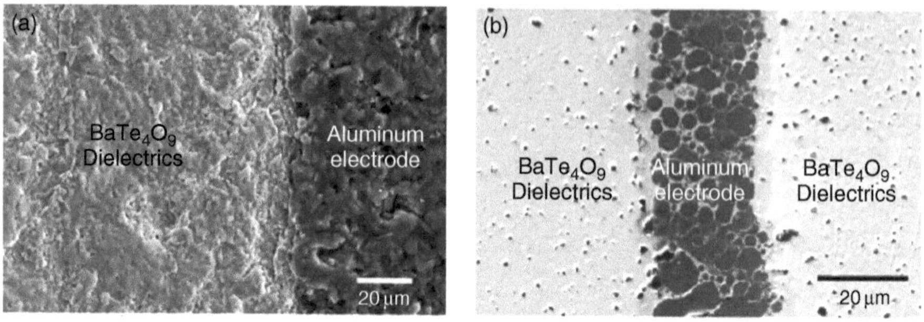

Figure 12.17 Cross sectional scanning electron micrographs of (a) cofired $BaTe_4O_9$ with Al top electrode (b) integrated and cofired $BaTe_4O_9$/Al inner electrode/$BaTe_4O_9$ sample (after Ref. [244]).

properties. Figure 12.17 shows the cross-sectional scanning electron micrographs of (a) cofired $BaTe_4O_9$ with Al top electrode (b) integrated and cofired $BaTe_4O_9$/Al inner electrode/$BaTe_4O_9$ sample. Kwon et al. [245] sintered $BaTiTe_3O_9$ at 625–700°C as a single-phase material and reported $\varepsilon_r = 29$, $Qf = 13\,000$ GHz and $\tau_f = -372$ ppm/°C. Udovic and Suvorov [246] reported low loss LTCC ceramics in the Bi_2O_3–TiO_2–TeO_2 system. They reported the dielectric properties of Bi_2TiTeO_8, $Bi_2Ti_3Te_2O_{12}$ and $Bi_6Ti_5TeO_{22}$ ceramics and are given in Appendix 2. The preparation of single-phase bismuth-based titanium tellurate is difficult due to the reduction of Te^{6+} and the evaporation of TeO_2. Hence to suppress the formation of secondary phase, Udovic and Suvorov sintered the ceramics by muffling the pellets with the same powder under an oxygen pressure of 10 bar. More recently, Subodh and Sebastian [247] reported $Zn_2Te_3O_8$ as very low temperature sinterable dielectric ceramic. They added TiO_2 to compensate for the high τ_f of $Zn_2Te_2O_8$. The $Zn_2Te_2O_8 + TiO_2$ sintered at 650°C showed ε_r of 19.3 with $Qf = 27\,000$ GHz and $\tau_f = -8$ ppm/°C.

12.7.6 ZnO–TiO_2 system

The ZnO–TiO_2 system contains $ZnTiO_3$ (hexagonal), Zn_2TiO_4 (cubic), $Zn_2Ti_3O_8$ (cubic) and the $Zn_2Ti_3O_8$ phase exists below 820°C [248]. The preparation of $ZnTiO_3$ from a mixture of ZnO and TiO_2 is difficult because the compound decomposes into Zn_2TiO_4 and rutile at about 945°C [249]. Haga et al. [250] and Golovochanski et al. [251] investigated the microwave dielectric properties of ZnO–TiO_2 ceramics. Later, it was found [252–255] that addition of glasses in $ZnTiO_3$ such as B_2O_3, ZnO–B_2O_3, and $B_2O_3 + LiF$ can effectively reduce the sintering temperature to about 875°C without degradation of the microwave dielectric properties. $ZnTiO_3$ has a negative τ_f, and addition of rutile improves the τ_f close to zero [252, 253]. Figure 12.18 shows the variation of microwave dielectric properties of $ZnTiO_3 + 0.25\ TiO_2$ with B_2O_3 content sintered at 875°C for 4 hours [252]. The dielectric properties considerably improved with B_2O_3 addition. The best properties, $\varepsilon_r = 30$, $Qf = 66\,000$ GHz and $\tau_f = 10$ ppm/°C, were obtained for 1 wt% B_2O_3 and sintered at 875°C/4 h [252]. It was found that the LTCC did not react with silver, and Zhang et al. proposed it as an ideal material for LTCC applications [253]. $ZnTiO_3$ and $MgTiO_3$ are both hexagonal and they can form solid solution. Mg is smaller (0.66 Å) than Zn (0.74 Å). Hence $Zn_{1-x}Mg_xTiO_3$ solid solution phases were prepared with a view of improving the

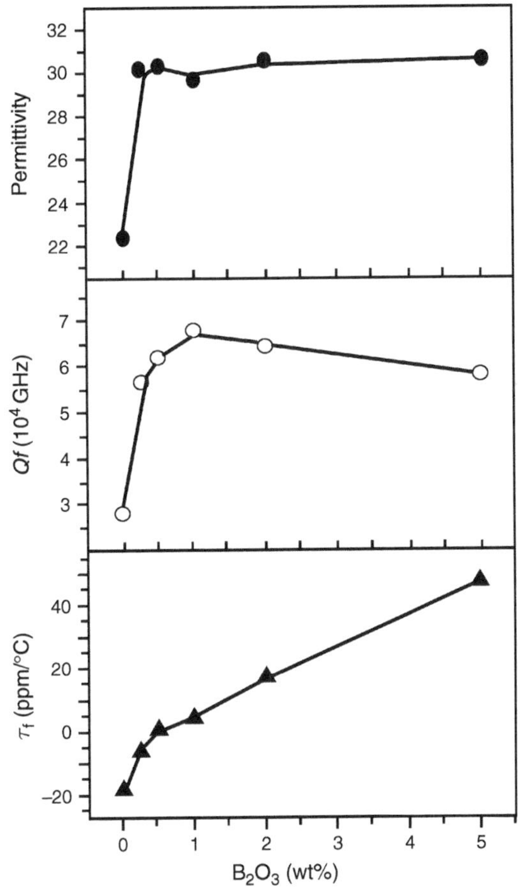

Figure 12.18 Variation of the microwave dielectric properties of the $ZnTiO_3 + 0.25$ TiO_2 mixture, relative to the B_2O_3 content. Specimens were sintered at 875°C for 4 hours (after Ref. [252]).

properties [256–258]. The thermal stability and unit cell parameters of the hexagonal phase varied with Mg content of the system [256]. The microwave dielectric properties are dependent on the phase composition, which was affected by the sintering conditions and the chemical composition. Figure 12.19 shows the variation of dielectric properties as a function of x in the range 0–0.1 for different sintering temperatures [258]. Zero τ_f can be obtained for $Zn_{1-x}Mg_xTiO_3$, x in the range 0–0.1, and it moved to a higher Mg containing composition as the sintering temperature increased. XRD and thermal analysis showed that the stability range of the hexagonal phase increased to higher temperatures as the Mg content increased. Sintering (Zn, Mg)TiO_3 above 950°C led to the formation of (Zn, Mg)$_2TiO_4$ + rutile. The presence of rutile is advantageous since $ZnTiO_3$ has negative τ_f. Lee et al. [257] studied the effect of addition of zinc borate to $Zn_{0.95}Mg_{0.05}TiO_3 + 0.25TiO_2$ in the temperature range 860–940°C. It was found that addition of about 1 wt% zinc borate significantly improved the density and microwave

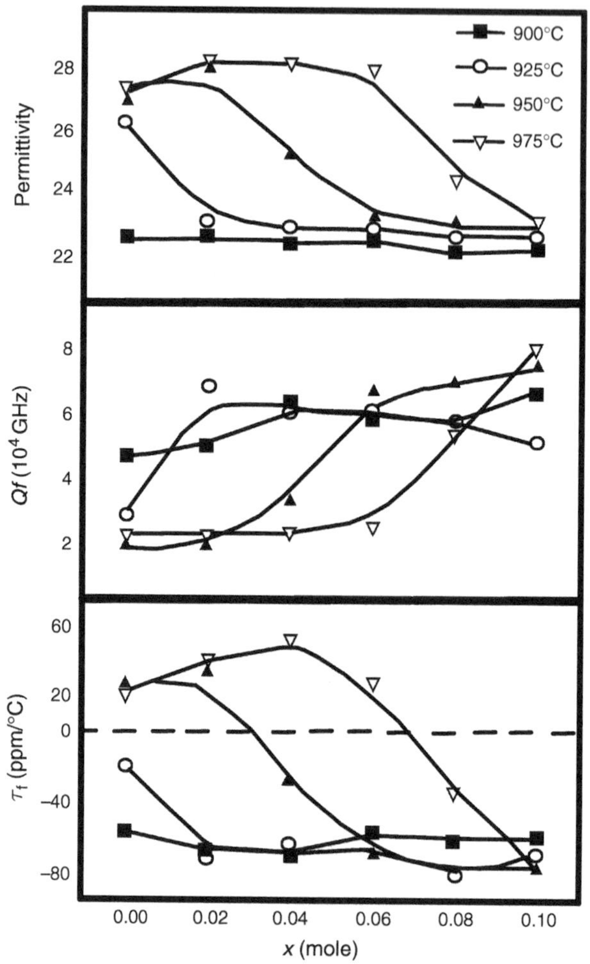

Figure 12.19 Microwave dielectric properties of $(Zn_xMg_{1-x})TiO_3$ ceramics (after Ref. [256]).

dielectric properties. The ceramic sintered at 900°C with 1 wt% zinc borate having $\varepsilon_r = 23.6$, $Qf = 30\ 990$ GHz and $\tau_f = -8$ ppm/°C.

12.7.7 $MgAl_2O_4$ and $ZnAl_2O_4$

The Spinels $MgAl_2O_4$ and $ZnAl_2O_4$ have low ε_r and high quality factors [259, 260]. The high sintering temperature of $MgAl_2O_4$ can be lowered by the addition of Li–Mg–Zn–B–Si–O glass [261, 262]. The glass-ceramic sintered at 900°C showed ε_r of 7.4 with high Qf of 48 000 GHz and τ_f of −90 ppm/°C. The LTCC showed a mechanical strength of 250 MPa and thermal conductivity of 5.59 W/mK. Although the composite has excellent quality factor, the temperature stability is poor which prevents its immediate application. However, recently, Surendran et al. [259] have reported that it is possible

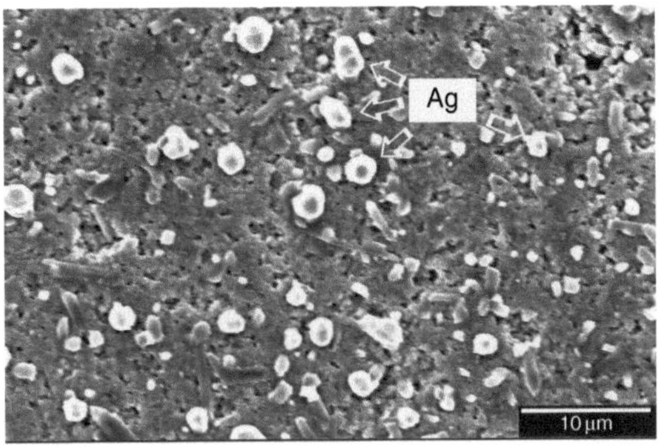

Figure 12.20 Microstructure of $0.83ZnAl_2O_4$–$0.17TiO_2$ + 10 wt% Bi_2O_3–B_2O_3–SiO_2–ZnO glass (after Ref. [263]).

to tune τ_f by the addition of TiO_2. Sherin and Sebastian [263] reported that $0.83ZnAl_2O_4$–$0.17TiO_2$ + 10 wt% Bi_2O_3–B_2O_3–SiO_2–ZnO ceramic-glass composite can be sintered at 950°C with ε_r=10.6, Qf> 10 000 GHz and τ_f= −23 ppm/°C. Addition of 0.3 wt% LiF further lowered the sintering temperature to 925°C/10 h with ε_r = 10.5, Qf> 14 500 GHz and τ_f= −28 ppm/°C. It was found from X-ray diffraction and SEM studies that Ag does not react with the $ZnAl_2O_4$–TiO_2-based glass-ceramic composite. Figure 12.20 shows the SEM picture of $0.83ZnAl_2O_4$–$0.17TiO_2$ + 10 wt% Bi_2O_3–B_2O_3–SiO_2–ZnO glass sintered with 20 wt% silver.

12.7.8 Tungsten bronze type LTCC ceramics

Several authors [37, 39, 44, 264–270] investigated the suitability of pseudo-tungsten bronze-type BaO–Ln_2O_3–TiO_2 ceramics (Ln = Nd, Sm) for LTCC applications. Ceramic materials in BaO–Ln_2O_3–TiO_2 (Sm, Nd) in 1:1:4 or 1:1:5 compositions are suitable for developing dielectric resonators in mobile phone handsets. These compositions are characterized by high relative permittivity, low dielectric loss and low temperature coefficient of resonant frequency [271]. Dernovsek et al. [44] reported the effect of addition of B_2O_3–Bi_2O_3–SiO_2–ZnO (BBSZ – barium–boron–silicon–zinc oxide). Addition of 10 vol% BBSZ and sintered at 900°C resulted in a composite with ε_r = 68, Qf> 6000 GHz and τ_f close to 0. Cheng et al. [268, 269, 272] studied the effect of addition of BaO–B_2O_3–SiO_2 (42:45:13 wt%), which has a softening temperature (T_s) of 619°C in Ba(Nd,Sm,Bi)$_2$$Ti_5O_{14}$. Addition of about 25 wt% glass resulted in a sintering temperature of about 900°C with Qf= 8500 GHz and ε_r = 40. Several authors [37, 39, 44, 264–266, 268, 272, 273] reported that addition of glasses such as B_2O_3–Bi_2O_3–SiO_2–ZnO (BBSZ), La_2O_3–B_2O_3–TiO_2 (LBT), BaO–B_2O_3–SiO_2, Li_2O–B_2O_3–SiO_2–Al_2O_3–CaO (LBSAC), B_2O_3, BaB_2O_4, $BaCu(B_2O_5)$ lower the sintering temperature of tungsten bronze-type ceramics to a level suitable for LTCC applications. Park et al. [39] investigated the effects of Li_2O–B_2O_3–SiO_2–Al_2O_3–CaO glass addition in MWF-38 and MBRT-90 [Ba(NdSm,Bi)$_2$$Ti_4O_{12}$ made by Fujitan] dielectric compositions . They reported the relative permittivity, dielectric loss and T_s and τ_f of several LBSAC glasses of varying compositions. MBRT-90 with 10 wt% of $52.45Li_2O$–$31.06B_2O_3$–$11.99SiO_2$–2CaO–$2.5Al_2O_3$ and sintered at 875°C resulted in excellent microwave dielectric properties of ε_r = 32 and Qf= 9000 GHz. Jung et al. [37] studied the effect of

60 wt% LBT (La_2O_3–B_2O_3–TiO_2 20:60:20) in $BaNd_2Ti_5O_{14}$ [BNT]. The composite sintered at 850°C showed $\varepsilon_r = 20$, $Qf = 8000$ and $\tau_f = 76.8$ ppm/°C. Low temperature form of LT-$LaBO_3$ second phase formed when sintered at 750°C and high-temperature HT–$LaBO_3$ was formed when sintered at temperatures in the range 800–900°C [37].

12.7.9 $Pb_{1-x}Ca_x(Fe_{1/2},Nb_{1/2})O_3$

Nakano et al. [274] modified the low-temperature sinterable ferroelectric $Pb(Fe_{2/3}W_{1/3})O_3$–$(Pb(Fe_{1/2}Nb_{1/2})O_3$ to a para electric at room temperature by the partial substitution of Pb by Ca. When Pb was partially substituted by Ca, a single phase (Pb, Ca) (W, Fe, Nb)O_3 (PCWFN) was formed with high relative permittivity and low dielectric loss. Nakano et al. [274] prepared $yPb(Fe_{2/3}W_{1/3})O_3$–$(1-y)(Pb_{1-x}Ca_x)(Fe_{1/2}Nb_{1/2})O_3$ (PFW–PFN) (for different values of x and y) solid solutions and studied their microwave dielectric properties. When sintered at temperatures in the range 930–1000°C, the ceramics have ε_r in the range 71–128, Qf up to 3800 GHz and τ_f of −30 to +90 ppm/°C, depending on the values of x and y and the sintering temperature. Kato et al. [275] reported the microwave dielectric properties of $(Pb, Ca)(Fe_{1/2}Nb_{1/2})O_3$ [PCFN] ceramics. Kucheiko et al. [276] found that partial substitution of Fe and Nb by Sn improves the microwave dielectric properties. The (Pb_{1-x},Ca_x) (Fe, Nb, Sn) ceramics have [276] $\varepsilon_r = 86$–90, $Qf = 7500$–8600 GHz and $\tau_f = 0$–9 ppm/°C when sintered at 1165°C. To lower the sintering temperature, Ha et al. [277] added CuO–Bi_2O_3 in $[(Pb_{0.45}Ca_{0.55})(Fe_{0.5}Nb_{0.5})_{0.9}Sn_{0.1}]O_3$ PCFNS ceramics. Addition of 0.2 wt% CuO and 0.1 wt% Bi_2O_3 lowered sintering temperature to 1000°C/3 h with $\varepsilon_r = 83$, $Qf = 6085$ GHz and $\tau_f = 8$ ppm/°C. Increase in Bi_2O_3 content led to the formation of secondary phases and increased relative permittivity but lowered the quality factor and τ_f. The PCLFN $[(Pb_{0.5}Ca_{0.5})_{1-x}La_x](Fe_{0.5}Nb_{0.5})O_3$ sinters at 1150°C with $\varepsilon_r > 90$, $Qf \sim 5500$ GHz and τ_f in the range 7–15 ppm/°C [278, 279]. Yang et al. [278] reported that addition of 1 wt% of PbO–B_2O_3–V_2O_5 (PBV) glass to PCLFN (Pb,Ca,La)(Fe,Nb)O_3 lowered the sintering temperature to 1050°C without degradation of the properties such as $\varepsilon_r = 101$, $Qf = 5400$ GHz and $\tau_f = 5.9$ ppm/°C. Pb_6BVO_{10} was detected as a secondary phase in the sintered specimens for PBV glass >0.5 wt%. Zhe et al. [280] found that addition of $Bi_2O_3 + MnO_2$ to $[(Pb_{0.5}Ca_{0.5})_{0.92}La_{0.08})(Fe_{0.5}Nb_{0.5})O_3$ (PCLFN) system lowered the sintering temperature to 1050°C with $\varepsilon_r = 91.1$, $Qf = 4870$ GHz and $\tau_f = 18.5$ ppm/°C.

12.7.10 $Ca(Li_{1/3}B_{2/3})O_{3-\delta}$ (B = Nb,Ta)

Choi et al. [281] found that complex perovskite $Ca(Li_{1/3}Nb_{2/3})O_{3-\delta}$ [CLN] is a useful low loss dielectric material but has a sintering temperature of 1150°C and that CLN need to be sintered in a Pt box to control the volatility of Li_2O at such high temperatures. Hence, to lower the sintering temperature several authors [282–287] added glasses to CLN. Liu et al. [282, 283] reported that addition of B_2O_3 significantly improved the density of non-stoichiometric CLN or $Ca(Li_{1/3}Ta_{2/3})O_{3-\delta}$. With the addition of 0.5–4 wt% B_2O_3, ε_r and Qf of CLN sintered at 990°C were as good as that sintered at 1150°C. The samples sintered at 1000°C with 4 wt% B_2O_3 showed $\varepsilon_r = 30.6$, $Qf = 31\,000$ GHz and $\tau_f = -17.5$ ppm/°C. However, addition of more than 4 wt% B_2O_3 deteriorated the dielectric properties due to the formation of $Li_2B_4O_7$. The sintering temperature was further reduced to 900°C by the addition of Bi_2O_3 but at the expense of the quality factor. The addition of B_2O_3 lowered the sintering temperature; however, the τ_f was high. To lower τ_f, Ti was partially substituted [228, 285–287] for Li-Nb. It was reported that addition of glass frit resulted in excellent microwave dielectric properties [287]. Substitution of Ti for Li–Nb increased ε_r, decreased Qf and changed τ_f from a negative to a positive value. $Ca[(Li_{1/3}Nb_{2/3})_{0.8}Ti_{0.2}]O_{3-\delta} + 12$ wt% glass frit sintered at 900°C for 3 hours has $\varepsilon_r = 40$, $Qf = 12\,500$ GHz, $\tau_f = -8$ ppm/°C [287]. Ha et al. [286] investigated the effect of Bi_2O_3 addition on lowering the sintering temperature, densification, and dielectric properties in $Ca[(Li_{1/3}Nb_{2/3})_{1-x}Ti_x]O_{3-\delta}$ (CLNT). As the amount of Bi_2O_3 increased, density and ε_r increased but the Qf decreased and the τ_f shifted to a positive value. The $Ca[(Li_{1/3}Nb_{2/3})_{0.95}Ti_{0.05}]O_{3-\delta} + 5$ wt% Bi_2O_3 sintered at 900°C has $\varepsilon_r = 20$, $Qf = 6500$ GHz, $\tau_f = -4$ ppm/°C and $Ca[(Li_{1/3}Nb_{2/3})_{0.8}Ti_{0.2}]O_{3-\delta} + 5$ wt% Bi_2O_3 sintered at 900°C has $\varepsilon_r = 35$, $Qf = 11\,000$ GHz and $\tau_f = 13$ ppm/°C.

Yoshida et al. [288] reported $(Ca_{1-x}Nd_{2x/3})TiO_3$ as a dielectric resonator material with $\varepsilon_r = 80$–100, $Qf = 150$–1000 GHz but has high sintering temperature of 1300°C. To lower the sintering temperature and to improve densification, Wei and Jean [279] added $3ZnO$–$2B_2O_3$ glass in $(Ca_{1-x}Nd_{2x/3})TiO_3$. Addition of more than 20 vol% of the glass lowered sintering temperature to 850–900°C. During firing, chemical reaction took place at the interface between the glass and the dielectric ceramics. The Ca in $(Ca_{1-x}Nd_{2x/3})TiO_3$ dissolved into the glass forming CaO–ZnO–B_2O_3 at 870–880°C, which improved the densification of the ceramics. The sample with 20–40 vol % glass sintered at 900°C showed a ε_r in the range 30–60, $Qf = 2000$–5000 GHz and $\tau_f = 20$–60 ppm/°C. Kim et al. [289] reported that addition of ZnO–H_3BO_3 from 1 to 4 wt% in $[Ca_{0.6}(Li_{0.5}Nd_{0.5})_{0.4}]_{0.45}Zn_{0.55}TiO_3$ (CLNZT) lowered sintering temperature from 1150°C to 900°C. The $[Ca_{0.6}(Li_{0.5}Nd_{0.5})_{0.4}]_{0.45}Zn_{0.55}TiO_3 + 2$ wt% $0.33ZnO$–$0.67H_3BO_3$ sintered at 875°C for 4 hours has ε_r of 42, $Qf = 10\,300$ GHz and $\tau_f = 19.5$ ppm/°C.

12.7.11 BaO—TiO_2-system

Several authors investigated [290–298] the effect of adding glasses such as B_2O_3, BaB_2O_4, BaO–B_2O_3–SiO_2, PbO–B_2O_3–SiO_2, ZnO–B_2O_3, on the microwave dielectric properties of $Ba_2Ti_9O_{20}$ ceramics. Lee et al. [294–296] studied the effect of addition of ZnO–B_2O_3 glass in ULF-280 dielectric powder (Ferro America) which contained $Ba_2Ti_9O_{20}$ and small quantities of ZrO_2, HfO_2, ZnO, SrO, B_2O_3, SiO_2. The samples sintered at 940°C/2 h with 1 wt% of $3ZnO$–B_2O_3 showed a Qf of 8300 GHz, $\varepsilon_r = 27$ and $\tau_f = 2.5$ ppm/°C. Addition of 3 wt% B_2O_3 to the $Ba_2Ti_9O_{20}$-based composite (ULF-280) and sintered at 940°C/2 h showed $\varepsilon_r = 28.3$, $Qf = 10\,800$ GHz, $\tau_f = -8.2$ ppm/°C [294]. It was found [290–292] that addition of 5 wt% B_2O_3 and sintered at 900°C gave single-phase $Ba_2Ti_9O_{20}$. However, addition of larger amount of B_2O_3 led to the formation of $BaTi(BO_3)_2$ and rutile secondary phases.

Choi et al. [299] studied the effect of addition of lithium borosilicate glass (10–35 wt% SiO_2, 23–43 wt% B_2O_3, 33–51 wt% Li_2O) in $BaTi_4O_9$. The glass has a relative permittivity of 7.5. Addition of 10 wt% glass frit and sintered at 900°C showed a density of 98% with $\varepsilon_r = 32$, $Qf > 9000$ GHz and $\tau_f = 10$ ppm/°C. Secondary phases of $BaTi_5O_{11}$ and $Ba_4Ti_{13}O_{30}$ were observed, which did not adversely affect the properties. Jhou and Jean [31] reported that with increasing BaO content in the barium zinc borate (BZB) glass, the softening and melting points of the resulting BZB glass decreased and the wetting between BZB and $BaTi_4O_9$ improved. For BZB glass with 0–20 mol% BaO content in 90 vol% $BaTi_4O_9$ + 10 vol% BZB, the ε_r varied in the range 28–33, Qf in the range 15 000–20 000 GHz. Kim et al. [300] and Lu et al. [301] studied the effect of zinc borate glass addition on the sintering behavior and microwave dielectric properties of $BaTi_4O_9$. Addition of small amount of glass lowered sintering temperature while maintaining good dielectric properties. Dense $BaTi_4O_9$ up to 96% relative density was obtained by sintering at 925°C/2 h with 5 wt% zinc borate glass [300]. The density increased sharply with increasing sintering temperature. Secondary phases of $Zn(BO_2)_2$ and $Zn_3(BO_3)_2$ were found in low-fired $BaTi_4O_9$ ceramics. The $BaTi_4O_9$ having 1 and 9 wt% zinc borate glass and sintered at 925°C and 875 °C has $Qf = 35\,200$ GHz and 27 900 GHz and $\varepsilon_r = 27$–30 respectively.

12.7.12 Vanadate system

The $Mg_3(VO_4)_2$ ceramic has an orthorhombic structure with space group *Cmca* [302]. Umemura et al. [303, 304] reported interesting microwave dielectric properties of $M_{3-x}Co_x(VO_4)_2$ (M = Mg, Ba). Figure 12.21 shows the variation of Qf as a function of sintering temperature in $Mg_3(VO_4)_2$ ceramics. The $Mg_3(VO_4)_2$ ceramics sintered at 1050°C showed the highest Qf of 64 000 GHz, with $\varepsilon_r = 9.1$ and $\tau_f = -93$ ppm/°C. They could also achieve the same properties by increasing the sintering duration to 50 hours at 950°C. X-ray diffraction studies showed that the ceramics do not react with silver. The $Mg_3(VO_4)_2$ decomposes at about 1074°C to form a liquid phase [305]. To lower the sintering temperature, Mg was substituted [303] partially by Co to form $Mg_{3-x}Co_x(VO_4)_2$. The partial substitution of Mg by Co lowered the sintering temperature from 1050°C to 850°C due to formation of $CoO–V_2O_5$ liquid phase. The $Mg_{3-x}Co_x(VO_4)_2$ with $x = 2$ sintered at 900°C for 5 hours showed a $Qf = 78\,000$ GHz, $\varepsilon_r = 9.4$ and $\tau_f = -95$ ppm/°C. Figure 12.22 shows the XRD patterns of $(Mg_{3-x}Co_x)(VO_4)_2$ ceramic sintered at 750°C/5 h. Figure 12.23 shows the effect of Co substitution and sintering temperature on ε_r and Qf in $Mg_{3-x}Co_x(VO_4)_2$. The $Ba_3(VO_4)_2$ ceramic has [304] a high sintering temperature of 1600°C/5 h has

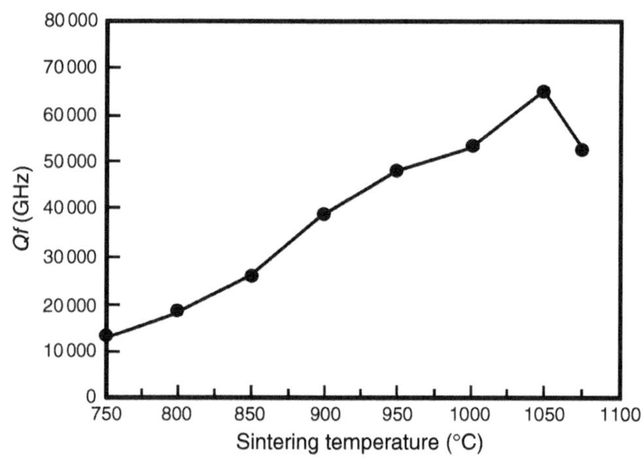

Figure 12.21 Relationship between Qf value and sintering temperature of $Mg_3(VO_4)_2$ ceramic sintered for 5 hours in air (after Ref. [303]).

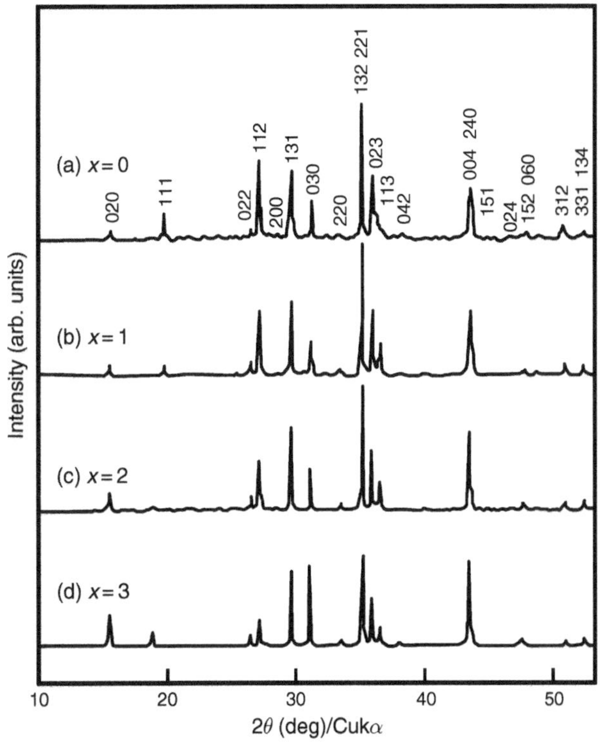

Figure 12.22 XRD pattens of $(Mg_{3-x}Co_x)(VO_4)_2$ ceramics sintered at 750°C for 5 hours (after Ref. [303]).

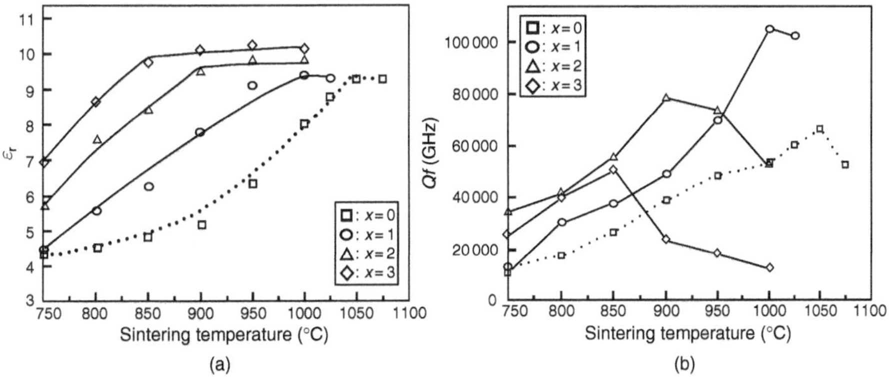

Figure 12.23 Effects of Co substitution for Mg on relative permittivity and *Qf* value in $(Mg_{3-x}Co_x)(VO_4)_2$ Ceramics (after Ref. [303]).

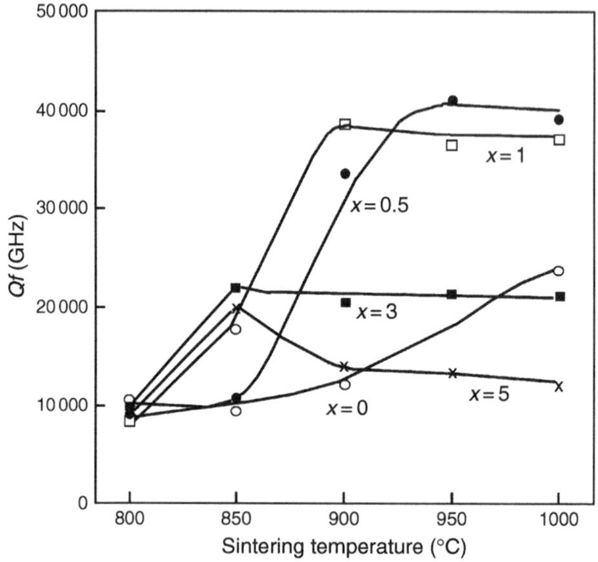

Figure 12.24 Influence of B_2O_3 additions on *Qf* value of $Ba_3(VO_4)_2$–xwt% B_2O_3 (x = 0–5) ceramics as a function of sintering temperature (after Ref. [304]).

$\varepsilon_r = 11$, $Qf = 62\,350$ GHz and $\tau_f = 28.8$ ppm/°C. Addition of 0.5–1 wt% B_2O_3 lowered the sintering temperature. Figure 12.24 shows the effect of B_2O_3 addition on the quality factor of $Ba_3(VO_4)_2$. Addition of 0.5 wt% B_2O_3 resulted in a *Qf* of 41 000 GHz, $\varepsilon_r = 12.5$, $\tau_f = 38.8$ ppm/°C when sintered at 950°C/5 h. Addition of larger amount of B_2O_3 led to formation of the secondary phase $Ba_2V_2O_7$. The $Ba_2V_2O_7$ sintered at 950°C has $Qf = 19\,000$ GHz, $\varepsilon_r = 7$, and τ_f of −74 ppm/°C.

12.7.13 Zinc and barium niobates

MNb_2O_6 (M = Ca, Co, Mn, Ni, Zn) ceramics are useful microwave dielectric material with medium ε_r values [242, 306]. However, their sintering temperatures are higher than that is needed for LTCC. The $ZnNb_2O_6$ sintered at 1150°C/2 h has $Qf = 83$ 700 GHz and $\varepsilon_r = 25$ [306] . Kim et al. [307] reported that addition of 5 wt% CuO to $ZnNb_2O_6$ decreases the sintering temperature to about 900°C. The presence of a CuO-rich intergranular phase was observed indicating liquid-phase sintering. The composition of the liquid phase was identified as $(ZnCu_2)Nb_2O_8$. This secondary phase has a low melting point with excellent dielectric properties such as $\varepsilon_r = 16.7$, $Qf = 41\,000$ GHz and $\tau_f = -76$ ppm/°C. Addition of 5 wt% CuO to $ZnNb_2O_6$ showed $\varepsilon_r = 22.1$, $Qf =$ 59 500 GHz and $\tau_f = -66$ ppm/°C. It was also found [308, 309] that addition or substitution of V_2O_5 to $ZnNb_2O_6$ lowers the sintering temperature to about 900°C. Addition of V_2O_5 led to the formation of a lossy secondary phase $V_3Nb_{17}O_{50}$, which degraded the quality factor. In contrast, the $Zn(Nb_{1-x}V_x)_2O_6$ samples with $x = 0.06$ sintered at 875°C/2 h showed a single-phase columbite structure with $\varepsilon_r = 23.9$, $Qf = 65\,000$ GHz and $\tau_f = -72.8$ ppm/°C. To lower the high negative τ_f, Kim et al. [310, 311] added TiO_2, which has a high positive τ_f. The $(1-x)ZnNb_2O_6$–x TiO_2 with $x = 0.58$ has a zero τ_f but its sintering temperature is relatively high. Hence to lower sintering temperature, Kim et al. [311] added CuO, which reduced preparation temperature to 875°C. However, addition of CuO decreased quality factor. XRD analysis showed that the sintered ceramics is a mixture of columbite $ZnNb_2O_6$, TiO_2 and ixiolite ($ZnTiNb_2O_8$). Secondary phases of $Cu_{0.85}Zn_{0.25}Nb_2O_6$ and $CuNb_2O_6$ were formed and they lowered the quality factor. The $(1-x)ZnNb_2O_6-xTiO_2$ ($x = 0.58$) with 10 wt% CuO sintered at 875°C showed $Qf = 17\,000$ GHz, $\varepsilon_r = 37$, and $\tau_f = -7$ ppm/°C.

The $Zn_3Nb_2O_8$ is another low loss material in the ZnO–Nb_2O_5-system [208, 312], but with a sintering temperature of about 1200°C. Addition of a small amount of V_2O_5, $0.29BaCO_3$–0.71CuO or $0.81MoO_3$–0.19CuO lowered the sintering temperature to about 850°C. The 2 mol% V_2O_5 added ceramic and sintered at 850°C/4 h resulted in $\varepsilon_r = 22.4$, $Qf = 67\,500$ GHz [312]. Pullar et al. [313] reported that $MCu_2Nb_2O_8$ (M = Zn, Co, Ni, Mg, Ca) are good candidates for LTCC and can be sintered in the temperature range 985–1010°C and $CaCu_2Nb_2O_8$ at 1100°C. The sintering temperature can be further lowered by adding 3 wt% of V_2O_5. However, doping V_2O_5 in general decreased the quality factor and the relative permittivity.

Sebastian et al. [314, 315] reported that $Ba_5Nb_4O_{15}$ is a suitable material for microwave applications, but its sintering temperature is quite high at about 1400°C with high positive τ_f. The sintering temperature of $Ba_5Nb_4O_{15}$ could be lowered to about 925°C [316] by the addition of B_2O_3. The ceramics contained hexagonal $BaNb_2O_6$, which has a high negative τ_f of −800 ppm/°C that compensated for the high positive τ_f of $Ba_5Nb_4O_{15}$. Addition of 3 wt% B_2O_3 to $Ba_5Nb_4O_{15}$ resulted a $\varepsilon_r = 39$ and $Qf = 18\ 700$ GHz with zero τ_f. Figure 12.25 shows the variation of dielectric properties as a function of B_2O_3 content. As the B_2O_3 content increases ε_r, τ_f and Qf decrease. Kim et al. [317, 318] reported addition of a mixture of 0.3 wt% B_2O_3 and 0.3 wt% V_2O_5 in $0.84Ba_5Nb_4O_{15}$–$0.16BaNb_2O_6$ composite further lowered the sintering temperature to 900°C/2 h with $\varepsilon_r = 42$, $Qf = 28\,000$ GHz and zero τ_f. The $Ba_5Nb_4O_{15}$ ceramic does not react with Ag and Cu electrode materials.

12.7.14 (Mg, Ca)TiO_3

Several authors [35, 319–325] investigated the effect of glass addition in $MgTiO_3$–$CaTiO_3$ (MCT) ceramics. Chen et al. [35, 319] studied the densification and microwave dielectric properties of RBS-(Mg, Ca)TiO_3, R = MgO, CaO, SrO, BaO, B = B_2O_3, S = SiO_2. The BaO–B_2O_3–SiO_2 (BBS) – $(Mg_{0.95}Ca_{0.05})TiO_3$ (MCT) (1:1 volume ratio) composite exhibited the highest ε_r and quality factor. Figure 12.26 shows the variation of microwave dielectric properties of the MCT ceramic glass composites as a function of sintering temperature. The BBS-MCT materials showed the best quality factor of 10 000 GHz when sintered at 900°C. XRD and SEM study of MCT with BBS (50:50 vol%) and sintered at 800–900°C indicated chemical reaction of MCT with glass and the formation of secondary phase of $BaTi(BO_3)_2$. The glass-ceramic composite and sintered tapes of the composite were found to be very porous. The MCT-BBS glass (50:50 vol%) possess very low shrinkage characteristics.

Zhang et al. [320] reported that Bi_2O_3–V_2O_5 addition in $MgTiO_3$ lowered sintering temperature from 1400°C to 875°C due to liquid-phase effect. With increasing V_2O_5 the ε_r decreased and the quality factor increased. This effect was attributed to the variation of the amount of different secondary phases such as $Bi_2Ti_2O_7$, $Bi_4V_{1.5}Ti_{0.5}O_{10.85}$ and $BiVO_4$. At 875°C, $MgTiO_3$ ceramics with 5 mol% Bi_2O_3 + 7 mol% V_2O_5 gave excellent microwave dielectric properties such as $\varepsilon_r = 20.6$,

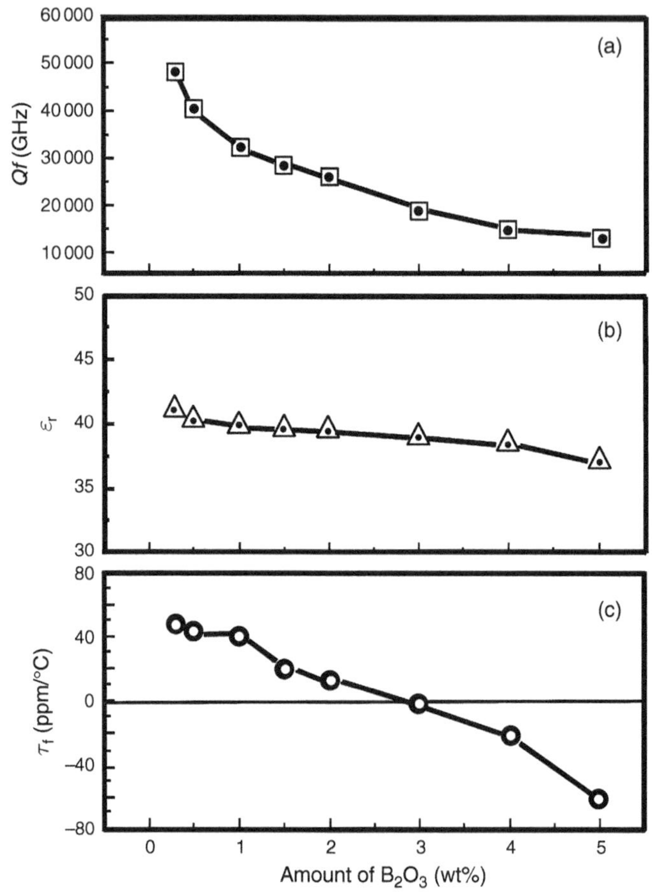

Figure 12.25 Microwave dielectric properties of $Ba_5Nb_4O_{15}$ ceramics sintered at 925°C/2 h as a function of the amount of B_2O_3 added (a) quality factor (b) relative relative permittivity (c) temperature coefficient of resonant frequency (after Ref. [316]).

$Qf = 10\,420$ GHz. However, Shin et al. [321] found that $MgTiO_3$-based dielectric decompose to $MgTi_2O_5$ and Mg_2TiO_4 during liquid-phase sintering using lithium borosilicate glass. However, this decomposition does not adversely affect the dielectric properties since $MgTi_2O_5$ has $\varepsilon_r = 17.4$ with $Qf = 47\,000$ GHz and Mg_2TiO_4 has $\varepsilon_r = 14.4$ and $Qf = 55\,000$ GHz. Jantunen et al. [322–325] made a detailed study of the effects of different glass compositions on the tape casting and the microwave dielectric properties. They [324] also investigated the sintering behavior and dielectric properties of mixtures of MMT-20 (MCT) with ZSB ($ZnO–SiO_2–B_2O_3$, 60.3:27.1:12.6) and BSB glasses ($BaO–SiO_2–B_2O_3$, 35:55:10). Jantunen et al. [322] prepared LTCC by mixing 30 wt% MCT (MMT-20) ceramic powder with 70 wt% of glass-forming oxides ZnO, SiO_2 and B_2O_3 in 60.3:12.6:27.1 mol%. The mixtures were ball-milled, dried and the cylindrical pucks made by sintering at 900°C. The samples prepared in this method was found to have better properties than prepared by mixing glass with ceramic powder. Hu et al. [325] reported

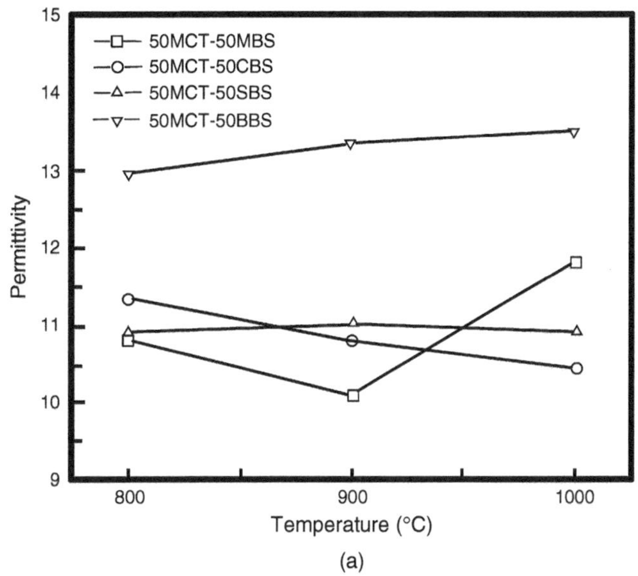

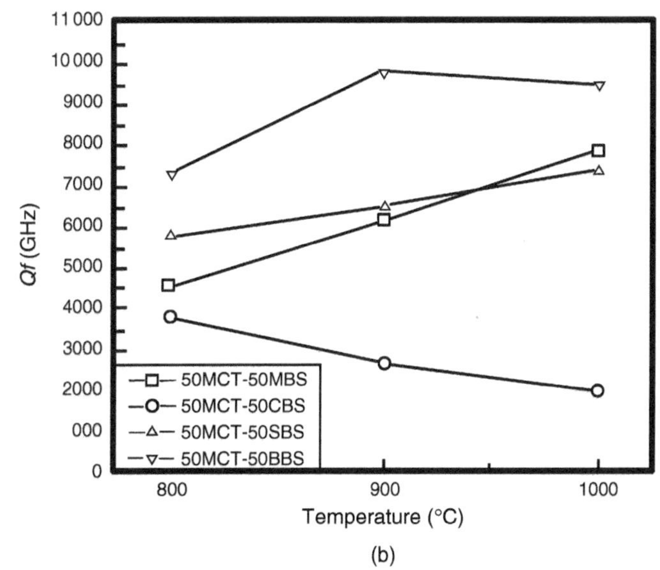

Figure 12.26 Variation of relative permittivity and quality factor of MCT-RBS glass-ceramic composites as a functon of the densification temperature (after Ref. [35]).

that if the $MgTiO_3$–$CaTiO_3$ powders contain free B_2O_3, then tape preparation is difficult regardless of the slurry system, whereas powders containing pre-reacted B_2O_3 did not cause any problem in making dense tapes with excellent properties. Choi et al. [326] reported that addition of lithium borosilicate glass to $CaZrO_3$–$CaTiO_3$ system lowered sintering temperature from 1450°C to 900°C. $CaTiO_3$ has positive τ_f and $CaZrO_3$ and glass frit have negative τ_f. When 15 wt% $CaTiO_3$ was mixed with 75 wt% of $CaZrO_3$ and 15 wt% glass frit (25–35 wt% of Li_2O, B_2O_3,-SiO_2 and small amounts of CaO, Al_2O_3, ZnO), sintered at 875°C showed $\varepsilon_r = 23$, *Qf* of 2400 GHz and zero τ_f. It is possible to shift the τ_f toward slightly positive or negative side by adjusting the $CaTiO_3$–$CaZrO_3$ composition to meet the circuit requirements.

12.7.15 $Mg_4(Nb/Ta)_2O_9$

The $Mg_4Nb_2O_9$ can be sintered at 850°C by the addition of 3 wt% LiF with excellent quality factor of 103 600 GHz and ε_r of 12.6 [327]. Substitution of a small amount of V for Nb in $Mg_4Nb_2O_9$ can considerably improve the quality factor with a decrease in the sintering temperature [328]. The limit of solid solution formation is close to $x = 0.125$. Secondary phases of $Mg_3(VO_4)_2$ formed for $x > 0.25$. The highest *Qf* of 160 000 GHz was obtained for $Mg_4(Nb_{2-x}V_x)O_9$ for $x = 0.0625$ sintered at 1025°C with ε_r=11.6. Small amount of V substitution is effective in lowering the sintering temperature without deterioration in the microwave dielectric properties. However, the τ_f is relatively high (−70 ppm/°C) for practical applications. To lower the τ_f, Yokoi et al. [329] added 6 wt% $CaTiO_3$ and sintered at 950°C for 10 hours to obtain $\varepsilon_r = 15.7$, *Qf* = 22 100 GHz and $\tau_f = -3.3$ ppm/°C. Although the microwave dielectric properties are useful, the compatibility with electrode materials needs to be investigated for practical use.

12.7.16 $Ba(Mg_{1/3}Nb_{2/3})O_3$

The complex perovskite $Ba(Mg_{1/3}Nb_{2/3})O_3$ (BMN) ceramic has good microwave dielectric properties with $\varepsilon_r = 32$, *Qf* up to 160 000 GHz and τ_f 33 ppm/°C [330, 331]. However, the sintering temperature of BMN is quite high at about 1450°C. Lim et al. [332] reported that addition of B_2O_3 can lower the sintering temperature of BMN to about 930°C but BaB_2O_4 was formed as a secondary phase. More recently, Lim et al. [333] reported that the sintering temperature can further be lowered to 875°C by the addition of 2 mol% B_2O_3 and 10 mol% CuO. The BMN sintered at 875°C for 2 hours showed a *Qf* value of 21 500 GHz with $\varepsilon_r = 31$ and τ_f= 21.3 ppm/°C. The addition of CuO suppressed the formation of BaB_2O_4. The authors believe that the CuO reacted with B_2O_3 to form CuO–B_2O_3 liquid phase, which assisted the sintering of BMN below 900°C.

12.7.17 $(Zr,Sn)TiO_4$ system

Several authors studied [334–340] the effect of additives on the low firing of $(Zr,Sn)TiO_4$ (ZST). Takada et al. [334] investigated the effect of addition of several glasses such as SiO_2, B_2O_3, 5ZnO–2B_2O_3 and commercial glasses on the lowering of the sintering temperature and microwave dielectric properties of (Zr, Sn)TiO_4. It is difficult to densify ZST when sintered at temperatures less than 1500°C as discussed in Chapter 4. Addition of 5 wt% glasses and sintering at 1100°C decreased the density considerably (less than 70% of theoretical density). Huang et al. [336–338] used ZnO, CuO, V_2O_5 and Bi_2O_3 in various combinations to lower the sintering temperature of ZST, whereas Zhang et al. [340] used La_2O_3–BaO to lower the sintering temperature. Jean and Lin [335] found that addition of 2.5–5 wt% $BaCuO_2$ + CuO enhances the densification kinetics of ZST but further increase in the additive content retard densification. Wang et al. [339] succeeded in sintering ZST in the temperature range 950–1150°C using ZnO–B_2O_3–SiO_2–Li_2O–CuO (ZBSLC), BaO–B_2O_3–SiO_2–Li_2O–CuO (BBSLC) and

BaO-SiO_2–TiO_2–CuO (BSTC) glasses. The samples with 10 wt% BBSLC and sintered at 950°C/4 h showed a relatively poor density with ε_r ~18, Qf = 12 700 GHz and τ_f = 1.5 ppm/°C. Addition of 10 wt% BBSLC glass to ZST and sintering at 1050°C for 4 hours resulted in $\varepsilon_r = 30$, Qf of 30 300 GHz and $\tau_f = -4$ ppm/°C. The ZBSLC and BSTC glasses were not effective in densifying the ceramics and the microwave properties were poor.

12.7.18 Ag(NbTa)O_3 ceramics

The $AgNbO_3$ and $AgTaO_3$ compounds undergo a series of structural phase transitions as they cool from the prototypic cubic perovskite phase [341]. $AgNbO_3$ exhibits a weak ferroelectric behavior at room temperature [341]. These materials have a high relative permittivity >400 and Qf in the range 600–900 GHz [342–344]. Composite microstructures consisting of 45 wt% $Ag(Nb_{0.65}Ta_{0.35})O_3$ and 55 wt% $Ag(Nb_{0.35}Ta_{0.65})O_3$ have $\varepsilon_r = 430$ and Qf = 700 GHz. Large grain size helps to minimize the reaction between these two phases during sintering [344]. Moreover, the evaporation and reduction of Ag_2O at high temperature and in oxygen-deficient atmospheres severely affect the densification and microwave properties [344]. To lower the sintering temperature, Sakabe et al. [345] added V_2O_5 and substituted Li for silver in $Ag(Nb_xTa_{1-x})O_3$. More recently, Kim et al. [346] succeeded in lowering the sintering temperature to below 950°C by liquid phase sintering with the addition of 1 wt% CuO. They adjusted [346] the temperature coefficient of capacitance by adjusting the Nb/Ta ratio in the solid solution and by creating composite microstructures. They prepared a two-phase assemblage consisting of $Ag(Nb_{3/4}Ta_{1/4})O_3$ (ANT-31) and $Ag(Nb_{1/4}Ta_{3/4})O_3$ (ANT-13) to get a temperature stable composite. The CuO added temperature stable composite had a relative permittivity of about 390 and Qf about 800 GHz. SEM and EDAX analysis revealed that CuO segregated at the grain boundaries. Figure 12.27 shows the variation of relative permittivity with temperature for $Ag(Nb_{3/4}Ta_{1/4})O_3$–$Ag(Nb_{1/4}Ta_{3/4})O_3$ (45:55 mol% mixing ratio) composite and the

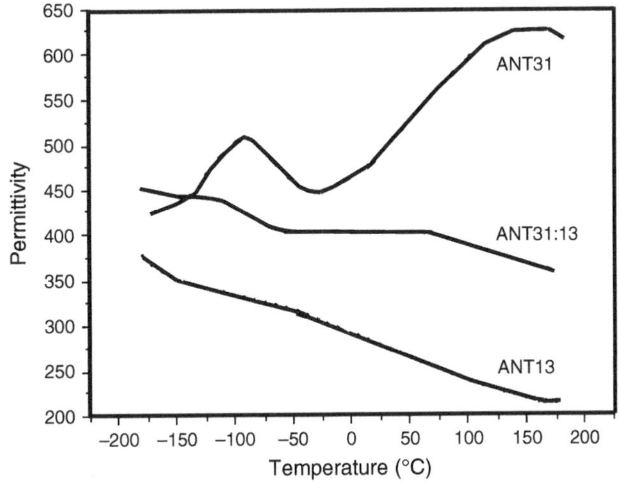

Figure 12.27 Temperature dependence of relative permittivity for ANT31:13 end members and composites (after Ref. [346]).

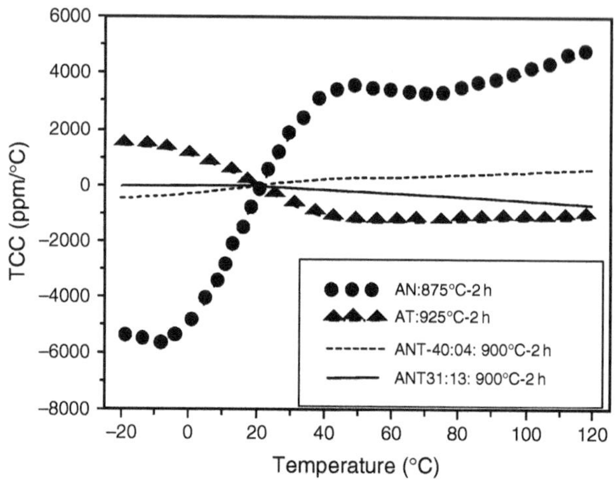

Figure 12.28 Temperature dependence of relative permittivity for ANT31:13 end members and composite (after Ref. [346]).

end members $Ag(Nb_{3/4}Ta_{1/4})O_3$ (ANT-31)and $Ag(Nb_{1/4}Ta_{3/4})O_3$ (ANT-13). The composite showed a very nominal variation in the relative permittivity with temperature ranging from −50°C to 100°C. Figure 12.28 shows the variation of temperature coefficient of capacitance for $Ag(Nb,Ta)O_3$ end members and composite. The ANT31:13 sintered at 900°C/2 h is nearly temperature stable. The $Ag(Nb_{2/4}Ta_{2/4})O_3$ (ANT-22) + 1 wt% CuO and sintered at 900°C showed a ε_r =398 and Qf= 400 GHz at 2.24 GHz. The ANT 31–13 + 1 wt% CuO and sintered at 875°C showed ε_r = 378 and Qf= 410 GHz. It was found that the material did not react with silver conductor. Guo et al. [347] reported that partial substitution of Sb for Nb/Ta lower the tan δ and improve the temperature stability of the dielectric ceramic. Kim et al. [346] proposed $Ag(Nb_{1-x}Ta_x)O_3$ solid solution and composites as promising candidates as embedded capacitors for high-frequency applications.

12.7.19 $A_2P_2O_7$ (A = Ca, Sr, Ba, Zn, Mg, Mn)

The crystal structure of $A_2P_2O_7$ has been extensively studied [348, 349]. The β-$Ca_2P_2O_7$ is tetragonal and α-$Ca_2P_2O_7$ is monoclinic. $Ca_2P_2O_7$ is also an important material in the field of luminescence and biomaterials [350, 351]. The $A_2P_2O_7$ (A = Ca, Sr, Ba, Mg, Zn, Mn) crystallizes in two forms [348, 349]. When the ionic radius of A is less than 0.97 Å it crystallizes in thortveitite form (A = Mg, Mn, Zn) and when the ionic radius of A is >0.97 Å it (Ca, Sr, Ba) crystallizes in the dichromate form. The $A_2P_2O_7$ except $Mn_2P_2O_7$ exist in allotropic forms. The thortveitite form undergoes a reversible phase transformation below 600°C from low temperature α-form to the high temperature β-form. The dichromate compounds undergo irreversible transformation at temperatures above 700°C. The $A_2P_2O_7$ (A = Mg, Mn, Zn) thortveitite forms are difficult to sinter into dense ceramics. Recently, Bian et al. and Cho et al. reported [352–354] the

Figure 12.29 Microstructure of $SrZnP_2O_7$ ceramics (after Ref. [353]).

microwave dielectric properties of $A_2P_2O_7$. Cho et al. [353] tailored the high negative τ_f of $A_2P_2O_7$ by the addition of TiO_2, which has a high positive τ_f forming a mixture with $A_2P_2O_7$. Bian et al. [352, 354, 361] reported that some of the $A_2P_2O_7$ such as $CaCuP_2O_7$, $SrCuP_2O_7$, $Mn_2P_2O_7$, α-$Zn_2P_2O_7$,$CaZnP_2O_7$, $SrZnP_2O_7$ are suitable glass-free LTCC materials. They can be sintered at temperatures of about 900°C with a low ε_r of about 7 and *Qf* up to 10 000 GHz and negative τ_f of about −70 ppm/°C. They all react with silver but $SrZnP_2O_7$ and $CaZnP_2O_7$ do not react with Cu. Figure 12.29a shows the microstructure of a typical $A_2P_2O_7$ sintered at 900°C.

12.7.20 ABO_4 (A = Ca, Sr, Ba, Mg, Mn, Zn: B = Mo, W)

Brower and Fang [273] prepared $CaMoO_4$ crystals by the Czocharlski method and reported it having $\varepsilon_r = 24 \pm 0.2$ along the *a*-axis and 20 ± 0.2 along the *c*-axis at 1.59 KHz with a tan δ of about 10^{-3}. The $CaMoO_4$ ceramic sinters at temperatures of about 1300°C but the sintering temperature can be lowered by changing the Ca/Mo ratio. Choi et al. [355] prepared $CaMo(x)O_4$ by sintering in the temperature range 900–1000°C for $x = 1.02$–1.08. The $CaMo(x)O_4$ with $x = 1.02$ sintered at 1000°C showed a $Qf = 70\,000$ GHz with $\varepsilon_r = 9.5$. Choi et al. [356] reported the microwave dielectric properties of $AMoO_4$ (A = Ca, Sr, Ba, Mg, Mn, Zn) ceramics sintered at temperatures in the range 800–1100°C. $ZnMoO_4$ has the lowest sintering temperature of 800°C. These ceramics have ε_r in the range 7–10.8, *Qf* up to 89 000 GHz and τ_f in the range −46 to −87 ppm/°C. The Ca-, Sr- and Ba-based ceramics have tetragonal sheelite structure and Mg, Mn and Zn based have wolframite structure. The $ZnMoO_4$ is triclinic, whereas $MgMoO_4$ and $MnMoO_4$ are monoclinic [356]. Several authors [357–360] reported the microwave dielectric properties of low temperature-sintered AWO_4 ceramics. The $MgWO_4$ sinters at 1050°C with $\varepsilon_r = 13.5$, $Qf = 69\,000$ GHz and $\tau_f = -58$ ppm/°C. $ZnWO_4$ and $MnWO_4$ sintered at 1100°C with high *Qf* and negative τ_f of about −60 ppm/°C. The sheelite ceramics have a relatively higher sintering temperature of about 1150°C with excellent dielectric properties.

12.8 Conclusion

There are about 400 low loss dielectric materials (see Appendix 2) reported with sintering temperature $\leq$1050°C. Within the low relative permittivity materials (<20), there are several temperature-compensated compositions like MMT-20 + $ZnO:B_2O_3:SiO_2$ glass, $Mg_3Sm_4Al_{44}O_{75} + B_2O_3–SiO_2–Al_2O_3$ and $NaAlSi_3O_8$ and several others needing only a slight adjustment fulfilling this requirement. Some of the materials like $K_xBa_{1-x}Ga_{2-x}Ge_{2+x}O_8$, $SrCuP_2O_7$, $Mg_4(Nb_{2-x}V_x)O_9$ ($x = 0.0625$), $CaWO_4$ + 1 wt% $MnSO_4$, $Cu_3Nb_2O_8$, $Mg_4Nb_2O_9$ + 3 wt% LiF have high Qf > 100 000 GHz. The $Mg_4(Nb_{2-x}V_x)O_9$ ($x = 0.0625$) sintered at 1025°C with $\varepsilon_r = 11.6$ has the highest Qf (> 160 000 GHz) among the LTCC materials. Although several authors reported success in reducing the sintering temperature to the level suitable for LTCC, very little attention was paid to know their chemical compatibility with electrode materials and silicon, thermal expansion, shrinkage, thermal conductivity, etc. Tapes of most of the materials are not being made and their sintering behavior, shrinkage, dielectric properties, chemical compatibility with electrode materials are not being investigated.

References

[1] L. Hongwei, H. L. Barnes, J. Laskar, and D. Estrich. Application of digital PGA technology to k band microcircuit and microwave subsystem package. IEEE Trans. Microw. Theory Technol. 48(2000)2644–2651.

[2] T. A. Midford, J. J. Wooldridge, and R. L. Sturdivant. The evolution of packages for monolithic microwave and millimeterwave circuits. IEEE Trans. Antennas Propagation 43 (1995)983–991.

[3] R. R. Tummala Rao. Ceramic and glass packaging in the 1990s. J. Am. Ceram. Soc. 74(1991)895–908.

[4] A. L. Eustice, S. J. Horowitz, J. J. Stewart, A. R. Travis, and H. T. Sawhill. Low temperature cofireable ceramics. A new approach for Electronic packaging. Proc. 36th Electronic Component Conference, IEEE, Seattle (1986) New York, 37–47.

[5] J. Muller, H. Thust, and K. H. Drue. RF design considerations for passive elements in LTCC material systems. Int. J. Microcircuits Electron. Packaging. 18(1995)200–206.

[6] M. A. Rodriguez, P. Yang, P. Kotula, and D. Dimos. Microstructure and phase development of buried resistors in low temperature cofired ceramics. J. Electroceram. 5(2000)217–223.

[7] H. T. Sawhil. Material compatibility and cosintering aspects of shrinkage control in low temperature cofired ceramic packages. Ceram. Trans. 26(1987)307–319.

[8] K. W. Long and K. M. Luk. Circular dielectric resonator antenna of high dielectric constant for low profile Application. IEEE Antenna Propagation 43(1995)517–519.

[9] K. Delaney, J. Barret, J. Barton, and R. Doyle. Characterisation and performance prediction of integral capacitors in low temperature cofired technology. IEEE Trans. Adv. Pack. 22 (1999)68–77.

[10] R. Ludwig and P. Bretchko. RF Circuit Design: Theory and Applications, Prentice Hall, NJ (2000).

[11] Y. Imanaka. Multilayers Low Temperature Cofired Ceramics (LTCC) Technology, Springer, New York (2005).

[12] R. E. Mistler and E. R. Twiname. Tape Casting, Theory and Practice. The American Ceramic Society, 735 Ceramic Place, Westerville, OH 43081 (2000).

[13] A. Dzuirdzia, S. Nowack, M. Ciez, and W. Gregorcyzk. Microwave resonant elements up to 20 GHz manufactured in photoimageable thin technology. Proc. European Microelectronic Packaging and Interconnection Symp., IMAPS, Cracow, Poland (2002) pp. 409–414.

[14] J. Hagberg, M. Pudas, M. Kittilä, and S. Leppävuori, Direct gravure printing (DGP) method for printing fine line electrical circuits. Proc. Intern. Symp. Microelectronics, IMAPS, Baltimore, MD, USA (2001) 681–687.

[15] H. Jantunen, T. Kangasvieri, J. Vähäkangas, and S. Leppävuori. Design aspects of microwave components with LTCC techniques. J. Eur. Ceram. Soc. 23(2003)2541–2548.

[16] H. Thust, K. H. Drue, J. Mueller, T. Thelemann, T. Tuschiek, J. Chilo, C. Golovanov, and F. Ndagijimana. Coupling behaviour between transmission lines with meshed ground planes in LTCC-MCMs. Microwave J. 41(1998)70–79.

[17] V. Gektin, A. Barcohen, and S. Witzman. Coffin-Manson based fatigue analysis of underfilled DCAS. IEEE Trans. Comp. Packaging and Manufacturing Technol. Part A. 21(1998)577–584.

[18] B. Schwartz. Microelectronic packaging II. Am. Ceram. Soc. Bull. 63(1984)577–581.

[19] H. Kagata, T. Inoue, and J. Kato. Low fire bismuth based dielectric ceramics for microwave use, Jpn. J. Appl. Phys. 31(1992)3152–3155.

[20] R. L. Whalers, S. J. Stein, Y. D. Huang, and M. R. Heinz. Lead Free Multilayer System for Telecommunications. Technical publications of ElectroScience Laboratories, Prussia, PA, USA (2001).

[21] L. Devlin, G. Pearson, and J. Pittock. RF and microwave component in LTCC. Proc. of the 38th IMAPS Nordic Conference, Oslo, Norway (2001)96–110.

[22] S. H. Hall, G. W. Hall, and J. A. Mc Call. A High Speed Digital System Design: A Hand Book of Interconnect Theory and Design Practices, John Wiley & Sons, New York (2000).

[23] D. I. Amey, M. Y. Keating, M. A. Smith, S. J. Horowitch, P. C. Donahue, and C. R. Needes. Low loss tape materials system for 10 to 40 GHz application. Proc. intern. symp. Microelectronics. IMAPS, Boston, USA (2000)654–658.

[24] R. Brown. LTCC bringing functional integration to RF and microwave products. Proc. RF Design 97. Conf. and Expo, Englewood Co, USA (1997)209–221.

[25] R. Kulke, W. Simon, C. Gunner, G. Mollenbeck, D. Kother, and M. Rittweger. R-F benchmark up to 40 GHz for various LTCC low loss tapes. IMAPS-Nordic, Stockholm, Sweden (2002)97–102.

[26] C. Wang and K. A. Zaki. Temperature compensation of combline resonators and filters. IEEE MMT-S Intern. Microwave Symp. Digest 3(1993)1041–1044.

[27] H. Jantunen and A. Turunen. Temperature compensated dielectric filter, US patent 5, 302 924 (1994).

[28] H. Jantunen. A novel low temperature cofiring ceramic (LTCC) material for telecommunication devices. Ph.D. Thesis, Department of Electrical Engineering, University of Oulu (2001).

[29] J. P. Sommer, B. Michel, U. Goebel, and J. Jelonnek. Thermally optimised millimetre wave package on an LTCC ceramic board. In the seventh intersociety Conference on Thermal and Thermomechanical Phenomena in Electronic systems. ITHERM, Las Vegas, NV, USA (2000).

[30] D. Antonio C. B. Bencoe, and K. G. Ewsuk. Proc. of SPIE-international Soc. for Optical Engn. 5231(2003)160.

[31] M.-Z. Jhou and J.-H. Jean. Low fire processing of $BaTi_4O_9$ dielectric with $BaO–ZnO–B_2O_3$ glasses. J. Am. Ceram. Soc. 89(2006)786–791.

[32] J. M. Wu and H. L. Huang. Microwave properties of zinc barium and lead borosilicate glasses. J. Non Cryst. Solids 260(1999)116–124.

[33] K. P. Surendran, P. Mohanan, and M. T. Sebastian. The effect of glass additives on the microwave dielectric properties of $Ba(Mg_{1/3}Ta_{2/3})O_3$ ceramics. J. Solid. State Chem. 177 (2004)4031–4046.

[34] T. Takada and K. Kageyama. Synthesis and microwave dielectric properties of $La_2O_3–xB_2O_3$ based melt mixtures for low temperature cofired ceramics. Jpn. J. Appl. Phys. 44(2005) 6629–6635.

[35] C.-S. Chen, C.-C. Chou, W.-J. Shih, K.-S. Liu, C.-S. Chen, and I.-N. Lin. Microwave dielectric properties of glass- ceramic composites for low temperature cofirable ceramics. Mater. Chem. Phys. 79(2003)129–134.

[36] H. Zhu, M. Liu, H. Zhou, L. Li, and A. Lv. Preparation, and properties of low temperature cofired ceramic of $CaO–SiO_2–B_2O_3$ system, J. Mater. Sci. Mater. Electron. 17(2006)637–641.
[37] B.-H. Jung, S.-J. Hwang, and H.-S. Kim. Glass-ceramic for low temperature cofired dielectric ceramic materials based on $La_2O_3–B_2O_3–TiO_2$ glass with BNT ceramics. J. Eur. Ceram. Soc. 25(2005)3187.
[38] Y.-J. Choi, J.-H. Park, W.-J. Ko, I.-S. Hwang, J.-H. Park, J.-G. Park, and S. Nahm. Cofiring and shrinkage matching in low and middle permittivity dielectric compositions for a low temperature cofired ceramics system. J. Am. Ceram. Soc. 89(2006)562–567.
[39] J. H. Park, Y.-J. Choi, and J.-H. Park. Low fire dielectric compositions with permittivity 20–60 for LTCC applications. Mater. Chem. Phys. 88(2004)308–312.
[40] H. Zhu, M. Lu, H. Zhou, L. Li, and A. Lv. Preparation and properties of low temperature cofired ceramic of $CaO–SiO_2–B_2O_3$ system. J. Mater. Sci. Mater. Electron. 17(2006) 637–641.
[41] J. H. Jean, T.-H. Kuan, and C.-R. Chang. Low temperature-fired low relative permittivity silica glass composite for substrate applications. Mater. Chem. Phys. 41(1995)123–127.
[42] A. Yang, H. Lin, L. Luo, and W. Chen. Microwave dielectric properties of low temperature cofired glass ceramic based on $B_2O_3–La_2O_3–MgO–TiO_2$ glass with $La(Mg_{.5}Ti_{.5})O_3$ ceramics. Jpn. J. Appl. Phys. 45(2006)1698–1701.
[43] S.-G. Kim, J.-S. Park, J.-S. An, K. S. Hong, H. Shin, and H. Kim. Effects of addition of different types of fillers on the properties of $BaO–ZnO–B_2O_3–SiO_2$ glass composites for application to barrier ribs of plasma display panels. J. Am. Ceram. Soc. 89(2006)902–907.
[44] O. Dernovsek, A. Naeini, G. Preu, W. Wersing, M. Eberstein, and W. A. Schiller, LTCC glass-ceramic composites for microwave application. J. Eur. Ceram. Soc. 21(2001)1693–1697.
[45] C.-C. Chiang, S.-F. Wang, V.-R. Wang, and W.-C. J. Wei. Densification and microwave dielectric properties of $CaO–B_2O_3–SiO_2$ system glass ceramics. Ceram. Int. 34(2008)599–604.
[46] S. Knickerbocker, A. H. Kumar, and L. W. Herron. Cordierite glass-ceramics for multilayer ceramic packaging. Am. Ceram. Soc. Bull. 72(1993)90–95.
[47] FERRO-TAPE-A6. Technical Publication, Ferro Corporation, Santa Barbara, CA (1996).
[48] K. Niwa, Y. Imanaka, N. Kamehara, and S. Aoki. Advances in Ceramics. M. P. Yan, K. Niwa, H. M. O. Bryan, and W. S. Young, (Eds), American Ceramic Society, Columbus, OH (1989), Vol. 26.
[49] A. H. Kumar, P. W. Mc Millan, and R. R. Tummala. Glass ceramic structures and sintered multilayered substrates thereof with circuit patterns of gold, silver or copper, US patent 4 301 324 (1981).
[50] O.-H. Kwon and G. L. Messing. Kinetics analysis of solution precipitation during liquid phase sintering of alumina. J. Am. Ceram. Soc. 73(1990)275–281.
[51] V. K. Singh. Densification of alumina and silica in the presence of liquid phase. J. Am. Ceram. Soc. 64(1981) C-133–C-135.
[52] K. G. Ewsuk and L. W. Harrison. Advances in Ceramics. C. Handwerker, J. Blendell, and W. Kayser (Eds), The American Ceramic Society, Westerville, OH (1999) Vol. 29, p. 356.
[53] W. D. Kingery, E. Niki, and M. D. Narasimhan. Sintering of oxide and carbide–metal compositions in the presence of liquid phase. J. Am. Ceram. Soc. 44(1961)29–34.
[54] R. M. German. Liquid Phase Sintering, Plenum Press, New York (1985).
[55] J.-H. Jean and T. K. Gupta. Liquid phase sintering in the glass cordierite system. J. Mater. Sci. 27(1992)1575–1584.
[56] J.-H. Jean and T. K. Gupta. Liquid phase sintering in the glass-cordierite system. Particle size effects. J. Mater. Sci. 27(1992)4967–4973.
[57] J.-H. Jean and T. K. Gupta. Densification kinetics and modelling of alumina filled alumina composites. J. Mater. Res. 9(1994)771–780.
[58] J.-H. Jean and T. K. Gupta. Densification kinetics of binary borosilicate glass composite. J. Mater. Res. 9(1994)486–492.
[59] J.-H. Jean and S.-C. Lin. Effects of borosilicate glass on densification and properties of borosilicate glass + TiO_2 ceramics. J. Mater. Res. 14(1999)1359–1363.

[60] M. Eberstein and W. A. Schiller. Development of high permittivity glasses for microwave LTCC tapes. Glass Sci. Technol. 76(2003)8–16.
[61] M. Eberstein and W. A. Schiller. New low-loss sintering microwave ceramics for applications in mobile communication technology. Keramische Zeitschrift 55(2003)344–348.
[62] A. L. Eustice, S. J. Horowitz, J. J. Stewart, A. R. Travis, and H. T. Sawhill. Low temperature co-fired ceramics: a new approach to electronic packaging. 36th Electronic Component Conf., Seattle, WA, May 5–7 (1986)37–67.
[63] T. Noro and H. Tozaki. Constituent materials for via conductor in multilayer glass-ceramic board, Japanese Patent No. 6070799A (1986).
[64] Y. Imanaka, S. Aoki, N. Kamehara, and K. Niwa. Crystallisation of low temperature fired glass-ceramic composite, J. Ceram. Soc. Jpn. 95(1987)1119–1121.
[65] Y. Imanaka, K. Yamazaki, S. Aoki, N. Kamehara, and K. Niwa. Effect of alumina addition on crystallization of borosilicate glass. J. Ceram. Soc. Jpn. 97(1989)309–313.
[66] S. Nishigaki, S. Yano, J. Fukuta, M. Fukuyama, and T. Fuwa. A new multilayered low temperature fireable ceramic substrate. Proc. 85 Int. Symp. Hybrid Microelectronics (ISHM) (1985) 225–234.
[67] G. C. Kuczynski and I. Zaplatynskyj. Sintering of glass. J. Am. Ceram. Soc. 39(1956) 349–350.
[68] B. Cutler and R. E. Henrichsen. Effect of particle shape on the kinetics of sintering of glass. J. Am. Ceram. Soc. 51(1968)604–605.
[69] Y. Imanaka and N. Kamehara. Influence of shrinkage mismatch between copper and ceramics on dimensional control of the multilayer ceramic circuit board. J. Ceram. Soc. Jpn. 100(1992)558–561.
[70] J. I. Steinberg, S. J. Horowitz, and R. J. Bacher. Low temperature cofired Tape dielectric material systems for multilayer interconnections. In: Advances in Ceramics, J. B. Blum and W. R. Cannon (Eds), American Ceramic Society, Westerville, OH (1998) Vol. 19.
[71] R. R. Tummala Rao, P. Garrou, T. Gupta, N. Kuramoto, K. Niwa, Y. Shimada, and M. Terasawa. Ceramic Packaging. In: Microelectronic Packaging Hand Book. Part II, R. R. Tummala Rao, E. J. Rymaszewski, and A. G. Klopfenstein (Eds), Chapman & Hall, New York (1997).
[72] Y. Shimada, Y. Shiozawa, M. Sizuki, H. Takimizawa, and Y. Tamashita. In: Proceedings of the 36th Electronic Components Conference (1984) pp. 395–405.
[73] K. H. Yoon, S. J. Yoo, W. S. Kim, J. B. Kim, and E. S. Kim. Effect of BaO/WO_3 on the microwave loss quality of $Ba(Mg_{1/3}Ta_{2/3})O_3$. Jpn. J. Appl. Phys. 38(1999)5616–5620.
[74] W. D. Kingery. Densification during sintering in the presence of liquid phase-I: Theory. J. Appl. Phys. 30(1959)301–306.
[75] J.-H. Jean and S.-S. Lin. Low fire processing of ZrO_2–SnO_2–TiO_2 ceramics. J. Am. Ceram. Soc. 83(2000)1417–1422.
[76] S. Kemethmuller, A. Roosen, F. Goetz-Neunhoffer, and J. Neubauer, Quantitative analysis of crystalline and amorphous phases in glass-ceramic composites like LTCC by the Rietveld method. J. Am. Ceram. Soc. 89(2006)2632–2637.
[77] M. Ebberstein, J. Moller, J. Wiegmann, and W. A. Schiller. Modifiation and simulation of dielectric properties of glass/crystal phase composites for LTCC applications. CFI-Ceramic Forum International 80(2003) E39–E46.
[78] N. Ichinose and H. Yamamota. Effect of additives on microwave dielectric properties in low temperature firing (Mg, Ca)TiO_3 based ceramics. Ferroelectrics 201(1997)255–262.
[79] H. Tamura, H. Kagata, T. Inoue, and I. Kameyama. Low fire bismuth based dielectric ceramics for microwave use. Jpn. J. Appl. Phys. 31(1992)3152–3155.
[80] L. Navias and R. L. Green. Dielectric properties of glasses at ultrahigh frequencies and their relation to composition. J. Am. Ceram. Soc. 29(1946)267–276.
[81] J. O. Israd. A study of the migration loss in glass and a generalised method of calculating the rise of dielectric loss with temperatures. Int. Conf. Comp. Mater. Electron. Pack. Eng. IEEE (1961)3636.
[82] E. B. Shand, (Ed). Properties of glass. Glass Engineering Hand Book, Mc Graw Hill, New York (1958) p. 16.

[83] K. Wakino. Proc. ISIF-92(1992)308–335.
[84] A. R. Von Hippel, (Ed). Dielectric Materials and Applications, Technology Press of MIT, John Wiley & Sons, Cambridge, MA (1954).
[85] D. L. Kinser. Electrical conduction in glass and glass ceramics. In: Physics of Electronic Ceramics, Part A, L. L. Hench and D. B. Dove (Eds), Marcel Dekker, New York (1971).
[86] A. I. Berezhoni, Properties of glass-ceramics. In: Glass Ceramics and Photo-Sitalls, A. G. Pincus (Ed.), Plenum Press, New York (1970).
[87] G. A. Pavlova and V. G. Chistoserdov. Structure of Glass, Consultants Bureau, New York, (1964).Vol. 3.
[88] H. Mandai, K. Sugo, K. Tsukamoto, H. Tani, and M. Murata. A low temperature cofired multilayer ceramic substrate containing copper conductors. Proc. IMC, Kobe (1986)61–64.
[89] H. Mandai. Proc Ceramic Science & Technology Congress (1990)83–86.
[90] H. Mandai and S. Okube. Low temperature fireable dielectric ceramic material. Ceram. Trans. 32(1992)91–101.
[91] T. Takada, H. Yammoto, and K. Kageyama. Synthesis and microwave dielectric properties of $x\mathrm{Re_2O_3}$–$y\mathrm{B_2O_3}$ (Re = La, Nd, Sm, Dy, Ho, Y) compounds. Jpn. J. Appl. Phys. 42(2003)6162.
[92] H. Kagata, R. Saito, and H. Katsumura. Al_2O_3–MgO–ReOx(Re = rare earth) based LTCC and its applications to multilayer Non-shrinkage substrate for microwave devices. J. Electroceram. 13(2004)277–280.
[93] C.-R. Chang and J.-H. Jean. Crystallisation kinetics and mechanism of low dielectric constant low temperature, cofirable CaO–B_2O_3–SiO_2 glass-ceramics. J. Am. Ceram. Soc. 82(1999)1725–1732.
[94] Y. Kobayashi and E. Kato. Low temperature fabrication of anorthite ceramics. J. Am. Ceram. Soc. 77(1994)833–834.
[95] C.-L. Lo, J.-G. Duh, B.-S. Chiou, and W.-H. Lee. Low temperature sintering and microwave dielectric properties of anorthite based glass-ceramics. J. Am. Ceram. Soc. 85(2002) 2230–2235.
[96] E. S. Lim, B. S. Kim, J.-H. Lee, and J.-J. Kim. Effect of BaO content on the sintering and physical properties of BaO–B_2O_3–SiO_2 glasses. J. Non-cryst. Solids 352(2006)821–826.
[97] J.-M. Wu and H.-L. Huang. Effect of crystallisation on microwave dielectric properties of stoichiometric cordierite glass containing B_2O_3 and P_2O_5. J. Mater. Res. 15(2000)222–227.
[98] T. Takada, S. F wang, S. Yoshikawa, S.-J. Jang, and R. E. Newnham. Effect of glass additions in BaO–TiO_2_WO_3 microwave ceramics. J. Am. Ceram. Soc. 77(1994)1909–1916.
[99] J. D. Breeze, X. Aupi, N. Mc. N. Alford. Ultra low loss polycrystalline alumina. Appl. Phys. Lett. 81(2002)5021–5023.
[100] W. C. J. Jei, C. L. Chen, and A. Roosen. La_2O_3–Al_2O_3–B_2O_3 based glass ceramics for LTCC application. Key Eng. Mater. 280–183(2005)929–993.
[101] M.-H. Lim, J.-H. Park, H.-G. Kim, and M.-J. Yoo. Fabrication and characterisation of glass-ceramic + Al_2O_3 composition for LTCC (low temperature cofired ceramic). Proc. IEEE Intl. Conf. Properties and Applications of Dielectric Materials. (2003), Vol. 2, pp. 757–760.
[102] J. J. Seo, D. J. Shin, and Y. S. Cho. Phase evolution and microwave dielectric properties of lanthanum borate based low temperature cofired ceramic materials. J. Am. Ceram. Soc. 89(2006)2352–2355.
[103] C.-L. Chen, W.-J. Wei, and A. Roosen. Wetting, densification and phase transformation of $La_2O_3/A_2O_3/B_2O_3$ based glass-ceramics. J. Eur. Ceram. Soc. 26(2006)59–65.
[104] N. Chakraborty J. E. Shelby, and R. A. Condrate. Properties and structure of lanthanum borate glasses, J. Am. Ceram. Soc. 67(1984)782–785.
[105] H. Jo, M. S. Kang, K. W. Chung, and Y. S. Cho. Chemical stability and dielectric properties of RO–La_2O_3–B_2O_3(R = Ca, Mg, Zn) based ceramics. Mater. Res. Bull. 43(2008)361–369.
[106] Y.-C. Fang and J.-H. Jean. Compositional design of lead free, low temperature cofired ceramic dielectric composite. Jpn. J. Appl. Phys. 45(2006)6357–6361.

[107] S. Dai, R.-F. Huang, and D. L. Wilcox. Use of titanates to achieve a temperature stable low temperature cofired ceramic dielectric for wireless applications. J. Am. Ceram. Soc. 85(2002)828–832.

[108] D. Wilcox, R.-F. Huang, and S. Dai. Enabling Materials for Wireless Multilayer Ceramic Integrated Circuit (MCIC) Applications. In: Ceramic Transaction, J.-H. Jean, T. K. Gupta, K. M. Nair, and K. Niwa (Eds), Multilayer Electronic Ceramic Devices. American Ceramic Society, Westerville, OH (1998) Vol. 97, pp. 201–213.

[109] S. Dai, R.-F. Huang, and D. Wilcox. Temperature stable low loss and low fire dielectric for consumer wireless Applications. Proc. of the 1st China International conference on High Performance Ceramics. Beijing Oct 31–Nov 3 (1998).

[110] Y. J. Seo, J. H. Jung, Y. S. Cho, J. C. Kim, and N. K. Kang. Influence of particle size of alumina filler in an LTCC system. J. Am. Ceram. Soc. (2006).

[111] E. A. Barringer and H. K. Bowen. Formation, packing and sintering of monodisperse TiO_2 powders. J. Am. Ceram. Soc. 65(1982)C199–C201.

[112] H. Hahn, J. Logas, and R. S. Averback. Sintering characteristics of nanocrystalline TiO_2. J. Mater. Res. 5(1990)609–614.

[113] D.-W. Kim, T. G. Kim, and K. S. Hong. Low firing of CuO doped anatase. Mater. Res. Bull. 34(1999)771–781.

[114] D.-W. Kim, B. Park, J.-H. Chung, and K. S. Hong. Mixture behaviour and microwave dielectric properties in the low fired TiO_2-CuO system. Jpn. J. Appl. Phys. 39(2000)2696–2700.

[115] S. Yano, N. Hirofumi, Y. Komaki, and T. Hirai. Distributed constant circuit board using ceramic substrate materials. US patent No. 5232 765. Aug. 1993.

[116] S.-H. Yoon, D.-W. Kim, S.-Y. Cho, and K. S. Hong. Phase analysis and microwave dielectric properties of LTCC-TiO_2 with glass system. J. Eur. Ceram. Soc. 23(2003) 2549–2552.

[117] J.-H. Jean and S.-C. Lin. Effects of borosilicate on densification and properties of borosilicate + TiO_2 ceramics. J. Mater. Res. 14(1999)1359–1363.

[118] X. M. Cui, J. Zhou, C. L. Miao, and J. H. Shen. The study on fabrication of dielectric materials for low temperature cofired ceramics. Rare Earth Mater. Eng. 34(2005)410–415.

[119] J.-H. Jean and T. H. Kuan. Compositional design and properties of a low k silica dielectric for multilayer ceramic substrate. Jpn. J. Appl. Phys. 34(1995)1901–1905.

[120] Y. Miyauchi and T. Arashi. The particle size effect of filler on sintering characteristics in glass-ceramics. J. Jpn. Soc. Powder Powder Metall. 52(2005)271–275.

[121] D. M. Mattox, S. R. Gurkovich, J. A. Olenik, and K. M. Mason. Low dielectric constant alumina compatible cofired multilayer substrate. Ceram. Eng. Sci. Proc. 9(1988)1567–1578.

[122] R. A. Gudla. Anorthite ceramic dielectrics. Am. Ceram. Soc. Bull. 50(1971)555–557.

[123] M. G. M. U. Ismail and H. Aria. Sol gel synthesis of B_2O_3 doped anorthite and its characterisation. J. Ceram. Soc. Jpn. 100(1992)1385–1389.

[124] C.-L. Lo, J.-G. Duh, B.-S. Chiou, and W.-H. Lee. Microstructure characteristics for anorthite composite glass with nucleating agents of TiO_2 under non-isothermal crystallisation. Mater. Res. Bull. 37(2002)1949–1960.

[125] C. Lo, J. G. Duh, B. S. Chiou, and W. H. Lee. Low temperature sintering and crystallisation behaviour of low loss anorthite based glass ceramics. J. Mater. Sci. 38(2002)693–698.

[126] V. M. F. Marques, D. U. Tulyaganov, S. Agathopoulos, V. K. Gataullin, G. P. Kothoyal, and J. M. F. Ferreira. Low temperature synthesis of anorthite based glass-ceramics via sintering and crystallisation of glass-powder compacts. J. Eur. Ceram. Soc. 26(2006)2503–2510.

[127] M. E. Villafuerte-Castrejon, A. Aragon-Pifia, R. Valenzuela, and A. R. West. Compound and solid solution formation in the system Li_2O-Nb_2O_5-TiO_2. J. Solid State Chem. 71(1987)103–108.

[128] A. Borosevich and P. K. Davies. Microwave dielectric properties of $Li_{1+x-Y}M_{1-x-3y}Ti_{x+4y}O_3$ ($M = Nb^{5+}, Ta^{5+}$) solid solutions. J. Eur. Ceram. Soc. 21(2000)1719–1722.

[129] A. Borisevich and P. K. Davies. Synthesis and dielectric properties of $Li_{1-x+y}Ta_{1-x-3y}Ti_{x+4y}O_3$ M-phase solid solutions. J. Am. Ceram. Soc. 85(2002)2487–2491.

[130] A. Borisevich and P. K. Davies. Crystalline structure and dielectric properties of $Li_{1+x-y}Nb_{1-x-3y}Ti_{x+4y}O_3$ M-phase solid solutions. J. Am. Ceram. Soc. 85(2002)573–578.
[131] Y. Borisevich and P. K. Davies. Effect of V_2O_5 doping on the sintering and dielectric properties of M-phase $Li_{1+x-y}Nb_{1-x-3y}Ti_{x+4y}$ ceramics. J. Am. Ceram. Soc. 87(2004) 1047–1052.
[132] D. H. Kang, K. C. Nahm and H. J. Cha. Effect of Li_2O–V_2O_5 on the low temperature sintering and microwave dielectric properties of $Li_{1.0}Nb_{0.6}Ti_{0.5}O_3$ ceramics. J. Eur. Ceram. Soc. 26(2006)2117–2121.
[133] Q. Zeng, W. Li, J.-L. Shi, J.-K. Guo, H. Chen, and M.-L. Liu. Effect of B_2O_3 on the sintering and microwave dielectric properties of M-phase $LiNb_{0.6}Ti_{0.5}O_3$ ceramics. J. Eur. Ceram. Soc. 27(2007)261–265.
[134] Q. Zeng, W. Li, J.-L. Shi, J.-K. Guo, M.-W. Zuo, and W.-J. Wu. A new microwave dielectric ceramic for LTCC applications. J. Am. Ceram. Soc. 89(2006)1733–1735.
[135] Q. Zeng, W. Li, J.-L. Shi, and J.-K. Guo. Fabrication and microwave dielectric properties of a new LTCC ceramic composite based on Li_2O–Nb_2O5–TiO_2 system. Mater. Lett. 60(2006)3203–3206.
[136] Q. Zeng, W. Li, J.-L. Shi, and J.-K. Guo. Microwave dielectric properties of $5Li_2O$–$0.583Nb_2O_5$–$3.248TiO_2$ ceramics with V_2O_5. J. Am. Ceram. Soc. 89(2006)3305–3307.
[137] Q. Zeng, W. Li, J.-L. Shi, X. L. Dong, and J.-K. Guo. Microwave dielectric properties and microstructures of the $11Li_2O$–$3Nb_2O_5$–$12TiO_2$ ceramics with B_2O_3 addition. Mater. Res. Bull. 43(2008)441–417.
[138] L. E. Depero and L. Sangaletti. Structural disorder and ionic conductiviton: The case of Bi_2O_3. J. Solid State Chem. 122(1996)439–443.
[139] H. Iwahara, T. Esaka, T. Sato, and T. Takahashi. Formation of high oxide ion conductive phases in sintered oxide of the system Bi_2O_3–Ln_2O_3(Ln=La–Yb). J. Solid State Chem. 39(1981)173–180.
[140] S. Esmaeilzadeh, S. Lundgren, U. Halenius, and J. Grins. $Bi_{1-x}Cr_xO_{1.5+1.5x}$ $0.05<x<0.15$: A new high temperature solid solution with a three dimensional incommensurate modulation. J. Solid State Chem. 156(2000)168–180.
[141] T. Takahashi, T. Esaki, and H. Iwahara. Electrical conduction in sintered oxides of Bi_2O_3–BaO. J. Solid State Chem. 16(1976)317–323.
[142] K. Z. Fung and A. V. Virkar. Phase stability, phase transformation kinetics and conductivity of Y_2O_3 –Bi_2O_3 solid electrolytes containing aliovalant dopants. J. Am. Ceram. Soc. 74(1991)1970–1980.
[143] T. Takahashi, H. Iwahara, and T. Esaka. High oxide ion conduction in sintered oxide of the system Bi_2O_3–M_2O_3. J. Electrochem. Soc. 124(1977)1563–1569.
[144] T. Takahashi and H. Iwahara. Oxide ion conductors based on bismuth sesquioxide. Mater. Res. Bull. 13(1978)1447–1453.
[145] A. A. Yaremchenko, V. V. Kharton, E. N. Naumovich, and A. Vechar. Oxygen ion transport in Bi_2O_3-based oxides. The solid solutions Bi_2O_3–Nb_2O_5. J. Solid State Electrochem. 2(1998)146–149.
[146] P. Samoukhina, S. Kamba, S. Santhi, J. Petzelt, M. Valant, and D. Suvorov. Infrared and terahertz of novel Bi_2O_3–Nb_2O_5 microwave ceramics. J. Eur. Ceram. Soc. 25(2005) 3085–3088.
[147] C. D. Ling, R. L. Withers, S. Schmidt, and J. G. Thomson. A review of bismuth rich binary oxides in system Bi_2O_3–Nb_2O_5, Bi_2O_3–Ta_2O_5, Bi_2O_3–MoO_3, Bi_2O_3–WO_3 systems. J. Solid State Chem. 137(1998)42–61.
[148] C. D. Ling. Structural relationships among bismuth rich phases and the Bi_2O_3–Nb_2O_5, Bi_2O_3–Ta_2O_5, Bi_2O_3–MoO_3, Bi_2O_3–WO_3 systems. J. Solid State Chem. 148(1999) 380–405.
[149] R. L. Withers, C. D. Ling, and S. Schimdt. Atomic modulation functions, perovskite nodal surfaces and three dimensional incommensurateley modulated $(1-x)Bi_2O_3$–xNb_2O_5 $0.06\leq x\leq 0.23$ solid solution. Z. Krist. 214(1999)296–304.

[150] C. D. Ling and M. Johnson. Modeling refinement and analysis of the type II δ-Bi_2O_3 related superstructures in the Bi_2O_3-Nb_2O_5 system. J. Solid State Chem 177(2004)1838–1846.
[151] R. S. Roth and J. L. Waring. Phase equilibrium relations in the binary system bismuth sesquioxide niobium pentoxide. J. Res. Natl. Bur. Stand (USA) 66a(1962)451–463.
[152] H. A. Harwig. On the structure of bismuth sesquioxide: The α, β, and δ-phases. Z. Anorg. Alleg. Chem. 444(1978)151–166.
[153] M. Valant and D. Suvorov. Bol. Soc. Esp. Ceram. Vidrio. 43(2004)634.
[154] C. D. Ling and M. Johnson. Modellig refinement and analysis of the type III δ-Bi_2O_3 relared superstructure in the Bi_2O_3–Nb_2O_5 system. J. Solid State Chem. 177(2004)1838–1846.
[155] C. D. Ling, R. L. Withers, S. Schmidt, and J. G. Thomson. A review of bismuth rich binary oxides in the system Bi_2O_3–Nb_2O_5, Bi_2O_3–Ta_2O_5, Bi_2O_3–MoO_3, and Bi_2O_3–WO_3. J. Solid State Chem. 137(1998)42–61.
[156] W. Zhou and D. A. Jefferson and J. M. Thomas. Defect fluorite structures containing Bi_2O_3-Nb_2O_5. Proc. Roy. Soc. London A 406(1986)173–182.
[157] A. Castor, E. Aguado, J. M. Rojo, P. Herrero, R. Enjalbert, and J. Galy. The new oxygen deficient fluorite Bi_3NbO_7: Synthesis, electrical behaviour and structural approach. Mater. Res. Bull. 33(1998)31–41.
[158] M. Valant, B. Jancar, U. Pirnat, and D. Suvorov. Order-disorder transition in Bi_2O_3–Nb_2O_5 fluorite like dielectrics. J. Eur. Ceram. Soc. 25(2005)2829–2834.
[159] U. Pirnat, M. Valant, B. Jancar, and D. Suvorov. Formation characteristics of the commensaurate fluorite type Bi_2O_3–Nb_2O_5 solid solutions. Chem. Mater. 17(2005)5155–5160.
[160] D. Liu, Y. Liu, S.-Q. Huang, and X. Yao. Phase structure and dielectric properties of Bi_2O_3–ZnO–Nb_2O_5 based dielectric ceramics. J. Am. Ceram. Soc. 76(1993)2129–2132.
[161] M. Valant and D. Suvorov. Dielectric properties of the fluorite like Bi_2O_3–Nb_2O_5 solid solution and the tetragonal Bi_3NbO_7. J. Am. Ceram Soc. 86(2003)939–944.
[162] N. Wang, M.-Y. Zhao, W. Li, and Z.-W. Yin. The sintering behavior and microwave dielectric properties of Bi(Nb,Sb)O_4 ceramics. Ceram. Int. 30(2004)1017–1022.
[163] Y.-C. Chen, W.-C. Tzou, C.-F. Yang, P.-S. Cheng. Low fire bismuth based dielectric ceramics for microwave use. Jpn. J. Appl. Phys. 40(2001)3252–3155.
[164] H. R. Lee, K. H. Yoon, and E. S. Kim. Low temperature sintering and microwave dielectric properties of $BiNbO_4$–$ZnNb_2O_6$ ceramics with CuV_2O_6 additive. Jpn. J. Appl. Phys. 42(2003)6168–6171.
[165] E. S. Kim and J. W. Choi. Intl. Conf. Microwave Materials and their Applications, Inuyama, Japan (2004).
[166] C.-L. Huang, M.-H. Weng, and C.-C. Yu. Low fired $BiNbO_4$ based microwave dielectric ceramics. Ceram. Int. 27(2001)343–350.
[167] C.-L. Huang and M.-H. Weng, and C.-C. Wu. The microwave dielectric properties and microstrucures of La_2O_3 modified $BiNbO_4$ ceramics. Jpn. J. Appl. Phys. 39(2000) 3506–3510.
[168] C.-L. Huang and M.-H. Weng. Low fired $BiTaO_4$ dielectric ceramics for microwave applications. Mater. Lett. 43(2000)32–35.
[169] N. Wang, M.-Y. Zhao, Z.-W. Yin, and W. Li. Effect of complex substitution of La and Nd for Bi on the microwave dielectric properties of $BiNbO_4$ ceramics. Mater. Res. Bull. 39(2004)439–448.
[170] C.-L. Huang, M.-H. Weng, and G.-M. Shan. Effect of V_2O_5 and CuO additives on sintering behavior and microwave dielectric properties of $BiNbO_4$ ceramics. J. Mater. Sci. 35(2000)5443–5447.
[171] D. Shihua, Y. Xi, and Y. Yang. Dielectric properties of B_2O_3 doped $BiNbO_4$ ceramics. Ceram. Int. 30(2004)1195–98.
[172] C.-F. Yang, and W. H. Weng. Improvement of quality value and shift of Tf value of $BiNbO_4$ ceramics with CuO–V_2O_5 mixture. Jpn. J. Appl. Phys. 38(1999)6797–6800.
[173] E. S. Kim and W. Choi. Effect of phase transition on the microwave dielectric properties of $BiNbO_4$. J. Eur. Ceram. Soc. 26(2006)1761–1766.

[174] W. C. Tzou C. F. Yang, Y. C. Chen, and P. S. Cheng. Improvement in the sintering and microwave dielectric properties of $BiNbO_4$ ceramics by V_2O_5 addition. J. Eur. Ceram. Soc. 20(2000)991–996.
[175] C.-L. Huang, M.-H. Weng, C.-C. Wu, and C.-T. Lion. Low fire $BiNbO_4$ microwave dielectric ceramics modified by Sm_2O_3 addition. Mater. Res. Bull. 35(2001)827–835.
[176] W. Choi and K. Y. Kim. Effect of Nd_2O_3 on the microwave dielectric properties of $BiNbO_4$ ceramics. J. Mater. Res. 13(1998)2945–2949.
[177] D. Shihua, Y. Xi, M. Yu, and L. Puling. Microwave dielectric properties of $(Bi_{1-x}R_x)NbO_4$ ceramics [R=Ce, Nd, Dy, Er]. J. Eur. Ceram. Soc. 26(2006)2003–2005.
[178] D. Shingua, Y. Xi, and Y. Yong. Dielectric properties of B_2O_3 doped $BiNbO_4$ ceramics. Ceram. Int. 30(2004)1195–1198.
[179] Y. Yang, S. Ding, and X. Yao. Influences of Fe_2O_3 additives on the dielectric properties of $BiNbO_4$ ceramics under different sintering conditions. Ceram. Int. 30(2004)1341–1345.
[180] H. R. Lee, B. D. Lee, and K. H. Yoon. Microstructures and dielectric properties of low temperature cofired $BiNbO_4$–$ZnNb_2O_6$ ceramics with Ag electrode. Jpn. J. Appl. Phys. 44(2005)133–1336.
[181] H. C. Ling, M. F. Yan, and W. W. Rhodes. High dielectric constant and small temperature coefficient bismuth based dielectric compositions. J. Mater. Res. 5(1990)1752–1762.
[182] W. F. Su and S. C. Lin. Interfacial behaviour between $Bi_{1.5}ZnNb_{1.5}O_7$–$0.02V_2O_5$ and Ag. J. Eur. Ceram. Soc. 23(2003)2593–2596.
[183] N. Wang, M.-Y. Zhao, W. Li, and Z.-W. Yin. Study of the relationship between the defect and dielectric properties of ZnO doped $BiNbO_4$ ceramics. Ceram. Int. 30(2004)1335–1339.
[184] Z. Wang, X. Yao, and L. Zhang. CeO_2 modified $BiNbO_4$ microwave ceramics sintered under atmosphere. Ceram. Int. 30(2004)1329–1333.
[185] C.-L. Huang and M.-H. Weng. Microwave dielectric properties and the microstructures of $Bi(Nb,Ta)O_4$ ceramics. Jpn. J. Appl. Phys. 38(1999)5949–4452.
[186] M.-H. Weng, and C.-L. Huang. Shifting of Tf value of $BiNbO_4$ ceramics by $BiTaO_4$ addition. J. Mater. Sci. Lett. 19(2000)375–376.
[187] K. Fukuda, R. Kitoh and I. Awai. Microwave characteristics of TiO_2–Bi_2O_3 dielectric resonaor. Jpn. J. Appl. Phys. 32(1993)4584–4588.
[188] A.-K. Axelsson, M. T. Sebastian, and N. Mc N. Alford. Microwave dielectric properties of Bi_2O_3–TiO_2 composite ceramics. J. Korean Ceram. Soc. 40(2003)340–345.
[189] A.-K. Axelsson and N. Mc N. Alford. Bismuth titanates candidates for high permittivity LTCC. J. Eur. Ceram. Soc. 26(2006)1933–1936.
[190] I. Levin, T. G. Amos, J. C. Nino, T. A. Vanderah, I. M. Reaney, C. A. Randall, and M. T. Lanagan. Crystal chemistry of the compound $Bi_2Zn_{2/3}Nb_{4/3}O_7$. J. Mater. Res. 17(2002)1406–1411.
[191] M. Valant and P. K. Davies. Crystal chemistry and dielectric properties of chemically substituted $Bi_{1.5}(Zn_{1.0}Nb_{1.5})O_7$ and $Bi_2(Zn_{2/3}\ Nb_{4/3})O_7$ pyrochlores. J. Am. Ceram. Soc. 83(2000)147–153.
[192] H.-J. Youn, T. Sogabe, C. A. Randall, T. R. Shrout, and M. T. Lanagan. Phase relations and dielectric properties of Bi_2O_3-ZnO-Ta_2O_5 system. J. Am. Ceram. Soc. 84(2001) 2557–2562.
[193] H. Du, X. Yao, L. Zhang. Structure, IR spectra and dielectric properties of Bi_2O_3-ZnO-SnO_2-Nb_2O_5 quaternary pyrochlore. Ceram. Int. 28(2002)231–234.
[194] Z. P. Wang and S. Z. Zhang. Electronic Components and Materials (in Chinese) 1(1979)11.
[195] X. Wang, H. Wang, X. Yao. Structures, phase transformations, and dielectric properties of pyrochlores containing bismuth. J. Am. Ceram. Soc. 80(1997)2745–2748.
[196] I. Levin, T. G. Anos, J. C. Nino, T. A. Vanderah, C. A. Rindall, and M. T. Lanagan. Structural study of unusual cubic pyrochlore $Bi_{1.5}Zn_{0.92}O_{6.92.}$ J. Solid State Chem. 168(2002)69–75.
[197] R. L. Withers, T. R. Welberry, A. K. Larsson, Y. Liu, L. Noren, H. Rundoff, and F. J. Brink. Local crystal chemistry induced strain and short range order in cubic pyroclhore $Bi_{1.5\alpha}Zn_{0.5\beta}(Zn_{0.5}2Nb_{1.5-\delta})O_7$-1-a(BZN). J. Solid State Chem. 177(2004)231–244.

[198] S. Kamba, V. Prokhonsky, A. Pashkin, V. Bovtun, and J. Petzelt. Anomalous broad dielectric relaxation in $Bi_{1.5}ZnNb_{1.5}O_7$ pyrochlore. Phys. Rev. B66(2002)054106.
[199] M.-C. Wu, S. Kamba, V. Bovtun, and W.-F. Su. Comparison of microwave dielectric behaviour between $Bi_{1.5}Zn_{0.92}Nb_{1.5}O_{6.92}$ and $Bi_{1.5}ZnNb_{1.5}O_7$. J. Eur. Ceram. Soc. 26(2006)1889–1893.
[200] J. C. Nino, M. T. Lanagan, C. A. Rindall and S. Kamba. Corelation between infrared phonon modes and dielectric relaxation in Bi_2O_3–ZnO–Nb_2O_5 cubic pyrochlore. Appl. Phys. Lett. 81(2002)4404–4406.
[201] W. Zhenping and Z. Shiying. No lead, low fired multilayer ceramic capacitor (MLS) dielectric material. J. Am. Ceram. Soc. 67(1984)413–419.
[202] E. A. Nenesheva and N. F. Kartenko. Magnesium, zinc-nickel substituted bismuth niobate. Russian Patent 2021207 Oct (1990).
[203] S.-Y. Chen, S.-Y. Lee, and Y. J. Lin. Phase transformation reactions kinetics and microwave characteristics of Bi_2O_3–ZnO–Nb_2O_5-ceramics. J. Eur. Ceram. Soc. 23(2003)873–881.
[204] M. F. Yan, H. C. Lig, and N. W. Rhodes. Low firing temperature stable dielectric composition based on bismuth nickel zinc niobates. J. Am. Ceram. Soc. 73(1990)1106–1107.
[205] Y. Hu and C.-L. Huang. Structure and dielectric properties of Bi_2O_3-ZnO-Nb_2O_5 based dielectric ceramics. Mater. Chem. Phys. 72(2001)60–65.
[206] S. L. Swartz and T. R. Shrout. Ceramic composition for BZN dielectric resonator. US patent 5449652 (1995).
[207] S.-Y. Chen and Y.-J. Lin. Effect of CuO based oxide additives on Bi_2O_3-ZnO-Nb_2O_5 microwave ceramics. Jpn. J. Appl. Phys. 40(2001)3305–3310.
[208] M.-C. Wu, K.-T. Huang, and W.-F. Su. Microwave dielectric properties of doped $Zn_3Nb_2O_8$ ceramics sintered below 950°C and their compatibility with silver electrode. Mater. Chem. Phys. 98(2006)406–409.
[209] G.-K. Choi, D.-W. Kim, S.-Y. Cho, and K. S. Hong. Influence of V_2O_5 substitution $Bi_2(Zn_{1/3}Nb_{2/3})_2O_7$ pyrochlore on sintering temperature and dielectric properties. Ceram. Int. 30(2004)1187–1190.
[210] M. F. Yan, H. C. Ling, and W. W. Rhodes. Low firing temperature stable dielectric compositions based on bismuth nickel zinc niobates. J. Am. Ceram. Soc. 73(1990)1106–1106.
[211] H. Wang, H. Du, Z. Peng, M. Zhang, and X. Yao. Improvement in sintering and dielectric properties of Bi_2O_3–ZnO–Nb_2O_5 pyrochlore ceramics by V_2O_5 addition. Ceram. Int. 30(2004)1225–1229.
[212] E. A. Nenesheva and N. F. Kartenko. Low sintering ceramic materials based on Bi_2O_3-ZnO-Nb_2O_5 compounds. J. Eur. Ceram. Soc. 26(2006)1929–1922.
[213] H. Wang, R. Elsebrock, T. Scheller, R. Waser, and X. Yao. Bismuth zinc niobate ($Bi_{1.5}ZnNb_{1.5}O_7$) ceramics derived from metallo organic decomposition precursor solution. Solid State Commun. 132(2004)481–486.
[214] H. Du and Y. Yao. Investigations on structural evolution and dielectric characteristics of high performance Bi based dielectrics. Mater. Res. Bull. 40(2005)1527–1535.
[215] H. Wang, S. Kamba, M. Zhang, X. Yao, S. Deniov, F. Kadlec, and J. Petzelt. Microwave and infrared dielectric response of monoclinic bismuth zinc niobate based pyrochlore ceramics with ion substitution in A site. J. Appl. Phys. 100(2006)034109.
[216] H. Wang, Z. Peng, H. Du, T. Yao, and X. Yao. Mn^{4+} and W^{6+} substitution on Bi_2O_3-ZnO-Nb_2O_5 based low firing ceramics. Ceram. Int. 30(2004)1219–1223.
[217] J. C. Nino, M. T. Lanagan, and C. Rindall. Phase formation and reactions in the Bi_2O_3-ZnO-Nb_2O_5-Ag pyrochlore system. J. Mater. Res. 16(2001)1460–1464.
[218] J. C. Nino, I. M. Reaney, M. T. Lanagan, and C. A. Rindall. Transmission electron microscopy investigation of Bi_2O_3–ZnO–Nb_2O_5 pyrochlore and related compounds. Mater. Lett. 57(2002)414–419.
[219] H. Du and X. Yao. Effects of Sr substitution on dielectric characteristics in $Bi_{1.5}ZnNb_{1.5}O_7$ ceramics. Mater. Sci. Eng. B. 99(2003)437–440.
[220] J. Wei, L. Zhang, and X. Yao. Melting properties of Bi_2O_3-ZnO—Nb_2O_5 based dielectric ceramics. J. Am. Ceram. Soc. 82(1999)2551–2552.

[221] S. Okawa, M. Suzuki, S. Okamoto, H. Uchida, S. Koda, and H. Funakubo. MOCVD growth of $Bi_{1.5}Zn_{1.0}Nb_{1.5}O_7$ (BZN) epitaxial thin films and their electrical properties. Jpn. J. Appl. Phys. 44(2005)6957–6959.
[222] W. Ren, S. Trolier-McKinsty, C. A. Rindall, and T. R. Shrout. Bismuth zinc niobate pyrochlore dielectric thin films for capacitative applications. J. Appl. Phys. 89(2001)767–774.
[223] J. Lu and S. Stemmer. Low loss tunable bismuth zinc niobate films deposited by rf magnetron sputtering. Appl. Phys. Lett. 83(2003)2411–2413.
[224] J.-Y. Lim, D.-W. Kim, H. S. Jung, and K.S. Hong. Voltage tunable dielectric properties of pyrochlore Bi–Zn–Nb–Ti–O solid solution thin films. Jpn. J. Appl. Phys. 44(2005) 6648–6653.
[225] Y. P. Hong, S. Ha, H. Y. Lee, Y. C. Lee, K. H. Ko, D.-W. Kim, H. B. Hong, and K. S. Hong. Voltage tuneable dielectric properties of rf sputtered Bi_2O_3–ZnO–Nb_2O_5 pyrochlore thin films. Thin Solid Films 419(2002)183–188.
[226] A. K. Tagantsev, J. Lu, and S. Stemmer. Temperature dependence of the dielectric tunability of pyrochore bismuth zinc niobate thin films. Appl. Phys. Lett. 86(2005)032901.
[227] A. K. Tagentsev, V. O. Sherman, K. F. Astafiev, J. Venkatesh, and N. Setter. Ferroelectrioc materials for microwave tunable applications. J. Electroceram. 11(2003)5–66.
[228] D. P. Cann, C. A. Randall, and T. R. Shrout, Investigation of the dielectric properties bismuth pyrochlores. Solid State Commun. 100(1996)529–534.
[229] H. J. Youn, C. A. Randall, A. Chen, T. R. Shrout, and M. T. Lanagan. Dielectric relaxation and microwave dielectric properties of Bi_2O_3–ZnO–Ta_2O_5 ceramics. J. Mater. Res. 17(2002)1502–1506.
[230] H. B. Hong, D.-W. Kim, and K.S. Hong, Microwave dielectric properties of $Bi_2(Zn_{1/3}Ta_{2/3})_2O_7$ polymorphs. Jpn. J. Appl. Phys. 42(2003)5172–5175.
[231] B. Shen, X. Yao, L. Kang, and D. Peng. Effect of CuO or and V_2O_5 oxide additives on Bi_2O_3–ZnO–Ta_2O_5 based ceramics. Ceram. Int. 30(2004)1203–1206.
[232] C. A. Randall, J. C. Nino, A. Baker, H.-J. Youn, A. Hitomi, R. Thayer, L. F. Edge, T. Sogabe, D. Anderson, T. R. Shrout, S. Trolier-McKinstry, and M. T. Lanagan. Bi-pyrohlore and zirconolite dielectrics for integrated passive component applications. Am. Ceram. Soc. Bull. 82(2003)644–652.
[233] A. Baker, M. Lanagan, C. Randall, E. Semouchkina, G. Semouchkin, K. Z. Rajab, R. Eitel, R. Mitra, S. Rhee, P. Geggier, C. Duschl, and G. Fuhr. Integration concepts for the fabrication of LTCC structures. Int. J. Appl. Ceram. Technol. 2(2005)514–520.
[234] E. Semouchkina, G. Semouchkin, M. Lanagan, and R. Mitra. Field simulation based strategy for designing microstrip filters. IEEE MTT-S. Intl. Microw. Symp. Dig. Philadelphia (2003) 1897–1900.
[235] M. Valant and D. Suvorov. Processing and dielectric properties of sillenite compounds $Bi_{12}MO_{20}$ (M = Si, Ge, Ti, Pb, Mn, $B_{1/2}$, $P_{1/2}$). J. Am. Ceram. Soc. 84(2001)2900–2904.
[236] A. D. Murray, C. R. A. Catlow, F. Beech, and J. Drennen. A neutron powder diffraction study of the low and high temperature structures of $Bi_{12}PbO_{19}$. J. Solid State Chem. 62(1986)290–296.
[237] G. Bayer, zur kristallchemie des Tellus: Telluroxide und oxidiverbindungen Fortschr. Miner. 46(1969)42–72.
[238] M. Udovic, M. Valant, and D. Suvorov. Dielectric characterisation of ceramics from the TiO_2-TeO_2 system. J. Eur. Ceram. Soc. 21(2001)1735–1738.
[239] M. Udovic, M. Valant, and D. Suvorov. Phase formation and dielectric characterisation of the Bi_2O_3-TeO_2 system prepared in an oxygen atmosphere. J. Am. Ceram. Soc. 87(2004) 591–597.
[240] S. Yamanaka and M. Miyake. Study of the ternary Ti-Te-O system. J. Less Common Metals 59(1990)179–189.
[241] G. Meunier and J. Galy. Sur une deformation inedite du reseau de type fuorine. Structure cristalline de phases MTe_3O_8 (M = Ti, Sn, Hf, Zr). Acta Cryst. 27B(1971)602–608.
[242] M. Maeda, T. Yamamura, and T. Ikeda. Dielectric characteristics of several complex oxide ceramics at microwave frequencies. Jpn. J. Appl. Phys. 26(1987)76–79.

[243] M. Valant and D. Suvorov. Glass-free low temperature cofired ceramics: calcium germanates, silicates and tellurates. J. Eur. Ceram. Soc. 24(2004)1715–1719.
[244] D.-K. Kwon, M. T. Lanagan, and T. Shrout. Microwave dielectric properties and low temperature cofiring of $BaTe_4O_9$ with Al metal electrode. J. Am. Ceram. Soc. 88(2005) 3419–3422.
[245] D.-K. Kwon, M. T. Lanagan, and T. R. Shrout. Synthesis of $BaTiTe_3O_8$ ceramics for LTCC application and its dielectric properties. J. Ceram. Soc. Jpn. 113(2005)216–219.
[246] M. Udovic and D. Suvorov. Sintering and dielectric characterization of pseudoternary compounds from the Bi_2O_3–TiO_2–TeO_2 system. J. Am. Ceram. Soc. 87(2007)591–597.
[247] G. Subodh and M. T. Sebastian. Glass free microwave ceramics for LTCC applications. J. Am. Ceram. Soc. 90(2007)2266–2268.
[248] O. Yamaguchi, M. Morimi, H. Kawabata, and K. Shimizu. Formation and transformation of $ZnTiO_3$. J. Am. Ceram. Soc. 70(1987)C97–C98.
[249] Y. S. Chang, Y. H. Chang, I. G. Chen, G. J. Chen, Y.L. Chai, S. Wu, and T. H. Fang. The structure and properties of zinc titanates doped with strontium. J. Alloys Compd. 354(2003) 303–309.
[250] K. Haga, T. Ishiji, J. Mashiyama, and T. Ikeda. Dielectric properties of two phase mixture ceramics composed of rutile and its compounds. Jpn. J. Appl. Phys. 31(1992)3156–3159.
[251] A. Golovochanski, H. T. Kim, and Y. H. Kim. Zinc titnate dielectric ceramics prepared by sol gel process. J. Korean Phys. Soc. 32(1998)S1167–S1169.
[252] H. T. Kim, S. H. Kim, S. Nahm, J. D. Byun, and Y. Kim. Low temperature sintering and microwave dielectric properties of zinc metatitanate-rutile mixtures using boron. J. Am Ceram. Soc. 82(1999)3043–3048.
[253] Q. L. Zhang, H. Yang, J. L. Zou, and H. P. Wang. Sintering and microwave dielectric properties of LTCC zinc titanate multilayer. Mater. Lett. 59(2005)880–884.
[254] Y.-L. Chai, Y.-S. Chang, Y.-J. Hsiao, and Y.-C. Lian. Effects of borosilicate glass addition on the structure and dielectric properties of $ZnTiO_3$ ceramics. Mater. Res. Bull. 43(2008) 257–263.
[255] A. Chaouchi, M. Aliouat, S. Marinel, S. D. Astorg, and H. Bourahla. Effects of additives on the sintering temperature and dielectric properties of $ZnTiO_3$ based ceramic. Ceram. Int. 33(2007)245–258.
[256] H. T. Kim, S. Nahm, J. D. Byun, and Y. Kim. Low fired $(ZnMg)TiO_3$ microwave dielectrics. J. Am. Ceram. Soc. 82(999)3476–3480.
[257] Y.-C. Lee, W.-H. Lee, and F.-T. Shiao. Microwave dielectric properties of $Zn_{0.95}Mg_{0.05}TiO_3 + 0.25\ TiO_2$ ceramics with $3ZnO$–B_2O_3 addition. Jpn. J. Appl. Phys. 43(2004)7596–7599.
[258] Y–C. Lee, and W.-H. Lee. Effect of glass addition on the microwave dielectric properties of $Zn_{0.95}Mg_{0.05}TiO_3 + 0.25TiO_2$ ceramics. Jpn. J. Appl. Phys. 44(2005)1838–1843.
[259] K. P. Surendran and M. T. Sebastian. (1–x)$MgAlO_4$-xTiO_2 dielectrics for microwave and millimeterwave applications. Appl. Phys. A. 81(2005)823–826.
[260] K. P. Surendran, M. T. Sebastian, M. V. Manjusha, and J. Philip. A low loss dielectric substrate in $ZnAl_2O_4$–TiO_2 for microelectronic applications. J. Appl. Phys. 98(2005)044101.
[261] N. Mori, Y. Sugimoto, J. Harada, and Y. Higuchi. Dielectric properties of new glass-ceramics of LTCC applied to microwave or millimeter wave frequencies. J. Eur. Ceram. Soc. 26(2006)1925–1928.
[262] N. Mori, Y. Sugimoto, J. Harada, and H. Takagi. Dielectric properties of new glass-ceramics for LTCC applied to microwave or millimeter wave frequencies. J. Ceram. Soc. Jpn. Supl. 112–1(2004)S1563–S1566.
[263] S. Thomas and M. T. Sebastian. Effect of B_2O_3–Bi_2O_3–SiO_2–ZnO glass on the sintering and microwave dielectric properties of $0.83ZnAl_2O_4$–$0.17TiO_2$. Mater. Res. Bull. 43(2008)843–851.
[264] J.-M. Yoon, J.-A. Lee, J.-H. Lee, J.-J. Kim, and S.-H. Cho. Sintering behaviour and microwave dielectric characteristics of BaO-Sm_2O_3-$4TiO_2$ ceramics with B_2O_3 and BaB_2O_4 additives. J. Eur Ceram. Soc. 26(2006)2129–2133.

[265] J.-B. Lim, K.-H. Cho, S. Nahm, J.-H. Paik, and J.-H. Kim. Effect of $BaCu(B_2O_5)$ on the sintering temperature and microwave dielectric properties of $BaO–Ln_2O_3–TiO_2$ (Ln = Sm, Nd) ceramics. Mater Res. Bull. 41(2006)1868–1874.
[266] I.-S. Cho, D.-W. Kim, J.-R. Kim, and K. S. Hong. Low temperature sintering and microwave dielectric properties of $BaO–(Nd_{1-x}Bi_x)_2O_3–4TiO_2$ by the glass additions. Ceram. Int. 30(2004)1181–1185.
[267] C. F. Yang, L. Wu, and T. S. Wu. Effect of CuO on the sintering and dielectric characteristics of $(Ba_{1-x}Sr_x)(Ti_{.9}Zr_{.1})O_3$ ceramics. J. Mater. Sci. 27(1992)6573–6578.
[268] C.-C. Cheng, T- Hsieh, and I.-N. Lin. Effects of composition on low temperature cofireable Ba-Nd-Sm-Ti-O microwave dielectric materials. J. Eur. Ceram. Soc. 24(2004) 1787–1790.
[269] C.-C. Cheng, T.-E. Hsieh, and I.-N. Lin. Microwave dielectric properties of glass ceramic composites for low temperature cofireable ceramics. J. Eur. Ceram. Soc. 23(2003) 2553–2558.
[270] S. Hwang and H. Kim. Effect of frit size on microwave properties of glass-ceramic in low temperature cofired ceramics. Thermochemica Acta 455(2007)119–122.
[271] H. Ohsato. Research and development of microwave dielectric ceramics for wireless communications. J. Ceram. Soc. Jpn. 113(2006)703–711.
[272] C.-C. Cheng, T.-E. Hsieh, and I.-N. Lin. The effect of composition on Ba-Nd-Sm-Ti-O microwave dielectric materials for LTCC application. Mater. Chem. Phys. 79(2003) 119–123.
[273] W. S. Brower and P. H. Fang. Dielectric constants of $PbMoO_4$ and $CaMoO_4$. Phys. Rev. 149(1966)646–646.
[274] M. Nakano, K. Suzuki, T. Miura, and M. Kobayashi. Low temperature fireable dielectric material $Pb(Fe_{2/3}W_{1/3})O_3$- $(PbCa)(Fe_{1/2}Nb_{1/2})O_3$ for microwave use. Jpn. J. Appl. Phys. 32(1993)4314–4318.
[275] J. Kato, H. Kagata, and K. Nishimoto. Dielectric properties of (Pb, Ca) $(MeNb)O_3$ at microwave frequencies. Jpn. J. Appl. Phys. 31(1992)3144–3147.
[276] S. Kucheiko, J. W. Choi, H. J. Kim, S. J. Yoon, and H. J. Jung. Microwave characteristics of (Pb, Ca) (Fe, Nb, Sn)O_3 dielectric materials. J. Am. Ceram. Soc. 80(1997)2937–2940.
[277] J.-Y. Ha, J.-W. Choi, H.-J. Kim, S.-J. Yoon, and K. H. Yoon. The effect of sintering aid $CuO-Bi_2O_3$ on microwave dielectric properties of $(Pb_{0.45}Ca_{0.55})[(Fe_{0.5}Nb_{0.5})_{0.9}Sn_{0.1}]O_3$ ceramics. Mater. Chem. Phys. 79(2003)261–265.
[278] Q.H. Yang, E. S. Kim, Y. J. Kim, and P. K. Kim. Effect of $PbO–B_2O_3–V_2O_5$ glass on the microwave dielectric properties of (Pb,Ca,La)(Fe,Nb)O_3 ceramics. Mater. Chem. Phys. 79(2003)236–238.
[279] C.-H. Wei and J.-H. Jean. Low fire processing $(Ca_{1-x}\,Nd_{2x/3})TiO_3$ microwave ceramics. J. Am. Ceram. Soc. 86(2003)93–98.
[280] H. M. Zhe, Z. D. Xiang, G. H. Shuang, W. Hão, Z. Tianjing, and Z. Bin. Influence of Bi_2O_3 and MnO_2 doping on microwave properties of [(Pb,Ca)La](Fe,Nb)O_3 dielectric ceramics. Mater. Sci. Eng. B. 117(2005)199–204.
[281] J.-W. Choi, C.-Y. Kang, S.-J. Yoon, H.-J. Kim, H.-J. Jung, K. H. Yoon, H. J. Kim, H. J. Jung, and K. H. Yoon. Microwave dielectric properties of $Ca[(Li_{1/3}Nb_{2/3})_{1-x}\,Mi_x]O_{3-d}$ [M = Sn, Ti] ceramics. J. Mater. Res. 14(1999)3567–3570.
[282] P. Liu, E. S. Kim, and K. H. Yoon. Low temperature sintering and microwave dielectric properties of $Ca(Li_{1/3}Nb_{2/3})O_{3-d}$ ceramics. Jpn. J. Appl. Phys. 40(2001)5769–5773.
[283] P. Liu, H. Ogawa, E. S. Kim, and A. Kan. Low temperature sintering and microwave dielectric properties of $Ca(Li_{1/3}Ta_{2/3})O_{3-d}$-$CaTiO_3$ ceramics. J. Eur. Ceram. Soc. 23(2003) 2417–2421.
[284] J. W. Choi, C.Y. Kang, S. J. Yoon, H. J. Kim, and K. H. Yoon. Microwave dielectric properties of B_2O_3 doped $Ca[(Ca_{1/3}Nb_{2/3})_{0.9}Ti_{0.1}]O_{3-\delta}$ ceramics. Ferroelectrics 262(2001)167–172.
[285] P. Liu, E. S. Kim, S. G. Kang, and H. S. Jang. Microwave dielectric properties of $Ca[(Li_{1/3}\,Nb_{2/3})_{1-x}Ti_x]O_{3-d}$ ceramics with B_2O_3. Mater. Chem. Phys. 79(2003)270–272.

[286] J.-Y. Ha, J.-W. Choi, S.-J. Yoon, D. J. Choi, K. H. Yoon, and H.-J. Kim. Microwave dielectric properties of Bi_2O_3 doped $Ca[(Li_{1/3}Ta_{2/3})_{1-x} Ti_x]O_{3-d}$ ceramics. J. Eur. Ceram. Soc. 23(2003)2413–2416.
[287] J.-Y. Ha, J.-W. Choi, C.-Y. Kang, S.-J. Yoon, D. J. Choi, and H. J. Kim. Low temperature sintering of $Ca[(Li_{1/3}Nb_{2/3})_{1-x}Ti_x]O_{3-\delta}$ based dielectric ceramics with glass frit. Jpn. J. Appl. Phys. 44(2005)1322–1325.
[288] Y. Yoshida, N. Hara, T. Takada, and A. Seki. Structure and dielectric properties of $Ca_{1-x}Nd_{2x/3}TiO_3$. Jpn. J. Appl. Phys. P 36(1997)6818–6823.
[289] E. S. Kim, B. S. Chun, J. D. Kim, and K. H. Yoon. Low temperature sintering and microwave dielectric properties of $[Ca_{0.6}(Li_{0.5}Nd_{0.5})_{0.4}]_{0.45}Zn_{0.65}TiO_3$ ceramics. Mater. Sci. Eng. B. 99(2003)243–246.
[290] S. Nishigaki, S. Yano, H. Kato, T. Hirai, and T. Nomura. BaO-TiO_2-WO_3 microwave ceramics and crystalline $BaWO_4$. J. Am. Ceram. Soc. 71(1988)C11–C16.
[291] S.-F. Wang, C.-C. Chiang, and C.-H. Wang. Effects of B_2O_3 on the phase formation of $Ba_2Ti_9O_{20}$ ceramics. Mater. Chem. Phys. 79(2003)256–260.
[292] W. Huang, K.-S. Liu, L.-W. Chu, G.-H. Hsiue, and I.-N. Lin. Microwave dielectric properties of glass-ceramic composites for low temperature cofireable ceramics. J. Eur. Ceram. Soc. 23(2003)2559–2564.
[293] J.-A. Lee, J.-H. Lee, and J.-J. Kim. Low temperature sintering and microwave dielectric characteristics of $Ba_2Ti_9O_{20}$ ceramics. J. Eur. Ceram. Soc. 26(2006)2111–2115.
[294] Y.-C. Lee, W.-H. Lee, and F.-S. Shieu. Effects of B_2O_3 addition on the microwave dielectric properties of $Ba_2Ti_9O_{20}$ ceramics. Jpn. J. Appl. Phys. 42(2003)1311–1314.
[295] Y.-C. Lee, W.-H. Leeb, and F.-S. Shiuc. Microwave dielectric properties and microstructures of $Ba_2Ti_9O_{20}$ based with $3ZnO$–B_2O_3 addition. J. Eur. Ceram. Soc. 25(2005) 3459–3468.
[296] Y.-C. Lee, W.-H. Lee, and F.-S. Shieu. Low temperature sintering and microwave dielectric properties of $Ba_2Ti_9O_{20}$ ceramics with $3ZnO$–B_2O_3 addition. Jpn. J. Appl. Phys. 41(2002)6049–6053.
[297] M.-H. Weng and C.-L. Huang. Single phase $Ba_2Ti_9O_{20}$ microwave dielectric ceramics prepared by low temperature liquid phase sintering. Jpn. J. Appl. Phys. 39(2000)3528–3529.
[298] C.-L. Huang, M.-H. Weng, C.-T. Lion, and C.-C. Wu. Low temperature sintering and microwave dielectric properties of $Ba_2Ti_9O_{20}$ ceramics using glass additions. Mater. Res. Bull. 35(2000)2445–2456.
[299] Y.-G. Choi, J.-H. Park, W.-J. Ko, J.-H. Park, S. Nahm, and J.-G. Park. Low temperature sintering of $BaTi_4O_9$ based middle k dielectric composition for LTCC applications. J. Electroceram. 14(2005)157–162.
[300] D. W. Kim, D.-G. Lee, and K. S. Hong. Low temperature firing and microwave dielectric properties of $BaTi_4O_9$ with Zn–B–O glass system. Mater. Res. Bull. 36(2001)585–595.
[301] S. G. Lu, K. W. Kwok, H. L. W. Chen, and C. L. Choy. Structural and electrical properties of $BaTi_4O_9$ microwave ceramics incorporated with glass phase. Mater. Sci. and Eng. B 99 (2003)491–494.
[302] N. Krishnamurti and C. Calvo. Refinement of the structure of $Mg_3(VO_4)_2$. Can. J. Chem. 49(1971)1629–1637.
[303] R. Umemura, H. Ogawa, H. Ohsato, A. Kan, and A. Yoko., Microwave dielectric properties of low temperature sintered $Mg_3(VO_4)_2$ ceramics. J. Eur. Ceram. Soc. 25(2005) 2865–2870.
[304] R. Umemura, H. Ogawa, A. Yokoi, H. Ohsato, and A. Kan. Low temperature sintering-microwave dielectric property relations in $Ba_3(VO_4)_2$ ceramic. J. Alloys Compd. 424 (2006) 388–393.
[305] R. C. Kerby and J. R. Wilson. Solid-liquid phase equilibria for the ternary system V_2O_5–Na_2O–, V_2O_5–Na_2O–Cr_2O_3, V_2O_5–Na_2O–MgO. Can. J. Chem. 51(1973)1032–1040.
[306] H. J. Lee, K. S. Hong, S. J. Kim, and I. T. Kim. Dielectric properties of MNb_2O_6 compounds (M = Ce, Mn, Co, Ni or Zn). Mater. Res. Bull. 32(1997)847–855.

[307] D.-W. Kim, K. H. Ko, and K. S. Hong. Influence of copper (II) oxide additions to zinc niobate microwave ceramics on sintering temperature and dielectric properties. J. Am. Ceram. Soc. 84(2001)1286–1290.

[308] S.-H. Wee, D.-W. Kim, S.-I. Yoo, and K. S. Hong. Low temperature sintering of V_2O_5 added and substituted $ZnNb_2O_6$ microwave ceramics. Jpn. J. Appl. Phys. 43(2004) 3511–3515.

[309] J. Wang, Z. Yue, Z. Gui, and L. Li. Low temperature sintered $Zn(Nb_{1-x}V_{x/2})_2O_{6-2.5x}$ microwave dielectric ceramics with high Q value for LTCC applications. J. Alloys Compd. 392(2005)263–267.

[310] D. W. Kim, D. Y. Kim, and K. S. Hong. Phase relations and microwave dielectric properties of $ZnNb_2O_6$–TiO_2. J. Mater. Res. 15(2000)1331–1335.

[311] D.-W. Kim, H. B. Hong, and K. S. Hong. Structural transitions and microwave dielectric properties of $ZnNb_2O_6$-TiO_2 sintered at low temperatures. Jpn. J. Appl. Phys. 41(2002) 1465–1469.

[312] Y.-C. Lee, C.-H. Lin, and I.-N. Lin. Bi–Zn–Nb–O microwave dielectric materials for multilayer filter applications. Mater. Chem. Phys. 79(2003)124–128.

[313] R. C. Pullar, C. Lai, F. Azough, R. Freer, and N. Mc N. Alford. Novel microwave dielectric LTCCs based on V_2O_5 doped $M^{2+}Cu_2Nb_2O_6$ compounds (M^{2+} = Zn, Co, Ni, Mg and Ca). J. Eur. Ceram. Soc. 26(2006)1943–1946.

[314] H. Sreemoolanathan, M. T. Sebastian, and P. Mohanan. High permittivity and low loss ceramics in the BaO–SrO–Nb_2O_5 system. Mater. Res. Bull. 30(1995)653–658.

[315] R. Ratheesh, M. T. Sebastian, P. Mohanan, M. E. Tobar, J. Hartnett, R. Woode, and D. G. Blair. Microwave characterisation of $BaCe_2Ti_5O_{15}$ and $Ba_5Nb_4O_{15}$ ceramic dielectric resonators using whispering gallery mode method. Mater. Lett. 45(2000)279–285.

[316] D.-W. Kim, J.-R. Kim, S.-H. Yoon, K. S. Hong, and C. K. Kim. Microwave dielectric properties of low fired $Ba_5Nb_4O_{15}$. J. Am. Ceram. Soc. 85(2002)2759–2762.

[317] D.-W. Kim, K. S. Hong, C.S. Yoon, and C. K. Kim. Low temperature sintering and microwave dielectric properties of $Ba_5Nb_4O_{15}$-$BaNb_2O_6$ mixtures for LTCC applications. J. Eur. Ceram. Soc. 23(2003)2597–2601.

[318] D.-W. Kim, H. J. Yoon, K. S. Hong, and C. K. Kim. Microwave dielectric properties of (1–x) $Ba_5Nb_4O_{15}$-x$BaNb_2O_6$ mixtures. Jpn. J. Appl. Phys. 41(2002)3812-3816.

[319] C.-S. Chen, C.-C. Chou, C.-S. Chen, and I.-N. Lin. Microwave dielectric properties of glass-MCT low temperature cofireable ceramics. J. Eur. Ceram. Soc. 24(2004)1795–1798.

[320] Q.-L. Zhang, H. Yang, and J.-X. Tong. Low temperature firing and microwave dielectric properties of $MgTiO_3$ ceramics with Bi_2O_3-V_2O_5. Mater. Lett. 60(2006)1188–1191.

[321] H.-K. Shin, H. Shin, H. S. Jung, S.-Y. Cho, J.-R. Kim, and K. S. Hong. Role of lithium borosilicate glass in the decomposition of $MgTiO_3$ based dielectric ceramic during sintering. Mater. Res. Bull. 41(2006)1206–1214.

[322] H. Jantunen, R. Rautioaho A. Uusumaki, and S. Leppavuori. Preparing low loss low temperature cofired ceramic material without glass addition. J. Am. Ceram. Soc. 83(2000)2855–2857.

[323] H. Jantunen, A. Uusumaki, R. Rautioaho, and S. Leppavuori. Temperature coefficient of microwave resonance frequency of a low temperature cofired (LTCC) system. J. Am. Ceram. Soc. 85(2002)697–699.

[324] H. Jantunen, A. Uusumaki, R. Rautioaho, and S. Leppavuori. Compositions of $MgTiO_3$-$CaTiO_3$ ceramic with two borosilicate glasses for LTCC technology. J. Eur. Ceram. Soc. 20(2000)2331–2336.

[325] T. Hu, R. Rautioaho, H. Jantunen, S. Leppavuori, K. Soponmanee, and S. Sirisoonthorn. Optimisation of $MgTiO_3$–$CaTiO_3$ based LTCC tapes containing B_2O_3 for use in microwave applications. Ceram. Int. 31(2005)85–93.

[326] Y.-J. Choi, J.-H. Park, J.-H. Park, and J.-G. Park. Middle permittivity LTCC dielectric composition with adjustable temperature coefficient. Mater. Lett. 58(2004)3102–3106.

[327] A. Yokoi , H. Ogawa, A. Kan, H. Ohsato, and Y. Higashida. Use of LiF to achieve a low temperature cofired ceramics (LTCC) with low dielectric loss. J. Ceram. Soc. Jpn. 112(2004)S1633–S1636.

[328] A. Kan , H. Ogawa, A. Yokoi, and H. Ohsato. Low temperature sintering and microstructure of $Mg_4((Nb_{2-x}V_x)O_9$ microwave dielectric ceramic by V substitution for Nb. Jpn. J. Appl. Phys. 42(2003)6154–6157.

[329] A. Yokoi , H. Ogawa, A. Kan, H. Ohsato, and Y. Higashida. Microwave dielectric properties of $Mg_4Nb_2O_9$-3 wt% LiF ceramics prepared with $CaTiO_3$ additions. J. Eur. Ceram. Soc. 25(2005)2871–2875.

[330] S. Nomura. Ceramics for microwave dielectric resonator. Ferroelectrics 49(1983)61–70.

[331] T. Kolodiazhnyi, A. Petric, A. Belous, V. Yunov, and O. Yancheevskij. Synthesis and dielectric properties of barium tantalates and niobates with complex perovskite structure. J. Mater. Res. 17(2002)3182–3189.

[332] J. B. Lim, J.-O. Son, S. Nahm, W. S. Lee, M. J. Yoo, N. K. Kang, H. J. Lee, and Y. S. Kim. Low temperature sintering of B_2O_3 added $Ba(Mg_{,1/3}Nb_{2/3})O_3$ ceramics. Jpn. J. Appl. Phys. 43(2004)5388.

[333] J.-B. Lim, D.-H. Kim, S. Nahm, J.-H. Paik, and H.-J. Lee. Effect of B_2O_3 and CuO additives on the sintering temperature and microwave dielectric properties of $Ba(Mg_{1/3}Nb_{2/3})O_3$ ceramics. Mater. Res. Bull. 41(2006)1199–1205.

[334] T. Takada, S.-F. Wang, S. Yoshikawa, S.-J. Jang, and R. E. Newnham. Effect of glass additions on $(Zr,Sn)TiO_4$ for microwave applications. J. Am. Ceram. Soc. 77(1994) 2485–88.

[335] J.-H. Jean and S.-S. Lin. Low fire processing of ZrO_2–SnO_2–TiO_2 ceramic. J. Am. Ceram. Soc. 83(2000)1417–1422.

[336] C.-L. Huang and M.-H. Weng. Liquid phase sintering of $(Zr,Sn)TiO_4$ microwave dielectric ceramics. Mater. Res. Bull. 35(2000)1881–1888.

[337] C.-L. Huang, M.-H. Weng, C.-C. Wu, and C.-C. Wei. Microwave dielectric properties and microstructure of V_2O_5-modified $(Zr_{.8}Sn_{.2})TiO_4$ ceramics. Jpn. J. Appl. Phys. 40 (2001)698–702.

[338] C.-L. Huang, M.-H. Weng, and H.-L. Chen. Effects of additives on microstructures and microwave dielectric properties of $(Zr,Sn)TiO_4$ ceramics. Mater. Chem. Phys. 71(2001) 17–22.

[339] Y.-H. Wang, S.-F. Wang, and C.-K. Wen. Low fire of $(Zr_{.8}Sn_{.2})TiO_4$ with glass additives. Mater. Sci. Eng. A 426(2006)143–146.

[340] S. X. Zhang, J. B. Li, and H. Z. Zhai. Low temperature sintering and dielectric properties of $(Zr_{.8}Sn_{.2})TiO_4$ microwave ceramics using La_2O_3–BaO additives. Mater. Chem. Phys. 77(2002)470–475.

[341] M. H. Framcombe and B. Lewis. Structural and electrical properties of silver niobate and silver tantalite. Acta Cryst. 11(1958)175.

[342] M. Valant, D. Suvorov, and A. Meden. New high permittivity $AgNb_{1-x}Ta_xO_3$ microwave ceramics. Part-1. Crystal structures and phase-decomposition relations. J. Am. Ceram. Soc. 82(1999)81–87.

[343] M. Valant, D. Suvorov, and A. Meden. New high permittivity $AgNb_{1-x}Ta_xO_3$ microwave ceramics. Part-2. Dielectric characteristics. J. Am. Ceram. Soc. 82(1999)88–93.

[344] M. Valant, D. Suvorov, C. Hoffmann, and H. Sommariva. $Ag(NbTa)O_3$ based ceramics with suppressed temperature dependence of permittivity. J. Eur. Ceram. Soc. 21(2001)2647–51.

[345] Y. Sakabe, T. Takeda, Y. Ogio, and N. Wada. Dielectric and ferroelectric properties of $(Ag,Li)(Nb,Ta)O_3$ ceramics. Proc. of 10th US-Japanese Seminar on Dielectric and Piezoelectric Ceramics, Providence, RI, Sept 26–29, (2001) pp 63–66.

[346] H. T. Kim, T. R. Shrout, C. Rindall, and M. Lanagan. Low temperature sintering and dielectric properties of $Ag(Nb,Ta)O_3$ composite ceramics. J. Am. Ceram. Soc. 85(2002) 2738–2744.

[347] X. Guo, N. Zhu, M. Xiao, and X. Wu. Structural and dielectric properties of $Ag(Nb_{0.8}\text{-}Ta_{0.2})_{1-x}Sb_xO_3$ ($x \leq 0.08$) ceramics. J. Am. Ceram. Soc. 90(2007)2467–2471.
[348] N. C. Webb. The crystal structure of β-$Ca_2P_2O_7$. Acta Cryst. 21(1966)942–948.
[349] C. Calvo. The crystal structure of α-$Ca_2P_2O_7$. Inorg. Chem. 7(1968)1345–1351.
[350] P. W. Ranby, D. H. Mash, and S. H. Henderson. The investigation of phosphor with particular reference to pyrophosphates. Br. J. Appl. Phys. 4 supl. (1955)18–25.
[351] F. H. Lin, J. R. Liaw, M. H. Hon, and C. Y. Wang. The efftects of $Na_4P_2O_7$–$10H_2O$ bioceramics on the mechanical properties of sintered $Ca_2P_2O_7$ bioceramics. Mater. Chem. Phys. 41(1995)110–116.
[352] J.-J. Bian, D.-W. Kim, and K.S. Hong. Glass free LTCC microwave dielectric ceramics. Mater. Res. Bull. 40(2005)2120–2129.
[353] I.-S. Cho, S.-K. Kang, D.-W. Kim, and K. S. Hong. Mixture behaviour and microwave dielectric properties of $(1-x)Ca_2P_2O_7$–$xTiO_2$, J. Eur. Ceram. Soc. 26(2006)2007–2010.
[354] J.-J. Bian, D.-W. Kim, and K. S. Hong. Mixture behaviour and microwave dielectric properties of $(Ca_{1-x}Zn_x)_2P_2O_7$. Mater. Lett. 59(2005)257–260.
[355] G.-K. Choi, S.-Y. Cho, J.-S. An, and K. S. Hong. Microwave dielectric properties and sintering behaviors of scheelite compound $CaMoO_4$. J. Eur. Ceram. Soc. 26(2006) 2011–2015.
[356] G.-K. Choi, J.-R. Kim, S. H. Yoon, and K. S. Hong. Microwave dielectric properties of Scheelite (A = Ca, Sr, Ba) and wolframite (A = Mg, Zn, Mn) AMO_4 compounds. J. Eur. Ceram. Soc. 27(2007)3063–3067.
[357] J. S. Kim, J. C. Lee, C. I. Cheon, and C. H. Lee. New type of wolframite $LiYW_2O_8$ for LTCC applications. International conference. Microwave Materials and their Applications (MMA-2004), Inuyama Japan. (2004) page 108.
[358] E. S. Kim, S.-H. Kim, and B. I. Lee. Low-temperature sintering and microwave dielectric properties of $CaWO_4$ ceramics for LTCC applications. J. Eur. Ceram. Soc. 26(2006) 2101–2104.
[359] R. C. Pullar, S. Farrah, and N. Mc N. Alford. $MgWO_4$, $ZnWO_4$, $NiWO_4$, and $CoWO_4$ microwave dielectric ceramics. J. Eur. Ceram. Soc. 27(2007)1059–1063.
[360] S. H. Yoon, D.-W. Kim, S.-Y. Cho, and K. S. Hong. Investigation of the relations between structure and microwave dielectric properties of divalent metal tungstate compounds. J. Eur. Ceram. Soc. 26(2006)2051–2054.
[361] J.-J. Bian, D.-W. Kim, and K. S. Hong. Microwave dielectric properties of $A_2P_2O_7$ (A = Ca, Sr, Ba, Zn, Mn). Jpn. J. Appl. Phys. 43(2004)3521–3525.

CHAPTER THIRTEEN

Tailoring the Properties of Low-Loss Dielectrics

13.1 Introduction

The microwave properties of dielectric materials are usually tuned by chemical methods like doping, slight nonstoichiometry, formation of solid solution or mixtures of dielectrics with opposite τ_f [1–7]. Santiago and co-workers [3, 8] developed a mechanical compensation technique where a resonator is constructed using two closely spaced identical cylindrical sapphire disks. The thermal expansion of the copper post causes a widening of the gap between the discs, which modulates the resonant frequency with opposite temperature dependence to the permittivity. However, this type of compensation is inherently sensitive to vibrations. Luiten et al. [5, 9] used paramagnetic effects of impurity ions to compensate the permittivity–temperature dependence. But this technique is not applicable at liquid nitrogen and room temperature due to the finite energy gap of a paramagnetic resonance. Hartnett et al. [10, 11] proposed a method of compensating the frequency–temperature dependence of high Q monolithic sapphire resonators near liquid nitrogen temperature by doping single-crystal sapphire with Ti^{3+} ions. Lim et al. [12] reported another method to design a coaxial ceramic resonator, whose resonant frequency is unchanged with temperature using positive τ_f and negative τ_f materials. High-temperature stability of the resonant frequency is realized by obtaining the lengths to be filled with respective materials. Breeze et al. [13] reported a new method of achieving temperature compensation by coating a film of TiO_2 over the surface of alumina disk. The composite resonators obtained by firing at 1400°C showed a temperature compensation depending on the volume fraction of TiO_2. More recently, Ohsato and coworkers reported [14] that it is possible to tune the permittivity and τ_f by texturing. The permittivity and τ_f depend on the crystallographic direction and by preparing grain-oriented ceramics, the ε_r and τ_f can be tuned.

13.2 Solid Solution Formation

The MW dielectric properties especially τ_f can be tailored by forming a solid solution between the positive τ_f and a negative τ_f material. The compatibility of ionic radius, ionic charge and structure are the conditions required for the formation of solid solutions without much degradation of the required properties. The properties can be tuned between that of the end compositions. For example, a solid solution can be prepared between $PrTiNbO_6$ and $GdTiNbO_6$. $PrTiNbO_6$ is orthorhombic with the space group *Pnma* while $GdTiNbO_6$ is orthorhombic with the space group *Pbcn*. $PrTiNbO_6$ has a

Dielectric Materials for Wireless Communications
ISBN-13: 978-0-08-045330-9

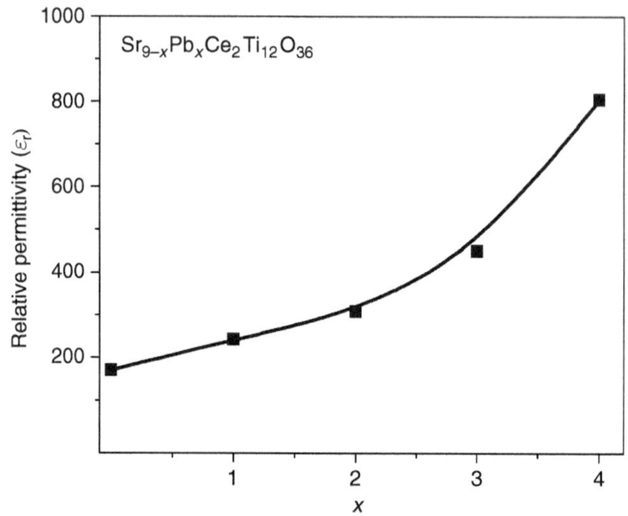

Figure 13.1 Variation of relative permittivity as a function of Pb substitution in $Sr_9Ce_2Ti_{12}O_{36}$ ceramics (after Ref. [16]).

$\varepsilon_r = 53$, $Qf = 12\ 300$ GHz and $\tau_f = 56$ ppm/°C. $GdTiNbO_6$ has $\varepsilon_r = 20$, $Qf = 9000$ GHz and $\tau_f = -52$ ppm/°C. Hence, it is possible to tailor the dielectric properties by forming a solid solution of $Pr_{1-x}Gd_xTiNbO_6$ [15]. The ε_r, Qf and τ_f varied in between those of the end members as shown in Figures 11.38 and 11.39 in Chapter 11. The ε_r, τ_f and Qf show a sudden change at $x = 0.8$ indicating structural transformation. The permittivity and τ_f decrease almost linearly up to 80 mol% substitution of Pr by Gd. Further substitution leads to sudden increase in ε_r and Qf and a decrease in τ_f. This sudden variation in ε_r, Qf and τ_f is due to the phase transformation from the space group *Pnma* to *Pbcn*. A linear variation in ε_r, Qf and τ_f may occur only if both the end member (positive and negative τ_f) materials have the same crystallographic structure.

Increasing the relative permittivity in dielectric ceramics is typically achieved by substitution of a cation of higher ionic polarizability and/or ionic radius. A good example is Pb substitution for Sr in Sr–Ce–TiO_2. Lead has a high dielectric polarizability and the permittivity of the material increases with increasing Pb substitution as shown in Figure 13.1 for the SrO–CeO_2–TiO_2 system [16]. Another example is the $BaZrO_3$–$BaTiO_3$ solid solution in which Zr substitution by Ti increases the B site polarizability with a commensurate increase in permittivity [17]. The ε_r depends on the dielectric polarizability and cell volume as discussed in Chapter 2.

13.3 Use of Additives

It is possible to tailor the MW properties of dielectric ceramics by adding suitable dopants. Addition of a small amount of suitable additives increases density, and leads to variations in the dielectric properties. Sebastian and coworkers doped $Ba(Mg_{1/3}Ta_{2/3})O_3$, $Ba(Zn_{1/3}Ta_{2/3})O_3$ (BZT) and $Ba(Zn_{1/3}Nb_{2/3})O_3$ with ions of different valance states and found that the quality factor increases when the dopant ionic radius is close to the ionic radius of B site ions [18–20] (see Figure 8.36 for $Ba(Mg_{1/3}Ta_{2/3})O_3$). The matching of the ionic size of the dopant with the ions of the parent material has significant influence than the valence state in improving the quality factor. Addition of small amount (<2 wt%) of low-loss glasses can not only improve densification but also improve the dielectric properties. However, addition of larger amount of glasses, although considerably lower the sintering temperature, degrade the MW dielectric properties. The glasses in general have a low permittivity and negative τ_f. Hence, addition of glasses lowers the ε_r and decreases the positive τ_f and alters the τ_f of negative τ_f materials. The effect of glass additives on sinterability and MW dielectric properties are described in detail in Chapter 12. Some of the materials have poor sinterability and such materials are usually sintered by adding sintering aids. The use of sintering aids also alters the MW dielectric properties. $MnCO_3$, CeO_2, ZnO, La_2O_3, Nb_2O_5, etc. are some of the common sintering aids for solid-state sintering, whereas CuO, B_2O_3, Bi_2O_3, V_2O_5, CuO–V_2O_5, $BaCu(B_2O_5)$, LiF, etc. are used as liquid-phase sintering aids.

13.4 Nonstoichiometry

The MW dielectric properties can also be tailored by allowing a slight nonstoichiometry [1, 21–26]. Historically, Desu and O'Bryan [21] made the first attempt to correlate the excellent MW dielectric properties of BZT ceramics with the B site cation nonstoichiometry. They reported that the escape of ZnO in BZT leads to an increase in the quality factor. They have found that the escape of ZnO leads to crystallographic distortion and improvement in the 1:2 ordering. The effect of a slight nonstoichiometry and chemical inhomogeneity on the order–disorder phase transformation was investigated in $Ba(Ni_{1/3}Nb_{2/3})O_3$ and $Ba(Zn_{1/3}Nb_{2/3})O_3$ [1]. Paik et al. [24] investigated the effect of Mg deficiency on the MW dielectric properties of $Ba(Mg_{1/3}Nb_{2/3})O_3$. They found an improvement in density and unloaded Q factor for $x=0.02$ in $Ba(Mg_{0.33-x}Nb_{0.67})O_3$ and was attributed to the enhanced grain boundary mass transport. Surendran et al. [22] found that a slight deficiency of Ba or Mg in $Ba(Mg_{1/3}Ta_{2/3})O_3$ increases the density, 1:2 ordering and the quality factor and are discussed in Chapter 8. They could also tune the ε_r and τ_f by Ba and Mg deficiency. Khalam and Sebastian [2] found that in $Sr(Eu_{1/2}Ta_{1/2})O_3$ a slight deficiency of Sr and a slight excess of Eu or Ta increase the quality factor and that the ε_r and τ_f also vary with nonstoichiometry. It may be noted that large nonstoichiometry leads to formation of secondary phases. Wu and Davies [25] found that a slight deficiency of Ba and Zn and excess of Nb in $Ba(Zn_{1/3}Nb_{2/3})O_3$ improve the quality factor. However, excess ZnO considerably lowers the *Qf* as discussed in Chapter 8. In zinc silicate, the nonstoichiometric composition $Zn_{1.8}SiO_{3.8}$ showed excellent properties such as $Qf=147\,000$ GHz, $\varepsilon_r=6.6$ and $\tau_f=-22$ ppm/°C when sintered at 1300°C [26]. The ε_r, *Qf* and τ_f vary as a function of x in $Zn_{2-x}SiO_{4-x}$ ceramics.

13.5 Stacked Resonators

Konishi et al. in 1971 [27] and Tsironis and Panker [28] outlined the possibility of obtaining temperature compensation by stacking two cylindrical resonators made of two different materials with opposite τ_f. Tobar and coworkers [4, 29, 30] showed that it is possible to annul the frequency–temperature dependence of single-crystal sapphire DRs using a dielectric of the opposite permittivity temperature dependence such as rutile or $SrTiO_3$. Tobar and coworkers developed a composite resonator structure for temperature compensation for WGMs in certain temperature ranges. In this, two very thin slices of rutile single crystal are clamped tightly against the upper and lower surfaces of a cylinder of sapphire [4, 29]. The rutile has permittivity and temperature coefficient of two orders of magnitude higher than that of sapphire and is of opposite sign. Thus the temperature compensation is achieved. The temperature compensation of the WG quasi-TM (transverse magnetic) modes were measured to be below 90 K with Q factors of the order of a few million depending on the mode. Using a piece of $SrTiO_3$ stacking over sapphire, Tobar et al. [30] succeeded in annulling the temperature dependence of resonant frequency in hybrid WGMs.

Recently, Sebastian et al. [31] tailored the MW dielectric properties by stacking resonators made from two different materials having τ_f of opposite sign. The ceramic pucks of one kind were placed over the other with axial symmetry. Figure 13.2 shows schematic sketch of stacking of resonators made of varying thickness. The DR samples of opposite τ_f were carefully prepared to have the same diameter and polished to minimize the air gap between the two DRs stacked together. They have used $Ba_5Nb_4O_{15}$ with $\varepsilon_r = 39$, $Qf = 25\ 000$ and $\tau_f = +78$ ppm/°C and $Sr(Y_{1/2}Nb_{1/2})O_3$ having $\varepsilon_r = 28$, $Qf = 40\,000$ GHz and $\tau_f = -58$ ppm/°C). The permittivity and τ_f varied with volume fraction of the positive and negative τ_f materials. Figure 13.3 shows the variation of effective permittivity and Qf as a function of volume fraction of $Sr(Y_{1/2}Nb_{1/2})O_3$. The permittivity decreased and Qf increased with volume fraction of $Sr(Y_{1/2}Nb_{1/2})O_3$. The τ_f decreased and became negative with increasing volume fraction of $Sr(Y_{1/2}Nb_{1/2})O_3$ as shown in Figure 13.4. The resonant frequency, permittivity and quality factor did not show any significant variation by the reversal of the stacking sequence (top and bottom). However, it was found that the DR material in the bottom of the stack has greater influence on the resultant τ_f.

In a similar way, Li et al. [32] stacked $MgTiO_3$ and $CaTiO_3$ which are having opposite sign of τ_f. With increasing thickness of $CaTiO_3$, the Qf decreased while effective permittivity and τ_f increased. By changing the thickness fraction of $CaTiO_3$, they could obtain a temperature-stable composite dielectric. They stacked two dielectric ceramics in different schemes such as $MgTiO_3/CaTiO_3$, $CaTiO_3/MgTiO_3$, $MgTiO_3/CaTiO_3/MgTiO_3$ and $CaTiO_3/MgTiO_3/CaTiO_3$. They also reported that finite element method could predict accurately the MW dielectric properties. The co-sintering of

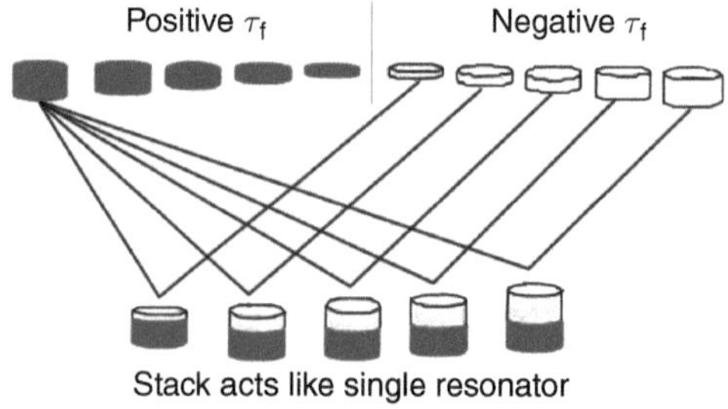

Figure 13.2 Schematic sketch of stacking of resonators of positive and negative τ_f made of varying thickness (after Ref. [31]).

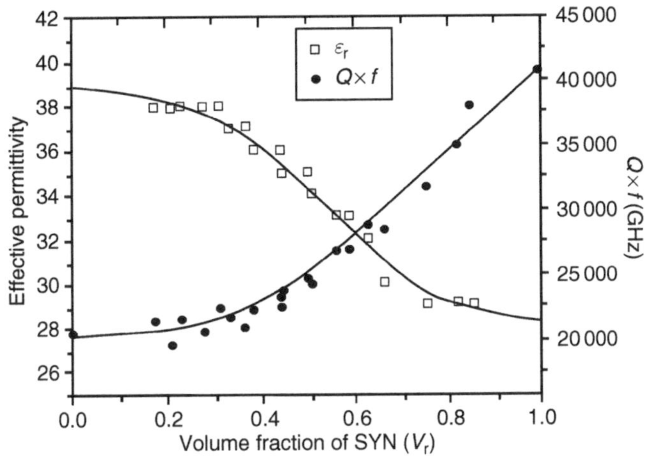

Figure 13.3 Variation of ε_r and Qf of $Ba_5Nb_4O_{15}$ ceramics as a function of volume fraction of stacked $Sr(Y_{1/2}Nb_{1/2})O_3$ (SYN) (after Ref. [31]).

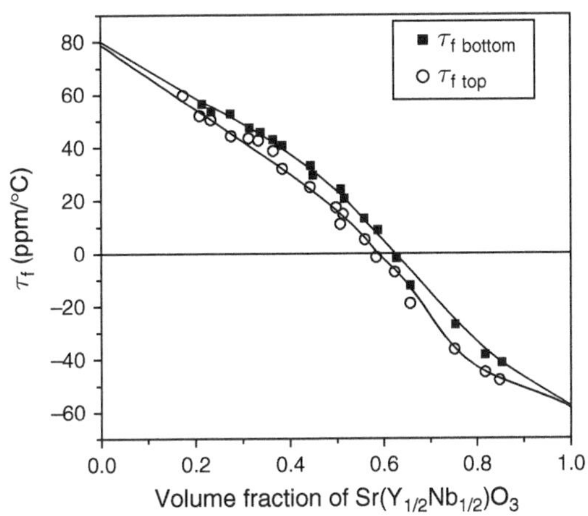

Figure 13.4 Variation of τ_f of $Ba_5Nb_4O_{15}$ ceramics as a function of volume fraction of stacked $Sr(Y_{1/2}Nb_{1/2})O_3$ (SYN) (after Ref. [31]).

MCT [$(Mg_{1-x}Ca_x)TiO_3$] leads to microcracking [33]. Li and Chen [34] reported the use of adhesive for bonding during stacking of resonators of opposite sign of τ_f considerably influence the quality factor, whereas the ε_r and τ_f are only slightly affected. The Qf decreases with increase in thickness of the adhesive. More recently, Zhou et al. [35] tailored the properties of $Bi_2(Zn_{2/3}Nb_{4/3})O_7$ by stacking with $BiNbO_4$ and then cofiring. To lower the sintering temperature, they partially substituted Nb with vanadium. The $BiNbO_4$ has $\varepsilon_r = 43$, $Q = 2500$ and $\tau_f = -3$ ppm/°C. $Bi_2(Zn_{2/3}Nb_{4/3})O_7$ has $\varepsilon_r = 76$, $Q \sim 900$ and $\tau_f = -78$ ppm/°C. The stacked composite structure was cosintered at 900°C. The sintered ceramics showed microcracks due to the unequal shrinkage of the two ceramic materials. The microstructure of the composite ceramics showed a narrow gap between the two ceramics. Although they could succeed in tailoring the ε_r, Qf and τ_f, the Qf degraded due to the presence of microcracks. The ε_r and τ_f varied in between the values of the end members as a function of the volume fraction of the two materials. The possibility of tuning the dielectric properties by stacking resonators of opposite τ_f is convenient for materials which do not form solid solution or a mixture but react to form new phases degrading the MW dielectric properties.

13.6 Tailoring the Properties by Mixture Formation

The dielectric properties can be conveniently tailored by forming mixture phases of two materials of different ε_r, Qf and τ_f. This method is possible only if the two materials do not react during heat treatment. Yoon et al. [36] added TiO_2 to $CaWO_4$ to tailor properties of $CaWO_4$ which has a negative τ_f of −53 ppm/°C. XRD as shown in Figure 13.5 reveals

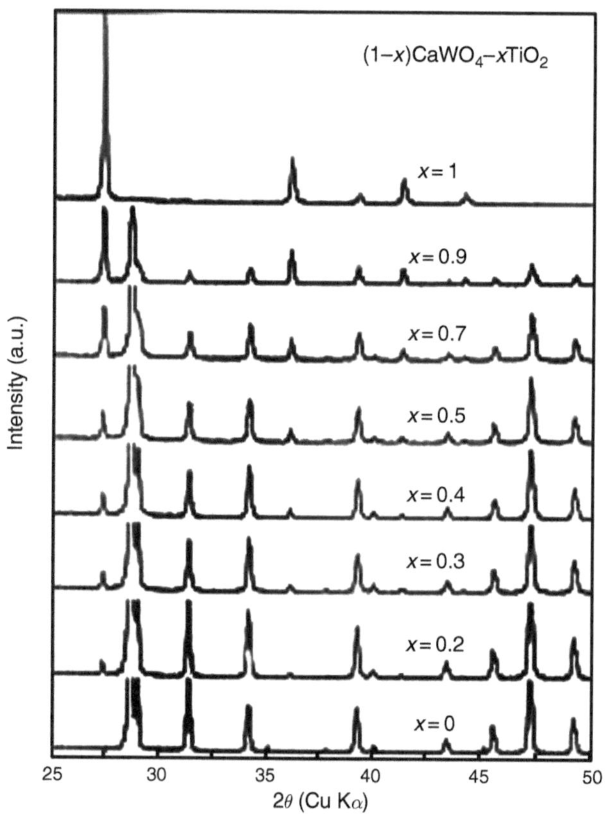

Figure 13.5 X-ray diffraction powder pattern recorded from sintered $(1-x)CaWO_4-xTiO_2$ mixture ceramics (after Ref. [36]).

the formation of mixtures of TiO_2 and $CaWO_4$. Figure 13.6 shows the backscattered image of a mixture of $0.5CaWO_4$–$0.5TiO_2$ mixture. The SEM picture shows the grains of both TiO_2 and $CaWO_4$. It is found that the τ_f varied linearly with the TiO_2 content as shown in Figure 13.7. The variation of ε_r is shown in Figure 13.8. The ε_r of the mixture varies in between the ε_r of the end compositions. The variation of the ε_r of the mixtures

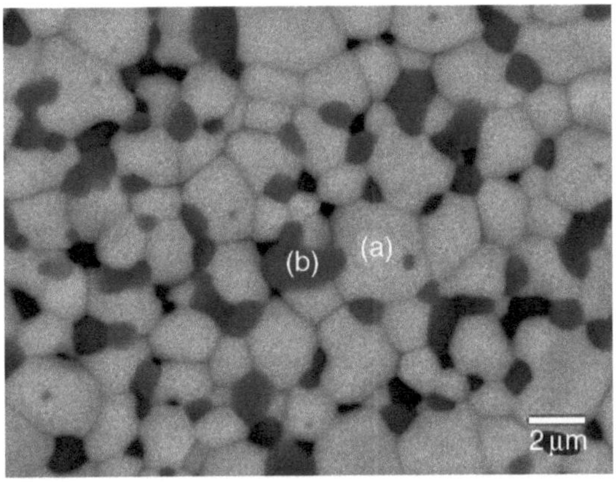

Figure 13.6 SEM backscattered image recorded from $0.5CaWO_4$–$0.5TiO_2$ ceramics. (a) $CaWO_4$ and (b) TiO_2 (after Ref. [36]).

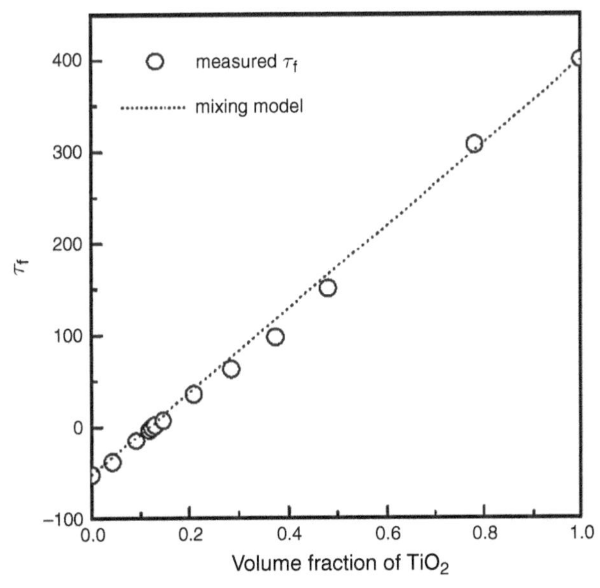

Figure 13.7 Variation of τ_f in $(1-x)CaWO_4$–$xTiO_2$ mixture ceramics as a function of volume fraction of TiO_2 (after Ref. [36]).

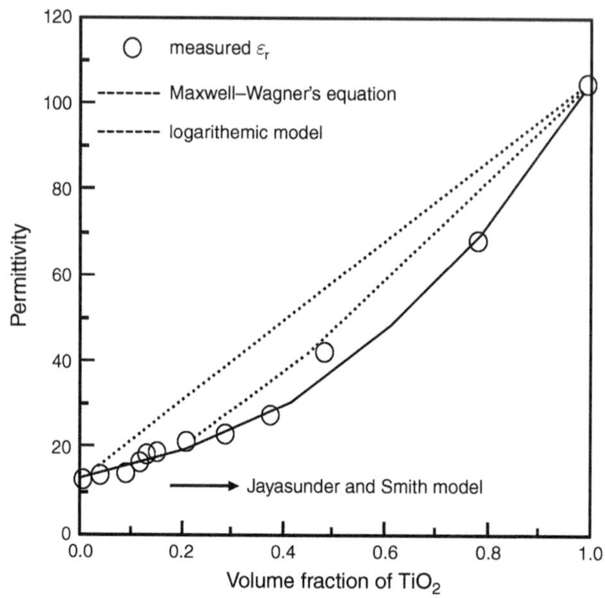

Figure 13.8 Variation of permittivity as a function of volume fraction of TiO_2 in (1–x) $CaWO_4$–$x TiO_2$ mixture ceramics (after Ref. [36]).

can be analyzed using different mixture rules. One of the commonly used mixture rule is the following Maxwell–Wagner equation [37]:

$$\varepsilon_c = V_1\varepsilon_1 + V_2\varepsilon_2$$

where V_1, V_2 and ε_1, ε_2 are the volume fractions and permittivities of phase 1 and phase 2, respectively. Lichtenecker suggested a logarithmic mixture rule model as given below [38]

$$\ln\varepsilon_c = V_1 \ln\varepsilon_1 + V_2 \ln\varepsilon_2$$

Recently, Jayasundere and Smith [39] and Poon and Shin [40] proposed another mixture rule model which considers the interactive effects between the fields of neighboring spheres. This model predicts the following variation in the permittivity of the mixed ceramic composition:

$$\varepsilon_c = \frac{\varepsilon_1 V_2 + \varepsilon_2 V_2[3\varepsilon_1/(\varepsilon_2+2\varepsilon_1)][1+3V_2(\varepsilon_2-\varepsilon_1)/(\varepsilon_2+2\varepsilon_1)]}{V_1 + V_2(3\varepsilon_1)/(\varepsilon_2+2\varepsilon_1)[1+3V_2(\varepsilon_2-\varepsilon_1)/(\varepsilon_2+2\varepsilon_1)]}$$

It is found [36] that the Jayasundere–Smith model gives better fit for the permittivity of the mixture ceramics as shown in Figure 13.8.

The variation of the τ_f with the volume fraction is shown in Figure 13.7, which is in agreement with that predicted by the mixture rule [38, 41]

$$\tau_f(mixture) = V_1\tau_{f1} + V_2\tau_{f2}$$

It is found that the variation in Qf does not follow any mixture rule. The composition $0.74CaWO_4$–$0.26TiO_2$ dielectric ceramic showed $\varepsilon_r = 17.5$, $Qf = 27\,000$ with $\tau_f \sim 0$ ppm/°C. The tailoring of dielectric properties by mixture formation has been reported in several systems such as Bi_2O_3–TiO_2 [41], $BaTi_4O_9$–CeO_2 [42], $ZnAl_2O_4$–TiO_2 [7], $MgAl_2O_4$–TiO_2 [43] and $MgSiO_4$–TiO_2 [44]. It may be noted that the MW dielectric properties will also vary with preparation conditions such as calcination and sintering temperatures and their durations as described in Chapter 1.

References

[1] K. S. Hong, I. T. Kim, and S. J. Yoon. Effect of non-stoichiometry and chemical inhomogeneity on the order-disorder phase transformation in the complex perovskite compounds $Ba(Ni_{1/3}Nb_{2/3})O_3$ and $Ba(Zn_{1/3}Nb_{2/3})O_3$. J. Mater. Sci. 30(1995)514–521.

[2] L. A. Khalam and M. T. Sebastian. Effect of cation substitution and non-stoichiometry on the microwave dielectric properties of $Sr(B'_{1/2}B''_{1/2})O_3$ [B′ = lanthanides] perovskites. J. Am. Ceram. Soc. 89(2007)3689–3695.

[3] S. G. Santiago, R. T. Wang, and G. J. Dick. Improved performance of a temperature compensated LN_2 cooled sapphire oscillator. Proc. IEEE Int. Freq. Control Symp. (1995)397–400.

[4] J. G. Hartnett, M. E. Tobar, E. N. Ivanov, and D. Cross. High frequency compensated sapphire/rutile resonator. Electron. Lett. 36(2000)726–727.

[5] A. N. Luitenen, A. G. Mann, and D. G. Blair. Paramagnetic susceptibility and permittivity measurements at microwave frequencies in cryogenic sapphire resonators. J. Phys. D. Appl. Phys. 29(1996)2082–2090.

[6] J. Krupka, D. Cross, A. N. Luitenen, and M. E. Tobar. Design of very high Q sapphire resonators. Electron. Lett. 32(1996)670–671

[7] K. P. Surendran, N. Santha, P. Mohanan, and M. T. Sebastian. Temperature stable low loss ceramic dielectrics in $(1-x)ZnAl_2O_4$–$xTiO_2$ system for microwave substrate applications. Eur. Phys. J. B 41(2004)301–306.

[8] G. J. Dick, D. G. Santiago, and R. T. Wang. Temperature compensated sapphire resonator for ultra-stable oscillator capability at temperatures above 77 Kelvin. Proc. IEEE Int. Freq. Control Symp. (1994)421–431.

[9] A. N. Luiten, A G Mann, N. J. Mc Donald, and D. G. Blair. Latest results of the UWA cryogenic sapphire oscillator. Proc. Freq. Control Symp. (1995)433–437.

[10] J. G. Hartnett, M. E. Tobar, A. G. Mann, J. Krupka, and E. N. Ivanov. Temperature dependence of Ti^{3+} doped sapphire whispering gallery mode resonator. Electron. Lett. 34(1998)195–196.

[11] J. G. Hartnett, M. E. Tobar, A. G. Mann, E. N. Ivanov, J. Krupka, and R. Geyer. Frequency-temperature compensation in Ti^{4+} doped sapphire whispering gallery mode resonators. IEEE Trans. Ultrason. Ferroelectr. Freq. Control 46(1999)993–999.

[12] S.-K. Lim, H.-Y. Lee, J.-C. Kim, and C. An. The design of a temperature stable stepped impedance resonator using composite ceramic materials. IEEE Microw. Guided Wave Lett. 9(1999)143–144.

[13] J. Breeze, S. J. Penn, M. Poole, and N. Mc N. Alford. Layered Al_2O_3–TiO_2 composite dielectric resonators. Electron. Lett. 36 (2000)883–884.

[14] K. Wada, K. -Kakimoto, and H. Ohsato. Anisotropic microwave dielectric properties of textured $Ba_4Sm_{9.33}Ti_{18}O_{54}$ ceramics. Key Eng. Mater. 269(2004)207–210.

[15] K. P. Surendran, M. R. Varma, P. Mohanan, and M. T. Sebastian. Microwave dielectric properties of $RE_{1-x}RE'_xTiNbO_6$ [RE = Pr, Nd, Sm; RE' = Gd, Dy, Y] ceramics. J. Am. Ceram. Soc. 86(2003)1695–1699.

[16] G. Subodh, M. T. Sebastian, S. Kamba, and J. Petzelt. (to be published).

[17] I. M. Reaney and D. Iddles. Microwave dielectric ceramics for resonators and filters in mobile phone networks. J. Am. Ceram. Soc. 89(2006)2063–2072.

[18] M. R. Varma, R. Reghunandan, and M. T. Sebastian. Effect of dopants on microwave dielectric properties of $Ba(Zn_{1/3}Ta_{2/3})O_3$ ceramics. Jpn. J. Appl. Phys. 44(2005)298–303.
[19] K. P. Surendran, M. T. Sebastian, P. Mohanan, and M. V. Jacob. The effect of dopants on the microwave dielectric properties of $Ba(Mg_{0.33}Ta_{0.67})O_3$ ceramics. J. Appl. Phys. 8(2005)084114.
[20] M. R. Varma and M. T. Sebastian. Effect of dopants on microwave dielectric properties of $Ba(Zn_{1/2}Nb_{2/3})O_3$ ceramics. J. Eur. Ceram. Soc. 27(2007)2827–2873.
[21] S. B. Desu and H. M. O'Bryan. Microwave loss quality of $BaZn_{1/3}Ta_{2/3}O_3$ ceramics. J. Am. Ceram. Soc. 68(1985)546–551.
[22] K. P. Surendran, P. Mohanan, M. T. Sebastian, R. L. Moreira and A. Dias. Effect of non-stoichiometry on the structure and microwave dielectric properties of $Ba(Mg_{1/3}Ta_{2/3})O_3$. Chem. Mater. 17(2005)142–151.
[23] S. Kawashima. Influence of ZnO evaporation on microwave dielectric loss and sinterablity of $Ba(Zn_{1/3}Ta_{2/3})O_3$ ceramics. Am. Ceram. Soc. Bull. 72(1993)120–126.
[24] J. H. Paik, S. Nahm, J. D. Byun, M. H. Kim, and H. J. Lee. The effect of Mg deficiency on the microwave dielectric properties of $Ba(Mg_{1/3}Nb_{2/3})O_3$ ceramics. J. Mater. Sci. Lett. 17(1998)1777–1780.
[25] Wu and P. K. Davies. Influence of non-stoichiometry on the structure and properties of $Ba(Zn_{1/3}Nb_{2/3})O_3$ microwave dielectrics. IV. Tuning Tf and the part size dependence of Qxf. J. Am. Ceram. Soc. 89(2007)2271–2278.
[26] N.-H. Nguyen, J.-B. Lim, S. Nahm, J.-H. Paik, and J. H. Kim. Effect of Zn/Si ratio on the microstructural and microwave dielectric properties of Zn_2SiO_4 ceramics. J. Am. Ceram. Soc. 90(2007)3127–3130.
[27] Y. Konishi. Microwave dielectric resonator (in Japanese). NHK (Nippon Hoso Kyokai), Tokyo. (Technical Report 1971).
[28] C. Tsironis and V. Panker. Temperature stabilization of GaAs MESFET oscillator using dielectric resonator. IEEE MTT-3 (1983)312–314.
[29] M. E. Tobar, J. Krupka, J. G. Hartnett, E. N. Ivanov, and R. A. Woode. IEEE Trans. Ultrason., Ferroelectr. Freq. Control 45(1998)830.
[30] M. E. Tobar, J. Krupka, E. N. Ivanov, and R. A. Woode. Dielectric frequency temperature compensated microwave whispering gallery mode resonators. J. Phys. D Appl. Phys. 30(1997)2770–2775.
[31] M. T. Sebastian, I. N. Jawahar, and P. Mohanan. A novel method of tuning the properties of microwave dielectric resonators. Mater. Sci. Eng. B 97(2003)258–264.
[32] L. Li, X. M. Chen, and X. C. Fan. Characterization of $MgTiO_3$–$CaTiO_3$ layered microwave dielectric resonators with TE01d mode. J. Am. Ceram. Soc. 89(2006)557–561.
[33] X. M. Chen, L. Li, X. Liu. Layered complex structures of $MgTiO_3$ and $CaTiO_3$ dielectric ceramics. Mater. Sci. Eng. B 99(2003)255–258, 557–561.
[34] L. Li and X. M. Chen. Adhesive bonded $Ca(Mg_{1/3}Nb_{2/3})O_3/Ba(Zn_{1/3}Nb_{2/3})O_3$ layered dielectric resonators with tunable temperature coefficient of resonant frequency. J. Am. Ceram. Soc. 89(2006) 544–549.
[35] D. Zhou, H. Wang, and X. Yao. Layered complex structures of $Bi_2(Zn_{2/3}Nb_{4/3})O_7$ and $BiNbO_4$ dielectric ceramics. Mater. Chem. Phys. 105(2007)151–153.
[36] S. H. Yoon, G.-K. Choi, D.-W. Kim, S.-Y. Cho, and K. S. Hong. Mixture behaviour and microwave dielectric properties of $(1-x)CaWO_4$–$xTiO_2$. J. Eur. Ceram. Soc. 27(2007)3087–3091.
[37] G. T. Micel and G. S. Frank. Validation of a novel dielectric constant simulation model and the determination of its physical parameter. Microelectronic J. 33(2002)627–632
[38] D. W. Kim, B. W. Park, J. H. Chung, and K. S. Hong. Mixture behaviour microwave dielectric properties in the low fired TiO_2–CuO system. Jpn. J. Appl. Phys. 39(2000)2696–2700.
[39] N. Jayasundere and B. V. Smith. Dielectric constant for binary piezoelectric 0-3 composites. J. Appl. Phys. 73(1993)2462–2466.

[40] Y. M. Poon and F. G. Shin. A Simple explicit formula for the effective dielectric constant of binary 0-3 composite. J. Mater. Sci. 39(2004) 1277–1281.
[41] K. Fukuda, R. Kitoh, and I. Awai. Microwave characteristics of TiO_2–Bi_2O_3 dielectric resonator. Jpn. J. Appl. Phys. 32(1993)4584–4588.
[42] P. S. Anjana and M. T. Sebastian. Microwave dielectric properties and low temperature sintering of $(1-x)CeO_2-xBaTi_4O_9$ ceramics. J. Appl. Ceram. Technol. 5(2008)84–93.
[43] K. P. Surendran, P. V. Bijumon, P. Mohanan, and M. T. Sebastian. $(1-x)MgAl_2O_4-xTiO_2$dielectrics for microwave and millimeter-wave applications. Appl. Phys. A 81(2005)823826.
[44] T. Tsunooka, M. Andou, Y. Higashida, H. Sugiura, and H. Ohsato. Effects of TiO_2 on sinterability and dielectric properties of high Q fosterite ceramics. J. Eur. Ceram. Soc. 23(2003)2573–2578.

CHAPTER FOURTEEN

CONCLUSION

Low-loss dielectric materials are extensively used in wireless communication systems as well as in several electronic devices. The important characteristics required for practical applications are low dielectric loss and good temperature stability of the resonant frequency and permittivity. For practical applications, these materials are usually prepared in the ceramic form. The dielectric properties depend considerably on the material, crystal structure, porosity and imperfections in the crystal lattice. The preparation conditions will also influence the material properties. Calcination conditions are important parameters which influence the structure, secondary phase formation and DR properties. Secondary phases play an important role in controlling the microstructure, density, grain growth and MW dielectric properties of the ceramic resonators. The origin and purity of the initial raw materials have considerable influence on the sinterability and sintering temperature, sintering duration, ordering, effect of annealing on ordering and dielectric properties. It may be noted that there may be small variations in calcination, sintering temperature and durations with the purity and origin of the initial raw materials. A survey of the literature shows that although the sintering temperature is lowered, the quality factor of the DRs made from wet chemical methods are in general no better than those made by conventional solid-state methods. Nanopowders often have poor sinterability and on increasing the sintering temperature, the volatile components may escape leading to nonstoichiometry. In many materials, it is found difficult to sinter the pucks due to the escape of more volatile components. For example, the Zn escapes on sintering pellets made of nanopowders of BZT leading to $BaTa_2O_6$ as the end product. The quality factor decreases abruptly as porosity increases. The porous ceramic traps moisture which increases the dielectric loss. These porous samples show higher quality factor when measured in vacuum or at slightly above room temperature due to elimination of moisture.

Although several manufacturers may produce similar components for the same application, there are subtle differences in circuit design, construction and packaging. Because frequency drift of a device is a consequence of the overall thermal expansion of its unique combination of construction materials, each design requires a slightly different τ_f for temperature compensation. Typically, ceramics with a specific τ_f in the range 15 to −15 ppm/°C are selected. The concept of ceramic DR is simple but it is difficult to control their dimensions and precise phase assemblage during processing. Any slight differences from batch to batch or within a batch may alter their resonant frequency and temperature stability.

The dielectric loss is caused by intrinsic and extrinsic factors. By careful preparation, the extrinsic losses can be minimized to a limited extent. The intrinsic loss is due to the interaction of the AC field with phonons. These intrinsic losses represent the lowest loss limit which can be achieved by material processing. The intrinsic loss can be estimated by

Dielectric Materials for Wireless Communications
ISBN-13: 978-0-08-045330-9

spectroscopic methods and the total loss can be obtained by MW methods. The intrinsic loss is mostly linearly frequency dependent. At ambient temperatures, they are mostly linearly temperature dependent [1], but generally the microscopic theory expects $\varepsilon'' \approx \omega^n T^m$ [2], whereas extrinsic loss is temperature dependent. By studying the loss factor as a function of temperature by both MW and spectroscopic methods, one can estimate the amount of loss due to extrinsic factors and thereby optimize the preparation conditions to lower the extrinsic loss. The ceramics belonging to different crystal symmetry classes have very different temperature and frequency dependencies of tan δ [2]. In real crystals, defects such as point defects, line defects and planar defects affect both the absolute value of the loss for a given temperature and also the temperature dependence of the loss. Losses due to different types of defects display different frequency and temperature dependence.

In the past, the search for better MW materials has been mainly empirical. One of the reasons for this is that it is difficult to find a correlation between microscopic material parameters and extrinsic dielectric loss at MW frequencies. In fact, the measured loss represents the overall loss, including extrinsic contributions such as porosity, second phase, grain boundaries, impurities and other crystal defects which obscure the view on the intrinsic effects. This can be overcome by dielectric spectroscopic study at frequencies $>10^{11}$ Hz, where predominantly intrinsic contributions (phonon) govern the complex permittivity. It has been demonstrated [1, 3–5] that FIR may help to characterize the MW properties of low-loss materials. It is generally accepted [6, 7] that intrinsic MW losses are determined by anharmonic phonon decay processes in the pure crystal lattice. The intrinsic losses increase at least proportionally to frequency and are expected to predominate in the FIR range over extrinsic losses. Some of the extrinsic loss mechanisms which may play an important role in the MW frequency range may not be able to follow the FIR frequencies.

The quality factor of the MW materials is very sensitive to a number of different processing and structural variables. The losses can vary from sample to sample even when they have same composition due to small differences in the intrinsic crystal structure, microstructure, density and impurity concentration, ordering, etc. The product *Qf* is a constant for a given sample over a limited frequency range. Different samples from the same compound (or material) can have different *Qf* value due to varying amount of defects. $Qf[(=\varepsilon'/\varepsilon'')f]$ should be constant, because ε' is frequency independent and ε'' is proportional to frequency. But, in practice, some samples measured at higher frequencies of 5–10 GHz can give higher *Qf* values compared with identical materials measured at lower frequencies of 1–3 GHz. This difference may be due to the difference in processing. The large samples used for measurement at lower frequencies contain more imperfections than a smaller ceramic puck. It was found that lowest losses in complex perovskites are observed when the two different B cations adopt a long-range 1:2 ordering, i.e., B′B″B″ repeat along the [111] direction of the cubic perovskite subcell. Clear correlations between the dielectric loss and degree of cation order and/or the size of the resultant ordered domains have been established in BZT, $Ba(Mg_{1/3}Ta_{2/3})O_3$, $Ba(Mg_{1/3}Nb_{2/3})O_3$, etc. as demonstrated in Chapter 8. The resonator materials, in general, show negligible aging effects and can be used for space applications for a very long period. Property deviations are typically less than measuring errors. This long-term stability of the properties is due to the high densification of the ceramic with negligible moisture absorption. The resonators made from ceramics using fully reacted powders are inert to environmental gases such as H_2O, N_2, O_2 and CO_2. This means that the bulk chemistry of the resonator material does not change.

Although a large number of ceramic dielectric materials have been developed, it has proved difficult to satisfy all these requirements in a single material at a reasonable cost. Figure 14.1 shows the variation of quality factor Qf as a function of permittivity. The Qf, in general, decreases with increase in permittivity. Figure 14.2 shows the variation of τ_f as a function of ε_r. In general, the low ε_r materials have negative τ_f and the high ε_r materials have positive τ_f. Harrop [8] predicted and demonstrated empirically that over a broad range of ε_r, τ_f is proportional to $-\tau_\varepsilon$ for dielectric materials. Therefore, neglecting

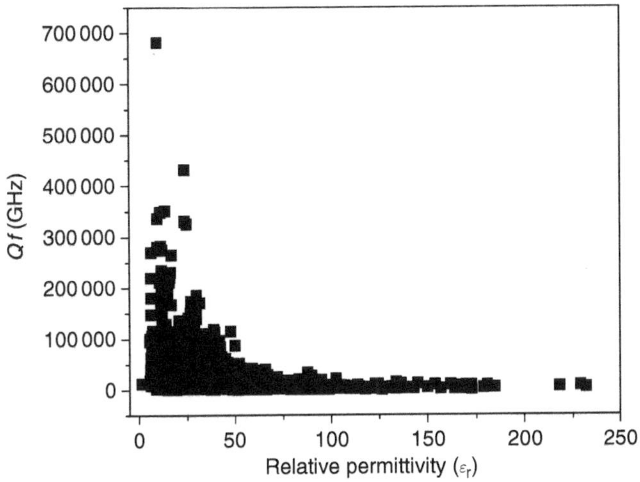

Figure 14.1 Variation of quality factor (Qf) versus permittivity (the data of ceramics with dopants or additives are excluded in the plot).

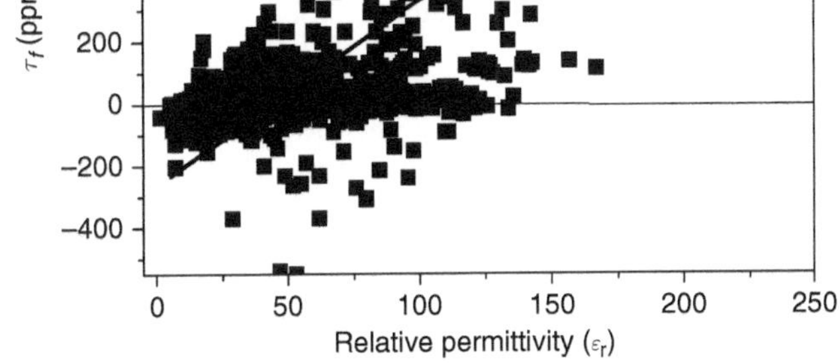

Figure 14.2 Variation of τ_f versus permittivity (the data of ceramics with dopants or additives are excluded in the plot).

the thermal expansion coefficient α_L which is typically 10–15 ppm/°C for electronic ceramics, it follows that ε_r is also directly proportional to τ_f as τ_f is proportional to $-\tau_\varepsilon$. This relationship has been confirmed for perovskites and related materials by several authors [9–11].

Among all the reported DR materials, $CaO–B_2O_3–SiO_2$ (29.3:9.3:61.4) [12] and $Li_2AlB_2O_6$ [13] have the lowest relative permittivities of 3.9 and 4.1, respectively. $Ba_{0.6}Sr_{0.4}TiO_3$ has the highest ε_r of 838 with a *Qf* of about 300 GHz. $SrTiO_3$ has $\varepsilon_r = 270$, $Qf = 3000$ and has the highest positive τ_f of 1500 ppm/°C. The hexagonal $BaNb_2O_6$ has $\tau_f = -800$ ppm/°C [14] and α-$Mg_2P_2O_7$ has $\tau_f = -746$ ppm/°C [15] which are the highest negative τ_f's reported. Alumina has the highest quality factor of $>10^6$ GHz [16] and $Ba(Mg_{1/3}Ta_{2/3})O_3$ has 430 000 GHz [17]. At cryogenic temperatures, the quality factor increases considerably mainly due to suppression of the damping of lattice vibrations. The $Ba(Mg,SnTa)O_3$ is a promising candidate for low-temperature MW dielectric applications as it has extremely high *Qf* value of about 1×10^6 GHz at 70 K [18, 19].

Hartnett et al. [20] reported a frequency–temperature compensated sapphire–rutile resonator with Q factors higher than 10^7. The high negative τ_f of sapphire was compensated by the use of two thin rings of rutile at the top and bottom of the sapphire resonator. The frequency–temperature dependence was annulled when a specific balance of electric energy in the rutile and sapphire was reached at 56 K in a WGM WGE_{900} mode at 13.1 GHz with a Q factor of 30 million. Thus, it is possible to annul the frequency–temperature dependence of sapphire DRs using a dielectric material of the opposite frequency–temperature dependence. Braginsky et al. [21] investigated the loss factor of sapphire crystals of different defect level over a wide temperature range and predicted a Q value of 2×10^{14} at 10 GHz at 4 K for an ideal crystal which is the upper limit of Q in sapphire. Luiten et al. [22] reported a *Qf* of 8.3×10^9 GHz at 12.7 GHz and 1.55 K, which is the highest quality factor reported in the literature. Hartnett et al. [23] reported a *Qf* factor of 1.7×10^9 at 11.2 GHz and 4.2 K for the $WGH_{16,0,0}$ sapphire resonator. The cryogenic sapphire oscillator fabricated with such high Q sapphire resonator showed an exceptionally high long-term frequency stability with a negative drift of about 2.2×10^{-15}/day [23]. An unloaded quality factor of 5×10^6 was reported [24] for sapphire DRs operating on a low-order TE mode at 77 K employing high-temperature superconducting films.

There are about 2300 low-loss DR materials reported in the literature (see table in Appendix 2 for a list of DR materials with properties). The DR table shows that about 60% of the resonator materials are based on alkaline earth metals Ba, Sr, Ca and Mg. About 46% of the resonator materials are titanates, 40% of the materials in the table contain rare earths, and 39% contain tantalates or niobates. Silicates and tungstates have also low-loss resonating properties. The silicates, in general, have predominantly covalent bonding. Hence the atoms in silicates cannot rattle around due to strong covalent bonding and this leads to low dielectric loss. The low dielectric polarizability of silicon and the strong covalent bond lead to low ε_r in silicates. The silicates and tungstates have low ε_r, niobates and tantalates have medium ε_r and titanates have relatively larger ε_r. A study of the DR table (Appendix 2) shows that majority of the resonator materials belong to the perovskite family. Moreover, most of the resonator materials in the table have an octahedral arrangement of atoms. Within a perovskite family of materials octahedral tilts accompanied by symmetry lowering occurs, depending on the size of the A and B site atoms. The tilting of the octahedra changes the MW dielectric properties.

A study of the DR table (Appendix 2) reveals that tantalates, niobates, titanates, silicates or tungstates based on alkali earth metal and rare earths have low dielectric loss. Now the question arises why these materials have low dielectric loss. Most of the low-loss resonator materials have an octahedral or tetrahedral arrangement of atoms. Further detailed investigations, especially spectroscopic studies, are needed to know about the relation between the chemical bonding, lattice vibrations, tetrahedral or octahedral arrangement of atoms and MW dielectric properties. Such studies will be useful in finding new materials for applications and in engineering the properties of known materials.

References

[1] J. Petzelt and S. Kamba. Submillimeter and infrared response of microwave materials. Extrapolation to microwave properties. Mater. Chem. Phys. 79(2003)175–180.

[2] G. L. Gurevich and A. K. Tagantsev. Intrinsic dielectric loss in crystals. Adv. Phys. 40(1991)719–767.

[3] K. Wakino, M. Murata, and H. Tamura. Far infrared reflection spectra of $Ba(Zn_{1/3}Ta_{2/3})O_3$–$BaZrO_3$ dielectric resonator material. J. Am. Ceram. Soc. 69(1986)34–37.

[4] H. Tamura, D. A. Sagala, and K. Wakino. Lattice vibrations of $Ba(Zn_{1/3}Ta_{2/3})O_3$ crystal with ordered perovskite structure. Jpn. J. Appl. Phys. 25(1986)787–791.

[5] J. Petzelt, S. Pacesova, J. Fousek, S. Kamba, V. Zelezny, V. Koukal, J. Schwarzbach, B. P. Gorshunov, G. V. Kozlav, and A. A. Volkov. Dielectric spectra of some ceramics for microwave applications in the range of 10^{10}–10^{14} Hz. Ferroelectrics 93(1989)77–85.

[6] W. Wersing. High frequency ceramic dielectrics and their applications for microwave components. In: Electronic Ceramics, B. C. H. Steele (Ed.), Elsevier Applied Science, London (1991) pp. 67–119.

[7] J. Petzelt and N. Setter. Far infrared spectroscopy and origin of microwave losses in low loss ceramics. Ferroelectrics 150(1993)89–102.

[8] P. J. Harrop. Temperature coefficient of capacitance of solids. J. Mater. Sci. 4(1964)370–374.

[9] P. L. Wise, I. M. Reaney, W. E. Lee, D. M. Iddles, D. S. Canel, and T. J. Price. Tunability of tauf in perovskites and related compounds. J. Mater. Res. 17(2002)2033–2040.

[10] M. P. Seabra, M. Avdeedv, V. M. Ferreira, R. C. Pullar, and N. Mc N. Alford, and I. M. Reaney. Structure property relations in $xBaTiO_3$–$(1-x)La(Mg_{1/2}Ti_{1/2})O_3$ solid solution. J. Am. Ceram. Soc. 87(2004)584–590.

[11] D. Cairns, I. M. Reaney, N. Otten, D. Iddles, and T. Price. Structural determination and microwave of $xReCo_{1/2}Ti_{1/2})O_3$–$(1-x)CaTiO_3$(Re = La, Nd) solid solution. J. Eur. Ceram. Soc. 26(2006)875–882.

[12] C.-C. Chiang, S. F. Wang, Y.-R. Wang, and W.-C. J. Wei. Densification and microwave dielectric properties of CaO-B_2O_3-SiO_2 system glass-ceramic. Ceram. Int. 94(2007) 599–604.

[13] M. Ohsahi, H. Ogawa, and A. Kan. Microwave dielectric properties of low temperature sintered $Li_3AlB_2O_6$ ceramics. J. Eur. Ceram. Soc. 25(2005)2877–2881.

[14] D.-W. Kim, H. B. Hong, K.-S. Hong, C. K. Kim, and D. J. Kim. The reversible phase transition and dielectric properties of $BaNb_2O_6$ polymorphs. Jpn. J. Appl. Phys. 41(2002) 6045–6048.

[15] J. J. Bian, D. W. Kim, and K. S. Hong. Microwave dielectric properties of $A_2P_2O_7$ (A = Ca, Sr, Ba, Mg, Zn, Mn). Jpn. J. Appl. Phys. 43(2004)3521–3525.

[16] J. D. Breeze, X. Aupi, and N. Mc N. Alford. Ultra low loss polycrystalline alumina. Appl. Phys. Lett. 81(2002)5021–5023.

[17] S. Nomura. Ceramics for microwave dielectric resonator. Ferroelectrics 49(1983)61–70.
[18] Y. Higuchi and H. Tamura. Recent progress on the dielectric properties of resonator materials with their applications from microwave to optical frequencies. J. Eur. Ceram. Soc. 23(2003)2683–2688.
[19] H. Tamura, H. Matsumoto, and K. Wakino. Low temperature properties of microwave materials. Jpn. J. Appl. Phys. Suppl. 28. 2(1989)21–23.
[20] J. Hartnett, M. E. Tobar, E. N. Ivanov, and D. G. Cros. High frequency temperature compensated sapphire/rutile resonator. Electron. Lett. 36(2000)726–727.
[21] V. B. Braginsky, V. S. Ilchenko, and S. Kh. Bagdassarov. Experimental observation of fundamental microwave absorption in high quality dielectric crystals. Phys. Lett. 120(1987)300–305.
[22] A. N. Luiten, A. G. Mann, and D. G. Blair. Ultra high Q factor cryogenic sapphire resonator. Electron. Lett. 29(1993)879–881.
[23] J. G. Hartnett, C. R. Locke, E. N. Ivanov, M. E. Tobar, and P. L. Stanwix. Cryogenic sapphire oscillator with exceptionally high long term frequency stability. Appl. Phys. Lett. 89(2006)203513.
[24] M. M. Driscoll, J. T. Haynes, R. A. Jelen, K. W. Weinert, J. R. Gavaler, J. Tavacchio, G. R. Wagner, K. A. Zaki, and X. P. Liang. Cooled, ultrahigh sapphire dielectric resonators for low noise microwave signal generation. IEEE Trans. Ultrasonic, Ferroelectric and Frequency control 39(1992)405–411.

Font: Bembo & Meta Size:10/11pt Margins:Top:3pc6 Gutter:4pc6 T.Area:29pc×44pc7 1 Color Lines: 45 Fresh Recto

Appendix 1

Ionic Radius

Ion	CN	Ionic radius
Ac^{3+}	6	1.12
Ag^{1+}	2	0.67
Ag^{1+}	4	1
Ag^{1+}	4	1.02
Ag^{1+}	5	1.09
Ag^{1+}	6	1.15
Ag^{1+}	7	1.22
Ag^{1+}	8	1.28
Ag^{2+}	4	0.79
Ag^{2+}	6	0.94
Ag^{3+}	4	0.67
Ag^{3+}	6	0.75
Al^{3+}	4	0.39
Al^{3+}	5	0.48
Al^{3+}	6	0.535
Am^{2+}	7	1.21
Am^{2+}	8	1.26

Ion	CN	Ionic radius
Am^{2+}	9	1.31
Am^{3+}	6	0.975
Am^{3+}	8	1.09
Am^{4+}	6	0.85
Am^{4+}	8	0.95
As^{3+}	6	0.58
As^{5+}	4	0.335
As^{5+}	6	0.46
At^{7+}	6	0.62
Au^{1+}	6	1.37
Au^{3+}	4	0.68
Au^{3+}	6	0.85
Au^{5+}	6	0.57
B^{3+}	3	0.01
B^{3+}	4	0.11
B^{3+}	6	0.27
Ba^{2+}	6	1.35

Dielectric Materials for Wireless Communications
ISBN-13: 978-0-08-045330-9

Ion	CN	Ionic radius
Ba^{2+}	7	1.38
Ba^{2+}	8	1.42
Ba^{2+}	9	1.47
Ba^{2+}	10	1.52
Ba^{2+}	11	1.57
Ba^{2+}	12	1.61
Be^{2+}	3	0.16
Be^{2+}	4	0.27
Be^{2+}	6	0.45
Bi^{3+}	5	0.96
Bi^{3+}	6	1.03
Bi^{3+}	8	1.17
Bi^{5+}	6	0.76
Bk^{3+}	6	0.96
Bk^{4+}	6	0.83
Bk^{+4}	8	0.93
Br^{1-}	6	1.96
Br^{3+}	4	0.59
Br^{5+}	3	0.31
Br^{7+}	4	0.25
Br^{7+}	6	0.39
C^{4+}	3	−0.08
C^{4+}	4	0.15
C^{4+}	6	0.16
Ca^{2+}	6	1
Ca^{2+}	7	1.06

Ion	CN	Ionic radius
Ca^{2+}	8	1.12
Ca^{2+}	9	1.18
Ca^{2+}	10	1.23
Ca^{2+}	12	1.34
Cd^{2+}	4	0.78
Cd^{2+}	5	0.87
Cd^{2+}	6	0.95
Cd^{2+}	7	1.03
Cd^{2+}	8	1.1
Cd^{2+}	12	1.31
Ce^{3+}	6	1.01
Ce^{3+}	7	1.07
Ce^{3+}	8	1.143
Ce^{3+}	9	1.196
Ce^{3+}	10	1.25
Ce^{3+}	12	1.34
Ce^{4+}	6	0.87
Ce^{4+}	8	0.97
Ce^{4+}	10	1.07
Ce^{4+}	12	1.14
Cf^{3+}	6	0.95
Cf^{4+}	6	0.821
Cf^{4+}	8	0.92
Cl^{1-}	6	1.81
Cl^{5+}	3	0.12
Cl^{7+}	4	0.08

Ion	CN	Ionic radius
Cl^{7+}	6	0.27
Cm^{3+}	4	0.97
Cm^{4+}	4	0.85
Cm^{4+}	8	0.95
Co^{2+}	4	0.58
Co^{2+}	5	0.67
Co^{2+}	6	0.65
Co^{2+}	6	0.745
Co^{2+}	8	0.9
Co^{3+}	6	0.545
Co^{3+}	6	0.61
Co^{4+}	4	0.4
Co^{4+}	6	0.53
Cr^{2+}	6	0.73
Cr^{2+}	6	0.8
Cr^{3+}	6	0.615
Cr^{4+}	4	0.41
Cr^{4+}	6	0.55
Cr^{5+}	4	0.345
Cr^{5+}	6	0.49
Cr^{5+}	8	0.57
Cr^{6+}	4	0.26
Cr^{6+}	6	0.44
Cs^{1+}	6	1.67
Cs^{1+}	8	1.74
Cs^{1+}	9	1.78

Ion	CN	Ionic radius
Cs^{1+}	10	1.81
Cs^{1+}	11	1.85
Cs^{1+}	12	1.88
Cu^{1+}	2	0.46
Cu^{1+}	4	0.6
Cu^{1+}	6	0.77
Cu^{2+}	4	0.57
Cu^{2+}	4	0.57
Cu^{2+}	5	0.65
Cu^{2+}	6	0.73
Cu^{3+}	6	0.54
D^{1+}	2	−0.1
Dy^{2+}	6	1.07
Dy^{2+}	7	1.13
Dy^{2+}	8	1.19
Dy^{3+}	6	0.912
Dy^{3+}	7	0.97
Dy^{3+}	8	1.027
Dy^{3+}	9	1.083
Er^{3+}	6	0.89
Er^{3+}	7	0.945
Er^{3+}	8	1.004
Er^{3+}	9	1.062
Eu^{2+}	6	1.17
Eu^{2+}	7	1.2
Eu^{2+}	8	1.25

Ion	CN	Ionic radius
Eu^{2+}	9	1.3
Eu^{2+}	10	1.35
Eu^{3+}	6	0.947
Eu^{3+}	7	1.01
Eu^{3+}	8	1.066
Eu^{3+}	9	1.12
F^{1-}	2	1.285
F^{1-}	3	1.3
F^{1-}	4	1.31
F^{1-}	6	1.33
F^{7+}	6	0.08
Fe^{2+}	4	0.63
Fe^{2+}	4	0.64
Fe^{2+}	6	0.61
Fe^{2+}	6	0.78
Fe^{2+}	8	0.92
Fe^{3+}	4	0.49
Fe^{3+}	5	0.58
Fe^{3+}	6	0.55
Fe^{3+}	6	0.645
Fe^{3+}	8	0.78
Fe^{4+}	6	0.585
Fe^{6+}	4	0.25
Fr^{1+}	6	1.8
Ga^{3+}	4	0.47
Ga^{3+}	5	0.55

Ion	CN	Ionic radius
Ga^{3+}	6	0.62
Gd^{3+}	6	0.938
Gd^{3+}	7	1
Gd^{3+}	8	1.053
Gd^{3+}	9	1.107
Ge^{2+}	6	0.73
Ge^{4+}	4	0.39
Ge^{4+}	6	0.53
H^{1+}	1	−0.38
H^{1+}	2	−0.18
Hf^{4+}	4	0.58
Hf^{4+}	6	0.71
Hf^{4+}	7	0.76
Hf^{4+}	8	0.83
Hg^{1+}	3	0.97
Hg^{1+}	6	1.19
Hg^{2+}	2	0.69
Hg^{2+}	4	0.96
Hg^{2+}	6	1.02
Hg^{2+}	8	1.14
Ho^{3+}	6	0.901
Ho^{3+}	8	1.015
Ho^{3+}	9	1.072
Ho^{3+}	10	1.12
I^{1-}	6	2.2
I^{5+}	3	0.44

Ion	CN	Ionic radius
I^{5+}	6	0.95
I^{7+}	4	0.42
I^{7+}	6	0.53
In^{3+}	4	0.62
In^{3+}	6	0.8
In^{3+}	8	0.92
Ir^{3+}	6	0.68
Ir^{4+}	6	0.625
Ir^{5+}	6	0.57
K^{1+}	4	1.37
K^{1+}	6	1.38
K^{1+}	7	1.46
K^{1+}	8	1.51
K^{1+}	9	1.55
K^{1+}	10	1.59
K^{1+}	12	1.64
La^{3+}	6	1.032
La^{3+}	7	1.1
La^{3+}	8	1.16
La^{3+}	9	1.216
La^{3+}	10	1.27
La^{3+}	12	1.36
Li^{1+}	4	0.59
Li^{1+}	6	0.76
Li^{1+}	8	0.92
Lu^{3+}	6	0.861

Ion	CN	Ionic radius
Lu^{3+}	8	0.977
Lu^{3+}	9	1.032
Mg^{2+}	4	0.57
Mg^{2+}	5	0.66
Mg^{2+}	6	0.72
Mg^{2+}	8	0.89
Mn^{2+}	4	0.66
Mn^{2+}	5	0.75
Mn^{2+}	6	0.67
Mn^{2+}	6	0.83
Mn^{2+}	7	0.9
Mn^{2+}	8	0.96
Mn^{3+}	5	0.58
Mn^{3+}	6	0.58
Mn^{3+}	6	0.645
Mn^{4+}	4	0.39
Mn^{4+}	6	0.53
Mn^{5+}	4	0.33
Mn^{6+}	4	0.255
Mn^{7+}	4	0.25
Mn^{7+}	6	0.46
Mo^{3+}	6	0.69
Mo^{4+}	6	0.65
Mo^{5+}	4	0.46
Mo^{5+}	6	0.61
Mo^{6+}	4	0.41

Ion	CN	Ionic radius
Mo^{6+}	5	0.5
Mo^{6+}	6	0.59
Mo^{6+}	7	0.73
N^{3-}	4	1.46
N^{3+}	6	0.16
N^{5+}	3	−0.104
N^{5+}	4	0.13
Na^{1+}	4	0.99
Na^{1+}	5	1
Na^{1+}	6	1.02
Na^{1+}	7	1.12
Na^{1+}	8	1.18
Na^{1+}	9	1.24
Na^{1+}	12	1.39
Nb^{3+}	6	0.72
Nb^{4+}	6	0.68
Nb^{4+}	8	0.79
Nb^{5+}	4	0.48
Nb^{5+}	6	0.64
Nb^{5+}	7	0.69
Nb^{5+}	8	0.74
Nd^{2+}	8	1.29
Nd^{2+}	9	1.35
Nd^{3+}	6	0.983
Nd^{3+}	8	1.109
Nd^{3+}	9	1.163

Ion	CN	Ionic radius
Nd^{3+}	12	1.27
Ni^{2+}	4	0.55
Ni^{2+}	4	0.49
Ni^{2+}	5	0.63
Ni^{2+}	6	0.69
Ni^{3+}	6	0.56
Ni^{3+}	6	0.6
Ni^{4+}	6	0.48
No^{2+}	6	1.1
Np^{2+}	6	1.1
Np^{3+}	6	1.01
Np^{4+}	6	0.87
Np^{4+}	8	0.98
Np^{5+}	6	0.75
Np^{6+}	6	0.72
Np^{7+}	6	0.71
O^{2-}	2	1.35
O^{2-}	3	1.36
O^{2-}	4	1.38
O^{2-}	6	1.4
O^{2-}	8	1.42
OH^{1-}	2	1.32
OH^{1-}	3	1.34
OH^{1-}	4	1.35
OH^{1-}	6	1.37
Os^{4+}	6	0.63

Ion	CN	Ionic radius
Os^{5+}	6	0.575
Os^{6+}	5	0.49
Os^{6+}	6	0.545
Os^{7+}	6	0.525
Os^{8+}	4	0.39
P^{3+}	6	0.44
P^{5+}	4	0.17
P^{5+}	5	0.29
P^{5+}	6	0.38
Pa^{3+}	6	1.04
Pa^{4+}	6	0.9
Pa^{4+}	8	1.01
Pa^{5+}	6	0.78
Pa^{5+}	8	0.91
Pa^{5+}	9	0.95
Pb^{2+}	4	0.98
Pb^{2+}	6	1.19
Pb^{2+}	7	1.23
Pb^{2+}	8	1.29
Pb^{2+}	9	1.35
Pb^{2+}	10	1.4
Pb^{2+}	11	1.45
Pb^{2+}	12	1.49
Pb^{4+}	4	0.65
Pb^{4+}	5	0.73
Pb^{4+}	6	0.775

Ion	CN	Ionic radius
Pb^{4+}	8	0.94
Pd^{1+}	2	0.59
Pd^{2+}	4	0.64
Pd^{2+}	6	0.86
Pd^{3+}	6	0.76
Pd^{4+}	6	0.615
Pm^{3+}	6	0.97
Pm^{3+}	8	1.093
Pm^{3+}	9	1.144
Po^{4+}	6	0.94
Po^{4+}	8	1.08
Po^{6+}	6	0.67
Pr^{3+}	6	0.99
Pr^{3+}	8	1.126
Pr^{3+}	9	1.179
Pr^{4+}	6	0.85
Pr^{4+}	8	0.96
Pt^{2+}	4	0.6
Pt^{2+}	6	0.8
Pt^{4+}	6	0.625
Pt^{5+}	6	0.57
Pu^{3+}	6	1
Pu^{4+}	6	0.86
Pu^{4+}	8	0.96
Pu^{5+}	6	0.74
Pu^{6+}	6	0.71

Ion	CN	Ionic radius
Ra^{2+}	8	1.48
Ra^{2+}	12	1.7
Rb^{1+}	6	1.52
Rb^{1+}	7	1.56
Rb^{1+}	8	1.61
Rb^{1+}	9	1.63
Rb^{1+}	10	1.66
Rb^{1+}	11	1.69
Rb^{1+}	12	1.72
Rb^{1+}	14	1.83
Re^{4+}	6	0.63
Re^{5+}	6	0.58
Re^{6+}	6	0.55
Re^{7+}	4	0.38
Re^{7+}	6	0.53
Rh^{3+}	6	0.665
Rh^{4+}	6	0.6
Rh^{5+}	6	0.55
Ru^{3+}	6	0.68
Ru^{4+}	6	0.62
Ru^{5+}	6	0.565
Ru^{7+}	4	0.38
Ru^{8+}	4	0.36
S^{2-}	6	1.84
S^{4+}	6	0.37
S^{6+}	4	0.12

Ion	CN	Ionic radius
S^{6+}	6	0.29
Sb^{3+}	4	0.74
Sb^{3+}	5	0.8
Sb^{3+}	6	0.76
Sb^{5+}	6	0.6
Sc^{3+}	6	0.745
Sc^{3+}	8	0.87
Se^{2-}	6	1.98
Se^{4+}	6	0.5
Se^{6+}	4	0.28
Se^{6+}	6	0.42
Si^{4+}	4	0.26
Si^{4+}	6	0.4
Sm^{2+}	7	1.22
Sm^{2+}	8	1.27
Sm^{2+}	9	1.32
Sm^{3+}	6	0.958
Sm^{3+}	7	1.02
Sm^{3+}	8	1.079
Sm^{3+}	9	1.132
Sm^{3+}	12	1.24
Sn^{4+}	4	0.55
Sn^{4+}	5	0.62
Sn^{4+}	6	0.69
Sn^{4+}	7	0.75
Sn^{4+}	8	0.81

Ion	CN	Ionic radius
Sr^{2+}	6	1.18
Sr^{2+}	7	1.21
Sr^{2+}	8	1.26
Sr^{2+}	9	1.31
Sr^{2+}	10	1.36
Sr^{2+}	12	1.44
Ta^{3+}	6	0.72
Ta^{4+}	6	0.68
Ta^{5+}	6	0.64
Ta^{5+}	7	0.69
Ta^{5+}	8	0.74
Tb^{3+}	6	0.923
Tb^{3+}	7	0.98
Tb^{3+}	8	1.04
Tb^{3+}	9	1.095
Tb^{4+}	6	0.76
Tb^{4+}	8	0.88
Tc^{4+}	6	0.645
Tc^{5+}	6	0.6
Tc^{7+}	4	0.37
Tc^{7+}	6	0.56
$Te^{(2-)}$	6	2.21
Te^{4+}	3	0.52
Te^{4+}	4	0.66
Te^{4+}	6	0.97
Te^{6+}	4	0.43

Ion	CN	Ionic radius
Te^{6+}	6	0.56
Th^{4+}	6	0.94
Th^{4+}	8	1.05
Th^{4+}	9	1.09
Th^{4+}	10	1.13
Th^{4+}	11	1.18
Th^{4+}	12	1.21
Ti^{2+}	6	0.86
Ti^{3+}	6	0.67
Ti^{4+}	4	0.42
Ti^{4+}	5	0.51
Ti^{4+}	6	0.605
Ti^{4+}	8	0.74
Tl^{1+}	6	1.5
Tl^{1+}	8	1.59
Tl^{1+}	12	1.7
Tl^{3+}	4	0.75
Tl^{3+}	6	0.885
Tl^{3+}	8	0.98
Tm^{2+}	6	1.03
Tm^{2+}	7	1.09
Tm^{3+}	6	0.88
Tm^{3+}	8	0.994
Tm^{3+}	9	1.052
U^{3+}	6	1.025
U^{4+}	6	0.89

Ion	CN	Ionic radius
U^{4+}	7	0.95
U^{4+}	8	1
U^{4+}	9	1.05
U^{4+}	12	1.17
U^{5+}	6	0.76
U^{5+}	7	0.84
U^{6+}	2	0.45
U^{6+}	4	0.52
U^{6+}	6	0.73
U^{6+}	7	0.81
U^{6+}	8	0.86
V^{2+}	6	0.79
V^{3+}	6	0.64
V^{4+}	5	0.53
V^{4+}	6	0.58
V^{4+}	8	0.72
V^{5+}	4	0.355
V^{5+}	5	0.46
V^{5+}	6	0.54
W^{4+}	6	0.66
W^{5+}	6	0.62
W^{6+}	4	0.42
W^{6+}	5	0.51
W^{6+}	6	0.6

Ion	CN	Ionic radius
Xe^{8+}	4	0.4
Xe^{8+}	6	0.48
Y^{3+}	6	0.9
Y^{3+}	7	0.96
Y^{3+}	8	1.019
Y^{3+}	9	1.075
Yb^{2+}	6	1.02
Yb^{2+}	7	1.08
Yb^{2+}	8	1.14
Yb^{3+}	6	0.868
Yb^{3+}	7	0.925
Yb^{3+}	8	0.985
Yb^{3+}	9	1.042
Zn^{2+}	4	0.6
Zn^{2+}	5	0.68
Zn^{2+}	6	0.74
Zn^{2+}	8	0.9
Zr^{4+}	4	0.59
Zr^{4+}	5	0.66
Zr^{4+}	6	0.72
Zr^{4+}	7	0.78
Zr^{4+}	8	0.84
Zr^{4+}	9	0.89

(Modified after R. D. Shannon, Acta Crystallogr. A 32(1976) 751–767.)

APPENDIX 2

LIST OF MICROWAVE DIELECTRIC RESONATOR MATERIALS AND THEIR PROPERTIES

The table lists the key property data of microwave dielectric materials available from published and to a far lesser extent from reputable unpublished sources. These data are the relative permittivity (ε_r), the product of the Q factor and the frequency (Qf), the frequency of measurement (f) and the temperature coefficient of the resonant frequency (τ_f). In tabulating these data, we make no judgement on the measurement method and the reliability of the result. It is known that the ceramic properties such as porosity, grain size, raw materials used, measurement methods and equipment used for measurements affect the dielectric properties and readers should be aware that exact comparison of data on materials of identical composition and manufactured in different laboratories using different processing conditions would be expected to lead to small variations in properties. The data of dielectric measurements carried out using impedance methods at low (MHz) frequency is excluded as the errors in these methods mean that a loss tangent less than 10^{-3} is unreliable.

Dielectric Materials for Wireless Communications
ISBN-13: 978-0-08-045330-9

No.		Sintering temp. (°C)	ε_r	*Qf* (GHz)	*f* (GHz)	τ_f	Reference
1	$CaO-B_2O_3-SiO_2$ (29.3:9.3:61.4 mol%)	900	3.9	1800	9.9	–	[1]
2	$CaO-B_2O_3-SiO_2$ (19.8:30.9:49.3 mol%)	900	4.1	2000	9.9	–	[1]
3	$CaO-B_2O_3-SiO_2$ (10.5:22.2:67.3 mol%)	900	4.1	2600	9.9	–	[1]
4	$Li_3AlB_2O_6$	650	4.2	12 460	16.8	−290	[2]
5	$Li_3AlB_2O_6$	700	4.9	12 609	16.9	−201	[2]
6	(cordierite) $Mg_2Al_4Si_5O_{18}$ + 7 wt% Yb_2O_3	1420	4.9	112 500	18		[3]
7	$Li_3AlB_2O_6$	775	5.4	20 450	17.4	−244	[2]
8	40 wt% Al_2O_3 + 60 wt% ($SiO_2-B_2O_3-Al_2O_3$)	875	5.4	8000	–	−50	[4]
9	$NaAlSi_3O_8$ (albite)	1025	5.5	11 200		−5	[5]
10	μ-cordierite + $B_2O_3-P_2O_5$	860	5.8	3000		−55	[6]
11	α-cordierite + $B_2O_3-P_2O_5$	950	5.8	6000		−15	[6]
12	$K_{0.67}Ba_{0.33}Ga_{1.33}Ge_{2.67}O_8$		5.9	94 100		−25	[7]
13	Al_2O_3 + $MgO-Al_2O_3-SiO_2-GeO_2$ + $ZnO-B_2O_3$	900	5.9	5590	8.4		[8]
14	$Na_{0.8}Ca_{0.2}Al_{1.2}Si_{2.8}O_8$		6	17 600		0	[9]
15	$MgO-Al_2O_3-B_2O_3-SiO_2-TiO_2$		6.1	4200			[10]
16	$\alpha-Mg_2P_2O_7$	1150	6.1	38 180		−746	[11]
17	$CaWO_4$ + 0.5 wt% B_2O_3	1050	6.1	38 100		−47	[12, 13]
18	$(Mg_{0.9}Ni_{0.1})_2Al_4Si_5O_{18}$	1440 for 2 hours	6.1	99 100		−32	[14]
19	$KGaGe_3O_8$	970	6.2	19 800		−21	[7]
20	$Mg_2Al_4Si_5O_{18}$	1440	6.2	40 000		−25	[6]
21	50 wt% Al_2O_3 + 50 wt% ($SiO_2-B_2O_3-Al_2O_3$)	875	6.2	11 400	–	−35	[4]
22	$AlSbO_4$	1100 for 3 hours	6.3	3200	4		[15]
23	45 wt% Al_2O_3 + 55 wt% ($SiO_2-B_2O_3-Al_2O_3$)	875	6.3	11 500	–	−33	[4]
24	55 wt% Al_2O_3 + 45 wt% ($SiO_2-B_2O_3-Al_2O_3$)	900	6.4	13 000	–	−58	[4]
25	$Mg_3(VO_4)_2$	950 for 5 hours	6.4	48 800	–	−83	[16]
26	$K_{0.9}Ba_{0.1}Ga_{1.1}Ge_{2.9}O_8$	1040	6.4	94 700	12	−23	[7]
27	$Y_2BaCu_{0.75}Ni_{0.25}O_5$		6.4	8350	13.5	−40	[17]
28	$K_{0.4}Ba_{0.6}Ga_{1.6}Ge_{2.4}O_8$	1040	6.4	94 700	12	−23	[7]
29	$Ca_{0.99}Mg_{0.01}SiO_3$	1290 for 2 hours	6.5	62 400		−43	[18]
30	Willemite (Zn_2SiO_4)	1340	6.6	219 000		−61	[19]
31	$CaWO_4$ + 1 wt% $MnSO_4$	1050	6.6	129 540		−56	[12, 13]
32	$K_{0.9}Ba_{0.1}Ga_{1.1}Ge_{2.9}O_8$	990	6.6	12 700		−21	[7]
33	$Zn_{1.8}SiO_{3.8}$	1300 for 3 hours	6.6	147 000		−22	[20]
34	$MgO-B_2O_3-SiO_2$ (42:45:13) glass		6.64	2130	6.9		[21]

No.		Sintering temp. (°C)	ε_r	Qf (GHz)	f (GHz)	τ_f	Reference
35	$ZnO-0.6\ SiO_2+Bi_2O_3-Li_2CO_3$	910 for 2 hours	6.65	33 000	11	−70	[22]
36	$CaO-SiO_2$	1320	6.69	25 400		–	[23]
37	50 wt% $(La_2O_3-B_2O_3)$ + 50 wt% Al_2O_3	850	6.7	2800	17.7		[24]
38	$Mg_{1.975}Mn_{.025}SiO_4$	1400	6.71	180 000		−71	[25]
39	Mg_2SiO_4 (fosterite)	1450	6.8	270 000		−67	[26]
40	$Mg_{1.93}Ca_{.07}SiO_4$	1400	6.87	105 000		−72	[25]
41	$ZnO:B_2O_3$ (50:50) glass	<800	6.88	1733	16.4	−10	[27]
42	$K_{0.67}Ba_{0.33}Ga_{1.33}Ge_{2.67}O_8$	1020	6.9	32 600	12	−27	[7]
43	$ZnO:B_2O_3:SiO_2$ (50:40:10) glass	<800	6.91	1710	15.8	−21	[27]
44	Fosterite + 1 wt% TiO_2	1300	7.0	230 000		−65	[28]
45	$Ba_2V_2O_7$	950	7.0	19 000		−74	[29]
46	$BaGa_2Ge_2O_8$	1100	7.0	10 600		−25	[7]
47	$\delta-Ba_2P_2O_7$	1150	7	12 300			[11]
48	$Ba_2MgSi_2O_7$	1350 for 10 hours	7	31 000		−60	[30]
49	$SrZnP_2O_7$	950	7.06	52 780		−70	[31]
50	$MgMoO_4$	900	7.07	79 100		−46	[32]
51	$ZnO:B_2O_3:SiO_2$ (50:30:20) glass		7.08	1670	15.9	−43	[27]
52	$\alpha-Sr_2P_2O_7$	1150	7.1	33 500		−23	[11]
53	$SrO-B_2O_3-SiO_2$ (32.85:52.09:15.05) glass		7.1	3608	6.7		[21]
54	$Mg_3B_2O_6$		7.2	150 400	16		[33, 34]
55	$Sr_xBa_{1-x}Al_2Si_3O_8$	1500 for 12 hours	7.2	40 000–77 000	10.5	−22 to −30	[35]
56	$BaAl_2Si_3O_8$	1500 for 12 hours	7.2	70 600	10.5	−22	[35]
57	$Sr_{0.05}Ba_{0.95}Al_2Si_3O_8$	1500 for 40 hours	7.2	77 000	10.5		[35]
58	$SrZnP_2O_7$	925 for 2 hours	7.25	71 520		−64	[31]
59	$BaO:B_2O_3:SiO_2$ (30:20:50) glass		7.28	1840	14.8	−62	[27]
60	$BaCu(B_2O_5)$	810	7.3	50 000		−32	[36]
61	$CaO-B_2O_3-SiO_2$(69.7:16.2:14.1 mol%)	900	7.3	2300	9.6		[1]
62	$CaO-B_2O_3-SiO_2$(38.3:31.5:30.2 mol%)	900	7.3	1800	9.6		[1]
63	$BaO:B_2O_3:SiO_2$ (30:40:30) glass		7.3	2700	15.4	−34	[27]
64	$BaO:B_2O_3:SiO_2$ (30:60:10) glass		7.3	3390	14.9	−25	[27]
65	$CaCuP_2O_7$	900 for 2 hours	7.33	71 620		−76	[31]
66	$Mn_2P_2O_7$	1150	7.34	23 850		−96	[11]
67	$Y_2BaCu_{0.6}Mg_{0.4}O_5$		7.4	25 320	12.9	−56	[17, 37]
68	$CaAl_2Si_2O_8$ (anorthite)	1500	7.4	12 000		−130	[5]

No.		Sintering temp. (°C)	ε_r	*Qf* (GHz)	*f* (GHz)	τ_f	Reference
69	$MgAl_2O_4$:Li–Mg–Zn–B–Si–O glass	1000	7.4	48 000	24	−90	[33]
70	$CaO-B_2O_3-SiO_2$ (42:45:13) glass		7.47	2380	6.2		[21]
71	$DyBO_3$, $HoBO_3$, YBO_3		7.5<	10 000<			[38]
72	α-$Zn_2P_2O_7$	1150	7.5	50 000		−204	[11]
73	$Mg_{2.05}SiO_4$	1550 for 3 hours	7.5	114 700	10.6	−59	[39]
74	$Ba_2ZnSi_2O_7$	1350	7.5	48 000		−74	[30]
75	$ZnO:B_2O_3:SiO_2$ (60:20:20) glass	<800	7.51	1410	15.4	−84	[27]
76	$ZnO:B_2O_3$ (60:40) glass	<800	7.51	1430	15.1	−3	[27]
77	$ZnO:B_2O_3:SiO_2$ (60:30:10) glass	<800	7.56	1440	15.5	−21	[27]
78	$CaZnP_2O_7$	900 for 2 hours	7.56	63 130		−82	[40]
79	$20CaO-20La_2O_3-60B_2O_3+40$ wt% Al_2O_3	850	7.6	5000	17.4		[41]
80	$BaO-B_2O_3-SiO_2$ (42:45:13) glass		7.63	4100	6.65		[21]
81	α−$Ca_2P_2O_7$	1290 for 4 hours	7.8	14 110		−97	[42]
82	$Mg_3Sm_4Al_{44}O_{75}+B_2O_3-SiO_2-Al_2O_3$	920	7.8	10 000		11	[43]
83	Yb_2BaCuO_5		7.9	7290		−44	[44]
84	$SrWO_4$	1150	7.9	56 000		−55	[45]
85	$Mg_3(VO_4)_2$	950 for 10 hours	7.9	53 000		−84	[16]
86	$CaO-B_2O_3-SiO_2$(50.1:22.2:67.3 mol%)	900	7.9	2100	9.6		[1]
87	CAS-T5 glass ($CaO:Al_2O_3:SiO_2:TiO_2:B_2O_3$)	950	8.0	22 500	10	−20	[46]
88	$CaAl_2Si_2O_8+5$ wt% TiO_2	900	8	22 500		−50	[46]
89	$Li_2O-B_2O_3-SiO_2-Al_2O_3-CaO$	550	8	2400		−48	[47, 48]
90	$La_2O_3-2B_2O_3-0.5ZnO$	900	8	72 000	13	–	[49]
91	$20MgO-20La_2O_3-60B_2O_3+40$ wt% Al_2O_3	950	8.1	19 000			[41]
92	$BaWO_4$	1150	8.1	56 000		−55	[45, 50]
93	$SrWO_4$	1150	8.1	57 500		−78	[45]
94	$MgZn_2(VO_4)_2$	800 for 5 hours	8.1	44 700		−108	[51]
95	60 wt% $La_2O_3-B_2O_3$ +40 wt% Al_2O_3	850	8.1	4500	17.5		[24]
96	$MgTiO_3-CaTiO_3$ (MMT-20) + $SiO_2-B_2O_3-BaO$	875	8.2	3000	7		[52]
97	$BaWO_4+0.5$ wt% B_2O_3	950	8.2	32 700		−85	[12]
98	$20ZnO-20La_2O_3-60B_2O_3+40$ wt% Al_2O_3	950	8.2	20 000	17.1		[41]
99	70 wt% $La_2O_3-B_2O_3+30$ wt% Al_2O_3	850	8.3	5500	17.2		[24]
100	Y_2BaCuO_5 (CIP)		8.3	53 300		−39.5	[37]
101	$20ZnO-20La_2O_3-60B_2O_3+40$ wt% Al_2O_3	950	8.3	18 600	17	–	[41]
102	80 wt% $La_2O_3-B_2O_3+20$ wt% Al_2O_3	850	8.4	9800	17.6		[24]

No.		Sintering temp. (°C)	ε_r	Qf (GHz)	f (GHz)	τ_f	Reference
103	$CaSiO_3$	1300	8.4	16 000	10		[53]
104	$Ca_3SnSi_2O_9$		8.4	93 300		−70	[53]
105	$\beta-Ca_2P_2O_7$	1150 for 2 hours	8.4	53 500		−53	[31]
106	$Ca_3SnSi_2O_9$	1400	8.4	54 800		−65	[53]
107	$Yb_2Ba(Cu_{0.5}Ni_{0.5})O_5$		8.5	13 300		−46.4	[54]
108	$TbPO_4$	1650 for 2 hours	8.5	20 100		−17.6	[55]
109	$MgAl_2O_4$	1650 for 3 hours	8.5	105 000		−63	[34, 56]
110	Li_2MgSiO_4		8.5	30 000	15		[34]
111	$Li_2O-B_2O_3-SiO_2$ frit glass	<800	8.5	1800		−157	[57]
112	$MgTiO_3-CaTiO_3$ (MMT)−20−$ZnO-B_2O_3-SiO_2$ (44.57:17.32:6.95:30.16)	875	8.5	7000	7	6.2	[52, 58]
113	$ZnAl_2O_4$	1375	8.5	56 000	12.3	−79	[59]
114	$ZnO-B_2O_3-SiO_2$−MMT-20 (44.57:17.32:6.95:30.16)	875	8.5	3000	7	6.2	[58]
115	$Mg_4Nb_2O_9$ (precipitation)	950	8.5	50 000		–	[60]
116	$0.84Al_2O_3-0.16TiO_2$ + 8 wt% MCAS glass	1250	8.5	9900		−2	[61]
117	$MnMoO_4$	900	8.5	54 100		−74	[32]
118	Mn_2SiO_4	1100/N_2	8.5	50 000		−90	[25, 54]
119	$Y_2BaCu_{0.25}Ni_{0.75}O_5$		8.6	31 290	12.5	−36	[17]
120	$0.88Al_2O_3-0.12TiO_2$ + 2 wt% MCAS glass	1250	8.63	9580		5	[62]
121	$ZnMoO_4$	800	8.67	49 900		−87	[32]
122	$CaWO_4$ + 0.5 wt% Bi_2O_3 + 9 wt% H_3BO_3	850	8.7	70 220		−15	[63]
123	$CaWO_4$	1200	8.7	75 000		−54	[45]
124	$YbPO_4$	1600 for 2 hours	8.7	71 600		−27.7	[55]
125	$(Al_{1/2}Ta_{1/2})O_2$	1600	8.7	60 800		−55	[64]
126	Ca−Al−B−Si−O + Al_2O_3 (K8)	870	8.7	900	3	–	[65]
127	$ZnO-B_2O_3-SiO_2$−MMT-0 (46.34:17.09:6.85:29.72)	900	8.9	7000	8	−24	[58]
128	$CaGeO_3$	1200	8.9	32 200	10		[66]
129	$ZnO-B_2O_3-SiO_2$−MMT-0 (44.97:17.2:6.9:29.93)	900	8.9	810	8	−15	[52, 58]
130	$ZnO-B_2O_3-SiO_2$−MMT-20 (46.34:17.09:6.85:29.72)	900	8.9	800	8	−24	[52, 58]
131	$ZnO-B_2O_3-SiO_2$−MMT-20 (49.21:16.15:6.49:28.15)	900	9	7000	8	−62	[58]
132	$Co_3O_4-Nb_2O_5-TiO_2$	1010	9	41 000	7.5	−59	[67]
133	$MgTiO_3-CaTiO_3-ZnO-B_2O_3-SiO_2$		9	7000	7	−55	[58]
134	$PbO:B_2O_3:SiO_2$ (30:60:10) glass		9	1700	13.5	−15	[27]

No.		Sintering temp. (°C)	ε_r	Qf (GHz)	f (GHz)	τ_f	Reference
135	$5ZnO-Ta_2O_5-TiO_2$	1150	9	9000	7.9	−45	[67]
136	Yb_2BaNiO_5		9.1	44 600		−37.5	[44]
137	$K_2O-B_2O_3-SiO_2-CaO-SrO-BaO$ (glass) + Al_2O_3	900	9.1	600	0.5	0	[68]
138	Yb_2BaZnO_5		9.1	44 600	−37.5		[44]
139	$DyPO_4$	1650 for 2 hours	9.1	28 600		−17.0	[55]
140	α-$Ca_3(PO_4)_2$	1500 for 10 minutes	9.1	22 000		−97	[69]
141	$BaO:B_2O_3:SiO_2$ (50:40:10) glass	<800	9.2	1220	13.1	−43	[27]
142	$Y_2BaCu_{0.9}Mg_{0.1}O_5$	1250	9.2	36 798		−36	[37]
143	$(1-x)LiYW_2O_8-xBaWO_4+y$ wt% B_2O_3 ($x=0.48$, $y=0.5$)	930	9.2	28 100		−52	[12]
144	$0.88CaMgSi_2O_6-0.12CaTiO_3$ + 1 wt% $Li_2CO_3-V_2O_5$	880 for 2 hours	9.2	46 200		1.3	[70]
145	$SmBO_3$		9.3	11 000			[38]
146	$0.84MgAl_2O_4-0.16TiO_2$	1410	9.3	82 200	10.1	−27	[56]
147	$BaMoO_4$	900	9.3	37 200		−79	[32]
148	$Zn_2SiO_4+TiO_2$	1250	9.3	113 000		1	[19]
149	$0.84Al_2O_3-0.16TiO_2$ + 4 wt% MCAS glass	1250	9.4	8200		10	[61]
150	Y_2BaCuO_5		9.4	3830	12	−35	[71]
151	$\alpha-Ca_3(PO_4)_2$	1500 for 8 hours	9.4	15 200		−97	[69]
152	$0.88CaMgSi_2O_6-0.12CaTiO_3$	1300 for 2 hours	9.4	50 800		6	[70]
153	$Mg_3(VO_4)_2$	1050	9.4	65 500		−90	[16]
154	$MgCo_2(VO_4)_2$	900 for 5 hours	9.4	78 900		−95	[72]
155	$SrMoO_4$	1050	9.49	61 000		−67	[32]
156	$La_2O_3-2B_2O_3-0.5ZnO+La_2O_3-3B_2O_3-0.5ZnO$	900	9–10	72 000	13		[49]
157	$BaO:B_2O_3:SiO_2$ (50:30:20) glass		9.5	1256	13.5	−95	[27]
158	$Y_2BaCu_{0.8}Mg_{0.2}O_5$		9.5	42 300	−38		[37]
159	$BaO:B_2O_3:SiO_2$ (50:20:30) glass		9.61	1310	14.3	−114	[27]
160	$ZnO-B_2O_3-SiO_2-MMT-20$ (44.77:17.59:7.05:30.59)	900	9.7	7000	8	8.8	[58]
161	$CrTaO_4$		9.7	1600	4		[15]
162	$Y_2BaCu_{0.6}Ni_{0.4}O_5$		9.7	36 000	−26.5		[73]
163	β-$Ca_3(PO_4)_2$	1125 for 8 hours	9.7	10 300		−47	[69]
164	$(Mg_{1/2}Ca_{1/2})WO_4$ + 1 wt% Li_2WO_4	950	9.9	30 150		−63	[12]
165	$MgWO_4$	950	9.9	5400			[74]

No.		Sintering temp. (°C)	ε_r	*Qf* (GHz)	*f* (GHz)	τ_f	Reference
166	$TiO_2-CaAl_2Si_2O_8$	960 for 0.5 hour	10	22 500		−10 to −60	[46]
167	$Y_2Ba(Cu_{0.8}Mg_{0.2})O_5$ (CIP)		10	49 180		−40	[37]
168	CAS-T10 glass ($CaO:Al_2O_3:SiO_2:TiO_2:B_2O_3$)	950	10	22 500	10	−15	[46]
169	Al_2O_3 + 0.5 wt% TiO_2	1550 for 5 hours	10	453 000	9		[75]
170	Al_2O_3	1550 for 5 hours	10	335 000		−60	[75]
171	$CaO-4ZnO-Ta_2O_5-TiO_2$	1225	10	15 000	8.4	−60	[67]
172	$Mg_4Nb_2O_9$ + 3 wt% LiF	950 for 10 hours	10	116 420		−72	[76]
173	$Mg_4Ta_2O_9$	1450	10	345 000		−70	[77]
174	Al_2O_3 + 500 ppm TiO_2		10	500 000			[78]
175	Mg_4NbSbO_9	1450 for 10 hours	10	280 000		−70	[79]
176	$BaTeO_3$	800	10	34 000		−54	[80]
177	Al_2O_3 + 0.015 vf TiO_2		10	300 000	10	0	[81]
178	$CaWO_4$	1150	10.0	75 000		−24	[45, 82]
179	Al_2O_3	1550	10.05	680 000		−60	[54]
180	$Y_2BaCu_{0.1}Ni_{0.9}O_5$		10.1	5830	12.3	−36	[17]
181	Al_2O_3		10.15	1 191 580			[83]
182	$(1-x)LiYW_2O_8-xBaWO_4$ + y wt% B_2O_3 (x = 0.46, y = 0.5)	900	10.2	24 300		−21	[12]
183	$Ce_2(WO_4)_3$	1025	10.2	10 500		−25	[84]
184	$0.88ZnAl_2O_4-0.12TiO_2$	1380	10.3	79 800	11.1	−21.5	[59]
185	$Mg_3Yb_4Al_{44}O_{75}$	1680	10.3	41 000		−57	[43]
186	$CaMo_{1..02}O_4$	1300 for 2 hours	10.3	71 000			[85]
187	$NdPO_4$	1300 for 2 hours	10.3	59 500		−47	[55]
188	$MgO-1.2Al_2O_3-2.8SiO_2-1.2TiO_2-0.8CeO_2$	1150 for 2 hours	10.4	15 300		−5	[86]
189	CAS-TB glass ($CaO:Al_2O_3:SiO_2:TiO_2:B_2O_3$)	950	10.5	14 200	10	−20	[46]
190	$0.83ZnAl_2O_4-0.17TiO_2$ + 10 wt% BBSZ + 0.3 wt% LiF	925 for 10 hours	10.5	14 500	5.5	−28	[87]
191	$MgTe_2O_5$	700 for 4 hours	10.5	61 000	5.3	−45	[88]
192	$Ca_3ZrSi_2O_9$	1400	10.6	93 300		−77	[53]
193	$MgTiO_3-CaTiO_3$ (MMT-20) + $SiO_2-B_2O_3-BaO$	900	10.6	6000	7		[52]
194	$0.83ZnAl_2O_4-0.17TiO_2$ + 10 wt% BBSZ glass	950 for 10 hours	10.6	10 000	5.5	−23	[87]
195	$0.75MgAl_2O_4-0.25TiO_2$	1450	10.7	105 400	7.5	−12	[56]
196	$CoWO_4$	1200	10.7	38 600			[74]

No.		Sintering temp. (°C)	ε_r	Qf (GHz)	f (GHz)	τ_f	Reference
197	$CaMoO_4$	1100	10.8	89 700		−57	[32]
198	$CaCu_2Nb_2O_8$ + 3 wt% V_2O_5	935	10.8	9300		−16	[89]
199	$CaGe_2O_5$	1180	10.9	39 000	10		[66]
200	$0.7Ca_2P_2O_7{-}0.3TiO_2$	1200 for 2 hours	10.9	44 000		−11	[90]
201	$0.76Mg_2SiO_4{-}0.24TiO_2$		11	85 000		0	[28, 91]
202	Sm_2BaCuO_5:Co		11	89 000		−7	[92]
203	$NdBO_3$		11	17 000			[38]
204	Ba_2MgTeO_6 + 0.2 wt% B_2O_3	1200	11	25 000	5.5	−16	[93]
205	Mg_2SiO_4 + 24 wt% TiO_2	1200	11	82 000		0	[26]
206	($TiO_2{-}B_2O_3{-}$:Ca−Al−Si−O) glass		11	1400	10		[46]
207	$Mg_5Ta_4O_{15}$ (MgO calcined)	1560	11	18 100	9.06	−54	[94]
208	$Mg_5Nb_4O_{15}$ (MgO calcined)	1450	11	37 400	8.3	−53	[94]
209	$Mg_4Nb_2O_9$	1300	11	210 000		−70	[77]
210	$Co_4Nb_2O_9$	1200	11	5000		−10	[77]
211	$(Mg_{4-x}Mn_x)Nb_2O_9$		11–16	21 000–50 000			[95]
212	PbO:B_2O_3 (40:60) glass		11.1	1320	12.22	−43	[27]
213	$LaPO_4$	1400 for 2 hours	11.1	64 500		−56	[55]
214	$SmPO_4$	1400 for 2 hours	11.1	60 500		−55	[55]
215	$NiCu_2Nb_2O_8$ + 3 wt% V_2O_5	935	11.2	5760		−11.7	[89]
216	$Cu_3Nb_2O_8$	910	11.2	25 560		−3.7	[89]
217	$MgCu_2Nb_2O_8$ + 3 wt% V_2O_5	935	11.3	2900		−27	[89]
218	$ZnCu_2Nb_2O_8$ + 3 wt% V_2O_5	935	11.4	10 200		−23	[89]
219	$Y_2BaCu_{0.4}Ni_{0.6}O_5$	1360	11.5	45 200	13.11	−20.3	[73]
220	$BaTi(BO_3)_2$	1000 for 2 hours	11.5	2300			[96]
221	$Mg_4Ta_2O_9$	1450	11.5	347 000		−70	[97]
222	$Y_3Ga_5O_{12}$	1500	11.56	104 700		−21	[98]
223	$Mg_4(Nb_{2-x}V_x)O_9$ ($x = 0.0625$)	1025	11.6	160 250		−75	[99]
224	$0.83ZnAl_2O_4{-}0.17TiO_2$ + 1 wt% BBSZ glass	1300 for 4 hours	11.6	49 000		−10	[87]
225	$ZnAlO_4{-}0.21TiO_2$	1500 for 3 hours	11.6	74 000	6.5	−0.4	[100]
226	$CaCu_2Nb_2O_8$	1110	11.6	2300		−17	[89]
227	$CePO_4$	1400 for 2 hours	11.6	68 300		−46	[55]
228	$0.6LiYW_2O_8{-}0.4BaWO_4$	900	11.7	19 750		14	[12]

No.		Sintering temp. (°C)	ε_r	*Qf* (GHz)	*f* (GHz)	τ_f	Reference
229	$0.83ZnAl_2O_4{-}0.17TiO_2 + 0.2$ wt% BBSZ glass	950 for 4 hours	11.7	120 000		−7.3	[87]
230	$Yb_2Ba(Cu_{0.75}Zn_{0.25})O_5$		11.7	11 200		−43.3	[44]
231	$CaMoO_4$ (hot pressed)	1100	11.7	55 000		−60	[85]
232	$La(Mg_{0.5}Ti_{0.5})O_3 + 0.6B_2O_3{-}0.12La_2O_3{-}0.28MgO$	850 for 2 hours	11.8	14 700		7.4	[101]
233	Mg_4NbTaO_9	1100	11.8	281 670		−66	[102]
234	$Mg_4Nb_{1.5}Ta_{.5}O_9$	1100	11.9	234 520		−67	[102]
235	$BaO{-}SrO{-}SiO_2{-}ZrO_2$	<1000	12	1000	5		[103]
236	$CoCu_2Nb_2O_8 + 3$ wt% V_2O_5	885	12	7530		−18	[89]
237	$CaO{-}4Co_3O_4{-}Nb_2O_5{-}TiO_2$	1100	12	28 000	6.9	−42	[67]
238	$Mg_4(TaNb_{1-x}V_x)O_9$ ($x = 0.025$)	1200	12	200 000		−73	[104]
239	$AlNbO_4$	1250	12	34 000		−56	[105]
240	$PbO{:}B_2O_3{:}SiO_2$ (40:20:40) glass		12.11	1420	12.2	−31	[27]
241	$Dy_3Ga_5O_{12}$	1450	12.15	42 100	14.6	−22	[98, 106]
242	$Mg_3CoNb_2O_9$	1150	12.3	34 560		−64	[102]
243	Yb_2BaZnO_5		12.3	27 000		−59.9	[44]
244	$0.72Ba(Mg_{1/2}W_{1/2})O_3{-}0.28BaTiO_3$	1500 for 6 hours	12.3	11 000		−5.3	[107]
245	$Sm_3Ga_5O_{12} + TiO_2$	1450	12.33	234 700	14.1	−16	[106]
246	$Sm_3Ga_5O_{12}$	1450	12.35	192 200	14	−19	[106]
247	$Nd_3Ga_5O_{12}$	1400	12.37	137 800	13.8	−33	[106]
248	$0.9Al_2O_3{-}0.1TiO_2$ Annealed at 1000 C	1350	12.4	117 000		1.5	[108]
249	$Al_2O_3{-}TiO_2{:}MnO$	1300	12.4	274 000	76	0.4	[109]
250	$0.9Al_2O_3{-}0.1TiO_2$	1300 for 2 hours	12.4	148 000		1.5	[110]
251	$(1-x)LiYW_2O_8{-}xBaWO_4 + y$ wt% B_2O_3 ($x = 0.2$, $y = 0$)	900	12.4	12 100		33	[12]
252	$Eu_3Ga_5O_{12}$	1400	12.48	169 100	14.6	−17	[106]
253	$LaBO_3$		12.5	53 000			[38]
254	$Ba_3ZnNb_{2-x}Sb_xO_9$ ($x = 1.875$)		12.5	2290	7.6		[111]
255	$Ba_3(VO_4)_2 + 0.5$ wt% B_2O_3	950	12.5	41 065		38.8	[29]
256	$Mg_4Nb_2O_9 + 3$ wt% LiF	950 for 10 hours	12.6	116 410		−72	[76]
257	$0.83ZnAl_2O_4{-}0.17TiO_2$	1410	12.6	100 200	10	0	[59, 112]
258	$Yb_2Ba(Cu_{0.25}Ni_{0.75})O_5$		12.6	50 040		−40.9	[44]
259	$YSmBaCuO_5$		12.6	25 130	11	−29.9	[113]
260	$Mg_3(VO_4)_2{-}0.5Ba_3(VO_4)_2 + 0.0625$ wt% Li_2CO_3	950 for 5 hours	12.6	74 400		−6	[114]
261	$Mg_4Nb_2O_9 + 3$ wt% LiF	850 for 10 hours	12.6	103 600		−70	[60]

No.		Sintering temp. (°C)	ε_r	*Qf* (GHz)	*f* (GHz)	τ_f	Reference
262	$PbO:B_2O_3:SiO_2$ (40:40:20) glass		12.74	1700	12	−69	[27]
263	Tm_2BaCuO_5	1250	12.8	14 400	9.77	−14.8	[113]
264	$NiCu_2Nb_2O_8$	985	12.8	4240		481	[89]
265	$Y_2Ba_{0.7}Sr_{0.3}CuO_5$		12.9	2960	10.7	+1.6	[115]
266	$0.89Al_2O_3-0.11TiO_2-0.5$ wt% ZnO	1350	12.9	187 000		−2	[116]
267	$Mg_4Nb_2O_9$	1300 for 10 hours	12.9	217 390	–	−70	[99, 102]
268	$Mg_3La_4Al_{44}O_{75}$	1680	13	7700		3	[43]
269	$2CaO-3ZnO-Ta_2O_5-TiO_2$	1300	13	20 000	6.6	−24	[67]
270	$Ba_3ZnNb_{2-x}Sb_xO_9$ ($x=2$)		13	1550	7.7		[111]
271	$Ba_3(VO_4)_2$	1200	13	46 700		17	[114]
272	$Mg_3(VO_4)_2-0.5Ba_3(VO_4)_2+0.0625$ wt% Li_2CO_3	950 for 5 hours	13	74 000		−6	[114]
273	$0.67Ba(Mg_{1/2}W_{1/2})O_3-0.33BaTiO_3$	1500 for 6 hours	13.1	35 000		−6.36	[107]
274	$MgWO_4$	1150	13.1	69 000		−58	[45]
275	$Al_2O_3-WO_3-TiO_2$	1150	13.18	3580	6.9	12	[117]
276	$Mg_2Co_2Nb_2O_9$		13.2	14 300		−51	[102]
277	$(Mg_{0.95}Ca_{0.05})TiO_3+BaO-B_2O_3-SiO_2$ (50:50 wt%)	900	13.2	10 000		–	[118]
278	$Ca_2MgTeO_6+0.2$ wt% B_2O_3	1250	13.2	81 000	5.5	−81	[93]
279	$Ba_2Ti_9O_{20}+50$ vol% BBS glass	900	13.2	1150			[96]
280	$0.5MgAl_2O_4-0.5TiO_2$	1460	13.2	88 000	6.8	−4.4	[56]
281	$YErBaCuO_5$		13.3	16 050	10.6	−34.2	[113]
282	$NiWO_4$	1200	13.3	24 900			[74]
283	$0.7CaWO_4-0.3LaNbO_4$	1150	13.3	50 000		−9	[82]
284	$Sm_2BaCu_{0.5}Zn_{0.5}O_5$	1280	13.4	65 740	10.6	−6.38	[119]
285	Er_2BaCuO_5		13.5	12 560	11	−26.1	[113]
286	$0.64Ba(Mg_{1/2}W_{1/2})O_3-0.36BaTiO_3$	1500 for 6 hours	13.5	14 000		−6.03	[107, 120]
287	$ZnWO_4$		13.5	62 800			[74]
288	$MgWO_4$	1050	13.5	69 000		−58	[45]
289	$ZnMnW_2O_8$	950	13.7	10 670		−17	[12]
290	$PbO:B_2O_3:SiO_2$ (50:40:10) glass		13.8	880	10.7	−98	[27]
291	$Y_2BaCu_{0.2}Ni_{0.8}O_5$	1390	13.8	87 200	12.8	−16.5	[73]
292	$Ba_2CeV_3O_{11}$	1025	13.8	10 000		−14	[84]
293	$YHoBaCuO_5$		13.9	12 056	10.7	−29.8	[113]
294	$Mg_4Ta_2O_9$	1200	14	350 000	–	−60	[97]
295	$YDyBaCuO_5$		14	42 600	10.8	−22.1	[113]
296	$LaMgAl_{11}O_{19}$	1700	14	28 000	7	−12	[121]

No.		Sintering temp. (°C)	ε_r	*Qf* (GHz)	*f* (GHz)	τ_f	Reference
297	$Y_2BaCu_{0.75}Zn_{0.25}O_5$	1270	14	56 230	10.8	−39	[71]
298	$Mg_5Nb_4O_{15}$	1475	14	14 600	7.3	−58	[94]
299	$5Co_3O_4-Ta_2O_5-TiO_2$	1150	14	48 000	6.5	−43	[67]
300	$Mg_3Ce_4Al_{44}O_{75}$	1680	14	9000		11	[43]
301	$Ba_3NiSb_2O_9$		14	41 840	7.8	−4.7	[111]
302	$YGdBaCuO_5$		14	14 300	10.9	−35.2	[113]
303	$0.80ZnAl_2O_4-0.20TiO_2$	1420	14	90 700	9.66	6.2	[59]
304	$Yb_2Ba(Cu_{0.5}Zn_{0.5})O_5$		14.2	20 630		−47.5	[44]
305	$Y_2BaCu_{0.5}Zn_{0.5}O_5$	1270	14.2	110 660	10.7	−41.5	[71]
306	Mg_2TiO_4	1500	14.2	160 000			[122, 123]
307	$0.8(Al_{1/2}Ta_{1/2})O_2-0.2(Mg_{1/3}Ta_{2/3})O_2$	1450	14.2	62 150		−40.5	[64]
308	80 wt% ($La_2O_3-B_2O_3-TiO_2$ in 20:60:20 mol%) + 20 wt% $BaNd_2Ti_5O_{14}$	850	14.2	9800	7.5	94	[124]
309	$Mg_{0.95}Co_{0.05}TiO_3$	1275 for 4 hours	14.3	128 000	7	−51	[125]
310	$Sr_2MgTeO_6 + 0.2$ wt% B_2O_3	1250	14.3	27 400	5.5	−60	[93]
311	$BaO-0.35MgO-0.33WO_3-0.32TiO_2$	1500 for 6 hours	14.4	74 000		−8.8	[107]
312	80 wt% ($La_2O_3-B_2O_3-TiO_2$ in 20:60:20 mol%) + 20 wt % $BaNd_2Ti_5O_{14}$	800	14.5	9100	7.5	86	[124]
313	$BaO-0.34MgO-0.32WO_3-0.34TiO_2$	1500 for 12 hours	14.5	107 000		−7.6	[107, 120]
314	$Mg_3Pr_4Al_{44}O_{75}$	1680	14.5	10 000		23	[43]
315	$Y_2BaCu_{0.6}Ni_{0.4}O_5$	1340	14.5	36 000	14.5	−26.5	[73]
316	$MnWO_4$	1000	14.5	32 000		−64	[45]
317	$Ba_3MgNb_{2-x}Sb_xO_9$ ($x = 0.5$)		14.7	81 300	6.3	4.8	[111]
318	$Zn_3Nb_2O_8 + 3$ wt% ($0.29BaCO_3 + 0.71CuO$)	950	14.7	8200	8.3		[126]
319	$LiYW_2O_8$	900	14.8	9550		−64	[12]
320	$Ba_3NiNb_{2-x}Sb_xO_9$ ($x = 1.875$)		14.8	38 380	6.8	−10	[111]
321	$Yb_2Ba(Cu_{0.25}Zn_{0.75})O_5$		14.9	52 810		−44.9	[44]
322	Dy_2BaCuO_5		14.9	31 610	10.56	−6.4	[113]
323	Sr_2TiO_4	1300 for 5 hours	15	1600	4		[15]
324	$5MgO-Nb_2O_5-TiO_2$	1325	15	59 000	6.8	−77	[67]
325	$Mg_3Nd_4Al_{44}O_{75}$	1680	15	11 000		35	[43]
326	$Ba_3MgNb_{2-x}Sb_xO_9$ ($x = 1.875$)		15	84 100	7.25	2.8	[111]
327	$Ba(Mg_{1/2}W_{1/2})O_3$	1550	15	57 300	12.7		[127]
328	$BaO-0.34MgO-0.33WO_3-0.33TiO_2$	1500 for 6 hours	15.1	72 000		−12.9	[107]
329	Al_2O_3 + Ca−Al−B−Si−O + Ba−(Sm,Nd)−Ti−O	870	15.1	2800	3		[65]
330	$MgCo_3Nb_2O_9$		15.2	14 290		−36	[102]

No.		Sintering temp. (°C)	ε_r	Qf (GHz)	f (GHz)	τ_f	Reference
331	$Y_2BaCu_{0.25}Zn_{0.75}O_5$	1270	15.2	70 080	9.95	−42	[71]
332	Ho_2BaCuO_5		15.3	9360	10.48	−19.3	[113]
333	$LaSrAlO_4$	1375	15.3	32 820		−17	[128]
334	$PbO:B_2O_3:SiO_2$ (60:20:20) glass		15.32	650	11.72	−124	[27]
335	Y_2BaZnO_5	1270	15.4	189 000	10	−41.5	[17, 71]
336	$BaO-0.35MgO-0.34WO_3-0.31TiO_2$	1500 for 6 hours	15.4	77 000		−7.6	[107]
337	$Ba(Mg_{1/3}Ta_{(2-2x)/3}W_{x/3}Ti_{/3})O_3$ ($x = 1$)	1550 for 4 hours	15.4	35 400		−25	[129]
338	$Sm_2Ba(Cu_{0.985}Co_{0.015})O_5$		15.5	59 300	10.3	−8.1	[92]
339	$Cu_3Nb_2O_8$	900 for 2 hours	15.6	48 400		−75	[130]
340	$BaO-0.33MgO-0.34WO_3-0.33TiO_2$	1500 for 6 hours	15.6	67 000		−10.3	[107]
341	$YAlO_3$		15.7	58 000	10	−59	[131]
342	$Mg_4Nb_2O_9$ + 3 wt% LiF + 6 wt% $CaTiO_3$	950 for 5 hours	15.7	22 100		−3	[132]
343	$YSmBaZnO_5$		15.85	63 210	9.9	−23	[17]
344	$Y_2Ba_{0.7}Sr_{0.3}Cu_{0.75}Zn_{0.25}O_5$		15.9	12 450		0.8	[115]
345	$MgCu_2Nb_2O_8$	1010	15.9	6780		−46	[89]
346	$Ba_3Ti_5Nb_6O_{28}$ + 5 wt% B_2O_3	900 for 2 hours	15.9	14 000		−13	[133]
347	$(0.4Bi_2O_3-La_2O_3-MgO-TiO_2)-0.6La(Mg_{0.5}Ti_{0.5})O_3$	900	15.9	14 300		35.3	[134]
348	$Zn_3Nb_2O_8$ + 3 wt% ($0.81MoO_3 + 0.19CuO$	950	15.9	10 200	8.2		[126]
349	75 wt% $ZnNb_2O_6-TiO_2$ + 25 wt% ($SiO_2-B_2O_3-Al_2O_3$)	875	15.9	15 000		−20	[4]
350	Gd_2BaCuO_5		16	3320	11.05	−27.7	[113]
351	$Mn_4Nb_2O_9$	–	16	50 000		–	[95]
352	$3CaO-2ZnO-Ta_2O_5-TiO_2$	1325	16	34 500	6.2	−49	[67]
353	$NdYBaZn_{0.45}Cu_{0.55}O_5$	1250 for 50 hours	16	100 270		–	[135]
354	$Ba_{10}Ta_{7.04}Sn_{1.2}O_{30}$		16	30 000		20	[136]
355	$(Mg_{0.95}Zn_{0.05})TiO_3$	1320 for 4 hours	16	210 000		−60	[137]
356	$Y_{1.5}Sm_{0.5}BaZnO_5$		16	120 000	−32		[138]
357	Tm_2BaZnO_5		16.1	8040	9.9	−20	[17]
358	$Sm_2Ba(Cu_{0.995}Co_{0.005})O_5$		16.1	87 800	10	−8.2	[92]
359	$Nd_2BaZn_{0.5}Cu_{0.5}O_5$		16.2	36 570		−13.2	[135]
360	$Zn_2Te_3O_8$	620	16.2	66 000	4.9	−60	[139]
361	$(Mg_{0.95}Ca_{0.05})TiO_3$ + 3 mol% B_2O_3	1100	16.2	62 000		50	[140]
362	Er_2BaZnO_5	1300	16.3	6836	9.9	−28	[17]

No.		Sintering temp. (°C)	ε_r	*Qf* (GHz)	*f* (GHz)	τ_f	Reference
363	$BaO-33MgO-0.35WO_3-0.32TiO_2$	1500	16.3	77 000		−9.5	[107]
364	$Ba_3MgNb_{2-x}Sb_xO_9$ ($x = 1$)		16.3	33 400	6.7	−4	[111]
365	$Co_4Nb_2O_9$	1100	16.4	5000		−11	[102, 141]
366	$Y_2Ba_{0.7}Sr_{0.3}Cu_{0.5}Zn_{0.5}O_5$		16.5	17 670	10.1	−1.6	[115]
367	$Y_2Ba_{0.7}Sr_{0.3}Cu_{0.15}Zn_{0..85}O_5$		16.5	23 640	10.5	−17.5	[115]
368	Sm_2BaCuO_5		16.5	53 200	9.9	−5.2	[92]
369	$Mg_3Sm_4Al_{44}O_{75}$	1680	16.5	11 000		93	[43]
370	$MgTiO_3$ (slow cooled 1°/minute)	1350	16.5	220 000		−55	[142]
371	$0.84Ba(Mg_{1/2}W_{1/2})O_3-0.16BaTiO_3$	1500	16.6	12 000		−11.3	[107]
372	$CoCu_2Nb_2O_8$	985	16.6	36 800		−37	[89]
373	$(Mg_{0.95}Ca_{0.05})TiO_{3+}$ 5 mol% V_2O_5	1000	16.6	13 700		49.9	[140]
374	$ErNbO_4$	1500	16.6	43 900		−64	[105]
375	$0.6Ba(Mg_{1/2}W_{1/2})O_3-0.4BaTiO_3$	1500	16.7	15 000		12.3	[107]
376	$Y_2Ba_{0.7}Sr_{0.3}ZnO_5$		16.7	4920	10.8	−35.3	[115]
377	$ZnCu_2Nb_2O_8$	900 for 2 hours	16.7	41000		−77	[143]
378	$Sm_2BaCu_{0.99}Co_{0.01}O_5$		16.8	90 700	9.9	−9.2	[17, 92]
379	70 wt% ($La_2O_3-B_2O_3-TiO_2$ in 20:60:20 mol%) + 30 wt% $BaNd_2Ti_5O_{14}$	800	16.8	5900	7.1	109	[124]
380	$(Mg_{0.95}Co_{0.05})TiO_3$		16.8	230 000	10	−54	[144]
381	$Sm_2BaCu_{0.25}Zn_{0.75}O_5$		16.9	42 200	−4.6		[119]
382	$3CaO-2ZnO-Ta_2O_5-TiO_2$	1400	17	30 000	6.6	−47	[67]
383	$CeO_2-0.5WO_3-0.5TiO_2$	1130	17	45 500		7	[145]
384	$Mg_5Ta_4O_{15}$	1550	17	14 400	7.2	−15	[94]
385	$BaNb_2O_6$		17	2600	7.01		[146]
386	$Ba(Mg_{1/2}W_{1/2})O_3$	1550 for 6 hours	17	57 000		−34	[107, 127]
387	$Ca_5Ta_2HfO_{12}$	1700	17	18 000	5.9	−32	[147]
388	$MgTiO_3$		17	166 400		−50	[148, 149]
389	$ErAlO_3$		17	44 200	10	−40	[131]
390	Dy_2BaZnO_5	1320	17.1	29 669	9.9	−1.5	[17]
391	$CuNb_2O_6$	1000	17.1	7100	7.4	−45	[146]
392	$BaO-0.34MgO-0.35WO_3-0.31TiO_2$	1500 for 6 hours	17.1	75 000		−8.1	[107]
393	Eu_2BaCuO_5		17.1	9820		−25.4	[150]
394	$LaSrAlO_4$	1450	17.1	30 770		3.4	[128]
395	$(Mg_{0.95}Zn_{0.05})TiO_3$	1300	17.1	264 000	7	−40	[151]
396	$NdYBaZn_{0.45}Cu_{0.55}O_5$		17.1	100 300	−30		[135]
397	$Ba(Ni_{1/2}W_{1/2})O_3$	1450	17.1	36 300	13.3		[127]
398	$In_2O_3-WO_3-TiO_2$	1175	17.15	5100	6.4	−68	[117]

No.		Sintering temp. (°C)	ε_r	Qf (GHz)	f (GHz)	τ_f	Reference
399	Gd_2BaZnO_5	1280	17.2	2580	9.8	−27	[17]
400	$Eu_2BaCu_{0.25}Zn_{0.75}O_5$		17.2	57 920		−28.8	[150]
401	Ho_2BaZnO_5	1300	17.2	6200	9.8	−23	[17]
402	$CaNb_2O_6$	1350	17.3	49 600	6.9	−53	[146]
403	$Mg_3Eu_4Al_{44}O_{75}$	1680	17.3	11 000		147	[44]
404	$LuNbO_4$	1500	17.4	56 600		−64	[105]
405	$Ba(M_{0.33}Ta_{0.13}Ti_{0.267}W_{0.267})O_3$	1560	17.4	43 780	7.1	−28.7	[129]
406	$CaTeO_3$	840	17.4	49 300	10		[66]
407	$BaTe_4O_9$	500	17.5	54 700		−90	[80, 152]
408	$0.96MgTiO_3-0.036SrTiO_3 + 4$ wt% CuO	1070 for 2 hours	17.5	25 100			[153]
409	$MgTi_2O_5$	1500 for 3 hours	17.4	47 000			[123]
410	$0.74CaWO_4-0.26TiO_2$	1250	17.5	27 000		0	[154]
411	$DyAlO_3$	1650 for 2 hours	17.6	38 000	10	−34	[131]
412	Nd_2BaCuO_5		17.6	2200		−18.4	[135]
413	$ZnWO_4$	1100	17.6	65 000		−60	[45]
414	MWF−38 + 10 wt% $Li_2O-B_2O_3-SiO_2-CaO-Al_2O_3$ (28:27:30:5:10)	875	17.7	3700		−15	[155]
415	$MgTiO_3 + 1$ mol% Nb_2O_5	1350	17.7	175 000		–	[156]
416	$Ce_2O_2-WO_3-TiO_2$	1025	17.8	13 100	6.2	85	[117]
417	$DyNbO_4$	1250	17.8	38 500		−66	[105]
418	$(Zr_{0.8}Sn_{0.2})TiO_4 + 10$ wt% $BaO-B_2O_3-SiO_2-Li_2O-CuO$	950 for 4 hours	17.8	12 700		1.1	[157]
419	$NiTiO_3$	1475 for 4 hours	17.8	13 900		−51	[158]
420	$0.96MgTiO_3-0.036SrTiO_3$	1170 for 2 hours	17.9	30 400	9	5	[153]
421	$Eu_2BaCu_{0.5}Zn_{0.5}O_5$		17.9	49 849		−29.5	[150]
422	$2/3LaCa_{0.5}Zr_{0.5}O_3-1/3CaTiO_3$	1575	18	26 000	5.3	−75	[159]
423	$5MgO-Ta_2O_5-TiO_2$	1325	18	114 000	6.61	−56	[67]
424	$Zn_{0.6}Mg_{0.4}TiO_3 + 5$ wt% $B_2O_3-SiO_2-ZnO-K_2O$	1100	18	29 400		–	[160]
425	$Sm_2BaCu_{0.5}Zn_{0.5}O_5$		18	65 700	−6.4		[161]
426	$(1-x)LaCa_{0.5}Zr_{0.5}O_3-xCaTiO_3$ $(x = 1/3)$		18	16 000		−75	[159]
427	$0.5CeO_2-0.5Sm_2O_3$	1650	18	90 000		−30	[162]
428	$5ZnO-Nb_2O_5-TiO_2$	1050	18	6000	5.9	−57	[67]
429	$5MgO-Ta_2O_5-TiO_2$	1325	18	114 000	6.6	−47	[67]
430	$CeO_2-0.5NiO-0.5TiO_2$	1200	18	25 300		−58	[145]
431	Eu_2BaZnO_5		18.1	23 360		−25.4	[150]

No.		Sintering temp. (°C)	ε_r	*Qf* (GHz)	*f* (GHz)	τ_f	Reference
432	$MgTiO_3 + 6\,wt\%\ CuO-Bi_2O_3-V_2O_5$	900 for 2 hours	18.1	20 300		−57	[163]
433	$Nd_2BaZn_{0.25}Cu_{0.75}O_5$	1250 for 10 hours	18.1	25 170		−18	[135]
434	$BaO-0.32MgO-0.28WO_3-0.4TiO_2$	1500 for 6 hours	18.1	48 000		−3	[120]
435	$Ba(Ni_{1/2}W_{1/2})O_3$	1450	18.1	52 000	8.22	−45	[164]
436	$CaSmAlO_4$		18.2	51 060		−3	[165]
437	$CaNdAlO_4$		18.2	17 980		−52	[165]
438	$0.995MgO-0.005BaO-TiO_2$	1320	18.3	18 500	9.83		[166]
439	$Mg_3Gd_4Al_{44}O_{75}$	1680	18.3	4800		175	[43]
440	$Mg_3Tb_4Al_{44}O_{75}$	1680	18.3	5900		200	[43]
441	$0.5ZnNb_2O_6-0.5Zn_3Nb_2O_8-12\,wt\%\ ZnCu_3B_2O_5$	875	18.3	39 750		−88	[167]
442	$Sr(Ni_{1/2}W_{1/2})O_3$	1570	18.3	56 000	8.4	−50	[164]
443	$0.8MgNb_2O_6-0.2CaTiO_3$	1300	18.35	73 700		−45	[168]
444	$GdAlO_3$	1650 for 2 hours	18.4	11 000	10	−54	[131]
445	$(Zr_{0.8}Sn_{0.2})TiO_4 + 10\,wt\%$ $BaO-B_2O_3-SiO_2-Li_2O-CuO$	950 for 8 hours	18.4	10 500		−0.3	[157]
446	$BaNd_2Ti_5O_{14} + La_2O_3-B_2O_3-TiO_2$	750	18.4	6100		3.6	[169]
447	Sm_2BaZnO_5		18.5	35 500		−6.4	[170]
448	$DyTiNb_xTa_{1-x}O_6\ (x = 0.05)$	1575	18.65	31 000	5.7	−28	[171]
449	$Nd_2BaZn_{045}Cu_{0.55}O_5$	1250 for 10 hours	18.8	44 100		−19.9	[135]
450	$SmNbO_4$	1250	18.8	56 300		−40	[105]
451	$CaNb_2O_6$	1350	18.8	49 600		−53	[172]
452	$CaYAlO_4$	1450 for 3 hours	18.9	39 980		6	[165]
453	Sr_2AlNbO_6 (oxygen atm)	1550	19	16 000	8.3	−5	[173]
454	$YTiNbO_6$	1400	19	8820	8.2	−45	[174]
455	$Sm_{0.1}Y_{0.9}TiNbO_6$	1420	19	11 700		−42	[175]
456	$Zn_{0.6}Mg_{0.4}TiO_3 + 5\,wt\%$ B−Si−Zn−K glass	950	19	18 950			[176]
457	$ZnTiO_3$	1100	19	30 000	10	−55	[177]
458	$0.96MgTiO_3-0.036SrTiO_3 + 2\,wt\%\ B_2O_3$	1170 for 2 hours	19	75 300	9	−9	[153]
459	$Tb(Ti_{1/2}W_{1/2})O_4$	1375	19.05	5900	6.6	−6	[117]
460	$0.99MgO-0.01BaO-TiO_2$	1320	19.1	21 500	9.53		[166]
461	$80\,wt\%\ ZnNb_2O_6-TiO_2 + 20\,wt\%$ $(SiO_2-B_2O_3-Al_2O_3)$	875	19.1	9600		9	[4]
462	$85\,wt\%\ ZnNb_2O_6-TiO_2 + 15\,wt\%$ $(CaO-B_2O_3-SiO_2)$	875	19.2	11 000		17	[4]
463	$YbTiTaO_6$	1560	19.3	31 800	6.2	−41	[178]

No.		Sintering temp. (°C)	ε_r	*Qf* (GHz)	*f* (GHz)	τ_f	Reference
464	$CaTe_2O_5$	780	19.3	13 400	10		[66]
465	$MgTi_2O_5 + 10\,wt\%$ LBS glass	950 for 2 hours	19.3	6800	–	−16	[47]
466	$Ba(Co_{1/2}W_{1/2})O_3$	1390	19.3	21 000	7.76	−55	[164]
467	TeO_2	640 for 15 hours	19.3	30 000	4	−119	[179]
468	$MgTi_2O_5 + 10\,wt\%$ LBS glass	950 for 2 hours	19.3	6800	–	−16	[47]
469	$Zn_2Te_3O_8 + 4\,wt\%\ TiO_2$	650	19.3	27 000		−8.7	[139]
470	$LaNbO_4$	1250	19.3	54 400		9	[105]
471	$Y(Ti_{1/2}W_{1/2})O_4$	1425	19.33	6200		−19	[117]
472	$0.94CaNb_2O_6{-}0.06CaTiO_3$	1300	19.54	69 500		−65	[168]
473	$Ca(La_{1/2}Ta_{1/2})O_3$	1500	19.5	30 000	9.8	−41	[180]
474	$CoTiO_3$	1375 for 5 hours	19.5	62 500		−49	[158]
475	Sm_2BaZnO_5		19.5	35 500	−6.4		[119]
476	$ZnTiO_3 + 0.25\,wt\%\ V_2O_5$	900	19.5	2700	7.4		[186]
477	$90\,wt\%\ ZnNb_2O_6{-}TiO_2 + 10\,wt\%$ $(SiO_2{-}B_2O_3{-}Al_2O_3)$	900	19.5	9200		18	4
478	$90\,wt\%\ (Mg,Ca)TiO_3 + 10\,wt\%\ Li_2O{-}B_2O_3{-}SiO_2$	950	19.5	26 700		−12	[182]
479	$PbO{:}B_2O_3{:}SiO_2$ (70:20:10) glass		19.57	500	10.3	−155	[27]
480	$NdNbO_4$	1250	19.6	33 000		−24	[105]
481	Sm_2BaCuO_5		19.6	3397	11.36	−8.7	[113]
482	$(Mg_{0.95}Ca_{0.05})TiO_3 + 5\,mol\%\ B_2O_3$	1200	19.6	86 000	10	−3	[140]
483	$0.92Ba(Mg_{1/2}W_{1/2})O_3{-}0.08BaTiO_3$	1500 for 6 hours	19.6	37 000		−18.7	[107]
484	$Nd_{0.5}La_{1.5}BaZnO_5$		19.6	16 320		−1.4	[183]
485	$Ca(Sm_{1/2}Ta_{1/2})O_3$	1500	19.6	26 500	9.8	−24	[180]
486	$LiYbW_2O_8$	900	19.7	8720		45	[12]
487	$Ba(Mg_{0.33}Ta_{0.33}Ti_{0.167}W_{0.167})O_3$	1580	19.7	58 200	6.5	−11.1	[129]
488	$Dy(Ti_{1/2}W_{1/2})O_4$	1425	19.85	6000	6.6	−5	[117]
489	$MgNb_2O_6 + 2\,wt\%\ CuO$	1170	19.9	110 000	10	−44	[184]
490	$BaNd_2Ti_5O_{14}$: $(20La_2O_3{-}60B_2O_3{-}20TiO_2)$ (60:40 wt%)	850	19.9	8200			[185]
491	$Ca(Al_{1/2}Ta_{1/2})O_3$		20	8,500	–	−90	[186]
492	$Ca[(Li_{1/3}Nb_{2/3})_{0.95}Ti_{0.05}]O_{3-\delta} + 5\,wt\%\ Bi_2O_3$	900 for 3 hours	20	6500		−4	[187]
493	$La_{0.9}Nd_{0.1}NbO_4$	1250	20.0	45 000		−1	[105]
494	$Ca(Nd_{1/2}Ta_{1/2})O_3$	1500	20	2400	9.7	−16	[180]
495	$0.95MgTiO_3{-}0.05CaTiO_3 + 0.25\,wt\%\ CuO$	1275 for 4 hours	20	51 000	7	−8.3	[188]

No.		Sintering temp. (°C)	ε_r	*Qf* (GHz)	*f* (GHz)	τ_f	Reference
496	$Ca(Nd_{1/2}Nb_{1/2})O_3$	1500	20	17 500	9.6	−33	[180]
497	$GdTiNbO_6$	1385	20	9050	7.27	−52	[174]
498	$MgO-SiO_2-TiO_2$ + 15 wt% $ZnO-B_2O_3$ + 2.4 wt% Co_2O_3	1160	20	100 000	10		[189]
499	$Sm_{0.3}Y_{0.7}TiNbO_6$	1420	20	19 200		−33	[175]
500	$Ba_{1-3x/2}La_x(Mg_{1/2}W_{1/2})O_3$	1450 for 2 hours	20	87 700	–	−1	[176]
501	$LaNbO_4$		20	15 000		50	[82]
502	$CaO-4ZnO-Nb_2O_5-TiO_2$	1125	20	9000	5.9	−47	[67]
503	$5NiO-Nb_2O_5-TiO_2$	1125	20	8200	5.9	−64	[67]
504	$CaO-4MgO-Ta_2O_5-TiO_2$	1360	20	50 000	5.6	−33	[67]
505	$5NiO-Ta_2O_5-TiO_2$	1300	20	14 000	5.9	−53	[67]
506	$CaO-4Co_3O_4-Ta_2O_5-TiO_2$	1210	20	26 000	5.8	−30	[67]
507	$Ba_{1-3x/2}La_x(Mg_{1/2}W_{1/2})O_3$ ($x = 0.02$)	1450 for 2 hours	20	87 680	–	−1	[190]
508	$xBa(Mg_{1/2}W_{1/2})O_3-(1-x)BaTiO_3$ ($x = 0.92$)	1500 for 6 hours	20	37 000		−19	[107]
509	$0.95MgTiO_3-0.05CaTiO_3$	1400	20	56 000		0	[191]
510	$0.94MgTiO_3-0.06CaTiO_3$ + 0.25 wt% CuO	1275	20	48 000		−3	[188]
511	$0.75MgAl_2O_4-0.25TiO_2$		20	10 500		0	[56]
512	$Ba(Mg_{1/3}Ta_{(2-2x)/3}W_{x/3}Ti_{x/3})O_3$ ($x = 0.15$)	1550 for 4 hours	20	90 000		0	[129]
513	$Ba[Ti_{1-x}(Ni_{1/2}W_{1/2})_x]O_3$ ($x = 0.6$)	1425	20	42 000		−9.9	[192]
514	$Zn_3Nb_2O_8$ + 2 wt% V_2O_5 + 0.5 wt% CuO	800	20	36 000	11.8	–	[193]
515	$La(MgSn)_{0.5}O_3$	1600 for 4 hours	20.1	63 000		−78	[194]
516	$SrNb_2O_6$	1300	20.1	16 900	6.5	–	[146]
517	$ZnTiO_4$	1300 for 2 hours	20.2	19 000		−55	[195]
518	$Ba_3NiNb_{2-x}Sb_xO_9$ ($x = 0.5$)		20.2	16 780	6	−28.5	[111]
519	$SrLaGaO_4$	1275 for 3 hours	20.3	16 200		−34	[196]
520	La_2BaZnO_5		20.3	17 800		−0.9	[183, 197]
521	$NdLaBaZnO_5$		20.3	7900		−5	[183]
522	$ZnTiO_3$ + 0.5 wt% V_2O_5	900	20.3	5200	7.8		[181]
523	90 wt% $ZnNb_2O_6-TiO_2$ + 10 wt% $(Li_2O-B_2O_3-SiO_2)$	875	20.3	8200		5	[4]
524	$0.95(Mg_{0.95}Co_{0.05})TiO_3-0.05CaTiO_3$	1275 for 4 hours	20.3	107 000	7	−22.8	[125]
525	CeO_2: (at 30K)	1675	20.3	600 000	5.5		[198]
526	$Pr(Ti_{1/2}W_{1/2})O_4$	1300	20.3	6900	6.53	−20	[117]
527	$MnTa_2O_6$	1350	20.3	16 500		−44	[199]

No.		Sintering temp. (°C)	ε_r	Qf (GHz)	f (GHz)	τ_f	Reference
528	$SmAlO_3$	1650 for 2 hours	20.4	65 000	10	−74	[131]
529	$MnTiO_3$	1350 for 2 hours	20.4	15 200		−56	[158]
530	$Ba_3NiNb_{2-x}Sb_xO_9$ ($x=1$)		20.4	43 880	6.3	−18	[111]
531	$CoNb_2O_6$	1300 for 4 hours	20.5	81 000		−70	[146, 200]
532	$Sm_{0.4}Y_{0.6}TiNbO_6$	1400	20.5	15 000		−30	[175]
533	$ErTiTaO_6$	1560	20.6	85 500		−29	[178]
534	$Ca(La_{1/2}Nb_{1/2})O_3$	1500	20.6	38 000	9.4	−50.6	[180]
535	$MgTiO_3 + 5$ mol% Bi_2O_3-7 mol% V_2O_5	875	20.6	10 420	6.3		[201]
536	$0.97MgO-0.03BaO-TiO_2$	1320	20.6	32 600	9.35		[166]
537	$ZnTiO_3 + 0.75$ wt% V_2O_5	900	20.6	8800	8.2		[182]
538	$Nd_2BaZn_{0.8}Cu_{0.2}O_5$		20.7	11 680		−1.6	[135]
539	$LaAlO_3 + 0.25$ wt% CuO	1460	20.7	48 000		−80	[202]
540	$0.964MgTiO_3-0.036SrTiO_3$	1270	20.8	71 000		−1.3	[203]
541	$La_{2-x/3}Na_x(Mg_{1/2}W_{1/2})O_3$ ($x=0.5$)	1450	20.8	5700		−47	[204]
542	$Nd_2BaZn_{0.7}Cu_{0.3}O_5$		20.8	19 793		−3.1	[135]
543	$0.94(Mg_{0.95}Co_{0.05})TiO_3-0.06CaTiO_3$	1275 for 4 hours	20.9	102 000	7	−9.5	[125]
544	$Ca(Sm_{1/2}Nb_{1/2})O_3$	1500	20.9	24 500	9.4	−28	[180]
545	$0.95MgO-0.05BaO-TiO_2$	1320	20.9	32 500	9.06		[166]
546	$0.96MgTiO_3-0.04SrTiO_3$	1300 for 4 hours	20.9	135 000	9	0	[205]
547	$Zn_{1.01}Nb_2O_6$	1300 for 4 hours	20.9	120 000		−74	[200]
548	$MnNb_2O_6$	1150	20.9	12 900	6.8	−74	[172]
549	$Li_{2.081}Ti_{0.676}Nb_{0.243}O_3 + 1.5$ wt% B_2O_3	880	20.9	34 100		8.3	[206]
550	$Mg_{1.03}Nb_2O_6$	1400 for 4 hours	21	121 000		−60	[200]
551	$0.964MgTiO_3\_0.036SrTiO_3$	1270	21	71 000		−1	[203]
552	$BaTe_2O_6$	650	21	50 300		−51	[80]
553	$Ca(Mg_{1/3}Ta_{2/3})O_3$		21	78 000	–	−61	[186]
554	$Zn_3Nb_2O_8$	1150 for 2 hours	21	83 300		−71	[130, 195]
555	$0.95MgTiO_3-0.05CaTiO_3$	1450	21	56 000	7	0	[188]
556	$5NiO-Nb_2O_5-TiO_2$	1125	21	8200	5.08	−64	[67]
557	$TbTiNbO_6$	1385	21	15 700	7.58	−45	[174]
558	$5ZnO-2Nb_2O_5$	1220	21	88 000	6.98	−73	[94]
559	$Sm_{0.6}Y_{0.4}TiNbO_6$	1400	21	11 500		−27	[175]
560	$Pr_{0.1}Gd_{0.9}TiNbO_6$	1385	21	3450		−50	[175]

No.		Sintering temp. (°C)	ε_r	*Qf* (GHz)	*f* (GHz)	τ_f	Reference
561	$Ca_5Ta_2ZrO_{12}$	1700	21	23 800	5.38	−27	[207]
562	$ZnTiO_3$	925	21	30 000		−90	[208]
563	$Ca[(Li_{0.33}Nb_{0.67})_{0.9}Ti_{0.1}]\ O_{3-\delta}+20\,wt\%\ LiF$	840	21	20 400		−18	[209]
564	$NiNb_2O_6$	1200	21	19 300	6.5	−71	[172]
565	$Mg_{0.95}Ca_{0.05}TiO_3+0.2\,mol\%\ Bi_2O_3$	1250	21	55 600	7	−12.4	[210]
566	$Mg_{0.95}Ca_{0.05}TiO_3+2\,wt\%\ B_2O_3$	1200	21.2	62 000	8	4	[191]
567	$Ca(Yb_{1/2}Ta_{1/2})O_3$	1500	21.2	24 000	9.6	−38	[180]
568	$SmZrTaO_6$	1650	21.2	24 190		−58.8	[211]
569	$CaTa_2O_6$	1600	21.2	11 600		1	[199]
570	$Sr(Co_{1/2}W_{1/2})O_3$	1450	21.2	14 000	7.72	−73	[164]
571	$Ca[Li_{0.33}Nb_{0.67}]_{0.9}Ti_{0.1}]O_{3-\delta}+20\,wt\%\ LiF$	840	21.3	20 450	4.59	−18	[209]
572	$ZnTiO_3+1\,wt\%\ V_2O_5$	900	21.3	8000	8.8		[181]
573	$Nd(Ti_{1/2}W_{1/2})O_4$	1285	21.3	10 600	5.5	−22	[117]
574	$0.5CeO_2-0.25MgO-0.25TiO_2$: 1 WO_3	1400	21.4	90 000	5.57	−50	[145]
575	$SrNdGaO_4$	1300 for 3 hours	21.4	16 600		7.1	[196]
576	$0.9ZnNb_2O_6-0.1(ZnO-V_2O_5)$	950	21.4	29 500			[212]
577	$MgNb_2O_6$	1300	21.4	93 800		−70	[199]
578	$MgTiO_3-CaTiO_3$ (MMT-20)	1360	21.4	26 000	7		[52]
579	$Sm(Ti_{1/2}W_{1/2})O_4$	1300	21.5	7100	5.5	−14	[117]
580	$BaO-0.32MgO-0.26WO_3-0.42TiO_2$	1500 for 6 hours	21.5	49 000		1	[120]
581	$MgNb_2O_6+0.25\,wt\%\ B_2O_3$	1260 for 3 hours	21.5	115 800		−48	[213]
582	$5Li_2O-0.583Nb_2O_5-3.248TiO_2+1\,wt\%\ V_2O_5$	920	21.5	32 950		6.1	[214]
583	85 wt% $Ba_5Nb_4O_{15}$ + 15 wt% $Li_2O-B_2O_3-SiO_2-CaO-Al_2O_3$	875	21.5	3400		−15	[182]
584	$NdAlO_3+0.25\,wt\%\ V_2O_5$	1410	21.5	64 000	9	−30	[215]
585	$ZnNb_2O_6$	1200	21.5	84 500	6.3	−75	[172]
586	$Ba_3MgSb_2O_9$		21.5	23 020	5.1	−5.5	[111]
587	$Zn_3Nb_2O_8$	1150 for 2 hours	21.6	83 300		−71	[195]
588	$0.93(Mg_{0.95}Co_{0.05})TiO_3-0.07CaTiO_3$	1275 for 4 hours	21.6	92 000	7	−1.8	[125]
589	$Ba[Ti_{1-x}(Ni_{1/2}W_{1/2})_x]O_3$ $(x=0.55)$	1425	21.6	38 400		−7.8	[192]
590	$Mg_{0.95}Ca_{0.05}TiO_3+0.5\,mol\%\ Bi_2O_3$	1250	21.7	52 400	7	−28.7	[210]
591	$0.56Ba(Mg_{1/2}W_{1/2})O_3-0.44BaTiO_3$	1500	21.8	13 000		43.8	[107]
592	90 wt% $CaZrO_3$ + 10 wt% $Li_2O-B_2O_3-SiO_2$	875	21.9	4700		−58	[182]
593	$Ca(Ca_{1/3}Ta_{2/3})O_3$		22	22 000	–	−41	[186]
594	$Li_{2.081}Ti_{0.676}Nb_{0.243}O_3+0.5\,wt\%\ B_2O_3$	880	22	32 000		9.5	[216]

No.		Sintering temp. (°C)	ε_r	Qf (GHz)	f (GHz)	τ_f	Reference
595	$Sr(Ca_{1/3}Ta_{2/3})O_3$		22	27 300	7	−91	[217]
596	$Ca(Ni_{1/3}Ta_{2/3})O_3$		22	21 000	–	−80	[186]
597	$CrNbO_4$	1300 for 5 hours	22	4000	4		[15]
598	$(Zn_{0.7}Mg_{0.3})TiO_3$	950	22	65 000		−80	[208]
599	$Sr(Mg_{1/3}Ta_{2/3})O_3$		22	5600	7	−50	[217]
600	$Y(Mg_{1/2}Ti_{1/2})O_3$	1650 for 2 hours	22	33 700	10	−46	[218]
601	$NdGaO_3$		22	85 000			[219]
602	Sr_4AlNbO_8	1525	22	3700	10.25		[173]
603	$(Zn_{0.3}Co_{0.7})TiO_3$	1150	22	80 000		−60	[220]
604	$DyTiNbO_6$	1385	22	19 100	7.76	−42	[174]
605	$YbTiNbO_6$	1400	22	11 000	7.4	−63	[174]
606	$Sm_{0.71}Y_{0.29}TiNbO_6$	1400	22	1400		−1.9	[175]
607	$Ba(Mn_{1/3}Ta_{2/3})O_3$	1600	32	58 200	11.4	34	[221]
608	$Ca_5Nb_2HfO_{12}$	1700	22	16 000	5.4	−29	[147]
609	$0.5CeO_2-0.25ZnO-0.25TiO_2$:4 Co_3O_4	1250	22	32 100	5.5	−48	[145]
610	$CeO_2-0.5CoO-0.5TiO_2$	1200	22	5000		−47	[145]
611	$CoNb_2O_6$	1150	22	41 700	6.7	−66	[146, 199]
612	$Zn_{0.5}Mg_{0.5}Nb_2O_6$	1150	22	33 100		−29	[222]
613	$(1-x)(Mg_{0.95}Zn_{0.05})TiO_3-xCa_{0.6}La_{0.8/3}TiO_3$ ($x=0.1$)	1320 for 4 hours	22	94 000		−20	[137]
614	$ZnNb_2O_6$ + 10 wt% V_2O_5	900 for 2 hours	22.1	10 300		−83	[223, 224]
615	$La_{2-x/3}Na_x(Mg_{1/2}W_{1/2})O_3$ ($x=0.4$)	1450	22.1	5500		−45	[204]
616	$YTiTaO_6$	1625	22.1	51 400		−20	[178]
617	$0.6(Al_{1/2}Ta_{1/2})O_2-0.4(Mg_{1/2}Ta_{2/3})O_2$	1450	22.1	90 930		−16.1	[64]
618	$0.92(Mg_{0.95}Co_{0.05})TiO_3-0.08CaTiO_3$	1275 for 4 hours	22.1	86 400	7	5.4	[125]
619	Nd_2BaCuO_5		22.1	4910		4.6	[183]
620	$Nd_2Ba(Zn_{1-x}Cu_x)O_5$ ($x=0.15$)		22.1	7700	2		[135]
621	$ZnNb_2O_6$ + 5 wt% CuO	925 for 2 hours	22.1	59 500		−65	[143]
622	$Ba(Yb_{1/2}Ta_{1/2})O_3$	1700	22.1	14 000	6.7	89.2	[180]
623	$Sm_2Ba_{0.95}Sr_{0.05}ZnO_5$		22.1	10 053		30.3	[170]
624	$Mg_{0.93}Ca_{0.07}TiO_3$	1350 for 3 hours	22.15	68 550		5.6	[225]
625	$Gd(Ti_{1/2}W_{1/2})O_4$	1375	22.18	5000	5.5	−16	[117]
626	$ZnTiO_3$ + 5 wt% $B_2O_3-SiO_2$	850	22.2	52 500		–	[226]
627	$NdAlO_3$	1650 for 2 hours	22.3	58 000	10	−33	[131]

No.		Sintering temp. (°C)	ε_r	Qf (GHz)	f (GHz)	τ_f	Reference
628	$Nd_2BaZn_{0.9}Cu_{0.1}O_5$	1250 for 10 hours	22.4	6340		1.6	[135]
629	$NdAlO_3 + 0.25$ wt% CuO	1420 for 2 hours	22.4	63 000	10	−35	[227]
630	$MnNb_2O_6$	1150	22.4	34 300		−60	[199]
631	$0.5CeO_2{-}0.25MgO{-}0.25TiO_2$	1400	22.4	17 500	5.5	−62	[145]
632	$Zn_3Nb_2O_8 + 2$ mol% V_2O_5	850−1000	22.4	67 500			[223]
633	$0.75ZnNb_2O_6{-}0.25TiO_2$	1200	22.5	15 000		−15	[168]
634	$0.91(Mg_{0.7}Zn_{0.3})TiO_3{-}0.09CaTiO_3$	1310 for 3 hours	22.5	86 000	7.5	3	[228]
635	$La_{5/3}MgTaO_6$	1500 for 6 hours	22.5	5000	7.2	−80	[229]
636	$NiNb_2O_6$	1150	22.6	40 100		−38	[230]
637	Nd_2BaZnO_5		22.6	12 451	8.9	4.6	[231]
638	$La_{2-x/3}Na_x(Mg_{1/2}W_{1/2})O_3$ ($x = 0.1$)	1400	22.6	19 700		−34	[204]
639	$La_{2-x/3}Na_x(Mg_{1/2}W_{1/2})O_3$ ($x = 0.2$)	1400	22.6	16 600		−27	[204]
640	$0.94MgTiO_3{-}0.06CaTiO_3 + 0.2$ mol% Bi_2O_3	1250	22.6	53 000	7	−2.9	[210]
641	$0.5ZnNb_2O_6{-}0.5Zn_3Nb_2O_8$	1200	22.7	95 500		−65	[167]
642	$LaAlO_3 + 10$ mol% $Sr_2Nb_2O_7$	1575 for 3 hours	22.8	18 610	7.6	46	[232]
643	$0.92CoNb_2O_6{-}0.08TiO_2$	1150	22.8	29 000		−12	[168]
644	$La_{2-x/3}Na_x(Mg_{1/2}W_{1/2})O_3$ ($x = 0.3$)		22.8	11 500		45	[204]
645	CeO_2 :1 mol% Nd_2O_3	1650	22.8	51 000		−63	[162]
646	$Ca(Cu_{1/3}Ta_{2/3})O_3$		23	5500	–	–	[186]
647	$0.93MgTiO_3{-}0.07CaTiO_3$ (Spark plasma sintering)	1150 for 10 minutes	23	7000		–	[233]
648	$Ca(Co_{1/3}Ta_{2/3})O_3$		23	12 000	–	−65	[186, 217]
649	$Sr(Ni_{1/3}Ta_{2/3})O_3$		23	49 000	7	−18	[217]
650	$Sr(Co_{1/3}Ta_{2/3})O_3$		23	17 500	7	−71	[217]
651	$Ba(Ni_{1/3}Ta_{2/3})O_3$		23	49 700	7	−18	[217]
652	$La(Co_{1/2}Ti_{1/2})O_3$		23	32 000		−57	[234]
653	$Dy(Mg_{1/2}Ti_{1/2})O_3$	1650 for 2 hours	23	36 800	10	−6	[218]
654	$0.7MgTiO_3{-}0.3MgTa_2O_6$	1460 for 3 hours	23	81 000		−2	[235]
655	$Ca(La_{1/2}Nb_{1/2})O_3$	1550 for 4 hours	23	31 000		−43	[236]
656	$Ca(Ho_{1/2}Nb_{1/2})O_3$	1550 for 4 hours	23	32 000		3	[236]
657	$Ca(La_{1/2}Ta_{1/2})O_3$	1600 for 4 hours	23	20 600		−32	[237]
658	$CaTiO_3{-}CaZrO_3{-}$frit glass (70:15:15)	875	23	2400		0	[57]

No.		Sintering temp. (°C)	ε_r	Qf (GHz)	f (GHz)	τ_f	Reference
659	$Sm_2Ba_{0.9}Sr_{0.1}ZnO_5$		23	8520		36.1	[170]
660	$CeO_2-CoO_4-TiO_2+0.5$ wt% CuO	1050	23	45 000		−55	[238]
661	$CaO-4NiO-Ta_2O_5-TiO_2$	1340	23	8500	5.5	−48	[67]
662	$CoNb_2O_6$	1150	23	40 000		−35	[230]
663	$CaTiO_3-CaZrO_3$−frit glass (70:15:15)	875	23	2400		0	[57]
664	$0.4(0.5ZnNb_2O_6-0.5Zn_3Nb_2O_8)-0.6ZnTa_2O_6$	1275	23	9300		−55	[167]
665	$CaO-4MgO-Nb_2O_5-TiO_2$	1340	23	52 000	5.5	−30	[67]
666	$2CaO-3ZnO-Nb_2O_5-TiO_2$	1140	23	15 000	5.5	−34	[67]
667	$Nd(Mg_{1/2}Ti_{1/2})O_3$	1650 for 2 hours	23	36 900	10	−49	[218]
668	$Ca[(Li_{1/3}Nb_{2/3})_{1-x}Sn_x]\ O_{3-\delta}$ $(x=0.3)$	1150 for 3 hours	23	46 300		−39	[239]
669	$0.93(Mg_{0.6}Zn_{0.4})_{0.95}Co_{0.05}TiO_3-0.07CaTiO_3$	1200	23.04	79 400		1.4	[240]
670	$HoTiTaO_6$	1550	23.1	46 900		−8	[178]
671	$Nd_{1.95}La_{0..5}BaZnO_5$		23.1	7165		1.6	[183]
672	$PrAlO_3$	1650 for 2 hours	23.2	51 000	10	−25	[131]
673	$ZnNb_2O_6$	1200	23.2	84 500	6.3	−76	[172]
674	$Zn(Nb_{1-x}V_{x/2})_2O_{6-2.5x}$ $(x=0.15)$	975 for 2 hours	23.3	37 000		−71	[223]
675	$Ca[(Li_{1/3}Nb_{2/3})_{1-x}Sn_x]O_3$ $(x=0.2)$	1150 for 3 hours	23.3	50 600		−30	[239]
676	$ZnNb_2O_6+5$ wt% $CuO+4B_2O_3$	900	23.3	46 800		−6.7	[241]
677	$Ca(Li_{1/3}Ta_{2/3})O_3-\delta$ +6 wt% B_2O_3	1100	23.33	27 900	10.99		[242]
678	$LaAlO_3$ +5 mol% $Sr_2Nb_2O_7$	1575 for 3 hours	23.4	20 790	10.81	−25	[232]
679	$LaAlO_3$	1650 for 2 hours	23.4	68 000	10	−44	[131]
680	CeO_2 :1 mol% Er_2O_3	1650	23.5	74 000		−60	[162]
681	$La_6Mg_4Ta_2W_2O_{24}$	1350 for 4 hours	23.5	13 600	5.4	−46	[243]
682	$Ba(Mg_{0.33}Ta_{0.53}Ti_{0.067}W_{0.067})O_3$	1590	23.6	75 900	5.7	−3.4	[129]
683	$La_{2/3}(Mg_{1/2}W_{1/2})O_3$	1250	23.6	32 500		−43	[204, 244]
684	$La_{2/3}(Mg_{1/2}W_{1/2})O_3$ +2 mol% TiO_2	1330	23.6	14 800		−10	[245]
685	$Zn_{0.95}Mg_{0.05}TiO_3$ +0.25 TiO_2 +1 wt% $3ZnO-B_2O_3$	940 for 2 hours	23.6	30 990	7.75	−8	[246]
686	$(Zn_{1-x}Mg_x)Nb_2O_6$	1150–1350	23.66–19.17	81 220–33 110		−71 to −29	[222]
687	$Zn(Nb_{1-x}V_{x/2})_2O_{6-2.5x}$ $(x=0.025)$	1000 for 2 hours	23.8	64 000		−50	[223]
688	$Zn_{1+x}Nb_2O_6$ $(x=0.01)$	1250	23.8	120 000		−73	[247]

No.		Sintering temp. (°C)	ε_r	Qf (GHz)	f (GHz)	τ_f	Reference
689	$Zn(Nb_{0.94}V_{0.06})_2O_6$	875 for 2 hours	23.9	65 000		−73	[224]
690	CeO_2 :1 mol% Sm_2O_3	1650	23.93	90 000		−50	[162]
691	CeO_2: 1 mol% CaO	1675	24	120 000	5.48	−60	[198]
692	CeO_2 + 1 mol% Sm_2O_3	1650 for 2 hours	24	90 000		−50	[162]
693	$2CaO-3CoO-Ta_2O_5-TiO_2$	1260	24	13 500	5.3	−19	[67]
694	CeO_2	1675	24	65 000		−55	[198]
695	$Ba_8Ta_6Ni_{0.25}Mg_{0.75}O_{24}$		24	93 000		25	[248]
696	$Ca(Li_{1/3}Ta_{2/3})\ O_{3-\delta}$	1200	24	42 300	10.8		[242]
697	$BaO-TiO_2-WO_3$ + 5 wt% $ZnO-2B_2O_3$	1100	24	13 000	9.4		[249]
698	$BaO-0.32MgO-0.25WO_3-0.43TiO_2$	1500 for 6 hours	24	19 500		34	[120]
699	$Ba(Mg_{1/3}Ta_{2/3})O_3$	1640 for 20 hours	24	430 000	10	5	[221]
700	$Ca(Pr_{1/2}Nb_{1/2})O_3$	1550 for 4 hours	24	31 500		−39	[236]
701	$Ca(Pr_{1/2}Ta_{1/2})O_3$	1600 for 4 hours	24	22 200		−31	[237]
702	$Ca(Nd_{1/2}Ta_{1/2})O_3$	1600 for 4 hours	24	22 400		−30	[237]
703	$Ca(In_{1/2}Ta_{1/2})O_3$	1600 for 4 hours	24	16 700		−35	[237]
704	$LiNb_3O_8$	1075	24	58 000		−96	[250]
705	$Nd_{0.3}Dy_{0.7}TiNbO_6$		24	27 750		−22	[175]
706	$Ca(Li_{1/3}Ta_{2/3})O_{3-\delta}$ + 3 wt% B_2O_3	1100	24	40 300	10.86		[242]
707	$0.75(Al_{1/2}Ta_{1/2})O_2-0.25(Ti_{1-x}Sn_x)O_2$		24–30	55 000–80 000		−25 to 15	[251]
708	$0.85Ba(Mg_{1/3}Ta_{2/3})O_3-0.15BaSnO_3$	1640 for 20 hours	24	330 000		−1	[252]
709	$La_{3/4}Mg_{2/4}Ta_{1/4}W_{1/4}O_3$	1350 for 4 hours	24	13 600		−46	[243]
710	$La_{2/3}(Mg_{1/2}W_{1/2})O_3$ + 2 mol% TiO_2		24	14 800	6	10	[245]
711	Sm_2SrZnO_5		24.1	19 283	8.1	−97	[170]
712	$Ca(Li_{1/3}Ta_{2/3})O_{3-\delta}$ + 1 wt% B_2O_3	1100	24.1	38 900	10.8		[242]
713	$0.20MgAl_2O_4-0.80TiO_2$	1460	24.1	48 900	5.5	11	[56]
714	$Ba(Sn,MgTa)O_3$		24.2	120 000	6.0		[253]
715	$Ba[Ti_{1-x}(Ni_{1/2}W_{1/2})_x]O_3$ ($x = 0.5$)	1425	24.2	35 000		−5.6	[192]
716	$Ba_8Ta_6(Ni_{1-x}Mg_x)O_{24}$ ($x = 0.75$)		24.2	93 100		26	[248]
717	$Ba(Mg_{1/3}Ta_{2/3})O_3$:0.5 mol% $Ba(Mg_{1/2}W_{1/2})O_3$		24.2	400 000	10		[254]
718	$Sm_2Ba_{0.75}Sr_{0.25}ZnO_5$		24.3	8670		30	[170]
719	$0.85(Mg_{0.95}Zn_{0.05})TiO_3-0.15Ca_{0.61}Nd_{0.26}TiO_3$	1300	24.3	112 000		−10	[255]

No.		Sintering temp. (°C)	ε_r	*Qf* (GHz)	*f* (GHz)	τ_f	Reference
720	$0.9ZnNb_2O_6-0.1(2ZnO-V_2O_5)$	950	24.3	72 800		–	[212]
721	$In_2O_3-TiO_2-Ta_2O_5$	1525	24.3	15 400		39	[178]
722	$LaTiTaO_6$	1530	24.4	45 300		−39	[178]
723	$Sm_2Ba_{0.15}Sr_{0.85}ZnO_5$		24.4	12 200		2.6	[170]
724	$Sm_2Ba_{0.1}Sr_{0.9}ZnO_5$		24.5	14 950		−35.5	[170]
725	$Ba(Mg_{0.33}Ta_{0.63}Ti_{0.017}W_{0.017})O_3$	1600	24.5	100 700	5.4	12.6	[129]
726	$Sm_2Ba_{0.05}Sr_{0.95}ZnO_5$		24.6	8690		81.5	[170]
727	$Zn_{0.95}Mg_{0.05}TiO_3 + 0.25TiO_2 + 1$ wt% $3ZnO-B_2O_3$	880	24.6	4000		−14	[256]
728	$Ba_{0.9925}(Mg_{0.33}Ta_{0.67})O_3$	1600	24.7	152, 00	5.7	1.2	[257]
729	$La(Mg_{2/3}Ta_{1/3})O_3$	1600	24.7	65 500	10	−65	[258]
730	$Mg_4Al_2Ti_9O_{25}$	1550	24.7	30 000			[259]
731	$Ca[Li_{0.33}Nb_{0.67}]_{0.9}Ti_{0.1}]O_{3-\delta} + 10$ wt% LiF	900	24.8	19 300	4.2	−15	[209]
732	$SmTaTi_{0.25}Zr_{0.75}O_6$		24.9	25 266		−44.5	[211]
733	$Ba[(Mg_{1-x}Zn_x)_{1/3}Ta_{2/3}]O_3$		24–26	200 000–300 000	8	−0.5 −(+)1.7	[260]
734	$CaO-ZrO_2-$glass		25	3500			[261]
735	$Ca(Nd_{1/2}Nb_{1/2})O_3$	1550 for 4 hours	25	31 800		−37	[236]
736	$Ca(Sm_{1/2}Nb_{1/2})O_3$	1550 for 4 hours	25	33 200		−34	[236]
737	$Ba(Co_{1/3}Ta_{2/3})O_3$	1500	25	71 400		−16	[217]
738	$Ca(Eu_{1/2}Nb_{1/2})O_3$	1550 for 4 hours	25	35 800		−30	[236]
739	$La(Co_{1/2}Ti_{1/2})O_3$	1550	25	67 000		−42	[234, 262]
740	$Ca(Ga_{1/2}Ta_{1/2})O_3$	1500 for 2 hours	25	80 000		−81	[263]
741	$Ca(Al_{1/2}Nb_{1/2})O_3$	–	25	7500		−87	[186]
742	Sr_2AlNbO_6	1600	25	4100		−3	[173]
743	$Ca(Zn_{1/3}Ta_{2/3})O_3$		25	25,000	–	−66	[186]
744	$ZnNb_2O_6$	1150 for 2 hours	25	83 700		−56	[230]
745	$NiTa_2O_6$	1600	25	31 000		35	[199]
746	$Ca[(Li_{1/3}Nb_{2/3})_{1-x}Sn_x]\ O_{3-\delta}$ $(x = 0.15)$	1150 for 3 hours	25	49 100		−25	[239]
747	$(Sr_{2/3}La_{1/3})(Li_{1/3}Ta_{2/3})O_3$		25	25 000			[264]
748	$Ba(Mg_{1/3}Ta_{2/3})O_3$ WGM mode	1650 for 4 hours	25	325 000		8	[265]
749	$La_6Mg_4Nb_2W_2O_{24}$	1400 for 4 hours	25	16 400	5.4	−56	[243]
750	$(Zn_{0.9}Mg_{0.1})TiO_3 + 4$ wt% Bi_2O_3	1000 for 4 hours	25	70 000		−10	[266]

No.		Sintering temp. (°C)	ε_r	Qf (GHz)	f (GHz)	τ_f	Reference
751	$Sm(Mg_{1/2}Ti_{1/2})O_3$	1650 for 2 hours	25	65 500	10	−26	[218]
752	$BaO-TiO_2-WO_3$ (N−35):5 wt% $PbO-SiO_2-B_2O_3$	1100	25	6500	6	–	[249]
753	$Ba_3MgNb_{2-x}Sb_xO_9$ ($x=0.25$)		25	96 290	5.6	6.4	[111]
754	$Sr_{2/3}La_{2/3}[Li_{1/3}Ta_{2/3}]O_3$	1350	25	25 200	10.2	−25	[264, 267]
755	$(1-x)(Mg_{0.95}Zn_{0.05})TiO_3-xCa_{0.6}La_{0.8/3}TiO_3$ ($x=0.15$)	1320 for 4 hours	25	86 000		0.5	[137]
756	$0.9LaAlO_3-0.1SrTiO_3$	1680	25.1	128 000	10	−51	[268]
757	$Sm_2Ba_{0.25}Sr_{0.75}ZnO_5$		25.1	1900		18	[170]
758	$Ba(Mg_{0..3183}Ta_{0.67})O_3$	1600	25.1	120 500	5.6	3.3	[257]
759	$Ca[(Li_{1/3}Nb_{2/3})_{1-x}Sn_x]O_3$ ($x=0.1$)	1150 for 3 hours	25.2	48 200		−14	[239]
760	$0.17Ba_5Nb_4O_{15}-0.83BaNb_2O_6$	1300	25.2	59 300		0	[269]
761	90 wt% $CoNb_2O_6$ + 10 wt% $CaTiO_3$	1150	25.2	21 700		2	[168]
762	$Sm_2Ba_{0.5}Sr_{0.5}ZnO_5$		25.3	10 075	8.1	29.6	[17, 170]
763	$0.9MgNb_2O_6-0.1TiO_2$	1300	25.36	19 000		−23	[168]
764	$Nd_2Ba_{0.5}Sr_{0.5}ZnO_5$		25.5	6120		26	[231]
765	$Sm(Co_{1/2}Ti_{1/2})O_3$	1360 for 4 hours	25.5	76 000	10	−16	[270]
766	$Ba[Mg_{1/3}(Nb_{1/4}Ta_{3/4})_{2/3}]O_3$		25.5	140 666		4.8	[271]
767	$Ba_{10}Ta_{7.04}Ti_{.045}Sn_{0.75}O_{30}$		25.6	59 100		30	[136]
768	$(Zr_{0.8}Sn_{0.2})TiO_4$ + 10 wt% $BaO-B_2O_3-SiO_2-Li_2O-CuO$	1000 for 4 hours	25.6	13 000		−7.8	[157]
769	Nd_2SrZnO_5		25.7	25 836	9.7	−80	[17, 231]
770	$Ba_5Nb_3TaO_{15}$	1500	25.7	21 600	4.93	16	[272]
771	$0.4(Al_{1/2}Ta_{1/2})O_2-0.6(Mg_{1/2}Ta_{2/3})O_2$	1450	25.8	111 230		−5	[64]
772	$La_6Mg_4Ta_2W_2O_{24}$	1400 for 4 hours	25.8	16 400		−56	[243]
773	$Ca(Ni_{1/3}Nb_{2/3})O_3$		26	11 000	–	−78	[186]
774	85 wt% $BaTi_4O_9$ + 15 wt% $Li_2O-B_2O_3-SiO_2-CaO-Al_2O_3$	875	26	10 200		0	[182]
775	$0.85(Mg_{0.95}Zn_{0.05})TiO_3-0.15Ca_{0.6}La_{0.8/3}TiO_3$	1320 for 4 hours	26	8600		0.5	[137]
776	$0.5LaCa_{0.5}Zr_{0.5}O_3-0.5CaTiO_3$	1575	26	13 500	4.5	−67	[159]
777	$Nd(Mg_{1/2}Ti_{1/2})O_3$	1650 for 2 hours	26	60 000		−72	[218, 273]
778	$Ca(Sm_{1/2}Ta_{1/2})O_3$	1600 for 4 hours	26	25 000		−25	[237]
779	$Ca(Er_{1/2}Ta_{1/2})O_3$	1600 for 4 hours	26	29 600		−12	[237]
780	$Ca(Yb_{1/2}Ta_{1/2})O_3$	1600 for 4 hours	26	59 200		−21	[237]

No.		Sintering temp. (°C)	ε_r	Qf (GHz)	f (GHz)	τ_f	Reference
781	$Ca(Gd_{1/2}Nb_{1/2})O_3$	1550 for 4 hours	26	11 000		−26	[236]
782	$Sr(Yb_{1/2}Ta_{1/2})O_3$	1600 for 4 hours	26	32 300		−79	[274]
783	$Sr(In_{1/2}Nb_{1/2})O_3$	1600 for 4 hours	26	32 700		−62	[275]
784	$BaO-TiO_2-WO_3$ (N-35):5 wt% $BaO-SiO_2-B_2O_3$	1100	26	8400	6.1	–	[276]
785	$Sr_3Ti_2O_7$	1300 for 5 hours	26	2400	4		[15]
786	$LaYbO_3$	1600 for 4 hours	26	20 600	7	−22	[277]
787	$0.75Ca_2AlNbO_6-0.25Ca_3Nb_2O_8$		26	13 200	6.97	−14	[288]
788	$2CaO-3NiO-Ta_2O_5-TiO_2$	1410	26	11 000	4.9	−41	[67]
789	$Ca_5Nb_2ZrO_{12}$	1690	26	22 800	4.8	−25	[207]
790	$3CaO-2ZnO-Nb_2O_5-TiO_2$	1325	26	22 000	5.3	−25	[67]
791	$2CaO-3MgO-Ta_2O_5-TiO_2$	1450	26	30 000	5.1	−28	[67]
792	$0.2CaTiO_3-0.8Sm(Mg_{0.5}Ti_{0.5})O_3$	1550	26	13 500	6	−33	[278]
793	MWF-38 + 10 wt% $Li_2O-B_2O_3-SiO_2-CaO-Al_2O_3$ (52.45:31.06:11.99:2:2.5)	875	26	10 200		−4	[155]
794	$Ba(Zn_{1/3}Ta_{2/3})O_3$ + 5 mol% B_2O_3 + 10 mol% CuO	870 for 2 hours	26	11 000		0	[279]
795	$0.5CeO_2-0.25ZnO-0.25TiO_2$	1250	26.1	24 100	5.1	−43	[145]
796	$Ca[(Li_{1/3}Ta_{2/3})_{0.95}Ti_{0.05}]O_{3-\delta}$ + 3 wt% B_2O_3	1050 for 4 hours	26.1	22 000	10.3	−97	[242]
797	$0.35(Al_{1/2}Ta_{1/2})O_2-0.65(Mg_{1/3}Ta_{2/3})O_2$	1450	26.1	112 500		0	[64]
798	$Ca_2Mg_3Nb_4TiO_{17}$	1525	26.2	13 750		−24	[280]
799	$MgTa_{1.3}Nb_{0.7}O_6$	1450	26.2	53 100		−4.1	[281]
800	$Ba(Mg_{0.30}Ta_{0.60}Ti_{0.10})O_3$	1600	26.3	100 000	5.2	14.4	[282]
801	$0.5CeO_2-0.25MnO-0.25TiO_2$	1200	26.3	17 100	5	−30	[145]
802	$Nd_2Ba_{0.5}Ca_{0.5}ZnO_5$		26.4	6185		24	[231]
803	$0.2(Al_{1/2}Ta_{1/2})O_2-0.8(Mg_{1/3}Ta_{2/3})O_2$	1450	26.5	103 190		25.3	[64]
804	$Ba_3NiNb_{2-x}Sb_xO_9$ ($x=0.1$)		26.5	31 110	5	−13	[111]
805	$Ba[Ti_{1-x}(Ni_{1/2}W_{1/2})_x]O_3$ ($x=0.45$)	1425	26.5	30 800		−3	[192]
806	$BaO-Sm_2O_3-4TiO_2$ + 10 wt% B_2O_3	1100 for 2 hours	26.5	11 800		−9.5	[283]
807	$Ba_3Ti_5Nb_6O_{28}$ + 5 wt% CuO	900 for 2 hours	26.6	14 100		21	[133]
808	$Ba_8Ta_6(Ni_{1-x}Mg_x)O_{24}$ (x = 0.5)		26.6	86 800		31	[248]
809	$Ba_8Ta_6MgO_{24}$		26.6	80 900		18	[248]
810	$Sr_6Ta_4ZrO_{18}$ + 3 wt% $Bi_2O_3-B_2O_3$	1625 for 2 hours	26.7	9100		−39	[284]

No.		Sintering temp. (°C)	ε_r	Qf (GHz)	f (GHz)	τ_f	Reference
811	$0.8LaAlO_3-0.2SrTiO_3$	1680	26.7	139 000	10	−50	[268]
812	$La(Mg_{2/3}Nb_{1/3})O_3$		26.8	52 000		−86	[285]
813	$(Zr_{0.8}Sn_{0.2})TiO_4$ + 10 wt% $BaO-B_2O_3-SiO_2-Li_2O-CuO$	1000 for 8 hours	26.8	21 900		−1.3	[157]
814	$Ca(Cu_{1/3}Nb_{2/3})O_3$		27	3300	–	–	[186]
815	$Sr(Ga_{1/2}Ta_{1/2})O_3$	1500 for 3 hours	27	91 000		−50	[286]
816	$La(Mg_{1/2}Ti_{1/2})O_3$ (sol–gel)	–	27	74 500		–	[287]
817	$Ca(Tb_{1/2}Nb_{1/2})O_3$	1550 for 4 hours	27	34 600		−13	[236]
818	$Ca(Eu_{1/2}Ta_{1/2})O_3$	1600 for 4 hours	27	23 600		−22	[237]
819	$Ca(Gd_{1/2}Ta_{1/2})O_3$	1600 for 4 hours	27	26 000		−16	[237]
820	$Ca(Y_{1/2}Ta_{1/2})O_3$	1600 for 4 hours	27	42 300		−9	[237]
821	$0.6Ca(Y_{1/2}Ta_{1/2})O_3-0.4\ Ba(Y_{1/2}Ta_{1/2})O_3$	1600 for 4 hours	27	42 000		−1	[237]
822	$Sr(Er_{1/2}Ta_{1/2})O_3$	1600 for 4 hours	27	22 100		−77	[274]
823	Ca_2AlNbO_6		27	14 000	7.02	−88	[288]
824	$Ba(Mg_{1/3}Ta_{2/3})O_3-Ba(Zn_{1/3},Ta_{2/3})O_3$		27	150 000	10	0	[289]
825	$BaO-TiO_2-WO_3$ (N-35) + 5 wt% $ZnO-B_2O_3-SiO_2$	1000	27	8400	7.0		[249, 276]
826	$Ca(Yb_{1/2}Nb_{1/2})O_3$	1500	27	7200	8.4	−30	[180]
827	$BaO-TiO_2-WO_3$ (N-35):5 wt% $PbO-Al_2O_3-SiO_2$	1100	27	8400	6.1	–	[249, 276]
828	$Ba_8Ta_6Ni_{0.25}Zn_{0.75}O_{24}$		27	91 730		35	[248]
829	$BaTi_4O_9$ + 20 wt% $B_2O_3-ZnO-La_2O_3$	900 for 3 hours	27	20 000		6.5	[290]
830	$Nd(Co_{1/2}Ti_{1/2})O_3$	1440 for 4 hours	27	140 000	9	−46	[291]
831	$Ba_6Ta_4TiO_{18}$	1625 for 2 hours	27	27 500		45	[284]
832	$CaZrO_3$		27	20 800			[292]
833	$Ba(Mn_{1/3}Ta_{2/3})O_3$	1600 in air	27	15 500		45	[293]
834	$Ba(Mn_{1/3}Ta_{2/3})O_3$	1600 in N_2	27	104 000		45	[293]
835	$Ba_6Ta_4TiO_{18}$ + 2 wt% $Bi_2O_3-B_2O_3$	1530 for 40 hours	27	27 500	4.7	45	[284]
836	$0.3CaTiO_3-0.7Sm(Mg_{0.5}Ti_{0.5})O_3$	1550	27	11 970	5.8	−29	[278]
837	$ZnO-TiO_2$−2 wt% $ZnO-B_2O_3-SiO_2$	930 for 3 hours	27	20 000		2	[294]
838	$ZnO-TiO_2$ + 2 wt% ZBS glass	900 for 3 hours	27	19 400	6	2	[294]

No.		Sintering temp. (°C)	ε_r	Qf (GHz)	f (GHz)	τ_f	Reference
839	$BaO-2CeO_2-4TiO_2$	1250	27	18 560		9	[295]
840	$Ba_5Nb_2Ta_2O_{15}$	1475	27	1064	4.7	22	[272]
841	$CaO-4NiO-Nb_2O_5-TiO_2$	1185	27	4000	4.6	58	[67]
842	$Ba_3NiNb_{2-x}Sb_xO_9$ (x = 0.25)		27	27 370	5	−20.6	[111]
843	$LaGaO_3$		27	97 000	5	−80	[296]
844	$Ba(Zn_{1/2}Ta_{2/3})O_3$ + 1 mol% CeO_2	1525 for 6 hours, 1350 for 5 hours	27	123 000		14	[297]
845	$Ba[(Mg_{0.4}Zn_{0.6})Ta_{2/3}]O_3$	1600 for 4 hours	27	109 900		4	[282]
846	$0.7Ba(Mg_{1/3}Ta_{2/3})O_3-0.3Ba(Co_{1/3}Nb_{2/3})O_3$	1530 for 5 hours	27	172 700		−1	[298]
847	$BaTi_4O_9 + B_2O_3-ZnO-La_2O_3$ glass	900	27	20 000		6.5	[299]
848	$Ba(La_{1/2}Ta_{1/2})O_3$	1450	27.1	18 000	8.7	51	[180]
849	$Sr(La_{1/2}Ta_{1/2})O_3$	1500	27.1	2600	8.4	−29	[180]
850	$Sr(Nd_{1/2}Ta_{1/2})O_3$	1500	27.1	25 000	8.3	−68	[180]
851	$Nd(Co_{1/2}Ti_{1/2})O_3$ + 0.75 wt% B_2O_3	1320 for 4 hours	27.2	153 000	9	0	[300]
852	$Ba_2Ti_9O_{20}$ + 1 wt% $ZnO-B_2O_3$	940 for 2 hours	27.3	8300	7.2	2.5	[301]
853	$Ba_8Ta_6(Ni_{1-x}Zn_x)O_{24}$ (x = 0.5)		27.4	83 800		36	[248]
854	$Ba_8Ta_6NiO_{24}$		27.5	81 750		33	[248]
855	$Sr(La_{1/2}Nb_{1/2})O_3$	1500	27.5	2000	8.3	−33	[180]
856	$Sr(Zn_{1/2}W_{1/2})O_3$	1360	27.5	51 000	7.0	−45	[164]
857	$0.8La(Mg_{1/2}Ti_{1/2})O_3-0.2La_{2/3}TiO_3$	1500 for 2 hours	27.5	16 600	7.9		[302]
858	$Ba(Mg_{1/3}Nb_{2/3})O_3 + B_2O_3$	900	27.5	8500		27	[303]
859	90 wt%(Zr,Sn)TiO_4 + 10 wt% $Li_2O-B_2O_3-SiO_2$	875	27.5	9000		14	[182]
860	$ZnTiO_3-0.25TiO_2$	925	27.5	14 000		−20	[304]
861	$Ba_8Ta_6(Ni_{1-x}Zn_x)O_{24}$ (x = 0.75)		27.6	91 700		37	[248]
862	$Ca[(Li_{1/3}Ta_{2/3})_{0.9}-Ti_{0.1}]O_{3-\delta}$ + 3 wt% B_2O_3	1000 for 4 hours	27.6	9800	10.2	-	[242]
863	$La(Mg_{1/2}Ti_{1/2})O_3$	1600	27.6	114 300	7.1	−81	[305, 306]
864	$Sr(Sm_{1/2}Ta_{1/2})O_3$	1500	27.7	59 000	8.5	−63	[180]
865	$Ba[(Mg_{0.4}Zn_{0.6})Ta_{2/3}]O_3$	1575	27.7	109 900	4.6	6.3	[282]
866	$Ba_6Ta_4ZrO_{18}$ + 2 wt% $Bi_2O_3-B_2O_3$	1525 for 2 hours	27.7	18 500		37	[284]
867	$Sm_{0.78}Y_{0.22}TiNbO_6$	1400	27.9	2300		11	[175]
868	$MgTa_{1.5}Nb_{0.5}O_6$	1450	27.9	33 100		−0.7	[307]
869	$0.8(Mg_{0.95}Co_{0.05})TiO_3-0.2(Ca_{0.6}La_{0.8/3}TiO_3$ + 1 wt% ZnO	1250	27.9	36 000	8	14.1	[308]

No.		Sintering temp. (°C)	ε_r	Qf (GHz)	f (GHz)	τ_f	Reference
870	$Ba_8Ta_6(Ni_{1-x}Mg_x)O_{24}$ ($x=0.25$)		27.9	81 500		32	[248]
871	$Sr_{1-x}Ca_x(Ga_{1/2}Ta_{1/2})O_3$		26–28	18 000–90 000	8	−50 to −97	[309]
872	$BaTi_4O_9$−10 mol% $BaO-ZnO-B_2O_3$ glass	925	28–33	20 000	6.6		[310]
873	$Nd(Mg_{1/2}Ti_{1/2})O_3$	1650 for 2 hours	28	36 900	10	−49	[218]
874	$Ba(Tb_{1/2}Ta_{1/2})O_3$	1625	28	28 200		−38	[311]
875	$Ca(Mg_{1/3}Nb_{2/3})O_3$		28	58,000	–	−48	[186, 312]
876	$Sr(Zn_{1/3}Ta_{2/3})O_3$	1500	28	21 700		−54	[217]
877	$Ba(Zn_{1/3}Ta_{2/3})O_3$	1350 for 20 hours	28	168 000		0.5	[289]
878	$Ca(Tb_{1/2}Ta_{1/2})O_3$	1600 for 4 hours	28	28 400		−10	[237]
879	$0.6Ca(Yb_{1/2}Ta_{1/2})O_3-0.4Ba(Yb_{1/2}Ta_{1/2})O_3$	1600 for 4 hours	28	48 000		2	[237]
880	$Ca(Yb_{1/2}Ta_{1/2})O_3$ + 4 mol% $CaTiO_3$	1600 for 4 hours	28	41 000		−2	[237]
881	$Ba_3[Zr_{0.0645}Ni_{0.1625}Zn_{0.816}Ta_{1..957}]O_3$	1520 for 48 hours	28	136 770		−3	[313]
882	$Ba(Zn_{1/3}Ta_{2/3})O_3$ + 1 mol% Cr_2O_3	1525 for 6 hours	28	125 500		−1.6	[297]
883	$Ca(Ho_{1/2}Ta_{1/2})O_3$	1600 for 4 hours	28	23 700		−8	[237]
884	$Sr(Dy_{1/2}Ta_{1/2})O_3$	1600 for 4 hours	28	34 200		−73	[274]
885	$Sr(Ho_{1/2}Ta_{1/2})O_3$	1600 for 4 hours	28	38 800		−75	[274]
886	$Sr(Y_{1/2}Ta_{1/2})O_3$	1600 for 4 hours	28	54 300		−77	[274]
887	$Ca(Ca_{1/3}Nb_{2/3})O_3$		28	17 000	–	−22	[186]
888	$Zn(Nb_{0.35}Ta_{0.65})O_6$	1300	28	50 000		0	[314]
889	$0.5Ca_2AlNbO_6-0.5Ca_3Nb_2O_8$		28	8900	6.86	36	[288]
890	$MgTa_2O_6$ + 0.5 wt% CuO	1400	28	58 000		18	[315]
891	$Pr(Mg_{1/2}Ti_{1/2})O_3$	1650 for 2 hours	28	27 800	10	−17	[218]
892	$Ba_5Ta_4O_{15}$	1550	28	31 600	5.55	12	[94, 272]
893	$3CaO-2CoO-Ta_2O_5-TiO_2$	1400	28	19 500	4.8	−14	[67]
894	$0.75(Al_{1/2}Ta_{1/2}O_2)-0.25(Ti_{0.85}Sn_{0.15})O_2$	1450 for 3 hours	28	68 000		0	[316]
895	$(1-x)LaMg_{1/2}Ti_{1/2}O_3-xLa_{2/3}TiO_3$ ($x=0.1$)		28	56 000	6.6	−66	[317]
896	$Ba_{10}Mg_{0.25}Ta_{7.9}O_{30}$	1600	28	3600	4.02	30	[318]
897	$(1-x)Sr(Li_{1/4}Nb_{3/4})O_3-xSr(Li_{2/3}W_{3/5})O_3$ ($x=0.283$)	1450	28	23 800	9.1	−29	[319]

No.		Sintering temp. (°C)	ε_r	Qf (GHz)	f (GHz)	τ_f	Reference
898	$La_5AlTi_3O_{15}$	1600	28.09	28 600	3.4	−39.2	[320]
899	$Ba(Zn_{1/2}W_{1/2})O_3$	1340	28.1	22 700	8		[127]
900	$Al_2O_3-TiO_2-Ta_2O_5$	1575	28.1	1000		20	[178]
901	$Ba_{10}Mg_{0.25}Ta_{7.9}O_{30}$	1600 for 12 hours	28.2	33 500		29	[318]
902	$La(Zn_{1/2}Ti_{1/2})O_3$ (sol–gel)		28.3	66 500			[321]
903	$Ba(Yb_{1/2}Nb_{1/2})O_3$	1700	28.3	16 500	8.7	118	[180]
904	$Ca[(Li_{1/3}Ta_{2/3})_{.8}Ti_{0.2}]O_{3-\delta}+3$ wt% B_2O_3	1050	28.4	12 900	9.9	−15	[242]
905	$Sr_6Ta_4ZrO_{18}+3$ wt% $Bi_2O_3-B_2O_3$ glass	1625 for 2 hours	28.4	9100		−39	[284]
906	$0.25Ba(Y_{1/2}Ta_{1/2})O_3-0.75Ba(Ca_{1/9}Y_{3/9}Ta_{5/9})O_3$		28.5	16 900	9.6	142	[322]
907	$Ba_3MgNb_{2-x}Sb_xO_9$ ($x=0.125$)		28.5	101 300	4.7	14	[111]
908	$(Pb_{1-3x/2}La_x)(Mg_{1/2}W_{1/2})O_3$ ($x=0.56$)	1200	28.7	18 100		−6	[323]
909	$Ba_8Ta_6ZnO_{24}$		28.85	85 000		40	[248]
910	$SmTaTi_{0.6}Zr_{0.4}O_6$		28.9	38 320		−11.6	[211]
911	$0.7LaAlO_3-0.3SrTiO_3$	1680	28.9	120 000	9.9	−44	[268]
912	$Ca(Co_{1/3}Nb_{2/3})O_3$		29	6200	–	−65	[186]
913	$BaO-TiO_2-WO_3$ (N-35): $ZnO-B_2O_3$	1100	29	7000	5.8	–	[249, 276]
914	$CoTa_2O_6$	1500	29	2300		23	[199]
915	$La(Mg_{1/2}Ti_{1/2})O_3$	1650 for 2 hours	29	114 000		−81	[218, 306]
916	$Ba(Zn_{1/2}Ta_{2/3})O_3+0.3$ mol% Ta_2O_5	1620 for 10 hours	29	152 000			[324]
917	0.95 $Ba(Zn_{1/2}Ta_{2/3})O_3-0.05Sr(Ga_{1/2}Ta_{1/2})O_3$	1550 for 2 hours, 1450 for 24 hours	29	162 000		0	[325, 326]
918	$Ba_3(Zr_{0.0645}Zn_{0.816}Ni_{0.1625}Ta_{1.957})O_9$	1510 for 24 hours	29	126 860		−2	[327]
919	$Sr(Tb_{1/2}Ta_{1/2})O_3$	1600 for 4 hours	29	34 200		−70	[274]
920	$3CaO-2NiO-Ta_2O_5-TiO_2$	1500	29	18 800	4.9	−33	[67]
921	$2CaO-3CoO-Ta_2O_5-TiO_2$	1175	29	18 500	4.9	−28	[67]
922	$La_{10}MgTi_9O_{34}$		29	13 000	5.9	−22	[317]
923	$BaTiTe_3O_9$	650	29	1700	7.6	−372	[328]
924	$(1-x)Ca(Li_{1/4}Nb_{3/4})O_3-xCa(Li_{2/3}W_{3/5})O_3$ ($x=0.333$)	1150	29	15 700	9.7	−35	[319]
925	$(Sr_{2/3}La_{1/3})(Li_{1/3}Nb_{2/3})O_3$	1300	29	6300	8.9	−76	[267]
926	$CeO_2+0.06CaTiO_3$	1650 for 2 hours	29	25 000		0	[162]
927	$0.5CeO_2-0.25MnO-0.25TiO_2$:0.4 Sb_2O_3	1200	29.05	7000	4.9221	−0.8	[145]
928	$Ba(Zn_{1/2}W_{1/2})O_3$	1330	29.1	36 000	6.8	−31	[164]

No.		Sintering temp. (°C)	ε_r	*Qf* (GHz)	*f* (GHz)	τ_f	Reference
929	$0.9La(Mg_{1/3}Ti_{1/3})O_3-0.1SrTiO_3$		29.2	14 500	7.33		[329]
930	$0.9Nd(Co_{1/2}Ti_{1/2})O_3-0.1SrTiO_3$	1440 for 4 hours	29.3	80 900		0	[330]
931	$Ba_6Ta_4TiO_{18}$	1550 for 2 hours	29.3	27 500		45	[284]
932	$Sr(Yb_{1/2}Nb_{1/2})O_3$	1500	29.4	50 000	8.1	−72	[180]
933	$0.8(Al_{1/2}Ta_{1/2})O_2-0.2TiO_2$	1450 for 3 hours	29.4	75 470		0	[331]
934	$Ca[(Li_{1/3}Ta_{2/3})_{.85}Ti_{0.15}]O_{3-\delta}+3$ wt% B_2O_3	1050 for 4 hours	29.4	20 700	10.47	−57	[242]
935	$0.94CoNb_2O_6-0.06TiO_2$	1150	29.6	20 300		4.4	[168]
936	$0.78ZnNb_2O_6-0.22TiO_2$	1200	29.6	27 700		22	[168]
937	$Ca[Li_{1/3}Nb_{2/3}]O_3$	1150 for 3 hours	29.6	40 000		−21	[239]
938	$La(Mg_{1/2}Ti_{1/2})O_3$ + 1 wt% CuO	1450	29.63	33 800		−68	[332]
939	$Sr(Nd_{1/2}Nb_{1/2})O_3$	1500	29.7	2500	8.1	−32	[180]
940	$SmTaTi_{0.5}Zr_{0.5}O_6$		29.7	32 173		−21.4	[211]
941	$Ba_{10}Co_{0.25}Ta_{7.9}O_{30}$	1600 for 24 hours	29.7	36 700		29	[318]
942	$0.75Ba(Y_{1/2}Ta_{1/2})O_3-0.25Ba(Ca_{1/9}Y_{3/9}Ta_{5/9})O_3$		29.8	34 200	9.6	153	[322]
943	$Ca[(Li_{1/3}Nb_{2/3})_{0.9}Zr_{0.1}]O_{3-\delta}$	1150	29.8	36 300		−4.9	[333]
944	$Ba[Ti_{1-x}(Ni_{1/2}W_{1/2})_x]O_3$ ($x=0.4$)	1425	29.8	26 700		6.5	[192]
945	$0.9La(Mg_{1/2}Ti_{1/2})O_3-0.1CaTiO_3$	1600	29.8	16 700	6.9	−70	[305]
946	$La(Co_{1/2}Ti_{1/2})O_3+0.25$ wt% CuO	1380	29.8	64 000	8	−56	[334]
947	$Ba(Zn,Ta)O_3-Ba(Zn,Nb)O_3$		30	164 000	12	0	[289]
948	$Ba(Ca_{1/3}Ta_{2/3})O_3$		30	27 400	7	145	[217]
949	$(Ca,Sr,Ba,)ZrO_3$		30	44 000	11	5	[337]
950	$BaNb_2O_6$ (orthorhombic)	1300	30	43 000		−45	[335]
951	$Pb_{0.5}Ca_{0.5}(Al_{1/2}Nb_{1/2})O_3$		30	1500	5.1	−23	[336]
952	$CaZrO_3$		30	26 400	11	−27	[337, 338]
953	$SrZrO_3$		30	13 600	11	−67	[337, 338]
954	$La(Co_{1/2}Ti_{1/2})O_3$	1440 for 6 hours	30	67 000	10	−64	[262]
955	$Ca[(Li_{1/3}Nb_{2/3})_{0.9}Zr_{0.1}]O_{3-\delta}$	1150	30	36 300		−5	[339]
956	$(1-x)Sr(Li_{1/4}Nb_{3/4})O_3-xSr(Li_{2/5}W_{3/5})O_3$ ($x=0.385$)	1450	30	21 200		−33	[319]
957	$(1-x)Ca(Li_{1/4}Nb_{3/4})O_3-xCa(Li_{2/5}W_{3/5})O_3$ ($x=0.238$)	1150	30	22 700		−33	[319]
958	$ZnTiO_3+0.25TiO_2+$ 1 wt% B_2O_3	875 for 4 hours	30	56 000		10	[340]
959	$Sm_{0.8}Y_{0.2}TiNbO_6$	1400	30	11 000		17	[175]

No.		Sintering temp. (°C)	ε_r	*Qf* (GHz)	*f* (GHz)	τ_f	Reference
960	$Ba_5Ta_4O_{15}$	1550 for 40 hours	30	31 600		12	[94]
961	$Bi_2ZnNb_2O_9 + ZnNb_2O_6 +$ 3wt% $PbO-Bi_2O_3-B_2O_3-ZnO-TiO_2$ glass	900	30	3500	6		[341]
962	$Ba_2Ti_9O_{20} + 9$ wt% B_2O_3	1050 for 2 hours	30	13 700		6	[342]
963	$Ba(Zn_{1/3}Ta_{2/3})O_3$ + 1 mol% Mn	1550	30	145 000		0	[221]
964	$Ba_3[Zr_{0.09}\ Ni_{0.125}Zn_{0.845}Ta_{1..94}]O_3$	1520 for 48 hours	30	138 710		−1	[313]
965	$Ba(Zr_{0.05}Zn_{0.32}Ta_{0.63})O_3$	1500 for 4 hours	30	148 000		8	[343]
966	$0.15TiTe_3O_8-0.85TeO_2$	700	30	22 000	5	0	[179]
967	$BaO-CeO_2-TiO_2 + 1.5$ wt% CuO	1050	30	32 000		−11	[238]
968	$Ca_{1-x}Nd_{2x/3})TiO_3 + 3ZnO-2B_2O_3$ glass (20–40 mol%)	880	30–60	200–5500		20–60	[344]
969	$3CaO-2MgO-Ta_2O_5-TiO_2$	1550	30	185 000	4.6	−24	[67]
970	$La(Zn_{1/2}Ti_{1/2})O_3$ (sol–gel)	1350	30	60 000		−71	[345]
971	$Ca(Yb_{1/2}Nb_{1/2})O_3$	1550 for 4 hours	30	32 500		−25	[236]
972	$Ca(In_{1/2}Nb_{1/2})O_3$	1550 for 4 hours	30	37 900		−33	[236]
973	$Sr(Eu_{1/2}Ta_{1/2})O_3$	1600 for 4 hours	30	45 500		−43	[274]
974	$Sr(Gd_{1/2}Ta_{1/2})O_3$	1600 for 4 hours	30	4000		−66	[274]
975	$Ca(Dy_{1/2}Ta_{1/2})O_3$	1600 for 4 hours	30	26 500		−6	[237]
976	$BaZn_2Ti_4O_{11}$	1200	30	68 000		−30	[346]
977	$Ba_5SrTa_4ZrO_{18} + 2$ wt% $Bi_2O_3-B_2O_3$ glass	1525 for 4 hours	30	18 500		37	[284]
978	$BaZn_{1.95}Ti_4O_{10.95}$	1200	30	110 000			[346]
979	$La(Co_{1/2}Ti_{1/2})O_3 + 0.25$ wt% B_2O_3	1350 for 6 hours	30	64 600	8	−48	[347]
980	$(1-x)Sr(Li_{1/4}Nb_{3/4})O_3-xSr(Li_{2/3}W_{3/5})O_3$ $(x = 0.385)$	1450	30	21 200	9.2	−33	[319]
981	$(Ca_{2/3}La_{1/3})(Li_{1/3}Nb_{2/3})O_3$	1250	30	26 500	8.7	−26	[267]
982	$BaZn_{2-x}Ti_4O_{11-x}$ $(x = 0-0.1)$	1250 for 4 hours	30	83 000		−30	[346]
983	$Ba_{10}Co_{0.25}Ta_{7.9}O_{30}$	1550	30	36 700	3.78	29	[318]
984	$Ba_6Ta_4ZrO_{18} + 2$ wt% $Bi_2O_3-B_2O_3$ glass	1625 for 2 hours	30.1	41 000		5	[284]
985	$La_2O_3-WO_3-TiO_2$	1350	30.1	9225	5.8	−17	[117]
986	MWF-38 + 10 wt% $Li_2O-B_2O_3-SiO_2$ (56.92:37.59:5.49)	875	30.2	9500		3	[155]

No.		Sintering temp. (°C)	ε_r	*Qf* (GHz)	*f* (GHz)	τ_f	Reference
987	$MgTa_2O_6$	1550	30.3	59 600		30	[199]
988	$ZnTa_2O_6$	1400	30.3	87 580		9.5	[348]
989	$0.5Ba(Y_{1/2}Ta_{1/2})O_3{-}0.5Ba(Ca_{1/9}Y_{3/9}Ta_{5/9})O_3$		30.4	36 000	9.56	160	[322]
990	$Eu(Zr_{1/3}Ti_{2/3})O_6$	1600	30.4	11 000	4.5	−4	[349]
991	$Ba(Y_{1/2}Ta_{1/2})O_3$		30.5	38 500	9.4	135	[350]
992	$Ba_8ZnTa_6O_{24}$	1400	30.5	62 000	8.9	36	[264]
993	$Ca(Li_{1/3}Nb_{2/3})O_{3-\delta}$ + 4 wt% B_2O_3	1000	30.6	31 000		−17.5	[351]
994	$Sr_6Ta_4ZrO_{18}$ + 3 wt% $Bi_2O_3{-}B_2O_3$	1625 for 2 hours	30.8	5600		−19	[284]
995	$0.1BaTiO_3{-}0.9La(Mg_{1/2}Ti_{1/2})O_3$		30.9	16 330	8.29	−72	[306]
996	$Li_{0.774}Zr_{0.057}NbO_3$	1150	30.9	3550		−16	[352]
997	$Ba_3ZnNb_{2-x}Sb_xO_9$ ($x = 0.75$)		30.9	23 700	5	0	[111]
998	$Ba_3ZnNb_{2-x}Sb_xO_9$ ($x = 0.5$)		30.9	35 620	5.3	−11	[111]
999	$Ba_{1-x}Ca_x(Nd_{1/2}Nb_{1/2})O_3$		30–42	25 000–5000			[350]
1000	$Ba_{1-x}Ca_x(Y_{1/2}Ta_{1/2})O_3$		30–22				[350]
1001	$(1-x)LaMg_{1/2}Ti_{1/2}{-}xLa_{2/3}TiO_3$ ($x = 0.2$)		31	43 000	6.3	−54	[317]
1002	$2CaO{-}3NiO{-}Nb_2O_5{-}TiO_2$	1275	31	7500	4.3	−49	[67]
1003	$(1-x)Sr(Li_{1/4}Nb_{3/4})O_3{-}xSr(Li_{2/3}W_{3/5})O_3$ ($x = 0.333$)	1450	31	27 400	8.71	−23	[319]
1004	$(1-x)Ca(Li_{1/4}Nb_{3/4})O_3{-}xCa(Li_{2/3}W_{3/5})O_3$ ($x = 0.238$)	1150	31	22 700	10.3	−33	[319]
1005	$(1-x)Ba(Li_{1/4}Nb_{3/4})O_3{-}xBa(Li_{2/3}W_{3/5})O_3$ ($x = 0.333$)	1470	31	19 000	7.8	18	[353]
1006	$Sm(Zn_{1/2}Ti_{1/2})O_3$	1310 for 2 hours	31	37 000	8	−19	[354]
1007	$Ba_3Zn_7Ti_{12}O_{34}$	1150 for 4 hours	31	4300		−25	[355]
1008	$BaO{-}TiO_2{-}WO_3$ (N-35):5 wt% $Al_2O_3{-}SiO_2{-}B_2O_3$	1100	31	5400	5.7	–	[249, 276]
1009	$Ba(Mg_{1/3}Nb_{2/3})O_3$ + 2 mol% B_2O_3 + 10 mol% CuO						[356]
1010	$0.5Ba(Mg_{1/2}W_{1/2})O_3{-}0.5BaTiO_3$		31	8200	–	125	[107]
1011	$(Zn_{0.5}Co_{0.5})TiO_3$	1150	31	60 000		75	[220]
1012	$0.75(Al_{1/2}Ta_{1/2})O_2{-}0.25(Ti_{1-x}Sn_x)O_2$	1450 for 3 hours	31–26	54 600–70 700		12.8 to −9.3	[316]
1013	$0.95Ba(Zn_{1/2}Ta_{2/3})O_3{-}0.05[Sr_{0.25}Ba_{0.75}](Ga_{1/2}Ta_{1/2})O_3$	1500	31	210 000			[326]
1014	$Ba_3Zn_7Ti_{12}O_{34}$	1150 for 4 hours	31	4300	10	−25	[355]
1015	$Ba_{0.85}Sr_{0.15}(Y_{1/2}Ta_{1/2})O_3$	1600	31	32 000		0	[274]
1016	$Sr(La_{1/2}Ta_{1/2})O_3$	1600 for 4 hours	31	4500		−42	[274]

No.		Sintering temp. (°C)	ε_r	Qf (GHz)	f (GHz)	τ_f	Reference
1017	$Sr(Sm_{1/2}Ta_{1/2})O_3$	1600 for 4 hours	31	45 200		−61	[274]
1018	$Sr(Al_{1/2}Nb_{1/2})O_3$	1600 for 24 hours	31	10 800		−27	[357, 275]
1019	$Ca(Y_{1/2}Nb_{1/2})O_3$	1550 for 4 hours	31	35 000		−13	[236]
1020	$Sr(Yb_{1/2}Nb_{1/2})O_3$	1600 for 4 hours	31	26 600		−73	[275]
1021	$Ca(Li_{1/3}Nb_{2/3})O_{3-\delta}$ + 4 wt% B_2O_3	1000	31	31 000		−18	[351]
1022	$0.4CaTiO_3-0.6Sm(Mg_{0.5}Ti_{0.5})O_3$	1550	31	12 000	5.3	−28	[278]
1023	$2CaO-3MgO-Nb_2O_5-TiO_2$	1325	31	33 000	4.6	−10	[67]
1024	$Ca_4MgTa_2TiO_{12}$	1625	31	43 000	4.7	−62	[67]
1025	$Ba(Li_{1/4}Nb_{3/4})O_3-Ba(Li_{2/3}W_{3/5})O_3$	1470	31	19 000	7.8	18	[353]
1026	$Ca_4NiTa_2TiO_{12}$	1625	31	40 000	4.7	−26	[67]
1027	$Ba(Ni_{1/3}Nb_{2/3})O_3$	1400	31	48 000		−18	[358]
1028	$Ba(Mg_{1/3}Nb_{2/3})O_3$	1350	31	46 000		18	[359]
1029	$Ba_4SrTa_4O_{15}$	1575	31.1	9500	5.2	8	[272]
1030	$Pr_{0.16}Gd_{0.84}TiNbO_6$	1400	31.1	3180		−7.9	[175]
1031	$SmTaTi_{0.7}Zr_{0.3}O_6$		31.1	37 481		−2.2	[211]
1032	$0.5Ba(Mg_{1/2}W_{1/2})O_3-0.5BaTiO_3$	1500 for 6 hours	31.2	8200		125.5	[107]
1033	$0.75(Al_{1/2}Ta_{1/2})O_2-0.25TiO_2$	1450	31.2	54 600		12.8	[316]
1034	$Nd(Zr_{1/3}Ti_{2/3})O_6$	1600	31.4	15 800	4.3	6	[349]
1035	$(Zr_{0.8}Sn_{0.2})TiO_4$ + 10 wt% $BaO-B_2O_3-SiO_2-Li_2O-CuO$	1050 for 4 hours	31.4	32 200		−1.4	[157]
1036	$0.9Ba(Zn_{1/3}Ta_{2/3})O_3-0.1BaTi_4O_9$	1320	31.5	68 500	6	4	[360]
1037	$Nd(Zn_{1/2}Ti_{1/2})O_3$	1330 for 4 hours	31.6	170 000	8.5	−42	[361]
1038	$SmTaTi_{0.8}Zr_{0.2}O_6$		31.7	30 654		6.1	[211]
1039	$(Zr_{0.8}Sn_{0.2})TiO_4$ + 10 wt% $BaO-B_2O_3-SiO_2-Li_2O-CuO$	1050 for 12 hours	31.7	29 700		−1.8	[604]
1040	$Ba_5NbTa_3O_{15}$	1500	31.7	21 500		16	[272]
1041	$BaSr_4Ta_4O_{15}$	1600	31.7	2800	5.34	−60	[272]
1042	90 wt% $BaTi_4O_9$ + 10 wt% $Li_2O-B_2O_3-SiO_2$	875	31.7	9000		10	[182]
1043	$Sr(Sm_{1/2}Nb_{1/2})O_3$	1500	31.8	41 000	8.1	−45	[180]
1044	MBRT-90 + 10 wt% $Li_2O-B_2O_3-SiO_2-CaO-Al_2O_3$ (28:27:30:5:10)	875	31.9	2200		20	[155]
1045	$Ba(Mg_{1/3}Nb_{2/3})O_3$		32	55 500	10	33	[221]
1046	$La_6Mg_{0.913}Ti_{4.04}O_{18}$		32	31 000	6.1	−46	[317]
1047	$Ca_5Nb_2Ti_{0.4}Hf_{0.6}O_{12}$	1675	32	22 000	4.5	± 0.5	[147]
1048	$Ba(Mg_{1/3-x}Nb_{2/3})O_{3-\delta}$ ($x = 0.02$)	1450	32	96 000		30	[362]

No.		Sintering temp. (°C)	ε_r	*Qf* (GHz)	*f* (GHz)	τ_f	Reference
1049	$Ba(Co_{1/3}Nb_{2/3})O_3$	1400	32	78 000		−12	[363, 364]
1050	$Ba(Yb_{1/2}Ta_{1/2})O_3$	1625 for 4 hours	32	35 900		112	[311]
1051	$Ca(Fe_{1/2}Ta_{1/2})O_3$	–	32	20 000		−61	[186]
1052	$Ca(Er_{1/2}Nb_{1/2})O_3$	1550 for 4 hours	32	31 800		−18	[236]
1053	$Ca(Dy_{1/2}Nb_{1/2})O_3$	1550 for 4 hours	32	32 500		5	[236]
1054	$Sr(Sm_{1/2}Ta_{1/2})O_3$ + 0.2 wt% TiO_2	1600	32	46 400		−46	[274]
1055	$Sr(Pr_{1/2}Ta_{1/2})O_3$	1600 for 4 hours	32	8400		−50	[274]
1056	$Sr(Nd_{1/2}Ta_{1/2})O_3$	1600 for 4 hours	32	38 500		−55	[274]
1057	$Sr(Ho_{1/2}Nb_{1/2})O_3$	1600 for 4 hours	32	20 400		−65	[275]
1058	$Sr(Y_{1/2}Nb_{1/2})O_3$	1600 for 4 hours	32	38 800		−66	[275]
1059	$Sr(Er_{1/2}Nb_{1/2})O_3$	1575 for 4 hours	32	36 100		−67	[275]
1060	$BaO{-}2CeO_2{-}5TiO_2$	1250	32	19 100		41	[295]
1061	$EuTiNbO_6$	1370	32	17 250	5.3	5	[174]
1062	$5CaO{-}2Nb_2O_5$	1500	32	6000	6.48	−37	[94]
1063	$BaTi_4O_9$ + 10 wt% glass frit	875	32	9000		10	[365]
1064	$0.1(Na_{0.5}La_{0.5})TiO_3{-}0.9CeO_2$	1400	32	8200		0	[366]
1065	$3CaO{-}2CoO{-}Nb_2O_5{-}TiO_2$	1400	32	15 000	4.3	−18	[67]
1066	$0.25Ca_2AlNbO_6 - 0.75Ca_3Nb_2O_8$		32	7500	6.34	64	[288]
1067	$Ba(Cd_{1/3}Ta_{2/3})O_3 + B_2O_3$	1350	32	50 000	2	80	[367]
1068	$Ca_5Nb_2Ti_{0.4}Hf_{0.6}O_{12}$	1675	32	22 000	4.458	≈0	[147]
1069	$BaO{-}TiO_2{-}WO_3$ (N-35):5 wt% $Al_2O_3{-}SiO_2$	1100	32	11 000	5.6	–	[276]
1070	$SnTe_3O_8$	700 for 15 hours	32	13 200	4		[15]
1071	$La_6Mg_{0.913}Ti_{4.04}O_{18}$		32	31 000	6.1	−46	[317]
1072	$Ba(Yb_{1/2}Ta_{1/2})O_3$	1625	32	35 800		112	[311]
1073	$Ba(Sn_{0.226}Zn_{0.258}Nb_{0.516})O_3$	1500	32	970 000		12	[368]
1074	$Ba(Zn_{1/3}Nb_{2/3})O_3$ + 5 mol% B_2O_3	900	32	3500		20	[369]
1075	$Ba_6Nb_4ZrO_{18}$	1625 for 2 hours	32.4	52 000		25	[284]
1076	$CaWO_4 + 0.005B_2O_3$	1050	32.7	70 600		−47	[13]
1077	$ZnNb_2O_6$ + 1.5 wt% $(CuO{-}V_2O_5{-}Bi_2O_3)$	870 for 2 hours	32.7	67 100		−33	[370]
1078	$Ba_4SrTa_4O_{15}$	1575	32.9	9000		8	[94]
1079	$[Ba_{1-z}Sr_zZn_{1/3}Ta_pNb_{1-p}]O_3{-}Sr_{1x}Ca_x(Ga_{1/2}Ta_{1/2})O_3$		32–34	180 000–80 000	7	0–10	[371]

No.		Sintering temp. (°C)	ε_r	Qf (GHz)	f (GHz)	τ_f	Reference
1080	$Sr(Mg_{1/3}Nb_{2/3})O_3$:Mn		33	23 700	10.3	−14	[221]
1081	$BaTi_4O_9$ + 5 wt% $ZnO-B_2O_3$ glass	900 for 2 hours	33	27 000		7	[372]
1082	$3CaO-2CoO-Nb_2O_5-TiO_2$	1400	33	15 000	4.3	−18	[67]
1083	$Sr(Dy_{1/2}Nb_{1/2})O_3$	1575 for 4 hours	33	32 700		−63	[275]
1084	$Ba[(Ni_{0.6}Zn_{0.4})_{1/3}Nb_{2/3}]O_3$ + 1 mol% B_2O_3	1300	33	39 700		−4	[373]
1085	$Ba_{0.95}Sr_{0.05}(Y_{1/2}Ta_{1/2})O_3$	1600	33	47 300		0	[274]
1086	$Ba(Y_{1/2}Ta_{1/2})O_3$	1625 for 4 hours	33	50 200		120	[311]
1087	$(1-x)LaMg_{1/2}Ti_{1/2}O_3-xLa_{2/3}TiO_3$ ($x=0.3$)		33	43 000	6.3	−54	[317]
1088	$Bi_6Te_2O_{15}$ (oxygen atm)	800 for 15 hours	33	41 000		−85	[374]
1089	$Ca[Li_{1/3}Nb_{2/3}]_{0.75}Ti_{0.25}]O_{3-\delta}$ + 5 wt% $Li_2O-B_2O_3-SiO_2$	950 for 4 hours	33	11 500		−5	[375]
1090	$Ba(Cd_{1/3}Ta_{2/3})O_3$ + 2 wt% ZnO	1550	33	37 500		80	[376]
1091	$La_5AlTi_3O_{15}$	1600 for 3 hours	33	28 600		−39.2	[320]
1092	$Li_{0.774}Zr_{0.057}NbO_3$	1150	33.01	4460		−28	[377]
1093	$Ba_3Sr_2Ta_4O_{15}$	1575	33.2	4300	5.2	−15	[272]
1094	$Sr_5Nb_2Ta_2O_{15}$	1575	33.2	2500	5.65	−2	[272]
1095	$Ba_3ZnNb_{2-x}Sb_xO_9$ ($x=0.375$)		33.2	44 940		−3	[111]
1096	$La_5CrTi_3O_{15}$	1625 for 2 hours	33.2	27 500	4.88	−34	[378]
1097	$La_4PrCrTi_3O_{15}$	1575 for 2 hours	33.2	23 700	4.7	−22	[378]
1098	$Ba_5SrTa_4TiO_{18}$	1550 for 2 hours	33.2	33 000		65	[284]
1099	$Pr(Zr_{1/3}Ti_{2/3})O_6$	1600	33.3	16 200	4.3	14	[349]
1100	$La_4NdCrTi_3O_{15}$	1600 for 2 hours	33.5	18 000	4.7	−36	[378]
1101	$Ba(Ho_{/2}Ta_{1/2})O_3$	1625	33.5	24 000		130	[311]
1102	$Ba_{1+x}[(Co_{0.7}Zn_{0.3})_{1/3}Nb_{2/3}]O_3$ ($x=-0.01$)	1450 for 10 hours	33.7	70 900		−4	[379]
1103	$Ba_2Sr_3Ta_4O_{15}$	1600	33.7	2400	5	−25	[272]
1104	$(Sr_{0.1}Ba_{0.9})(Ti_{0.1}Zn_{0.3}Ta_{0.6})O_3$		33.7	36 000	7	23	[343]
1105	$ZnTa_2O_6$	1400 for 10 hours	33.7	79 310	8.5	9	[380]
1106	93 wt% $BaTi_4O_9$ + 10 wt% $Li_2O-B_2O_3-SiO_2$	950	33.8	12 700		25	[182]
1107	$Sr_6Nb_4ZrO_{18}$ + 2 wt% $Bi_2O_3-B_2O_3$	1625 for 2 hours	33.9	21 000		−8	[284]
1108	$ZnTiNb_2O_8$	1250 for 2 hours	34	42 500		−52	[195]

No.		Sintering temp. (°C)	ε_r	Qf (GHz)	f (GHz)	τ_f	Reference
1109	$0.2TiTe_3O_8-0.8TeO_2$	670	34	22 000		24	[179]
1110	$BaO-TiO_2-WO_3$ (N-35): B_2O_3	1200	34	70 500	8.5	–	[249, 276]
1111	$La(Zn_{1/2}Ti_{1/2})O_3$	1550	34	59 000	10	−52	[381, 382]
1112	$Ba_8Ta_{4+0.8x}Ti_{3-x}O_{24}$ ($x=0.6$)	1400 for 40 hours	34	30 820		57	[383]
1113	$Ca_5Ta_2Ti_{0.6}Hf_{0.4}O_{12}$	1675	34	26 000	4.4	± 0.5	[147]
1114	$Ca_5Nb_2Ti_{0.2}Zr_{0.8}O_{12}$	1670	34	24 000	4.4	± 0.5	[207]
1115	$0.25Ba(Zn_{1/3}Nb_{2/3})O_3-0.75Ba(Mg_{1/3}Nb_{2/3})O_3+B_2O_3-LiF$	1350	34	76 700	7.6	−4	[384]
1116	$La_5Mg_{0.5}Ti_{3.5}O_{15}$		34	31 000	6	−16	[317]
1117	$La_4SmCrTi_3O_{15}$	1575 for 2 hours	34	15 900	4.89	−38	[378]
1118	$0.5CaTiO_3-0.5Sm(Mg_{0.5}Ti_{0.5})O_3$	1550	34	10 400	4.91	−24	[278]
1119	$(1-x)(Mg_{0.95}Zn_{0.05})TiO_3-xCa_{0.6}La_{0.8/3}TiO_3$ ($x=0.3$)	1320 for 4 hours	34	61 000		66	[137]
1120	$Ca[(Li_{1/3}Nb_{2/3})_{0.84}Ti_{0.16}]O_{3-\delta}$ + 2 wt% LiF + 3 wt% B_2O_3	900 for 2 hours	34	17 400		−5	[239]
1121	$Ba(Dy_{1/2}Ta_{1/2})O_3$	1625	34	20 600		−48	[311]
1122	$Sr(Gd_{1/2}Nb_{1/2})O_3$	1575 for 4 hours	34	8800		−56	[275]
1123	$0.99Ba(Co_{1/3}Nb_{2/3})O_3-0.01Ba(Y_{1/2}Nb_{1/2})O_3$	1380	34	38 690		#	[385]
1124	$0.95Ba(Yb_{1/2}Nb_{1/2})O_3-0.05Ca(Y_{1/2}Nb_{1/2})O_3$	1600	34	47 500		1	[236]
1125	$Sr(Tb_{1/2}Nb_{1/2})O_3$	1575 for 4 hours	34	36 300		−61	[275]
1126	$Ca_5Nb_2Ti_{0.2}Zr_{0.8}O_{12}$	1670	34	24 000	4.4	≈0	[207]
1127	$Ca_5Ta_2Ti_{0.6}Hf_{0.4}O_{12}$	1675	34	26 000	4.4	0	[147]
1128	$xBa(Zn_{1/3}Nb_{2/3})O_3-(1-x)Ba(Mg_{1/3}Nb_{2/3})O_3$ ($x=0.25$)	1500	34	76 700		−4	[384]
1129	$Ba[(Ni_{0.6}Zn_{0.4})_{1/3}Nb_{2/3}]O_3$ + 0.5 mol% B_2O_3	1340	34	42 100		−8	[373]
1130	$LiNb_3O_8$	1075 for 3 hours	34	58 000		−96	[250]
1131	$Zr_{0.034}Hf_{0.966}TiO_4$		34.1	34 000	9.9		[386]
1132	$Ba_8Ta_4Ti_3O_{24}$		34.2	23 050		75.8	[383]
1133	$Ba_5Nb_2Ta_2O_{15}$	1475	34.2	10 500		22	[387]
1134	$Ba(Sm_{1/2}Ta_{1/2})O_3$	1500	34.3	27 000	7.7	4.9	[180]
1135	$Ba[Ni_{0.6}Zn_{0.4}]_{0.33}Nb_{0.67}]O_3$ + 0.5 mol% B_2O_3	1350	34.3	42 100		−2.7	[373]
1136	$0.5ZnNb_2O_6-0.5TiO_2$	1250 for 2 hours	34.3	42 500		−52	[388]
1137	$ZnTiNb_2O_8$	1250	34.3	42 500		−52	[195]
1138	$Ca[(Li_{1/3}Nb_{2/3})_{0.84}Ti_{0.16}]O_{3-\delta}$ + 2 wt% LiF + 3 wt% $ZnO-B_2O_3-SiO_2$	900 for 2 hours	34.3	17 400		−5	[389]
1139	$Ba_3Sr_2Ta_4O_{15}$	1575	34.3	4000		−15	[387]

No.		Sintering temp. (°C)	ε_r	*Qf* (GHz)	*f* (GHz)	τ_f	Reference
1140	$0.5La_{2/3}TiO_3-0.5LaAlO_3$	1425	34.4	45 000	6.7	−23	[390]
1141	$Ba_3Co_{0.7}Zn_{0.3}Nb_2O_9+$ +0.4 wt% CeO_2	1450 for 4 hours	34.5	84 000	4	0	[391]
1142	$0.7Ba(Co_{1/3}Nb_{2/3})O_3-0.3Ba(Zn_{1/3}Nb_{2/3})O_3$	1400 for 20 hours	34.5	97 000	6.5	0	[391, 392]
1143	$0.5LaAlO_3-0.5SrTiO_3$ + 0.25 wt% B_2O_3	1430 for 2 hours	34.5	43 200	7	−10.7	[393]
1144	$Ba_3Co_{0.7}Zn_{.3}Nb_2O_9+V_2O_5$	1450	34.5	85 000	4	0	[391]
1145	$La_4SmCrTi_3O_{15}$	1650	34.5	17 300	4.7	−38	[378]
1146	$ZnTa_2O_6$ + 0.5 wt% CuO	1230	34.6	65 500		5	[394]
1147	$DyTiTaO_6$	1500	34.6	40 100		7	[178]
1148	$BaTi_4O_9$ +3 wt% MCAS glass	1200	34.6	42 050	7	14	[395]
1149	$La_4PrCrTi_3O_{15}$	1575	34.6	23 700	4.8	−22	[378]
1150	$0.09(0.5ZnNb_2O_6-0.5Zn_3Nb_2O_8)-0.91ZnTa_2O_6$	1350	34.7	41 950		0	[167]
1151	$La_5CrTi_3O_{15}$	1650	34.8	34 000	4.8	−35	[378]
1152	$Sr_6Ta_4TiO_{18}$ + 3 wt% $Bi_2O_3-B_2O_3$ glass	1625 for 2 hours	34.8	5600		−19	[284]
1153	$BiNbO_4$ + 0.03 wt% CuV_2O_6	1050	34.9	9870		−3.4	[396]
1154	$Ba_5SrTa_4TiO_{18}$	1550 for 4 hours	34.9	33 000		65	[284]
1155	$Ca[(Li_{1/3}Ta_{2/3})_{.7}Ti_{0.3}]O_{3-\delta}$ + 3 wt% B_2O_3	1050	35	22 800	9.45	−4	[242]
1156	$Ca[(Li_{1/3}Nb_{2/3})_{.9}Ti_{0.1}]O_{3-\delta}$ + 0.7 wt% B_2O_3	1000	35	22 100		−5	[242]
1157	$Ca[(Li_{1/3}Nb_{2/3})_{1-x}Ti_x]\ O_{3-\delta}$ ($x=0.1$)	1150 for 3 hours	35	27 200		−2	[239]
1158	$0.3TeO_2-0.5SnTe_3O_8$	650	35	8500		176	[397]
1159	$BaTi_4O_9-0.1WO_3$	1400	35	52 000	8	−0.5	[50]
1160	$0.46LaAlO_3-0.54SrTiO_3$ + 2 wt% B_2O_3	1460	35	38 000	7	−1	[398]
1161	$0.5LaAlO_3-0.5SrTiO_3$		35	27 000		−18	[399]
1162	$Ca[(Li_{1/3}Nb_{2/3})_{0.8}Ti_{0.2}]O_{3-\delta}$ + 5 wt% Bi_2O_3	900 for 3 hours	35	11 000		13	[187]
1163	$BaZrO_3$		35	8800			[292]
1164	$Ba_{10}Ta_{8-0.8x}Ti_xO_{30}$ ($x=1.2$)	1400 for 40 hours	35	25 760		64	[383]
1165	$Ca(Zn_{1/3}Nb_{2/3})O_3$		35	16,000	–	−43	[186]
1166	$0.42Zn_3Nb_2O_8-0.58TiO_2$	1250	35	48 000		0	[388]
1167	$Pb_{0.75}Ca_{0.25}(Al_{1/2}Nb_{1/2})O_3$		35	1100	4.7	133	[336]
1168	$Ca_{4.75}Ni_{0.25}Ta_2TiO_{12}$	1625	35	34 000	4.5	± 0.5	[67]
1169	$0.5CeO_2-0.25CaO-0.25TiO_2{:}6.5Cr_2O_3$	1550	35	4300	4.4	0	[145]
1170	$Sr(Cr_{1/2}Nb_{1/2})O_3$	1600 for 4 hours	35	6400		−80	[275]
1171	$Sr(Eu_{1/2}Nb_{1/2})O_3$	1575 for 4 hours	35	44 000		−52	[275]

No.		Sintering temp. (°C)	ε_r	*Qf* (GHz)	*f* (GHz)	τ_f	Reference
1172	$0.9Ba(Zn_{0.6}Co_{0.4})_{0.33}Nb_{0.67}]O_3-0.1Ba(Ga_{1/2}Ta_{1/2})O_3$		35	93 550	3.06	0	[400]
1173	$BaO-4TiO_2-0.1WO_3$		35	52 400	6	0	[50]
1174	$Ba[(Zn_{0.3}Co_{0.7})_{1/3}Nb_{2/3}O_3 + 0.25$ wt% V_2O_5	1450 for 4 hours	35	85 000		0	[401]
1175	$Ba[(Co_{0.7}Zn_{0.3})_{1/3}Nb_{2/3}]O_3 + 0.4$ wt% CeO_2	1450	35	84 000		0	[391]
1176	$0.9Ba[(Zn_{0.6}Co_{0.4})_{1/3}Nb_{2/3}]O_3-0.1Ba(Ga_{1/2}Ta_{1/2})O_3$		35	97 600		0	[402]
1177	$0.35Ba(Ni_{1/3}Nb_{2/3})O_3-0.65Ba(Zn_{1/3}Nb_{2/3})O_3$	1450 for 4 hours	35	25 000		1	[403]
1178	$BaO-Al_2O_3-4TiO_2$		35	5000	–	−15	[404]
1179	$Ca_{4.75}Ni_{0.25}Ta_2TiO_{12}$	1625	35	34 000	4.49	0	[67]
1180	$BaO-4TiO_2-0.1WO_3$	1400 for 2 hours in O_2	35	50 400		−0.5	[50]
1181	$0.5LaAlO_3-0.5SrTiO_3 + 0.25$ wt% B_2O_3	1430	35	43 200		−11	[393]
1182	$La_5GaTi_3O_{15}$	1550	35.01	30 300	3.09	−54.5	[320]
1183	$0.17Ba_5Nb_4O_{15}-0.83BaNb_2O_6$ (hexagonal)	1250 for 2 hours	35.2	59 300		0	[405]
1184	$BaTi_{4.35}Zn_{0.55}O_{10.25}$	1260 for 6 hours	35.2	5000		36	[406]
1185	$Ba_2Sr_3Ta_4O_{15}$	1575	35.2	2400		−25	[387]
1186	$0.6Ba(Zn_{1/3}Nb_{2/3})O_3-0.4Ba(Co_{1/3}Nb_{2/3})O_3$	1450	35.5	86 000	2	0	[407]
1187	$Ba(Tb_{1/2}Ta_{1/2})O_3$	1625	35.5	31 900		−38	[311]
1188	$La_4NdCrTi_3O_{15}$	1650	35.6	19 400	4.7	−34	[378]
1189	$Ba(Sc_{1/2}Nb_{1/2})O_3$	1700 for 12 hours	35.7	20 000			[408]
1190	$Ba_3ZnNb_{2-x}Sb_xO_9$ ($x = 0.125$)		35.7	56 980	4.8	14	[111]
1191	$SmTaTi_{0.9}Zr_{0.1}O_6$		35.8	27 730		14.9	[211]
1192	$Ba_3ZnNb_{2-x}Sb_xO_9$ ($x = 0.25$)		35.8	35 090	5.4	6	[111]
1193	$0.615BaTi_4O_9-0.35ZnO-0.3Nb_2O_5 + 0.3$ wt% Mn	1280 for 2 hours	35.8	50 800		1.1	[409]
1194	$0.5LaAlO_3-0.5SrTiO_3$	1680	35.9	108 800	9.7	−21	[268]
1195	$(Zr_{0.8}Sn_{0.2})TiO_4 + 0.2$ wt% NiO	1280	35.9	56 700	9.2		[410]
1196	$Ba_6Nb_4ZrO_{18}$	1625 for 2 hours	35.9	52 000		25	[284]
1197	$BaO-TiO_2-WO_3$ (N-35)		36	50 400	6	0	[276]
1198	$Ba_{1-x}Ca_x(Sc_{1/2}Nb_{1/2})O_3$		36–55	20 000–55 000			[408]
1199	$Ba(Nd_{1/2}Ta_{1/2})O_3$	1500	36	18 000	7.3	2.9	[180]
1200	$Ba_2Ti_9O_{20} + 9$ wt% BaB_2O_4	1050 for 2 hours	36	12 600		−2	[342]
1201	$Ba(Zn_{1/3}Nb_{2/3})O_3 +$ 5 mol% $B_2O_3 + CuO$	875	36	19 000		21	[745]

No.		Sintering temp. (°C)	ε_r	Qf (GHz)	f (GHz)	τ_f	Reference
1202	Bi_2TiTeO_8	840 for 10 hours	36	4700	5.7	41	[411]
1203	$BaTi_4O_9$ (citrate route)	1250 for 10 hours	36	50 470		16	[412]
1204	$Ba_{10}Ta_{7.04}Ti_{1.2}O_{30}$		36	30 000		52	[136]
1205	$BaTi_4O_9-ZnO-Ta_2O_5+0.1$ wt% Mn	1280	36	45 000	4.5	0	[413]
1206	$Nd(Zn_{1/2}Ti_{1/2})O_3$		36	42 300		−47	[414]
1207	$TiTe_3O_8$	700 for 5 hours	36	13 600	4		[15]
1208	$0.25Zn_3Nb_2O_8-0.75TiO_2$	1200 for 2 hours	36	5160		4	[195]
1209	$BaO-TiO_2-WO_3$ (N-35) + 5 wt% SiO_2	1200	36	4500	8.5		[249]
1210	$ZrTiO_4$ (polymer route) + 0.5 wt% Hf	1600	36	29 700	5.3		[415]
1211	$Ba(Yb_{1/2}Nb_{1/2})O_3$	1600	36	38 100		2	[416]
1212	$Nd_2Ti_2O_7$	1300	36	16 400		−118	[417, 418]
1213	$Ba(Gd_{1/2}Ta_{1/2})O_3$	1625 for 4 hours	36	3200		−18	[311]
1214	$Ba(Tb_{1/2}Ta_{1/2})O_3$	1625 for 4 hours	36	31 900		−38	[311]
1215	$(Pb_{0.2}Ca_{0.8})(Ca_{1/3}Nb_{2/3})O_3$	1350	36	12 500		−27	[419]
1216	$0.6Ba(Zn_{1/3}Nb_{2/3})O_3-0.4Ba(Co_{1/3}Nb_{2/3})O_3$	1400	36	86 000		0	[407]
1217	$Sr(Sm_{1/2}Nb_{1/2})O_3$	1575 for 4 hours	36	32 300		−47	[275]
1218	$Ca_4NiNb_2TiO_{12}$	1550	36	31 500	4.1	−30	[67]
1219	$Ca_{4.88}Co_{0.12}Ta_2TiO_{12}$	1625	36	35 000	4.49	0	[67]
1220	$Ca_5Ta_2Ti_{0.7}Zr_{0.3}O_{12}$	1650	36	28 000	4.4	0	[207]
1221	$La_4MgTi_3O_{12}$		36	25 000	5.8	−39	[317]
1222	$0.2CaTiO_3-8Ca(Mg_{1/3}Nb_{2/3})O_3$	1450	36	39 000	7.7	−30	[312]
1223	$Li_2O-Nb_2O_5-TiO_2$ + 1 wt% B_2O_3	1100	36	10 450	5.9	12.2	[420]
1224	$Ba_2Ti_{9-x}[La_{0.5}Ta_{.5}]_xO_{20}$ ($x=0.05$)	1350	36	60 000		12	[421]
1225	$Ba_8Ta_{4+0.8x}Ti_{3-x}O_{24}$($x=0-0.4$)	1400 for 40 hours	36	12 000		–	[421]
1226	$BaO-ZnO-TiO_2+0.52$ mol% SnO_2	1160	36.1	40 000		6	[406]
1227	$Sr_5NbTa_3O_{15}$	1575	36.2	6900	5.14	31	[272]
1228	$La_5GaTi_3O_{15}$	1600 for 30 hours	36.2	30 300		−54.5	[320]
1229	$BaTi_4O_9$ + 5 mol% CuO + 2 mol% B_2O_3	900 for 2 hours	36.3	30 500		28.1	[422]
1230	$La_6MgTi_4O_{18}$	1625 for 2 hours	36.3	27 350		−39	[284]
1231	$CaTi_{.3}(Al_{1/2}Nb_{1/2})_{0.7}O_3$ + 1 wt% Li_3NbO_4	1300 for 5 hours	36.4	38 900	7	−57	[423]

No.		Sintering temp. (°C)	ε_r	Qf (GHz)	f (GHz)	τ_f	Reference
1232	$Sr_6Nb_4ZrO_{18}$ + 2 wt% $Bi_2O_3-B_2O_3$ glass	1625 for 2 hours	36.4	21 000		−8	[284]
1233	$Nd_2Ti_2O_7$		36.5	16 400	9.1	−118	[417]
1234	$Ba_2Ti_9O_{20}$ + 5 wt% B_2O_3	1200	36.5	40 200		38	[422]
1235	$BaO-TiO_2-ZnO$	1160 for 6 hours	36.5	42 000	4.7	1.8	[406]
1236	$La_5Zn_{0.5}Ti_{3.5}O_{15}$	1500	36.5	23 000	5.3	−38.6	[320]
1237	$BaTi_{4.35}Zn_{0.55}O_{10.25}$ + 0.5 mol% SnO_2	1160 for 6 hours	36.5	42 000		1.8	[406]
1238	$Ba_3Ti_5Nb_6O_{28}$ + 3 wt% ZnB_2O_4	925	36.6	19 100		5	[424]
1239	$(5-x)BaO-xMgO-2Nb_2O_5$ ($x=0.5$) + 1 wt% CuO	1200	36.7	20 000		61	[425]
1240	$(Zr_{0.8}Sn_{0.2})TiO_4$ + 0.2 wt% MgO	1320	36.7	60 000	6.5		[426]
1241	$CaTi_{0.3}(Al_{1/2}Nb_{1/2})_{0.7}O_3$	1500 for 5 hours	36.8	29 800	7	−61	[423]
1242	$Ca(Fe_{1/2}Nb_{1/2})O_3$		36.8	15 800	8.2	−93	[427]
1243	$TbTiTaO_6$	1525	36.8	32 300		10	[178]
1244	$BaO-ZnO-TiO_2$ + 0.5 mol% $MnCO_3$	1250	36.8	39 000		−6.5	[406]
1245	$La_4Ti_9O_{24}$	1350	37	24 800	8.1	15	[417, 418]
1246	$Ba(La_{1/2}Ta_{1/2})O_3$	1625 for 4 hours	37	20 950		−36	[311]
1247	$Ba(Y_{1/2}Nb_{1/2})O_3$	1600	37	49 600		15	[416]
1248	$Ba(Eu_{1/2}Ta_{1/2})O_3$	1625 for 4 hours	37	41 200		−16	[311]
1249	$Sr(La_{1/2}Nb_{1/2})O_3$	1575 for 4 hours	37	4000		−20	[275]
1250	$Sr(Nd_{1/2}Nb_{1/2})O_3$	1575 for 4 hours	37	20 100		−40	[275]
1251	$0.9Ba(Co_{1/3}Nb_{2/3})O_3-0.1Ba(Y_{1/2}Nb_{1/2})O_3$	1380	37	25 560		#	[385]
1252	$0.35CaTiO_3-0.65LaAlO_3$	1600	37	47 000		−2	[428]
1253	$0.42ZnNb_2O_6-0.58TiO_2$ + 10 wt% CuO	875	37	17 000		−7	[429]
1254	$Ba_2Ti_9O_{20}$ (citrate route)	1300 for 2 hours	37	57 000	10.7	−6	[412]
1255	$0.9Ba(Zn_{1/3}Nb_{2/3})O_3-0.1Ba(Ga_{1/2}Ta_{1/2})O_3$		37	93 500	2.9	15	[400]
1256	$Ba(Y_{1/2}Nb_{1/2})O_3$	1600	37	49 600		15	[416]
1257	$Zr_{0.8}Sn_{0.2}TiO_4$ + 1 mol% Sb_2O_5 + 0.35 wt% $B_2O_3-Li_2O$ + slow cooled	1300 for 5 hours	37	62 000		–	[430]
1258	$Ca_{4.18}Co_{0..82}Nb_2TiO_{12}$	1550	37	30 000	4.31	≈0	[67]
1259	$Ca_{4.85}Zn_{0.15}Ta_2TiO_{12}$	1625	37	35 000	4.15	≈0	[67]
1260	$Ca_{4.82}Mg_{0.18}Ta_2TiO_{12}$	1625	37	36 000	4.356	≈0	[67]
1261	$Zr_{0.8}Sn_{0.2}TiO_4$ + 2 wt% La_2O_3, 1 wt% NiO	1370 for 2 hours	37	62 000		−9	[431]

No.		Sintering temp. (°C)	ε_r	*Qf* (GHz)	*f* (GHz)	τ_f	Reference
1262	$Zr_{0.648}Sn_{0.332}TiO_4 + La_2O_3 + NiO$	1370 for 20 hours	37	41 500		–	[432]
1263	$3CaO-2MgO-Nb_2O_5-TiO_2$	1340	37	19 000	4.2	−20	[67]
1264	$Ba(La_{1/2}Ta_{1/2})O_3$	1625	37.1	18 200		−35	[311]
1265	$0.6CaTiO_3-0.4NdAlO_3$	1450 for 10 hours	37.2	40 750		114	[433]
1266	$0.7La(Mg_{1/2}Ti_{1/2})O_3-0.3CaTiO_3$	1600	37.2	15 300	6.3	−54	[305]
1267	$Ba_5Nb_3TaO_{15}$	1435	37.2	4500		35	[387]
1268	$Ba(In_{1/2}Ta_{1/2})O_3$	1625	37.2	35 500		25	[311]
1269	$Ba(Y_{1/2}Ta_{1/2})O_3$	1625	37.3	45 900		120	[311]
1270	$SnTe_3O_8$	660	37.3	10 000		223	[397]
1271	$Bi_{12}(B_{0.5}P_{0.5})O_{20}$	780	37.4	850		−19	[434]
1272	$Ba(La_{1/2}Nb_{1/2})O_3$	1650	37.4	8000	7.2	−7.6	[180]
1273	Sr_2TiO_4		37.4	8160	3.1	137	[435]
1274	$Nd_4Ti_9O_{24}$	1300	37.5	24 100	8	65	[417, 418]
1275	$Ca_3Mg_2Nb_4TiO_{17}$	1225	37.5	22 500	3.9	−4	[280]
1276	$ZnTa_2O_6$	1350	37.6	65 200		9	[199]
1277	$Ba(Sm_{1/2}Ta_{1/2})O_3$	1625	37.6	15 000		−10	[311]
1278	$Ba_5SrNb_4ZrO_{18}$	1600 for 2 hours	37.6	36 000		68	[284]
1279	$Bi_{12}SiO_{20}$	850	37.6	8100		−20	[434]
1280	$La_6ZnTi_4O_{18}$	1600 for 4 hours	37.7	21 850		−37	[284]
1281	$(Zr_{0.8}Sn_{0.2})TiO_4$ 1 wt% ZnO, 0.25 wt% WO_3	1340	37.8	61 000	7	−4	[436]
1282	$GdTiTaO_6$	1540	37.9	12 900		11	[178]
1283	$0.95Ba(Zn_{1/3}Nb_{2/3})O_3-0.05Ba(Ga_{1/2}Ta_{1/2})O_3$		38	102 950	2.9	19	[400]
1284	$Bi_{12}GeO_{20}$	850	38	7800		−31	[434]
1285	$Ba(Sm_{1/2}Ta_{1/2})O_3$	1625 for 4 hours	38	15 000		−10	[311]
1286	$Ba(Ho_{1/2}Nb_{1/2})O_3$	1600	38	21 600		−11	[416]
1287	$Sr(Pr_{1/2}Nb_{1/2})O_3$	1575 for 4 hours	38	3300		−34	[275]
1288	$Ca_3Nb_2O_8$		38	7100	5.9	113	[288]
1289	$Zr_{0.8}Sn_{0.2}TiO_4$		38	62 000	4	0	[437]
1290	$Zr_{0.8}Sn_{0.2}TiO_4$ (sol−gel derived)	1300	38	55 000	6	0.5	[438]
1291	$0.5LaCa_{0.5}Zr_{0.5}O_3-0.5SrTiO_3$	1575	38	7000	3.8	8	[159]
1292	$BaO-2CeO_2-3TiO_2$	1250	38	7200		159	[295]
1293	$Ca_5Ta_2TiO_{12}$	1625	38	33 000	4.2	10	[439, 440]
1294	$Ca_5Ta_2TiO_{12} + 0.2$ wt% $Al_2O_3-B_2O_3-SiO_2$	1550	38	38 000		8	[440]
1295	$Ca_5Ta_2TiO_{12} + 0.1$ wt% $2MgO-Al_2O_3-5SiO_2$	1550	38	40 000		5	[440]

No.		Sintering temp. (°C)	ε_r	*Qf* (GHz)	*f* (GHz)	τ_f	Reference
1296	$Ba(Zn_{1/3}Nb_{2/3})_{0.9}Zr_{0.1}O_3$	1400	38	61 000	10	15	[441]
1297	$La_{0.42}Ca_{0.58}[Ca_{0.05}Mg_{0.16}Ti_{0.79}]O_3$		38	20 000	5.3	25	[317]
1298	$Ca_{(1-x)}Y_xTi_{1-x}Al_xO_3$ $(x=0.3)$		38	14 200		−14	[442]
1299	$Ba_{0.2}Sr_{0.71}(Zr_{0.951}Ti_{0.039}Ta_{0.01})O_3$		38	1700		0	[443]
1300	$Ba_{0.29}Sr_{0.71}(Zr_{0.973}Ti_{0.027})O_3$		38	2000		−40	[443]
1301	$Ba(Zn_{1/3}Nb_{2/3})O_3$ + 1 mol% WO_3	1450	38	95 150		39	[444]
1302	$0.95Ba(Zn_{1/3}Nb_{2/3})O_3-0.05Ba(Ga_{1/2}Ta_{1/2})O_3$		38	102 955		19	[400]
1303	$Ba_4NdTiNb_3O_{15}$	1430 for 3 hours	38.2	18 700	5.4	12	[445]
1304	$BaO-ZnO-TiO_2$	1250	38.2	5000		36	[406]
1305	$Sr_{1-x}Ca_x[(Li_{1/4}Nb_{3/4})_{1-y}Ti_y]O_3$		38.2–45.8	35 000		30 to −70	[446]
1306	$0.7La(Mg_{1/2}Ti_{1/2})O_3-0.3SrTiO_3$		38.3	10 550	6.6		[329]
1307	$Ba(Pr_{1/2}Ta_{1/2})O_3$	1625	38.5	42 800		−8	[311]
1308	$Sm(Nb_{0.25}Ta_{0.75})TiO_6$		38.5	22 100		26	[211]
1309	$(Sm_{0.5}Y_{0.5})(Ti_{1.5}W_{0.5})O_6$	1400 for 10 hours	39.4	36 900		−6.3	[447]
1310	$0.5Ba(Y_{1/2}Nb_{1/2})O_3-0.5Ba(Ca_{1/9}Y_{3/9}Nb_{5/9})O_3$		38.6	17 400	8.1		[448]
1311	MWF-38	1360	38.6	44 500		1.3	[155]
1312	$Bi_{12}PbO_{19}$		38.6	2900		−84	[434]
1313	$Ca[(Li_{1/3}Nb_{2/3})_{1-x}Ti_x]O_3$ $(x=0.2)$	1150 for 3 hours	38.6	26 100		0	[239]
1314	$Ba_3Ti_5Nb_6O_{28}$ + 3 wt% B_2O_3 + 1 wt% CuO	900 for 2 hours	38.6	29 800		5	[449]
1315	$Ba(Nd_{1/2}Ta_{1/2})O_3$	1625 for 3 hours	38.7	12 000		−4	[322]
1316	$Ba(Eu_{1/2}Ta_{1/2})O_3$	1625	38.8	36 200		−10	[311]
1317	$Ba(Dy_{1/2}Nb_{1/2})O_3$	1600	38.9	20 600		−4	[416]
1318	$Zr_{0.8}Sn_{0.2}TiO_4$	1600 for 4 hours	38.9	51 500		0.7	[450]
1319	$Ba(In_{1/2}Nb_{1/2})O_3$	1600	39	30 700		17	[416]
1320	$Ba_5Nb_4O_{15}$	1380	39	23 700	4.7	78	[94]
1321	$Ba_5Nb_4O_{15}$ + 6.3 vol% $BaNb_2O_6$ + 3 wt% B_2O_3	925 for 2 hours	39	18 700		0	[451]
1322	$(1-x)Ba_3(ZnNb_2)O_9-xBa_3W_2O_9$ $(x=0.007)$	1380	39	118 000		21	[452]
1323	$Ca[(Li_{1/3}Nb_{2/3})_{1-x}Ti_x]\ O_{3-\delta}$ $(x=0.15)$	1150 for 3 hours	39	26 100		0	[239]
1324	$Ba(Tb_{1/2}Nb_{1/2})O_3$	1600	39	52 400		−2	[416]
1325	$Ba_3Ti_5Ta_6O_{28}$	1430	39	4000	5.3	30	[453]
1326	$Ca_4SrTa_2TiO_{12}$	1625	39	21 000	3.59	12	[454]
1327	$Ba(Mn_{1/3}Nb_{2/3})O_3$		39	9300	9.3	27	[221]

No.		Sintering temp. (°C)	ε_r	*Qf* (GHz)	*f* (GHz)	τ_f	Reference
1328	$Ba_4LaSnNb_3O_{15}$	1480 for 4 hours	39	14 800	5.9	−29	[455]
1329	$(1-x)LaMg_{1/2}Ti_{1/2}-xLa_{2/3}TiO_3$ ($x = 0.48$)		39	3800	4.6	23	[317]
1330	$Li_{0.774}Zr_{0.057}NbO_3$	1150	39	4500	6	−17	[377]
1331	$SrTiO_3-LaAlO_3$		39	60 000		0	[456]
1332	$BaMg_6Ti_6O_{19}$	1450	39	20 000	2	370	[457]
1333	$Ba_2Ti_9O_{20}$	1350 for 3 hours	39	32 000	2		[458, 459]
1334	$Ba_5Nb_4O_{15}$ + 3 wt% B_2O_3	925	39	18 700		0	[451]
1335	$Bi_2Te_2O_8$ (oxygen atm)	650 for 10 hours	39	23 000		−43	[746]
1336	$0.9BiNbO_4-0.1ZnNb_2O_6 + 0.8CuV_2O_6$	900	39	31 000		−10	[460]
1337	$Ba(Dy_{1/2}Ta_{1/2})O_3$	1625	39.1	18 200		−48	[311]
1338	$Sr_{1.6}Ca_{0.4}TiO_4$	1600	39.2	8100	3	195	[461]
1339	$(5-x)BaO-xMgO-2Nb_2O_5$ ($x = 1$) + 1 wt% CuO	1200	39.2	43 800		38	[425]
1340	$Sr_2La_4Ti_5O_{18}$	1625 for 2 hours	39.2	27 350		20	[284]
1341	$Sm(Nb_{0.5}Ta_{0.5})TiO_6$		39.3	19 600		33	[211]
1342	$Ba_2Ti_9O_{20}$ + 1.64 mol% SnO_2	1390 for 6 hours	39.3	38 400		–	[462]
1343	$Ba_3LaTa_3O_{12}$	1500	39.4	26 800	6	−46	[463]
1344	$Bi_{12}MnO_{20-\delta}$	720	39.4	800		−35	[434]
1345	$Sm(Ti_{1.5}W_{0.5})O_6$	1400 for 10 hours	39.4	35 400		−0.7	[447]
1346	$Ba_{1-x}La_x[Zn_{(1+x)/3}Nb_{(2-x)/3}]O_3$ ($x = 0$)	1350 for 4 hours	39.5	112 280		18.5	[464]
1347	$Ba_2Ti_9O_{20}$ + 1.64 mol% ZrO_2	1390 for 6 hours O_2	39.5	41 700		2.1	[465]
1348	$Ba_5SrNb_4ZrO_{18}$	1600 for 4 hours	39.5	36 000		68	[284]
1349	$Ba[Zn_{1/4}Ti_{1/4}Ta_{1/2}]O_3$	1500	39.6	15 000	7.4		[466]
1350	$0.6La_{2/3}TiO_3-0.4LaAlO_3$	1400 for 33 hours	39.6	42 200	6	−15	[390]
1351	$Ba(Ho_{1/2}Ta_{1/2})O_3$	1625	39.6	21 900		130	[311]
1352	$0.95Ba(Zn_{1/3}Nb_{2/3})O_3-0.05BaZrO_3$ + 1 wt% CuO	1360 for 2 hours	39.7	70 000	7	17	[467]
1353	$Ba(Sm_{1/2}Nb_{1/2})O_3$	1500	39.7	21 500	7.1	21	[180]
1354	$Ba(Yb_{1/2}Ta_{1/2})O_3$	1625	39.7	31 700		112	[311]
1355	$GdTiTaO_6$	1540	39.9	12 900		11	[178]
1356	$Ba(Gd_{1/2}Nb_{1/2})O_3$	1600	40	5700		4.6	[416]
1357	$Ca_5Nb_{0.5}Ta_{1.5}TiO_{12}$	1600	40	31 500		19	[469]
1358	$Ba(Eu_{1/2}Nb_{1/2})O_3$	1600	40	40 200		6.7	[416]

No.		Sintering temp. (°C)	ε_r	Qf (GHz)	f (GHz)	τ_f	Reference
1359	$Ba_3Ti_5Nb_3Ta_3O_{28}$	1375	40	8000	4.8	9	[453]
1360	$Ba_5Nb_4O_{15}$	–	40	53 000		78	[470–472]
1361	$Ca_4ZnNb_2TiO_{12}$	1550	40	30 500	4.2458	−37	[67]
1362	$Sr_5Nb_4O_{15}$	1400	40	19 400	4.84	55	[94]
1363	$Ca(Fe_{1/2}Nb_{1/2})O_3$	1500 for 6 hours	40	20 000		−76	[186]
1364	$Ca[(Li_{1/3}Nb_{2/3})_{0.8}Ti_{0.2}]O_{3-\delta}$	920	40	20 500	8	4.7	[473]
1365	$Ca[(Li_{1/3}Nb_{2/3})_{0.8}Ti_{0.2}]O_{3-\delta}$ + 12 wt% $B_2O_3-ZnO-SiO_2-PbO$ frit glass	900	40	12 500		−8	[468]
1366	$Sr(Zn_{1/3}Nb_{2/3})O_3$		40	36 800	9.2	−39	[221]
1367	$Ba[(Zn_{0.8}Co_{0.2})_{1/3}Nb_{2/3}]O_3$	1410	40	61 000		18	[745]
1368	$Ba(Zn_{1/3}Nb_{2/3})O_3$	1390	40	87 000		30	[221, 474]
1369	$Ba_{0.3}Sr_{0.7}(Zn_{1/3}Nb_{2/3})O_3$	1500 for 1 hour	40	30 500	10	−5	[475]
1370	$0.7Ca(Mg_{1/3}Nb_{2/3})O_3-0.3CaTiO_3$		40	27 900	3.8	−15	[476]
1371	$(Zr,Sn)TiO_4$	1600	40	53 000	10	0	[477]
1372	$Ba_8Ta_{4+0.8x}Ti_{3-x}O_{24}$ ($x = 0$)	1400 for 40 hours	40	12 960			[383]
1373	$0.6La_{2/3}TiO_3-0.4LaAlO_3$	1400 for 33 hours (oxygen)	40	50 800	6	−15	[390]
1374	$Ba_{0.75}Sr_{0.25}(Zn_{1/3}Ta_{2/3})_{0.94}Ti_{0.06}O_3$	1400	40	65 000	10	−13	[441]
1375	$Ba_5Nb_4O_{15}$ + 0.3 wt% ZnB_2O_4 glass	900	40	12 100		48	[478]
1376	Ba−Nd−Sm−Bi−Ti−O + 9 wt% $BaO-B_2O_3-SiO_2$	950 for 2.5 hours	40	3000			[479]
1377	$Ca_5Ta_4TiO_{17}$	1525	40.1	24 000	4.2	−55	[280]
1378	$Ba_3Ti_5Nb_6O_{28}$ + 2 wt% B_2O_3 + 2 wt% CuO	900 for 2 hours	40.2	32 200		5	[133]
1379	$La_6MgTi_4O_{18}$	1625 for 2 hours	40.2	35 000		−39	[284]
1380	$Ba_3Ti_5Nb_6O_{28}$ + 1 wt% B_2O_3 + 3 wt% CuO	900 for 2 hours	40.3	32 500		9	[133]
1381	$Ca_4MgNb_4TiO_{17}$	1250	40.6	18 250		1.5	[280]
1382	$BiNb_{0.6}Sb_{0.4}O_4$	920	40.7	9500		−31	[480]
1383	$Zr_{0.8}Sn_{0.2}TiO_4$ + 1 wt% ZnO + 1 mol% Sb_2O_5	1400 for 5 hours	40.8	60 900			[481]
1384	$La_6ZnTi_4O_{18}$	1600 for 4 hours	40.8	21 900		−37	[284]
1385	$Ba_2Ti_9O_{20}$:Mn		41	45 000	9	2	[482]
1386	$Ca_{4.35}Mg_{0.65}Nb_2TiO_{12}$	1550	41	33 000	4.1	0	[67]
1387	$Ba(Zn_{1/3}Nb_{2/3})O_3$ (annealed in N_2)	1500	41	90 000		4	[221]
1388	$(Ti_{0.8}Sn_{0.2})Te_3O_8$	700 for 5 hours	41	22 000	4		[15]

No.		Sintering temp. (°C)	ε_r	Qf (GHz)	f (GHz)	τ_f	Reference
1389	$Bi_{0.95}Sm_{0.05}NbO_4$	1040	41	5200		−200	[483]
1390	$Bi_{12}TiO_{20}$		41	3300		−32	[434]
1391	$0.65CaTiO_3-0.35SmAlO_3$	1450 for 12 hours	41	42 000		−18	[433]
1392	$0.9BiNbO_4-0.12ZnNb_2O_6 + 1.2$ wt% CuV_2O_6	850	41	28 120		4	[460]
1393	$BaTi_5O_{11}$ (hot pressed)	1050 for 48 hours	41	46 000	10	40	[484]
1394	$0.65CaTiO_3-0.35LaAlO_3$	1450 for 12 hours	41	33 000		−17	[433]
1395	$Sr_5Ta_4O_{15}$	1610	41	2400	5.99	–	[94]
1396	$Ca_3Nb_2O_8$		41	8700	8.6	123	[288]
1397	$Ba_3Ti_5Nb_6O_{28}$	1300	41	4500	5.4	8	[453]
1398	$Zr_{0.513}Hf_{0.487}TiO_4$	1600	41	20 400		13	[386]
1399	$5CaO-2Ta_2O_5$	1550	41	5900	5.9	140	[94]
1400	$CaLa_4Ti_4O_{15}$	1550 for 24 hours	41.1	50 240		−25	[485]
1401	$CaTi_{0.4}(Al_{1/2}Nb_{1/2})_{.6}O_3$ + 1 wt% Li_3NbO_4	1300 for 5 hours	41.1	36 200	7	−36	[423]
1402	$Ba(Zn_{1/3}Nb_{2/3})O_3$	1390	41.1	86 900	9.5	31	[221, 475]
1403	$Zr_{0.513}Hf_{0.487}TiO_4$		41.1	43 000	8.5	13	[386]
1404	$CaTi_{0.4}(Al_{1/2}Nb_{1/2})_{.6}O_3$	1500 for 5 hours	41.3	27 100	7	−44	[423]
1405	$EuTiTaO_6$	1525	41.3	59 500		19	[178]
1406	$Li_2O-Nb_2O_5-TiO_2$ (5:1:5) + 1 wt% B_2O_3	900	41.3	9320			[420]
1407	$CaTi_{0.5}(Al_{1/2}Ta_{1/2})_{0.5}O_3$	1500 for 15 hours	41.4	26 100	8	−20	[486]
1408	$Ba_{0.9}Ca_{0.1}(Y_{.285}Nb_{1/2})O_{3+\delta}$		41.5	48 860	7.85	258	[448]
1409	$BiNbO_4$:0.4 wt% B_2O_3	960 for 2 hours	41.5	21 000		−2.4	[487]
1410	$0.7CaTiO_3-0.3(La_{0.5}Nd_{0.5})AlO_3$		41.5	37 000	8	4	[488]
1411	$Ca[Ti_{1-x}(Mg_{1/3}Nb_{2/3})_x]O_3$ ($x = 0.7$)	14 504 h	41.56	29 450		−12	[489]
1412	$Sm(Nb_{0.75}Ta_{0.25})TiO_6$		41.6	18 900		36	[211]
1413	$5Li_2O-Nb_2O_5-5TiO_2 + 1$ wt% V_2O_5	900 for 6 hours	41.7	7800		45	[490]
1414	$SmTiTaO_6$	1500	41.8	24 500		24	[178]
1415	$(Ni_{1/3}Ta_{2/3})_{1-x}Ti_xO_2$ ($x = 0.3$)	1300	41.8	20 600		35.3	[491]
1416	$BaNb_{1-x}Mo_xO_4$ ($x = 0.01$)	950	41.8	3500		−15	[492]
1417	$Pr_{0.2}Gd_{0.8}TiNbO_6$	1400	41.9	9500		35	[175]
1418	$Ba(Nd_{1/2}Nb_{1/2})O_3$	1500 for 96 hours	41.9	15 000	6.8	13.3	[180]
1419	$0.67CaTiO_3-0.33NdAlO_3$	1450 for 10 hours	41.98	42 900		45	[433]

No.		Sintering temp. (°C)	ε_r	Qf (GHz)	f (GHz)	τ_f	Reference
1420	$ZrTiO_4$	1400	42	31 000	7	58	[437]
1421	$Ca_{4.38}Ni_{0.62}Nb_2TiO_{12}$	1550	42	28 200	4	≈0	[67]
1422	$Sr_4Ti_3O_{10}$	1300 for 5 hours	42	960	4		[15]
1423	$0.95Ba(Zn_{1/3}Nb_{2/3})O_3{-}0.05BaZrO_3$	1450 for 2 hours	42	96 000	7	27	[493]
1424	$0.84Ba_5Nb_4O_{15}{-}0.16BaNb_2O_6$ + 0.3 wt% B_2O_3	900	42	28 000		0	[269]
1425	$Ba_xLa_4Ti_{3+x}O_{12+3x}$ ($x = 0.2$)	–	42	86 000		−17	[494]
1426	$BaTi_5O_{11}$	1120 for 24 hours	42	61 100	–	40	[495–497]
1427	$La_{0.43}Ca_{0.57}[Ca_{0.08}Mg_{0.14}Ti_{0.78}]O_3$		42	18 000	5.0	9	[317]
1428	$La_9Mg_{0.5}Ti_{8.5}O_{31}$		42	15 000	8.4	−11	[317]
1429	$0.6CaTiO_3{-}0.4SmGaO_3$	1450 for 12 hours	42	35 000		−11	[433]
1430	$(1-x)LaMg_{1/2}Ti_{1/2}O_3{-}xLa_{2/3}TiO_3$ ($x = 0.45$)		42	4500	4.8	−30	[317]
1431	$BaNb_2O_6$ (hexagonal)	1050	42	4000		−800	[269, 335]
1432	$Ba_{0.79}Sr_{0.21}Ti_5O_{11}$ (hot pressed)	1050 for 72 hours	42	39 000	10	44	[484]
1433	$Ca_{4.38}Ni_{0.62}Nb_2TiO_{12}$	1550	42	28 200	4	±0.5	[67]
1434	$Ca_3Sr_2Ta_2TiO_{12}$	1600	42	16 000	3.5	14	[454]
1435	$0.6CaTiO_3{-}0.4Sm(Mg_{0.5}Ti_{0.5})O_3$	1550	42	9200	4.8	6.2	[278]
1436	$Ba_{0.8}Sr_{0.2}(Zn_{1/3}Ta_{2/3})_{0.94}Ti_{0.06}O_3$	1400	42	82 000	10	−13	[441]
1437	$(Ca_{0.6}(Li_{0.5}Nd_{0.5})_{0.4})_{0.45}Zn_{.55}TiO_3$ + 2 wt% $0.33ZnO{-}0.67H_3BO_3$	900 for 4 hours	42	10 300		19. check	[498]
1438	$0.1CaTiO_3{-}0.9Nd(Mg_{1/2}Ti_{1/2})O_3$	1400	42	35 000		−10	[499]
1439	$Sr(Sm_{1/2}Ta_{1/2})O_3$ + 3 wt% TiO_2	1600	42	8800		3	[274]
1440	$0.76ZrTi_2O_6{-}0.24ZnNb_2O_6$	1300 for 4 hours	42	22 976			[500]
1441	$0.4LaAlO_3{-}0.6SrTiO_3$	1680	42.1	83 000	9.5	8.4	[268]
1442	$Ca[La_{0.5}Nd_{0.5})_4Ti_4O_{15}$	1525	42.3	15 200	8.3	−5.8	[501]
1443	$Ba_{11}TiNb_8O_{33}$	1400	42.3	27 000		47	[502, 503]
1444	$CaLa_{0.5}Nd_{0.5}Ti_4O_{15}$	1525	42.3	15 200	8.3	−5.8	[501]
1445	$Ba_4LaTiTa_3O_{15}$	1540 for 6 hours	42.3	28 800		33	[504]
1446	$Sr_{0.92}[Li_{1/4}Nb_{3/4}]_{0.92}Ti_{0.08}O_3$	1350 for 2 hours	42.3	31 500	9	–	[446]
1447	$ZnO{-}Nb_2O_5{-}TiO_2{-}SnO_2$ + 1.5 wt% $CuO{-}V_2O_5$	860	42.3	9000		8	[505]
1448	$Ba_{1-x}La_x[Zn_{(1+x)/3}Nb_{(2-x)/3}]O_3$ ($x = 0.05$)	1350 for 4 hours	42.4	46 530		35.1	[464]
1449	$BaLa_3Ti_2NbO_{12}$	1460 for 6 hours	42.4	33 600		6	[506]
1450	$Zr_{0.7}(Zn_{1/3}Ta_{2/3})_{0..3}TiO_4$	1300	42.5	40 200	5	1.1	[507]

No.		Sintering temp. (°C)	ε_r	Qf (GHz)	f (GHz)	τ_f	Reference
1451	$0.3BaTiO_3-0.7La(Mg_{1/2}Ti_{1/2})O_3$		42.5	14 225	7.23	−34	[306]
1452	$Ba_2La_2TiNb_2O_{12}$	1440	42.7	31 130		−4	[508]
1453	$Ba_{21}Nb_{16}TiO_{63}$		42.7	19 000		25.3	[502]
1454	$Ba_2La_3Ti_3NbO_{15}$	1460	42.8	21 700	5.85	−8	[509]
1455	$Ba_{16}Nb_{12}TiO_{48}$		42.9	29 000		25	[502]
1456	$Ba_{0.9}Ca_{0.1}(Y_{0.315}Nb_{1/2})O_{3+\delta}$		42.9	63 500	7.78	235	[448]
1457	$Pb_{0.5}Ca_{0.5}(Cr_{1/2}Nb_{1/2})O_3$		43	3800	4.6	293	[336]
1458	$BiNbO_4$	875	43	15 700	4.3	38	[510, 511]
1459	$0.7CaTiO_3-0.3NdAlO_3$		43	47 000	–	0	[512]
1460	$Sr(Zn,Nb)O_3-SrTiO_3$		43	25 000	5	−5 to 5	[513]
1461	$La_{0.57}Ca_{0.43}[Ca_{0.11}Mg_{0.18}Ti_{0.71}]O_3$		43	26 000	3.5	−19	[317]
1462	$Ca_4BaTa_2TiO_{12}$	1575	43	5000	3.9	14	[514]
1463	$0.16BaNb_2O_6-0.84Ba_5Nb_4O_{15}$ + 0.3 wt% B_2O_3 + 0.3 wt% V_2O_5	900	43	19 500		0	[269]
1464	$Ca_5NbTaTiO_{12}$	1580	43	30 000	–	28	[454, 469]
1465	$Ba(Nd_{0.8}Sm_{0.2})_2Ti_4O_{12}$ + 1 wt% B_2O_3	1020	43	5500			[515]
1466	$Ca_{4.36}Zn_{0.64}Nb_2TiO_{12}$	1550	43	29 000	4.0	0	[67]
1467	$Ba_6Nb_4TiO_{18}$	1625 for 2 hours	43	9500		53	[284]
1468	$Zr_{0.7}(ZnTa)_{0.3}TiO_4$	1300 for 3 hours	43	40 200		1.1	[507]
1469	$Ba(Sm_{1/2}Nb_{1/2})O_3$	1600	43	18 400		9	[416]
1470	$Sr_6Nb_4TiO_{18}$	1625 for 2 hours	43	6700		26	[284]
1471	$Sm_{0.9}Y_{0.1}TiNbO_6$	1560	43	10 230		47	[175]
1472	$0.52Nd(Co_{1/2}Ti_{1/2}O_3-0.48CaTiO_3$	1550	43	4000		0	[516]
1473	$Ba_xLa_4Ti_{3+x}O_{12+3x}$ $(x=2.3)$		43	23 480		−17	[517]
1474	$La_2Ti_2O_7$		43	2200	5.5	−6	[317]
1475	$0.7CaTiO_3-0.3La(Ga_{0.5}Al_{0.5})O_3$	1540	43	40 000		13	[488]
1476	$BiNbO_4$ + 0.4 wt% V_2O_5 + 0.1 wt% CuO	900	43	20 400		8	[518]
1477	$0.5La(Mg_{1/2}Ti_{1/2})O_3-0.5CaTiO_3$	1600	43	28 000	5.5	−13	[519]
1478	$Ca_{4.5}Mg_{0.5}Nb_4TiO_{17}$	1250	43	17 850		−33	[280]
1479	$Ba(Sm_{1/2}Nb_{1/2})O_3$	1600	43	18 400		9	[416]
1480	$NdTiTaO_6$	1550	43.1	26 400		30	[178]
1481	$Ca[(Li_{1/3}Nb_{2/3})_{0.7}Ti_{0.3}]O_{3-\delta}$ + 6 wt% Bi_2O_3, 2 wt% B_2O_3	920	43.1	10 600	7.68	10	[473]
1482	$Ca[(Li_{1/3}Nb_{2/3})_{0.7}Ti_{0.3}]O_{3-\delta}$ + 3 wt% Bi_2O_3: 2 wt% B_2O_3	940	43.1	12 900	7.73	53.7	[473]
1483	$CaLa_{0.875}Nd_{0.125}Ti_4O_{15}$	1550	43.1	29 820	7.85	−9	[501]
1484	$Zr_{0.752}Hf_{0.248}TiO_4$		43.2	20 000	8.5	–	[386]

No.		Sintering temp. (°C)	ε_r	*Qf* (GHz)	*f* (GHz)	τ_f	Reference
1485	$BiNbO_4$ + 0.5 wt% CuO	900	43.3	13 000	6.3	15	[520]
1486	$Ca[La_{0.875}Nd_{0.125}]_4Ti_4O_{15}$	1550	43.4	32 900	7.52	−13	[501]
1487	$Ba_3LaNb_3O_{12}$	1500	43.5	9000		−100	[521, 522]
1488	$Sr_5Ta_4O_{15}$	1510	43.5	2400		–	[387]
1489	$Ba_{11}Nb_8TiO_{33}$		43.5	12 000		33.2	[502]
1490	$BiTaO_4$	950	43.5	12 000		−40	[523]
1491	$0.7CaTiO_3-0.3NdAlO_3$		43.5	30 000	8	−2.1	[488]
1492	$Bi_{0.99}(La_{0.38}Nd_{0.62})_{0.01}NbO_4$	820	43.5	12 300		12.8	[524]
1493	$0.66CaTiO_3-0.34(La_{0.5}Nd_{0.5})GaO_3$		43.6	43 000	8	−9.5	[488]
1494	$CaLa_4Ti_4O_{15}$	1550	43.6	33 850	7.8	−17	[501]
1495	$0.7CaTiO_3-0.3NdAlO_3$	1450 for 10 hours	43.7	34 800		14	[433, 488]
1496	$SrLa_4Ti_4O_{15}$		43.8	50 200	4.15	−14	[485, 525]
1497	$Ba_{1-x}La_x[Zn_{(1+x)/3}Nb_{(2-x)/3}]O_3$ ($x = 0.35$)	1350 for 4 hours	43.8	2180		−10.3	[464]
1498	$Bi_{0.992}Gd_{.0.008}NbO_4$	900 for 3 hours	43.8	16 850	4.3	0	[526]
1499	$Ca[(Li_{1/3}Nb_{2/3})_{0.7}Ti_{0.3}]O_{3-\delta}$ + 1 wt% Bi_2O_3 + 1 wt% B_2O_3	960	43.9	16 600	7.6	35	[473]
1500	$Bi_2Ti_3TeO_{12}$	900 for 10 hours	44	12 500		146	[411]
1501	$ZrTiO_4-ZnNb_2O_6$		44	48 000		0	[500]
1502	$Ca_2Sr_3Ta_2TiO_{12}$	1575	44	8500	3.5	18	[454]
1503	$BaTiNb_4O_{13}$	1250	44	9000	4.7	15	[453]
1504	$Sr_2Zn_4Ti_{15}O_{36}$	1150 for 8 hours	44	3600	10	160	[355]
1505	$Ba_8Ta_{4+0.8x}Ti_{3-x}O_{24}$ ($x = 0.8$)	1400 for 40 hours	44	9720			[383]
1506	$Ba(Nd_{1/2}Nb_{1/2})O_3$	1600	44	11 700		10	[416]
1507	$Ca_{0.7}Nd_{0.3}T_{0.7}Al_{0.3}O_3$		44	40 000		0	[527]
1508	$Ba(La_{1-y}Al_y)_4Ti_4O_{15}$		44	47 000		1.3	[528]
1509	$0.7CaTiO_3-0.3LaAlO_3$	1450 for 12 hours	44	30 000		−3	[433]
1510	$Ba_{0.9}Ca_{0.1}(Y_{0.33}Nb_{1/2})O_{3+\delta}$		44	41 210	7.7	234	[448]
1511	$0.7Ca(Li_{1/4}Nb_{3/4})O_3-0.3CaTiO_3$	1250	44	12 000		−9	[529]
1512	$Ca_{0.7}Nd_{0.3}Ti_{0.7}Al_{0.3}O_3$	1600	44	33 000		0	[527]
1513	$Sr_2La_4Ti_5O_{18}$ + 0.3 wt% $Bi_2O_3-B_2O_3$	1625 for 2 hours	44	23 000		22	[284]
1514	$0.6CaTiO_3-0.4NdGaO_3$	1450 for 12 hours	44	30 000		−18	[433]
1515	$BiNbO_4$ + 0.5 wt% V_2O_5	895	44	15 800	7	18	[518]
1516	$BiNbO_4$ + 0.25 wt% CuO + V_2O_5	900	44	18 660		−7.8	[518]

No.		Sintering temp. (°C)	ε_r	*Qf* (GHz)	*f* (GHz)	τ_f	Reference
1517	$Bi_{0.95}Sm_{0.05}NbO_4 + 0.5$ wt% CuO	900	44	12 900		−4.2	[530]
1518	$BiNb_{0.4}Ta_{0.6}O_4$	940	44	21 000		−30	[531]
1519	$Sr_2Zn_4Ti_{15}O_{36}$	1150 for 8 hours	44	36 000		160	[355]
1520	$0.5La(Mg_{1/2}Ti_{1/2})O_3$−0/5$CaTiO_3$ + 1 wt% B_2O_3	1400	44.	28 000	7	−8	[519]
1521	$Ba_8Nb_4Ti_3O_{24}$	1450	44.1	22 000		115	[532]
1522	$Ba_xLa_4Ti_{3+x}O_{12+3x}$ $(x = 1)$		44.1	11 580		7	[517]
1523	$0.66CaTiO_3-0.34(La_{0.5}Nd_{0.5})(GaO_3)$		44.1	43 000	8	0.7	[488]
1524	$Ba_2La_2TiNb_2O_{12}$	1350 for 6 hours	44.2	31 660	6.9	−5	[522]
1525	$Ca[Ti_{1-x}(Mg_{1/3}Nb_{2/3})_x]O_3$ $(x = 0.65)$	1450 for 4 hours	44.2	28 340		−2	[533]
1526	$BaLa_4Ti_4O_{15}$	1600 for 2 hours	44.4	41 000		−26	[485, 528, 534]
1527	$Bi_{0.95}Sm_{0.05}NbO_4$	950	44.4	13 000	7.2	−4.2	[535]
1528	$BiNb_{0.95}Sb_{.05}O_4$	880	44.5	14 278		−5.2	[480]
1529	$BiNb_{0.88}Ta_{0.12}O_4$ + 0.5 wt% CuO	920	44.5	14 000		−0.2	[520]
1530	$Ba(Pr_{1/2}Nb_{1/2})O_3$	1600	44.5	28 500		−22	[416]
1531	$Ba(Y_{0.3}Bi_{0.2}Nb_{.5})O_3$	1300	44.6	2000		6	[536]
1532	$Ba_4Nd_2Ti_3Nb_2O_{18}$	1450 for 8 hours	44.6	13 100		18	[537]
1533	$Ba_{1-x}La_x[Zn_{(1+x)/3}Nb_{(2-x)/3}]O_3$ $(x = 0.3)$	1350 for 4 hours	44.7	1990		7.7	[464]
1534	$Ca_2La_4Ti_5O_{18}$		44.7	20 100	4.19	6	[525]
1535	$0.67CaTiO_3-0.33(La_{0.5}Nd_{0.5})(GaO_3)$		44.7	41 000	8	6.3	[488]
1536	$Ca[(Li_{1/3}Nb_{2/3})_{1-x}Ti_x]O_3$ $(x = 0.3)$	1150 for 3 hours	44.7	22 500		20	[239]
1537	$Ba_6Nb_4TiO_{18}$	1450	44.9	12 000		33.2	[284, 502]
1538	$0.7La_{2/3}TiO_3-0.3LaAlO_3$	1400 for 33 hours	44.9	33 000	6	7	[390]
1539	$Ca_5Nb_4TiO_{17}$	1475	44.9	17 600		−113	[280]
1540	$BiNbO_4$ + 0.03 wt% $CuV_2O_{6\ (ortho)}$	1000	44.9	16 100		−3.4	[396]
1541	$0.73CaTiO_3-0.27NdAlO_3$	1450 for 10 hours	45	31 000		−15	[433]
1542	$Ca[(Li_{1/3}Ta_{2/3})_{.5}Ti_{0.5}]O_{3-\delta}$ + 3 wt% B_2O_3	1050	45	12 300	8	75	[242]
1543	$0.58ZnNb_2O_6-0.42TiO_2$	1250 for 2 hours	45	6000		0	[388]
1544	$0.42ZnNb_2O_6-0.58TiO_2$	1250	45	48 000	8	0	[388]
1545	$0.45Nd_3Ga_5O_{12}-0.65CaTiO_3$	1450	45	46 000	5	−2	[98]
1546	$0.7CaTiO_3-0.3SmAlO_3$	1450 for 12 hours	45	42 000		1	[433]

No.		Sintering temp. (°C)	ε_r	*Qf* (GHz)	*f* (GHz)	τ_f	Reference
1547	$0.7CaTiO_3-0.3NdAlO_3$	1450 for 10 hours	45	44 000		3	[433]
1548	$Ba(La_{1/2}Nb_{1/2})O_3$	1600	45	5700		7	[416]
1549	$Sr(Fe_{1/2}Nb_{1/2})O_3$	1450 for 4 hours	45	4800		−24	[427]
1550	$CaSr_4Ta_2TiO_{12}$	1550	45	15 500	3.4220	21	[454]
1551	$Ba_xLa_4Ti_{3+x}O_{12+3x}$ $(x=0.4)$	–	45	60 000		−15	[494]
1552	$Ba_xLa_4Ti_{3+x}O_{12+3x}$ $(x=0.6)$	–	45	50 000		−13	[494]
1553	$BaTiTa_2Nb_2O_{13}$	1350	45	3500	5.2	96	[453]
1554	$BaSr_4Nb_4O_{15}$	1400	45	23 300	4.57	82	[94]
1555	$0.2CaTiO_3-0.8Sr(Mg_{1/3}Nb_{2/3})O_3$	1600	45	9000		0	[538]
1556	$Ba_2La_3Ti_3TaO_{15}$	1520	45	26 800		1	[539]
1557	$0.6CaTiO_3-0.4LaGaO_3$	1450 for 12 hours	45	34 000		−20	[433]
1558	$0.65CaTiO_3-0.35SmGaO_3$	1450 for 12 hours	45	34 000		1	[433]
1559	$SmTiNbO_6$		45	18 000	4.89	50	[174]
1560	$0.7CaTiO_3-03NdAlO_3$	1450 for 10 hours	45	44 000		0	[540]
1561	$0.65CaTiO_3-0.35NdGaO_3$	1450	45	46 000		−2	[433]
1562	$Ca[(Li_{1/3}Nb_{2/3})_{1-x}Ti_x]$ $O_{3-\delta}$ $(x=0.3)$	1150 for 3 hours	45	22 500		20	[239]
1563	$La_{2/3}TiO_3-LaAlO_3$		45	33 000		7	[390]
1564	$0.48La(Co_{1/2}Ti_{1/2}O_3-0.52CaTiO_3)$	1550	45	5000	7.8	0	[516]
1565	$0.5Nd(Zn_{1.2}Ti_{1/2})O_3-0.5CaTiO_3$	1300 for 4 hours	45	56 000		0	[541]
1566	$Ba(Mn_{1/2}Ti_{1/2})O_3$	1450 for 2 hours	45	11 600	5.75	−4	[542]
1567	$Ba_3La_2Ti_2Ta_2O_{15}$	1540	45.1	31 000		−13	[543]
1568	$0.71CaTiO_3-0.29NdAlO_3$	1450 for 10 hours	45.1	38 450		6	[433]
1569	$0.7CaTiO_3-0.3LaGa_{0.5}Al_{0.5})O_3$		45.2	40 000	8	13.4	[513]
1570	$0.7CaTiO_3-0.3$ $(La_{.5}Nd_{.5})(Ga_{0.5}Al_{0.5})O_3$		45.2	43 000	8	9.3	[488]
1571	$0.7CaTiO_3-0.3Nd(Ga_{0.5}Al_{0.5})O_3$		45.3	38 000	8	11.5	[488]
1572	$Ba_8Nb_4Ti_3O_{24}$	1400	45.3	23 500	5.6	115	[532]
1573	$Ba_2La_3Ti_3TaO_{15}$	1520	45.4	26 800		−1	[539]
1574	$Ba_{1-x}Sr_x$ $La_4Ti_4O_{15}$ $(x=0.6)$	1550	45.4	47 500		−7.3	[544]
1575	$Ba_{0.2}Ca_{0.8}(Fe_{1/2}Nb_{1/2})O_3$		45.5	2300	7.4	−34	[427]
1576	$(Ba_{1-x}Sr_x)La_4Ti_4O_{15}$ $(x=0.4)$	1450 for 4 hours	45.7	44 200		−5.5	[544]
1577	$PrTiTaO_6$	1500	45.8	32 300		33	[178]
1578	$Pr_{0.5}Gd_{0.5}TiNbO_6$	1400	45.9	9500		41	[175]

No.		Sintering temp. (°C)	ε_r	*Qf* (GHz)	*f* (GHz)	τ_f	Reference
1579	$Zr_{0.992}Hf_{0.008}TiO_4$		45.9	13 000	8.5	53	[386]
1580	$Pb_{0.25}Ca_{0\cdot75}(Mg_{1/3}Nb_{2/3})O_3$		46	8700	3.7	−34	[336]
1581	$BaLa_4Ti_4O_{15}$		46	47 000		−11	[528, 534]
1582	$Ca_3Ti_2O_7$		46	2600	2.69	50	[435]
1583	$0.1La_2Ti_2O_7{-}0.9La_4Ti_9O_{24}$	1300	46	5500		0	[417, 418]
1584	$BaTiTa_4O_{13}$	1450	46	6000	4.6	145	[453]
1585	$La_{0.39}Ca_{0.61}[Ca_{0.11}Mg_{0.08}Ti_{0.81}]O_3$		46	17 000	4.7	36	[317]
1586	$Ba_2La_4Ti_5O_{18}$	1575 for 10 hours	46	31 850		−36	[472]
1587	$Ca_5Nb_{.1.5}Ta_{05}TiO_{12}$	1560	46	28 400		35	[469]
1588	$CeTiTaO_6$	1540	46	33 300		41	[178]
1589	$0.3SrTiO_3{-}0.7Ca(Mg_{1/3}Nb_{2/3})O_3$	1475 for 3 hours	46	29 300	6.8	2	[545]
1590	$Ca_{0.7}Ti_{0.7}La_{0.3}O_3$ + 0.25 wt% Al_2O_3	1500	46	38 200	4	12	[546]
1591	$7Bi_2O_3{-}2TeO_2$ (oxygen atm)	750 for 15 hours	46	1100		−144	[374]
1592	$Ba(Er_{03}Bi_{0.2}Nb_{0.5})O_3$	1300	46.1	1500		−27	[536]
1593	$(Ba_{1-x}Sr_x)La_4Ti_4O_{15}$ ($x = 0.8$)	1600 for 4 hours	46.1	52 800		−3.0	[544]
1594	$0.65LiNb_3O_8{-}0.35TiO_2$	1100 for 2 hours	46.2	5800		0	[250]
1595	$Sr_6Nb_4TiO_{18}$	1625 for 2 hours	46.2	6700		26	[284]
1596	$BaLa_4Ti_4O_{15}$		46.3	16 200	5.15	−13	[525]
1597	$(Bi_{0.95}Ce_{0.05})NbO_4$	950 for 2 hours	46.4	15 000			[547]
1598	$0.64CaTiO_3{-}0.36LaGaO_3$		46.5	48 000	8	−2.9	[488]
1599	$Bi_{0.95}Ce_{0.05}NbO_4$ + 0.4 mol% CuO	950	46.5	3000			[547]
1600	$Ba_3La_2Ti_2Nb_{2-x}Ta_xO_{15}$ ($x = 1$)	1500	46.5	27 140	–	−3.7	[543]
1601	$CaTi_{0.54}(Al_{1/2}Ta_{1/2})_{0.46}O_3$	1500 for 15 hours	46.5	27 300	8	0	[486]
1602	$Ba_3Nd_2Ti_2Nb_2O_{15}$	1450 for 3 hours	46.8	19 500	5.1	28	[445]
1603	$(Ba_{1-x}Sr_x)La_4Ti_4O_{15}$ ($x = 0.2$)	1450 for 4 hours	46.8	24 500		−7.5	[544]
1604	$Ba_{1-x}Ca_x(Sc_{1/2}Nb_{1/2})O_3$ ($x = 0.5$)	1650	46.9	28 000			[548]
1605	$La_2Ti_2O_7$		47	8500	7.8	−10	[417]
1606	$0.75CaTiO_3{-}0.25LaAlO_3$	1450 for 12 hours	47	36 000		13	[433]
1607	$Bi_2Ti_4O_{11}$		47	4800		−540	[549]
1608	$Ca_2Zn_4Ti_{15}O_{36}$	1150 for 8 hours	47	41 200	10	120	[355]

No.		Sintering temp. (°C)	ε_r	*Qf* (GHz)	*f* (GHz)	τ_f	Reference
1609	$BaTi_{0.3}Ga_{0.35}Nb_{0.35}O_3$	1500 for 4 hours	47	2470	5.5		[550]
1610	$0.65CaTiO_3-0.35LaGaO_3$	1600	47	40 000		0	[296]
1611	$Pr_{.6}Gd_{.4}TiNbO_6$	1400	47.1	9500		44.3	[175]
1612	$0.76ZrTi_2O_6-0.24ZnNb_2O_6$	1260 for 4 hours in oxygen	47.1	34 200		0	[500]
1613	$Ca[Ti_{1-x}(Mg_{1/3}Nb_{2/3})_x]O_3$ (x = 0.6)	1450 for 4 hours	47.3	25 630		8.2	[489]
1614	$Ba_3La_3Ti_4NbO_{18}$	1480 for 6 hours	47.4	17 330	5.18	35	[551]
1615	$0.66CaTiO_3-0.34LaGaO_3$		47.5	46 000		3.6	[488]
1616	$0.5MgTiO_3-0.5CaTiO_3-0.25(Nd_2O_3-TiO_2)$		47.6	30 000		7.9	[552]
1617	$(Ba_{1-x}Ca_x)La_4Ti_4O_{15}$ ($x = 0.2$)	1575 for 4 hours	47.7	47 100		−7.9	[544]
1618	$Ba_5SrNb_4TiO_{18}$	1450 for 2 hours	47.9	7000		83	[284]
1619	$Ba(Gd_{0.3}Bi_{0.2}Nb_{0.5})O_3$	1300	47.9	2100		−3	[536]
1620	$Ba_xLa_4Ti_{3+x}O_{12+3x}$ ($x = 2.5$)		47.9	19 480			[517]
1621	$Ba_3La_2Ti_2Nb_{2-x}Ta_xO_{15}$ ($x = 0.5$)	1480	47.9	25 300	–	1.8	[543]
1622	$Pb_{0.75}Ca_{0.25}(Cr_{1/2}Nb_{1/2})O_3$		48	3600	4.3	7.6	[33]
1623	$0.65CaTiO_3-0.35LaGaO_3$	1450 for 12 hours	48	32 000		2	[433]
1624	$Ca[(Li_{1/3}Ta_{2/3})_{1-x}Ti_x]O_{3-\delta}+3$ wt% B_2O_3 ($x = 0.5$)	1150	48	21 000		–	[242]
1625	$Ca_5Nb_2TiO_{12}$	1550	48	26 600	3.7	40	[514]
1626	$Ca_3Ba_2Ta_2TiO_{12}$	1540	48	3000	3.8	+18	[514]
1627	$Ba_3Ti_4Ta_4O_{21}$	1380	48	7000	4.3	50	[453]
1628	$Pr_{0.7}Gd_{0.3}TiNbO_6$	1400	48	4500		47	[175]
1629	$Ba_4SrNb_4O_{15}$	1400	48	14 600	4.7	140	[94]
1630	$0.34CaTiO_3-0.66Ca(Mg_{1/3}Nb_{2/3})O_3$	1450	48	32 500		−2	[553]
1631	$CaTi_{0.5}(Al_{1/2}Nb_{1/2})_{.5}O_3$	1500 for 5 hours	48	26 100	7	−4	[423]
1632	$CaTi_{0.5}(Al_{1/2}Nb_{1/2})_{.5}O_3$ + 1 wt% Li_3NbO_4	1300 for 5 hours	48	32 100	7	−2	[423]
1633	$Ba_8Nb_4Ti_3O_{24}$	1450	48	23 500	5.5	115	[554]
1634	$Sr_2La_4Ti_5O_{18}$	1625 for 2 hours	48	27 350		20	[284]
1635	$Ba_3LaNb_3O_{12}$	1350 for 6 hours	48.3	38 000	6.76	−40	[521]
1636	$Ca_2Zn_4Ti_{16}O_{38}$	1100 for 4 hours	48.4	31 600	6.7	48	[555]
1637	$CaLa_8Ti_9O_{31}$		48.6	19 350	3.65	−6	[525]
1638	$Ba(Dy_{0.3}Bi_{0.2}Nb_{0.5})O_3$	1300	48.6	2000		6	[536]

No.		Sintering temp. (°C)	ε_r	*Qf* (GHz)	*f* (GHz)	τ_f	Reference
1639	$Sr_2La_4Ti_5O_{18}+0.3$ wt% $Bi_2O_3-B_2O_3$ glass	1625 for 2 hours	48.7	23 000		22	[284]
1640	$(Ba_{1-x}Ca_x)La_4Ti_4O_{15}$ $(x=0.4)$	1575 for 4 hours	48.9	42 400		−6.6	[544]
1641	$0.9Bi_2O_3-0.1Nb_2O_5$	900 for 2 hours	49	800		−234	[556]
1642	$Ca(Zr_{0.8}Ti_{0.2})O_3$		49	10 800			[292]
1643	$0.7CaTiO_3-0.3NdGaO_3$	1450 for 12 hours	49	32 000		35	[433]
1644	$0.66CaTiO_3-0.34(La_{0.5}Nd_{0.5})$ GaO_3	1540	49	43 000		0	[488]
1645	$Ca_2La_4Ti_5O_{18}$		49.3	20 100		6	[525]
1646	$0.7CaTiO_3-0.3LaGaO_3$		49.4	29 000	8	21.5	[488]
1647	$CaTi_{0.53}Al_{1/2}Nb_{1/2})_{0.47}O_3$ + 1 wt% Li_3NbO_4	1300 for 5 hours	49.4	31 400	7	12	[423]
1648	$Pr_{0.8}Gd_{0.2}TiNbO_6$	1400	49.5	9500		51	[175]
1649	$(Ba_{1-x}Ca_x)La_4Ti_4O_{15}$ $(x=0.8)$	1575 for 4 hours	49.5	42 400		−5.0	[544]
1650	$(Ni_{1/3}Ta_{2/3})_{1-x}Ti_xO_2$ $(x=0.4)$	1300	49.6	17 600		40	[491]
1651	$CaTi_{0.53}(Al_{1/2}Nb_{1/2})_{0.47}O_3$	1500 for 5 hours	49.8	26 000	7	7	[423]
1652	$Ba_3La_2Ti_2Nb_2O_{15}$	1460	49.8	22 000	5	6.9	[509]
1653	$Li_{1+x+y}Ta_{1-x-3y}Ti_{x+4y}O_3$ $(x=0.1,\ y=0.175)$	1175 for 1 hour	49.8	10 528		32	[557, 558]
1654	$Ba_6Ti_{14}Nb_2O_{39}$	1260 for 4 hours	50	2600	4	165	[559]
1655	$Ca[(Li_{1/3}Nb_{2/3})_{0.9}Ti_{0.3}]O_{3-\delta}+1$ wt% B_2O_3	940	50	6500		−7.6	[560]
1656	$Pb_{0.7}Ca_{0.3}La_{0.5}(Mg_{1/2}Nb_{1/2})O_3$	1350 for 2 hours	50	86 000		0	[561]
1657	$0.5La(Mg_{1/2}Ti_{1/2})O_3-0.5La_{2/3}TiO_3$	1400 for 2 hours	50	10 000		5	[562]
1658	$0.5CaTiO_3-0.5La(Zn_{1/2}Ti_{1/2})O_3$	1550 for 3 hours	50	38 000 ??	7	0	[563]
1659	$Ba_2Sr_3Nb_4O_{15}$	1400	50	16 500	4.7	232	[94]
1660	$La_{0.33}TaO_3$	1525	50	8000	3.8	144	[564]
1661	$Ca_5Nb_2TiO_{12}+0.1$ wt% $2MgO-Al_2O_3-5SiO_2$	1520 for 2 hours	50	30 000	4	38	[565]
1662	$TiTe_3O_8$	720	50	30 600	5	133	[179]
1663	$Ca_5Nb_2TiO_{12}+0.1$ wt% $2MgO-Al_2O_3-5SiO_2$	1520	50	30 000	–	38	[565]
1664	$(1-x)(Mg_{0.95}Zn_{0.05})TiO_3-xCa_{0.6}La_{0.8/3}TiO_3$ $(x=0.5)$	1320 for 4 hours	50	43 500		122	[137]
1665	$(1-x)La_{2/3}TiO_3-xNiTiO_3$ $(x=0.2)$	1340	50.7	13 900	3.5	23.5	[566]
1666	$Ba(Sm_{0.3}Bi_{0.2}Nb_{0.5})O_3$	1300	50.8	1600		14	[536]
1667	$Ba_5SrNb_4TiO_{18}$	1450 for 4 hours	50.8	7000		83	[284]

No.		Sintering temp. (°C)	ε_r	Qf (GHz)	f (GHz)	τ_f	Reference
1668	$0.75CaTiO_3-0.25SmAlO_3$	1450 for 12 hours	51	31 000		31	[433]
1669	$0.7CaTiO_3-0.3SmGaO_3$	1450 for 12 hours	51	18 000		41	[433]
1670	$Ba_3Sr_2Nb_4O_{15}$	1400	51	21 200	4.6	117	[94]
1671	$Ca_{0.6}(Li_{0.5}Nd_{0.5})_{.4})_{0.45}Zn_{0.55}TiO_3$	1150	51	12 700		17	[498]
1672	$Pr_{0.9}Gd_{0.1}TiNbO_6$	1400	51	8400		53	[175]
1673	$0.3La(Mg_{1/2}Ti_{1/2})O_3-0.7CaTiO_3$	1500	51.2	43 800	5.8	–	[305]
1674	$(1-x)La_{2/3}TiO_3-xNiTiO_3$ (0.01)	1380	51.4	3.55	1563	24.7	[566]
1675	$Ba_3La_2Ti_2Nb_2O_{15}$	1450	51.5			15	[567]
1676	$(Pb_{0.2}Ca_{0.8})[(Ca_{1/3}Nb_{2/3})_{0.8}Ti_{0.2}]O_3$	1350 for 3 hours	51.7	7270		0	[568]
1677	$Ba_{0.5}Ca_{0.5}(Fe_{1/2}Nb_{1/2})O_3$		51.7	960	6.9	45	[427]
1678	$Bi_2(Zn_{1/3}Ta_{2/3})_2O_7$	850	51.8	2600		−26	[569]
1679	$0.3LaAlO_3-0.7SrTiO_3$	1680	52	50 800	9.3	56.4	[268]
1680	$(Li_{1/2}Sm_{1/2})TiO_3$	1300	52	2290	3	−266	[570, 571]
1681	$Ca_4SrNb_2TiO_{12}$	1550	52	15 000	3.59	42	[454]
1682	$Ce_{0.33}TaO_3$	1525	52	10 000	3.58	159	[564]
1683	$NdTiNbO_6$	1370	52	4480	4.93	46	[174]
1684	$Pr_{0.95}Gd_{0.05}TiNbO_6$	1370	52	18 500		54	[175]
1685	$CaTi_{0.6}(Al_{1/2}Ta_{1/2})_{0.4}O_3$	1500 for 15 hours	52	13 200	8	37	[486]
1686	$0.7CaTiO_3-0.3LaGaO_3$	1450 for 12 hours	52	27 000		40	[433]
1687	$Ba_4LaTiNb_3O_{15}$	1450	52	15 600	4.47	93	[455]
1688	$Ba_4Sm_{9.33}Ti_{18}O_{54}$ + 10 wt% $BaCu(B_2O_5)$	950	52	4000		−29	[572]
1689	$Ba_{0.9}Ca_{0.1}(Fe_{1/2}Nb_{1/2})O_3$		52.1	620	6.8	45	[427]
1690	$Pb_{0.4}Ca_{0.6}(Mg_{1/3}Nb_{2/3})_{1-x}Sn_x]O_3$ ($x = 0.01$)	1280	52.2	8150	5	−3.4	[573]
1691	$CaTi_{0.5}$ $(Fé_{0.5}Nb_{0.5})_{.5}O_3$ + 3 wt% B_2O_3	900 for 2 hours	52.3	2930		13	[574]
1692	$Pb_{0.2}Ca_{0-8}(Fe_{1/2}Nb_{1/2})O_3$		53	10 000	4.1	−69	[336]
1693	$PrTiNbO_6$	1370	53	12 300	4.85	56	[174]
1694	$BaLa_4Ti_4O_{15}$ (textured)	1600 for 2 hours	53	41 400		−1	[575]
1695	$0.8TiO_2-0.2Bi_2O_3$		53.2	4500		−550	[576]
1696	$(1-x)La_{2/3}TiO_3-xNiTiO_3$ (x = 0.15)	1340	53.3	12 950	3.4	21.3	[566]
1697	$CaLa_4Ti_5O_{17}$		53.7	17 400	3.7	−26	[525]
1698	$0.8La_{2/3}TiO_3-0.2LaAlO_3$	1400	53.9	29 000	5.4	35	[390]
1699	$CeTiNbO_6$	1360	54	6530	4.4	67	[174]
1700	$0.4CaTiO_3-0.6Ca(Mg_{1/3}Nb_{2/3})O_3$	1450	54	32 000	6.7	18	[312]
1701	$0.8La_{2/3}TiO_3-0.2LaAlO_3$	1400	54	29 000		35	[390]

No.		Sintering temp. (°C)	ε_r	*Qf* (GHz)	*f* (GHz)	τ_f	Reference
1702	$Ca_3Sr_2Nb_2TiO_{12}$	1540	54	10 000	3.5	45	[454]
1703	$(Ca_{0.85}Nd_{0.1})[Ti_{0.5}(Mg_{0.33}Nb_{0.67})_{0.5}]O_3$		54.1	7660		0.8	[577]
1704	$Ca[Ti_{1-x}(Mg_{1/3}Nb_{2/3})_x]O_3$ $(x=0.5)$	1450 for 4 hours	54.3	22 900		39	[489]
1705	$Ba_{0.6}Ca_{0.4}(Fe_{1/2}Nb_{1/2})O_3$		54.5	600	6.6	55	[427]
1706	$Li_{1+x+y}Ta_{1-x-3y}Ti_{x+4y}O_3$ $(x=0.1,\ y=0.15)$	1175 for 1 hour	54.8	10 400		34	[558]
1707	$Li_{1+x-y}Nb_{1-x-y}Ti_{x+4y}O_3$ $(x=0.1,\ y=0.175)$	1100 for 1 hour	54.9	8890	6.2	28	[558, 578]
1708	$Ba_3Ti_4Nb_4O_{21}$	1270	55	9500	5.5	100	[453]
1709	$Sm_{(2-x)/3}Li_xTiO_3$ $(x=0.5)$	1350	55	2000		−260	[579]
1710	$Ba[Zn_{1/4}Ti_{1/4}Nb_{1/2}]O_3$	1400	54.8	13 200	6.1		[466]
1711	$CaLa_8Ti_9O_{31}$		54.9	19 300		−6	[525]
1712	$0.2PbZrO_3{-}0.8Ca(Fe_{1/2}Nb_{1/2})O_3$		55.1	450	5.3	−51.5	[580]
1713	$Ba_4La_2Ti_3Nb_2O_{18}$	1450 for 6 hours	55.1	21 270	5.1		[581]
1714	$Ca[(Li_{1/3}Nb_{2/3})_{1-x}Ti_x]O_3$ $(x=0.5)$	1150 for 3 hours	55.2	18 600		83	[239]
1715	$CaLa_4Ti_5O_{17}$		55.2	17 400		−20	[525]
1716	MBRT-90 + 10 wt% $Li_2O{-}B_2O_3{-}SiO_2$−(56.92:37.59)	875	55.3	2500		26	[155]
1717	$CaTi_{.6}(Al_{1/2}Nb_{1/2})_{.4}O_3$	1500 for 5 hours	55.7	21 800	7	47	[423]
1718	$0.64BaTi_4O_9{-}0.35BaPr_2Ti_4O_{19}$		56	1000			[582]
1719	$(Sr_{0.1}Ca_{0.9})_3Ti_2O_7$		56	3000	2.5	141	[435]
1720	$BaTi_3Nb_4O_{17}$	1310 for 4 hours	56	8400	4	86	[559]
1721	$BaTi_{0.95}Ni_{0.05}O_{3-\delta}$	1450 for 2 hours	56	2400			[583]
1722	Bi_2TeO_6 (oxygen atm)	720 for 15 hours	56	10 400		−49	[374]
1723	$Li_{1+x-y}Nb_{1-x-y}Ti_{x+4y}O_3$ $(x=0.1,\ y=0.15)$	1100 for 1 hours	56.2	8350	6	15	[558, 578]
1724	$0.5MgTiO_3{-}0.5CaTiO_3{-}0.25(Nd_2O_3{-}2TiO_2)$		56.3	23 500		67.8	[552]
1725	$Ca_{1-x}Zn_xLa_4Ti_5O_{17}$ $(x=0)$	1500 for 4 hours	56.5	12 500	6.6	3.7	[584]
1726	$Li_{1+x-y}Nb_{1-x-3y}Ti_{x+4y}O_3$ $(x=0.1,\ y=0.1)$	1150 for 10 hours	56.5	4500		−7	[585]
1727	$CaTi_{0.6}(Al_{1/2}Nb_{1/2})_{0.4}O_3$ + 1 wt% Li_3NbO_4	1300 for 5 hours	56.6	28 000	7	53	[423]
1728	$Ba_xLa_4Ti_{3+x}O_{12+3x}$ $(x=3)$		56.6	13 380		−191	[517]
1729	$0.7CaTiO_3{-}0.3Sm(Mg_{0.5}Ti_{0.5})O_3$	1550	57	11 150	4.1	54	[278]
1730	$La_{0.4}Ba_{0.6}Ti_{0.6}Y_{0.4}O_3$	1600 for 4 hours	57	750		12	[586]

No.		Sintering temp. (°C)	ε_r	*Qf* (GHz)	*f* (GHz)	τ_f	Reference
1731	$Bi_2O_3-CaO-Nb_2O_5$ (46.15:23.08:30.77)	950	57	470	3.7	24	[587]
1732	$Ca_{1-x}Zn_xLa_4Ti_5O_{17}$ ($x=0.025$)	1500 for 4 hours	57	17 400	6.7	7.5	[584]
1733	$Ba_5LaTi_2Nb_3O_{15}$	1420 for 6 hours	57.3	18 450	4.7		[581]
1734	$Pb_{0.4}Ca_{0.6}(Mg_{1/3}Nb_{2/3})_{1-x}\,Sn_x]O_3$ ($x=0.05$)	1280	57.4	8120	5	−3.6	[573]
1735	$Ca_{1-x}Zn_xTi_5O_{17}$ ($x=0.01$)	1450	57.6	17 100		4.9	[584]
1736	$Ba_{0.7}Ca_{0.3}(Fe_{1/2}Nb_{1/2})O_3$		57.7	830	7.8	101	[427]
1737	$0.85La_{2/3}TiO_3-0.15LaAlO_3$	1375	57.7	27 900	5.2	65	[390]
1738	$Sr_3Ti_2O_7$		57.9	18 850	2.5	317	[435]
1739	$Sr_{2.4}Ca_{0.6}Ti_2O_7$	1600	57.9	25 700	2.5	359	[461]
1740	$(Sr_{0.8}Ca_{0.2})_3Ti_2O_7$		58	2500	2.5	359	[435]
1741	$0.8Ca_{0.85}Nd_{0.1}TiO_3-0.2SmAlO_3$	1400	58	14 000		13	[588]
1742	$4CaO-BaO-Nb_2O_5-TiO_2$	1490	58	4000	3.4	44	[514]
1743	$Bi_2O_3-CaO-Nb_2O_5$ (45.75:21.75:32.5)	1050	58	1060	3.8	20	[587]
1744	$(Ni_{1/3}Ta_{2/3})_{1-x}Ti_xO_2$ ($x=0.5$)	1300	58.3	13 900		85.5	[491]
1745	$Li_{1+x-y}Nb_{1-x-y}Ti_{x+4y}O_3$ ($x=0.05$, $y=0.1$)	1100 for 1 hour	58.4	6230	6.3	−31	[558, 578]
1746	$0.3La(Mg_{1/2}Ti_{1/2})O=-0.7CaTiO_3$	1500	58.8	40 390	5.5	71	[305]
1747	$Li_{1+x+y}Ta_{1-x-3y}Ti_{x+4y}O_3$ ($x=0.1$, $y=0.1$)	1175 for 1 hour	58.9	7720		26	[558, 578]
1748	$0.42(La_{1/2}Na_{1/2})TiO_3-0.58Ca(Fe_{1/2}Nb_{1/2})O_3$	1300 for 10 hours	58.9	14 070	6.6	0	[589]
1749	$Pb_{0.4}Ca_{0.6}(Ni_{1/3}Nb_{2/3})O_3$		59	7100	4.2	6.2	[336, 590]
1750	$(Pb_{0.4}Ni_{0.6})(Mg_{1/3}Nb_{2/3})O_3$		59	7100		6.2	[336]
1751	$Bi_{18}Ca_8Nb_{12}O_{65}$	950	59	610	3.7	25	[587]
1752	$Li_{1+x-y}Nb_{1-x-y}Ti_{x+4y}O_3$ ($x=0.1$, $y=0.125$)	1100 for 1 hour	59.2	7560	6	22	[558, 578]
1753	$Ba_{0.8}Ca_{0.2}(Fe_{1/2}Nb_{1/2})O_3$		59.5	550	6.3	80	[427]
1754	$(1-x)La_{2/3}TiO_3-xNiTiO_3$ (0.05)	1360	59.6	14 860	3.2	21.5	[566]
1755	$Li_{1+x+y}Ta_{1-x-3y}Ti_{x+4y}O_3$ ($x=0.15$, $y=0.1$)	1175 for 1 hour	59.6	9100		42	[558, 578]
1756	$Pb_{0.25}Ca_{0.75}[(Mg_{1/3}Nb_{2/3})_{0.75}Ti_{0.25}]O_3$		60	11 000		0	[590]
1757	$Ba(Ti_{0.85}Mn_{0.15})O_3$	1400	60	12 000		225	[591]
1758	$BaSm_2Ti_4O_{12}+16$ mol% $BaCuB_2O_5$	875	60	4500		−30	[592]
1759	$(Sr_{0.2}Ca_{0.8})_3Ti_2O_7$		60	2630	2.5	232	[435]
1760	$0.5CaTiO_3-0.5Sr(Mg_{1/3}Nb_{2/3})O_3$	1600	60	14 000		60	[538]
1761	$BaNd_2Ti_3O_{10}$		60	5300	4.2	140	[593]
1762	$Ca_2Sr_3Nb_2TiO_{12}$	1530	60	6000	3.5	48	[454]
1763	$(Ca_{1-x}Nd_{2x/3})TiO_3$ ($x=0.3$) + 25 vol% $3ZnO-2B_2O_3$	900	60	3700		62	[344]

No.		Sintering temp. (°C)	ε_r	*Qf* (GHz)	*f* (GHz)	τ_f	Reference
1764	$[Ca_{0.4}(Mg_{1/3}Ta_{2/3})_{0.6}]TiO_3$	1350	60.2	36 900		−10	[594]
1765	$Li_{1+x+y}Ta_{1-x-3y}Ti_{x+4y}O_3$ ($x=0.1$, $y=0.075$)	1175 for 1 hour	60.5	5014		−5	[557, 558]
1766	$Ba(Pr_{0.3}Bi_{.02}Nb_{0.5})O_3$	1300	60.7	1500		15	[536]
1767	$0.5BaTiO_3-0.5La(Mg_{1/2}Ti_{1/2})O_3$		60.9	9600	5.2	−2	[306]
1768	$Ca_2Ba_3Ta_2TiO_{12}$	1525	61	1800	3.4	21	[514]
1769	$x(Ba_4Nd_{9.33}Ti_{18}O_{54})-(1-x)BaLa_4Ti_4O_{15}$ ($x=0.75$) $+Bi_2O_3-B_2O_3-ZnO-SiO_2$ glass	1140	61	2300		38	[595]
1770	MBRT-90 + 10 wt% $Li_2O-B_2O_3-SiO_2-CaO-Al_2O_3$ (52.45::31.06:11.99:2:2.5)	875	61.6	2500		18	[155]
1771	$(Pb_{0.4}Ca_{0.6})(Fe_{1/2}Ta_{1/2})O_3$	1050 for 3 hours	62	9000		−15	[596]
1772	$0.83Bi_2O_3-0.25Nb_2O_5$	900 for 3 hours	62	560		−372	[556]
1773	$0.9Bi_2O_3-0.1Nb_2O_5$	900 for 3 hours	62	800		−234	[556]
1774	$CaSr_4Nb_2TiO_{12}$	1530	62	11 500	3.4	51	[454]
1775	$Li_{1+x+y}Ta_{1-x-3y}Ti_{x+4y}O_3$ ($x=0.15$, $y=0.075$)	1175 for 1 hour	62.1	6190		13	[557, 558]
1776	$Li_{1+x-y}Nb_{1-x-y}Ti_{x+4y}O_3$ ($x=0.1$, $y=0.05$)	1100 for 1 hour	62.4	3750	6.	−53	[557, 558]
1777	$Ba_{0.725}Ca_{0.275}(Fe_{1/2}Nb_{1/2})O_3$		62.4	640	6.3	82	[427]
1778	$Ca[Ti_{1-x}(Mg_{1/3}Nb_{2/3})_x]O_3$ ($x=0.4$)	1450 for 4 hours	62.9	12 200		92	[489]
1779	$Pb_{0.4}Ca_{0.6}(Mg_{1/3}Nb_{2/3})_{1-x}Sn_x]O_3$ ($x=0.03$)	1280	63	7540	5	−3.8	[573]
1780	$(1-x)BaLa_4Ti_4O_{15}-xBa_4Nd_{9.333}Ti_{18}O_{54}$ ($x=0.55$)		63	10 000		−20	[597]
1781	$0.9La_{2/3}TiO_3-0.1LaAlO_3$	1350	63	26 100		82	[390]
1782	$Bi_2Zn_{2/3}Ta_{4/3}O_7$ + 0.05 wt% CuO + 0.05 wt% V_2O_5	930	63	6800	5.35		[598]
1783	$Ba_3La_4Ti_6O_{21}$		63	9100		198	[599]
1784	$(1-x)La_{2/3}TiO_3-xNiTiO_3$ ($x=0.02$)	1380	63.3	6210	3.3	21.8	[566]
1785	$0.4Ba(Mg_{1/2}W_{1/2})O_3-0.6BaTiO_3$	1500 for 6 hours	63.9	3800		303.5	[107]
1786	$CaBa_4Ta_2TiO_{12}$	1500	64	1400	3.6	+24	[514]
1787	$Pb_2Ta_{1.5}Nb_{0.5}O_7$		64	12 800		223	[600]
1788	$Ba_{6-3x}Sm_{8+2x}Ti_{18}O_{54x-2/3}$ + 3 wt% $Al_2O_3-B_2O_3-SiO_2$	1175	64	8500	–	0	[601]
1789	$Li_{1+x-y}Nb_{1-x-y}Ti_{x+4y}O_3$ ($x=0.15$, $y=0.075$)	1100 for 1 hour	64.	4610	5.9	−15	[578]
1790	$Pb_{.4}Ca_{.6}(Mg_{1/3}Nb_{2/3})_{1-x}Sn_x]O_3$ ($x=0.01$)	1280	64.7	7130	5	0	[573]
1791	$LiNb_{0.6}Ti_{0.5}O_3$ + 0.5 wt% 0.17 $Li_2O-0.83V_2O_5$	850	64.7	5900		9	[585]
1792	$Li_{1+x-y}Nb_{1-x-y}Ti_{x+4y}O_3$ ($x=0.1$, $y=0.1$)	1100 for 1 hour	64.8	6385	5.7	8	[557, 558]

No.		Sintering temp. (°C)	ε_r	*Qf* (GHz)	*f* (GHz)	τ_f	Reference
1793	$La_{2/3}TiO_3$		65	15 700			[602]
1794	$La_{0.4}Ba_{0.6}Ti_{0.6}Yb_{0.4}O_3$	1600 for 4 hours	65	4500		1	[586]
1795	$Ba_3Ti_4Nb_4O_{21}$ + 3 wt% CuO + 1 wt% B_2O_3	900 for 2 hours	65	16 000		101	[603]
1796	$(Pb_{0.4}Ca_{0.6})[(Mg_{1/2}Nb_{1/2})O_3Sn_x]$ ($x = 0.01$)	1350	65	7100		136	[573]
1797	$Ba_3Ti_4Nb_4O_{21}$ + 1 wt% B_2O_3 + 3 wt% CuO	900 for 2 hours	65	16 000		101	[603]
1798	$0.1Pb(Fe_{2/3}W_{1/3})O_3-0.9Pb_{0.2}Ca_{0.8}(Fe_{1/2}Nb_{1/2})O_3$	1000	65.3	2270		−24	[605]
1799	$CaTi_{0.7}(Al_{1/2}Ta_{1/2})_{0.3}O_3$	1500 for 15 hours	65.4	20 000	8	113	[486]
1800	$0.5CeO_2-0.25CaO-0.25TiO_2$	1550	65.5	9500	3.2	+399	[145]
1801	$Ca_3Ba_2Nb_2TiO_{12}$	1475	66	2600	3.3	48	[514]
1802	$(1-x)(Mg_{0.95}Zn_{0.05})TiO_3-xCa_{0.6}La_{0.8/3}TiO_3$ ($x = 0.7$)	1320 for 4 hours	66	39 000		168	[137]
1803	$BaTi_{0.4}Ga_{0.3}Nb_{0.3}O_3$	1500 for 4 hours	66	3720	4.7		[550]
1804	$LiNb_{0.6}Ti_{0.5}O_3)$ + 2 wt%V_2O_5	900 for 1 hour	66	3800	5.6	11	[585]
1805	$Bi_2O_3-CaO-Nb_2O_5$ (52.5:17.5:30)	925	66	330	3.6	35	[587]
1806	$Pb_{0.4}Ca_{0.6}(Mg_{1/3}Nb_{2/3})O_3$	1280	66.3	6940	5	3.1	[573]
1807	$Bi_2(Zn_{1/3}Ta_{2/3})_2O_7$	850	66.3	6200		−8.8	[598]
1808	$(Ni_{1/3}Ta_{2/3})_{1-x}Ti_xO_2$ ($x = 0.6$)	1300	66.4	2180		147	[491]
1809	$0.92La_{2/3}TiO_3-0.08LaAlO_3$	1350	66.9	28 350	4.8	82	[390]
1810	$(Ba_{4.2}Sm_{9.2})Ti_{16.6}Al_{1.4}O_{54}$	1440	67	1543	5.4	−90	[606]
1811	$BaNd_2Ti_4O_{12} + B_2O_3-Bi_2O_3-SiO_2-ZnO$ glass + $La_2O_3-B_2O_3-TiO_2$	900	67	6000	6	4	[607]
1812	$ZrTe_3O_8$	760 for 15 hours	67.5	1800	4	362	[15, 397]
1813	$CaTi_{0.7}(Al_{1/2}Nb_{1/2})_{0.3}O_3$	1500 for 5 hours	67.8	18 700		138	[423]
1814	$0.8CaTiO_3-0.2Sm(Mg_{0.5}Ti_{0.5})O_3$	1550	68	12 400	3.5	147	[278]
1815	$Ba_{6-3x}(Sm_{1-y}Nd_y)_{8+2x}(Ti_{1-z}Sn_z)O_{54}$ ($x = 2/3$, $y = 0$, $z = 0.1$)		68	4020		20	[608]
1816	$0.6CaTiO_3-0.4Ca(Mg_{1/3}Nb_{2/3})O_3$	1450	68	17 000	5.4	108	[312]
1817	$0.3La(Mg_{1/2}Ti_{1/2})O_3-0.7SrTiO_3$		68.3	4500	5.41		[329]
1818	$(Pb_{1-x}Ca_x)[Fe_{1/2}Nb_{1/2}]_{1-y}Zr_y]O_3$ ($y = 0.01$, $x = 0.6$)	1150	68.7	6800	4.2	−17	[609]
1819	$(Ni_{1/3}Nb_{2/3})_{1-x}Ti_xO_2$ ($x = 0.3$)	1200	68.7	19 300		56.6	[491]
1820	$Ba_4Gd_9Ti_{18}O_{54}$	1350 for 10 hours	69	3300	1	−60	[610]
1821	$CaTi_{0.7}(Al_{1/2}Nb_{1/2})_{0.3}O_3$ + 1 wt% Li_3NbO_4	1300 for 5 hours	69.0	21 500	7	145	[423]

No.		Sintering temp. (°C)	ε_r	Qf (GHz)	f (GHz)	τ_f	Reference
1822	$(1-x)La_{2/3}TiO_3-xNiTiO_3$ (0.03)	1375	69.4	16 960	3.1	18.4	[566]
1823	$Ba_{2-x}Sm_{(4+2/3x)}Ti_9O_{26}$ ($x=0.3$)	1360 for 4 hours	69.4	9700		5.8	[611]
1824	$(Ba_{4.2}Sm_{.9.2})_aTi_{17.}AlO_{54}$	1440	70	4360	5.22	−57	[606]
1825	$Bi_2(Zn_{1/3}Nb_{2/3})_2O_7$ + 1 wt% of $0.15CuO-0.85MoO_3$	900	70	4800	3		[612]
1826	$Ba_{6-3x}Sm_{8+2x}Ti_{18}O_{54}$ ($x=2/3$) + 0.5 wt% ($50Al_2O_3-50\ SiO_2$)	1220	70	8500	–	−21	[601]
1827	$0.97La_{2/3}TiO_3-0.03NiTiO_3$	1350	70	17 000		18	[566]
1828	$Ba_{6-3x}Sm_{8+2x}Ti_{18}O_{54}$ ($x=2/3$) + 0.5 wt% ($44Al_2O_3-30B_2O_3-26SiO_2$)	1220	70	8600	–	−12	[601]
1829	$LiNb_{0.6}Ti_{0.5}O_3$ + 1 wt% B_2O_3	880	70	5400		−6	[613]
1830	$(Ba_{4.2}Sm_{9.2})_\alpha Ti_{18-y}Al_yO_{54}$ ($y=1$, $\alpha=1+y/36$, $x=0.6$)	1440	70.2	4350		−57	[606]
1831	$Pb_{0.4}Ca_{0.6}[(Fe_{1/2}Nb_{1/2})_{0.9}Sn_{0.1}O_3$	1150 for 3 hours	70.3	8200		−19	[614]
1832	$0.1CaTiO_3-0.5(Li_{1/2}Nd_{1/2})TiO_3-0.4(Dy_{1/3}Nd_{1/3})TiO_3$	1350 for 3 hours	70.6	1470		−156	[615]
1833	$Ba_{6-3x}\ Sm_{8+2x}Ti_{18}O_{54}$ ($x=2/3$) + 0.5 wt% ($22MgO-22Al_2O_3-56SiO_2$)	1200	71	5890	–	−19	[601]
1834	$Ba_{6-3x}\ Sm_{8+2x}Ti_{18}O_{54}$ ($x=2/3$) + 0.5 wt% ($35Bi_2O_3-32ZnO-6SiO_2-27B_2O_3$)	1200	71	8900	–	−10	[601]
1835	$0.14(BaO-Nd_2O_3-4TiO_2)+0.86(BaO-Al_2O_3-4TiO_2)$		71	8200	–	0	[404]
1836	$Ba\{Ti_{0.95}Mn_{0.05}\}O_{3-\delta}$	1450 for 2 hours	71.1	7700	4.8		[583]
1837	$0.2Pb(Fe_{2/3}W_{1/3})O_3-0.8Pb_{.2}Ca_{.8}(Fe_{1/2}Nb_{1/2})O_3$	1000	71.4	1520		−29	[605]
1838	$Ba_{(2-x)}Sm_{(4+2/3x)}Ti_9O_{24}$ ($x=0.25$)	1370	71.5	10 700	5.1	4.3	[616]
1839	$Pb_{0.5}Ca_{0.5}(Na_{1/4}Nb_{3/4})O_3$		72	1500	3.5	230	[336]
1840	$Pb_{0.5}Ca_{0.5}(Zr_{0.95}Ti_{0.05})O_3$	1350	72	4100	4	2.4	[617]
1841	$0.96La_{2/3}TiO_3-0.04LaAlO_3$	1325	72	24 000		123	[390]
1842	$Ba_{6-3x}Sm_{8+2x}Ti_{18}O_{54}$ ($x=2/3$) + 0.5 wt% ($60ZnO-30B_2O_3-10SiO_2$)	1200	72	4530	–	−17	[601]
1843	$Ba_{6-3x}Sm_{8+2x}Ti_{18}O_{54}$ ($x=2/3$) + 0.5 wt% ($40MgO-40B_2O_3-20SiO_2$)	1200	72	4450	–	−16	[601]
1844	$Ba_{6-3x}\ Sm_{8+2x}Ti_{18}O_{54}-1.9TiO_2$ ($x=2/3$)	1350 for 2 hours	72	10 300		7.2	[618]
1845	$(Ba_{4.2}Sm_{9.2})_\alpha Ti_{18-y}Al_yO_{54}$ ($y=0.8$, $\alpha=1+y/36$, $x=0.6$)	1440	72.1	4600		−42	[606]
1846	$Pb_{0..5}Ca_{0..5}(Ni_{1/3}Nb_{2/3})O_3$		73	5100	3.5	52	[336]
1847	$Pb_{0.4}Ca_{0.6}(Mg_{1/3}Nb_{2/3})O_3$		73	4100	3.1	3.7	[336]
1848	$Ba_{6-3x}\ Sm_{8+2x}Ti_{18}O_{54}$ ($x=2/3$) + 0.5 wt% ($71ZnO-29B_2O_3$)	1200	73	4830	–	−14	[601]
1849	$(Pb_{0.5}Ni_{0.5})(Mg_{1/3}Nb_{2/3})O_3$		73	4900		52	[336]

No.		Sintering temp. (°C)	ε_r	*Qf* (GHz)	*f* (GHz)	τ_f	Reference
1850	$Ba_{6-3x}Sm_{8+2x}Ti_{18}O_{54}$ ($x = 2/3$) + 0.5 wt% (40 B_2O_3–60SiO_2)	1200	73	7900	–	−16	[601]
1851	$Ba_{0.98}Sr_{0.02})Sm_2Ti_4O_{12}$	1375 for 6 hours	73	7920		−6	[619]
1852	$Ca_4La_2Ti_5O_{17}$		73	16 000	3.3	127	[620]
1853	$Ba_4Nd_{3.33}Eu_6Ti_{18}O_{54}$	1460	73.9	8898		−11	[621]
1854	$BaTi_{0.95}Co_{0.05}O_{3-\delta}$	1450 for 2 hours	74	1300			[583]
1855	$Pb_{0.5}Ca_{0.5}ZrO_3$	1500	74	3900	3.7	−16.9	[496]
1856	$Ba_{6-3x}Sm_{8+2x}Ti_{18}O_{54}$ ($x = 2/3$) + 0.5 wt% (50ZnO–50B_2O_3)	1220	74	5330	–	−17	[601]
1857	TiO_2 + zinc borosilcate glass	900	74	8000		340	[622]
1858	0.88TiO_2–0.12$Bi_2Ti_4O_{11}$	1200	74	9500		3	[549]
1859	BaO–Sm_2O_3–TiO_2		74	12 000		10	[623]
1860	$BaTi_{0.92}Ga_{0.08}O_{2.96}$	1450	74	7810	5.5		[624]
1861	$Ba_{(2-x)}Sm_{(4+2/3x)}Ti_9O_{24}$ ($x = 0.2$)	1370	74.8	10 900	5.78	2.4	[616]
1862	$Ca_2Ba_3Nb_2TiO_{12}$	1500	75	1600	3.04	53	[514]
1863	Ba_{6-3x} $Sm_{8+2x}Ti_{18}O_{54}$ ($x = 2/3$) + 0.5 wt% (40PbO–40B_2O_3–20SiO_2)	1200	75	6500	–	−17	[601]
1864	$(Pb_{0.5}Ca_{0.5})(Mg_{1/3}Nb_{2/3})O_3$		75	1400		16	[336]
1865	$(Bi_{1.92}Zn_{0.08})(Zn_{0.64}Nb_{1.36})O_7$	1000	75	1800			[625]
1866	$Ba_{0.98}Sr_{0.02}Sm_2Ti_4O_{12}$		75	7920		−6	[619]
1867	$Pb_{0.5}Ca_{0.5}(Co_{1/3}Nb_{2/3})O_3$		75	1450	3.7	16	[336]
1868	$Li_{1/2}Nd_{1/2}TiO_3$		75	2000		−274	[615]
1869	0.6$(Sm_{1/2}Li_{1/2})TiO_3$–0.4$(Sm_{1/2}Na_{1/2}Ti)O_3$	1350	75	2000	10		[626]
1870	$Ba_4Sm_{9.33}Ti_{18}O_{54}$ + 0.5 wt% GeO_2	1200	75.2	5200		−12	[627]
1871	$BaSm_2Ti_4O_{12}$ + 1 wt% CuO	1160	75.8	4900		−7.7	[628]
1872	$Ba_{6-3x}(Sm_{1-y}Nd_y)_{8+2x}Ti_{1-z}Sn_z)_{18}O_{54}$ ($x = 2/3$, $y = 0$, $z = 0.05$)		76	6260	4.1	2	[621]
1873	$Ba_{6-3x}Sm_{8+2x}Ti_{18}O_{54}$ ($x = 2/3$) + 0.5 wt% (30BaO–40B_2O_3–30SiO_2)	1220	76	9100	–	−7	[601]
1874	$Ba_{6-3x}(Sm_{1-y}Nd_y)_{8+2x}Ti_{1-z}Sn_z)_{18}O_{54}$ ($x = 2/3$, $y = 0.1$, $z = 0.05$)		76	7130	4.2	6	[629]
1875	$(Bi_{1.92}Ca_{0.08})(Zn_{0.64}Nb_{1.36})O_7$	960	76	3900			[625]
1876	$Ba_{6-3x}(Sm_{1-y}Nd_y)_{8+2x}(Ti_{1-z}Sn_z)O_{54}$ ($x = 2/3$, $y = 0$, $z = 0.05$)		76	6260		2	[629]
1877	$(Bi_{1.92}Sr_{.08})(Zn_{0.64}Nb_{1.36})O_7$	940	76	1100			[625]
1878	$Ba_4Sm_{9.33}Ti_{18}O_{54}$ + 0.5 wt% B_2O_3	1200	76.1	10 500		−19	[627]
1879	$Ba_{4.5}Gd_9Ti_{18}O_{54}$		76.1	2050		−35	[630]
1880	$Sr_4Ti_3O_{10}$		76.1	12 700	2.2	576	[435]
1881	$Ba_{(2-x)}Sm_{(4+2/3x)}Ti_9O_{28}$ ($x = 0.15$)	1370	76.1	12 800	5.2	0.8	[616]

No.		Sintering temp. (°C)	ε_r	*Qf* (GHz)	*f* (GHz)	τ_f	Reference
1882	$(Ba_{4.2}Sm_{9.2})_{\propto}Ti_{18-y}Al_yO_{54}$ ($y=0.4$, $\propto=1+y/36$, $x=0.6$)	1440	76.1	3800		−33	[606]
1883	$Bi_2(Zn_{1/3}Nb_{2/3})_2O_7$	950 for 2 hours	76.2	2980			[631]
1884	$Li_{1+x-y}Nb_{1-x-y}Ti_{x+4y}O_3$ ($x=0.15$, $y=0$)	1100 for 1 hour	76.2	1000	5.3	−62	[558, 578]
1885	$Ba_4Sm_{7.33}Eu_2Ti_{18}O_{54}$	1460	76.4	8990		−16.2	[621]
1886	$Ba_4Nd_{1.33}Eu_8Ti_{18}O_{54}$	1400	76.6	8590		−21	[621]
1887	$Ba(Mg_{0.167}Ta_{0..333}Ti_{0.50})O_3$	1550	76.6	10 000	4.1	113	[282]
1888	$Ba_4Eu_{9.33}Ti_{18}O_{54}$	1400	77	6580		34	[621]
1889	$Ba_{6-3x}(Sm_{1-y}Nd_y)_{8+2x}Ti_{1-z}Sn_z)_{18}O_{54}$ ($x=2/3$, $y=0.3$, $z=0.05$)		77	7850	4.2	9	[629]
1890	$(Ti_{0.9}Zr_{0.1})O_2$	1400 for 5 hours	77	14 000	4		[15]
1891	24 wt% $BaTiO_3$−76 wt% Nd_2O_3−$3TiO_2$	1220	77	11 000		123	[632]
1892	$(Ba_{1-\propto}Sm_{\propto})_{4.2}Sm_{9.2}Ti_{18}O_{54}$ ($\propto=0.1$)	1450 for 2 hours	77	6680		−11.4	[633]
1893	$BaSm_2Ti_5O_{14}$		77	9300		12	[634, 635]
1894	$Ba_{6-3x}(Sm_{1-y}Nd_y)_{8+2x}(Ti_{1-z}Sn_z)O_{54}$ ($x=2/3$, $y=0.3$, $z=0.05$)	1360 for 12	77	8185	4	1	[629]
1895	$Ba_4Sm_{9.33}Ti_{18}O_{54}$ + 0.5 wt% GeO_2	950	77.3	8900		−19	[627]
1896	Ba_{6-3x} $Sm_{8+2x}Ti_{18}O_{54}$ + 0.5 wt% GeO_2 + 0.5 wt% B_2O_3 ($x=2/3$)	1150	77.3	8900		−12.6	[627]
1897	$Ba_{(2-x)}Sm_{(4+2/3x)}Ti_9O_{24}$ ($x=0$)	1360	77.5	11 200	5.2	−3.4	[616]
1898	$Li_{1+x-y}Nb_{1-x-y}Ti_{x+4y}O_3$ ($x=0.05$, $y=0.05$)	1100 for 1 hours	77.8	2180	5.2	−42	[558, 578]
1899	$Ba_4Nd_{5.33}Eu_4Ti_{18}O_{54}$	1460	78	10 460		10.4	[621]
1900	$(Ba_{4.2}Sm_{.9.2})_aTi_{17.8}Al_{.0.2}O_{54}$	1460	78	8233	4.8	−18	[606]
1901	$Pb_{0.5}Ca_{0.5}(Li_{1/4}Nb_{3/4})O_3$		78	2000	3.7	460	[336]
1902	$Pb_{0.4}Ca_{0.6}[(Fe_{1/2}Nb_{1/2})_{0.95}Sn_{0.05}O_3$	1150 for 3 hours	78	6000		−9	[614]
1903	$Ca(Zr_{0.6}Ti_{0.4})O_3$		78	7840			[292]
1904	$(Ni_{1/3}Nb_{2/3})_{1-x}Ti_xO_2$ ($x=0.4$)	1200	78	17 060		97.7	[491]
1905	$(Ba_{4.2}Sm_{9.2})_{\propto}Ti_{18-y}Al_yO_{54}$ ($y=0.2$, $\propto=1+y/36$, $x=0.6$)	1460	78	8200		−18	[606]
1906	$Ba_4(Nd_{28/3-y}Yb_y)Ti_{18}O_{54}$ ($y=1$)	1480 for 2 hours	78.4	6780		53.1	[636]
1907	$Ba_{(2-x)}Sm_{(4+2/3x)}Ti_9O_{24}$ ($x=0.05$)	1360	78.5	11 900	5.2	−2.1	[616]
1908	$Bi_2(Zn_{1/3}Nb_{2/3-x}V_x)_2O_7$ ($x=0.001$)	850 for 2 hours	78.5	3780			[631]
1909	$Ba_4Nd_{8.33}DyTi_{18}O_{54}$	1480	78.6	10 040	4.8	33.8	[636]
1910	$Ba_{6-3x}Sm_{8+2x}Ti_{18}O_{54}$ ($x=0.75$)		78.6	8700		–	[634]

No.		Sintering temp. (°C)	ε_r	*Qf* (GHz)	*f* (GHz)	τ_f	Reference
1911	$Bi_2(Zn_{1/3}Nb_{2/3-x}V_x)_2O_7$ ($x=0.003$)	850 for 2 hours	78.6	3140			[631]
1912	$Ba_4Sm_{8.33}EuTi_{18}O_{54}$	1460	78.7	9560		−10.5	[621]
1913	$Ba_4Sm_{9.33}Ti_{18}O_{54}$ + 0.3 mol% TiO_2	1350	78.8	10 750	4.8	−21	[618]
1914	$Ba_{6-3x}Sm_{8+2x}Ti_{18}O_{54}$ ($x=0.5$)		78.9	8400		−19.2	[634, 637]
1915	$Pb_{0.5}Ca_{0.5}(Na_{1/4}Nb_{3/4})O_3$		79	400	3.7	550	[336]
1916	$(Ba_{0.97}Ca_{0.03})-Sm_2O_3-4.5TiO_2$		79	10 500	10	−5	[634]
1917	$Bi_{18}(Ca_{1-x}Zn_x)_8Nb_{12}O_{65}$ ($x=0.725$)	925	79	1000	3.2	1	[587]
1918	$Ba_4Nd_{8.33}HoTi_{18}O_{54}$	1480	79.3	9690	4.7	31.1	[636]
1919	$Ba_4Nd_{8.33}ErTi_{18}O_{54}$	1480	79.5	8290	4.1	32.5	[636]
1920	$Ba_{6-3x}Sm_{8+2x}Ti_{18}O_{54}-0.1TiO_2$ ($x=2/3$)	1350 for 2 hours	79.8	9880		−17.6	[618]
1921	$Ba_4Sm_{9.33}Ti_{18}O_{54}$	1450	80	10 700	4.7	−15	[627]
1922	$0.19TiO_2-0.81Bi_2O_3$		80	9000		21	[576]
1923	$0.58(Sm_{1/2}Li_{1/2})TiO_3-0.42(Sm_{1/2}Na_{1/2}Ti)O_3$	1350	80	2000	10		[626]
1924	$0.8Bi_2O_3-0.3Nb_2O_5$	920 for 3 hours	80	420		−306	[556]
1925	$(Ti_{0.9}Ge_{0.1})O_2$	1400 for 5 hours	80	24 000	4		[15]
1926	$(Sm_{1/2}Na_{1/2}Ti)O_3$	1350	80	13 000	10		[626]
1927	$(La_{0.44}Sr_{0.33})TiO_3$	1350	80	7500	3	70	[638]
1928	$(Li_{1/2}Nd_{1/2})TiO_3$		80	3100	4	−310	[639]
1929	$0.15(Ba_{0.95}Sr_{0.05})-0.15Sm_2O_3-0.7TiO_2$	1380	80	11 000	3	0	[640]
1930	$Pb_{0.4}Ca_{0.6}(Fe_{1/2}Nb_{1/2})O_3$		80	6100	3.2	−25	[336]
1931	$Ba_{6-3x}(Sm_{1-y}Nd_y)_{8+2x}Ti_{18}O_{54}$ ($x=2/3$, $y=0.1$)	1340 for 6 hours	80	9620	3.75		[641]
1932	$Ba_{6-3x}(Sm_{1-y}Nd_y)_{8+2x}(Ti_{1-z}Sn_z)O_{54}$ ($x=2/3$, $y=0.8$, $z=0.05$)	1360 for 12 hours	80	10 600	3.9	11	[629]
1933	$Ba_{6-3x}(Sm_{1-y}Nd_y)_{8+2x}(Ti_{1-z}Sn_z)O_{54}$ ($x=2/3$, $y=0.5$, $z=0.05$)	1360 for 12 hours	80	10 050	4	5	[629]
1934	$BaO-Bi_2O_3-Nd_2O_3-TiO_2$ + 0.4 wt% $Mn(CH_3COO)_2+WO_3$	1320	80	7000		0	[642]
1935	$(Ba_{1-\alpha}Sr_\alpha)_{4.2}Sm_{9.2}Ti_{18}O_{54}$ ($\alpha=0.01$)	1450 for 2 hours	80	8890		−11.3	[633]
1936	$(Ba_{1-\alpha}Sr_\alpha)_{6-3x}Sm_{8+2x}Ti_{18}O_{54}$ ($\alpha=0.06$, $x=0.6$)		80	10 075		−7	[633]
1937	$(Pb_{1-x}Ca_x)ZrO_3$		80–120	2000–4000		–	[590]
1938	$Sr(Bi_{1-x}Nd_x)_8Ti_7O_{27}$	1250	80–120	120–2100			[643]
1939	$(Ba_{0.952}Sr_{0.048})_{4.2}Sm_{9.2}Ti_{18}O_{54}$		80.1	10 205	4.9	−9	[633]
1940	$(Ba_{1-\alpha}Sm_\alpha)_{4.2}Sm_{9.2}Ti_{18}O_{54}$ ($\alpha=0.06$)	1450 for 2 hours	80.2	10 075		−7.4	[633]

No.		Sintering temp. (°C)	ε_r	Qf (GHz)	f (GHz)	τ_f	Reference
1941	$(Ba_{1-\alpha}Sm_{\alpha})_{4.2}Sm_{9.2}Ti_{18}O_{54}$ ($\alpha=0.0$)	1450 for 2 hours	80.3	9500		−8.6	[633]
1942	$(Ba_{1-\alpha}Sm_{\alpha})_{4.2}Sm_{9.2}Ti_{18}O_{54}$ ($\alpha=0.04$)	1450 for 2 hours	80.6	9590		−11.9	[633]
1943	$Ba_{6-3x}Sm_{8+2x}Ti_{18}O_{54}$ ($x=2/3$)	1360 for 3 hours	80.8	11 330	4	−11	[618, 644, 645]
1944	$Ba_{6-3x}(Sm_{0.2}Nd_{0.8})_{8+2x}Ti_{18}O_{54}$ ($x=2/3$)	1330 for 3 hours	80.8	8100		35.6	[646]
1945	$BaO-Pr_6O_{11}-TiO_2$		81	9000	5	130	[623]
1946	$(Ba_{1-x}Ca_x)O-Sm_2O_3-4.5TiO_2$ ($x=0.05$)		81	9500		2	[647]
1947	$Pb_{1-x}Ca_x[(Fe_{1/2}Nb_{1/2})_{1-y}Sn_y]O_3$ ($x=0.6$, $y=0.05$)	1150 for 3 hours	81	4830		3	[614]
1948	$Bi_2O_3-TiO_2$ (1:11.3) + 0.112 wt% CuO	915 for 2 hours	81	8900		0	[648]
1949	$Ba_{6-3x}(Sm_{1-y}Nd_y)_{8+2x}Ti_{18}O_{54}$ ($x=2/3$, $y=0.3$)	1340 for 12 hours	81	9630	3.9		[641]
1950	$(Ba_{0.95}Ca_{0.05})-Sm_2O_3-4.5TiO_2$		81	11 000	10	2	[647]
1951	$Ba_4Sm_{5.33}Eu_4Ti_{18}O_{54}$	1460	81.1	7109		−20	[621]
1952	$Ba_4Nd_{7.33}Eu_2Ti_{18}O_{54}$	1460	81.1	10 660		30.9	[621]
1953	$Ba_4Sm_{3.33}Eu_6Ti_{18}O_{54}$	1400	81.2	8604		−26	[621]
1954	$Ba_4Sm_{9.33}Ti_{18}O_{54}$ (SPS)	1200 for 5 minutes	81.2	10 099	5	−17.2	[649]
1955	$(Ba_{0.98}Sr_{0.02})_{4.2}Sm_{9.2}Ti_{18}O_{54}$		81.4	9661	4.8	−11	[633]
1956	$Ba_4Sm_{9.33}Ti_{18}O_{54}$ + 14 mol% TiO_2	1350	81.5	10 415	5	−0.3	[618]
1957	$Ba_{6-3x}Sm_{8+2x}Ti_{18}O_{54}-0.1TiO_2-1.4TiO_2$ ($x=2/3$)	1350 for 2 hours	81.5	10 400		−0.3	[618]
1958	$Ba_{6-3x}(Sm_{1-y}Nd_y)_{8+2x}Ti_{18}O_{54}$ ($x=2/3$, $y=0.84$)	1400 for 10 hours	81.7	10 500		2.1	[650]
1959	$(Li_{1/2}Nd_{1/2})TiO_3$		82	2220		292	[571]
1960	$Ba_{6-x}Sm_{8+2x}Ti_{18}O_{54}$ ($x=0.5$)	1300	82	10 150		−17	[650]
1961	$(Pb_{0.45}Ca_{0.55})[(Fe_{0.5}(Nb_{0.96}Ta_{0.04})_{0.5})]O_3$	1150 for 3 hours	82	7650		−5	[651]
1962	$Ba_{6-3x}(Sm_{1-y}Nd_y)_{8+2x}Ti_{18}O_{54}$ ($x=2/3$, $y=0.5$)	1340 for 16 hours	82	9500	3.75		[641]
1963	$(Ba_{0.8}Ca_{0.2})_{6-3x}Sm_{8+2x}Ti_{18}O_{54}$ ($x=1.5$)	1350	82	10 000		−20	[653]
1964	$Ba_{6-3x}(Sm_{1-y}Nd_y)_{8+2x}(Ti_{1-z}Sn_z)_{18}O_{54}$ ($x=2/3$, $y=0.8$, $z=0.05$)	1360 for 3 hours	82	1000	4.1	17	[629]
1965	$Ba_4Sm_{8.08}Li_{0.25}Ti_{18}O_{54}$	1400	82.1	5620	4.7	−2	[652]
1966	$Ba\{Ti_{0.95}Fe_{0.05}\}O_{3-\delta}$	1450	82.1	4800	4		[583]
1967	$Ba_{6-3x}(Sm_{0.2}Nd_{0.8})_{8+2x}Ti_{18}O_{54}$ ($x=2/3$) + 1 wt% Bi_2O_3	1200 for 3 hours	82.1	8530		17.3	[646]
1968	$Ba_4La_4Ti_7O_{24}$		82.2	500		317	[599]
1969	$Ba_{6-3x}[Nd_{(8+2x)-y}Bi_y]Ti_{18}O_{54}$ ($x=2/3$, $y=0.05$)	1380	82.2	9760		62.2	[630]

No.		Sintering temp. (°C)	ε_r	*Qf* (GHz)	*f* (GHz)	τ_f	Reference
1970	$(Ba_{0.8}Sr_{0.2})_{4.2}Sm_{9.2}Ti_{18}O_{54}$		82.3	2860		0.4	[633]
1971	$Ba_4(Sm_{0.95}Bi_{0.05})_{9+1/3}Ti_{18}O_{54}$	1420	82.3	8810		−16.5	[654]
1972	$Ba_{6-3x}Sm_{8+2x}Ti_{18}O_{54}$ $(x=0.6)$	1450	82.5	10 500	4.6	−12	[644]
1973	$Ba_4Nd_{9.33}Ti_{18}O_{54}$	1460	82.5	10 060		71.1	[623]
1974	$(Pb_{1-x}Ca_x)[Fe_{1/2}Nb_{1/2}]_{1-y}Zr_y]O_3$ $(y=0.01,\ x=0.55)$	1150	82.5	6800	3.8	−3	[609]
1975	$Ba_4Nd_{8.33}EuTi_{18}O_{54}$	1480	82.6	10 400	4.8	47.3	[636]
1976	$Ba_{6-3x}(Sm_{1-y}Nd_y)_{8+2x}Ti_{18}O_{54}$ $(x=2/3,\ y=0.72)$	1400 for 10 hours	82.7	10 500		3.8	[650]
1977	$Ba_{4.2}Sm_{9.2}Ti_{18}O_{54}$	1500	83	8950	3.5	−13	[655]
1978	$BaO-Nd_2O_3-5TiO_2$	1450 for 2 hours	83	10 500		70	[623]
1979	$Ba_{4.2}(Sm_{0.9}Nd_{0.1})_{9.2}Ti_{18}O_{54}$	1500	83	8936	3.5	−6	[655]
1980	$CaBa_4Nb_2TiO_{12}$	1470	83	1200	2.9	60	[514]
1981	$(Pb_{0.45}Ca_{0.55})[(Fe_{0.5}Nb_{0.5})_{0.9}Sn_{0.1}]O_3$ + 0.2 wt% CuO + 0.1 wt% Bi_2O_3	1000 for 3 hours	83	6080		8	[656]
1982	$(Ni_{1/3}Nb_{2/3})_{1-x}Ti_xO_2$ $(x=0.5)$	1200	83.1	19 300		165.7	[491]
1983	$Ba_{6-3x}(Sm_{1-y}Nd_y)_{8+2x}Ti_{18}O_{54}$ $(x=2/3,\ y=0.6)$	1400 for 10 hours	83.4	10 700	4	11	[641]
1984	$Ba(Ti_{0.92}Ga_{0.08})O_{3-\delta}$	1450	83.7	4200	4.2		[583]
1985	$Ba_{6-3x}(La_{1-y-z}Sm_yBi_z)_{8+2x}Ti_{18}O_{54}$ $(x=2/3,\ y=0.7,\ z=0)$	1350 for 3 hours	83.8	8000	4.0	33	[657]
1986	$(Pb_{1/2}Ca_{1/2})(Fe_{1/2}Ta_{1/2})O_3$	1250	83.9	6680	7		[658]
1987	$Ba_4(Nd_{.95}Bi_{.05})_{9.33}Ti_{18}O_{54}$	1360	83.9	8330		32	[654]
1988	$Ba_{4.2}(Sm_{0.9}La_{0.1})_{9.2}Ti_{18}O_{54}$	1500	84	9050	3.5	1.6	[655]
1989	$Ba_{6-3x}(Sm_{1-y}Nd_y)_{8+2x}Ti_{18}O_{54}$ $(x=0.6,\ y=0.2)$	1500 for 2 hours	84	9000		0	[655]
1990	$Ba_{6-3x}Sm_{8+2x}Ti_{18}O_{54}$ $(x=0.7)$ hotpressed	1200	84	9960		−14	[659]
1991	$0.95TiO_2-0.05Bi_2Ti_4O_{11}$	1200	84	12 500		230	[549]
1992	$BaPr_2Ti_5O_{14}$		84	9000	5	150	[660]
1993	$(Ba_{0.9}Ca_{0.1})-Sm_2O_3-4.5TiO_2$		84	9500	10	25	[647]
1994	$Ba_4(Nd_{0.9}Bi_{0.1})_{9+1/3}Ti_{18}O_{54}$	1380	84.1	7840		−21	[654]
1995	$0.4PbZrO_3-0.6Ca(Fe_{1/2}Nb_{1/2})O_3$		84.7	3000	4.2	40.5	[580]
1996	$Ba_{4.2}(Sm_{0.7}Nd_{0.3})_{9.2}Ti_{18}O_{54}$	1500	85	9160	3.5	8.6	[655]
1997	$Ba_{6-3x}(Sm_{1-y}Nd_y)_{8+2x}Ti_{18}O_{54}$ $(x=2/3,\ y=0.8)$	1340 for 12 hours	85	9460	3.9	0.8	[641]
1998	$0.77Bi_2O_3-0.23Nb_2O_5$	900 for 3 hours	85	350		−215	[556]
1999	$Pb_{1-x}Ca_x[(Fe_{1/2}Nb_{1/2})_{1-y}Sn_y]O_3$ $(x=0.55,\ y=0.1)$	1150 for 3 hours	85	8600		0	[614]
2000	$(Pb_{1-x}Ca_x)[(Fe_{1/2}Nb_{1/2})_{1-y}Zr_y]O_3$ $(y=0.1,\ x=0.55)$	1200	85	8600		−1	[609]

No.		Sintering temp. (°C)	ε_r	Qf (GHz)	f (GHz)	τ_f	Reference
2001	$Ba_4Sm_{8.33}LiTi_{18}O_{54}$	1400	85.4	5045	4.5	45	[652]
2002	$Ba_{4.2}(Sm_{.5}Nd_{..5})_{9.2}Ti_{18}O_{54}$	1500	86	9170	3.4	25	[65]
2003	$Pb_{0.5}Ca_{0.5}(Mg_{1/3}Nb_{2/3})O_3$		86	4600	3.0	34	[336]
2004	$(Pb_{0.45}Ca_{0.55})[(Fe_{0.5}Nb0._{0.5})_{0.9}Sn_{0.1}]O_3$ + 0.2 wt% CuO + 0.4 wt% Bi_2O_3	1000 for 3 hours	86	4340		8	[661]
2005	$0.74Bi_2O_3-0.26Nb_2O_5$	900 for 2 hours	86	1000		120	[556]
2006	$BaTi_{0.5}Ga_{0.25}Nb_{0.25}O_3$	1500 for 4 hours	86	3050	4		[550]
2007	$Ba_{1.37}Na_{0.63}Nd_2TiO_{10}$	1250	86	1500		270	[662]
2008	$Ba_{6-3x}Nd_{8+2x}Ti_{18}O_{54}$ ($x=0.75$)		86	10 450		–	[634]
2009	$Ba_{6-3x}(Sm_{1-y}Nd_y)_{8+2x}Ti_{18}O_{54}$ ($x=0.5$, $y=0.67$)	1400 for 10 hours	86	7850		23.5	[650]
2010	$0.15(Ba_{0.93}Sr_{0.07})O-0.15(Sm_{0.4}La_{0.6})_2O_3-0.7TiO_2$	1370	86.2	16 700		95	[663]
2011	$Pb_{0.45}Ca_{0.55}[(Fe_{1/2}Nb_{1/2})_{0.95}Sn_{0.05}]O_3$	1150 for 3 hours	86.3	6250		2	[614]
2012	$(Pb_{1-x}Ca_x)[(Fe_{1/2}Nb_{1/2})_{1-y}Zr_y]O_3$ ($y=0.01$, $x=0.5$)	1150	86.3	6800	3.7	25	[609]
2013	$Pb_{0.45}Ca_{0.55}[(Fe_{1/2}Nb_{1/2})_{0.9}Sn_{0.1}]O_3$	1150 for 3 hours	86.7	7900		0	[614]
2014	$Ba_{6-3x-z}Sr_zNd_{8+2x-y}Bi_yTi_{18}O_{54}$ ($y=0$, $z=0.9$, $x=0.5$)		86.7	7200		63	[664]
2015	$Ba_{6-3x}(La_{1-y-z}Sm_yBi_z)_{8+2x}Ti_{18}O_{54}$ ($x=2/3$, $y=0.5$, $z=0$)	1350 for 3 hours	86.9	7360	3.8	83	[657]
2016	$0.9CaTiO_3-0.1Sm(Mg_{0.5}Ti_{0.5})O_3$	1550	87	9500	3.2	285	[278]
2017	$(Pb_{1-x}Ca_x)[(Fe_{1/2}Nb_{1/2})_{1-y}Zr_y]O_3$ ($y=0.05$, $x=0.55$)	1200	87	8500		−10	[609]
2018	$Sr(Bi_{1-x}Nd_x)_8Ti_7O_{27}$ ($x=0.05$)		87	190			[665]
2019	$(Ba_{1-z}Pb_z)_{6-x}Nd_{8+2/3x}Ti_{18}O_{54}$ ($x=2/3$, $z=0.4$)	1400 for 2 hours	87	4000		−32	[666]
2020	$Pb_{0.46}Ca_{.54}[(Fe_{1/2}Nb_{1/2})_{0.9}Sn_{0.1}]O_3$	1150 for 3 hours	87.8	7870		5	[614]
2021	$Ba_{4.2}(Sm_{0.1}Nd_{0..9})_{9.2}Ti_{18}O_{54}$	1500	88	9500	3.44	64	[655]
2022	$Ba_{6-x}Nd_{8+2x}Ti_{18}O_{54}$ ($x=0.7$) hot pressed	1300	88	4920		55	[659]
2023	$Ba_{4.2}(Sm_{.7}La_{.3})_{9.2}Ti_{18}O_{54}$	1500	88	8050	3.44	44	[655]
2024	$Ba_{6-3x}(Sm_{1-y}Nd_y)_{8+2x}Ti_{18}O_{54}$ ($x=0.6$, $y=0.9$)	1500 for 2 hours	88	8500		64.2	[655]
2025	$Ba_{6-3x}(Sm_{1-y}Nd_y)_{8+2x}Ti_{18}O_{54}$ ($x=0.6$, $y=1.0$)	1500 for 2 hours	88	8300		76	[655]
2026	$BaO-Bi_2O_3-TiO_2-Nd_2O_3$		88	5500		8	[632]
2027	$(Ba_{1-z}Pb_z)_{6-x}Nd_{8+2/3x}Ti_{18}O_{54}$ ($x=2/3$, $z=0.22$)	1400 for 2 hours	88	5500		0	[666]
2028	$BaO-Nd_2O_3-4TiO_2$ + 0.5 wt% Al_2O_3 + 8 wt% Bi_2O_3		88	8000	–	0	[404]

No.		Sintering temp. (°C)	ε_r	*Qf* (GHz)	*f* (GHz)	τ_f	Reference
2029	$(1-x)(Mg_{0.95}Zn_{0.05})TiO_3-xCa_{0.6}La_{0.8/3}TiO_3$ ($x=0.9$)	1320 for 4 hours	88	32 800		205	[137]
2030	$Ba_{4.2}Nd_{9.2}Ti_{18}O_{54}$	1500	88	8315	3.4	76	[655]
2031	$Pb_{0.46}Ca_{0.54}[(Fe_{1/2}Nb_{1/2})_{0.95}Sn_{0.05}\ O_3$	1150 for 3 hours	88.2	6100		7	[614]
2032	$Ba_{6-3x}(La_{1-y-z}Sm_yBi_z)_{8+2x}Ti_{18}O_{54}$ ($x=2/3$, $y=0.7$, $z=0.04$)	1350 for 3 hours	88.4	6690	4.4	1	[657]
2033	$0.05Pb(Fe_{2/3}W_{1/3})O_3-0.95Pb_{0.4}Ca_{0.6}(Fe_{1/2}Nb_{1/2})O_3$	1000	88.4	3800		−6	[605]
2034	$Ba_4(Nd_{0.85}Bi_{0.15})_{9+1/3}Ti_{18}O_{54}$	1360	88.9	6620		−20	[654]
2035	$Ba(Nd_{0.82-z}Sm_zBi_{0.18})Ti_4O_{12}$ ($z=0.7$)	1300 for 2 hours	89	6880	4.5		[667]
2036	$(Pb_{0.45}Ca_{0.55})[(Fe_{0.5}Nb_{0.5})_{0.9}Sn_{0.1}]O_3 + 5$ wt% BiO_3-LiF	950	89	800		−15	[668]
2037	$0.2CaTiO_3-0.5(Li_{1/2}Nd_{1/2})TiO_3-0.3(Dy_{1/3}Nd_{1/3})TiO_3$	1350 for 3 hours	89.4	4650		−87	[615]
2038	$(Ni_{1/3}Nb_{2/3})_{1-x}Ti_xO_2$ ($x=0.6$)	1200	89.4	12 800		193	[491]
2039	$Ba_{6-3x}(Nd_{1-y}Bi_y)_{8+2x}Ti_{18}O_{54}$ ($x=2/3$, $y=0.04$)	1340 for 3 hours	89.6	7700	4	21	[669]
2040	$Ba_{6-3x}Nd_{8+2x}Ti_{18}O_{54}+PbO/Bi_2O_3$		90	9000	–	0	[655]
2041	$BaO-(Nd_{0.95}Bi_{0.05})_2O_3-4TiO_2$	1300	90	7600		33	[48]
2042	$0.96La_{2/3}TiO_3-0.04CaTiO_3$		90	27 000	10	190	[670]
2043	MBRT-90	1300 for 2 hours	90	6100		6	[155]
2044	$0.75Bi_2O_3-0.25Nb_2O_5$	900 for 3 hours	90	630		60	[556]
2045	$0.5Sm_{1/2}Li_{1/2}TiO_3-0.5Sm_{1/2}Na_{1/2}TiO_3$	1300	90	1500		−140	[626]
2046	$BaNd_2Ti_5O_{14}$ + 25 wt% Nd_2O_3 + 0.5 mol% PbO	1250 for 2 hours	90	6000		−20	[671]
2047	$(Pb_{1-x}Ca_x)[Fe_{1/2}Nb_{1/2}]_{1-y}Zr_y]O_3$ ($y=0.01$, $x=0.45$)	1150	90.6	2500	3.6	41	[609]
2048	$BaSm_{1.8}La_{.2}Ti_5O_{14}$		90.7	8900	–	4.2	[672]
2049	$Ba_{6-3x}(Nd_{(8+2x)-y}Bi_y)Ti_{18}O_{54}$ ($x=2/3$, $y=0.1$)	1360	90.71	7020		24.4	[654]
2050	$Ba_{6-x}Sm_{8+2x}Ti_{18}O_{54}$ ($x=0.5$) hot pressed	1300	91	10 870		3.2	[659]
2051	Bi_3NbO_7		91	730		100	[673]
2052	$Ba_4Sm_{6.33}Li_3Ti_{18}O_{54}$	1400	91.3	3990	4.4	111	[652]
2053	$0.3Pb(Fe_{2/3}W_{1/3})O_3-0.7Pb_{0.2}Ca_{0.8}(Fe_{1/2}Nb_{1/2})O_3$	1000	91.3	1650		7.3	[605]
2054	$(Li_{1/2}Pr_{1/2})TiO_3$		92	1010		403	[571]
2055	$0.15Ba_4Gd_9Ti_{18}O_{54}+0.85Ba_4Nd_9Ti_{18}O_{54}$		92	5000	1	0	[610]
2056	$Ba_{4.5}Nd_9Ti_{18}O_{54}$ + 15 mol% $Ba_{4.5}Gd_9Ti_{18}O_{54}$	1350 for 10 hours	92	5000		0	[674]
2057	$0.75Bi_2O_3-0.25Nb_2O_5$	850 for 3 hours	92	720		96	[556]

No.		Sintering temp. (°C)	ε_r	Qf (GHz)	f (GHz)	τ_f	Reference
2058	$Ba_{6-3x}(Nd_{1-y}Bi_y)_{8+2x}Ti_{18}O_{54}$ ($x=2/3$, $y=0.08$)		92.3	6460	4	10	[669]
2059	$Ba_4Sm_{3.33}Li_6Ti_{18}O_{54}$	1350	92.4	3580	4.4	303	[652]
2060	$(Ca_{0.3}Li_{0.14}Sm_{0.42})TiO_3$	1350	92.12	8292		2.6	[225]
2061	$Ba_4(Nd_{0.8}Bi_{0.2})_{9+1/3}Ti_{18}O_{54}$	1360	92.4	5680		−12	[654]
2062	$Ca_{2/5}Sm_{2/5}TiO_3-Li_{1/2}Sm_{1/2}TiO_3-0.8TiO_2$	1300 for 2 hours	92.5	4900	4	9	[675]
2063	$(Pb_{0.45}Ca_{0.55})(Fe_{0.5}Nb_{0.5})O_3$	1100 for 3 hours	92.55	5970			[336]
2064	$Ba_4Sm_{7.33}Li_2Ti_{18}O_{54}$	1400	92.7	3720	4.4	89	[652]
2065	$Pb_{0..5}Ca_{0.5}(Li_{1/4}Nb_{3/4})O_3$		93	2000	3.2	630	[336]
2066	$Ba_{4.2}(Sm_{0.5}La_{0.5})_{9.2}Ti_{18}O_{54}$	1500	93	1300	3.3	118	[655]
2067	$Ca_{1-x}Nd_{2x/3}TiO_3$ ($x=0.42$)	1400	93	6940	7	228	[676]
2068	$BaO-(Nd_{1-x}Bi_x)_2O_3-4TiO_2$ ($x=0.1$)	1300	93	5900		15	[48]
2069	$Ba_{6-3x-z}Sr_zNd_{8+2x-y}Bi_yTi_{18}O_{54}$ ($y=0.5$, $x=0.5$)		93.4	5700		40	[664]
2070	$(Pb_{0.48}Ca_{0.52})Fe_{1/2}Nb_{1/2})_{0.9}Sn_{0.1}O_3$	1150 for 3 hours	93.6	7100		18	[614]
2071	$Ba_4(Nd_{0.85}Bi_{0.15})_{9.33}Ti_{18}O_{54}$	1380	93.7	6350		17	[654, 677]
2072	$(Pb_{0.48}Ca_{0.52})(Fe_{1/2}Nb_{1/2})O_3$ + 2.2 mol% CeO_2	119 for 2.5 hours	93.7	6770		2	[678]
2073	$Ba_{6-3x}[Nd_{(8+2x)-y}Bi_y)]Ti_{18}O_{54}$ ($x=2/3$, $y=0.15$)	1360	93.7	6350		17.3	[654]
2074	$Ba_{6-3x}(La_{1-y-z}Sm_yBi_z)_{8+2x}Ti_{18}O_{54}$ ($x=2/3$, $y=0.5$, $z=00.4$)	1340 for 3 hours	93.9	4337	4	29	[657]
2075	$Pb_{0.6}Ca_{0.4}ZrO_3$	1450	94	3600	3	−9.8	[496]
2076	$Pb_{0.6}Ca_{0.4}(Ni_{1/3}Nb_{2/3})O_3$		94	3800	3.2	130	[336]
2077	$3SrO-Ta_2O_5-3TiO_2$	1400 for 5 hours	94	5200			[15]
2078	$(Pb_{0.6}Ni_{0.4})(Mg_{1/3}Nb_{2/3})O_3$	#	94	3800		130	[336]
2079	$Pb_{0.48}Ca_{0.52}[(Fe_{1/2}Nb_{1/2})_{0.95}Sn_{0.05}O_3$	1150 for 3 hours	94.3	5950		24	[614]
2080	$Ca_{1-x}Sm_{2x/3}TiO_3$ ($x=0.6$)	1450	94.5	14 900	5		[679]
2081	$CaO-BaO-Li_2O-Sm_2O_3-TiO_2$ (14:4:9:12:63)	1325	94.5	7400		3	[680]
2082	$Ba_{6-3x}Nd_{8+2x}Ti_{18}O_{54}$ ($x=0.5$ + 10 wt% $Bi_4Ti_3O_{12}$)	1300 for 3 hours	94.9	5620		21.4	[681]
2083	$Ba_{6-3x}Pr_{8+2x}Ti_{18}O_{54}$		95	6000	–	200	[682, 683]
2084	$Ba_4Sm_{5.33}Li_4Ti_{18}O_{54}$	1350	95	1000	4.4	142	[652]
2085	$Ba_{6-3x}(La_{1-y-z}Sm_yBi_z)_{8+2x}Ti_{18}O_{54}$ ($x=2/3$, $y=0.5$, $z=0.08$)	1320 for 3 hours	95	3510	3.8	3	[657]
2086	$0.3(Sm_{1/2}Li_{1/2})TiO_3-0.7(Sm_{1/2}Na_{1/2}Ti)O_3$	1350	95	1000	10		[626]
2087	$0.98TiO_2-0.02Bi_2Ti_4O_{11}$	1200	95	18 000		351	[549]
2088	$(Ba_{0.6}Pb_{0.4})_{6-3x}La_{8+2x}Ti_{18}O_{54}$ ($x=1.5$)	1380	95	6000		200	[653]
2089	$0.3Sm_{1/2}Li_{1/2}TiO_3-0.7Sm_{1/2}Na_{1/2}TiO_3$	1300	95	1000		−240	[626]

No.		Sintering temp. (°C)	ε_r	Qf (GHz)	f (GHz)	τ_f	Reference
2090	Pb,Ca(Fe, W, Nb)O_3	1000	95.7	3840		9.6	[605]
2091	$0.67Ca_{2/5}Sm_{2/5}TiO_3{-}0.33Li_{1/2}Sm_{1/2}TiO_3$	1300 for 3 hours	95.5	7200		0	[639]
2092	$Ba_{6-3x}(Nd_{1-y}Bi_y)_{8+2x}Ti_{18}O_{54}$ ($x=2/3$, $y=0.12$)		95.8	5820	4	8	[669]
2093	$Ba_{6-x}Sm_{8+2x}Ti_{18}O_{54}$ ($x=0.3$) hot pressed	1300	96	1440		−14	[659]
2094	$CaO{-}BaO{-}Li_2O{-}Sm_2O_3{-}TiO_2$ (14:4:8:12:63)	1325	96	7580		−6	[684]
2095	$0.92Ba_{4.5}(Nd_{1-y}Bi_y)_9Ti_{18}O_{54}{-}0.08BaTi_4O_9$ ($y=0.12$)		96	5590		27	[685]
2096	$CaO{-}SrO{-}Li_2O{-}0.83Sm_2O_3{-}0.17Yb_2O_3$ TiO_2		96.4	2690	–	36	[686]
2097	$Ba_{6-3x-z}Sr_zNd_{8+2x-y}Bi_yTi_{18}O_{54}$ ($y=1$, $x=0.5$)		97	5500		22	[664]
2098	$Sr(Bi_{1-x}Nd_x)_8Ti_7O_{27}$ ($x=0.1$)		97	740			[665]
2099	$0.3CaTiO_3{-}0.4(Li_{1/2}Nd_{1/2})TiO_3{-}0.3(Dy_{1/3}Nd_{1/3})TiO_3$	1350 for 3 hours	97.6	5150		0.4	[615]
2100	$0.98TiO_2{-}0.019Bi_2O_3$		97.8	3700		354	[576]
2101	TiO_2 + 2 wt% CuO	900 for 2 hours	98	14 000		374	[687]
2102	$0.92Ba_{4.5}(Nd_{1-y}Bi_y)_9Ti_{18}O_{54}{-}0.08BaTi_4O_9$ ($y=0.145$)		98	5500		17	[685]
2103	$Ca_{1-x}Nd_{2x/3}TiO_3$ ($x=0.39$)	1400	98	8560	7	247	[676]
2104	$(1\_x)Ca_{2/5}Sm_{2/5}TiO_3{-}xLi_{1/2}Nd_{1/2}TiO_3$ ($x=0.3$)	1300 for 3 hours	98	5100	5		[688]
2105	$0.75Bi_2O_3{-}0.25Nb_2O_5$	930 for 3 hours	98	300		−154	[556]
2106	$0.3CaTiO_3{-}0.4(Li_{1/2}Nd_{1/2})TiO_3{-}0.3Dy_{1/3}Nd_{1/3}TiO_3$	1350	98	5100		0	[615]
2107	$(Nd_{1/2}Na_{1/2})TiO_3$		98	2700		190	[689]
2108	$(Ca_{0.275}Sm_{0.4}Li_{0.25})TiO_3$ + 0.5 wt% $B_2O_3{-}Li_2O$	1200 for 3 hours	98.7	5930		−3.7	[690]
2109	$CaO{-}BaO{-}Li_2O{-}Sm_2O_3{-}TiO_2$ (14:2:9:12:63)	1325	98.7	6180		7.85	[680]
2110	$Ba_4Sm_{(28-y)/3}Li_yTi_{18}O_{54}$ ($y=8$)	1300	98.8	280		515	[652]
2111	$Ba_{6-3x}(La_{1-y-z}Sm_yBi_z)_{8+2x}Ti_{18}O_{54}$ ($x=2/3$, $y=0.7$, $z=0.08$)	1325 for 3 hours	99	4920	4.38	−18	[657]
2112	$0.7Ca_{2/5}Sm_{2/5}TiO_3{-}0.3Li_{1/2}Nd_{1/2}TiO_3$	1200 for 3 hours	99	6200		9	[688, 691]
2113	$Ba_{6-x}Nd_{8+2x}Ti_{18}O_{54}$ ($x=0.3$) hot pressed	1300	99	3680		110	[659]
2114	$(Ca_{1-x}Nd_{2x/3})TiO_3$ ($x=0.6$)	1300	99	3500	3.2		[692]
2115	$Ba_{6-3x}(Nd_{1-y}Bi_y)_{8+2x}Ti_{18}O_{54}$ ($x=2/3$, $y=0.15$)		99.1	5290		−5.5	[669]
2116	$CaO{-}SrO{-}Li_2O{-}0.83Sm_2O_3{-}0.17Dy_2O_3{-}TiO_2$		99.5	5930	–	30	[686]
2117	$Pb_{0.5}Ca_{0.5}[(Fe_{1/2}Nb_{1/2})_{0.9}Sn_{0.1}]O_3$	1150 for 3 hours	99.6	6570		32	[614]
2118	$Sr_5Ti_4O_{13}$		99.8	4000	1.9	801	[435]
2119	$(Pb_{1/2}Ca_{1/2})_{0.94}(La_{1/2}Nd_{1/2})_{0.06}$ $[Fe_{1/2}Nb_{1/2}]O_{3+d}$	1200 for 3 hours	99.9	5822	5.5	0	[693]

No.		Sintering temp. (°C)	ε_r	Qf (GHz)	f (GHz)	τ_f	Reference
2120	$(Ca_{1-x}Nd_{2x/3})TiO_3$ ($x=0.5$)	1300	100	14 600	3.2		[692]
2121	$Ba(Nd_{0.82-z}Sm_zBi_{0.18})Ti_4O_{12}$ ($z=0.1$)	1300 for 2 hours	100	3950	4.7		[667]
2122	$[(Pb_{0.5}Ca_{0.5})_{0.95}Nd_{0.05}](Fe_{1/2}Nb_{1/2})O_3$		100	5800		0	[694]
2123	$CaO-SrO-Li_2O-(1-x)Sm_2O_3-xDy_2O_3-TiO_2$ ($x=0.17$)		100	5900		30	[686]
2124	$0.45CaTiO_3-0.35(Li_{1/2}Nd_{1/2})TiO_3-0.2(Dy_{1/3}Nd_{1/3})TiO_3$	1350 for 3 hours	100.1	6430		118	[615]
2125	$0.05Pb(Fe_{2/3}W_{1/3})O_3-0.95Pb_{0.45}Ca_{0.55}(Fe_{1/2}Nb_{1/2})O_3$	1000	100.8	3250		20	[605]
2126	$(Pb,Ca)ZrO_3$		>100	1000	–	–	[695]
2127	$Ba_{6-3x-z}Pb_zNd_{8+2x-y}Bi_yTi_{18}O_{54}$ ($y=1$, $z=1.0$, $x=0.5$)		101	4000		−4	[664]
2128	$Bi_{1.733}(Zn_{0.733}Nb_{4/3})O_{6.67}$	1000	101	4800	4.8		[696]
2129	$0.55Ca_{0.61}Nd_{0.91}TiO_3-0.45Li_{1/2}Nd_{1/2}TiO_3$	1400 for 4 hours	101	5300	7.2	13	[676]
2130	$Ca_{1-x}Sm_{2x/3}TiO_3$ ($x=0.4$)	1450	101	14 090	5		[679]
2131	$Pb_{0.5}Ca_{0.5}[(Fe_{1/2}Nb_{1/2})_{0.95}Sn_{0.05}O_3$	1150 for 3 hours	102	4900		38	[614]
2132	$Ba_{6-x}La_{8+2x}Ti_{18}O_{54}$ ($x=0.7$) hot pressed	1300	102	2380		399	[659]
2133	$Ba(Nd_{0.82-z}Sm_zBi_{0.18})Ti_4O_{12}$ ($z=0.03$)	1300 for 2 hours	102	3650	4.74		[667]
2134	$0.1La(Mg_{1/2}Ti_{1/2})O_3-0.9CaTiO_3$	1350	102.2	20 200	4.3	395	[305]
2135	$Ba_{6-3x}(Nd_{1-y}Bi_y)_{8+2x}Ti_{18}O_{54}$ ($x=2/3$, $y=0.18$)		102.6	4400	4	−16.5	[669]
2136	$CaO-BaO-Li_2O-Sm_2O_3-Nd_2O_3-TiO_2$ (14:4:8:10:2:63)	1350	103	7200		2	[684]
2137	$0.4CaTiO_3-0.5(Li_{1/2}Nd_{1/2})TiO_3-0.1(Dy_{1/3}Nd_{1/3})TiO_3$	1350 for 3 hours	103	4214		146	[615]
2138	$Ba_4(Nd_{0.7}Bi_{0.3})_{9+1/3}Ti_{18}O_{54}$	1320	103.3	2980		8.6	[654]
2139	$(Pb_{1/2}Ca_{1/2})_{0.95}La_{0.05}[Fe_{1/2}Nb_{1/2}]O_{3+\delta}$	1150	103.4	5640		7.1	[693]
2140	$Pb_{0.5}Ca_{0.5}(Fe_{1/2}Nb_{1/2})O_3$		104	4000	2.8	26	[336]
2141	TiO_2	1200	104	44 000		–	[15, 697]
2142	$Sr(Bi_{1-x}Nd_x)_8Ti_7O_{27}$ ($x=0.3$)		104	350			[665]
2143	TiO_2 + 0.05 mol% Fe	1500	104	50 300			[697]
2144	$CaO-Li_2O-Sm_2O_3-TiO_2$ (16:9:12:63)	1325	104.1	4320		13	[571, 680]
2145	$Ba_{6-3x}La_{8+2x}Ti_{18}O_{54}$		105	2000	–	450	[682, 683]
2146	$5BaO-Ta_2O_5-3TiO_2$	1400 for 5 hours	105	800			[15]
2147	$0.5Ca_{0.6}La_{0.2667}TiO_3-0.5Li_{1/2}Nd_{1/2}TiO_3$	1400 for 4 hours	105	7000		4.5	[698]
2148	$Ba(Nd_{0.82-z}Sm_zBi_{0.18})Ti_4O_{12}$ ($z=0.12$)	1300 for 2 hours	105	4150	3.64		[667]

No.		Sintering temp. (°C)	ε_r	*Qf* (GHz)	*f* (GHz)	τ_f	Reference
2149	$Ba_{6-3x}(La_{1-y-z}Sm_yBi_z)_{8+2x}Ti_{18}O_{54}$ ($x=2/3$, $y=0.7$, $z=0.12$)	1325 for 3 hours	105.1	4170	4.59	−15	[615]
2150	$0.45CaTiO_3-0.45(Li_{1/2}Nd_{1/2})TiO_3-0.1(Dy_{1/3}Nd_{1/3})TiO_3$	1350 for 3 hours	105.1	5160		155	[615]
2151	$0.7BaTiO_3-0.3La(Mg_{1/2}Ti_{1/2})O_3$		105.8	6940	4.55	525	[306]
2152	$BaO-(Nd_{0.8}Bi_{0.2})_2O_3$-$4TiO_2$	1300	106	4200		8	[48]
2153	$CaO:BaO:Li_2O(Sm_{1-y}Nd)_2O_3:TiO_2$ (14:4:8:12:63) ($y=0.33$)	1400 for 3 hours	106	6600		22	[684]
2154	$0.3Ca_{2/5}Sm_{2/5}TiO_3-0.7Li_{1/2}Nd_{1/2}TiO_3$		106	3100			[688]
2155	$0.5Ca_{2/5}Sm_{2/5}TiO_3-0.5Li_{1/2}Nd_{1/2}TiO_3$	1300 for 3 hours	106	3710	5		[688]
2156	$Ca_{1-x}Nd_{2x/3}TiO_3$ ($x=0.3$)	1400	107	6590	7	316	[676]
2157	$Ca_{2/5}Sm_{2/5}TiO_3-Li_{1/2}Nd_{1/2}TiO_3-0.8TiO_2$	1300 for 2 hours	107	3300	4	0	[675]
2158	$CaO-Sm_2O_3-Li_2O-TiO_2$ (11:8:5:40)	1250	107.2	5700		0	[699]
2159	$0.2Pb(Fe_{2/3}W_{1/3})O_3-0.8Pb_{0.4}Ca_{0.6}(Fe_{1/2}Nb_{1/2})O_3$	930	107.2	3790		48	[605]
2160	$CaO-SrO-Li_2O-0.83Sm_2O_3-0.17Sm_2O_3-TiO_2$		108	5480	–	15	[686]
2161	$(Ca_{1-x}Nd_{2x/3})TiO_3$ ($x=0.39$)	1300	108	17 200	3		[692]
2162	$Sr(Bi_{0.6}Nd_{0.4})_8Ti_7O_{27}$	1260	108	2000		–	[665]
2163	$Ba_{3.75}La_{9.5}Ti_{18}O_{54}$		109.2	1800	3		[634]
2164	$0.1Pb(Fe_{2/3}W_{1/3})O_3-0.9Pb_{0.45}Ca_{0.55}(Fe_{1/2}Nb_{1/2})O_3$	930	109.4	3500		52	[605]
2165	$Pb_{0.63}Ca_{0.37}ZrO_3$	1450	110	3000	2.8	13	[496]
2166	$(1-x)Ca_{2/5}Sm_{2/5}TiO_3-xLi_{1/2}Nd_{1/2}TiO_3$ ($x=0.6$)	1300 for 3 hours	110	3400	5		[688]
2167	$[(Pb_{0.5}Ca_{0.5})_{0.98}Nd_{0.02}](Fe_{1/2}Nb_{1/2})O_3$		110	5800		17	[700]
2168	$0.15CaO-0.01SrO-0.09Li_2O-0.12Sm_2O_3-0.63TiO_2$		110	4500	3	7	[571]
2169	$0.3CaTiO_3-0.4(Li_{1/2}Nd_{1/2})TiO_3-0.3La_{1/3}Nd_{1/3}TiO_3$	1350	110	1400		−93	[615]
2170	$Ba_{6-x}La_{8+2x}Ti_{18}O_{54}$ ($x=0.5$) hot pressed	1300	110	2460		340	[659]
2171	$0.2CaTiO_3-0.5(Li_{1/2}Nd_{1/2})TiO_3-0.3(La_{1/3}Nd_{1/3})TiO_3$	1350 for 3 hours	110.4	1460		−93	[615]
2172	$Ba_2Sr_2Sm_2Ti_{4+x}Ta_{6-x}O_{30-x/2}$ ($x=3$)	1340 for 2 hours	111	200	3.3		[701]
2173	$Ba_{6-3x}(La_{1-y-z}Sm_yBi_z)_{8+2x}Ti_{18}O_{54}$ ($x=2/3$, $y=0.5$, $z=0.12$)	1320 for 3 hours	111.3	2470	3.7	−30	[657]
2174	$Ba_{6-3x}(La_{1-y-z}Sm_yBi_z)_{8+2x}Ti_{18}O_{54}$ ($x=2/3$, $y=0.7$, $z=0.16$)	1300 for 3 hours	111.4	2530	4.3	−21	[657]
2175	$0.6(Na,La)TiO_3-0.4(Li,Sm)TiO_3$		112	1060		18	[570]
2176	$Pb_{0.95}Ca_{.05}ZrO_3$	1250	112	720	2.8	–	[496]
2177	$Ba_{6-3x}Nd_{8+2x-y}Bi_yTi_{18}O_{54}$ ($y=2$, $x=0.5$)		112	3000		25	[664]

No.		Sintering temp. (°C)	ε_r	Qf (GHz)	f (GHz)	τ_f	Reference
2178	$0.05Pb(Fe_{2/3}W_{1/3})O_3-0.95Pb_{0.5}Ca_{0.5}(Fe_{1/2}Nb_{1/2})O_3$	1000	112.2	2730		52	[605]
2179	$CaO-SrO-Li_2O-0.83Sm_2O_3-0.17Nd_2O_3-TiO_2$		112.5	4900	–	13	[686]
2180	$[Ca_{0.4}(Li_{1/2}Nd_{1/2})_{0.6}]TiO_3$	1350	112.6	4480		8	[594]
2181	$Sr_2Ce_2Ti_5O_{16}$ ($Sr_{1-3x/2}Ce_xTiO_3$, $x=0.4$)	1300 for 2 hours	113	8000		306	[702]
2182	$0.3CaTiO_3-0.7Li_{1/2}Sm_{1/2}TiO_3$	1300 for 3 hours	114	3700		11.5	[703]
2183	$Ba_2Sr_2Sm_2Ti_{4+x}Ta_{6-x}O_{30-x/2}$ ($x=2$)	1340 for 2 hours	114	150	3.63		[701]
2184	$Ba_2Sr_2Sm_2Ti_{4+x}Ta_{6-x}O_{30-x/2}$ ($x=2.5$)	1340 for 3 hours	114	140		–	[701]
2185	$Ba_4(Nd_{0.7}Bi_{0.3})_{9.33}Ti_{18}O_{54}$	1320	114.1	2706		44	[654]
2186	$CaO-SrO-Li_2O-0.83Sm_2O_3-0.17Pr_6O_{11}O_3-TiO_2$		114.3	4850	–	14	[686]
2187	$BaO-(Nd_{0.7}Bi_{0.3})_2O_3-4TiO_2$	1275	115	2100		26	[48]
2188	$Ba_{6-3x}(La_{1-y-z}Sm_yBi_z)_{8+2x}Ti_{18}O_{54}$ ($x=2/3$, $y=0.5$, $z=0.15$)	1300 for 3 hours	115.4	1884	3.58	−22	[657]
2189	$0.3CaTiO_3-0.4(Li_{1/2}Nd_{1/2})TiO_3-0.3(La_{1/3}Nd_{1/3})TiO_3$	1350 for 3 hours	116	1675		23	[615]
2190	$0.3(Na_{1/2}La_{1/2})TiO_3-0.7(Li_{1/2}Sm_{1/2})TiO_3$	1300	117	2280	3	−19	[570]
2191	$Ba_{6-3x}(La_{1-y-z}Sm_yBi_z)_{8+2x}Ti_{18}O_{54}$ ($x=2/3$, $y=0.7$, $z=0.2$)	1275 for 3 hours	117	1780	4.29	−36	[657]
2192	$0.6CaTiO_3-0.1(Li_{1/2}Nd_{1/2})TiO_3-0.3(La_{1/3}Nd_{1/3})TiO_3$	1350 for 3 hours	117	3950		258	[615]
2193	$0.4CaTiO_3-0.3(Li_{1/2}Nd_{1/2})TiO_3-0.3(La_{1/3}Nd_{1/3})TiO_3$	1350 for 3 hours	117	2070		119	[615]
2194	$CaO-SrO-Li_2O-0.83Sm_2O_3-0.17La_2O_3-TiO_2$		118	410	–	15	[686]
2195	$Ca(Zr_{0.4}Ti_{0.6})O_3$		118	6400			[292]
2196	$Pb_{0.65}Ca_{0.35}ZrO_3$	1450	118	1260	2.8	29	[496]
2197	$Bi_{1.5}Zn_{0.92}Nb_{1.5}O_{6.92}$ +3 wt% $(0.81MoO_3-0.19CuO)$	900 for 4 hours	118.2	1000	2.3		[704]
2198	$Ca_{1-x}Sm_{2x/3}TiO_3$ ($x=0.2$)	1450	119.3	12 330	5		[679]
2199	$Bi_{1.5}Zn_{0.92}Nb_{1.5}O_{6.92}$ +3 wt% $(0.21BaCO_3-0.79CuO)$	950 for 4 hours	120.1	1050	2.3		[704]
2200	$Ba_{6-3x}(La_{1-y-z}Sm_yBi_z)_{8+2x}Ti_{18}O_{54}$ ($x=2/3$, $y=0.5$, $z=0.18$)	1300 for 3 hours	120.2	1571	3.8	−15	[657]
2201	$0.4CaTiO_3-0.5(Li_{1/2}Nd_{1/2})TiO_3-0.1(La_{1/3}Nd_{1/3})TiO_3$	1350 for 3 hours	121.3	3040		113	[615]
2202	$0.45CaTiO_3-0.25(Li_{1/2}Nd_{1/2})TiO_3-0.3(Dy_{1/3}Nd_{1/3})\ TiO_3$	1350 for 3 hours	121.3	3650		109	[615]
2203	$(La_{1/2}Na_{1/2})TiO_3$	1300	122	9800	3	480	[570]
2204	$0.15CaO-0.011SrO-0.09Li_2O-13\ 820.06Sm_2O_3-0.63TiO_2$		123	4150		10.8	[686]

No.		Sintering temp. (°C)	ε_r	*Qf* (GHz)	*f* (GHz)	τ_f	Reference
2205	$0.5CaTiO_3-0.4(Li_{1/2}Nd_{1/2})TiO_3-0.1(La_{1/3}Nd_{1/3})TiO_3$	1350 for 3 hours	123	4148		136	[615]
2206	$Sr_3Ce_2Ti_6O_{19}$ ($Sr_{1-3x/2}Ce_xTiO_3$, $x=0.333$)	1350 for 2 hours	123	10 000		392	[702]
2207	$Ba_{6-3x}(La_{1-y-z}Sm_yBi_z)_{8+2x}Ti_{18}O_{54}$ ($x=2/3$, $y=0.5$, $z=0.2$)	1300 for 3 hours	124.5	1430	3.58	−9	[657]
2208	$0.4CaTiO_3-0.6Li_{1/2}Nd_{1/2}TiO_3$		126	2600	2.1	127	[705]
2209	$Bi_{1.5}Zn_{0.92}Nb_{1.5}O_{6.92}$	1050 for 4 hours	126.2	520	2.4		[612]
2210	$Ba_3La_3Ti_5Ta_5O_{30}$	1425	126.6	110	3.1	100	[706]
2211	$(Pb_{1-x}Ca_x)[(Fe_{1/2}Nb_{1/2})_{1-y}Zr_y]O_3$ ($y=0.01$, $x=0.4$)	1150	126.7	3630	3	118	[609]
2212	$0.2Pb(Fe_{2/3}W_{1/3})O_3-0.8Pb_{0.45}Ca_{0.55}(Fe_{1/2}Nb_{1/2})O_3$	930	127.2	2300		96	[605]
2213	$0.6CaTiO_3-0.3(Li_{1/2}Nd_{1/2})TiO_3-0.1(La_{1/3}Nd_{1/3})TiO_3$	1350 for 3 hours	128.2	4460		256	[615]
2214	$0.5(Li_{1/2}Nd_{1/2})TiO_3-0.5(Na_{1/2}Nd_{1/2})TiO_3$		130	2000			[707]
2215	$0.2CaTiO_3-0.8Li_{.5}Nd_{.5}TiO_3 + 5$ wt% $Bi_2Ti_2O_7$	1300	130	2400		20	[708]
2216	$(La_{0.44}Pb_{0.33})TiO_3$	1300	130	5000	3	300	[638]
2217	$Ba_4La_2Ti_4Ta_6O_{30}$	1425	131.8	540	3.47	–	[706]
2218	$Pb_{0.7}Ca_{0.3}ZrO_3$	1400	132	1800	2.8	86	[496]
2219	$0.2CaTiO_3-0.68(Li_{1/2}Nd_{1/2})TiO_3-0.12(La_{1/3}Nd_{1/3})TiO_3$	1350 for 3 hours	132.6	1450		−17.2	[615]
2220	$Sr_4Ce_2Ti_7O_{22}$	1325	133	11 100	2.3		[702]
2221	$Sr_{0.8}Ca_{0.2}TiO_3$	1400	133.9	3950	1.62	1534	[461]
2222	$5CaO-2Nb_2O_5-3TiO_2$	1300 for 5 hours	134	1500			[15]
2223	$0.8CaTiO_3-0.2(Li_{1/2}Nd_{1/2})TiO_3$		134	13 800		200	[707]
2224	$0.5(Ca_{0.7}Nd_{0.2})TiO_3-0.5(Li_{1/2}Nd_{1/2})TiO_3$	1150 for 4 hours	134	2200		20	[709]
2225	$Sr_4Ce_2Ti_7O_{22}$ ($Sr_{1-3x/2}Ce_xTiO_3$, $x=0.286$)	1350 for 2 hours	136	10 800		428	[702]
2226	$0.4CaTiO_3-0.48(Li_{1/2}Nd_{1/2})TiO_3-0.12(La_{1/3}Nd_{1/3})TiO_3$	1350 for 3 hours	136.4	2220		122	[615]
2227	$Pb_{0.6}Ca_{0.4}[(Fe_{1/2}Nb_{1/2})_{0.95}Sn_{0.05}]O_3$	1150 for 3 hours	139.4	2450		140	[614]
2228	$PbZrO_3-CeO_2$	1250 for 4 hours	140	2500	3	−1080	[710]
2229	$0.6PbZrO_3-0.4Ca(Fe_{1/2}Nb_{1/2})O_3$		140.7	1776	3.02	120	[580]
2230	$(Ca_{1-x}Nd_{2x/3})TiO_3$ ($x=0.15$)	1300	141	11 300	2.77		[692]
2231	$(Ca_{1-x}Nd_{2x/3})TiO_3$ ($x=0.27$)	1300	141	10 350	3.07		[692]
2232	$0.6CaTiO_3-0.28(Li_{1/2}Nd_{1/2})TiO_3-0.12(La_{1/3}Nd_{1/3})TiO_3$	1350 for 3 hours	142	3327		283	[615]
2233	$Sr_5Ce_2Ti_8O_{25}$	1325	142	11 100	2.3		[702]

No.		Sintering temp. (°C)	ε_r	Qf (GHz)	f (GHz)	τ_f	Reference
2234	$Pb_{0.6}Ca_{0.4}[(Fe_{1/2}Nb_{1/2})_{0.9}Sn_{0.1}O_3$	1150 for 3 hours	142.6	2520		130	[614]
2235	$Sr_5Ce_2Ti_8O_{25}$ ($Sr_{1-3x/2}Ce_xTiO_3$, $x=0.25$)	1375 for 2 hours	143	11 000		478	[702]
2236	$Sr_{0.8}Ca_{0.2}TiO_3$		145	4050	1.62	1534	[435]
2237	$Bi_{1.5}Zn_{0.92}Nb_{1.5}O_{6.92}+0.6$ wt% V_2O_5	850 for 1 hour	148.0	120			[612]
2238	$0.4CaTiO_3-0.6Li_{1/2}Nd_{1/2}TiO_3$ + 15 wt% $Bi_2O_3-2TiO_2$		150	2200		65	[705]
2239	$0.4CaTiO_3-0.6Li_{0.5}Nd_{0.5}TiO_3$ + 10 wt% $Bi_2Ti_2O_7$	1175	150	2400		70	[705]
2240	$Bi_{1.5}ZnNb_{1.5}O_7$		150	300	2.9		[711, 712]
2241	$Sr_6Ce_2Ti_9O_{28}$ ($Sr_{1-3x/2}Ce_xTiO_3$, $x=0.222$)	1375 for 2 hours	150	9600		497	[702]
2242	$Ca(Zr_{0.2}Ti_{0.8})O_3$		153	4400			[292]
2243	$Pb_{0.6}Ca_{0.4}(Fe_{1/2}Nb_{1/2})O_3$		154	1700	2.3	135	[336]
2244	$Sr_7Ce_2Ti_{10}O_{31}$ ($Sr_{1-3x/2}Ce_xTiO_3$, $x=0.2$)	1375 for 2 hours	157	9300		544	[702]
2245	$CaTiO_3$	1400	162	1290	1.5	859	[292, 435]
2246	$0.1Pb(Fe_{1/2}Nb_{1/2})O_3-0.9CaTiO_3$	1200 for 3 hours	164	6180	2.8	583	[713]
2247	$Pb_{0.75}Ca_{0.25}ZrO_3$	1300	167	960	2.4	111	[496]
2248	$Sr_8Ce_2Ti_{11}O_{34}$ ($Sr_{1-3x/2}Ce_xTiO_3$, $x=0.182$)	1375 for 2 hours	167	8000		601	[702]
2249	$PbZrO_3-CeO_2-SrTiO_3-NiO-Nb_2O_5$	1250 for 4 hours	170	3600	3	800	[710]
2250	$Sr_{0.1}Ca_{0.9}TiO_3$		170	8320	1.	931	[435]
2251	$0.3Pb(Fe_{1/2}Nb_{1/2})O_3-0.7CaTiO_3$	1200 for 3 hours	172.7	810	2.7	392	[713]
2252	$Sr_9Ce_2Ti_{12}O_{37}$ ($Sr_{1-3x/2}Ce_xTiO_3$, $x=0.167$)	1375 for 2 hours	173	3000		637	[702]
2253	$Sr_{10}Ce_2Ti_{13}O_{40}$ ($Sr_{1-3x/2}Ce_xTiO_3$, $x=0.154$)	1400 for 2 hours	179	8000		724	[702]
2254	$Sr_{0.2}Ca_{0.8}TiO_3$		181	3900	1.4	991	[435]
2255	$Sr_{11}Ce_2Ti_{14}O_{43}$ ($Sr_{1-3x/2}Ce_xTiO_3$, $x=0.154$)	1400 for 2 hours	185	6000		789	[702]
2256	$Sr_{0.4}Ca_{0.6}TiO_3$		218	7180	1.3	1164	[435]
2257	$Sr_8Ce_2PbTi_{12}O_{36}$	1300	229	4400	2	950	[714]
2258	$0.5Pb(Fe_{1/2}Nb_{1/2})O_3-0.5CaTiO_3$	1200 for 3 hours	232.1	870	2.3	433	[713]
2259	$Sr_{0.5}Ca_{0.5}TiO_3$		236	4120	1.2	1234	[435]
2260	$Sr_{0.8}Ca_{0.2}TiO_3$		255	3800	1.6	1534	[435]
2261	$Pb_{1.5}Nb_2O_{6.5}$		259	3010		1239	[715]
2262	$SrTiO_3$		270	3000	2	150	[435, 716]

No.		Sintering temp. (°C)	ε_r	*Qf* (GHz)	*f* (GHz)	τ_f	Reference
2263	$Ag(Nb_{1/3}Ta_{2/3})O_3$ + 1 wt% CuO	875	271	800			[717]
2264	$BaTi_{0.7}Ga_{0.15}Nb_{0.15}O_3$	1500 for 4 hours	275	100	2.4		[550]
2265	$Ag(Nb_{2/4}Ta_{2/4})O_3$	1200	285	300	2.4		[717]
2266	$Ag(Nb_{1/4}Ta_{3/4})O_3$	925	295	600	2.6		[717]
2267	$Sr_7Ce_2Pb_2Ti_{12}O_3$	1250	301	4300	1.8	1287	[714]
2268	$0.8PbZrO_3-0.2Ca(Fe_{1/2}Nb_{1/2})O_3$		335.8	314	2	386	[580]
2269	$Bi_6Ti_5TeO_{22}$	1010 for 10 hours	350	220		2600	[411]
2270	$Ba_{0.2}Sr_{0.8}TiO_3$	1450 for 3 hours	363	2400	2.3		[718]
2271	$Ba_{0.3}Sr_{0.7}TiO_3$−10 mol% $MgTiO_3$	1350	365	1500			[719]
2272	$AgTa_{0.57}Nb_{0.43}O_3$	1200	380	800			[720]
2273	$Ag(Nb_{2/4}Ta_{2/4})O_3$ + 1 wt% CuO	900	398	400	2.3		[717]
2274	$Ag_{0.52}Ta_{0.48}O_3$	1250 for 20 hours	415	430	2		[721]
2275	$Ba_{0.5}Sr_{0.5}TiO_3$		420	2250			[435]
2276	$Sr_6Ce_2Pb_3Ti_{12}O_{36}$	1200	430	2300	1.7	2218	[714]
2277	$Ag(Nb_{3/4}Ta_{1/4})O_3-Ag(Nb_{1/4}Ta_{3/4})O_3$ (5:55)	925	463	200	1.97		[717]
2278	$Ag(Nb_{3/4}Ta_{1/4})O_3$	925	487	200	1.89		[717]
2279	$Ba_{0.8}Sr_{0.2}TiO_3$	1450 for 3 hours	560	850	1.9		[718]
2280	$0.7Pb(Fe_{1/2}Nb_{1/2})O_3-0.3CaTiO_3$	1150 for 3 hours	566	120	1.42	1075	[713]
2281	$Ba_{0.4}Sr_{0.6}TiO_3$	1450 for 3 hours	672	1600	1.7		[718]
2282	$Ba_{0.6}Sr_{0.4}TiO_3$	1450 for 3 hours	838	300	1.6		[718]

Data not available.

Dielectric Properties of Single Crystals

	Material	Temperature of measurement (K)	f (GHz)	ε_r perpendicular	Qf (GHz)	τ_ε (ppm/K)	ε_r parallel	$Qf_{\parallel}$ (GHz)	τ_ε (ppm/K)	Reference
1	Quartz	295	17	4.443	1.4×10^6	9	4.644	2×10^6	28.7	[722]
2	MgF_2	300	10–30	5.48	4.9×10^5	210	4.765	2.2×10^5		[723]
3	MgF_2	15 K	39 GHz	5.4	3.9×10^7		4.6			[724]
4	SrF_2	300	8.5	6.45	7.3×10^4	230				[723]
5	CaF_2	300	8.1	6.8	9.2×10^4	238				[723, 725]
6	CaF_2	50	17.2	6.45	2.0×10^5					[726]
7	BaF_2	300	7.9	7.35	5.7×10^4	204				[723]
8	BaF_2	14	7	10.4	1×10^5					[727]
9	LiF	300	7.1	9.02	1.92×10^5	257				[723, 725]
10	LiF	20	8	8.42	2.22×10^5	224				[728]
11	LiF	290	16.5	8.42	4.21×10^6					[729]
12	LiF	20	16.5	8.33	7.66×10^5	239				[728]
13	YVO_4	300	16.3	9.36	2.8×10^5	84				[730]
14	Sapphire	297	20, (6–16)	9.935	1.17×10^6	85	11.59	1.89×10^6	121	[722]
15	Sapphire	1.55	12.7	–	10.54×10^{10}					[731]
16	Sapphire	4.55			1.2×10^{10}					[732]
17	YAG	289	20.2	10.6	8.67×10^5	108				[722, 733]
18	$SrLaAlO_4$	297	12	16.85	6.28×10^5	50	19.8	1.8×10^5	115	[722]

19	Lanthanum gallium silicate	300	6.9	18.15	1.72×104	107	50	1.13×10^4	−810	[734]
20	$NdGaO_3$	297	18.5	21.9	4.4×10^4	183				[735]
21	$(La_2Sr)(Al,Ta)O_3$ [LSAT]	290	15.5	23.13	8.6×10^4	−0.37				[736]
22	$LaAlO_3$	300	10	24	2×10^5					[737]
23	$LiTaO_3$	300	11.4	41	1.75×10^4	275				[738]
24	Rutile	299	5.5	85.7	6.46×10^4	−760	163.2	5.6×10^4	−1200	[739]
25	$SrTiO_3$	297	7.2	318	1.4×10^4	−3380				[739]
26	$KTaO_3$	297	3	238	2.27×10^4	−3300				[740]
27	$CaNdAlO_4$			18	1×10^4					[741]
28	$LaSrGaO_{4(100)}$		18	21.3	7.5×10^4					[742]
29	$LaSrGaO_{4(001)}$		18	21.8	5.42×10^4					[742]
30	$LaAlO_3$ (100top)		10	24.1	2.94×10^5					[743]
31	$LaAlO_3$ (110top)		10	24	5.09×10^5					[743]
32	Si		300	11.6	3.0000					[744]
33	$LaGaO_3$		500	26	6×10^5					[744]
34	MgO	300	10 GHz	9	1×10^6					[78]
35	MgO	80	8 GHz	9.8	1.75×10^7					[78]
36	TiO_2	300	10 GHz	89	7×10^4					[78]

REFERENCE

[1] C.-C. Chiang, S. F. Wang, Y.-R. Wang, and W.-C. J. Wei. Densification and microwave dielectric properties of $CaO–B_2O_3–SiO_2$ system glass-ceramics. Ceram. Int. 34(2008)599–604.
[2] M. Ohsahi, H. Ogawa, A. Kan, and E. Tanaka. Microwave dielectric properties of low temperature sintered $Li_3AlB_2O_6$ ceramic. J. Eur Ceram. Soc. 25(2005)2877–2881.
[3] T. Okamura and T. Kishina. Dielectric properties of rare earth added cordierite at microwave and millimeterwave frequencies. Jpn. J. Appl. Phys. 37(1998)5364–5366.
[4] Y.-J. Choi, J.-H. Park, W.-J. Ko, I.-S. Hwang, J.-H. Park, J.-G. Park, and S. Nahm. Cofiring and shrinkage matching in low and middle permittivity dielectric compositions for a low temperature cofired ceramics system. J. Am. Ceram. Soc. 89(2006)562–567.
[5] M. M. Krzmanc, M. Valant, and D. Suvorov. A structural and dielectric characterization of $Na_xCa_{1-x}Al_{2-x}Si_{2+x}O_8$ ($x = 0,1$) ceramics. J. Eur. Ceram. Soc. 25(2005)2835–2838.
[6] J.-M. Wu and H. L. Huang. Effect of crystallization on microwave dielectric properties of stoichiometric cordierite glasses containing B_2O_3 and P_2O_5 glasses. J. Mater. Res. 15(2000)222–227.
[7] M. M. Krzmanc, A. Meden, and D. Suvorov. The correlation between the structure and the dielectric properties of $K_xBa_{1-x}Ga_{2-x}Ge_{2+x}O_8$ ceramics. J. Eur. Ceram. Soc. 27(2007)2957–2961.
[8] M.-H. Lim, J.-H. Park, H.-G. Kim, and M.-J. Yoo. Proc. IEEE Int. Conf. Properties and Applications of Dielectric Materials. 2(2003)757–760.
[9] M. M. Krzmanc, M. Valant, B. Jancar, and D. Suvorov. Subsolidus synthesis and microwave dielectric characterisation of plagioclase feldspars. J. Am. Ceram. Soc. 88(2005)2472–2479.
[10] Wakino et al. Proc. ISIF. 92(1992)308.
[11] J. J. Bian, D.-W. Kim, and K. S. Hong. Microwave dielectric properties of $A_2P_2O_7$ ($A = Ca$, Sr, Ba, Mg, Zn, Mn). Jpn. J. Appl. Phys. 43(2004)3521–3525.
[12] M.-H. Kim, J. C. Lee, C. I. Cheon, and H.-J. Kang. Crystal structure and low temperature cofiring ceramic property of $(1-x)(Li, Re)W_2O_8–xBaWO_4$ ceramics. ($Re = Y$,Yb). Jpn. J. Appl. Phys. 45(2006)7397–7400.
[13] J. S. Kim, J. C. Lee, C. I. Cheon, and C. H. Lee. New type of wolframite $LiYW_2O_8$ for LTCC applications. MMA, Inuyama, Japan (2004)108.
[14] M. Terada, K. Kuwamura, I. Kagomiya, K. Kakimoto, and H. Ohsato. Effect of Ni substitution on the microwave dielectric properties of cordierite. J. Eur. Ceram. Soc. 27(2007)3045–3048.
[15] M. Maeda, T. Yamamura, and T. Ikeda. Dielectric characteristics of several complex oxide ceramics at microwave frequencies. Jpn. J. Appl. Phys. Suppl. 26-2(1987)76–79.
[16] R. Umemura, H. Ogawa, H. Ohsato, A. Kan, and A. Yoko. Microwave dielectric properties of low temperature sintered $Mg_3(VO_4)_2$ ceramics. J. Eur. Ceram. Soc. 25(2005)2865–2870.
[17] A. Kan. Ph.D Thesis, Meijo University, Japan.
[18] H. Sun, Q. Zhang, H. Yang, and J. Tzou. $(Ca_{1-x}Mg_x)SiO_3$: a low permittivity microwave dielectric ceramic system. Mater. Sci. Eng. B138(2007)46–50.
[19] Y. Guo, H. Ohsato, and K. Kakimoto. Characterization and dielectric behaviour of willemite and TiO_2 doped willemite at millimeterwave frequency. J. Eur. Ceram. Soc. 26(2006)1827–1830.
[20] N.-H. Nguyen, J.-B. Lim, S. Nahm, J.-H. Paik, and J.-H. Kim. Effect of Zn/Si ratio on the microstructural and microwave dielectric properties of Zn_2SiO_4 ceramics. J. Am. Ceram. Soc. 90(2007)3127–3130.
[21] C.-S. Chen, C.-C. Chou, W.-J. Shih, K.-S. Liu, C.-S. Chen, and I.-N. Lin. Microwave dielectric properties of glass ceramic composites for low temperature cofireable ceramics. Mater. Chem. Phys. 79(2003)129–134.
[22] J. L. Tzou, Q.-L. Zhang, H. Yang and H.-P. Sun. A new system of low temperature sintering $ZnO–SiO_2$ dielectric ceramics. Jpn. J. Appl. Phys. 45(2006)4143–4145.
[23] H. Wang, Q. Zhang, H. Yang, and H. Sun. Synthesis and microwave dielectric properties of $CaSiO_3$ nanopowder by the sol-gel process. Ceram. Int. (2008) in press.

[24] J. J. Seo, D. J. Shin, and Y. S. Cho. Phase evolution and microwave dielectric properties of lanthanum borate based low temperature cofired ceramic materials. J. Am. Ceram. Soc. 89(2006)2352–2355.
[25] T. Sugiyama, T. Tsunooka, K. Kakimoto, and H. Ohsato. Microwave dielectric properties of forsterite based solid solutions. J. Eur. Ceram. Soc. 26(2006)2097–2100.
[26] T. Tsunooka, H. Sugiyma, K. Kakimoto, H. Ohsato, and H. Ogawa. Zero temperature coefficient τ_f and sinterability of forsterite ceramics by rutile addition. J. Ceram. Soc. Jpn. 112(2004)S1637–S1640.
[27] J.-M. Wu and H.-L. Huang. Microwave properties of zinc, barium and lead borosilicate glasses. J. Non-Cryst. Solids. 260(1999)116–124.
[28] T. Tsunooka, M. Andou, Y. Higashita, H. Sugiura, and H. Ohsato. Effects of TiO_2 on sinterability and dielectric properties of high Q forsterite ceramics. J. Eur. Ceram. Soc. 23(2003)2573–2378.
[29] R. Umemura, H. Ogawa, A. Yokoi, H. Ohsato, and A. Kan. Low temperature sintering-microwave dielectric property relations in $Ba_3(VO_4)_2$ ceramic. J. Alloys Compd. 424(2006)388–393.
[30] A. Yokoi, H. Ogawa, A. Kan, and Y. Nakamura. Relationship between crystal structure and microwave dielectric properties of melilite type ceramics. J. Eur. Ceram. Soc. 27(2007) 2989–2993.
[31] J.-J. Bian, D.-W. Kim, and K. S. Hong. Glass free LTCC microwave dielectric ceramics. Mater. Res. Bull. 40(2005)2120–2129.
[32] G.-K. Choi, J.-R. Kim, S. H. Yoon, and K. S. Hong. Microwave dielectric properties of scheelite (A = Ca, Sr, Ba) and wolframite (A = Mg, Zn, Mn) $AMoO_4$ compounds. J. Eur. Ceram. Soc. 27(2007)3063–3067.
[33] N. Mori, Y. Sugimoto, J. Harada, and H. Takagi. Dielectric properties of new glass-ceramics for LTCC applied to microwave or millimeter-wave frequencies. J. Ceram. Soc. Jpn. Suppl. 112–1 (2004)S1563–S1566.
[34] N. Mori, Y. Sugimoto, J. Harada, and Y. Higuchi. Dielectric properties of new glass-ceramic for LTCC applied to microwave or millimeter-wave frequencies. J. Eur. Ceram. Soc. 26(2006)1925–1928.
[35] M. M. Krzmanc, M. Valant, and D. Suvorov. The synthesis and microwave dielectric properties of $Sr_xBa_{1-x}Al_2Si_2O_8$ and $Ca_yBa_{1-y}Al_2Si_2O_8$ ceramics. J. Eur. Ceram. Soc. 27(2007)1181–1185.
[36] S. Nahm, J.-B. Lim, J.-H. Paik, J.-H. Kim, and H.-J. Lee. Synthesis of $BaCu(B_2O_5)$ ceramics and their effect on the sintering temperature and microwave dielectric properties of $Ba(Zn_{1/3}Nb_{2/3})O_3$. J. Amer. Cer. Soc. 89(2006)3124–3128.
[37] K. Mori, H. Ogawa, A. Kan, and H. Ohsato. Microwave dielectric-microstructure relationships in $Y_2Ba(Cu_{1-x}Mg_x)O_5$ solid solutions. J. Eur. Ceram. Soc. 24(2004)1749–1753.
[38] T. Takada, H. Yamamoto, and K. Kageyama. Synthesis and microwave dielectric properties of xRe_2O_3–yB_2O_3 (Re = La, Nd, Sm, Dy, Ho and Y) compounds. Jpn. J. Appl. Phys. 42(2003)6162–6167.
[39] K. X. Song and X. M. Chen. Effects of Mg/Si ratio on microwave dielectric characteristics of forsterite ceramics. J. Am. Ceram. Soc. 90(2007)1808–1811.
[40] Q. L. Zhang, H. Yang, J. L. Zou, and H. P. Wang. Microwave dielectric properties of $(Ca_{1-x}Zn_x)P_2O_7$. Mater. Lett. 59(2005)257–260
[41] Y. H. Jo, M. S. Kang, K. W. Chung, and Y. S. Cho. Chemical stability and dielectric properties of RO–La_2O_3–B_2O_3 (R = Ca,Mg,Zn) based ceramics. Mater. Res. Bull. 43(2008)361–369.
[42] J.-J. Bian, D.-W. Kim, and K. S. Hong. Microwave dielectric properties of $Ca_2P_2O_7$. J. Eur. Ceram. Soc. 23(2003)2589–2592.
[43] H. Kagata, R. Saito, and H. Katsumura. Al_2O_3-MgO-ReOx (Re = rare earth) based LTCC and its applications to multilayer Non-shrinkage substrate for microwave devices. J. Electroceram. 13(2004)277–280.
[44] A. Kan, H. Ogawa, H. Ohsato, and J. Sugishita. Influence of M (M = Zn, Ni) substitution for Cu on microwave dielectric characteristics of $YB_2Ba(Cu_{1-x}M_x)O_5$ solid solutions. Jpn. J. Appl. Phys. 40(2001)5774–5778.

[45] S. H. Yoon, D.-W. Kim, S.-Y. Cho, and K. S. Hong. Investigation of the relations between structure and microwave dielectric properties of divalent metal tungstate compounds. J. Eur. Ceram. Soc. 26(2006)2051–2054.
[46] C.-L. Lo, J.-G. Duh, B.-S. Chou, and W.-H. Lee. Low temperature sintering and microwave dielectric properties of anorthite based glass-ceramics. J. Am. Ceram. Soc. 85(2002) 2230–2235.
[47] H. Shin, H.-K. Shin, H. S. Jung, S.-Y. Cho, and K. S. Hong. Phase evolution and dielectric properties of $MgTi_2O_5$ ceramic sintered with lithium borosilicate glass. Mater. Res. Bull. 40(2005)2021–2028.
[48] I.-S. Cho, D.-W. Kim, J.-R. Kim, and K. S. Hong. Low temperature sintering and microwave dielectric properties of $BaO–(Nd_{1-x}Bi_x)_2O_3–4TiO_2$ by the glass addition. Ceram. Int. 30(2004)1181–1185.
[49] T. Takada and K. Kageyama. Synthesis and microwave dielectric properties of $La_2O_3–xB_2O_3$ based melt mixtures for low temperature cofired ceramics. Jpn. J. Appl. Phys. 44(2005)6629–6635.
[50] S. Nishigaki, S. Yano, H. Kato, T. Hirai, and S. Nomura. $BaO–TiO_2–WO_3$ microwave ceramics and crystalline $BaWO_4$. J. Am. Ceram. Soc. 71(1988)C11–C17.
[51] R. Umemura, H. Ogawa, and A. Kan. Low temperature sintering and microwave dielectric properties of $(Mg_{3-x}Zn_x)(VO_4)_2$ ceramics. J. Eur. Ceram. Soc. 26(2006)2063–2068.
[52] H. Jantunen, A. Uusumaki, R. Rautioaho, and S. Leppavuori. Compositions of $MgTiO_3$–$CaTiO_3$ ceramic with two borosilicate glasses for LTCC technology. J. Eur Ceram. Soc. 20(2000)2331–2336.
[53] A. Kan, H. Ogawa, and H. Ohsato. Synthesis and crystal structure–microwave dielectric property relation in Sn substituted $Ca_3(Zr_{1-x}Sn_x)Si_2O_9$ solid solution with cuspidine structure. Jpn. J. Appl. Phys. 46(2007)7108–7111.
[54] H. Ohsato, T. Tsunooka, Y. Ohishi, Y. Miyauchi, M. Ando, and K. Kakimoto. Millimeterwave dielectric ceramics of alumina and fosterite with high quality factor and low dielectric constant. J. Korean. Ceram. Soc. 40(2003)350–353.
[55] I.-S. Cho, G.-K. Choi, and J.-S. An, J.-R. Kim, and K. S. Hong. Sintering, microstructure and microwave dielectric properties of rare earth orthophospahates $RePO_4$ (Re = La, Ce, Nd, Sm, Tb, Dy, Y, Yb). J. Mater. Res. (2007).
[56] K. P. Surendran, P. V. Bijumon, P. Mohanan, and M. T. Sebastian. $(1-x)\ MgAl_2O_4–xTiO_2$ dielectrics for microwave and millimeter wave applications. Appl. Phys. A81(2005)823–826.
[57] Y.-J. Choi, J.-H. Park, J.-H. Park, and J.-G. Park. Middle permittivity LTCC dielectric composition with adjustable temperature coefficient. Mater. Lett. 58(2004)3102–3106.
[58] H. Jantunen, A. Uusumaki, R. Rautioaho, and S. Leppavuori. Temperature coefficient of microwave resonance frequency of a low temperature cofired (LTCC) system. J. Am. Ceram. Soc. 85(2002)697–699.
[59] K. P. Surendran, N. Santha, P. Mohanan, and M. T. Sebastian. Temperature stable low permittivity ceramic dielectrics in $(1-x)ZnAl_2O_4–xTiO_2$ system for microwave substrate applications. Eur. Phys. J. B41(2004)301–306.
[60] A. Kan and H. Ogawa. Low temperature synthesis and microwave dielectric properties of $Mg_4Nb_2O_9$ ceramics synthesized by a precipitation method. J. Alloys Compd. 364(2004) 247–249.
[61] C.-F. Fang, Y.-C. Chen, W.-C. Tzou, and S.-L. Chang. Sintering and microwave dielectric characteristics of MCAS glass-aaded $0.8Al_2O_3–0.16TiO_2$. Mater. Lett. 57(2003)2945–2949.
[62] W.-C. Tzou, S.-L. Chang, C.-F. Yang, and Y.-C. Chen. Sintering and dielectric properties of $0.88Al_2O_3–0.12TiO_2$ microwave ceramics by glass addition. Mater. Res. Bull. 38(2003)981–989.
[63] E. S. Kim, S.-H. Kim, and B. I. Lee. Low temperature sintering and microwave dielectric properties of $CaWO_4$ ceramics for LTCC applications. J. Eur. Ceram. Soc. 26(2006) 2101–2104.
[64] J. W. Choi, S. Kucheiko, S.-J. Yoon, J. Jai, and K.-H. Yoon. Microwave dielectric properties of $(1-x)\ (Al_{1/2}Ta_{1/2})O_2$-$x(Mg_{1/3}Ta_{2/3})O_2$ ceramics. J. Am. Ceram. Soc. 84(2001)2570–2572.

[65] Y. Higuchi, Y. Sugimoto, J. Harada, and H. Tamura. LTCC system with new high ϵ_r and high Q material cofired with conventional low ε_r base material for wireless communications. J. Eur. Ceram. Soc. 27(2007)2785–2788.
[66] M. Valant and D. Suvorov. Glass free low temperature fired ceramics: calcium germanates, silicates and tellurates. J. Eur. Ceram. Soc. 24(2004)1715–1719.
[67] P. V. Bijumon, M. T. Sebastian, and P. Mohanan. Experimental investigations and three dimensional transmission line matrix simulation of $Ca_{5-x}A_xB_2TiO_{12}$ (A = Mg, Zn, Ni and Co: B = Nb and Ta) ceramic resonators. J. Appl. Phys. 98(2005)124105.
[68] S. X. Dai, R.-F. Huang, and D. L. Wilcox. Use of titanates to achieve temperature stable low temperature cofired dielectrics for wireless applications. J. Am. Ceram. Soc. 85(2002)828–832.
[69] I.-S. Cho, H.-S. Ryu, J.-R. Kim, D.-W. Kim, and K. S. Hong. Sintering behaviour and microwave dielectric properties of tricalcium phosphate polymorphs. Jpn. J. Appl. Phys. 46(2007)2999–3003.
[70] Q.-L. Zhang, H. Yang, and H.-P. Sun. A new microwave ceramic with low permittivity for LTCC applications. J. Eur. Ceram. Soc. 28(2008)605–609.
[71] M. Watanabe, H. Ogawa, H. Ohsato, and C. Humphreys. Microwave dielectric properties of $Y_2Ba(Cu_{1-x}Zn_x)$ solid solution. Jpn. J. Appl. Phys. 37(1998)5360–5363.
[72] J. Bian. Microwave dielectric properties of $MgCo_2(VO_4)_2$ ceramics synthesized by a sol-gel method. J. Eur. Ceram. Soc., 27(2007)3855–3859.
[73] H.-C. Chen, M.-H. Weng, C. I. Hung, M.-P. Houng, and Y.-H. Wang. Effect of Ni substitution for Cu on the dielectric properties of $Y_2Ba(Cu_{1-x}Ni_x)O_5$ solid solutions. Physica B. 349(2004)304–309.
[74] R. C. Pullar, S. Farrah, and N. Alford. $MgWO_4$, $ZnWO_4$, $NiWO_4$, and $CoWO_4$ microwave dielectric ceramics. J. Eur. Ceram. Soc. 27(2007)1059–1063.
[75] N. Mc N. Alford and S. J. Penn. Sintered alumina with low dielectric loss. J. Appl. Phys. 80(1996)5895–5898.
[76] A. Yokoi, H. Ogawa, A. Kan, and H. Ohsato. Use of LiF to achieve a low temperature co-fired ceramics (LTCC) with low dielectric loss. J. Ceram. Soc. Jpn. Suppl. 112(2004) S1633–S1636.
[77] A. Kan, H. Ogawa, A. Yokoi, and Y. Nakamura. Crystal structural refinement of corundom-structured $A_4M_2O_9$ (A = Co and Mg, M = Nb,Ta) microwave dielectric ceramics by high temperature X-ray powder diffraction. J. Eur. Ceram. Soc. 27(2007)2977–2981.
[78] N. Mc N. Alford, J. Breeze, X. Wang, S. J. Penn, S. Dalla, S. J. Webb, N. Ljepojevic, and X. Aupi. Dielectric loss of oxide single crystals and polycrystalline analogues from 10 to 320K. J. Eur. Ceram. Soc. 21(2001)2605–2611.
[79] H. Ogawa, H. Taketani, A. Kan, A. Fujita, and G. Zouganelis. Evaluation of electronic state of $Mg_4(Nb_{2-x}Sb_x)O_9$ microwave dielectric ceramics by first principal calculation method. J. Eur. Ceram. Soc. 25(2005)2859–2863.
[80] D.-W. Kwon, M. T. Lanagan, and T. R. Shrout. Microwave dielectric properties of BaO–TeO_2 binary compounds. Mater. Lett. 61(2007)1827–1831.
[81] J. Breeze, X. Aupi, and N. Mc N. Alford. Layered Al_2O_3–TiO_2 composite dielectric resonators. Electron. Lett. 36(2000)883–884.
[82] E. S. Kim and S. H. Kim. Effects of structural characteristics on microwave dielectric properties of $(1-x)CaWO_4$-$xLaNbO_4$. J. Electroceram. 17(2006)471–477.
[83] J. Breeze, X. Aupi, and N. Mc N. Alford. Ultralow polycrystalline alumina. Appl. Phys. Lett. 81(2002)5021–5023.
[84] P. S. Anjana and M. T. Sebastian (to be published).
[85] G.-K. Choi, S.-Y. Cho, J.-S. An, and K.-S. Hong. Microwave dielectric properties and sintering behaviours of Scheelite compound $CaMoO_4$. J. Eur. Ceram. Soc. 26(2006)1969–1971.
[86] J. Wang, W. Chen, and L. Luo. Effect of CeO_2 on microwave dielectric properties of MgO–Al_2O_3–SiO_2–TiO_2 glass ceramic. Jpn. J. Appl. Phys. 46(2007)5218–5220.
[87] S. Thomas, and M. T. Sebastian. Effect of B_2O_3–Bi_2O_3–SiO_2–Zno glass on the sintering and microwave dielectric properties of. 0. $83ZnAl_2O_4$–$0.17TiO_2$. Mater. Res. Bull. 43(2008)843–851.

[88] G. Subodh, R. Ratheesh, M. V. Jacob, and M. T. Sebastian. Microwave dielectric properties and vibrational spectroscopic analysis of $MgTe_2O_5$ ceramics. J. Materials Research (2008) in press.

[89] R. C. Pullar, F. Azough, R. Freer, and N. Mc N. Alford. Novel microwave dielectric LTCCs based upon V_2O_5 doped $M^{2+}CuNbO_6$ compounds (M = Zn, Co, Ni, Mg and Ca). J. Eur. Ceram. Soc. 26(2007)1943–1946.

[90] I.-S. Cho, S.-K. Kang, D. W. Kim, and K. S. Hong. Mixture behaviour and microwave dielectric properties of $(1-x)Ca_2P_2O_7-xTiO_2$. J. Eur. Ceram. Soc. 26(2006)2007–2010.

[91] H. Ohsato, T. Tsunooka, T. Sugiyama, K. Kakimoto, and H. Ogawa. Forsterite ceramics for millimeterwave dielectrics. J. Electroceram. 17(2006)445–450.

[92] K. Mori, H. Ogawa, A. Kan, and H. Ohsato. Dielectric property-microstructure relations in Co-O doped $(Y_{2-x}Sm_x)BaCuO_5$ ceramics. J. Eur. Ceram. Soc. 23(2003)2603–2606.

[93] G. Subodh, and M. T. Sebastian. Microwave dielectric properties of A_2MgTeO_6 (A = Ba, Ba, Sr) double perovskites. (to be published).

[94] I. N. Jawahar, P. Mohanan, and M. T. Sebastian. $A_5B_4O_{15}$ (A = Ba, Sr, Ca, Mg: B = Nb,Ta) microwave dielectric ceramics. Mater. Lett. 57(2003)4043–4048.

[95] A. Kan, H. Ogawa, and A. Yokoi. Electric state and chemical bonding of $(Mg_{4-x}Mn_x)Nb_2O_9$ microwave dielectrics. J. Mater. Res. 20(2005)2456–2461.

[96] W. Huang, K.-S. Liu, L.-W. Chu, G.-H. Hsiue, and I. N. Lin. Microwave dielectric properties of LTCC materials consisting of glass-$Ba_2Ti_9O_{20}$ composites. J. Eur. Cer. Soc. 23(2003)2559–2563.

[97] H. Ogawa, A. Kan, S. Ishihara, and Y. Higashida. Crystal structure of corundom type $Mg_4(Nb_{2-x}Ta_x)O_9$ microwave dielectric ceramics with low dielectric loss. J. Eur. Ceram. Soc. 23(2003)2485–2488.

[98] J. C. Kim, M.-H. Kim, J.-B. Kim, S. Nahm, J. H. Paik, and J.-H. Kim. Synthesis and microwave dielectric properties of $Re_3Ga_5O_{12}$ (Re = Nd, Sm, Eu, Dy, Yb, and Y) ceramics. J. Am. Ceram. Soc. 90(2007)641–664.

[99] A. Kan, H. Ogawa, A. Yokoi, and H. Ohsato. Low temperature sintering and microstructure of $Mg_4(Nb_{2-x}V_x)O_9$. microwave dielectric ceramic by V substitution for Nb. Jpn. J. Appl. Phys. 42(2003)6154–6157.

[100] W.-Z. Lu, W. Lei, J.-H. Zhu, and F. Liang. Calcining temperature dependence of microwave dielectric properties of $(1-x)ZnAl_2O_4-xTiO_2$ (x = 0.21) ceramics. Jpn. J. Appl. Phys. 46(2007)L724–L726.

[101] H. Lin, A. Yang, L. Luo, and W. Chei. Microwave dielectric properties of low temperature cofired glass- ceramic based on B_2O_3–La_2O_3–MgO glass with $La(Mg_{0.5}Ti_{0.5})O_3$ ceramics. Mater. Lett. 62(2008)611–614.

[102] A. Kan, H. Ogawa, and H. Ohsato. Relationship between bond strength and microwave dielectric properties of corundom type $(Mg_{4-x}Co_x)Nb_2O_9$ and $Mg_4(Nb_{2-y}Ta_y)O_9$ solid solutions. J. Ceram. Soc. Jpn. Supl. 112- 1(2004)S1622–S1626.

[103] H. Mandai et al. Proc. Ceramic Science & Technlogy congress. (1990)83.

[104] A. Kan, H. Ogawa, and A. Yokoi. Sintering temperature dependence of microwave dielectric properties in $Mg_4(TaNb_{1-x}V_x)O_9$ compounds. Mater. Res. Bull. 41(2006) 1178–1184.

[105] D.-W. Kim, D.-K. Kwon, S. H. Yoon, and K. S. Hong. Microwave dielectric properties of rare earth ortho- niobates with ferroelasticity. J. Am. Ceram. Soc. 89(2006)3861–3864.

[106] J. C. Kim, M.-H. Kim, S. Nahm, J.-H. Paik, J.-H. Kim, and H.-J. Lee. Microwave dielectric properties of $Re_3Ga_5O_{12}$ (Re = Nd, Sm, Eu, Dy and Yb) ceramics and effect of TiO_2 on the microwave dielectric properties of $Sm_3Ga_5O_{12}$ ceramics. J. Eur. Ceram. Soc. 27(2007)2865–2870.

[107] H. Takahashi, K. Ayusawa, and N. Sakamoto. Microwave dielectric properties of $Ba(Mg_{1/2}W_{1/2})O_3$-$BaTiO_3$ ceramics. Jpn J. Appl. Phys. 36(1997)5597–5599.

[108] Y. Ohishi, Y. Miyauchi, H. Ohsato, and H. K. Kakimoto. Controlled temperature coefficient of resonant frequency of Al_2O_3–TiO_2 ceramics by annealing treatment. Jpn. J. Appl. Phys. 43(2004)L749–L751.

[109] M. Kono, H. Takagi, T. Tatekawa, and H. Tamura. High Q dielectric resonator material with low dielectric constant for millimeterwave application. J. Eur. Ceram. Soc. 26(2006)1909–1912.
[110] Y. Miyauchi, Y. Ohishi, S. Miyake, and H. Ohsato. Improvement of the dielectric properties of rutile-doped Al_2O_3 ceramics by annealing treatment. J. Eur. Ceram. Soc. 26(2006)2093–2096.
[111] M. W. Lufasso, E. Hopkins, S. M. Bell, and A. Llobet. Crystal chemistry and microwave dielectric properties of $Ba_3MNb_{2-x}Sb_xO_9$ (M = Mg, Ni, Zn). Chem. Mater. 17(2005) 4250–4255.
[112] K. P. Surendran, P. Mohanan, M. V. Manjusha, J. Philip and M. T. Sebastian. A low loss dielectric substrate in $ZnAl_2O_4$–TiO_2 system for microelectronic applications. J. App. Phys. 98(2005) 044101–044109.
[113] A. Kan, H. Ogawa, H. Ohsato, and S. Ishihara. Effects of variations in crystal structure on microwave dielectric properties of Y_2BaCuO_5 system. J. Eur. Ceram. Soc. 21(2001)2593–2598.
[114] H. Ogawa, A. Yokoi, R. Unemura, and A. Kan. Microwave dielectric properties of $Mg_3(VO_4)_2{-}xBa_3(VO_4)_2$ ceramics for LTCC with near zero temperature coefficient of resonant frequency. J. Eur. Ceram. Soc. 27(2007)3099–3104.
[115] A. Kan, H. Ogawa, H. Ohsato, and S. Ishihara. Microwave dielectric properties and crystal structure $Y_2(Ba_{1-x}Sr_x)(Cu_{1-y}Zn_y)O_5$ solid solutions synthesized by a solid state reaction method. Jpn. J. Appl. Phys. 39(2000)5654–5657.
[116] Y. Ohishi, Y. Miyamauchi, K. Kakimoto, and H. Ohsato. Proceeding AMF. (2004).
[117] P. V. Bijumon, S. Solomon, M. T. Sebastian, and P. Mohanan. Preparation characterization and properties of rare earth titanium tungstate dielectric ceramics. J. Mater. Sci. Mater. Electr. 14(2003)5–8.
[118] C.-S. Shen, C.-C. Chou, W.-J. Shih, K.-S. Liu, C.-S .Chen, and I.-N. Lin. Microwave dielectric properties of glass- ceramic composites for low temperature cofireable ceramics. Mater. Chem. Phys. 79(2003)124–128.
[119] A. Kan, H. Ogawa, M. Watanabe, S. Hatanaka, and H. Ohsato. Microwave dielectric properties of $Sm_2Ba(Cu_{1-x}Zn_x)O_5$ solid solutions. Jpn. J. Appl. Phys. 38(1999)5629–5632.
[120] H. Takahashi, K. Ayusawa, and N. Sakamoto. Ba-Mg-W-Ti-O ceramics with temperature stable low microwave loss. Jpn. J. Appl. Phys. 37(1998)908–911.
[121] P. V. Bijumon, P. Mohanan, and M. T. Sebastian. Microwave dielectric properties of $LaMgAl_{11}O_{19}$. Mater. Res. Bull. 37(2002)2129–2133.
[122] A. Belous, O. Ovchar, D. Durylin, M. Valant, M. M. Krzmanc, and D. Sunilov. Microwave composite dielectrics based on magnesium titanates. J. Eur. Ceram. Soc. 27(2007)2963–2966.
[123] H.-K. Shin, H. S. Shin, H. S. Jung, S.-Y. Cho, J.-R. Kim, and K. S. Hong. Role of lithium borosilicate glass in the decomposition of $MgTiO_3$ based dielectric ceramics during sintering. Mater. Res. Bull. 41(2005)1206–1214.
[124] S. J. Hwang, Y.-J. Kim, and H. S. Kim. La_2O_3–B_2O_3–TiO_2 glass/BaO–Nd_2O_3–TiO_2 ceramic for high quality factor low temperature cofired ceramic dielectric. J. Electroceram. 18(2007)121–128.
[125] S. J. Hwang, Y.-J. Kim, and H. S. Kim. Dielectric properties of $(1-x)(Mg_{0.95}Co_{0.05})TiO_3{-}xCaTiO_3$ ceramic system at microwave frequency. Mater. Res. Bull. 37(2002)2483–2490.
[126] M.-C. Wu, K.-T. Huang, and W.-F. Su. Microwave dielectric properties of doped $Zn_3Nb_2O_8$ ceramics sintered below 950°C and their compatibility with silver electrode. Mater. Chem. Phys. 98(2006)406–409.
[127] D. D. Khalyavin, J. Han, A. M. R. Senos, and P. Q. Mantas. Synthesis and dielectric properties of tungsten based complex perovskites. J. Mater. Res. 18(2003)2600–2607.
[128] X. Q. Liu, X. M. Chen, and Y. Xiao. Preparation and characterisation of $LaSrAlO_4$ microwave dielectric ceramics. Mater. Sci. Eng. B. 103(2003)276–280.
[129] K. P. Surendran, P. Mohanan, and M. T. Sebastian. Microwave dielectric properties of $Ba(Mg_{1/32-2x/3})W_{x/3}Ti_{x/3})O_3$ ceramics. Mater. Res. Bull. 41(2006)784–790.

[130] D.-W. Kim, I.-T. Kim, B. Park, K. S. Hong, and J. H. Kim. Microwave dielectric properties of $(1-x)Cu_3Nb_2O_8$-$xZn_3Nb_2O_8$ ceramics. J. Mater. Res. 16(2001)1465–1470.

[131] S.-Y. Cho, I.-T. Kim, and K. S. Hong. Microwave dielectric properties and applications of rare earth aluminates. J. Mater. Res. 14(1999)114–119.

[132] A. Yokoi, H. Ogawa, A. Kan, H. Ohsato, and Y. Higashida. Microwave dielectric properties of $Mg_4Nb_2O_9-3$ wt% LiF ceramics prepared with $CaTiO_3$ additions. J. Eur. Ceram. Soc. 25(2005)2871–2875.

[133] R. Kim, D.-W. Kim, S. H. Yoon, and K. S. Hong. Low temperature sintering and microwave dielectric properties of $Ba_3Ti_5Nb_6O_{28}$ with B_2O_3 and CuO additives. J. Electroceram. 17(2006)439–443.

[134] A. Yang, H. Lin, and W. Chen. Microwave dielectric properties of low temperature cofired glass-ceramic based on B_2O_3–La_2O_3–MgO–TiO_2 with $La(Mg_{0.5}Ti_{0.5})O_3$ ceramics. Jpn. J. Appl. Phys. 45(2006)1698–1701.

[135] H. Ogawa, A. Kan, H. Taketani, and H. Ohsato. Phase transition and microwave dielectric properties of $Nd_2Ba(Zn_{1-x}Cu_x)O_5$ (x = 0 to 0.55) by Cu substitution for Zn. J. Ceram. Soc. Jn. Suppl. 112−1(2004)S1627–S1632.

[136] A. Yokoi, H. Ogawa, and A. Kan. Microwave dielectric properties of BaO–Ta_2O_5–TiO_2 system. J. Eur. Ceram. Soc. 26(2006)2069–2074.

[137] J.-J. Wang, C.-L. Huang, P.-H. Li. Microwave dielectric properties of (1−x) $(Mg_{0.95}Zn_{0.05})TiO_3$–$Ca_{0.6}La_{0.8/3}Ti_{O3}$ ceramic system. Jpn. J. Appl. Phys. 45(2006)6352–6356.

[138] Y. Tsuji, A. Kan, H. Ogawa, and S. Ishihara. Relationship between microwave dielectric properties and chemical bonding in R_2BaMO_5 [R = rare earth, M = Cu and Zn] compounds. J. Eur. Ceram. Soc. 25(2005)2883–2887.

[139] G. Subodh and M. T. Sebastian. Glass free microwave ceramics for LTCC applications. J. Am. Ceram. Soc. 90(2007)2266–2268.

[140] N. Ichinose and H. Yamamoto. Effect of additives on microwave dielectric properties in low temperature firing $(Mg,Ca)TiO_3$ based ceramics. Ferroelectrics. 201(1997)255–262.

[141] A. Kan, H. Ogawa, A. Yokoi, and Y. Nakamura. Crystal structural refinement of corundom structured $A_4M_2O_9$ (A = Co and Mg, M = Nb and Ta) microwave dielectric ceramics by high temperature X-ray powder diffraction. J. Eur. Ceram. Soc. 27(2007)2977–2981.

[142] M.-L. Hsieh, L.-S. Chen, S.-M. Wang, C.-H. Sun, M.-H. Weng, M.-P. Houng, and S.-L. Fu. Low temperature sintering of microwave dielectrics $(Zn,Mg)TiO_3$. Jpn. J. Appl. Phys. 44(2005)5045–5048.

[143] D.-W. Kim, K. H. Ko, and K. S. Hong. Influence of copper (II) oxide additions to zinc niobate microwave ceramics on sintering temperature and dielectric properties. J. Am. Ceram. Soc. 84(2001)1286–1290.

[144] J.-H. Sohn, Y. Inaguma, S.-O. Yoon, M. Itoh, T. Nakamura, S.-J. Yoon, and H.-J. Kim. Microwave dielectric characteristics of illmenite type titanates with high Q values. Jpn. J. Appl. Phys. 33(1994)5466–5470.

[145] P. S. Anjana, M. T. Sebastian, A.-K. Axelsson, and N. Mc N. Alford. Microwave dielectric properties of CeO_2- 0.5AO-0.5TiO_2 (A = Ca, Mg, Zn, Mn, Co, Ni, W) ceramics. J. Eur. Ceram. Soc. 27(2007)3445–3452.

[146] R. C. Pullar, J. D. Breeze, and N. Mc N. Alford. Microwave dielectric properties of columbite structure niobate ceramics $M^{2+}Nb_2O_6$. Key Eng. Mater. 224–226(2002)1–4.

[147] P. V. Bijumon, P. Mohanan, and M. T. Sebastian. Microwave dielectric properties of temperature stable $Ca_5A_2Ti_{1-x}Hf_xO_{12}$ (A = Nb,Ta) ceramics. J. Electroceram. 16(2006) 239–245.

[148] K. Wakino. Recent developments of dielectric resonator materials and filters in Japan. Ferroelectrics. 91(1989)69–86.

[149] V. M. Ferreira, F. Azough, R. Freer, and J. L. Baptista. The effect of Cr and La on $MgTiO_3$ and $MgTiO_3$–$CaTiO_3$ microwave dielectric ceramics. J. Mater. Res. 12(1997)3293–3299.

[150] A. Kan, H. Ogawa, and H. Ohsato. Role of Zn substitution on the microwave dielectric properties and crystal structure of $Eu_2Ba(Cu_{1-x}Zn_x)O_5$ solid solutions. Physica B. 322(2002)403–407.
[151] C. L. Huang and S.-S. Liu. Characterization of extremely low loss dielectrics $(Mg_{0.95}Zn_{0.05})TiO_3$ at microwave frequency. Jpn. J. Appl. Phys. 46(2007)283–285.
[152] D.-W. Kwon, M. T. Lanagan, and T. R. Shrout. Microwave dielectric properties and low temperature cofiring of $BaTe_4O_9$ with aluminium metal electrode. J. Am. Ceram. Soc. 88(2005)3419–3422.
[153] W. W. Cho, K. Kakemoto, and H. Ohsato. Microwave dielectric properties and low temperature sintering of $MgTiO_3$– $SrTiO_3$ with B_2O_3 or CuO. Mater. Sci. Eng. B. 121(2005)48–53.
[154] S. H. Yoon, G.-K. Choi, D.-W. Kim, S.-Y. Cho, and K. S. Hong. Mixture behaviour and microwave dielectric properties of $(1-x)CaWO_4$-$xTiO_2$. J. Eur. Ceram. Soc. 27(2007)3087–3091.
[155] J. H. Park, Y.-J. Choi, J.-H. Park, and J.-G. Park. Low fire dielectric composition with permittivity 20-60 for LTCC applications. Mater. Chem. Phys. 88(2004) 308–312.
[156] V. M. Ferreira and J. L. Baptista. The role of niobium in magnesium titanate microwave dielectric ceramics. J. Am. Ceram. Soc. 79(1996)1697–1698.
[157] Y.-R. Wang, S.-F. Wang, and C.-K. Wen. Low fire $(Zr_{0.8}Sn_{0.2})TiO_4$ with glass additives. Mater. Sci. Eng. A426(2006)143–146
[158] P. S. Anjana and M. T. Sebastian. Synthesis, characterization and microwave dielectric properties of $ATiO_3$ (A = Co, Mn, Ni) ceramics. J. Am. Ceram. Soc. 89(2006)2114–2117.
[159] I. Levin, T. A. Vanderah, and R. Coutts. Phase equlibria and dielectric properties in perovskite like $(1-x)LaCa_{0.5}Zr_{0.5}O_3-xATiO_3$ (A = Ca,Sr) ceramics. J. Mater. Res. 17(2002)1729–1734.
[160] Y.-R. Wang, S.-F. Wang, and Y.-M. Lin. Low temperature sintering of $(Zn_{1-x}Mg_x)TiO_3$ microwave dielectrics. Ceram. Int. 31(2005)905–909.
[161] S. Kawaguchi, H. Ogawa, A. Kan, and E. Tanaka. Effect of Sr substitution for Ba on the microwave dielectric properties of $Sm_2(Ba_{2-x}Sr_x)(Cu_{0.5}Zn_{0.5})O_5$ ceramics. J. Eur. Ceram. Soc. 25(2005)2853–2858.
[162] M. T. Sebastian, N. Santha, P. V. Bijumon, A.-K. Axelsson, and N. Mc N. Alford. Microwave dielectric properties of $(1-x)CeO_2$-$xCaTiO_3$ and $(1-x)CeO_2$-xSm_2O_3 ceramics. J. Eur. Ceram. Soc. 24(2004)2583–2589.
[163] Q.-L. Zhang, and H. Yang. Low temperature sintering and microwave dielectric properties of $MgTiO_3$ ceramics. J. Mater. Sci.: Mater. Electr. 18(2007)967–971.
[164] F. Zhao, Z. Yue, Z. Gui, and L. Li. Preparation characterization and microwave dielectric properties of A_2BWO_6 (A = Sr, Ba: B = Co, Ni, Zn) double perovskites. Jpn. J. Appl. Phys. 44(2005)8066–8070.
[165] Y. Xiao, X. M. Chen, and X. Q. Liu. Microstructures and microwave dielectric characteristics of $CaRAlO_4$ (R = Nd, Sm, Y) ceramics with tetragonal K_2NiF_4 structure. J. Am. Ceram. Soc. 87(2004)2143–2146.
[166] H.-C. Chen, M.-H. Weng, C. I. Huang, and M.-P. Houng. Microstructure and microwave dielectric properties of $(1-x)MgO$–$xBaO$–TiO_2 ceramics. Jpn. J. Appl. Phys. 43(2004)2564–2567.
[167] P. S. Anjana and M. T. Sebastian. Low loss temperature stable dielectric ceramics in $ZnNb_2O_6$–$Zn_3Nb_2O_8$ system for LTCC applications. J. All. Comp. (to be published).
[168] R. C. Pullar, K. Okereme, N. Mc N. Alford. Temperature compensated niobate microwave ceramics with the columbite structure, $M^{2+}Nb_2O_6$. J. Eur. Ceram. Soc. 23(2003)2479–2483.
[169] B.-H. Jung, S.-J. Hwang, H.-S. Kim. Glass-ceramic for low temperature cofired dielectric ceramic materials based on La_2O_3–B_2O_3–TiO_2 glass with BNT ceramics. J. Eur. Ceram. Soc. 25(2005)3187–3193.
[170] K. Mori, H. Ogawa, A. Kan, and H. Ohsato. Relationship between Sr substitution for Ba and dielectric characteristics in Sm_2BaZnO_5 ceramics. J. Eur. Ceram. Soc. 24(2004) 1745–1748.

[171] J. T. Joseph, H. Padmakumar, M. R. Varma, J. K. Thomas, and S. Solomon. Effect of addition Nb_2O_5 on the dielectric characteristics of $DyTiTaO_6$ microwave ceramics. Mater. Lett. 62(2008)911–913.

[172] R. C. Pullar, J. Breeze, and N. Mc N.Alford. Characterisation and microwave dielectric properties of $M^{2+}Nb_2O_6$ ceramics. J. Am. Ceram. Soc. 88(2005)2466–2471.

[173] J. Y. Chen, I. Levin, T. A. Vanderah, K. G. Geyer, and R. S. Roth. Subsolidus phase relations and dielectric properties in the $SrO–Al_2O_3–Nb_2O_5$ systems. Int. J. Inorg. Mater. 2(2000)107–114.

[174] M. T. Sebastian, S. Solomon, R. Ratheesh, J. George, and P. Mohanan. Preparation characterization and microwave dielectric properties of $RENbTiO_6$ (RE = Ce, Pr, Nd, Sm, Eu, Gd, Tb, Dy, Y and Yb) dielectric ceramics. J. Am. Ceram. Soc. 84(2001) 1487–1489.

[175] K. P. Surendran, M. R. Varma, and M. T. Sebastian. Microwave dielectric properties of $RE_{1-x}RE'_xTiNbO_6$ [RE = Pr, Nd, Sm; Re' = Gd, Dy, Y] ceramics. J. Am. Ceram. Soc. 86(2003)1695–1699.

[176] Y.-R. Wang, S.-F. Wang, and Y.-M. Lin. Low temperature sintering of $(Zn_{1-x}Mg_x)TiO_3$ microwave dielectrics. Ceram. Int. 31(2005)905–907.

[177] A. Golovchanski, H. T. Kim, and Y. H. Kim. Zinc Titanate dielectric ceramics prepared by sol gel process. J. Kor. Phys. Ceram. Soc. 32(1998) S1167–S1169.

[178] K. P. Surendran, S. Solomon, M. R. Varma, P. Mohanan, and M. T. Sebastian. Microwave dielectric properties of $RETiTaO_6$ [RE = La, Ce, Pr, Nd, Sm, Eu, Gd, Tb, Dy, Ho, Y, Yb and In] ceramics. J. Mater. Res. 17(2002) 2561)2566.

[179] M. Udovic, M. Valant, and D. Suvorov. Dielectric characterisation of ceramics from the TiO_2-TeO_2 system. J. Eur. Ceram. Soc. 21(2001)1735–1738.

[180] M. Tagata and K. Kageyama. Microwave charateristics of $A(B^{3+}{}_{1/2}B^{5+}{}_{1/2})O_3$ ceramics (A = Ba, Sr, Ca; B^{3+} = La, Nd, Sm, Y; B^{5+} = Nb,Ta). J. Am. Ceram. Soc. 72(1989)1955–1959.

[181] X. Liu, F. Gao, L. Zhao, and M. Zhao. Effects of V_2O_5 addition on the phase structure and dielectric properties of zinc titanate ceramics. J. Electroceram. 18(2007)103–109.

[182] Y.-C. Choi, J.-H. Park, J.-H. Park, S. Nahm, and J.-G. Park. Middle and high permttitivity dielectric compositions for low temperature cofired ceramics. J. Eur. Ceram. Soc. 27(2007)2017–2024.

[183] A. Kan, H. Ogawa, K. Mori, and J. Sugishita. Effects of La substitution for Nd in $(Nd_{2-x}La_x)BaZnO_5$ solid solution on microwave dielectric properties and crystal structure. Mater. Res. Bull. 37(2002)1509–1518.

[184] C.-S. Hsu, C.-L. Huang, J.-F. Tseng, and C.-Y. Huang. Improved high Q microwave dielectric resonator using CuO doped $MgNb_2O_6$ ceramics. Mater. Res. Bull. 38(2003) 1091–1099.

[185] S. Hwang and H. Kim. Effect of frit size on microwave properties of glass ceramic in low temperature cofired ceramics. Thermochemica Acta A45(2007)119–122.

[186] H. Kagata and J. Kato. Dielectric propereties of Ca based complex perovskite at microwave frequencies. Jpn. J. Appl. Phys. 33(1994)5463–5465.

[187] J.-Y. Ha, J.-W. Choi, S.-J. Yoon, D.-J. Choi, K. H. Yoon, and H.-J. Kim. Microwave dielectric properties of Bi_2O_3 doped $Ca[Li_{1/3}Nb_{2/3}]_{1-x}Ti_x]O_{3-\delta}$ ceramics. J. Eur. Ceram. Soc. 23(2003)2413–2416.

[188] C.-L. Huang, C.-L. Pan, S.-J. Shim. Liquid phase sintering of $MgTiO_3$–$CaTiO_3$ microwave dielectric ceramics. Mater. Chem. Phys. 78(2003)111–115.

[189] L. X. Li, X. Wu, and Z. Zhu. (to be published).

[190] J. J. Bian, K. Yan, and Y. F. Dong. Microwave dielectric properties of $A_{1-3x/2}La_x(Mg_{1/2}W_{1/2})O_3$ (A = Ba, Sr, Ca) double perovskites. Mater. Sci. Eng. B. 147(2008)27–34.

[191] C.-L. Huang and M.-H. Weng. Improved high Q values of $MgTiO_3$–$CaTiO_3$ microwave dielectric ceramics at low sintering temperatures. Mater. Res. Bull. 36(2001)2741–2750.

[192] F. Zhao, Z. Yue, J. Pei, H. Zhuang, Z. Gui, and L. Li. Structure and microwave dielectric properties of hexagonal $Ba[Ti_{1-x}(Ni_{1/2}W_{1/2})_x]O_3$ ceramics. J. Am. Ceram. Soc. 90(2007)2461–2466.

[193] O. A. Shlykhtin and Y.-J. Oh. Low temperature sintering of $Zn_3Nb_2O_8$ ceramics from fine powders. J. Am. Ceram. Soc. 89(2006)3366–3372.

[194] G. S. Babu, V. Subramanian, and V. R. K. Murthy. Structure determination and microwave dielectric properties of $La(Mg,Sn)O_3$ ceramics. J. Eur. Ceram. Soc. 27(2007)2973–2976.

[195] D. W. Kim, J.-H. Kim, J.-R. Kim, and K. S. Hing. Phase constitutions and microwave dielectric properties of $Zn_3Nb_2O_8$–TiO_2. Jpn. J. Appl. Phys. 40(2001)5994–5998.

[196] X. Q. Liu and X. M. Chen. Microwave dielectric characteristics of $SrLaGaO_4$ and $SrNdGaO_4$ ceramics. J. Eur. Ceram. Soc. 26(2006)1969–1971.

[197] K. Mori, H. Ogawa, A. Kan, H. Ohsato, and S. Ishihara. Crystal structural characterization of Nd_2BaZnO_5 type microwave dielectric ceramics with rare earth substitution for Nd. Mater. Chem. Phys. 79(2002)273–275.

[198] N. Santha, M. T. Sebastian, P. Mohanan, N. Mc N. Alford, K. Sarma, and R. C. Pullar. Effect of doping on the dielectric properties of CeO_2 in the microwave and sub-millimeter frequency range. J. Am. Ceram. Soc. 87(2004)1233–1237.

[199] H.-J. Lee, I.-T. Kim, and K. S. Hong. Dielectric properties of AB_2O_6 comopunds at microwave frequencies (A = Ca, Mg, Mn, Co, Ni, Zn and B = Nb,Ta). Jpn. J. Appl. Phys. 36(1997)L1318–L1320.

[200] A. Belous, O. Ovchar, B. Jancar, and J. Bezjak. The effect of non-stoichiometry on the microstructure and microwave dielectric properties of the columbite $A^{2+}Nb_2O_6$. J. Eur. Ceram. Soc. 27(2007)2933–2936.

[201] Q.-L. Zhang, H. Yang, and J.-X. Tong. Low temperature sintering and microwave dielectric properties of $MgTiO_3$ ceramics with Bi_2O_3–V_2O_5. Mater. Lett. 60(2005)1188–1191.

[202] C.-S. Hsu, and C.-L. Huang. Effect of CuO additive on sintering and microwave dielectric behavior of $LaAlO_3$ ceramics. Mater. Res. Bull. 36(2001)1939–1947.

[203] W. W. Cho, K. Kakemoto, and H. Ohsato. High Q microwave dielectrics $SrTiO_3$ doped $MgTiO_3$ material with near zero temperature coefficient of resonant frequency. Jpn. J. Appl. Phys. 43(2004)6221–6224.

[204] J. J. Bian, K. Yan, and J. Li. Structure and microwave dielectric properties of $La_{(2-x/3}$ $Na_x(Mg_{1/2}W_{1/2})O_3$. J. Eur. Ceram. Soc. 26(2006)1957–1960.

[205] C. L. Huang, J.-J. Wang, and Y.-P. Chang. Dielectric properties of low loss (1−x) $(Mg_{0.95}Zn_{0.05})TiO_3$-$xSrTiO_3$ ceramic system at microwave frequency. J. Am. Ceram. Soc. 90(2007)858–862.

[206] Q. Zeng, W. Li, J.-L. Shi, J.-K. Guo, M.-W. Zuo, and W.-J. Wu. A new microwave dielectric ceramic for LTCC applications. J. Am. Ceram. Soc. 89(2006)1733–1735.

[207] P. V. Bijumon and M. T. Sebastian. Temperature stable microwave dielectric ceramics in the $Ca_5A_2Ti_{1-x}Zr_xO_{12}$ (A = Nb, Ta) system. J. Mater. Res. 19(2004)2922–2928.

[208] H. T. Kim, S. Nahm, J. D. Byun, and Y. Kim. Low fired (Zn,Mg,)TiO_3 microwave dielectrics. J. Am. Ceram. Soc. 82(1999)3476–3480.

[209] J. X. Tong, Q. L. Zhang, H. Yang, and J. L. Zou. Low temperature firing and microwave dielectric properties of $Ca[(Li_{0.33}Nb_{0.67})_{0.9}Ti_{0.1}]O_{3-\delta}$ ceramics with LiF addition. Mater. Lett. 59(2005)3252–3255.

[210] C.-L. Huang and C.-L. Yu. Low temperature sintering and microwave dielectric properties of (1−x) $MgTiO_3$–$CaTiO_3$ ceramics bismuth addition. Jpn. J. Appl. Phys. 41(2002)707–711.

[211] T. Oishi, A. Kan, H. Ohsato, and H. Ogawa. Crystal structure microwave dielectric property relations in $Sm(Nb_{1-x}Ta_x)(Ti_{1-y}Zr_y)O_6$ ceramics. J. Eur. Ceram. Soc. 26(2006) 075–2079.

[212] J. Wang, Z. Yue, Z. Gui, and L. Li. Low temperature sintered $ZnNb_2O_6$ microwave dielectric ceramics doped with ZnO–V_2O_5 additives. J. Mater. Sci. Lett. 40(2005) 6581–6583.

[213] C.-L. Huang and K.-H. Chiang. Improved high Q microwave dielectric material using B_2O_3 doped $MgNb_2O_6$ ceramics. Mater. Sci. Eng. A. 474(2008)243–246.

[214] Q. Zeng, W. Li, J.-L. Shi, and J.-K. Guo. Microwave dielectric properties of $5Li_2O$-$0.583Nb_2O_5$-$3.248TiO_2$ ceramics with V_2O_5. J. Am. Ceram. Soc. 89(2006)3305–3307.

[215] C.-L. Huang and Y.-C. Chen. Influence of V_2O_5 additions to $NdAlO_3$ ceramics on sintering temperature and microwave dielectric properties. J. Eur. Ceram. Soc. 23(2003)167–173.

[216] Q. Zeng, W. Li, J. L. Shi, J.-K. Guo, M.-W. Zuo, and W.-J. Wu. A new microwave dielectric ceramics for LTCC applications. J. Am. Ceram. Soc. 89(2006)1733–1637.

[217] H. Tamura, T. Konoike, Y. Sakabe, and K. Wakino. Improved high dielectric resonator with complex perovskite structure. J. Am. Ceram. Soc. 67(1984)C59–C61.

[218] S.-Y. Cho, C.-H. Kim, D.-W. Kim, and K. S. Hongs. Dielectric properties of $Ln(Mg_{1/2}Ti_{1/2})O_3$ as substrates for high Tc superconducting thin films. J. Mater. Res. 14(1999)2484–2487.

[219] M.-H. Kim, S. Nahm, C.-K. Choi, H.-J. Lee, and H.-M. Park. Dielectric properties of $(1-x)$ $NdGaO_3$-$xCaTiO_3$ solid solution at microwave frequencies. Jpn. J. Appl. Phys. 41(2002)717–721.

[220] H. T. Kim and M. T. Lanagan. Structure and microwave dielectric properties of $(Zn_{1-x}Co_x)TiO_3$ ceramics. J. Am. Ceram. Soc. 86(2003)1874–1878.

[221] S. Nomura. Ceramics for microwave dielectric resonator. Ferroelectrics. 49(1983)61–70.

[222] Y.-C. Zhang, Z. X. Yue, Z. L. Gui, and L. T. Li. Microwave dielectric properties of $(Zn_{1-x}Mg_x)Nb_2O_6$ ceramics. Mater. Lett. 57(2003)4531–4534.

[223] J. Wang, Z. Yue, Z. Gui, and L. Li. Low temperature sintered $Zn(Nb_{1-x}V_{x/2})_2O_{6-2.5x}$ microwave dielectric ceramics with high Q value for LTCC applications. J. All. Comp. 392(2005)263–267.

[224] S.-H. Wee, D.-W. Kim, S.-I. Yoo, and K. S. Hong. Low temperature sintering of V_2O_5 added and substituted $ZnNb_2O_6$ microwave ceramics. Jpn. J. Appl. Phys. 43(2004)3511–3515.

[225] J. Y. Cho, K. H. Yoon, and E. S. Kim. Effect of stress on microwave dielectric properties of $Mg_{0.93}Ca_{0.07}TiO_3$–$(Ca_{0.0.3}Li_{0.14}Sm_{0.42})TiO_3$ ceramics. Mater. Chem. Phys. 79(2003)286–288.

[226] Y.-L. Chai, Y.-S. Chang, Y.-J. Hsiao, and Y.-C. Lian. Effects of borosilicate glass addition on the structure and dielectric properties of $ZnTiO_3$ ceramics. Mater. Res. Bull. 43(2008)257–263.

[227] C.-L. Huang and Y.-C. Chen. Liquid phase sintering and microwave dielectric properties of $NdAlO_3$ ceramics. Jpn. J. Appl. Phys. 41(2002)1459–1464.

[228] H. Su and S. Wu. Studies on the $(Mg,Zn)TiO_3$–$CaTiO_3$ microwave dielectric ceramics. Mater. Lett. 59(2005)2337–2341.

[229] D. D. Khalyavin, A. M. R. Senos, P. Q. Mantas, D. N. Argyriou, I. T. Gomes, L. G. Viera, and J. L. Ribeiro. Structure and dielectric characterization of a new A site deficient $La_{5/3}MgTaO_6$ perovskite. J. Sol. St. Chem. 180(2007)41–48.

[230] H.-J. Lee, K. S. Hong, S.-J. Kim, and I.-T. Kim. Dielectric properties of MNb_2O_6 compounds (where M = Ca, Mn, Co, Ni or Zn). Mater. Res. Bull. 32(1997)847–855.

[231] H. Ogawa, A. Kan, S. Yokola, H. Ohsato, and S. Ishihara. Microwave dielectric properties and refinements in M (M = Sr,Ca) doped $Nd_2Ba_{1-x}M_xZnO_5$ solid solutions. J. Eur. Ceram. Soc. 21(2001)1731–1734.

[232] X. Q. Liu and X. M. Chen. Dielectrics and mechanical characteristics of lanthanum aluminate ceramics with strontium niobate addition. J. Eur. Ceram. Soc. 24(2004)1999–2004.

[233] S. H. Shim, B. G. Choi, S. H. Kim, and K. B. Shim. Microwave characteritics of $MgTiO_3$–$CaTiO_3$ dielectric ceramics fabricated using spark plasma sintering. Jpn. J. Appl. Phys. 44(2005)5073–5075.

[234] D. L. Cairns, I. M. Reaney, H. Zheng, D. Iddles, and T. Price. Synthesis and characterisation of $La(Co_{1/2}Ti_{1/2})O_3$. J. Eur. Ceram. Soc. 25(2005)433–439.

[235] C.-L. Huang and K.-H. Chiang. Structures and dielectric properties of a new dielectric material system $xMgTiO_3$—$(1-x)MgTa_2O_6$ at microwave frequency. J. All. Comp. 431(2007)326–330.

[236] L. A. Khalam and M. T. Sebastian. Low loss dielectrics in the $Ca(B'_{1/2}Nb_{1/2})O_3$ [B' = lanthanides, Y] system. J. Am. Ceram. Soc. 90(2007)1467–1474.

[237] L. A. Khalam, S. Thomas, and M. T. Sebastian. Temperature stable and low loss microwave dielectrics in the $Ca(B'_{1/2}Ta_{1/2})O_3$ [B' = lanthanides, Y, In] system. J. Am. Ceram. Soc. 90(2007)2476–2486.

[238] P. S. Anjana et al. (to be published).
[239] J. W. Choi, C.-Y. Kang, S.-J. Yoon, H.-J. Kim, H.-J. Jung, and K. H. Yoon. Microwave dielectric properties of $Ca[Li_{1/3}Nb_{2/3}]_{1-x}M_x]O_3$. [M = Sn, Ti]. J. Mater. Res. 14(1999) 3567–3570.
[240] H. J. Cha, D. H. Kang, and Y. S. Cho. Optimised microwave dielectric properties of Co- and Ca-substituted $Mg_{0.06}Zn_{0.04}TiO_3$. Mater. Res. Bull. 42(2007)265–273.
[241] C.-L. Huang, R. J. Lin, and J. H. Wang. Effect of B_2O_3 additives on sintering and microwave dielectric behaviors of CuO doped $ZnNb_2O_6$ ceramics. Jpn. J. Appl. Phys. 41(2002)758–762.
[242] P. Liu, H. Ogawa, E. S. Kim, and A. Kan. Low temperature sintering and microwave dielectric properties of $Ca(Li_{1/3}Ta_{2/3})O_{3-\delta}$–$CaTiO_3$ ceramics. J. Eur. Ceram. Soc. 23(2003)2417–2421.
[243] R. Rajamma and M. T. Sebastian. Microwave dielectric properties of $La_6Mg_4A_2W_2O_{24}$ [A = Ta and Nb] ceramics. J. Am. Ceram. Soc. 90(2007)2472–2475.
[244] J. J. Bian, H. B. Gao, and X. W. Wang. Microwave dielectric properties of $(Pb_{1-3x/2}La_x)(Mg_{1/2}W_{1/2})O_3$. Mater. Res. Bull. 39(2004)2127–2135.
[245] J. J. Bian, K. Yan, and H. B. Gao. Effect of TiO_2 addition on the microwave dielectric properties of $La_{2/3}(Mg_{1/2}W_{1/2})O_3$. Mater. Chem. Phys. 96(2006)349–352.
[246] Y.-C. Lee, W.-H. Lee, and F.-T. Shiao. Microwave dielectric properties of $Zn_{0.95}Mg_{0.05}TiO_3$ + 0.25 TiO_2 ceramics with 3 wt% ZnO–B_2O_3 addition. Jpn. J. Appl. Phys. 43(2004)7596–7599.
[247] A. G. Belous, O. V. Ovchar, A. V. Kramarenko, B. Jancar, J. Bezjak, and D. Surov. Synthesis and microwave dielectric properties of $Zn_{1+x}Nb_2O_{6+x}$. Inorg. Mater. 43(2007)277–280.
[248] S. Kawaguchi, H. Ogawa, A. Kan, and S. Ishihara. Microwave dielectric properties of $Ba_8Ta_6(Ni_{1-x}M_x)O_{24}$ (M = Zn and Mg) ceramics. J. Eur. Ceram. Soc. 26(2006)2045–2049.
[249] T. Takada, S. F. Wang, S. Yoshikawa, S.-J. Jang, and R. E. Newnham. Effect of glass additions on BaO–TiO_2–WO_3 microwave ceramics. J. Am. Ceram. Soc. 77(1994)1909–1916.
[250] S.-O. Yoon, J. H. Yoon, S. H. Shim, and Y. K. Pyon. Microwave dielectric properties of $LiNb_3O_8$ ceramics with TiO_2 additions. J. Eur. Ceram. Soc. 26(2006)2031–2034.
[251] J.-W. Choi, J. Y. Ha, S.-J. Yoon, H.-J. Kim, and K. H. Yoon. Microwave dielectric characteristics of $0.75(Al_{1/2}Ta_{1/2})O_2$–$0.25(Ti_{1-x}Sn_x)O_2$ ceramics. J. Eur Ceram. Soc. 23(2003)2507–2510.
[252] H. Matsumoto, H. Tamura, and K. Wakino. $Ba(MgTa)O_3$–$BaSnO_3$ high Q dielectric resonator. Jpn. J. Appl. Phys. 30(1991)2347–2349.
[253] H. Tamura, H. Matsumoto, and K. Wakino. Proc. 7th Meeting on Ferroelectric Materials and their Applications, Kyoto. Jpn. J. Appl. Phys. Suppl. 28-2(1989)21.
[254] M. Furuya and A. Ochi. Microwave dielectric properties of $Ba(Mg_{1/3}Ta_{2/3})O_3$–$A(Mg_{1/2}W_{1/2})O_3$(A = Ba,Sr and Ca) ceramics. Jpn. J. Appl. Phys. 33(1994)5482–5487.
[255] C.-L. Huang, C.-F. Tasi, Y.-B. Chen, and Y.-C. Cheng. New dielectric material system of $(Mg_{0.95}Zn_{0.05})TiO_3$–$Ca_{0.61}Nd_{0.26}TiO_3$. J. All Comp. 453(2008)337–346.
[256] Y.-C. Lee and W.-H. Lee. Effect of glass addition on the microwave dielectric properties of $Zn_{0.95}Mg_{0.5}TiO_3$ + $0.25TiO_2$ ceramics. Jpn. J. Appl. Phys. 44(2005)1838–1843.
[257] K. P. Surendran, M. T. Sebastian, P. Mohanan, R. L. Moreira, and A. Dias. Effect of stoichiometry on the microwave dielectric properties of $Ba(Mg_{1/3}Ta_{2/3})O_3$ dielectric ceramics. Chem. Mater. 17(2005)142–151.
[258] K.-J. Youn, I.-T. Kim, K. S. Hong, and H. Kim. Crystal structure and microwave dielectric properties of $La(Mg_{2/3}Ta_{1/3})O_3$ ceramics. Ferroelectrics. 193(1997)167–174.
[259] S. Jayanthi and T. R. N. Kutty. Microwave dielectric properties of $Mg_4Al_2Ti_9O_{25}$ ceramics. Mater. Lett. (2007).
[260] T. Shimada. Dielectric loss and damping constants of lattice vibrations in $Ba(Mg_{1/3}Ta_{2/3})O_3$ ceramics. J. Eur. Ceram. Soc. 23(2003)2647–2652.
[261] H. Mandai and S. Okube. Ceram. Trans. 32(1992)91.

[262] C.-L. Huang and J.-F. Tseng. Dielectric Characteristics of $La(Co_{1/2}Ti_{1/2})O_3$ ceramics at microwave frequencies. Mater. Lett. 58(2004)3732–3736.
[263] K. Kageyama. Microwave dielectric properties of $CaO–Ga_2O_3–Ta_2O_5$ ceramics. Ferroelectrics. 109(1990)173–178.
[264] P. K. Davies, A. Borosevich, and M. Thirumal. Communicating with wireless perovskites: cation order and zinc volatilization. J. Eur. Ceram. Soc. 23(2003)2461–2466.
[265] R. Ratheesh, M. T. Sebastian, P. Mohanan, M. E. Tobar, J. Hartett, and D. G. Blair. Whispering Gallery mode microwave characterization of $Ba(Mg_{1/3}Ta_{2/3})O_3$ dielectric resonators. J. Phys. D. Appl. Phys. 32(1999)2821–2826.
[266] M.-H. Hsieh, L.-S. Chen, S.-M. Wang, C.-H. Sunand, and M. H. Weng. Low temperature sintering of microwave dielectrics $(Zn,Mg)TiO_3$. Jpn. J. Appl. Phys. 44(2005)5045–5048.
[267] A. Boriosevic and P. Davies. 1:2 cation order in $A(Li_{1/3}(Nb,Ta_{2/3})O_3$ microwave ceramics. Appl. Phys. Lett. 84(2004)1347–1349.
[268] T. Shimda, and S. Ohtsuki. Dielectric properties and farinfrared reflectivity of lanthanum aluminate- strontium titanate system., Interntl. Conf. Microwave Materials and Their Applications, Inuyama, Japan (2004).
[269] D.-W. Kim, H. J. Youn, K. S. Hong, and C. K. Kim. Microwave dielectric properties of$(1-x)Ba_5Nb_4O_{15}$-$xBaNb_2O_6$ mixtures. Jpn. J. Appl. Phys. 41(2002)3812–3816.
[270] H.-T. Song, C.-S. Hsu, M.-T. Kuo, and C.-L. Huang. Microwave characteristics of $Sm(Co_{1/2}Ti_{1/2})O_3$ dielectric resonators. Mater. Lett. 58(2004)2829–2833.
[271] C.-F. Lin, H.-H. Lu, T.-I. Chang, and J.-L. Huang. Microstructural characteristics and microwave dielectric properties of $Ba[(Mg_{1/3}(Nb_{x/4}Ta_{(4-x)/4})_{2/3}O_3$ ceramics. J. All. Comp. 407(2006)318–325.
[272] I. Jawahar, M. T. Sebastian, and P. Mohanan. Microwave dielectric properties of $Ba_{5-x}Sr_xTa_4O_{15}$, $Ba_5Nb_xTa_{4-x}O_{15}$ and $Sr_5Nb_xTa_{4-x}O_{15}$ ceramics. Mater. Sci. Eng. B. 106(2004)207–212.
[273] J. B. Kim, K. H. Yoon, and Y. S. Cho. Compositional dependence of microwave dielectrics in $(1-x)(Na_{1/2}Nd_{1/2})TiO_3$-$xNd(Mg_{1/2}Ti_{1/2})O_3$ ceramics. J. Am. Ceram. Soc. 86(2003)1159–1161.
[274] L. A. Khalam and M. T. Sebastian. Effect of cation substitution and non-stoichiometry on the microwave dielectric properties of $Sr(B'_{0.5}Ta_{0.5})O_3$ [B' = lanthanide] perovskites. J. Am. Ceram. Soc. 89(2006)3689–3695.
[275] L. A. Khalam and M. T. Sebastian. Microwave dielectric properties of $Sr(B'_{1/2}Nb_{1/2})O_3$ [B' = La, Pr, Nd, Sm, Eu, Gd, Tb, Dy, Ho, Y, Er, Yb and In] ceramics. Int. J. Appl. Ceram. Tech. 3(2006)364–374.
[276] T. Takada, S. F. Wang, S. Yoshikawa, S.-J. Jang, and R. E. Newnham. Proc. 9th International Symp. on Applications of Ferroelectrics, IEEE (1994)626.
[277] A. Feteira, L. J. Gillie, R. Elsebrock, and D. C. Sinclair. Crystal structure and dielectric properties of $LaYbO_3$. J. Am. Ceram. Soc. 90(2007)1475–1482.
[278] N. Santha, I. N. Jawahar, P. Mohanan, and M. T. Sebastian. Microwave dielectric properties of $(1-x)CaTiO_3$-$xSm(Mg_{1/2}Ti_{1/2})O_3$ ceramics. Mater. Lett. 54(2002)318–322.
[279] M.-H. Kim, S. Nahm, W. S. Sung, M.-J. Yoo, N.-K. Kang, H.-T. Kim, and H.-J. Lee. Effect of B_2O_3 and CuO on the sintering temperature and microwave dielectric properties of $Ba(Zn_{1/3}Ta_{2/3})O_3$ ceramics. Jpn. J. Appl. Phys. 44(2005)3091–3094.
[280] P. S. Anjana, T. Joseph, and M. T. Sebastian. Microwave dielectric properties of $Ca_{5-x}Mg_xNb_4TiO_{17}$ (x = 0–3) ceramics. Proc. IEEE Applied Electromagnetic Conference, (Kolkatta) (2007) to be published.
[281] C.-M. Cheng, Y.-C. Chen, C.-F. Yang, and C.-C. Chen. Sintering and compositional effects on the microwave dielectric characteristics of $Mg(Ta_{1-x}Nb_x)_2O_6$ ceramics with $0.2 \leq x \leq 0.35$. J. Electroceram. 18(2007)155–160.
[282] K. P. Surendran and M. T. Sebastian. Low loss dielectrics in $Ba[(Mg_{1/3}Ta_{2/3})_{1-x}Ti_x]O_3$ and $Ba[(Mg_{1-x}Zn_x)_{1/3}Ta_{2/3}]O_3$ systems. J. Mater. Res. 20(2005)2919–2928.
[283] J.-M. Yoon, J.-A. Lee, J.-H. Lee, J.-J. Kim, and S.-H. Cho. Sintering behaviour and microwave dielectric characteristics of $BaO–Sm_2O_3–4TiO_2$ ceramics with B_2O_3 and BaB_2O_4 additives. J. Eur Ceram. Soc. 26(2006)2129–2133.

[284] N. Santha and M. T. Sebastian. Microwave dielectric properties of $A_6B_5O_{18}$ type perovskites. J. Am. Ceram. Soc. 90(2007)496–501.
[285] J. H. Park, C. Choi, S. Nahm, J. D. Byun, and H. J. Lee. Structural and microwave dielectric properties of $La(Mg_{2/3}Nb_{1/3})O_3$ ceramics. J. Mater. Sci. Lett. 18(1999)889–993.
[286] J. Takahashi, K. Kageyama, T. Fuji, Y. Yamada, and K. Kodaira. Formation and microwave dielectric properties of $Sr(Ga_{0.5}Ta_{0.5})O_3$ based complex perovskites. J. Mater. Sci. Mater. Electr. 8(1997)79–84.
[287] M. P. Seabra and V. M. Ferreira. Synthesis and characterisation of microwave $La(Mg_{1/2}Ti_{1/2}) O_3$ ceramics. Key Eng. Mater. 206–213(2001)1501–1504.
[288] T. A. Vanderah, W. Febo, J. Y. Chan, R. S. Roth, J. M. Loezos, L. D. Totter, R. G. Geyer, and D. B. Minor. Phase equlibria and dielectric behaviour in the $CaO-Al_2O_3-Nb_2O_5$ system. J. Sol. St. Chem. 155(2000)78–85.
[289] S. Kawashima, N. Nishida, I. Ueda, and H. Ouchi. $Ba(Zn_{1/3}Ta_{2/3})O_3$ ceramics with low dielectric loss at microwave frequencies. J. Am. Ceram. Soc. 66(1983)421–423.
[290] E. Guan, W. Chen, and L. Luo. Low firing and microwave dielectric properties of $BaTi_4O_9$ with B_2O_3-ZnO- La_2O_3 glass addition. Ceram. Int. 33 (2007) 1145–1148.
[291] C.-T. Tseng, C.-L. Huang, and C. H. Hsu. Low dielectric characteristics of $Nd(Co_{1/2}Ti_{1/2}) O_3$ ceramics at microwave freqncies. J. Am. Ceram. Soc. 90(2007)1619–1622.
[292] V. Subramanian, V. R. K. Murthy, and B. Viswanathnan. Microwave dielectric properties of certain simple alkaline earth perovskite compounds as a function of tolerance factor. Jpn. J. Appl. Phys. 36(1997)194–197.
[293] S. Nomura, and K. Kaneta. $Ba(Mn_{1/3}Ta_{2/3})O_3$ ceramic with ultra low loss at microwave frequency. Jpn. J. Appl. Phys. 33(1984)507–508.
[294] Q. L Zhang, H. Yang, J. L. Zou, and H. P Wang. Sintering and microwave dielectric properties of LTCC zinc titanate multilayers. Mater. Lett. 59(2005)880–884.
[295] H. Sreemoolanathan, M. T. Sebastian, R. Blachnik, M. Woechlecke, B. Schneider, M. Newmann, and P. Mohanan. Microwave dielectric properties of $BaO-2CeO_2-nTiO_2$ ceramics. J. Sol. St. Chem. 177(2004)3995–4000.
[296] H. Zheng, G. D. C. Csette De Gyorgyfalva, and I. M. Reaney. Microstructure and microwave properties of $CaTiO_3-LaGaO_3$ solid solutions. J. Mater. Sci. 40(2005) 5207–5214.
[297] M. R. Varma, R. Reghunathan, and M. T. Sebastian. Effect of dopants on the microwave dielectric properties of $Ba(Zn_{1/3}Ta_{2/3})O_3$ ceramics. Jpn. J. Appl. Phys. 44(2005)298–303.
[298] M. K. Lee, J.-H Chung, K. W. Ryu, and Y. H. Lee. Microwave dielectric properties of $Ba(Mg,Ta)O_3$–$Ba(Co,Nb)O_3$ ceramics. Proceedings of International Symposium. Insulating Materials, Toyahashi, Japan, (1998) pp. 139–142.
[299] E. Guan, W. Chen, and L. Luo. Low firing and microwave dielectric properties of $BaTi_4O_9$ with B_2O_3–ZnO–La_2O_3 glass addition. Ceram. Int. 33(2006)1145–1148.
[300] C.-F. Tseng, and C.-L. Huang. Improved microwave dielectric properties of B_2O_3 doped $Nd(Co_{1/2}Ti_{1/2})O_3$ ceramics with near zero temperature coefficient of resonant frequency. Mater. Res. Bull. 42(2006)9–16.
[301] Y.-C Lee, W.-H Leeb, and F.-S Shiuc. Microwave dielectric properties and microstructure of $Ba_2Ti_9O_{20}$ based ceramics with $3ZnO–B_2O_3$ addition. J. Eur. Ceram. Soc. 25(2005) 3459–3468.
[302] A. Salak, D. D. Khalyavavin, P. Q. Mantas, A. M. Senos, and V. M. Ferreira. Structure-depedent microwave dielectric properties of $(1-x)La(Mg_{1/2}Ti_{1/2})O_3$-$xLa_{2/3}TiO_3$ ceramics. J. Appl. Phys. 98(2005)034101.
[303] J. B. Lim, J.-O. Son, S. Nahm, W. S. Lee, M. J. Yoo, N. K. Kang, H. J. Lee, and Y. S. Kim. Low temperature sintering of B_2O_3 added $Ba(Mg_{1/3}Nb_{2/3})O_3$ ceramics. Jpn. J. Appl. Phys. 43(2004)5388–5391.
[304] H. T. Kim, S. H. Kim, S. Nahm, J. D. Byun, and Y. Kim. Low temperature sintering and microwave dielectric properties of zinc metatitanate-rutile mixtures using boron. J. Am. Ceram. Soc. 82(1999)3043–3048.

[305] M. P. Seabra, M. Avdeev, V. M. Ferreira, R. C. Pullar and N. Mc N Alford. Structure and microwave dielectric properties of $La(Mg_{0.5}Ti_{0.5})O_5$–$CaTiO_3$ system. J. Eur. Ceram. Soc. 23(2003)2403–2408.

[306] M. P. Seabra, M. Avdeev, V. M. Ferreira, R. C. Pullar, N. Mc N Alford, and I. M. Reaney. Structure property relations in $xBaTiO_3$–$(1-x)La(Mg_{1/2}Ti_{1/2})O_3$ solid solutions. J. Am. Ceram. Soc. 87(2004)584–590.

[307] C.-F Yang, C.-C Chan, C.-M Cheng, and Y.-C Chen. Sintering and compositional effects on the microwave dielectric characteristics of $Mg(Ta_{1-x}Nb_x)_2O_6$ ceramics with $0.25 \leq x \leq 0.35$. J. Eur. Ceram. Soc. 25(2005)2849–2852.

[308] Y.-B. Chen, C.-L. Huang, and S.-H. Lin. Influence of ZnO additions to $0.8(Mg_{0.95}Co_{0.05})TiO_3$–$0.2Ca_{0.6}La_{0.8/3}TiO_3$ ceramics on sintering behaviour and microwave dielectric properties. Mater. Lett. 60(2006)3591–3595.

[309] T. Shimada. Far infrared reflection of $(Sr_{1-x}Ca_x)(Ga_{1/2}Ta_{1/2})O_3$ ceramics. J. Eur. Ceram. Soc. 21(2001)2617–2620.

[310] M.-Z. Jhou, and J.-H. Jean. Low fire processing of $BaTi_4O_9$ dielectric with BaO–ZnO–B_2O_3 glasses. J. Amer. Ceram. Soc. 89(2006)786–791.

[311] R. L. Moreira, L. A. Khalam, M. T. Sebastian and A. Dias. Raman spectroscopy investigations on the crystal structure and phonon modes of $Ba(RE_{1/2}Ta_{1/2})O_3$ microwave ceramics. J. Eur. Ceram. Soc. 27(2007)2803–2810.

[312] C.-L. Huang, R.-Y. Yang, and M.-H. Weng. Dielectric properties of $CaTiO_3$–$Ca(Mg_{1/3}Nb_{2/3})O_3$ ceramic system at microwave frequency. Jpn. J. Appl. Phys. 39 (2000) 6608–6611.

[313] G. Rong, N. Newman, B. Shaw, and D. Cronin. Role of Ni and Zr doping on the electrical optical magnetic and structural properties of barium zinc tanatalite ceramics. J. Mater. Res. 14(1999)4011–4019.

[314] Y. C. Zhang, Z. X. Yue, X. Qi, B. Li, Z. Gui, and L. T. Li. Microwave dielectric properties of $Zn(Nb_{1-x}Ta_x)_2O_6$ ceramics. Mater. Lett. 58(2004)1392–1395.

[315] C.-L. Huang, K.-H. Chiang, and C.-Y. Huang. Microwave dielectric properties and microstructure of $MgTa_2O_6$ ceramics with CuO addition. Mater. Chem. Phys. 90(2005) 373–377.

[316] J.-W. Choi, J.-Y. Ha, S.-J. Yoon, H.-J. Kim, and K. H. Yoon. Microwave dielectric characteristics of $0.75(Al_{1/2}Ta_{1/2})O_2$–$0.25(Ti_{1-x}Sn_x)O_2$ ceramics. J. Eur. Ceram. Soc. 23(2003)2507–2510.

[317] T. A. Vanderah, V. L Miller, I. Levin, S. M. Bell, and T Negas. Phase relations, crystal chemistry and dielectric properties in sections of the La_2O_3–CaO–MgO–TiO_2 system. J. Sol. St. Chem. 177(2004)2023–2038.

[318] P. Mallinson, J. B. Clridge, D. Iddles, T. Price, R. M. Ibberson, M. Allix, and M. J. Rossiensky. New 10- layer hexagonal perovskites: Relationship between cation valency and ordering and microwave dielectric loss. Chem. Mater. 18(2006)6227–6238.

[319] H. Wu and P. Davies. Ordered perovskites in the $A^{2+}(Li_{1/4}Nb_{3/4})O_3$–$A^{2+}(Li_{2/5}W_{3/5})O_3$ (A^{2+} = Sr, Ca) systems. J. Sol. St. Chem. 177(2004)4305–4315.

[320] X. Kuang, M. M. B. Allix, J. B. Claridge, H. J. Niu, M. J. Rossiensky, R. M. Ibberson, and D. Iddles. Crystal structure, microwave dielectric properties and AC conductivity of B-cation deficient hexagonal perovskites $La_5M_xTi_{4-x}O_{15}$ ($x = 0.5, 1$: M = Zn, Mg, Ga, Al). J. Mater. Chem. 16(2006)1038–1045.

[321] S. Kucheiko, H.-J. Kim, D.-H. Yeo, and H.-J. Jung. Microwave dielectric properties of $LaZn_{0.5}Ti_{05}O_3$ ceramics prepared by sol-gel process. Jpn. J. Appl. Phys. 35(1996)668–672.

[322] Y. Itou, Y. Takai, K. Tsumoto, M. Takemoto, and H. Ikawa. 11 th US-Japan seminar on dielectric and piezoelectric materials, Sapporo (2003).

[323] B. J. Jiang, G. H. Bo, and W. Xiaowu. Microwave dielectric properties of $(Pb_{1-3x/2}La_x)(Mg_{1/2}W_{1/2})O_3$. Mater. Res. Bull. 39(2004)2127–2135.

[324] Y.-H. Jeong, M.-H. Kim, S. Nahm, W.-S. Lee, M.-J. Yoo, N.-K. Kang, and H.-J. Lee. Effect of Ta_2O_5 on microstructure and microwave dielectric properties of $Ba(Zn_{1/3}Ta_{2/3})O_3$ ceramics. Jpn. J. Appl. Phys. 44(2005)956–960.

[325] I. M. Reaney, P. L. Wise, I. Qazi, C. A. Miller, T. J. Price, D. S. Cannell, D. M. Idles, M. J. Rosienky, S. M. Moussa, M. Bieringer, L. D. Noailles, and R. M. Ibberson. Ordering and quality factor in $0.95Ba(Zn_{1/3}Ta_{2/3})O_3$– $0.05Sr(Ga_{1/2}Ta_{1/2})O_3$ production resonators. J. Eur. Ceram. Soc. 23(2003)3021–3034.

[326] K. Kageyama. Crystal structure and microwave dielectric properties of $Ba(Zn_{1/3}Ta_{2/3})O_3$–$(Sr,Ba)(Ga_{1/2}Ta_{1/2})O_3$ ceramics. J. Am. Ceram. Soc. 75(1992)1767–1771.

[327] P. K. Davies, J. Tong and T. Negas. Effect of ordering induced domain boundaries of low loss $Ba(Zn_{1/3}Ta_{2/3})O_3$–$BaZrO_3$ perovskite microwave dielectrics. J. Am. Ceram. Soc. 80(1997)1727–1740.

[328] D.-W. Kwon, M. T. Lanagan, T. R. Shrout. Synthesis of $BaTiTe_3O_9$ Ceramics for LTCC Application and Its Dielectric Properties J. Ceram. Soc. Jpn. 113(2005)216–219.

[329] M. P. Sebra, A. N. Salak, V. M. Ferreira, J. L. Ribeiro, and L. G. Viera. Dielectric properties of $(1-x)La(Mg_{1/2}Ti_{1/2})O_3$-$xSrTiO_3$ ceramics. J. Eur. Ceram. Soc. 24(2004)2995–3002.

[330] C.-F. Tseng and C.-L. Huang. Microwave dielectric properties of $(1-x)Nd(Co_{1/2}Ti_{1/2})$ O_3-$xSrTiO_3$ ceramics with near zero temperature coefficient of resonant frequency. Jpn. J. Appl. Phys. 46(2007)1571–1573.

[331] J.-W. Choi, S. Yoon, H.-J. Kim, and K. H. Yoon. Microwave dielectric characteristics of $(1-x)$ $(Al_{1/2}Ta_{1/2})O_2$-$xTiO_2$ ceramics. Jpn. J. Appl. Phys. 41(2002)3804–3807.

[332] Y.-B. Chen, C.-L. Huang, and C.-W Lo. Microwave dielectric properties and microstructure of $La(Mg_{1/2}Ti_{1/2})O_3$ with CuO doped. Mater. Sci. Eng. B 128(2006)98–102.

[333] J.-W. Choi, J.-Y. Ha, S.-J. Yoon, H.-J. Kim, and K. H. Yoon. Microwave dielectric characteristics of $Ca[(Li_{1/3}Nb_{2/3})]_{1-x}Zr_x]O_{3-d}$ ceramics. Jpn. J. Appl. Phys. 43(2004)223–225.

[334] C.-F. Tseng and C.-L. Huang. Microwave dielectric properties of CuO doped $La(Co_{1/2}Ti_{1/2})O_3$ ceramics. Jpn. J. Appl. Phys. 45(2006)826–830.

[335] D.-W. Kim, H. B. Hong, K.-S. Hong, C. K. Kim, and D.J. Kim. The reversible phase transition and dielectric properties of $BaNb_2O_6$ polymorphs. Jpn. J. Appl. Phys. 41(2002) 6045–6048.

[336] J. Kato, H. Kagata, and K. Nishimoto. Dielectric properties of $(Pb,Ca)(Me,Nb)O_3$ at microwave frequencies. Jpn. J Appl. Phys. 31(1992)3144–3147.

[337] Y. Yamaguchi, Y. Komatsu, T. Otobe, and Y. Murakami. Newly developed ternary (Ca,Sr,Ba) Zirconate ceramic system for microwave resonators. Ferroelectrics 27(1980)273–276.

[338] Y.-J. Choi, J.-H. Park, and J.-G.Park. Middle permittivity LTCC composition with adjustable temperature coefficient. Mater. Lett. 58(2004)3102–3106.

[339] J. W. Choi, J.-Y. Ha, S.-J. Yoon, H.-J. Kim, K. H. Yoon. Microwave dielectric characteristics of $[Ca(Li_{1/3}Nb_{2/3})_{1-x}Zr_x]O_{3-\delta}$ ceramics. Jpn. J. Appl. Phys. 43(2004)223–225.

[340] H. T. Kim, S. H Kim, S. Nahm, and J. D. Kim. Low temperature sintering and microwave dielectric properties of zinc metatitanate –rutile mixture using boron. J. Am. Ceram. Soc. 82(1999)3043–3048.

[341] H. Wang, Z. Peng, H. Du, T. Yao, and X. Yao. Mn^{4+} and W^{6+} substitution on Bi_2O_3–ZnO–Nb_2O_5 based low firing ceramics. Ceram. Int. 30(2004)1219–1223.

[342] J.-A. Lee, J.-H. Lee, and J.-J. Kim. Low temperature sintering and microwave dielectric characteristics of $Ba_2Ti_9O_{20}$ ceramics. J. Eur. Ceram. Soc. 26(2006)2111–2115.

[343] K. Wakino, M. Murata, and H. Tamura. Far infrared reflection spectra of $Ba(Zn,Ta)O_3$-$BaZrO_3$ dielectric resonator material. J. Am. Ceram. Soc. 69(1986)34–37.

[344] C.-H. Wei and J.-H. Jean. Low fire processing $(Ca_{1-x}Nd_{2x/3})TiO_3$ microwave ceramics. J. Am. Ceram. Soc. 86(2003)93–98.

[345] S. Kucheiko, H.-J. Kim, D.-H. Yeo, and H.-J. Jung. Microwave dielectric properties of $LaZn_{0.5}Ti_{0.5}O_3$ ceramics prepared by sol-gel process. Jpn. J. Appl. Phys. 35(1966)668–675.

[346] A. G. Belous, O. V. Ovchar, M. Macek-Krzmanc, and M. Valant. The homogeneity range and the microwave dielectric properties of the $BaZn_2Ti_4O_{11}$ ceramics. J. Eur. Ceram. Soc. 26(2006)3733–3739.

[347] C.-F. Tseng and C.-L. Huang. Effects of B_2O_3 addition to $La(Co_{1/2}Ti_{1/2})O_3$ ceramics on sintering behaviour and microwave dielectric properties. Jpn. J. Appl. Phys. 45(2006)8766–8768.

[348] A. Kan, H. Ogawa, and H. Ohsato. Influence of microstructure on microwave dielectric properties of $ZnTa_2O_6$ ceramics with low dielectric loss. J. All. Comp. 337(2002)303–308.

[349] S. Solomon, H. Padmakumar, L. Jacob, J. Thomas, and M. R. Varma. $Ln(Zr_{1/3}Ti_{2/3})O_6$ (Ln = Ce, Pr, Nd, and Eu): A novel group of microwave ceramics. J. All. Comp. (2008) (in press).

[350] H. Ikawa and M. Takemoto. Products and microwave dielectric properties of ceramics with nominal composition $(Ba_{1-x}Ca_x)(B_{1/2}B'_{1/2})O_3$ ($B = Y^{3+}$, Nd^{3+},Gd^{3+}, B" = Nb^{5+},Ta^{5+}). Mater. Chem. Phys. 79(2004)222–225.

[351] P. Liu, E. S. Kim, and K. H. Yoon. Low temperature sintering and microwave dielectric properties of $Ca(Li_{1/3}Nb_{2/3})O_{3-\delta}$ ceramics. Jpn. J. Appl. Phys. 40(2001)5769–5773.

[352] A. Lu, P. Lin, and S.-F. Wang. Microstructures and microwave dielectric properties of Li_2O–Nb_2O_5–ZrO_2 ceramics. Ceram. Int. 33(2007)1389–1393.

[353] H. Wu and P. K. Davies. Non-stoichiometric 1:2 ordered perovskites in the $Ba(Li_{1/4}Nb_{3/4})O_3$–$Ba(Li_{2/5}W_{3/5})O_3$ system. J. Sol. St. Chem. 177(2004)3469–3478.

[354] C.-S. Hsu, H.-T. Soong, C.-C. Yu, C.-L. Huang, and M.-T. Ku. The effect of sintering temperature and time on microwave properties of $Sm(Zn_{1/2}Ti_{1/2})O_3$ ceramic resonators. Ceram. Int. 33(2006)951–955.

[355] H. T. Kim, J. D. Byun, and Y. Kim. Microstructure and microwave dielectric properties of modified zinc titanates. Mater. Res. Bull. 33(1998)975–986.

[356] J. B. Lim, D. H. Kim, S. Nahm, J. H. Paik, and H.-J. Lee. Effect of B_2O_3 and CuO additions on the sintering temperature and microwave dielectric properties of $Ba(Mg_{1/3}Nb_{2/3})O_3$ ceramics. Mater. Res. Bull. 41(2006)1199–1205.

[357] R. Guo, S. Bhalla, J. Sheen, F. W. Ainger, S. Edei, E. C. Subbarao, and L. E. Cross. Strontium aluminum tantalum oxide and strontium aluminum niobium oxide as potential substrates for high Tc thin films. J. Mater. Res. 10(1995)18–25.

[358] I.-T. Kim, Y. H. Kim, and S. J. Chung. Ordering and microwave dielectric properties of $Ba(Ni_{1/3}Nb_{2/3})O_3$ ceramics. J. Mater. Res. 12(1997)518–525.

[359] Y.-W. Kim, J.-H. Park, and J.-G. Park. Local cationic ordering behaviour in $Ba(Mg_{1/3}Nb_{2/3})O_3$ ceramics. J. Eur. Ceram. Soc. 24(2004)1775–1779.

[360] W.-C. Tzou, Y.-S. Yang, C.-F. Yang, H.-H. Chung, C.-C. Huang, and C.-C. Diao. Microstructure and microwave dielectric characteristics of $(1-x)Ba(Zn_{1/3}Ta_{2/3})O_3$–$xBaTi_4O_9$ ceramics. Mater. Res. Bull. 42(2007)1897–1904.

[361] C.-F. Tseng, C.-L. Huang, W.-R. Yang, and C.-L. Hsu. Dielectric characterization of $Nd(Zn_{1/2}Ti_{1/2})O_3$ ceramics at microwave frequencies. J. Am. Ceram. Soc. 89(2006)1465–1470.

[362] J. H. Paik, S. Nahm, J.-D. Byulin, M. H. Kim, and H. J. Lee. The effect of Mg deficiency on the microwave dielectric properties of $Ba(Mg_{1/3}Nb_{2/3})O_3$. J. Mater. Sci. Lett. 17(1998)1777–1780.

[363] C.-W. Ahn, S. Nahm, Y.-S. Lim, W. Choi, H.-M. Park, and H.-J. Lee. Microstructure and microwave dielectric properties of $Ba(Co_{1/3}Nb_{2/3})O_3$ ceramics. Jpn. J. Appl. Phys. 41(2002)5277–5280.

[364] F. Azough, C. Leach, R. Freer. Effect of non-stoichiometry on the structure and microwave dielectric properties of $Ba(Co_{1/3}Nb_{2/3})O_3$ ceramics. J. Eur. Ceram. Soc. 26(2006)2877–2884.

[365] Y.-G. Choi, J.-H. Park, W.-J. Ko, J.-H. Park, S. Nahm, and J.-G. Park. Low temperature sintering of $BaTi_4O_9$ based middle k dielectric composition for LTCC applicaions. J. Electroceram. 14(2005)157–162.

[366] M.-Z Hu, J. Qian, H.-S. Gu, X. Yu, and G. Xiong. A novel temperature compensated microwave dielectric based on $(Na_{0.5}La_{0.5})TiO_3$–CeO_2 system. Mater. Sci. Eng. B 134 (2006) 89–93.

[367] S. Li, J. Sun, R. Taylor, D. J Smith, and N. Newmann. Microstructure and dielectric properties of $Ba(Cd_{1/3}Ta_{2/3})O_3$ microwave ceramics synthesized with boron oxide sintering aid. J. Mater Res. 19(2004)3526–3533.

[368] J.-S. Sun, J.-J. Yu, J.-C You, and C. C. Wei. Microwave dielectric properties of $Ba(Sn_xZn_{(1-x)/3}Nb_{(1-x)2/3})O_3$ ($0 \leq x \leq 0.32$) ceramics. J. Mater. Sci. 28(1995)2163–2168.

[369] C. W. Ahn, H-J. Jang, S. Nahm, H-M. Park, H-J. Lee. Effect of microstructure on the microwave dielectric properties of $Ba(Co_{1/3}Nb_{2/3})O_3$ and (1-x)$Ba(Co_{1/3}Nb_{2/3})O_3$−x $Ba(Zn_{1/3}Nb_{2/3})O_3$ ceramics. J. Eur. Cer. Soc. 23(2003)2473–2478.
[370] Y.-C. Zhang, Z. X. Yue, L. Gui, and L. Li. Microwave dielectric properties of CuO–V_2O_5–Bi_2O_3 doped $ZnNb_2O_6$ ceramics with low sintering temprature. J. Electroceram. 14(2005)67–74.
[371] T. Shimada, US Patent 6,413896B1.
[372] D.-W. Kim, D.-G. Lee, and K. S. Hong. Low temperature firing and microwave dielectric properties of $BaTi_4O_9$ with Zn-B-O glass system. Mater. Res. Bull. 36(2001)585–595.
[373] J.-H. Lee, Y. I. Jang, H. J. Youn, J. W. Jang, and B. K. Kim. Low firing and microwave dielectric properties of $Ba[(Ni_{0.6}Zn_{0.4})_{0.33}Nb_{0.64}]O_3$ ceramics doped with Sb_2O_5 and B_2O_3. J. Mater. Sci. 34(1999)625–628.
[374] M. Udovic, M. Valant, and D. Suvorov. Phase formation and dielectric characterisation of the Bi_2O_3-TeO_2 system prepared in an oxygen atmosphere. J. Am. Ceram. Soc. 87(2004)591–597.
[375] S. George and M. T. Sebastian. Effect of lithium based glass addition on the microwave dielectric properties of $Ca[Li_{1/3}Nb_{2/3}]_{1-x}Ti_x]O_{3-\delta}$ LTCC ceramics. (to be published).
[376] S. Liu, R. Taylor, N. S. Petrovic, L. Budd, M. V. Schilfgaarde, and N. Newmann. Experimental and theoretical investigation of the structural, chemical, electronic and high frequency dielectric properties of barium cadmium tantalate based ceramics. J. Appl. Phys. 97(2005)014105.
[377] C. -A. Lu, P. Lin, and S. -F. Wang. Microstructures and microwave dielectric properties of Li_2O–Nb_2O_5–ZrO_2 ceramics. Ceram. Int. 33(2007)1389–1393.
[378] R. Rejini and M. T. Sebastian. Microwave dielectric properties of $La_5CrTi_3O_{15}$ and La_4MCr-Ti_3O_{15} (M = Pr, Nd, and Sm) ceramics. Int. J. Appl. Ceram. Tech. 3(2006)230–235.
[379] J. J. Bian, G. X. Song, and K. Yan. Structure and microwave dielectric properties of $Ba_{1+x}[(Co_{0.7}Zn_{0.3})_{1/3}Nb_{2/3}]O_3$ ($-0.015 \leq x \leq -0.015$). J. Eur. Ceram. Soc. 27(2007) 2817–2821.
[380] A. Kan, H. Ogawa, and H. Ohsato. Influence of microstructure on microwave dielectric properties of $ZnTa_2O_6$ ceramics with low dielectric loss. J. All. Comp. 337(2002)303–308.
[381] S.-Y. Cho, M.-K Seo, K. S Hong, S. J Park, and I.-T Kim. Influence of ZnO evaporation on the microwave dielectric properties of $La(Zn_{1/2}Ti_{1/2})O_3$. Mater. Res. Bull. 32(1997)725–735.
[382] R. Ubic, Y. Hu, K. Khamoudshi, and I. Abrahams. Structure and properties of $La(Zn_{1/2}Ti_{1/2})O_3$. J. Eur. Ceram. Soc. 26(2006)1787–1790.
[383] N. Baranov and Y.-J. Oh. Microwave frequency dielectric properties of hexagonal perovskites in the $Ba_5Ta_4O_{15}$–$BaTiO_3$ system. J. Eur. Ceram. Soc. 25(2005)3451–3457.
[384] A. Veres, S. Marinel, and F. Roulland. Dielectric properties of $Ba(Mg,Zn)_{1/3}Nb_{2/3})O_3$ and effect of B_2O_3 and LiF addition. J. Eur. Ceram. 25(2005)2759–2762.
[385] I. Melodetsky and P. K. Davies. Effect of $Ba(Y_{1/2}Nb_{1/2})O_3$ and $BaZrO_3$ on the cation order and properties of $Ba(Co_{1/3}Nb_{2/3})O_3$ microwave ceramics. J. Eur. Ceram. Soc. 21(2001)2587–2591.
[386] H. Ikawa, H. Harita, and O. Fukanaga. Microwave dielectric properties of solid solutions in the system $ZrTiO_4$–$HfTiO_4$. J. Ceram. Soc. Jpn. 98(1990)860–863.
[387] I. N. Jawahar, M. T. Sebastian, and P. Mohanan. Microwave dielectric properties of $Ba_{5-x}Sr$-Ta_4O_{15}, $B_5Nb_xTa_{4-x}O_{15}$ and $Sr_5Nb_xTa_{4-x}$ ceramics. Mater. Sci Eng. B 106(2004)207–212.
[388] D.-W. Kim, D.-Y. Kim, and K. S. Hong. Phase relations and microwave dielectric properties of $ZnNb_2O_6$–TiO_2. J. Mater. Res. 15(2000)1331–1335.
[389] J.-X. Tong, Q.-L. Zhang, H. Yang, and J.-L. Zou. Low temperature firing and microwave dielectric properties of $Ca[(Li_{1/3}Nb_{2/3})_{0.84}Ti_{0.16}]O_{3-\delta}$ ceramics for LTCC applications. J. Am. Ceram. Soc. 90(2007)845–849.
[390] D. Suvorov, M. Valant, S. Skapin, and D. Kolar. Microwave dielectric properties of ceramics with composition along the $La_{2/3}TiO_3$–$LaAlO_3$ tie line. J. Mater. Sci. 33(1998)85–89.

[391] F. Azough, C. Leach, and R. Freer. Effect of CeO_2 on the sintering behaviour cation order and properties of $Ba_3Co_{0.7}Zn_{0.3}Nb_2O_9$ ceramics. J. Eur. Ceram. Soc. 26(2006)1883–1887.

[392] C.-W. Ahn, H.-J. Jang, S. Nahm, H.-M. Park, and J.-J. Lee. Effects of microstructure on the microwave dielectric properties of $Ba(Co_{1/3}Nb_{2/3})O_3$ and $(1-x)Ba(Zn_{1/3}Nb_{2/3})O_3$-$xBa(Co_{1/3}Nb_{2/3})O_3$ ceramics. J. Eur Ceram. Soc. 23(2003)2473–2478.

[393] C.-L. Huang and C.-S. Hsu. Improved high Q value $0.5LaAlO_3$-$0.5SrTiO_3$ microwave dielectric ceramics at low sintering temperature. Mater. Res. Bull. 36(2001)2677–2687.

[394] C.-L. Huang and K.-H. Chiang. Characterization and dielectric behaviour of CuO doped $ZnTa_2O_6$ ceramics at microwave frequency. Mater. Res. Bull. 39(2004)1701–1708.

[395] C.-F. Yang. The microwave characteristics of glass-$BaTi_4O_9$ ceramics. Jpn. J. Appl. Phys. 38(1999)3576–3579.

[396] E. S. Kim and W. Choi. Effect of phase transition on the microwave dielectric properties of $BiNbO_4$. J. Eur Ceram. Soc. 26(2006)1761–1766.

[397] G. Subodh and M. T. Sebastian. Microwave dielectric properties of ATe_3O_8 (A = Sn, Zr) ceramics. Jpn. J. Appl. Phys. (2008) (in press).

[398] C.-L. Huang and K.-H. Chiang. Dielectric properties of B_2O_3 doped $(1-x)LaAlO_3$-$xSrTiO_3$ ceramic system at microwave frequency. Mater. Res. Bull. 37(2002)1941–1948.

[399] S.-Y. Cho, K. S. Hong, and K.-H. Ko. Mixture like behavior in the microwave dielectric properties of $(1-x)LaAlO_3$-$xSrTiO_3$ system. Mater. Res. Bull. 34(1999)511–516.

[400] H. Hughes, D. M. Iddles, and I. M. Reaney. Niobate based dielectrics suitable for third generation mobile phone base stations. Appl. Phys. Lett. 79(2001)2952–2954.

[401] F. Azough, C. Leach, and R. Freer. Effect of V_2O_5 on the sintering behavior, cation order and properties of $BaCo_{0.7}Zn_{0.3}Nb_2O_9$ ceramics. J. Eur. Ceram. Soc. 25(2005)2839–2841.

[402] S. Kamba, H. Hughes, D. Noujini, S. Santha, R. C. Pullar, P. Samoukhim, J. Petzelt, F. Freer, N. Mc N. Alford, and D. Iddles. Relationship between microwave and lattice vibration properties in $Ba(Zn_{1/3}Nb_{2/3})O_3$ based microwave dielectric ceramics. J. Phys. D Appl. Phys. 37(2004)1980–1986.

[403] M. Barwick, F. Azough, and R. Freer. Structure and dielectric properties of perovskite ceramics in the system $Ba(Ni_{1/3}Nb_{2/3})O_3$–$Ba(Zn_{1/3}Nb_{2/3})O_3$. J. Eur. Ceram. Soc. 26(2006)1767–1773.

[404] F. Azough, T. Lowe, and R. Freer. Control of microwave dielectric properties in the system $BaO.Nd_2O_3.4TiO_2$–$BaO.Al_2O_3.4TiO_2$. J. Electroceram. 15(2005)183–192.

[405] D.-W. Kim, K. S. Hong, C. S. Yoon, and C. K. Kim. Low temperature sintering and microwave dielectric properties of $Ba_5Nb_4O_{15}$–$BaNb_2O_6$ mixtures for LTCC applications. J. Eur. Ceram. Soc. 23(2003)2597–2601.

[406] W. Guoquing, W. Shunhua, and S. Hao. Microwave dielectric ceramics BaO–TiO_2–ZnO systemdoped with $MnCO_3$ and SnO_2. Mater. Lett. 59(2005)2229–2231.

[407] R. J. Scott, M. Thomas, and C. Hampson. Development of low cost, high performance $Ba(Zn_{1/3}Nb_{2/3})O_3$ based materials for microwave resonator applications. J. Eur. Ceram. Soc. 23(2003)2467–2471.

[408] H. Ikawa, M. Ohara, T. Iida, M. Takayama, and M. Takemoto. Products and microwave dielectric properties of ceramics with nominal composition $(Ba_{1-x}Ca_x)(Sc_{1/2}Nb_{1/2})O_3$. J. Ceram. Soc. Jpn. Supl. 12(2004) S1614–S1616.

[409] M. J. Lee, B. D. You, and D. S. King. Properties of Mn doped $BaTi_4O_9$–ZnO–Ta_2O_5 ceramics. Part 1. reaction sequence, microstructural analysis and microwave dielectric properties. J. Mater. Sci. Mater. Electr. 6(1995)165–172.

[410] A. Ioachim, M. G. Banciu, M. I. Toacsan, L. Nedelcu, D. Ghetu, H. V. Alexandru, G. Stoica, G. Annino, M. Cassattari, and M. Martinnelli. Nickel doped $(Zr_{0.8}Sn_{0.2})TiO_4$ for microwave and millimeter wave applications. Mater. Sci. Eng. B. 118(2005)205–209.

[411] M. Udovic and D. Suvorov. Sintering and dielectric characterization of Psuedoternary compounds from the Bi_2O_3–TiO_2–TeO_2 system. J. Am. Ceram. Soc. 90(2007)2404–2408.

[412] J. H. Choy, Y. S. Han, J. H. Sohn, and M. Itoh. Microwave characteristics of BaO-TiO_2 ceramics prepared via a citrate route. J. Am. Ceram. Soc. 78(1995)1169–1172.

[413] T. Negas, G. Yeager, S. Bell, N. Coats, and I. Minis. $BaTi_4O_9/Ba_2Ti_9O_{20}$ based ceramics resurrected for modern microwave applications. Am. Ceram. Soc. Bull. 72(1993)80–89.

[414] R. Ubic, K. Khomouski, and A. C. Spray. Structure and properties of rare earth zinc and magnesium titanates. Interntl. Conference on Microwave Materials and their Applications. MMA, Inuyama, Japan (2004).

[415] A. Bianco, G. Gusmano, R. Freer, and P. Smith. Zirconium titanate microwave dielectrics prepared via polymeric precursor route. J. Eur. Ceram. Soc. 19(1999)959–963.

[416] L. A. Khalam, H. Sreemoolanathan, R. Ratheesh, P. Mohanan, and M. T. Sebastian. Preparation characterisation and microwave dielectric properties of $Ba(B'_{1/2}B''_{1/2})O_3$ [B' = La, Pr, Nd, sm, Eu, Gd, Tb, Dy, Ho, Y, Yb, In] ceramics. Mater. Sci. Eng. B. 107(2004)264–270.

[417] J. Takahasi, K. Kageyama, and K. Kodaira. Microwave dielectric propewrties of lanthanide titanate ceramics. Jpn. J. Appl. Phys. 32(1993)4327–4331.

[418] J. Takahashi, K. Kageyama, and T. Hayashi. Dielectric properties of double oxide ceramics in the system Ln_2O_3–TiO_2 (Ln = La, Nd, Sm). Jpn. J. Appl. Phys. 30(1991)2354–2358.

[419] E. S. Kim, Y. H. Kim, J. H. Chae, D. W. Kim, and K. H. Yoon. Dielectric properties of $[Pb_{0.2}Ca_{0.8}]$ $(Ca_{1/3}Nb_{2/3})O_3$ eramics at microwave frequencies. Mater. Chem. Phys. 79(2003)230–232.

[420] Q. Zeng, W. Li, J.-L. Shi, and J.-K. Guo. Fabrication and microwave dielectric properties of a new LTCC ceramic composite based on Li_2O_3–Nb_2O_5–TiO_2 system. Mater. Lett. 60(2006)3203–3206.

[421] V. Parvanova and I. Iliev. Synthesis and microwave properties of substituted $Ba_2Ti_9O_{20}$. J. Mater. Sci. Mater. Electr. 17(2006)815–818.

[422] J.-J. Lim, S. Nahm, H.-T. Kim, J.-H. Kim, J.-J. Paik, and H.-J. Lee. Effect of B_2O_3 and CuO on the sintering temperature and microwave dielectric properties of the $BaTi_4O_9$ ceramics. J. Electroceram. 17(2005)393–397.

[423] S. Kucheiko, D.-H. Yeo, J.-W. Choi, S.-J. Yoon, and H.-J. Kim. Microwave dielectric properties of $CaTiO_3$–$Ca(Al_{1/2}Nb_{1/2})O_3$ ceramics doped with Li_3NbO_4. J. Am. Ceram. Soc. 85(2002)1327–1329.

[424] J.-R. Kim, D.-W. Kim, I.-S. Cho, B. S. Kim, J.-S. An, and K. S. Hong. Low temperature sintering and microwave dielectric properties of $Ba_3Ti_5Nb_6O_{28}$ with ZnO-B_2O_3 glass additions for LTCC applications. J. Eur. Ceram. Soc. 27(2007)3075–3079.

[425] Y.-C. Liou, C.-Y. Shih, and W.-H. Shiu. (5–x)BaO–xMgO–2Nb_2O_5 ceramics prepared using a reaction sintering process. Mater. Res. Bull. 42(2007)1104–1110.

[426] A. Ioachim, M. G. Banciu, M. I. Toacsen, L. Nedelcu, D. Ghetu, H. V. Alexandru, C. Berbecaru, A. Dutu, and G. Stoica. High k Mg doped ZST for microwave applications. Appl. Surf. Sci. 253(2006)335–338.

[427] H. Ikawa, T. Tsunneyama, and M. Takemoto. Microwave Materials and their Applications, MMA, Inuyama, Japan, (2004).

[428] J. H. Moon, H. M. Jung, H. S. Park, J. Y. Shin, and H. S. Kim. Sintering behaviour and microwave dielectric properties of $(Ca,La)(Ti,Al)O_3$ ceramics. Jpn. J. Appl. Phys. 38(1999)6821–6827.

[429] D.-W. Kim, H. B. Hong, and K. S. Hong. Structural transition and microwave dielectric properties of $ZnNb_2O_6$-TiO_2 sintered at low temperatures. Jpn. J. Appl. Phys. 41(2002)1465–1469.

[430] Y. S. Ahn, K. H. Yoon, and E. S. Kim. Effect of cooling rate on loss quality of $(Zr_{.8}Sn_{.2})TiO_4$ ceramics with additives. J. Eur. Ceram. Soc. 23(2003)2519–2523.

[431] D. Houivet, J. El. Fallah, and J. M. Haussone. Phases in La_2O_3 and NiO doped (Zr, Sn)TiO_4 microwave dielectric ceramics. J. Eur. Ceram. Soc. 19(1999)1095–1099.

[432] D. Houivet, J. El Fallah, B. Lamagenere, and J. M. Haussonne. Effect of annealing on the microwave properties of $(Zr,Sn)TiO_4$ ceramics. J. Eur. Ceram. Soc. 21(2001)1727–1730.

[433] D. Suvorov, M. Valant, B. Jancar, and S. D. Skapin. $CaTiO_3$-based ceramics: Microstructural development and dielectric properties. Acta Chim. Slov. 48(2001)87–99.

[434] M. Valant and D. Suvorov. Processing and dielectric properties of sillenite compounds $Bi_{12}MO_{20}$ (M = Si, Ge, Ti, Pb, Mn, $B_{1/2}$, $P_{1/2}$). J. Am. Ceram. Soc. 84(2001)2900–2904.

[435] P. L. Wise, I. M. Reaney, W. E. Lee, T. J. Price, D. M. Iddles, and D. S. Cannell. Structure-microwave property relations in $(Sr_{1-x}Ca_x)_{n+1}Ti_nO_{3n}$. J. Eur. Ceram. Soc. 1(2001)1723–1726.

[436] C.-L. Huang, C.-S. Hsu, and R.-J. Lin. Improved high Q microwave dielectric resonatorusing ZnO-WO_3 doped $Zr_{0.8}Sn_{0.2}TiO_4$ ceramics. Mater. Res. Bull. 36(2001) 1985–1993.

[437] H. Tamura. Microwave loss quality of $(Zr_{0.8}Sn_{0.2})TiO_4$. Am. Ceram. Soc Bull. 73(1994)94–97.

[438] Y.-S. Ho, M. H. Weng, B.-T. Dai, and S.-S. Wang. Nanopowder and microwave dielectric properties of sol-gel derived $Zr_{0.8}Sn_{0.2}TiO_4$ ceramics. Jpn. J. Appl. Phys. 44(2005) 6152–6155.

[439] P. V. Bijumon, P. Mohanan, and M. T. Sebastian. Synthesis, characterization and properties of $Ca_5A_2TiO_{12}$ (A = Nb, Ta) ceramic dielectric materials for applications in microwave communication systems. Jpn. J. Appl. Phys. 41(2002)3834–3835.

[440] P. V. Bijumon and M. T. Sebastian. Tailoring the microwave dielectric properties of $Ca_5Ta_2TiO_{12}$ ceramics through glass addition. J. Am. Ceram. Soc. 88(2005)3433–3439.

[441] V. D. Parnova and S. K. Andreev. Microwave properties of the systems $Ba_{1-x}Sr_x(Zn_{1/3}Ta_{2/3})_{0.94}Ti_{0.06}O_3$ and $Ba(Zn_{1/3}Nb_{2/3})_{1-x}Zr_xO_3$. J. Mater. Sci. Mater. Electr. 16(2005) 249–252.

[442] A. Feteira D.C. Sinclair and M. T. Lanagan. Structure and microwave dielectric properties of $Ca_{1-x}Y_xTi_{1-x}Al_xO_3$(CYTA) ceramics. J. Mater. Res. 20(2005)2391–2399.

[443] J. Joseph, T. M. Vimala, K. C. J. Raju, and V. R. K. Murthy. Structural investigations on $(Ba_xSr_{1-x})(Zr_yTi_{1-y+\delta}Ta_\delta)O_3$ dielectric resonator used for microwave applications. Jpn. J. Appl. Phys. 35(1996)179–181.

[444] M. R. Varma and M. T. Sebastian. Effects of dopant addition on the microwave dielectric properties of $Ba(Zn_{1/3}Nb_{2/3})O_3$ ceramics. J. Eur. Ceram. Soc. 27(2007)2827–2833.

[445] L. Fang, C. L. Diao, H. Zhang, R. Z. Yuan, R. Dronskaoski, and H. X. Liu. Preparation and characterization of two new dielectric ceramics $Ba_4NdTiNb_3O_{15}$ and $Ba_3Nd_2Ti_2Nb_2O_{15}$. J. Mater. Sci. Mater. Electr. 15(2004)803–805.

[446] N. Ichinose, T. Mizutani, H. Hiraki, and H. Ookumapp.Microwave dielectric materials in the system $(Sr_{1-x}Ca_x)[(Li_{1/4}Nb_{3/4}]_{1-y}Ti_y]O_3$. Ceramuraga Int. 3(1977)100–102.

[447] T. Oishi, H. Ogawa, and A. Kan. Effect of sintering time on crystal structure and microwave dielectric properties of aeschenite-structured $(Sm_{1-x}Y_x)(Ti_{1.5}W_{0.5})O_6$ (x = 0 and 0.5) ceramics. Mater. Res. Bull. 42(2007)2072–2081.

[448] H. Ikawa, Y. Itou, Y. Takei, K. Tsumoto, and M. Takemoto.11 th US-Japan seminar on dielectric and piezoelectric ceramics Sapporo, Japan, (2003) pp. 93–96.

[449] J.-R. Kim, D.-W. Kim, S. H. Yoon, and K. S. Hong. Low temperature sintering and microwave dielectric properties of $Ba_3Ti_5Nb_6O_{28}$ with B_2O_3 and CuO additions. J. Electroceram. 17(2006)439–443.

[450] D.-J. Kim, J.-W. Hahn, G.-P. Han, S.-S. Lee, and T.-G. Choy. Effects of Alkaline-Earth-Metal Addition on the Sinterability and Microwave Characteristics of (Zr,Sn)TiO_4 Dielectrics. J. Am. Ceram. Soc. 83(2000)1010–1012.

[451] D.-W. Kim, J. R. Kim, S.-H. Yoon, K. S. Hong, and C. K. Kim. Microwave dielectric properties of low fired $Ba_5Nb_4O_{15}$. J. Am. Ceram. Soc. 85(2002)2759–2762.

[452] H. Wu and P. K. Davies. Influence of non-stoichiometry on the structure and properties of $Ba(Zn_{1/3}Nb_{2/3})O_3$ microwave dielectrics. I. Substitution of $Ba_3W_2O_9$. J. Am. Ceram. Soc. 89(2006)2239–2249.

[453] M. T. Sebastian. New loss microwave dielectric ceramics in the BaO-TiO_2-Nb_2O_5/Ta_2O_5 system. J. Mater. Sci. Mater. Electr. 10(1999)475–478.

[454] P. V. Bijumon, M. T. Sebastian, A. Dias, R. L. Moreira, and P. Mohanan. Low loss $Ca_{5-x}Sr_xA_2TiO_{12}$ [A = Nb,Ta] ceramics. Microwave dielectric properties and vibrational spectroscopic analysis. J. Appl. Phys. 97(2005)104108.

[455] L. Fang, H. Zhang, T. H. Huang, R. Z. Yuan, and R. Dronkawski. Preparation, structure and dielectric properties of $Ba_4LaMNb_3O_{15}$ (M = Ti, Sn) ceramics. Mater. Res. Bull. 39(2004)1649–1654.
[456] M. Reaney and D. Iddles. Microwave dielectric ceramics for resonators and filters in mobile phone networks. J. Am. Ceram. Soc. 89(2006)2063–2072.
[457] S. Jayanthi and T. R. N. Kutty. Dielectric properties of baium mgnesium titanate ceramics $BaMg_6Ti_6O_{19}$. Mater. Lett. 60(2006)796–800.
[458] J. K. Plourde, D. F. Lim, H. M. O'Bryan, and J. Thomson Jr. $Ba_2Ti_9O_{20}$ as a microwave dielectric resonator. J. Am. Ceram. Soc. 58(1975)418–420.
[459] S. F. Wang, Y. F. Hsu, T.-H. Weng, C.-C. Chiang, J. P. Chu, and C-Y. Huang. Effects of additives on the microstructure and dielectric properties of $Ba_2Ti_9O_{20}$ microwave ceramics. J. Mater. Res. 18(2003)1179–1187.
[460] H. R. Lee, K. H. Yoon, and E. S. Kim. Low temperature sintering and microwave dielectric properties of $BiNbO_4$–$ZnNb_2O_6$ ceramics with CuV_2O_6 additives. Jpn. J. Appl. Phys. 42(2003)6168–6169.
[461] P. L. Wise, I. M. Reaney, R. Ubic, and W. Lee. Inst phys conf ser 161. Proc. Conference on Electron Microscopy and Analysis EMA 99, Sheffield, (1999) 161. IOP publishing House. Bristol.
[462] K. H. Yoon, J. B. Kim, and E. S. Kim. Effect of $BaSnO_3$ on the microwave dielectric properties of $Ba_2Ti_9O_{20}$. J. Mater. Res. 11(1996)1996–2001.
[463] L. Fang, Y. Wu, C.-Z. He, P. Chen, and H. Zhang. $Ba_3LaTa_3O_{12}$: A new microwave dielectric of $A_4B_3O_{12}$- type cation deficient perovskites. Mater. Lett. 62(2008)594–596.
[464] C.-T. Lee, Y.-C. Lin, C.-Y. Huang, C.-Y. Su, and C.-L. Hu. Structural and dielectric characteristics of barium lanthanum zinc niobate. J. Am. Ceram. Soc. 89(2007)3662–3668.
[465] W.-Y. Lin, R. F. Speyer, W. S. Hackenberger, and T. R. Shrout. Microwave properties of $Ba_2Ti_9O_{20}$ doped with zirconium and tin oxide. J. Am. Ceram. Soc. 82(1999)1207–1211.
[466] S. Marinel, and F. Rolland. Synthesis and characterisation of the double perovskite $Ba_2(Zn_{05}Ti_{0.5}X)O_6$ (X = Nb,Ta) ceramics. Mater. Res. Bull. 40(2005)962–969.
[467] J.-J. Weng, C.-H. Hsu, C.-L. Huang, and R.-J Lin. Effect of CuO additives on sintering and microwave dielectric behaviors of $0.95Ba(Zn_{1/3}Nb_{2/3})O_3$–$0.05BaZrO_3$ ceramics. Jpn. J. Appl. Phys. 44(2005)8039–8042.
[468] J.-Y. Ha, J.-W. Choi, C.-Y. Kang, S.-J. Yoon, D. J. Choi, and H.-J. Kim. Low temperature sintering of $Ca(Li_{1/3}Nb_{2/3})_{1-x}Ti_x)O_{3-d}$ based microwave dielectric ceramics with glass frit. Jpn. J. Appl. Phys. 44(2005)1322–1325.
[469] P. V. Bijumon, P. Mohanan, and M. T. Sebastian. High dielectric constant low loss microwave dielectric ceramics in the $Ca_5Nb_{2-x}Ta_xTiO_{12}$ system. Mater. Lett. 57(2003)1380–1384.
[470] R. Ratheesh, H. Sreemoolanathan, and M. T. Sebastian. Vibrational analysis of $Ba_{5-x}Sr_xNb_4O_{15}$ microwave dielectric ceramic resonators. J. Sol. St. Chem. 131(1997)2–8.
[471] H. Sreemoolanathan, M. T. Sebastian, and P. Mohanan. High permittivity and low loss ceramics in the BaO–SrO–Nb_2O_5-system. Mater. Res. Bull. 30(1995)653–658.
[472] C. Vineis, P. K. Davies, T. Negas, and S. Bell. Microwave dielectric properties of hexagonal perovskites. Mater. Res. Bull. 31(1996)431–437.
[473] P. Liu, H. Ogawa, E. S. Kim, and A. Kan. Microwave dielectric properties of low temperature sintered $Ca[(Li_{1/3}Nb_{2/3}),Ti]O_{3-d}$ ceramics. J. Eur. Ceram. Soc. 24(2004)1761–1764.
[474] Z. Yue, F. Zhao, Y. Zhang, Z. Gui, and L. Li. Microwave dielectric properties of $Ba[(Zn_{1-x}Co_x)_{1/3}Nb_{2/3}]O_3$ ceramics. Mater. Lett. 58(2004)1830–1834.
[475] M. Onoda, J. Kuwata, K. Kanaeta, K. Toyoma, and S. Nomura. $Ba(Zn_{1/3}Nb_{2/3})O_3$–$Sr(Zn_{1/3}Nb_{2/3})O_3$ solid solution ceramics with temperature stable high dielectric constant and microwave loss. Jpn. J. Appl. Phys. 21(1982)1707–1710.
[476] W. Lu, F. Azough, E. R. Kipkoch, and R. Freer.The structural and microwave dielectric properties of CMN- based ceramics. International Conf on Microwave Materials and their Applications. MMA, Inuyama, Japan (2004,).

[477] S. Hirano, T. Hayashi, and A. Hattai. Chemical processing and microwave characteristics of (Zn,Sn)TiO_4 microwave dielectrics. J. Am. Ceram. Soc. 74(1991)1320–1324.

[478] J.-R. Kim, D.-W. Kim, H. S. Jung, and K. S. Hong. Low temperature sintering and microwave dielectric properties of $Ba_5Nb_4O_{15}$ with ZnB_2O_4 glass. J. Eur. Ceram. Soc. 26(2006)2105–2109.

[479] C.-C. Cheng, T.-E. Hsieh, and I.-N. Lin. Microwave dielectric properties of glass ceramic composites for low temperature cofireable ceramics. J. Eur. Ceram. Soc. 23(2003)2553–2558.

[480] N. Wang, M.-Y. Zhao, W. Li, and Z-W. Yin. The sintering behaviour and microwave dielectric properties of Bi(Nb,Sb)O_4 ceramics. Ceram. Int. 30(2004)1017–1022.

[481] K. H. Yoon, Y. S. Kim, and E. S. Kim. Microwave dielectric properties of $(Zr_{0.8}Sn_{0.2})TiO_4$ ceramics with pentavalent additives. J. Mater. Res. 10(1995)2085–2090.

[482] S. Nomura, K. Tomaya, and K. Kaneta. Effect of Mn doping on the dielectric properties of $Ba_2Ti_9O_{20}$ ceramics microwave frequencies. Jpn. J. Appl. Phys. 22(1983)1125–1128.

[483] Y.-C. Chen, W.-C. Tzou, C.-F. Yang, and P.-S. Cheng. The effect of the crystalline phase on the sintering and microwave dielectric properties of CuO doped $(Bi_{0.95}Sm_{0.05})NbO_4$ ceramics. Jpn. J. Apl. Phys. 40(2001)3252–3255.

[484] S. Hirano, T. Hayashi, K. Kikuta, and J. Otsuka. Chemical processing and properties of $BaTi_5O_{11}$ microwave dielectrics. Ceram. Trans. 12(1990)733–740.

[485] Y. Tohdo, K. Kakimoto, H. Ohsato, H. Yamada, and T. Okawa. Microwave dielectric properties and crystal structure of homologous compounds $ALa_4Ti_4O_{15}$ (A = Ba, Sr, and Ca) for base station applications. J. Eur. Ceram. Soc. 26(2006)2039–2043.

[486] S. Kucheiko, and J-w. Choi, H.-J. Kim, H.-J. Jung. Microwave dielectric properties of $CaTiO_3$-$Ca(Al_{1/2}Ta_{1/2})O_3$ ceramics. J. Am. Ceram. Soc. 79(1996)2739–2743.

[487] D. Shihara, Y. Xi, and Y. Yong. Dielectric properties of B_2O_3 doped $BiNbO_4$ ceramics. Ceram. Int. 30(2004)1195–1198.

[488] E. A. Nenesheva, L. P. Mudroliuba, and N. F. Kartenko. Microwave dielectric properties of ceramics based on $CaTiO_3$-$LnMO_3$ system (Ln = La, Nd,; M = Al, Ga). J. Eur. Ceram. Soc. 23(2003)2443–2448.

[489] F. Zhao, X. Yue, Y. Zhang, Z. Giu and L. Li. Microstructure and microwave dielectric properties of $Ca[Ti_{1-x}(Mg_{1/3}Nb_{2//3})_x]O_3$ ceramics. J. Eur. Ceram. Soc. 25(2005)3347–3352.

[490] Q. Zeng, W. Lei, J.-l. Shi, X-l. Dong and J.-K.Guo. Influence of V_2O_5 additions to $5Li_2O$–$1Nb_2O_5$–$5TiO_2$ ceramics on sintering temperature and microwave dielectric properties. J. Am. Ceram. Soc. 90(2007)2262–2263.

[491] E. S. Kim, D. H. Kang, and S. J. Kim. Effect of crystal structure on microwave dielectric properties of $(Ni_{1/3}B_{2/3})_{1-x}Ti_xO_2$ (B = Nb and Ta). Jpn. J. Appl. Phys. 46(2007)7101–7104.

[492] Y. Mao, S. H. Ding, X. Yao, and L. Y. Zhang. Microwave dielectric property of $BiNb_{1-x}Mo_xO_4$ ceramics. Mater. Chem. Phys. 98(2006)198–202.

[493] C.-L. Huang, C.-S. Hsu, and S.-J. Lu. Dielectric properties of $0.95Ba(Zn_{1/3}Nb_{2/3})O_3$–$0.05BaZrO_3$ ceramics at microwave fequency. Mater. Lett. 57(2003)3602–3605.

[494] H. Yamada, T. Okawa, Y. Tohdo, and H. Ohsato. Microwave dielectric properties of $BaLa_xTi_{3+x}O_{12+3x}$ (x = 0.0–1) ceramics. J. Eur. Ceram. Soc. 26(2007)2059–2062.

[495] K. Okuyama, Y. Yokotani, S. Kawashima, K. Kugimiya, K. Nishimoto, T. Ishizaki, J. Kato, and Y. Hakotani. Proc. Asia Pacific Microwave Conf. The Institute of Electronics Information and Communication Engineers, (1994) Vol. II, p. 751.

[496] J. Kato. Material produces small resonators with high dielectric constant. JEE Sept. (1991)114–118.

[497] T. Fukui, C. Sakurai, and M. Okuyama. Effect of heating rate on sintering of alkoxide drived $BaTi_5O_{11}$ powder. J. Mater. Res. 7(1992)192–196.

[498] E. S. Kim, B. S. Chun, J. D. Kim, and K. H. Yoon. Low temperature sintering and microwave dielectric properties of $[Ca_{0.6}(Li_{0.5}Nd_{0.5})_{0.4}]_{05}Zn_{0.55}TiO_3$. Mater. Sci. Eng. B99(2003)243–246.

[499] C.-L. Huang, Y.-B. Chen, and S.-H. Lin. New dielectric material system of $Nd(Mg_{1/2}Ti_{1/2})O_3$–$CaTiO_3$ at microwave frequency. Solid St. Ionics. 49(2005)1921–1924.

[500] P. Riazikhoei, F. Azough, and R. Freer. The influence of $ZnNb_2O_6$ on the microwave dielectric properties of $ZrTi_2O_6$ ceramics. J. Am. Ceram. Soc. 89(2006)216–223.
[501] Z. Yue, F. Zhao, Z. Gui, and Z. Li. Phases, microstructure and microwave dielectric properties of hexagonal perovskite $Ca(La_{1-x}Nd_x)_4Ti_4O_{15}$ ceramics. J. All. Comp. 395(2005)126–131.
[502] F. Zhao, Z. Yue, J. Pei, H. Zhuans, Z. Gui, and L. Li. Improvement on the temperature coefficient of resonant frequency of hexagonal perovskites through intergrowth structures. Appl. Phys. Lett. 89(2006)202901.
[503] I.-S. Cho, J.-R. Kim, D. W. Kim, and K. S. Hong. Microwave dielectric properties and far infra red spectroscopic analysis of $Ba_{5+n}Ti_nNb_4O_{15+3n}$ $(0.3 < n < 1.2)$ ceramics. J. Eur. Ceram. Soc. 27(2007)3081–3086.
[504] L. Fang, G. Zhang, C. Hu, and H. Zhang. Preparation and microwave dielectric properties of a new $A_5B_4O_{15}$ type cation deficient perovskites: $Ba_4LaTiTaO_{15}$. Materl. Lett. (2008) (in press).
[505] Q.-L. Zhang and H. Yang. Low temperature firing and microwave dielectric properties of $ZnO–Nb_2O_5–TiO_2–SnO_2$ ceramics with $CuO–V_2O_5$. Mater. Res. Bull. 40(2005)1891–1898.
[506] L. Fang, H.-F. Liu, H. Zhang, and Z.-Q. Liu. Preparation and microwave dielectric properties of a new $A_4B_3O_{14}$ type cation deficient perovskite $BaLa_3Ti_2NbO_{12}$. J. Mater. Res. 41(2006)1281–1284.
[507] W. S. Kim, J. H. Kim, J. H. Kim, K. H. Hur, and J. Y. Lee. Microwave dielectric properties of the $ZrO–ZnO–Ta_2O_5–TiO_2$ systems. Mater. Chem. Phys. 79(2003)204–207.
[508] L. Fang, S.-S. Meng, H. Zhang, and Z. Liu. $Ba_2La_2TiNb_2O_{12}$: A new microwave dielectric of $A_4B_3O_{12}$ type cation deficient perovskites. Mater. Lett. 60(2006)1147–1150.
[509] L. Fang, H. Zhang, X. K. Hong, F. C. Meng, J. F. Yang, and Y. H. Huang. Preparation and characterization of two new dielectric ceramics $Ba_3La_2Ti_2Nb_2O_{15}$ and $Ba_2La_3Ti_3NbO_{15}$. Mater. Lett. 58(2004)3884–3887.
[510] Kagata, I. Inoue, J. Kato, and I. Kameyana. Low fire bismuth based dielectric ceramics for microwave use. Jpn J. Appl. Phys. 31 (1992) 3152–3255.
[511] E. S. Kim, and W. Choi. Effect of phase transition on the microwave dielectric properties of $BiNbO_4$. Int. Conf. Microwave Materials and their Applications. Inuyama, Japan (2004).
[512] S. Hirahara, N. Fujikawa, S. Enami, and T. Nishi. US Patent 5,356,844 (1994).
[513] M. Takasugi and R. Kitoh. Natl. Conv. Rec. of IECE Japan, SC-96, 1988.
[514] A. Dias, P. V. Bijumon, M. T. Sebastian, and R. L. Moreira. Vibrational spectroscopy and microwave dielectric properties of $Ca_{5-x}Ba_xNb_2TiO_{12}$ and $Ca_{5-x}Ba_xTa_2TiO_{12}$ ceramics. J. Appl. Phys. 98(2005)084105.
[515] L.-C. Chang and B.-S. Chiou. Effect of B_2O_3 nano coating on the sintering behaviour and electrical microwave properties of $Ba(Nd_{2-x}Sm_x)Ti_4O_{12}$ ceramics. J. Electroceram. 13(2004)829–837.
[516] D. L. Cairns, I. M. Reaney, N. Otten, D. Iddles and T. Price. Structural determination and microwave dielectric properties of $(x)Re(Co_{1/2}Ti_{1/2})O_3–(1-x)CaTiO_3$ (Re = La and Nd) solid solutions. J. Eur Ceram. Soc. 26(2006)875–882.
[517] H. Ohsato, A. Harada, T. Okawa, and H. Okabe. US patent 6,905,994 (2005).
[518] C.-L. Huang, M.-H. Weng, and G.-M. Shan. Effect of V_2O_5 and CuO additiives on sintering behavior and microwave dielectric properties of $BiNbO_4$ ceramics. J. Mater. Sci. 35(2000)5443–5447.
[519] C.-L. Huang, Y.-B. Chen, and C.-W. Lo. Microwave dielectric properties and micostructures of $0.5La(Mg_{1/2}Ti_{1/2})O_3–0.5CaTiO_3$ ceramics with B_2O_3 addition. Jpn. J. Appl. Phys. 44(2005)6706–6708.
[520] C.-L. Huang, M.-H. Weng, and C.-C Yu. Low fireable $BiNbO_4$ based microwave dielectric ceramics. Ceram. Int. 27(2001)343–350.
[521] R. Rawal, A. Feteira, N. C. Hyatt, D. C. Sinclair, K. Sarma and N. Mc N. Alford. Microwave dielectric properties of hexagonal 12R -$Ba_3LaNb_3O_{12}$ ceramics. J. Am. Ceram. Soc. 89(2006)332–335.
[522] L. Fang, S.-S. Meng, H. Zhang, and B.-I. Wu. Preparation and characterization of new microwave dielectric ceramics $Ba_{3-n}La_{1+n}Nb_{3-n}O_{12}$. Trans. Nonferrous. Met. Soc. China. 16(2006)S521–S523.

[523] C.-L. Huang and M.-H. Weng. Low fire $BiTaO_4$ dielectric ceramics for microwave applications. Mater. Lett. 43(2000)32–35.
[524] N. Wang, M.-Y. Zhao, Z-W. Yin, and W. Li. Effect of complex substitution of La and Nd for Bi on the microwave dielectric properties of $BiNbO_4$ ceramics. Mater. Res. Bull. 39(2004)439–448.
[525] I. N. Jawahar, N. Santha, M. T Sebastian, and P. Mohanan. Microwavedielectric properties of $MO–La_2O_3–TiO_2$ (M = Ca, Sr, Ba) ceramics. J. Mater. Res. 17(2002)3084–3089.
[526] Y. Pang, C. Zhong, and S. Zhang. Effects of Gd doping on the sintering and microwave dielectric properties of $BiNbO_4$ ceramics. J. Mater. Sci. 42(2007)7052–7055.
[527] S. Suvorov, M. Valant, B. Jancar, and S. D. Skapin. $CaTiO_3$ based ceramics: Microstructural development and dielectric properties. Acta Chim. Slov. 48(2001)87–89.
[528] H. Okawa, K. Kiuchi, H. Okabe, and H. Ohsato. Microwave dielectric properties of $Ba_nLa_4Ti_{3+n}O_{12+3n}$ homologous compounds and substitution of trivalent cations for La. Ferroelectrics 272(2002)345–350.
[529] S.-O. Yoon, D.-M. Kim, S. H. Kim, J.-K. Park, and K.-S. Kang. Microwave dielectric properties of $Ca(Li_{1/4}Nb_{3/4})O_3–CaTiO_3$ ceramic system. J. Eur. Ceram. Soc. 26(2006)2023–2026.
[530] C.-L Huang, M.-H. Weng, C.-C. Wu, and C.-T. Lion. Low fire $BiNbO_4$ microwave dielectric ceramics modified by Sm_2O_3 addition. Mater. Res. Bull. 35(2001)827–835.
[531] M.-H. Weng and C.-L. Huang. Shifting of Tf value of $BiNbO_4$ ceramics by $BiTaO_4$ addition. J. Mater. Sci. Lett. 19(2000)375–376.
[532] R. Rawal, A. Feteira, A. A. Flores, N. C. Hyatt, A. R. West, D. C. Sinclair, K. Sarma, and N. Mc N. Alford. Dielectric properties of the twinned 8H-hexagonal perovskite Ba_8Nb_4-Ti_3O_{24}. J. Am. Ceram. Soc. 89(2006)336–339.
[533] F Zhao, Z. Yue, Y. Zhang, Z. Gui, and L. Li. Microstructure and microwave dielectric properties of $Ca[Ti_{1-x}(Mg_{1/3}Nb_{2//3})_x]O_3$ ceramics. J. Eur. Ceram. Soc. 25(2005)3347–3352.
[534] T. Okawa, K. Kiuchi, H. Okabe, and H. Ohsato. Microwave dielectric properties of $Ba_nLa_4Ti_{3+n}O_{12+3n}$ homologous series. Jpn. J. Appl. Phys. 40(2001)5779–5782.
[535] C.-L. Huang, M.-H. Weng, and C.-C Wu. Low fire $BiNbO_4$ microwave dielectric ceramic modified by Sm_2O_3 addition. Mater. Res. Bull. 35(2001)827–835.
[536] S. Solomon, H. S. Nathan, M. T. Sebastian, and P. Mohanan. Microwave dielectric resonators based on $Ba[(Bi_{0.2}D_{0.3})Nb_{0.5}]O_3$ (D = Y, Pr, Sm, Gd, Dy, Er). Mater. Lett. 28(1996)107–111.
[537] H. Zhang, L. Fang, R. Elsebrocke, R. Z. Yuan, and H. X. Liu. Characterization of materials and microwave dielectric properties of a new $A_6B_5O_{18}$ type cation deficient perovskite $Ba_4Nd_2Ti_3Nb_2O_{18}$. J. Mater. Sci. 40(2005)4427–4429.
[538] H. Badshaw, D. Iddles, R. Quinby, and I.-M. Reaney. Structure-property relations in x $CaTiO_3–(1-x)SrMg_{1/3}Nb_{2/3}O_3$ based microwave dielectric ceramics. J. Eur. Ceram. Soc. 23(2003)2435–2441.
[539] L. Fang, Q. Yu, H. Zhang, C. Z. Hu, and B. L. Lu. Microwave dielectric properties of a new $A_5B_4O_{15}$ type cation deficient perovskite $Ba_2La_3Ti_3TaO_{15}$. J. Am. Ceram. Soc. 90(2007)1626–1628.
[540] B. Jancar, D. Suvorov, M. Valant, and G. Drazic. Characterization of $CaTiO_3–NdAlO_3$ dielectric ceramics. J. Eur. Ceram. Soc. 23(2003)391–400.
[541] C.-F. Tseng, C.-L. Huang, and W. R. Yang. Microwave dielectric properties of $xNd(Zn_{1/2}Ti_{1/2})O_3–(1-x)CaTiO_3$ ceramics. Mater. Lett. 61(2007)4054–4057.
[542] G. M. Keith, A. A. Krik, K. Sarma, N. Mc N. Alford, E. J. Cussen, M. J. Rosiensky, and D. K. Sinclair. Synthesis, crystal structure and characterization of $Ba(Ti_{1/2}Mn_{1/2})O_3$: A high permittivity 12R-type hexagonal perovskite. Chem. Mater. 16(2004)2007–2015.
[543] L. Fang, Q. Yu, H. Zhang, and C-Z. Hu. Microwave dielectric properties of $Ba_3La_2Ti_2Nb_{2-x}Ta_xO_{15}$ ceramics. Mater. Lett 61(2007)3093–3095.
[544] J. Pei, Z. Yue, F. Zhao, Z. Gui, and L. Li. Microwave dielectric ceramics of hexagonal $(Ba_{1-x}A_x)La_4Ti_4O_{15}$ (A = Sr,Ca) for base station applications. J. All. Comp. (2008) (in press).

[545] J.-J Wang and C.-L Huang. New dielectric materials of $SrTiO_3$–$Ca(Mg_{1/3}Nb_{2/3})O_3$ ceramic system at microwave frequency. Mater. Lett. 60(2006)1280–1283.
[546] G. A. Ravi, F. Azough, and R. Freer. Effect of Al_2O_3 on the structure and microwave dielectric properties of $Ca_{0.7}Ti_{0.7}La_{0.3}Al_{0.3}O_3$. J. Eur. Ceram. Soc. 27(2007)2855–2859.
[547] Z. Wang, X. Yao, and L. Zhang. CeO_2 modified $BiNbO_4$ microwave ceramics sintered under atmosphere. Ceram. Int. 30(2004)1329–1333.
[548] H. Ikawa, M. Ohara, T. Iida, M. Takayama, and M. Takemoto. Products and microwave dielectric properties of ceramics with nominal composition $Ba_{1-x}Ca_x(Sc_{1/2}Nb_{1/2})O_3$. J. Ceram. Soc. Jpn. Suppl. 112-1. Special issue (2005) S1614–S1616.
[549] A. K. Axelsson, M. T. Sebastian, and Mc N. Alford. Microwave dielectric properties of Bi_2O_3–TiO_2 composite ceramics. J. Kor. Ceram. Soc. 40(2003)340–345.
[550] A. Feteiro, D. C. Sinclair, I. M. Reaney, and M. T. Lanagan. Structure property relationship of $BaTi_{1-2y}Ga_yNb_yO_3$ ($0\leq y\leq 0.35$) ceramics. J. Am. Ceram. Soc. 88(2005) 3055–3062.
[551] L. Fang, H. Zhang, R. Z. Yuan, and H. X. Liu. $Ba_3La_3Ti_4NbO_{18}$: a new microwave dielectric of $A_6B_5O_{18}$ type cation deficient perovskite. Mater. Lett. 59(2005)535–538.
[552] T. Takado, K. Kageyama, M. Yonemura, N. Hara, and S. Takao. Microwave dielectric properties of mixed ceramics $Ba(Zn_{1/3}Ta_{2/3})O_3$–$xCaTiO_3$ and $xMgTiO_3$–$yCaTiO_3$–$z(Nd_2O_3,wTiO_2)$. J. Mater. Sci. Mater. Electr. 14(2003) 205–214.
[553] C.-L. Huang, H.-L. Huang, and C.-C. Wu. Improved high Q value of $CaTiO_3$–$Ca(Mg_{1/3}Nb_{2/3})O_3$ solid solution with near zero temperature coefficient of resonant frequency. Mater. Res. Bull. 36(2001)1645–1652.
[554] A. R. Rawal, A. Feteiro, A. A. Flores, N. C. Hyatt, A. R. West, D. C. Sinclair, K. Sharma, and N. Mc N. Alford. Dielectric properties of the twinned 8H hexagonal perovskite $Ba_8Nb_4Ti_3O_{24}$. J. Am. Ceram. Soc. 89(2005)336–339.
[555] F. Zhao, Z. Yue, J. Pei, Z. Gui, and L. Li. A novel microwave dielectric ceramic $Ca_2Zn_4Ti_{16}O_{38}$. Preparation and dielectric properties. J. Sol. St. Chem. 179(2006)1719–1725.
[556] P. Samaoukhima, S. Kamba, S. Santhi, J. Petzelt, M. Valant, and D. Suvorov. Infrared and terahertz dielectric spectra of novel Bi_2O_3–Nb_2O_5 microwave ceramics. J. Eur. Ceram. Soc. 25(2005)3085–3088.
[557] A. Y. Borisevich and P. K. Davies. Synthesis and dielectric properties of $Li_{1-x+y}Ta_{1-x-3y}Ti_{x+4y}O_3$ M-phase solid solutions. J. Am. Ceram. Soc. 85(2002)2487–2491.
[558] A. Borisevich and P. K. Davies. Microwave dielectric properties of $Li_{1+x-y}M_{1-x-3y}Ti_{x+4y}O_3$ ($M = Nb^{5+}, Ta^{5+}$) solid solutions. J. Eur. Ceram. Soc. 21(2001)1719–1722.
[559] R. Ratheesh, H. Sreemoolanathan, S. Suma, M. T. Sebastian, K. A. Jose, and P. Mohanan. New high permittivity and low loss ceramics in the BaO–TiO_2–Nb_2O_5 system. J. Mater. Sci. Mater. Electr. 9(1998)291–294.
[560] P. Liu, E. S. Kim, S.-G. Kang, and H. S. Jang. Microwave dielectric properties of $Ca(Li_{1/3}Nb_{2/3})_{1-x}Ti_{3x})O_{3-\delta}$ ceramics with B_2O_3. Mater. Chem. Phys. 79(2003)270–272.
[561] C.-L. Liu and T.-B. Wu. Effect of calcium substitution on the structural and microwave dielectric characteritstics of $[(Pb_{1-x}Ca_x]_{1/2}La_{1/2}](Mg_{1/2}Nb_{1/2})O_3$ ceramics. J. Am. Ceram. Soc. 84(2001)1291–1295.
[562] A. N. Salak, M. P. Seabra, and V. M. Ferreira. $La(Mg_{1/2}Ti_{1/2})O_3$–$La_{2/3}TiO_3$ microwave dielectric ceramics. J. Eur. Ceram. Soc. 23(2003)2409–2412.
[563] D.-H. Yeo, J. B. Kim, J. H. Moon, S. J. Yoon, and H. J. Kim. Dielectric properties of $(1-x)CaTiO_3$-$xLa(Zn_{1/2}Ti_{1/2})O_3$ ceramics at microwave frequencies. Jpn. J. Appl. Phys. 35(1996)663–667.
[564] L. Gowri. M. Sc Project Report, Regional Research Laborator (2004).
[565] P. V. Bijumon and M. T. Sebastian. Influence of glass additives on the microwave dielectric properties of $Ca_5Nb_2TiO_{12}$ ceramics. Mater. Sci. Eng. B123(2005)31–40.
[566] D. Houivet, J. El Fallah, J. Bernad, F. Roulland, and J. M. Haussonne. Microwave properties and microstructure of $La_{2/3}TiO_3$ stabilised with NiO. J. Eur. Ceram. Soc. 21(2001)1715–1718.

[567] Y.-C. Liou and C.-Y. Liu. A simple and effective process for $Ba_3La_2Ti_2Nb_2O_{15}$ microwave ceramics. Mater. Sci. Eng. A448(2007)351–355.
[568] E. S. Kim, Y. H. Kim, B. S. Chun, Y. T. Kim, and K. H. Yoon. Effect of Ti^{4+} substitution on microwave dielectric properties of $(Pb_{0.2}Ca_{0.8})$ $[(Ca_{1/3}Nb_{2/3})_{1-x}Ti_x]O_3$ ceramics. Mater. Chem. Phys. 79(2003)233–235.
[569] H. B. Hong, D. W. Kim, and K. S. Hong. Microwave dielectric properties of $Bi_2(Zn_{1/3}Ta_{2/3})_2O_7$ polymorphs. Jpn. J. Appl. Phys. 42(2003)5172–5175.
[570] N. Ichinose and K. Mutoh. Microwave dielectric properties in the (1−x) $(Na_{1/2}La_{1/2})TiO_3$-x$(Li_{1/2}Sm_{1/2})TiO_3$ ceramic system. J. Eur Ceram. Soc. 23(2003)2455–2459.
[571] K. Ezaki, Y. Baba, H. Takahashi, H. Shibata, and S. Nakano. Microwave dielectric properties of $CaO–Li_2O–Ln_2O_3–TiO_2$ ceramics. Jpn. J. Appl. Phys. 32(1993)4319–4322.
[572] N. Santha and M. T. Sebastian. Low temperature sintering and microwave dielectric properties of $Ba_4Sm_{9.33}Ti_{18}O_{54}$. Mater. Res. Bull. (2008) (in press).
[573] J.-W. Choi, J.-Y. Ha, C.-Y. Kang, S.-J. Yoon, H.-J. Kim, and K. H. Yoon. Microwave dielectric properties of $(Pb,Ca)(Mg,Nb,Sn)O_3$ ceramics. Jpn. J. Appl. Phys. 39(2000) 5923–5926.
[574] K.-W. Wang, H. T. Kim, M. Lanangan, and T. Shrout. Low temperature sintering and microwave dielectric properties of $CaTi_{1-x}(Fe_{.5}Nb_{.5})_xO_3$ ceramics with B_2O_3 addition. Mater. Res. Bull. 41(2006)1385–1391.
[575] K. Wada, Y. Fukami, K.-I. Kakemoto, and H. Ohsato. Microwave dielectric properties of textured $BaLa_4Ti_4O_{15}$ ceramics. With layered perovskite structure. Jpn. J. Appl. Phys. 44(2005)7094–7097.
[576] K. Fukuda, R. Kitoh, and I. Awai. Microwave dielectric characteristics of $Bi_2O_3–TiO_2$ dielectric resonator. Jpn. J. Appl. Phys. 32(1993)4584–4588.
[577] E. S. Kim, S. H. Kim, S. G. Kang, and K. H. Yoon. Microwave dielectric properties of $(Ca_{0.85}Nd_{0.1})$ $[Ti_{1-x}(Mg_{0.33}Nb_{0.67})_x]O_3$ ceramics. J. Ceram. Soc. Jpn. Suppl. 112–1 (2004)S1641–S1644.
[578] A.-Y. Borisevich and P. K. Davies. Crystalline structure and dielectric properties $Li_{1+x-y}Nb_{1-x-3y}Ti_{x+4y}$ M-phase solid soltions. J. Am. Ceram. Soc. 85(2002)573–578.
[579] J. J. Bian, K. Yan, and R. Ubic. Structure and microwave dielectric properties of $Sm_{(2-x)/3}Li_xTiO_3$. J. Electroceram. 18(2007)283–288.
[580] X. Hu and X. M. Chen. Microstructure and dielectric properties of $Ca(Fe_{1/2}Nb_{1/2})O_3$ modified $PbZrO_3$ ceramics. Jpn. J. Appl. Phys. 43(2004)215–218.
[581] H. Zhang, L. Fang, R. Dronskowski, P. Muller, and R. Z. Yuan. Some $A_6B_5O_{18}$ cation deficient perovskites in the $BaO–La_2O_3–TiO_2–Nb_2O_5$ system. J. Sol. St. Chem. 177(2004)4007–4012.
[582] K. Fukuda, R. Kitoh, and I. Awai. Microwave characteristics of mixed phases of $BaTi_4O_9–BaPr_2Ti_5O_{14}$ ceramics. J. Mater. Sci. 30(1995)1209–1216.
[583] G. M. Keith, M. J. Pampling, K. Sarma, N. Mc N. Alford, and D. C. Sinclair. Synthesis and characterization of doped 6H $BaTiO_3$ ceramics. J. Eur. Ceram. Soc. 24(2004)1721–1724.
[584] F. Zhao, Z. Yue, Z. Gui, and L. Li. Effects of Zn substitution on crystal structure and microwave dielectric properties of $CaLa_4Ti_5O_{17}$ ceramics. J. Am. Ceram. Soc. 89(2006)3421–3423.
[585] D. H. Kang, K. C. Nahm, and H. J. Cha. Effect of $Li_2O–V_2O_5$ on the low temperature sintering and microwave dielectric properties of $LiNb_{0.6}Ti_{0.5}O_3$ ceramics. J. Eur. Ceram. Soc. 26(2006)2117–2121.
[586] A. Feteira, R. Slsebrock, A. Dias, R. L. Mmoreira, M. T. Lanagan, and D. C. Sinclair. Synthesis and characterization of $La_{0.4}B_{0.6}Ti_{0.6}RE_{0.4}O_3$ (RE = Y,Yb) ceramics. J. Eur. Ceram. Soc. 26(2006)1947–1951.
[587] H. Kagata, T. Inoue, and J. Kato. Low fire bismuth based dielectric ceramics for microwave use. Jpn. J. Appl. Phys. 31(1992)3152–3155.
[588] E. S. Kim, B. S. Chun, and D. H. Kang. Effects of structural characteristics on microwave dielectric properties of (1−x)$Ca_{0.85}Nd_{0.1}TiO_3$-x$LnAlO_3$ (Ln = Sm, Er, and Dy) ceramics. J. Eur. Ceram. Soc. 27(2007)3005–3010.

[589] H.-J. Kim, S. Kucheiko, S.-J. Yoon, and H. J. Jung. Microwave dielectrics in the $(La_{1/2}Na_{1/2})TiO_3$–$Ca(Fe_{1/2}Nb_{1/2})O_3$ system. J. Am. Ceram. Soc. 80(1997)1316–1318.
[590] H. Kagata, J. Kato, K. Nishimoto, and T. Inoue. Dielectric properties of Pb based perovskite substituted by Ti for B site at microwave frequencies. Jpn. J. Appl. Phys. 32(1993) 4332–4334.
[591] S.-F. Yang, Y.-C. Chuan, Y.-C. Hsu, J. P. Chu, and C.-H. Wu. Properties of hexagonal $Ba(Ti_{1-x}Mn_x)O_3$ ceramics. Effects of sintering temperature and Mn content. Jpn. J. Appl. Phys. 46(2007)2978–2983.
[592] M.-H. Kim, J.-B. Lim, S. Nahm, J.-H. Paik, and H.-J. Lee. Low temperature sintering of $BaCu(B_2O_5)$-added BaO–Re_2O_3–TiO_2 (Re = Sm, Nd) ceramics. J. Eur. Ceram. Soc. 27(2007)3033–3037.
[593] R. Ratheesh, H. S. Nathan, M. T. Sebastian, and P. Mohanan. Preparation characterization and dielectric properties of ceramics in the BaO-Nd_2O_3-TiO_2 system. Ferroelectrics. 211(1998)1–8.
[594] T. Liu, X.-Z. Zhao, and W. Chen. A/B site modified $CaTiO_3$ dielectric ceramics for microwave application. J. Am. Ceram. Soc. 89(2006)1153–1155.
[595] H. Zheng, I. M. Reaney, D. Muir, T. Price, and D. Iddles. Effect of glass additions on the sintering and microwave properties of composite dielectric ceramics based on BaO–Ln_2O_3–TiO_2 (Ln = Nd, La). J. Eur. Ceram. Soc. 27(2007)4479–4487.
[596] H. Yoon, H. S. Park, J. Y. Cho, and E. S. Kim. Microwave dielectric properties of $(Pb_{.4}Ca_{.6})(Fe_{.5}Ta_{.5})O_3$ ceramics prepared by mechanochemical processing. J. Eur. Ceram. Soc. 23(2003)2579–2582.
[597] H. Zheng, I. M. Reaney, D. Muir, T. Price, and D. Iddles. Composite dielectric ceramics based on BaO–Ln_2O_3–TiO_2 (Ln = Nd,La). Jpn. J. Appl. Phys. 44(2005)3087–3090.
[598] B. Shen, X. Yao, L. Kang, and D. Peng. Effect of CuO or and V_2O_5 oxide additives on Bi_2O_3–ZnO–Ta_2O_5 based ceramics. Ceram. Int. 30(2004)1203–1206.
[599] H. Ohsato, A. Harada, T. Okawa, and H. Okabe. Microwave dielectric composition and method for producing the same. Eur. Pat. EP1422207A1.
[600] R. Ubic and I. M. Reaney. Microwave properties of doped lead pyroniobates. J. Eur. Ceram. Soc. 21(2001)2659–2662.
[601] N. Santha, M. T. Sebastian, V. George, and J. Philip. Microwave dielectric properties of glass-added $Ba_{6-3x}R_{8+2x}Ti_{18}O_{54}$ (x = 2/3). Mater. Chem. Phys. 100(2006)423–429.
[602] J. Petzelt, E. Buixaderas, G. Komandin, A. V. Pronin, M. Valant, and D. Suvorov. Infrared dielectric response of the $La_{2/3}TiO_3$–$LaAlO_3$ microwave ceramic system. Mater. Sci. Eng. B 57(1998)40–45.
[603] D. K. Yim, J.-R. Kim, D.-W. Kim, and K. S. Hong. Microwave dielectric properties and low temperature sintering of $Ba_3Ti_4Nb_4O_{21}$ ceramics with B_2O_3 and CuO additions. J. Eur. Ceram. Soc. 27(2007)3053–3057.
[604] Y-R. Wang, S-F. Wang, C-K. Ken. Lowfire $(Zr_{0.8}Sn_{0.2})TiO_4$ with glass additives. Mater. Sc. Eng. A 426(2006)143–146.
[605] M. Nakano, K. Suzuki, T. Miura, and M. Kobayashi. Low temperature fireable dielectric material $Pb(Fe_{2/3}W_{1/3})O_3$–$(PbCa)(Fe_{1/2}Nb_{1/2})O_3$ for microwave use. Jpn. J. Appl. Phys. 32(1993)4314–4318.
[606] M. Mizuta, K. Uenoyama, H. Ohsato, S. Nishigaki, and T. Okuda. Formation of tungsten bronze type $(Ba_{6-3x}Sm_{8+2x})_{\alpha}Ti_{18-y}Al_yO_{54}$(= 1 + y/36)solid solutions and microwave dielectric properties. Jpn. J. Appl. Phys. 35(1996)5065–5068.
[607] O. Dernovsek, A. Naeini, G. Preu, W. Wersing, M. Eberstein, and W. A. Schiller. LTCC m glass-ceramic composites for microwave application. J. Eur. Ceram. Soc. 21(2001) 1693–1697.
[608] X. M. Chen and Y. Li. A and B site cosubstituted $Ba_{6-3x}Sm_{8+2x}Ti_{18}O_{54}$ microwave dielectrics ceramics. J. Am. Ceram. Soc. 85(2002)579–584.
[609] M. Hu, D. Zhou, D. Zhang, W. Lu, B. Li, J. Huang, and S. Gong. Microwave dielectric properties of $(Pb,Ca)(Fe,Nb,Zr)O_3$ ceramics. Mater. Sci. Eng. B99(2003)403–407.

[610] M. Valant, D. Suvorov, and D. Kolar. High permittivitty microwave ceramics-the inglu-ence of $Ba_{4.5}Gd_9Ti_{18}O_{54}$ addition on the temperature stability of dielectric properties. In: Proc. Conf. Euroceramics. Wasser et al. (Ed.), RWTH (1994) pp. 146–148.
[611] Y.-C. Chen and C.-L. Huang. Microwave dielectric properties of $Ba_{2-x}Sm_{4+2/3}$ xTi_9O_{26} ceramics with zero temperature coefficient. Mater. Sci. Eng. A334(2002) 250–256.
[612] S.-Y. Chen and Y.-J. Lin. Effect of CuO based oxide additives o Bi_2O_3–ZnO–Nb_2O_5 microwave ceramics. Jpn. J. Appl. Phys. 40(2001)3305–3310.
[613] Q. Zeng, W. Li, J.-l. Shi, J.-K. Guo, H. C. Chen, and M.-l. Liu. Effect of B_2O_3 addition on the sintering and microwave dielectric properties of of M-phase $Li_{0.6}Nb_{0.5}TiO_3$ ceramics. J. Eur. Ceram. Soc. 27(2007)261–265.
[614] S. Kucheiko, J.-W. Choi, H.-J. Kim, S.-J. Yoon, and H.-J. Jung. Microwave characteristics of (Pb,Ca)(Fe,Nb,Sn)O_3 dielctric materials. J. Am. Ceram. Soc. 80(1997)2937–2940.
[615] J.-S. Kim, C. I. Cheon, H.-J. Kang, C.-H. Lee, K.-Y. Kim, S. Nahm, and J.-D. Byun. Crystal structure and microwave dielectric properties of $CaTiO_3$–$(Li_{1/2}Nd_{1/2})TiO_3$–$(Ln_{1/3}Nd_{1/3})TiO_3$(Ln = La, Dy) ceramics. Jpn. J. Appl. Phys. 38(1999)5633–5637.
[616] Y.-C. Chen and C.-L. Huang. Microwave dielectric properties of $Ba_{2-x}Sm_{4+2/3x}Ti_9O_{28}$ ceramics with zero temperature coefficient. Mater. Sci. Eng. A334(2002)250–256.
[617] C.-L. Huang and M.-H. Weng. The effect of PbO loss on the microwave dielectric of (Pb,Ca)(Zr,Ti)O_3 ceramics. Mater. Res. Bull. 36(2001)683–691.
[618] H. Ohsato, A. Komura, Y. Takagi, S. Nishigaki, and T. Okuda. Microwave dielectric properties of and sintering of $Ba_{6-3x}R_{8+2x}Ti_{18}O_{54}$ (R = Sm, x = 2/3) solid solutions with added rutile. Jpn. J. Appl. Phys. 37(1998)5357–5359.
[619] C.-F. Yang. Microwave properties of $(Ba_{1-x}Sr_x)Sm_2Ti_4O_{12}$ ceramics and its application as dielectric resonator antenna. Jpn. J. Appl. Phys. 39(2000)1430–1434.
[620] R. Rejini, G. Subodh, and M. T. Sebastian. $Ca_4La_2Ti_5O_{17}$: A novel low loss dielectric ceramics in the CaO–La_2O_3–TiO_2 system. J. Mater. Sci. Mater. Electr. (2008) (in press).
[621] Y. Futamata, H. Ohsato, K. Kakimoto, and S. Nishigaki. Microwave Dielectric Properties of $Ba_{6-3x}Eu_{8+2x}Ti_{18}O_{54}$ (x = 2/3) with Sm and Nd Substituted for Eu. Ferroelectrics. 272(2002)213–218.
[622] S.-H. Yoon, D.-W. Kim, S.-Y. Cho, and K. S. Hong. Phase analysis and microwave dielectric properties of LTCC TiO_2 glass system. J. Eur. Ceram. Soc. 23(2003)2549–2552.
[623] K. Fukuda, I. Fijui, R. Kitoh, Y. Cho, and I. Awai. Influence of rare earth ions on BaO–TiO_2-rare earth oxide ceramics for microwave applications. Jpn. J. Appl. Phys. 32(1993)1712–1715.
[624] A. Fetira, K. Sarma, N. Mc N. Alford, I. M. Reaney, and D. C. Sinclair. Microwave dielectric properties of gallium doped hexagonal barium titanate ceramics. J. Am. Ceram. Soc. 86(2003)511–513.
[625] H. Wang, S. Kamba, M. Zhang, X. Yao, S. Denisov, F. Kadlec, and J. Petzelt. Microwave and infrared dielectric response of monoclinic bismuth zinc niobate based pyrochlore ceramics with ion substitution in A site. J. Appl. Phys. 100(2006)034109.
[626] A. G. Belous and O. V. Ovchar. Temperature compensated microwave dielectrics based on lithium containing titanates. J. Eur. Ceram. Soc. 23(2003)2525–2528.
[627] Y. Ota, K. Kakemoto, H. Ohsato, and T. Okawa. Low temperature sintering of $Ba_{6-3x}Sm_{8+2x}Ti_{18}O_{54}$ microwave dielectric ceramics by B_2O_3 and GeO_2 addition. J. Eur. Ceram. Soc. 24(2004)1755–1760.
[628] M. W. Zuo, W. Li, J.-L. Shi, and Q. Zeng. Influence of CuO addition to $BaSm_2Ti_4O_{12}$ microwave ceramics on sintering behavior and dielectric properties. Mater. Res. Bull. 41(2006)1127–1132.
[629] X. M. Chen and Y. Li. Raman spectra of Nd/Sn substituted $Ba_{6-3x}Sm_{8+2x}Ti_{18}O_{54}$ microwave dielectric ceramics. J. Appl. Phys. 96(2004)5683–5686.
[630] M. Valant and D. Suvorov. Incorporation of bismuth into $Ba_{6-x}R_{8+2x}Ti_{18}O_{54}$ (R = Nd,Gd). J. Mater. Sci. 36(2001)2991–2997.

[631] G.-K. Choi, D.-W. Kim, S.-Y. Cho, and K. S. Hong. Influence of V_2O_5 substitutions to $Bi_2(Zn_{1/3}Nb_{2/3})_2O_7$ pyrochlore on sintering temperature and dielectric properties. Cer Int. 30(2004)1187–1190.

[632] J. M. Durand and J. P. Boilot. Microwave dielectric characteristics of $BaO–Bi_2O_3–TiO_2–Nd_2O_3$ dielectric resonators. J. Mater. Sci. Lett. 6(1987)134–136.

[633] M. Imaeda, K. Ito, M. Mizuta, H. Ohsato, S. Nishigaki, and T. Okuda. Microwave dielectric properties of $Ba_{6-3x}Sm_{8+2x}Ti_{18}O_{54}$ solid solutions with Sr substituted for Ba. Jpn. J. Appl. Phys. 36(1997)6012–6015.

[634] H. Ohsato, S. Nishigaki, and T. Okuda. Superlattice and dielectric properties of $BaO–R_2O_3–TiO_2$ (R = La, Nd, Sm) microwave dielectric compounds. Jpn. J. Appl. Phys. 31(1992)3136–3138.

[635] P. Laffezz, G. Desgardin, and B. Raveau. Influence of calcinations, sintering and composition upon microwave properties of the $Ba_{6-x}Sm_{8+2x/3}Ti_{18}O_{54}$ type oxide. J. Mater. Sci. 27(1992)5229–5238.

[636] H. Ohsato, J. Sugino, A. Komura, S. Nishigaki, and T. Okuda. Microwave dielectric properties of $Ba_4(Nd_{28/3-y}R_y)Ti_{18}O_{54}$ (R = Eu, Dy, Ho, Er, and Yb) solid solutions. Jpn. J. Appl. Phys. 38(1999)5625–5628.

[637] P.-S. Cheng, C.-F. Yang, Y.-C. Chen, and W.-C. Tzou. The reaction sequence and dielectric properties of $BaSm_2Ti_4O_{12}$ ceramics. Ceram. Int. 26(2000)877.

[638] Y.-W. Liu and P. Lin. Phase formation and microwave dielectric properties of Pb^{2+} and Sr^{2+} doped $La_4Ti_9O_{24}$ ceramics. Mater. Res. Bull. 41(2006)1845–1853.

[639] K. H. Yoon, Y. H. Chang, W. S. Kim, J. B. Kim, and E. S. Kim. Dielectric properties of $Ca_{1-x}Sm_{2x/3}TiO_3–Li_{1/2}Ln_{1/2}TiO_3$ ceramics. Jpn. J. Appl. Phys. 35(1996)5145–5149.

[640] S. Nishigaki, H. Kato, S. Yano, and R. Kamimura. Microwave dielectric properties of $(Ba,Sr)O–Sm_2O_3–TiO_2$ ceramics. Am. Ceram. Soc. Bull. 66(1987)1405–1410.

[641] Y. Li and X. M. Chen. Effects of sintering conditions on microstructure and microwave dielectric properties of $Ba_{6-3x}(Sm_{1-y}Nd_y)_{8+2x}Ti_{18}O_{54}$ ceramics (x = 2/3). J. Eur. Ceram. Soc. 22(2002)715–719.

[642] M. Valant, D. Suvorov, and D. Kolar. Microwave Ceramics of High Dielectric Constant. 3rd Euroceramics. P. Duram and J. F. Fernandez (Eds), Faenza Editrice Iberics S.L. (Spain) Vol. 2, (1993) pp. 235–240.

[643] Z. X. Xiong, C. Fang, J. R. Huang, H. Qiu, and M. Lin. Preparation and properties of $Sr(Bi_{1-x}Nd_x)Ti_2O_{27}$ ceramics for wireless communication. Mater. Sci. Eng. B99(2000) 138–142.

[644] H. Ohsato, M. Imaeda, Y. Takagi, A. Komura, and T. Ohuda. Microwave quality factor improved by ordering of Ba and rare earth on the tungsten bronze type $Ba_{6-3x}R_{8+2x}$-$Ti_{18}O_{54}$ (R = La, Nd, Sm) solid solutions. Proc. Symp. Applications of ferroelectrics XI, Switzerland, August. (1998) pp. 509–512.

[645] Y. Xu, G. Huang, and Y. He. Sol gel preparation of $Ba_{6-3x}Sm_{8+2x}Ti_{18}O_{54}$ microwave dielectric ceramics. Cer Int. 31(2005)21–25.

[646] Y. Zheng, X. Zhao, W. Lei, and S. Wang. Effects of Bi_2O_3 addition on the microstructures and microwave dielectric characteristics of $Ba_{6-3x}(Sm_{0.2}Nd_{0.8})_{8+2x}Ti_{18}O_{54}$ (x = 2/3) ceramics. Mater. Lett. 60(2005)459–463.

[647] A. G. Belous and O. V. Ovchar. MW dielectrics with perovskite like structure based on Sm containing system. J. Eur. Ceram. Soc. 19(1999)1119–1122.

[648] A.-K. Axelsson and N. Mc N. Alford. Bismuth titanates candidates for high permittivity LTCC. J. Eur. Ceram. Soc. 26(2006)1933–1936.

[649] Y. Guo, K. Kakemoto, and H. Ohsato. Microwave dielectric properties of $Ba_{6-3x}Sm_{8+2x}$-$Ti_{18}O_{54}$ ceramics (x = 2/3) prepared by spark plasma sintering. Jpn. J. Appl. Phys. 42(2003)7410–7413.

[650] A. Webhoffer and A. Feltz. Microwave dielectric properties of ceramics of the system $Ba_{6-3x}(Sm_yNd_{1-y})_{8+2x}Ti_{18}O_{54}$. J. Mater. Sci. Lett. 18(1999)719–721.

[651] K. H. Yoon, E. S. Kim, and J. S. Jeon. Understanding the microwave dielectric properties of $(Pb_{0.45}Ca_{0.55})[Fe_{0.5}(Nb_{1-x}Ta_x)_{0.5}]O_3$. J. Eur. Ceram. Soc. 5(2003)2391–2396.

[652] H. Ohsato, A. Komura, and S. Nishigaki. Occupation in the trigonal columns of tungsten bronze like $Ba_{6-3x}Sm_{8+2x}Ti_{18}O_{54}$ solid solutions. J. Ceram. Soc. Jpn. Suppl. 112(2004)S1616–S1621.

[653] A. G. Belous, O. Ovchar, M. Valant, D. Suvorov, and D. Kolar. The effect of partial isovalent substitution in the A sublattice on MW properties of materials based on $Ba_{6-x}Ln_{8+2x/3}Ti_{18}O_{54}$ solid solutions. J. Eur. Ceram. Soc. 21(2001)2723–2730.

[654] T. Okawa, M. Imaeda, and H. Ohsato. Microwave dielectric properties of Bi added $Ba_4Nd_{9+1/3}Ti_{18}O_{54}$ solid solutions. Jpn. J. Appl. Phys. 39(2000)5645–5649.

[655] H. Ohsato, H. Kato, M. Mizuta, S. Nishigaki, and T. Okuda. Microwave dielectric properties of the $Ba_{6-3x}(Sm_{1-y}R_y)_{8+2x}Ti_{18}O_{54}$ (R = Nd and La) solid solutions with zero temperature coefficient of the resonant frequency. Jpn J. Appl. Phys. 34(1995) 5413–5417.

[656] J.-Y. Ha, J.-W. Choi, H.-J. Kim, S.-J. Yoon, and K. H. Yoon. The effect of sintering aid $CuO–Bi_2O_3$ on microwave dielectric properties of $(Pb_{0.45}Ca_{0.55})[(Fe_{0.5}Nb_{0.5})_{0.9}Sn_{0.1}]O_3$ ceramics. Mater. Chem. Phys. 79(2003)261–265.

[657] N. Qin and X. M. Chen. Effect of Sm//Bi cosubstitution on microstructures and microwave dielectric characteristics of $Ba_{6-3x}La_{8+2x}Ti_{18}O_{54}$ (x = 2/3) solid solution. Mater Sci. Eng. B111(2004)90–94.

[658] E. S. Kim, H. S. Park, and K. H. Yoon. Porosity dependence of microwave dielectric properties of $(Pb_{0.5}Ca_{0.5})(Fe_{0.5}Ta_{0.5})O_3$ ceramics. Mater. Chem. Phys. 79(2003)213–217.

[659] C. Hoffmann and R. Waser. Hotforging of $Ba_{6-3x}RE_{8+2x}Ti_{18}O_{54}$ ceramics (RE = La, Ce, Nd, Sm). Ferroelectrics. 201(1997)127–135.

[660] K. Fukuda, R. Kitoh, and I. Awai. Microwave characteristics of $BaPr_2Ti_4O_{12}$ and $BaPr_2Ti_5O_{14}$ ceramics. J. Mater. Res. 10(1995)312–319.

[661] W. Choi, H. J. Kim, S. J. Yoon, and J. Y. Ha. The effect of sintered aid $CuO–Bi_2O_3$ on microwave dielectric properties of $(Pb,Ca)(Fe,Nb)O_3$. Mater. Chem. Phys. 79(2003)261–265.

[662] Y. Yamada, K. Wada, I. Kagomiya, K. Kakimoto, and H. Ohsato. Soft chemical reaction of layered perovskite $Na_2Nd_2Ti_3O_{10}$ and its microwave dielectric properties. J. Eur. Ceram. Soc. 27(2007)3049–3052.

[663] J.-S. Sun, C.-C. Wei, and L. Wu. Dielectric properties of $(Ba,Sr)O–(Sm,La)_2O_3–TiO_2$ ceramics at microwave frequencies. J. Mater. Sci. 27(1992)5818–5822.

[664] T. Negas and P. K. Davies. Influence of chemistry and processing on the electrical properties of $Ba_{6-3x}Ln_{8+2x}Ti_{18}O_{54}$ solid solutions. In: Materials and Processes for Wireless Communication. T. Negas and H. Ling (Eds), American Ceram. Soc, Westerville, OH (1995) pp. 179–196.

[665] Z. X. Xiong, C. Fang, J. R. Huang, H. Qiu, and M. Lin. Preparation and properties of $Sr(Bi_{1-x}Nd_x)_8Ti_7O_{27}$ ceramics for wireless technologies. Mater. Sci. Eng. B99(2003)138–142.

[666] M. Podlinik, M. Valant, and D. Suvorov. The influence of isovalent doping on the microwave dielectric properties of $Ba_{6-3x}Nd_{8+2x}Ti_{18}O_{54}$ based ceramics. Key Engn. Mater. 132–136(1997)1211–1214.

[667] Y. J. Wu and X. M. Chen. Bismuth/samarium cosubstituted $Ba_{6-3x}Nd_{8+2x}Ti_{18}O_{54}$ microwave dielectric ceramics. J. Am. Ceram. Soc. 83(2000)1837–1839.

[668] C. Qin, Z. Yue, Z. Gui, and L. Li. Low fired $(Pb,Ca)(FeNb)O_3$ ceramics for multilayer filter applications. Mater. Sci. Eng. B99(2003)286–289.

[669] Y. J. Wu, and X. M. Chen. Structures and microwave dielectric properties of $Ba_{6-3x}(Nd\text{-}Bi)_{8+2x}Ti_{18}O_{54}$(x = 2/3) solid solutions. J. Mater. Res. 16(2001)1734–1738.

[670] I. Kim, W. H. Jung, Y. Inaguma, T. Nakamura, and M. Itoh. Dielectric properties of a site deficient perovskite type lanthanum-calcium- titanium oxide solid solution. Mater. Res. Bull. 30(1995)307–316.

[671] A. Ioachim, M. I. Toacsan, M. G. Banciu, L. Nedeku, H. Alexandru, C. Berbecaru, D. G. Hetur, and G. Stoica. BNT ceramics synthesis and characterisation. Mater. Sci. Eng. B109(2004)183–187.

[672] N. Ichinose, and H. Amada. Preparation and microwave dielectric properties of BaO–$(Sm_{1-x}La_x)_2O_3$–$5TiO_2$ ceramic system. J. Eur. Ceram. Soc. 21(2001)2751–2753.
[673] M. Valant and D. Suvorov. Dielectric properties of fluorite like Bi_2O_3–Nb_2O_5 solid solution and Bi_3NbO_7. J. Am. Ceram. Soc. 86(2003)939–944.
[674] M. Valant, D. Suvorov, and D. Kolar. High permitivitty microwave ceramics-the influence of $Ba_{4.5}Gd_9Ti_{18}O_{54}$ addition on the temperature stability of dielectric properties. Proc. Conf. Euroceramics, Aachen, Waser. (Ed.) (1994) pp. 146– 147 Ed.
[675] W. S. Kim, E. S. Kim, and K. H. Yoon. Effect of excess TiO_2 on the microwave dielectric properties of the $Ca_{2/5}Sm_{2/5}TiO_3$–$Li_{1/2}Ln_{1/2}$ TiO_3 (Ln = Sm, Nd) ceramics. Ferroelectrics. 223(1999)277–284.
[676] H.-L. Chen and C.-L. Huang. Microwave dielectric properties and microstructure $Ca_{1-x}Nd_{2x/3}TiO_3$–$Li_{1/2}Nd_{1/2}TiO_3$ ceramics. Jpn. J. Appl. Phys. 41(2002)5650–5653.
[677] T. Okawa, M. Imaeda, and H. Ohsato. Microwave dielectric properties of $Ba_4(Sm_{1-y}Bi_y)_{9+1/3}Ti_{18}O_{54}$. Key Eng. Mater. 214–215(2002)37–42.
[678] Z. D. Xiang, H. M. Zue, J. S. Lin, and H. Jing. Microwave properties of CeO_2 modified (Pb,Ca)(Fe,Nb) dielectric ceramics. Mater. Lett. 58(2004)2684–2687.
[679] W. S. Kim, E. S. Kim, and K. H. Yoon. Effects of Sm^{3+} substitution on dielectric properties of $Ca_{1-x}Sm_{2x/3}TiO_3$ ceramics at microwave frequencies. J. Am. Ceram. Soc. 82(1999)2111–2115.
[680] Y. C. Chen, P.-S. Cheng, C.-F. Yang, and W.-C. Tzou. The microwave dielectric properties of CaO–BaO–Li_2O–Sm_2O_3–TiO_2 ceramics. J. Mater. Sci. Lett. 20(2001)863–865.
[681] Y. J. Wu and X. M. Chen. Modified $Ba_{6-3x}Nd_{8+2x}Ti_{18}O_{54}$ microwave dielectric ceramics. J. Eur. Ceram. Soc. 19(1999)1123–1126.
[682] H. Ohsato. Science of tungsten-bronze type $Ba_{6-3x}R_{8+2x}Ti_{18}O_{54}$ (R = rare earth) microwave dielectric solid solutions. J. Eur. Ceram. Soc. 21(2001)2703–2711.
[683] H. Ohsato and M. Imaeda. The quality factor of the microwave dielectric materials based on crystal structure –as an example: $Ba_{6-3x}R_{8+2x}Ti_{18}O_{54}$ (R = rare earth) solid solutions. Mater. Chem. Phys. 79(2000)208–212.
[684] C.-L. Huang, J.-T. Tsai, and J.-L. Li. Microwave dielectric properties of $(1-x)$CaO–xBaO–Li_2O–$(1-y)Sm_2O_3$–yNd_2O_3–TiO_2 ceramic system. J. Mater. Sci. 35(2000)4901–4905.
[685] D. Suvorov, M. Valant, and D. Kolar. Chemistry and microwave properties of Bi_2O_3 doped $Ba_{4.5}Nd_9Ti_{18}O_{54}$ based ceramics. Materials and Processes for Wireless Communication, Ed. T. Negas and H. Ling, Ceram. Trans. 53(1995)197–206.
[686] H. Takahashi, Y. Baba, K. Ezaki, and K. Shibata. Microwave dielectric properties and crystal structure of CaO–Li_2O–$(1-x)Sm_2O_3$–xLn_2O_3–TiO_2 (Ln = lanthanide) ceramic system. Jpn. J. Appl. Phys. 35(1996)5069–5073.
[687] D.-W. Kim, B. Park, J.-H. Chung, and K. S. Hong. Mixture behavior and microwave dielectric properties of low fired CuO–TiO_2. Jpn. J. Appl. Phys. 39(2000)2696–2700.
[688] W. S. Kim, K. H. Yoon, and E. S. Kim. Microwave dielectric characterisctic of $Ca_{2/5}Sm_{2/5}$ TiO_3–$Li_{1/2}Nd_{1/2}TiO_3$ ceramics. Jpn. J. Appl. Phys. 39(2000)5650–5653.
[689] H. Takahashi, Y. Baba, and K. Ezaki. Dielectric characteristics of $A^{1+}{}_{1/2}A^{3+}{}_{1/2}TiO_3$ ceramics at microwave frequencies. Jpn. J. Appl. Phys. 30(1991)2339–2342.
[690] K. H. Yoon, M. S. Park, J. Y. Cho, and E. S. Kim. Effect of B_2O_3–Li_2O on microwave dielectric properties of $(Ca_{0.275}Sm_{0.4}Li_{0.25})TiO_3$ ceramics. J. Eur. Ceram. Soc. 23(2003)2423–2427.
[691] G. Huang, D. Zhou, J. Yu, Z. Zheng, and S. Gong. Synthesis of $(1-x)Ca_{2/5}Sm_{2/5}TiO_3$-x $Li_{1/2}Nd_{1/2}TiO_3$ ceramic powder via ethylenediaminetetracacetioc acid precursor. Mater. Res. Bull. 40(2005)13–19.
[692] M. Yoshida, N. Hara, T. Takada, and A. Seki. Structure and dielectric properties $Ca_{1-x}Nd_{2x/3}TiO_3$. Jpn. J. Appl. Phys. 36(1997)6818–6823.
[693] Q. Yang, E. S. Kim, J. Xu, and Z. Meng. Effect of La^{3+} and Nd^{3+} on the microwave dielectric properties of $(Pb_{0.5}Ca_{0.5})(Fe_{0.5}Nb_{0.5})O_3$ ceramics. Mater. Sci. Eng. B99(2004)259–261.

[694] Q. H. Yang, E. S. Kim, and J. Xu. Effect of Nd^{3+} on the microwave dielectric properties of $(Pb_{0.5}Ca_{0.5})(Fe_{0.5}Nb_{0.5})O_3$ ceramics. Mater. Sci Eng. B113(2004)224–227.
[695] J. Kato, H. Kagata, and K. Nishimoto. Dielectric properties of lead alkaline earth zirconate at microwave frequencies. Jpn. J. Appl. Phys. 30(1991)2343–2346.
[696] S. L. Schwartz, and T. R. Shrout. US Patent, 5449652 (1995).
[697] A. Templeton, X. Wang, S. J. Penn, S. J. Webb, L. F. Cohn, and N. Mc N. Alford. Microwave dielectric loss of titanium oxide. J. Am. Ceram. Soc. 83(2000)95–100.
[698] C. L. Huang, J.-T. Tsai, and Y. B. Chan. Dielectric properties of $(1-y)Ca_{1-x}La_{2x/3}TiO_3$-y $Li_{1/2}Ti_{1/2}O_3$ ceramic system at microwave frequency. Mater. Res. Bull. 36(2002)547–556.
[699] E. R. Kipokoch, and W. Lu. Effect of Li_2O–Bi_2O_3–TiO_2 additions on the microwave dielectric properties of CaO–Sm_2O_3–Li_2O–TiO_2 Ceramics. Int. Conf. Microwave Materials and their Applications, MMA, Inuyama, Japan (2004).
[700] O. H. Yang, E. S. Kim, and J. Xu. Effect of Nd^{3+} on the microwave dielectric properties of $(Pb_{0.5}Ca_{0.5})(Fe_{0.5}Nb_{0.5})O_3$ceramics. Mater. Sci. Eng. B113(2004)224–227.
[701] J. Li, X. M. Chen, and Y. J. Wu. Strontium/titanium cosubstituted $Ba_4Sm_2Ti_4Ta_6O_{30}$ dielectric ceramics. J. Eur. Ceram. Soc. 22(2002)87–91.
[702] G. Subodh, J. James, M. T. Sebastian, R. M. Paniago, A. Dias, and R. L. Moreira. Structure and microwave dielectric properties of $Sr_{2+n}Ce_2Ti_{5+n}O_{15+3n}$ ($n \leq 10$) homologous series. Chem. Mater. 19(2007)4077–4082.
[703] E. S. Kim and K. H. Yoon. Microwave dielectric properties of $(1-x)CaTiO_3$-$xLi_{1/2}Sm_{1/2}TiO_3$ ceramics. J. Eur. Ceram. Soc. 23(2003)2397–2401.
[704] M.-C. Wu, Y. C. Huang, and W.-F. Su. Silver cofireable $Bi_{1.5}Zn_{0.92}Nb_{1.5}O_{6.92}$ microwave ceramics containing CuO based dopants. Mater. Chem. Phys. 100(2006)391–394.
[705] T. Lowe, F. Azough, and R. Freer. The microwave dielectric properties of $CaTiO_3$–$Li_{0.5}Nd_{0.5}TiO_3$ ceramics prepared with $Bi_2Ti_2O_7$ additions. J. Eur. Ceram. Soc. 23(2003)2429–2433.
[706] X. M. Chen, Y. H. Sun, and X. H. Zheng. High permittivity and low loss dielectric ceramics in the BaO–La_2O_3–TiO_2–Ta_2O_5 system. J. Eur. Ceram. Soc. 23(2003)1571–1575.
[707] N. Ichinose and N. Chida.Applications of ferroelectriics. Proc IEEE Intl Symp (1998) p. 513.
[708] T. Lowe, F. Azough, and R. Freer. Control of $0.2CaTiO_3$–$0.8Li_{0.5}Nd_{0.5}TiO_3$ microwave dielectric ceramics by the addition of $Bi_2Ti_3O_7$. J. Electroceram. 16(2006)257–261.
[709] E. S. Kim, B. S. Chun, D. W. Yoon, and K. H. Yoon. Microwave dielectric properties of $(1-x)(Ca_{0.7}Nd_{0.2})TiO_3$–$x(Li_{0.5}Nd_{0.5})TiO_3$. Mater. Sci. Eng. B99(2003)247–251.
[710] K. Tatsui, K. Murano, T. Kawamura, H. Kato, S. Yano, and S. Nishigaki. Dielectric properties of PbO–ZrO_2–REO_x ceramics at microwave frequencies. Jpn. J. Appl. Phys. Suppl. 26-2(1987)80–82.
[711] W.-F. Su, and S.-C. Lim. Interfacial behaviour between $Bi_{1.5}ZnNb_{1.5}O_7$–$0.02V_2O_5$ and Ag. J. Eur. Ceram. Soc. 23(2003)2593–2596.
[712] J. C. Nino, M. T. Lanagan, and R. C. Randall. Dielectric relaxation in Bi_2O_3–ZnO–Nb_2O_5 cubic pyrochlore. J. Appl. Phys. 89(2001)4512–4516.
[713] X. Hu, X. M. Chen, and S.-Y. Wu. Preparation properties and characterization of $CaTiO_3$ modified $Pb(Fe_{1/2}Nb_{1/2})O_3$ dielectrics. J. Eur. Ceram. Soc. 23(2003)1919–1924.
[714] G. Subodh, S. Kamba and M. T. Sebastian (to be published).
[715] R. Ubic, I. M. Reaney, W. E. Lee, and N. Mc. Alford, Conference dielectrics for emerging technologies, Institute of Physics, Edinburgh, (2003).
[716] N. Mc N. Alford. London South Bank University (unpublished results).
[717] H. T. Kim, T. Shrout, C. Rindall, and M. Lanagan. Low temperature sintering and dielectric properties of $Ag(Nb,Ta)O_3$ composite ceramics. J. Am. Ceram. Soc. 85(2002) 2738–2744.
[718] L. Wu, Y.-C. Chen, L.-J. Chen, Y.-P. Chou, and Y.-T. Tsai. Preparation and microwave dielectric characterization of $Ba_{1-x}Sr_xTiO_3$ ceramics. Jpn. J. Appl. Phys. 38(1999)5612–5615.

[719] N. Lin, J. P. Chu, and S. F. Wang. Structures and properties of $Ba_{0.3}Sr_{0.7}TiO_3$:$MgTiO_3$ ceramic composites. Mater. Lett. 59(2005)2786–2789.
[720] J. Petzelt, S. Kamba, E. Buixaders, V. Boutin, Z. Zikmund, A. Kanica, V. Koukal, J. Pokormy, J. Polivka, V. Pashkov, G. Komandin, and A. Volkov. Infrared and microwave dielectric response of disordered and antiferroelectric Ag(Ta,Nb)O_3 system. Ferroelectrics. 223(1999)235–246.
[721] M. Valant, and D. Suvorov. New high permittivity $AgNb_{1-x}Ta_xO_3$ microwave ceramics: Part II- Dielectric characteristics. J. Am. Ceram. Soc. 82(1999)88–93.
[722] J. Krupka, K. Derzakowski, M. E. Tobar, J. Hartnett, and R. G. Geyer. Complex permittivity of some ultra low loss dielectric crystals at cryogenmnic temperature. Meas. Sci. Tech. 10(1999)387–392.
[723] R. G. Geyer, J. Baker-Jarvis, and J. Krupka, Annual Report, Conference On Electrical Insulation And Dielectric Phenomena, October. 17–20, Boulder, Colorado (2004) pp.493–497.
[724] M. V. Jacob, J. Krupka, J. Mazierska, and G. S. Woods. Cryogenic complex anisotropic permittivity of magnesium fluoride. Mater. Sci. Eng. A427(2006)175–180.
[725] A. Fowler, J. Hartnett, M. Tobar, and J. Krupka. Proc. (2003) IEEE Intnl Freq. Control Symp., PDA Exhibition, and 17th EFTF (2003) p. 391.
[726] M. V. Jacob, J. Mazierska, and J. Krupka. Ceramic Eng & Science Proceeding 26. American Ceramic Society. 26(2005)161–168.
[727] M. V. Jacob, J. G. Hartnett, J. Mazierska, J. Krupka, and M. E. Tobar. Dielectric characterization of barium fluoride at cryogenic temperatures. Using TE01 and quasi TEomn mode dielectric resonator. Cryogenics. 46(2006)730–735.
[728] M. V. Jacob. Low temperature microwave characterization of lithium fluoride at different frequencies. Sci. Techn. Adv. Mater. 6(2005)944–949.
[729] M. V. Jacob. Low temperature characterization of lithium fluoride at different temperatures. Tech. Adv. Mater. 6(2005)944–949.
[730] M. V. Jacob, J. Mazierska, J. Krupka, and S. Takeuchi. Dielectric properties of ytterium vanadate crystal from 15–295K. Interntl. Conf. Microwave Materials and Their Applications. MMA, Inuyama. Japan (2004).
[731] A. N. Luiten, A. G. Mann, and D. G. Blair. Ultra high Q factor cryogenic sapphire resonator. Electronic Lett. 29(1993)879–881.
[732] J. G. Hartnett, C. R. Locke, E. N. Ivanov, M. E. Tobar, and P. L. Stanwix. Cryogenic sapphire oscillator with exceptionally high long term frequency stability. Appl. Phys. Lett. 89(2006)203513–203515.
[733] J. G. Hartnett, A. N. Luiten, J. Krupka, M. E. Tobar, and P. Bilski. Influence of paramagnetic chromium ions on crystalline YAG at microwave frequencies. J. Phys. D. Appl. Phys. 35(2002)1459–1466.
[734] V. Giordano, J. Hartnett, J. Krupka, Y. Kersalé, P. Y. Bourgeois, and M. Tobar. Whispering gallery mode technique applied to the measurement of dielectric properties of langasite between 4 and 300 K. IEEE UFFC. 51(2004)484–490.
[735] J. Krupka, R. G. Geyer, M. Kuhn, and J. Hinken. Dielectric properties of single crystal Al_2O_3, $LaAlO_3$, $NdGaO_3$, $SrTiO_3$ and MgO at cryogenic temperature. IEEE MTT. 42(1994)1886–1890.
[736] M. V. Jacob, J. Mazierska, and J. Krupka. Trans. Mater. Res. Soc. Jpn. 9(4)(2004)1093.
[737] Mazierska, J. Krupka, M. V. Jacob, and D. Ledenyov. Complex permittivity measurements at variable temperatures of low loss dielectric substrate employing split post and single post dielectric resonators. IEEE MTT Symp. Dig. (2004) 1825 Forth-Worth.
[738] M. V. Jacob, J. G. Hartnett, J. Mazierska, V. Giordano, J. Krupka, and M. E. Tobar. Temperature dependence of permittivity and loss tangent of lithium tantalate at microwave frequencies. IEEE MTT. 52(2004)536–541.

[739] M. E. Tobar, J. Krupka, E. N. Ivanov, and R. A. Woode. Anisotropic complex permittivity measurement of monocrystalline rutile between 10 and 300K. J. Appl. Phys. 83(1998) 1604–1609.

[740] R. G. Geyer, J. Krupka, B. Riddle, and L. A. Boatner. (unpublished).

[741] J. Konopka. Dielectric properties of $CaNdAlO_4$ at microwave frequencies. J. Appl. Phys. 72(1992)218–223.

[742] S. Erdei, M. Mc Neal, J. S. Jang, L. E. Cross, A. S. Bhalla, F. W. Ainger, A. Dabkowski, O. I. Sirokiski, H. A. Dabkoeska. Microwave dielectric property measurement of $LaSrGaO_4$ single crystal having possible HTSC substrate applications. J. Crystal Growth. 174(1997) 324–327.

[743] T. Shimada. Microwave dielectric properties of lanthanum aluminate ceramics and single crystal. J. Eur Ceram. Soc. 25(2005)2901–2905.

[744] R. Sobdewski, P. Gierlowski, W. Kula, S. Zarembinski, S. J. Lewanbski, M. Berkowski, and O. I. Sirokiskipp. High Tc thin films on low microwave loss alkaline rae earth aluminate crystals. IEEE Trans. Magn. 27(1991)876–879.

[745] M. H. Kim, J. H. Jeong, H. T. Kim and H. J. Lee. Effect of B_2O_3 and additives on the sintering temperature and microwave dielectric properties of $Ba(Zn_{1/3}\,Nb_{2/3})O_3$ ceramics. J. Eur. Cer. Soc. 26(2006)2139–2142.

[746] M. Udovic, M. Valant, D. Suvorov. Phase formation and dielectric characterization of Bi_2O_3-TeO_2 system prepared in an oxygen atmosphere. J. Amer. Cer. Soc. 87(2004) 591–597.

Index

FSC
www.fsc.org
MIX
Papier | Fördert
gute Waldnutzung
FSC® C083411

Kolibri 360 GmbH
Ferdinand-Jühlke-Straße 7
99095 Erfurt, Deutschland
produktsicherheit@kolibri360.de